THE OCCUPATIONAL ERGONOMICS
SECOND EDITION

FUNDAMENTALS AND ASSESSMENT TOOLS FOR OCCUPATIONAL ERGONOMICS

THE OCCUPATIONAL ERGONOMICS HANDBOOK
SECOND EDITION

FUNDAMENTALS AND ASSESSMENT TOOLS FOR OCCUPATIONAL ERGONOMICS

INTERVENTIONS, CONTROLS, AND APPLICATIONS IN OCCUPATIONAL ERGONOMICS

THE OCCUPATIONAL ERGONOMICS HANDBOOK
SECOND EDITION

FUNDAMENTALS AND ASSESSMENT TOOLS FOR OCCUPATIONAL ERGONOMICS

Edited by

William S. Marras
The Ohio State University
Columbus, Ohio, U.S.A.

Waldemar Karwowski
University of Louisville
Louisville, Kentucky, U.S.A.

CRC Press
Taylor & Francis Group
Boca Raton London New York

CRC Press is an imprint of the
Taylor & Francis Group, an **informa** business

A TAYLOR & FRANCIS BOOK

CRC Press
Taylor & Francis Group
6000 Broken Sound Parkway NW, Suite 300
Boca Raton, FL 33487-2742

First issued in paperback 2019

© 2006 by Taylor & Francis Group, LLC
CRC Press is an imprint of Taylor & Francis Group, an Informa business

No claim to original U.S. Government works

ISBN-13: 978-0-8493-1937-2 (hbk)
ISBN-13: 978-0-367-86498-9 (pbk)
Library of Congress Card Number 2005050633

Library of Congress Cataloging-in-Publication Data

Fundamentals and assessment tools / [edited by] W.S. Marras and W. Karwowski.-- 2nd ed.
    p. cm.
  Includes bibliographical references and index.
  ISBN 0-8493-1937-4
  1. Human engineering--Handbooks, manuals, etc. 2. Industrial hygiene--Handbooks, manuals, etc. I. Marras, William S. (William Steven), date. II. Karwowski, Waldemar, date.

TA166.O258 2005
620.8'2--dc22
                                                2005050633

**Visit the Taylor & Francis Web site at
http://www.taylorandfrancis.com**

**and the CRC Press Web site at
http://www.crcpress.com**

Preface

Development of the 2nd edition of the *Occupational Ergonomics handbook* was motivated by our desire to facilitate a wide application of ergonomics knowledge to work systems design, testing, and evaluation in order to improve the quality of life for millions of workers around the world. Ergonomics (or human factors) is defined by the International Ergonomics Association (www.iea.cc) as the scientific discipline concerned with the understanding of interactions among humans and other elements of a system, and the profession that applies theory, principles, data, and methods to design in order to optimize human well-being and overall system performance. Ergonomists contribute to the design and evaluation of tasks, jobs, products, environments, and systems in order to make them compatible with the needs, abilities, and limitations of people.

The ergonomics discipline promotes a holistic approach to the design of work systems with due consideration of the physical, cognitive, social, organizational, environmental, and other relevant factors. The application of ergonomics knowledge should help to improve work system effectiveness and reliability, increase productivity, reduce employee healthcare costs, and improve the quality of production processes, services, products, and working life for all employees. In this context, professional ergonomists, practitioners, and students should have a broad understanding of the full scope and breadth of knowledge of this demanding and challenging discipline.

Fundamentals and Assessment Tools contains a total of 50 chapters divided into two parts.

Part I introduces the discipline and profession of ergonomics, including the systems approach and human-centered design, quality management, risk theory in human-machine systems, legal issues, cost justification for implementing ergonomics interventions, as well as professional certification and education issues. The fundamental ergonomics knowledge covered also includes the areas of epidemiology, engineering anthropometry, biomechanics, motor control, human strength evaluation, cumulative spine loading, application of basic knowledge for the assessment of loading on the human back, shoulders, legs, and feet; rehabilitating low back disorders, low-level exertions, pathomechanics and musculoskeletal injury pathways; understanding of individual factors for musculoskeletal disorders and adaptation. Other important topics include consideration of cognitive factors, design of information devices and controls, cognitive processing, multimodal information processing, tolerances and variation in human performance, the effects of personality, psychosocial work factors, as well as the aging processes. Finally, the work environment issues, including vision and visual and tactile performance, noise and auditory effects, vibrometry, and shiftwork are also discussed.

Part II focuses on ergonomics assessment methods and tools and their validity. These comprise tools for the assessment of physical and cognitive work demands and efforts. In the physical domain, the selected topics include methods for evaluating working postures and assessment of the entire body (REBA, RULA, and LUBA), methods for analysis of upper extremity loading and exposure (such as

PLIBEL, HAL, or SHARP method), wrist posture assessment and back assessment (NIOSH Lifting Equation, 3DSSPM, Industrial Lumbar Motion Monitor, TLVs). Methods that are focused on the psychophysical assessment techniques, cognitive task analysis, application of subjective scales of effort and workload, and determination of rest allowances, are also presented.

We hope that the fundamental knowledge presented in this book will help the readers to improve their understanding of the nature of complex human-artifacts interactions that occur in a variety of working environments, especially from the perspective of the design, testing, evaluation, and management of human-compatible systems.

We also hope that this book will be useful to a large number of professionals, practitioners, and students who strive everyday to optimize the design of systems, products and processes, manage the workers' health and safety, and improve the overall quality and productivity of contemporary businesses.

William S. Marras
The Ohio State University

Waldemar Karwowski
University of Louisville

About the Editors

William S. Marras, Ph.D., D.Sc. (Hon), C.P.E., holds the Honda endowed chair in transportation in the department of industrial, welding, and systems engineering at The Ohio State University. He is the director of the biodynamics laboratory and holds joint appointments in the departments of orthopedic surgery, physical medicine, and biomedical engineering. He is also the codirector of The Ohio State University Institute for ergonomics. Dr. Marras received his Ph.D. in bioengineering and ergonomics from Wayne State University in Detroit, Michigan. He is also a certified professional ergonomist (CPE).

His research is centered around occupational biomechanics. Specifically, his research includes workplace biomechanical epidemiologic studies, laboratory biomechanic studies, mathematical modeling, and clinical studies of the back and wrist. His findings have been published in over 170 refereed journal articles, 7 books, and over 25 book chapters. He also holds several patents, including one for the Lumbar Motion Monitor (LMM). Professor Marras has been selected by the National Academy of Sciences to serve on several committees investigating causality and musculoskeletal disorders. He also serves as the chair of the Human Factors Committee for the National Research Council within the National Academy of Sciences.

His work has attracted national and international recognition. He has been twice winner (1993 and 2002) of the prestigious Swedish Volvo Award for low back pain research as well as Austria's Vienna Award for physical medicine. He recently won the Liberty Mutual Prize for injury prevention research. Recently, he was awarded an honorary doctor of science degree from the University of Waterloo for his work on the biomechanics of low back disorders.

In his spare moments, Dr. Marras trains in Shotkan karate (a black belt), enjoys playing and listening to music, sailing, and fishing.

Waldemar Karwowski, Sc.D., Ph.D., P.E., C.P.E., is professor of industrial engineering and director of the center for industrial ergonomics at the University of Louisville, Louisville, Kentucky. He holds an M.S. (1978) in production engineering and management from the Technical University of Wroclaw, Poland, and a Ph.D. (1982) in industrial engineering from Texas Tech University. He was awarded the Sc.D. (dr hab.) degree in management science by the Institute for Organization and Management in Industry (ORGMASZ), Warsaw, Poland (June 2004). He is also a board certified professional ergonomist (BCPE). He also received doctor of science honoris causa from the South Ukrainian State K.D. Ushynsky Pedagogical University of Odessa, Ukraine (May 2004). His research, teaching, and consulting activities focus on human system integration and safety aspects of advanced manufacturing enterprises, human–computer interaction, prevention of work-related musculoskeletal disorders, workplace and equipment design, and theoretical aspects of ergonomics science.

Dr. Karwowski is the author or coauthor of more than 300 scientific publications (including more than 100 peer-reviewed archival journal papers) in the areas of work systems design, organization, and management; macroergonomics; human–system integration and safety of advanced manufacturing; industrial ergonomics; neuro-fuzzy modeling in human factors; fuzzy systems; and forensics. He has edited or coedited 35 books, including the *International Encyclopedia of Ergonomics and Human Factors*, Taylor & Francis, London (2001).

Dr. Karwowski served as a secretary-general (1997–2000) and president (2000–2003) of the International Ergonomics Association (IEA). He was elected as an honorary academician of the *International Academy of Human Problems in Aviation and Astronautics* (Moscow, Russia, 2003), and was named the alumni scholar for research (2004–2006) by the J. B. Speed School of Engineering of the University of Louisville. He has received the Jack A. Kraft Innovator Award from the Human Factors and Ergonomics Society, USA (2004), and serves as a corresponding member of the European Academy of Arts, Sciences and Humanities.

Contributors

W. Gary Allread
Industrial, Welding and
  Systems Engineering
The Ohio State University
Columbus, Ohio

Thomas J. Armstrong
University of Michigan
Ann Arbor, Michigan

Stephen Bao
SHARP, Washington State
  Department of Labor
  and Industries
Olympia, Washington

Jack P. Callaghan
Department of Kinesiology
University of Waterloo
Waterloo, Ontario, Canada

Pascale Carayon
Ecole des Mines
Nancy, France

John G. Casali
Virginia Polytechnic Institute
  and State University
Grado Department of Industrial
  and Systems Engineering
Blacksburg, Virginia

Don B. Chaffin
Industrial Engineering
University of Michigan
Ann Arbor, Michigan

Donald C. Cole
Institute for Work and Health
Toronto, Ontario, Canada

Robert G. Cutlip
NIOSH Health Effects
  Laboratory Division
Morgantown, West Virginia

Kermit G. Davis, III
Department of Environmental
  Health
The University of Cincinnati
Cincinnati, Ohio

Patrick G. Dempsey
Liberty Mutual Research
  Institute for Safety
University of Louisville
Louisville, Kentucky

Colin G. Drury
Department of Industrial
  Engineering
State University of New
  York at Buffalo
Buffalo, New York

Mircea Fagarasanu
Department of Physical
  Therapy
University of Alberta
Edmonton, Alberta, Canada

Arthur D. Fisk
Georgia Institute of Technology
School of Psychology
Atlanta, Georgia

John M. Flach
Psychology Department
Wright State University
Dayton, Ohio

Andris Freivalds
Pennsylvania State University
University Park, Pennsylvania

Ash Genaidy
Department of Mechanical
  Industrial and Nuclear
  Engineering
University of Cincinnati
College of Engineering
Cincinnati, Ohio

Krystyna Gielo-Perczak
Liberty Mutual Research
  Institute for Safety
Hopkinton, Massachusetts

Damian Green
Defence Technology Centre
  for Human Factors
  Integration
BIT Lab, School of
  Engineering & Design
Brunel University
Uxbridge, Middlesex, UK

Chris Hamrick
Ohio Bureau of Workers'
  Compensation
Pickerington, Ohio

Sue Hignett
Nottingham City Hospital
Loughborough University
Nottingham, UK

Sven Hinrichsen
Institute of Industrial
   Engineering and
   Ergonomics, RWTH
   Aachen University
Aachen, Germany

Ninica Howard
SHARP, Washington State
   Department of Labor
   and Industries
Olympia, Washington

Susan J. Isernhagen
DSI Work Solutions, Inc.
Duluth, Minnesota

Julie A. Jacko
Wallace H. Coulter Department
   of Biomedical Engineering
Georgia Institute of Technology
   and Emory University School
   of Medicine
Atlanta, Georgia

Richard J. Jagacinski
The Ohio State University
Columbus, Ohio

Dieter W. Jahns
Board of Certification in
   Professional Ergonomics
Bellingham, Washington

Bente Rona Jensen
Department of Human
   Physiology, Institute of
   Exercise and Sports Sciences
University of Copenhagen
Copenhagen, Denmark

Waldemar Karwowski
Center for Industrial
   Ergonomics
University of Louisville
Louisville, Kentucky

Dohyung Kee
Department of Industrial and
   Systems Engineering
Keimyung University
Taegu, Korea

Kristina Kemmlert
Psychology, Social Affairs
   and Ergonomics Division
Medical and Social Department
National Board of Occupational
   Safety and Health
Solna, Sweden

Stephan Konz
Department of Industrial
   Engineering
Kansas State University
Manhattan, Kansas

Melichar Kopas
Technical University
Kosice, Slovakia

Karl H.E. Kroemer
Industrial Ergonomics Lab
Virginia Polytechnic Institute
   and State University
Blacksburg, Virginia

Shrawan Kumar
Department of Physical
   Therapy
University of Alberta
Edmonton, Alberta, Canada

Steven A. Lavender
Industrial, Welding and
   Systems Engineering
The Ohio State University
Columbus, Ohio

V. Kathlene Leonard
Wallace H. Coulter Department
   of Biomedical Engineering
Georgia Institute of Technology
   and Emory University School
   of Medicine
Atlanta, Georgia

Soo-Yee Lim
NIOSH
Atlanta, Georgia

Holger Luczak
Forschungsinstitut für
   Rationalisierung (FIR)
an der RWTH Aachen University
Aachen, Germany

Pepe Marlow
Ergonomics Consultant
Concord, NSW, Australia

William S. Marras
Department of Industrial, Welding
   and Systems Engineering
The Ohio State University
Columbus, Ohio

Christopher B. Mayhorn
Department of Psychology
North Carolina State University
Raleigh, North Carolina

Lynn McAtamney
National Occupational Health
   and Safety Commission
Nottingham City Hospital
Nottingham, UK

Stuart M. McGill
Department of Kinesiology
University of Waterloo
Waterloo, Ontario, Canada

Gary Mirka
Department of Industrial
   Engineering
North Carolina State University
Raleigh, North Carolina

Tracy L. Mitzner
School of Psychology
Georgia Institute of Technology
Atlanta, Georgia

Kevin P. Moloney
Center for Interactive Systems
   Engineering
Institute for Health Systems
   Engineering and School of
   Industrial and Systems
   Engineering
Georgia Institute of Technology
Atlanta, Georgia

Timothy H. Monk
Clinical Neuroscience
   Research Center
Western Psychiatric Institute
   and Clinic
University of Pittsburgh
   Medical Center
Pittsburgh, Pennsylvania

Max Mulder
Delft University of
 Technology
Delft, Netherlands

Susane Mütze-Niewöhner
Institute of Industrial
 Engineering and Ergonomics,
 RWTH Aachen University
Aachen, Germany

Milan Oravec
Technical University
Kosice, Slovakia

Maurice Oxenburgh
Ergonomics Consultant
Concord, NSW, Australia

Hana Pačaiová
Technical University
Kosice, Slovakia

Barbara J. Peters
Peters and Peters
Santa Monica, California

George A. Peters
Peters and Peters
Santa Monica, California

Gary S. Robinson
Virginia Polytechnic
 Institute and State
 University
Blacksburg, Virginia

Wendy A. Rogers
Virginia Polytechnic
 Institute and State
 University
Blacksburg, Virginia

Kris Rightmire
Board of Certification in
 Professional Ergonomics
Bellingham, Washington

Irina Rivilis
Department of Public
 Health Sciences
University of Toronto
Toronto, Canada

Paul Salmon
Defence Technology Centre for
 Human Factors Integration
Brunel University
BIT Lab, School of
 Engineering and Design
Uxbridge, Middlesex, UK

Nadine Sarter
Department of Industrial and
 Operations Engineering
University of Michigan
Ann Arbor, Michigan

James Sheedy
College of Optometry
The Ohio State University
Columbus, Ohio

B. Sherehiy
Department of Industrial
 Engineering
University of Louisville
Louisville, Kentucky

Gwanseob Shin
Department of Industrial
 Engineering
North Carolina State
 University
Raleigh, North Carolina

Barbara Silverstein
SHARP, Washington State
 Department of Labor
 and Industries
Olympia, Washington

Juraj Sinay
Technical University
Kosice, Slovakia

Gisela Sjøgaard
Department of Physiology
National Institute of
 Occupational Health
Copenhagen, Denmark

Philip J. Smith
Industrial, Welding and
 Systems Engineering
The Ohio State University
Columbus, Ohio

Amy L. Spencer
Cognitive Systems
 Engineering, Inc.
Columbus, Ohio

Peregrin Spielholz
SHARP, Washington State
 Department of Labor
 and Industries
Olympia, Washington

Neville Stanton
Defence Technology Centre
 for Human Factors
 Integration
Brunel University
BIT Lab, School of Engineering
 and Design
Uxbridge, Middlesex, UK

R. Brian Stone
Department of Design
The Ohio State University
Columbus, Ohio

Setenay Tuncel
Department of Mechanical,
 Industrial and Nuclear
 Engineering
University of Cincinnati
Cincinnati, Ohio

Marinus M. Van Paassen
Control and Simulation
 Engineering Department
Delft University of Technology
Delft, Netherlands

Guy Walker
Defence Technology Centre
 for Human Factors
 Integration
Brunel University
BIT Lab, School of
 Engineering and Design
Uxbridge, Middlesex, UK

Donald E. Wasserman
D.E. Wasserman, Inc.
Cincinnati, Ohio

Thomas R. Waters
Division of Applied Research
and Technology
National Institute for
Occupational Safety and Health
Cincinnati, Ohio

David G. Wilder
Department of Biomedical
Engineering
University of Iowa
Iowa City, Iowa

Beth A. Winkelstein
Departments of Bioengineering
and Neurosurgery
University of
Pennsylvania
Philadelphia, Pennsylvania

Charles B. Woolley
Industrial Engineering
University of Michigan
Ann Arbor, Michigan

**National Research
Council Panel on
Musculoskeletal
Disorders and the
Workplace**
Department of Industrial,
Welding and Systems
Engineering
The Ohio State University
Columbus, Ohio

Contents

II Assessment Tools

I

Fundamentals of Ergonomics

1

A Guide
to Certification
in Professional
Ergonomics*

Dieter W. Jahns
*Board of Certification in
Professional Ergonomics*

1.1　Introduction

Some form of "quality assurance" effort is natural to most professions. This generally involves development of credentialing in educational programs and/or of individuals. Three types of processes are most common: *Accreditation* is established for the regulation of instructional programs. It is voluntary and generally developed and administered by an association of professionals within the field. *Certification* involves a voluntary process of evaluation and measurement of individuals, which can then indicate whether they have achieved a professional level of qualifications as judged by professional peers. It is developed and administered by a professional association or a group specifically established for professional development purposes. *Licensure*, while it does credential individuals, is a mandatory process and is administered by a political or governing body. When laws are implemented "to protect the public" from unprofessional practices, it becomes illegal to practice one's profession without a license. Thus, these processes are distinguishable by three aspects: (a) the recipient of the credential, (b) the credentialing body, and (c) the degree of volunteerism involved in obtaining the credential (Jahns, 1991).

In 1994, Dr. Carol Slappendel reviewed nine ergonomics certification/registration programs in operation around the world. Her findings are summarized in Table 1.1 and Table 1.2. Since International Ergonomics Association (IEA) Federated Societies are more oriented towards "information dissemination," and not so much towards "control" of the profession as a guild structure, there is an increasing trend for cooperative, yet independent credentialing agencies. In "open-market" societies there are also opportunities for sham operators, which makes a supervisory role by IEA Federated Societies desirable. Examples of such efforts include the Association of Canadian Ergonomists (ACE), which prior to developing and launching certification processes and criteria for Canadian ergonomists, recognized the

---

*With updates from Kris Rightmire, Executive Administrator.

TABLE 1.1   Certification of the Ergonomist: Programs in Operation as of May 1994[a]

Certification/ Registration	Authority	Designation	Acronyms
Nonsociety	Board of Certification in Professional Ergonomics (BCPE), U.S.A.	Certified Professional Ergonomist	CPE
		Certified Human Factors Professional	CHFP
		[Certified Ergonomics Associate]	[CEA]
		Ergonomist in Training:	
		Associate Ergonomics Professional	AEP
		Associate Human Factors Professional	AHFP
	Center for Registration of European Ergonomists (CREE) European Union	European Ergonomist	Eur.Erg.
	Stichting Registratie ergonomen (SRe), Netherlands	Registered Ergonomist	R.e.
Society	[Board for Certification of New Zealand Ergonomists (BCNZE)]	[Certified New Zealand Ergonomist]	[CNZErg]
	[Canadian College for the Certification of Professional Ergonomists (CCCPE)]	[Canadian Certified Professional Ergonomist]	[CCPE]
		[Ergonomist in Training:]	
		[Associate Ergonomist]	[AE]
	Professional Affairs Board (PAB) of the Ergonomics Society, U.K.	Registered Member of the Ergonomics Society (Professional Member)	M.Erg.S.
		Fellow of the Ergonomics Society Practitioner of the Professional Register	F.Erg.S.
	Professional Affairs Board (PAB) of the [Human Factors and] Ergonomics Society of Australia [(HFESA)]	Certified Professional [Ergonomist]	[CPE]
	Membership Subcommittee of the New Zealand Ergonomics Society	Professional Member	M.NZ.Erg.S

[a]Programs are also in operation in France, Belgium, and Sweden, but information on these was unavailable at the time of original publication.

Note: Brackets indicate updations.

Source: From Slappendel, C. 1994. *Proceedings of the 12th Triennial Congress of the IEA*, Toronto, ON, Canada. D. Jahns. A Guide to Certification in Professional Ergonomics. CRC Press 1998. With permission.

Board of Certification in Professional Ergonomics (BCPE) as a valid and reliable certification organization. BCPE has also served informally as a consultant for certification efforts underway in Japan and South Africa.

Similarly in Europe, the center for Registration of European Ergonomists (CREE) works with the ergonomics societies of member countries in the European Union in evaluating and registering applicants for the "Eur.Erg." designation. The BCPE and CREE have a "reciprocity" agreement in place. As former CREE President E. N. Corlett (personal communication, December 11, 1996) wrote: "Our policy at the moment is to be linked with only one Registering body in each country. Because of our constitution, this body has to have certain requirements, as laid out in the European Standard 45013 to which we adhere. We have confirmed that BCPE fulfills these requirements."

Further, the IEA, in an effort to establish professional practice standards for ergonomists around the world, as well as harmonize ergonomic credentialing organizations on an international scale, has developed criteria and procedures for endorsing professional certifying bodies and programs. In 2001, the BCPE became the first certification organization endorsed by the IEA, in accordance with the following criteria (Criteria for IEA Endorsement, 2001):

1. The certifying body is acceptable to any relevant Federated Society.
2. The certifying body is national or international in scope.
3. The certifying body operates independently of any educational body or institute.

TABLE 1.2 Criteria Applied in Certification Programmes

Designation	Criteria	Recertification
[Certified Ergonomics Associate (BCPE)]	[Bachelor's degree from an accredited university, *plus* 200 h of ergonomics training, *plus* 2 yr of full-time ergonomics practice, *plus* a passing score on the CEA written certification examination]	[To be implemented in 2005; annual maintenance fee]
[Canadian Certified Professional Ergonomist (CCCPE)]	[Based on education and four years experience, including one mentored year, *or* education and five years experience, *or* 10 yr of experience (mature candidates)]	[Annual maintenance fee plus the cost of ACE "member" category membership]
[Certified New Zealand Ergonomist (BCNZE)]	[At least 3 yr of academic formation in any field of which the total amount of education in ergonomics is at least 1 year, *plus* at least 1 yr of training, *plus* at least 2 yr of experience]	[Not required]
Certified Professional Ergonomist/ Certified Human Factors Professional (BCPE)	Master's degree in ergonomics (human factors) or equivalent, *plus* 3 yr of full-time professional practice in ergonomics with emphasis on ergonomic design, *plus* submission of a work product, *plus* a passing score on a written certification examination	[To be implemented in 2005]; annual [maintenance] fee
[Certified Professional Ergonomist (HFESA)]	[Active membership in HFESA (or other IEA Affiliated Society) for the past 2 yr, *plus* completed an education program which provides a comprehensive set of ergonomics competencies, *plus* expertise in ergonomics demonstrated through the provision of at least one major work sample or supported by one or more work samples or products of smaller magnitude, *plus* a minimum of four years of full-time practice in HF/E or the part-time equivalent]	[Not required]
European Ergonomist (CREE)	At least 3 yr of academic formation in any field of which the total amount of education in ergonomics is at least 1 yr, *plus* at least 1 yr of training, plus at least 2 yr of experience	Registration is for a 5-yr period
Registered Ergonomist (SRe)	Not specified, but are in line with CREE criteria	Every 3 yr
Registered Member of the Ergonomics Society (a.k.a. Professional Member)	At least 3 yr (or part-time equivalent) in the practice of ergonomics, or teaching or research of ergonomics relevance since admission to the Society, *plus* evidence of academic achievements	Not required

(Table continued)

TABLE 1.2 *Continued*

Designation	Criteria	Recertification
Fellow of the Ergonomics Society	Registered Member for at least 6 yr plus significant contribution to the practice of, teaching of, or research in ergonomics for a period of 10 yr since becoming an Ordinary Member *plus* substantial contribution to the activities of the Society	Not required
Practitioner on the Professional Register of the Ergonomics Society	Must be a Registered Member of the Society *plus* a minimum of 3 yr in active practice during the preceding year	Every 3 yr
Certified professional member of the Ergonomics Society of Australia	A suitable qualification *plus* 3 yr full-time equivalent experience in the practice of ergonomics	Required
Professional member of the New Zealand Ergonomics Society	A tertiary qualification in ergonomics, or a qualification of which ergonomics made up a substantial portion of the course content, *plus* experience in the practice of ergonomics, or teaching or research of ergonomics relevance	Not required

Note: Brackets indicate updations.

Source: Slappendel, C. 1994. *Proceedings of the 12th Triennial Congress of the IEA*, Toronto, ON, Canada. D. Jahns. A Guide to Certification in Professional Ergonomics. CRC Press 1998. With Permission.

4. The certifying body has a governing body comprised of certified ergonomists, which is impartial and reflects the range of interests practiced by ergonomists.
5. The certifying body has a governing body responsible for the formulation of policy matters relating to the operation of the certifying body.
6. The certifying body clearly demonstrates the line of responsibility, the reporting structure and the relationship between the assessment and certification functions.
7. The certifying body has the financial resources to conduct certification efficiently.
8. The certifying body is operated on a nonprofit basis.
9. The certifying body is explicit about its legal status.
10. The certifying body is staffed by personnel who are knowledgeable about ergonomics and competent in the functions for which they are employed to carry out.
11. The eligibility criteria used by the certifying body are clearly defined and include: specific reference to qualifications, supervised experience, professional experience in ergonomics and any forms of evidence required for the certification process; are independent of whether a person is a member of a relevant ergonomics society; are nondiscriminatory in terms of gender, ethnicity, religion or physical status; are related to contemporary ergonomics theory and practice; and, refer to requirements for recency of an individual's practice.
12. The procedural information provided by the certifying body to applicants includes: literature clearly outlining the formal procedures to be followed by the applicant in seeking certification; the deadlines for applying for certification in any year; information on all fees relevant to the process; the process used by the organization in evaluating the suitability of the applicant for certification; and, the standards of competency to be applied in all aspects of the review.

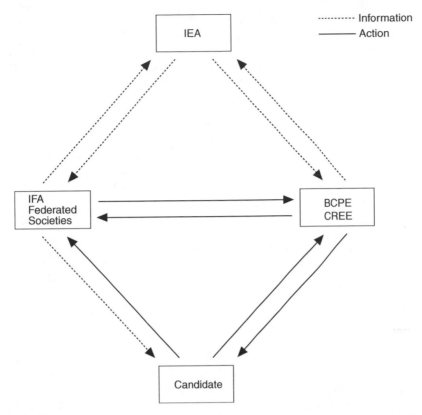

FIGURE 1.1 Communication and actions among ergonomics societies and certification agencies. (From D. Jahns. A Guide to Certification in Professional Ergonomics. CRC Press 1998. With permission.)

13. The processes followed by the certifying body are documented properly in accordance with the minimum IEA criteria for certification. They include: statements and rules relating to the current process of certification and policies relating to the granting of certification and are reviewed regularly to ensure their currency in relation to ergonomics practice; and, include a documented appeal mechanism.

14. For those certifying bodies where an examination forms part of the review, the standards should be relevant to current practice and should be clearly defined; mechanisms should exist to ensure confidentiality of the examination and its outcomes; the form of evaluation should be a valid test of the competencies assessed; and, methods used by the certifying body to test the reliability of the assessment should be described.

15. When appointing certification personnel, the certifying body must have access to a pool of qualified and competent certification personnel and to other facilities to carry out a certification review initially and for recertification purposes; the certifying personnel must be competent in the areas where they will make evaluations; maintain up to date information on relevant qualifications, training and experience of certifying personnel; and, provide clear guidelines relating to duties and responsibilities of certifying personnel.

16. The certifying body should have established processes for giving adequate feedback about deficiencies to applicants who have not attained certification.

17. The certifying body should keep a record of all policies and regulations relating to its process; keep a confidential record of details of each certification procedure followed for individual applicants; publish an annual report, including reference to numbers of applicants and outcomes of

the process; maintain an up to date register of those who have been certified, and make it accessible to public; and, publish its financial statements annually.

18. The certifying body must have already established, or be developing, a recertification process. That process should define the period of currency for any certification awarded and address criteria relevant to the applicant's competence in relation to contemporary practice in ergonomics.

The candidates for certification usually follow the pathways shown in Figure 1.1 (solid lines) by contacting either the certification agency directly or by making inquiry to one of the IEA Federated Societies, which then coordinates the certification procedures. Both BCPE and CREE have highly coordinated "information exchanges" (dashed lines in Figure 1.1) with the IEA and selected, regionally-active Federated Societies to harmonize the professional development of ergonomists. Interested readers can contact the organizations listed in "For Further Information." A general overview of BCPE certification criteria and procedures follows for interested individuals.

1.2   BCPE Certification Requirements

1.2.1   Criteria for Certification

1. CPE(Certified Professional Ergonomist)/CHFP(Certified Human Factors Professional). The BCPE job/task analyses led to the following minimum criteria for certification at the CPE/ CHFP level:
   - A master's degree in ergonomics or human factors, or an equivalent educational background in the life sciences, engineering sciences, *and* behavioral sciences to comprise a professional level of ergonomics education.
   - Three years of full-time professional practice as an ergonomist with emphasis on design involvement (derived from ergonomic analysis or ergonomic testing/evaluation).
   - Documentation of education, employment history and ergonomic project involvement by means of the BCPE "Application for Certification."
   - A passing score on the CPE/CHFP written examination. (*Note*: A person who has graduated from a HF/E degree program accredited by an IEA Federated Society, e.g., HFES, ES, is not required to take Part-I of the exam).
   - Payment of all fees levied by the BCPE for processing and maintenance of certification.
2. CPE/CHFP by portfolio review. On September 5, 1996, the BCPE reinstated the portfolio review process to accommodate senior ergonomics/human factors professionals. Individuals with at least 15 yr of ergonomic work experience may apply for designation as a CPE or CHFP via portfolio review if he or she:
   - Has a master's degree in ergonomics or human factors, or an equivalent educational background in the life sciences, engineering sciences *and* behavioral sciences to comprise a professional level of ergonomics education.
   - Has at least 15 yr of ergonomic work experience, which emphasizes design involvement (derived from ergonomic analysis or ergonomic testing/evaluation).
   - Documents his or her education, employment history, publications, work experience, and ergonomic project involvement by means of the BCPE "Application for CPE/CHFP by Portfolio."
   - Pays all fees levied by BCPE for processing and maintenance of certification.
3. AEP(Associate Ergonomics Professional)/AHFP(Associate Human Factors Professional). On March 26, 1995, the BCPE created an interim or transient Associate category of certification. An individual with the AEP [AHFP] designation is considered to be an "ergonomist in training" who has 6 yr to complete 3 full-time yr of work experience *and* transition to the higher CPE [CHFP] designation. A person takes the basic knowledge portion (Part-I) of the CPE [CHFP] exam immediately after fulfilling the education requirement for CPE [CHFP] certification (*Note*: A person who has graduated from a HF/E degree program accredited by an IEA Federated

Society, e.g., HFES, ES, is not required to take Part-I of the exam). Parts II and III of the exam may be taken after fulfilling the other CPE [CHFP] requirements.

4. CEA(Certified Ergonomics Associate). On April 17, 1998, the BCPE established a technical level of certification to meet the growing need for certified ergonomists who use commonly accepted tools and techniques for analysis and enhancement of human performance in existing systems, but they are not required to solve complex and unique problems, develop advanced analytic and measurement technologies, provide a broad systems perspective, or define design criteria and specifications. The minimum criteria for designation as a CEA are:
   - A bachelor's degree from an accredited university.
   - At least 200 contact hours of ergonomics training.
   - Two years of full-time practice in ergonomics.
   - Documentation of education, employment history and work experience by means of the BCPE "Application for CEA Certification."
   - A passing score on the CEA written examination.
   - Payment of all fees levied by the BCPE for processing and maintenance of certification.

1.2.2 Procedures for Certification

1. Request application materials from the Online Store of the BCPE web site at www.bcpe.org, by phone at 888-856-4685, or by sending a check of U.S. $10 to:
   BCPE
   PO Box 2811
   Bellingham WA 98227
   Application materials consist of three or four pages of instructions and four or seven pages of forms to be filled out by the applicant. These include:
   Section A — Personal data
   Section B — Academic qualifications
   Section C — Employment history (CPE/CHFP); ergonomics training hours (CEA)
   Section D — Work experience in ergonomic analysis, design, and testing/evaluation (CPE/CHFP); employment history (CEA)
   Section E — Work product description (CPE/CHFP)
   Section F — Signature and payment record
2. The candidate completes the application and submits it with (1) the appropriate processing and examination fee, (2) an official academic degree transcript, and (3) CPE/CHFP only, a work product (article/technical report/project description/patent/application, etc.).
3. A review panel evaluates all submitted materials and makes a recommendation whether or not the applicant qualifies to take the written examination.
4. The qualified applicant may take the written examination anytime during his or her 2-yr eligibility period. Certification is awarded upon receipt of a passing score on the examination.

The nonqualified applicant has up to 2-yr after his or her application is received to correct any deficiencies or missing elements and to take the written examination. If certification is not awarded during the 2-yr eligibility period, the applicant is required to reapply for certification.

1.2.3 Examination Administration

Applicants who have demonstrated eligibility for the examination will be notified regarding the date and location of the next examination approximately 2 months prior to the testing date. The examination, requiring a full day for CPE [CHFP] candidates and 4 hr for CEA candidates, will generally be scheduled for the spring and fall, and is usually offered as an adjunct to the meetings of ergonomics-related professional societies and associations. Qualified applicants needing accommodations in compliance with the Americans with Disabilities Act (ADA) are asked to specify their accommodation needs to the BCPE prior to signing up for the examination.

1.2.4   Scoring Methods

A panel of BCPE certificants with expertise in psychometrics determines the method of establishing passing scores to be used for the examination. Passing scores are established to ensure the candidate's mastery of the knowledge and skills required for the related level of ergonomics practice. The BCPE will periodically review, evaluate, and, as necessary, revise the examination and scoring to assure that valid and reliable measures of requisite performance capability for ergonomics practice are maintained.

1.2.5   Retaking the Examination

Candidates who do not pass the examination may retake the examination up to three times during their 2-yr eligibility period. A 6-month waiting period is required between successive exam attempts. A reduced examination fee is assessed for each retake attempt.

1.2.6   Written Examination Approximate Weighting of Subject Areas

1. Methods and Techniques (M&T)   30%
2. Design of Human–Machine Interface (DHMI)   25%
3. Humans as Systems Components (Capabilities/Limitations) (HSC)   25%
4. Systems Design and Organization (SDO)   15%
5. Professional Practice (PP)   5%

References

Criteria for IEA Endorsement of Certifying Bodies. 2001. Retrieved May 26, 2004, from http://www.iea.cc/events/edu_criteria.cfm

Jahns, D. W. 1991. Certification of professional ergonomists: a status report. *Proceedings of the 24th Annual Conference of the HFAC/ACE*, Vancouver, BC, Canada.

Slappendel, C. 1994. Harmonising the different approaches to the certification of the ergonomist. *Proceedings of the 12th Triennial Congress of the IEA*, Toronto, ON, Canada.

For Further Information

The Board of Certification in Professional Ergonomics (BCPE) — PO Box 2811, Bellingham, WA 98227-2811, U.S.A., Phone: 888-856-4685, Fax: 866-266-8003.
E-mail: bcpehq@bcpe.org
Web: www.bcpe.org

Center for Registration of European Ergonomists (CREE) — Prof. Philippe Mairiaux, President, Dept. of Occupational Health & Health Education, University of Liege, CHU Sart Tilman-B23, B-4000 Liege, Belgium
E-mail: ph.mariaux@ulg.ac.be
Web: www.eurerg.org

International Ergonomics Association (IEA) — Prof. Sebastiano Bagnara, Secretary General, ISTC-CNR, Viale Marx 15, 00137 Roma, Italy, Phone: +39-06-86090281, Fax: +39-06-824737
E-mail: ieasecr@ip.rm.cnr.it
Web: www.iea.cc

2

Magnitude of Occupationally-Related Musculoskeletal Disorders*

National Research Council Panel on Musculoskeletal Disorders and the Workplace
The Ohio State University

The panel's effort to evaluate the scientific basis for a relationship between work factors and musculo-skeletal disorders of the back and upper extremities required comprehensive reviews of the epidemiologic literature. For each of the two anatomical regions, reviews of the physical and the psychosocial factors were undertaken. Referring back to Figure 1.2, the review of the epidemiologic evidence addresses several components. The workplace factors considered include all three main elements and their relationship to the person. The person is considered in terms of the several outcomes reported in these studies, while adjusting or stratifying for the individual factors that are relevant.

2.1   Methods

2.1.1   Criteria for Selection and Review of Articles

In planning for this process, the panel set a number of criteria specific to the task of selecting articles for the epidemiology review:

- Both the exposed and the nonexposed (or comparison) populations are clearly with explicit inclusion and exclusion criteria. It is evident why subjects who were studied were eligible and why those not studied were ineligible.
- The participation rate was 70% or more.
- Health outcomes relate to musculoskeletal disorders of the low back, neck, and upper extremities and were measured by well-defined criteria determined before the study. The health outcomes

*Reprint from NRC book.

studied are carefully defined so that it is evident how an independent investigator could identify the same outcome in a different study population. Outcomes are measured either by objective means or by self-report. For self-reported outcomes, however, there are explicit criteria for how the data were collected and evidence that the collection method would permit another investigator to repeat the study in another population.

- The exposure measures are well defined. Self-report of exposure is acceptable so long as the method of collecting self-reports was well specified and there was evidence that the self-reports were reliable reflections of exposures. Job titles as surrogates for exposure were acceptable when the exposure of interest was inherent in the job (e.g., vibration exposure for those operating pneumatic chipping hammers).
- The article was published in English.
- The article was peer reviewed.
- The study was done within the last 20 yr (preferably).

No specific limitations were placed on study designs acceptable for consideration. The advantages of prospective studies, however, were recognized. For example, there were sufficient prospective studies of low back pain to examine these separately among the studies of physical factors and exclusively among the studies of psychosocial factors.

2.1.2   Literature Search Methods

The literature reviews were conducted using computer-based bibliographic databases, with MEDLINE (National Library of Medicine, U.S.A) a component of all searches. Additional databases included: NIOSHTIC (National Institute for Occupational Safety and Health, U.S.A), HSELINE (Health and Safety Executive, U.K), CISDOC (International Labour Organization, Switzerland), Ergoweb (Internet site of the University of Utah), Psychinfo, Oshrom, Ergonomics Abstracts, and ArbLine (National Institute for Working Life, Sweden).

The bibliographies of articles (particularly review articles) and the NIOSH comprehensive review (Bernard, 1997b) were examined to identify additional relevant articles.

Using these sources, a candidate list of articles were established and then systematically screened to determine which ones met the strict criteria, described above, for inclusion in the review. Each process reduced the list substantially. For physical work factors studied in association with back disorders, 255 studies were initially identified as relevant and 41 met the selection criteria and were reviewed. For psychophysical factors and back disorders, the search resulted in 975 references, which were then reduced to 21 work-related risk factor studied and 29 individual risk factor studies. For work-related physical factors and upper extremity disorders, the initial list of 265 references was reduced to 13 that provided direct and 29 that provided indirect measures of exposure. For psychophysical factors and upper extremity disorders, the initial 120 references were reduced to 28.

2.1.3   Analysis of Study Results

2.1.3.1   Definition of Measure: Relative Risk

In epidemiology, the *relative risk* is a measure of the strength of an association, here meaning the relationship between the frequency of an exposure and the occurrence of an outcome (e.g., amount of vibration and incidence of back pain). Because human populations typically have a variety of exposures occurring in near proximity, relative risk is typically measured as the incidence of disease in the exposed (e.g., helicopter pilots who experience vibration) and the incidence of disease in the unexposed (similar people, like ground crews, who are considered to share nearly the same other exposures as the exposed, such as recreational activities, diet, and living conditions). The ratio of incidence provides a measure of association, and the higher this ratio of incidence (the relative risk), the stronger the association, the more confidence we can place in a conclusion that the association is meaningful.

Because incidence is a rate calculated by following people over time, and many studies are cross-sectional or retrospective (case-control), other measures, such as the *prevalence ratio* and the *odds ratio*, have been developed to summarize the association between exposure and outcomes for these other study designs. Our analysis focused on associations expressed by such risk estimates as the odds ratio and the relative risk. These estimates were retrieved from the original article or calculated when sufficient raw data were presented.

2.1.3.2 Definition of Measure: Attributable Risk

The *attributable risk* is another measure used to help generate inferences. In its simplest form, it is the difference between the incidence in those exposed and those unexposed — a risk difference. This risk difference is thought of as attributable risk in that, in theory, removing this exposure entirely would reduce the frequency of the outcome to the level of those who are unexposed. Rotham and Greenland (1998a, b) discuss some of the limitations of this simple assumption. Attributable risk is often calculated as a ratio rather than a difference: risk in the exposed is divided by risk in the unexposed, producing an attributable fraction. The attributable fraction is the proportion by which the rate of the outcome among the exposed would be reduced if the exposure were eliminated. This fraction is calculated as the ratio of $(RR - 1)/RR$, where RR is the relative risk or the prevalence ratio of risk in the exposed compared with the unexposed:

$$AF_e = (RR - 1)/RR \tag{2.1}$$

The attributable fraction helps scientists and policy makers recognize that in many cases a variety of factors contribute to the total incidence of a disease or other outcome, so that removal of an exposure typically does not reduce the outcome rate to zero. However, in its simplest form, the attributable risk is a measure that suggests that if the offending exposure were removed (by intervention or regulation), then the amount of disease outcomes would be estimated to be reduced by the calculated amount. As is noted in the following para, this simple summary is enmeshed in caveats.

It is important to recognize in this calculation that the result depends on what is included. This is, if one considers a calculation of one factor as it relates to an outcome and then performs a separate calculation for another factor for the same outcome, there is overlapping (correlation) between factors that could make the sum of the two separate factors sum to more than 100%. Attributable fraction, then, represents a crude but important estimation of the impact of control of risk factors. An estimate of the attributable fraction for a multifactorial disease such as a musculoskeletal disorder provides only an estimate of the relative importance of the various factors studied. It is not, and cannot be, considered a direct estimate of the proportion of the disease in the population that would be eliminated if only this single factor was removed (Rotham and Greenland, 1998a). Rather it provides guidance to the relative importance of exposure reduction in those settings in which the exposure under study is prevalent. Consequently, we have not attempted to rank or further interpret the findings for attributable fractions and have chosen only to report them as a rough guide to the relative importance of the factors in the study settings in which they have been examined.

In this review, the relative risk in longitudinal studies and the prevalence or odds ratio in cross-sectional surveys were used to calculate the attributable fraction for the risk factors studied. For example, if workers exposed to frequent bending and twisting have a prevalence of low back pain that is three times that of those not exposed, then among the exposed the attributable fraction will be:

$$AF_e = (3 - 1)/3 = 0.67 \tag{2.2}$$

By this hypothetical calculation, 67% of low back pain in the exposed group could be prevented by eliminating work that requires bending and twisting.

2.1.3.3   Confounding

None of the musculoskeletal disorders examined in this report is uniquely caused by work exposures. They are what the World Health Organization calls work-related conditions. "Work-related diseases may be partially caused by adverse working conditions. They may be aggravated, accelerated, or exacerbated by workplace exposures, and they may impair working capacity. Personal characteristics and other environmental and socio-cultural factors usually play a role as risk factors in work-related diseases, which are often more common than occupational diseases" (World Health Organization, 1985).

In Chapter 3 we note that the epidemiologic study of causes related to health outcomes such as musculoskeletal disorders requires careful attention to the several factors associated with the outcome. The objective of a study will determine which factor or factors are the focus and which factors might "confound" the association. In the case of musculoskeletal disorders, a study may have as its objective the investigation of individual risk factors. Such a study, however, cannot evaluate individual risk factors effectively if it does not also consider relevant work exposures; the work exposures are potential confounders of the association with individual risk factors. Conversely, a study that evaluates work exposures cannot effectively evaluate these factors if it does not also consider relevant individual risk factors; the individual risk factors are potential confounders of the association with work exposures.

Therefore, when studying the relationship of musculoskeletal disorders to work, it is necessary to consider the other known factors that cause or modify the likelihood that the disorder will occur, such as individual factors and nonwork exposures. For example, the frequency of many musculoskeletal disorders is a function of age, so age has to be taken into account before attributing a musculoskeletal disorder to a work exposure. Another common concern is whether a recreational exposure accounts for an outcome that otherwise might be attributed to work.

In every epidemiologic study, confounders need to be measured and, when relevant, included in the data analysis. The confounders selected for consideration in the analysis of data from a specific study depend on the types of exposures studied, the types of outcomes measured, and the detail on potential confounders that can be accurately collected on a sufficient number of the study subjects. As a consequence, our approach to reviewing epidemiologic studies of work and musculoskeletal disorders documented the attention given to a wide range of potential confounders (see the panel's abstract form in Figure 2.1). No study can measure every possible confounder; however, the papers included by the panel were judged to have given adequate attention to the primary individual factors that might have confounded the work exposures under study. These include in particular age and gender, as well as, when necessary and possible, such factors as obesity, cigarette smoking, and comorbid states.

The role of potential confounders in epidemiologic studies and their proper management is often confusing to the nonepidemiologist. The difficulty stems from the fact that the potential confounder is often known to be associated with the disease, in this case musculoskeletal disorders. The association of a risk factor such as age with the disease, however, does not make it a true confounder of the study's examination of a separate risk factor such as work exposures. True confounding occurs only when, for example, both the risk factors being studied (age and work exposures) are associated with the outcome (musculoskeletal disorders) *and* the two risk factors are also correlated (e.g., those with more work exposure are also older). Fortunately, as noted in Chapter 3, there are statistical methods available to manage confounding that provide a way to "separate," in this example, the effects of the work exposure from the effects of age.

The panel recognizes that a number of nonwork factors are associated with or also cause the musculoskeletal disorders under study. These were not separately studied, but they were considered, as necessary, to evaluate the significance of the work factors that were studied. In our judgment, it is evident that confounding alone is highly unlikely to explain the associations of musculoskeletal disorders with work that are noted. More detailed consideration of confounding in future studies, however, should further improve the precision and accuracy of risk estimates.

	Described	Used in Analysis	Does Not Vary
☐ Age	☐	☐	☐
☐ Gender	☐	☐	☐
☐ Body mass index	☐	☐	☐
☐ Weight	☐	☐	☐
☐ Height	☐	☐	☐
☐ Smoking	☐	☐	☐
☐ Marital status	☐	☐	☐
☐ Income	☐	☐	☐
☐ Educational status	☐	☐	☐
☐ Comorbid states	☐	☐	☐
☐ Hormone-related conditions (e.g., pregnancy)	☐	☐	☐
☐ Strength or capacity	☐	☐	☐
☐ Race	☐	☐	☐
☐ Workers' compensation policies	☐	☐	☐
☐ Nonoccupational exposure factors	☐	☐	☐

Methods used to control confounding:
- ☐ Matching
- ☐ Stratification
- ☐ Standardization
- ☐ None
- ☐ Regression
- ☐ Other:——

Consideration of interactions:
- ☐ Interaction between different types of work exposures
- ☐ Interaction between work exposures and nonwork exposures/cofactors

FIGURE 2.1   Individual Factors Considered in Analyses Form Used in Describing Studies Included in the Review.

2.1.4   Measures of Workplace Exposures

2.1.4.1   Physical Exposures

The measures of physical exposures investigated include force, repetition, posture, vibration, and temperature. Available approaches for estimating exposure to these physical stressors include worker self-report, bioinstrumentation, and direct observation. The optimal choice among methods depends on characteristics of the methods as well as of the jobs under study. Job exposure can be considered as a weighted sum of the different task-specific exposures that make up the job, with weights coming from task distributions (Winkel and Mathiassen, 1994). Each of the two components — exposures in each task and the relative frequency of each task — must be estimated. Workers with the same job title may have different exposure levels because of between-worker variability in either the duration and distribution of task within jobs or the exposures with tasks. Furthermore, job title may indicate homogenous exposure groups for some stressors, such as repetitiveness and force demands, while other features such as posture may vary widely among workers in the same job (e.g., Punnett and Keyserling, 1987; Silverstein et al., 1987). In highly routinized or cyclical work, such as that at a machine-paced assembly line, without job rotation there is only one task, the short duration and regularity of which make the exposure determination a relatively simple problem. In contrast, in nonroutinized work, such as construction and maintenance, determination of task distributions over an extended period of time may be a more difficult undertaking. As jobs become less routinized, that is, less predictably structured, valid estimation of both task distributions and task-specific exposures becomes increasingly challenging.

Typically, both observational and direct measurement techniques generate highly detailed, accurate exposure analyses for a relatively short period of elapsed time in each job. Most protocols for these methods assume that the work is cyclical, with little variability over time, so that it is reasonable to measure exposures for a short period and extrapolate them to the long term. But many jobs do not fit this model: they are not comprised of work cycles, or the cycles are highly variable in their total duration or content (the number or sequence of steps that comprise each cycle) and do not account for all of the work performed by an individual with any given job title. For these jobs, it would be infeasible to undertake continuous measurements for entire cycles as an exposure assessment strategy, because either there are no cycles, or a very large number of (long) cycles would have to be recorded in order to quantify accurately the total and average duration of exposures. With short measuring times, the data collected are of uncertain representativeness because these time periods do not match the duration of exposures that are thought to be relevant for the development of musculoskeletal disorder.

A versatile alternative for estimating physical exposures is the use of data collected directly from workers. Such reports may address both task-specific exposures within jobs and the distributions of tasks performed by each worker. In addition to being time-efficient, self-reports permit assessment of exposures in the past as well as the present and may be structured with task-specific questions or organized to cover the job as a whole. Some researchers have explicitly recommended a composite approach to the analysis of nonroutine jobs, in which task-specific exposures are measured directly and the temporal distribution (frequency and duration) of each task is obtained from self-report. Self-reported data can take various forms, including duration, frequency, and intensity of exposure. In some studies, absolute ratings have agreed well with observations or direct measurements of the corresponding exposures, while others have diverged significantly, especially with use of continuous estimates or responses that required choices among a large number of categories (e.g., Burdorf and Laan, 1991; Faucett and Rempel, 1996; Lindström, et al., 1994; Rossignol and Baetz, 1987; Torgén et al., 1999; Viikari-Juntura, 1996; Wiktorin et al., 1993).

Retrospective recall of occupational exposures has been frequently employed in studies of musculoskeletal disorders, but there are few data on the reproducibility of such information. Three studies have examined the potential for differential error (i.e., information bias) in self-reported exposure with respect to musculoskeletal disorders with mixed results; some risk estimates were biased away from the null value, some toward it, and others not at all (Torgén et al., 1999; Viikari-Juntura, 1996; Wiktorin et al., 1993). In the REBUS[1] study follow-up population, Toomingas et al. (1997a) found no evidence that individual subjects systematically overrated or underrated either exposures or symptoms in the same direction. Self-reported exposures have promise, but their validity depends on the specific design of the questions and response categories.

A variety of instrumentation methods exist for direct measurement of such dimensions as muscle force exertion (electromyography), joint angles and motion frequency (e.g., electrogoniometry), and vibration (accelerometers). For example, the goniometer has been used in a variety of studies of wrist posture, including field assessments of ergonomic risk factors (Moore et al., 1991; Wells et al., 1994), comparisons of keyboard designs (Smutz et al., 1994), and clinical trials (Ojima et al., 1991). Hansson et al. (1996) evaluated the goniometer for use in epidemiologic studies, and Marras developed a device for measuring the complex motion of the spine (Marras, 1992). While many consider these methods to represent collectively the standard for specific exposures, each instrument measures only one exposure, and usually only at one body part. When multiple exposures are present simultaneously and must be assessed at

---

[1]In the original REBUS study conducted in 1969, participants were asked to complete a questionnaire regarding health status — all selected were given a medical examination. A diagnosis of musculoskeletal disorder required signs and symptoms. The follow-up study, conducted in 1993, asked the younger participants in the original REBUS study to participate in a re-examination.

multiple body parts, the time required to perform instrumented analyses on each subject may limit their applicability to epidemiologic research (Kilbom, 1994). Another practical concern is the potential invasiveness that may interfere with job performance, alter work practices, or reduce worker cooperation. Thus, there is a trade-off between the precision of bioinstrumentation and the time efficiency and flexibility of visual observation and worker self-report. As discussed in Chapter 6, gross categorical exposure measures (e.g., >10 kg vs. <10 kg) used in epidemiologic studies may limit the possibility of observing an exposure–risk relationship; a continuous measure based on bioinstrumentation might make such a relationship more apparent. Thus, their high accuracy (for the period of measurement) gives these methods utility for validating other methods on population subsets and added value when they can be applied in epidemiologic studies.

A large number of observational methods for ergonomic job analysis have been proposed in the last two decades (see Kilbom, 1994). These include checklists and similar qualitative approaches to identify peak stressors (e.g., Keyserling et al., 1993; Stetson et al., 1991). The limitation with checklists is that they provide little information beyond the presence or absence of an exposure, with a possibly curde estimate of the exposure duration. The qualitative approaches are not likely to provide sufficient detail to effectively assess exposure for epidemiologic studies.

The most common observational techniques used to characterize ergonomic exposures are based on either time study or work sampling. Both of these techniques require a trained observer to characterize the ergonomic stressors. Methods based on time study (e.g., Armstrong et al., 1982; Keyserling, 1986) are usually used to create a continuous or semi-continuous description of posture and, occasionally, force level. Therefore, changes in the exposure level, as well as the proportion of time a worker is at a given level, may be estimated. Because methods based on time study tend to be very time intensive, they are better suited to work with fairly short and easily definable work cycles. A different approach, work sampling, involves observation of worker(s) at either random or fixed, usually infrequent, time intervals and is more appropriate for nonrepetitive work (e.g., Karhu et al., 1977; Buchholz et al., 1996). Observations during work sampling provide estimates of the proportion of time that workers are exposed to various stressors, although the sequence of events is lost. Though less time intensive than time study, work sampling still requires too much time for use in an epidemiologic study, especially one that employs individual measures of exposure.

There are also a few highly detailed, easily used observational analyses for use as an exposure assessment tool in an epidemiologic study. These methods employ subjective ratings made by expert observers. For example, Rodgers (1988, 1992) has developed methods based on physiological limits of exposure that rate effort level, duration, and frequency. The method developed by Moore and Garg (1995) employs ratings similar to those of Rodgers and adds posture and speed of work ratings. Moore and Garg's strain index is designed to estimate strain for the distal upper extremity. It is the weighted product of six factors placed on a common five-point scale (subjective ratings of force, hand/wrist posture, and speed of work and measurement of duration of exertion, frequency of exertion, and duration of task per day). The strain index is a single priority score designed to represent risk for upper extremity musculoskeletal disorders and is conceptually similar to the lift index for low back disorders. The lift index was developed as part of the revised NIOSH lifting equation (Waters et al., 1993) and is the ratio of the load lifted and the recommended weight limit.

Recently, Latko et al. (1997) developed a method employing visual analog scales for expert rating of hand activity level (called HAL). The method has also been generalized to assess other physical stressors, including force, posture, and contact stress (Latko et al., 1997, 1999). The HAL employs five verbal anchors, so that observers can rate the stressors reliably. In an evaluation, a team of expert observers comes to a consensus on ratings for individual jobs. These ratings correlated well with two quantitative measures, recovery time/cycle and exertions/second, and are found to be reliable when compared with ratings of the same jobs 1.5–2 yr later (Latko et al., 1997).

In sum, there are many methods for assessment of ergonomic exposures. The challenge for ergonomists and epidemiologists is to determine a method for characterizing the level of exposure that is efficient enough to permit analysis of intersubject and intrasubject variability across hundreds of subjects and that can also produce exposure data at the level of detail needed to examine etiologic relationships with musculoskeletal disease. The HAL, as developed by Latko, is easy to apply and has proven to be predictive of the prevalence of upper extremity musculoskeletal disorders in cross-sectional studies.

2.1.4.2 Psychosocial Exposures

Measures of psychosocial exposures reported in the literature are obtained through the use of various self-report surveys. These surveys are typically presented to subjects in a paper format in which the subject is requested to complete a series of questions. These survey tools typically comprise multiple scales used to assess psychosocial risk factors. Many of these measures assess the construct of interest using a continuous scale of measurement, by which it is possible to provide a measure of exposure in terms of degree, and not simply whether it was present or absent. Response items vary depending on the scale and typically range from 0 to 5, 0 to 7, or 0 to 10, with options anchored so that the respondent has a frame of reference for various responses.

Some measures are standardized, well-developed, self-report tools whose psychometric properties (reliability and validity) have been established based on past research, while other items or scales were developed for the purposes of single study. Currently, all scales used are self-report. Depending on the length of the survey, the time for completion can range from 10 min to several hours. It is rare that the perceptions reported by the respondent are corroborated by an independent assessment tool or process (e.g., supervisor or coworker evaluations or direct observation of a workplace). Although it can be helpful to assess such independently collected information to support workers' reports of their sense or opinions of their environments, perceptions are, by their nature, best collected through self-report.

The most common work-related psychosocial constructs measured in the epidemiologic literature include: job satisfaction, mentally demanding work, monotony, relationships at work that include co-worker and supervisor support, daily problems at work, job pressure, hours under deadline per week, limited control over work, job insecurity, and psychological workload (a composite of a number of sub-items that include stress at work, workload, extent of feeling tired, feeling exhausted after work, rest break opportunities, and mental strain),

The job content questionnaire (JCQ) is an example of a workplace psychosocial measure whose measurement properties are well defined; it has been used frequently in the psychosocial epidemiology literature. The JCQ comprises three key measures of job characteristics: mental workload (psychological job demands), decision latitude, and social support (Karasek, 1985). Decision latitude is based on the worker's decision authority and the worker's discretion over skill use — that is, the worker's ability to control the work process and to decide which skills to utilize to accomplish the job. Psychological job demands reflect both physical pace of work and time pressure in processing or responding to infor-mation. In the Karasek and Theorell model (1990), high psychological job demands in combination with low decision latitude result in residual job strain and, over time, chronic adverse health effects. The JCQ, as an instrument for measuring such strain, has been shown to be highly reliable and has been validated as a predictor, in numerous countries and industrial sectors, of increased risk of cardio-vascular morbidity (Karasek and Theorell, 1990; Karasek et al., 1998; Kawakami et al., 1995; Kawakami and Fujigaki, 1996; Kristensen, 1996; Schwartz et al., 1996; Theorell, 1996).

2.1.5   Measures of Musculoskeletal Disorder Outcomes

The epidemiologic literature on the relationship between exposure to physical and psychosocial risk factors and the development of musculoskeletal disorders in the workplace focuses on four major types of outcomes. Two outcomes rely on patient self-report (symptoms and work status), and two rely on sources independent of the patient (evaluation by a clinician and review of workplace or insur-ance records). Table 2.1 summarizes the outcomes assessed in 132 epidemiologic studies. These do not include the 29 upper extremity studies that provided indirect measures of exposure.

Self-report symptom measures were the most common outcomes, with 61 studies assessing presence of symptoms (usually nonstandardized questionnaires asking about prevalence or incidence), 19 studies

TABLE 2.1 Outcome Measures in Epidemiologic Studies of Work and Back and Upper Extremity Musculoskeletal Disorders

Risk Factor and Body Region	Number of Studies	Self-Report Symptoms			Self-Report Work Status		Clinical Evaluation[a]			Records		
		Present	Severity[b]	Disability	Sick Days	Return to Work	Visit Only	Physical Exam	Tests	Claim	Sick Days	Return to Work
Phychosocial — back												
Work-related factors (longitudinal)	21	6	5	2	2	4		2		3	4	1
Individual factors (longitudinal; not including studies above)	29	9	8	6	2	6	1	1				1
Psychosocial — upper extremities												
All factors (cross-sectional)	25	13	6	1				8				
All factors (longitudinal)	3	1						2				
Physical — back												
Workers only (cross-sectional)	21	21					1	1				
Community (cross-sectional)	9	7										
Workers (longitudinal)	7	2								4	1	
Workers (case-control)	4						1	1		2		
Physical — upper extremities												
Workers (cross-sectional)	13	2						7	4			
Total[c]	132	61	19	9	4	10	3	22	4	9	5	2

[a]Studies are counted only once regarding clinical evaluation; some studies simply noted that a clinical visit occurred; some further specified that a physical examination was performed; and some also noted that diagnostic tests were done.

[b]Severity usually measured with standardized pain or symptom severity measure.

[c]The total number of specific outcomes exceeds the number of studies (i.e., 132), since some studies assessed multiple outcomes.

Source: Reprinted with permission from NACS (Musculoskeletal Disorders and the workplace. Low Back and Upper Extremities © 2001 by the National Academy of Sciences, Courtesy of the National Academies Press, Washington, D.C.)

assessing symptom severity (often with standardized pain and symptom questionnaires), and 9 studies assessing symptom-related disability. A total of 14 studies assessed the self-reported effect of the musculoskeletal disorder on work status, either as number of sick days ($n = 4$) or return (or nonreturn) to work ($n = 14$). Formal clinical evaluation constituted an outcome in 29 studies, most of which relied on a physical examination by a physician or other health care professional (e.g., physical therapist). Diagnostic tests such as x-rays or nerve conduction studies were a standard outcome in only a few studies. Information obtained from records constituted an outcome in 16 studies, including claims data, sick days, or return to work. The predominance of symptoms as an outcome is inherent in the nature of musculoskeletal disorders, which are primarily defined by pain or other symptoms. Indeed, the results of physical examination and diagnostic tests may be normal in a large proportion of individuals with musculoskeletal disorders.

There were a greater number of high-quality studies related to back pain than to upper extremity musculoskeletal disorders. More of the back pain studies were longitudinal rather than cross-sectional, providing stronger evidence for a potentially causal relationship between particular risk factors and back disorders. A greater proportion of upper extremity musculoskeletal disorder studies used clinical evaluation as an outcome.

2.2   Results

2.2.1   Work-Related Physical Factors

2.2.1.1   Back Disorders

The scientific literature on work-related back disorders was reviewed to identify those risk factors of physical load that are consistently shown to be associated with back disorders and to determine the strength of their associations. A total of 43 publications were selected that provided quantitative information on associations between physical load at work and the occurrence of back disorders. These risk factors were found significant in almost all of the studies: lifting and carrying of loads in 24 of the 28 in which it was studied, whole-body vibration in 16 of the 17, frequent bending and twisting in 15 of the 17, and heavy physical work in all 8 in which this factor was studied. The following significant findings are summarized from these studies: for lifting and carrying of loads, risk estimates varied from 1.1 to 3.5, and attributable fractions were between 11 and 66% for whole-body vibration, risk estimates varied from 1.3 to 9.0, with attributable fractions between 18 and 80%; for frequent bending and twisting, risk estimates ranged from 1.3 to 8.1, with attributable fractions between 19 and 57%; and for heavy physical work, risk estimates varied from 1.5 to 3.7, with attributable fractions between 31 and 58%. Appendix Table 2.1, Appendix Table 2.2, Appendix Table 2.3, and Appendix Table 2.4 provide the detailed findings in the 43 publications selected in this review. Three publications are not included in these tables because they did not present any significant association (Hansen, 1982; Lau et al., 1995; Riihimäki et al., 1994).

The evidence on static work postures and repetitive movements is not consistent. The characteristics of the studies have some impact on the magnitude of the risk estimate, but these characteristics do not explain the presence or absence of an association. Table 2.2 provides a compilation of results from all studies in terms of the importance of each general type of exposure.

Study designs affect these findings. Studies with small samples tend to have higher risk estimates, which may be an indication of publication bias. Due to power considerations, in smaller studies the effect of a risk factor needs to be larger in order to reach the level of statistical significance. Hence, the evaluation of the magnitude of a particular risk factor should take into account the sample size.

Case-control studies (Appendix Table 2.4) reported higher risk estimates than cross-sectional studies (Appendix Table 2.1 and Appendix Table 2.2) for manual material handling and frequent bending and twisting. An explanation may be that in case-control study design, recall bias (by subjects of exposure) is stronger than in cross-sectional studies, since there was usually a long period between exposure and recall. However, the case-control study with the highest risk estimate was based on observations at the workplace.

TABLE 2.2 Summary of Epidemiologic Studies with Risk Estimates of Null and Positive Associations of Work-Related Risk Factors and the Occurrence of Back Disorders

| | Risk Estimate | | | | Attributable Fraction (%) | |
| | Null Association[a] | | Positive Association | | | |
Work-Related Risk Factor	n	Range	n	Range	n	Range
Manual material handling	4	0.90–1.45	24	1.12–3.54	17	11–66
Frequent bending and twisting	2	1.08–1.30	15	1.29–8.09	8	19–57
Heavy physical load	0		8	1.54–3.71	5	31–58
Static work posture	3	0.80–0.97	3	1.30–3.29	3	14–32
Repetitive movements	2	0.98–1.20	1	1.97	1	41
Whole-body vibration	1	1.10	16	1.26–9.00	11	18–80

Notes: n = number of associations presented in epidemiologic studies. Details on studies are presented in Appendix Table 2.1 Appendix Table 2.2, Appendix Table 2.3 and Appendix Table 2.4.

[a]Confidence intervals of the risk estimates included the null estimate (1.0). In only 12 of 16 null associations was the magnitude of the risk estimate presented.

Source: Reprinted with permission from NACS (Musculoskeletal Disorders and the workplace. Low Back and Upper Extremities © 2001 by the National Academy of Sciences, Courtesy of the National Academics Press, Washington, D.C.)

In general, risk estimates in community-based surveys (Appendix Table 2.2) were smaller than those in cross-sectional studies in occupational populations (Appendix Table 2.1). A reasonable explanation is that contrast in exposure is less in community-based studies that survey a large variety of jobs. In various cross-sectional studies, contrast in exposure has played a role in the selection of subjects.

Multivariate analyses with more than two confounders showed smaller risk estimates (see, e.g., the longitudinal study by Smedley et al., 1997) than statistical analyses with just one or two confounders (see, e.g., the longitudinal studies by Gardner et al., 1999; Kraus et al., 1997; Strobe et al., 1988 and Venning et al., 1987). For lifting as a risk factor, this difference was statistically significant, with average risks of 1.42 and 2.14. Most studies have adjusted only for a limited number of potential confounders.

In addition to study design issues, some of the differences in findings appear related to the different ways exposure was measured. For manual material handling, the seven studies with observations and direct measurements showed a significantly higher risk estimate than the 21 studies based on questionnaires, with average risk estimates of 2.42 and 1.86, respectively. This finding may be explained by larger misclassification of exposure in questionnaire studies, or by larger contrast in exposure in studies that used actual workplace surveys to determine exposure levels. In general, questionnaire studies showed associations between physical load and back disorders similar to those shown in studies that represented much more detailed exposure characterization. Therefore, the information from these questionnaire studies provides useful corroborating evidence.

The magnitude of the risk estimate could not be evaluated in relation to the contrast in exposure, since exposure parameters were not very comparable. Some studies have used reference groups (low exposure) that may nonetheless have had measurable exposure to physical load in other studies.

This review concludes that there is a clear relationship between back disorders and physical load imposed by manual material handling, frequent bending and twisting, physically heavy work, and whole-body vibration. Although much remains to be learned about exposure-outcome relationships (see Chapter 3), the epidemiologic evidence presented suggests that preventive measures may reduce the exposure to these risk factors and decrease the occurrence of back disorders (see Chapter 6). However, the epidemiologic evidence itself is not specific enough to provide detailed, quantitative guidelines for design of the workplace, job, or task. This lack of specificity results from the absence of exposure measurements on a continuous scale, as opposed to the more commonly used dichotomous (yes/no) approach. Without continuous measures, it is not Possible to state the "levels" of exposure associated with increased risk of low back pain.

2.2.1.2 Upper Extremity Disorders

A variety of disorders of the upper extremity were studied in the selected literature. Primary among these was carpal tunnel syndrome, identified by symptoms and physical examination alone or in combination with nerve conduction testing. A second important outcome was hand-arm vibration syndrome (Raynaud's disease or other vibration-related conditions of the hand). There were also a number of operationally defined but les well-specified outcomes (defined for epidemiologic, not clinical, purposes) such as musculoskeletal disorders of the wrist, tendinitis, and bone- or joint-related abnormalities. Studies that met the most stringent criteria were not based on self-report alone. The anatomical areas with the greatest number of studies were the hand and the wrist, although a number of studies focused more generally on the upper extremities. Although a number of studies of the neck/shoulder region were considered, only two were included. The neck, shoulders, and upper arms operate as a functional unit, which makes it difficult to estimate specific exposure factors for the neck/shoulder region at a level beyond that of job or job tasks. Further complicating study of the region is the fact that most of the reported musculoskeletal problems of this region are nonspecific, without well-defined clinical diagnoses.

Table 2.3 provides a compilation of point estimates of risk from all studies across the major types of work-related physical exposure that were studied. Appendix Table 2.5 presents the risk ratios for various exposures; these ratios cover a very wise range (2–84), depending on how specifically the exposure and the outcome were defined. With the exception of the few studies of bone- and joint-related abnormalities, most of the results demonstrate a significant positive association between upper extremity musculoskeletal disorders and exposure to repetitive tasks, forceful tasks, the combination of repetition and force, and the combination of repetition and cold. A number of good studies demonstrated that there is also an important role for vibration.

There were nine studies in which carpal tunnel syndrome were defined by a combination of a history of symptoms and physical examination or nerve conduction testing (Appendix Table 2.5 and Appendix Table 2.6). In these studies, there were 18 estimates of risk based on various specificities of carpal tunnel syndrome diagnosis and varying degrees of work exposure. Of these, 12 showed significant odds ratios greater than 2.0 (range 23–39.8), 4 showed nonsignificant odds ratios of greater than 2.0 and 2 showed nonsignificant odds ratios between 1.7 and 2.0. These findings were supported when less specific outcomes were examined. In most instances (8 out of 10), conditions classified as "wrist cumulative trauma disorders" or "nonspecific upper extremity musculoskeletal disorders" were found to be significantly associated with work-related physical risk factors with a similar range of elevated

TABLE 2.3   Summary of Epidemiologic Studies with Risk Estimates of Null and Positive Associations of Specific Work-Related Physical Exposures and the Occurrence of Upper Extremity Disorders

Work-Related Risk Factor	Risk Estimate				Attributable Fraction (%)	
	Null Association[a]		Positive Association			
	n	Range	*n*	Range	*n*	Range
Manual material handling	4	0.90–1.45	24	1.12–3.54	17	11–66
Repetition	4	2.7–3.3	4	2.3–8.8	3	53–71
Force	1	1.8	2	5.2–9.0	1	78
Repetition and force	0	—	2	15.5–29.1	2	88–93
Repetition and cold	0	—	1	9.4	1	89
Vibration	6	0.4–2.7	26	2.6–84.5	15	44–95

Notes: n = number of associations presented in epidemiologic studies. Details on studies are presented in Appendix Table 2.5.

[a]Confidence intervals of the risk estimates included the null estimate (1.0).

Source: Reprinted with permission from NACS (Musculoskeletal Disorders and the workplace. Low Back and Upper Extremities © 2001 by the National Academy of Sciences, Courtesy of the National Academics Press, Washington, D.C.)

risk. Hand-arm vibration syndrome and other vibration disorders were significantly associated with vibration exposures in 12 of 13 studies, with risk elevated 2.6–84.5 times that of nonexposed or low-exposed comparison workers.

It should be noted that the majority of studies were cross-sectional. Therefore, it is important to consider the temporal direction of the findings. It is likely that the occurrence of upper extremity symptoms or disorders contributes to increased work-related and nonwork-related stress. If this is the case and a reciprocal relationship exists, it does not preclude the need to reduce the impact of stress (as either cause or consequence) on these disorders, given the potential health effects of repeated or prolonged stress. A second limitation in cross-sectional studies is the healthy-worker effects. This effect refers to the observation that healthy workers tend to stay in the workforce, and unhealthy workers tend to leave it. Those who may have left the workforce due to the health condition being studied will be absent from the study group, resulting in an underestimation of an effect if one is present.

The findings from the studies reviewed indicate that repetition, force, and vibration are particularly important work-related factors associated with the occurrence of symptoms and disorders in the upper extremities. Although these findings are limited by the cross-sectional nature of the research designs, the role of these physical factors is well supported by a number of other studies in which exposure assessment was less specific (Appendix Table 2.6). Despite indirect objective exposure information, the jobs studied appeared to represent conspicuously contrasting ergonomic exposures. These articles were not used to estimate exposure–response relationships for specific physical hazards (e.g., repetition, force, and posture), but they do provide a foundation for demonstrating a hazard (Appendix Table 2.6). Only three studies included in the review examined the effects of computer keyboard work (Bernard et al., 1994; Murata et al., 1996; Sauter et al., 1991). In two, significant associations were found with pain or discomfort in the upper extremity, and the third found association with slowed median nerve velocity in subclinical carpal tunnel syndrome.

The attributable fractions related to the physical risk factors that were found to be important provide additional useful information. They suggest that, when present, each of the physical factors listed in Table 2.3 is an important contributor to upper extremity disorders. The studies for which attributable fractions are reported explored associations primarily with hand/wrist disorders such as carpal tunnel syndrome and hand–arm vibration syndrome. Study of these physical factors in each of the other upper extremity disorders is indicated to further explore how strong an influence these same factors might have specifically on the other disorders. Even given the limitations on generalizing from specific studies, the estimates suggest that substantial benefit could result from reducing the most severe of these physical risk factors (Table 2.3 and Appendix Table 2.5).

As with other epidemiology study reviews, there are limitations in the available literature. Characterization of exposure with sufficient specification to segregate and adequately describe exposure to the different physical factors for such regions as the neck/shoulder area provides an important example. Literature reviews by Anderson (1984), Hagberg and Wegman (1987), Sommerich et al. (1993), Bernard (1997a), and Ariens et al. (2000) provide support for the view that physical work factors are associated with neck and shoulder musculoskeletal disorders. Had the review of the literature presented in this chapter been less restrictive regarding study specifications of exposure, it is likely that much stronger conclusions would have been drawn for each of the upper extremity musculoskeletal disorders. Our review, along with the substantial literature that has used less well-specified exposures, demonstrates the high priority to be placed on developing better exposure measures for study of the neck/shoulder as well as the other upper extremity disorders.

An equally important need is for more prospective studies to address individual physical risk factors and their combination as these relate to each of the upper extremity musculoskeletal disorders. The cross-sectional findings demonstrating a strong interaction between repetition and force and between repetition and cold indicate combinations that should be priorities for future study. Given the findings on work-related psychosocial risk factors and upper extremity disorders (follows later), it will be particularly important to carry out studies that examine the combined effects of physical and psychosocial factors.

2.2.2   Psychosocial Factors

Psychosocial risk factors for work-related musculoskeletal disorders can be separated into two major categories: those that are truly specific to the workplace (job satisfaction, poor social support at work, work pace, etc.) and those that are individual psychosocial factors (such as depression). Both types of factors are important to review for several reasons. First, there is an abundance of literature regarding the relationship between both types, particularly for back pain. Second, individual psychosocial factors such as depression are typically present both at work and outside it, making it nearly impossible to distinguish which aspects of depression are work-related and which are nonwork-related. As a result, we summarize the literature on both types of risk factors, describing each separately. For research on back pain, separate tables are provided. For upper extremity disorders, fewer studies examining individual psychosocial factors were identified. Therefore, the two types of risk factors are distinguished but included in the same table.

2.2.2.1   Back Disorders

2.2.2.1.1   Work-Related Psychosocial Factors

A relatively large number of work-related psychosocial factors have been suggested as related to back pain and the resultant disability. These range from general conceptualizations, such as "job satisfaction," to more specific variables, such as "decision latitude" or "work pace." A great many measurement techniques and research designs have been employed, making direct comparison among studies difficult.

The robustness of the association between work-related psychosocial factors and back pain is suggested by two facts. First, the findings are relatively consistent in this literature despite vastly different methodologies. Second, the relationship remains and sometimes becomes stronger when possible biasing factors are controlled.

When discrepancies are found, it may be necessary to call on several factors to help explain them. These include the sample composition and size, severity of the injury/disease, measures of predictors, time of outcome, outcome criteria, study design, and possible treatment received between initial assessment and outcome. It is difficult to calculate the exact size of the effects observed, even though many of the psychosocial variables prove to be better predictors than biomedical or biomechanical factors.

Taken as a whole, the body of research provides solid evidence that work-related psychosocial factors are important determinants of subsequent back pain problems (Table 2.4 and Appendix Table 2.7). The studies produced strong evidence (i.e., at least three studies showing a positive association) for six factors, including low job satisfaction, monotonous work, poor social support at work, high perceived stress, high perceived job demands (work pace), and perceived ability to return to work. In addition, moderate evidence was found for linking low back pain to low job control, an emotionally demanding job, and the

TABLE 2.4   Summary of Work-Related Psychosocial Factors and Back Pain: 21 Prospective Studies

Work-Related Psychosocial Factor	Null Association (*n*)	Positive Association (*n*)	Attributable Fraction (%) *n*	Range
High job demands	1	5	2	21–48
Low decision latitude/control	0	2		
Low stimulus from work (monotony)	2	4	1	23
Low social support at work	0	7	3	28–48
Low job satisfaction	1	13	6	17–69
High perceived stress	0	3	1	17
High perceived emotional effort	0	3		
Perceived ability to return to work	0	3		
Perceived work dangerous to back	0	2		

Note: Details on studies are presented in Appendix Table 2.7.

Source: Reprinted with permission from NACS (Musculoskeletal Disorders and the workplace. Low Back and Upper Extremities © 2001 by the National Academy of Sciences, Courtesy of the National Academics Press, Washington, D.C.)

perception that the work could be dangerous for the back. Genéral measures, such as job satisfaction and stress, showed a very distinct relationship. However, such general measures may reflect other aspects of the psychological work environment, such as relationships at work or job demands. Therefore, the studies provide relatively little information about the mechanisms or processes involved. Despite huge differences in study design and some problems outlined below, the general methodological quality of these studies is relatively high, and participation rates are good. Few studies employed a theoretical framework, and a consequence has been difficulty in specifying which predictor variables should be measured.

The relationships examined involve a large number of parameters that may influence the strength of the association. A given risk factor may, for instance, interact with the outcome variable employed. The belief that work is dangerous would seem to be relevant for the outcome variable of return to work, but possibly not for the onset of back pain. Similarly, some risk factors may be relevant only for certain types of work. As an illustration, for assembly line employment, work pace may be strongly related to future back pain complaints, but for professionals, such as nurses, it may have a weaker relationship.

The general quality of the studies was high. By selecting prospective investigations, a minimum standard was set. Nevertheless, there is great diversity in the methodology and this causes several prominent problems. One concern is that the same concept has been measured in many different ways. Since reliability and validity are generally not specified, it is possible that two studies claiming to measure the same entity may in fact be measuring quite different ones. There was also substantial variation from study to study in the definition and measurement of the outcome variable, and this may have had considerable consequences on the results obtained. There is, for example, a difference between a simple report of having had back pain during the past year with dysfunction, with health care visits, or with sick leave.

2.2.2.1.2 *Individual Psychosocial Factors*

The results demonstrate that individual psychosocial factors are related to back pain from its inception to the chronic stage (Table 2.5 and Appendix Table 2.8). Indeed, these variables were shown to be important in the development of pain and disability. Nonetheless, since psychosocial factors account for only a portion of the variance, and since other factors are known to be of importance, the present findings may underscore the necessity of a multidimensional view in which psychological factors interact with other variables. Although psychological factors are considered to be of particular importance in chronic pain, the data reviewed show distinctly that psychosocial factors are also pivotal in the transition from acute to chronic pain as well as being influential at onset. Moreover, the results suggest that psychosocial factors are not simply an overlay, but rather an integral part of a developmental process that includes emotional, cognitive, and behavioral aspects.

Considerable research has examined the relationship between psychosocial variables and back pain, but few have penetrated the reasons why these variables may be important. A challenge for future

TABLE 2.5   Summary of Individual Psychosocial Factors and Back Pain: 38 Prospective Studies

Individual Psychosocial Factor	Null Association (*n*)	Positive Association (*n*)	Attributable Fraction (%)	
			n	Range
Depression or anxiety[a]	5	17	6	14−53
Psychological distress[b]	0	11	4	23−63
Personality factors	3	4	4	33−49
Fear-avoidance-coping	1	8	1	35
Pain behavior/function[c]	1	6	1	38

Note: Details on studies are presented in Appendix Table 2.8.

[a]Seventeen studies assessed depression only, two studies anxiety only, and three studies both depression and anxiety.

[b]Nine studies assessed psychological distress, and two assessed stress.

[c]Four studies assessed pain behavior, and three assessed pain-related functioning.

Source: Reprinted with permission from NACS (Musculoskeletal Disorders and the workplace. Low Back and Upper Extremities © 2001 by the National Academy of Sciences, Courtesy of the National Academics Press, Washington, D.C.)

research is therefore to devise studies that include a theoretical perspective. Too often, studies have simply employed a convenience measure of a "psychological" variable, without considering why or how the variable might work. With a theoretical model, stronger designs could be used that would provide answers to specific questions.

Few investigations have amply treated the temporal aspects of the problem. The data reviewed suggest that certain factors are important very early, while others may be important at first consultation or a recurrence. Moreover, the reciprocal nature of pain and psychological variables was almost always treated as unidirectional, such as depression causing pain rather than pain affecting depression.

Even though all studies were prospective, methodological shortcomings ranged from selection bias and inappropriate use of statistical tests to failure to account for the intercorrelation of measures. The use of self-ratings as both the dependent and independent variables is a particular problem that may inflate risk estimates. It is difficult to summarize some results, because different terminology and measurement methods have been used to assess similar concepts (e.g., reluctance to participate in activities such as "fear-avoidance," "disability," or "somatic anxiety"). There is a need to improve the quality of prospective studies in this area and to foster the use of a more structured terminology.

Some prominent psychological factors do emerge, however. First, a cognitive component represented by attitudes, beliefs, and thoughts concerning pain, disability, and perceived health seems to be a central theme. A second theme is an emotional dimension in which distress, anxiety, and depression are central. Third, a social aspect appears, in which family and work issues seem to be relevant, even if the data are less convincing. Finally, a behavioral domain emerges, in which coping, pain behaviors, and activity patterns are consequential elements.

It is tempting to conclude that since the studies included in Appendix Table 2.7 and Appendix Table 2.8 have prospective designs, the observed relationships are causal; however, this may be incorrect. Although the relationships may be temporal, they need not be causal in nature. Caution in drawing conclusions concerning causality does not lessen the value of the reviewed findings, but points to the need for experimental or other designs to advance understanding.

An important implication is how this knowledge may be incorporated into clinical practice. First, considerable psychosocial information that could be of the utmost importance in conjunction with medical examinations may be overlooked if proper assessment of these variables is not conducted. Second, if psychosocial elements play a central role in back pain, then better interventions could be designed to deal with these factors to provide better care and prevention.

2.2.2.1.3 *Summary of Work-Related and Individual Psychosocial Factors*

Based on the studies reviewed here, there is ample evidence that both work-related and individual psychosocial factors are related to subsequent episodes of back pain (Table 2.4 and Table 2.5; Appendix Table 2.7 and Appendix Table 2.8). *Strong* evidence for a risk factor was defined by at least three studies demonstrating a positive association and a distinct majority (i.e., at least 75%) of the studies examining that risk factor showing a positive association. *Moderate* evidence for a risk factor was defined by two studies showing a positive association and none showing a negative association. *Inconclusive* evidence for a risk factor meant neither strong nor moderate evidence was demonstrated. Of the nine types of work-related psychosocial risk factors, six had strong evidence for an association with back pain (low job satisfaction, monotonous work, poor social support at work, high perceived stress, high perceived job demands, and perceived ability to return to work), and three had moderate evidence (low job control, emotionally demanding job, and perception that work could be dangerous). Of the five types of individual psychosocial risk factors, four had strong evidence, while one was inconclusive. Conclusions regarding psychosocial risk factors are further strengthened by the fact that a main criterion for selection of back pain studied for review was a prospective design, thus ensuring that the psychosocial factor was measured before the outcome. Nonetheless, the studies do not elucidate the mechanisms or the developmental process whereby "normal" acute back pain becomes chronic.

The attributable fractions related to work-related psychosocial risk factors suggest that improvement in job satisfaction may reduce risk for back disorders by 17 to 69%, while improved social support at work might reduce risk by 28 to 48%. Acknowledging the limitations associated with the interpretation of attributable fractions (as discussed earlier in the chapter) we conclude that these results point to the potential for structural changes in job supervision, teamwork structures, and the ways in which work may be organized to reduce risk. The most consistent evidence related to individual psychosocial risk factors suggests that reduction in depression and anxiety symptoms could reduce the risk for back disorders by 14 to 53%, and reduction in psychological distress could reduce risk by 23 to 63%. This is important because a number of effective treatments are available for depression, anxiety, and psychological distress. In a number of studies, the attributable risk associated with a particular psychosocial factor could not be estimated, because although the factor was significantly associated with back disorders in multivariate models, the exact data sufficient to calculate relative risk were not provided.

2.2.2.2 Upper Extremity Disorders

Exposure measures investigated among the 28 reviewed studies of the impact of psychosocial factors on upper extremity disorders included specific work demands (e.g., number of hours on deadline), perceptions of the degree of support from supervisors and coworkers; perceived control over high work demands; and reports of symptoms that may be stress-related (e.g., stress-related abdominal distress), which is a measure of response to stressors rather than a stressor itself. Such a measure is used as a proxy to stress exposure (assuming the response is indicative of exposure to stress) and is not therefore a direct measure of exposure to a stressor. This type of measure was found only in studies of nonwork-related psychosocial exposures. Table 2.6 provides a compilation of results from all studies across all anatomic areas, as well as for each specific anatomic location. Detailed summaries can be found in Appendix Table 2.9 and Appendix Table 2.10.

The most frequently studied outcome was the report of symptoms (pain, numbness, tingling, aching, stiffness, or burning) in a specific anatomical area over the past week, month, or year, measured by self-report survey. Of 28 studies, seven included confirmation of symptoms by physical examination. The anatomical areas with the greatest number of studies were the shoulder and the neck, although a number of studies focused on the hand and the elbow.

The tables indicate that the risk ratios for work-related exposures ranged from 1.4–4.4. The majority of the findings were below 2.0. Considering all upper extremity sites, this table indicates that the number of studies reporting a positive association for high job demands, high perceived stress, and nonwork-related worry and distress was greater than those reporting no significant effect for these exposures. This table also indicates that a number of potential psychosocial risk factors were not shown to be associated with the onset of work-related upper extremity symptoms or disorders. Specifically, the majority of studies that met the methodological criteria for inclusion did not report a significant effect for low decision latitude, work-related and nonwork-related (friends and family) social support, or few rest break opportunities. A similar pattern of results was observed for each of the specific anatomical locations. It should be noted that the majority of studies were cross-sectional; therefore, it is difficult to determine the direction of the findings.

The findings from the review of psychosocial work factors indicate that high job stress and high job demands are work-related factors that are consistently associated with the occurrence of symptoms and disorders in the upper extremities. The review also indicated that nonwork-related worry, tension, and psychological distress were consistently associated with work-related upper extremity symptoms and disorders. Although these findings are limited by the cross-sectional nature of the research designs, the role of job stress as a risk for upper extremity disorders was also supported by one large-scale prospective study (Bergqvist, 1995). These findings are also consistent with a prospective study in a community sample of recently diagnosed workers with a number of work-related upper extremity diagnoses (Feuerstein et al., 2000). This study indicated that level of perceived job stress predicted a composite index of outcomes (symptoms, function, lost time from work, mental health) at 3 months after diagnosis.

The attributable fractions related to these risk factors suggest that modification of the high job demands could potentially reduce the risk for upper extremity disorders and symptoms by 33 to 58%.

TABLE 2.6 Summary of Epidemiologic Studies: Psychosocial Risk Factors and Work-Related Upper Extremity Disorders

| | Risk Estimate | | | | | |
| | Null Association[a] | | Positive Association | | Attributable Fraction (%) | |
Work-Related Risk Factor	n	Range	n	Range	n	Range
Wrist/Forearm						
High job demands	4	1.2–1.4	5	1.6–2.3	4	37–56
Low decision latitude; low control and low stimulus from work	8	1.0–1.7	3	1.6–6.3	3	37–84
Low social support	4	—	3	1.4–2.1	3	28–52
Low job satisfaction	4	1.4	0	—	—	—
High perceived stress	1	1.5	3	—	—	—
Few rest break opportunities	5	2.7	2	1.5	1	33
Low support nonwork-related	4	—	0	—	—	—
Worry, tension, psychological distress, nonwork-related	0	—	2	2.3–3.4	2	56–71
Shoulder/Upper Arm						
High job demands	6	1.1	6	1.5–1.9	3	33–47
Low decision latitude; low control and low stimulus from work	8	1.1	6	1.6–1.9	3	37–47
Low social support	7	1.2	5	—	—	—
Low job satisfaction	2	—	0	—	—	—
High perceived job stress	3	1.5	3	—	—	—
Few rest break opportunities	3	—	1	3.3	1	70
Low support nonwork-related	3	—	0	—	—	—
Worry, tension, psychological distress, nonwork-related	1	—	1	4.8	—	79
Elbow/Arm						
High job demands	3	1.1	6	2.0–2.4	2	50–58
Low decision latitude; low control and low stimulus from work	5	1.0–3.0	1	2.8	1	64
Low social support	5	1.2–1.7	0	—	—	—
Low job satisfaction	2	—	0	—	—	—
High perceived job stress	1	1.4	2	2.0	1	50
Few rest break opportunities	1	—	1	3.1	1	67
Low support nonwork-related	1	—	0	—	—	—
Worry, tension, psychological distress, nonwork-related	0	—	1	1.4–1.8	1	28–44
All Upper Extremity						
High job demands	6	1.1–1.4	10	1.5–2.4	6	33–58
Low decision latitude; low control and low stimulus from work	10	1.1–1.7	6	1.6–2.8	4	37–64
Low social support	7	1.2	7	1.4–2.1	3	28–52
Low job satisfaction	4	1.1–1.4	0	—	—	—
High perceived job stress	2	1.4	5	2.0	1	50
Few rest break opportunities	3	1.4–1.5	3	1.5–3.3	2	33–70
Low support nonwork-related	3	—	0	—	—	—
Worry, tension, psychological distress, nonwork-related	1	—	3	1.4–4.8	3	28–79

Notes: n = number of associations presented in epidemiologic studies. Details on studies are found in Appendix Table 2.9.
   [a]Confidence intervals of the risk estimates included the null estimate (1.0). The magnitude of the risk estimate often was not presented.
   Source: Reprinted with permission from NACS (Musculoskeletal Disorders and the workplace. Low Back and Upper Extremities © 2001 by the National Academy of Sciences, Courtesy of the National Academics Press, Washington, D.C.)

Reduction in perceived levels of job stress could reduce the risk for upper extremity disorders and symptoms by 50%, and reduction in nonwork-related worry, tension, and distress has the potential to reduce risk by 28 to 79%. These findings highlight the potential impact of modifying both work-related and nonwork-related sources of stress; however, they must be considered within the limitations presented earlier in this chapter on the interpretation of attributable fractions. The observation that no study that considered both psychosocial and physical risk factors met review inclusion criteria is important, since many models assume a complex interaction among medical, physical/ergonomic, and workplace and individual psychosocial factors (e.g., Armstrong et al., 1994).

There is a need for more prospective studies. Unlike the area of back pain, there are very few prospective studies of psychosocial risk factors in work-related upper extremity disorders. There is also a need for more consistent use of measures that assess specific psychosocial exposures. These measures should have sound psychometric properties (e.g., reliability and validity) that justify their use. The inclusion of various measures should also be based on well-conceived hypotheses based on working models of how these factors may affect the occurrence of these symptoms and disorders (Chapter 7 discusses such models). The case definitions used in studies should be carefully delineated, and a more consistent use of outcome measures of symptoms, disorders, and functional limitations should be implemented. The criteria used be select studies for review may have been too restrictive, given the relative level of sophistication of the psychosocial literature in this area. Nevertheless, despite this rigor, an association among perceived job stress, high job demands, nonwork-related distress, and upper extremity disorders was noted. These findings highlight the importance of conducting additional studied to identify specific factors that contribute to the identified risk factors and to explain how these interact to influence the development, exacerbation, or maintenance of work-related upper extremity disorders. It is also important to determine how these psychosocial factors interact with medical and ergonomic risk factors to modify risk. It is possible that the psychosocial factors that were not found to be consistently associated with the occurrence of work-related upper extremity symptoms and disorders may influence the recovery process following onset. It is also possible that these factors may impact other outcomes, such as functional limitation or the ability to sustain a full day's work. The role of psychosocial factors in the exacerbation and maintenance of these disorders requires further investigation.

This review highlights the potential utility of increased efforts directed at understanding the mechanisms by which job stress may impact work-related upper extremity disorders and the biological basis for such an association. The review also supports the need to investigate approaches that eliminate or reduce work- and nonwork-related sources of stress in prevention efforts.

2.3   Conclusion

A number of general and specific reviews were identified in which physical and psychosocial factors were examined in relation to musculoskeletal disorders of the upper extremities and back (see review references). These reviews served as a resource to supplement the panel's efforts to identify relevant epidemiologic studies. They also were examined to determine whether conclusions when drawn from the panel's review were consistent with previous review efforts. The objectives of the reviews differed; some focused on specific industries, job, or exposures, but others were more general. As a whole, the findings from these other reviews are consistent with those arrived at in the panel's review and provide additional support for the conclusions.

The approach for considering causal inferences described in Chapter 3 is useful for summarizing our review of the data from epidemiologic studies. As the tables in this chapter show, a number of studies were judged to be of sufficient quality for inclusion in this review, and these vary in terms of the types of designs and measurement approaches. While this variety complicates the generalization of causal inferences, the summary tables indicate meaningful associations between work-related physical and psychosocial exposures and musculoskeletal disorders. The tables show not only a preponderance of evidence for some exposures (e.g., 26 of 32 studies found a significant association between vibration and upper extremity musculoskeletal disorders), but also a consistency of association for many of the exposures and outcomes. Although the literature contains mostly cross-sectional surveys, some work to establish temporality; combined with the available prospective studies, evidence for temporal association has been included in this chapter.

Most studies reviewed here also show a meaningful strength of association measured by both estimates of the relative risk and calculation of attributable risk. The attributable risk provides an estimate of the proportion of musculoskeletal disorders that might be prevented if effective interventions were implemented; the calculations are appreciable for most of the exposures summarized here.

While the measure of attributable risk is meaningful for conceptualizing public health impact, the calculations are presented for one factor at a time and do not account for other factors. As noted in this chapter, many studies did account for potential confounders that could provide alternative explanations for the observed findings, but the number of confounders examined in each study tends to be limited. While this is due to multiple factors (including expense associated with satisfying sample size requirements), the fact that the associations persist after accounting for the confounders measured to date supports the fundamental association, but it also justifies more detailed investigation.

The joint effect of exposures is another element of the risk estimation suggested in Chapter 3 and illustrated in this chapter. The attributable fraction summarizes the impact of a single exposure. However, scant attention has been paid to the joint effect, or interaction, of two (or more) exposures, increasing risk beyond the level of either alone. As noted in Chapter 3, some combinations of exposures might work jointly, although their individual actions may or may not be significant. The studies by Silverstein (e.g., Silverstein et al., 1987) showed an interaction between high force and high repetition for upper extremity disorders among industrial workers. Further investigation for joint effects of exposures is indicated from the current review. The effect of joint exposures can be investigated within physical (vibration, force, load, etc.), and psychosocial (job strain, job demand, etc.) domains. This review indicated the utter lack of studies that were found to be of sufficient quality and that examine both physical and psychosocial factors together. Because evaluation of each has shown important effects on the development of musculoskeletal disorders, and some of the current evidence (although modest) suggests that one does not explain the other, it is unlikely that more detailed investigation will demonstrate that the association of either with musculoskeletal disorders is due to confounding with the other. However, additional studies are needed to understand the degree to which each contributes to the overall incidence of musculoskeletal disorders, and the extent to which both work synergistically in selected work settings.

While the results presented in this chapter are consistent with one another, it is important to examine the degree to which they are consistent with the results from the basic science and the biomechanics studies (Chapter 5 and 6). Some of these studies have been mentioned in this chapter; their results are generally consistent, providing here some suggestion of biological plausibility for the association between physical forces and musculoskeletal disorders. The degree of consistency across different levels of study will be discussed in more detail in the integration chapter.

Most epidemiologic studies have been summarized as having exposure and outcome measures dichotomized. The ability to make inferences about dose–response relationship is limited in this context. While there are step-wise differences in dichotomous measures across studies (e.g., see Boshuizen et al., 1992: Bovenzi and Zadini, 1992) that make cross-comparisons tantalizing, the differences in comparison groups and other design features hinder the combining of results for generating inferences on dose–response relationships. Future studies can help generate strong inferences by paying greater attention to more refined levels of measurement. While this is a challenge, the strength of the current studies justifies this effort.

In conclusion, the epidemiologic evidence provides support for associations between workplace physical and psychosocial exposures and both back and upper extremity musculoskeletal disorders.

Panel on Musculoskeletal Disorders and the Workplace

Jeremiah A. Barondess (*Chair*), New York Academy of Medicine

Mark R. Cullen, Occupational and Environmental Medicine Program, School of Medicine, Yale University

Barbara de Lateur, Department of Physical Medicine and Rehabilitation, School of Medicine, Johns Hopkins University

Richard A. Deyo, Department of Medicine, Division of General Internal Medicine, University of Washington, Seattle

Sue K. Donaldson, School of Nursing and School of Medicine, Johns Hopkins University

Colin G. Drury, Department of Industrial Engineering, State University of New York, Buffalo

Michael Feuerstein, Departments of Medicial/Clinical Psychology and Preventive Medicine/Biometrics, Uniformed Services, University of the Health Sciences, Bethesda, Maryland, and Division of Behavioral Medicine, Georgetown University Medical Center

Baruch Fischhoff, Department of Social and Decision Sciences and Department of Engineering and Public Policy, Carnegie Mellon University

John W. Frymoyer (retired), McClure Musculoskeletal Research Center, Department of Orthopaedics and Rehabilitation, University of Vermont, Burlington

Jeffrey N. Katz, Division of Rheumatology, Immunology, and Allergy, Brigham and Women's Hospital, Harvard University Medical School

Kurt Kroenke, Regenstrief Institute for Health Care and School of Medicine, Indiana University, Indianapolis

Jeffrey C. Lotz, Orthopaedic Bioengineering Laboratory and Department of Orthopaedic Surgery, University of California, San Francisco

Susan E. Mackinnon, Division of Plastic and Reconstructive Surgery, School of Medicine, Washington University, St. Louis

William S. Marras, Institute for Ergonomics and Department of Industrial, Welding, and Systems Engineering, The Ohio State University, Columbus

Robert G. Radwin, Department of Biomedical Engineering, University of Wisconsin, Madison

David M. Rempel, School of Medicine, University of California, San Francisco

Robert M. Szabo, Department of Orthopaedic Surgery, School of Medicine, University of California, Davis

David Vlahov, Center for Urban Epidemiologic Studies, New York Academy of Medicine, and Johns Hopkins University School of Public Health

David H. Wegman, Department of Work Environment, University of Massachusetts, Lowell

3

Legal Issues in Occupational Ergonomics

George A. Peters
Barbara J. Peters
Peters and Peters

3.1 Introduction

3.1.1 Objectives of the Law

The law serves a vital function in complex social settings. It provides the standards or guidelines that define acceptable or unacceptable human behavior. Human conduct that transgress the limits is considered a violation of the law and the transgressor could face civil or criminal penalties. The legal system has been established by society in a form thought best to meet the needs for a harmonious, efficient, evolving, and beneficial social order.

Modern technology has imposed profound social demands or forced radical transitions on a human civilization that now interacts on a worldwide basis and has become more socially interdependent. The legal system that exists in every governmental jurisdiction and at every level of organized society must constantly adapt to changing demands, so there has been and will be a continuing effusion of controlling laws. Yet, there is an old dictum that ignorance of the law is no excuse. Without the dictum most of the laws would be conveniently ignored. With the dictum, there is presumed awareness and an incentive to learn and comply with the relevant applicable laws pertaining to a human endeavor.

3.1.2 Human Fault

In essence, the legal system is based on concepts of human fault. Without human wrongdoing, there is no need or logical reason for punishment or legal redress. For example, the concept of *negligence* relates to the failure to exercise due care (reasonable or ordinary care), which persons of ordinary prudence would use to avoid injury or damage. The behavioral standard is only that of conformance with ordinary everyday care and that is not a very high standard of conduct. For example, *strict liability* requires a defect and a remedy that was technically and economically feasible. The defect may be defined as excessive preventable risks, failure to meet reasonable consumer expectations, or some other term or phrase that connotes fault or wrongdoing. Only reasonable efforts to eliminate defects or unsafe conditions are generally required. The term *foreseeability* simply means that the harm was predictable at the time of the design, construction, or sale of the product, component, or system. In other words, the law is not unreasonable in its requirements. However the legal terminology is defined, it is intended to balance the interests of all parties. The jury criterion is what conduct is reasonable under the circumstances, unless a statute defines it specifically (such as a posted speed limit on a roadway).

3.1.3 Liability Prevention

Liability prevention or mitigation is the avoidance of fault by investigation, analysis, evaluation, risk assessment, corrective action, and preventive remedies. The objective for the human factors specialist is to reduce human error or human performance variance to that which is acceptable, tolerable, or within prescribed levels. Rather than gross speculative estimates of variance, an objective detailed approach is desirable. To identify or determine excessive uncontrolled human performance deficiencies, there must be an understanding of the process at all stages. In terms of human error, what are the measurements, what are the critical parameters, and what are the tolerable limits? Is the system robust and relatively free from unanticipated variance? The foundational question is whether or not there is a formal implemented program to reduce relevant human performance variance in design, manufacture, logistics, and actual customer applications and use.

3.1.4 (Your) Involvement with the Law

It is highly probable that many scientists and most engineers, during their lifetime, will have direct or indirect personal involvement with the legal system. It is wise to accumulate some knowledge and become somewhat familiar about this seemingly foreign discipline, sphere of activity, or prospective area of entanglement. Any sudden introduction to the legal process may be a disconcerting, anxiety provoking, error producing, and frustrating experience. The contacts with the law may be varied. A university professor may want to be an expert witness and to teach the jury about his specialty and its applications to a defined fact situation. The research specialist may be asked to explain his findings as either an expert witness or as a consultant to a law firm. An engineer may be asked by his company's Legal Department to assist in furnishing information for the completion of interrogatories (written questions), the production of records, to arrange possible inspections of the accident site or product, and help in the technical instruction of the lawyers and their expert witnesses. The engineer may be asked to review related accident or incident records, warranty claims, adjustment data, test reports, quality inspection data, and possibly help to develop or revise a liability prevention program. The engineer may be asked to perform or work with others on special research or test projects related to ongoing litigation. There are a myriad of roles that can be played in state or federal common law actions, worker's compensation cases, allegations of statutory violations, arbitrations or mediations, government agency hearings, or citations involving OSHA, NHTSA, EPA, CPSC, DOT, and FTC or other legal assessment, conflict, or dispute resolution entities (Vinal, 1999). Finally, the scientist or engineer may be the subject of a lawsuit or be subpoenaed as an eye witness to an investigated or litigated event. Thus, early preparation would seem desirable rather than to wait for a deluge of litigation-related information, quick forced learning, and accommodation to the unfamiliar.

3.2 Human Factors Testimony

The primary focus of this chapter is the critically important subject of the expert's interaction with lawyers, the court, and others directly or indirectly involved in the legal system. The expert is not expected to be a legal specialist, but some familiarity with the customary and anticipated role of the expert, consultant, or advisor is expected. The expert witness should inquire, from the attorneys, about unique applicable law, procedure, or factual issues pertaining to the case that might effect his research hypotheses, case preparation, or trial testimony. This is because the relevant law varies in each state, may be changing, the applicaion of the law to the facts is in the province of the practice of law, and legal issues may be unique. The suggestions, in this chapter, are equally applicable to other roles and functions assumed by the human factors/ergonomics specialist that are not directly involved in litigation, but within his occupational sphere of activity and personal interest.

3.2.1 Typical Testimony

The human factors specialist may act as a paid expert witness in a civil lawsuit under the common law advocacy system. The specialist may participate in a worker's compensation claim process on behalf of his occupational employer. It may be within the criminal law system, a statutory or governmental agency legal process, or simply as a consultant, advisor, reviewer, research engineer, test specialist, or providing documents and reference material to a lawyer. There are many diverse roles that can be played by the human factors specialist, but court testimony is a critically important role. Did the manual controls on a machine induce human error? Was the weight to be lifted excessive under the conditions that existed in a given workplace? Could repetitive motion, on a particular piece of equipment, cause injury?

There may be cases involving rather simple and narrow human factors issues. For example, was there a blind intersection for the driver of a truck who was working within the course and scope of his employment? Could a warning have been sufficiently effective as to have prevented an accident on a machine tool located in a factory workplace? Would more adequate illumination or better stairway markings have prevented a trip and fall accident in an office records repository? Was a high voltage line appropriately identified during a building construction project? Were work procedures confusing to a worker?

A member of the Human Factors and Ergonomics Society was, with his authorization, designated as an expert witness in a lawsuit involving a conveyor system. His area of expertise was human factors and safety. During his oral deposition, he was asked questions about his accident reconstruction to determine whether or not he understood the broader issues and the context within which he formulated his opinions and answers. The expert relied upon the injured worker's deposition for a basic description of the behavior involved and the work tasks being performed. He reconstructed the events by adding his expert knowledge to achieve a more detailed human factors description and conclusionary explanation of the accident scenario. The worker had testified in deposition that he was cleaning a conveyor roller (pulley) and moving belt with a hand tool (a scraper), when suddenly he was pulled in and crushed between the roller and the belt. The expert was asked about the careless and unsafe behavior of the worker. The expert explained the human reaction time in such circumstances and how it contributed to the accident. The tool was pulled into the in-running nip-point so quickly that the worker could not release his grip on the hand tool quickly enough to prevent the accident. The expert described the long history of such accidents with hand tools, cleaning rags, human hair, and work uniforms. He indicated that most workers do not understand why their reaction time is too long under such circumstances that they believe they are acting in a safe manner and doing their work as prescribed by their employer. The expert had old trade standards and articles describing the hazard and the various precautionary measures that should be taken on all conveyor systems (such as barrier guards, access by work platforms, cable emergency cut-offs, and warnings). The expert was questioned about appropriate training, instruction, violation of work rules, fear of losing a job if there was a refusal to do unsafe work, possible human error, and the allocation of human fault. The expert's deposition took two full days to cover all of the human factors issues that the opposing counsel thought was relevant to the accident.

Another expert witness testified about a vehicle control or handling problem that may have led to an accident. The vehicle manufacturer had some written design objectives about driver handling parameters that were used as a reference. In essence, the vehicle should compensate for driver overreactions or excessive steering wheel inputs, and remain stable under all operational conditions including accident avoidance maneuvers. In addition, the vehicle should signal or give the driver some perceptible signals when the vehicle handling limits are approached. There should be sufficient stability and predictability that the driver will not lose control of the vehicle and initiate an accident situation. There were questions as to how human factors considerations, in a vehicle stability index, could assure controllability and have a reasonable margin of safety in resisting rollover, tipover, and side slip or slide. The expert was asked about reasonable design parameters and limits, given various risk-benefit balances, the technical feasibility at the time of manufacture, the cost implications, and the effects on customer satisfaction. Also, what constitutes an adequate perceptual signal to the driver and what is likely in terms of the driver reaction? Many questions relied upon the subjective judgment of the expert, others upon extrapolation of known test results, and others on specific published data. Different experts have varying success in quick, decisive, and subjective answers. The more experience, the greater credibility of the answer. The more decisive, the more the jury believes and remembers.

3.2.2 The Prediction of Behavior

The human factors specialist has an occupational focus on human behavior that rejects ill-informed speculation. Thus, there should be some factual objective basis for any conclusions, recommendations, opinions, or interpretations as to the cause, characteristics, and predictability of human behavior. The best approach is ethical targeted research, conducted for a lawsuit, but such testing is often just a form of advocacy to prove a predetermined point and it can be very deceptive. Independent experimental research findings can accumulate and permit a conditional relevant application. Probably, the most understandable for a judge and jury is controlled observational data and conclusions that pertain to individual, group, or team performance. In terms of predictive indices or equations, a best fit for the available data may be helpful. Whatever the analytic process, there should be some objective data that serves as a foundation for an opinion.

The human factors psychologist, by training and experience, may provide informed guidance on the critical question of the foreseeability of certain behavior, including intentional error, misuse, abuse, destructive behavior, and the relative predictability of certain acts of commission or omission. However, these are ultimate issues of law and fact that pervade many cases and are interpreted differently by various participants in the legal system. If the prediction of a certain behavior is a question for the jury, the human factors specialist may assist the jury by defining the factors to be considered in the prediction of behavior.

The human operator may be just one variable in a machine system. The ergonomic issues may include the learning and adaptability of a selected or unselected, trained or untrained, machine operator from a diverse pool of prospective talent and skill. But, all machines are increasingly variable over time, usage, maintenance, repair, and environmental conditions. How is the operator informed or signaled as to an approach to decreasing and undesired machine limits? How will the operator respond to out-of-tolerance, emergency, or panic situations after long-term familiarity with the machine or process? In driving a company vehicle, is there some warning that the vehicle dynamics are close to the handling instability limit? This might be an understeer–oversteer gradient of such a character as to permit the driver to maintain safe control or for an electronic stability control system to actuate. In other words, questions of human behavior are not isolated, they are intertwined with a specific machine function or a process control.

3.3 Forensic Issues

3.3.1 Business Records

It may be wise for a participant in the legal process (such as an expert witness) to carefully segregate business records, pertaining to litigation, from other business or personal records. This includes financial records that document payments made for expert witness services that might be subpoenaed in some legal jurisdiction. It includes a list of lawsuits maintained by case name and number, the court and location, the attorneys, the subject matter, and the result if known. This information may be required to be produced in some cases in many jurisdictions. Any documents that are relied upon in formulating an expert opinion should be identified and retained until the lawsuit is fully resolved. A list of personal publications should be maintained along with an updated resume covering past education, employment, patents, awards and honors, seminars attended, or other continuing education.

3.3.2 Description of the Discipline

It may be desirable to formulate a short and simple statement that describes your occupational discipline as you perceive it. Those involved in the legal process may not have heard about the discipline in specifics or may have misconceptions generated by other incorrect, improper, or superficial use. The written statement could include the purpose or objective of the discipline, the typical qualifications of its practitioners, its standard methods or procedures, relevant college curricula and textbooks, the date of its founding, its relative size and sophistication, and examples of its accomplishments. Its general acceptance by other disciplines, where there is professional interaction, may be important for a judge acting as a gatekeeper and attempting to assess its substantive value and reliability. This statement should be consistent with similar statements published by relevant professional organizations, peer certification boards, and state licensing boards.

3.3.3 The Failure of Daubert

There have been many lawsuits where the Daubert doctrine (Daubert, 1993) was strictly applied and expert witness testimony was excluded as "unreliable." The expert's opinions must be based on *sufficient facts* or data, the testimony must be based on the product of reliable principles and *methods*, and the application to the facts of the case must be considered *reliable* (Rule 702).

If a judge is assigned the role of gatekeeper, it is assumed that the judge is sufficiently knowledgeable to perform that function as it relates to many technical, engineering, scientific, and medical specialties. This is often an unfair, unwise, and burdensome role. For example, one judge stated that a designated expert, a neuropsychologist, was just a technician who applied electrodes to the skull of a patient. The judge as a gatekeeper, decided that the neuropsychologist could not discuss brain injury in front of the jury. Yet, this highly qualified individual had performed extensive neurospychological testing on that patient, provided treatment, and did his own brain mapping (EEG). His work was in conjunction with a neurologist (MD) who deferred to the neuropsychologist in the precise aspects of the closed head injury. The judge misunderstood the actual function and credibility of the expert witness.

Similarly, the presumed testimony of an expert may be the subject of a motion to strike (to exclude that testimony from consideration or prevent it, in whole or in part, to be given to a jury). Thus, the potential expert witness should consider how to convince a judge as well as a jury as to the objectives, methods, substance, "reliability," general acceptability, and role of the discipline. This is in addition to providing an adequate foundation (justification) for the individual expert's testimony and opinions. The weight given the testimony, by a jury, is dependent upon the believability of the data or assertations, the personal credibility of the witness, the judge's admonitions, if any, and how the opinions are used during the advocacy of final argument.

There is always the possibility of having a double translation; that is, to first convince the judge, in his language, of the "reliability" of the proferred testimony and, then, convince a jury in the "everyday language" that they understand. There is always the question of inequality, that different judges in the same courthouse may have different standards as to the acceptability of evidence.

However, in most courts, the ultimate question is whether the expert witness testimony will help the trier of fact (judge or jury) understand the evidence or to determine a fact. Qualifications of the expert generally go to the weight of the evidence presented, not to its admissibility (Campbell, 2001; Goodstein, 2000).

The issue for the expert witness is to determine whether the judge favors information helpful to the jury or whether the judge may be excessively strict and politically motivated. The lawyer who retains the expert should be able to indicate, from past rulings, whether the judge will be zealous in applying Daubert or its progeny. Daubert applies to all federal courts and is of considerable interest in many state courts.

Daubert is considered a failure, by many scientists and engineers, because it has erected barriers to what they believe is valid and relevant information that may not be familiar to the evidentiary gatekeeper. They may believe that efficient judicial administration may conflict with facilitating the delivery of pertinent or illuminating information to the jury. If just one key expert has his testimony curtailed or excluded, it may torpedo or signal the end of the plaintiff's or defense's case. In fact, it may be quick judicial resolution of a case based on what seems to be a legal technicality unrelated to the merit of the claim or defense.

3.3.4   Junk Science

The expert witness should expect, during deposition or trial, to receive some incisive questioning about key research or testing that was performed by the expert or that the expert relied upon in formulating an opinion or conclusion. The purpose is to undercut the justification so that the opinion falls, becomes somewhat questionable, or uncertain to some degree. The inevitable opposing lawyer's interpretation will be that the conclusions are not supported by the data, there are serious questions about flawed methodology, and that it appears to be junk science. The expert witness should respond in a civil manner, citing other supporting data and peer investigators, also indicating that the findings are not unexpected given the logic of contemporary science or engineering, and should give the reasons why the findings are to be considered truthful and accurate in comparison with other studies. The junk science allegation may provoke an uncalm defensive personal reaction, just as inferences regarding possible violations of professional ethics may upset the witness, but this may be the intent of the opposing advocate. While truthful statements should be admitted and not argued, the expert is an expert and should hold firm and strong where justified.

3.3.5   Differential Diagnosis

Under Daubert, proferred scientific testimony must be relevant and reliable. It is reliable if the principles and methodology are grounded in the methods of science. The factors to be considered are whether the theory or technique can be *tested*, have been subjected to peer review and *publication*, whether there is an *error rate*, and whether the theory or technique is *generally accepted* in the scientific community. The issue might be whether the expert opinion was developed for purposes of testimony or was developed from research conducted *independent* of litigation. The research may not have been published because it is too recent, too specific, or of too limited interest. One universally accepted "scientific" method of establishing root cause is differential diagnosis or differential etiology. It is the systematic elimination of likely causes until the most probable cause remains isolated and cannot be excluded or ruled out. For example, a comprehensive list of hypotheses that might explain a finding is compiled, then a process of elimination occurs with an explanation why each alternative cause was ruled-in or ruled-out. Precise information may not be available or necessary to provide a basis for an expert's opinion (Clausen, 2003).

3.3.6 Root Cause

There are various forms of technical or engineering analyses that are very similar to the differential diagnosis method used by scientists. One generally accepted and utilized method is root cause analysis. It is a step-by-step procedure that is defined, rigorous, detailed, standardized, and repeatable.

During attempts to identify and correct problems by normal trouble shooting methods, it was found that proximate causes were being identified rather than the inherent deeper true causes or root causes. Correction of only the superficial proximate causes often resulted in the reoccurrence of the problem. To get to a more fundamental understanding of cause, the questioner keeps asking why did the proximate cause occur, why did the intermediate cause occur, and so on. The widely used Kepner-Tregoe Methodology (Kepner, 1965) requires that at least five rounds of asking why may be necessary to reach the root cause (The Five Whys Technique). Such techniques are considered organizational procedures for the expression of logic and the completeness of an inquiry.

One example of a root cause technique is to bring together a group of skilled persons (Root Cause Team). Then, have them evaluate whatever evidence is available pertaining to an undesired event from their own perspectives and knowledge bases (what happened). Then, have them attempt to determine as many different causes of the event as possible, however remote (the whys). Through discussions with knowledgeable people, all noncontributing causes are gradually eliminated until only the most likely contributory causes remain. At that point, a process or flow diagram and a failure mode or fault tree may be constructed for clarification, assurance, and verified isolation of the true root cause.

There is an *8D process* (The Eight Disciplines of Problem Solving), widely utilized in the automotive industry, that extends the root *cause* problem-solving to a formal plan to implement the recommended *changes*. It is directed at the "owner" of the problem, calls for both emergency and permanent corrective action, reviews the priority of the problem, determines whether the problem is inherent in the process or is somehow unique, and insures that there is a true team consensus in the implementation process.

Another example is accident investigation and accident reconstruction, if following clearly established procedures such as those contained in a published Collision Investigation Manual. The manual may list report headings and subheadings, the content in each section, where and how measurements are to be taken (such as pacing, steel tape, rollmeter, or laser), the facts, the parties, the equipment, the tools, the scene, the time, the place, the weather, evidence collected, witness statements, conclusions, and so on. Investigation is evidence gathering. The reconstruction may be even more detailed, if necessary, to recreate the specific elements of the story or to reconstruct the overall event. The words used in the narrative may be specified to assure common and correct communication. The inferences, assumptions, and logic that are required should be indicated, explained, and their foundation basis mentioned.

Other potentially credible and valuable methodologies include system safety analyses (fault tree) and reliability engineering (failure mode and effect) or some derivative, tailored, or supplemental application that has general peer approval and usage. The human factors investigation and reconstruction may use detailed task analysis, perhaps combined or interpreted with other extrinsic (independent) research findings in support of an opinion, conclusion, recommendation, or proffered testimony.

3.3.7 Cognitive Impairment

Lawyers and jurors may believe that it is reasonable to expect that the human factors expert, as a specialist in human behavior, should be able to recognize human cognitive deficits and those behavioral functions adversely effected by brain injury. Such human performance impairments may be a causal or contributing factor in an analysis of an undesired incident or an accident involving personal injury, property damage, or process interruption. It is reasonable that the human factors specialist should be aware of such conditions and include term in his analysis? The key word is recognition, not the precise diagnosis of a licensed neuropsychologist or neurologist. The human factors specialist may detect impairment, but must rely on others for court testimony establishing the particular cognitive impairment.

Some of the symptoms indicative of specific brain injury include memory problems (forgetfulness in personal and occupational activities), problems in vigilance (maintenance of selective attention), poor divided attention (on concurrent tasks), distractibility (from a perceptual set), speed and accuracy of information organization and processing, and personality changes such as episodic hyperirritability, aggressive outbursts, mood swings, emotional blunting, and socially inappropriate behavior. There are many possible symptoms of deviant behavior (from some norm), good tests to assess various mental functions, various combinations of localized and general brain insults or damage, generally accepted diagnostic categories, and many treatment modalities. Similarly, over-the-counter and prescribed medications affect the brain and may produce undesired behavioral side effects that may or may not contribute to a human factors problem in a particular situation (Price, 1988).

The critical question may be whether or not the cognitive impairment existed before, during, or after a particular event. Conversely, did the impairment result from an accident in which there was head impact, some sudden acceleration or deceleration, or an unusual head rotation? Was there a head concussion (loss of consciousness), a medical diagnosis of a closed head injury, or an unusual change in work performance?

This suggests that the human factors specialist should be on the alert for unusual behavior, if appropriate, consult with other team members of different specialties, and where available review available documentation including medical records. The opposing counsel may do likewise and ask the expert pertinent questions at depositions or at trial.

3.3.8   Complexity

An expert may find that it is fairly easy to discuss complex issues in complex language. The expert may be familiar with the concepts, symbols, equations, and specialized definitions of terminology used within the specialty. It may be far more difficult to simplify, be direct, avoid unconditional qualifications that add ambiguity and uncertainty, and reduce the key concepts to demonstrable analogies and graphic representations. The expert should effectively communicate with the jurors, judges, and lawyers at a reasonable cost in terms of time, money, and effort. Remember that rather complex issues are resolved every day by judges determining the applicable law and by jurors deciding the factual issues. If the system did not work, it would have been modified or replaced a long time ago. What this means is that it is up to the expert to translate complex issues, in his specialty, to a form that can be understood by the average layperson (juror) who can rise to the occasion. If they can decide narrow issues in neurosurgery, nuclear engineering, chemical processing, patent infringement, cost accounting, and pharmacology, it suggests that they are able to learn, understand, and decide in a relatively short time, under judicial guidance, if there is effective communication of the complexities of a specialty. Look up, do not look down on the jurors or others in the legal system. Establish an equality-based rapport with all those in the legal system who are functioning under time limitations, cost restraints, and often high stress levels. They are exercising considerable personal responsibility, so should the expert as a member of the litigation team who still exercises the independent discretion of a professional.

An example of complexity is the human input (control) as it effects vehicle dynamics. The input may be accomplished by attempting to achieve a desired vehicle direction by steering wheel movements (rate of movement and excursion angle limits), by depressing the throttle (force, position, and resulting vehicle acceleration or deceleration), and braking (when, how much, and as effected by weather and road conditions). Is the human input modified by perceptions that the lateral stability of the vehicle is approaching its limits (requiring a precautionary input) or has it exceeded its limits (requiring corrective action for an out-of-control vehicle)? Has there been a panic reaction by the driver with excessive, untimely, insufficient, or inappropriate steering maneuvers? Can the driver be expected to exert a timely and effective human input, given the handling characteristics of the vehicle? Steer angle changes may be monitored to prevent overshoot, excessive lateral acceleration, and dampen system oscillations in steady-state cornering maneuvers. The human input may be modified with active steering (transient steering torque under automatic control) where the steering ratio varies with vehicle speed (Triggs, 1988). The suspension may modify steering response if there is air suspension or variable torque anti-sway bars. There may be roll stabilization, lateral body movement, or other active chassis, suspension, or steering devices. The human input may be overridden, for a brief period of time, when electronic stabilization systems control quick impulses to various brakes, effect throttle position or power dynamics, and have active steering to restore an out-of-control vehicle to a straight ahead position. A yaw velocity and roll velocity sensor may institute proactive inputs to an electronic control system that completely overrides human input where the driver's reaction time is inadequate to meet the needs. There are numerous peer accepted technical terms that can be used. There are numerous technical devices that can assist in maintaining vehicle stability or act as automatic driver support systems. They serve to recognize problems inherent in those human characteristics important to safe vehicle handling. In short, describing the interactions between driver, the vehicle, and the roadway can be very complex, confusing, and difficult. They may be the subject of both the human factors specialist and the vehicle dynamics engineer acting in a cooperative fashion.

Complexity is often the result of the use of precise terminology that is appropriate where brevity of communication between peers is desirable or for those in research where exact replication is important. However, it may be just bureaucratic clutter or dress up that is unnecessary. The question is always what purpose does the complexity serve and how can it be truthfully simplified for a lay audience such as a jury? The university professor is often seen as a person who converts the complex into something that can be understood and retained by a select group of motivated students. But, the background of those students is homogenous and elevated compared to jurors.

Lawyers in their pre-trial preparation gather, analyze, and determine the implications of a considerable body of evidence. As they study the evidence they attempt to narrow issues, condense the key facts, select among the witnesses, and emphasize certain evidence. It is a process of gradual simplification, not unknown among other professional disciplines.

A treating physician may be obligated to inform and to explain to the patient something about the diagnosis of a disease, the treatment options, the prognosis, and the various risks. The physician must simplify the complexities, tailor the discussion to the needs and inherent level of understanding of the patients, and secure actual informed consent where necessary. A reasonably direct and honest approach requires simplification and truth for effective communication to the recipient, if there is to be mutual trust created or affirmed. The proponent of any discipline must engender personal trust if there is to be reliance on the analysis and opinion in a complex subject area.

Jury instructions are of particular importance to an expert witness, since they are the operative guidelines for the trier-of-fact (usually a jury). These are the landmarks around which all of the testimony is oriented (i.e., there must be relevancy as to the contested issues of fact in the case before the court). There have been attempts to simplify the complex legal language of some jury instructions into "plain English" instructions. The advocacy aspects of a trial, such as final argument, permits the lawyer to use plain English to explain the meaning of the admitted evidence.

Thus, the simplification of complexities is a continual ongoing process for all those involved in the litigation process. It may not appear to be simplification with numerous and lengthy depositions, many motions and declarations, and endless discovery in the form of interrogatories and requests for production or admission that may or may not be seen by the expert witness, consultant, coordinator, or remote employee. But, after any search for possible evidence, the simplification process must take place for all those involved. The court may impose strict limits as to time, both for preparation and in-court testimony.

3.4 Court Appearances

3.4.1 Full and Timely Disclosure

A potential expert witness should complete his analysis as early as possible, subject to modification as additional facts become known. The expert should ascertain, from the lawyer, all relevant deadlines for his work. The Court may require an expert to fully disclose his opinions to opposing counsel at or before his deposition is taken. A written report may or may not be required that states all opinions and the basis for those opinions. An early oral report to the retaining attorney may be helpful for him in preparing or responding to interrogatories, the production of documents, and requests for admission. It is not wise to hold back information on key issues for use at trial or to somehow attempt to supplement or revise statements made at depositions or in reports. Thus, full and timely disclosure of all opinions should be made before the deadlines and the start of the trial.

3.4.2 Organized Files

All documents should be carefully organized so that deposition or trial questions can be quickly answered. Three ring binders may be used for reference material that supports or justifies expert opinions. If a publication is from a peer reviewed journal, there may be a form of presumption that it would not have been published if it did not meet high standards, was not peer acceptable, or was not trustworthy. Voluminous records may be impressive, but only if they are easily accessed. Causation is often the prime reason for supporting documentation. The analysis should be focused and efficient, but the expert should understand other aspects of the case to assure compatability and the respect of the jury. The testimony that follows should be direct and to the point. Communication may be enhanced by unconditional opinions, demonstrative evidence, and quickly available references.

3.4.3 Criticism

Some lawyers like to encourage one expert to criticize another for some perceived mistakes, some violations of rules, some possible misapplications or misinterpretations, or for omissions or failures to meet some standard of conduct. For example, the questions might start as follows "what would you have done under the same circumstances" (a hypothetical question) or "what should be done to achieve a reliable and fair opinion or conclusion" (a general proposition or standard of care). The lawyer might believe that his case is strengthened by discrediting the opponent's expert. The expert may believe that his profession could be harmed by accusations of sloppy work or deceptive tests. The experts may differ only because each formulates an opinion on different set of facts, each gives different weight to some key facts, or one lets advocacy influence his perception. The cardinal rule is for the experts to stay within and conform to the code of ethics of their profession. Civility should be paramount, despite the actions of others. There may be ethical rules that encourage a challenge to improper, invalid, or immoral testing, an insufficient or inappropriate basis for conclusions and opinions, and illegal or unethical practices.

3.4.4 Records Requested from Expert Witnesses

At the time of deposition notice, the retained or designated expert may be requested to bring certain specified documents to the deposition. Some requests are minimal, some reasonable, and some are unconscionably burdensome. The retaining lawyer may be able to provide advice as to how to respond to the request, may want to review the records in advance of the deposition, or may be able to provide some documents for production at deposition. It would simplify matters if, at an early stage, prospective experts would gradually add to a computer disk a list of personal publications, academic and employment history, and cases on which testimony has been given. A similar disk could be developed for particular subject areas; for example, a list of all warning references, reaction time references, or visibility references that might be relied upon, depending on the questions asked in various cases. A few updated disks could convert the burdensome to the minor task of making a copy of the appropriate disk. The following example of a rather comprehensive, perhaps overbroad, list of records suggests that early prior planning is advisable.

1. A current curriculum vitae (i.e., a detailed *biographical* statement).
2. Each *report* prepared by you with respect to this case, including draft copies submitted to others for review or comments.

3. All *photographs* in your possession pertaining to the accident, the accident scene, or the specified equipment and its component parts, including any associated products involved in the accident.

4. The entire contents of your file with respect to the subject lawsuit, including all *documents* received from counsel or any of their representatives and any documents, which you have compiled independent of counsel.

5. True and correct copies of any and all *product analyses* and derivative charts, diagrams, reports, computer disks, computer programs, and journal articles in your possession upon which you relied or will rely in forming your opinions.

6. True and correct copies of all videotapes, audio recordings, computer disks, and photographs of any *testing* on *the product* involved in this case.

7. All *correspondences*, which were prepared, signed, sent, received, drafted, or delivered by you to any other person, which pertains to or refers to your involvement in this litigation matter.

8. Any and all reports, memoranda, graphs, drawings, work papers, calculations, images, photographs, moving pictures, video tapes, computer disks, and correspondences in your possession concerning the *testing* done on any *similar* or identical products, which provide information that supports any of your opinions.

9. Any and all reports, memoranda, graphs, drawings, work papers, calculations, images, photographs, moving pictures, videotapes, computer disks, and correspondences in your possession concerning *testing done* on any product *other than* the product in this case that provides information regarding any of your opinions.

10. A listing of any *other lawsuits* in which you have testified as an expert, either at trial or by deposition, within the preceding 5 yr.

11. *All references*, articles, publications, presentations, books, book chapters, lecture materials, and other documents relating to any publication authored, program attended, or any presentation in which you (the deponent) participated, which in any way relates or refers to the subject matter and opinions that you (the deponent) may offer, or the area of your expertise as an expert in this lawsuit.

12. All *billings*, fee agreements, time records, financial statements, contracts involving fees and costs, and all correspondence and other documents relating to your retention (the deponent) and those, which show time and charges incurred by you (the deponent) in connection with your activities in this lawsuit.

13. All *medical records* evaluations, neuropsychological tests, consult reports, raw data, x-rays, CT scans, MRIs, EEGs, SPEC scans, electrodiagnostic findings, and all other files, documents, and reports relied upon or used by you (the deponent) in connection with your activities in this lawsuit, arbitration, or mediation.

14. A list of all cases or projects in which you participated as an *advisor*, consultant, or employee that related to the design and development of this product, function, service, or system.

15. The names, occupational designations, and addresses of each and every person from whom *information* was obtained that could be utilized by you in this case.

16. A list of all meetings, conferences, or discussions *with other experts* retained in this matter, and pertaining to this case, including dates, locations, and the names of those involved.

17. A list of written or oral *statements* of all witnesses that pertain in any way to this litigation seen or reviewed by you, including a copy of each statement and any notes or reviews made by you of the statements.

18. A list of all witnesses *interviewed*, questioned, heard, or observed that pertain to the accident, the scene of the accident, or the products, objects, or materials involved in the accident or injury-causing event.

19. All maps, diagrams, sketches, measurements, and material analyses, related to the injury-causing event that were made by you or directed by you, which may serve as foundation or demonstrative evidence supporting your opinions in this case.

3.4.5 Fees

The fees to be charged for professional services should be reasonable under the circumstances. The fees may vary widely for consultation, analysis, evaluation, travel, deposition time, trial appearances, preparation of demonstrative evidence, the assistance of associates or other professionals, and administrative support services. There should be a printed and up-to-date fee schedule that is applicable to all kinds of work. It should not differ for various projects or parties. It is important that the expectations of both the retaining party and the expert or consultant be clearly known. The expectations include what is to be achieved within the agreed time and cost estimates. There should be approval of any extraordinary expenditures such as those for special research, laboratory testing, and special exhibits. Equality of performance on different projects or cases suggest the need for budgeting sufficient calendar time and avoiding schedule conflicts, since a crowded or disorganized schedule could disrupt the schedules of other participants. The knowledge of the project or case schedule is important to avoid last minute efforts and shortcomings in preparation; for example, "I did not know of that or thought of it" in the middle of court testimony. In essence, the timely and cost-effective performance should meet or exceed the comparable accomplishments of peers within the specialty.

The fees should not be excessive. High fees do not suggest that there will be a high level of performance. Experts are evaluated on past performance and reputation. The past is prologue. Some of the very best and well-known consultants and experts charge rather moderate fees, divide the costs of travel among several projects, and share test costs that can be applied to multiple projects. It is the total final cost that is important, not the hourly rate. Excessive fees may suggest "purchased testimony" and possible ethical problems.

It is generally assumed that the expert is already familiar and prepared, in a general sense, with the content of the specialty that is applicable to the project. But, time should be allocated to learn recent developments in the field, to refine what may be said about possible conflicts on key issues, and to prepare a list of supporting publications. In other words, an informed estimate of the overall cost should be made. This is to avoid insufficient preparation that could result in court testimony that is mistaken or inadequate, since a trial error can be costly and irremediable. Poor advice given during an urgent effort to correct a liability problem, within a company, could have serious consequences in terms of monetary costs and human lives. In essence, there may be a fleeting window of opportunity, in a competitive enterprise, for informed relevant information, rather than "old hat" opinions. Some specialists do choose the path of least effort with conceptions of consultant-only project aspirations, but most projects related to litigation are an intellectual challenge requiring "best efforts." Never underestimate what is known about your specialty by those in other disciplines, so provide something exceptional for the purchasers of your services.

Retainers also should be reasonable under the circumstances. Before charging any fees, there should be full disclosure of any possible conflicts-of-interest and of any potential problems that could render the professional services impaired, useless, void, or excludable. If problems arise during the job performance, they should be disclosed early enough to permit repair or replacement of the expert. Do not wait until after there is a formal disclosure of experts in litigation, since after naming the expert witnesses they may not be replaceable and a big hole could be left in the trial presentation. It is advantageous to have everything that relates to fees and your professional activities in writing, in anticipation of fee disputes, but such writings should be reviewed by legal counsel for meaning, effect, legality, and possible interpretation by other parties. Care is required where money motivates action by others.

3.4.6 Personal Opinion Testimony

Some state court judges distinguish between science-based testimony and personal opinion testimony. It is the scientific type testimony, with its deductive reasoning (from general to specifics) that has resulted in the erection of judicial limitations on that type of opinion testimony (Frye, 1923; Daubert, 1993). The attempt is to limit where there is new, novel, experimental, or exotic testimony that might be misleading to a jury (Davis, 2001). Limitations exist where testimony is based on scientific principles, formulas, discoveries, or procedures developed by others (Rickgauer, 2001).

There is another form of testimony, based on inductive reasoning (from particulars to the whole) (Holy Cross, 2001). This includes pure opinion based on the expert's own training, experiences, observations, and research (Ronnie Jones, 2003). For example, tire expert opinion may not be scientific testimony (Kumho, 1999).

The logic involved may be a derivative of both hearsay and speculation objections. There is more credibility when opinions are based on direct personal knowledge. There is less credibility when expert opinions may involve some speculation in just applying someone else's results, beliefs, procedures, and conclusions. A juror might not be able to distinguish between the weight that should be given to direct knowledge as opposed to indirect knowledge, so the judge acts to balance the scales and assure that only competent testimony reaches the jury.

3.4.7 Proffered Testimony

3.4.7.1 The Most Common Scenario

The vast majority of litigated cases involve expert testimony in which the expert witness's basic qualifications are quickly established and the expert testimony in court is generally unimpeded (with few objections and restrictions). In general, the testimony must be relevant (a tendency to prove or disprove, in some way, the veracity of one or more of the basic issues contested by one or more of the parties). It should be in a form acceptable to the court. Experts are generally held in high regard, depending on the scope and depth of their qualifications, their publications, their occupational history, and any honors they have received that have some connection with the issues of the case before the court. They may be given special privileges such as utilization of otherwise hearsay or objectionable evidence if that is the practice of their profession. They may be permitted to draw conclusions, whereas the lay witness may be restricted to what they personally observed, heard, felt, or did.

3.4.7.2 The Defined Purposes of the Testimony

In complex or vigorously contested lawsuits, each segment of the expert's testimony may be related to a defined objective. There may be pretrial or in-trial hearings, before the trial judge, to determine the expert's qualifications and the admissibility of the proffered testimony. Such hearings may go far beyond proof of general fault or its absence. It may focus on whether the proof relates to claims of a product *defect* or an unsafe *condition*, inferences of knowledge or *notice* of a dangerous situation, the *causes* of an accident, a failure to *remedy* or *warn*, evidence of mandated or secret *recalls*, records of *prior accidents* that are substantially similar and not too remote, historical statistics on reasonable human *behavior* under the circumstances, or the absence of prudent or *due care*. The judge may admit evidence on one objective and deny it on other objectives.

For example, is there sound evidence, based on adequate foundation, concerning the specific (in this case) tire failure? Does the proffered evidence relate directly to the tire *defect* (an unexpected steel belt-from-belt peel or tire failure)? Does it relate to *causation* (wedge cracking and circumferential belt edge failure due to the fatigue reversion of the skim stock holding the belts together)? Does it relate to predictable human *driver reactions* to unexpected tire failure? Was a recall, customer *notification* program, or safety improvement program campaign necessary, timely, or sufficient? The trial judge may go further on specific issues, to determine whether or not an opinion is justified or consists of mere speculation. As a general rule, if the testimony is admissible, it is credibility that effects the weight given to the testimony by the trier of fact.

3.4.7.3 Conflict in Governing Law

There may be a conflict as to which case precedents govern the admissibility of expert testimony. In California, in 2004, there were two different appellate decisions (Roberti, 2003; Jennings, 2003). One case stated that an expert witness "does not possess a carte blanche to express any opinion within the area of expertise." An opinion has no evidentiary value without a "reasoned explanation connecting the factual predicates to the ultimate conclusion." If the testimony is that it "could have been a cause-in-fact" that is insufficient, because it is a mere possibility, and then it becomes "the duty of the court to direct a verdict." In essence, the court held that the expert opinion must explain why the facts convinced the expert and, therefore, should convince a jury. A conclusion is not an explanation. If not more likely (probable) than not, it is speculation that must be excluded.

In the other case, the expert testimony was challenged on the basis of the "possibility of causation" opinions and that they were "unsupported by peer reviewed scientific and medical literature." The appellate court held that this was a matter of the "credibility and weight of the expert testimony," and that "jurors may temper the acceptance of his testimony with a healthy skepticism born of their knowledge that all human beings are fallible." In essence, it was not a question of admissibility (by the judge) but of the weight to be given (by the jury) to the "underlying bases" for the opinions. The opinions were based on generally accepted methods, tests, interview techniques, and procedures. In other words, the first case emphasized a strict role for the judge concerning the admissibility of evidence (by excluding expert opinions). Whereas the second case de-emphasized admissibility and instead relied upon the role of the jury in determining the weight to be given to expert testimony.

3.4.7.4   Cross Examination

Opposing counsel may have very little cross examination if they believe their side could be hurt by an expert merely reinforcing his opinions, using the questioning and explanations in a soap-box fashion, or introducing evidence when the door is opened by opposing counsel. In contrast, opposing counsel may go on the attack by calling the testimony simply that of a hired gun or highly paid testifier for only one side who wants repeat business. Other less aggressive counsel may focus on what seems to be flaws in the logic or reasoning, some apparent bias or mistakes, or the neglect of what he believes are the real issues in the case. They may infer that all facts were not considered, that certain records or photographs could change the opinions, or that the seemingly authoritative tests or specialists should not have been relied upon in formulating an opinion. The expert should "stay the course" where reasonable and correct or the cross examination will feed upon itself and be prolonged.

3.4.7.5   On-the-Stand Behavior

The verbal behavior of a witness (the spoken words), during trial testimony, will be observed and interpreted by the jurors in an unfavorable, favorable, or neutral fashion. Does the witness, on the stand, argue with the lawyers, provide evasive answers, refuse to provide short answers, or engage in boring responses that are not very informative? Is there respect for the jury and the court?

Nonverbal behavior (nonspoken communication) is equally important. Does the witness assume a posture that seems arrogant, aggressive, or indifferent? Does he lean back in a comfortable position or lean forward in an attentive position? Does he act in a condescending manner, a relaxed manner, or a friendly manner? What impression is created? Does it relate to the credibility or believability of the testimony being provided?

The expert witness may observe the reactions and behavior of the jurors as the testimony is given. The juror's nonverbal behavior may consist of signs of attentive listening or inattention, writing down the important details or complexities in the testimony, some jurors may nod in agreement or shrink back in disagreement, they may fold their arms in front of them, gaze away in disinterest, or dress in a fashion that suggests socioeconomic status. These nonverbal cues are often given far too much weight or interpretation, but the expert witness should observe whether rapport and receptivity have been established.

The trial judge is often well experienced in providing nonverbal signals to the jury, particularly if there is an attempt to avoid the court reporter's written record of what has been said to the jury. The judge may make facial expressions of acceptance, disbelief, hostility, or agreement with what the expert says. The judge may be attempting to control, steer, or advise the jurors by direct or indirect signals. The expert should attempt to discern whether the judge understands the technical concepts, since the jurors may be in the same position in terms of effective communication.

The trial is often a diverse mixture of law and facts, lawyers and judges, jurors and witnesses, cues and instructions, and the colorization or expression of different life experiences, cultures, and value systems. All of these elements interact to make prediction of a trial outcome difficult. It is the uncertainty that acts as a compromising force in arbitrations, mediations, and settlements before, during, and after trials. Unfortunately, the expert witness is privy to a rather small portion of the litigation process. The trial consultant may advise only on the selection of the jurors as the trial lawyer attempts to reduce the uncertainties. The uncertainty inherent in the trial process should be understood by the expert witness, since the expert should be capable of adaptation to variations in format, interpretations of content, and unexpected or novel issues.

3.5 Bias

Each occupational "discipline" has its own shared beliefs or value systems. They are intended to guide or influence personal decisions and judgments. This is a form of a known bias. It can easily serve as a factual filter and may provide an extraneous perceptual flavoring about a fact situation. People can see what they want to see and perceive what is consistent with their expectations and beliefs. This can result in misunderstanding or difficulty when what is appropriate in one discipline (such as ergonomics) is applied, utilized, or translated for use in another discipline (such as the law). The translator (between disciplines) needs to know something about both worlds or errors of communication, understanding, and comprehension should be created. For example, the research scientist may believe that absolute certainty does not exist for any conclusion and further research is always necessary, but the law may define *scientific certainty* as something that might be only above the 50% level of present belief.

The term *peer approval* may elicit statements that there is and always should be constructive criticisms by fellow scientists, appropriate questions as to better methodology, and productive doubts by fellow scientists. Scientists may believe that attaining full peer approval is just an illusionary concept. This questioning attitude, by peer group scientists, may be particularly present for research *applications* (to the "real world") or for *generalizing* research findings (going beyond the original research context). However, despite rejection of the concepts of full peer approval or the need to attempt to attain such a approval from other scientists, some proof of peer approval may be a minimum requirement for permitting an opinion to be expressed in a court of law. No peer approval may mean no helpful testimony for the trier-of-fact. The translator (between disciplines) provides meaning to words and concepts by using the message recipient's own language. Obviously, there should be effective, truthful, and meaningful communication by those persons who testify or offer scientific opinions in a societal context. If there is bias, it should be disclosed.

It is the trial court judge who has "the power" to make decisions regarding whether or not the expert witness's testimony will be presented to the jury. If the testimony is admissible, the judge may then decide on what issues, in what depth or scope, and what limiting jury instructions or judge's cautionary comments may be given to the jurors. The discretion of the trial judge is considerable. How the judges control the courtroom and testimony varies widely within a particular jurisdiction or courthouse. They vary widely in interpreting the law and applying it to the facts of a particular case. This is because of biases created by strong personal beliefs, basic value systems, their strengths and limits of knowledge in specialized areas, the techniques they employ to control the courtroom, the kind of evidence and advocacy being presented, the possible outcome of the case and its consequences, and possible local reelections or judicial advancement concerns. Therefore, the expert witness should not be offended, displeased, shocked, chagrined, or elated at the admissibility of the testimony or the judge's attitude toward his or her testimony or that of any other expert witness in the case. Perceptions of bias in the courtroom should not encourage insertion of responsive bias or altered demeanor. The expert witness should be prepared to comply and to adjust to the trial judge's rulings. It is the trial judge who controls the legal process by pretrial, in-trial, and posttrial rulings. The expert's role is not to respond to perceived bias, but to remain a calm, independent, neutral person who can assist the trier-of-fact (judge or jury), while maintaining personal integrity, intellectual rigor, and fairness. Bias may be everywhere, but it should not prevail.

References

Campbell v Metro. Prop. & Gas Ins. Co., 239 F.3d 179, 184 (2d Cir., 2001). Expert qualifications.

Clausen v M/V New Carissa, 339 F.3d 1049 (9th Cir., 2003).

Daubert v Merrell Dow Pharms. Inc., 509 U.S. 579 (1993); *Gen. Elec. Co. v Joiner*, 522 U.S. 136 (1997); and *Kumho Tire Co., Ltd. v Carmichael*, 526 U.S. 137 (1999). The Daubert doctrine is codified in Federal Rule of Evidence 702. It states: "If scientific, technical, or other specialized knowledge will assist the trier of fact to understand the evidence or to determine a fact in issue, a witness qualified as an expert by knowledge, skill, experience, training, or education, may testify thereto in the form of an opinion or otherwise, if (1) the testimony is based upon sufficient facts or data, (2) the testimony is the product of reliable principles and methods, and (3) the witness has applied the principles and methods reliably to the facts of the case."

Davis v Caterpillar, Inc., 787 So. 2d 894 (Fla. 3d DCA, 2001).

Frye v United States, 293 F.1013 (D.C. Cir, 1923).

Goodstein, David, How science works, in *Reference Manual on Scientific Evidence*, 2nd ed., 2000, 70.

Holy Cross Hops., Inc v Marrone, 816 So. 2d 1113, 1117 (Fla. 4th DCA, 2001).

Jennings v Palomar Pomerado Health Systems, Inc., 114 Cal. App. 4th 1108 (Dec. 2003), 4th Dist., Review denied.

Kepner, C.H. and Tregoe, B.B., The rational manager: a systematic approach to problem solving and decision making. New York: McGraw-Hill, 1965.

Kumho Tire Co. v Carmichael, 526 U.S. 137, 141, 142 (1999).

Price, Dennis L., Effects of alcohol and drugs, in Peters, George A. and Peters, Barbara J., Eds., *Automotive Engineering and Litigation*, Vol. 2, New York and London: Garland Law Publishing, 1988, pp. 489–334.

Rickgauer v Sarkar, 804 So. 2d 502, 504 (Fla. 5th DCA, 2001).

Roberti v Andy's Termite & Pest Control Inc., 113 Cal. App. 4th 893 (Nov. 2003), 2nd Dist., Review denied 2-18-04.

Ronnie Jones and Sylvia Jones v Goodyear Tire & Rubber Co., DCA, 3d Dist., Fla., July 2003 (Case 3 DO1-3583).

Triggs, Thomas J., Speed estimation, in Peters, George A. and Peters, Barbara J., Eds., *Automotive Engineering and Litigation*, Vol. 2, New York and London: Garland Law Publishing, 1988, pp. 569–598.

Vinal, Robert W., Criminal liability of contractors, engineers, and building owners regarding asbestos projects, in Peters, George A. and Peters, Barbara J., Eds., *Sourcebook on Asbestos Diseases: Medical, Legal, and Engineering Aspects*, Vol. 5, Asbestos Abatement. Salem, NH: Butterworth Legal Publishers, 1991, pp. 45–62.

Wang, C. Julius., Product improvement from integrated quantitative techniques, in Peters, George A. and Peters, Barbara J., Eds., *Automotive Engineering and Litigation*, Vol. 5, New York: Wiley Law Publications, 1993, pp. 121–148.

Recommended Further Reading

Peters, George A. and Peters, Barbara J., *Automotive Vehicle Safety*, London and New York: Taylor & Francis, 2002. *Note*: This book illustrates many of the *technical considerations* involved in a lawsuit that may be present, in some form, in a diverse range of products, processes, systems, and industrial equipments. Includes accident reconstruction, human error control, and design safety.

Peters, George A. and Peters, Barbara J., *Warnings, Instructions, and Technical Communication*, Tucson, AZ: Lawyers & Judges Publishing Co., 1999. *Note*: This book illustrates the factors to be considered when *warnings* and instructions become a part of a lawsuit.

Peters, George A. and Peters, Barbara J., *Handling Soft Tissue Injury Cases: Legal Aspects*, 2nd ed., Vol. 1, Charlottesville, VA: Lexis Law Publishing, 1999. *Note*: This book illustrates many of the *legal considerations* involved in lawsuits, from the viewpoint of the lawyers. Includes the general procedures utilized by lawyers in the preparation of their cases.

Kazan, Steven and Moscowitz, Ellyn, The role of the plaintiff's attorney in asbestos litigation, in Peters, George A. and Peters, Barbara J., Eds., *Sourcebook on Asbestos Diseases: Medical, Legal, and Engineering Aspects*, Vol. 5, Asbestos Abatement. Salem, NH: Butterworth Legal Publishers, 1991, pp. 1–24. *Note*: This chapter and this book illustrate the *legal history* aspects of lawsuits or how the information accumulates on a particular problem area.

Watson, Donald, Ed., *Architectural Design Data*, 7th ed., New York: McGraw-Hill, 1997. *Note*: This book contains a range of useful topics, such as lighting, stair design, elevators, door and windows, construction, fire safety, units of measurement, and human figure dimensions. *Note*: Illustrative of books that provide background information for human factors analysis.

Peters, George A., Product liability and safety, in Kreith, Frank, Ed., *The CRC Handbook of Mechanical Engineering*, Boca Raton, FL and London: CRC Press, 1998, pp. 20.11–20.15. *Note*: This book provides valuable background and reference data with discussions of subjects ranging from engineering design and mathematics to project management and mechanical systems. It may be considered a supplement to this chapter in this Handbook.

National Research Council and the Institute of Medicine, *Musculoskeletal Dirsorders and the Workplace*, Washington, D.C.: National Academy Press, 2001. *Note*: Illustrates the kind of peer group authored book given great weight in lawsuits. Deals with a common human factors problem.

4

Cost Justification for Implementing Ergonomics Intervention

Maurice Oxenburgh
Pepe Marlow
Ergonomics Consultant

4.1 Introduction

4.1.1 Productivity and Profit for an Enterprise Is Compatible with Good Working Conditions

Ergonomists and finance managers bring to health, safety and production problems different assumptions about what is acceptable. For instance, from a health and safety point of view it is not considered acceptable to have dusty, dirty and unsafe working conditions. However, from the point of view of an economic rationalist dusty, dirty and unsafe working conditions can be totally acceptable on the basis that (a) someone has to do the work and (b) the worker has agreed to work under these conditions for the wages paid. Accounting systems that underlie enterprise financial structures are based on this economic premise.

In this article we will consider the "economics of ergonomics" and show that good working conditions are compatible with profit. We will discuss:

- The economic reasoning behind cost-benefit analysis in enterprises
- A cost-benefit model that ergonomists and others may use to support their economic arguments for good working conditions
- Case studies to illustrate that "good ergonomics is good economics"

4.1.2   Why Should Economics be Part of Ergonomics?

In these times where expenditure is considered in terms of its impact on enterprise profits, the so-called "bottom-line," it is not enough for ergonomists simply to imply that they told management about the safety problem but management chose to ignore their advice. Ergonomists need to use the language of accountants and finance managers to argue for the ergonomics interventions required. Those concerned with health and safety have a responsibility and need to use all reasonable means to get their ideas accepted and implemented.

There have been many literature references discussing the finances (benefits as well as costs) of workplace interventions and, to support our argument, some are mentioned below.

Dr. Arne Aarås has done some remarkable work in controlling and reducing musculoskeletal illnesses in a telephone wiring company and in his paper (Spilling et al., 1986), he has also discussed the cost-benefit of this intervention. Comparing a 7-yr period prior to the ergonomics intervention with a 7-yr post intervention period he showed that there was a saving of about nine times the outlay for ergonomics improvements, measuring only injury and labor turnover reduction. The length of the study and its results, both in injury prevention and cost effectiveness, has been a highlight in ergonomics.

It is commonplace nowadays that, before governments introduce safety legislation, they make a cost-benefit prediction. This is often in response to industry saying that it cannot afford whatever is proposed — almost a knee-jerk reaction to any thought of expenditure. In an early paper on the economics of the introduction of regulation or code on safe manual handling (Oxenburgh and Guldberg, 1993) unsafe lifting was determined in terms of the code and thus what the code would do in preventing injury. A statistically determined cross section of workplaces was surveyed by a team of investigators who measured the number of unsafe lifting practices (i.e., lifts addressed and potentially "corrected" by the code). The cost to correct each unsafe lift was determined and the cost expanded to an estimate of the entire industry and related to the estimated savings in back injuries. Although it was shown that the industry-wide costs outweighed the financial benefits the difference was considerably less than that estimated by industry. No estimate was made of the reduction in public expenditure due to fewer injuries. An interesting side issue in this study was that a considerable portion of industry was already reducing injury potential for financial reasons, independently of whether the code was introduced or not.

Professor Guy Ahonen surveyed 340 small- and medium-sized companies in Finland in order to develop their occupational health and safety practices. He compared the individual companies with the best in its industry group based on sickness absence; a "best practice" methodology. Ahonen was then able to estimate the financial benefit to an enterprise of reducing the absence rate both in actual savings and in productivity increase (Ahonen, 1998).

Connon et al. (2003) have taken a slightly different approach to Ahonen to measure the value of good working conditions and the company finances. Through site visits, they were able to show a positive relationship between the organizational health aspects and low workers' compensation claims (safe workplaces) in the furniture manufacturing industry.

On a broad front, there is literature regarding the financial results from good working conditions but there has been very little development of generic models available for the ergonomist to use in any given situation.

Dr. Paula Liukkonen of Stockholm University was instrumental in helping Swedish industry examine the costs of poor working conditions. We believe that she was the first to codify injury and low productivity costs in a generic form, an arrangement that enterprises could use to improve working conditions. Most of her work was published in Swedish but she has summarized some of her ideas in English (Kupi et al., 1993).

A Special Issue of Applied Ergonomics (Vol. 34, 5, September 2003) was devoted to cost effectiveness. Unfortunately, none of the articles in this Special Issue proposed a generic method that could be used by ergonomists and others wishing to implement ergonomics solutions. Each author took their case study and showed its financial effectiveness, but the methodology was unique to that case and could not easily be used in other workplaces.

It was a similar situation at the European Conference on cost effectiveness (Mossink and Licher, 1997) in that very few papers indicated useful working models that could be used by ergonomists.

It seems unfortunate that more models are not developed for occupational health and safety on the market so that users would be able to choose the model which best fits their needs. Cost-benefit models that are *not* suitable are, for example, engineering ones that only derive from the technology of the design or process and do not express the affect on, and the effect of, the workers and other persons concerned. It is, at least partly, to fill this gap that the Productivity Assessment Tool (and its predecessor, the Productivity Model) has been developed and will be discussed later in this article.

4.1.3 Using Financial Arguments

By not using financial arguments, ergonomists and other occupational health and safety practitioners are not presenting a balanced view when it comes to arguing for funds to be expended; to just present the cost of a project does not reflect its value to the enterprise. Cost-benefit analysis is one tool that can be used to rectify this absence of balance.

Cost benefit analysis is in its infancy for ergonomists but engineers have long made use of the tool to look at, for example, the value of replacement versus maintenance of existing plant and buildings. This analysis may be given names such as "return on investment," but essentially it is an attempt to look at both costs and benefits of a projected change, hence a cost-benefit analysis. Over time engineers and others have gained more confidence and have developed a way of putting a monetary value on what is, essentially, a prediction about future performance.

It is our hope that using the concepts of cost-benefit analysis will give ergonomists and other occupational health and safety practitioners a similar level of confidence to that of engineers, accountants, managers, etc., with whom they need to argue for funds.

By these means ergonomists will not just present the costs for implementing ergonomics solutions but also quantify the benefits. The costs derive from the implementation of better working conditions while the benefits come from a safer workplace with lower injury costs, higher productivity, reduced warranty costs, improved staff retention, multi-skilling, etc.

4.2 Financial Modeling

Ergonomists commonly assert that their projects will benefit an enterprise citing these benefits qualitatively while presenting the costs of the project quantitatively. When it comes to arguing for funds to be expended, to just present the cost of a project does not reflect its value. If several projects are competing for funding then, in financial terms, the project that provides the greatest value to the enterprise should be funded first. Financial modeling can be used to calculate the value of a future project, incorporating both its costs and benefits quantitatively. This requires access to the enterprise's costs data, selection of those items that the project will affect and estimation of the size of the effect.

In the engineering field data are selected, costed and benefits estimated on the basis of custom and practice. What can ergonomists learn from the engineers' approach?

Ergonomists are often cautious about making predictions on the benefits of a project; epidemiology tells us that we cannot know, for instance, that there will be injuries among a particular team of workers, just the likelihood of injuries. How then is it possible to say that an intervention will prevent injury and thus include reduced injury or other costs in a cost-benefit analysis? When we make predictions about the future we are limited in the accuracy of our predictions. Enterprises must use historical data to make predictions about the future and hence can only estimate costs and benefits.

We suggest that ergonomists follow the lead of engineers in making use of cost-benefit analysis tools. Engineers have developed cost-benefit models that suit the questions they want to answer such as "should we buy that new machine or should we continue to maintain the old one?" Engineers are also comfortable with the need to make estimates and how to go about estimation on the basis of their experience. Think of cost-benefit analysis as just a financial tool to assist in asserting one's point of view about the need for, and worth of, one's ergonomics interventions.

Cost-benefit analysis is actually an economic model and to use it effectively in situations where one's proposal is competing for funds against other projects, it is necessary to understand at least some of the economics behind it.

4.2.1   How is Work Valued in Economics?

Ergonomists need a way not just to present the costs for implementing ergonomics solutions but also a way to quantify the benefits from this implementation. Many of the benefits from ergonomics interventions flow to workers, and so, how do we quantify this?

Accounting systems are based on financial concepts but are narrow in focus, usually constructed for the purposes of taxation, regulation and financial accounting. Simplistically within the accounting system all financial transactions are recorded to be either as assets (money-in), or expenses (money-out). Traditional accounting practice considers workers merely as an expense and hence safety measures are just another expense, while ergonomists and human resource professionals consider workers and their safety and well-being to be the paramount. In many cases this has led to disputation about the value of ergonomics interventions with enterprises arguing that they cannot afford the expense of the intervention.

The debate centers on the meaning of the term "asset". In an accounting sense an asset is something that the organization owns and has the right to sell. The theory is that the value of an enterprise (its asset base) is equal to the money that someone would pay for the components of the enterprise, which includes buildings, machinery and products waiting to be sold. As the enterprise has no right to sell its workers they are not considered assets. Hence any intervention that increased the value of the workers does not result in an increase in the asset base of the enterprise.

This accounting concept of an asset was developed early last century in manufacturing industries. The nature of work is changing with an increasing proportion of people working in service-based industries, where the product is intangible, or working from home or out of a car where there is little investment in buildings or machinery. The value of these enterprises is based on their people, which economics terms as "human capital," and accounting is grappling with ways of placing a value on these types of industries.

4.2.2   What Models Suit Ergonomists?

Ergonomics is person focused (Oborne et al., 1993) and hence cost-benefit analysis models that also focus on people are needed. As it is not possible to capture the value of an ergonomics intervention to workers directly within enterprise financial systems, ergonomists need to measure the benefits indirectly, that is, through changes to the enterprise costs or the value of physical assets.

For example, implementation of multi-skilling could render work more interesting, make workers feel valued and reduce work stress be it physical or mental. If a cost-benefit analysis was undertaken in support of funding for this project these benefits may be reflected in lower injury costs, higher productivity, reduced warranty costs, reduced waste, improved staff retention, etc. Section 4.4 provides other examples of ways enterprises have applied a cost-benefit model to ergonomics projects.

4.2.3   What Do We Need to Measure?

We need to measure any changes to the enterprise's production costs or the value of physical assets that will reflect the impact of the ergonomics intervention. Any factors that do not change as a result of the intervention will measure the same before and after, and so do not affect the cost-benefit calculation; hence they are not included in the calculation.

It is assumed within economic modeling that the enterprise will be in compliance with all pertinent regulations, including OHS regulations. If this is not the case then the *costs* and *benefits* of becoming compliant with regulations should also be included.

4.2.3.1   Production Costs

Production costs are the costs for an enterprise to produce a product or service ready to sell and they are typically grouped into three categories: labor, raw material costs and capital (equipment, machinery, computer systems, etc.). For each area of production costs consider if that factor will change in value as a result of the project and estimate the size of that change.

The costs that most directly reflect the benefits of projects that have an impact on people are labor costs. The costs of labor that may need to be included are:

- Wages costs and on-costs
- Overtime
- Turnover costs, which include recruitment, training and any extra supervision of workers as they are learning
- Training
- Injury costs (workers' compensation)
- Absenteeism costs
- Supervision

Raw material costs may be influenced through projects that impact on the quality of the finished product. If there is less waste during the production process, then raw materials may be reduced and this saving should be included in the cost-benefit analysis.

The capital costs of production (equipment, machinery, computer systems, etc.) are usually replaced only periodically. It is assumed in economics that enterprises will monitor the efficiency of their production processes and use the cheapest and most efficient means of production available to produce their products and services. However, often enterprises continue using old and less efficient machinery or systems, for a variety of reasons, and they may choose not to replace old equipment or systems. This may lead to increased production costs that are not accounted for. Ergonomists and human resource managers are then called upon to assist in making these older production systems safer and more productive, for example, through improved work practices or training.

A full cost-benefit analysis may show that updated equipment or technology is cheaper in the long run and safer for the workers. The cost-benefit analysis should compare the costs of the short-term solution, for example, training program, plus the costs of the inefficiencies in the system against the cost of the replacement machinery or system. The following inefficiency factors may need to be included in one's cost-benefit analysis:

- Efficiency of equipment reflected in breakdown costs
- Maintenance of equipment
- Warranty or product defect costs
- Delays in supply

4.2.3.2   Changes in Product or Service Sales

The argument that investing in workers adds to production costs and pushes up the price of products and services is often given as the reason that training and ergonomics interventions are not funded.

This follows from an economic philosophy that says increased production costs will push up the price and so less people will buy the product, which is not always true. For instance, motor vehicle manufacturers in Sweden during the last several decades found that improving the skills of their workforce and giving workers the responsibility for quality has resulted in improved quality of vehicles manufactured. This has not just saved warranty costs but there is more demand for these higher quality motor vehicles.

Estimating the impact of an ergonomics intervention on future product sales is very complex and necessarily imprecise. The best estimates of the impact of such projects come from past experience. Look for case studies of a similar intervention to the one proposed and, if necessary, adjust the estimated impact up or down, in one's cost-benefit analysis, to better match one's enterprise.

4.2.3.3 Legislation Compliance Costs

Cost-benefit analysis models assume that the enterprise will already be compliant with the law. If this is not the case both the cost of becoming compliant and the penalties for noncompliance should be included in the cost-benefit analysis. Consider the enterprise's compliance with the following standards:

- Safety
- Environment
- Employment practices

4.2.4 What Are the Limitations of Cost-Benefit Analysis?

The critical limitation of cost-benefit analysis is that social costs are not included. Micro-economic models consider only costs borne by the enterprise and are limited to consideration of the effects of manufacture or sale of a product or service.

The Productivity Assessment Tool, described in the following section, is a micro-economic model and does not take into account the effect of social and other outside costs. Although this is a limitation as the assumptions made within the model are relatively few it means that the model can be more easily and simply used.

Where social costs are significant to the enterprise (e.g., environmental pollution) and need to be included in one's analysis, one will need to use another tool, macro-economic models, for the analysis.

The following three areas are examples of social costs that may need to be considered.

4.2.4.1 Injury Costs to the Person

Where a worker is injured at work the direct cost to the enterprise is usually limited to the workers' compensation premium cost. This is the case even where a worker is permanently disabled as a result of a work injury.

The cost to the worker will include the social costs of that worker's pain and suffering and restrictions on activities outside of work, expenses not covered by compensation, reduced earnings, etc. These factors are not included in financial analysis.

For example, if a worker is permanently disabled, a decision may need to be made whether to re-train the worker for another role or terminate the worker's employment. Research has shown that workers with work-related disabilities have greatly reduced chances of finding employment on the open job market, a personal cost not considered by financial analysis. If, on the basis of the enterprise's costs (workers compensation premium), the best option is to terminate the worker's employment the case for retraining the worker might need to include an argument about social justice.

4.2.4.2 Responsibility to Provide Workers with a Living Wage

Under economic theory it is an enterprise's responsibility to minimize production costs but there is no mention of any reciprocal responsibility to provide workers with a living wage. For example, increasing service industries such as call centers and hotels use casual workers who can be rostered to work on relatively short notice and for short periods of time. This is not necessarily the best situation for the workers who are involved. The enterprise can capture the cost to itself by placing a value on the costs of training, recruitment and worker quality and morale but these will not include the social or personal disruptive effects on the concerned workers. An argument on the basis of social justice may be needed.

4.2.4.3 Occupational Diseases

Due to the long lag time between exposure and the development of occupational diseases such as asbestosis, silicosis and hearing loss, it can be difficult to prove that a particular employer was responsible and the cost of this illness is borne by the individual and the society. From the enterprise's perspective, if workers are at risk of occupational disease in 15 or 20 years time, the enterprise may not include an allowance for the costs of future treatment in the current costs of production. Compared with the potential short-term gains, the long-term risks may seem too far off to worry about. This has two effects:

- It is more difficult to argue in support of expenditure to prevent exposure to the hazardous substance within the workplace
- The costs of diseases that occur will be borne by the worker, their family and social services, that is, the whole community

These long-term projects need a different model and are the most difficult to cost as they need to consider the impact of factors outside the enterprise including changes in interest rates. For information about long-run projects see Oxenburgh et al. (2004).

4.2.5 What Are the Benefits of Enterprise-Based Cost-Benefit Analysis?

Used judiciously cost-benefit analysis can be a powerful tool to present one's argument for funding any ergonomics project. Cost-benefit analysis enables benefits of an ergonomics intervention for that enterprise to be quantified, presenting the value of the intervention to the decision makers of the enterprise. Cost-benefit analysis also provides a logical framework to assess the impact of an ergonomics intervention across all aspects of the enterprise.

4.3 Cost-Benefit Analysis

4.3.1 Cost-Benefit Analysis as a Checklist

Most ergonomists will be familiar with checklists to assist in determining work tasks, work organization or work stations that may lead to injury; a cost-benefit analysis model can act as a generic checklist and, in addition, identify economic parameters with which the ergonomist will be less familiar.

One essential feature of any checklist is the *relevant* questions it poses. In our experience, finding the data is not usually too difficult; asking the *relevant* questions is the crux of the matter. A cost-benefit analysis model must direct the user to the relevant questions although it allows the user to determine the appropriateness of the individual questions. A difference between a specific ergonomics and a cost-benefit analysis checklist is that the former is usually directed to the health and safety of the worker and the latter to the productivity of the worker. Actually they will be measuring similar parameters; a cost-benefit analysis does not replace the ergonomics checklist and subsequent intervention but is designed to support it through a different mechanism.

4.3.2 Cost-Benefit Analysis Assumptions

A cost-benefit analysis assumes that the present work situation is not optimal and that changes (an intervention) may be made to improve worker productivity and other cost factors, including injury and absence costs. That is, the cost-benefit analysis assesses a work place at a particular point in time and compares it with future or alternative situations, which are the test cases.

The data collected is centered around the people and not the product which, to some extent, is a point of departure from the organizational concepts originating from the work of, for example, Deming (1982). The major determination of this cost-benefit analysis is of the workers who produce the goods or services and not of the goods or services themselves. Although there are programs that take the starting point as the equipment and processes of the manufacturing or service systems, this cost-benefit analysis concentrates on the people's side. In the long run both approaches need to be integrated but, for the purposes of this article, we will concentrate on the people's side.

Certain assumptions need to be made for a cost-benefit analysis model and these include:

- The important and/or critical data relating to employment costs
- The costs for implementing the intervention in the work place
- The financial benefits due to the intervention

These assumptions are little different from the assumptions made when the warehouse manager wants a new fork lift truck, the call center manager wants new computer software or the hotel manager requires more cleaning staff. The initial costs may be known but the benefits are only assumptions. Will the fork lift truck reduce loading time and goods damage? Will the new software bring in more telephone customers? Will the extra staff increase the standard and the guest fees of the hotel? Determining the future can only be based on the best guess even if based on experience.

Once the assumptions are agreed upon then data can be collected. Some of the employment data required are straightforward and usually easy to obtain. These include the direct labor costs of hours worked, wages, social costs, training, absenteeism, etc. but may also include a portion of the organizational costs of supervision, management and head office costs.

Other data may be less easy to obtain but, in our experiences, give the greatest return from ergonomics investment. These include:

- Productivity (gross output and quality)
- Labor turnover
- Error and warranty costs

One can also add equipment and material costs (equipment failures, waste, errors, etc.) particularly if these are expected to change, for better or worse, due to the intervention. The costs for all these items, and the eventual savings, are related mainly to the labor costs.

4.3.3 A Cost-Benefit Analysis Model: The Productivity Assessment Tool

As we have noted earlier, there is a dearth of cost-benefit analysis models specifically directed to occupational ergonomics, health, and safety. In this article we will describe one generic cost-benefit analysis model (the Productivity Assessment Tool) developed for use by ergonomists and other occupational health and safety practitioners. The model is equally applicable for service or manufacturing industries.

A basic software version of the Productivity Assessment Tool has been published in Oxenburgh et al. (2004), which allows one employee or an employee group and one test case for comparison with the initial or present situation.

The full version of the Productivity Assessment Tool (ProductAbility, 2004) allows the user to have upto five individual employees or five employee groups in the selected workplace. That is, the employees can be individuals or groups, each working for a different number of hours per week, different rates of pay, overtime, absenteeism and degrees of productivity. With the reduction in full-time or permanent employment and the concomitant rise in precarious or casual employment (Bohle and Quinlan, 2000), such employment variations are common.

In addition, the present or basic state of the workplace (the initial case) can be compared with upto four possible interventions (the test cases). The advantage is that one can propose several possible solutions to any problem and directly compare their economic effectiveness. Needless-to-say, these test cases must be comparably effective in terms of injury prevention or other worker benefits and not solely economic comparisons.

This particular cost-benefit analysis model can be used to answer several economic questions, which follow.

4.3.3.1 Productivity Measurement

The pay-back period and yearly savings are usually the major determinants needed by management when considering financial investment; that is, how effective in cost terms will the investment be?

Most ergonomics interventions at workplaces give a short pay-back period, frequently 6 months or less. Longer pay-back periods are usually associated with redesign of specific areas, new building or products. If the pay-back period is greater than one year, then other factors may need to be taken into account; for instance, the cost of borrowing money, the discounted cash flow, etc.

The savings speak for themselves releasing finance for other investments or simply be considered as profit.

4.3.3.2 Of Several Alternatives, which Intervention Will Be the Most Cost-Effective?

If, in any work situation, there are alternative interventions for ergonomics improvement, then a cost-benefit analysis allows a comparison between these alternative ways. This comparison does not determine the most effective in health terms but can be a guide between the alternative financial costs and can guide these aspects of the intervention.

For example, if, in a warehouse, the alternatives considered to be similar to reduce back injuries include improving shelving or purchasing an extra fork lift truck then, by entering in the employment cost data for the two test cases, the benefits will be calculated (savings and pay-back) and a direct comparison may be made.

4.3.3.3 Sensitivity Analysis

If one is to make an ergonomics intervention it is just as well to identify those parameters that, by their alteration, will lead to the greatest cost benefit. For example, if overtime is a major employment cost factor it will be advantageous if the proposed intervention will reduce this cost factor.

By varying the appropriate data, the sensitivity to changes in the employment cost factors can be determined. For example, by reducing overtime by half, the cost-benefit analysis will show how this affects the yearly employment costs.

4.3.3.4 Rehabilitation

As a cost-benefit analysis can be used for individual employees it is suitable for determining the cost of rehabilitation of an injured worker. The data used assume that the injured employee will come back to work but with various restrictions on his/her work tasks and the medical and rehabilitation costs can be added to give a complete financial picture. By these means the financial investment needed to complete a rehabilitation program can be determined.

The following case studies illustrate the use of cost-benefit analysis in implementing ergonomics intervention and are restricted to illustrating cost-benefit analysis in productivity measurement (see Section 4.3.1.1). For more details on these and studies illustrating other uses of cost-benefit analysis, the reader is referred to Oxenburgh et al. (2004).

4.4 Applications

The following three case studies are "traditional" in that the problems were musculoskeletal injuries in manual workers with the solutions within the realm of standard ergonomics interventions.

They illustrate three types of interventions, all of which were costed by the use of cost-benefit analysis.

The ergonomics interventions (cases) are:

- Section 4.4.1 — almost no-cost changes to the work methods where the management and men experimented to find the best solution
- Section 4.4.2 — an ergonomics intervention where an ergonomist altered working heights to overcome poor posture
- Section 4.4.3 — large capital investment where engineers designed new equipment for injury prevention

We intend that these case studies are illustrative of cost-benefit analysis rather than ergonomics solutions and are simplified examples of the ones given by Oxenburgh et al. (2004). For a more detailed analysis, please see this reference.

4.4.1 Warehouse Work — Truck Loading

The manager of a warehousing and transporting enterprise became concerned when a number of his warehouse staff and drivers began reporting of neck and shoulder pain. A risk assessment identified the source of the shoulder and neck pains as arising from loading bulky packages into the delivery trucks.

The packages were about 1.5 m wide by 0.4 m in diameter, weighing about 8 to 10 kg, soft and flexible and covered by a slippery plastic and, with no handles, the packages were awkward to handle. Although the weight of the individual packages was not high their soft, flexible character made the packages awkward to lift and, as the rows filled, it required pressure to push the last packages into position on each row.

The simple ergonomics solution was to provide a platform within the truck, which enabled the warehouse staff to work at a height that gave them better mechanical efficiency. This eliminated most of the loading above head height so that the top layers could be packed with minimal stress to the shoulders. The warehouse staff have found the method more comfortable and they are not getting neck and shoulder pains any longer.

Although the new packing system has been successful interms of injury prevention, has it been successful in terms of cost-benefit analysis?

The new system increased the time required to load each truck from 35 to 45 min thus increasing the cost to load each truck by about 30%. However, using the original loading system each truck was packed to 89% volume capacity but, with the new system, the capacity was increased to 95%. This made such a difference that the drivers were able to load the entire day's deliveries of product into one less truck.

Table 4.1 shows the cost-benefit analysis of the improved loading system. In this analysis only the direct wage costs and truck running costs are used. Despite the increased loading/employment costs, the improved loading of the trucks led to a net savings of about 45,000 "units" per year. The pay-back period was less than 1 week.

TABLE 4.1 Cost-Benefit Analysis for Loading and Delivering Bulk Packages

	Initial Case	Improved Loading
Employment costs for the loading work (units/year)[a]	50,200	55,700
Truck costs (units/year)	501,400	451,300
Total yearly costs for loading and delivery	551,600	507,000
Intervention costs: management and warehouse staff time (units)[b]	—	625
Savings (units/year)	—	44,600
Pay-back period	—	1 week

[a]For reasons of confidentiality, "units" are used for costing.
[b]A "one-off" cost.

4.4.2   Manual Handling Made Easy: Barrel Handling

The work station was the wash line in a brewery where empty aluminium beer casks were received from public houses and washed ready for reuse. Preparatory to washing the plastic keystones, the top caps, were removed from the casks by levering out with a chisel and the shives, wooden bungs on the sides of the casks, were removed with a hammer and chisel.

The barrels travelled along a conveyor at floor level so that the operators had to bend their backs when manually removing the casks off the conveyor. The force and repetition needed to break and remove the wooden shives using the hammer and chisel led to musculoskeletal shoulder injuries. This work situation led to back and shoulder injuries and lost time for the two employees who worked in this section of the wash line. Other employees had to work overtime to cover this lost time.

Additionally, when the shives were removed, the wooden pieces fell to the floor forming both a tripping hazard and getting trapped in the conveyor belt causing damage to the belt. When the belt was damaged it had to be stopped. A mechanic repaired the damage and this took about 15 min for each stoppage; on an average, a belt stoppage occurred three times each day. When this happened the rest of the wash line, an additional nine men, were also idle.

The ergonomist measured the average elbow height of the men working there and, after allowing for the heights of the casks, determined the most suitable conveyor belt height. Of course, the height of the conveyor belt had to be low enough so that the keystone cap on top of the barrel could be removed and so bending could not be completely eliminated. Other means (improved hand tools) reduced the stress on the shoulders. Although the solutions were compromises (not ergonomically "ideal"), they have been effective in reducing, and in fact eliminating, absence due to bad backs and shoulders. Reduced injury resulted in reduced overtime (Table 4.2).

During the belt breakdowns not only were the two deshiving men idle but so were the nine men of the wash line. The cost-benefit analysis model (the Productivity Assessment Tool) used here, calculated the productive employment cost of the 11 men on the wash line and the mechanic to repair the conveyor belt at approximately £43.00 per breakdown of 15 min. By expansion to the full year, at a rate three of breakdowns per day and taking into account factory closures, the total cost of breakdowns is approximately £32,000.

The ergonomics intervention resulted in improvements both in safety and productivity. Reductions in overtime and conveyor breakdowns paid for themselves in 3 months although the benefit continued beyond this period, which is an increase in profit.

In this case study, as well as the previous one, there were unexpected improvements in productivity beyond the immediate reason for the ergonomics intervention. In the first case, the improved loading of the trucks and, in the second, the reduction in line breakdowns. These were not the reasons for the intervention and were not forecast in the original assumptions made by the management. We believe that this is a common occurrence as "good ergonomics is also good economics."

TABLE 4.2   Reduction in Injury Absence, Overtime, and Conveyor Belt Breakdowns

	Initial Case	Improved Work Case
Total cost of employment for the deshiving area, including overtime (£/year)	56,400	50,400
Cost of breakdowns (wash line idle) (£/year)	32,000	650
Total cost (£/year)	88,400	51,050
Intervention costs (£)[a]	—	8,830
Savings (£/year)	—	37,350
Pay-back Period (months)	—	3

[a] A "one-off" cost.

TABLE 4.3   Cost-Benefit Analysis for the Introduction of a Cable Handling Machine Including Direct
Wages and Supervisory Costs

	Manual Handling of the Cable	Design and Use of the Cable Handling Machine
Miners' wage cost to move the cable ($ per year)	36,600	5,200
Direct supervisory costs for total hours on cable handling ($ per year)	6,800	980
Net labor costs for total hours on cable handling ($ per year)	43,400	6,180
Intervention costs ($)[a]		65,300
Savings in labor costs ($ per year)	—	37,220
Pay-back period (months)	—	21

[a]A "one-off" cost.

4.4.3 Manual Handling in Coal Mines

In longwall excavation, as the coal seam is progressively removed the longwall mining machinery has to
be moved forward. Although heavy machinery and trucks are used to move the longwall equipment there
are still some tasks requiring manual labour and one of these tasks is moving and rehanging 11 kV cables.

The cables, which are heavily armoured, weigh nearly 12 kg/m. Due to the stiffness of the cable, the six
men on the task would be holding about 20 m length above the ground at shoulder or head height at any
one time (about 240 kg weight of cable) and it would take 3 to 4 h to put up a 300 m run. This task is
repeated weekly.

Clearly, this task is a back and shoulder stress problem and there had been about six injuries per year
accounting for 1200 h of lost time (an average of nearly 6 weeks lost per injured miner).

The colliery engineering and safety staff visited several coal mines in the area to see if there were any
better systems and came to the conclusion that the various methods they saw were not good enough and
decided that they would have to design a system themselves.

In collaboration with a mining equipment manufacturing company, a cable handling machine was
designed and constructed at a unit cost of $50,000. All the wages of the 80 underground miners who
were trained in the use of the machine as well as the direct wages of the colliery staff who assisted in
the design of the machine were included in the intervention costs. This added an extra $15,300 to the
intervention costs.

After installation of the cable handling machine the injuries had been halved to three per year with a
remarkable reduction in injury severity; only a total of 22 h was lost in each year or about 1 day lost per
injured miner per year.

Table 4.3 illustrates the cost-benefit analysis of the introduction of the cable handling machine but it
only includes the direct wages of the miners and the underground supervisors (deputies) and the direct
costs of the cable handling unit and associated training.

For major machinery purchases in this industry a pay-back period of less than 4 yr can be considered
to be financially satisfactory. In any cost-benefit analysis it is only necessary to include enough information to provide a reasonable estimate of the costs and benefits. Additional costs would include the
administration staff, head office costs and so on but the extra effort required to include the other
costs may not be necessary unless the project is of marginal benefit.

References

Ahonen, G., The nation-wide programme for health and safety in SMEs in Finland. Economic evaluation
and incentives for the company management. Protection to Promotion. Occupational Health and
Safety in Small-scale Enterprises. People and Work. Research Reports 25, Finnish Institute of
Occupational Health, 1998, 151–156.

Applied Ergonomics, Special Issue: Cost Effectiveness. Vol. 34, number 5, September 2003.

Bohle, P. and Quinlan, M., *Managing Occupational Health and Safety*, 2nd ed., MacMillan, Melbourne, Australia, 2000.

Connon, C., Reeb-Whitaker, C., and Curwick, C., Healthy Workplaces Technical Report 67-3-2003, 2003, Washington State Department of Labor and Industries, Olympia, WA.

Deming, W.E., *Out of the Crisis*, 1982, Cambridge University Press, Cambridge, UK.

Kupi, E., Liukkonen, P., and Mattila, M., Staff use of time and company productivity, *Nordisk Ergonomi*, 4, 9–11, 1993.

Mossink, J. and Licher, F., Proceedings of the European Conference on Costs and Benefits of Occupational Safety and Health, *Proceedings of the Conference*, Amsterdam, NIA TNO, 1997.

Osborne, D.J., Branton, R., Leal, F., Shipley, P., and Stewart, T., *Person-Centred Ergonomics: A Brantonian View of Human Factors*, Taylor & Francis, London, 1993.

Oxenburgh, M.S. and Guldberg, H.H., The economic and health effects on introducing a safe manual handling code of practice, *Int. J. Ind. Ergon.*, 12, 241–253, 1993.

Oxenburgh, M., Marlow, P., and Oxenburgh, A., *Increasing Productivity and Profit through Health & Safety: The Financial Returns from a Safe Working Environment*, 2nd ed., Boca Raton, FL, CRC Press, 2004.

ProductAbility, 2004. *Software for the Productivity Assessment Tool*. See www.productAbility.co.uk or e-mail maurice_oxenburgh@compuserve.com

Spilling, S., Eitrheim, J., and Aarås, A., Cost-benefit analyses of work environment investment at STK's plant at Kongsvinger, in *The Ergonomics of Working Postures*, Corlett, Wilson, and Manencia, Eds., Taylor & Francis, London, 1986, pp. 380–397.

5

Humans in Work System Environment

Holger Luczak
Sven Hinrichsen
Susane
Mütze-Niewöhner
RWTH Aachen University

5.1   Objectives and Structure of the Article

System theories are closely connected to the development of individual branches of science. Thus, system-theoretical approaches can be found in numerous scientific fields like biology, ecology, pedagogy, business administration, organization theory, engineering science, and political science. System-theoretical realizations are integrated into the own scientific field — with the concept of the work system.

The starting point of various system-theoretical developments was formed by the system-theoretical-cybernetic approach. This mathematical approach contributed, considerably, to the fact that the system idea is applied to a set of scientific subjects. Individual scientific fields, in turn, provided impulses for the advancement of system theories. Due to various system-theoretical realizations an overview of system theories and approaches — how it is given in the first part of this article — can, inevitably, be only one interpretation of the manifold literature. The depiction of individual system-theoretical approaches in the first part of this article orients itself primarily towards several branches of science. Thus, the socio-technical system approach originating from social psychology is introduced as it corresponds with its own conception of a work system. The theory of social systems — originating from sociology — provides several impulses for other scientific fields. As a result, the individual development stages of the theory of social systems are outlined. This theory and other branches of science like business administration in turn took up realizations from the evolution theory. For this reason, this theory — originating from biology — will also be explained. Furthermore, this part exemplifies how system-theoretical realizations were integrated into engineering science. The single system-theoretical approaches are constituted in conjunction with the appropriate idea of man in each case. This should help when comparing the individual approaches with the own normative requirement of designing work in a human and economic way.

In another part of the article the work system concept is described in detail as an entry to the systematic analysis, common in ergonomic literature. This part deals with the term "work system," the components of a typical work system, and its substantial reciprocal effects.

The last part of the article is dedicated to the systematic design of work systems. First of all general strategies of work system design are explained. Subsequently, models and methods are presented, exemplarily, based on a level scheme for the systematical order of work-referred realizations. These models and methods are able to support a systematical proceeding with the anticipating and prospective work system design. These approaches and methods are the result of research projects, which were accomplished by the Institute of Industrial Engineering and Ergonomics (IAW) and by the Research Institute for Operations Management (FIR) at the RWTH Aachen. The exemplarily cited research results are supposed to show that the use of the work system approach is possible and reasonable on different levels of ergonomic research.

5.2   Overview of Selected System Approaches

5.2.1   General System Theory and Cybernetics

Ludwig von Bertalanffy (1949) is considered as an important founder of the "General System Theory." He defines systems as whole units, which consist of interconnected parts. He explains the proportion of whole units to their parts with the help of a holistic axiom. Referring to this, the characteristics and behaviors of higher levels are not explainable by the summation of the characteristics and behaviors of their components, as long as these are regarded in isolation. Only if the ensemble of the components and the relations, which exist between them are realized — so the axiom — the higher levels are derivable from their components. The whole — so the central theorem — is more than the sum of its parts.

During the investigation of the development of individual fields of science, Ludwig von Bertalanffy (1951) discovered "the remarkable phenomenon that similar general problems and viewpoints appear in the different branches of science." Among others, he cites the exponential law as an example. This law "applies to radioactive decay, to the breakdown of a chemical compound in monomolecular reaction, to the death of bacteria under the influence of light or disinfectants, to the consumptions of an animal by starvation, and to the decrease of an animal or human population where death rate is higher than birth rate." The discovery of this phenomenon of "isomorphy of laws" in the different scientific fields caused Ludwig von Bertalanffy to aim for a "Unity of Science" with help of the "General System Theory." It was apparent to him that this ambitious goal could only be achieved — if ever — in the inscrutable future. His proceeding is marked mathematically. A standardization of the sciences is finally seen by him as the "reduction of all science to physics, in the final resolution of all phenomena into physical events."

The system approach of cybernetics (Wiener, 1961) was developed at about the same time as the "General System Theory." This approach is likewise mathematically formed. Wiener and others understand cybernetics (Greek: "kybernetes" = steermanship) as the "entire field of control and communication theory, whether in the machine or in the animal" (Wiener, 1961). The cybernetics takes up concepts of feedback, regulation, and control and interprets social systems in a cybernetic way. Methods, procedures, and realizations of the (automatic) control engineering are generalized and applied to nontechnical concepts. A process is described with the control loop, which functions autonomously on the basis of the exact given premises. A desired value is given to the system. If it comes to a disturbance, the system implements prior defined corrections autonomously, in order to achieve the programmed specified condition again. The system compensates any environmental changes with self change. It experiences information about required changes via feedback of the results of its procedures. Figure 5.1 shows an example of a control loop with reference to the handling of a boat (Frank, 1964). The boat can be interpreted as a socio-technical system. The captain formulates the destination. The pilot determines the particular location (current status) and composes a program to transfer the current status to the desired status. He has to "save" the desired status, to measure the current status, to compare the two values and to derive a program. This program has to be conveyed to the steersman by the pilot in terms of individual decisions (so-called "determined" decisions). The steersman transfers these orders into navigation positions.

Self-regularization, adaptation, and learning aptitude are, therefore, system specifications, which are examined by the cybernetics. The cybernetics discusses for the first time the relationship between system and environment as a problem of constancy and change. In the center of the considerations the question is formed, how system constancy can be maintained in a changing environment. For the

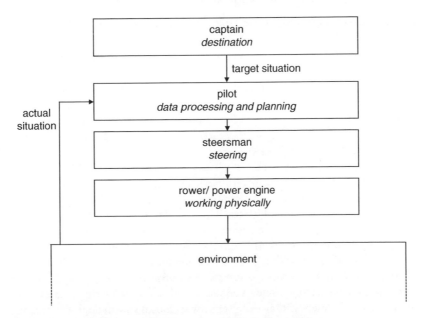

FIGURE 5.1 The socio-technical system "boat" as an example of a control loop.

first time, the stability of a system was not defined as a trait of systems (ontological approach) but as a problem, which has to be solved perpetually (Luhmann, 1973).

An integration of the "General System Theory" and the cybernetics to a system-theoretical-cybernetic concept began for the first time with Ashby (1956). He uses the "Black Box theory" to describe the relations between an experimenter and his environment with special attention to the flow of information. "The primary data of any investigation of a Black Box consists of a sequence of values of the vector with two components: input state, output state" (comparable to cybernetics). Referring to the "General System Theory," two machines are isomorphic, "if a one–one transformation of the states (input and output) of the one machine into those of the other can convert the one representation to the other." Thus, Ashby explains the fundamentals of cybernetics: the processes within the system and the processes of communication between system and system. Those are important to study the central theme of cybernetics — regulation and control. Regulation is essentially related to the flow of variety; without regulation the variety is large — with regulation it is small.

In the meantime it emerged that various impulses for individual scientific fields proceeded from the system-theoretical-cybernetic concept. However, their requirement to promote a standardization of science in different material systems with the help of a discipline-spreading terminology and the assumed existence of a more formal structure, that is, isomorphic laws, can only be partly fulfilled. In fact system approaches and theories were developed by the single branches of science. These consider the specific questions of the individual scientific field and contain the respective specialized terminology. The system-theoretical-cybernetic approach proceeds from a "shortened" idea of man. Humans are understood less as socio-emotional nature, which would like to carry out individual motives and needs. Rather, humans are regarded as a self-regulating system, which compensates for environmental changes by self changes. Despite this "shortened" idea of man, the view of humans is of importance as automatic controller for the understanding of human data processing. Related to man–machine systems a regulation-technical model can appear in such a way that humans function as an automatic controller and the object, which can be regulated is represented by the machine. The goal value is given from the "outside." The task of the automatic control loop consists of adapting the output quantity (actual value) as precisely as possible to the target value.

5.2.2   Socio-Technical System Approach

The socio-technical system approach — originating from social psychology — dates back particularly to a set of studies of the London Tavistock Institute of Human Relations. A study accomplished in English coal mines forms the starting point for the development of the socio-technical system approach (Trist and Bamforth, 1951). The Tavistock Institute was assigned to determine the causes for the low work motivation of the workers, for the high absence and fluctuation rates, as well as for a high number of accidents and labor disputes in a coal pit. A cause for the problems formed the introduction of the "long-wall method of coal-getting." The researchers found out that in the course of the introduction of this half-mechanized digging method existing social structures were destroyed. Thus, before the introduction of this "mass-production engineering" the complete coal-getting task — consisting of digging, loading, transport etc — was accomplished by a "single, small, face-to-face group, which experiences the entire cycle of operations within the compass of its membership." These shift-spreading groups divided their wages in the same proportion among themselves and were responsible also for their security. "Leadership and 'supervision' were internal of the group, which had a quality of responsible autonomy." With the introduction of the longwall method the groups were dissolved and the holistic task was divided into subtasks, transferred to specialists and spread over shifts. Trist and Bamforth (1951) could show that these changes had a negative effect on work motivation. In doing so the joint founders of the socio- technical system approach stress the dependence of social and structural aspects of an organization on the applied technology.

In consequence further investigations were accomplished concerning the socio-technical system approach. The fundamental system-theoretical theses of the socio-technical system approach can be summarized as the following (Emery, 1972; Rohmert and Weg, 1976; Alioth, 1980; Antoni, 1996; Ulich, 2001):

Organizations are open, dynamic, and goal-oriented systems, which consist of a social and a technical subsystem. The social subsystem of an organization contains the employees with their knowledge and abilities, as well as their individual and group-specific needs. The technical subsystem contains the entirety of the technical and spatial conditions of work (Figure 5.2).

In order to avoid suboptimal results of system design, technical and social subsystems have to be optimized together. The socio-technical system approach acts on the assumption that technology has a crucial influence on the organization. However, the organization is not completely determined by the technology. Also with a given technical system organizational options exist. This clearance of system design, obviously, increases, if the technical system and the work organization are planned together. When planning, the needs and requirements of the coworkers in reference to their work are to be considered.

The system "enterprise" is subject to fluctuations. These are caused, on the one hand, by the system environment (e.g., changes of demand). On the other hand, system fluctuations also have internal causes (e.g., disturbance at a machine; necessary reworking measures due to errors during the work execution). A central thesis of the socio-technical system approach proves that enterprises with small decentralized, self-regulating organizational units are more able to adapt to changes and fluctuations than central-controlled systems.

With the design of self-regulating organizational units there are three principles to consider (Ulich, 2001): (1) The formation of organizational units, which are relatively independent from each other should prevent the fluctuations and disturbances propagating uncontrollably. (2) An internal task coherent of an organizational unit makes it possible for the coworkers to determine their operational procedure on their own. This design principle stresses the motivational aspects of holistic tasks. (3) The design of the organization should be product-orientated, if possible. Thus, the formation of independent organizational units is supported. Furthermore, the work task shows a stronger connection with the product.

The socio-technical system approach raises the claim to provide a theoretical reference framework for the analysis of practical problems in organizations. Sydow (1985) shows that the socio-technical system approach only partly comes up to its claim. He justifies his animadversion on the approach among other things with the fact that central constructs like the one of the technical and that of the social system, their connection to the task system and to the feeling-orientated system are formulated imprecisely. Beyond that he shows that, many a time, technology is accepted as fixed although the necessity for a common optimization of the technical and social system is continually stressed.

The idea of man based on the socio-technical system approach refers particularly to the motive structure of humans (Sydow, 1985). The focus of attention is the effect of the work task on humans and their motivation. The socio-technical system approach focuses on intrinsic motives of humans. It shows that

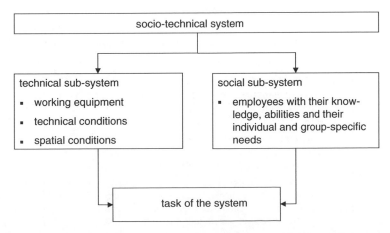

FIGURE 5.2 Elements of the socio-technical system (in modification of Ulich, E., *Arbeitspsychologie*, 5th ed., Schäeffer-Poeschel, Stuttgart, 2001. With Permission).

humans are regarded in a social nature. Expectations and needs of humans are, therefore, to be considered with the design of work organization and technical system. The economic orientation of humans, which manifests itself in their interest in a maximum payment is played down despite differing project experiences of the Tavistock researchers (Kelly, 1978; Sydow, 1985).

Despite the aforementioned animadversion on the socio-technical system approach it is very important from an ergonomic view as the principle of optimizing the social and technical system together corresponds with ergonomic aims. Beyond that, it has to be emphasized that small, decentralized organizational units, according to the socio-technical system approach, in particular correspond with today's requirements of high flexibility and short response and lead times in production.

5.2.3 Evolution-Theoretical Approach

The evolution - theoretical approach — originating from biology — closely follows the evolution theory of Darwin. The subject matter of the evolution theory is the adaptation of organisms to their environment after periods of time.

According to Maturana and Varela (1987), reproduction is the characteristic of an organism and is differentiated into heredity transmission and reproductive variation. Heredity transmission means that structural similarities arise, which already belong to the preceding generations. Variations in transmissions are the differences, which show up in the course of the development and can explain the diversity of species. The process of structural changing of a unit without a loss of their basic organization is called ontogenesis. This structural change takes place in each instant. It is released either by interactions with the surrounding environment or as result of the internal dynamics of a unit. This change ends only with the dissolution of a unit.

During the interactions between the organism and its environment the disturbances in the environment do not determine what happens to the organism. Rather, the structure of the organism determines what kind of change happens within it as a result of the disturbances.

As long as the unit does not interact destructively with the environment, compatibility can be found between the environment and the unit. In this case, environment and unit act as sources of disturbances in a mutual status and cause changes in each other. This constant process is called structural coupling. The basis of the continuity of organisms is the structural compatibility of the organism with the environment. This is also called adaptation. The adaptation of a unit to its environment is a necessary consequence of the structural coupling of this unit with its environment.

Necessary conditions for the existence of an organism are preservations of autopoiesis and adaptation. Autopoiesis in this context means that an organism strives for producing itself continuously. Lines of descent threatened by extinction can, therefore, be explained by not having the ability to adapt sufficiently. Basic principles of the evolution-theoretical approach are depicted in Figure 5.3.

1. *Heredity transmission*: structural similarities, which already belong to the preceding generations of a system arise

2. *Variation in transmission*: differences, which show up in the course of the development and can be explained by the adaptation and diversity of species

3. *Selection*: resources are rare; a lack of adaptation of a system to their environment results in finishing of system existence. The more resources in system environment exist the greater is the chance to survive.

FIGURE 5.3 Basic principles of the evolution-theoretical approach.

Based on this biological point of view, Malik (1993) transferred the evolution-theoretical approach to the management of enterprises. The main issue of importance for enterprises is their adaptability to unforeseeable changes. This ability becomes particularly important in times of difficult economic situations, in order to increase the probability of survival. On the one hand, the goal is to avoid behavior, which disturbs or endangers a system. On the other hand, the goal is to take over useful changes. The evolution-theoretical approach was also transferred to the theory of social systems (Section 5.2.5). Autopoiesis in the context of social systems means that they strive continuously to maintain their sense (Krieger, 1996).

Critics of the evolution-theoretical approach provide examples, which this theory does not apply to (Krieger, 1996). On the level of individuals, political or religious martyrs of all kinds show that for humans the preservation of an idea (thus, a certain form of sense) can be more important than the preservation of the own life. The same applies to believers who do not touch animals and plants because of religious taboo, although they are threatened by starvation.

Humans are to be considered primarily of a biological, rather, than of a socio-emotional nature in the evolution-theoretical approach. Humans are part of the evolution and their task consists of the best possible adaptability to their environment. According to this theory, humans act because of self-preservation.

5.2.4 Engineering-Scientific System Approaches

The system theory in engineering science developed, on the one hand, a firmly established basic subject. On the other hand, the system approach was integrated into single engineering-scientific subjects in different ways, which is shown through examples of measurement and control engineering, design engineering, manufacturing technique, thermodynamics, and rationalization research.

In measurement and control engineering methodology is usually based upon the preposition of mathematical models, in order to win insights into technical connections with different applications and to obtain quantitative results. The models mentioned represent mathematical pictures for the interaction of the physical features, which are the basis for the technical procedures. The advantage of the application of the system theory is regarded by the fact that a multiplicity of features in measurement and control engineering can be explained as a consequence of a few system-theoretical basic concepts (Unbehauen, 1990).

In design engineering, technical objects such as facilities, apparatuses, machines, devices, and individual parts are considered as artificial systems. These consist of an entirety of elements, which are arranged and due to their characteristics linked with one another by relations. Technical systems serve a process in which energies, materials, and signals are led and/or changed (Pahl, 1995). Design engineering proceeds from a conventional system understanding. Such a system is distinguished from its environment. Connections to the environment are formed by input and output variables. A system can be subdivided into subsystems. A clutch, for example, represents a component within a machine while it can be divided into the two subsystems "flexible clutch" and "shift clutch" again as independent components. The subsystems can be divided again into individual elements of the system.

A manufacturing system has the purpose to fulfill a fixed task of manufacturing (Warnecke, 1995). Therefore, coordinated processes are necessary. These are generated by subsystems of the manufacturing system (Figure 5.4). Among these subsystems are the control system, the power-supply-system, the handling-of-work-pieces-system, the tool-handling-system, the measuring-and-control-system, the auxiliary-supplies-system, as well as the disposal system. The basic task of manufacturing is fulfilled in the subsystem "work system."

A thermodynamic system is understood as a material formation whose thermodynamic characteristics are examined (Stephan, 1995). Examples of thermodynamic systems are a mass of gas, a liquid and its steam, a mixture of several liquids, or a crystal. A thermodynamic system has a system border, which can shift, however, during the procedure, for example, by expansion of gas. Furthermore, the distinction between open and closed systems is of importance in thermodynamics. Closed systems are impermeable for matter. Open systems can change their mass, if masses flowing in and leaking out are unequal.

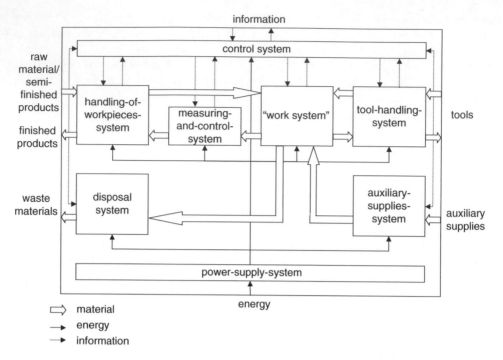

FIGURE 5.4 Functional structure of the manufacturing system (in modification of Warnecke, H.-J., *Dubbel-Taschenbuch für den Maschinenbau*, 18th ed., Springer, Berlin, 1995. With permission.)

In the rationalization research, concepts of the cybernetics, synergetic, and complexity theory were taken up (Luczak and Fricker, 1996). The situation of the rationalization is formed by increasing the dynamics of changes, which makes shorter response times necessary and causes an increasing complexity of operational settings of tasks. For the operational rationalization in principle the question is derived whether the solution is to be searched in the decrease and avoidance of complexity or in better methods of its handling and control (Warnecke and Hueser, 1995). Approaches, which aim at excluding a decrease of complexity of operational structures are not regarded as sufficient, however, for a successful complexity management since usually only short-term efficiency advantages are achieved for the disadvantage of long-term stability (Bullinger et al., 1993). In terms of the complexity accomplishment, on the one hand, the connection between operational structures and the system behavior is regarded (Malik, 1992). The question arises how operational structures are to be arranged. On the other hand, the behavior of the involved people during problem release processes is examined. For the design of problem solution processes various realizations and experiences are present. Regarding the modeling and design of order structures, however, few approaches are present. Therefore, an instrument was developed and evaluated for the modeling and design of complex production structures (Luczak and Fricker, 1996; Fricker, 1996). As degrees of complexity, variety (Ashby, 1956), entropy (Shannon, 1976), and the effective complexity (Gell-Mann, 1994) are used. The developed instrument enables the following:

- Representation and quantitative evaluation of interlaced operational coherences and order structures in the sense of monitoring the structural development
- Evaluation of the operational complexity, the partial equilibrium, and the adaptability to defined and internal measuring points
- Identification of operational complexity drivers and of starting points for reorganization measures
- Quantification of order and organizational structures.

5.2.5   Sociologically Formed System Approaches

Willke (2000) differentiates between several development stages of the social system theory. The structural-functional system theory represents the first fundamental delineation of a sociological system theory. Structures consist of elements that are in some way related to one another. On the one hand, structures limit the options. On the other hand, they ensure a certain assurance that expectations are fulfilled. It is not only the term of structure, but also the term of function that is of great importance to the structural-functional system theory. For example, the political system has the function to achieve collectively binding decisions. The structural-functional system theory is characterized by the fact that the term of structure is paramount compared to the term of function. The initial point is the assumption that all social systems necessarily show certain structures. The research-leading question is: which functional achievements have to be performed by the system, so that this system remains with its given structures. Structures are accepted to a large extent as given. How these develop remains unconsidered in the theory.

The system-functional approach — as a second development stage of the theory of social systems — regards social systems as complex, adaptive, and goal-seeking entities. Social systems are characterized by the fact that they are able to change or elaborate their structure to modified environmental conditions, in order to ensure their survival or viability. "We might generally say that the system has 'mapped' parts of the environmental variety and constraints into its organization as structure and/or 'information'" (Buckley, 1968). The functional theory does not circumscribe its view to the internal state of the system — unlike, for example, the classic organization theory, which only investigates the organization itself and the jurisprudence, which only deals with legal norms. It incorporates the environment as far as it is relevant for the stabilization of the system (Luhmann, 1971). The progress of the theory formation lies in the fact that structures are regarded as variables. The research question is: which structural adjustments must social systems be able to accomplish under certain changing environmental conditions, in order to be able to fulfill their substantial system functions.

The relationship of system and environment takes center stage of the functional-structural approach — as the third development stage of social systems theory. The environment is no longer regarded as a causing factor, but as a constituting factor for the formation of a system. The sense of the system formation is observed by the fact that excluded areas are created. This exclusion of areas, the creation of an environment, makes it possible for humans with their limited absorption capacity to seize and process complexity. In Luhmann's opinion there is one conclusion about the form of how a system solves its problems: with the help of structure formation (Luhmann, 1971). Structure formation leads to behavior expectations, referring to operation results. Systems form a regulator between incidental and processable complexity. The progress of the theory formation is to be seen in the fact that the emergence of certain characteristics of social systems is no longer attributed to individual internal or external conditions. Social systems are, according to the theory, reasonable constituted units, which have to solve the problem of processing complexity. The solution of this problem is regarded as a condition for reaching all of the other goals of a system.

It is emphasized in a further development step of the sociological system theory that the fixing of the boundaries between system and environment is an achievement, which is performed by the social system itself. The boundary is a social construction. The process of fixing the boundaries consists in establishing a difference between system and environment. In doing so, certain action patterns are created. These patterns make it possible to order the variety of possibilities, that is, processing and reducing the complexity. For the theory of self-referential systems a radical reorientation is suggested (Luhmann, 1984). It considers the theory of the autopoiesis originating from biology (Maturana, 1985). "Autopoieseis" means "self-creating" (griech.: auto = self; poiein = make, bring out). If the operations of a system only consist of producing the elements and relations that the system itself is made up of, then the system is an autopoietic one. That means it is a system whose goal is nothing else than itself (Krieger, 1996). The substantial idea of the theory is that a system does not only produce its structures itself, but also the elements of which it consists. Similar to cell biology, elements are understood as temporal operations, which decay continuously and are reproduced sequentially by the elements of the system itself. An autopoietic system appears now as entirety contrary to the system-theoretical basic postulate of the necessary candidness of complex systems. These entireties are closed in their core region, their internal control structure.

In the theory of social systems, humans are regarded as a system as well. On the one hand, this personal system can be part of a superordinate system or part of a system environment. On the other hand, the system "human" consists of subsystems like the genetic, organic, neuro-physiological, or mental systems. The idea of man is characterized by a high degree of abstraction. An entire, human-orientated point of view is not undertaken.

5.3 Work System Approach as an Analytical Framework

5.3.1 Overview of Work System Approach

5.3.1.1 Purpose and Characteristics of Work System Approach

The realizations of systems theory particularly were introduced into ergonomics via the work system approach. A work system is understood as an open, socio-technical system, which consists of elements. Candidness is expressed by the fact that the system receives an input from its environment, which is transformed by purposeful cooperating of the system elements into an output, in order to fulfill the system purpose. Synonymously to the term of work system, the terms "man–machine system" and "socio-technical system" are also used (Heeg, 1988).

The work system approach acts particularly as an analysis framework. It is conducive to systemic thinking, because it does not regard individual elements in isolation, but considers the various interdependences between the elements. Besides, the application of this analysis framework forces the ergonomist to draw up a system limitation and, thus, make a clear problem definition.

Ergonomic literature refers to different concepts of work systems (e.g., Rohmert and Rutenfranz, 1975; Karwowski, 1991; REFA, 1993; Hettinger and Wobbe, 1993; Luczak, 1998; Carayon and Smith, 2000; Hendrick, 2001). These differ from each other regarding individual system components. Most of the system concepts have the following in common, humans in work systems change work pieces coming from the system environment in a purposeful way and dispense them into the system environment again. This happens with the help of working equipment and operating resources. Hereby, the work environment affects the working person.

In addition to this general work system approach, job order, and work task should be differentiated in accordance with Figure 5.5. In order to be able to execute an order, the work system receives different system inputs from its environment. Among these are work object, energy, and information. The information concerning the job order contains rules for the work process. This process determines the spatial arrangement and temporal sequence of cooperation between inputs, human, and techniques and is designated as working method. The information about the job order is interpreted by the working person. The person takes over the job order as a task in a redefined form. In accordance with the task, the work object is converted to a nominal condition by activities of the working person with the help of the working equipment and operating resources. Energies are converted and information is produced so that the essential output of a work system consists of a changed work piece, as well as information and energy. During the process of transformation of inputs to outputs the physical environmental influences affect humans. These are known as the work environment.

The term "work system" is widely interpreted in the literature, that is, parts of an individual workplace can be constituted as a work system like a whole enterprise (Luczak, 1998). Zülch (1992) differentiates between work systems on a macro- and micro-level. Micro-work systems refer on the one hand to individual jobs and on the other hand to working methods. The macro-level of a work system corresponds with the level of working groups. On this level the work organization and the technical process are regarded.

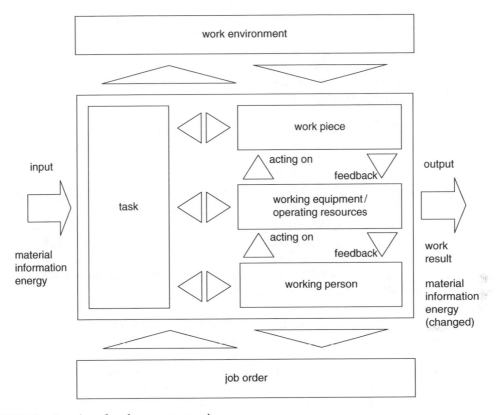

FIGURE 5.5 Overview of work system approach.

The differentiation between job order and work task provides a first reference point for the fact that the system purpose is dependent on the point of view on the work system. From the systems environment point of view, the main purpose of a work system consists of purposeful handling of job orders. Work systems are regarded by their environment under purpose-rational and economic criteria. From the point of view of a human who is working in the system, the purpose of the work system is the realization of personal motives. On the one hand, his work accounts for the assurance of his subsistence via the payment. On the other hand, the purpose of the system can be seen as an interesting and varied task.

5.3.1.2 Goals of Work System Design and the Idea of Man in Ergonomics

Basic goal of work system design is the optimization of the entire work system. In the sense of the main objective of ergonomics work systems are to be suitably arranged for both human and economical purposes. Thereby, economy is determined by the best possible proportion of inputs (e.g., raw materials, energy) to outputs (e.g., finished products, service). Human abilities and technical conditions are to be adjusted, in order to arrange work systems to be suitable for humans. Ergonomics is based on the assumption that rationalization and humanization goals work complementary to each other. On the one hand, humane conditions of work lead to effectiveness (in the sense of result reaching) and efficiency (in the sense of small resources inputs). On the other hand, effectiveness and efficiency form an important basis, in order to be able to create humane conditions of work. The consideration of the "resources coworker," the "human factors," has become more important. A one-sided pursuit of one or another goal clearly leads to suboptimal results.

Ergonomics considers the technology, the organization, and the personnel in the context of human work. In doing so, ergonomics distinguishes itself from predominantly human being-referred disciplines by including technical aspects into the working process. On the other hand, a differentiation takes place towards economic-technical disciplines, which exclusively aims at the optimization of the work result. Therefore, the requirement of ergonomics must proceed from a human conception, which regards humans, on the one hand, as a performance generator in work systems and, on the other hand, as a complex nature whose needs are to be considered during the work system design. Complexity results from the various, inter-individual performance rates and motives of humans, which can change depending on the situation and the course of time.

5.3.1.3 Limitations of the Application of Work System Approach

Even if the work system approach can be applied in principle to different levels — from a single execution via the work place and the entire enterprise to interplant cooperation — the level of workplaces is emphasized. Hereby, the workplace is understood as the spatial range in a work system in which the work task is performed by an employee. This focus on the level of the workplace in the application of the work system approach may be connected above all to the fact that the system approach was developed in a time, which was influenced by industrial mass production and high division of labor. Accordingly, work systems were arranged in such a way that an individual working person had to execute simple, short-cyclic, and repetitive activities. The work was determined to a great extent by working methods specified in the job order and by technology. Ergonomic efforts for the improvement of labor situation for industrial workers with the help of humane design of workplaces came to the fore.

Today, however, industrial manufacturing is characterized by a high variety of products and variants. Mass production was replaced gradually by a series production with small- and middle-sized batches in many industries. Work organization was also developed further because of these changed requirements. Thus, decentralized team and process-orientated work systems, which regulate themselves exceedingly and are flexible to react to changing requirements can be found in many enterprises. The borderlines of work systems are shifted because of the implementation of teamwork. Groups do not only realize executive tasks, but also tasks of regularization, preservation, and optimization within the work system. Rules are no longer set only by the environment of a work system, but are also the result of group-dynamic processes. A control of work systems at group level is carried out increasingly today through target agreements between groups and supervisor.

Moreover, the importance of the service sector in the western industrial nations increases. Services are characterized by the integration of the customers or of a factor, which are imported by the last-mentioned. The work system approach is to be modified to the extent that the "work object" in accordance with the work system conception outlined earlier can also concern a subject, in this case, the customer. During the following description of the individual elements of a work system the changes in the general conditions are considered.

5.3.2 Elements of a Work System

5.3.2.1 Working Person

Referring to the working person, several dimensions of human work capacity can be distinguished theoretically, even if it is difficult to account for the differentiation in dimensions in a metrological way concerning the concrete application. These dimensions are called, on the one hand, ability and, on the other hand, willingness to perform. In regards to ability the characteristics, which cause the achievement structure of a working person are considered physiologically as achievement capacity of the organs and/or organ systems and psychologically as achievement potential of psychological functions and/or appropriate components. Willingness to perform is determined physiologically by the excitation level of organs and/or organ systems in the psychological sense by achievement attitudes and motives such as needs, interests, intentions, or convictions. Components of the willingness to perform are, thus, a necessary condition, in order to utilize the existing potential of abilities. This means that only people who are physiologically above a certain excitation level (e.g., muscle tonus) and those who are additionally motivated (psychological dimension) can achieve the performance level, which they are enabled to achieve due to their physiological and psychological characteristics.

The characteristics of a person who has the ability to perform is influenced by four categories. These are the constitution-, disposition-, qualification-, and adaptation characteristics (Figure 5.6). Constitution characteristics are unchangeable ratios in the life cycle of humans. Among these characteristics are gender, somatotype, and ethnical origin. Disposition characteristics are variable, however, the working person has no direct influence on this. Disposition characteristics cover the age, body weight, the state of health, and the intelligence of a working person. Beyond that, the biorhythm belongs to the disposition characteristics. This means that the human efficiency in the course of a day is subject to fluctuations. The entirety of all knowledge, skills, and experiences of a working person are understood as qualification characteristics, which a working person must have for carrying out activities at the workplace. Characteristics, which affect the human ability and are changeable at short notice are called adaptation characteristics. Adaptation characteristics refer to the fatigue and recovery, as well as the strain of a working person. With the beginning of an activity numerous physical and psychological conversion procedures start, whereby humans adapt increasingly to the workload caused by the activity. Manual labor, for example, leads to a rise of the heart rate and the muscle blood circulation, in order to be able to meet the demanded achievement. The realization became generally accepted that an optimal achievement of working humans is only ensured in the long term, if the fatigue caused by the work stays within limits.

Most of the determinants mentioned for the ability are mutually dependent on each other, for example, somatotype and body weight, age and experience, skills and strain. Differences between people or changes within a person, however, do not have to lead inevitably to differences in the achievement. Measurable effects on the achievement cannot be determined due to the existence of mutual compensation mechanisms.

Constitution characteristics are solely influenced by the selection of personnel. Disposition characteristics can also be affected by other measures. Specifications of some characteristics can be changed within a limited range also through measures of work system design. For example, realizations of the influence of biorhythm on the human ability effect the shift work models. Qualification characteristics can be

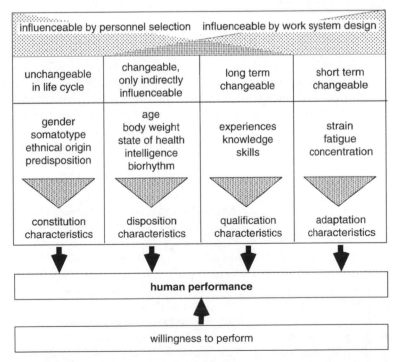

FIGURE 5.6 Individual determinants of human performance (From Luczak, H., 1989, Wesen Menschlicher Leistung. In: Institut für angewandte. Arbeitswissenschaft (ed.): Arbeitsgestaltung in Produktion und Verwaltung. Cologne: Bachem, pp. 39–60, With Permission).

determined by both, a purposeful selection of personnel and by measures of the work system design and personnel development. For example, the introduction of a continuous improvement process (CIP) can lead to the fact that employees exchange knowledge and experiences during operational problem solving, causing the qualification level within the CIP-team to increase. Adaptation characteristics are not or only indirectly affected in their developments by a selection of personnel. Measures of the work system design can exert influence on variables at short notice such as strain and fatigue. For example, organizational rules of recovery periods can help to avoid an excessive demand and a possible harm to working persons.

The willingness to perform, the work motivation of a person, results from the reciprocal effects between personal motives of the working person and the motivation potential of the work. Neither the motives nor the motivation potential of the work alone are sufficient, in order to explain the development of work motivation. The motivation potential of work results from the expectations of employees to be able to achieve their motives. For example, motivation potentials regarding achievement, power, and integration can be differentiated from each other, since motivation potential always refers to specific motives (Kleinbeck, 1996). A work situation with a high achievement-related motivation potential is, for example, characterized by establishing a range of decision and activity for employees in which a complete and in their view important task with adequate difficulty can be worked on. Motivation potential regarding power always contains work activities, whenever the work activity is implemented by several employees simultaneously and the task makes it necessary that a person (e.g., guidance person) affects other people in the sense of achieving a certain goal. High motivation potential regarding integration exhibit work activities, which give many opportunities to the working person to achieve social contacts and maintain them. If motivation potentials are highly developed, these do not lead inevitably to a high motivation. Motivation potentials are regarded, therefore, as a condition of work motivation. For example, if a work system is spatially arranged in such a way that it permits interactions between working persons, this leads to a high motivation for the employees with a strong integration motive. For people with low integration motive, however, the possibility of exchanging itself with colleagues does not lead to increased motivation.

5.3.2.2 Job Order and Work Task

The objective job order — or in the sense of a recognized obligation leading to a self-identified order — usually would be redefined by the employee in the form of a task. A distinction between job order and work task appears essential, as employees rarely adopt job orders accurately or how they are intended, but follow their own subjective interpretation of the job order (Hackman, 1969; Hacker, 1995). In Hackman's (1969) perception, the job order — termed by him as "objective task input" — consists of "stimulus materials" and instructions. Regarding instructions, there is a differentiation between "instructions about goals" and "instructions about operations." Mostly orders consist of a combination of goals and rules concerning the work method, that is, "Minimize the wavering in the tone you hear (instructions about goals) by adjusting the four knobs on this panel (instructions about operations)." "Stimuli are the actual physical materials with which subjects work." "Think about this picture and tell me what it means" is an example of an objective task input, which is conveyed by stimulus materials.

During the instructions about goals, realizations from the theory of goal setting should be considered (Locke and Latham, 1990; Kleinbeck, 1996). These show that from goals, a performance-enhancing effect can proceed. Thus, on the one hand, ambitious goals and, on the other hand, specific, preferable quantified goals lead to a higher performance than low set and nonspecific, qualitatively formulated goals as, for example, "do your best" or "work at a moderate pace." A condition for these positive effects on the performance is, however, the fact that a working person assesses a given or declared goal as realistic (Locke and Latham, 1990). The extent to which the goals affect the performance, additionally, depends on variables like the goal commitment and feedback. The goal commitment expresses how willing employees are to commit themselves to a certain goal or how high the resistance is to give up the goal or to downgrade the emphasis of the goal. The higher the goal commitment is, the larger the action intensity and perseverance and, thus, the achievement. A possibility of strengthening the goal commitment and increasing the associated performance exists in setting monetary incentives (Kleinbeck, 1996).

Thus, coworkers pursue goals whose reaching or reaching degree has effects on their payment substantially more consistently than goals, which are not coupled to their payment. Thus, the probability of achieving the objectives rises. Feedbacks, which give information to the employees about the goal reaching degree during the work process positively influence the effects of goal setting on performance. Due to the knowledge about existing discrepancies between the goal and the current performance level, the employee can increase action intensity and duration and, thus, eliminate these discrepancies (Locke and Latham, 1990).

The valid rules of the work process (instructions about operations) are also called work methods. A work method usually applies to all people who are active in a work system. By work manner the individual execution of the work method is understood (REFA, 1993). If the work method is given and the dispersion of the individual work manner is small, then a high method level is referred to. The method level is affected, on the one hand, by the level of organization of a work system and, on the other hand, by the abilities and skills of the employees. These two influencing variables depend again on the repeating frequency of job orders. A high method level — usual in line and mass production — offers only small interpretation clearance to the employee concerning the redefinition of the job order. Beyond that, the factors including the comprehensibility and acceptance of the job order, typical expectations and values of the working person, and existing experiences with considerably similar orders affect the redefinition process (Hackman, 1969).

Work tasks can be differentiated into primary and secondary tasks. "Each system or sub-system has, however, at any given time, one task which may be defined as its primary task — the task which it is created to perform" (Rice, 1958). Value generating activities are a direct result of a primary task. They create value for the internal and external customers. The processing of secondary tasks is not directly valuable, however, it forms a condition for fulfilling the primary task and achieving the goals of a work system (e.g., annual cost reduction around 3%). Secondary tasks can be separated into tasks of system regularization, preservation, and optimization (Miller and Rice, 1967; Antoni, 1996). The tasks of system regularization include, for example, the job order planning and production control. Typical tasks of system preservation are preventive maintenance and repair related to the technical subsystem of a work system. With reference to the social subsystem, consisting of the employees, for example, training measures can be added to the tasks of system preservation. Among tasks of system optimization is, for example, the implementing of the CIP in a work system. The detailed fixing of the boundaries cannot be made solely by formal rules set by the enterprise, as the work system, on the one hand, is subject to changes in the time response (e.g., coworker turnover) and the environment of the work system, on the other hand, changes its demands (e.g., due to changed customer's requests) on the system. According to the theory of social systems (Section 5.2.5), the fixing of the boundaries rather takes place by the work system itself. Creating boundaries between system and environment is an achievement, which generates the system partially itself. Even if processes creating the boundaries do not take place consciously in every case, the creating of the boundaries can be interpreted as another secondary task. The subject of this task clarifies whether certain activities are to be implemented in the work system or its environment (e.g., a work system, which is preliminary in the process). Furthermore, the work system and its environment (e.g., other work systems, supervisor) have to clarify how certain tasks have to be fulfilled. This clarifying process, thereby, primarily refers to those activities, which were not specified clearly by organizational rules or the work method.

The type and range of the primary and secondary tasks determine the range for decision and activity of employees in a work system. Range for decision and activity makes an individual and collective self-regularization possible and forms a condition for the increase in work motivation of the employees.

5.3.2.3 Work Object and/or Recipient of Services (Subject)

The purposeful change of the work object is regarded (on the point of view of the system environment), as a rule, as the main objective of the work system. Regarding a work system in a production plant, the work object is changed by manufacturing methods as, for example, metal forming, cutting, and coating. Concerning service enterprises — as aforementioned — the "work object" can be the customer who takes up, for example, an advisory service. The substantial output of the work system can be, in this case, immaterial and exist in additional knowledge, which the customer can acquire by the usage of the advisory service.

5.3.2.4 Working Equipment and Operating Resources

A meaningful classification of technical systems, which are also called working equipment and operating resources can be considered in accordance with Figure 5.7 on the basis of the system function and the system output (Ropohl, 1979).

The categories "transformation," "transport," and "storage" can be differentiated regarding the system function. Transformation means that the outputs of a system are quantitatively and/or qualitatively different from the system inputs. Transport signifies that only the spatial and temporal coordinates between inputs and outputs change while the material, energetic, and informational attributes remain constant in quality and quantity. The function class "storage" is concerned; if only the temporal coordinates of the output differ from those of the input while all other attributes (material, energy, information, space) are not subject to change. These function classes can be assigned to the following three terms: production technique, transport technique, and storage technique. The term of production is to be understood in close technical sense.

The output of a technical system can be differentiated regarding material, energy, and information. Material is everything that possesses mass and takes up physical space. Energy is defined as the ability to carry out work in a physical sense. By information, a configuration or a series of symbols is understood to which a certain meaning can be attached and which can lead to certain behavior.

Single technical subjects can be assigned to the individual fields of the pattern represented in Figure 5.7. Thus, for example, the process engineering and the manufacturing technique are located in the field of "material/transformation." While process-engineering systems produce materials with exactly defined chemical and physical characteristics, the output of manufacturing systems consists of products with exactly defined material and geometrical characteristics. Complex technical systems, which contain many subsystems in relation with each other can be assigned to a field regarding their

FUNCTION CATEGORY OUTPUT	TRANSFORMATION production technique	TRANSPORT transport technique	STORAGE storage technique
MATERIAL / material technique	process engineering manufacturing technique	materials handling and conveying	warehouse engineering
ENERGY / energy technique	energy transformation technique	energy transmission technique	energy storage technique
INFORMATION / information technique	information processing technique metrology control and feedback control technique	information transfer technique	information memory technique

FIGURE 5.7 Classification of technical systems (in modification of Ropohl, G., *Eine Systemtheorie der Technik — zur Grundlegung der allgemeinen Technologie.* Hanser, München, 1979. With Permission).

primary function and their substantial output, but apart from their characteristic technique most of the other techniques are also used in subordinated functions. A manufacturing system, for example, apart from its main technical function covers handling- and storage-technical subfunctions, as well as energy- and information-technical subfunctions (Section 5.2.4).

Systematics of technical systems — like Ropohl's scheme — can serve as a framework for the design of the technical subsystem of a work system. Further technical systematics can be allocated to the individual fields, referring to Ropohl's scheme. For example, in reference to the manufacturing technique (field "material/transformation" in the scheme) different manufacturing processes (referring to DIN 8580) can be distinguished (Figure 5.8). These individual manufacturing processes can in turn be differentiated further. Thus, cutting processes are divided into cutting with geometrically welldefined tool edges and cutting with nondefined tool edges. The first-mentioned category consists, for example, of turning, milling, roaching, drilling, and boring. These individual processes can in turn be classified according to different criteria.

5.3.2.5 Work Environment/Environmental Influences

On the one hand, social and cultural factors, which affect the work system are assigned to the work environment. On the other hand, the term "environment" means the spatially surrounding fields, from which physical and chemical, and in addition, biological (e.g., bacteriological) influences affect humans and their physical ability. The physical–chemical work environment can be differentiated according to the kind of influences it has on the working person:

- Working materials
- Radiation
- Climate
- Noise and sound
- Mechanical vibrations
- Lighting

On the basis of the fact that individual work environment factors rarely appear isolated, but in combination with each other a consideration of effects is only permissible for the entirety of all environment factors in combination with the work-specific types of stress. So far, these inter-relationships have remained widely unexplored so that in practice the consideration of effects for each individual variable of stress seems appropriate.

The following proceeding has proven useful for the analysis, evaluation, and design of single work environment factors (Luczak, 1998):

1. Knowledge of the scientific bases
2. Measurement of the work environment variable

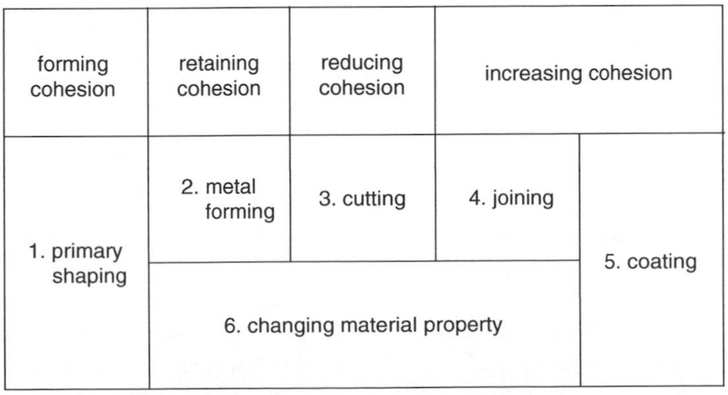

forming cohesion	retaining cohesion	reducing cohesion	increasing cohesion	
1. primary shaping	2. metal forming	3. cutting	4. joining	5. coating
	6. changing material property			

FIGURE 5.8 Systematics of manufacturing processes referring to DIN 8580.

3. Evaluation of the results of measurement
4. Estimation and deducing rules

A condition for the measurement of the environmental influences is a realization of the physical and chemical, and (with noise, climate, and lighting) the physiological variables and regularities. If these scientific fundamentals are well-known, the quantitative value of individual environment variables can be identified with the help of specific measuring methods and devices. In the next step the question arises, which effects different levels of load (as a consequence of the variation of single environment variables) have on people. These effects can be the detriment of the working person, the influence of physiological variables, and, in addition, the influence of the person's feelings or the work behavior (e.g., error frequency). Furthermore, the factors and individual characteristics of the working person on which the effects depend must be indicated. In the sense of the stress-strain-concept the goal of this step is to indicate the strain and/or detriment associated with the environmental load. If sufficient knowledge is available, desired or limit values for the single environment variables can be derived in a fourth step, in order to estimate the endangerments and strains of environmental load. In each case, these values are coupled with design goals. The goals, freedom from impairment and risk avoiding, are accepted for all environmental exposure. An avoidance of disturbance applies to substances (dirtiness, unpleasant odor), sound (noise), and electrical and magnetic fields, only concerning the population while they often are considered as substantially tolerable for the workforce.

Even if the inter-related effects of individual work environment factors are practically unexplored, a "bottleneck observation" seems adequate. First the influence of each load variable on working persons is evaluated and the specific organic system, which is loaded is identified. If the same organic system is loaded several times, a "bottleneck observation" should be made. This proceeding was applied successfully, for example, when climatic factors affect humans in connection with a high energetic load (heat work; Section 5.5.2). Both load factors lead to an increased utilization of the cardiovascular system, which is to be regarded in this case as a shortage (Luczak, 1979; Wenzel and Piekarski, 1980).

5.4   Systematic Design of Work Systems

5.4.1   Process of Work System Design

The implementation of ergonomic insights in concrete solutions of work system design is carried out in different operational functions. The individual divisions approach their task of work system design with different areas of interest. Regarding the systematization of such design interests Fürstenberg (1983) developed a conception from a personal view. He differentiates interests of preservation, interests of design, and interests of utilization of human work. In reference to the interest of preservation of the human work ability the occupational safety was developed, which in turn endues own systematic procedures, for example, the TOP model (Compes, 1970). Work technology is referred to, regarding the general interest of the function of work systems. The work technology is essentially affected by systematics of the technological determinism in its design procedure. In direction of the interest of utilization of the human work, the labor economics and the industrial organization are to be identified, which developed their systematics, for example, towards types of work processes (concerning humans, equipment, and work object) and towards types of requirements (e.g., for the job evaluation and remuneration).

The framework of a comprehensive ergonomic design process contains design interests, which can be related to Fürstenberg and proceeds from the system ergonomic procedure (Kirchner, 1972; Döring, 1986; Kraiss, 1986). The approach proceeds from a hierarchical-sequential procedure with feedback possibilities during the design process. Thereby, the design areas of the technological, technical, organizational, and ergonomic design are covered (Figure 5.9).

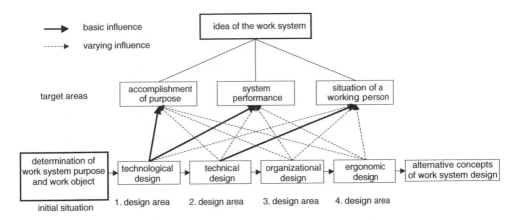

FIGURE 5.9 System ergonomic procedure according to the principle of sequential designing.

In this context, technology should be understood as a describing level of technique. Technology deals with the systematics and analysis of concrete techniques. Technologies in this sense are, for example, manufacturing processes (Section 5.2.4 and Section 5.3.2). For instance, the manufacturing process "turning" (technology) is to be carried out with a corresponding technique (turning machine, control program, etc.).

5.4.1.1 Technological Design

The technological design is mainly determined with the construction process of a product. For instance, necessary exactitudes (e.g., fits), stabilities (basic material, geometry, heat treatment), and surface properties (surface roughness, corrosion protection, etc.) in this process are determined in consideration of existing technologies (Section 5.3.2: Ropohl's scheme). From an ergonomic point of view the technological design must consider the effects a decision for a specific technology can have on the conditions of work (e.g., stress caused by certain work environmental conditions, degree of the simplicity of assembly task).

5.4.1.2 Technical Design

In a second stage in the context of the technical design the degree of mechanization and automation, that is, the division of labor between working person, on the one hand, and working equipment and operating resources, on the other hand, are discussed. Thereby, human weaknesses and evolution-caused limitations of humans are adjusted by mechanization and automatization. Besides, remainder functions for people, for example, as automation gaps, are to be equally avoided.

Referring to the degree of mechanization and automatization, according to Figure 5.10, three levels can roughly be distinguished, namely, the manual activities, the mechanized fulfillment, and the automated fulfillment (Kirchner, 1972).

The manual fulfillment of a task may take place with the help of tools. The energy, however, which is necessary for the purposeful modification of the work object is mustered with the physical strength of a working person. Thereby, the human being poses as a regulator (Section 5.2.1), that is, he gathers information concerning the actual condition of the work object and processes it as long as the desired work result is available. The informational effort of the working persons can be reduced with the help of working equipments like templates or apparatuses. Mechanization means the substitution of human energy with technical forms of energy. In comparison with the manual fulfillment, however, the informational effort can be further reduced, but no complete substitution of human information processing in the sense of a technical control or regulation takes place. The automation is characterized by the fact that — beyond the mechanization — the regulation of the process also takes place within the technical

degree of mechanization and automatization	subfunctions of a work system		
	acting on work object	steering	controlling
manual fulfillment	human	human	human
mechanized fulfillment	technique	human	human
automated fulfillment	technique	technique	human

FIGURE 5.10   Levels of functional division between human and technique in a work system.

system. The working person solely has a monitoring function. The individual levels of mechanization and automation can be distinguished further (Kirchner, 1972).

The concept of supervisory control supplies basic design indications, referring to the degree of mechanization and automation. "The term supervisory control derives from the close analogy between a supervisor's interaction with subordinate people in a human organization and a person's inter-action with intelligent automated subsystems" (Sheridan, 2002). A supervisor gives instructions to the subordinates who have to understand and transform them into detailed actions. The subordinates, again, present their results to the supervisor. The supervisor has to compare his goals — in analogy to the cybernetic system approach (Section 5.2.1) — with the results presented by the subordinates. Then he has to decide on the further action to be taken. The same sort of interaction occurs between a human supervisor and the automation (Sheridan, 1992). Thus, the automation can be compared to a subordinate, but with less intelligence. Five roles can be cited for the human supervisor: he has to plan, to teach, to monitor, to intervene, and to learn.

To find out whether it is better to carry out a task manually or supervisory, Figure 5.11 can be used. The dashed line in the figure represents the time required for direct manual control. It is assumed that the more complex a task is, the more time it takes to carry out. The thin curve in the figure shows how much time someone needs to plan and teach a task. If the task execution time is required, a slice between the thin curve and the heavy solid curve is needed. The sum of these two values is shown by the heavy solid curve. The heavy solid curve intersects the dashed line in two places. On the one hand, the left end of the scale demonstrates that it is better to do simple tasks on your own, because this is much quicker than to explain it to a computer or another person. On the other hand, the right end represents that very complex tasks are too hard to figure out how to program and so it is faster to do them manually. If a task is to be repeated many times and the environmental conditions do not change, it is better to program it. In this case this automation (and supervisory control) is the fastest way.

5.4.1.3   Organizational Design

In the third stage the organizational design mainly deals with the division of labor between humans. Different activity elements are combined to a whole task, which supports the motivation of the working person. The principle of the completeness of activities should have priority over principles of Taylorism and Fordism. Basic principles, which are formulated in the context of the socio-technical

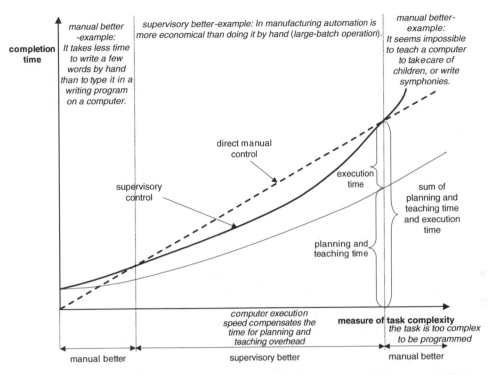

FIGURE 5.11 Range in which supervisory control outperforms manual control (in modification of Sheridan, T.B., *Humans and Automation — System Design and Research Issues*, John Wiley & Sons, Santa Monica, 2002. With Permission).

system approach (Section 5.2.2) are to be considered with the task design. Moreover, regarding the implementation of group work, important organizational design fundamentals are described in Section 5.5.6.

5.4.1.4 Ergonomic Design

In the fourth stage of the ergonomic design the "division of labor between segments of the organism" is designed. On the basis of anthropometric and physiological and also information- and cognition-psychological realizations, activity elements and working equipment are improved for the functions of energy and power generation and the functions of the information absorption, processing, and delivery.

The model of user–computer interface, which is described in the following exemplifies the stage of ergonomic design. This model is based on the language model, which derives from Foley and van Dam (1982). The model supports the process of designing the user–computer interface and consists of four steps (Figure 5.12).

1. In the first step (*conceptual design*), the user has to clarify how a task can be carried out with the help of a computer. At this level the structure of the fulfilment of a task is partially determined by the analysis of the task.
2. In a second step, the software functions are defined. The functions influence the activities of the working person (semantic design).
3. In the next level, the user has to find a way to solve the function. Therefore, the *syntactic design* defines the sequence of input and output. "For input, sequence is grammar — the rules by which sequences of tokens (words) in the language are formed into proper . . . sentences." These input sentences are, for example, commands, names, coordinates, etc. The output tokens are often symbols and drawings.

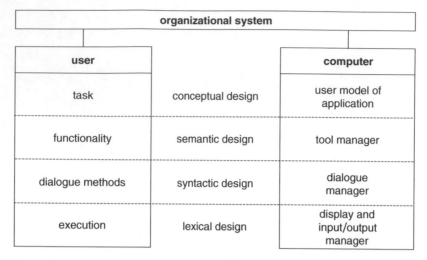

FIGURE 5.12   Model of user–computer interface (in modification of Foley, J.D., and Van Dam, A., *Fundamentals of Interactive Computer of graphics*, Addison Wesley, Menlo Park, CA, 1982. With Permission).

4. In the last step (*lexical design*), it is determined how these input and output tokens are formed from the available hardware components. "We see, then, that lexical design represents the binding of hardware capabilities to the hardware — independent tokens of the input and output languages."

General strategies of the work system design, apart from this systematic, sequential proceeding, can be differentiated further and are described in the following.

5.4.2   Corrective and Conceptive Work System Design

In general, the design process of a work system has two variables:

- The changes (modernization, extension, etc.) of existing work systems
- The development of new work systems

From an ergonomic view the meaning of former type of design process is important, because existing work systems are frequently adapted to the (changed) requirements of human work (humanization measures). Such corrective measures of the work system design usually concentrate on ergonomic and/or organizational aspects (e.g., change of control elements, additional sound insulation, job enrichment, job enlargement).

The importance of the conceptive work system design increases in connection with the introduction and establishment of the concurrent engineering approach or related concepts for the parallelism and/or integration of a product-, process-, and production system design (Hacker refers also to the term of "activity project planning," [1991]). One of the objectives is to consider the requirements of human work in the draft stage and/or to anticipate the effects of design decisions.

Ulich (1993) differentiates between two kinds of conceptive work designs: "Preventive work design" as the mental anticipation of possible harming of the health and psychosocial impairment, and "prospective work design" as the conscious anticipation of possibilities of the personality development by the creation of scopes of activity, which can be used by employees in different ways. A prospective work design should result in motivation and enable learning through activities and tasks, which offer the potential for personal expansion and development.

It has been indicated that although the conceptive (anticipated) work design can noticeably reduce the time and cost for the corrective work design, it does not eliminate these completely. An evaluation of the realized work system, however, is still necessary; the correctness of the obtained assumptions and forecasts must be verified with consideration of the individual characteristics of the working persons (see also for this stress-strain concepts in Luczak and Rohmert, 1997).

5.4.3 Sequential and Integrated Work System Design

Concerning the temporal reference with which the individual design range of technology, technique, ergonomics, and organization are processed two further strategies can be differentiated: the sequential and the integrated work system design (see Kirchner, 1972). The characteristic of the sequential work system design mentioned earlier — is that the design ranges are worked on gradually in a phase conception. (A conceptive design presupposes naturally that all phases will go through the thought process first before the work system is realized.) A fundamental problem consists, however, of the fact that the design levels are not independent of each other. Generally the decision made in a planning phase for a design condition limits the decision range of the following planning phases. The advantage of the transparent planning process, which limits the complexity is always faced with the disadvantage that interdependences between the design ranges are not considered necessary. It is criticized further that sequential concepts usually begin with the technological work system design and the following decisions are a subordinate of the selected technological concept (technological determinism).

This determinism can at least be partly avoided by iterative procedures, that is, by the installation of feedbacks and presumptions:

- *Feedbacks:* if no satisfying solutions can be found in a design phase, completed phases are repeated in the form of a loop and new insights can be integrated
- *Presumptions:* in each design phase the consequences are anticipated, which leads to a respective decision in the following stages of the design

The iterative design represents the transition towards integrated work system designs. The concept of the integrated work system design provides for a simultaneous processing of the four design stages, in contrast to the sequential work system design (Kirchner, 1972, 1993, 1997). Ergonomic problems and aspects of the work organization are to be considered in particular with reference to technological and technical determinations. A goal is to consider reciprocal effects from the beginning and avoid the compensatory measures, which are accomplished in the sequential design process. Compensatory measures abolish unwanted secondary conditions of proceeding decisions (Luczak, 1998). Due to the complexity of the planning process interdisciplinary planning teams are necessary for the realization of such an integrated approach (Luczak, 1996).

5.4.4 Technocentric and Anthropocentric Work System Design

Every decision made concerning the mechanization degree of a work system also means a determination of the function division between humans and operating resources. This allocation process can orientate itself towards different criteria. The spectrum of design strategies is clarified on the basis of two typical conceptions: the technocentric and the anthropocentric design strategies (Brödner, 1986).

The aim of a technocentric design is to attain an operating goal (production of goods or services) possibly independent of the requirements of manpower. The intention of the approach is to reach a high degree of mechanization and automation. The remaining work carried out by humans is characterized by a large amount of division in labor, in order to reach a high exchangeability in the workforce. Human work itself is a subordinate of technical requirements (e.g., cycle time) and is regarded in principal as a potential interference factor. A goal, in the long run, is a fully automated production.

In contrast, an anthropocentric design proceeds from the requirements of the workforce. Humans are perceived as a complex nature whose needs are to be considered during the work system design. Activity elements, which are helpful to this idea of man are carried out by humans, all others are transferred to the technical system. The goal is that the technology subordinates to the human (in particular also the individual human) requirements. Both strategies are faced with limitations in technical and economic basic conditions, as long as they are represented in a true form.

5.5   Approaches to Anticipatory Work System Design

5.5.1   Ordering Model

The basic philosophy of successful ergonomic design seems to be the anticipatory thinking in terms of design processes and their consequences. Thus, anticipation means that the designer is aware of the consequences of his actions. A generalized approach to the problem of anticipation in ergonomics must fail, because different disciplinary concepts propose design solutions of different scopes and different logical basis (Luczak, 1995). So the question arises what might be a structure to differentiate between different approaches to the problem of "anticipation," between different concepts of anticipatory design, and evaluation of work systems respectively?

The concept of macroergonomics (Hendrick, 1993) implies at least a differentiation between two levels: a microergonomic level dealing with specific tasks and related human–machine interfaces, and a macroergonomic level with a top-down socio-technical systems approach to organizational and work systems design.

Combining the knowledge for anticipatory work design with clear ideas for successful design procedures and structural design alternatives means to acknowledge that his two-level model only roughly structures the scope of ergonomics. A concept that differentiates the analysis of human work and working conditions, according to levels of integration (and the degree of abstraction), was developed by Luczak et al. (1989). The structural levels of this concept of analysis and design of working processes can be used to define or to find out what kind of models of the human or the working situation are available to anticipate design procedures and respective solutions. The procedural levels of this concept, however, allow a separation of the work-related, time-line oriented specificity, or depth of analysis. Coordinating this approach with the unquestioned philosophy of ergonomics (macro and micro), namely, to adapt humans and working situations means to identify models of the human or models of the work situation, which are able to predict design necessities, design approaches, and design solutions to a considerable extent, but only on the level that they are designed for and on the processes that they are validated for. Each level has a relative autonomy in the whole concept, but it can be combined with design recommendations of the other levels (Figure 5.13).

The concept divides working processes and working structures into seven levels. The highest level represents work related to its societal impacts and to cooperative processes between companies (level S7). The lowest (level S1) represents elementary physiological processes.

Level S1 deals in general with the physiology of autonomous organic systems in relation to physical and chemical environments, but through the perspective of anticipation and simulation models of control theory in particular can be used to find out stability limitations and stability reserves of the human body under environmental settings or to predict the combined effect of different components of an environmental factor in one summarized figure by previous experimental investigations; for example, to combine climatic factors to the result of a "basic normal temperature" (psychophysical model), a "heat stress index" (thermodynamical model), or a "P4SR — predicted four hour sweat rate" (physiological model).

Level S2 concerns the actions of willfully steered organic systems — the biological and psychological basis for limb operation — in relation to operations and movements with tools, work means, interfaces, displays, and controls, etc. the predictive concepts on this level focus on anthropometrical man models, for instance, that combine design data in a human-oriented arrangement for body segments and the whole body, or they imply specific design technologies for tools and machines such as video-somatography or CAD systems.

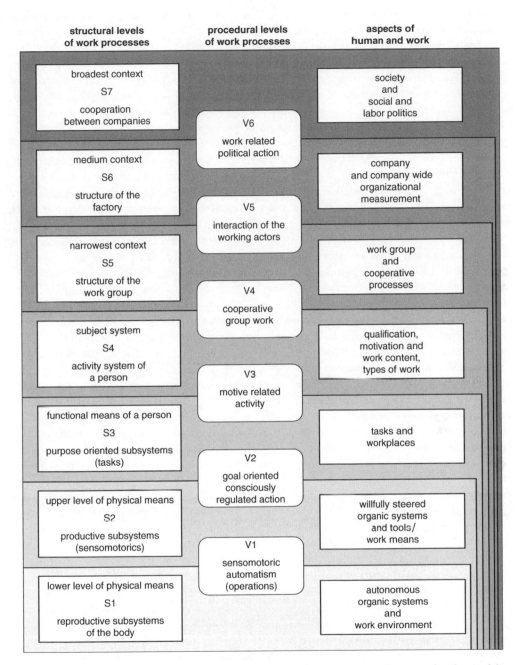

FIGURE 5.13 Structural and procedural levels of working processes and assigned aspects for the modeling of humans and work.

The focus of level S3 is on human tasks embedded in motive-related execution procedures in combination with systems analysis of workplaces. The concept of systems analysis and task analysis leads to a variety of models — the goals operations methods selection rules (GOMS) model for human–computer interaction, the work-factor-mento approach to visual inspection and, last but not least, the different systems of predetermined times for manual assembly. The analytical and constructive "power" of these models leads to a broad application in industry with millions of workplaces designed and operated according to an optimization of time consumption in highly repetitive tasks.

The central level, level S4, concerns the human being as an individual working person: the typical approach on this level is a holistic view of human work as an entity of motivational, qualification-based, and social-interactive elements, which together result in forms or types of work with characteristic sets of work content, work demands, and work-induced stressors. The models to anticipate design results on this level are multifarious according to the specific anthropological perspective of the human: man as a personality regulating his actions in an environmental context of self-set or accepted tasks in social interaction with others or man as a person, underlying, withstanding, and influencing his workload (by a set of stressors) according to his abilities to cope with them in stress-strain concepts. Limitations of use and abuse of human resources are frequently based upon those concepts, and, thus, standards of acceptable and endurable working conditions can be designed anticipatorily for a person or a group of persons after stressor analysis.

Level S5 brings into focus the working groups and group work, which means the cooperation of individual working persons with their functions in the network of interactions to other persons that is determined by division of labor, hierarchy, behavioral traits, participation in decision processes, as well as questions of communication and information transfer with respect to human relations. Mostly models of function allocation and simulation models for crew design and crew operations in specific goal-oriented and task-oriented settings are used to find out the effects of independent variables on crew performance, crew workload, crew behavior, etc. before implementing a specific design configuration.

Level S6 describes company organization with special reference to the personnel and to the industrial relations of employers' and employees' representatives. Economic and social aspects in the design of company structure and functions, related strongly to job design and evaluation, are combined in industrial engineering or operations research models for the preparation of decisions about a management of human resources. Cost-benefit analysis of design solutions and economical optimization criteria following cost-structure models, quantitative production output models or, recent quality-oriented modeling of production behavior are the tools for the prognostic or anticipatory approach. Whereas level S5 implies clearly a human-oriented approach by the macroergonomic perspective, level S6 derives design intentions and goals for level S5 or sets limitations for design efforts on this level.

Level S7 is oriented to comprehensive socio-political and societal contexts of work. Typical questions on this level deal with work regulation and standardization, work in the national economy, structural and economical components of employment, the labor market, activities of employers, and unions in socio-politics and cooperations between companies. Econometric models and growth models for national/international economies try to anticipate the effects of political decisions in complex situations, obviously with limited success regarding the world-wide problem of unemployment in many national economies.

The application of the seven-level concept of work sciences to the problem of "anticipation" shows that the microergonomic models for predictive and prescriptive design reach from level S1 to level S3, whereas macromodels of work sciences can be assigned to levels S5 to S7. An intermediate position in between macro and micro, named meso, can be given to level S4 where the working person is seen in a reacting position to environmental conditions caused by work, as well as in an acting position in coping with the work environment and, thus, influencing work itself by individual and collective efforts.

5.5.2 Inter-relations between Ordering Model and Work System Approach

Systematic considerations are always expedient, if simple cause-and-effect chains do not exist between variables but, however, there are more than two variables in relation to each other. The analysis, evaluation, and design frames of a work system are arranged by the determination of system borders. The foreclosure of fields and, therefore, the creation of a system environment enables the ergonomist to acquire and process complexity with his limited intake capacity (Section 5.2.5). Envisioning the central aim of ergonomics — that is, to say the humane and economic design of work systems (Section 5.3.1) — the task to reach this aim is complex to such an extent that a focus on individual subsystems (e.g., the effect of certain environmental variables on humans) is normally inalienable in research projects. Thus, the individual elements of work system approach are regarded as systems in each case and — if necessary — are to be divided in subsystems. Different aspects of individual elements of a work system are the center of consideration at every level of ordering model. The system border is displaced in each case with the change from one level to another (Figure 5.14).

On level S1, the human is divided into physical "subsystems," for example, the cardiovascular system, muscles and sinews, etc. The physiological systems of people and their interaction concerning the

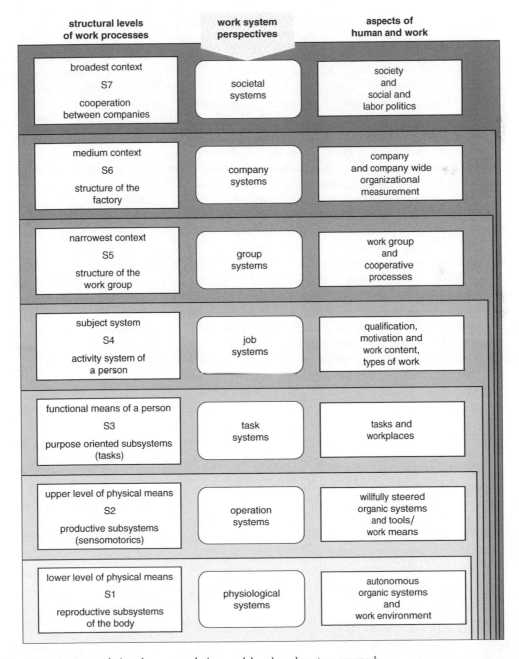

FIGURE 5.14 Inter-relations between ordering model and work system approach.

dependence on the work environment and the work task take center stage in these considerations. On level S2 elementary physical and psychological functions are examined regarding the work person — mostly in cooperation with working equipment. From these investigations, design recommendations for the technical system are derived (e.g., references to the design of displays and manual controls). The central focus of work systems exists, on the one hand, concerning the subsystems of humans (e.g., muscles and skeletal) and, on the other hand, concerning the technical system (e.g., workplace design). A connection between human and technical systems is produced by the individual operations of the people. Therefore, at level S2 a work system is considered through the perspective of an "operator," so that at this level it can be spoken from an "operation system." It is typical for this level of the ordering model that only subsystems of the work system element "task," namely operations and activities, are regarded. At level S3 the work system is not considered from the perspective of operators or individual operations. The system border shifts in such a way that individual subtasks or activities are no longer regarded in relation to the work system element "task," however, one or more total tasks are. Therefore, at this level "task systems" should be taken into consideration. Questions concerning action regulation in connection with task systems are examined in relation to humans. With this, the center of attention is the humane design of the task system. Beyond that, questions of functional and temporal cooperation of humans and technique are discussed. Primarily, the goal of the utilization of human work is pursued (Section 5.4.1). Jobs are considered at level S4. These develop through the combination of many tasks. The job description usually takes place independent of the individual working person. From these descriptions the individual tasks and their cooperation in the "job system" can be derived to side the requirements of the working person and a quantified job evaluation. Requirement determinations serve, therefore, as a basis for a choice of personnel, a personnel development and for the design of remuneration. The center of attention is on the adjustment of humans to the work. On the other hand, it is the design of "job system" that affects the personality and motivation promoting of the working person. The central focus of considerations is the adjustment of humans to the "job system." At level S5 of the model, a work system is considered in the perspective of "group systems." Level S5 is not similar to level S4 in terms of just one working person or only one job is considered rather a majority of people, that is, jobs. Groups are characterized by the fact that they consist of several coworkers who work together over a longer period with direct interaction. Members of a "group system" fulfill a whole task together, but to a certain degree role-differentiated. Apart from the execution of this primary group task, they also take over secondary tasks (Section 5.3.2). At level S6 such system components of the complex "company system" are examined, which concern the human work. The system borderline shifts in such a way so that in reference to the working person all people or a large part of the staff are also included into the views of the system. Ergonomical questions of the design of the "company system" refer for example to operational work time regulations, the control of job execution processes, and the organization design. If the system is regarded as societal, the system borders of an enterprise and its staff shift all enterprises of a society or a majority of enterprises.

In the following, models and methods — to each level of the scheme — are presented exemplarily. These models and methods are able to support a systematical proceeding with the anticipating and prospective work system design. These approaches and methods are the result of research projects, which were accomplished by the IAW and FIR at the RWTH Aachen. The exemplarily cited research results are supposed to show that the use of the work system approach is possible and reasonable on different levels of ergonomic investigation. Beyond that, the exemplarily represented models and methods make clear how the system limitations of one level shift ordering models to the next.

5.5.3 Identification of Recovery Times in Heat Work as an Example of Work System Design at Level S1

In the level S1 of the ordering model the human is subdivided into physical subsystems, for example, the cardiovascular system, muscles, and metabolic system. Subject of the level S1 is the physiological work system design, which includes all measures that have a direct influence on human physiology.

According to the work system approach, different influences of work environment affect a working person (Figure 5.15). These influences hardly ever appear isolated, but mostly in combination (Section 5.3.2). Concerning heat work, for example, unfavorable climatic conditions take effect in combination with hard dynamic muscle work (Luczak, 1979). In doing so, the climate conditions are allocated to environmental influences of a work system, whereas, muscle work is considered as an action resulting from the work task.

The subjects of investigation concerning heat work are the questions, which "human subsystems" are especially loaded, how these loads interact, and when "bottlenecks" arise in the human body. For example, the cardiovascular and the muscle system cooperate, because the amplitude and the frequency of respiration depend on the muscular load. In the same way, the cardiovascular and the metabolic system depend on each other due to the perspiration under heat work conditions — that is, an increasing loss of water and salt — these materials urgently have to be taken in. Therefore, interferences caused by salt deficiency (e.g., heat convulsions, which effect the cardiovascular system and, therefore, also the muscle system) can be anticipated.

At the time of publication of the following described study (Luczak, 1979), numerous scientific realizations were present concerning "pure" muscle work and especially the endurance limit and the recovery time. However, combinations with environmental influences were examined only sporadically, referring to a few points of the entire load continuum.

A goal of the investigation was the obtaining of physiologically justified realizations for the determination of recovery time for heat work. These realizations should cover a wide range of relevant load continuums. In addition recovery times should be assigned to measurable load values.

The general objective of the investigation is based on the following partial goals:

- Development of a model for the coupling of the thermal regularization system and the cardio vascular system
- Examination of the model on the basis of individual results from the literature
- Determination of recovery times and the superposition principles of energetic effects and climatic loads in the fast-time-simulation based on the model

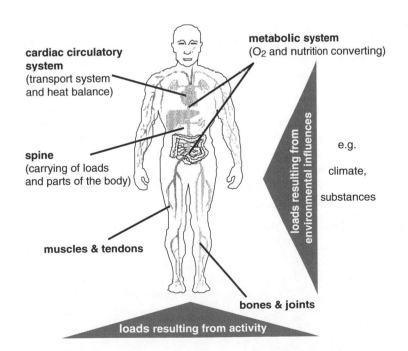

FIGURE 5.15   Physical subsystems of humans and influences on these subsystems.

- Experimental evaluation of the model results concerning the recovery times at selected test conditions

In the following, physiological bases of the thermal regularization and selected results of the study are presented.

5.5.3.1 Physiological Bases of the Thermal Regulation

The core body temperature is kept constant within a wide range independent of outside climatic conditions. The desired value of the core body temperature is 37°C. High and low outside temperatures, as well as manual labor lead to measurable deviations in the target value. The temperatures are measured by central and peripheral thermoreceptive cells.

Smaller climatic loads, warming and cooling stimuli, can be compensated by regulation of the heat transfer resistance between core areas and body surface. Larger thermal loads are compensated by increased perspiration, whereby a multiplicity of theoretical and experimental results regarding the relationship between different, locally distinguishable temperatures and the perspiration level are indicated.

The core body temperature rises during combined muscular and thermal load. The temperature can reach a steady degree, as long as heat production and heat dissipation are balanced. The circulatory system is a limiting factor of the heat dissipation and in supplying the musculature with high-energy substances. A rise of the body core temperature is usually accompanied with a rise of the heart frequency. The heat dissipation becomes insufficient, if an increase in blood circulation of the skin is no longer possible. Other reasons for the insufficient heat dissipation are loss of water and a conversion of thermal balance (environmental temperature is more than core body temperature) under the conditions of increasing duration of work load and very high outside temperatures.

5.5.3.2 Model for the Description of the Thermal Regularization System and the Cardiovascular System

Regarding the determination of the recovery times and because of the interconnection of thermo-regulatory and cardio-respiratory indicators it appears necessary to consider these both, several multiple connecting physiological systems. The model describes the regulation and rhythm of the momentary heart frequency, the respiration, the blood pressure, the core temperature and the skin temperature, that is, those physiological functions, which are predominantly affected by a change in heat work and interfere in a quantitatively describable way.

In terms of this model different approaches described in the literature were examined. In principle all the available approaches at the time were similar so that easiest correlation indicated by Behling (1971) has been transferred into the model. Thereby, a correction factor (Scarperi et al., 1972) was considered. The factor contains the loss of water in the organism in a long-term test. Behling adopts two blocks, body core and body surface. A defined heat exchange takes place between these blocks. Perspiration rate, oxygen intake, and heat development are determined by a weighted sum of core and skin temperatures. The model was compared and examined with the measured values from physiological experiments regarding its reactions to thermal and energetic-effective load both for the steady state and for the courses of the most important model parameters.

5.5.3.3 Test Run to Determine the Recovery Time Diagrams

With the successful examination of the model, the requirements were given to superimpose the load factors "hard dynamic muscle work" and "thermal environment" for defined work durations and subsequently to pursue the recovery time processes. The climatic conditions of the test runs were varied effectively within the range of 0–35°C, since the model is appropriate for a temperature range, which does not include the conversion of thermal balance with the rise in the human body temperature by external thermal influences.

The energetic load was varied upto 150 W, whereby the fictitious endurance level of the model was focused at 100 W. A hard dynamic muscle work with high efficiency is presupposed, since the model equations based on the investigations were accomplished with such work forms. The fictitious endurance level of 100 W corresponds to the continuous endurance level of a young, healthy, male employee. It is realistic, because only suitable employees are involved in heat work because of personal selection and adaptation processes. The values for the work duration were applied in five geometrical stages: 15, 30 min, 1, 2, and 4 h. These values appear meaningful regarding their delimitations, since a continuous work duration under 15 min represents a special case in operational practice, and a work duration over 4 h may not be accepted due to the laws of working time regulation.

5.5.3.4 Experimental Examination of the Model Results

Four test subjects were selected. These subjects differed broadly in terms of size, weight, and physical efficiency and covered, therefore, a wide range of observable continuum. Three of these persons were heat-adapted, that is, the test subjects had to generate a performance (on an ergometer) between 50 and 100 W for 2 h daily in a climatic chamber at an effective temperature of 33°C.

5.5.3.5 Conclusion

Both model results and experimental results were recorded in diagrams. The diagrams indicate that the recovery time increases exponentially starting from a certain energetic load value and from a certain effective temperature. The influence of the effective temperature begins with a certain temperature level. Until this level the thermal regularization system of the human body is able to regulate the body core temperature independent from the work duration due to a reconciliation of heat production and heat dissipation.

It is evident in all test results that the four factors "effective temperature, muscular load, work duration, and individual characteristics of the working person" determine the characteristic curves. The influence of the effective temperature begins, thereby, from a threshold value. This value is again supposed to be dependent on the energetic-effective load, that is, an endurance level for the superposition of these two types of loads. Beyond this point, the slope of the course of the curve is determined by the energetic-effective load. The recovery time increases to the power of the energetic load value and the effective temperature.

If the model results are compared with the experimental results, then it shows up that the tendency of the courses of the curves exhibits large agreements (Figure 5.16).

5.5.4 Coordination of Movements as an Example of Work System Design at Level S2

The combination of power-, speed-, and acceleration development of muscles and skeletal elements in a purposeful process with spatial and temporal coaction is denominated coordination of movements. Working persons coordinate the movement of their extremities and their body to act on the work object indirectly and (with the help of work and operational funds) directly. Thus, they fulfill their work tasks. Powers, which are raised by individual muscles are inter-connected regarding amount and direction. Thereby, the movements of the extremities and the body are spatially and temporally controlled.

The movement of the human body and its extremities in its entirety is a complicated mechanical process. The principles of this process can quantitatively and causally are conceived by the use of biomechanical analysis. Apart from the proposition concerning movement coordination based on a predominately energetic-effective view, however, informational-mental principles are also important for the work system design. Furthermore, movements are also temporal-operative processes, that is, the adjustment quality to an optimal course concerning spatial approximation and movement economy, as well as concerning speed-oriented, temporal aspects is achieved by continuous iteration of movements. This process is denominated as an exercise, especially with short-cyclic-repetitive sensorimotor activities.

Selected realizations concerning biomechanic analyses, valuate-informatical processes, and exercise processes are presented in the following (Luczak, 1983).

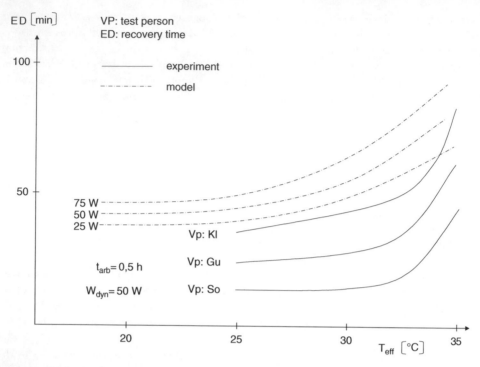

FIGURE 5.16   Comparison between experimental results and model/test results concerning the recovery time.

5.5.4.1   Biomechanical Analysis and Functional Anatomy

The presentation of the human body in mechanical systems of equations (deductive model systems) is a characteristic for the biomechanical analysis. It allows to type exactly the quality and quantity of a movement and, thus, to estimate the mechanical load of the human body in a certain circumstance of load. The biomechanical analysis does not deal with those corporal structures in which these powers are produced and absorbed. These are the topics of the functional anatomy. In this context, the coordination of movements is an organized interaction between different skeletal elements and muscles.

Isolated individual movements are rather uncommon. The consideration of normal movements reveals that entire units of muscles are involved in the carrying out of even simple movements. More or less all muscles of the full arm collaborate in the rather simple process of outstretching the arm. Furthermore, the mustered power spreads out through the entire brace of the torso. So this movement, which seems to be so easy demands a reasonable interaction of a great amount of muscles of the entire body. The analysis of such muscle units as a complementary view of the biomechanical analysis is quite useful for the practical work system design, in order to identify the bottlenecks in building up power and moving. This applies especially for all forms of posture work, whereby the powers released to the outside are negligible compared to the inner powers.

Realizations from physiology, which are useful for the coordination of movements mainly refer to the minimization of the energetic and physiological effort in consideration of the complex functional system. Movements do not only consist of a summation of tetanic twitches of several muscles, but are determined by the inertia powers, the attenuation powers, and the elastic powers of the sinews, muscles, and ligaments of the extremities.

5.5.4.2   Movement Informational Processes

An idea of the complexity of movement informational processes can only be achieved once an attempt is made to accomplish a movement that is familiar to others but unfamiliar to you. The informational difficulty of a movement becomes clear in the observation of the first clumsy trial of playing tennis, ice-skating, or chipping. This difficulty cannot be handled until the brain is able to store the movement programs in a way as subprograms. This means that only the order to retrieve the subprogram has to become conscious and, thus, engage the upper nerve center, after a movement is once available as a subprogram.

One of this automated work takes place with very little mental load. One possibility to judge the degree of automation of an activity is to examine the performance proportion, which is possible to carry out besides the automated activity. This idea is based on the model of a single-channeled information processing with defined maximal channel capacity. According to this, information processing degrees originating from different activities can be summarized additively to the maximal channel capacity.

The absence of the consciousness in automated work makes a wide division of labor with its continuous repetitions bearable, since the "disencumbered" consciousness is able to deal with other things but work.

Methods originating from information theory and control theory are the appropriate measures, if one tries to quantify the entirety of information processing with the movement coordination instead of trying to fractionate only the conscious part. In the majority of cases it is only possible to describe manual aim movements with the help of information-theoretical and control-theoretical models. These manual aim movements are interspersed to a certain extent between the individual "stationary" movement elements, for example, grabbing, assembling, or joining. Regarding the entirety of the motor activity, the signaling can be analyzed with the help of control technical systems by the use of actuators handling elements, that is, the motor information transfer in general, which is also called signal-motor work. Thereby, predications concerning work design can result.

5.5.4.3 Operations of Exercise

Each exercise starts with the comprehension of the rules, which determine the movements. These rules are also called work methods (see also the remarks concerning work tasks in Section 5.3.2.2). A condition for the perfection of the movement course by conscious and continuous repetition is this view of the informational and motor operations in which a movement is integrated. Repetitions lead to the fact that a consecution of operations eventually transfers to an automated component of a conscious activity.

The quantitative and qualitative increase of performance, the decrease of mistakes, the decrease of the required time, the decrease of energy effort and, furthermore, the decrease of different central-physiological parameters of load, for example, the heart rate and electrical activity of working muscles are all criteria of the exercise success.

Examinations concerning the contribution of individual movement elements to the exercise course showed that an increase of performance could only be achieved by the use of temporal shortening of the difficult movement elements (e.g., grabbing, assembling). Thus, it seems reasonable only to intensely exercise the complex movements of a work course. This method is suggested for complicated and long-term cyclic activities.

The exercise, according to the entirety method, is the most popular. Thereby, the whole task is carried out from the beginning to the end before it is repeated. This method is used especially in situations where the workers have already started producing for the manufacturing while still exercising (training on the job). Financial incentives turned out to be effective for increasing the speed of practice.

The entirety method is a kind of active exercise. Thereby, the movement, which is to be accomplished is orderly and actually repeated. Besides, these possibilities of coordination exercise, training forms are possible, which are based on the fact that working persons store the movement as an "internal model," that is, that the working persons import, consolidate, or correct the informational program of this movement. Thereby, it is possible to distinguish between observative, mental, and verbal training. With the observative training the practicing person watches a very proficient person while carrying out the movement. With mental training the trainees imagine the movement course. With verbal training the affected people talk about the movement course.

5.5.5 User-Centered Design of an Autonomous Production Cell (APC) as an Example of Work System Design at Level S3

The design of a user interface of an APC is a typical example for work system design on the third level, because APC are based on effective interaction of highly skilled operators as well as complex CNC-based machinery and industrial robots linked with an automated materials handling system. Its operation is controlled through a supervisory computer (Pfeifer et al., 2000). An APC that is very closely related to flexible manufacturing systems is dedicated to a single part family, so that each part family can be completely produced within the cell, and surrounding cells have minimum interaction with each other. APC are next generation manufacturing systems where human operator and technical subsystems develop a cellular organizational unit within the organism of flexible production plants. A further characteristic of APC is that planning, set-up, numerical control programming, process control, quality inspection, and fault management tasks are integrated locally in the manufacturing cell. In the light of human–machine interaction, APCs are highly automated manufacturing systems. Hence, the task spectrum of the APC operator is shifted from manual process control to planning, programming, and monitoring tasks (Luczak and Schlick, 2000; Reuth et al., 2001). Figure 5.17 shows the APC as a work system (Luczak et al., 2003).

Especially human performance models (see, e.g., Pew and Baron, 1983) are arranged on level S3 from a subject-related point of view. These describe behavior patterns of agents with the machine interaction. Models of informational user performance (cognitive models) are a subset of this model family. In ergonomics the approach of cognitive modeling is crucial for the design of complex and adaptive human–machine systems. Because of the variety of cognitive models available in the literature this area of expertise seems to be a good demonstrator for the solution of the question how a specific cognitive modeling approach can be assigned to a specific design problem, namely that of an user-centered design of an APC (Luczak and Schlick, 2001).

5.5.5.1 Requirements for Cognitive Models

As a basis for the selection of the appropriate cognitive model, Schlick (1999) has defined a total of 11 requirements. Due to the application context of flexible manufacturing, the following nine requirements

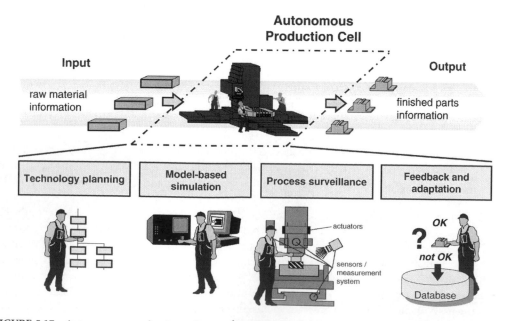

FIGURE 5.17 Autonomous production cell as work system.

must be taken into account (Luczak and Schlick, 2000, 2001). First, it is required to model human information processing in complex task settings. That means the model must be able to represent input–output transformations, which significantly extend Sternberg's (1969) classical stimulus-response approaches. Therefore, the theoretical construct of a "mental model" must be considered, because tasks of production planning and diagnosis often require an appropriate reasoning space for the operator. Second, it is required to represent concurrent threads of reasoning and, therefore, cope with aspects of mental resource allocation. Therefore, a symbiosis of stage- and resource-oriented models of human information processing is preferred (Kahnemann, 1973). Third, there is a need to model different time bases of human information processing, because some tasks, for example, process control, need synchronous information processing, while other tasks, for example, production planning, are completely self-paced. Fourth, especially synchronous human information processing involves different sensory modalities, which must be considered. Fifth, the autonomy of the manufacturing system requires to model aspects of human learning and adaptation. Sixth, human errors play an important role for performance prediction and evaluation of system reliability should be taken into account. Seventh, a strict modeling formalism is required to anticipate time consumption of relevant cognitive functions. Eighth, it is required to represent cognitive stress and strain, because equal performance levels may be assessed differently with regard to "mental costs." Ninth, it is required to integrate formal aspects of human communication in terms of semiotic modeling, so that different levels of information exchange between human and machine can be differentiated (physical, syntactical, semantical, pragmatical).

In addition to these application-driven requirements, two utility-driven requirements must be taken into account: first, it is useful to have software tools for supporting cognitive analysis, modeling, and evaluation. Second, widespread models are preferred, because they ease data collection and comparative assessments.

5.5.5.2 Preset Catalog of Cognitive Models

It is necessary to distinguish between normative and explicative approaches of cognitive modeling. Normative models provide an answer to the question "what should be?," that means the model structure is developed with a system of design goals in mind. These models usually stem from the domain of cognitive engineering (Rasmussen, 1986) or the domain of naturalistic decision-making (Zsambok and Klein, 1997).

The following normative models were considered:

1. Caccibue (1998): cognitive simulation model — COSIMO
2. Card et al. (1983): model human information processor, which is the basis for the well-known approach of goals, operators, methods, and selection rules — GOMS (e.g., Kieras, 1997)
3. Rasmussen (1983, 1986): "model of skills, rules, knowledge, signals, signs, and symbols" — SRK

In contrast, explicative models provide an answer to the question "what is the cause?." That means the model structure is developed through inductive reasoning steps regarding cause-and-effect relationships. Therefore, explicative models have their roots in cognitive science or artificial intelligence research (Newell, 1990).

The following explicative models were considered:

1. Anderson (1993): ACT-R theory
2. Newell (1990, 1992): unified theories of cognition — UCT

5.5.5.3 Model Evaluation by Multi-Criteria Value Functions

The goal of further investigation was to select the model with the best goodness-of-fit concerning the fulfillment level of the cited requirements. Therefore, a decision theoretic approach with multi-criteria value functions was preferred (Eisenführ and Weber, 1994). In a first step, the fulfillment level of the requirements was ranked model-by-model using a three-level ordinal scale (requirement fully fulfilled = value of 1, requirement fulfilled partially = value of 0.7, requirement not fulfilled = value of 0). In a

TABLE 5.1   Evaluation of Cognitive Models (Cells of Matrix Refer to Requirement Fulfillment Level, the Goodness-of-Fit is the Weighted Sum of Fulfillment Levels)

Model/Requirement	COSIMO	GOMS	SRK	UCT	ACT-R	Priority
Complex tasks	2	2	1	2	3	1
Resource	3	2	3	1	1	2
Different time bases	1	2	1	1	3	1
Multiple modalities	3	1	3	2	3	1
Learning	2	2	1	1	1	2
Human error	2	3	1	3	3	1
Time consumption	1	1	2	1	1	1
Stress/Strain	3	3	1	3	1	3
Semiotics	3	3	1	3	3	1
Software support	2	1	3	2	2	3
Spreading	2	1	2	2	3	2
Goodness-of-fit	0.53	0.61	0.73	0.62	0.30	

second step, the corresponding weights of the requirements were defined with the help of an exponential priority scale (first priority has twice the weight of second priority, second priority has twice the weight of third priority). The goodness-of-fit for each model was calculated as the sum of the weighted fulfillment levels (Table 5.1).

With regard to human information processing in complex task situations, the SRK approach represents the construct of mental model in terms of a detailed means-ends abstraction hierarchy (Rasmussen, 1985) and, therefore, fully fulfills the first requirement. However, COSIMO, UCT, and GOMS define different levels of cognitive abstraction and, therefore, partially fulfill this requirement. ACT-R uses the action-rules monolithically and, therefore, does not fulfill this requirement. Concerning the allocation of mental resources, the UCT, as well as ACT-R provide features to activate and schedule cognitive rules in parallel and, therefore, cope well with limited cognitive resources. Also GOMS-extensions, for example, Gray et al. (1993), can model parallel execution threads, but only with a restricted horizon of resource allocation. The other models offer limited features only and, therefore, do not fulfill this requirement.

Multiple time bases concerning task coupling are represented well in SRK, COSIMO, and UCT, which rely on a multi-layered architecture of cognition. GOMS has restricted abilities when considering different partial models (like unit task and keystroke level). Multiple sensory modalities can be modeled with GOMS appropriately. UCT differentiates between sensory input and output channels, but is lacking specific operators. The other models do not fulfill this requirement. Learning processes are represented well with SRK, ACT-R, and UCT. The SRK-framework interprets human learning as a shift towards lower levels of cognitive control while ACT-R and UCT include chunking mechanisms. The other approaches only partially fulfill this requirement.

Aspects of human error and reliability are represented well with extensions of SRK such as Reason (1987) and Hannaman et al. (1985). COSIMO also copes with human reliability, but is restricted to error detection and recovery. The other models are lacking sufficient mechanisms for human error modeling. COSIMO, UCT, GOMS, and ACT-R provide strict formalisms for modeling of mental time consumption and, therefore, fulfill this requirement completely. SRK offers limited functions in terms of rules of thumb. Aspects of mental stress and strain are a fundamental part of ACT-R, which controls rational behavior. Therefore, this model fulfills this requirement completely. Moreover, the work of Moray et al. (1988a, b) must be taken into account, which enriches SRK with workload modeling. The other models are lacking such evaluation mechanisms and do not fulfill this requirement. Semiotic aspects of human–machine communication are represented well in SRK, which differentiates among signals, signs, and symbols for information exchange. The other models do not integrate semiotic aspects and, therefore, do not fulfill this requirement.

Concerning the first utility-driven requirement, the GOMS approach offers a variety of software tools for analysis, modeling and evaluation and, therefore, fulfills this requirement completely. However,

COSIMO, UCT, and ACT-R also offer specialized software tools for model simulation and, therefore, fulfill this requirement partially. Finally, GOMS is the most widespread model of human information processing and is a "de facto" standard for the investigation of human–computer interaction (Gugerty, 1993) and fulfills this requirement completely. COSIMO, SRK, and UCT are well known in their corresponding scientific communities and partially fulfill this requirement. ACT-R is a highly specialized approach.

In conclusion, Rasmussen's (1983, 1986) model of skills, rules, knowledge, and the abstraction hierarchy for knowledge representation (Rasmussen, 1985) in conjunction with additional work on human reliability modeling (Reason, 1987; Hannaman, 1985), as well as workload assessment (Moray et al., 1988a, b) represent the framework for cognitive modeling with the highest overall goodness-of-fit in the application domain of flexible manufacturing. Therefore, this framework was used extensively to investigate operator requirements for human–machine interfaces of APCs (Schlick et al., 1995), to design the human–machine interface on a conceptual (Schlick et al., 1996), as well as a detailed level (Schlick et al., 1997), and to simulate and assess the production system as a whole, which includes aspects of human error and labor division (Schlick, 1999).

5.5.6 Computer-Based Prospective Job Design and Evaluation with Space+ as an Example of Work System Design at Level S4

The concept of concurrent engineering (CE) aims at shortening time to market, reducing processing cost, and improving product quality. It requires, besides, a simultaneous product and process design also a simultaneous design of the production system (Clausing, 1993; Prasad, 1996). For such an early job design, tools are necessary, which support the iterative process of design and evaluation of jobs and tasks in production. In this context criteria of personality development are — among others, for example, safety and feasibility — considered to be important.

However, the repertoire of methods in CE research lacks instruments, which are especially developed to allow prospective job design with respect to these criteria within a CE environment. The demand for such supporting methods and instruments is based on time and cost arguments. In conjunction with other design instruments (e.g., for industrial organization) they should reveal requirements and potentialities of production as early as possible and assure a fast and smooth start of production process. On the one hand, there is a need to initiate time and cost-intensive work organizational- and personnel-related measures in an early stage (in this case the job design is the basis for determination of KSA profiles for personnel planning; KSA = knowledge, skills, and abilities). On the other hand, there is a strong need to keep existing staff — especially in times with a shortage of skilled engineers and technicians — and to ease personnel recruitment (this means the use of human-oriented criteria to create attractive jobs and to lower staff turnover and absenteeism). The applicability of such methods crucially relies on the capability to tolerate CE-specific circumstances such as incompleteness of information, frequency of iteration and modification, as well as progressively increasing level of detail (main CE requirements).

5.5.6.1 Basic Concepts and Development of Space+

Starting from a criteria-based analysis of existing job analysis methods and job design systems and the ascertained deficits, the system Space+ (system for prospective job design and evaluation in CE) has been developed. The basic concepts of Space+ are a modeling concept, a KSA model, and nine assessment models for evaluating designed jobs and tasks with respect to criteria of personality development. Realized as a software prototype (operating system Linux, programming language C++, development toolkit QT from Trolltech), it supplies an instrument designed to meet the earlier mentioned CE requirements.

The modeling of jobs in Space+ is based on Harel's Hi-graphs and his "blob"-notation (Harel, 1988). Taking up the job descriptor concept of McCormick et al. (1989), tasks and jobs can be described using "job elements, process elements, and cooperation elements," provided in so-called "descriptor libraries." All elements are linked with sets of KSA requirements. Job elements describe work behavior on a general level and refer to general traits in terms of abilities. Process elements allow a more detailed description of task execution (concrete operations, tools, etc.) and are directly linked with knowledge and skills. Taking into account the specific nature of process elements, the corresponding library is open and can be modified by users to their needs. A detailed description of the modeling concept and its advantages for the use in a CE environment can be found in Stahl (1998) and Stahl et al. (2000). The KSA model and the way in which job descriptors and KSAs are linked are presented in Stahl and Luczak (2000). Furthermore, a reference catalog gives access to predefined tasks, relevant for manufacturing processes. Figure 5.18 shows an extract of the graphical user interface (GUI) of Space+ and contains an example to clarify the modeling syntax and the hierarchical structure of job models.

The main goal of Space+ is to assist the user-designing jobs, which offer motivation potentialities and opportunities for learning and personal development to (potential) employees. For assessing the created job models nine characteristic values are determined concerning the following nine criteria:

1. Regulation requirements (requirements for planning, cogitating, decision-making)
2. Hierarchical completeness
3. Cyclical (or sequential) completeness
4. Cooperation and communication requirements
5. Autonomy (degrees of freedom)
6. Responsibility
7. Feedback
8. Variety of KSA requirements
9. Opportunities for learning and further qualification

Additionally, the job designer or rather the job design team is supported in continuously improving the modeled jobs by recallable design recommendations.

5.5.6.2   Assessment Models

During the system development, special emphasis was laid on the development of the assessment models and algorithms. The challenge was to meet the CE-specific requirements and the demands of prospective job design. The basis for the development had been an analysis of condition-related instruments, based

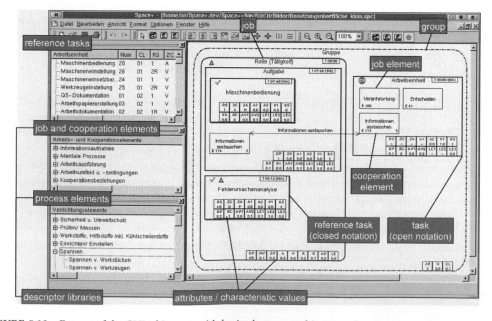

FIGURE 5.18   Extract of the GUI of Space+ with basic elements and its exemplary use.

on the action regulation theory (e.g., Tätigkeitsbewertungssystem, TBS, from Hacker et al., 1995). Taking these assessment concepts into account, the characteristics of humanitarian jobs had been systematically worked out, sorted, and selected. Each of the criteria mentioned earlier has been operationalized in an assessment model in terms of subcriteria, relative characteristic values, equations, and instructions, necessary for determining attributes. Two of the nine assessment models are presented in Figure 5.19. It becomes obvious that not only the jobs are assessed and made comparable by the characteristic values, but also the tasks and subtasks, which is another specialty of Space+ to handle the increasing level of detail during design.

The implemented assessment algorithms evaluate the used descriptors, respectively, the linked KSA requirements. In case, an evaluation based on these entities is not possible, the algorithms utilize the pre-attributed reference tasks. For the analytical description of the criteria, a deliberate limitation to the designable aspects of jobs and working conditions had been made, in order to accommodate for prospectiveness and incompleteness of information. Such characteristics were left out, which result only after the complete realization of the work system from the interaction of all or at least some system elements and which can, therefore, not be influenced directly by the job designer (e.g., feedback from superiors, working atmosphere).

With the exception of set memberships, represented by the blob-notation, and cooperation relationships, no relations between tasks and jobs are evaluated. The inclusion of control-, material-, and information-flows into the assessment concept would already necessitate a very detailed modeling in the early stages and would therewith reduce the tolerance towards incomplete information. In contrast to other methods, Space+ does not require a complete description of the work system. The assessment algorithms linked directly to the design allow modifying, gradually completing, and recognizing interdependencies of different jobs within the work system.

5.5.6.3 Catalog of Reference Tasks

A catalog of reference tasks has been developed together with eight experts (all engineers with wide experience in industrial engineering and ergonomics). The catalog consists of 54 tasks typical for manufacturing, for example, machine setup, work documentation, error analysis, and is based on the works of Schumann (1995). The reference tasks have been described by assigning descriptors (job and cooperation elements). The quality of the description and the capability to differentiate the reference tasks by means of the descriptors had been proven by a cluster analysis. Furthermore, the reference tasks have been rated by the experts with regard to certain criteria (e.g., regulation requirements, feedback) necessary for the assessment of jobs and tasks at a higher hierarchical level, which cannot be determined automatically by evaluating the descriptors and the linked KSA requirements. By integrating

Criterion Level	Regulation Requirements (RR)	Hierarchical Completeness (HC)
Job (J)	$RR_J = \max(\{RS_{ST_n} \mid n = 1, 2, \cdots, N\})$	$HC = \sum_{rs=1R}^{5} (\frac{h_{rs}}{L_{rs}} \cdot \min(\frac{T_{rsJ}}{T_J}; L_{rs}))$ with $T_{rsJ} = \sum_{n=1}^{N} T_{STn} \mid RS_{ST_n} = rs, \ rs = 1R, 1, 2R, \cdots, 5$ (visualized as a histogram)
Task (Tk)	$RR_{Tk} = \max(\{RS_{ST_n} \mid n = 1, 2, \cdots, N\})$	$T_{rsTk} = \sum_{n=1}^{N} T_{STn} \mid RS_{ST_n} = rs, \ rs = 1R, 1, 2R, \cdots, 5$ (visualized as a histogram)
Subtask (ST)	Determination of the regulation stage (RS) using the 10-stage-model of Oesterreich/Volpert 1991 (not necessary when using preattributed reference tasks)	
N = number of subtasks within the job / task; L_{rs} = limit of the allocated time per regulation stage h_{rs} = weighing factors; $T_{rsJ,Tk}$ = time allocated for sub tasks related to total time of the job / task		

FIGURE 5.19 Two of the nine assessment models for determining the characteristic values in Space+.

this reference catalog into Space+, the process of design and redesign had been significantly shortened and simplified, which is especially important for the efficient use of the system in a CE environment.

A special methodology for attributing newly designed tasks and subtasks or modified reference tasks ensures the applicability of the assessment algorithms even in later phases of the design process and hence allows the user to adapt his hierarchical structure of jobs to the increasing maturity level of information within the CE process. The system allows, therewith, the representation of different levels of information quality and quantity.

5.5.6.4 System Evaluation

Standard methods of validation could not be used for the evaluation of Space+, as Space+ focuses on the assessment of incompletely or vaguely described jobs that have not been realized. To get an indication of reliability, validity, and usability of Space+ and, furthermore, to identify shortcomings and sources of errors, a controlled experiment with 20 experts, from industry and research in the field of industrial engineering and ergonomics, had been conducted. The results of the study were used in the sense of a formative evaluation to optimize and improve the system.

The evaluation study has shown that Space+ (despite its prototype-character) is a suitable and easy-to-use instrument for prospective, human-oriented job design. The hierarchical modeling, the atomistic approach of description, the adapted and linked assessment models, the catalog of described and preattributed reference tasks, and the methodology for attributing self-designed tasks are the characteristical features and concepts of Space+, making it appropriate for the use in a CE environment. Space+ is able to support an early production system design and, thus, can help to reap the benefits outlined in the introduction.

Due to the uncertainty in a CE environment and, therefore, the required simplifications, the assessment results should not be interpreted as absolute values, but rather should be seen as decision guidance and be used as such within a cooperative design process. The Space+ system should not dictate, but assist. Further research refers to the extension of the system in terms of other kinds of jobs as well as further assessment criteria (e.g., safety, feasibility).

5.5.7 Criteria-Based Identification of Areas for Group Work as an Example of Work System Design at Level S5

If group work is implemented, the boundaries of existing work systems change. In addition to the primary task a group assumes further secondary tasks (work task: Section 5.3.2), which have been performed by the supervisors and indirect departments (e.g., industrial engineering, maintenance) before. As advantages of group-oriented forms of work organization, for example, self-regulation, flexibility, and motivated employees are denominated. A basic success factor concerning the implementation of group work consists in the "right" setting of the system boundaries of group work systems.

In the following, criteria are presented for the identification of areas for group work in production systems taking into account the scientific and practical knowledge derived from many industry-oriented projects. With the introduction of group work the role of the worker should change from an individual maximization strategy of the input−output relationship "performance to individual result" to a holistic optimization of the groups' output or the company's production. To achieve a respective change in behavior by the workers, it is necessary to select group-oriented production areas, according to the following variables, formulated mostly as quantitative indices (Luczak and Metz, 1996; Metz, 1997).

- *Group size:* number of workers in the shift. The flexible reaction to disturbances in terms of order size, absenteeism, etc. requires a sufficient group size. Groups of less than five people become critical, especially, when the production island is equipped with tool machines that have to be surveyed permanently.
- *Bottle-neck machines:* proportion of workplaces in an island that cause limits of output in terms of quantity and time. Bottle-neck machines in an island lead to idle time in sequentially depending workplaces. The risk increases the fact that external disposition of the bottle-neck limits group responsibility as well as planning flexibility.

- *Completeness of the task:* proportion of task elements in the group that can be taken over by a coworker. The design of jobs in an island should be oriented to a mixture of direct and indirect tasks: inspection of tools and machines, quality control, documentation or results, etc. give improved scope for self-determined actions and for building up a feeling of responsibility.

In particular, the analytical and simulation models clearly indicate how important these variables are for a quantitative ex-ante evaluation of a situation: person-oriented job design depends on labor-partition and cooperation between persons (group size), the availability and demands of tools and machines, and the scope of tasks to be performed by a more or less qualified working person.

- *Spatial extension:* proportion of workplaces that have no direct communication (visual or auditory) with the others. Group coherence and the direct help in solving problems will suffer with spatial extension.
- *Homogeneity of workplaces:* differences in qualification demands between workplaces. A high homogeneity of workplaces eases an exchange of workforce and increases the flexibility of manning
- *Continuity in personnel:* proportion of works that are not directly bound to the process and, thus, facilitate disposition of personnel capacities (over $\sim$ and under $\sim$). A continuous manning of the island with a well-defined crew is important for the group coherence. People without a permanent group membership do not share the responsibility for quality, production times, etc. In cases of overcapacity in personnel, additional workplaces in the production island should be available to keep persons within their group.
- *Stability of the orders:* variations in the time series of loading by a different order quantity. Different order quantities lead to increased efforts in disposition of personnel. If better workplaces are not available in the production island, personnel deficits or overflow have to be regulated by external workplaces. Negative consequences for group coherence and indirect tasks (planning efforts) arise.

These variables determine to what extent procedures of introduction of *new forms of work organization*, as well as the procedures in the group can be anticipated with respect to their success and their stability. Thus, they are derived from ideas stemming from organizational development in theory and combined in a practicable manner in the procedural and participative approach.

- *Continuity of flows:* number of interrupts in process chains by external operations (not performed on the production island), inspections, or intermediate storage. Problems for group work arise, because of internal responsibility for delivery due date with external (uncontrolled) operations.
- *Length of process chains:* relation between the number of workplaces in a production island, which are bound to sequential operations at parts, components, or products to the total of workplaces in the island. To implement "group thinking" in the production island, a reciprocal interdependency between the workplaces is useful. This is introduced by process chains in the sequential operations. Self-control is improved, when a person knows about the working processes before and after him and, thus, the reciprocal responsibility for quality and deadliness.
- *Scope in job shop and order scheduling:* proportion of orders (in terms of quantity and time) for which a planning of the sequence and time of operations and orders is possible. An optimal job shop scheduling and disposition of times and sequences of orders is possible only near to production. Responsibility for delivery dates implies scope, in order scheduling for the group.
- *Purity of the production island:* relation between completely manufactured parts in the island to the total sum of parts in the island. Islands should be designed in a way such that the manufacturing process is complete. External operations cause increased efforts of coordination, the responsibility for the completed accounting quantity and date becomes weak, and quality deficits are not corrected immediately, but delayed by weak feelings of responsibility.

The combination of work-structuring and technological criteria in the programmatical approach becomes visible in these variables. A determination of the variables on a positive/neutral/negative scale for different production areas gives indications for possibilities to introduce group work. In

Figure 5.20, profile in the form of a polar diagram is outlined for a manufacturing area of rotors for pumps, which signalize a status quo that can be developed to group work with a good prognosis of successful design efforts.

The development of the variables is an experience- and knowledge-based process to the same extent as a systematical research process. Nevertheless, the variables themselves show a considerable amount of face validity for suitability for group work. The scaling and evaluation itself is done normally in a discussion process between company experts and university staff with the help of visual aids. Thus, the identification of design possibilities is a premise for the identification of applicable design solutions.

5.5.8 Aachener PPC-Model as an Example of Work System Design at Level S6

If a whole production enterprise is understood as a work system, then production planning and control (PPC) provides a substantial design field. PPC supports the entire technical job execution including mass-, time-, and capacity planning, controlling, and monitoring of the manufacturing process. Due to the high complexity of the PPC and because of its central meaning for production enterprises, in addition to the structure and sequence organization in the PPC systems, the information-technical support is of great importance. In operational practice, it is evident that the firm-specific design of the PPC and the exhaustion of the optimizing potential can only be reached efficiently, if adequate practical reference models are provided.

On this basis, the Aachener PPC model was developed (Laakmann and Much, 1995; Luczak and Kees, 1998; Luczak and Eversheim, 1998). With the Aachener PPC model the goal of selecting and arranging PPC systems, according to the specific requirements of an enterprise, is pursued. Thus, in the analysis phase of a project the existing PPC structures are illustrated first with the help of the model. On the basis of the results of this analysis an organization of the PPC system, which meets the companies' requirements takes place with the help of the model.

In order to reach this objective, the Aachener PPC model offers the following views for the operational organization:

- Tasks of the PPC (task view)
- Job execution processes (process view)

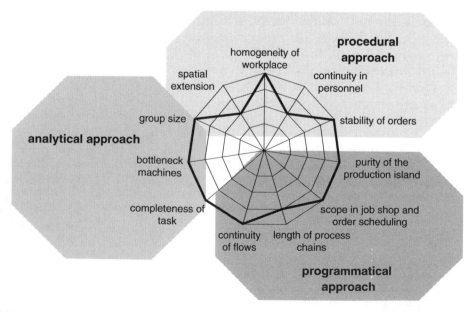

FIGURE 5.20 Polar diagram for the anticipation of design success in the introduction of group work in production systems.

- IT system functions of the PPC (function view)
- IT system data of the PPC (data view)

Task and process views primarily serve the purpose of organizational design and are, thus, subsumed under the term organization views.

Function and data views above all provide the basis for the selection, design, and introduction of PPC systems and, therefore, are named IT-System-views (Figure 5.21). In the following the individual reference views are briefly outlined.

5.5.8.1 Task View

The task view specifies and particularizes the tasks of the PPC in a generally accepted, hierarchical abstraction, that is, the task view is structured independently concerning structure-organizational solutions and determines no special process organization. In principle, tasks are differentiated between core and cross-sectional tasks. The subject of the core tasks is the completion of orders. The core tasks include the production program planning and the production capacity planning, as well as the self-manufacturing planning and the external procurement planning and control. Again, individual subtasks are assigned to these tasks. The cross-division integration and optimization of the PPC is subject of the cross-sectional task. This task includes order coordination, stock control, as well as PPC Controlling. The task model is suitable for the analysis and design of operational tasks.

5.5.8.2 Process View

Within the framework of the process view of the Aachener PPC model, the operational-specific tasks, which are identified with the help of the task view are arranged into a time-related and sequence-logical order. Additionally, contents of individual tasks are concretized. The heterogeneity of the task fulfillment in production enterprises does not permit a comprehensive processing, model which is available for every enterprise. Such a model would become complex and unclear due to the very high number of cases, which would have to be considered. The Aachener PPC model differentiates, therefore, between the following four operating types:

- Order manufacturer
- Frame order manufacturer

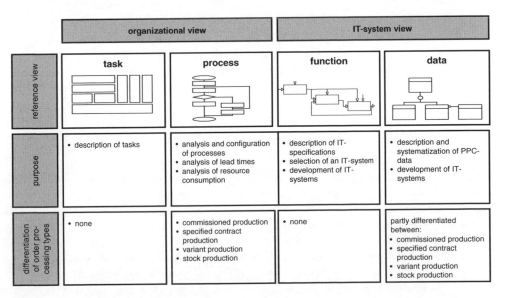

	organizational view		IT-system view	
reference view	**task**	**process**	**function**	**data**
purpose	• description of tasks	• analysis and configuration of processes • analysis of lead times • analysis of resource consumption	• description of IT-specifications • selection of an IT-system • development of IT-systems	• description and systematization of PPC-data • development of IT-systems
differentiation of order processing types	• none	• commissioned production • specified contract production • variant production • stock production	• none	partly differentiated between: • commissioned production • specified contract production • variant production • stock production

FIGURE 5.21 A reference to the Aachener PPC model.

- Variant manufacturer
- Stock manufacturer

A morphologic characteristic pattern assigns specific characteristics to the respective job execution types. For instance, one characteristic of an order manufacturer is that the production is initiated by a customer-specific order (make-to-order production). Whereas, the stock manufacturer can fulfill all customer orders ex stock, since he produces order-neutrally and exclusively program-dependent. A goal of the usage of this typology is to be able to quickly create an expressive, operation-specific process model by the allocation of a certain production enterprise to one of the job execution types.

5.5.8.3   Function View

The function view of the Aachener PPC model has the task to describe the requirements to an EDP system, which supports the PPC activities. Therefore, individual functions are described. The structure of the function view corresponds to the arrangement of the task model. Thus, those functions can quickly be identified, which can support certain tasks. For example, the functions "order quantity calculation, soliciting and evaluating quotations, selection of suppliers, and order monitoring" are assigned to the task "external procurement planning and control." The individual PPC functions are described again by characteristics. Possible specifications are assigned in turn to the individual characteristics.

5.5.8.4   Data View

PPC data comprises all data, which are relevant and able to format for the execution of the PPC. The data describe the product as well as the manufacturing and order structure. The data reference view is to offer support with the development of relationally developed PPC systems. Furthermore, the systematic collection of quantitative parameters must be supported for the dimensioning of PPC hardware.

The basis for the structure of the data view is the major task defined in the task view and the job execution types defined in the process view. Within the major task ranges (e.g., order coordination, stock control) partial models are provided for the individual job execution types. The individual partial models are based on the master data model, which contains the long-term valid information. It illustrates the structure of the enterprise and the products, which are relevant for the PPC. The task- and type-specific partial models cover, in addition to the master data those, the so-called transaction data, which are relevant at short notice, usually, in the case of an order.

5.5.8.5   Outlook

Due to the increasing cross-linking of enterprises, the design range of in-plant aligned PPC expanded to the inter-plant management of complete supply chains. In this context the questions arise how inter-plant processes are to be arranged and how the coordination of the distributed production has to take place. In order to accommodate for the changed requirements, the Research Institute for Operations Management is currently working on an extension of the Aachener PPC model (Schiegg and Lücke, 2004).

5.5.9   Cooperation between Companies as an Example of Work System Design at Level S7

Work systems do not only exist within an enterprise but can also be constituted inter-organizational. Thus, tasks are handled in the context of cooperation of companies. In this case work systems are designed for the duration of the cooperation. Although the goals, which are aimed at cross-plant cooperations can be different the procedures for obtaining the goals are quite similar. In the context of a research project, a process-model and a method were developed, which support companies with the setting up and the operation of inter-organizational cooperation (Killich and Luczak, 2003).

The steps for the build-up and the operation of company cooperation can be allocated to four phases (Figure 5.22). In general, these phases are not passed through completely, but represent the "maximal case" of a cross-plant cooperation. It is possible, for example, that because of operational-specific

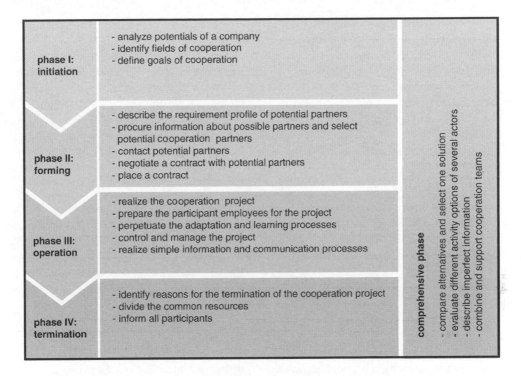

FIGURE 5.22 Phases of cooperation processes (Killich, S. and Luczak, H., *Unternehmenskooperation für kleine und mittelständische Unternehmen-Lösungen für die Praxis*, Springer, Berlin, 2003. With Permission).

reasons only regional business associates are chosen as project associates. In this case, the purchase of information is not that important, since numerous information is already available because of the long-lasting business connections. The following Sections describe the single phases of the process model in detail.

5.5.9.1 Phase of Initiation

The strategic arrangement of an organization is the initial point of the initiation of a cross-plant cooperation. In the context of the development of a corporate strategy, potential fields of cooperation can already be identified. Starting from the corporate strategy and/or in the context of a strategy development it has to be decided, whether the implementation of the strategy should be carried out by the company single-handedly (make), with cooperation of other organizations (cooperate), or should take place in the open market (buy). This decision should be based on several criteria, which do not only consider the related costs. It is possible, for example, that a new, not yet developed product can be of such an important and strategic interest that the autonomous generation of the know-how (make) represents a better choice for the organization, although the cooperative alternative would be cheaper.

The phase of initiation is very important, since the basis of a purposeful cooperation, which refers to the own interests created without violating the interests of the cooperation partners. The fact that this basis for a common coordination should not only be created by the own enterprise is critical for success, since the decisions of the partners also influence the cooperation project. The involvement of the individual organizations in the design of the cooperation project and the project goals depends on the point of time, when the organization is included in the project. Two levels of involvement can be distinguished:

- The active initiation is characterized by the fact that the cooperation project is generated out of own interests, which leads to more freedom with the design of the project. As a first step, an analysis of the organizational potential should take place. Hence, fields of cooperation can be deduced and the cooperation project can be defined.

- With the passive initiation the own organization has already been identified as a potential business associate and is asked whether it is interested in participating in the cooperation project. In this context, the following question arises: "Is the predefined cooperation project important for the own organization, or not?." In answering this question, again the strategic orientation of the own organization is highly important. The procedure of identifying and/or judging a cooperation is identical for the active and passive initiation. Only the actuators of the planning of the cross-plant cooperation differ.

5.5.9.2 Phase of Forming

The cooperation projects derived from the analysis of the organizational potentials contain implicit requirements towards the cooperation partner. Thus, the requirements towards the cooperation partner, which add to the success of the cooperation project have to be identified in the phase of forming. Therefore, a requirement profile for the partner(s) is created. This profile specifies the characteristics an organization has to possess to be seen as a "perfect" cooperation partner. By means of this requirement profile organizations that meet as many of the requirements as possible and can, therefore, be seen as a potential cooperation partner have to be identified.

If one or several organizations can be chosen with the requirement profile, it is to be clarified which possibilities of approach exist. Concerning the potential cooperation partner, interest should be evoked with this approach. The cooperation intention, however, at least partially discloses the strategical orientation of the own company. Therefore, the information that can be passed on should be determined beforehand.

The cooperation contract is an important result of the "phase of forming." The contractual arrangement of individual aspects of the inter-working helps to safeguard the organization against opportunistic action on the part of the cooperation partner.

5.5.9.3 Phase of Operation

The execution of a cross-plant cooperation project includes — like inner-organizational projects — the design fields project management, organizational development, and personnel management.

Concerning the project management the same methods are used as in inner-organizational projects, like, for example, bar charts or flow charts. Even the election of common support instruments can be problematic, since every organization tries to use their "own" software, if possible. Furthermore, the determination of binding milestones of the cooperation project can complicate the project management of a cross-plant cooperation.

The organizational development contains the design of the operational, as well as the organizational structure of a cooperation project. The operational structure of the cooperation project has a great influence on the collaboration of the cooperation partners. It determines the communication channels that should be used in case of appearing problems. A further characteristic of the cross-plant cooperation results from the fact that the respective operational structures of the cooperating organizations can be very different. The organizational concept of the project should consider this fact.

With the process organization, particularly, the interfaces between the organizations are of importance. One basic task in this context is to choose and apply appropriate communicative devices on the basis of the cross-plant processes. At this juncture the problem often arises that every organization intends to use those instruments, which are available in the own enterprise.

Concerning the personnel management, it has to be clarified which employees participate in the cooperation and which tasks are fulfilled by these employees. These tasks also refer to the cooperation-specific activities, which often occur in inter-organizational teams. Similarly, it is important that an appropriate qualification of the employees is provided for.

5.5.9.4 Phase of Termination

Experience shows that cross-plant cooperations often run out without a defined ending. According to this, the "phase of termination" of a cross-plant cooperation usually does not take place structured and often only refers to parts of the cooperation project. In this context, common investments and/or the benefits and disprofits are divided between the individual organizations. Thereby, important hints for future cooperation projects can be identified through the analysis of the reasons for the ending of a cooperation. Moreover, it is often not apparent to the involved employees why the collaboration failed. Thus, it becomes difficult to win these employees for future inter-organizational projects.

5.6 Conclusion

The realizations of systems theory particularly were introduced into ergonomics via the work system approach. This conception acts as an analysis framework. It is conducive to systemic thinking, because it does not regard individual elements in isolation, but considers the various interdependences between the elements. Besides, the application of this analysis framework forces the ergonomist to draw up a system limitation and, thus, make a clear problem definition.

The focus of the application of the work system concept is in operational practice at levels S3 and S4 of the ordering model. The work system approach can also be applied to the other levels, which are shown in the project examples in the previous sections. The following example of a professor reveals that the system borderlines shift from one level to another.

A professor performs his lectures usually standing or walking, while the students follow while sitting. The professor's physical load during the lecture is, therefore, higher than that of his students. The professor uses, in addition, media during his lecture like an overhead projector, which produces heat. Thus, the professor's core body temperature possibly increases, so that his "physiological systems" require a lower room temperature compared to his students (level S1). During his lecture, the professor performs a series of coordinated movements. For example, he moves the mouse of a computer to begin his presentation or writes formulas on the board with chalk. The more these coordinations are performed, the less is the time taken and the quality of the movement (e.g., better typeface on the board) increases to a certain level. Work resources are also considered at level S2 (operation systems), as well as psychological and physical functions like the coordination of movements. Referring to the example of the professor, this means, for example, that the height of desk in the lecture room is to be designed corresponding to anthropometric measures. Considering the teaching work order of a professor, or the redefined task of teaching ("task system," level S3), it does not consist of only performing lectures. The professor has to prepare his lectures, for example, to read up on the adequate literature. Furthermore, the professor applies the continuous improvement process to the concept of teaching, in integrating the latest research results and takes out the dated contents. As well as the teaching tasks, further tasks belong to the "job system" of a professor (level S4); primarily, the research and tasks, which result from the leadership of an institute. Referring to the leadership, the professor has scientific and nonscientific coworkers at his disposal, in order to cope with the various activities and subtasks, which are related to the "job system" of a professor. These employees work together in "group systems" (level S5). The individual groups represent the main research fields of the institute. The members of the individual groups work together, so that the typical group characteristics like role differentiation and sense of togetherness are developed. Besides, an institute can be compared to a "company system" (level S6). The professor takes on the role of a manager, for example, in setting goals, formulating strategies, and coordinating the research processes. In doing so, the professor interacts with his employees and represents the institute. Therefore, his tasks and activities are imbedded in a "societal system" (level S7). In this context the rough contents of public research are worked out by the ministries. Furthermore, the institute, for instance, is imbedded in a legal framework (e.g., industrial law, salary law, university law).

Bibliography

Alioth, A., *Entwicklung und Einführung alternativer Arbeitsformen*, in *Schriften zur Arbeitspsychologie*, Ulich, E., ed., Vol. 27, Huber, Bern, 1980.

Anderson, J.R., *Rules of the Mind*. Lawrence Erlbaum Assoc, Hillsdale, 1993.

Antoni, C.A., *Teilautonome Arbeitsgruppen - Ein Königsweg zu mehr Produktivität und einer menschengerechten Arbeit?* Psychologie Verlags Union, Weinheim, 1996.

Ashby, W.R., *An Introduction to Cybernetics*. Chapman & Hall, London, 1956.

Behling, K., *Ein analoges Modell der Thermoregulation des Menschen bei Ruhe und Arbeit aufgrund experimenteller Daten*. Dissertation Universität Hamburg, 1971.

Bertalanffy, L. von, General systems theory — a new approach to unity of science, in *Human Biology*, Vol. 23, Winsor, Ch. et al., eds., Baltimore, Maryland, 1951, pp. 306–361.

Brödner, P., *Fabrik 2000 - Alternative Entwicklungspfade in die Zukunft der Fabrik*, 2nd ed., Sigma, Berlin, 1986.

Buckley, W., Society as a complex adaptive system, in: Modern Systems Research for the Behavioral Scientist, Buckley, W., ed., Aldine, Chicago, 1968, pp. 490–513.

Bullinger, H.-J., Fähnrich, K.-P., and Niemeier, J., Informations- und Kommunikationssysteme für "schlanke Unternehmungen," in *Office Management*, 41(1–2), 1993, pp. 6–19.

Caccibue, P.C., *Modelling and Simulation of Human Behaviour in System Control*, Berlin, Springer, 1998.

Carayon, P., and Smith, M.J., Work organization and ergonomics, in *Appl. Ergon.*, 31, 2000, pp. 649–662.

Card, S.K., Moran, T.P., and Newell, A., *The Psychology of Human–Computer Interaction*, Lawrence Erlbaum Assoc., Hillsdale, 1983.

Clausing, D.P., World-class concurrent engineering, in *Concurrent Engineering: Tools and Technologies for Mechanical System Design*, Haug, E.J., ed., NATO ASI Series, Series F: Computer and Systems Sciences, Vol. 108, Springer, Berlin, 1993, pp. 3–40.

Compes, P., *Sicherheitstechnisches Gestalten - Gedanken zur Methodik und Systematik im projektiven und konstruktiven Maschinenschutz*, Habilitationsschrift RWTH Aachen, 1970.

Döring, B., Systemergonomie bei komplexen Arbeitssystemen, in *Arbeitsorganisation und Neue Technologien*, Hackstein, R., Heeg, F.-J., and Below, F.v., eds., Springer, Berlin, 1986, pp. 399ff.

Eisenführ, F., and Weber, M., *Rationales Entscheiden*, 2nd ed., Springer, Berlin, 1994.

Emery, F.E., Characteristics of socio-technical Systems, in *Design of Jobs*, Davis, L.E. and Taylor, J.C., eds., Penguin, Harmondsworth, 1972, pp. 177–198.

Foley, J.D., and van Dam, A., *Fundamentals of Interactive Computer Graphics*, Addison Wesley, Menlo Park, CA, 1982.

Frank, H., Was ist Kybernetik? in *Kybernetik - Brücke zwischen den Wissenschaften*, Frank, H., ed., 4th ed., Umschau, Frankfurt a.M., 1964, pp. 9–20.

Fricker, A. R., *Eine Methodik zur Modellierung, Analyse und Gestaltung komplexer Produktionsstrukturen*, Augustinus, Aachen, 1996.

Fürstenberg, F., Humane Arbeitsgestaltung als Partizipations problem, in *Menschengerechte Gestaltung der Arbeit*, Fürstenberg, F., Hanau, P., Kreikelbaum, H., and Rohmert, W., eds., BI Wissenschaftsverlag, Mannheim, 1983.

Gell-Mann, M., Das *Quark und der Jaguar - Vom Einfachen zum Komplexen - Die Suche nach einer neuen Erklärung der Welt*, Piper, München, 1994.

Gray, W.D., John, B.E., and Atwood, M.E., Project Ernestine: a validation of GOMS for prediction and explanation of real-world task performance, in *Hum.–Comput. Interact.*, 8(3), 1993, pp. 237–259.

Grochla, E., *Einführung in die Organisationstheorie*, Poeschel, Stuttgart, 1978.

Gugerty, L., The use of analytical models in human–computer interface design, in *Int. J. Man–Machine Stud.*, 38(1), 1993, pp. 625–660.

Hacker, W., Projektieren von Arbeitstätigkeiten: Möglichkeiten-Probleme-Grenzen, in *Zeitschrift für Arbeitswissenschaft*, 45(4), 1991, pp. 193–198.

Hacker, W., *Arbeitstätigkeitsanalyse: Analyse und Bewertung psychischer Arbeitsanforderungen*, Asanger, Heidelberg, 1995.

Hacker, W., Fritsche, B., Richter, P., and Iwanowa, A., Tätigkeitsbewertungssystem (TBS) - Verfahren zur Analyse, Bewertung und Gestaltung von Arbeitstätigkeiten. *Schriftenreihe Mensch-Technik-Organisation*, Vol. 7, Ulich, E., ed., Zürich: vdf, 1995.

Hackman, J.R., Toward understanding the role of tasks in behavioral research, in *Acta Psychol.*, 31, 1969, pp. 97–128.

Hannaman, G.W., Spurgin, A.J., and Lukic, Y., A Model for Assessing Human Cognitive Reliability in PRA Studies. Third IEEE Conference on Human Factors and Nuclear Power Plants, Monterey, CA, 1985, pp. 343–353.

Harel, D., On visual formalisms, *Commun. ACM*, 31(5), 1988, pp. 514–530.

Heeg, F.J., *Moderne Arbeitsorganisation - Grundlagen der organisatorischen Gestaltung von Arbeitssystemen bei Einsatz neuer Technologien*, München, Hanser, Wien, 1988.

Hendrick, H.W., Organizational design and management (ODAM) in HCI, in Luczak, H., Cakir, A., and Cakir, G. eds., *Work with Display Units 92*, Amsterdam, North-Holland: 1993, pp. 30–34.

Hendrick, H.W., Sociotechnical systems theory: the sociotechnical systems model of work systems, in *International Encyclopedia of Ergonomics and Human Factors*, Karwowski, W., ed., Taylor & Francis, New York, 2001, pp. 1720–1722.

Hettinger, T., and Wobbe, G., *Kompendium der Arbeitswissenschaft - Optimierungsmöglichkeiten zur Arbeitsgestaltung und Arbeitsorganisation*, Kiehl, Ludwigshafen, 1993.

Hill, W., Fehlbaum, R., and Ulrich, P., *Organisationslehre 2 - Ziele, Instrumente und Bedingungen der Organisation sozialer Systeme*, 3rd ed., Paul Haupt, Bern, 1981.

Kahnemann, D., *Attention and Effort*, Prentice Hall, Englewood Cliffs, 1973.

Kaminsky, G., *Praktikum der Arbeitswissenschaft - Analytische Untersuchungen beim Studium der menschlichen Arbeit*, 2nd ed., Hanser, Darmstadt 1979.

Karwowski, W., Complexity, fuzziness and ergonomic incompatibility issues in the control of dynamic work environments, in *Ergonomics*, 34(6), 1991, pp. 671–686.

Kelly, J.E., A reappraisal of sociotechnical systems theory, in *Hum. Relations*, 31(12), 1978, pp. 1069–1099.

Kieras, D.E., A guide to GOMS model usability evaluation using NGOMSL, in *Handbook of Human Computer Interaction*, 2nd ed., Helander, M., Landauer, T.K., and Prabhu, P., eds., Elsevier Science, Amsterdam 1997.

Killich, S., and Luczak, H., *Unternehmenskooperation für kleine und mittelständische Unternehmen - Lösungen für die Praxis*, Springer, Berlin, 2003.

Kirchner, J.H., *Arbeitswissenschaftlicher Beitrag zur Automatisierung - Analyse und Synthese von Arbeitssystemen*, Beuth, Berlin, 1972.

Kirchner, J.H., Arbeitswissenschaft - Entwicklung eines Grundkonzeptes, in *Zeitschrift für Arbeitswissenschaft*, 47, 1993, pp. 85–92.

Kirchner, J.H., Integrative Arbeitssystemgestaltung, in *Handbuch Arbeitswissenschaft*, Luczak, H., and Volpert, W., eds., Schäffer-Poeschel, Stuttgart, 1997, pp. 805–810.

Kleinbeck, U., *Arbeitsmotivation: Entstehung, Wirkung und Förderung*, Weinheim, Juventa, München, 1996.

Kraiss, K.-F., Rechnergestützte Methoden zum Entwurf und zur Bewertung von Mensch-Maschine-Systemen, in *Arbeitsorganisation und Neue Technologien*, Hackstein, R., Heeg, F.-J., and Below, F.V., eds., Springer, Berlin, 1986, pp. 435f.

Krieger, D.J., *Einführung in die allgemeine Systemtheorie*, Fink, München, 1996.

Laakmann, J., and Much, D., Das Aachener PPS-Modell - FIR entwickelt Grundlage für organisatorische und informationstechnische PPS-Gestaltung, in *fir+iaw-Mitteilungen*, 27(1), 1995, pp. 14/15.

Locke, E.A., and Latham, G.P., *A theory of goal setting and task performance*, Prentice-Hall, Englewood Cliffs, 1990.

Luczak, H., Arbeitswissenschaftliche Untersuchungen von maximaler Arbeitsdauer und Erholzeiten bei informatorisch-mentaler Arbeit nach dem Kanal- und Regler-Mensch-Modell sowie superponierten Belastungen am Beispiel Hitzearbeit, in *VDI-Z*, Vol. 10, No. 6, VDI, Düsseldorf, 1979.

Luczak, H., Koordination der Bewegungen, in *Praktische Arbeitsphysiologie*, 3rd ed., Rohmert, W., and Rutenfranz, J. eds., Thieme, Stuttgart, 1983.

Luczak, H., Wesen menschlicher Leistung, in Institut für angewandte Arbeitswissenschaft, ed., *Arbeitsgestaltung in Produktion und Verwaltung*, Bachem, Cologne, 1989, pp. 39–65.

Luczak, H., Volpert, W., Raeithel, A., and Schwier, W., *Arbeitswissenschaft - Kerndefinition, Gegenstandskatalog, Forschungsgebiete*, 3rd ed., RKW, TÜV Rheinland, Cologne, 1989.

Luczak, H., Macroergonomic anticipatory evaluation of work organization in production systems, in *Ergonomics*, 38(8), 1995, pp. 1571–1599.

Luczak, H., Arbeitssysteme und Arbeitstechnologien. in *Produktion und Management - Betriebshütte*. Part 2, 7th ed., Eversheim, W., and Schuh, G., eds., Springer, Berlin, 1996, chap. 12, pp. 11–39.

Luczak, H., and Fricker, A.R., Komplexitätsbewältigung - Bedingung zur ganzheitlichen Rationalisierung, in *Wie rational ist Rationalisierung heute? Ein öffentlicher Diskurs anlässlich des 75-jährigen Jubiläums des RKW e.V.*, Hoß, D. ed., Raabe, Stuttgart, 1996.

Luczak, H., and Metz, A., Development of a decision support system for the selection of a pilot section to implement self-directed work teams in manufacturing and assembling, in *Human Factors in Organizational Design and Management — V*. Proceedings of the Fifth International Symposium on Human Factors in Organizational Design and Management held in Breckenridge, CO., U.S.A., Brown, Jr., ed., Elsevier Science B.V., Amsterdam, 1996, pp. 383–389.

Luczak, H., Kerndefinition und Systematiken der Arbeitswissenschaft, in *Handbuch Arbeitswissenschaft*, Luczak, H., and Volpert, W., eds., Schäffer-Poeschel, Stuttgart, 1997, pp. 11–19.

Luczak, H., and Rohmert, W., Belastungs-Beanspruchungs-Konzepte, in *Handbuch Arbeitswissenschaft*, Luczak, H., and Volpert, W., eds., Schäffer-Poeschel, Stuttgart, 1997, pp. 326–332.

Luczak, H., *Arbeitswissenschaft*, 2nd ed., Springer, Berlin, 1998.

Luczak, H., and Eversheim, W., eds., *Produktionsplanung und -steuerung - Grundlagen, Gestaltung und Konzepte*, Springer, Berlin, 1998.

Luczak, H., and Kees, A., Das Aachener PPS-Modell, in Referenzmodellierung '98 - Anwendungsfelder in Theorie und Praxis, Conference, Aachen, 1998.

Luczak, H., and Schlick, C., Utility of Cognitive Models, in Proceedings of the IEA 2000/HFES 2000 Congress, eds., *Human Factors and Ergonomics Society,*. San Diego, 2000, S. 1/585–1/588.

Luczak, H., and Schlick, C., An essay about the theory-to-practice relationship of ergonomics, in International Congress on Humanizing Work and Work Environment, Bombay, 2001, pp. 1–10.

Luczak, H. Reuth, R., and Schmidt, L., Development of error-compensating UI for autonomous production cells, in *Ergonomics*, 46(1–3), 2003, pp. 19–40.

Luhmann, N., *Soziologische Aufklärung I*, 2nd ed., Westdeutscher Verlag, Opladen, 1971.

Luhmann, N., *Zweckbegriff und Systemrationalität - Über die Funktion von Zwecken in sozialen Systemen*, Suhrkamp, Frankfurt a.M., 1973.

Luhmann, N., *Soziale Systeme - Grundriß einer allgemeinen Theorie*, Suhrkamp, Frankfurt a. M., 1984.

Malik, F., *Strategie des Managements komplexer Systeme. Ein Beitrag zur Management-Kybernetik evolutionärer Systeme*, 4th ed., Haupt, Bern, 1992.

Malik, F., *Systemisches Management, Evolution, Selbstorganisation: Grundprobleme, Funktionsmechanismen und Lösungsansätze für komplexe Systeme*, Haupt, Bern, 1993.

Maturana, H.R., *Erkennen - Die Organisation und Verkörperung von Wirklichkeit*, 2nd ed., Vieweg, Wiesbaden, 1985.

Maturana, H., and Varela, F., *Der Baum der Erkenntnis - Die biologischen Wurzeln des menschlichen Erkennens*, Scherz, Bern, 1987.

McCormick, E.J., Jeanneret, P.R., and Mecham, R.C., Position Analysis Questionnaire (PAQ) Form C. Consulting Psychologists Press, Palo Alto, 1989.

Metz, A., Entscheidungshilfe für die Auswahl von Pilotbereichen zur Einführung von Gruppenarbeit in der Produktion. Dissertation RWTH Aachen, 1997.

Miller, E.J., and Rice, A.K., *Systems of Organization*, London: Tavistock, 1967.

Moray, N., Eisen, P., Money, L., and Turksen, I.B., Fuzzy analysis of skill- and rule-based workload, in *Human Mental Workload*, Hancock, P., and Meshkati, L., eds., North-Holland, Amsterdam, 1988a.

Moray, N., Kruschelnicky, E., Money, L., and Turksen, I.B., On measuring the combined workload of skill-, rule- and knowledge-based workload, Proceedings of the Human Factors Society — 32nd Annual Meeting — Anaheim (CA), 1988b.

Mütze-Niewöhner, S., and Luczak, H., Prospective job design and evaluation in early stages of production system design, in *Human Factors in Organizational Design and Management — VII*, Proceedings of the Seventh International Symposium on Human Factors in Organizational Design and Management held in Aachen, Luczak, H. and Zink, K.J., eds., IEA Press, Santa Monica, 2003, pp. 323–328.

Newell, A., *Unified Theories of Cognition*, Harvard University Press, Cambridge, 1990.

Newell, A., Precis of unified theories of cognition, *Behav. Brain Sci.*, 15, 3, 1992, pp. 425–492.

Pahl, G., Grundlagen der Konstruktionstechnik, in *Dubbel - Taschenbuch für den Maschinenbau*, 18th ed., Beitz, W., and Küttner, K.-H., eds., Springer, Berlin, 1995.

Pew, R.W., and Baron, S., Perspectives on human performance modeling, *Automatica*, 19(6), 1983, pp. 663–676.

Pfeifer, T., Sack, D., Stemmer, M., and Orth, A., Sensor/Actuator Network — The Nervous System of a New Production Concept: The Autonomous Production Cells, Congresso Brasileiro de Automática, Florianópolis, Brasil, 2000.

Prasad, B., *Concurrent Engineering Fundamentals — Integrated Product and Process Organization*, Vol. 1, Prentice Hall, Upper Saddle River, 1996.

Rasmussen, J., Skills, rules, knowledge: signals, signs, and symbols and other distinctions in human performance models, *IEEE Trans. Sys., Man and Cybern.*, 13(3), 1983, pp. 257–267.

Rasmussen, J., The role of hierarchical knowledge representation in decision making and system management, *IEEE Trans. Syst., Man and Cybern.*, 15(2), 1985, pp. 234–243.

Rasmussen, J., *Information Processing and Human-Machine Interaction: An Approach to Cognitive Engineering*, North-Holland, New York, 1986.

Reason, J., Generic error-modeling system: a cognitive framework for locating common human error forms, in Rasmussen, J., Duncan, K., and Leplat, J., eds., *New Technology and Human Error*, John Wiley & Sons, New York, 1987.

REFA — Verband für Arbeitsstudien und Betriebsorganisation e.V. ed., *Methodenlehre der Betriebsorganisation. Part 1 Grundlagen der Arbeitsgestaltung*, 2nd ed., Hanser, München, 1993.

Reuth, R., Schlick, C., and Luczak, H., A simulation approach: comparative assessment of knowledge, skill and abilities in autonomous production cells, in Proceedings of HCI International 2001, Vol. 2, Systems, Social and Internationalization Design Aspects of Human–Computer Interaction, 2001 in New Orleans, Smith, J.M., and Salvendy, G., eds., Lawrence Erlbaum Associates, Mahwah, NJ, pp. 207–211.

Rice, A.K., *Productivity and Social Organization — The Ahmedabad Experiment*, Tavistock, London, 1958.

Rohmert, W., and Rutenfranz, J., Arbeitswissenschaftliche Beurteilung der Belastung und Beanspruchung an unterschiedlichen industriellen Arbeitsplätzen. Der Bundesminister für Arbeit und Sozialordnung ed., Referat Öffentlichkeitsarbeit, Bonn, 1975.

Rohmert, W., and Weg, F.J., Organisation teilautonomer Gruppenarbeit - betriebliche Projekte, Leitregeln zur Gestaltung, ed., RKW (Beiträge zur Arbeitswissenschaft, Reihe 1, Bd. 1), Hanser, München, 1976.

Ropohl, G., *Eine Systemtheorie der Technik - zur Grundlegung der allgemeinen Technologie*, Hanser, München, 1979.

Scarperi, M., Scarperi, S., Behling, K., Bleichert, A., and Kitzing, J., Antriebe und effektorische Maßnahmen der Thermoregulation bei Ruhe und während körperlicher Arbeit. III. Über den Einfluss des Wasserhaushaltes auf die Thermoregulation bei Ruhe und Arbeit, *Int. Z. f. angew. Physiol.* 1972, 30, pp. 186–192.

Schiegg, P., and Lücke, T., Referenzmodell der Produktionsplanung und -steuerung, in *Betriebsorganisation im Unternehmen der Zukunft*, Luczak, H., and Stich, V., eds., Springer, Berlin, 2004.

Schlick, C., Modellbasierte Gestaltung der Benutzungsschnittstelle autonomer Produktionszellen. Dissertation RWTH Aachen, 1999.

Schlick, C., Springer, J., and Luczak, H., Operator requirements for man–machine interfaces of CNC-machine tools, in *Automated Systems Based on Human Skill*, Brandt, D., and Martin, T., eds., Elsevier Science, Oxford, 1995.

Schlick, C., Springer, J., and Luczak, H., Support system design for man–machine interfaces of autonomous production cells, in Proceedings of the 13th World Congress of IFAC, Gertler, J.J., Cruz, J.B., and Peshkin, M., eds., Elsevier Science, Oxford, 1996.

Schlick, C., Daude, R., Luczak, H., Weck, M., and Springer, J., Head-mounted display for supervisory control in autonomous production cells, *Int. J. Displays*, 17(3/4), 1997, pp. 199–206.

Schumann, R., Entwicklung eines modellgestützten Systems zur Gestaltung teilautonomer Arbeitsgruppen in der Fertigung von Automobilzulieferbetrieben, Dissertation RWTH Aachen, 1995.

Shannon, C.E., Die mathematische Theorie der Kommunikation, in *Mathematische Grundlagen der Informationstheorie*, Shannon, C.E., and Weaver, W., eds., Oldenbourg, München, 1976.

Sheridan, T.B., *Telerobotics, Automation, and Human Supervisory Control*, The MIT Press, Cambridge, 1992.

Sheridan, T.B., *Humans and Automation — System Design and Research Issues*, John Wiley & Sons, Santa Monica, 2002.

Stahl, J., Entwicklung einer Methode zur Integrierten Arbeitsgestaltung und Personalplanung im Rahmen von Concurrent Engineering, Dissertation RWTH Aachen, 1998.

Stahl, J., and Luczak, H., Personnel planning in concurrent engineering: a case study, in *Hum. Factors Ergon. Manufacturing*, 10(1), 2000, pp. 23–44.

Stahl, J., Mütze, S., and Luczak, H., A method for job design in concurrent engineering, in *Hum. Factors Ergon.Manufacturing*, 10(3), 2000, pp. 291–307.

Stephan, K., Thermodynamik, in *Dubbel - Taschenbuch für den Maschinenbau*, 18th ed., Beitz, W., and Küttner, K.-H., eds., Springer, Berlin, 1995.

Sternberg, S., The discovery of processing stages: extension of Donders' method, in *Acta Psychol.*, 30(1), 1969, pp. 276–315.

Sydow, J., *Der soziotechnische Ansatz der Arbeits- und Organisationsgestaltung - Darstellung, Kritik, Weiterentwicklung*, Campus, Frankfurt, 1985.

Trist, EL., and Bamforth, K.W., Some social and psychological consequences of the longwall method of coal getting, in *Hum. Relations*, 4, 1951, pp. 3–38.

Ulich, E., Gestaltung von Arbeitstätigkeiten, in Lehrbuch Organisationspsychologie, Schuler, H., ed., Huber, Bern, pp. 189–208, 1993.

Ulich, E., *Arbeitspsychologie*, 5th ed., vdf; Zürich; Schäffer-Poeschel, Stuttgart, 2001.

Unbehauen, R., *Systemtheorie - Grundlagen für Ingenieure*, 5th ed., Oldenbourg, München, 1990.

Warnecke, H.-J. and Fertigungsverfahren, in *Dubbel — Taschenbuch für den Maschinenbau*, Beitz, W., and Küttner, K.-H., eds., 18th ed., Springer, Berlin, 1995.

Warnecke, H.-J., and Hüser, M., Selbstorganisation im Produktionsbetrieb, in *ZWF Zeitschrift für wirtschaftlichen Fabrikbetrieb*, 90(1–2), 1995, pp. 12–16.

Wenzel, H.-G., and Piekarski, C., *Klima und Arbeit*, Parcus, München, 1980.

Wiener, N., *Cybernetics — or control and communication in the animal and the machine*, 2nd ed., The MIT Press, John Wiley & Sons, New York, 1961.

Willke, H., *Systemtheorie I: Grundlagen - Eine Einführung in die Grundprobleme der Theorie sozialer Systeme*, 6th ed., Lucius und Lucius, Stuttgart, 2000.

Zsambok, C.E., and Klein, G.A., eds., *Naturalistic Decision Making*, Lawrence Erlbaum Assoc., Mahwah 1997.

Zülch, G., Ansätze und Defizite einer arbeitsorganisatorischen Methodenlehre - Teil 1: Bezugsrahmen der Arbeitsorganisation, in *Zeitschrift für Arbeitswissenschaft*, 46(3), 1992, pp. 133–138.

6

Human Factors and TQM

Colin G. Drury
State University of New York
 at Buffalo

6.1 Introduction: TQM and Human Factors Programs in Industry

Over the past decade the pace of change in industry has been remarkable. Whether in manufacturing or service, industry has moved on an unprecedented scale to new technologies, new forms of organization, and new programs (Mize, 1992). It has not done this from an innate love of change, but because of strategic imperatives. The removal of tariff barriers and creation of trading blocs (EU, NAFTA, etc.) in the 1980s and 1990s has exposed even the smaller companies to unprecedented competition. One response in industrial countries has been to join the competition rather than fight it, for example by using manufacturing (and service) facilities in areas of relatively low labor costs. Thus, European and Japanese automobile plants have appeared in the U.S., while American apparel plants have been built in Central and South America. Even service operations, such as data entry and computer programming have been moved "offshore" using modern communication links.

However, many companies have chosen to remain in their traditional locations and compete by application of more advanced knowledge to their business. For example, Kleiner and Drury (1996) show how a number of companies in a rust-belt region chose to remain and expand by exploiting regional knowledge and skills.

One area in which global competition has benefited companies has been the free flow of ideas, matching the freer flow of goods and services. Thus, developments in microprocessor based technology, productivity software, organizational change, cellular manufacturing, and quality solutions arising in one county have been rapidly emulated throughout the developed (and now the developing) world. A major movement within this has been the quality imperative — the realization that without high quality, products will not sell, and the simultaneous realization that organizing for quality will produce benefits in productivity, efficiency, and safety (Crosby, 1979; Dobyns and Crawford-Mason, 1991; Krause, 1993; Deming, 1986).

Through the quality imperative in particular, companies have realized the importance of process control, i.e., ensuring that the process produces its intended output in a highly reliable manner. As more is learned about the process through quality methodologies (e.g., Statistical Process Control: Grant and Leavenworth, 1995; Designed Experiments: Taguchi, 1986), so the process can change from closed loop control using performance feedback, to open loop control using valid prediction models (Drury and Prabhu, 1994). Predictive control allows the process to operate with minimum setup time after a product change, thus facilitating moves toward just-in-time manufacturing.

The other company response to the quality imperative is the active management of quality. This has comprised both the realization that managerial leadership is important (Witcher, 1995) and specific policies for managing quality (Deming, 1986). An obvious policy is the use of teams as both a change agent (Blest, Hunt, and Shadle, 1992) and as a natural group for controlling a process (Brennan, 1990).

At the same time that these strategic-driven changes in quality have been taking place, there have been simultaneous programs at other levels. Thus, new technology has been introduced to reduce labor costs and improve process capability. In response to both rising costs and public/government pressure, there has been a movement toward managing the costs associated with human errors and injuries. Company responses here have been injury reduction/safety programs (Rahimi, 1995), ergonomics programs (Liker, Joseph and Armstrong, 1984) focusing on workforce injury reduction, and similar programs for reduction of the consequences of human error (e.g., Taylor, 1990). These latter are usually termed "human factors" programs but have many characteristics in common with "industrial ergonomics" programs. Indeed, programs incorporating both injury and error reduction are possible (Drury, 1995). In this chapter the terms "ergonomics" and "human factors" will be used interchangeably.

With few exceptions, programs arising from the quality movement and the human factors movement have been simultaneous but unrelated in industry. They have many similarities and some obvious differences, but there is no *a priori* reason for them to be separate. The remainder of this chapter takes up the managerial challenge of integrating these largely parallel programs so as to gain additional benefits. For a more detailed comparison and discussion of their linkages, see Drury (1996). In particular, that paper looks at many facets of the quality movement, such as TQM, Quality awards, the ISO-9000 series, and justin-time manufacturing, while in the current chapter we concentrate on just one (TQM) for simplicity.

6.2   Fundamentals: The Basic Tenets of TQM and Human Factors

Before we can discuss interactions between the quality movement (e.g., TQM) and human factors/ergonomics, we must at least review their basic beliefs. Both programs "work" in the sense of improving industrial performance. For example, see Larson and Sinha (1995) for TQM and Drury (1992) for ergonomics.

TQM is not a monolithic philosophical structure, but rather a set of beliefs built upon the largely parallel efforts of a number of early practitioners. Rather than debate the merits of including each tenet, a recent review paper will be used to provide a convenient synopsis. Hackman and Wageman (1995) provide a thoughtful review of TQM so that we will use their structure of TQM. Table 6.1 provides this in outline form and does not appear to contradict the writings of most TQM practitioners, e.g., Deming's fourteen points (Deming, 1986). Hackman and Wageman (1995) also note two enhancements routinely used by (at least) U.S. practitioners of TQM: competitive benchmarking and employee involvement (EI). Benchmarking is the measurement of the level of performance of equivalent parts of other organizations to provide goals for those processes in your own organization. Goals are typically seen as being the best available rather than the average. Employee involvement is a generic title for a movement common in at least the larger companies to extend the employee voice in organizational affairs beyond the traditional union/management bargaining and beyond the roles specified in TQM (Russell, 1991). In fact, many companies see benchmarking and EI as integral parts of the TQM process.

Finding an equivalent set of basic beliefs or tenets of ergonomics has proven rather more difficult. As with TQM, human factors has been a largely empirical discipline, which defines what it *does* rather than

TABLE 6.1   Tenets of TQM

Assumptions	1. Good quality is less costly to an organization than is poor workmanship.
	2. Employees naturally care about quality and will take initiatives to improve it.
	3. Organizations are systems of highly interdependent parts: problems cross functional lines.
	4. Quality is viewed as ultimately the responsibility of top management.
Change Principles	1. Focus on the work processes.
	2. Uncontrolled variability is the primary cause of quality problems: it must be analyzed and controlled.
	3. Management by fact: use systematically collected data throughout the problem-solving cycle.
	4. The long-term health of the organization depends upon learning and continuous improvement.
Interventions	1. Explicit identification and measurement of customer requirements.
	2. Creation of supplier partnerships.
	3. Use of cross-functional teams to identify and solve quality problems.
	4. Use scientific methods to monitor performance and identify points for process improvement.
	5. Use process-management heuristics to enhance team effectiveness.

Source: Adapted from Hackman, J. R. and Wageman, R. 1995. *Administrative Science Quarterly*, 40: 308–342.

its basic beliefs. Thus, societies within the International Ergonomics Association have their own definitions of ergonomics, as do textbooks and journals. Although some authors have begun to consider the underpinnings of the discipline (e.g., Karwowski, Marek, and Noworol, 1988; Meister, 1996), there is no simple list of tenets similar to Table 6.1. As a working list, Table 6.2 is proposed, keeping the structure of the equivalent TQM list to facilitate comparison. Note that this listing is biased toward design ergonomics, rather than more overtly sociotechnical systems approaches (e.g., Taylor and Felten, 1993).

As is obvious from Table 6.2, ergonomics is a human-oriented process, using detailed knowledge of human functioning as a basis for designing high-performance, safe systems. Indeed, the current book

TABLE 6.2   Tenets of Ergonomics/Human Factors

Assumptions	1. Errors and stress arise when task demands are mismatched.
	2. In any complex system, start with human needs and system needs, and allocate functions to meet these needs.
	3. Honor thy user: use measurements and models to provide the detailed technical understanding of how people interact with systems.
	4. Changing the system to fit the operator is usually preferable to changing the operator to fit the system. At least develop personnel criteria and training systems in parallel with equipment, environment, and interface.
	5. Design for a range of operators rather than an average; accommodate those beyond the design range by custom modifications to equipment.
	6. Operators are typically trying to do a good job within the limitations of their equipment, environment, instructions, and interfaces. When errors occur, look beyond the operator for root causes.
Change Principles	1. Begin design with an analysis of system and human needs using function and task analysis.
	2. Use the task analyses to discover potential as well as existing human/system mismatches.
	3. Operators have an essential role in designing their own jobs and equipment, and are capable of contributing to the design process on equal terms with professional designers.
	4. Optimize the job via equipment, environment, and procedures design before optimizing the operator through selection, placement, motivation, and training.
	5. Use valid ergonomic techniques to measure human performance and well-being before and after the job change process.
Interventions	1. Prepare well for any technical change, especially at the organizational level.
	2. Involve operators throughout the change process, even those in identical jobs and on other shifts.
	3. Use teams comprising operators, managers, and ergonomists (at least) to implement the change process.

gives many examples of both the detailed human knowledge, and its use in design. We now need to consider the linkages between TQM and ergonomics explicitly, to find how to manage both programs together.

6.3　Applications of TQM and Ergonomics to Each Other

As a point of departure, a comparison of Table 6.1 and Table 6.2 is useful in considering the matches between TQM and ergonomics. It is immediately obvious that there are both similarities and differences between the two lists. Under assumptions, both consider the complexity of the system (#3 for TQM, #2 for ergonomics) as an explicit element in design and analysis. Both also have a belief in the integrity of the human operator in the system (#2 in TQM, #6 in ergonomics).

At the level of change principles, TQM starts with a focus on the work process, typically the *existing* work process (#1 in TQM). Ergonomics similarly starts from the work process (#1 and #2 in ergonomics), but this is typically advocated at a function level in the sense of all possible processes which *could* perform the task, rather than in the sense of the current process.

Neither ergonomics nor TQM advocates starting from the existing *jobs*, that is what individuals currently do in the process. This does not stop much of current ergonomics practice being oriented toward small changes in existing jobs. Also within the Change Principles, both TQM and ergonomics advocate a measurement-based approach (#3 in TQM, #5 in ergonomics).

For Interventions, the main point of similarity is in the use of small teams to control the change process (#3 in TQM, #3 in ergonomics). In a similar vein, specific measurements are stressed as #4 in TQM corresponding to #5 in ergonomics, listed, however, under Change Principles.

In those tenets where TQM and ergonomics differ, it is primarily due to differences in level of application. TQM is concerned with the company as a whole, its customers, and suppliers. It advocates a managerial approach, emphasizing responsibility, overall costs, continuous improvement, and managerial heuristics. Human factors/ergonomics in contrast deals with the system defined in narrower mission-oriented technical terms. It advocates particular solutions (hardware before training), detailed task analysis, user involvement and use of specific data on human functioning. Ergonomics is still a technical discipline, perhaps more at level equivalent to statistical quality/process control in TQM, than at the level of management intervention. As a single glaring example, Drury (1996) notes the almost complete absence of leadership considerations in the ergonomics/human factors literature.

With these similarities and differences in mind, we can explore some of the reported interactions between ergonomics and TQM. Practitioners of TQM have been noticeably silent on ergonomics/human factors. In the management literature many papers examine the impact of TQM on their discipline, for example Waldman (1994) on a theory of work performance, Grant, Shari, and Krishnan (1994) on management theory, or Costigen (1995) on human resource management, but nothing on ergonomics. However, the human factors literature has reported on how the quality movement affects the human factors profession, for example Zink, Hauer, and Schmidt (1994) on quality awards or Wilson, Neely, and Chew (1993) on the effects of modern manufacturing on worker well-being. This latter is taken further by Bjorkman's (1996) analysis of the similarities and differences between various quality movements and the tenets of modern work organization design. However, there are two relatively well-developed sets of ergonomics/TQM studies. The first uses TQM ideas in safety, while the second provides evaluations of joint TQM/ergonomics programs.

In the traditional safety area, a number of authors have pointed out the similarities between safety and quality (Rahimi, 1995; Roughton, 1993; Smith and Larson, 1991; Krause, 1993). One similarity is that both have departments (safety department, quality department) which may no longer be needed with a TQM approach (Rahimi, 1995). Krause (1993) shows their similarity in the measures they use. Traditionally, both have focused on downstream measures (defects reaching the customer, injury-producing accidents), but both should move toward process measures (SPC charts, behavioral measures) to help control the downstream events nearer to their source.

Two recent papers show how these ideas have progressed in Finland. Vainio and Mattila (1996) integrated safety concerns within the TQM system for an electrical utility. They made safety and health an integral part of TQM largely by addressing safety and health issues within the total quality handbook. More evaluation data were provided by Saari and Laitinen (1996) in a manufacturing setting. They set up continuous improvement teams for safety, defining best work practices in each area. Then, using the measurement-based TUTTAVA system, the teams set goals, made continuous improvements, and validated the results. A posture survey across the whole plant showed considerable improvement over the course of the project. In addition, injury and illness days lost were reduced by about 90% over three years.

Beyond safety is the safety role of ergonomics, typically designing to avoid injury. Here also a considerable literature is developing. Stuebbe and Houshmand (1995) characterize the production system as an inadvertent injury-producing system and advocate applying quality control approaches such as control charting, Pareto analysis, etc., to an "integrated ergonomic-quality system." This consists of analysis of the task, worker, and environment using these quality control techniques. Getty, Abbott, and Getty (1995) link quality initiatives to ergonomic projects, showing how an intervention to control cumulative trauma disorder in a panel drilling task also had a substantial effect on quality and productivity.

A major program in Sweden, the Quality, Working Environment and Productivity (QPEP) project (Axelsson, 1994; Eklund, 1995) examined specifically the linkages between quality and ergonomics in a car assembly plant. In eight departments, they produced an inventory of ergonomically demanding jobs, both those which were physically demanding and those causing production problems. Two different measures of quality showed significant differences between ergonomically good and ergonomically poor tasks, indicating the close link between ergonomics and quality.

One of the most integrated quality ergonomics efforts so far appears to be the implementation of ergonomic change within the TQM philosophy at the mail order clothing manufacturer and distributor, L. L. Bean (Rooney and Morency, 1992; Rooney, Morency, and Herrick, 1993). Their ergonomic objective was initially to eliminate the cumulative trauma disorder exposures of repetitive sewing production in a 400-person manufacturing plant. TQM was seen as defining the mission, objectives, and responsibility for safety with line management. Ergonomics moved over a six-year period from reacting to employee injuries, through proactive job design using teams, to now become part of the management and employee performance expectations and rewards.

In a follow-on paper, Rooney et al. (1993) were able to tackle some of the more deep-seated problems of repetitive work. They redesigned payment systems (with active operator involvement), replacing direct piece-rates with an annual appraisal system in which units produced were only 35 to 33% of the weighting. More complexity was built into jobs, by using cross-training and team work. Management and supervisor commitment for the ergonomics program was shown by their active support. Rooney et al. (1993) see these changes as a way of incorporating the musculoskeletal injury reduction aspects of ergonomics into a wider framework based upon macroergonomics (Hendrick, 1992) and TQM principles.

We can, however, go beyond these examples to provide managerial advice on TQM/human factors interactions, making use of similarities where they exist and exploiting differences to enhance each program. Starting with the similarities between the tenets of TQM and ergonomics (Table 6.1 and Table 6.2) we can suggest, with some combining of categories:

1. *Study and Measure the Process.* Start from a systems focus rather than the current process (also advocated in Business Process Reengineering, Hammer and Champy, 1990). Use this as the basis for a detailed quantitative understanding of the process. Standard quality techniques should be used to measure process parameters, and models of human performance and well-being to measure and understand the role of the operator in the system. Use these measurements as the basis for directing and quantifying continuous improvement.
2. *Honor Thy User.* (To quote Kantowitz and Sorkin, 1987). Respect the operators in the system as people trying to do their best, and having an inherent stake in performing well. Do not necessarily blame the operator alone for poor quality/productivity/safety. Tap the potential for

operatorempowered improvement by giving real power to small teams which include operators. The rewards will be improvements in performance, safety, and job satisfaction.

From differences between TQM and ergonomics we can show first how ergonomics can learn from TQM practice. These represent largely a shift from a technical process level of intervention to a more strategic, managerial level. (Longer discussions of each issue are presented in Drury, 1996.)

3. *Consider the Strategic Level.* Understand the forces beyond the process within the factory, such as requirements of the ultimate customer, and active management of the supply chain. Ensure that ergonomic interventions are truly customer-driven by explicitly measuring customer needs.

4. *Understand Leadership.* Any change activity needs responsibility of managers, up to the highest level. Do not take the mechanistic view of an organization which defines each manager by function. Understand the principles of leadership, recognize leaders, and practice leadership. All change projects need a powerful champion.

5. *Use Well-Developed Team Skills.* TQM, and many other change disciplines have standard methods of starting, organizing, and running successful teams. Use these methods where they are appropriate. At least understand these methods so that you can build on the teamwork training existing within the organization from TQM programs.

Where TQM can learn from ergonomics is in the area of technical knowledge of the human operator, and how to incorporate this in process design.

6. *Use Allocation of Function Techniques.* A basic building block of human factors is the concept of function allocation, i.e., permanent or flexible assignment of logical functions between human and machine. This has been used by ergonomists at levels ranging from the whole complex system (Older, Clegg, and Waterson, 1996) to a single human–machine system (Drury, 1994). Without an explicit treatment of function allocation, technology can easily fail. For example, consider the baggage handling system at the Denver International Airport.

7. *Error-Free Manufacturing/Service.* While TQM is calling for drastic reductions in error rates, human factors is coming to grips with the causes of human error (e.g., Reason, 1990). In airline flight operations (Wiener and Nagel, 1988) and maintenance (Wenner and Drury, 1997) we have classified errors and derived logical interventions, moving from a consideration only of the accident-precipitating event to a study of root causes and latent pathogens.

8. *Interface Design.* From physical workplace layout to reduce injuries (e.g., Kroemer, Kroemer, and Kroemer-Elbert, 1994) to the interface between software and the user (Helander, 1988), human factors engineers have been designing less error-prone interfaces between people and systems. This set of techniques is largely ignored in the TQM literature, despite the latter's emphasis on error reduction, parts per million, and six-sigma processes.

6.4 Summary

In this chapter we have examined the relationship between quality programs, specifically TQM, on the one hand and ergonomics/human factors programs on the other. Simple listing of their tenets, although these may still be arguable, led to recognition of the similarities and differences between the programs. Examples of use of ergonomics within a TQM context showed that sensible linkages had already been reported.

The aim of the chapter was to find prescriptions which would help the manager exploit the similarities and differences, so as to find new linkages between human factors and TQM. Seven prescriptions are given which can lead to greater integration between the two programs in the future. Readers who *do* use these for successful integration of the human factors and the quality imperative are urged to continue to report their work in the open literature and continue the integration process for the benefit of all.

References

Axelsson, J. R. C. 1994. Ergonomic aspects on design and quality, *IEA'94*, Vol. 4: 18–21.

Bjorkman, T. 1996. The rationalisation movement in perspective and some ergonomic implications, *Applied Ergonomics*, 27(2): 71–77.

Blest, J. P., Hunt, R. G., and Shadle, C. C. 1992. Action teams in the total quality process: experience in a job shop, *National Productivity Review/Spring 1992*, 195–202.

Brennan, L. 1990. The human dimension to statistical process control within advanced manufacturing systems, in *Ergonomics of Hybrid Automated Systems II*, Eds. W. Karwowski and M. Rahimi, pp. 527–534, Elsevier Science Publishers, London.

Costigan, R. D. 1995. Adaptation of traditional human resources processes for total quality environments, *Quality Management Journal 95 Spring*, 7–23.

Crosby, 1979. *Quality is Free, The Art of Making Quality Certain*, McGraw-Hill, New York.

Deming, W. E. 1986. *Out of the Crisis*, Massachusetts Institute of Technology, Cambridge, Mass.

Dobyns, L. and Crawford-Mason, C. 1991. *Quality or Else. The Revolution in World Business*, Houghton Mifflin Company, Boston.

Drury, C. G. 1992. Ergonomics of job and equipment design, *Impact of Science on Society*, 165: 41–52.

Drury, C. G. 1994. Function allocation in manufacturing, in *Proceedings of the Ergonomics Society 1994 Annual Conference*, Ed. S. A. Robertston, University of Warwick, 19–22 April 1994, pp. 2–16.

Drury, C. G. 1995. Work design, in *Human Factors Guide for Aviation Maintenance*, Ed. M. Maddox, Chapter 6, Federal Aviation Administration/DOT, Washington, D.C.

Drury, C. G. 1996. Ergonomics and the quality movement. The 1996 Ergonomics Society Lecture, Leicester, U.K., 10–12 April 1996.

Drury, C. G. and Prabhu, P. V. 1994. Human factors in test and inspection, in *Design of Work and Development of Personnel in Advanced Manufacturing*, Eds. G. Salvendy and W. Karwowski, Chapter 13, pp. 355–492, John Wiley & Sons, New York.

Eklund, J. A. E. 1995. Relationships between ergonomics and quality in assembly work. *Applied Ergonomics*, 26(1): 15–20.

Evans, J. R. and Lindsay, W. M. (1995). *The Management and Control of Quality*, Third Edition, West Publishing Company, Minneapolis/St. Paul, MN, 143.

Getty, R. L., Abbott, W. L., and Getty, J. M. 1995. ISO 9000 methodology enhances ergonomics effort: ergonomics becomes a tool for continuous improvement, *ASQC 49th Annual Quality Congress Proceedings*, 904–913.

Grant, E. L. and Leavenworth, R. S. 1995. *Statistical Quality Control*, McGraw-Hill, 1984, New York.

Grant, R. M., Shari, R. and Krishnan, R. 1994. TQM's challenge to management theory and practice, *Sloan Management Review/Winter*, 25–35.

Hackman, J. R. and Wageman, R. 1995. Total quality management: empirical, conceptual, and practical issues, *Administrative Science Quarterly*, 40: 308–342.

Hammer, M. and Champy, J. 1993. *Reengineering the Corporation*, Harper Business, New York.

Helander, M. Ed. 1988. *Handbook of Human-Computer Interaction*, Elsevier Science Publishers B.V., Amsterdam, The Netherlands.

Hendrick, H. W. 1992. A macroergonomic approach to work organization for improved safety and productivity, in *Advances in Industrial Ergonomics and Safety IV*, Ed. S. Kumar, pp. 3–10, Taylor & Francis, Hampshire.

Kantowitz, B. H. and Sorkin, R. D. 1987. Allocation of function, in *Handbook of Human Factors*, Ed. G. Salvendy, pp. 355–369, John Wiley & Sons, New York.

Karwowski, W., Marek, T., and Noworol, C. 1988. Theoretical basis of the science of ergonomics. *Ergonomics International 88, Proceedings of the 10th Congress of the International Ergonomics Association*, Sydney, Australia, 1–5 August 1988, Eds. A. S. Adams, R. R. Hall, B. J. McPhee and M. S. Oxenburgh, pp. 756–758, Taylor & Francis, London.

Kleiner, B. M. and Drury, C. G. 1996. Macroergonomics in regional planning and economic development in O. Brown, Jr. and H.W. Hendrick (Eds.), *Human Factors in Organizational Design and Management*, 523–528.

Krause, T. R. 1993. Safety and quality: two sides of the same coin, *Occupational Hazards*, April 1993, 47–50.

Kroemer, K.H.E., Kroemer, H.J., and Kroemer-Elbert, K.E. 1994. *Ergonomics: How to Design for Ease and Efficiency*, Prentice Hall, Englewood Cliffs, NJ, 430–441.

Larson, P. D. and Sinha, A. 1995. The TQM impact: a study of quality managers' perceptions, *Quality Management Journal 95 Spring*, 53–66.

Liker, J. K., Joseph, B. S., and Armstrong, T. J. 1984. From ergonomic theory to practice: Organizational factors affecting the utilization of ergonomic knowledge, in *Human Factors in Organizational Design and Management*, Eds. H. W. Hendrick and O. Brown, Proc. 1st Symp., Honolulu, HI, August 1984, North-Holland, Amsterdam.

Meister, D. 1996. A new theoretical structure for developmental ergonomics. Paper to *4th Pan-Pacific Conference on Occupational Ergonomics*, Taipei 1996.

Mize. J. H. 1992. Constant change, constant challenge, in *Manufacturing Systems, Foundations of World-Class Practice*, Eds. J. A. Heim and W. D. Compton, pp. 196–203, National Academy Press, Washington, D.C.

Older, M., Clegg, C. W. and Waterson, P. E. 1996. Task allocation in complex systems, in *Advances in Applied Ergonomics*, Eds. A. F. Ozok and G. Salvendy, pp. 471–474, U.S.A. Publishing Corp, West Lafayette, IN.

Rahimi, M. 1995. Merging strategic safety, health and environment into total quality management, *International Journal of Industrial Ergonomics*, 16: 83–94, Elsevier, London.

Reason, J. 1990. *Human Error*, Cambridge University Press, Cambridge, U.K.

Rooney, E. F. and Morency, R. R. 1992. A practical evaluation method for quantifying ergonomic changes at L. L. Bean, in *Advances in Industrial Ergonomics and Safety IV*, Ed. S. Kumar, pp. 475–481, Taylor & Francis, New York.

Rooney, E. F., Morency, R. R., and Herrick, D. R. 1993. Macroergonomics and total quality management at L. L. Bean: a case study, in *Advances in Industrial Ergonomics and Safety V*, Eds. R. Nielsen and K. Jorgensen, pp. 493–498, Taylor & Francis, New York.

Roughton, J. 1993. TQM Integrating a total quality management system into safety and health programs. *American Society of Safety Engineers*, June 1993, 32–37.

Russell, S. 1991. Employee involvement aspects of total quality management, *P+ European Participation Monitor*, (2): 29–32.

Saari, J. and Laitinen, H. 1996. Towards continuous improvement of workplace, in *Advances in Applied Ergonomics*, Eds. A. F. Ozok and G. Salvendy, pp. 82–87, U.S.A. Publishing Corp, West Lafayette, IN.

Smith, T. J. and Larson, T. L. 1991. Integrating quality management and hazard management: a behavioral cybernetic perspective, *Proceedings of the Human Factors Society 35th Annual Meeting — 1991*, 903–907.

Stuebbe, P. A. and Houshmand, A. A. 1995. Quality and ergonomics, *Quality Management Journal 95 Winter*, 52–64.

Taguchi, G. 1986. *Introduction to Quality Engineering*, UNIPUB, White Plains, NY.

Taylor, J. C. and Felten, D. F., 1993. *Performance by Design*, Prentice Hall, NJ.

Taylor, J.C. 1990. Organizational context for aircraft maintenance and inspection, in *Proceedings of the Human Factors Society 34th Annual Meeting, Volume 2*, 1176–1180.

Vainio, P. and Mattila, M. 1996. Development of a safety and ergonomics oriented total quality system for an electricity company, in *Advances in Applied Ergonomics*, Eds. A. F. Ozok and G. Salvendy, pp. 43–46, U.S.A. Publishing Corp, West Lafayette, IN.

Waldman, D. A. 1994. The contributions of total quality management to a theory of work performance, *Academy of Management Review 1994*, 19(3): 510–536.

Wenner, C. and Drury, C. G. 1997. Deriving targeted interventions for ground damage, *Proceedings of the 1997 SAE Airframe Engine Maintenance & Repair Conference* (AEMC '97) August 1997, SAE Technical Paper Series 972591, Warrendale, PA.

Wiener, E. L. and Nagel, D. C. Eds. 1988. *Human Factors in Aviation.* Academic Press, Inc., San Diego.

Wilson, J. R., Neely, A. D., and Chew, T. 1993. Human and production requirements in modern manufacturing: complementary or contradictory? *Journal of Design and Manufacturing (1993)*, 3: 167–175.

Witcher, B. 1995. The changing scale of total quality management, *Quality Management Journal 95 Summer*, 9–29.

Zink, K. J., Hauer, R., and Schmidt, A. 1994. Quality assessment: instruments for the analysis of quality concepts based on EN 29000, the Malcolm Baldridge Award and the European Quality Award, in *Total Quality Management*, Ed. G. K. Kanji, 5(5): 329–343, Carfax Publishing Company, UK.

For Further Information

The ideas in this chapter were based on the rather fuller treatment in Drury (1996), and the concept of tenets of the discipline reported by Hackman and Wageman (1995). The latter is a good and thoughtful review of TQM from a management viewpoint.

Standard works on TQM are Deming (1986), Evans and Lindsay (1995), and Taguchi (1986).

Excellent evaluations of the social role of TQM can be found in Wilson, Neely, and Chew (1993) and Bjorkman (1996). Comments on TQM from a sociotechnical systems viewpoint are given by Taylor and Felten (1993).

7

User-Centered Design of Information Technology

V. Kathlene Leonard
Kevin P. Moloney
Julie A. Jacko
Georgia Institute of Technology

7.1 Primer: The Need for User-Centered Design (UCD)

Increasingly, terms such as *Easy to use, User-friendly, Usable,* and similar others, pervade the way in which systems are described in their marketing to users and organizations. While the use of these terms in the advertizing and sales vernacular are becoming more and more (if not already) cliché, the attention which they are paid is not unsubstantiated. In a business sense, it is intuitive — if the systems in place are not accepted by the employees, or if they are inefficient, ineffective, and inconsistent with the organization's goals, then the financial consequences can be devastating. Poor design can take a tremendous toll on productivity and employee satisfaction as well as increase the costs of maintenance. Attention must be paid to the user interface to maintain a competitive edge in the marketplace (Shneiderman, 1998).

Several examples of the severe implications of poor design fill the headlines of the popular press. Perhaps the most recently infamous is the butterfly ballot design from Palm Beach County, FL that caused much turmoil in the 2000 U.S. presidential election. Using this punch card ballot system, voters were instructed to punch out the hole in the center next to the arrow, which points from the candidate of choice. This setup is depicted in Figure 7.1. From an UCD perspective, the design of this ballot violates the principles of mapping and alignment (Lidwell et al., 2003). The actual punch cards, which are slid into a frame, have no candidate names written on them, so if improperly inserted, they have a high probability for error. The misalignment of the holes and candidate names has been identified as the source of error. Several ballots cast in Palm Beach contained two votes with the holes punched for candidates listed adjacently on the ballot. A surprisingly high number of votes were also cast for Pat Buchanan. Several voters reported that they inadvertently voted for Buchanan because the placement of his punch hole was in between Bush's and Gore's names (Van Natta and Candey, 2000).

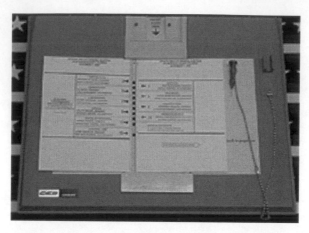

FIGURE 7.1 The punch card voting system, as experienced by voters in Palm Beach County, FL, for the 2000 presidential election. (From Psephos Corporation, 2005. With permission.)

Cranor (2001) outlines the difficulties with the two-page layout, even for someone without vision impairment with healthy hand-eye motor skills and cognition. The population of Palm Beach County is comprised of mostly adults over 50. The population commonly experiences age-related declines in their sensory, motor, and cognitive abilities. Thus, it is easy to understand how the insufficient spatial mapping resulting from this design could prove highly problematic for users. In addition to the poor layout, there were several problems with the actual reading of the cards to count votes. Often at times the hole was not punched out completely leaving a ballot that was "dimpled" or not completely perforated through (i.e., the infamous hanging chads), and henceforth not readable by the machine. This led to thousands of work hours dedicated to manual sorting and classifying these paper cards, and ultimately delayed the outcome of the election.

While hindsight is 20/20, had the decision to implement the butterfly ballot taken into consideration abilities of the actual voters, the demands presented by the task, and the costs of recounting these ballots, officials may not have been inclined to use such a system, or (at least) they may have been more deliberate in the layout. The core problem with the butterfly ballot system was its narrow focus on the automation and functionality of counting votes. The users of the system (the voters) and all of their inherent variability were not considered as an operational component of this system. Instead, they were given tasks for which the vote counting system was operationally incapable of completing. That is, the people in this system took on the leftover work, such as inserting the cards into the butterfly ballot, selecting the candidate, and in case of the 2000 election, manually counting chads and dimples.

Critical features implicating the voting system/voter interaction were overlooked, such as the cognitive and physical abilities of the users, voter trust in the system, and verification that the intended vote was cast. Because the design of the system was remiss in addressing these critical components, impediments emerged to the ultimate goal of electing a president. It is critical in the design process to consider how to allocate the functions of work between people and systems, in order to more readily facilitate when the user is considered a component of that system (Noyes and Barber, 1999).

Quality in systems and products is not an accident; it's an outcome of judicious design (Ostroff, 2001). Furthermore, good design does not happen by chance. While the tools and systems integrated into the workplace can be effective and beneficial to work goals and a pleasure to use, potential always exists for these tools to be ineffective and cumbersome to the goals and needs of the user, work, and organization. Additionally, the impact of an ill-fit system in the workplace can propagate throughout an organization to seriously implicate total quality, worker satisfaction, worker retention, productivity, and downtime.

In extreme cases, users abandon technology intended to augment their work, or abandon the job out of frustration when the use of a poorly designed system is mandatory. This can ultimately influence the bottom line and long-term organizational endurance.

New tools and systems introduce change from many perspectives (Karat et al., 2000). Underestimating the impact of a new system can have negative implications. For example, the introduction of electronic purchasing reports may be aimed at improving productivity by removing the need for face-to-face meetings with a purchasing manager. This organizational productivity may come at the cost of individual productivity, as the purchasing manager may obtain information beyond what the electronic system records provide, such as the immediacy of the needs of different orders. Furthermore, the elimination of personal contact with other employees may prompt a feeling of isolation and a significant decrease in job satisfaction. Providing tools that enable users through directing and facilitating activity and not by restricting and controlling their activity should always be a consideration (Karat et al., 2000).

The design process entails innumerable assumptions and decisions by the designer(s) about the work and the user. Good designs are more likely to emerge when those assumptions and decisions are well-informed. The quality, quantity, and accuracy of the information gathered directly impact the efficacy of the final design product. Good designs represent a symphonic integration of characteristics of both the users and the technology. That is, when the decision makers who orchestrate the design process possess an adequate understanding of the people using the system, the technology, and what drives the interactions between the two, the resulting designs are more liable to promote harmonious interactions. Good designs incorporate consideration of the user(s), the organization, the environment, and the technology at several points throughout the design and implementation process. This process should incorporate communication, analyses, and design techniques, which are creative, yet grounded in knowledge (Hackos and Redish, 1998).

According to Karwowski (2003), the ergonomics approach to information technology (IT) "advocates the systematic use of knowledge concerning relevant human characteristics to achieve compatibility in the design of interactive systems among people, computers, and outside environments ... to achieve other specific goals, such as system effectiveness, safety, ease of performance and to contribute to overall user well-being" (p. 1227). This *compatibility* is representative of the overarching goal of human factors and ergonomics, and is to be considered at all levels in the design (e.g., physical, emotional, perceptual, cognitive, social, organization, environmental, etc.) to optimize the user-interface system, and its subsequent output. The integration of IT into the workplace creates a paradox. As technologies evolve to include more sophisticated features and provide richer information to the users, an increase in the complexity of the system results, as does a corresponding propensity for usability problems (Hackos and Redish, 1998; Karwowski, 2003; Norman, 1988). Norman termed this concept *creeping featurism* and suggested that the square additional functionality, which is incorporated in a design materializes as additional complexity is imparted upon then required interaction (Norman, 1988). This is also known as the *complexity-compatibility paradigm*, which states that the added complexity of a system decreases the potential for efficacy in ergonomic interventions and overall design efficacy (Karwowski, 2003). These things considered, a deliberate approach to the incorporation of user factors is essential for the design of usable, useful systems.

7.2 Introduction to UCD

This chapter provides an overview of UCD — the practice of eliciting and incorporating user characteristics throughout the design process. While the UCD process is appropriately applied to systems embodying a range of technical sophistication (including no technology) with which people interact, this chapter will focus on UCD in the context of IT systems in organizations. The underlying principles of UCD are propagated through a variety of documented and tested methods that involve the user throughout the design and implementation of a new system. Readers will be provided with a justification for the importance of UCD and some common strategies for incorporating accurate information about users and their work.

The term user-centered design first emerged in the mid 1980s and is often attributed to Norman's (1988) renowned book, *The Psychology of Everyday Things* (later reissued with the title: *The Design of Everyday Things*) (Norman, 1988, 2002). UCD was embraced by the human–computer interaction (HCI) discipline as the foundation for usable, useful, and successful interfaces. *The Berkshire Encyclopedia of Human Computer Interaction*, more recently, defines UCD as a "broad term, used to describe the process in which end users influence how a design takes shape" (Bainbridge, 2004, p. 763). UCD directs focus of the design process to the users' role in the interactions between man and machine, and the resulting dynamics that ensue with the work, environment, and organization.

Changes and modifications are introduced to the UCD process by designers in reaction to an evolving network of users, goals, environments, technologies, and associated constraints with which the design must operate. This improvisational approach aims to achieve a balanced harmony of interactions between users, goals, context and constraints, and to avoid dissonance within the network that could impede productivity. UCD embodies a set of design strategies and information gathering techniques and tools that are perceived as personalized rather than controlling. The vehicle to achieve this level of usability and usefulness is the multifaceted task of understanding the users, the work context, and the tasks to be accomplished (Karat et al., 2000).

In truth, the specific set of techniques used for UCD depends on the specific combination of circumstances and constraints related to the users, work, and organization associated with the given project (Mao et al., 2001; Marcus, 2005). It is only through practice in developing and hands on experience in implementing these techniques that designers acquire their own combinations of tools and techniques optimal for the constraints and requirements of the target problem and goals. Those who are highly skilled in UCD demonstrate the ability to adapt techniques to the constraints of a given domain and situation. UCD is an exercise in improvization of the well-established, tested techniques used to acquire and integrate user characteristics, in order to adapt these techniques concurrently as the requirements of the design evolve. As such, this chapter provides a high-level roadmap for the UCD process, and introduces readers to fundamental methods for incorporating the user into the UCD process. An operational definition is provided of UCD, followed by two sections, which emphasize common methods of user involvement in the design process. The first section details requirements gathering, and the second underscores user involvement in the testing and evaluation of designs and prototypes.

7.2.1   Use of UCD

UCD has been applied in a variety of domains from aviation to HCI. The general UCD guidelines and techniques are constantly revised to sufficiently account for the domain and context-specific goals, characterizations of the user, and organization, beyond those accounted for the general "good design" principles developed and introduced by Nielsen (1994), Norman (1988, 2002), and Shneiderman (1998). While UCD initially focused on the design of computer interfaces and technical interfaces, more recently it has been given much attention to the fields of bioinformatics, health systems, children's applications, and designing for individuals with limited abilities.

Table 7.1 features examples of several domains that have applied UCD. Often, the result of extending UCD to other domains is a new set of guidelines aimed at standardizing the UCD approach across a given domain. The negative outcome of this includes the influx of guidelines, standards, and requirements in both literature and legal initiatives, which creates an additional burden on the design team. Rarely (if ever) do the guidelines specify, for practical situations *how* design teams should apply them in the design process. Largely, designers are just given several lists of "rules" for the design, but no means by which to obey these rules (Cassim and Dong, 2003; Koubek et al., 2003; Law, 2001; Lebbon and Coleman, 2003).

TABLE 7.1 Examples of Recent Application Domains of the UCD Process

Domain	Title	UCD Notions
Bioinformatics	Beyond power: making bioinformatics tools user-centered (Javahery et al., 2004)	The efficacy of bioinformatics in health care is influenced by the degree to which tools are adopted, and their long-term reliability. UCD methods can inform systems that are more easily used by novices, while enhancing the skills of subject matter experts. In a domain with a notoriously high cost of personnel, UCD can help to optimize how efficiently tasks are accomplished
Intra-organizational learning	Intranets and intra-organizational learning, (Emery et al., 2004)	The evolution and matriculation of knowledge through an organization is often indicative of an organization's long-term sustainability. Intranets are a central repository for this knowledge, and can facilitate intra-organizational learning. UCD processes, when followed in the design of IT such as intranets, can result in tools that best champion the learning styles of a given organization, and the individual knowledge of workers as well
Health care	A user-centered framework for redesigning health care interfaces (Johnson et al., 2005)	The rapid proliferation of health care interfaces into the field has resulted in a considerable number of usability problems. Therefore, current interfaces need to be redesigned, in consideration of UCD processes in order to retrofit these interfaces to better meet the user and task requirements
Accessible technology	User-sensitive inclusive design: in search of a new paradigm (Newell and Gregor, 2000)	UCD needs to make special consideration for users with extraordinary characteristics (a.k.a. disabilities) and users who are ordinary, but the context within which they operate is extraordinary (e.g., noisy, poor lighting, stressful). Through UCD, designers can determine the potential user population to fit into these special circumstances. Incorporating UCD and inclusive design can result in systems that are usable by a greater number of people
Design for children	KidReporter: a user requirements gathering technique for designing for children (Bekker et al., 2003)	UCD techniques for working with children need to be motivating and stimulating, yet appropriate for the children's cognitive skill level. Special consideration is given to the consistency of answers solicited from children during the UCD process. Strategies for overcoming unpredictability of the children involved in UCD include the collection of information from several sources (or children) to increase reliability

7.2.1.1 The UCD Process

For practical purposes, the discussion of UCD practices will be couched within the generally-accepted vision of iterative software-based system development, the usability engineering lifecycle (Mayhew, 1999, 2003). The usability engineering lifecycle represents a structured and systematic approach to applying usability and UCD principles during the system (or interface or product) development process.

Readers unfamiliar with this vision of product development should refer to Mayhew (1999, 2003) for further details.

UCD focuses on incorporating the user, task, and environment into the design process. While the user lies at the center of the design process, they are part of an intricate network consisting of workplace-, environment-, and organization-related factors. Critical information about the users is gathered and applied in the formulation of design requirements, as well as the continued evaluation of design alternatives and final implementation. Figure 7.2 illustrates this high level conceptualization of UCD.

Eason (1995) describes three overarching goals of UCD, which include: (1) the translation of articulated user/context/work needs into overall product requirements and design specifications; (2) the production of several design options; and (3) the evaluation of the degree to which the design options fulfill the requirements. With this view of the system design (or redesign) and development process in mind, the primary activities of UCD can be divided into two phases. The first phase involves the front-end development of an appropriate understanding and representation of the system users, their goals, their need, their work tools, and their work (i.e., "understanding the work system"). The second phase involves the back-end testing and evaluation of proposed design solutions to examine and validate this front-end work and its interpretation (i.e., "testing and evaluation"). These two primary phases of the UCD process are shaded in Figure 7.2.

UCD mandates that inquiries be made into user and work conditions. In order to create appropriate systems for users, the needs of the users should be at the forefront of the process (Sugar, 2001). This

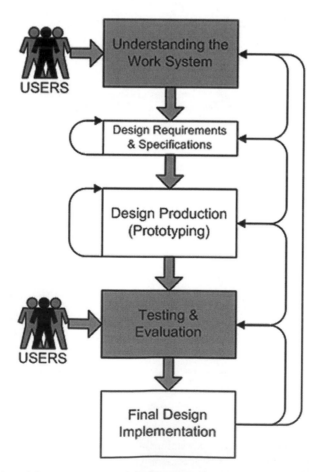

FIGURE 7.2   Illustration of the UCD process and the key steps of user input or feedback.

facilitates the derivation of design goals, creates a formal account of the design's requirements, and translates these into a tangible list of specifications, or action items to inform design. Intuitively, the next step in UCD is the application of the specifications to the actual system design. Once system designs have been developed through a process of prototyping, these potential designs will be tested and evaluated for their appropriateness, utility, and usability. Finally, when a good design results, a full working system will be implemented and released.

Additionally, as can be seen in Figure 7.2, this process is highly iterative with continual verification and validation of decisions and progress along the path to implementing designs. As has been commonly outlined in the usability engineering or design "lifecycles" (see Mayhew, 1999; e.g., Nielsen, 1993; Samaras and Horst, 2005), there are several feedback loops in the process, often heavily involving users as sources for validation and testing. While this concept of iterative design will not be heavily discussed in this chapter, it is one of the cornerstone principles of UCD and the software and usability engineering lifecycles.

7.2.1.2 User Involvement in UCD: The Fundamental Principle

As shown in Figure 7.2, users should be heavily involved directly in the collection of information (i.e., understanding the work system) to inform the design requirements and specifications, and again in the evaluation and testing of the design. Depending on the preferences of the UCD team and the UCD methods they utilize, along with other practical considerations (e.g., time, resources, organizational priorities, etc.), users may also be involved, albeit to a lesser extent, in directly participating in the design of the system (e.g., participatory design) and through user feedback after the final implementation.

Involvement of the users in the formulation of system design requirements and the subsequent testing and evaluations of the system leads to more effective, efficient, and safe systems and contributes to the perceived success and acceptance by the actual users. Noyes and Barber (1999) summarized these steps, which are the fundamental components of UCD: (1) designing to accommodate the physical and cognitive capabilities of the users; and (2) designing to accommodate for the work that is required. That said, the scope of the UCD process is constantly evolving during the design and evaluation activities, soliciting user opinions, understanding work needs and expectations, designing to accommodate the work activities, and designing to accommodate user abilities. The involvement of users helps to assess the impact of a new design with respect to not only the system or interface, but also to the related job aspects such as staffing levels, allocation of tasks, new policies, and training (Kontogiannis and Embrey, 1997).

The inclusion of users in the design process should not be accomplished arbitrarily. It requires deliberate coordination and well-calculated timing. Intuitively the quality of the information collected (and the subsequent quality of the design) is a function of the timing in the design process at which the user is involved and the extent of their involvement. Consideration of UCD factors too late in the design process can prove costly. Adjusting for problems in the development phase is estimated to cost ten times more than during the design phase, and the cost of fixing a problem in the postimplementation phase can cost 100 times more than during the design phase (Johnson et al., 2005). A poorly implemented system can be costly in terms of productivity — it has been estimated that a poorly designed system, when implemented can cause employees to spend upto 40% of their time asking each other how to use the system to do tasks (Hackos and Redish, 1998).

In terms of the degree to which users should be involved in the UCD process, it is important to remember that users are not necessarily a part of the design and development team, but rather a rich source of information, by which decisions in UCD are influenced. It is upto the design team to integrate the information collected during observations of the users with additional information retrieved about the organization, tasks, and stakeholders. However, using only what end-users convey as important is likely to introduce substantial bias and incomplete assumptions in the design, as the users tend to be insufficiently articulate in the explicit communication of their needs and typically have a restricted view of their work within the perspective of the entire organization, often proposing solutions to problems, which can be solved by more effective means (Newell and Gregor, 2000).

The UCD approach just looks beyond principles of physical design — beyond traditional ergonomics — to consider more fundamental issues such as the structure of information presented, the degree of process automation, implications of skill transfer and retention, and training requirements. If operators and supervisors are involved only at the later phases of the design process, then the final product may be difficult to learn, and incompatible with the existing well-established working practices, while also triggering increases in vigilance tasks and diminished user acceptance and trust (Kontogiannis and Embrey, 1997).

The result is a set of integrated techniques, which combine users' assertions with observations across several levels of the organization, and several classes of users and stakeholders. Damodaran (1996) has suggested a well-defined taxonomy of the roles of individuals involved in the design process and the identification and selection of the most representative users from the organization. In fact, users should be identified at several levels from within an organization, ranging from end users to middle- and top-level management. This focus on all stakeholders is because the outcomes of technical design decisions may have profound implications on job design, and subsequently working life (Damodaran, 1996; Damodaran et al., 1980).

7.2.2   Summary

This chapter will focus on the inclusion of users at the two phases in the design process during which user feedback can have the greatest influence on its eventual success. This includes, as illustrated by Figure 7.2, understanding the work system and the testing and evaluation of designs and prototypes. The integration of specifications into tangible designs will briefly be touched on, as a more complete explanation is out of the scope of a manuscript of this length. The actual design production step is best learnt through practice and case studies of the experiences of others. In a study of the efficacy of novice designers to UCD, Sugar (2001) investigated the challenges that novice designers face in the application of UCD principles and guidelines. These designers were observed to have incomplete mental models of the translation of requirements into creative design solutions and tended to focus on overt observations of user needs.

In this chapter, readers will be provided with guidance on how to incorporate users into and leverage their knowledge and insight during the UCD process. UCD is partly a structured methodology and partly a skillful improvization. Readers should recognize the importance of practical experience and reviewing actual design case studies, both grounded in empirically validated guidelines and underlying philosophies of UCD in order to achieve effective designs. In UCD, those designing the system are responsible for: (1) facilitating the task or work for the user; (2) ensuring that the user can use the product as it was intended to be; and (3) making certain that the training and learning required to use the product is minimized.

7.3   Common Tools and Techniques for UCD

7.3.1   Overview

There are, in both literature and practice, several well-documented and commonly applied methods for gathering user information and including users in the design and evaluation processes to achieve UCD. The International Organization for Standardization (ISO) in an attempt to consolidate these methods has published principles of UCD and steps required in the UCD development cycle (ISO 13407, 1999). In his article entitled "Methods to support Human-Centered Design," Maguire (2001) provides an overview of the main UCD principles and an extensive list of methodologies appropriate for each step in the UCD process.

There are a seemingly infinite number of approaches and frameworks for UCD. These range from general principles (e.g., active user participation or iterative design and testing) to specific tools [e.g., the Unified Modeling Language (Booch et al., 1998; Rumbaugh et al., 2004) or hierarchical task analysis (Shepherd, 2000, 2001)] or techniques [e.g., contextual design (Beyer and Holtzblatt, 1998; Holtzblatt,

2003) or ISO 13407 (ISO, 1999) or 18529 (ISO, 2000)] to amalgamations of any and all approaches that suit the needs and purposes of the UCD team. However, despite this considerable variability in the techniques and approaches one might use or the design philosophy one might subscribe to, there are common threads linking all of these methods and techniques to one common goal. This goal, of course, is ensuring that the customers', users', and stakeholders' needs and goals are met in the design of a system or tool.

One of the reasons that the UCD process is such a nebulous topic to summarize is the variability among approaches and the differences imposed by context disparities. As is well known by all those who practice UCD, the ideal circumstances for UCD, such as complete user buy-in, unlimited resources, unlimited time, and unlimited access to users and their work never actually exist in practice. As such, UCD practitioners often pick their preferred techniques or tools and go about the business of designing or redesigning work systems for users. This being said, this chapter will discuss the UCD process in fairly broad terms, focusing more on the general ideas, principles, purposes, and methods or tools of UCD.

As a preface to this discussion, readers should note that UCD practitioners (e.g., ergonomists, human factors engineers, psychologists, ethnographers, designers, developers, etc.) often deal with a whole host of constraints, limitations, and obstacles to overcome in real-world domains. This being said, we begin the discussion of UCD and the process by which practitioners can translate an understanding of users, their needs, and their work into appropriate and successful system design solutions.

7.3.2 Understanding Users, Their Needs, and Their Work

7.3.2.1 Setting the Stage(s)

As previously discussed, the UCD process can be depicted as a complex, iterative process with two principle phases involving the users — the process of understanding the work system, which is used to define the needs and develop design solutions, and then the testing and evaluation of the formulated design solutions (see Figure 7.2). The present section will discuss this initial phase of the UCD process, in which the users, their needs, and their work will be examined in order to develop an understanding of the goals and requirements to be met by the system, application, or tool, which will culminate in design solutions.

Figure 7.3 summarizes this initial front-end work of the UCD process including stages of defining the needs, examining the work system, and then representing the work system and its needs. This information and acquired knowledge is then used in the definition and development of system requirements and specifications, which are used to develop system design alternatives. As discussed, these potential design solutions (i.e., prototypes) must finally be tested and evaluated (to be discussed later). This general conception of the UCD process has been the basis for many other models of systems engineering, which have been neatly summarized in a recent paper by Samaras and Horst (2005).

This section on understanding users, their needs, and their work will focus on the steps of this process from the work system analysis through the development of system requirements, ending with some discussion on the translation of these requirements and specifications into design prototypes. It should be noted that this representation is extremely simplified, as it does not display the iterative nature of the process or the numerous feedback loops that occur throughout the process to ensure the verification and validation of the previous steps (see, e.g., Mayhew, 1999, 2003; Samaras and Horst, 2005).

7.3.2.1.1 The Work System

A *work system* refers to the collective set of components that define a context under which goals are set and actions are executed in an attempt to accomplish these goals. Work systems are dynamically evolving networks of components operating under a shared set of goals and constraints (although components, they may not share *all* goals and constraints). An organization, or learning organization, is itself a work system, but also comprises of several smaller work systems. The components of the work system consist of all related facets of the users, stakeholders, environments, tools, artifacts, and facets of the tasks that have been identified as the means to attain goals. This holistic notion of the user, tools, and context, which implicate the work system stems from contextual computing, HCI, and cognitive engineering (Dourish, 2004; Kirlik, 1998; Oulasvirta, 2004; Sears et al., 2003).

The work system, as it relates to UCD, involves: (1) the human users and all of their capabilities, limitations, demographics, knowledge, experience, and skills; (2) the actual work and all related issues including allocation of functions, roles, responsibilities, complexity, and sequences; (3) the work tools and all related issues including hardware attributes, physical device attributes, nontechnical artifacts; and (4) the work context and all related issues including organizational needs, environmental variables, social interaction, and communication.

In the context of UCD we must draw an important distinction between a *work system* versus a *system*. While a *work system* refers to the comprehensive view of components operating under similar constraints to achieve shared goals, a *system* is often used in UCD in reference to a general tool, application, or interface that the user interacts with to complete a task. Systems are also commonly referred to as tools, designs, interfaces, and ITs. In other words, the work system is what is being examined, while the system is what is being designed and developed.

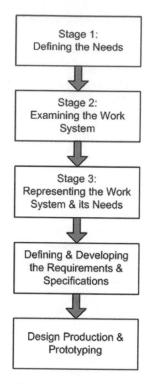

FIGURE 7.3   The process of translating an understanding of the work system into design. This figure is expanded from Figure 7.2 to fully illustrate more of the initial steps in the overall UCD process.

Noyes and Barber (1999) present a similar view of the work system, within the context of the UCD of systems and tools.

This clarification of the work system is important, because it is the focus of UCD. The work system comprises of the users, their needs, and their work. It is the focus of the front-end work in the UCD process, which provides the boundaries and inspirations for the designs and system prototypes. As can be seen in Figure 7.3, the resulting outcome of the front-end work is a set of system requirements, which will ultimately serve as the guide for system design. This process, sometimes referred to as requirements engineering (RE) (Kotonya and Sommerville, 1998; Nuseibeh and Easterbrook, 2000; Sommerville and Sawyer, 1997), is the manifestation of users, their needs, and their work. The following discussion will be dedicated to the process of RE and the common UCD methods and techniques used to collect and represented the user-centered data that will inform system design.

7.3.2.1.2   RE: The Overall Goal

Before getting into the details of the methods and techniques used, the underlying purpose of this process should be discussed. Whether taking a linear or iterative approach, or starting from a clean slate or a well-established work system, the process of UCD generally begins with the examination and formulation of the scope, nature, and needs of the work system. This process has been commonly referred to as the engineering, analysis, gathering, or specification of requirements. RE is the process of discovering, documenting, and maintaining a set of requirements, or descriptions of system properties, attributes, behaviors, or constraints, for a (computer-based) work system (Kotonya and Sommerville, 1998; Sommerville and Sawyer, 1997).

RE is the process of eliciting, analyzing, defining, documenting, validating, and managing requirements for a system (Kotonya and Sommerville, 1998; Sommerville and Sawyer, 1997). This process involves a number of activities, such as discovering the purpose for which the system is intended, identifying the system stakeholders and their needs, and (finally) translating and documenting these needs into a form that can be communicated, validated, and implemented (Nuseibeh and Easterbrook, 2000). RE is, arguably the seminal activity in the UCD process, because it represents the initial step in involving the users (and their needs) into the design process and, ultimately, serves as the blueprints for the designs that are to be tested and evaluated in later phases of the system development process.

As alluded to, RE is a process, meaning that it is (necessarily) somewhat ill-defined, fluid, and complex. This process involves any number of activities, ranging from simply defining the purpose of the system or goals of the customer or organization to diligently observing and modeling the actual work of the end-users. As all of the experts agree, there is no best way of conducting the RE process, just as there is no single or best way to define and document the requirements that result from the process.

While there are considerable issues regarding the practice of RE, such as requirements validation, change management, documentation styles, and traceability, this discussion falls outside of the scope of this chapter. Readers can be directed to any number of books and reports that deal with these delicate issues (e.g., Christel and Kang, 1992; Ransley, 2003; Rosenberg, 1999; Young, 2001). Some of these issues, such as insuring that representative users are chosen, will be discussed in relation to the activities at the various stages depicted in Figure 7.3. There are a number of good texts detailing the process of and considerations associated with the RE process, including those by Wiegers (2003), Robertson and Robertson (1999), and Young (2001, 2004).

The real purpose of the RE process is to minimize problems resulting from costly and destructive misunderstandings or misinterpretation between the stakeholders or users and the design and development team (Ransley, 2003; Sommerville and Sawyer, 1997). The RE process is able to limit these types of problems by utilizing systematic and structured process from the start of the system development process before making definitive judgments or assumptions about design alternatives or work system needs. Overall, RE is able to provide a better understanding of the system to be developed, help communication between users and the design and development team, minimize risk and maximize acceptance through good structuring, clear language, clear linkages between requirements and documented work system needs, and direct user involvement.

In more concrete terms, RE is the process of translating knowledge of users, their needs, and their work, and so on into specific, tangible requirements of the system to be designed and developed. As a simple summary, the RE process is utilized to produce system requirements in such a way that representative, accurate, and complete work system data are provided in a feasible, affordable, and ethical manner. We will now turn to the primary component or outcome of the RE process — the system and design requirements.

7.3.2.1.3 The Nature and Importance of Requirements

As suggested by popular models of design lifecycles [e.g., the usability engineering lifecycle (Mayhew, 1999, 2003)], requirements are defined during the early stages of system development, prior to any efforts in the design of system prototypes or implementation efforts. Requirements are generally described as descriptions or statements of system services, functions, properties, attributes, or constraints typically detailing how a system should behave within its context of use to help perform work (Kotonya and Sommerville, 1998; Nuseibeh and Easterbrook, 2000; Ransley, 2003; Sommerville and Sawyer, 1997). Requirements are typically a mixture of some problem information, statements of system behavior and properties, and design and manufacturing constraints (Sommerville and Sawyer, 1997).

There is no best way to generate or communicate requirements. The process of creating requirements from the collected and modeled data of the work system is nebulous to characterize, often requiring considerable interpretation, rationalization, and translation of the raw data into concise, accurate, and communicable ideas. There is also no set way to communicate system requirements. The requirements document, the comprehensive list of requirements, is the communication tool between the users and stakeholders, the UCD team, and the systems engineers and developers. Depending on the audience, organizational culture, and work domain (e.g., medicine versus commercial logistics versus fast-food), the nature of the requirements and the language used to express them can vary considerably.

Regardless of the method of creation or language, there are some established criteria for system requirements. Good system requirements possess the qualities or properties of specificity, feasibility, testability, clarity, comprehensiveness, consistency, exclusivity, traceability, accuracy, and so on (Ransley, 2003). The properties comprise the basis for good systems requirements as they specify the guidelines to ensure that the requirements are appropriate and effective for guiding UCD. It should also be mentioned that there are also a number of different types of requirements.

Design requirements can range from general requirements, which set out what the system should do in broad terms, to performance requirements, which specify a minimum acceptable performance level of the system (Kotonya and Sommerville, 1998; Ransley, 2003; Sommerville and Sawyer, 1997). Traditionally, system requirements govern functional capabilities rather than nonfunctional requirements such as issues of human cognition and capabilities, usability, individual differences and preferences, and social and organizational concerns (see Dix et al., 1998, for a good overview of this philosophy and appropriate techniques).

Not surprisingly, there is little consensus on the number and nomenclature regarding the classification of requirements. Generally speaking, there are two major classes of requirements: (1) functional requirements, which outline the things that the system must do (e.g., services and functionality to provide); and (2) nonfunctional requirements, which outline the properties that the system must have (e.g., usability, efficiency, reliability, etc.) (Robertson and Robertson, 1999; Young, 2004). Table 7.2 outlines some of these broad classes of system requirements. For a more comprehensive examination of requirement classification and delineation, readers can refer to texts by Young (2001, 2004) and Kovitz (1998).

As can be seen, requirements represent the collective set of issues, problems, needs, and constraints of the work system that need to be met by the new design or tool. With the concepts of the RE process and the requirements established, it is appropriate to discuss the specific activities of this process and how to equate it to requirements. These activities effectively supply the data on which the requirements are constructed.

7.3.2.1.4 UCD Methods and Techniques

The RE process consists of two basic classes of activities: (1) those that focus primarily on collecting data on the work system and (2) those that focus primarily on representing this data. The former approaches can be considered as requirement elicitation techniques, while the latter can be considered as requirements analysis and modeling techniques (Nuseibeh and Easterbrook, 2000).

TABLE 7.2   Summary of the Different Classes of System Requirements

Requirement Class	General Description
User requirements	Based on the facets of the users — often in terms of user expertise, training needs, limitations, and capabilities
Task requirements	Based on the facets of the work — often in terms of artifact requirements and sequence of steps
Work environment requirements	Based on the facets of the nature of the work context, based on environmental conditions such as communication needs and physical constraints
Organizational requirements	Based on the facets of the organization including cultural, technological, and legal needs and standards
Usability requirements	Based on the usability needs and guidelines to be met by the design — often in terms of efficiency, effectiveness, satisfaction, learnability, and acceptability
Functional requirements	Based on the services and functions provided by the design — often in terms of tasks, actions, and activities that the system must be able to accomplish or support the user in accomplishing (sometimes accounted for under user requirements)
Performance requirements	Based on the expectations of system functionality — often in terms of timeliness, reliability, and accuracy

While this division of approaches is not absolute, it does distinguish that there are clear differences between these two types of approaches. For example, elicitation techniques are generally (arguably) quick preliminary techniques, which involve direct interaction with the users to yield rough data that often requires considerable interpretation, refinement, and validation. On the other hand, analysis and modeling techniques are often more extensive follow-up efforts that are used to represent this rough data in a manner that can be more easily communicated and interpreted and applied. For these reasons, the following discussion will consider elicitation techniques (Stage 2: Examining the work system) and analysis and modeling techniques (Stage 3: Representing the work system and its needs) as independent stages in the RE process as shown in Figure 7.3.

Before talking about these approaches and techniques in greater detail, Table 7.3 outlines many of the popular techniques of requirements elicitation and analysis/modeling commonly employed by UCD professionals. For greater detail on these (and other) methods and techniques, readers should examine some of the literature (e.g., Beyer and Holtzblatt, 1998; Maguire, 2001; Sommerville and Sawyer, 1997) and Websites (http://www.usabilitynet.org; http://usability.jameshom.com/index.htm) that provide detailed insight into different UCD techniques.

7.3.2.1.5 Selection of UCD Methods and Techniques

It is well-known that the UCD approach taken and the methods and tools chosen heavily depend on the constraints of a given project-time, money, staffing, and accessibility to users. Readers will find that even expert practitioners adopt their own unique combination of the techniques, as suits the needs and work within the constraints of a project. In a recent article, Marcus (2005) summarized the factors that mostly affect collection of UCD techniques used. These factors include: (1) availability of testing labs and equipment; (2) availability of usability professionals; (3) availability of users; (4) budget; and (5) calendar schedule.

A survey of UCD professionals showed that the cost-benefit tradeoff was instrumental in the consideration of which UCD method, if any would be implemented in a given design process (Mao et al., 2001).

TABLE 7.3 Common Requirements Elicitation and Requirements Analysis and Modeling Techniques

Requirements Elicitation	Requirements Analysis and Modeling
RE stage	
Stage 2: Examining the work system	Stage 3: Representing the work system and its needs
Approach summary	
Gathering or capturing system requirements through direct observation or interaction with stakeholders or end-users	Defining system requirements through the development of models based on the data collected using elicitation techniques and documentation
Characteristics	
Ethnographic in nature	Analytical in nature
Direct interaction with users	Direct involvement with data
Results in rough data	Refined representations of users, needs and tasks
Relatively quick and simple	Time and resource intensive
Requires more interpretation	Requires less interpretation
Preliminary activities	
Common techniques	
Surveys and questionnaires	Use cases/scenarios
Interviews	Visioning
Observation/ethnography	Personas/user profiles
Contextual inquiry/user interview	Error and critical incident analysis
Focus groups and brainstorming	Affinity diagramming
Think aloud	Work modeling
Laddering and card sorting	Decision/action diagrams
Documentation and product review	Cognitive work/task analysis

According to the 103 survey respondents, the top five UCD methods easily applied in practice included: (1) informal usability testing, (2) user analysis and profiling, (3) evaluations of existing systems, (4) low-fidelity prototyping, (5) heuristic evaluations, (6) task identification, (7) navigation design, and (8) scenario-based design. To summarize, practitioners tend to use those methods that are most flexible, informal, and least structured. The use of these "low hanging fruit" UCD methods reflects that in practice, actives are in fact driven by time, money, and the attention, which the process is given by key stakeholders.

7.3.2.1.6 Summary

While not all of these techniques and tools will be discussed in detail, representative methods that are particularly popular with UCD practitioners will be chosen for review. These tools and techniques will be summarized, followed by a summary section discussing how ergonomics, human factors, and HCI researchers and practitioners can get from this initial phase of understanding users and their work to leveraging and incorporating this new knowledge into the actual design of work tools.

As will be seen, all of these methodologies and techniques, while somewhat interrelated and in some ways redundant, comprise the initial activities of the UCD process (i.e., understanding the work system). All of these methods can be used to help examine, understand, and model the users, their needs, and their work. It is this intimate knowledge of the work system that helps to drive ergonomic design. These user-centered methods, techniques, and activities directly concern the documentation or modeling of the user and their work. At the heart of all of the aforementioned methods and techniques is the focus on human users — their abilities, needs, work context, and work.

7.3.2.2 Contextual Design: A Summary Example

7.3.2.2.1 Why Contextual Design?

The process of contextual design has been selected as an example of the UCD (and RE) process, because it represents a comprehensive approach to the RE process including both requirements elicitation and requirements analysis and modeling techniques as well as a defined process for how to use this data to define system requirements and spark design alternatives. Contextual design, a holistic and customer -focused design process popularized by Beyer and Holtzblatt (1998, 2003), or variants on the process, is one commonly used approach to UCD. The choice to give this procedure a more detailed summary stems from its inclusion of variants of all major RE activities wrapped up in one prescribed procedure. While contextual design (Beyer and Holtzblatt, 1998) is by no means the only solution to RE or UCD approach, it contains all the integral elements of the process of taking information about users, their needs, their work, and so on and translating into a data-driven design. According to Holtzblatt, "contextual design is a full front-end design process that takes a cross-functional team from collecting data about users in the field, though interpretation and consolidation of that data, to the design of product concepts and a tested product structure . . ." (2003, p. 942).

7.3.2.2.2 Procedural Overview

The process of contextual design begins with, albeit abbreviated, the process of defining the overall problem or goal being addressed, doing background and market research, setting the project focus and scope, and establishing parameters of the process (e.g., selecting a team, tailoring method to resources, project planning, and scope).

Additionally, the contextual design process stresses the importance of selecting an appropriate and representative user population keeping in mind the possible variability with their working environment, their work role and importance as a stakeholder, and considers the nature of the work practices to be studied (Stage 1). While the *Contextual Design Manual* (Beyer and Holtzblatt, 1998) provides relatively little detail on these stages of the process, there is considerable detail in laying out the rules and procedures for collecting data from users (Stage 2). In fact, the fundamental component of the contextual design process is the contextual inquiry. Contextual inquiry is the process of interviewing and interacting with the customer or user within the actual context of their work (Beyer and Holtzblatt, 1995; Holtzblatt and Jones, 1993).

7.3.2.2.3 Process Outcomes

This structured observation and probing process is the basis for collecting the necessary data to build a solid and comprehensive understanding of the user, their needs, and their work. In order to present this picture of work in a comprehensible and concise manner (Stage 3), the contextual design process incorporates the process of developing graphical work models including the flow, sequence, artifact, culture, and physical work models (Beyer and Holtzblatt, 1998). Through a well-prescribed process of consolidating data and models, the contextual design team then "walks the data" to create a team focus for the process of group brainstorming a series of visions, or large-scale, rough representations of the work system components, structure, and work tasks. These visions are then used to direct work and organizational re-engineering, use scenarios, storyboards, and technical/hardware requirements.

From these consolidated work models, the vision diagrams, storyboards, and the like, a unique contextual design work system model is developed — the user environment diagram (UED; Beyer and Holtzblatt, 1998). The UED represents the key concepts for the work system model and reveals how a system can support work in terms of the fundamental activities of work tasks, the links and interdependency between these tasks and activities, and the purposes or goals that direct the work tasks. Finally, by using the structural UED and sequential storyboards and some nifty reverse engineering, requirements and specifications of the system can be developed from examining the coherent tasks and activities that need to be supported, the types of data and artifacts needed to support the tasks, and the sequence and interdependence of the task activities.

7.3.2.2.4 Summary

Arguably, the contextual design process will not always be the best or most appropriate method of RE and UCD. For example, contextual design is particularly well suited for software-focused, computer-based systems, which are fairly radical upgrades to the previous legacy systems that exist within the organization. However, it is a good example of a work system design process that stresses the need to incorporate the user into the design process and ensure that the needs and work processes of the user will be appropriately addressed in design solutions.

7.3.2.3 Stage 1: Defining the Needs

The first step in any endeavor of designing and developing a new system to be used to support work is to understand the problem domain that the system is being designed and developing for. We refer to this stage of the UCD process as *defining the work system*. Activities involved in this stage include identification of the stakeholders, context of use analysis, and competitor analysis, all of which help to both define the work system and help direct the choice and planning of RE techniques (i.e., activities of Stages 2 and 3). Many of these activities are often associated with the preliminary planning and scope, although a slight distinction can be made.

7.3.2.3.1 Project Planning

As practitioners can attest to, the first step in the system development lifecycle is planning and scoping the project. Project planning and scope refer to those activities involved with determining the high-level business goals and requirements, establishing the scope and boundaries of the project, determining the UCD methods to use, researching the market and domain for the future system, and determining logistics (i.e., resources, costs, timeline). Building executive buy-in and agreement on the overall business goals and requirements is also necessary. While these activities are particularly involved for UCD practitioners working in the industry, these details are often glazed over in academic and technical publications assuming that these activities have already been completed before starting up the RE process.

In keeping with tradition, no detailed discussion of these activities, and their associated issues, will be provided here. It is necessary to acknowledge both the inclusion of these activities in the UCD process and the importance of these activities — especially for practitioners working in the industry. However, readers can refer to texts by Vredenburg et al. (2002) or Rouse (1991) or Mayhew (1999) for more details on these preparatory activities and high-level issues related to the usability engineering or system development lifecycles. It is important to note, however, that some of these activities help to define some classes of system requirements as discussed in Table 7.2, primarily including organizational requirements, general constraints on the system or process, and definition of the overall purpose and business goals to be met by the system. Thus, the project planning activities are important not only for ensuring project success, but also help to define the requirements.

7.3.2.3.2   Activities that Help Define the Work System

The distinction between these preparatory, planning activities and those that are more directly involved with defining the work system is cloudy, at best. However, these activities tend to focus more on setting the stage for the main requirements gathering processes (Stages 2 and 3), either through involvement with stakeholders and representative users or through examination of relevant domain knowledge, documentation, or competitive systems. This is the "context and groundwork" (Nuseibeh and Easterbrook, 2000) that serves as the basis for project assessment, selection of RE methods, and where to direct these methods. Some of the more unique or interesting work system defining activities will now be discussed.

Stakeholder Analysis   Stakeholder analysis, or the stakeholder meeting refers to the general process of identifying those individuals within the work system that are affected by the development of a new system and getting their input and buy- in on the development process. In the systems development (or RE) process, stakeholders generally refer to any individuals who affect or are affected by the system of interest (see Sharp et al., 1999, for a good review of stakeholder theory). In practice, this often includes business managers, project managers, user representatives, training and support staff, developers, etc. Stakeholder analysis is a technique used to identify, assess, and prioritize the needs, goals, and requirements of the stakeholders (Damodaran et al., 1980). As stakeholders, by definition, have the ability to significantly influence the success of the project, stakeholder analysis is also an opportunity to establish buy- in and agreement at the onset of the project.

Stakeholder analysis helps to define the work system through the identification of the overall goals, requirements, constraints, and needs related to the system to be developed. This process helps to define some of the overarching organizational goals and business requirements that will help to scope and define the system requirements. Additionally, usability analysis and the establishment and agreement of usability priorities and goals, which are translated into usability requirements, can also be conducted during this time (Maguire, 2001). The overall goal of stakeholder analysis is to bring together all relevant parties to create a common vision of the overall purpose, scope, constraints, and high-level requirements of the system to be developed.

Context and Domain Analysis   Context or domain analysis refers to the general process of collecting information about the context of use in which the system will be used. This *context of use* idea is similar to our idea of the work system, as described previously. The basic goal is to gather stakeholders and domain experts together to identify information such as identifying the users and work tasks, as well as laying out the technical and environmental constraints of the work system (Bevan et al., 1996; Maguire, 2001). Context (of use) and domain analysis is also helpful in familiarizing the UCD team with important background knowledge of application domain considerations and terminology. As previously discussed, a firm understanding of domain-specific issues is necessary for the development of meaningful system requirements.

More specifically with respect to the development and delineation of usability needs and requirements, Bevan et al. (1996) and Bevan and Macleod (1994) have developed a structured method, called Usability Context Analysis, which is used to elicit details from stakeholders about a potential system and how it will be used in context. The method results in a list of important characteristics of systems' users, their work tasks, and the context of work. This list serves as a framework to ensure that all factors affecting system usability have been identified. Context, or context of use, analysis is also often the spawning ground for associated RE activities such as analysis of competitive or existing system (Preece et al., 1994) and usability planning (Mayhew, 1999).

7.3.2.3.3 Summary

As suggested by the term UCD, the key to incorporating the needs of users into the design of systems, tools, or interfaces lies in developing an understanding of the needs of the users being designed for. This includes examination of their work, their objectives, their activities, their (cap)abilities, their differences, their environment, their constraints, etc. In essence, this is developing a picture of the work system as previously discussed. While the techniques discussed in the initial stage of defining the work system needs do serve to information system requirements, this is not the complete picture. As is well-acknowledged in the scientific and practical knowledge base, there are limitations to merely collecting a group of experts or stakeholders together and asking questions, as users cannot necessarily vocalize or make their understanding of their own work explicit — especially out of context. Karen Holtzblatt, cocreator of the popular contextual design approach to system design, elucidates this point well: "... requirements gathering is not simply a matter of asking people what they need. A product is always part of a larger work practice. It is used in the context of other tools and manual processes. Product design is fundamentally about the redesign of work or life practice, given the technological possibility. Work practice cannot be designed well if it is not understood in detail" (2003, p. 944). In response to this phenomenon, researchers and practitioners in ergonomics, human factors, psychology, cognitive science, systems engineering, and HCI have developed knowledge elicitation techniques to help augment this understanding of the work system.

An important outcome of this stage is defining the problem that is to be solved, which is often done through the definition of boundaries. This is an important activity within Stage 1 efforts as the definition of the problem space with boundaries helps to define the scope of the design to be implemented. As noted by Nuseibeh and Easterbrook (2000), the nature of the boundaries chosen affects all subsequent requirements elicitation efforts (Stage 2), as it affects the identification and selection of stakeholders and user cases, the identification of applicable goals, and the choice of tasks to observe. Once the boundaries are drawn, various goals and constraints settled upon by the primary stakeholders and the general model of the work system (including user, context, and work) envision, the process of examining the work system and collecting data begins. Generally, this process begins with the use of requirements elicitation techniques.

7.3.2.4 Stage 2: Examining the Work System

While there may have been previous contact with users (or stakeholders) or other work system components (e.g., documentation, artifacts) upto this point, as part of the work system's needs definition activities (Stage 1), Stage 2 marks the true beginning of the work system examination. Work system examination methods and techniques are the collection of activities involving the investigation, either directly or indirectly, of the end-users, the work context, and the work itself. These techniques commonly include traditional methods such as surveys and questionnaires, interviews, user observations, and more involved user-guided activities including protocol analysis, card sorting, and affinity diagramming or brainstorming. These UCD methods comprise the requirements elicitation tools of the RE process.

7.3.2.4.1 Some Examples of Methods and Tools

As there are innumerable requirements elicitation methods and techniques available to UCD practitioners, only a few interesting examples of these tools are discussed here. As may be recalled, requirements elicitation techniques refer to activities used to capture knowledge and data through (often) direct interaction with the stakeholders or end-users. For a more detailed discussion of the philosophy and practice of requirements elicitation, readers should be directed to relevant texts (e.g., Christel and Kang, 1992; Kotonya and Sommerville, 1998; Young, 2004). These requirements elicitation techniques can be thought of as UCD knowledge solicitation tools. These methods and tools represent the primary activity of the RE process (Nuseibeh and Easterbrook, 2000) focusing on the *collection* of data on the work system.

Focus Groups and Brainstorming   Focus groups and brainstorming are two cheap, easy, and popular ways that UCD professionals directly interact with stakeholders or users to collect data that can be used to understand the work system and define appropriate requirements. The basic concept of focus groups and brainstorming is to gain insight into the real events of work by providing a loosely moderated forum in which stakeholders or users can collectively discuss aspects of their work. Focus groups tend to be more rigidly designed and moderated to help keep the participants on the focus (Caplan, 1990; Kontio et al., 2004), while brainstorming techniques tend to allow for more leeway in the scope of the discussion — as long as it remains on the focus. Additionally, these techniques are also good for achieving internal buy-in and support by providing stakeholders with the forum to provide input and collectively agree on what aspects of the work system are important.

Another derivative of focus groups and brainstorming, which is commonly used, are requirements workshops. These techniques use the "whole is more than the sum of its parts" principle of collecting stakeholders and the design team together in a shared space with a shared purpose of defining design requirements (Gottesdiener, 2002). The output of these meetings includes use cases, business rules, system criteria, and various models of the work system. These requirements meetings and workshops are often very efficient and useful methods of data collection and requirements gathering.

Contextual Inquiry: User Interviews in Context   It is likely that the majority of readers will be familiar with the concept of contextual inquiry, or will have been introduced already with the previous discussion of contextual design. Contextual inquiry is a well-formulated interview technique popularized by Beyer and Holtzblatt (1998) as the primary data collection tools of their Contextual Design approach to system design. As the authors explain, "The core premise of Contextual Inquiry is very simple: go where the customer works, observe the customer as he or she works, and talk to the customer about the work" (1998, p. 41).

While this method is effectively a user interview (another common requirements elicitation technique), it is actually more akin to ethnography (Blomberg et al., 2003; Saferstein, 1998) applying the principles of being anchored in natural settings, focusing on the holistic context, providing descriptive (rather than prescriptive) understanding of events, and using the point-of-view of the individual being observed. While there is interaction between the interviewer and the interviewee, this is largely to clarify the interpretations of the interviewee's words and actions and to help keep the interviewee vocal and focused on the project scope.

In addition, contextual inquiry is both structured and unstructured as the creators have outlined several principles and procedures to follow, yet the bulk of the process is the naturalistic observation of user work. As has been outlined by its creators, contextual inquiry has defined philosophies including the "Master/Apprentice" model of the relationship between the customer and the designer as well as the four steps of the interview process — the conventional interview, the transition, the contextual interview proper, and the wrap-up, wherein each has its own prescribed methods and suggestions (Beyer and Holtzblatt, 1995, 1998; Holtzblatt and Jones, 1993). The end result of the contextual inquiry process is a very rich set of notes and observations, which are then used to build models to represent this interpretation of the work system. These analyses and modeling techniques along with other popular tools for representing the work system will be discussed in the next section.

7.3.2.4.2   *Summary*

As can be understood from the discussion, requirements elicitation techniques are used to gather information on the work system including the users, their goals, their tasks, their work context, their abilities, their needs, etc. However, as discussed, the information gathered through requirements elicitation techniques is often rough, verbose, and overbearing for immediate use. As such, this data needs to be arranged, interpreted, analyzed, and modeled to achieve the end goal of creating system requirements (Nuseibeh and Easterbrook, 2000). Thus, researchers and practitioners developed analysis and modeling methods to help wrangle this work system data into a form that was more amenable to communication, study, interpretation, and translation. We refer to these techniques as requirements analysis and modeling tools.

7.3.2.5 Stage 3: Representing the Work System and Its Needs

As there are innumerable requirements analysis and modeling methods and techniques available to UCD practitioners, only a few of the more interesting and popular requirements analysis and modeling techniques will be discussed here. As may be recalled, requirements analysis and modeling techniques refer to those activities used to build representations of the work system data gathered through the requirements elicitation processes. This data comes in a number of forms ranging from the raw observational notes of an ethnographic study to the more clearly defined results of user surveys and questionnaires. This data coming from a variety of sources is then used to create more consolidated and concise views of the work system.

7.3.2.5.1 *Some Examples of Methods and Tools*

The variability of analysis and modeling techniques and the output of these techniques is as diverse as that of the requirements elicitation techniques. For example, the highly popular unified modeling language approach (UML; Booch et al., 1998; Rumbaugh et al., 2004) alone contains nine different viewpoints of the work system including activity diagrams, class, diagrams, collaboration diagrams, component diagrams, deployment diagrams, object diagrams, sequence diagrams, statechart diagrams, and use case diagrams. In terms of more traditional work and system modeling techniques, the U.S. Department of Defense (DOD; 1999) published a handbook of human engineering processes and procedures that outlines an extensive list of analysis and modeling techniques ranging from cognitive task analysis to decision/action diagrams to operational sequence diagrams (1999). Despite the variability, all of these analyses and modeling tools are used to *represent* the collected work system data.

Contextual Design Work Models and the User Environment Design (UED)   As previously mentioned, contextual design includes a fairly well-defined set of analysis of modeling tools, which are used to represent the free-form data and observations that are collected during the contextual inquiry (Beyer and Holtzblatt, 1998). According to the contextual design protocol, five work models are generated during an interpretation session from the collective observation data. These models include:

- The artifact model, which represents the composition, structure, and usage of physical work artifacts
- The cultural model that represents the influences on the user including all organizational partnerships, the hierarchical structure of power, business policies, and standards, etc
- The flow model that represents the users' responsibilities and duties and the communication and collaboration related to carrying out these duties
- The physical model that represents the actual work environment including the physical layout, the environmental constraints, the organization of business objects, etc
- The sequence model that represents the procedures and actions as well as their sequence required to perform specific work tasks

In addition to these work models, the raw data from the contextual inquiry is also used to build an affinity diagram, which uses a bottom-up approach to link all of the insights gained during the interview into one hierarchical representation that can be used to reveal common issues and themes to be addressed (Beyer and Holtzblatt, 1998). This affinity diagram and the work models are then used to develop an idea of the work redesign necessary for the work system and, finally, through a process of visioning and storyboards, develops the UED (Beyer and Holtzblatt, 1998). The UED is composed of focus areas representing the system components that support specific work activities (the purpose), in which the system functions and links to other focus areas are linked. Using the UED and the previous models and materials constructed, the team can derive system requirements and specifications through linking the sequence and structure of the new work system model (the UED and other models) to needed system functions and behaviors.

Personas and Use Case Models   Two more recent requirements analysis and modeling techniques that are commonly used are personas and use cases (or use case models). Personas, which were first introduced in a remarkable book by Alan Cooper (2004), are becoming a wildly popular technique for UCD. Personas are typically built from data collected in ethnographic interviews (e.g., contextual inquiry) and are expressed in 1 to 2 page descriptions that include a general profile, behavior patterns, goals, skills, attitudes, and environment. A persona is a user archetype that is based on a collection of job responsibilities and goals (Pruitt and Grudin, 2003). Personas help to enhance the utility of scenarios and other user-centered techniques such as participatory design by strengthening the focus on realistic users and work contexts — albeit through a fictionalized setting (Grudin and Pruitt, 2002; Pruitt and Grudin, 2003). In this way, personas are really user models, which can then be used to guide and enhance other techniques such as use case models.

Use case models are largely popularized through their use in the UML method (Booch et al., 1998; Rumbaugh et al., 2004) and are used to capture the system requirements through a narrative-like format of the steps to perform the tasks, the tasks to be performed, and the system's behavior for each task (Cockburn, 2000; Kulak and Guiney, 2004). The use cases are generated from work scenarios created through the requirements elicitation processes such as user interviews and brainstorming. Use cases essentially represent a goal-oriented set of interactions between the users (actors) and the system to be used and describe the sequence of the user's actions and the system's reactions. In this way, each use case presents the system functionality needed to support user work for a given task and goal.

7.3.2.5.2 *Summary*

If any reader happens to already know UCD or system engineering practitioners, there are a lot of other popular methods, many of which include structural diagramming methods such as task analysis or hierarchical task analysis (Hackos and Redish, 1998; Shepherd, 2000, 2001) and work domain or cognitive work analysis (Rasmussen, 1986; Vicente, 1999), just to name a few. Researchers and practitioners have developed analysis and modeling methods and tools to represent nearly all imaginable aspects of the work system from models representing the organizational culture and policies of the work system [the cultural model of contextual design (Beyer and Holtzblatt, 1998)] to models representing complex tasks in terms of goals and subgoals with associated plans for carrying out those subgoals [hierarchical task analysis (Shepherd, 2001)].

However, a discussion of all of these work system analyses and modeling techniques could itself produce a multi-volume handbook. As can be seen, work system analysis and modeling techniques are used to create fairly concise, yet complex representations of the work system. While each of these analysis and modeling tools might represent completely different aspects of the work system, they are all used to provide a representation of the work system that can be used to generate ideas of (primarily) what the system is supposed to do and (to some extent) how the system should go about supporting these goals and needs. In other words, work system analyses and modeling tools are used to provide a basis to form system requirements.

7.3.2.6 Defining Requirements and Specifications to Drive Design

As discussed, the overall purpose of investing all this time, money, and trouble in eliciting knowledge, opinions, and ideas from the stakeholders and end-users and then analyzing and modeling this data is to feed the process of developing representative, appropriate, and useful requirements for the work system. These requirements are vital to UCD as they are the foundation for the engineering design specification, which will be used as the blueprint in the design and development (Samaras and Horst, 2005). The system requirements are used to define the engineering design specifications. These specifications are then used to guide design and development. This basic process is outlined in Figure 7.3.

7.3.2.6.1 *More on Requirements*

From the previous discussions, RE in actuality is a process — involving sequential activities typically proceduralized and have been validated through previous use, and building the knowledge about and understanding of the work system. The culmination of this process is the requirements document — the list of requirements. The translation of this collected information into an actual list of requirements is very complex and nuanced. It is often an art as much as it is a science (although RE theoreticians may cringe at the thought). There is no prescribed or preferred methodology for this process, rather several useful summaries of this process (Nuseibeh and Easterbrook, 2000; Ransley, 2003; Wiegers, 2003; Young, 2004).

Fortunately, however, even Karl Wiegers (a noted RE heavyweight) acknowledges that RE is primarily a communication activity rather than a technical activity (Wiegers, 2000). A number of guidelines can help make the requirements formulation considerably less random and more manageable. Many of these guidelines lay out general principles and processes for generating requirements (Alexander and Stevens, 2002; Robertson and Robertson, 1999), while others actually propose standardized guidelines for the process and specification (IEEE, 1998a, b).

As a reminder, requirements are descriptions or statements of system services, functions, properties, attributes, or constraints laying out how a system should behave within its context of use to help perform work (Kotonya and Sommerville, 1998; Nuseibeh and Easterbrook, 2000; Ransley, 2003; Sommerville and Sawyer, 1997). Requirements are typically a mixture of some problem information, statements of system behavior and properties, and design and manufacturing constraints (Sommerville and Sawyer, 1997). In a somewhat sensible, yet magical way, these requirements naturally "drop out" of the RE process as the requirements elicitation and analysis and modeling tools are used.

For example, the project planning and stakeholders' meetings generally set out the overall business goals to be met by the system as well as any business requirements and constraints and general usability requirements. The actual nascence of a requirement lies in an idea that is generated from the study of the work system (e.g., activities of Stages 1 and 2). Once the goals and needs of the work system are elicited and modeled, the defining goals (e.g., the new system should be usable) or needs (e.g., the new system needs to interface with legacy systems) should be broken down into atomic, defined problems, put in a natural language format, described, and referenced back to the actual data or model that spawned the requirement. For example, in the contextual design process (Beyer and Holtzblatt, 1998; Holtzblatt, 2003), it is suggested that the requirements be referenced back to the part of the UED or storyboard that outlines the need for a system function, component, or property and also attaches a list of any relevant data used (e.g., sections of the affinity diagram or pieces of the consolidated models).

These statements or descriptions serve as specifications of what should be implemented. As the RE process, or even how the results are communicated, is often a function of the nature and culture of the audience of the RE results (i.e., the requirements) and the domain in which the system is being implemented (Kotonya and Sommerville, 1998; Sommerville and Sawyer, 1997), the establishment of the actual system requirement is a highly variable and flexible process. However, as Ransley (2003) notes, having an actual process is prerequisite to control and repeatability allowing the ability to refine the process, improve the knowledge, and increase the skill in turning the identified problems into appropriate solutions. The RE process also provides a clear auditing trail allowing the traceability of design decisions back to system requirements and, ultimately, to the identified problems that needed to be addressed. Arguably, this concept of traceability is of utmost importance as it allows the design decisions, which have been based on system requirements and specifications to be traced back to actual work system needs and goals, which were explained and enumerated in the system requirements and specifications.

7.3.2.6.2 *Creating Design Alternatives*

Once the systems requirements and specifications have been agreed upon and settled, then the design process can begin. The truth is, especially in practice, that there is considerable interpretation that goes into transforming the data collected into a set of well-formulated system requirements or specifications. However, what happens next — the development of design alternatives — is even less concrete. Fortunately, researchers and practitioners have developed a number of tools and techniques for translating an abstract understanding of what the system needs to do into a more tangible representation of how the system will meet these requirements.

Along with the requirements document, which explicitly lays out the system specifications, the models and representations of the work system are used to guide the potential design alternatives. For example, personas, story boards, use cases, and decision/action diagrams could be used to help envision the order and organization in which the system provides content or functionality in order to support the natural workflow of the end-user. However, this process requires a clear and accurate interpretation of the data and creative ingenuity to meet all of the functional and nonfunctional requirements, while working within the constraints of the organization or environment. In fact, the boundaries and limitations created by constraints on the system help design by reducing the number of possible design alternatives (e.g., if the system needs to be implemented in Java^{TM}). If a design solution meets all of these requirements and can be elegant and appealing to users, then designers have truly earned their keep.

As mentioned previously, the transition from requirements to tangible design is somewhat nebulous. Design solutions arise from the requirements in many ways, but progress and mature through iterative development. Designers and their teams should accept a process of iterative prototyping, developing first a low-fidelity mockup, and iteratively integrating feedback from evaluations of the prototype into redesigns incrementally until a final working simulation is produced. Ideally, the design process should gradually iterate on several prototypes, although this may necessarily be realistically feasible or affordable given the constraints and resources associated with the project. At the minimum, however, the process should incorporate at least one low-cost low- fidelity prototype in the design of a final operational system. In fact, the value of design information garnered from the testing and evaluation of low-fidelity, low-cost prototypes lies in the considerable positive influence of this collected insight on the ultimate production of usable, useful, attractive final designs (Hall, 2001).

7.3.2.6.3 *Design Production: Prototyping*

In his summary of UCD methods and techniques, Maguire (2001) provides a comprehensive review of nine methods, which facilitate the design production step of UCD. Figure 7.4 illustrates different methods for prototyping according to the number of design alternatives typically generated juxtaposed with the level of fidelity the prototypes exude. The level of fidelity integrated into a design option is directly proportional to the number of prototypes typically generated with such techniques and related to the cost and time involved in the prototype building. Additionally, any prototype developed should be done so with the foresight of realistic evaluations that the design and development team

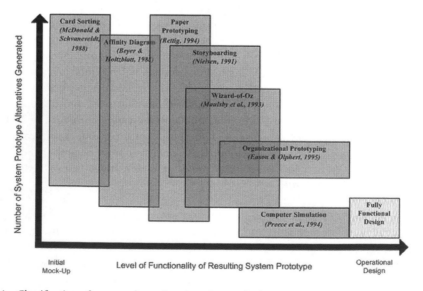

FIGURE 7.4 Classification of commonly used prototyping methods.

have resources available for and, which will generate the type of information needed to inform the iterative design.

The other methods mentioned by Maguire (2001), which relate more to processes for generating initial prototypes include brainstorming, parallel design, and the use of design guidelines and standards. Table 7.4 provides operational definitions for these nine design activities.

7.3.2.6.4 For Consideration in UCD: Standards and Guidelines

Perhaps more clear than the typical interpretation of the work system models to guide design decisions, designers can also use established standards and guidelines to help constrain and define design decisions. Standards are documented agreements on technical specifications, aimed at ensuring quality control and abiding legislation or regulations related to design (Sherehiy et al., 2005). Standards may be international, national and regional with international standards being the most expansive (Sherehiy et al., 2005). Sherehiy et al. (2005) in their extensive account of human factors and ergonomics standards, summarize ergonomic standards put forth by the ISO, the European Committee for Standardization (CEN), the International Labor Organization, (ILO), the U.S. Government, the Occupation Safety and Health Administration (OSHA), the American National Standards Institute (ANSI), and other standards created at the state level as well as by individual organizations.

These standards, while all focus on human factors and ergonomics also deal with a range of human capabilities and environments providing instructions for dealing with the physical world and interface design. For example, in terms of user interface standards, a number of conventions have been developed.

TABLE 7.4 Methods for Transitioning between RE and Design Production

Method	Description
	Design Initiators
Brainstorming	The collective generation of creative design ideas by design and task experts useful for the initial design when little information is available
Parallel design	Groups separately generate design solutions to the same problem to give more variety to the solutions that are considered
Design guidelines and standards	Review of legal requirements and style guides in the generation of design alternatives
	Design to Prototype Catalysts
Affinity diagrams	A technique suggested in contextual design, sticky notes, create representation of the organization of screens or functions from the user's perspective
Card sorting	Users arrange card with relevant design terms written on them into grouping analyzed for common patterns for the hierarchical organization of information and functions in a design
	Types of Prototypes
Storyboarding	Images/Illustrations depict a scenario of design use explaining input and output exchanges between the user and the design
Paper prototyping	Paper-based simulation of interface for evaluation with the user
Wizard-of-Oz prototyping	The user interacts with the front-end or interface of a design and a hidden developer acts as a "wizard", providing the relevant output to the user's actions with the prototype for realistic design scenarios
Organizational prototyping	Simulation of prototyped design within the context of the actual user environment to determine the level of integration within this context
Computer simulation	Software developed for realistic representation of the design focused on different aspects of structure, screen layouts, and navigation

The ISO has also fairly recently developed checklists based on standards and principles of user guidance and dialog as subsets of the overarching ISO 9241 Standard on Ergonomics Requirements for Visual Display Terminals (ISO, 1998). Sample checklists associated with these standards are provided by Stewart and Travis (2003) in their chapter on "Guidelines, Standards, and Style Guides" in Jacko and Sears' (2003), *The Handbook of Human–Computer Interaction.*

In addition to these international standards, there are several other sets of guidelines that have been established by developers, HCI researchers, and usability professionals (e.g., Apple Computer, 2003; Microsoft Corporation, 2004; W3C, 2004). With so many standards and legal requirements promoting accessibility, designers must have guidance in how to incorporate these requirements into the mainstream UCD process. A comprehensive review of accessibility, universal design standards, and techniques for how to meet these requirements can be found in work by Preiser and Ostroff (2001).

However, unless the design team has a substantial portfolio of experience in a specific design domain, the translation of specifications into tangible systems is typically a challenge, for which there is little guidance (Ostroff, 2001). That being said, a more effective approach is the generation of several design alternatives and creating low-fidelity mock-ups of these alternatives. Through testing and evaluation, the design alternatives are evaluated to verify the degree to which they address the initially derived requirements. Based on the evaluation, designs are altered and iteratively evaluated with prototypes of increased fidelity until the final optimal design is ready for implementation.

7.3.2.6.5 *For Consideration in UCD: Principles of Usable Design*

Even after the implementation of a design, the process of UCD continues. Alterations and adjustments are continually integrated into the system based on the need to optimize the compatibility of the user– system interaction as both evolve relative to the other over time (Karwowski, 2003). Seemingly endless variations of this conceptualization of UCD exist in the literature under several guises such as the usability engineering lifecycle (Mayhew, 1999), contextual design (Beyer and Holtzblatt, 1998), compatible design (Karwowski, 2003), and others. Nonetheless, at the core of each approach lies an ideology of establishing a conscious awareness of users, context, and work.

That said, the system design and its implementation must place emphasis on derived usability goals, the environment, task, workflow, and characteristics of the user (cognitive and physical). As the users and system are evolving and dynamically interrelated, UCD must reflect this propensity for change. This is accounted for in the cyclical pattern of UCD and the iterative appraisal of the systems, users, and work to best preserve productivity and success.

Many, including Norman (1988, 2002) formulated general design guidelines that are universal to any usable design and serve as the underlying goals of UCD. Examples include Norman's seven principles of design, Shneiderman's eight golden rules of design (Norman, 2002; Shneiderman, 1998), and Nielson's design heuristics (Nielsen, 1994), summarized in Table 7.5. Additionally, there are several additional principles of design spanning across the disciplines of art, design, natural science, psychology, and more (see Lidwell et al., 2003, for an extensive review). Common to all of these guidelines is the purported need to consider these rules in light of the actual user's needs, desires, and individual differences taking into account who the users are, what they do, and where they do it.

However, since the preliminary introduction of these guidelines in the mid 1980s, there are innumerable guidelines published as the outcome of research, practice, and the emergence of new technologies. Furthermore, these guidelines typically do not become obsolete. Recently, Nielsen (2005) reviewed the present-day relevance of 944 usability guidelines published in an 1986 Air Force documentation on usability knowledge and guidelines for the design of military command and control centers. In his appraisal, Nielsen determined that 70% of these guidelines were fully applicable to present-day systems. The majority of those irrelevant guidelines were less important due to the anticipated obsolescence of certain design elements and technologies.

For designers this poses a problem of information overload. One cannot possibly fully address all guidelines as it takes a vast amount of time and money to accomplish and it is also likely that there will be some contradiction between guidelines. Through UCD, designers must apply knowledge about the

TABLE 7.5 Summary of Three Popular Collections of Design Principles and Guidelines

Eight Golden Rules of Design (Shneiderman, 1998)	Design Heuristics (Nielsen, 1994; Nielsen and Molich, 1990)	Seven Principles of Design (Norman, 2002)
Strive for consistency	Visibility of system status	Use of knowledge in the world and knowledge in the head
Enable frequent users to use shortcuts	Match between system and the real world	Use mental aides and consistency in information presentation
Offer informative feedback	User control and freedom	Make things visible, enabling the user to figure out the use of an object by seeing the buttons/devices required for operations
Design dialog to yield closure	Consistency and standards	Get mappings right and easily understood
Offer simple error handling	Error prevention	Use constraints to direct users' actions
Permit easy reversal of actions	Recognition rather than recall	Design for error, planning for the recovery from any possible error
Support internal locus of control	Flexibility and efficiency of use	Standardize (when everything else fails) in order to avoid arbitrary mappings
Reduce short-term memory load	Aesthetic and minimalist design help users to recognize, diagnose, and recover from errors help and documentation	

user, work, organization, and context to generate a prioritization of usability goals, which then informs the relative importance of guidelines and leads to the integration of the most appropriate guidelines for the situation at hand.

7.3.2.7 Summary

As has been discussed, the process of translating the understanding of the work system and its needs into tangible system designs can be best characterized as highly variable or nebulous. A considerable amount of interpretation goes into this process of first creating the system requirements and specifications along with considerable creativity and ingenuity to translate these specifications into actual designs. While there are several documented methods for this procedure along with design standards and guidelines and design principles to help guide this process, it is still an art in further need or standardization through advances in research and science.

This section has provided a cursory overview of some of the popular methods and techniques for the elicitation, representation, and development of system requirements. It is from this collected knowledge represented in lists, diagrams, and models that UCD emerges. All of the methods and techniques discussed, whether they be ethnographic in nature (e.g., contextual inquiry) or modeling tools (e.g., hierarchical task analysis), have one common link — an implicit focus on users, their needs, and their work. All of these methods and techniques also have a common goal — the development of the system requirements, which will be used to guide the development of UCD alternatives.

As discussed, these methods and techniques used to understand the work system fall within the preliminary phase of the UCD lifecycle (i.e., defining, examining, and representing the work system and its needs). The knowledge gained from these initial efforts is the fuel used to elicit, understand, and develop the functional and nonfunctional system requirements. These requirements are then translated into system engineering specifications, which represent the blueprints for the design and development process of work tools (i.e., systems, applications, interfaces, etc.). However, it is this understanding of users and their work that ensures that the system requirements generated are appropriate and useful. If these requirements are incorrect, then the specifications used to guide the system design and development will be incorrect and the system or tool will ultimately fall short of good UCD. Thus, the importance of these initial stages of UCD should be clear.

However, simply performing the user and work analysis and modeling stages is not enough. For example, the "requirements analysis" stage (Mayhew, 2003) is a necessary, but not sufficient step to ensure effective and useful designs within the usability engineering lifecycle. In a way, this initial process of UCD is focused on *solving the correct problem*. That is, the RE process is directed at identifying the needs and problems within the work system that the new system is being designed to solve. The following section, which presents some representative and popular testing and evaluation methods for UCD is focused on *solving the problem correctly*.

7.3.3 Testing and Evaluation of Proposed Design Solutions

7.3.3.1 Overview

In the overall conception of the UCD process, the previously discussed methods and tools are the precursors to design development yielding design alternatives and prototypes. The second major phase of UCD to be discussed in this section includes methods and tools to iteratively test and evaluate the prototypes (low-fidelity mock-ups through functional systems) in light of how well the prototype meets the documented requirements and standards. As the RE process generates various work system models and lists of system requirements and specifications, which in turn generate design alternatives (e.g., prototypes), designers must ensure sure that the interpretation and prediction of the system requirements and specifications generate an appropriate design.

If the design passes this inspection, the collection and interpretation of the work system data yield accurate and appropriate system requirements. This aptitude assessment of design appropriateness is carried out through testing and evaluation techniques. The landscape of evaluation and testing methodologies are as numerous as requirements gathering methods introduced earlier in this chapter; some grounded in theory, empirical research, practical experience, or some combination. In this section, basic theory and principles of design testing and evaluation will be covered along with a review of the several of the most frequently employed testing and evaluation methods used by researchers and practitioners. Readers through practical experience are likely to craft their own unique set of evaluation methodologies appropriate to their context and domain of focus.

7.3.3.1.1 *Testing and Evaluation Basics*

In reaction to the increased demand for usable systems, there is an amplified emphasis on conducting tests to assure the generation of usable, useful design selection. Testing and evaluation provides verification that the requirements of the system established in the initial phase of UCD (i.e., understanding the work system and its needs) are in fact fulfilled. This is key, because the process of design production can never be free of uncertainty; requirements and specifications merely assist in making well-informed decisions. Testing and evaluation provides checks and balances to UCD, as shown in Figure 7.2, it is not uncommon to revisit stages in the process as through feedback loops in order to better clarify or correct requirements and their interpretation. The necessity of evaluating designs iteratively using representative users and tasks within specified work environments should not be misjudged (Hall, 2001). Performing testing and documenting the results aids the establishment of a more trusting relationship between the designers and the users (Shneiderman, 1998), which means an improved user acceptance rate of the final design. Testing and evaluation methods serve as critical catalysts in completing the UCD process. They aid in assessing the integrity of a design by motivating incremental, iterative improvements and the eventual implementation (if deemed worthy) (Leonard et al., 2005).

The specific testing and evaluation methodologies used are dependent on the level of prototype fidelity and contextual constraints required for implementation of the assessment method(s) (e.g., budget, time, manpower, etc.). Typically, as a prototype's fidelity increases from low to high, so does the number of applicable evaluation techniques. Stanton and Young (1999) provide a summary of specific interface evaluation techniques appropriate at different levels of design fidelity from concept through implementation. From this study, it is presumed that the ease with which an evaluation method is applied is interrelated with the level of abstraction required of those conducting the assessment to conceptualize the design and tasks interactions. Those prototypes that have a physical presence and are highly tangible maintain more flexibility in applicable evaluation techniques than prototypes of a more conceptual nature (e.g., Wizard-of-Oz methods or paper-based prototypes).

7.3.3.1.2 *Testing and Evaluation Considerations*

In their summary of human factors methods, Leonard et al. (2005) summarized both resource- and method-specific criteria for the selection of evaluation techniques, summarized in the form of questions in Table 7.6. The authors defined these criteria in the context of all human factors methods and not just UCD. In the context of UCD, the purpose of the evaluation is the assessment of the design in terms of the requirements and specification that emerged in the first step of UCD. The intended outcome of the evaluation in terms of UCD is to iteratively improve upon the design until it is appropriate for implementation.

Evaluation methods can serve three roles (Stanton and Young, 1999):

1. *Functional analyses*, to understand the scope of functions that a given design supports
2. *Scenario analyses*, to evaluate the scope and sequence of activities users must step through with the design to achieve the desired outcome(s)
3. *Structural analyses*, to evaluate the efficacy and opinions of a design from the users' perspectives

The generalizability of evaluation outcomes is of key concern in the selection of evaluation studies. Namely, the information gathered in this testing process must be easily translated into actionable design changes relative to the users, tasks, and environments of the actual proposed context of use. That said, those directing the testing process must be sensitive to the trade-offs assumed in the selection of an evaluation method, setting (field or lab), functionality and tasks assessed, evaluation participants, and sample size. For example, in the decision to collect evaluation information in the field through observations in lieu of a more formal usability testing method, designers should question if more is gained from watching the interactions in context than what is lost from the lack of structure and control. A lack of generalizable results from an evaluation can translate into longer lead-time on final development or the recommendation of erroneous design changes that are not truly advantageous for actual design use in the target work system.

Three levels of inclusion of users in the evaluation process have been identified (Maguire, 2001):

1. *Participative*, to directly educe user opinions and internal thoughts upon interaction with the design without evaluator/designer prompting
2. *Assistive*, which employs a think-aloud protocol methodology in which the user talks out loud while working with a design to explain their motivation for actions and train of thought. The designer only intervenes when the user hits an obstacle

TABLE 7.6 Questions for Consideration in the Selection of UCD Testing and Evaluation Techniques

Resource-Specific Criteria	Method-Specific Criteria
What is the cost-benefit ratio of using this method?	What is the purpose of the evaluation?
How much time is available for the study?	What is the state of the product/system?
How much money is available for the study?	What is the intended outcome of the evaluation?
How many staffs are available for the implementation and analysis of the study?	
How can the users be involved in the evaluation?	
How can the designers be involved in the evaluation?	

3. *Controlled*, taking place in a test environment controlling external factors to assess effectiveness, efficiency, and satisfaction with defined qualitative and quantitative metrics

7.3.3.2 Summary

Evaluations are typically *user-based* or *expert-based* in reference to who conducts the assessment of the design(s). Again, the generalizability trade-offs of using either class should be considered. User-based methods have been found to derive more practically realistic and legitimate problems. Expert-driven evaluations, however, provide a means to extract valid, unbiased issues more accurately than a user group sampling that is insufficiently small (Maguire, 2001). It is not uncommon in evaluations to have a shortage of actual users to work with.

Recently, Bainbridge (2004) identified *user-based evaluations* as consisting of the identification of representative users and tasks and the observation of problems arising out of the use of a design to create a task. The evaluations can also be *formative* in their role in design selection and prototype iteration or alternatively *summative* in the documentation of efficacy, efficiency, and satisfaction at the completion of the design iterations. Applicable methods, prototype robustness, measures, and purpose are decidedly different between summative and formative methods (Bainbridge, 2004). As formative methods are intimately linked to the iterative portion of UCD, evaluations must be timely, in order not to delay the overall design approach.

7.3.3.3 Some Examples of Methods and Tools

The testing used in UCD is typically categorized under the umbrella of methods classified as *usability evaluation methods*. However, there are several categories of methods that inform the evaluation of the design process and usability testing is one of those categories. Rubin (1994) asserts ten basic techniques relevant to the UCD process and specifically concerning the evaluation of potential design choices. Table 7.7 summarizes several of the methods mentioned by Rubin (1994) including focus group research, surveys, design walk-through, paper-and-pencil evaluations, expert evaluations, usability audits, usability testing, field studies, and follow-up studies. Again, note that usability testing is just one of the several classes of evaluation methodologies.

Even so, usability testing is probably the most common approach used in implementing UCD and has been a practice since the early 1980s (Dumas, 2003). It has received significant attention in the last 15 yr as a valid, reliable, and efficient means of assessment. Several texts portray usability testing methods in great detail providing extensive instructions for the planning, design, and management of these evaluations (e.g., Dix et al., 1998; Nielsen and Mack, 1994; Rubin, 1994). These texts and others alike provide step-by-step instructions for specific types of evaluations. Like the entire UCD process, the use of evaluation techniques is affected by several contextual factors, to which the technique must be adapted. In this section, three approaches to testing and evaluation are discussed. Two of these fall under the category of expert-based, inspection methods and the other technique discussed is user-observation.

Heuristic evaluation and cognitive walkthroughs are two usability evaluation methods that are classified as expert-based inspection methods. These techniques represent two of the original usability inspection methods introduced to the HCI community. As a result, several of the subsequently emerging methods were grounded on heuristic evaluation and cognitive walkthrough (Nielsen and Mack, 1994). Inspection methods are often lauded for their quick turnaround, limited training requirements, and ability to generate suitable usability problems without a great deal of involvement from actual users (Sears and Hess, 1999). To contrast, user-based observation techniques will also be introduced. Intuitively, a class of techniques requires significant participation from actual users. User-based observations may be anchored in usability evaluations, but can in fact range from basic observations to highly controlled empirical methods using special equipment to measure interactions. Typically, because they involve actual users (or at least representative users) they yield a higher degree of validity in their assessment of the ability of a design to match the identified requirement. User-based methods can sometimes provide surprising design flaws and strengths not observable through expert-based methods.

TABLE 7.7 UCD Assessment Methods

Method	Summary	Advantage	Disadvantage
Focus groups	Appropriate for low-fidelity prototypes to assess the suitability of basic design concepts derived form requirements gathering	In depth exploration of a few people's opinion and rationale of their opinions	Opinionated members of a focus group can bias other members
Surveys	Used at any stage in the process but most advantageous in early design to obtain generalizations for a large population	Ability to gather large samples of opinions to generalize a large user population	The interpretations of the questions by the user can differ if not carefully worded and verified
Design walk-throughs	Early prototypes/concept is used to verbally/pictorially guide the user through the tasks supported with a moderator	Provide insight into navigation or acceptability of tasks by user	The level of abstraction required by the user in envisioning the tasks can be problematic
Paper-and-pencil evaluation	An aspect of the design represented on page and users are asked questions about it	Inexpensive, fast way to collect information that is critical to the design	The system cannot be assessed in its entirety
Usability audit	Evaluation of a design against checklists of standards	Checklists are readily available in research literature and usability criteria of previous projects	Information overload in the selection of specific standards and lack of guidance in the implementation of the standards
Usability testing	Observations of representative end-users with the design	Quick turnaround on problem identification	Decisions on representative tasks and users can be challenging
Field study	Assessment of a design's use in its natural context in the last stages of design	Design is used in the field with an actual user to refine product prior to release	Despite the richness of data collected significant changes to the design are unlikely; lack of control in these studies limit the reproducibility of the results
Follow-up study	A field study completed after implementation of the final design	Reliable and valid, because they assess actual users within the actual context	Similar to field studies, a loss of control is experienced in the collection of data

7.3.3.3.1 *Heuristic Evaluation*

Heuristic evaluations, introduced by Nielsen and Molich (1990), direct several usability experts to generate a collaborative design critique. Their evaluations are structured according to general usability heuristics or rules of thumb. The experts independently evaluate the design using the rules to trigger their appraisal. When first conceived, there were nine heuristics; today the heuristics have evolved into the following ten (Nielsen, 1994):

- Visibility of system status
- Match between system and real world
- User control and freedom
- Consistency and standards
- Error prevention
- Recognition rather than recall
- Flexibility and efficiency of use
- Aesthetic and minimalist design

Nielsen derived these usability principles from a factor analysis of 249 potential usability problems. Additionally, evaluators can consider other usability principles and it is not uncommon to develop category/domain specific heuristics to supplement the original ten (Nielsen, 1994).The true power in the heuristic evaluation method is the involvement of several evaluators. Evaluators tend to find different usability problems and when their findings are aggregated (and duplicates reconciled), they can account for the majority of the usability flaws. Nielsen typically recommends 3 to 5 evaluators because there is a point of diminishing returns with a large number of evaluators (Nielsen, 1994).

Evaluators may require a training/familiarization session depending on the target user group and domain and other aspects of the work system (a definite minus to this method as there is no documented means by which to train the experts). For example, to evaluate an information kiosk, a walk-up and use system to be used by the general public, an evaluator would not typically receive training or prompting to explain the use of the system. In contrast, in the evaluation of a nurses' scheduling tool, a system for which nurses receive specialized training evaluators should really be provided by giving training for that system as well as preparation on the subject matter expertise possessed by the nurses. In addition, it may be useful to provide the evaluators with typical usage scenarios for the design to help them antici-pate the realistic demands on the design's functionality. To formally account for these factors, Muller et al. (1995) introduced three additional heuristics to encourage evaluators to consider the context of use (arguably ignored in the introductions to heuristic evaluation). These heuristics include:

1. Respect the user and his or her skills
2. Promote a pleasurable experience with the system
3. Support quality of work

First, each evaluator steps themselves through the interface several times independently, record any potential and any usability problems they identify with the interface for target users and tasks. Each usability problem identified should be accompanied by a sufficiently explicatory description. The more specific evaluators are in the vindication of the issues observed, the better, as it isolates target pro-blems and their priority in the subsequent redesign activities.

Following the independent evaluation, evaluators' problems and descriptions are aggregated and duplicates are accounted for. Then evaluators discuss their findings eliminating any duplicate problems and resolving any issues that may be contradictory. In this discussion, the evaluators often work to form consensus on the severity levels for each issue. The outcome of the heuristic evaluation is a list of specific problems and reference to which heuristics are violated and a severity level, which provides guidance as to which issues take priority in redesign. While the outcomes of heuristic evaluations are not recipes that explicitly direct redesign activities to achieve "correct" design, solutions that emerge from the evaluation are often intuitive, because of the heuristics' connection to fundamental usability principles.

A shortcoming of heuristic evaluations is that evaluators are not fully prepared for inspection in terms of applying the heuristics and relative to the target domain. Direction is typically not provided for the specific approaches taken up in the validation of design in terms of each heuristic (and even experienced evaluators may have inconsistent approaches to this). An additional limitation, some argue, is the pro-pensity for heuristic evaluation to generate several false positive usability issues. This is especially true in circumstances when the evaluators have inconsistent and unreliable knowledge of the domain and context (Cockton et al., 2003). In their assessment of heuristic evaluation, Cockton et al. (2003) noted a trend for evaluators to underestimate users' capabilities in display interactions. Still, the convenience and efficiency afforded by the heuristic evaluation technique prompts its use in situations requiring a quick turnaround in the iterative design process and in situations proving impractical or impossible for the inclusion of end-users directly in the evaluation and iteration stages of UCD.

The following key points summarize the heuristic evaluation technique:

- *Prototype requirements*: Compatible with a range of fidelity — from paper through operational systems
- *Number of evaluators*: 3 to 5, but possibly more in complex design situations.
- *Testing environment:* Flexible
- *Time involved*: 2 to 3 h, but longer if the system is highly complex

- *Output*: A document delineating specific design problems and explanations of why they are problems in terms of usability principles
- *Special equipment needed*: None

7.3.3.3.2 Cognitive Walkthrough

The cognitive walkthrough evaluation method (Polson et al., 1992) was introduced by Polson et al. around the same time as Nielsen and Molich introduced heuristic evaluation. While heuristic evaluations focused on several usability goals with little attention to actual users' knowledge, the cognitive walkthrough method pays particular attention to the *learnability* of the system for the target user population. The method was inspired by exploratory interactions of users new to a system or the acquisition of knowledge and features as the work mandates (Wharton et al., 1994). As such, this method has grounded in psychological theory, but incorporates aspects of a more informal walk-through and has been revised to be usable by designers and developers (Dix et al., 1998).

Cognitive walkthrough considers the sequence of actions required by a design to achieve some goal. Like heuristics evaluations, usability and human factors professionals direct cognitive walkthroughs. Evaluators are provided with a prototype with at least some detail on layout, wording and control placement, and a description of a representative task(s) that apply to the system. The selection of the task should be compelled by what tasks the users would actually be driven to accomplish in the context of the work system. Along with the prototype and task, the evaluators must be given a comprehensive list of steps required to complete the identified task(s) with the prototype. Finally, evaluators are provided with a profile of the users including a summary of the anticipated experience and knowledge of target users (Dix et al., 1998). When performing the walkthrough, the evaluator records for each sequential action, a story that advocates or dispels the usability of the design in accommodating each required action.

Answering several questions at each sequential step in the completion of the task shapes a story of interaction and highlights any usability and requirements issues. These questions are summarized in Table 7.8. In this process, the evaluator considers what the user must know before and during the performance of each action. The story should contain information that is critical for accomplishing interactions such as knowledge/information required of the user, assumptions made about the user population through the design, notes about side issues that may need resolution, and examples when the system did and did not support the intended actions or user requirements. The output of the cognitive walkthrough is a detailed account including an estimation of the severity of each problem in relation to impeding task goals.

The limitations of cognitive walkthrough are influenced by the method's emphasis on learnability (Wharton et al., 1994). Also, similar to heuristic evaluations, the level of detail provided about the task and system before the assessment can alter significantly the types of problems that are identified. Sears and Hess (1999) reported a study on this effect. They concluded that detailed descriptions of the task retrieved more problems with the system feedback. Conversely, nondetailed descriptions uncovered more problems with the controls in the interface. Wharton et al. (1994) also have conjectured that a rise in the number of tasks evaluated leads to an increased number of identified problems related to general design issues.

The following key points summarize the cognitive walkthrough:

- *Prototype requirements*: Can be completed with any fidelity, but the more detailed the prototype, the better the representation of the user's sequence of actions and subsequent extraction of usability issues

TABLE 7.8 Questions Used in the Cognitive Walkthrough Evaluation Method

Will the user try to achieve the right effect?
Will the user notice that the correct action is available?
Will the user associate the correct action with the effect that the user is trying to achieve?
If the correct action is performed, will the user recognize that progress is being made toward completion of the task?

- *Number of evaluators*: one or more
- *Testing environment*: Flexible
- *Time involved*: Varies greatly, dependent on prototype fidelity and complexity of the design — In other words, the number of steps required for consideration in completing each task
- *Output*: A document delineating specific design problems and explanations of why they are problems, in terms of usability principles
- *Special equipment needed*: Task instructions and users' profiles

7.3.3.3.3 User Observations

Simply put, user observations are a collection of techniques and tools for summarizing what happens when real users interact (and react) to the design. These summaries can be generated watching the users with the system design in the field (which requires a highly functional prototype), but typically involves the assessment of a prototype with incomplete functionality in a laboratory environment. Three major considerations exist in the planning and execution of observational techniques: (1) the level of mediation with the user; (2) the methods of data collection; and (3) the measures used to summarize the interactions.

The first consideration is the amount of mediating activity required to allow the user to operate the prototype and the degree of contact needed with the users in order to elicit user opinions, motivations, and internal dialog. Those leading the evaluation can sit behind a two-way mirror without speaking to the user at all during the task (allowing for a highly natural/representative interaction), or they can sit beside the user, walking them through the prototype, educing their reactions through pointed questions. While those are two extreme scenarios, methods vary along a continuum of intervention with the user.

Second, the specific methods of data collection must be determined. For example, the observation session may be video taped and retrospectively reviewed and classified. Alternatively, those leading the evaluation may take copious notes of user verbalizations and observations about the interface. In some instances, users take "diaries" or notebooks with their own observations on the iteration, which is useful in longitudinal evaluations of a design (Dix et al., 1998). It may also be advantageous to have the system automatically log user information during their interactions. Third, the metric generated to analyze the degree to which the design meets the design's requirements must be clearly defined. Both quantitative and qualitative data are collected during observations and each informs assessments of the design's quality. While the measurement of design quality is a seemingly vague proposition, there are several initiatives aimed at the standardization of assessment metrics. For example, the ESPRIT measuring usability in context (MUSIC) project has introduced a performance measurement method for software evaluation (Bevan and Macleod, 1994). A useful review of international usability standards and metrics has been produced by Bevan (2001) and human factors and a comprehensive account of ergonomic standards has been assembled in work by Sherehiy et al. (2005).

Dumas (2003) recently pointed out the following shortcomings of user-observations, citing: (1) the difficulty in inferring causality of observed behavior; (2) the inability to control the rate at which events occur or if they occur at all; (3) participants may change their behavior when they are being watched; and (4) Observers can become narrow-minded and only see what they want to and challenge the validity of what is captured. Truly these issues impose doubt on the validity of the resultant conclusions and recommendations from user-observations. Rubin (1994) makes the following suggestions to inhibit potentially destructive variations in the evaluation process: (1) use scripts so that the instructions between subjects are consistent in their goal direction; (2) using checklists to assure all activities were completed by the user and that the information has been collected; and (3) consistency in the person(s) conducting the tests, which can influence the behaviors of the users.

A final word of advice when organizing user-based evaluations is to always run a pilot test prior to the actual assessment session(s). This should include running through actual protocol using any surveys and working with the data types that are collected. This dress rehearsal allows the evaluator to make fine adjustments to the protocol and ensures that the time with the users is spent gathering data and not in recovering from unforeseen mishaps with equipment or protocols. As in the design process, it is much more feasible to make changes prior to the full implementation of the assessment and the implications of any "hiccups" in the protocol will have limited impact. Human behavior is by nature variable and hard to predict; pilot testing can control some of the negative impacts of this irregularity.

Outcomes of user-observation based studies are typically an indication of usability problems, what triggers them, and sometimes recommendations for solutions. While it used to be a standard protocol to develop a formal written report on the finding, a more common trend is a debriefing, during which the problems, possible solutions, and prioritizations are communicated (sometimes in combination with a report). Additionally, video footage collected can truly supplement the degree to which the development team is on board with the suggested problems and fixes (Dumas, 2003; Rubin, 1994). In some cases, just having the development and design team observe the interactions proved to be sufficient to inspire the necessary requirements (Wixon, 2003).

The focus of UCD evaluation is to include the target end-users or potential end-users in the protocol. Working with actual users is both a challenging and rewarding venture. The evaluation team has to decide in many cases the most relevant population of users to test. Dumas (2003) recommends that this should be based on the priorities of management (which was identified in requirements gathering), not the ease with which subjects are recruited. Because usability studies typically require 5 to 10 participants to uncover the majority of problems with a design, subject recruitment is typically not difficult (Dix et al., 1998; Dumas, 2003). While there is some contention in the literature regarding the optimal number of participants (Wixon, 2003) there is a point of diminishing return in including additional participants.

Despite the challenges facing user-observations, the discerning power of the results, when gathered with attention to these details can transform a problematic design into a product that is usable and pleasurable to use. Table 7.9 introduces specific methods for user observation and a description of each along with advantages and disadvantages.

The following key points summarize the user evaluations:

- *Prototype requirements*: Can be completed with prototypes of any level, but the more detailed the prototype, the better the representation of the user's sequence of actions and subsequent extraction of usability issues
- *Number of evaluators*: Varies depending on the types of analyses and statistical power that is mandated; some suggest that between 8 to 25 users be evaluated depending on the power requirements and user variability (Maguire, 2001)
- *Testing environment*: Typically controlled laboratory setting, but can sometimes be in the field (or both in the case of remote testing)
- *Time involved*: Varies greatly, dependent on prototype fidelity, complexity of the design, and the number of subjects to run; evaluators coordinating assessment sessions should be sensitive to time required of the users
- *Output*: A document delineating specific design problems with justification from observation and recommended solutions
- *Special equipment needed*: Video cameras, one-way mirrors/observation area, tape recorders, video editing equipment, information collection forms, software for logging interaction dialogs, and even eye tracking to assess the gaze paths of the user

7.3.3.4 Summary

In this section, we discussed both advantages and limitations to the fundamental methods and provided classification of types of methods for consideration in the evaluation process. Several breeds of testing methods have spawned from the ones mentioned in this section. The set of methods presented just scratches the surface on the available UCD methods. Researchers and practitioners are continually developing new techniques, assessing and improving upon the current steadfast favorites (e.g., heuristic evaluation). As is the case with the overall UCD process, the selection and utilization of testing and evaluation

TABLE 7.9 Methods Used in User-Observations to Assess Usability

Observation Method	Summary	Advantage	Challenges
Empirical evaluation in controlled user testing (Dix et al., 1998)	No contact between evaluator and user except the provision of instructions; collects user verbalization, interaction events, post test evaluations of the system to support a particular hypothesis	These techniques allows for the most natural interaction setting for the user without interruption; time measurements are valid	A session can end prematurely if the user cannot accomplish the task
Assisted evaluation (Maguire, 2001)	Contact is made only in situations where the user is baffled in the use of the system and the possible actions are not clear	Hiccups in the prototype will not end the evaluation; prototype fidelity can be relatively incomplete	Can induce unintended workload on the user to have someone seated; it can be easy to include encouraging or discouraging statements to the user by mistake
Think aloud protocol (Rubin, 1994)	Participants are encouraged to provide their own verbal commentary about their interactions. This includes feelings, questions, and motivations; evaluators do not ask questions, but do remind the users to think aloud if they forget	Simultaneous collection of preference and performance data; users may be more focused and directed in their role; can identify what triggers confusion before it implicates into a bigger problem	Unnatural and some users may not be highly articulate; interferes with their performance (in negative or positive ways); distraction or fatigue is likely after a few hours
Remote testing (Dumas, 2003)	Evaluator and user are geographically separated from each other	Testing occurs in a familiar environment to the user; testing occurs with the actual equipment and contextual cues; reduced costs from not traveling	Unreliable conferencing software; company firewalls can prevent live tests

methods is subject to changes — based on the given constraints of a study. In a perfect world, evaluators would apply a combination of these techniques to account for as many design problems as possible. What has become more feasible, however, has been the hybridization of different techniques to improve on the benefits generated by each [e.g., "heuristic walkthroughs"; see Cockton et al. (2003); Cockton and Woolrych (2001). The challenge for evaluators, designers, and developers is how to best match the evaluation method with the identified goals of the design within the organizational constraints of funding, lead-time, and availability of evaluators.

7.4   Conclusions

UCD is just not a methodology for design; its a philosophy under which the entire organization and design/development process must operate in order to realize the goals of the new system. There are organizational-wide implications in terms of the commitment to, support for, and acceptance of the UCD process. Rubin (1994) has identified common characteristics of organizations, which successfully carry out UCD. These include:

- Use a phase-based approach for development that integrates incremental evaluations from user and expert feedback at critical stages in the design process
- Utilize multidisciplinary teams to provide the variety of skills, knowledge, and information about the target users and their activities, including engineering, marketing, training, interface design, human factors, and multimedia

- Facilitate an involved and concerned management committed to following and promoting the UCD activities

As discussed, within the realm of UCD there are countless variations of techniques and tools to include end-users and their needs in the design process, to generate data and knowledge of the work system, and to test and evaluate the usability, appropriateness, and of system design alternatives. This chapter provided a high level overview of the most prevailing methods as well as guidance in the selection and application of UCD methods and tools validated to be effective through both research and practice.

Overall, UCD is the principle of including users and stakeholders in the design process, an idea that was unheard of (or ignored) just a few decades ago. However, with the steadily increasing ubiquity of computer-based technologies within the personal and work lives of the general populace, UCD is becoming not only a "good practice," but a legal mandate and a practical necessity to ensure the success and acceptability of technologies. Despite the relatively long history and continual development of UCD tools and procedures, the UCD philosophy is still far from reaching a saturation point in today's world of product and technology design. Champions of UCD are needed to propel this ideology and techniques in a variety of domains to produce systems that uphold high standards of safety, quality, and efficiency and support a high level of user satisfaction. Users and the various aspects of the work system must be deliberately and judiciously woven into the design process to emerge a usable, useful appealing design.

7.5 Message from the Authors

As introduced in this chapter, the introduction of UCD in the mid 1980s resulted in several initiatives that correspond with the underlying fundamentals of this approach. Specifically, these include approaches for integrating the needs of users with disabilities in design referred to as universal or inclusive design. The inclusive approach to the design of products, technologies, and systems entails designing for as expansive and varied user population as possible (Preiser and Ostroff, 2001). Problems arise, however, as these design philosophies can result in tension when considered in the context of UCD. They place greater demand on the requirements gathering process (Newell and Gregor, 2000).

Inclusive design is not wholly inclusive. This approach does not account for end-users who are not included in the scope of the product and task requirements (Coleman et al., 2003). Simply put, and "designing for all" is not always appropriate in all situations (Norman, 1988). Still, there is a noted lack of progress in the application of inclusive design even in mainstream products and systems. This lack of progress has been attributed to shortcomings in the guidance offered in the practical application of the inclusive design strategies and requirements in terms of real world design constraints and priorities (Cassim and Dong, 2003; Lebbon and Coleman, 2003). As previously discussed, designers and developers of products and systems have to work within several practical and limiting constraints when creating or modifying designs. As such, a more practical approach to the UCD process is needed to help facilitate its actual use in practice.

Inclusive design is a challenging venture in its own and imposes additional requirements on the UCD process. Newell and Gregor (2000) have identified these additional considerations and challenges, which inclusive design impose on the UCD process. They include:

- A much greater spectrum of user characteristics and functionality to be considered
- Developing precise specification of user group characteristics and functionality
- Defining, finding, and recruiting representative users
- Resolution of conflicts between accessibility and ease of use for less disabled individuals and varieties of impairments
- Identifying and justifying situations in which universal design may not be appropriate
- The need to provide additional components for a system to afford access to

In truth, considering the needs of users with disabilities merits integration into the priorities of UCD in order to create opportunity for users of all abilities and operate under a variety of contexts to be gainfully employed and be productive members of society. This is especially true in light of the significant population of aging adults, who will experience the normally anticipated age-related declines in physical, cognitive, and sensory functions. If systems and tools are not usable by a percentage of the population, then this percentage of individuals can never be trained to serve the roles typically associated with a given set of tasks. The design itself may become an impediment to the successful completion of related tasks.

The initiatives driving accessibility agendas such as the Americans with Disabilities Act clearly states the minimum requirements that must be met in the design of systems and products. However, the actual challenge lies in the improvement of existing UCD methods and development of new UCD method that will allow UCD researchers and practitioners to meet these needs and expectations of a highly variable and important population of end-users. This directive should guide the future of UCD and, will hopefully, result in more clearly defined, systematic, and acceptable ways for designers and developers to incorporate the principles of UCD and inclusive design into tangible, practical design solutions.

References

Alexander, I.F. and Stevens, R., *Writing Better Requirements*. Addison-Wesley Professional, Boston, MA, 2002.

Apple Computer, *Introduction to the Apple Human Interface Guidelines*. Retrieved March 8, 2004, from http://developer.apple.com/documentation/UserExperience/Conceptual/OSXHIGuidelines/index.html#//apple_ref/doc/uid/20000957.

Bainbridge, W.S., 2004 User–centered design, in *Berkshire Encyclopedia of Human–Computer Interaction*, Bainbridge, W.S., Ed., Vol. 2, Berkshire Publishing Group, Great Barrington, 2004, pp. 763–768.

Bekker, M., Beusman, J., Keyson, D., and Lloyd, P., KidReporter: a user requirements gathering technique for designing with children. *Interact. Comput.*, 15 (2), 187–202, 2003.

Bevan, N., International standards for HCI and usability. *Int. J. Hum.–Comput. Stud.*, 55 (4), 533–552, 2001.

Bevan, N., Bowden, R., Corcoran, R., Curson, I., Macleod, M., Maissel, J. et al., *Usability Context Analysis—A practical guide, v4.02.* Middlesex: NPL Usability Services, National Physical Laboratory, 1996.

Bevan, N. and Macleod, M., Usability measurement in context. *Behav. Inf. Technol.*, 13, 132–145, 1994.

Beyer, H. and Holtzblatt, K., Apprenticing with the customer. *Commun. ACM*, 38 (5), 45–52, 1995.

Beyer, H. and Holtzblatt, K., *Contextual Design: Defining Customer-Centered Systems.* Morgan Kaufmann Publishers, San Francisco, CA, 1998.

Blomberg, J., Burrell, M., and Guest, G., An ethnographic approach to design, in *The Human–Computer Interaction Handbook: Fundamentals, Evolving Technologies and Emerging Applications*, Jacko, J.A. and Sears, A. Eds., Lawrence Erlbaum Associates, Inc., Mahwah, NJ, 2003, pp. 964–986.

Booch, G., Rumbaugh, J., and Jacobson, I., *The Unified Modeling Language User Guide*, 1st ed., Addison-Wesley Professional, Boston, MA, 1998.

Caplan, S., Using focus groups methodology for ergonomic design. *Ergonomics*, 33 (5), 527–533, 1990.

Cassim, J. and Dong, H., Critical users in design innovation, in *Inclusive Design: Design for the whole Population.* Clarkson, J., Coleman, R., Keates, S., and Lebbon, C. Eds., Springer, New York, 2003.

Christel, M.G. and Kang, K.C., *Issues in Requirements Elicitation* (Technical Report; CMU/SEI-92-TR-12; ESC-TR-92–012). Carnegie Mellon University, Pittsburgh, PA, 1992.

Cockburn, A., *Writing Effective Use Cases* (1st ed.). Addison-Wesley Professional, Boston, MA, 2000.

Cockton, G., Lavery, D. and Woolrych, A., Inspection-Based Evaluation, in *The Human–Computer Interaction Handbook: Fundamentals, Evolving Technologies and Emerging Applications* Jacko, J. A. and Sears, A. Eds., Lawrence Erlbaum Associates, Inc, Mahwah, NJ, 2003, pp. 1118–1138.

Cockton, G. and Woolrych, A., Understanding inspection methods: Lessons from an assessment of heuristic evaluation, in *People and Computers XV*, Blandford, A. and Vanderdonckt, J. Eds., Springer-Verlag, Berlin, 2001, pp. 171–192.

Coleman, R., Clarkson, J., Lebbon, C., and Keates, S., Introduction: from margin to mainstream, in *Inclusive Design: Design for the Whole Population*, Clarkson, J., Coleman, R., Keates, S., and Lebbon, C. Eds., Springer, New York, 2003.

Cooper, A., *The Inmates Are Running the Asylum: Why High Tech Products Drive Us Crazy and How to Restore the Sanity* (2nd ed.). SAMS Publishing, Pearson Education, Indianapolis, IN, 2004.

Damodaran, L., User involvement in the systems design process–a practical guide for users. *Behav. Inf. Technol.*, 15 (6), 363–377, 1996.

Damodaran, L., Simpson, A., and Wilson, P., *Designing systems for people*. NCC Publications, Manchester, 1980.

Dix, A.J., Finlay, J.E., Abowd, G.D., and Beale, R., *Human–Computer Interaction* (2nd ed.). Prentice Hall, London, UK, 1998.

Dourish, P., What we talk about when we talk about context. *Personal Ubiquitous Comput.*, 8 (1), 19–30, 2004.

Dumas, J.S., User-Based Evaluations, in *The Human-Computer Interaction Handbook: Fundamentals, Evolving Technologies and Emerging Applications*, Jacko, J. A., and Sears, A. Eds., Lawrence Erlbaum Associates, Inc., Mahwah, NJ, 2003, pp. 1093–1117.

Eason, K.D., User-centred design: For users or by users? *Ergonomics*, 38 (8), 1667–1673, 1995.

Emery, V.K., Jacko, J.A., Sainfort, F. and Moloney, K., Intranet and intra-organizational communication, in *The Handbook of Human Factors in Web Design*, Proctor, R.W. and Vu, K.P.L. Eds., Lawrence Erlbaum Associates, Inc., Mahwah, NJ, 2004, pp. 528–550.

Gottesdiener, E., 2002 *Requirements by Collaboration: Workshops for Defining Needs* (1st ed.). Addison-Wesley Professional, Boston, MA, 2002.

Grudin, J. and Pruitt, J., Personas, participatory design and product development: An infrastructure for engagement. *Proceedings of the Participatory Design Conference (PDC 2002)* Computer Professionals for Social Responsibility (CPSR), Malmö, pp. 144–161, 2002.

Hackos, J.T. and Redish, J.C., *User and Task Analysis for Interface Design*. John Wiley & Sons, Inc., New York, NY, 1998.

Hall, R., Prototyping for usability of new technology. *Int. J. Hum.–Comput. Stud.*, 55, 485–501, 2001.

Holtzblatt, K., Contextual Design, in *The Human-Computer Interaction Handbook: Fundamentals, Evolving Technologies and Emerging Applications*, Jacko, J. A., and Sears, A. Eds., Lawrence Erlbaum Associates, Inc., Mahwah, NJ, 2003, pp. 941–963.

Holtzblatt, K. and Jones, S., Contextual inquiry: A participatory technique for system design, in *Participatory design: Principles and practices*, Schuler, D. and Namioka, A. Eds., Lawrence Erlbaum Associates, Inc., Hillsdale, NJ, 1993, pp. 177–210.

IEEE., *EEE Recommended Practice for Software Requirements Specifications (IEEE Std 830–1998)*. IEE, New York, 1998a.

IEEE., *IEEE Guide for Developing System Requirements Specifications (IEEE Std 1233)*. IEEE, New York, 1998b.

International Organization for Standardization, ISO 9241–11: Ergonomic requirements for office work with visual display terminals (VDTs). Part 11: Guidance on Usability, 1998.

International Organization for Standardization (ISO), *ISO 13407: Human-Centred Design Processes for Interactive Systems*. International Organization for Standardization, Geneva, 1999.

International Organization for Standardization (ISO), *ISO/TR 18529: Ergonomics—Ergonomics of Human–System Interaction—Human-Centred Lifecycle Process Descriptions*. International Organization for Standardization, Geneva, 2000.

Jacko, J.A. and Sears, A., *The Handbook of Human-Computer Interaction Handbook: Fundamentals, Evolving Technologies and Emerging Applications*. Lawrence Erlbaum Associates, Mahwah, NJ, 2003.

Javahery, H., Seffah, A., and Radhakrishnan, T., Beyond power: Making bioinformatics tools user-centered. *Commun. ACM*, 47 (11), 58–63, 2004.

Johnson, C.M., Johnson, T.R. and Zhang, J., A user-centered framework for redesigning health care interfaces. *J. Biomed. Informatics*, 38 (1), 75–87, 2005.

Karat, J., Karat, C.-M. and Ukelson, J., Affordances, motivations, and the design of the user interface. *Commun. ACM*, 43, 49–51, 2000.

Karwowski, W., Achieving compatibility in human-computer interface design and evaluation, in *The Human–Computer Interaction Handbook*, Jacko, J.A. and Sears, A. Eds., Lawrence Erlbaum Associates, Inc., Mahwah, NJ, 2003, pp. 1226–1238.

Kirlik, A., The design of everyday life environments, in *A Companion to Cognitive Science*, Bechtel, W. and Graham, G. Eds., Blackwell Publishers, Oxford, UK, 1998, pp. 702–712.

Kontio, J., Lehtola, L., and Bragge, J., Using the Focus Group Method in Software Engineering: Obtaining Practitioner and User Experiences. *Proceedings of the ACM-IEEE International Symposium on Empirical Software Engineering (ISESE 2004)*, Redondo Beach, CA, pp. 271–280, 2004.

Kontogiannis, T. and Embrey, D., A user-centred design approach for introducing computer-based process information systems. *Appl. Ergon.*, 28 (2), 109–119, 1997.

Kotonya, G. and Sommerville, I., *Requirements Engineering: Processes and Techniques.* John Wiley & Sons, Inc., New York, 1998.

Koubek, R., Benysh, D., Buck, M., Harvey, C., and Reynolds, M., The development of a theoretical framework and deisgn tool for process usability assessment. *Ergonomics*, 46 (1–3), 220–241, 2003.

Kovitz, B.L., *Practical Software Requirements: A Manual of Content and Style.* Manning Publications Co., Greenwich, CT, 1998.

Kulak, D. and Guiney, E., *Use Cases: Requirements in Context* (2nd ed.). Pearson Education, Inc., (Addison-Wesley), Boston, MA, 2004.

Law, C., Keeping designers up with the times . . . projects for enabling the rapid development of universally accessible IT by industry. *Proceedings of the Inclusion by Design — Planning a Barrier — Free World, An International World Congress* Montreal, Canada, 2001.

Lebbon, C. and Coleman, R., A design-centered approach, in *Inclusive design: Design for the Whole Population*, Clarkson, J., Coleman, R., Keates, S. and Lebbon, C. Eds., Springer, New York, 2003, pp. 500–518.

Leonard, V.K., Jacko, J.A., Yi, J., and Sainfort, F., Human Factors & Ergonomic Methods, in *Handbook of Human Factors & Ergonomics*, Salvendy, G. Ed., (3rd ed., pp. in press). John Wiley and Sons, Inc., New York, 2005.

Lidwell, W., Holden, K., and Butler, J., *Universal Principles of Design: 100 Ways to Enhance Usability, Influence Perception, Increase Appeal, Make Better Design Decisions, and Teach through Design.* Rockport Publishers, Inc., Gloucester, MA, 2003.

Maguire, M., Methods to support human-centred design. *Int. J. Hum.–Comput. Stud.*, 55 (4), 587–634, 2001.

Mao, J.-Y., Vrendenberg, K., Smith, P.W., and Carey, T., User-centered design methods in practice: A survey of the state of the art. *Proceedings of the 2001 Conference of the IBM Centre for Advanced Studies on Collaborative Research (CASCON '01)*, IBM Press, Toronto, pp. 12–24, 2001.

Marcus, A., User-centered design in the enterprise. *Interactions*, 12, 18–23, 2005.

Mayhew, D.J., *The Usability Engineering Lifecycle: A Practitioner's Handbook for User Interface Design* (1st ed.). Morgan Kaufmann Publishers, San Francisco, CA, 1999.

Mayhew, D.J., Requirements Specifications Within the Usability Engineering Lifecycle, in *The Human-Computer Interaction Handbook: Fundamentals, Evolving Technologies and Emerging Applications*, Jacko, J.A. and Sears, A. Eds., Lawrence Erlbaum Associates, Inc., Mahwah, NJ, 2003, pp. 913–921.

Microsoft Corporation, *Microsoft Tablet PC — Visual Design Guidelines (Microsoft Tablet PC)*. Retrieved March 8, 2004, from http://msdn.microsoft.com/library/default.asp?url = /library/en-us/tpcsdk10/html/whitepapers/designguide/tbcontpcvisualdesignguidelines.asp, 2004.

Muller, M.J., McClard, A., Bell, B., Dooley, S., Meisky, L., Meskill, J.A. et al., Validating an extension to participatory heuristic evaluation: Quality of work and quality of life. *Proceedings of the ACM SIGCHI Conference on Human Factors in Computing Systems (CHI '95)*, ACM Press, Denver, CO, pp. 115–116, 1995.

Newell, A.F. and Gregor, P., "User sensitive inclusive design" — In search of a new paradigm. *Proceedings of the ACM Conference on Universal Usability (CUU '00)*, ACM Press, Arlington, VA, pp. 39–44, 2000.

Nielsen, J., *Usability Engineering* (1st ed.). Morgan Kaufmann, San Francisco, CA, 1993.

Nielsen, J., Heuristic Evaluation, in *Usability Inspection Methods* Nielsen, J. and Mack, R.L. Eds., John Wiley & Sons, New York, 1994, pp. 25–62.

Nielsen, J., *Durability of Usability Guidelines*. Retrieved February 20, 2005, from http://www.useit.com, 2005.

Nielsen, J. and Mack, R.L., *Usability Inspection Methods*. John Wiley & Sons, Inc., New York, 1994.

Nielsen, J. and Molich, R., Heuristic evaluation of user interfaces. *Proceedings of the ACM SIGCHI Conference on Human Factors in Computing Systems (CHI '90)*, ACM Press, Seattle, WA, pp. 249–256, 1990.

Norman, D.A., *The Psychology of Everyday Things* (reprint ed.). Basic Books (Perseus), New York, 1988.

Norman, D.A., *The Design of Everyday Things* (1st ed.). Basic Books (Perseus), New York, 2002.

Noyes, J. and Barber, C., *User-Centred Design of Systems*. Springer-Verlag, London, UK, 1999.

Nuseibeh, B. and Easterbrook, S., Requirements Engineering: A Roadmap. *Proceedings of the International Conference on Software Engineering (ICSE 00)*, ACM Press, Limerick, pp. 35–46, 2000.

Ostroff, E., Universal Design: The New Paradigm, in *Universal Design Handbook*, Preiser, W.F.E. and Ostroff, E. Eds., McGraw Hill, Boston, MA, 2001, pp. 1.3–1.12.

Oulasvirta, A., Finding Meaningful Uses for Context-Aware Technologies: The Humanistic Research Strategy. *CHI Lett.*, 6 (1), 247–254, 2004.

Polson, P., Lewis, C., Rieman, J. and Wharton, C., Cognitive walkthroughs: A method for theory-based evaluation of user interfaces. *Int. J. Man–Machine Stud.*, 36, 741–773, 1992.

Preece, J., Rogers, Y., Sharp, H., Benyon, D., Holland, S. and Carey, T., *Human-Computer Interaction: Concepts And Design* (1st ed.). Addison-Wesley, Reading, MA, 1994.

Preiser, W.F.E. and Ostroff, E., *Universal Design: The New Paradigm*. McGraw Hill, MA, Boston, 2001.

Pruitt, J. and Grudin, J., Personas: Practice and Theory. *Proceedings of the 2003 Conference on Designing for User Experiences (DUX2003)*, ACM Press, San Francisco, pp. 1–15, 2003.

Psephos Corporation., 2005 *Butterfly ballot voting machine*. Retrieved February 22, 2005, from http://www.xeniashopper.com/psephos_voting_machine.htm

Ransley, P., *White Paper: A Primer on Requirements Engineering*, 2003.

Introducing Beavers Requirements Engineering Services Capability, Beaver Computer Consultants, Ltd, Essex, UK.

Rasmussen, J., *Information Processing and Human–Machine Interaction: An Approach to Cognitive Engineering*. North-Holland, Amsterdam, 1986.

Robertson, S. and Robertson, J., *Mastering the Requirements Process*. Addison-Wesley Professional, Boston, MA, 1999.

Rosenberg, L., *Requirements Engineering: A Methodology for Writing High Quality Requirement Specification and for Evaluating Existing Ones*. Retrieved February 17, 2005, from http://www.stc-online.org/cd-rom/1999/slides/MethWrit.pdf, 1999.

Rouse, W.B., *Design for Success: A Human-Centered Approach to Designing Successful Products and Systems*. John Wiley & Sons, Inc., New York, 1991.

Rubin, J., *Handbook of Usability Testing: How to Plan, Design, and Conduct Effective Tests*. John Wiley & Sons., New York, 1994.

Rumbaugh, J., Jacobson, I. and Booch, G., *The Unified Modeling Language Reference Manual* (2nd ed.). Addison-Wesley Professional, Boston, MA, 2004.

Saferstein, B., Ethnomethodology, in *A Companion to Cognitive Science*, Bechtel, W. and Graham, G. Eds., Blackwell Publishers, Oxford, UK, 1998, pp. 391–401.

Samaras, G.M. and Horst, R.L., A systems engineering perspective on the human-centered design of health information systems. *J. Biomed. Informatics*, 38(1), 61–74, 2005.

Sears, A. and Hess, D., Cognitive walkthroughs: Understanding the effect of task description detail on evaluator performance. *Int. J. Hum.–Comput. Interact.*, 11 (3), 185–200, 1999.

Sears, A., Lin, M., Jacko, J.A., and Xiao, Y., When Computers Fade. . . Pervasive Computing and Situationally-Induced Impairments and Disabilities, in *Human–Computer Interaction: Theory and Practice (Part II)*, Stephanidis, C. and Jacko, J.A. Eds., Lawrence Erlbaum Associates, Inc., Mahwah, NJ, 2003, pp. 1298–1302.

Sharp, H., Finkelstein, A. and Galal, G., Stakeholder Identification in the Requirements Engineering Process. *Proceedings of the 10th International Workshop on Database & Expert Systems Applications (DEXA)*, IEEE Computer Society, Florence, pp. 387–391, 1999.

Shepherd, A., HTA as a framework for task analysis, in *Task Analysis*, Annett, J. and Stanton, N.A. Eds., Taylor and Francis, New York, 2000, pp. 9–24.

Shepherd, A., *Hierarchical Task Analysis.* Taylor & Francis, Inc., New York, 2001.

Sherehiy, B., Karwowski, W., and Rodrick, D., Human Factors and Ergonomics Standards, in *Handbook of Human Factors and Ergonomics*, Salvendy, G. Ed., (3rd ed., pp. in press). John Wiley & Sons, Inc., New York, 2005.

Shneiderman, B., *Designing the User Interface: Strategies for Effective Human-Computer Interaction* (3rd ed.). Addison-Wesley Longman, Inc., Reading, MA, 1998.

Sommerville, I. and Sawyer, P., *Requirements Engineering: A Good Practice Guide.* John Wiley & Sons, Inc., New York, 1997.

Stanton, N.A. and Young, M.S., *A Guide to Methodology in Ergonomics.* Taylor & Francis., New York, 1999.

Stewart, T. and Travis, D., Guidelines, Standards, and Style Guides, in *The Human–Computer Interaction Handbook: Fundamentals, Evolving Technologies and Emerging Applications*, Jacko, J.A. and Sears, A. Eds., Lawrence Erlbaum Associates, Inc., Mahwah, NJ, 2003, pp. 991–1005.

Sugar, W.A., What is so good about user-centered design: Documenting the effect of usability sessions on novice software designers. *J. Res. Comput. Educ.*, 33 (3), 235–250, 2001.

U.S. Department of Defense (DOD), Section 8.3: Methods, in *MIL-HDBK-46855A: Department of Defense Handbook: Human Engineering Program Processes and Procedures* 1999, pp. 113–174.

Van Natta, D., Jr. and Canedy, D., The 2000 Elections: The Palm Beach Ballot. Florida Democrats Say Ballot's Design Hurt Gore. *The New York Times*, p. A1, 2000.

Vicente, K.J., *Cognitive work analysis: Toward safe, productive, and healthy computer-based work.* Lawrence Erlbaum Associates, Inc, Mahwah, NJ, 1999.

Vredenburg, K., Isensee, S., and Righi, C., *User-Centered Design: An Integrated Approach.* Prentice-Hall, Inc., Upper Saddle River, NJ, 2002.

W3C., 2004 *Authoring Tool Accessibility Guidelines 2.0 (W3C Working Draft).* Retrieved March 8, 2004, from http://www.w3.org/TR/2004/WD-ATAG20–20040224/

Wharton, C., Rieman, J., Lewis, C. and Polson, P., The Cognitive Walkthrough Method: A Practitioner's Guide, in *Usability Inspection Methods*, Nielsen, J. and Mack, R.L. Eds., John Wiley & Sons, New York, 1994, pp. 105–141.

Wiegers, K.E., 2000 *When Telepathy Won't Do: Requirements Engineering Key Practices.* Retrieved February 22, 2005, from http://www.processimpact.com/articles/telepathy.pdf

Wiegers, K.E., *Software Requirements* (2nd ed.). Microsoft Press, Redmond, WA, 2003.

Wixon, D., Evaluating usability methods: Why the current literature fails the practitioner. *ACM Interact.*, 10 (4), 28–34, 2003.

Young, R.R., *Effective Requirements Practices.* Addison-Wesley Professional, Boston, MA, 2001.

Young, R.R., *The Requirements Engineering Handbook.* Artech House, Inc., Norwood, MA, 2004.

8

Application of Risk Theory in Man–Machine–Environment Systems

Juraj Sinay
Hana Pačaiová
Melichar Kopas
Milan Oravec
Technical University of Kosice

All activities in the man–machine–environment system also create the risk state. Risk state influences health and safety, for example, health injury due to accident or due to long-time influences to the health. At the same time risk causes economical loses due to failures of machines or due to mechanical accidents/explosions, fire, destructions. The main goal of all activities of risk management is to control the risk, to minimize all negative influences, for example, illness, health injury, death and also technical consequences or major industrial accidents.

All necessary demands on the technical safety are integrated into two basic legislation products of EU:

- Direction 391/89/EU:[1] Increasing safety and health protection in working conditions
- Direction 392/89/EU:[2] together with 93/44/EU, 93/68/EU — Machine safety, approximation of member states legislation

According to the first Direction, it is necessary to evaluate all possible risks in the working process to perform measures for protection. This is the duty of the employer, who must take into consideration all-important influences and conditions, that is, organization of working process, working conditions, hazardous situations and many others. For such a process, one must know all methods for the identification, analysis and classification of risk, which is an integrated part of the complex system of risk.

In the Slovak Republic these directions are integrated into the Law "Safety and Health Protection at Work" (Law No. 330/1996).

8.1   Relation between Safety and Technical Risk

Technical safety is a very important ability of subject/machine, technology, operation to perform its functions without any hazardous situations for persons or environment. Hazardous situation is such circumstance in which a person is exposed to a hazard. Hazard is a potential source of harm. Harm is physical injury or damage to the health of people either directly or indirectly as a result of damage to property or to the environment.

All methods used for analysis of global safety must take into consideration not only technical aspects of machines safety and devices safety but also protection of health and environment.

Technical risk is defined as a combination of the probability of occurrence of harm and the severity of that harm/weight factor. Risk is generally defined as follows:

$$R = P \cdot D \tag{8.1}$$

where P is a probability of risk situation occurrence, D is a consequence after risk situation occurrence (damage in the whole man — machine — environment system).

Also it is possible to write:

$$R = P \cdot D \cdot W \cdot E \tag{8.2}$$

where W is a prohibition possibility of the observed phenomenon, E is an exposure time of the phenomenon.

It is the task of statistical methods to define probability P. But it is not so easy to define the consequence D. We can say that all damages are most frequently expressed in terms of economical effects.

For the computation of risk level, we must know the real values of the probability of occurrence of harm and values of severity of harm. If the final computed risk value is greater than the acceptable level, the necessary measures for risk reduction or elimination must be performed. This shortly described process is *risk management* and at the same time it is one subsystem of global management of work safety. Safety, in this meaning, is a freedom from unacceptable risk. Risk management or in other words control of risk is based on three points:

1. Identification of hazards
2. Risk analysis
3. Reduction or elimination of risk

8.2   Evaluation of Technical Risk

8.2.1   Example 1

The example considered here are big gearboxes with a power output of about 3.6 to 5.4 MW. We analyzed two basic conditions for failure in risk evaluation of parts of gearboxes:

- Current operation with all suitable parameters
- Operation with improper maintenance

8.2.1.1   Description of the Mill Train Set

The observed gearboxes (six gearboxes with constant speed and three with variable speed) are working in the hot mill train set and they transmit power from 3.6 to 5.4 MW. Gearboxes are one- or two-stage types with helical gears and with a synchronous motor (Figure 8.1). According to the last analysis, it is just the gearbox that is the part with the highest number of failures.

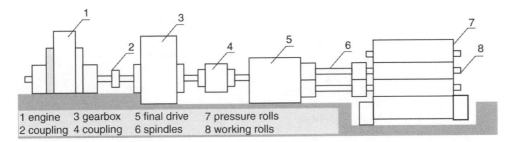

FIGURE 8.1 Gearboxes are one- or two-stage types with helical gears and with a synchronous motor.

8.2.1.2 Method of Solution

To evaluate the technical risk of the above-mentioned gearboxes, one needs to determine the probability P and the weight factor or consequence D.

There are three information sources available for the determination of P: statistics, expert system, individual estimate of an expert-worker.

We chose the way of an expert system, which is suitable for our purposes (Figure 8.2). Description of this flexible observation system is too complicated.

The software is based on the fuzzy groups and this method enables one to perform condition-based maintenance — CBM. It is a well-known fact that the CBM is the best maintenance method at the present time.

On the other hand, the consequence D is most frequently expressed in terms of money. Only the primary consequences are taken into consideration here, that is, price of spare parts and price of work. But the secondary consequences (losses due to lost time) are often much higher. There is a list of gearbox parts shown in Table 8.1. Two states of the gearboxes were simulated by means of the expert system:

1. Current operation with optimal parameters
2. Operation with insufficient maintenance

The result of the simulation is the final comparison of risk values for individual parts of the gearbox (Figure 8.3).

8.2.2 Example 2

8.2.2.1 Risk Evaluation in the Pipe Mill Hall

In the next part the example considered is the pipe mill hall. It is a big technological complex from the technical risk point of view. Products of this pipe mill are pipes with diameters from 500 to 1420 mm,

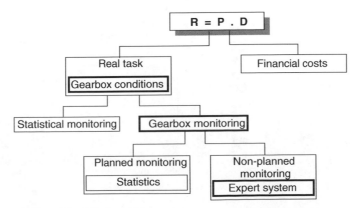

FIGURE 8.2 The complicated flexible observation system.

TABLE 8.1

Part of Gearbox	Sign
Pinion bearing	Rlp
Pinion	Rp
Bearing and pinion	Rl + p
Countershaft bearing	Rlpr
Countershaft gearing	Rpr
Countershaft shaft	Rhpr
Whole block	Rcb
Big gear	Rlvk
Output gear	Rvk
Bearing and output gear	Rlvk + vk

with thickness from 5.6 to 12.5 mm and with length from 9 to 12 m. The arrangement of pipe mill hallis shown in Figure 8.4.

8.2.2.2   Risk Level Determination

To obtain the values of technical risk levels in this technological complex we need to know the values of probability P and consequence D. Number of failures or the frequency of failures equals to the probability of risk state occurrence. The kind of accident equals to the range of health-injury, that is, equals to the consequence or to the weight factor. We can see all the important parts of pipe mill hall in the next table. For each part, the frequency of failures and the consequence of accident are determined. Combination of these two parameters is the risk level. Operation is with three levels of failure frequencies: *low, middle* and *high*. Three kinds of consequences are taken into consideration: *accident light, accident difficult* and *deadly accident*. Hence three risk levels are also obtained: low, middle and high (Table 8.2). It is possible to determine from this table the so-called weak points in the whole technological complex. From the analysis of maintenance costs and the technical risk it is evident that the most important subject of interest is the *ultrasonic welding 1*. Therefore, it is very useful to perform the technical risk analysis for all components of this machine for ultrasonic welding 1 (Figure 8.5).

The result is shown in the graph in Figure 8.5. In this graph the risk level becomes equal to the point evaluation: 1–5 is unacceptable risk; 6–9 is undesirable risk; 10–17 is risk acceptable with control and 18–20 is risk acceptable without control.

It is necessary to take into consideration that during the process of risk determination two aspects were combined together: technical factor — frequency of failures and human factor — kind of accident.

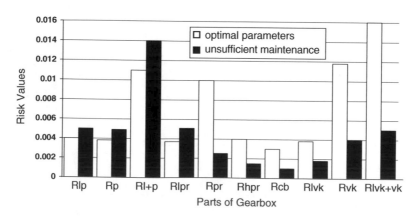

FIGURE 8.3   Comparison of risk values.

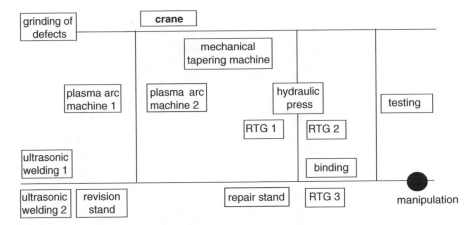

FIGURE 8.4 The arrangement of pipe mill hallis.

Therefore, we obtained the technical–human combined risk values or levels. These levels are only the primary and preliminary information for the designer. But this information is very important for all the successive steps of detailed technical risk analysis.

8.3 Complex Method of Risk Evaluation in the Workplace

This method of the risk evaluation directly in the workplace was developed in the section Machines Safety IVSS — working group for mechanical, physical and chemical risk [3]. Current praxis in small and medium companies needs such methods for risk evaluation, which are not expensive from both time and knowledge point of view. Such a method must be very simple and easy. Several components of the whole system can be neglected and the other components are very important and necessary. System in this meaning is one complex of components, which performs various functions. The principle of this method consists in point evaluation of each important factor of the final risk level. After the point evaluation of each component is computed, the final risk value is compared with the value of acceptable

TABLE 8.2 Risk Level Determination

Machine	Frequency of Failures	Consequence	Risk Level
Ultrasonic welding 1	Middle	Difficult accident	Middle
Ultrasonic welding 2	Middle	Difficult accident	Middle
Revision stand	Low	Deadly accident	Middle
Plasma arc tapering machine 1	Middle	Light accident	Low
Plasma arc tapering machine 2	Middle	Light accident	Low
Grinding of defects	Low	Light accident	Low
Crane	Middle	Deadly accident	High
Mechanical tapering machine	High	Light accident	Low
Hydraulic press	Low	Light accident	Low
RTG 1	Low	Light accident	Low
RTG 2	Low	Light accident	Low
Repair stand	High	Light accident	Low
RTG 3	Low	Light accident	Low
Testing	Middle	Light accident	Low
Binding	Middle	Light accident	Low
Manipulation	Low	Deadly accident	High

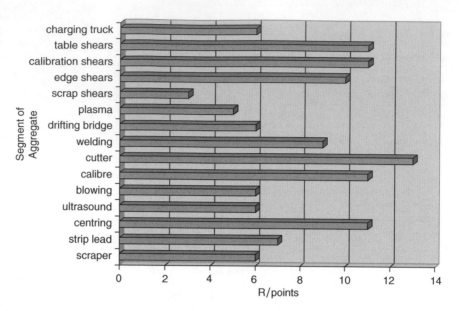

FIGURE 8.5 Technical risk analysis for all components of machine for ultrasonic welding 1.

risk. It is recommended to perform this evaluation of technical risk always by the same person/team, to reduce the human subjectivity factor. This method is based on the inductive principles.

8.3.1 Application and Purpose

The main application area for this method is human risks. This method integrates several basic components of human factor analysis and also enables the classification of risk level of the working area and the working subject. It is applicable in all stages of technical life of the observed system. This method is suitable for determination of immediate measures but not for global complex solutions.

8.3.2 Procedure for Risk Evaluation

This is a very suitable method for the evaluation of technical risk, which includes and takes into consideration not only technical but also ergonomic criteria. It is based on the point evaluation of all-important risk factors.

This method consists of the following steps:

- Computation of risk caused by technical device, that is, risk factor of machine M
- Computation of environment influences, that is, risk factor of environment U
- Computation of human factor, that is, risk factor of person P
- Computation of final risk value R
- Decision about the risk acceptability
- Suggestion and realization of necessary measures

Using the above-mentioned designation — M, U, P, R — the final risk value is given as:

$$R = M \cdot U - P \cdot (M/30) \tag{8.3}$$

where M-factor is determined as: $M = S \cdot Ex \cdot Wa \cdot Ve$

S	Value of possible damage, it is chosen from interval	$\langle 1; 10 \rangle$
	1 Accidents with light consequences	
	10 Accidents with very hard consequences, including death	
Ex	Exposure to hazard/frequency and duration/ from interval	$\langle 1; 2 \rangle$
	1 Temporary, seldom exposure	
	2 Frequent or permanent exposure	
Wa	Probability of accident combined with the correction factor of device	$\langle 0.5; 1.5 \rangle$
	0.5 Low level of probability, proper technical protection	
	1.5 High level of probability without any technical protection	
Ve	Possibility of prevention, from interval	$\langle 0.5; 1 \rangle$
	0.5 High prevention	
	1 Low prevention	

U-factor is determined as: U = Ua + Ub + Uc,

Ua	Arrangement of working area, from interval	$\langle 0.5; 1 \rangle$
	0.5 Suitable, well-arranged working place	
	1 Distressed surroundings	
Ub	Ergonomical conditions, from interval	$\langle 0.3; 0.6 \rangle$
	0.3 Pleasant working conditions	
	0.6 Hard conditions	
Uc	Other negative influences, from interval	$\langle 0.2; 0.4 \rangle$
	0.2 Almost none	
	0.4 Heavy loads carrying	

P-factor is determined as: P = Q + Psi + O

Q	Degree of personal qualification, from interval	$\langle 0; 10 \rangle$
	0 Person without qualification, inexperienced	
	10 Person fully qualified	
Psi	Personal psychological ability to perform given work, from interval	$\langle 0; 3 \rangle$
	0 Very bad ability	
	3 Very good ability	
O	Level of work organization, from interval	$\langle 0; 5 \rangle$
	0 Zero-level of organization	
	5 High level	

8.3.3 Example 3

Determination of acceptable risk level is the most important step. If it is not exactly determined, it will limit the value for the acceptance of technical risk. It depends on various influences, for example, state of science, working relations, legislation and social policy. Acceptable risk is such a risk, which can be accepted from the human and technical point of view.

Metallurgy is always a very important factor of economical development of each economy. Branches like chemistry, electrotechnics, power engineering and metallurgy create the base for the fast, successful and continuous economical growth. Therefore, it is necessary to analyze the technical risk in the metallurgical technologies.

There is one most serious problem in the day-to-day metallurgical operation from the technical risk point of view: this problem is the transport of liquid metal and their manipulation.

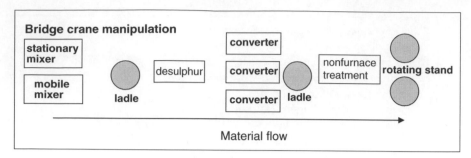

FIGURE 8.6   Representation of one chosen material flow in the steel plant.

Dangerous manipulations in the new and well-developed technologies are quite reduced or eliminated. But there are always lot of very dangerous manipulations with the liquid iron and liquid steel in many old steel plants.

The main source of the technical risk in this part of metallurgical technology is the improper manipulation with the liquid metal. There is a representation of one chosen material flow in the steel plant shown in Figure 8.6. The whole transport of material is performed by means of bridge crane and transport truck.

In connection with the hazard identification it is necessary to distinguish two criteria above all:

1. Exposition on risk
2. Cause of risk situation

After determination of all necessary values we obtain the final result: risk values for each element, which takes part in the working process in the chosen segment of metallurgical technology. Values and results are given in Table 8.3.

If we determine that the acceptable risk level is $R \leq 10$, we can discuss the results from various points of view.

This process of technical risk analysis may be very useful if computerized. The flow diagram developed for this purpose is shown in Figure 8.7.

8.4   Application of Risk Management in Maintenance

Risk management is a complex of activities, which includes requirements to control, to eliminate or to minimize risk possibility of death, injury, illness, damage of technical device or environmental impact. Legal requirements are also to demonstrate systematic implementation of safety requirements according to valid legislative directions.

There are also involved steps, which lead to identification of dangers. However, it is often a problem during finding of a solution to collect data about the state of a given system or device. Illustrated in Figure 8.8 is the method of risk theory application and FMEA (failure mode and effect analysis)

TABLE 9.3   Values and Results

	S	Ex	Wa	Ve	M	Ua	Ub	Uc	U	Q	φ	O	P	Risk
I. Smelter	10	1.5	1.3	1	19.5	1	0.6	0.4	2	6	1.5	2.5	10	32.5
II. Smelter	10	1.5	1.3	1	19.5	1	0.6	0.4	2	6	1.5	2.5	10	32.5
III. Smelter	10	1.5	1.5	1	22.5	0.7	0.6	0.4	1.7	5	1.5	2.5	9	31.5
Operator — liquid steel	10	1.5	1.2	1	18	0.8	0.6	0.3	1.7	8	1.5	3	13	23.1
Charging crane	10	1.1	0.7	1	7.7	0.7	0.5	0.3	1.5	8	2	3	13	8.213
Casting crane 220 t	7	1.1	0.7	1	5.39	0.7	0.5	0.3	1.5	8	2	3	13	5.749
Casting crane 240 t	7	1.1	0.7	1	5.39	0.7	0.5	0.3	1.5	8	2	3	13	5.749

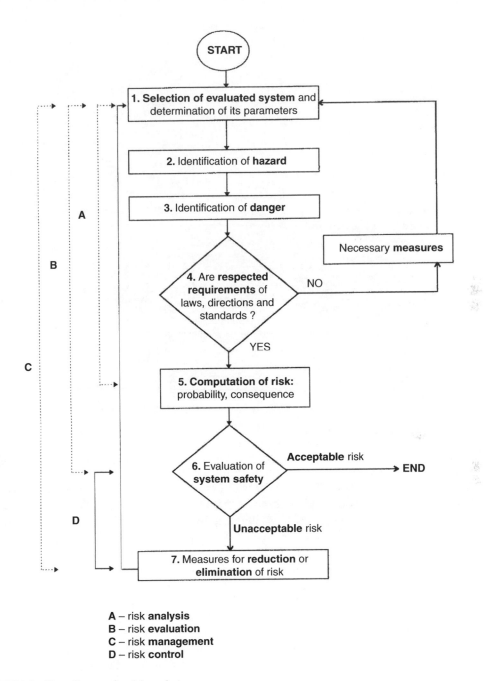

FIGURE 8.7 Flow diagram for risk analysis.

application for data collection and evaluation for crane 250 t. The first step was creation of crane function structures. On the base of these structures, possible failures were defined — causes and consequences, predominately from the point of view of safety, environment influence, time losses of individual system components (also taken into consideration were failures, which process quality).

By means of creation of such database it was enabled to analyze and evaluate risks. Applying Pareto-analysis, such failures (their causes) were specified, whose risk value was unacceptable (for safety).

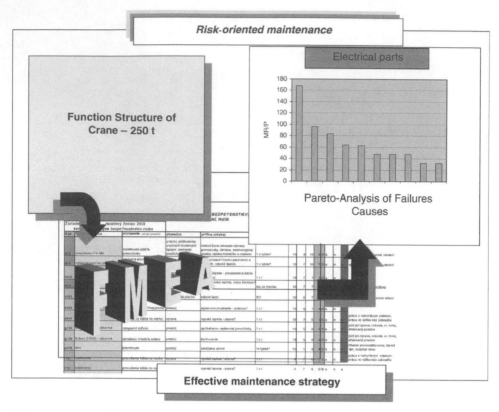

FIGURE 8.8 RCM method for risk evaluation — 250 t crane.

Suggestion of new measurements led to minimization of safety failure influences towards residual or acceptable risk levels. Of course, failures failures with impact such as great time losses of device were also taken into consideration. Evaluation of the present maintenance strategy and suggestion of new steps for increasing efficiency were the result of the analysis.

After the introduction of device into operation, it was evident that the value of operation costs was dependent on the maintenance quality. RCM (reliability centered maintenance) helps to master problems of incomplete data collection about machines and devices, which leads to incorrect decisions and the result is often not profit able, but is a unreasonable loss. The flexible maintenance strategy requires elimination of the most risk factors to apply new methods of personal management such as the basic principle of total production maintenance. It enables flexible response to "requirements" of own operation (devices).

All changes require initial investment. But predominately in the area of maintenance, it is important to reduce costs that lead to output profit because of time loss and higher production quality.

Bibliography

1. Direction 391/89/EU.
2. Direction 392/89/EU.
3. Sinay, J., Pačaiová, H., Kopas, M., Tomková, M., Application of technical risk theory for evaluation of gearbox damaging processes, in *IEA 97*, Helsinki, s.559–561.
4. Sinay, J., Oravec, M., Kopas, M., Juris, M., New development trends in the application of risk theory of man–machine–environment systems, *in Advances in Occupational Ergonomics and Safety*, IOS Press, Washington, 1997, s.573–575.

5. Sinay, J., Oravec, M., Pačaiová, H., Kopas, M., Risk evaluation and working conditions – ergonomical point of view, in *HAAMAHA 98*, Hong-Kong, 1998, s.547–550.
6. Sinay, J., Pačaiová, H., Kopas, M., Maintenance and risks during maintenance operation, in *Int. Conf. on Occupational Risk Prevention – ORP 2000*, Tennerife, 2000, CD.
7. Sinay, J., Kopas, M., Tomková, M., Safety and risk analysis of steel wire ropes, in *CAES 99*, Barcelona, 1999, CD-ROM.

9

Engineering Anthropometry

Karl H.E. Kroemer
*Virginia Polytechnic Institute
and State University*

9.1 Overview

People come in a great variety of sizes and the proportions of their body parts are not the same. Thus, devising tools, gear, and workstations to fit their bodies requires careful consideration; design for the statistical "average" will not do. Instead, for each body segment to be fitted, the designer must determine what dimension(s) is (are) critical: this may be a minimal or a maximal value, or a range. Often, a series of such decisions is necessary to accommodate body segments or the whole body by clothing, workspace, and equipment.

The following text describes the steps involved, provides statistical tools, and supplies anthropometric data.

9.2 Terminology

While all humans have heads and trunks, arms and legs, the body parts come in various sizes, assembled in different proportions. Anthropometry is the name of the science of measuring human bodies. Table 9.1 lists special terms often used in anthropometry. Together with the reference planes shown in Figure 9.1, they describe major aspects of anthropometric information useful to designers and engineers.

TABLE 9.1   Terms Used in Engineering Anthropometry

Anthropometry — measure of the human body. The term is derived from the Greek words "anthropos" human and "metron" (measure)

Breadth — straight line, point-to-point horizontal measurement running across the body or a body segment

Circumference — closed measurement following a body contour, hence this measurement is usually not circular

Curvature — point-to-point measurement following a body contour; this measurement is neither closed nor usually circular

Depth — straight line, point-to-point horizontal measurement running fore-aft the body

Distance — straight line, point-to-point measurement, usually between landmarks of the body

Height — straight line, point-to-point vertical measurement

Reach — point-to-point measurement following the long axis of an arm of leg

Coronal — in a plane that cuts the body into fore-aft (anterior – posterior) sections; same as frontal

Frontal — in a plane that cuts the body into fore-aft (anterior – posterior) sections; same as coronal

Medical — in a plane that cuts the body into left and right halves; same as mid-sagittal (also see below)

Mid-sagittal — in a plane that cuts the body into left and right halves; same as medical

Sagittal — in a plane parallel to the medical plane (occasionally used as medical)

Transverse — in a plane that cuts the body into upper and lower (superior and inferior) sections

Terms related to location

Anterior — in front of, toward the front of the body

Deep — away from, below, the surface; opposite of superficial

Distal[a] — away from the center of the body; opposite of proximal

Dorsal — toward the back or spine; opposite of ventral

Inferior — below, toward the bottom; opposite of superior

Medical — near or toward the middle (also see above)

Lateral — to the side, away from the middle

Posterior — behind, toward the back of the body; opposite of anterior

Proximal[a] — toward or near the center of the body; opposite of distal

Superficial — on or near the surface of the body; opposite of deep

Superior — above, toward the top; opposite of inferior

Ventral — toward the abdomen (occasionally used like anterior)

Note: For terms related to body reference planes, see Figure 9.1.

[a]Distal and proximal usually refer to limbs with the point of reference at the attachment to the next larger section of the body.

9.3   Designing to Fit the Body

A few common statistic terms can describe the results of anthropometric surveys because body data usually appear, statistically speaking, in a *normal (Gaussian)* distribution. If the sample size is large enough, the statistical descriptors *mean* (same as *average*), *standard deviation*, and *range* completely define a normal distribution of data (see below for more detail).

Misunderstanding and misuse has led to the false idea that one could "design for the average"; yet, the mean value is larger than half the data, and smaller than the other half. Consequently, the "average" does not describe the ranges of different statures, arm lengths, or hip breadths. Furthermore, one is unlikely ever to encounter a person who displays mean values in several, many, or all dimensions. The mythical "average person" is nothing but a statistical phantom.

STEPS IN DESIGN THAT FIT CLOTHING, TOOLS, WORKSTATIONS, AND EQUIPMENT TO THE BODY
    (*Adapted from Kroemer et al., 1997, 2001*)
    Step 1: Select those anthropometric measures that directly relate to defined design dimensions. Examples are: hand length related to handle size; shoulder and hip breadth related to escape-hatch diameter; head length and breadth related to helmet size; eye height related to the heights of windows and displays; knee height and hip breadth related to the leg room in a console.

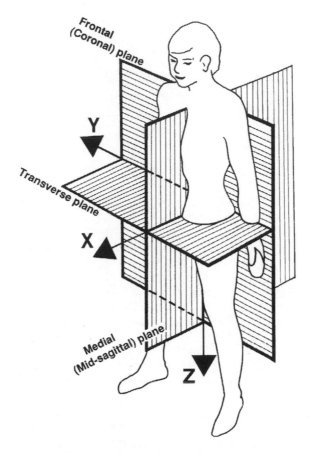

FIGURE 9.1   Reference planes used in conventional anthropometry.

Step 2: For each of these pairings, determine whether the design must fit only one given percentile (minimal or maximal) of the body dimension, or a range along that body dimension. Examples are: the escape hatch must be big enough to accommodate the very largest value of shoulder breadth and hip breadth, with clothing and equipment worn; the handle size of pliers is probably selected to fit the smallest hand; the leg room of a console must accommodate the tallest knee heights; the height of a seat should be adjustable to fit persons with short and with long lower legs. (The explanation on how to use and calculate percentiles follows.)

Step 3: Combine all selected design values in a careful drawing, computer model or mock-up to ascertain that they are compatible. For example, the required height of the leg-room clearance that sitting persons with long lower legs need, may be very close to or may even overlap with the height of the working surface determined from the elbow height.

Step 4: Determine whether one design will fit all users. If not, several sizes or adjustment features must be provided to fit all users. Examples are: one extra-large bed size fits all sleepers; gloves and shoes must come in different sizes; seat heights should be adjustable.

9.3.1   Using Percentiles

Most body dimensions are normally distributed. A plot of their individual measures falls inside the well-known bell curve, shown in Figure 9.2. Only a few persons are very short, or very tall, but many cluster around the center of the distribution, the mean. Figure 9.2 shows an approximate stature distribution of

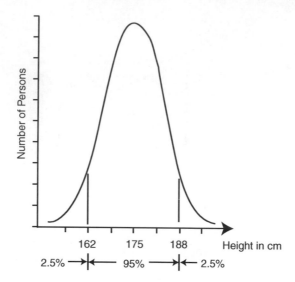

FIGURE 9.2 Frequency distribution of body height (stature) in Americans. About 95% of all males are between 162 and 188 cm tall, about 2.5% are shorter, another 2.5% taller. (From Kroemer et al., *Engineering Physiology. Bases of Human Factors/Ergonomics*, 3rd ed., Van Nostraud Reinhold — John Wiley & Sons, New York, NY 1997. With permission.)

male Americans; only 2.5% are shorter than approximately 1,620 mm, and another about 2.5% are taller than 1,880 mm. In other words, about 95% of all men are in the height range of 1,620 to 1,880 mm, because the 2.5th percentile value is at 1,620 mm and the 97.5th percentile is at 1,880 mm. The 50th percentile is at 1,750 mm.

(In a normal data distribution, mean [m], average, median, and mode coincide with the 50th percentile. The standard deviation [S] describes the peak or the flatness of the data set. More details on these statistical descriptors are discussed later in this chapter under "Estimation by Probability Statistics.")

There are two ways to determine the given percentile values. One is simply to take a distribution of data, such as that shown in Figure 9.2, and to determine from the graph (measure, count, or estimate) critical percentile values. This works whether the distribution is normal, skewed, binomial, or in any other form. Fortunately, most anthropometric data are normally distributed, which allows the second, even easier (and usually more exact) approach to calculate percentile values. This involves the standard deviation, S. If the distribution is flat (the data are widely scattered), the value of S is larger than when the data cluster is close to the mean.

To calculate the percentile value, one simply multiplies the standard deviation S by a factor k, selected from Table 9.2. Then one adds the product to the mean m:

$$p = m + k \times S \tag{9.1}$$

If the desired percentile is above the 50th percentile, the factor k has a positive sign and the product $k \times S$ is added to the mean; if the p-value is below average, k is negative and hence the product $k \times S$ is subtracted from the mean m.

Examples

1st percentile is at $m + k \times S$	with $k = -2.33$ (see Table 9.2; note the negative value of k)	
2nd percentile is at $m + k \times S$	with $k = -2.01$	
2.5th percentile is at $m + k \times S$	with $k = -1.96$	
5th percentile is at $m + k \times S$	with $k = -1.64$	
10th percentile is at $m + k \times S$	with $k = -1.28$	

TABLE 9.2 Percentile Values with their *k* factors

Below Mean		Above Mean	
Percentile	Factor *k*	Percentile	Factor *k*
0.001	−4.25	**50**	**0**
0.01	−3.72	51	0.03
0.1	−3.09	52	0.05
0.5	−2.58	53	0.08
1	−2.33	54	0.10
2	−2.05	**55**	**0.13**
2.5	−1.96	56	0.15
3	−1.88	57	0.18
4	−1.75	58	0.20
5	**−1.64**	59	0.23
6	−1.55	**60**	**0.25**
7	−1.48	61	0.28
8	−1.41	62	0.31
9	−1.34	63	0.33
10	**−1.28**	64	0.36
11	−1.23	**65**	**0.39**
12	−1.18	66	0.41
13	−1.13	67	0.44
14	−1.08	68	0.47
15	**−1.04**	69	0.50
16	−0.99	**70**	**0.52**
17	−0.95	71	0.55
18	−0.92	72	0.58
19	−0.88	73	0.61
20	**−0.84**	74	0.64
21	−0.81	**75**	**0.67**
22	−0.77	76	0.71
23	−0.74	77	0.74
24	−0.71	78	0.77
25	**−0.67**	79	0.81
26	−0.64	**80**	**0.84**
27	−0.61	81	0.88
28	−0.58	82	0.92
29	−0.55	83	0.95
30	**−0.52**	84	0.99
31	−0.50	**85**	**1.04**
32	−0.47	86	1.08
33	−0.44	87	1.13
34	−0.41	88	1.18
35	**−0.39**	89	1.23
36	−0.36	**90**	**1.28**
37	−0.33	91	1.34
38	−0.31	92	1.41
39	−0.28	93	1.48
40	**−0.25**	94	1.55
41	−0.23	**95**	**1.64**
42	−0.20	96	1.75
43	−0.18	97	1.88
44	−0.15	98	2.05
45	**−0.13**	99	2.33
46	−0.10	99.5	2.58
47	−0.08	99.9	3.09
48	−0.05	99.99	3.72
49	−0.03	99.999	4.26
50	**0**		

Note: Any percentile value *p* can be calculated from the mean *m* and the standard deviation SD (normal distribution assumed) by $p = m + kD$.

50th percentile is at m with $k = 0$
60th percentile is at $m + k \times S$ with $k = 1.28$
95th percentile is at $m + k \times S$ with $k = 1.64$

Percentiles serve the designer in several ways.

First, they help to establish the portion of a user population that will be included in (or excluded from) a specific design solution. For example, a certain product may need to fit everybody who has more than the 5th percentile, but less than the 60th percentile in hand reach. Thus, only the 5% having values smaller than the 5th percentile, and the 40% having values larger than the 60th percentile will not be fitted, while 55% (60 − 5%) of all the users will be accommodated.

Second, percentiles help to select subjects for fit tests. For example, if the product needs to be tested, persons having the 5th or 60th percentile values in the critical dimensions can be employed for tests.

Third, any body dimension, design value, or score of a subject can be exactly located. For example, a certain foot length can be described as a given percentile value of that dimension, or a certain seat height can be described as fitting a certain percentile value of lower leg length (e.g., popliteal height), or a test score can be described as falling at a certain percentile value.

Fourth, the use of percentiles helps in the selection of persons who can use a given product. For example, if a cockpit of an airplane is designed to fit the 5th to 95th percentiles, one can select cockpit crews whose body measures are at or between the 5th and 95th percentiles in the critical design dimensions.

9.3.2 To Determine a Single (Distinct) Percentile Point

(a) Select the desired percentile value
(b) Determine the associated k value from Table 9.2
(c) Calculate the p value from $p = m$ plus k times S (note that k and hence the product may be negative)

9.3.3 To Determine a Range

1(a) Select upper percentile p_{max}
1(b) Find related k value in Table 9.2
1(c) Calculate upper percentile value: $p_{max} = m + k \times S$
2(a) Select lower percentile p_{min}

(Note that the two percentile values need not be at the same distance from the 50th percentile, i.e., the range does not have to be "symmetrical to the mean.")

2(b) Find related k value in Table 9.2
2(c) Calculate lower percentile value $p_{min} = m + k \times S$
3. Determine range $R = p_{max} - p_{min}$

9.3.4 To Determine Tariffs

You may want to divide a distribution of body dimensions into certain sections, for instance, to establish clothing tariffs. A common example is the use of neck circumference to establish selected collar sizes for men's shirts. The first step is to establish the ranges (as seen earlier), which shall be covered by the tariff sections. The next step is to associate other body dimensions with the primary one, such as chest circumference, or sleeve length, with collar (neck) circumference. This can become a rather complex procedure, because the combinations of body dimensions (and their derived equipment dimensions) depend on correlations among these dimensions. More details on correlations follow but for detailed information see McConville's chapter in NASA (1978) and the book by Roebuck in 1995.

9.4 Body Postures

To standardize measurements, the subjects holds the body in defined static postures:

Standing: the instruction is "stand erect; heels together; (often, rears of heels, buttock, and shoulders touching a vertical wall); head erect; look straight ahead; arms hang straight down (or upper arms hang, forearms horizontal and extended forward); fingers extended."

Sitting: a plane, horizontal, hard surfaced plane adjusted in height so that the thighs are horizontal; "sit with lower legs vertical; feet flat on the floor; trunk and head erect; look straight ahead; arms hang straight down (or upper arms hang, forearms horizontal and extended forward); fingers extended."

Whether in the standing or sitting measurement postures, the *head* (including the neck) is held erect (or "upright"): *in the front view*, the pupils are aligned horizontally, and, *in the side view*, the ear-eye (EE) line is angled about 15° above the horizon (see Figure 9.3). The EE line runs through the whole ear and the outside juncture of the eyelids.

People do not stand or sit in these standard measurement postures naturally. Thus, the designer must convert dimensions taken on the immobile body in the standardized postures to reflect natural stances and motions. Of course, these vary widely at work or leisure. Therefore, it is impossible to give fixed "conversion factors" that apply to all conditions, and it remains the designer's task to select corrections that are appropriate for the given condition. For this, Table 9.3 offers guidelines.

9.5 Available Body Size Data

Traditional anthropometry follows well-established rules for selecting subjects, measuring the dimensions, and reporting the results (Roebuck, 1995). An outstanding example for the use of proper procedure is the survey of U.S. Army personnel conducted in 1988 (Gordon et al., 1989). Table 9.4 contains a selection from their data. Figure 9.4 illustrates the measurements.

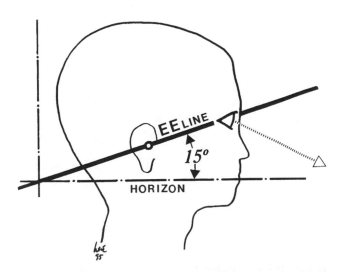

FIGURE 9.3 The EE line serves as a reference to describe head posture and the angle of the line-of-sight. The EE line is easier to use than the older "Frankfurt Plane," which is about 15° more declined. (From Kroemer, K.H.E., *Ergonomics in Design*, pp. 7–8, 1993; p. 40, 1994. With permission.)

TABLE 9.3 Guidelines for the Conversion of Standard Measuring Postures to Functional Stances and Motions

To consider	Do the following
Slumped standing or sitting	Deduct 5 to 10% from appropriate height measurements
Relaxed trunk	Add 5 to 10% to trunk circumferences and depths
Wearing shoes	Add approximately 25 mm to standing and sitting heights: more for "height heels"
Wearing light clothing	Add about 5% to appropriate dimensions
Wearing heavy clothing	Add 15% or more to appropriate dimensions (note that heavy clothing may severely reduce mobility)
Extended reaches	Add 10% or more for extensive motions of the trunk
Use of hand tools	Center of handle is at about 40% hand length, measured from the wrist
Forward bent head (and neck) posture	EE line close to horizontal
Comfortable seat height	Add or subtract upto 10% to or from standard seat height

Source: Adapted from Kroemer et al., *Engineering Physiology. Bases of Human Factors/Ergonomics*, 3rd ed., Van Nostrand Reinhold — John Wiley & Sons, New York, NY 1997. With permission.

In spite of the military sampling bias, the data in Table 9.4 are the best available information on the civilian North American adult population as Kroemer et al. argued in 1997. Their main reservation concerns body weight, which is obviously more variable in the civilian population than in the military and, given the current trend toward general obesity, on average is likely to be larger among the civilians. Head, foot, and hand sizes should not differ appreciably between soldiers and civilians.

Table 9.4 gives the mean (50th percentile), as well as the standard deviation, for 36 body segments and for body weight. This allows calculation of any percentile value of interest, as discussed earlier.

Table 9.5, Table 9.6, and Table 9.7 present, in similar fashion, descriptive data from Taiwan and Japan, Great Britain and France, Germany and Russia.

Reliable and comprehensive information on body sizes is at hand, unfortunately, for only the few populations listed previously. For other groups, some limited anthropometric information are available, which are compiled in Table 9.8. However, these data resulted from widely varying techniques of data gathering, often on only small samples. This makes it doubtful that the listed statistics for mean and standard deviation truly represent the underlying population. For most humans on earth, only gross estimates exist that are listed in Table 9.9. The data in Table 9.8 and Table 9.9 may serve as rough approximations of regional anthropometry, but they cannot replace exact measurements.

9.6   How to Get Missing Data

Often we design products for users about for whom we lack exact body size or strength information. This does not pose a great problem if the product is similar to items already in use, and if we know the users fairly well, such as our colleagues or at least people from our own country. In this case, we can probably take a few measurements on our acquaintances and make a "rough guestimate" of what the needed dimensions might be.

For exact and comprehensive information, however, more than such informal information gathering is necessary. Two avenues are open: one is to conduct a formal anthropometric survey, which is a major enterprise and best done by qualified anthropometrists (Bradtmiller, 2000; Robinette and Daanen, 2003; Roebuck, 1995). The other option is to deduce from existing data those that we need to know. For such approximations, several approaches are at hand as follows:

9.6.1   Estimation by "Ratio Scaling"

"Ratio scaling" (used by Pheasant in 1986 and 1996 to establish the British data partly shown in Table 9.6) is one technique to estimate data from known body dimensions (Pheasant, 1982). It relies on the

TABLE 9.4 Anthropometric Measured Data in mm of U.S. Army personnel, 17 to 51 yr of Age, According to Gordon et al. (1989) with Their Reference Numbers in Square Brackets. These Data also Describe the U.S. Civilian Population — see text

Dimension	Men		Women		Applications
	Mean	Std. Dev.	Mean	Std. Dev.	
1. Stature The vertical distance from the floor to the top of the head when standing [99]	1756	67	1629	64	A main measure for comparing population samples. Reference for the minimal height of overhead obstructions Add height for more clearance, hat, shoes stride
2. Eye height, standing The vertical distance from the floor to the outer corner of the right eye when standing [D19]	1634	66	1516	63	Origin the visual field of a standing person. Reference location of visual obstructions and of targets such as displays; consider slump and motion
3. Shoulder height (acromion), standing The vertical distance from the floor to the tip (acromion) of the shoulder when standing [2]	1443	62	1334	58	Starting point for arm length measurements; near the center of rotation of the upper arm. Reference point for hand reaches; consider slump and motion
4. Elbow height, standing The vertical distance from the floor to the lowest point of the right elbow when standing with the elbow flexed at 90° [D16]	1073	48	998	45	Reference for height and distance of the work area of the hand and location of controls and fixtures; consider slump and motion
5. Hip height (trochanter), standing The vertical distance from the floor to the trochanter landmark on the upper side of the right thigh when standing [107]	928	48	862	45	Traditional anthropometric measure, indicator of leg length and the height of the hip joint. Used for comparing population samples
6. Knuckle height, standing The vertical distance from the floor to the tip of the extended index finger of the right thigh when standing [107]	nd	nd	nd	nd	Reference for low locations of controls, handles, and handrails; consider slump and motion of the standing person
7. Fingertip height, standing The vertical distance from the floor to the tip of the extended index finger of the right hand when standing [D13]	653	40	610	36	Reference for lowest location of controls, handles, and handrails; consider slump and motion of the standing person
8. Sitting height The vertical distance from the sitting surface to the top of the head when sitting [93]	914	36	852	35	Reference for the minimal height of overhead obstructions. Add height for more clearance, hat, trunk motion of the seated person
9. Sitting eye height The vertical distance from the sitting surface to the outer corner of the right eye when sitting [49]	792	34	739	33	Origin of the visual field of a seated person. Reference point for the location of visual obstructions and of target such as displays; consider slump and motion
10. Sitting shoulder height (acromion) The vertical distance from the sitting surface to the tip (acromion) of the shoulder when sitting [3]	598	30	556	29	Starting point for arm length measurements; near the center of rotation of the upper arm. Reference for hand reaches; consider slump and motion

(Table continued)

TABLE 9.4 *Continued*

Dimension	Men		Women		Applications
	Mean	Std. Dev.	Mean	Std. Dev.	
11. Sitting elbow height The vertical distance from the sitting surface to the lowest point of the right elbow when sitting with the elbow flexed at 90° [48]	231	27	221	27	Reference for the height of an armrest, of the work area of the hand, and of keyboard and controls; consider slump and motion of the seated person
12. Sitting thigh height (clearance) The vertical distance from the sitting surface to the highest point on the top of the horizontal right thigh with the knee flexed at 90° [104]	168	13	160	12	Reference for the minimal clearance needed between see pan and the underside of a structure such as a table or desk; add clearance for clothing and motions
13. Sitting knee height The vertical distance from the floor to the top of the right kneecap when sitting with the kness flexed at 90° [73]	559	28	515	26	Traditional anthropometric measure for lower leg lengt. References for the minimal clearance needed below the underside of a structure, such as a table or desk; add height for shoe
14. Sitting popliteal height The vertical distance form the floor to the underside of the thigh directly behind the right knee; when setting with the kness flexed at 90° [86]	434	25	389	24	Reference for the height of a seat; add height for shoe
15. Shoulder-elbow length The vertical distance from the underside of the right elbow to the right acromion with the elbow flexed at 90° and upper arm hanging vertically [91]	369	18	336	17	Traditional anthropometric measure for comparing population samples
16. Elbow-fingertip length The distance from the back of the right elbow to the tip of the extended middle finger with the elbow flexed at 90° [54]	484	23	443	23	Traditional anthropometric measure. Reference for fingergrip reach when moving the forearm in the elbow
17. Overhead grip reach, sitting The vertical distance from the standing surface to the center of a cylindrical rod firmly held in the palm of the right hand [D45]	1310	55	1212	51	Reference for the height of overhead controls operated by a seated person. Consider ease of motion, reach, and finger/hand/arm strength
18. Overhead grip reach, standing The vertical distance from the standing surface to the center of a cylindrical rod firmly held in the palm of the right hand [D42]	2107	92	1947	87	Reference for the height of overhead controls operated by a standing person. Add shoe height. Consider ease of motion, reach, and finger/hand/arm/strength
19. Forward grip reach The horizontal distance from the back of the right shoulder blade to the center of a cylindrical rod firmly held in the palm of the right hand [D21]	751	37	686	34	Reference for forward reach distance. Consider ease of motion, reach, and finger/hand/arm strength
20. Arm length, vertical The vertical distance from the tip of the right middle finger to the right acromion with the arm hanging vertically [D3]	790	39	724	38	A traditional measure for comparing population sample Reference for location of controls very low on the side of the operator. Consider ease of motion, reach and finger/hand/arm/strength

					Comments
21. Downward grip reach The vertical distance from the right acromion to the center of a cylindrical rod firmly held in the palm of the right hand with the arm hanging vertically [D43]	666	33	700	33	Reference for the location of controls low on the side of the operator. Consider ease of motion, reach, and finger/hand/arm/strength
22. Chest depth The horizontal distance from the back to the right nipple [36]	243	22	239	21	A traditional measure for comparing population sample. Reference for the clearance between seat backrest and the location of obstructions in front of the trunk
23. Abdominal depth, sitting The horizontal distance from the back to the most protrucing point on the abdomen [1]	236	28	219	26	A traditional measure for comparing population sample. Reference for the clearance between seat backrest and the location of obstructions in front of the trunk
24. Buttock-knee depth, sitting The horizontal distance from the back of the buttocks to the most protrucing point on the right knee when setting with the kness flexed at 90° [26]	616	30	589	30	Reference for the clearance between seat backrest and the location of obstructions in front of the kness
25. Buttock-popliteal, sitting The horizontal distance from the back of the buttocks to back of the right knee just below the thigh when sitting with the knees flexed at 90° [27]	500	30	589	30	Reference for the depth of the seat
26. Shoulder breadth (biacromial) The distance between the right and left acromion [10]	397	18	363	17	A traditional measure for comparing population sample. Indicator of the distance between the centers of rotation of the two upper arms
27. Shoulder breadth (bideltoid) The maximal horizontal breath across the shoulders between the lateral margins of the right and left deltoid muscles [12]	492	26	433	23	Reference for the lateral clearance required at shoulder level. Add space for ease of motion and tool use
28. Hip breadth, sitting The maximal horizontal breath across the hips or thighs, whatever is greater, when sitting [66]	367	25	385	27	Reference for seat width. Add space for clothing and ease of motion
29. Span The distance between the tips of the middle fingers of the horizontally outstretched arms and hands [98]	1823	82	1672	81	A traditional measure for comparing population sample. Reference for sideway reach
30. Elbow span The distance between the tips of the elbows of the horizontally outstretched upper arms when the elbows are flexed so that the fingertips of the hands meet in front of the trunk	nd	nd	nd	nd	Reference for the lateral space needed at upper body level for ease of motion and tool use

(Table continued)

TABLE 9.4 *Continued*

Dimension	Men		Women		Applications
	Mean	Std. Dev.	Mean	Std. Dev.	
31. Head length The distance from the glabella (between the browridges) to the most rearward protrusion (the occiput) on the back, in the middle of the skull [62]	197	7	187	6	A traditional measure for comparing population sample. Reference for hear gear size
32. Head breadth The maximal horizontal breath of the head above the attachment of the ears [60]	152	5	144	5	A traditional measure for comparing population sample. Reference for head gear size
33. H and length The length of the right hand between the crease of the wrist and the tip of the middle finger with the hand [59]	194	10	181	10	A traditional measure for comparing population sample. Reference for hand tool and gear size. Consider manipulation, gloves tool use
34. Hand breadth The breath of the right hand, across the knuckles of the four fingers [57]	90	4	79	4	A traditional measure for comparing population sample. Reference for hand tool and gear size and for the opening through which a hand may fit. Consider manipulations, gloves, tool use
35. Foot length The maximal length of the right foot when standing [51]	270	13	244	12	A traditional measure for comparing population sample. Reference for Shoe and pedal size
36. Foot breadth The maximal breadth of the right foot, at right angle to the long axis of the foot, when standing[50]	101	5	90	5	A traditional measure for comparing population sample. Reference for shoe size, spacing of pedals
37. Weight (in kg) Nude body weight taken to the nearest tents of a kilogram	79	11	62	8	A traditional measure for comparing population sample. Reference for body size, clothing, strength, health, etc. Add weight for clothing and equipment worn on the body

Note: Measurements were taken in 1987/8 on U.S. Army soldiers, 1774 men and 2208 women. nd: no data available.

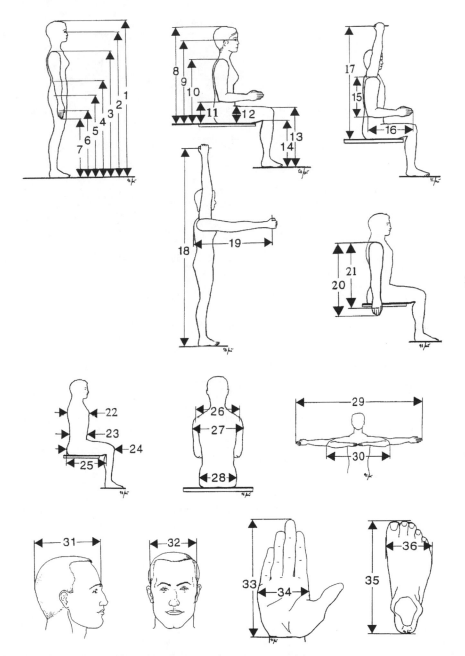

FIGURE 9.4 Illustrations of standard anthropometric measurements. (From Kroemer et al., *Engineering Physiology. Bases of Human Factors/Ergonomics*, 3rd ed., Van Nostraud Reinhold — John Wiley & Sons, New York, NY 1997. With permission.)

assumption that, though people vary greatly in size, they are likely to be relatively similar in proportions. This premise holds true for body components that are related to each other in size. For example, many body "lengths" highly correlate with each other; also, groups of body "breadths" are interrelated, as are "circumferences." However, it is not true that *all* body lengths (or breadths, or circumferences) are tightly associated with each other, and certainly many lengths are rather independent of breadths and depths and circumferences. Thus, one should better be very careful in deriving one set of data from another.

TABLE 9.5   Anthropometric Data (mm) of Adults in Taiwan and Japan

	Chine (Taiwan)				Japan			
	Men		Women		Men		Women	
Dimension	Mean	Std. Dev.	Mean	Std. Dev.	Mean	Std. Dev.	Mean	Std. Dev.
1. Stature	1705	59	1572	53	1688	55	1584	50
2. Eye height, standing	nda	nda	nda	nda	1577	53	nda	nda
3. Shoulder height (acromion), standing	1396	53	1285	50	1370	50	1279	48
4. Elbow height standing	1059	40	978	38	1035	39	967	37
5. Hip height (trochanter)	860	48	802	41	834	38	787	35
6. Knuckle height, standing	757	32	708	33	nda	nda	nda	nda
7. Fingertip height, standing	659	30	618	32	644	30	608	27
8. Sitting height	910	30	846	32	910	30	855	28
9. Sitting eye height	791	29	732	31	790	29	733	27
10. Sitting shoulder height (acromion)	602	26	561	27	591	26	551	24
11. Sitting elbow height	264	24	252	25	254	23	236	20
12. Sitting thigh height (clearance)	nda	nda	nda	nda	156	12	143	10
13. Sitting knee height	521	29	471	24	509	22	475	20
14. Sitting popliteal height	411	19	379	18	402	19	372	17
15. Shoulder-elbow length	338	19	309	18	337	18	315	15
16. Elbow-fingertip length	427	27	384	27	448	18	416	17
17. Overhead grip reach, sitting	1208	49	1105	44	nda	nda	nda	nda
18. Overhead grip reach, standing	2002	79	1831	67	nda	nda	nda	nda
19. Forward grip reach	710	36	651	33	nda	nda	nda	nda
20. Arm length, vertical	738	33	669	31	nda	nda	nda	nda
21. Downward grip reach	nda	nda	nda	nda	nda	nda	nda	nda
22. Chest depth	217	19	213	19	217	18	215	19
23. Abdominal depth, sitting	nda	nda	nda	nda	208	20	188	17
24. Buttock-knee depth, sitting	558	31	530	26	567	23	550	22
25. Buttock-popliteal depth, sitting	nda	nda	nda	nda	nda	nda	nda	nda
26. Shoulder breadth (biacromial)	369	28	324	25	395	17	367	14
27. Shoulder breadth (bideltoid)	460	23	406	24	nda	nda	nda	nda
28. Hip breadth, sitting	360	27	353	23	349	19	358	17
29. Span	1738	69	1571	62	1690	63	1579	62
30. Elbow span	894	45	801	39	nda	nda	nda	nda
31. Head length	197	7	187	6	190	7	177	6
32. Head breadth	167	8	161	9	161	6	151	6
33. Hand length	183	10	167	8	nda	nda	nda	nda
34. Hand breadth	86	5	75	4	85	4	75	3
35. Foot length	nda	nda	nda	nda	251	11	232	9
36. Foot breadth	nda	nda	nda	nda	104	5	96	4
37. Weight (in kg)	67.0	8.8	52.1	7.2	66	8	54	6

Note: Chinese (Taiwan) Adults, 25 to 34 yr of age, according to Wang et al (2000a,b). Data measured during 1996-2000 on civilians. Japanese Adults 18 to 35 yr of age, according to Kagimoto (1990). Data measured in 1988. nda: no data available.

A basic rule for ratio scaling is to use only pairings of data that are closely related in a statistical way to each other. Even if the coefficient of correlation is 0.7, for example, the variability of the predictor determines the variability of the derived information by only about 50% (the correlation coefficient 0.7 squared is 0.49). Ratio scaling should never be used if one has to assume that the sample to be scaled has body proportions different from those of the other set. For example, many Asian populations have proportionally shorter legs and longer trunks than the Europeans or North-Americans.

TABLE 9.6 Anthropometric Data (mm) of Adults in Great Britain and France

| | Britain | | | | France | | | |
| | Men | | Women | | Men | | Women | |
Dimension	Mean	Std. Dev.	Mean	Std. Dev.	Mean	Std. Dev.	Mean	Std. Dev.
1. Stature	1740	70	1610	62	1747	nda	1620	nda
2. Eye height, standing	1630	69	1505	61	nda	nda	nda	nda
3. Shoulder height (acromion), standing	1425	66	1310	58	1434	nda	1331	nda
4. Elbow height standing	1090	52	1005	46	nda	nda	nda	nda
5. Hip height (trochanter)	920	50	810	43	918	nda	844	nda
6. Knuckle height, standing	755	41	720	36	nda	nda	nda	nda
7. Fingertip height, standing	655	38	625	38	nda	nda	nda	nda
8. Sitting height	910	36	850	35	918	nda	867	nda
9. Sitting eye height	790	35	740	33	819	nda	772	nda
10. Sitting shoulder height (acromion)	595	32	555	31	nda	nda	nda	nda
11. Sitting elbow height	245	31	235	29	nda	nda	nda	nda
12. Sitting thigh height (clearance)	160	15	155	17	nda	nda	nda	nda
13. Sitting knee height	545	32	500	27	533	nda	487	nda
14. Sitting popliteal height	440	29	400	27	nda	nda	nda	nda
15. Shoulder-elbow length	365	20	330	17	365	nda	332	nda
16. Elbow-fingertip length	475	21	430	19	472	nda	427	nda
17. Overhead grip reach, sitting	1245	60	1150	53	nda	nda	nda	nda
18. Overhead grip reach, standing	2060	80	1905	71	nda	nda	nda	nda
19. Forward grip reach	780	34	705	31	nda	nda	nda	nda
20. Arm length, vertical	780	36	705	32	nda	nda	nda	nda
21. Downward grip reach	665	32	600	29	nda	nda	nda	nda
22. Chest depth	250	22	250	27	nda	nda	nda	nda
23. Abdominal depth, sitting	270	32	255	30	nda	nda	nda	nda
24. Buttock-knee depth, sitting	595	31	570	30	595	nda	569	nda
25. Buttock-popliteal depth, sitting	495	32	480	30	nda	nda	nda	nda
26. Shoulder breadth (biacromial)	400	20	355	18	382	nda	340	nda
27. Shoulder breadth (bideltoid)	465	28	395	24	457	nda	410	nda
28. Hip breadth, sitting	360	29	370	38	342	nda	346	nda
29. Span	1790	83	1605	71	nda	nda	nda	nda
30. Elbow span	945	47	850	43	nda	nda	nda	nda
31. Head length	195	8	180	7	155	nda	148	nda
32. Head breadth	155	6	145	6	142	nda	134	nda
33. Hand length	190	10	175	9	190	nda	173	nda
34. Hand breadth	85	5	75	4	86	nda	76	nda
35. Foot length	265	14	235	12	264	nda	237	nda
36. Foot breadth	95	6	90	6	101	nda	91	nda
37. Weight (in kg)	75	12	63	11	70	nda	58	nda

Note: British Adults, 19 to 35 yr of age, according to Pheasant (1986, 1996) who estimated the data in or before 1986. French Adults, 18 to 51 yr of age, according to Coblentz, A (1997), personal communication of 22 April 1997 regarding ERGODATA taken on 1015 French Soldiers (687 males, 328 females). Data measured in or before 1997. nda: no data available.

For sets of highly correlated data, one can establish an estimate E of a ratio scaling factor for a desired dimension (d_y) in the population sample Y if:

- The value of that dimension in sample X (d_x) is known
- The values of a reference dimension D in both samples X and Y (D_x and D_y) are known

TABLE 9.7   Anthropometric Data (mm) of Adults in Germany and Russia

| | Germany (East) | | | | Russia | | | |
| | Men | | Women | | Men | | Women | |
Dimension	Mean	Std. Dev.	Mean	Std. Dev.	Mean	Std. Dev.	Mean	Std. Dev.
1. Stature	1715	66	1608	59	1736	61	1606	53
2. Eye height, standing	1601	64	1504	57	1613	58	1404	52
3. Shoulder height (acromion), standing	1414	60	1319	53	1425	58	1303	48
4. Elbow height Standing	nda	nda	nda	nda	1070	49	974	39
5. Hip height (trochanter)	nda	nda	nda	nda	nda	nda	nda	nda
6. Knuckle height, standing	748	42	703	37	763	41	nda	nda
7. Fingertip height, standing	652	39	616	35	655	39	606	28
8. Sitting height	903	34	854	31	909	32	859	30
9. Sitting eye height	775	34	733	30	791	32	729	31
10. Sitting shoulder height (acromion)	601	31	562	29	nda	nda	nda	nda
11. Sitting elbow height	244	29	234	28	241	26	228	24
12. Sitting thigh height (clearance)	151	15	148	15	146	16	145	14
13. Sitting knee height	531	27	497	24	550	25	494	26
14. Sitting popliteal height	452	26	416	23	450	22	405	19
15. Shoulder-elbow length	nda	nda	nda	nda	nda	nda	nda	nda
16. Elbow-fingertip length	465	20	425	19	nda	nda	nda	nda
17. Overhead grip reach, sitting	nda	nda	nda	nda	1268	52	nda	nda
18. Overhead grip reach, standing	2121	89	1973	79	nda	nda	nda	nda
19. Forward grip reach	763	37	706	35	753	36	nda	nda
20. Arm length, vertical	762	35	703	33	nda	nda	nda	nda
21. Downward grip reach	nda	nda	nda	nda	nda	nda	nda	nda
22. Chest depth	nda	nda	nda	nda	256	28	252	29
23. Abdominal depth, sitting	nda	nda	nda	nda	nda	nda	nda	nda
24. Buttock-knee depth, sitting	603	27	585	27	601	31	565	28
25. Buttock-popliteal depth, sitting	486	25	479	26	nda	nda	nda	nda
26. Shoulder breadth (biacromial)	399	20	365	17	408	17	361	18
27. Shoulder breadth (bideltoid)	471	24	437	27	450	22	418	24
28. Hip breadth, sitting	369	22	401	35	348	24	379	23
29. Span	1760	75	1616	70	1787	74	1624	80
30. Elbow span	895	39	817	38	936	37	868	36
31. Head length	190	7	181	6	nda	nda	nda	nda
32. Head breadth	158	6	151	6	nda	nda	nda	nda
33. Hand length	189	9	174	9	184	8	165	8
34. Hand breadth	88	5	78	44	88	4	79	4
35. Foot length	264	13	241	12	270	12	241	12
36. Foot breadth	102	6	93	6	98	5	91	5
37. Weight (in kg)	nda	nda	nda	nda	72	10	61	8

Note: East German Adults, 18 to 59 yr of age, according to Fluegel et al. (1986). Data measured between 1979 (some 1967) and 1982. Ethnic Russian Factory Workers: 192 males between 18 and 29 yr old and 205 females between 20 and 29 yr old, all from Moscow, according to Strokina and Pakhomova (1999). Data measured during 1895 and 1986. nda: no data available.

In this case, the scaling factor E is calculated from

$$E = d_x/D_x \qquad (9.2)$$

Since the basic assumption is that the two samples are similar in proportion, the same scaling factor applies to both samples X and Y:

$$E = d_x/D_x = d_y/D_y \qquad (9.3)$$

TABLE 9.8 International Anthropometry: Adults, Measured; Averages (and Standard Deviations)

	Sample Size	Stature (mm)	Sitting Height (mm)	Knee Height Sitting (mm)	Weight (kg)
ALGERIA					
Females (Mebarki and Davies, 1990)	666	1576 (56)	795 (50)	487 (36)	61 (1)
AUSTRALIA					
Females, 77 (8) yr old	138	1521 (70	775 (40)	—	61 (13)
Males 76 (7) yr old (Kothiyal and Tettey, 2000)	33	1658 (79)	843 (56)	—	72 (11)
BRAZIL					
Males (Ferreira, 1988; cited by Al-Haboubi, 1991)	3076	1699 (67)	—	—	—
CHINA					
Females (Hong Kong)	69	1607 (54)	838 (45)	510 (31)	—
Females (Taiwan) (Huang and You, 1994)	300	1582 (49)	—	—	51 (7)
Males (Hong Kong) (Chan, So and Ng, 2000)	286	1737 (49)	884 (42)	552 (29)	—
Males (Canton) (Evans, 1990)	41	1720 (63)	—	—	60 (6)
EGYPT					
Females (Moustafa, Davies, Darwich and Ibraheem, 1987)	4960	1606 (72)	838 (43)	499 (25)	63 (4)
FRANCE					
Females	328	1620	867	487	58
Males (Coblentz, personal communication, 1997)	687	1747	918	533	70
GERMANY (East)					
Females	123	1608 (59)	854 (31)	497 (24)	—
Males (Fluegel, Greil and Sommer, 1986)	30	1715 (66)	903 (34)	531 (27)	—
INDIA					
Females	251	1523 (66)	775 (39)	483 (28)	50 (10)
Males (Chakarbarti, 1997)	710	1650 (70)	937 (45)	520 (30)	57 (11)
East-Ctr. India male farm workers (Victor, Nath and Verma, 2002)	300	1638 (56)	775 (40)	—	57 (7)
Central India male farm workers (Gite and Yadav, 1989)	39	1620 (50)	739 (26)	509(30)	49 (6)
South India male workers (Fernandez and Uppugonduri, 1992)	128	1607 (60)	791 (40)	542 (38)	57 (5)
East Indian male farm workers (Yadav, Tewari and Prasad, 1997)	134	1621 (58)	809 (22)	515 (29)	54 (67)
INDONESIA					
Females	468	1516 (54)	719 (34)	—	—
Males (Sama'mur, 1985; cited by Intaranont, 1991)	949	1613 (56)	872 (37)	—	—
IRAN					
Female students	74	1597 (58)	861 (36)	488 (23)	56 (10)
Male students (Mououdi, 1997)	105	1725 (58)	912 (26)	531 (24)	66 (10)
IRELAND					
Males (Gallwey and Fitzgibbon, 1991)	164	1731 (58)	911 (30)	508 (28)	74 (9)
ITALY					
Females*	753	1610 (64)	850 (34)	495 (30)	58 (8)
Females**	386	1611 (62)	—	—	58 (9)
Males*	913	1733 (71)	896 (36)	541 (30)	75 (10)
Males**	410	1736 (67)	—	—	73 (11)

*(Coniglio, Fubini, Masali et al., 1991)
**(Robinette, Blackwell, Daanen et al., 2002)

(*Table continued*)

TABLE 9.8 *Continued*

	Sample Size	Stature (mm)	Sitting Height (mm)	Knee Height Sitting (mm)	Weight (kg)
	JAMAICA				
Females	123	1648	832	—	61
Males (Lamey, Aghazadeh, and Nye, 1991)	30	1749	856	—	68
	JAPAN				
Females	240	1584 (50)	855 (28)	475 (20)	54 (6)
Males (Kajimito, 1990)	248	1688 (55)	910 (30)	509 (22)	66 (8)
	KOREA (South)				
Female workers (Fernandez, Malzahn, Eyada, and Kim, 1989)	101	1580 (57)	833 (32)	460 (22)	54 (7)
	MALASIA				
Females (Ong, Koh, Phoon, and Low, 1988)	32	1559 (66)	831 (39)	—	—
	NETHERLANDS				
Females, 20–30 yr old*	68	1686 (66)	—	—	67 (10)
Females (18–65 yr old)**	691	1679 (75)	—	—	73 (16)
Males (20–30 yr old)*	55	1848 (80)	—	—	81 (14)
Males (18–65 yr old)**	564	1813 (90)	—	—	84 (16)
*(Steenbekkers and Beijsterveldt, 1998)					
**(Robinette, Blackwell, Daanen et al., 2002)					
	RUSSIA				
Female herders (ethnic Asians)	246	1588 (55)	—	—	—
Female students (ethnic Russians)	207	1637 (57)	859 (32)	527 (24)	61 (8)
Female students (ethnic Usbeks)	164	1578 (49)	839 (28)	487 (25)	56 (7)
Female factory workers (ethnic Russians)	205	1606 (53)	849 (30)	494 (26)	61 (8)
Female factory workers (ethnic Usbeks)	301	1580 (54)	845 (31)	484 (26)	58 (9)
Male students (ethnic Russians)	166	1757 (56)	912 (32)	562 (25)	71 (9)
Male students (ethnic Usbeks)	150	1700 (52)	905 (29)	531 (23)	65 (7)
Male factory workers (ethnic Russians)	192	1736 (61)	909 (32)	550 (25)	72 (10)
Male factory workers (ethnic mix)	150	1700 (59)	896 (32)	541 (24)	68 (8)
Male farm mechanics (ethnic Asians)	520	1704 (58)	902 (31)	530 (25)	64 (8)
Male coal miners (ethnic Russians)	150	1801 (61)	978 (33)	572 (25)	—
Male construction workers (ethnic Russians) (Strokina and Pakhomova, 1999)	150	1707 (69)	—	—	—
	SAUDI ARABIA				
Males (Dairi, 1986; cited by Al-Haboubi, 1991)	1440	1675 (61)	—	—	—
	SINGAPORE				
Females (Ong, Koh, Poon and Low, 1988)	46	1598 (58)	855 (31)	—	—
Males (pilot trainees) (Singh, Peng, Lim, and Ong, 1995)	832	1685 (53)	894 (32)	—	—
	SRI LANKA				
Females	287	1523 (59)	774 (22)	—	—
Males (Abeysekera, 1985; cited by Intaranont, 1991)	435	1639 (63)	833 (27)	—	—
	SUDAN				
Males					
Villagers*	37	1687 (63)	—	—	57 (8)
City dwellers*	16	1704 (72)	—	—	62 (13)
City dwellers**	48	1668	—	—	51
Soldiers*	21	1735 (71)	—	—	71 (8)
Soldiers**	104	1728	—	—	60
*(ElKarim, Sukkar, Collins and Doré, 1981)					
**(Ballal et al., 1982; cited by Intaranont, 1991)					

(*Table continued*)

TABLE 9.8   *Continued*

	Sample Size	Stature (mm)	Sitting Height (mm)	Knee Height Sitting (mm)	Weight (kg)
THAILAND					
Females*	250	1512 (48)	—	—	—
Females**	711	1540 (50)	817 (27)	—	—
Males*	250	1607 (20)	—	—	—
Males**	1478	1654 (59)	872 (32)	—	—
*(Intaranont, 1991)					
**(NICE; cited by Intaranont, 1991)					
TURKEY					
Females					
Villagers	47	1567 (52)	792 (38)	486 (27)	69 (14)
City dwellers (Goenen, Kalinkara, and Oezgen, 1991)	53	1563 (55)	786 (05)	471 (05)	66 (13)
Male Soldiers (Kayis and Oezok, 1991)	5108	1702 (60)	888 (34)	513 (280	63 (7)
U.S.A.					
Midwest workers with shoes and light clothes					
Females	125	1637 (62)	—	—	65 (12)
Males (Marras and Kim, 1993)	384	1778 (73)	—	—	84 (16)
U.S. male miners (Kuenzi and Kennedy, 1993)	105	1803 (65)	—	—	89 (15)
U.S. Army soldiers					
	2208	1629 (64)	852 (35)	515 (26)	62 (8)
Females					
Males (Gordon, Churchill, Clauser et al., 1989)	1774	1756 (67)	914 (36)	559 (28)	76 (11)
North Americans (Canada and U.S.A.)					
Females (18–26 yr old)	1255	1640 (73)	—	—	69 (18)
Males (18–65 yr old)(Robinette, Blackwell, Daanen et al., 2002)	1120	1778 (79)	—	—	86 (18)
Vietnamese, living in the USA					
Females	30	1559 (61)	—	—	49
Males (Imrhan, Nguyen and Nguyen, 1993)	41	1646 (60)	—	—	59

Note: Last updated 15 January 2004. Contact the author for source references.

With $E = d_y/D_y$ (Equation [9.3a]) thus known, one can calculate

$$d_y = E \times D_y \tag{9.4}$$

in a stepwise fashion, as follows:

Step 1: If the shoulder height has to be estimated for sample Y, and the value in sample X is known then calculate using stature values as reference D, known in both samples: E = (shoulder height in sample X)/ (stature in sample X), see Equation 9.2.

Step 2: With E now known, the desired unknown dimension in population sample Y equals E times the reference parameter in sample Y. In this example, shoulder height in sample Y = $E \times$ (stature in sample Y), see Equation 9.4.

The technique of ratio scaling usually serves to estimate the mean of a required dimension and its standard deviation. The common parameter is often stature. Note, however, that while stature generally correlates well with other heights, it is not necessarily associated with depths, breadths, circumferences,

TABLE 9.9   Estimates of Average Anthropometric Data (mm) for 20 Regions of the Earth

	Stature		Sitting Height		Knee Height, Sitting	
	Females	Males	Females	Males	Females	Males
North America	1650	1790	880	930	500	550
LATIN AMERICA						
Indian Population	1480	1620	800	850	445	495
European and Negroid Population	1620	1750	860	930	480	540
EUROPE						
North	1690	1810	900	950	500	550
Central	1660	1770	880	940	500	550
East	1630	1750	870	910	510	550
Southeast	1620	1730	860	900	460	535
France	1630	1770	860	930	490	540
Iberia	1600	1710	850	890	480	520
AFRICA						
North	1610	1690	840	870	500	535
West	1530	1670	790	820	480	530
Southeast	1570	1680	820	860	495	540
Near east	1610	1710	850	890	490	520
INDIA						
North	1540	1670	820	870	490	530
South	1500	1620	800	820	470	510
ASIA						
North	1590	1690	850	900	475	515
Southeast	1530	1630	800	840	460	495
South China	1520	1660	790	840	460	505
Japan	1590	1720	860	920	395	515
AUSTRALIA						
European extraction	1670	1770	880	930	525	570

Source: Adapted from Juergens et al., *International Data on Anthropometry*, International Labour Office, Geneva, Switzerland, 1990.

or weight (as discussed earlier). Thus, ratio scaling requires careful consideration of the circumstances, especially taking into account statistical correlations.

9.6.2   Estimation by Regression Equation

Another way of estimating the relations among dimensions is through regression equation. Most regression equations are bivariate in nature, meaning two variables are involved, and that the two variables vary linearly with each other (explicit confirmation of that linear relationship is usually neglected). The general form is:

$$y = a + b \times x \tag{9.5}$$

where x is the known mean value and y the predicted mean. The constants a (the "intercept") and b (the "slope") must be determined (known) for the data set of interest. A recent example of this procedure is the estimation of body dimensions of American soldiers by Cheverud et al. (1990).

If mean value of y is predicted (for any value of x) using the regression Equation 9.5, the actual values of y are scattered about the mean in a normal (Gaussian) probability distribution. The standard error (SE) of the estimate depends on the correlation r between x and y, and on the standard deviation of y (S_y) according to

$$SE_y = S_y(1 - r^2)^{1/2} \tag{9.6}$$

Roebuck (1995) discussed the implications in some detail, including its extension to develop multi-variate regression equations, principal component analyses, and boundary description analyses.

9.6.3 Estimation by Probability Statistics

In most cases, we are unable to measure every person with respect to body size or strength. If we were able to do so, we would describe the parameters of that total population by the mean (average) and standard deviation, designated by the Greek letters mu (μ) and sigma (σ). (This is the terminology convention used in most statistics books.) In reality, we can measure only a subgroup (sample), and from its parameters we infer or estimate what the actual population would have yielded. Using roman letters to describe the sample data, we say that

$$m = (\Sigma x)/n \tag{9.7}$$

where m is the mean (average), x is the individual measurement, and n is the number of measured individuals. The distribution of the data is described by the equation

$$S = [\Sigma(x - m)^2/n]^{1/2} \tag{9.8}$$

with S called the standard deviation of the sample (often designated SD in statistics texts). If the sample size is small (conventionally, 30 or less) one usually makes an arbitrary correction by using $(n - 1)$ instead of n:

$$S = [\Sigma(x - m)^2/(n - 1)]^{1/2} \tag{9.8a}$$

The smaller the n in the sample, the larger is the standard error SE. The standard error SE of the mean is determined from

$$SE \text{ of the mean} = (S/n)^{1/2} \tag{9.9}$$

The standard error of the standard deviation is calculated from

$$SE \text{ of the standard deviation} = S/(2n)^{1/2} \tag{9.10}$$

As the number n increases, the mean m and the standard deviation S become more reliable estimates of the underlying general population, mu and sigma.

A useful way to describe the variability of a sample is to divide the standard deviation S by its mean m (and multiplying the result by 100). This yields the coefficient of variation

$$CV \text{ (in percent)} = 100 \ S/m \tag{9.11}$$

This expression is independent of the magnitude of the measurement and of the unit of measurement.

Groups of human dimensions show characteristic variabilities. Table 9.10 lists typical coefficients of variation.

TABLE 9.10   Typical Coefficients of Variation

Variable Measured	CV (%)
Body heights such as stature, sitting height, elbow height	3 to 5
Body breadth such as of shoulder or hip	5 to 9
Body depths such as chest or abdominal	6 to 9
Reaches	4 to 10
Joint mobility ranges	7 to 30
Muscular static strength	10 to 85
Total body weight	10 to 20

Source: Adapted from Kroemer et al., *Engineering Physiology: Bases of Human Factors/Ergonomics*, 3rd ed., van Nostrand Reinhold — John Wiley & Sons, New York, NY 1997. With permission.

Use of the CV is often of great help in assessing the credibility of data published in the literature; unusually large or small CV values indicate that, either, the distribution of the measured population is indeed different from the other populations, or that irregularities occurred in measuring, or in data treatment or reporting. The CV is also useful for estimating the standard deviation of an unknown data set.

9.7   Combining Anthropometric Data Sets

Occasionally one might may encounter rather involved considerations; one example is information about composite populations such as a% females and b% males. Another set of design issues concerns link lengths and mass properties of body segments including locations of mass centers and related dynamic body characteristics. Related information and solutions for such challenges can be found in publications by Annis and McConville (1996), Kroemer et al. (1997), or Roebuck (1995).

Often one must add or subtract anthropometric values, for example, total arm length is the sum of the upper and lower arm lengths. If two measures such as leg length and torso (with head) length have to be added, a new combined distribution should be generated in this case of stature. In doing so, one must take into account the covariation COV between the measures to be added (or subtracted). For example, usually (but not always) a taller torso is associated with a taller head and neck. The correlation coefficient r expresses the COV between two data sets, x and y, and their standard deviations, S_x and S_y:

$$\text{COV}(x,y) = r_{x,y} \times S_x \times S_y \tag{9.12}$$

After calculating the *sum* of the two mean values, m_{sum}, of the x and y distributions from

$$m_{\text{sum}} = m_x + m_y \tag{9.13}$$

one can determine the associated standard deviation of m_{sum} from

$$S_{\text{sum}} = [S_x + S_y + 2(r \times S_x \times S_y)]^{1/2} \tag{9.14}$$

The *difference* between two mean values, m_{diff}, is, of course,

$$m_{\text{diff}} = m_x - m_y \tag{9.15}$$

and the standard deviation of m_{diff} derives from

$$S_{\text{diff}} = [S_x + S_y - 2(r \times S_x \times S_y)]^{1/2} \tag{9.16}$$

Examples

Example 1: What is the 95p shoulder-to-fingertip length? Assume the mean lower arm (LA) link length (with the hand) to be 442.9 mm with a standard deviation of 23.4 mm. Also assume the mean upper arm (UA) link length to be 335.8 mm and its standard deviation 17.4 mm.

The multiplication factor of $k = 1.64$ (from Table 9.2) leads to the 95th percentile. But one cannot calculate the sum of the two 95p lengths because this would disregard their covariance; instead, one should calculate the sum of the mean values first, using Equation (9.13):

$$m = m_{\text{LA}} + m_{\text{UA}} = 442.9 + 335.8 = 778.7 \text{ mm.}$$

Inserting the standard deviations for LA and UA into Equation (9.14), and using an assumed coefficient of correlation of 0.4, one can obtain:

$$S = \{23.4^2 + 17.4^2 + 2 \times 0.4 \times 23.4 \times 18.4\}^{1/2} \text{ mm}$$
$$S = 34.6 \text{ mm}$$

Now, one can calculate the 95p total arm length (AL), using Equation (9.1):

$$AL_{95} = 778.7 \text{ mm} + 1.64 \times 34.6 \text{ mm} = 835.7 \text{ mm.}$$

Example 2: What is the average arm (acromion to wrist) length of an American pilot? it is known that for a standing pilot, the 90th percentile acromial (shoulder) height is 1532.0 mm and the wrist height is 905.6 mm; for the 10th percentile, the values are 1379.5 and 808.6 mm, respectively. The correlation between shoulder and wrist heights is estimated to be 0.3.

One must first calculate the mean acromion (A) and wrist (W) heights to be able to estimate the standard deviations. For the 90th percentile this is, as per Equation (9.7):

$$m_{\text{A90}} = (1532.0 + 1379.5) \text{ mm}/2 = 1455.75 \text{ mm.}$$

Using Equation (9.1), with $k = 1.28$ taken from Table 9.2, one obtains:

$$S_A = (1532.0 - 1455.75) \text{ mm}/1.28 = 59.6 \text{ mm.}$$

The same result is obtained on using the 10th percentile.

$$S_A = [(1455.75 - 1379.50) \text{ mm}/1.28] = 59.6 \text{ mm.}$$

Likewise,

$$m_{\text{W90}} = (905.6 + 808.6) \text{ mm}/2 = 857.1 \text{ mm.}$$
$$S_W = (905.6 - 857.1) \text{ mm}/1.28 = 37.9 \text{ mm.}$$
$$[\text{or: } S_W = (857.1 - 808.6) \text{ mm}/1.28 = 37.9 \text{ mm.}]$$

Now Equation (9.15) can be used to calculate the average arm length (acromion to wrist, AW):

$$m_{AW} = m_A - m_W = 1455.75 \text{ mm} - 857.1 \text{ mm} = 598.65 \text{ mm}.$$

According to Equation (9.16), the standard deviation of that arm length is

$$S_{AW} = (59.6^2 + 37.9^2 - 2 \times 0.3 \times 59.6 \times 37.9)^{1/2} \text{ mm} = 60.3 \text{ mm}.$$

Example 3: What is the mass of the head of a 75p Japanese female? The mass (weight) of the total body has a mean of 54.0 kg and a standard deviation of 6.0 kg — see Table 9.5. The estimated mass of the head is 6.2% of the total body mass (from data compiled by Kroemer et al. 1997). One can assume that the correlation coefficient between the masses of head and total body to be 1. (That assumption may be challenged, though.)

The mean head mass is, of course,

$$\text{Mean mass } (m_{head}) = 0.062 \times 54.0 \text{ kg} = 3.35 \text{ kg}.$$

The scaling factor E is, according to Equation (9.2):

$$E = S_{total\ body}/m_{total\ body} = 6.0 \text{ kg}/54.0 \text{ kg} = 0.11.$$

The standard deviation of the head mass is calculated from Equation (9.4):

$$S_{head} = 3.35 \text{ kg} \times 0.11 = 0.37 \text{ kg}.$$

According to Equation (9.1), the mass of a 75th percentile head is (with $k = 0.67$ taken from Table 9.2)

$$\text{mass}_{head75} = 3.35 \text{ kg} + 0.67 \times 0.37 \text{ kg} = 3.6 \text{ kg}.$$

9.8　The "Normative" Adult versus "Real Persons"

Without formally stating so, even without consciously being aware of it, we commonly design for a group of "regular" people who are in the 25- to 45-yr age bracket; who are of "normal" anthropometry, that is, have body dimensions such as stature, hand reach, or weight close to the 50th percentile; who are "healthy" in their metabolic, circulatory, and respiratory subsystems; whose nervous control, sensory capabilities, and intelligence are all "near average," and who are able and willing to perform "normally." Thus, by default or for reasons, the normative stereotype of many human factor engineers is the "regular" adult woman or man. (This prototype seems to be a close relative of the mysterious "average person" who appears in speeches of politicians, in newspapers, and in some flawed design models.)

This normative adult has become our prototype to which we compare other subgroups, such as children, temporarily or permanently impaired persons, women during their pregnancy, or aging people (Kroemer, 2006). Yet, most individuals and whole population subgroups deviate in size, strength, or other performance capabilities from the adult norm. In reality, hardly a person exists who is average in most or all respects; consequently products or processes "designed for the average" fit nobody well (Kroemer et al., 2001). To achieve ease, efficiency, and safety, it is mandatory to consider the ranges of, the variations in, and the combinations of physiologic and psychologic traits; the foregoing discussions showed ways to accommodate anthropometric variability.

9.9 Posture versus Motions

A similarly simplistic approach incorporates the idea of designing for a presumed body "posture." In part, this false concept may have been provoked by the standardized erect posture, sitting or standing, utilized in measuring body size. Unfortunately, the "erect" (or "upright") posture has often served as a design model, probably because it is easy to visualize and make into a drafting template. Orthopedists of the late nineteenth century promoted the upright idol as better than the hunched and bent bodies they saw in workshops and offices (Kroemer and Kroemer, 2001). Yet, over extended time, the human is unable to maintain *any* given posture, upright or otherwise. Standing still, immobile sitting, even lying stiffly, quickly become uncomfortable and then, with time; physically impossible to maintain; if enforced by injury or sickness, circulatory and metabolic functions become impaired, bed sores appear. The human body is made to move.

Our bodies allow for movement especially in the arms with shoulder and elbow joints providing extensive angular freedom. The strong legs are able to propel the body on the ground with major motions occurring in the knee and hip joints. Movements of the trunk occur mostly in flexion and extension

TABLE 9.11 Comparison of Mobility Data (in degrees) for Females and Males

Joint	Movement	5th percentile		50th percentile		95th percentile		Difference[a]	
		Female	Male	Female	Male	Female	Male	Female	Male
Neck	Ventral flexion	34.0	25.0	51.5	43.0	69.0	60.0	+8.5	
	Dorsal flexion	47.5	38.0	70.5	56.5	93.5	74.0	+14.0	
	Right rotation	67.0	56.0	81.0	74.0	95.0	85.0	+7.0	
	Left rotation	64.0	67.5	77.0	77.0	90.0	85.0	NS	
Shoulder	Flexion	169.5	161.0	184.5	178.0	199.5	193.5	+6.5	
	Extension	47.0	41.5	66.0	57.5	85.0	76.0	+8.5	
	Adduction	37.5	36.0	52.5	50.5	67.5	63.0	NS	
	Abduction	106.0	106.0	122.5	123.5	139.0	140.0	NS	
	Medical rotation	94.0	68.5	110.5	95.0	127.0	114.0	+15.5	
	Lateral rotation	19.5	16.0	37.0	31.5	54.5	46.0	+5.5	
Elbow-forearm	Flexion	135.5	122.51	148.0	138.0	160.5	150.0	+10.0	
	Supination	87.0	86.0	108.5	107.5	130.0	135.0	NS	
	Pronation	63.0	42.5	81.0	65.0	99.0	86.5	+16.0	
Wrist	Extension	56.5	47.0	72.0	62.0	87.5	76.0	+10.0	
	Flexion	53.5	50.5	71.5	67.5	89.5	85.0	+4.0	
	Adduction	16.5	14.0	26.5	22.0	36.5	30.0	+4.5	
	Abduction	19.0	22.0	28.0	30.5	37.0	40.0	−2.5	
Hip	Flexion	103.0	95.0	125.0	109.5	147.0	130.0	+15.5	
	Adduction	27.0	15.5	38.5	26.0	50.0	39.0	+12.5	
	Abduction	47.0	38.0	66.0	59.0	85.0	81.0	+7.0	
	Medical rotation (prone)	30.5	30.5	44.5	46.0	58.5	62.5	NS	
	Lateral rotation (prone)	29.0	21.5	45.5	33.0	62.0	46.0	+12.5	
	Medical rotation (sitting)	20.5	18.0	32.0	28.0	43.5	43.0	+4.0	
	Lateral rotation (sitting)	20.5	18.0	33.0	26.5	45.5	37.0	+6.5	
Knee	Flexion (standing)	99.5	87.0	113.5	103.5	127.5	122.0	+10.0	
	Flexion (prone)	116.0	99.5	130.0	117.0	144.0	130.0	+143.0	
	Medical rotation	18.5	14.5	31.5	23.0	44.5	35.0	+8.5	
	Lateral rotation	28.5	21.0	43.5	33.5	58.5	48.0	+10.0	
Ankle	Flexion	13.0	18.0	23.0	29.0	33.0	34.0	−6.0	
	Extension	30.5	21.0	41.0	35.5	51.5	51.5	+5.5	
	Adduction	13.0	15.0	23.5	25.0	34.0	38.0	NS	
	Abduction	11.5	11.0	24.0	19.0	36.5	30.0	+5.0	

[a] Listed are only differences at the 50th percentile, and if significant ($\alpha < 0.5$).

Source: Adapted from Houy, D.A., *Proceedings of the Human Factors Society's 27th Annual meeting*, Human Factors Society, Santa Monica, CA, pp. 374–378, 1983; Staff, K.R., *A Comparison of Range of Joint Mobility in College Females and Males*, Unpublished Master's Thesis, Texas A&M University, College Station, TX, 1983. With permission.

at the lower back. However, these bending and unbending lumbar motions (in the medial plane) are rather limited, and often lead to overexertions, especially if combined with lateral twisting of the torso; low back pain has been reported throughout the history of humans (Snook, 2000). Wrist problems have been associated with excessive motion requirements since the early 1700s (Kroemer, 2001). Our head and neck have limited mobility in bending and twisting. Our thumbs and fingers, as well as the eyes, have limited but finely controlled motion capability.

Ranges of motion (also called mobility or flexibility) depend much on age, health, fitness, training, and skill. Diverse measuring instructions and techniques have been applied to dissimilar groups of people to assess their mobility; hence, there is much diversity in reported ranges of motion. However, one set of mobility measurements has been taken on groups of 100 females and of 100 males each by the same researchers using the same techniques (Houy, 1983; Staff, 1983). Excerpts from these data appear in Table 9.11. Note that the differences in mobility between males and females are generally negligible.

Designing to fit motion ranges, instead of fixed postures, is not difficult. The articulations in the human body have varying degrees of freedom for movement. These are shown in Figure 9.5 for major body joints and the motion ranges are listed in Table 9.11. These maximal ranges were measured on students of physical education; hence, many people will have slightly less mobility than shown. "Convenient" mobility is within the range of maximal values shown in Table 9.11, but not always in the middle of the ranges. Occasionally, convenient motions are near the limits of mobility. Habits and skill as well as strength requirements may make different ranges preferred.

Design for motions starts by establishing the actual movement ranges. Convenient motions may cluster around the mean of mobility in a body joint or may be close to the limits of flexibility. For example, a person walking about, or standing, has the knees most of the time nearly extended, that is, the knee angle — in the sagittal view — is close to its extreme value of about 180°. The sagittal hip angle (between trunk and thigh) is also in the neighborhood of 180°. Of course, when sitting, both hip and knee angles change and cluster to about 90° when sitting — see Table 9.12. While sitting or moving about, the trunk is normally nearly erect, as are the neck and head. In most work situations, the upper arm hangs from the shoulder while the elbow angle tends to be near 90°; but the wrist is best held straight i.e., at about 180°.

DESIGN FOR MOTION IN THESE STEPS

Step 1: Determine which major body joints are involved.

Step 2: Adjust body dimensions reported for standardized postures (see Tables 10.4 through 10.9) to accommodate the real work conditions. Use Table 9.3 for guidance.

Step 3: Select appropriate motion ranges in the body joints. The range is between two positions such as knee angles ranging between 60 and 105° for a comfortably sitting person; or it is a motion envelope, such as circumscribed by combined hand-and-arm movement, or by the clearance envelope under (through, within, beyond) which body parts must fit. Use Table 9.12 for guidance.

Basic work space design faults should be avoided. These include:

1. Avoid twisted body positions, especially of the trunk and neck; often caused by bad locations of work objects, controls, and displays
2. Avoid forward bending of trunk, neck, and head; frequently provoked by improperly positioned controls and visual targets, and work surfaces that are too low.
3. Avoid postures that must be maintained for a long time, especially at the extreme limits of the range of motion. This is particularly important for the wrist, neck, and the back.
4. Avoid holding the arms raised. This results commonly from locating controls or objects too high, higher than the elbow when the upper arm hangs down. The upper limit for regular manipulation tasks is about chest height.

Preferred work areas of the hands and feet are in front of the body, within curved envelopes that reflect the mobility of the forearm in the elbow joint, or of the total arm in the shoulder joint, of the lower leg in the knee joint, and of the total leg in the hip joint. Thus, these envelopes may be described as (partial) spheres around the presumed locations of the body joints. However, the preferred ranges within the possible motion zones differ depending on whether the main requirements are strength, or speed, or accuracy, or vision — as discussed in some detail, by Kroemer et al. (1997, 2001). Thus, there is not one reach envelope, but different preferred envelopes, depending on the task.

For each job situation, the ergonomic designer determines the dominant requirements of the task, for example, whether the operator

- Works while sitting or walking (standing)
- Performs wide-ranging or specialized work

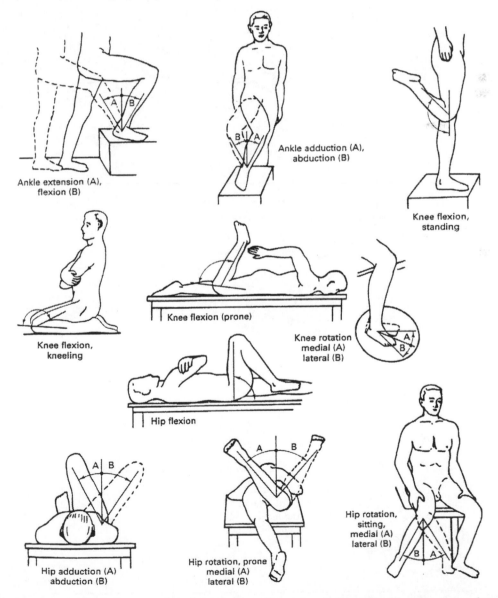

FIGURE 9.5 Maximal displacements in body joints. (From Kroemer et al., *Engineering Physiology. Bases of Human Factors/Ergonomics*, 3rd ed., Van Nostraud Reinhold — John Wiley & Sons, New York, NY 1997. With permission.)

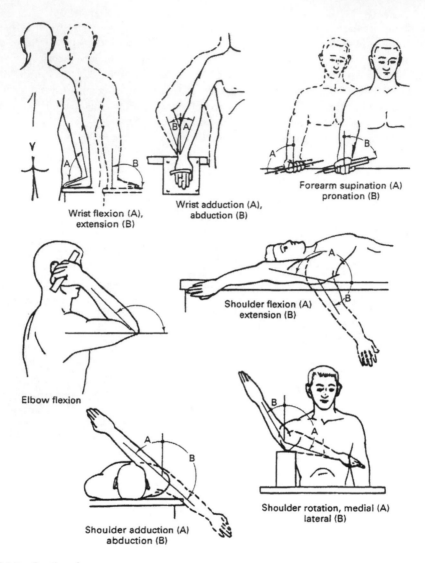

FIGURE 9.5   Continued

TABLE 9.12   Mobility Ranges at Work

Angels at	Walking about, Standing	Sitting
Knee	Near extreme stretch: 180° or slightly less	Mostly mid-range: about 90°
Hip (lateral view)	Near extreme stretches: about 180°	Mostly mid-range: about 90°
Shoulder	Mostly mid-range: upper arm often hanging down	
Elbow	Mostly mid-range: about 90°	
Wrist	Mostly mid-range: about straight	
Neck/Head	Mostly mid-range: about straight	
Back	Near extreme stretch: about erect	

Source: Adapted from Juergens et al., *International Data on Anthropometry*, International Labour Office, Geneva, Switzerland, 1990.

TABLE 9.13   Guidelines for Workspace Design

Consider	In order to
Human strength—	Facilitate exertion of strength (work, power) by object location and orientation
Human speed—	Place items so that they can be reached and manipulated quickly
Human effort—	Arrange work so that it can be performed with least effort
Human accuracy—	Select and position objects so that they can be manipulated and seen with ease
Importance—	Place the most important items in the most accessible locations
Frequency of use—	Place the most frequently used items in the most accessible locations
Function—	Group items with similar functions together
Sequence of use—	Lay out items, which are commonly used in sequence

- Must exert large or small forces
- Executes fast and gross or slow and exact motions
- Needs high or low visual control

Such circumstances affect the selection of the specific work envelope.

General criteria for workspace layout relate to human strength, speed, effort, accuracy, importance, frequency, function, and sequence of use as listed in Table 9.13. Achieving the task while assuring safety for the humans, avoiding overuse and unnecessary effort, and assuring ease and efficiency, are the primary design goals.

9.10   Summary

It is inexcusable to design tasks, tools, or workstations for the phantom of "the average person" in a static position. No such persons exist and design for the average fits nobody well. Instead, ranges of body sizes, of motions, and of strengths (see elsewhere in this book) establish the design criteria. This is easy to do for the designer and engineer who starts with proper anthropometric information and applies it ergonomically, that is, with "ease and efficiency" as the guiding principles.

References

Annis, J.F. and McConville, J.T., Anthropometry, in *Occupational Ergonomics*, Bhattacharya, A. and McGlothlin, J.D. Eds., Dekker, New York, NY, 1996, chap. 1, pp. 1–46.

Bradtmiller, B., *Anthropometry for Persons with Disabilities: Needs in the Twenty-First Century*. Paper presented at RESNA 2000 Annual Conference and Research Symposium, 28–30 June 2000, Orlando, FL. Rehabilitation Engineering and Assistive Technology Society of North America, Arlington, VA, 2000.

Cheverud, J., Gordon, C.C., Walker, R.A., Jacquish, C., Kohn, L., Moore, A., and Yamashita, N., *1988 Anthropometric Survey of U.S. Army Personnel: Correlation Coefficients and Regression Equations* (NATICK TR 90/032-6), U.S. Army Research, Development and Engineering Center, Natick, MA, 1990.

Fluegel, F., Greil, H., and Sommer, K., *Anthropologischer Atlas*, Tribuene, Berlin, Germany, 1986.

Gordon, C.C., Churchill, T., Clauser, C.E., Bradtmiller, B., McConville, J.T., Tebbetts, I., and Walker, R.A., *1988 Anthropometric Survey of U.S. Army Personnel: Summary Statistics Interim Report* (Natick-TR-89/027), U.S. Army Natick Research, Development and Engineering Center, Natick, MA, 1989.

Houy, D.A., Range of joint motion in college males, in *Proceedings of the Human Factors Society's 27th Annual Meeting*, Human Factors Society, Santa Monica, CA, pp. 374–378, 1983.

Juergens, H.W., Aune, I.A., and Pieper, U., *International Data on Anthropometry*. (Occupational Safety and Health Series No. 65), International Labour Office, Geneva, Switzerland, 1990.

Kagimoto, Y., ed., *Anthropometry of JASDF Personnel and its Applications for Human Engineering*, Aeromedical Laboratory, Air Development and Test Wing JASDF, Tokyo, Japan, 1990.

Kroemer, K.H.E., Locating the computer screen: how high, how far? *Ergon. Design*, October issue, 7–8, 1993; and January issue, p. 40, 1994.

Kroemer, K.H.E., Keyboards and keying: an annotated bibliography of the Literature from 1878 to 1999. *Int. J. Universal Access Inf. Soc. UAIS*, 1/2, pp. 99–160, 2001.

Kroemer, K.H.E., *Extra-ordinary Ergonomics: Designing for Small and Big Persons, The Disabled and Elderly, Expectant Mothers and Children*, CRC Press, Boca Raton, FL, 2006.

Kroemer, K.H.E. and Kroemer, A.D., *Office Ergonomics*, Taylor & Francis, London, UK, 2001.

Kroemer, K.H.E., Kroemer, H.B., and Kroemer-Elbert, K.E., *Ergonomics: How to Design for Ease and Efficiency*, 1st ed. Prentice Hall, Englewood Cliffs, NJ, 1994.

Kroemer, K.H.E., Kroemer, H.B., and Kroemer-Elbert, K.E., *Ergonomics: How to Design for Ease and Efficiency*, 2nd ed., Prentice Hall, Upper Saddle Rivert, NJ, 2001.

Kroemer, K.H.E., Kroemer, H.J., and Kroemer-Elbert, K.E., *Engineering Physiology. Bases of Human Factors/Ergonomics*, 3rd ed., Van Nostrand Reinhold — John Wiley & Sons, New York, NY, 1997.

McConville, J.T., Anthropometry in sizing and design, in *Anthropometric Sourcebook*, Vol. 1, NASA/Webb, eds., NASA Reference Publication 1024, LBJ Space Center, Houston, TX, chap. 8, 1978, pp. 8.1–8.23.

Pheasant, S.T., A technique for estimating anthropometric data from the parameters of the distribution of stature, *Ergonomics*, 25, pp. 981–992, 1982.

Pheasant, S., *Bodyspace: Anthropometry, Ergonomics and Design*, Taylor & Francis, London, UK., 1986.

Pheasant, S., *Bodyspace: Anthropometry, Ergonomics and the Design of Work*, 2nd ed., Taylor & Francis, London, UK, 1996.

Robinette, K.M., Blackwell, S., Daanen, H., Boehmer, M, Fleming, S., Brill, T., Hoeferlin, D., and Burnsides, D., *Civilian American and European Surface Anthropometry Resource (CAESAR) Final Report, Volume 1: Summary* (AFRL-HE-WP-TR-2002-0169), United States Air Force Research Laboratory, Wright-Patterson AFB, OH, 2002.

Robinette, K.M. and Daanen, H., Lessons learned from CAESAR: a 3-D anthropometric survey, in *Proceedings of the 15th Triennial Congress of the International Ergonomics Association*, August 24–29, Paper No. 00730, 2003.

Roebuck, J.A., *Anthropometric Methods*, Human Factors and Ergonomics Society, Santa Monica, CA, 1995.

Snook, S.H., Back risk factors: an overview, in Violante, F., Armstrong, T., and Kilbom, A., eds., Muskuloskeletal Disorders of the Upper Limb and Back, Taylor & Francis, London, UK, 2000, chap. 11, pp. 129–148.

Staff, K.R., *A Comparison of Range of Joint Mobility in College Females and Males*, Unpublished Master's Thesis, Texas A&M University, College Station, TX, 1983.

Strokina, A.N. and Pakhomova, B.A., *Anthropo-Ergonomic Atlas* (in Russian), Moscow State University Publishing House, Moscow, Russia, 1999.

Wang, M.J.J., Wang, E.M.Y., and Lin, Y.C., *Anthropometric Data Book of the Chinese People in Taiwan*, The Ergonomics Society of Taiwan, Hsinchu, ROC, 2002.

10

Human Strength Evaluation

Karl H.E. Kroemer

Virginia Polytechnic Institute and State Univesity

10.1 Overview

Skeletal muscles are able to move body segments with respect to each other against internal and external resistances. Muscle components can shorten dynamically, statically retain their length, or be lengthened. Various methods and techniques are available for assessing muscular strength. The engineering application of data on available body strength requires the determination of whether minimal or maximal exertions, static or dynamic, are critical. Data on body strength are available for the design of tools, equipment, and work tasks.

10.2 Background and Terminology

Muscular efforts have been of special interest to physiological science; therefore, there is a long tradition of philosophical and experimental approaches and use of terminology. Of particular importance are the "Newton's Three Laws:"

1. The first explaining that unbalanced force acting on a mass changes its motion condition
2. The second stating that force f equals mass m multiplied by acceleration a: $f = m \times a$
3. The third making it clear that force exertion requires the presence of an equally large counter force

Physiology books published until the middle of the twentieth century often divided muscle activities into either dynamic efforts lasting for minutes or hours, with work, energy, and endurance typical topics; or short bursts of contractile exertion. Much research on muscle effort concerned the "isometric" condition in which muscle length (and hence body segment position) did not change. Consequently, much information on muscle strength applies to such static exertion. All other muscle activities were typically called "anisometric," often even falsely labeled "isotonic" or "kinetic," meant to cover all the many possible dynamic muscle uses. Chaffin et al. (1999), Marras et al. (1993), Kroemer (1999), and Kumar (2004) discuss proper terminology: Table 10.1 lists and explains terms that correctly describe muscular events.

For the engineer, skeletal muscles are of primary interest since they pull on segments of the human body and generate energy for exertion to outside objects. Skeletal muscles connect two body links across their joint, as shown in Figure 10.1; in some cases muscles cross even two joints. Muscles are usually arranged in "functional pairs" so that contracting muscles counteract each other. The muscle, or the group of synergistic muscles, pulling in the intended direction is the agonist (also called protagonist) and the opposite is the antagonist. Cocontraction, the simultaneous activation of paired opposing muscles, serves to control speed and strength exertion.

There are several hundred skeletal muscles in the human body, known by their Latin names. Connective tissue (fascia) enwraps them; it imbeds nerves and blood vessels. At the ends of the muscle, the tissues combine to form tendons, which usually attach to bones.

Thousands of individual muscle fibers run, more or less parallel to the length of the muscle. Seen via a microscope, skeletal muscle fibers appear striped (striated) crosswise: thin and thick, light and dark bands run across the fiber in regular patterns, which repeat along the length of the fiber. One such thick dark stripe appears to penetrate the fiber like a membrane or disc: this is the so-called z-disk (from the German *zwischen*, between). The distance between two adjacent z-lines defines the sarcomere. Its length at rest is approximately 250 Å ($1 \text{ Å} = 10^{-10}$ m), meaning that there are about 40,000 sarcomeres in series within 1 mm of muscle fiber length.

Within each muscle fiber, thread-like myofibrils (from the Greek *mys*, muscle) lie in parallel by the hundreds or thousands. Each of these, in turn, consists of bundles of myofilaments. A network of tubular channels, sacs and cisterns, which connect with a larger tubular system in the z-disks, fill the spaces between the filaments. All of this is part of the networks of blood vessels and nerves in the fascia. This is the "plumbing and control" system of the muscle, the sarcoplasmic reticulum. It provides fluid transport between the cells inside and outside the muscle and carries chemical and electrical messages.

Two of the myofibrils, myosin and actin, have the ability to slide along each other; this is the source of muscular contraction. Small projections, called cross-bridges, protrude from the myosin filaments towards neighboring actins. The actin filaments are twisted double-stranded protein molecules, wrapped in a double helix around the myosin molecules. This is the "contracting microstructure" of the muscle.

The only *active* action that a muscle can take is contraction; external forces that stretch the muscle bring about passive elongation. According the "sliding filament theory," the heads of adjacent actin rods moving toward each other cause contraction. This pulls the z-disks closer together: sarcomeres in series (and those parallel) shorten, and as a result, the whole muscle shortens. After a contraction, the muscle returns to its resting length, primarily through a recoiling of its shortened filaments, fibrils, fibers, and other connective tissues. Force external to the muscle can stretch the muscle beyond its resting length, either by gravity or other force acting from outside the body, or by the action of antagonistic muscle. (Refer to texts by Asimov, 1963; Astrand and Rodahl, 1977, 1986; Chaffin et al., 1999; Enoka, 2002; Kroemer et al., 1997, 2001; Schneck 1990, 1992; and Winter, 1990, among others, for more information.)

10.3   Relation Between Muscle Length and Tension

Stimulation from the central nervous system (CNS) causes the muscle to contract to its smallest possible length, which is about half the resting length with no external load. In this condition, the actin proteins

TABLE 10.1 Glossary of Muscle Terms

Activation of muscle — See contraction

Cocontraction — Simultaneous contraction of two or more muscles

Concentric (muscle effort) — Shortening of a muscle against a resistance

Contraction[a] — Literally, "pulling together" the z lines delineating the length of a sarcomere, caused by the sliding action of actin and myosin filaments. Contraction develops muscle tension only if the shoutening is resisted.

Distal — Away from the center of the body

Dynamics — A subdivision of mechanics that deals with forces and bodies in motion

Eccentric (muscle effort) — Lengthening of a resisting muscle by external force

Fiber — See muscle

Fibril — See muscle fibers

Filament — See muscle fibers

Force — As per Netwton's Third Law, the product of mass and acceleration. The proper metric unit is the Newton, with $1\,N = 1\,kg\,m\,s^{-2}$. on earth 1 kg applies a (weight) force of 9.81 N (1 lb exerts 4.44 N) to its support. Muscular force is defined as muscle tension multiplied by the transmitting cross-sectional area

Free dynamic — In this context, an experimental condition in which neither displacement and its time derivatives, nor force are controlled as independent variables.

Iso — A prefix meaning constant or the same

Isoacceleration — A condition in which the acceleration is kept constant

Isoforce — A condition in which the muscular force (tension) is constant, that is, isokinetic. This term is equivalent to isotonic

Isoinertial — A condition in which muscle moves at a constant mass

Isojerk — A condition in which the time derivative of acceleration, jerk, is kept constant

Isokinetic — A condition in which muscle tension (force) is kept constant, See isoforce and isotonic; compare with isokinematic

Isokinematic — A condition in which the velocity of muscle shortening (or lengthening) is constant. (Depending on the given biomechanical conditions, this may or may not coincide with a constant angular speed of a body segment about its articulation.) Compare with isokinetic

Isometric — A condition in which length of the muscle remains constant

Isotonic — A condition in which muscle tension(force) is kept constant — see isoforce (in the past, this term was occasionally falsely applied to any condition other than isometric.)

Kinematics — A subdivision of dynamics that deals with the motions of bodies, but not the causing forces

Kinetics — A subdivision of dynamics that deals with forces applied to masses

Mechanical advantage — In this context, the lever arm (moment arm, leverage) at which a muscle pulls about a bony articulation.

Mechanics — The branch of physics that deals with forces applied to bodies and their ensuring motions

Moment — The product of force and the length of the (perpendicular) lever arm at which it acts. Mechanically equivalent to torque

Motor unit — All muscle filaments under the control of one efferent nerve axon

Muscle — A bundle of fibers, able to contract or be lengthened. In this context striated (skeletal) muscle that moves body segments about each other under voluntary control

Muscle contraction — The result of contractions of motor units distributed through a muscle so that the muscle is shortened. See contraction

Muscle fibers — Elements of muscle containing fibrils, which consist of filaments

Muscle fibrils — Elements of muscle fibers containing filaments

Muscle filaments — Muscle fibril elements, especially actin and myosin (polymerized protein molecules) capable of sliding along each other, thus shortening the muscle and, if doing so against resistance, generating tension

Muscle force — the product of tension within a muscle multiplied by the transmitting muscle cross-section

Muscle strength — the ability of a muscle to generate and transmit tension in the direction of its fibers. See also body strength

Muscle tension — the pull within a muscle expressed as force divided by the transmitting cross-section

Myo — A prefix referring to muscle (Greek mys. muscle)

Mys — A prefix referring to muscle (Greek mys, muscle)

Proximal — Towards the center of the body

Repetition — Performing the same activity more than once (one repetition indicates two exertions)

Statics — A subdivision of mechanics that deals with bodies at rest

Strength — See body strength and muscle strength

Tension — Force divided by the cross-sectional area through which it is transmitted.

Torque — The product of force and the length of the (perpendicular) level arm at which it acts. Mechanically equivalent to moment.

[a]Note that during an isometric "contraction" no change in sarcomere length occurs and that in an eccentric "contraction" the sarcomere is actually lengthened. To avoid such contradiction in terms, it is often better to use the terms activation, effort, or exertion.

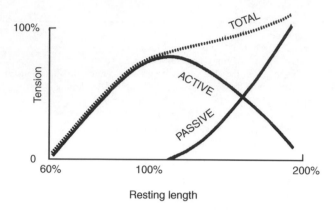

FIGURE 10.1   The biceps muscle reduces the elbow angle as agonist, counteracted by the triceps muscle and antagonist. Note the simplification of the actual conditions in modeling: in addition to the biceps, two other muscles (radialis and brachioradialis) also contribute to flexion about the elbow joint.

curl completely around the myosin rods. This is the shortest possible length of the sarcomeres, below which the muscle cannot develop any additional active contraction force.

Near resting length, the cross-bridges between the actin and myosin rods are in an optimal position to generate contact for contractile pull force. If the muscle is elongated further, the actin and myosin fibrils slide along each other, reducing the cross-bridge overlap between the protein rods. At about 150% of resting length, very little overlap remains that the proteins cannot actively resist the elongation anymore and only passive stretch resistance remains. Thus, the curve of *active contractile tension* developed within a muscle in isometric twitch is near zero at approximately half resting length, then rises to a maximum at about resting length, and finally falls back to a minimum at about 150% resting length. Figure 10.2 indicates this schematically while Figure 10.5 in Marras' chapter on "Occupational Biomechanics" in this book shows more detail.

The muscle also *passively resists stretch* like a rubber band. This passive resistance becomes stronger as the muscle is pulled more beyond its resting length and is strongest near the point of muscle or tendon (attachment) breakage. Thus, above the resting length, the tension in the muscle is the summation of active and passive strains. The summation effect explains why we stretch muscles for a strong exertion, like in bringing the arm behind the shoulder before throwing a rock. This "preloading" tenses the muscle for a strong exertion – see Figure 10.2 and Marras' Figure 10.5.

In engineering terms, the muscles exhibit "viscoelastic" qualities. They are viscous in that their behavior depends both on the amount by which they are deformed, and on the rate of deformation. They are elastic in that they return to the original length and shape after deformation. These behaviors help to explain why the tension that can be developed isometrically ("statically," especially in a state of eccentric stretch) is the highest possible one, while in active shortening (in a "dynamic" concentric movement) muscle tension is decidedly lower (Schneck, 1990, 1992).

10.4   Muscle Endurance and Fatigue

Sufficient supply of arterial blood to the muscle and its unimpeded flow through the capillary bed into the venules and veins are crucial, because they determine the ability of the metabolic contractile and contractile processes of the muscle to continue. Blood brings needed energy carriers and oxygen and it removes metabolic by-products, particularly lactic acid and potassium as well as heat, carbon dioxide, and water liberated during metabolism (Rodgers, 1997).

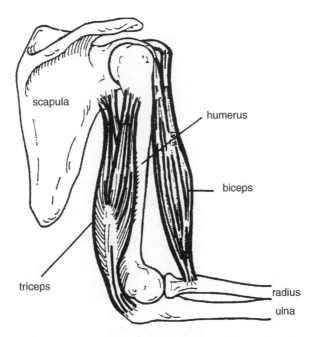

FIGURE 10.2 Active, passive and, total tension within a muscle at different lengths. (With permission by the publisher from Kroemer et al., *Engineering Physiology. Bases of Human Factors/Ergonomics*, 3rd ed., Van Nostrand Reinhold — John Wiley & Sons, New York, 1997. All rights reserved.)

Paradoxically, by contracting the muscle can compress its own fine blood vessels that permeate the muscle tissues. A strongly contracting muscle generates pressure within itself, as can be felt by touching a tightened biceps or calf muscle. By such strong pressure, the muscle reduces the cross-section of the vessels, diminishing or even shutting off its own blood circulation. The interruption of blood flow through a muscle stops the metabolic energy conversion and allows metabolic by-products to be accumulated in the muscle tissues; this quickly leads to muscle fatigue, forcing relaxation. Experiencing such fatigue is painful as it occurs slowly when the muscle is not contracting to its maximum, when working overhead with raised arms, for example, while fastening a screw in the ceiling of a room. Muscle fatigue in the shoulder muscles makes it impossible to keep one's arms raised even after a minute or so, even though nerve impulses from the CNS still arrive at the neuromuscular junctions, and the resulting action potentials continue to spread over the muscle fibers.

The operational definition of muscle fatigue is "a state of reduced physical ability that can be restored by rest." Figure 10.3 shows the relation between static exertion and muscle endurance schematically: a maximal exertion can be maintained for just a few seconds; 50% of the tension is present for about 1 min; but less than 20% can be applied for long endurance periods.

10.5 Muscle Tension and Its Internal Transmission to the Point of Application

In common use, the term "strength" may refer to any or all of the

- Tension *within* a muscle
- *Internal transmission* via body links across joints
- *External application* of force or torque by a body segment to an outside object

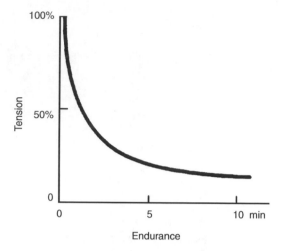

FIGURE 10.3   Muscle exertion and endurance. (With permission by the publisher from Kroemer et al., *Engineering Physiology. Bases of Human Factors/Ergonomics*, 3rd ed., Van Nostrand Reinhold — John Wiley & Sons, 1997. New York, All rights reserved.)

10.5.1   Muscle Tension — "Muscle Strength"

Within the muscle, all filament pulls combine to a resultant tension in the muscle. Its magnitude depends mostly on the involved number of parallel muscle fibers, that is, on the cross-sectional thickness of the muscles. Maximal tensions reported on human skeletal muscles are within the range of 16 to 61 N/cm^2. Enoka (1988) used 30 N/cm^2 as a typical value, calling it "specific (human muscle) tension." If the muscle cross-section area is known (such as from cadaver measurements or MRI scans), one can calculate a resultant muscle force.

From the muscle, one tendon extends proximally to the origin and, in the opposite direction, another tendon extends outward to the insertion — see Figure 10.4. Like cables, tendons transmit the muscle tension, usually to the surface of a bone but in some cases, such as in the fingers, to the strong connective tissue. The distal tendon may be quite long, for example, the tendons that reach from the muscles in the forearm to the digits of the hand can be 20 cm in length.

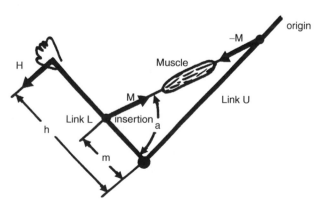

FIGURE 10.4   The muscle-tendon unit exerts pull forces *M* to links L and U at origin and insertion. Note the simplification in modeling the actual conditions at the elbow: in reality, the biceps muscle has its origin proximal to the shoulder joint: two other muscles (radialis and brachioradialis) contribute to flexion about the elbow joint. Torque $T = m \cdot M \cdot \sin \alpha$ must be transmitted across the wrist joint to generate hand force *H*. (With permission by the publisher from Kroemer et al., *Engineering Physiology. Bases of Human Factors/Ergonomics*, 3rd ed., Van Nostrand Reinhold — John Wiley & Sons, New York, 1997. All rights reserved.)

10.5.2   Internal Transmission

The tension inside a muscle-tendon unit tries to pull together the origin and insert the muscle-tendon unit across a joint. This generates torque about the articulation with the long bones, which are the lever arms at which muscle pulls. The torque developed depends, hence, not only on the strength of the muscle, but also on the effective lever arm, and on the pull angle. Figure 10.4 depicts these conditions as a simplified scheme of the lower arm link L attached to the upper arm link U, articulated in the elbow joint. The muscle has its origin at the proximate end of U. Its insertion point is at the distance m from the joint on the link L. It generates a muscle force M, pulling at an angle α. The torque T generated by M depends on the lever arm m and the pull angle α according to

$$T = m \times M \times \sin\alpha \tag{10.1}$$

This torque T then counteracts an external force H, acting perpendicularly at its lever arm h, according to

$$T = m \times M \times \sin\alpha = h \times H. \tag{10.2}$$

10.5.3   External Application — "Body (Segment) Strength"

The final output of the biomechanical system is the torque or force (H in Figure 10.4) available at the hand, foot, or other body segments for exertion to a resisting object. This object is usually outside the body, but the resistance may be from an antagonistic muscle. The "body (segment) strength" available for application to an outside object is of primary importance to the engineer, designer, and manager.

The model depicted in Figure 10.4 shows that the amount of force (H) available at the body interface with an external object depends on muscle force (M); lever arms (m and h); pull angle (α) which, in turn, depends on the angle between the two links. If all of these are known, the body segment force can be calculated from Equation (10.2)

$$H = (m/h) \times M \times \sin\alpha \tag{10.3}$$

DEFINITIONS

To help distinguish among muscle tension, its internal transmission, and the final exertion to an outside object, it is useful to define terms as follows:

Muscle strength is the maximal tension (or force) that muscle can develop voluntarily between its origin and insertion.

The best word to refer to this is "muscle tension" (in N/mm^2 or N/cm^2) but the term strength (force in N) is commonly used. If the variables m, h, α and H in Equation (10.2) are known, one can solve for muscle force as follows:

$$M = (h/m) \times H / \sin\alpha \tag{10.4}$$

Internal transmission is the manner in which muscle tension transfers in the form of torque inside the body along links and across joint(s) to the point of application to a resisting object.

If several link-joint systems in series constitute the internal path of torque (in N m or N cm) transmission, each transfers the arriving torque by the existent ratio of lever arms (m and h in the example shown earlier) until resistance is met, which is usually the point where the body interfaces with an outside object. This transfer of torques is more complicated under dynamic conditions than in the static case because of changes in muscle functions with motion, changes in lever arms and pull angles, and because of the effects of accelerations and decelerations of masses.

> *Body segment strength* is the force or torque that a body segment can apply to an object that is external to the body.
>
> Hand, shoulder, back, and foot are the body segments with which we commonly apply our "strength" as force (in N) or torque (in N m or N cm) to an object.

The quality and quantity of the force or torque transmitted to an outside object depends on many biomechanical and physical conditions, which are as follows:

- Body segment employed, for example, hand or foot
- Type of body object attachment, for example, by a touch or grasp
- Coupling type, for example, by friction or interlocking
- Direction of force/torque vector
- Static posture or body motions with dynamic exertion

Within the field of ergonomics (aka human factors aka human engineering)

- *Muscle strength* draws in particular, the attention of the engineering physiologist
- *Internal transmission* is of concern to both the biomechanist and the designer because of the implications for body segment posture and motion
- *Body segment strength* is of great practical interest to the designer of tools, equipment, and work tasks

Figure 10.5 shows, in the form of a flow diagram, the feed-forward of excitation signals sent from the CNS to the muscle in order to generate tension. The diagram also shows that one can record associated signals (EEGs, electroencephalograms), muscle activation (EMGs, electromyograms), and calculate muscle tension and developed torques (via biomechanical modeling). Resulting torque and force that are available for application by a body segment to a resisting object (often a handle or pedal) can be measured together with body posture and motion.

Figure 10.5 also identifies three feedback paths, although at present they do not provide convenient avenues for measurements. The first is a reflex loop F_1 that originates at interoceptors. The other two loops start at exteroceptors and lead to a comparator, where they modify the input to the CNS. F_2 provides kinesthetic signals related to touch, body position, and motion; F_3 is similar but relates specifically to task execution. F_3 commonly involves sound and vision. (Note that the experimenter can easily manipulate these feedback signals, for example, via verbal exhortation, or with an instrument that shows the magnitude of the applied force, or a desired value.)

10.6   Assessment of Body Segment Strength

In physiological terms, generation of muscle strength is a complex procedure of myofilament activation through nervous feed-forward and feed-back control. It may involve substantial shortening or lengthening of the muscle, that is, either a concentric or eccentric effort, or there may be no perceptible change in length, that is, the effort is isometric. Mechanically, the main distinction between muscle actions is whether they are "dynamic" or "static."

10.6.1   Static Strength

When there is no perceptible change in muscle length during an isometric effort, then the involved body segments do not move; in physical terms, all forces acting within the system are in static equilibrium, as stated by Newton's First Law. Therefore, the physiological "isometric" case is equivalent to the "static" condition in physics.

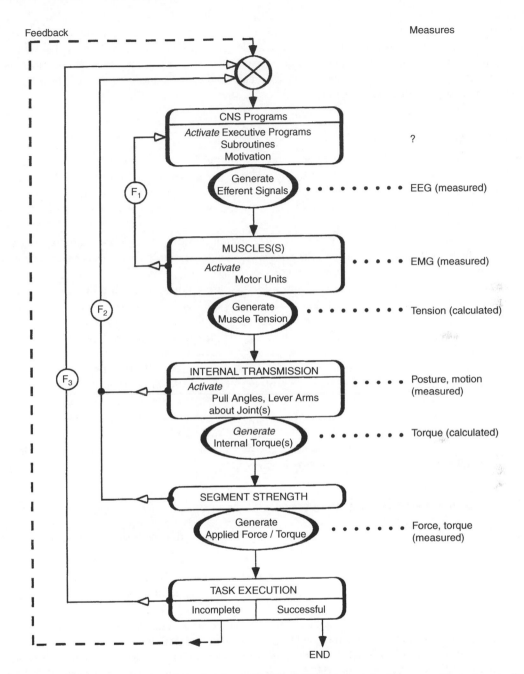

FIGURE 10.5 A conceptual model of the feed-forward and feed-back loops employed in generating and controlling muscle strength exertion. (Modified from Kroemer K. H. E. In *Proceedings, Annual Conference of the Human Factors Society,* Santa Monica, CA, pp. 19–20, 1979; Kroemer et al. *Ergonomics: How to Design for Erase and Efficiency,* Prentice Hall, Englewood cliffs, NJ, 1994; Kroemer K. H. E. et al. *Bases of Human Factors/Ergonomics,* 3rd ed., Van Nostrand Reinhold — John Wiley & Sons, New York, 1997. With permission.)

The static condition is theoretically simple and experimentally well controllable. It allows a rather easy measurement of muscular effort. Therefore, most of the information currently available on "human strength" describes the outcomes of static (isometric) testing. Accordingly, most of the tables on body segment strength in this chapter, and in other human engineering or physiologic literature, contain static data.

Besides the simple convenience of dealing with statics, measurement of isometric strength yields for most cases of practical design interest a reasonable estimate of the maximally possible exertion. That estimate also applies to slow body segment movements, especially if they are eccentric. However, the data do not estimate fast exertions well, especially if they are concentric and of the ballistic-impulse type, such as throwing or hammering.

10.6.2 Dynamic Strength

Dynamic muscular efforts are more difficult to describe and control than static contractions. In dynamic activities, muscle length changes, and therefore the involved body segments move. This results in displacement. The amount of travel is relatively small at the muscle but usually amplified along the links of the internal transmission path to the point of application to the outside, for example, at the hand or foot.

The time derivatives of displacement (velocity, acceleration, and jerk) are of importance for both the muscular effort (as discussed earlier) and the external effect, for example, change in velocity determines impact and force, as per Newton's Second Law.

Definition and experimental control of dynamic muscle exertions are much more complex tasks than static testing. Various new classification schemes for independant and dependant experimental variables can be developed (Kroemer, 1999; Kroemer et al., 1989, 1997; Kumar, 2004, Marras et al., 1993). Table 10.2 shows one such a system that includes the traditional isometric and isoinertial approaches.

Table 10.2 shows that, indeed, dynamic strength tests require more effort to describe and control than static (isometric) measurements. This complexity explains why, in the past, dynamic measurements (other than isokinematic and isoinertial testing) have been rare. In the most practical "free dynamic" test, common in sports, the experimenter can exert very little control (if any, then usually over mass and repetition). The independant variables, force and displacement (and their time derivatives), are the free choice of the subject. The dependant output is likely to be some measure of performance, such as the distance a discus is thrown.

10.7 Designing for Body Strength

The engineer or designer wanting to consider operator strength has to make a series of decisions. These include:

1. Is the exertion mostly static or dynamic? If static, we can use information about isometric capabilities, listed later. If dynamic, other considerations apply in addition, concerning, for example, physical (circulatory, respiratory, metabolic) endurance capabilities of the operator, or prevailing environmental conditions. Physiologic and ergonomic texts (e.g., Astrand and Rodahl, 1986; Kroemer et al., 1997, 2001; Winter, 1990) provide such information.
2. Is the exertion by hand, by foot, or with other body segment? For each, specific design information is available. If we can chose, we follow physiologic and ergonomic considerations to achieve the safest, least strenuous, and most efficient performance. For example, foot movements in comparison to hand movements over the same distance, consume more energy, are less accurate and slower — but they are stronger.
3. Is a maximal or a minimal strength exertion the critical design factor?
   - *Maximal* user strength usually relates to the structural strength of the object, so that even the strongest operator cannot break a handle or pedal. Accordingly, we set the design value with a safety margin above the highest perceivable strength application.
   - *Minimal* user strength is that expected from the weakest operator, which still yields the desired result, so that a door handle or brake pedal can be operated successfully or a heavy object be moved.

TABLE 10.2 Techniques to Measure Muscle Performance by Selecting Specific Independent and Dependent Variables

Names of Technique Variables	Isometric (Static) Indep. Dep.	Isovelocity (Dynamic) Indep. Dep.	Isoacceleration (Dynamic) Indep. Dep.	Isojerk (Dynamic) Indep. Dep.	Isoforce (Static or Dynamic) Indep. Dep.	Isoinertia (Static or Dynamic) Indep. Dep.	Free Dynamic Indep. Dep.
Displacement, linear/angular	constant*(zero)	C or X	C or X	C or X	C or X	C or X	X
Velocity, linear/angular	0	constant	C or X	C or X	C or X	C or X	X
Acceleration, linear/angular	0	0	constant	C or X	C or X	C or X	X
Jerk, linear/angular	0	0	0	constant	C or X	C or X	X
Force, torque	C or X	C or X	C or X	C or X	constant	C or X	X
Mass, moment of inertia	C	C	C	C	C	constant	C or X
Repetition	C or X	C or X	C or X	C or X	C or X	C or X	C or X

Note: Indep: independent; Dep: dependent; C: variable can be controlled; *: set to zero; 0: variable is not present (zero); X: can be dependent variables. The boxed constant variable provides the descriptive name.

Source: [With permission by the publisher from Kroemer, et al., *Engineering Physiology. Bases of Human Factors/Ergonomics*. 3rd ed., Van Nostrand Reinhold — John Wiley & Sons, New York, 1997. All rights reserved.]

- A *range* of expected strength exertions is, obviously, that between the considered minimum and maximum. The infamous "average user" strength is usually of no design value — see the Chapter 9 on "Engineering Anthropometry" in this handbook.

Most data on body segment strength apply to static (isometric) exertions. They provide reasonable guidance also for slow motions, although they are probably too high for concentric motions and a bit too low for eccentric motions. Of the little information available for dynamic strength exertions, much is limited to isokinematic (constant velocity) cases. As a rule, strength exerted in motion is less than that measured in static positions located on the path of motion.

The usual statistical treatment of measured strength data assumed that they fall into a normal distribution, which allows describing them in terms of averages (means) and standard deviations. This also allows the use of common statistical techniques to determine data points of special interest to the designer — as discussed in detail in the Chapter 9 on "Engineering Anthropometry" in this book. In reality, data describing body segment strength often appear in a skewed rather than in a bell-shaped distribution. The actual shape of the distribution is not of great concern, however, if the data points of special interest are the extremes. We can determine the maximal forces or torques, which the equipment must be able to bear without breaking, as those above the strongest measured data points. We can identify the minimal exertions, which even "weak" persons are able to generate at the low end of the distribution; again, see the "Anthropometry" Chapter 9 for procedures to calculate or estimate the design values.

10.7.1 Designing for Hand Strength

The human hand is able to perform a large variety of activities, ranging from those that require fine control to others that demand large forces. (However, the feet and legs are capable of more forceful exertions than the hand.)

One may divide hand tasks in this manner:

- Fine manipulation of objects with little displacement and force. Examples are writing by hand, assembly of small parts, adjustment of controls.
- Fast movements to an object requiring moderate accuracy to reach the target, but there is fairly a small force exertion. An example is the movement to a switch and its operation.
- Frequent movements between targets, usually with some accuracy but little force; such as in an assembly task, where parts must be taken from bins and assembled.
- Forceful activities with little or moderate displacement (such as with many assembly or repair activities, for example, when turning a hand tool against resistance) and forceful activities with large displacements (e.g., when hammering).

Texts book by Karwowski (2001), Kroemer et al. (2001), and Salvendy (1997) contain guidelines for design for accuracy and for displacement. For strength exertion, the designer keeps in mind:

Of the digits of the hand, the thumb is the strongest and the little finger the weakest. Finger forces depend on the finger joint angles, as listed in Table 10.3 and Table 10.4. Table 10.5 provides detailed information about manual force capabilities of male students and machinists. Female students developed between 50 and 60% digit strength of their male peers, but achieved 80 to 90% in "pinches."

Gripping and grasping strengths of the whole hand depend on the coupling between the hand and the handle — see Figure 10.6. The forearm can develop considerable twisting torques. Large force and torque vectors are available with the elbow at about the right angle, but the extended arm can exert the strongest pulling/pushing forces toward/away from the shoulder, especially if the trunk braces against a solid structure. Torque about the elbow depends on the elbow angle as depicted in Figure 10.7 and, in more detail, in Figure 10.8. Obviously, body posture and body support have great effects on the strength exerted with the arm and shoulder muscles.

TABLE 10.3 Average Forces and Standard Deviations (N) Exerted by Nine Subjects with their Fingertips, Depending on the Angle of the Proximal Interphalangeal Joint (PIP)

| | PIP at 30° | | | PIP at 60° | | |
| | Direction | | | Direction | | |
Digit	Fore	Aft	Down	Fore	Aft	Down
2, index finger	5.4 (2.0)	5.5 (2.2)	27.4 (13.0)	5.2 (2.4)	6.8 (2.8)	24.4 (13.6)
2, non-preferred	4.8 (2.2)	6.1 (2.2)	21.7 (11.7)	5.6 (2.9)	5.3 (2.1)	25.1 (13.7)
3, middle finger	4.8 (2.5)	5.4 (2.4)	24.0 (12.6)	4.2 (1.9)	6.5 (2.2)	21.3 (10.9)
4, ring finger	4.3 (2.4)	5.2 (2.0)	19.1 (10.4)	3.7 (1.7)	5.2 (1.9)	19.5 (10.9)
5, little finger	4.8 (1.9)	4.1 (1.6)	15.1 (8.0)	3.5 (1.6)	3.5 (2.2)	15.5 (8.5)

Source: [With permission by the publisher from Kroemer, et al., *Ergonomics: How to Design for Ease and Efficiency*, 2nd ed., Prentice Hall, Upper Saddle River, NJ, 2001. All rights reserved.]

10.7.2   The Use of Tables of Exerted Torques and Forces

There are many sources for data on body strengths that operators can apply, see the tables and figures in this handbook or in related human engineering texts (such as by Karwowski, 2001; Kroemer et al., 1997, 2001; Kumar, 2004; Peebles and Norris, 1998, 2000, 2003). Such data indicate "orders of magnitude" of forces and torques and the exact numbers should be viewed with great caution because they stem from measurements on various subject groups of often rather small numbers under widely varying circumstances. Therefore, it is good practice to take body strength measurements on a sample of the intended user population to verify that a new design is operable.

Note that the thumb and finger forces, for example, depend decidedly on "skill and training" of the digits as well as the posture of the hand and wrist. Hand forces (and torques) also depend on wrist position and on arm and shoulder postures. Exertions with arm, leg, and "body" (shoulder, backside) depend much on the posture of the body and on the support provided to the body (i.e., on the "reaction force" in the sense of Newton's Third Law) in terms of friction or bracing against solid structures. Table 10.6 and Figure 10.9 illustrate this; both rely on the same set of empirical data but their extrapolation shows the effects of:

- Location of the point of force exertion
- Body posture
- Friction at the feet

on horizontal push (and pull) forces applied by male soldiers.

It is obvious that the amount of strength available for exertion to an object outside the body depends on the weakest part in the chain of strength-transmitting body parts. Hand pull force, for example, may be limited by finger strength, or shoulder strength, or low back strength, and by the reaction force available to the body, as per Newton's Third Law. Figure 10.10 helps in determining where the "critical body

TABLE 10.4 Mean Poke Forces, and Standard Deviations (N) Exerted by 30 Subjects in the Direction of the Straight Digits

Digit	10 Male Mechanics	10 Male Students	10 Female Students
1, Thumb	83.8 (25.2) A	46.7 (29.2) C	32.4 (15.4) D
2, index finger	60.4 (25.8) B	45.0 (30.0) C	25.4 (9.6) DE
3, middle finger	55.9 (31.9) B	41.3 (21.6) C	21.5 (6.5) E

Note: Entries followed by different letters are different from each other ($p \leq 0.05$).

Source: [With permission by the publisher from Kroemer, et al., *Ergonomics: How to Design for Ease and Efficiency*, 2nd ed. Prentice Hall, Upper Saddle River, NJ, 2001. All rights reserved.]

TABLE 10.5 Mean Forces and Standard Deviations (N) Exerted by 21 Male students* and by 12 Male Machinists

Couplings (see Figure 8–26)	Digit 1 (Thumb)	Digit 2 (Index)	Digit 3 (Middle)	Digit 4 (Ring)	Digit 5 (Little)	
Push with digit tip in direction of the extended digit ("Poke")	91 (39)*	52 (16)*	51 (20)*	35 (12)*	30(10)*	See also Table 11.4
	138 (41)	84 (35)	86 (28)	66 (22)	52 (14)	
Digit touch (Coupling #1) perpendicular to extended digit	84 (33)*	43 (14)*	36 (13)*	30 (13)*	25 (10)*	—
	131 (42)	70 (17)	76 (20)	57 (17)	55 (16)	
Same, but all fingers press on one bar	—	Digits 2, 3, 4, 5 combined: 162 (33)				—
Tip force (like in typing; angle between distal and proximal phalanges about 135°)	—	30 (12)*	29 (11)*	23 (9)*	19 (7)*	—
		65 (12)	69 (22)	50 (11)	46 (14)	
Palm touch (Coupling #2) perpendicular to palm (arm, hand, digits extended and horizontal)	—	—				233 (65)
Hook force exerted with digit tip pad (Coupling #3, "Scratch")	61 (21)	49(17)	48 (19)	38 (13)	34 (10)	All digits combined: 108 (39)*
	118 (24)	89 (29)	104 (26)	77 (21)	66 (17)	252 (63)
Thumb — fingertip grip (Coupling #4, "Tip Pinch")		1 on 2	1 on 3	1 on 4	1 on 5	—
		50 (14)*	53 (14)*	38 (7)*	28 (7)*	
		59 (15)	63 (16)	44 (12)	30 (6)	
Thumb — finger palmer grip (Coupling #5, "Pad Pinch")	1 on 2 and 3	1 on 2	1 on 3	1 on 4	1 on 5	—
	85 (16)*	61 (12)*	61 (16)*	41 (12)*	31 (9)*	
	95 (19)	34 (7)	70 (15)	54 (15)	34 (7)	
Thumb — forefinger side grip (Coupling #6, "Side Pinch")	—	1 on 2	—	—	—	—
		98 (13)*				
		112 (16)				
Power grasp (Coupling #10, "Grip Strength")	—	—	—	—	—	318 (61)*
						366 (53)

Source:

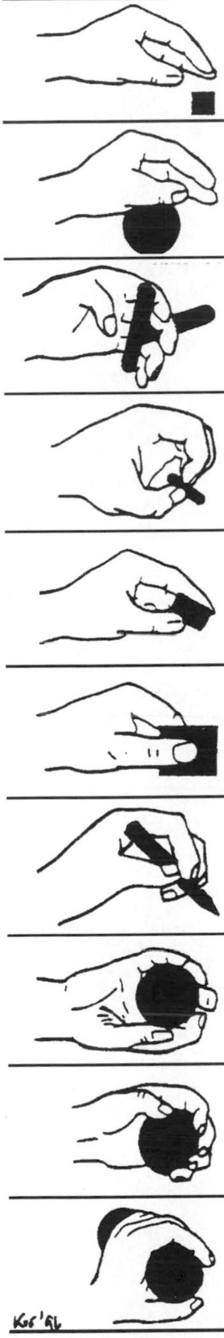

	Coupling #1	Digit Touch:
		One digit touches an object.
	Coupling #2	Palm Touch:
		Some part of the palm (or hand) touches the object.
	Coupling #3	Finger Palmar Grip (Hook Grip):
		One finger or several fingers hook(s) onto a ridge, or handle. This type of finger action is used where thumb counterforce is not needed.
	Coupling #4	Thumb-Fingertip Grip (Tip Pinch):
		The thumb tip opposes one fingertip.
	Coupling #5	Thumb-Finger Palmar Grip (Pad Pinch or Plier Grip):
		Thumb pad opposes the palmar pad of one finger (or the pads of several fingers) near the tips. This grip evolves easily from coupling #4.
	Coupling #6	Thumb-Forefinger Side Grip (Lateral Grip or Side Pinch):
		Thumb opposes the (radial) side of the forefinger.
	Coupling #7	Thumb-Two-Finger Grip (Writing Grip):
		Thumb and two fingers (often forefinger and middle finger) oppose each other at or near the tips.
	Coupling #8	Thumb-Fingertips Enclosure (Disk Grip):
		Thumb pad and the pads of three of four fingers oppose each other near the tips (object grasped does not touch the palm). This grip evolves easily from coupling #7.
	Coupling #9	Finger-Palm Enclosure (Collet Enclosure):
		Most, or all, of the inner surface of the hand is in contact with the object while enclosing it. This enclosure evolves easily from coupling #8.
	Coupling #10	Power Grasp:
		The total inner hand surfaces is grasping the (often cylindrical) handle which runs parallel to the knuckles and generally protrudes on one or both sides from the hand. This grasp evolves easily from coupling #9.

FIGURE 10.6   Couplings between hand and handle. (Adapted from Kroemer, *Hum. Factors*, 28, pp, 337–339. With permission. All rights reserved.)

segment" is present in the sequence of torques about body joints. Starting at the point of external exertion, for instance, at the hand, one assesses the strength requirements joint-by-joint along the arm, shoulder, and back. Often, the lumbar back area is the "weak link."

A walking or standing person receives all support from the ground, of course, and not seldom hip or knee joint strength limits the ability to do hard efforts, such as lifting a load on the back. A slippery

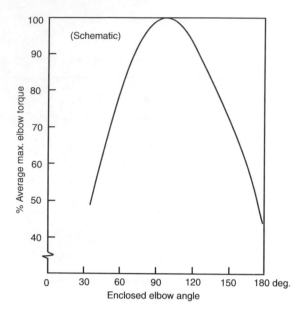

FIGURE 10.7 Relation of elbow angle and elbow torque. (With permission by the publisher from Kroemer et al., *Ergonomics: How to Design for Ease and Efficiency*, Prentice Hall, Upper Saddle River, NJ, 2001. All rights reserved.)

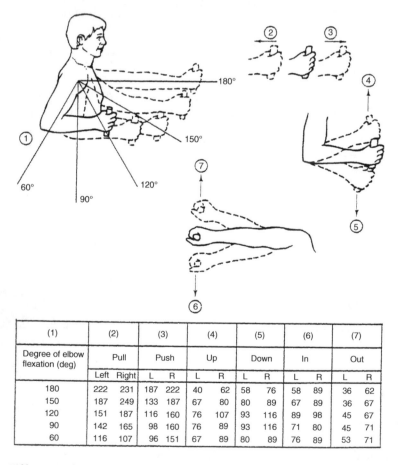

(1)	(2)		(3)		(4)		(5)		(6)		(7)	
Degree of elbow flexation (deg)	Pull		Push		Up		Down		In		Out	
	Left	Right	L	R	L	R	L	R	L	R	L	R
180	222	231	187	222	40	62	58	76	58	89	36	62
150	187	249	133	187	67	80	80	89	67	89	36	67
120	151	187	116	160	76	107	93	116	89	98	45	67
90	142	165	98	160	76	89	93	116	71	80	45	71
60	116	107	96	151	67	89	80	89	76	89	53	71

FIGURE 10.8 Fifth-percentile arm forces (N) exerted by sitting men. (Adapted from MIL HDBK 759.)

TABLE 10.6 Horizontal Push and Pull Forces (N) that Male Soldiers can Exert Intermittently or for Short Periods of Time

Horizontal Force[a] at least	Applied with[b]	Condition (μ is the coefficient of friction at the floor)
100 N push or pull	Both hands, or one shoulder, or the back	With low traction, μ between 0.2 and 0.3
200 N push or pull	Both hands, or one shoulder, or the back	With medium traction, μ about 0.6
250 N push	One hand	If braced against a vertical wall, 50 to 150 cm from the push panel and parallel to it
300 N push or pull	Both hands, or one shoulder, or the back	With high traction, μ above 0.9
500 N push or pull	Both hands, or one shoulder, or the back	If braced against a vertical wall, 50 to 150 cm from the push panel and parallel to it, or if anchoring the feet on a perfectly nonslip ground, such as at footrest
750 N push	The back	If braced against a vertical wall, 60 to 110 cm from the push panel and parallel to it, or if anchoring the feet on a perfectly non slip ground, such as at footrest

[a]May be nearly doubled for two and less than tripled for three operators pushing simultaneously. For the fourth and each additional operator, add about 75% of their push capabilities.
[b]See Figure 11.9 for example.
Source: Adapted from MIL-STD 1472.

surface may make it impossible to push a heavy object sideways with the shoulder; one can experience this in the winter on an icy ground when trying to push a car out of the ditch. To a sitting person, the seat provides most of the reaction that counters the forces actively exerted through the upper body and arms, although some support may be gathered from the floor via the legs.

10.7.3 Designing for Foot Strength

If a person must stand at work, fairly little force and only infrequent operations of foot controls should be required because, during these exertions, the operator has to support the body solely on the other leg. For a seated operator, however, operation of foot controls is much easier because the seat supports the body. Thus, the feet can move more freely and, under suitable conditions, can exert large forces and energies. A typical example for such an exertion is pedaling a bicycle where all energy transmits from the leg muscles through the feet to the pedals. For normal use, these should be located underneath the body, so that the body weight above them provides the reactive force to the force transmitted to the pedal. Placing the pedals forward makes body weight less effective for generation of reaction force to the pedal effort; but if a suitable backrest is present against which the buttocks and the low back rest, the feet can push forward on the pedals.

Foot movements are relatively slow compared to hand motions, because rather large leg messes are involved. Yet, feet can generate small forces in nearly all directions, such as for the operation of switches,

Force-plate[1] height	Distance[2]	Force, N Mean	Force, N SD
 50	80	664	177
50	100	772	216
50	120	780	165
70	80	716	162
70	100	731	233
70	120	820	138
90	80	625	147
90	100	678	195
90	120	863	141
Percent of shoulder height		Both hands	
 60	70	761	172
60	80	854	177
60	90	792	141
70	60	580	110
70	70	698	124
70	80	729	140
80	60	521	130
80	70	620	129
80	80	636	133
Percent of shoulder height			
 70	70	623	147
70	80	688	154
70	90	586	132
80	70	545	127
80	80	543	123
80	90	533	81
90	70	433	95
90	80	448	93
90	90	485	80
Percent of shoulder height		Both hands	
 100 percent of shoulder height		Both hands	
	50	581	143
	60	667	160
	70	981	271
	80	1285	398
	90	980	302
	100	646	254
		Preferred hand	
	50	262	67
	60	298	71
	70	360	98
	80	520	142
	90	494	169
	100	427	173
	Percent of thumb-tip reach*		
 100 percent of shoulder height	50	367	136
	60	346	125
	70	519	164
	80	707	190
	90	325	132
	Percent of span**		

[1]Height of the center of the force plate – 20 cm high by 25 cm long – upon which force is applied.
[2]Horizontal distance between the vertical surface of the force plate and the opposing vertical surface (wall or footrest, respectively) against which the subjects brace themselves.

*Thumb-tip reach – distance from backrest to tip of subject's thumb as arm and hand are extended forward.
**Span – the maximal distance between a person's fingertips when arms and hands are extended to each side.

FIGURE 10.9   Mean horizontal push forces and standard deviations (N) exerted by standing men with their hands, the shoulder, and the back. (Adapted from NASA STD. 3000 A, 1989.)

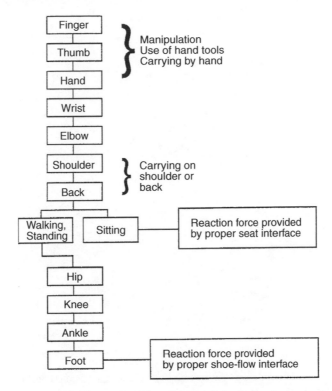

FIGURE 10.10 Determining the critical body segment strength for manipulating and carrying. (With permission by the publisher from Kroemer et al., *Engineering Physiology. Bases of Human Factors/Ergonomics*, 3rd ed., Van Nostrand Reinhold — John Wiley & Sons, New York, 1997. All rights reserved.)

but the downward or down-and-fore directions are preferred, especially for forceful exertions. Nearly fully extended legs can generate very large forces roughly in the direction of the lower leg. A fully extended knee angle (β in Figure 10.11), body inertia, and buttock and back support all limit leg force. Current automobile designs illustrate these principles. For example, under normal conditions, the driver an easily operate clutch and brake pedals with a knee angle of about 90°. However, a failure of the power-assist system unexpectedly requires very large forces from the feet; in this case, the driver must thrust the backside against a strong backrest and extend the legs to generate the needed pedal force.

Figure 10.11 through Figure 10.15 provide information about the forces that can be applied with legs and feet to a pedal. Of course, the forces depend on body support and hip and knee angles. The largest forward thrust force can be exerted with the nearly extended legs, which leaves very little room to move the foot control further away from the hip.

Of course, the strength that a foot can exert to an object such as a pedal, depends on the existent muscle strength, the means by which it can be transmitted along the "joint chain" ankle-knee-hip, and the way in which the seat provides the needed reactive support — see Figure 10.16. The diagram shows that an improper seat, a frail hip, knee, or ankle, or bad coupling of the shoe with the object may all make for a "weak kick."

Information on body strengths has been compiled in NASA and U.S. Military Standards; for example, by Chengular et al. (2003), Kroemer et al. (1997, 2001); Peebles and Norris (1998, 2000, 2002), Weimer (1993), and Woodson et al. (1991). However, caution is necessary when applying these data because they were measured with various techniques on different populations under varying conditions.

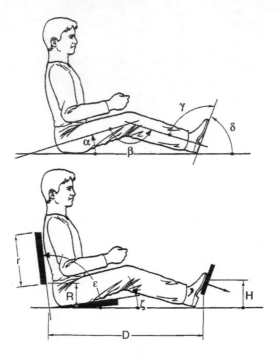

FIGURE 10.11   Body segment angles, seat dimensions, and pedal location affecting foot force exertion. (With permission by the publisher from Kroemer et al., *Ergonomics: How to Design for Ease and Efficiency*, Prentice Hall, Upper Saddle River, NJ, 2001. All rights reserved.)

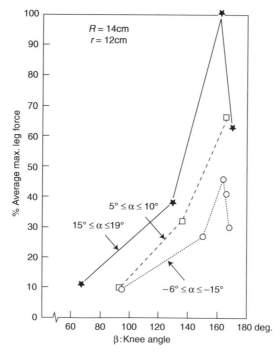

FIGURE 10.12   Effects of thigh angle α and knee angle β — see Figure 10.11 — on pedal push force. (With permission by the publisher from Kroemer et al., *Ergonomics: How to Design for Ease and Efficiency*, Prentice Hall, Upper Saddle River, NJ, 2001. All rights reserved.)

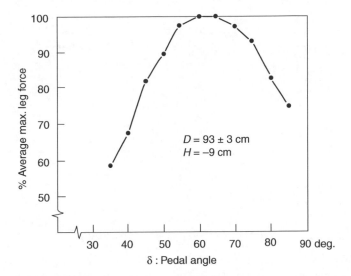

FIGURE 10.13   Effects of ankle (pedal) angle δ — see Figure 10.11— on foot force generated by ankle rotation. (With permission by the publisher from Kroemer, et al., *Ergonomics: How to Design for Ease and Efficiency,* Prentice Hall, Upper Saddle River, NJ, 2001. All rights reserved.)

10.8   Summary

Muscle contraction is brought about by active shortening of muscle substructures. Elongation of the muscle is due to external forces. Maximal muscle tension depends on the individual's muscle size and exertion skill.

Prolonged strong contraction leads to muscular fatigue, which hinders the continuation of the effort and finally cuts it off. Hence, maximal voluntary contraction can be maintained only for a few seconds.

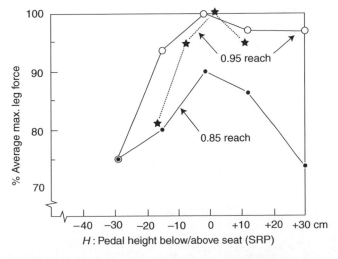

FIGURE 10.14   Effects of pedal height *H* and leg extension — see Figure 10.11 — on pedal push force. (With permission by the publisher from Kroemer et al., *Ergonomics: How to Design for Ease and Efficiency,* Prentice Hall, Upper Saddle River, NJ, 2001. All rights reserved.)

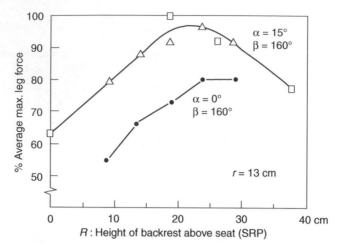

FIGURE 10.15   Effects of backrest height *R* — see Figure 10.11 — on pedal push force. (With permission by the publisher from Kroemer et al., *Ergonomics: How to Design for Ease and Efficiency,* Prentice Hall, Upper Saddle River, NJ, 2001. All rights reserved.)

In isometric contraction, muscle length remains constant, which establishes a static condition for the body segments affected by the muscle. In an isotonic effort, the muscle tension remains constant, which usually coincides with a static (isometric) effort.

Dynamic activities result from changes in muscle length, which bring about motion of body segments. In an isokinematic effort, speed remains unchanged. In an isoinertial test, the mass properties remain constant.

Human body (segment) strength is measured routinely as the force (or torque) exerted to an instrument external to the body. This is information of great importance to the ergonomic designer/engineer.

Design of equipment and work tasks to match human body strength capabilities considers these aspects:

- Determine whether the exertion is static or dynamic
- Establish the part of the body on which the force or torque is exerted
- Select the segment strength percentile (minimum and maximum) that is critical for the operation
- Follow the chain of strength vectors through the involved body segments to find the "weak link" and to improve and rearrange the conditions if possible

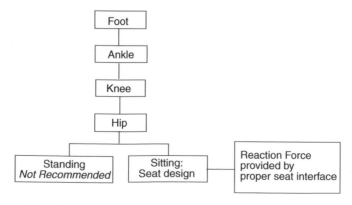

FIGURE 10.16   Determining the "critical body segment strength" for foot operation.

References

Asimov, I., *The Human Body. Its Structure and Operation*, The New American Library/Signet, New York, 1963.

Astrand, P.O. and Rodahl, K., *Textbook of Work Physiology*, 2nd, 3rd ed., McGraw-Hill, New York, 1977, 1986.

Chaffin, D.B., Andersson, G.B.J., and Martin, B.J., *Occupational Biomechanics*, 3rd ed., John Wiley & Sons, New York, 1999.

Chengular, S.N., Rodgers, S.H., and Bernard, T.E., *Kodak's Ergonomic Design for People at Work*, 2nd ed., John Wiley & Sons, New York, 2003.

Enoka, R.M., *Neuromechanical Basis of Kinesiology*, Human Kinetics Books, Champaign, IL, 1988.

Enoka, R.M., *Neuromechanics of Human Movement*, 3rd. ed., Human Kinetics Books, Champaign, IL, 2002.

Karwowski, W., ed., *International Encyclopedia of Ergonomics and Human Factors*, Taylor & Francis, London, UK, 2001.

Kroemer, K.H.E., A new model of muscle strength regulation, in *Proceedings, Annual Conference of the Human Factors Society*, Human Factors Society Santa Monica, CA, pp. 19–20, 1979.

Kroemer, K.H.E., Assessment of human muscle strength for engineering purposes: basics and definitions, *Ergonomics*, 42, pp. 74–93, 1999.

Kroemer, K.H.E., Kroemer, H.B., and Kroemer-Elbert, K.E., *Ergonomics: How to Design for Ease and Efficiency*, 2nd ed., Prentice Hall, Upper Saddle River, NJ, 2001.

Kroemer, K.H.E., Kroemer, H.J., and Kroemer-Elbert, K.E., *Engineering Physiology. Bases of Human Factors/Ergonomics*, 3rd ed., Van Nostrand Reinhold — John Wiley & Sons, New York, 1997.

Kroemer, K.H.E., Marras, W.S., McGlothlin, J.D., McIntyre, D.R., and Nordin, M., *Assessing Human Dynamic Muscle Strength* (Technical Report, 8-30-89), Virginia Tech, Industrial Ergonomics Laboratory Blacksburg, VA, 1989. Also published in *Int. J. Indus. Ergon.*, 6, pp. 199–210, 1990.

Kumar, S., Ed., *Muscle Strength*, CRC Press — Taylor & Francis, Boca Raton, FL, 2004.

Marras, W.S., McGlothlin, J.D., McIntyre, D.R., Nordin, M., and Kroemer, K.H.E., *Dynamic Measures of Low Back Performance*, American Industrial Hygiene Association, Fairfax, VA, 1993.

Peebles, L. and Norris, B., *Adultdata. The Handbook of Adult Anthropometric and Strength Measurements — Data for Design Safety.* (DTI/Pub 2917/3k/6/98/NP), Department of Trade and Industry, London, UK, 1998.

Peebles, L. and Norris, B., *Strength Data* (DTI/URN 00/1070), Department of Trade and Industry, London, UK, 2000.

Peebles, L. and Norris, B., Filling "gaps" in strength data for design, *Appl. Ergon.*, 34, pp. 73–88, 2003.

Rodgers, S.H., Work physiology — fatigue and recovery, in G. Salvendy, Ed., *Handbook of Human Factors and Ergonomics*, 2nd ed., John Wiley & Sons, New York, Chap. 10, pp. 268–297, 1997.

Salvendy, G., Ed., *Handbook of Human Factors and Ergonomics*, 2nd ed., John Wiley & Sons, New York, 1997.

Schneck, D.J., *Engineering Principles of Physiologic Function*, New York University Press, New York, 1990.

Schneck, D.J., *Mechanics of Muscle*, 2nd ed., New York University Press, New York, 1992.

Weimer, J., *Handbook of Ergonomic and Human Factors Tables*, Prentice Hall, Englewood Cliffs, NJ, 1993.

Winter, D.A., *Biomechanics and Motor Control of Human Movement*, 2nd ed., John Wiley & Sons, New York, 1990.

Woodson, W.E., Tillman, B., and Tillman, P., *Human Factors Design Handbook*, 2nd ed., McGraw-Hill, New York, 1991.

11

Biomechanical Basis for Ergonomics

William S. Marras
The Ohio State University

11.1 Biomechanic Analyses and Ergonomics

In 1978, E.R. Tichauer published a book entitled *The Biomechanical Basis of Ergonomics* (Tichauer, 1978). This book introduced much of the world to the concept of applying engineering techniques to the human body so that the limits of exposure could be identified. Since this time much has changed in the fields of ergonomics and biomechanics. However, his approaches to addressing occupationally related musculoskeletal problems remain the same to this day. Dr. Tichauer's book serves as the motivation for this chapter.

11.1.1 Definitions

Biomechanics can be defined as an interdisciplinary field in which information from both the biological sciences and engineering mechanics is used to assess the function of the body. A major assumption of occupational biomechanics is that the body behaves according to the laws of Newtonian mechanics. By definition, "mechanics is the study of forces and their effects on masses" (Kroemer, 1987). The object of interest in an occupational ergonomics context is most often a quantitative assessment of mechanical loading that occurs within the musculoskeletal system. The goal of an occupational biomechanics assessment is to quantitatively describe the musculoskeletal loading that occurs during work so that one can derive an appreciation for the degree of risk associated with an occupationally related task. The characteristic that distinguishes occupational biomechanics analyses from other types of ergonomic analyses is that the comparison is quantitative in nature. The quantitative nature of occupational biomechanics permits ergonomists to address the question of "how much exposure to the occupational risk factors is too much exposure?"

The portion of biomechanics dealing with ergonomics issues is often labeled industrial or occupational biomechanics. Chaffin et al. (1999) have defined occupational biomechanics as "the study of the physical interaction of workers with their tools, machines, and materials so as to enhance the worker's performance while minimizing the risk of musculoskeletal disorders." This chapter will address occupational biomechanical issues exclusively in this ergonomics framework.

11.1.2   Occupational Biomechanics Approach

The approach to biomechanical assessment is to characterize the human-work system situation through a mathematical representation or model. The idea behind such models is to represent the various underlying biomechanical concepts through a series of rules or equations in a "system" or model that helps us understand how the human would be affected by exposure to work. One can think of a biomechanical model as the "glue" that holds our logic together when considering the various factors that would affect risk in a specific work situation.

An advantage of representing the worker in a biomechanical model is that the model permits one to quantitatively consider the *trade-offs* associated with risk to various parts of the body in the design of a workplace. When one considers biomechanical rationale, one finds that it is difficult to accommodate all parts of the body in an ideal biomechanical environment. It is often the case that in attempting to accommodate one part of the body, the biomechanical situation at another body site is compromised. Therefore, the key to the proper application of biomechanical principles is to consider the appropriate biomechanical trade-offs associated with various parts of the body as a function of the work requirements and the various workplace design options. For this reason, this chapter will focus upon the information required to develop proper biomechanical reasoning when considering a workplace. The chapter will first present and explain a series of key concepts that constitute the underpinning of biomechanical reasoning. Next, these concepts will be applied to different parts of the body. Once this reasoning is developed an attempt will be made to examine how the various biomechanical concepts must be considered collectively in terms of trade-off, when designing a workplace from an ergonomic perspective under realistic conditions. This chapter will demonstrate that one *cannot* successfully practice ergonomics by simply memorizing a set of "ergonomic rules" (e.g., keep the wrist straight or don't bend from the waist when lifting). These types of rule-based design strategies ultimately result in sub-optimizing the workplace ergonomic conditions.

11.2   Biomechanical Concepts

11.2.1   The Load — Tolerance Construct

The fundamental concept in the application of occupational biomechanics to ergonomics is that one should design workplaces so that the load imposed upon a structure does not exceed the tolerance of the structure. This basic concept is illustrated in Figure 11.1. The figure illustrates the traditional concept of biomechanical risk in occupational biomechanics (McGill, 1997). A loading pattern is developed on a body structure that is repeated as the work cycles recur during a job. The structure tolerance is also shown in this figure. If the magnitude of the load imposed on a structure is far less than tissue tolerance, then the task is considered safe and the magnitude of the difference between the load and the tolerance is considered the safety margin. Also implicit in this figure, is the idea that risk occurs when the imposed load exceeds the tissue tolerance. While tissue tolerance is defined as the force that

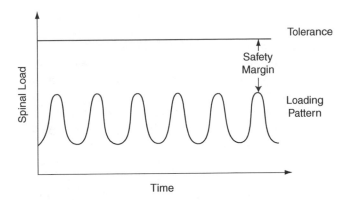

FIGURE 11.1   Traditional concept of biomechanical risk.

results in tissue damage, some ergonomists are beginning to expand the concept of tolerance to include not only mechanical tolerance of the tissue, but also the point at which the tissue exhibits an inflammatory reaction.

Industrial tasks are becoming more repetitive involving lighter loads. The conceptual load tolerance model can also be adjusted to also account for this type of risk exposure. As shown in Figure 11.2, occupational biomechanics logic can account for the fact that with repetitive loading the tolerance of the structure tissue may decrease over time to the point where it is more likely that the structure loading will exceed the structure tolerance and result in injury or illness. Thus, occupational biomechanical models and logic are moving towards systems that consider manufacturing and work trends in the workplace and attempt to represent these observations (such as cumulative trauma disorders) in the model logic.

11.2.2   Acute vs. Cumulative Trauma

It is well recognized that in occupational settings two types of trauma can affect the human body and lead to musculoskeletal disorders. First, *acute* trauma can occur, which refers to an application of force that is so large that it exceeds the tolerance of the body structure during an occupational task. Thus, acute trauma is typically associated with large exertions of force that would occur infrequently. For example, an acute trauma can occur when a worker is asked to lift an extremely heavy object as when moving a heavy part. This situation would relate to a peak load pattern that exceeded the load tolerance

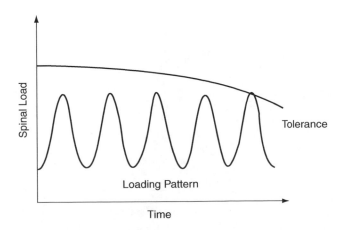

FIGURE 11.2   Realistic scenario of biomechanical risk.

in Figure 11.1. *Cumulative* trauma, on the other hand, refers to the repeated application of force to a structure that tends to wear down a structure, thus, lowering the structure tolerance to the point where the tolerance is exceeded through a reduction of the tolerance limit. This situation was illustrated in Figure 11.2. Cumulative trauma represents more of a "wear and tear" on the structure. This type of trauma is becoming far more common in the workplace since more repetitive jobs are becoming common in industry and is the mechanism of concern for many ergonomics evaluations.

Cumulative trauma can initiate a process that can result in a tissue reactive cycle that is extremely difficult to break. This process is illustrated in Figure 11.3. The cumulative trauma process begins by exposing the worker to manual exertions that are either frequent (repetitive) or prolonged. This repetitive application of force can affect either the tendons or the muscles of the body. If the tendons are affected, the following sequence occurs. The tendons are subject to mechanical irritation when they are repeatedly exposed to high levels of tension and groups of tendons may rub against each other. The physiologic reaction to this mechanical irritation can result in inflammation and swelling of the tendon. This swelling will stimulate the nociceptors surrounding the structure and signal the central control mechanism (brain) via pain perception that a problem exists. In response to this pain, the body will attempt to control the problem via two mechanisms. First, the muscles surrounding the irritated area will coactivate in an attempt to stabilize the motion of the tendons or stiffen the structure. Since motion will further stimulate the nociceptors and result in further pain, motion avoidance is often indicative of the start of a cumulative trauma disorder. Second, in an attempt to reduce the friction occurring within the tendon, the body will increase its production of synovial fluid within the tendon sheath. However, given the limited space available between the tendon and the tendon sheath the increased production of synovial fluid often exacerbates the problem by further expanding the tendon sheath and, in thus, further stimulating the surrounding nociceptors. As indicated in the figure, this initiates a viscous cycle where the response of the tendon to the increased friction results in a reaction (inflammation and the increased production of synovial fluid) that exacerbates the problem. Once this cycle is initiated it is very difficult to stop and often anti-inflammatory agents are required. This process results in chronic joint pain and a series of musculoskeletal reactions such as reduced strength, reduced tendon motion, and reduced mobility. Collectively, these reactions result in a functional disability.

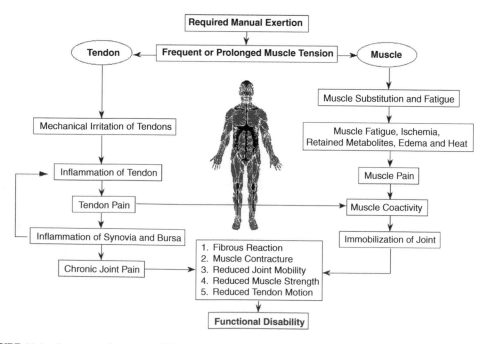

FIGURE 11.3   Sequence of events in CTDs.

A similar process occurs if the muscles are affected by cumulative trauma as opposed to the tendons. Muscles can be easily overloaded when they become fatigued. Fatigue can lower the tolerance to stress and can result in microtrauma to the muscle fibers. This microtrauma typically means that the muscle is partially torn and the tear will cause capillaries to rupture and result in swelling, edema, or inflammation near the site of the tear. This process can stimulate nociceptors and result in pain. As with cumulative trauma to the tendons, the body reacts by cocontracting the surrounding musculature and thereby minimizing the motion of the joint. Since the tendons are not involved with cumulative trauma to the muscles there is no increased production of synovial fluid. However, the end result is the same series of musculoskeletal reactions resulting from tendon irritation (i.e., reduced strength, reduced tendon motion, and reduced mobility). The ultimate result of this process is once again a functional disability.

Even though the stimulus associated with the cumulative trauma process is somewhat similar between tendons and muscles there is a significant difference in the time required to heal from the damage to a tendon compared to a muscle. The mechanism of repair for both the tendons and muscles is dependent upon blood flow to the damaged structure. Blood provides nutrients for repair as well as dissipates waste materials. However, the blood supply to a tendon is just a fraction (typically about 5% in an adult) of that supplied to a muscle. Thus, given an equivalent strain to a muscle and a tendon, the muscle will heal rapidly (in about 10 days if not reinjured), whereas the tendon could take months (20 times longer) to accomplish the same level of repair. For this reason, ergonomists must be particularly vigilant in the assessment of workplaces that could pose a danger to the tendons of the body. This lengthy repair process also explains why most ergonomics processes place a high value on identifying potentially risky jobs prior to a lost time incident through mechanisms such as discomfort surveys.

11.2.3 Moments and Levers

Biomechanical loads are *not* defined solely by the magnitude of weight supported by the body. The position of the weight relative to the axis of rotation of the joint of interest defines the imposed load on the body and is referred to as a *moment*. Thus, a moment is defined as the product of force and distance. For example, a 50 N mass held at a horizontal distance of 75 cm (0.75 meters) from the shoulder joint imposes a moment of 37.5 Nm (50 N × 0.75 m) on the shoulder joint, whereas, the same weight held at a horizontal distance of 25 cm from the shoulder joint imposes a moment or load of only 12.5 Nm (50 N × 0.25 m) on the shoulder. Thus, the load on a joint is a function of where the load is held relative to the joint and the mass of the weight held. Joint load is not simply a function of weight.

As implied by this example, moments are a function of the mechanical lever systems of the body. The musculoskeletal system can be represented by a system of levers and these are the lever systems that are used to describe the tissue loads with a biomechanical model. Three types of lever systems are present in the human body. First-class levers are those that have a fulcrum in the middle of the system, an imposed load on one end of the system and an opposing (*internal*) load imposed on the opposite end of the system. As will be discussed later, the trunk is an example of a first-class lever. In this example, the spine serves as the fulcrum. As the worker lifts, a moment is imposed anterior to the spine due to the object weight times the distance of the object from the spine. This moment is counterbalanced by the activity of the back musculature, however, back muscles are at a mechanical disadvantage since the distance between the back muscles and the spine is much less than the distance between the object lifted and the spine. A second-class lever system can be found in the lower extremity. In a second-class lever system the fulcrum is on one end of the lever, the opposing force is on the other end of the system and the applied load is in between these two. In the body, the foot is a good example of this lever system. In this example, the ball of the foot acts as the fulcrum, the load is applied through the tibia or bone of the lower leg. The restorative force is applied through the gastrocnemius or calf muscle. In this manner the muscle activates and causes the body to move about the fulcrum or ball of the foot and move the body forward. Finally, a third-class lever system is one where the fulcrum is located at one end of the system, the applied load acts at the other end of the system and the opposing

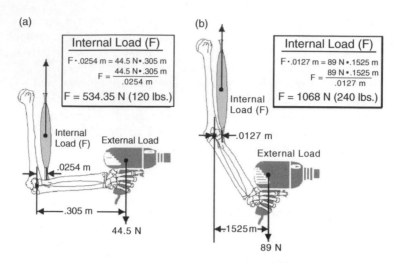

FIGURE 11.4　An example of an anatomical third-class lever (a) demonstrating how the mechanical advantage changes as the elbow position changes (b).

force acts in between the two. An example of this system in the human body is the elbow joint and is shown in Figure 11.4.

11.2.4　External and Internal Loading

Two types of forces can impose loads on a tissue during work. *External* loads refer to those forces that are imposed on the body as a direct result of gravity acting upon an external object being manipulated by the worker. For example, in Figure 11.4a, the tool held in the worker's hand is subject to the forces of gravity, which impose a 44.5 N (10 lb) external load at a distance from the elbow joint of 30.5 cm (12 inches). However, in order to maintain equilibrium, this external load must be counteracted by an *internal* load that is supplied by the muscles of the body. Figure 11.4a also shows that the internal load (muscle) acts at a distance relative to the elbow joint that is much closer to the fulcrum than the external load (tool). Thus, the internal load or force is at a biomechanical disadvantage and must be much larger (534 N or 120 lb) than the external load (44.5 N or 10 lb) in order to keep the musculoskeletal system in equilibrium. As shown in this example it is not unusual for the magnitude of the internal load to be much greater (typically 10 times greater) than the external load. Thus, it is typically the internal loading that contributes mostly to the cumulative trauma of the musculoskeletal system during work. The sum of the external load and the internal load define the total loading experienced at the joint. When evaluating a workstation the ergonomist must not only consider the externally applied load but must be particularly sensitive to the magnitude of the internal forces that can load the musculoskeletal system.

11.2.5　Factors Affecting Internal Loading

The previous discussion has discussed the importance of understanding the relationship between the external loads imposed upon the body and the internal loads generated by the force generating mechanisms within the body. The key to proper ergonomic design involves designing workplaces so that the internal loads are minimized. One can consider the internal forces as both the component that loads the tissue as well as a structure that can be subject to over-exertion. Muscle strength or capacity can be considered as a tolerance measure. If the forces imposed on the muscles and tendons as a result of the task exceed the strength of the muscle or tendon an injury is possible. In general, three components of the physical work environment (biomechanical arrangement of the musculoskeletal lever system, length–strength relationships, and temporal relationships) can be manipulated in order to facilitate this goal and serve as the basis for many ergonomic recommendations.

11.2.5.1 Biomechanical Arrangement of the Musculoskeletal Lever System

The posture required by the design of the workplace can affect the arrangement of the body's lever system, and thus, can greatly affect the magnitude of the internal load required to support the external load. The arrangement of the lever system can influence the magnitude of the external moment imposed upon the body as well as dictate the magnitude of the internal forces and the subsequent risk of cumulative trauma. Consider the biomechanical arrangement of the elbow joint that is shown in Figure 11.4. In Figure 11.4a, the mechanical advantage of the internal force generated by the biceps muscle and tendon is defined by a posture that keeps one's arm bent at a 90° angle. If one palpates the tendon and inserts the index finger between the joint center and the tendon, one can gain an appreciation for the internal moment arm distance. One can also appreciate how this internal mechanical advantage can change with posture. With the index finger still inserted between the elbow joint and the tendon and if the arm is slowly straightened one can appreciate how the distance between the tendon and the joint center of rotation is significantly reduced. If the imposed moment about the elbow joint is held constant (as shown in Figure 11.4b by a heavier tool) under these conditions, the mechanical advantage of the internal force generator is significantly reduced. Thus, the internal moment must generate greater force in order to support the external load. This greater force is transmitted through the tendon and can increase the risk of cumulative trauma. Therefore, the positioning of the mechanical lever system (which can be accomplished though work design) can greatly affect the internal load transmission within the body. The same task can be performed in a variety of ways but some of these positions are much more costly in terms of loading of the musculoskeletal system than others.

11.2.5.2 Length–Strength Relationship

Another important relationship in defining the load on the musculoskeletal system is the length–strength relationship of the muscles. Figure 11.5 shows this relationship. The active portion of this figure refers to structures that actively generate force such as muscles. The figure indicates that when muscles are close to their resting length (generally seen in the fetal position), they have the greatest capacity to generate force. However, when the muscle length deviates from this resting position the capacity to generate force is greatly reduced because the cross-bridges between the components of the muscle proteins become inefficient. Hence, when a muscle stretches or when a muscle attempts to generate force while at a short length the ability to generate force is greatly diminished. Note also, as indicated in Figure 11.5 that passive tissues in the

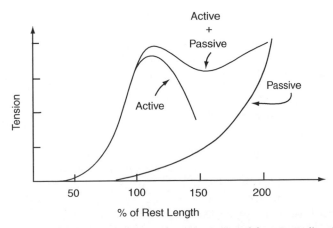

FIGURE 11.5 Length–tension relationship for a human muscle. (Adapted from Basmajian, J.V. and De Luca, C.J., *Muscles Alive: Their Functions Revealed by Electromyography*, 5th ed., Williams and Wilkins, Baltimore, MD, 1985. With permission.)

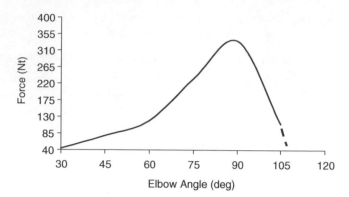

FIGURE 11.6 Position-force diagram produced by flexion of the forearm in pronation. "Angle" refers to included angle between the longitudinal axes of the forearm and upper arm. The highest parts of the curve indicate the configurations where the biomechanical lever system is most effective. (Adapted from Chaffin, D.B. and Andersson, G.B., *Occupational Biomechanics*, John Wiley & Sons, Inc. New York, 1991. With permission.)

muscle (and ligaments) can generate tension when muscles are stretched. Thus, the orientation of the muscle fibers during a task can greatly influence the force available to perform work and can, therefore, influence risk by altering the available internal force within the system. A given tension on a muscle can either tax the muscle greatly or be a minimum burden on the muscle. What might be considered a moderate force for a muscle at the resting length can become the maximum force a muscle can produce when it is in a stretched or contracted position, thus, increasing the risk of muscle strain. When this relationship is considered in conjunction with the mechanical load placed on the muscle and tendon via the arrangement of the lever system, the position of the joint arrangement becomes a major factor in the design of the work environment. It is typically the case that the length–strength relationship interacts synergistically with the lever system. Figure 11.6 shows the effect of elbow position on the force generation capability of the elbow. This figure indicates that position can have a dramatic effect on force generation. As already discussed this position can also have a great effect on internal loading of the joint and the subsequent risk of cumulative trauma.

11.2.5.3 Force–Velocity Relationship

Motion can profoundly influence the ability of a muscle to generate force and, therefore, load the biomechanical system. Motion can be a benefit to the biomechanical system if momentum is properly used or it can increase the load on the system if the worker is not taking advantage of momentum. The relationship between muscle velocity and force generation is shown in Figure 11.7. This figure indicates that, in general, the faster the muscle is moving the greater the reduction in force capability of the muscle. As with most of the biomechanical principles discussed in this chapter, this reduction in muscle capacity can result in the muscle strain that can occur at a lower level of external loading and a subsequent increase in the risk of cumulative trauma. In addition, this effect is considered in many dynamic ergonomic biomechanical models.

11.2.5.4 Temporal Relationships

11.2.5.4.1 Strength Endurance

Strength can be considered both an internal force as well as a tolerance limiter, but it is important to realize that strength is transient. A worker may generate a great amount of strength during a one-time exertion. However, if the worker is required to exert his strength either repeatedly or for a prolonged period of time, the amount of force that the worker can generate is reduced dramatically. Figure 11.8 demonstrates this relationship during an isometric exertion. The dotted line in this figure indicates the maximum force generation capacity of a static exertion of force over time. Maximum force output is only generated for a very brief period of time. As time increases, strength output decreases exponentially and levels off at about 20% of maximum after about 7 min. Similar trends occur under

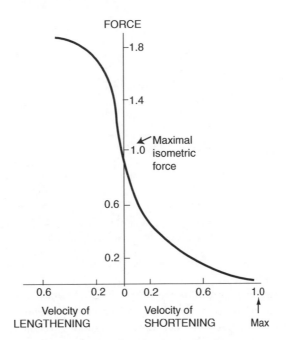

FIGURE 11.7   Influence of velocity upon muscle force (Adapted from *The Textbook of Work Physiology*, McGraw-Hill, 1977. With permission.)

repeated dynamic conditions. This indicates that if it is determined that a task requires a large portion of a workers' strength, one must consider how long that portion of the strength is required in order to ensure that the work does not strain the musculoskeletal system.

11.2.5.4.2   *Rest Time*

As mentioned previously, the risk of cumulative trauma increases when the capacity to exert force is exceeded by the force requirements of the job. Another factor that can affect this strength capacity (and tolerance to muscle strain) is rest time. Rest time has a profound effect on the ability to exert force. Figure 11.9

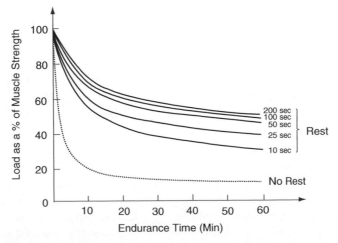

FIGURE 11.8   Forearm flexor muscle endurance times in consecutive static contractions of 2.5 sec duration with varied rest periods. (Adapted from Chaffin, D.B. and Andersson, G.B., *Occupational Biomechanics*, John Wiley & Sons, Inc. New York, 1991. With permission.)

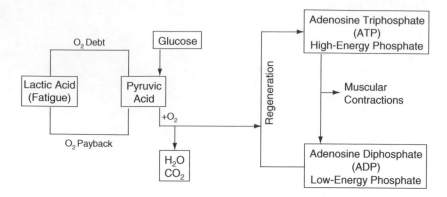

FIGURE 11.9　The body's energy system during work. (Adapted from Grandjean, E., *Fitting the Task to the Man: An Ergonomic Approach*, Taylor & Francis, Ltd., London, 1982. With permission.)

shows how energy for a muscular contraction is regenerated during work. Adenosine triphosphate (ATP) is required to produce a significant muscular contraction. ATP changes to adenosine diphosphate (ADP) once a muscular contraction has occurred. This ADP must be converted to ATP in order to enable another muscular contraction. This conversion can occur with the addition of oxygen to the system. If oxygen is not present, then the system goes into oxygen debt and there is insufficient ATP for a muscular contraction. Thus, this flow chart indicates that oxygen is a key ingredient to maintain a high level of muscular exertion. Oxygen is delivered to the target muscles via the blood flow. However, under static exertions the blood flow is reduced and there is a subsequent reduction in the blood flow to the muscle. This restriction of blood flow and subsequent oxygen deficit are responsible for the rapid decrease in force generation over time as shown in Figure 11.8. The solid lines shown in Figure 11.8 indicate how the force generation capacity of the muscles increase when different amounts of rest are permitted in a fatiguing exertion. As more and more rest time is permitted, increases in force generation are achieved when more oxygen is delivered to the muscle and more ADP can be converted to ATP. This relationship also shows that any more than about 50 sec of rest, under these conditions, does not result in a significant increase in force generation capacity of the muscle. Practically, this indicates that in order to optimize the strength capacity of the worker and minimize the risk of muscle strain, a schedule of frequent and brief rest periods would be more beneficial than lengthy infrequent rest periods.

11.2.6　Load Tolerance

As mentioned previously, occupational biomechanical analyses must consider not only the loads imposed upon a structure but also the ability of the structure to withstand or tolerate a load during work. This section will briefly review the knowledge base associated with body structure tolerances.

11.2.6.1　Muscle, Ligament, Tendon, and Bone Capacity

The exact tolerance characteristics of human tissues such as muscles, ligaments, tendons, and bones loaded under various working conditions are difficult to estimate. Tolerances of the structures in the body vary greatly under similar loading conditions. In addition, tolerance depends upon many other factors such as strain rate, age of the structure, frequency of loading, physiologic influences, heredity, conditioning, and many unknown factors. Furthermore, it is not possible to measure these tolerances under human *in vivo* conditions. Therefore, most of the published estimates of tissue tolerance have been derived from various animal and/or theoretical sources.

11.2.6.2 Muscle and Tendon Strain

Muscle appears to be the structure that has the lowest tolerance in the musculoskeletal system. The ultimate strength of a muscle has been estimated at 32 MPa (Hoy et al., 1990). In general, it is believed that the muscle will rupture prior to the tendon in a healthy tendon (Nordin and Frankel, 1989), since tendon stress has been estimated at between 60 and 100 MPa (Nordin and Frankel, 1989; Hoy et al., 1990). Hence, as indicated in Table 11.1, there is a safety margin between the muscle failure point and the failure point of the tendon of about twofold (Nordin and Frankel, 1989) to threefold (Hoy et al., 1990).

11.2.6.3 Ligament and Bone Tolerance

Ligaments and bone tolerances within the musculoskeletal system have also been estimated. Ultimate ligament stress has been estimated at approximately 20 MPa. The ultimate stress of bone varies depending upon the direction of loading. Bone tolerance can range from as low as 51 MPa in transverse tension to over 190 MPa in longitudinal compression. Table 11.1 also indicates the ultimate stress of bone loaded in different loading conditions.

A strong temporal component to ligament recovery appears to exist. Solomonow has found that ligaments require long periods of time to regain structural integrity and compensatory muscle activities are recruited (Solomonow et al., 1998, 1999, 2000, 2002; Stubbs et al., 1998; Gedalia et al., 1999; Wang et al., 2000; Solomonow, 2004). Recovery time has been found to be several fold the loading duration and can easily exceed the typical work-rest cycles observed in industry.

11.2.6.4 Disc/Endplate and Vertebrae Tolerance

The mechanism of cumulative trauma in the disc is thought to be related to repeated trauma to the vertebral endplate. The endplate is a very thin (about 1 mm thick) structure that facilitates nutrient flow to the disc fibers (anulus fibrosis). Repeated microfracture of this vertebral endplate is thought to impair the nutrient flow to the disc fibers and thereby lead to atrophy and degeneration of the fiber. It is believed that if one can determine the level at which the endplate experiences a microfracture, one can then minimize the effects of cumulative trauma and disc degeneration within the spine. Several studies of disc endplate tolerance have been performed. Figure 11.10 shows the levels of endplate compressive loading tolerance that have been used to establish safe lifting situations at the worksite (NIOSH, 1981). This figure shows the compressive force mean (column value) as well as the compression force distribution (thin line and normal distribution curve) that would result in vertebral endplate failure (microfracture). This figure indicates that for those under 40 years of age endplate microfracture damage begins to occur at about 3432 N, of compressive load on the spine. If the compressive load is increased to 6375 N, approximately 50% of those exposed to the load will experience vertebral endplate microfracture. When

TABLE 11.1 Tissue Tolerance of the Musculoskeletal System

Structure	Estimated Ultimate Stress (σ_u) (MPa)
Muscle	32–60
Ligament	20
Tendon	60–100
Bone longitudinal loading	
Tension	133
Compression	193
Shear	68
Bone transverse loading	
Tension	51
Compression	133

Source: Adapted from Ozkaya and Nordin, *Fundamentals of Biomechanics, Equilibrium, Motion and Deformation*, Van Nostrand Reinhold, New York, 1991. With permission.

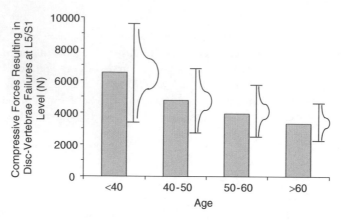

FIGURE 11.10 Mean and range of disc compression failures by age. (Adapted from National Institute for Occupational Safety and Health (NIOSH) Work practices guide for manual lifting, Department of Health and Human Services (DHHS), NIOSH, Cincinnati, OH, 81–122, 1981. With permission.)

the compressive load on the spine reaches a value of 9317 N, almost all of those exposed to the loading will experience a vertebral endplate microfracture. It should also be noted that the tolerance distribution shifts to lower levels with increasing age (Adams et al., 2000). In addition, it should be emphasized that this tolerance is based upon compression of the vertebral endplate alone. Shear and torsional forces in combination with compressive loading would be expected to further lower the tolerance of the end plate.

This distribution of risk has been widely used as the tolerance limits of the spine. However, it should be noted that others have identified different limits of vertebral endplate tolerance. Jager et al. (1991) have reviewed 13 studies of spine compressive strength and suggested different compression value limits. Their summary of these spine tolerance limits are shown in Table 11.2. These researchers have also been able to describe the vertebral compressive strength based upon an analysis of 262 values collected from 120 samples. They have related the compressive strength of the lumbar spine according to a regression equation:

$$\text{Compressive Strength (kN)} = (7.26 + 1.88\,\text{G}) - 0.494 + 0.468\,\text{G}) \cdot A$$
$$+ (0.042 + 0.106\,\text{G}) \cdot C - 0.145 \cdot L - 0.749 \cdot S,$$

TABLE 11.2 Investigations into Static Lumbar Compressive Strength

Reference	n	Strength in kN Mean	s.d.
Wyss and Ulrich, 1954	8	5.89	2.24
Brown et al., 1957	5	5.20	0.54
Perey, 1957	142	5.15	2.10
Decoulx and Rienau, 1958	9	4.41	1.14
Evans and Lissner, 1959	11	3.51	1.22
Roaf, 1960	3	4.83	2.06
Eie, 1966	16	3.70	1.60
Farfan, 1973	39	3.84	1.22
Hutton et al., 1979	23	5.35	2.67
Hansson et al., 1980	109	3.85	1.71
Hutton and Adams, 1982	33	7.83	2.87
Brinckmann and Horst, 1983	22	6.42	2.00
Brinckmann et al., 1989	87	5.35	1.76
Female	132	3.97	1.50
Male	174	5.81	2.58
Total	507	4.96	2.20

Source: Adapted from Jager, Luhman, and Laurig, Int. J. Indust. Ergo., 1991. With permission.

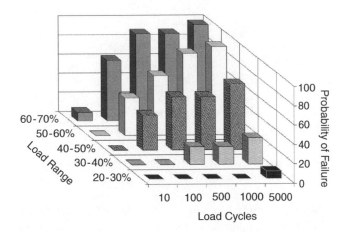

FIGURE 11.11 Probability of a motion segment to be fractured in dependence on the load range and the number of load cycles. (Adapted from Brinckmann, et al., *Clin. Biomech.*, 3(Suppl. 1), S1–S23, 1988. With permission.)

where *A* is the age in decade; G is the gender coded as 0 for female or 1 for male; *C* is the cross-sectional area of the vertebrae in cm^2; *L* is the the lumbar level unit where 0 is the L5/S1 disc, 1 represents the L5 vertebrae, etc. through 10, which represents the T10/L1 disc; S is the structure of interest where 0 is a disc and 1 is a vertebra.

This analysis suggests that the decrease in strength within a lumbar level is about 0.15 kN of that of the adjacent vertebrae and that the strength of the vertebrae is about 0.8 kN lower than the strength of the discs (Jager et al., 1991). Using this equation these researchers were able to account for 62% of the variability among the samples.

It has also been suggested that the tolerance limits of the spine varies as a function of frequency of loading (Brinkmann et al., 1988). Figure 11.11 indicates that spine tolerance varies as a function of spine load level and frequency of loading.

11.2.6.5 Pain Tolerance

It is believed that there are numerous pathways to pain perception associated with musculoskeletal disorders (Cavanaugh, 1995; Cavanaugh et al., 1997; Khalsa, 2004). It is important to understand these pathways since they are the basis for the structure and tissue limits employed in ergonomic logic. One can consider the quantitative limits above which a pain pathway is initiated as a tolerance limit for ergonomic purposes. While none of these pathways have been defined quantitatively, they are appealing since they represent biologically plausible mechanisms that complement the view of injury association derived from the epidemiologic literature.

In general, several broad categories of pain pathways are believed to exist that may affect the design of the workplace. These categories are associated with: (1) structural disruption, (2) tissue stimulation and pro-inflammatory response, (3) physiologic limits, and (4) psychophysical acceptance. Each of these pathways is expected to have different tolerance limits to mechanical loading of the tissue. Although many of these limits have yet to be quantitatively defined, future biomechanical assessments are expected to compare tissue loads to these limits when the dose–response relationship becomes better defined.

11.3 The Application of Biomechanics to the Workplace

11.3.1 Biomechanics of Commonly Affected Body Structures

Now that the basic concepts and principles of biomechanics relevant to ergonomics situations have been established we can apply these principles to various work situations. This section will show how one can apply these principles to various regions of the body that are typically affected by occupational tasks.

11.3.1.1 Shoulder

Shoulder pain is suspected of being one of the most under-recognized musculoskeletal disorders in the workplace. Second only to low back injury and neck pain, shoulder disorders are increasingly being recognized as a major workplace problem by those organizations that have reporting systems sensitive enough to detect such trends. The shoulder is one of the more complex structures of the body with numerous muscles and ligaments crossing the shoulder joint-girdle complex. Because of its biomechanical complexity surgical repair of the shoulder can be problematic. During many shoulder surgeries it is often necessary to damage much of the surrounding tissue in an attempt to reach the structure in need of repair. Often the target structure is small in size and difficult to reach. Thus, often at times, more damage is done to surrounding tissues than the benefits derived to the target tissue. Therefore, the best course of action is to ergonomically design work stations so that the risk of initial injury is minimized.

Since the shoulder joint is so biomechanically complex, much of our biomechanical knowledge is derived from empirical evidence. The shoulder represents a statically indeterminate system in that we can typically measure six external moments and forces acting about the point of rotation, yet, there are far more internal forces (over 30 muscles and ligaments) that must counteract the external moments. Thus, quantitative estimates of shoulder joint loading are rare.

With respect to the shoulder, optimal workplace design is typically defined in terms of preferred posture during work. Shoulder *abduction*, defined as the elevation of the shoulder in the lateral direction, is of concern when work is performed overhead. Figure 11.12 indicates shoulder performance measures in terms of both available strength and perceived fatigue while the shoulder is held in varying degrees of abduction. This figure indicates that shoulder can produce a considerable amount of strength throughout shoulder abduction angles of between 30 and 90°. However, when comparing fatigue characteristics at these same abduction angles it is apparent that fatigue increases rapidly as the shoulder is abducted above 30°. Thus, even though strength is not a problem at shoulder abduction angles upto 90°, fatigue becomes the limiting factor. Therefore, the only position of the shoulder that is acceptable from both a strength and fatigue standpoint is a shoulder abduction of at most 30°.

Shoulder *flexion* has been examined almost exclusively as a function of fatigue. Chaffin (1973) has shown that even slight shoulder flexion can influence fatigue characteristics of the shoulder musculature. Figure 11.13 and Figure 11.14 indicate the effects of vertical height of the work and horizontal distance, respectively, during shoulder flexion while seated upon fatigability of the shoulder musculature. During vertical flexion/extension (Figure 11.13), fatigue occurs more rapidly as the workers' arm becomes more elevated. This trend is most likely due to the fact that the muscles are farther from the neutral position as

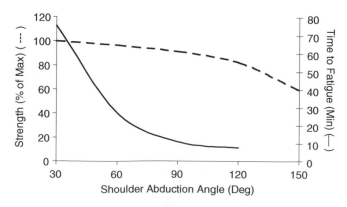

FIGURE 11.12   Shoulder abduction strength and fatigue time as a function of shoulder abducted from the torso. (Adapted from Chaffin, D.B. and Andersson, G.B., *Occupational Biomechanics*, John Wiley & Sons, Inc., New York, 1991. With permission.)

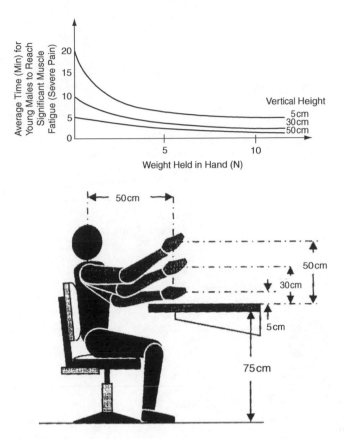

FIGURE 11.13 Expected time to reach significant shoulder muscle fatigue for varied arm flexion postures. (Adapted from Chaffin, D.B. and Andersson, G.B., *Occupational Biomechanics*, John Wiley & Sons, Inc., New York, 1991. With permission.)

the shoulder becomes more elevated thus affecting the length–strength relationship (Figure 11.5) of the shoulder muscles. Figure 11.14 shows that as the horizontal distance between the work and the body is increased, the time to reach significant fatigue is decreased. This trend is due to the fact that as a load is held further from the body, more of the external moment (force · distance) must be supported by the shoulder. Thus, the shoulder muscles must produce a greater internal force when the load is held further from the body and they fatigue quicker. Elbow supports have been shown to significantly increase the endurance time in these postures. In addition an elbow support has the effect of changing the bio-mechanical situation by providing a fulcrum at the elbow. Thus, the axis is rotation becomes the elbow instead of the shoulder and this makes the external moment much shorter. As shown in Figure 11.15, this not only increase the time one can maintain a posture, but also significantly increases the external load one can hold in the hand.

11.3.1.2   Neck

Neck disorders can also be associated with sustained work postures. In general, the more upright posture of the head, the less muscle activity and neck strength is required to maintain the posture. Upright neck postures also have the advantage of reducing the extent of fatigue perceived in the neck region. This relationship is shown in Figure 11.16. This trend indicates that when the head is tilted forward by 30° or more from the vertical position, the time to experience significant neck fatigue decreases rapidly. From a biomechanical standpoint, as the head is flexed forward the center of mass of the head moves forward relative to the base of support of the head (spine). Therefore, as the head is moved forward,

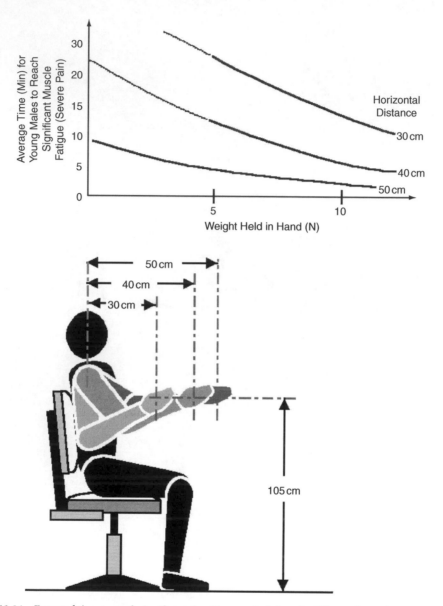

FIGURE 11.14 Expected time to reach significant shoulder muscle fatigue for different forward arm reach postures. (Adapted from Chaffin, D.B. and Andersson, G.B., *Occupational Biomechanics*, John Wiley & Sons, Inc., New York, 1991. With permission.)

more of a moment is imposed about the spine, which necessitates increased activation of the neck musculature and greater risk (probability of fatigue) since a static posture is maintained by the neck muscles. When the head is not flexed forward and is relatively upright, the neck can be positioned in such a way that minimal muscle activity is required of the neck muscles and thus fatigue is minimized.

11.3.1.3 Trade-Offs in Work Design

The key to optimal ergonomic workplace design, from a biomechanical standpoint, is to consider the biomechanical trade-offs associated with a given work situation. Trade-off considerations are necessary because it is often the case that a situation that is advantageous for one part of the body is

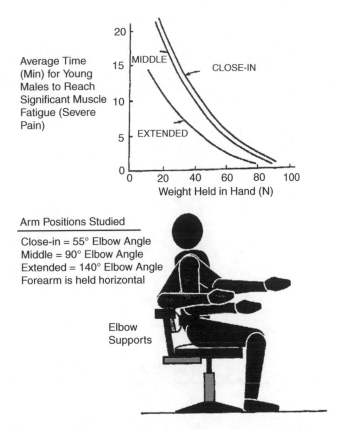

FIGURE 11.15 Expected time to reach significant shoulder and arm muscle fatigue for different arm postures and hand loads with the elbow supported. The greater the reach, the shorter the endurance time. (Adapted from Chaffin, D.B., and Andersson, G.B., *Occupational Biomechanics*, John Wiley & Sons, Inc., New York, 1991. With permission.)

disadvantageous for another part of the body. Thus, ergonomic design of the workplace requires one to consider the various trade-offs and rationales for various design options.

One common trade-off encountered in ergonomic design is the trade-off between accommodating the shoulders and accommodating the neck. This trade-off is resolved by considering the hierarchy of needs required by the task. Figure 11.17 illustrates this reasoning. The recommended height of the work is a function of the type of work that is to be performed. Precision work requires a high level of visual acuity, which becomes the greatest need in order to perform the work task. However, if the work is performed at too low of a level the head must be flexed in order to accommodate the visual requirements of the job and this becomes a problem for the neck. Therefore, in this circumstance, visual accommodation is at the top of the hierarchy of task needs, so that the work is raised to a relatively high level (95 to 110 cm above the floor) in order to accommodate vision and the neck posture. This posture accommodates the neck but creates a problem for the shoulders since they must be abducted when the work level is high. Thus, a trade-off should be considered. In this instance, ideal shoulder posture is sacrificed in order to accommodate the neck since the visual requirements of the job represent the greater priority for work performance, whereas, the minimal shoulder strength is required for precision work and, thus, represents a lower priority. Thus, visual accommodation is given a higher priority in the hierarchy of task needs and this criterion must be given priority over any other criteria. Besides, the shoulder problems can be minimized by providing wrist or elbow supports at the workplace.

The other extreme example of the working height situation involves heavy work. The greatest demand on the worker during heavy work involves a high degree of arm strength, whereas, visual requirements in

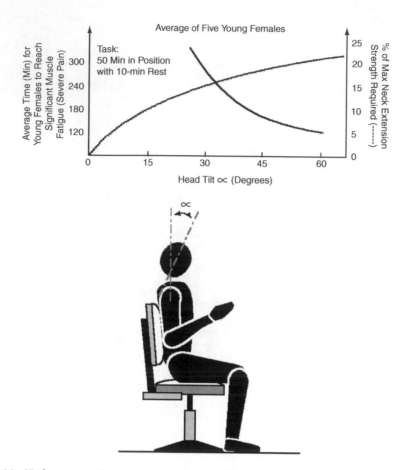

FIGURE 11.16 Neck extensor fatigue and muscle strength required versus head tilt angle. (Adapted from Chaffin, D.B. and Andersson, G.B., *Occupational Biomechanics*, John Wiley & Sons, Inc., New York, 1991. With permission.)

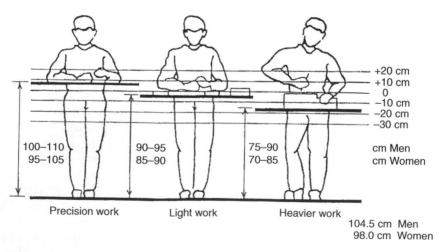

FIGURE 11.17 Recommended heights of bench for standing work. The reference line (+0) is the height of the elbows above the floor. (Adapted From Grandjean, E., *Fitting the Task to the Man: An Ergonomic Approach*, Taylor & Francis, Ltd., London, 1982. With permission.)

this type of work are often minimal. Thus, shoulder position represents a higher priority in the hierarchy of task needs in this situation. In this situation, ideal neck posture is typically sacrificed in favor of more favorable shoulder and arm postures. For this reason, heavy work is performed at a height of 70 to 90 cm above floor level. With the work set at this height, the position wherein the elbows are close to 90° maximizes strength (Figure 11.6). In addition, the shoulders are close to 30° of abduction, which minimizes fatigue. In this situation, the neck is not in an optimal position but the hierarchy logic dictates that the visual demands of a heavy task would not be substantial and thus the neck would not be flexed for prolonged periods of time and, therefore, do not pose much of a risk.

The third work height situation involves light work. Light work is a mix of moderate visual demands with moderate strength requirements. In this situation, work is a compromise between shoulder position and visual accommodation and neither visual nor strength demands dominate the hierarchy of work needs. Thus, the height of the work is set at a height between those of the precision work height level and the heavy work height level. In this manner, a compromise between the benefits and costs associated with accommodating the neck versus the shoulder is resolved. This situation dictates that the work is performed at a level of between 85 and 95 cm off the floor under light work conditions.

11.3.1.4   The Back

Low back disorders (LBD) have been identified as one of the most common and significant musculoskeletal problems in the U.S. that results in substantial amounts of morbidity, disability, and economic loss (Hollbrook et al., 1984; Praemer et al., 1992). LBD are one of the most common reasons for workers to miss work. Back disorders were responsible for the loss of over 100 million lost workdays in 1988 with 22 million cases reported that year (Guo, 1993; Guo et al., 1999). Among those under 45 years of age, LBD is the leading cause of activity limitations and can affect upto 47% of workers with physically demanding jobs (Andersson, 1997). The prevalence of LBD has also been observed to increase by 2700% since 1980 (Pope, 1993). The costs associated with LBD are significant with health care expenditures incurred by individuals with back pain in the U.S. exceeding $90 billion in 1998 (Luo et al., 2004).

It is clear that the risk of LBD can be associated with industrial work (NRC, 1999, 2001). Thirty percent of occupation injuries in the U.S. are caused by overexertion, lifting, throwing, holding, carrying, pushing, and or pulling objects that weigh 50 lb or less. Twenty percent of all workplace injuries and illnesses are back injuries, which account for upto 40% of compensation costs. Estimates of occupational LBD prevalence vary from 1 to 15% annually depending upon occupation and, over a career, can seriously affect 56% of workers.

Manual materials handling (MMH) activities, specifically lifting, dominate occupationally related LBD risk. It has been estimated that lifting and MMH account for upto two-thirds of work-related back injuries (NRC, 2001). From a biomechanical standpoint, we assume that most serious and costly back pain is discogenic in nature and has a mechanical origin (Nachemson, 1975). Studies have found increased degeneration in the spines of cadaver specimens who had previously been exposed to physically heavy work (Videman, et al., 1990). This suggests that occupationally related LBDs are closely associated with spine loading.

11.3.1.4.1   Significance of Moments

The most important concept associated with occupationally related LBD risk is that of the external moments imposed about the spine (Marras et al., 1993, 1995). As with most structures, the loading of the trunk is influenced greatly by the external moment imposed about the spine. However, because of the geometric arrangement of the trunk musculature relative to the trunk fulcrum during lifting, very large loads can be generated by the muscles and imposed upon the spine. Figure 11.18 shows this biomechanical arrangement of lever system. As indicated here, the back musculature is at a severe biomechanical disadvantage in many manual materials handling situations. Supporting an external load of 222 N (about 50 lb) at a distance of 1 m from the spine imposes a 222 Nm external moment about the spine. However, since the spine supporting musculature are at a relatively close proximity relative to the external load, the trunk musculature must exert extremely large forces (4440 N or 998 lb) to simply hold the external load in equilibrium. These internal loads can be far greater if dynamic

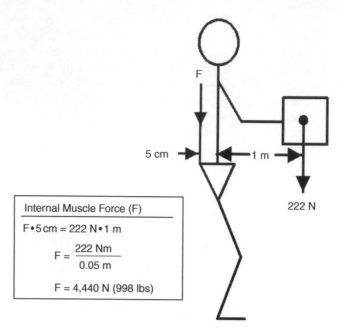

FIGURE 11.18   Internal muscle force required to counterbalance an external load during lifting.

motion of the body is considered (since force is a product of mass and acceleration). Thus, the most important concept to consider in workplace design from a back protection standpoint is to keep the moment arm at a minimum.

11.3.1.4.2   Lifting Style

The external moment concept has major implications for lifting styles or the best "way" to lift. Since the externally applied moment significantly influences the internal loading, the lifting style is of far less concern compared to the magnitude of the applied moment. Some have suggested that proper lifting involves lifting by "using the legs" as opposed to "stoop" lifting (bending from the waist). However, spine loading has also been found to be a function of anthropometry as well as lifting style. Biomechanical analyses (Park and Chaffin, 1974; van Dieen et al., 1999) have demonstrated that no one lift style is correct for all body types. For this reason the National Institute of Occupational Safety and Health (NIOSH, 1981) has concluded that liftstyle need not be a consideration when assessing the risk of occupationally related LBD. Some have suggested that the internal moment of the trunk has a greater mechanical advantage when lumbar lordosis is preserved during the lift (NIOSH, 1981; Anderson et al., 1985; McGill et al., 2000; McGill, 2002a,b). Thus, from a biomechanical standpoint, the primary indicator of spine loading and, thus, the correct lifting style is whatever style permits the worker to bring the center of mass of the load as close to the spine as possible.

11.3.1.4.3   Seated vs Standing Workplaces

Seated workplaces have become more prominent of late, especially with the aging of the workforce and the introduction of service-oriented and data processing jobs. It has been well documented that loads on the lumbar spine are always greater when one is seated compared to a standing posture (Andersson et al., 1975). This is due to the tendency for the posterior (bony) elements of the spine to form an active load path when one is standing. When seated, these elements are disengaged and more of the load passes through the intervertebral disc. Thus, work performed in a seated position puts the worker at greater risk of loading and therefore damaging the disc. Given this situation, it is important to consider the design features of a chair since it may be possible to influence disc loading through chair design. Figure 11.19 shows the results of pressure measurements made in the intervetebral disc of workers as

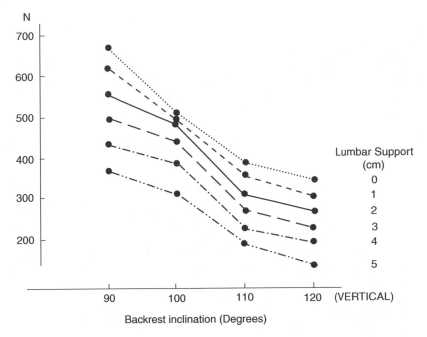

FIGURE 11.19 Disc pressures measured with different backrest inclinations and different size lumbar supports. (Adapted from Chaffin, D.B. and Andersson, G.B., *Occupational Biomechanics*, John Wiley & Sons, Inc., New York, 1991. With permission.)

the back angle of the chair and magnitude of lumbar support are varied. Since it is infeasible to directly measure the forces in the spine *in vivo*, disc pressure measures have traditionally been used as a rough approximation of loads imposed upon the spine. This figure indicates that both the seat back angle and lumbar support features have a significant effect on disc pressure. Disc pressure is observed to decrease as the backrest angle is increased. However, increasing the backrest angle in the workplace is often not practical, since it also has the effect of moving the worker away from the work and thereby increasing external moment. The figure also indicates that increasing lumbar support can also significantly reduce disc pressure. This reduction in pressure is most likely due to the fact that as lumbar curvature (lordosis) is reestablished (with lumbar support) the posterior elements play more of a role in providing an alternative load path as is the case when standing in the upright position.

Less is known about risk to the low back associated with prolonged standing. It is known that the muscles experience low level static exertions and may be subject to the static overload through the muscle static fatigue process described in Figure 11.9. This fatigue can result in lowered muscle force generation capacity and can, thus, initiate the cumulative trauma sequence of events (Figure 11.3). It has been demonstrated that this fatigue and cumulative trauma sequence can be minimized by two actions. First, foot rails provide a mechanism to allow relaxation of the large back muscles and thus increased blood flow to the muscle. This reduces the static load and fatigue in the muscle by the process described in Figure 11.9. When a leg is lifted and rested on the foot rest the large back muscles are relaxed on one side of the body and the muscle can be supplied with oxygen. Alternating legs on the foot rest provides a mechanism to minimize back muscle fatigue throughout the day. Second, floor mats have been shown to decrease the fatigue in the back muscles provided that the mats have proper compression characteristics (Kim et al., 1994). Floor mats are believed to induce body sway, which facilitate the pumping of blood through back muscles, thereby, minimizing fatigue.

Our knowledge of when standing workplaces are preferable is dictated mainly by work performance criteria. In general, standing workplaces are preferred when: (1) the task required a high degree of mobility (reaching and monitoring in positions that exceed the reach envelope or when performing tasks at different heights or different locations), (2) precise manual control actions are not required, (3) leg room is not available (when leg room is not available, the moment arm distance between the external load and the back is increased and thus greater internal back muscle force and spinal load result), and (4) heavy weights are handled or large forces are applied. When jobs must accommodate both sitting and standing, it is important to ensure that the positions and orientations of the body, especially the upper extremity, are in the same location under both standing and sitting conditions.

11.3.1.5 Wrists

The wrist has been of increased interest to ergonomists in the past three decades. The Bureau of Labor Statistics reports that repetitive trauma has increase from 18% of occupational illnesses in 1981 to 63% of occupational illnesses in 1993. Based upon these figures, repetitive trauma has been described as the *fastest growing* occupational problem. Even though these numbers and statements appear alarming one must acknowledge that occupational illnesses represent 6% of all occupational injuries and illnesses. Furthermore, these figures for illness include illnesses unrelated to musculoskeletal disorders such as noise-induced hearing loss. Thus, the magnitude of the cumulative trauma problem must not be over-stated. Nonetheless, there are specific industries (i.e., meat packing, poultry processing, etc.) where cumulative trauma to the wrist is a major problem and this problem has reached epidemic proportions within these industries.

11.3.1.5.1 Wrist Anatomy and Loading

In order to understand the biomechanics of the wrist and how cumulative trauma occurs in this structure one must appreciate the anatomy of the upper extremity. Figure 11.20 shows a simplified anatomical drawing of the wrist. This figure shows that few power-producing muscles reside in the hand itself. The thenar muscle, which activates the thumb is one of the few power producing muscles in the hand. The vast majority of the hand's power-producing muscles are located in the forearm. Force is trans-mitted from these forearm muscles to the fingers through a network of tendons (tendons attach muscles to bone). These tendons originate at the muscles in the forearm traverse the wrist (with many of them passing through the carpal canal), pass through the hand, and culminate at the fingers. These tendons are secured or "strapped down" at various points along this path with ligaments that keep the tendons in close proximity to the bones forming a sort of pulley system. This system results in a hand that is very small and compact, yet capable of generating large amounts of force. The price the musculoskeletal system pays for this design is friction. The forearm muscles must transmit force over a very long distance in order to supply internal forces to the fingers. Thus, a great deal of tendon travel must occur and this tendon travel can result in significant tendon friction under repetitive motion conditions thereby initi-ating the events outlined in Figure 11.3. Thus, the key to controlling wrist cumulative trauma is rooted in an understanding of those workplace factors that adversely affect the internal force generating (muscles) and transmitting (tendons) structures.

11.3.1.5.2 Biomechanical Risk Factors for the Wrist

A number of risk factors for wrist cumulative trauma have been documented in the literature. First, deviated wrist postures are known to reduce the volume of the carpal tunnel and, thus, increase tendon friction. In addition, grip strength is dramatically reduced by deviations in the wrist posture. Figure 11.21 indicates that any deviation from the wrist's neutral position significantly decreases the grip strength of the hand. This reduction in strength is caused by a change in the length–strength relationship (Figure 11.5) of the forearm muscles once the wrist is bent. Hence, the muscles are working at a level that is greater than necessary. This reduced strength potential associated with deviated wrist positions can, therefore, more easily initiate the sequence of events associated with cumulative trauma (Figure 11.3). Thus, deviated wrist postures not only increase tendon travel and friction, but also increase the amount of muscle strength necessary to perform the gripping task.

Second, increased frequency or repetition of the work cycle has been identified as a risk factor for cumulative trauma disorders (CTD; Silverstein et al., 1996, 1997). Studies have indicated that increased frequency of wrist motions increases the risk of developing a cumulative trauma disorder. Repeated

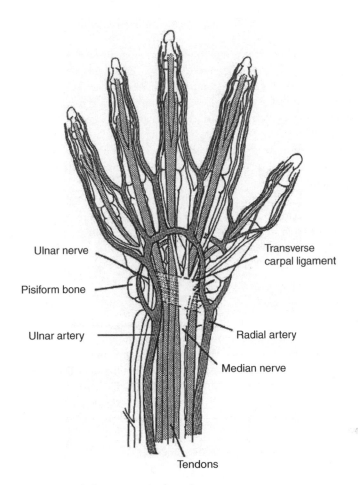

FIGURE 11.20 Important anatomical structures in the wrist.

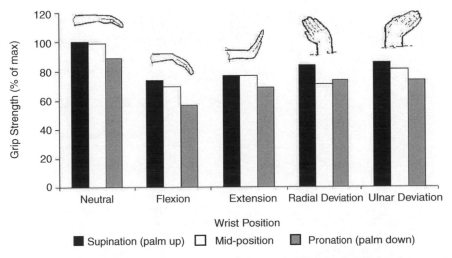

FIGURE 11.21 Grip strength as a function of wrist and forearm position. (Adapted from Sanders, M.S. and McCormick, E.F., *Human Factors in Engineering and Design*, McGraw-Hill Inc., New York, 1993. With permission.)

motions requiring a cycle time of less than 30 sec is considered a candidate for cumulative trauma disorder risk.

Third, the force applied by the hands and fingers during a work cycle has been identified as a risk factor. In general, the greater the force required by the work the greater the risk of CTD. Greater hand forces result in greater tension within the tendons and result in greater tendon friction and tendon travel. Another factor related to force is wrist acceleration. Industrial surveillance studies have reported that repetitive jobs resulting in greater wrist acceleration are associated with greater CTD incident rates (Marras and Schoenmarklin, 1993; Schoenmarklin et al., 1994). Since force is a product of mass and acceleration, jobs that increase the angular acceleration of the wrist joint result in greater tension and force transmitted through the tendons. Thus, wrist acceleration can be another mechanism of imposing force on the wrist structures.

Fourth, as shown in Figure 11.20, the anatomy of the hand is such that the median nerve becomes very superficial at the palm. Direct impact to the palm of the hand through pounding or striking an object with the palm, as is done often in assembly work, can directly stimulate the median nerve and initiate symptoms of cumulative trauma even though the work may not be repetitive.

11.3.1.5.3 Grip Design

The design of a tool's gripping surface can dramatically affect the activity of the internal force transmission system (tendon travel and tension). The grip opening and shape have a major influence on the available grip strength. Figure 11.22 shows how grip strength capacity changes as a function of the separation distance of the grip opening. This figure indicates that maximum grip strength occurs within a very narrow range of grip openings. If the grip opening deviates from this ideal range by as little an inch (a couple of centimeters), then grip strength is dramatically reduced. This change in strength is also due to the length–strength relationship of the forearm muscles. Also indicted in Figure 11.22 are the effects of hand anthropometry. The workers hand size as well as hand preference

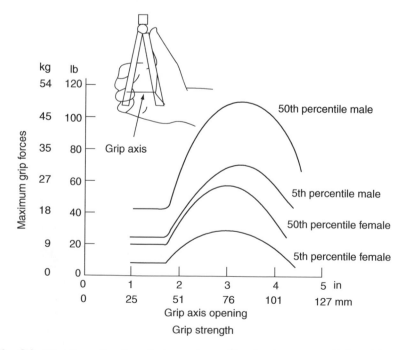

FIGURE 11.22 Grip strength as a function of grip opening and hand anthropometry. (Adapted from Sanders, M.S. and McCormick, E.J., *Human Factors in Engineering and Design*, McGraw-Hill Inc., New York, 1993. With permission.)

can influence grip strength and risk. Therefore, proper design of the handles is crucial in ergonomic workplace design.

Handle shape can also affect the strength of the wrist. Figure 11.23 shows how changes in the design of screwdriver handles can affect the maximum force that can be exerted. The biomechanical origin of these differences in strength capacity is most likely related to the length–strength relationship of the forearm muscles as well as contact area with the tool. The handle designs that result in less strength permit the wrist to twist or permit the grip to slip resulting in a deviation from the ideal length–strength position in the forearm muscles.

11.3.1.5.4 Gloves

The use of gloves can significantly influence the generation of grip strength and may play a role in the development of CTDs. When gloves are worn during work three effects must be considered. First, the grip strength generated is often reduced. There is typically a 10 to 20% reduction in grip strength when gloves are worn. When using gloves the coefficient of friction between the hand and the tool can be reduced which, in turn, permits some slippage of the hand upon the tool surface. This slippage can result in a deviation from the ideal muscle length and thus a reduction in available strength. The degree of slippage and the degree of strength loss depends upon how well the gloves fit the hand and the type of material used in the glove. Poorly fitting gloves result in greater strength loss. Figure 11.24 indicates how the glove material and glove fit can dramatically influence grip force application.

Second, when wearing gloves, even though the externally applied force (grip strength) is often reduced, the internal forces are often very large compared to not using a glove. For a given grip strength the muscle activity is significantly greater when using gloves compared to a bare-handed condition (Kovacs et al., 2002). Thus, the musculoskeletal system is less efficient when wearing a glove.

Third, the ability to perform a work task is affected negatively when wearing gloves. Figure 11.25 shows the increase in time required to perform work tasks when wearing gloves composed of different materials compared to performing the task bare-handed. The figure indicates that task performance can increase upto 70% when wearing gloves.

These effects have indicated that there are bio-mechanical costs associated with glove usage. Less strength capacity is available to the worker, more internal force is generated, and worker productivity is affected. These negative effects of gloves do not mean that gloves should never be worn at work. When hand protection is needed gloves should be considered as a potential solution. However, protection should only be provided to the parts of the hand that require protection. For example, if the palm of the hand requires protection, fingerless gloves might provide an acceptable solution. If the fingers require protection, but there is little risk to the palm of the hand, then grip tape wrapped around the fingers might be considered. In addition, different styles, materials, and sizes of gloves will fit workers differently. Thus, gloves produced by various manufacturers and of different sizes should be available to the worker to minimize the negative effects mentioned before.

11.3.1.5.5 Design Guidelines

This discussion has indicated that there are many factors that can affect the biomechanics of the

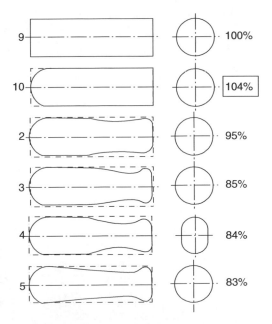

FIGURE 11.23 Maximum force, which could be exerted on a screwdriver as a function of handle shape. (From Konz, S.A., *Work Design: Industrial Ergonomics*, 2nd ed., Grid Publishing, Inc., Columbus, OH, 1983. With permission.)

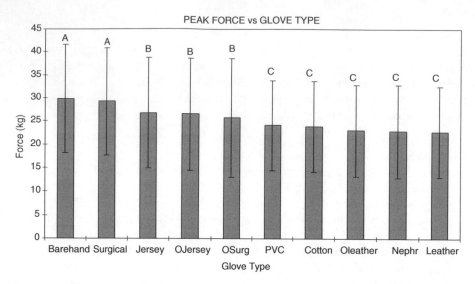

FIGURE 11.24   Peak grip force shown as a function of type of glove. Different letters above the columns indicate statistically significant differences.

wrist and the subsequent risk of CTDs. This suggests that proper ergonomic design of a work task cannot be accomplished by simply providing the worker with an "ergonomically designed" tool. Since ergonomics is associated with matching the workplace design to the workers' capabilities it is not possible to design an "ergonomic tool" without considering the workplace design and task requirements simultaneously. What might be an "ergonomic" tool for one work situation may be improper for use while a worker is assuming another work posture. For example, using an *in-line* tool may keep the wrist straight when inserting a bolt into a horizontal surface. However, if the bolt is to be inserted into a vertical surface a *pistol grip* tool may be more appropriate. Using the in-line tool in this situation (inserting a bolt into a vertical surface) may cause the wrist to be significantly deviated. Hence, there are no ergonomic tools.

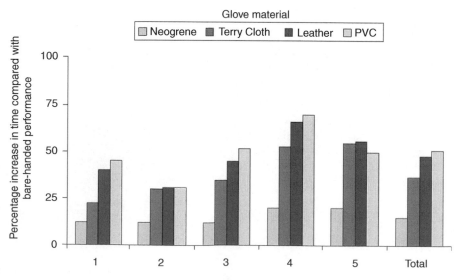

FIGURE 11.25   Performance (time to complete) on a maintenance-type task while wearing gloves constructed of five different materials. (From Sanders, M.S. and McCormick, E.J., *Human Factors in Engineering and Design*, McGraw-Hill Inc., New York, 1993. With permission.)

There are just ergonomic *situations*. What may be an ergonomically correct tool in one situation may not be ergonomically correct in another work situation.

Workplace design should be performed with care and trade-offs between different parts of the body must be considered by taking into consideration the various biomechanical trade-offs. Given these considerations, the following components of the workplace should be considered when designing a workplace so that cumulative trauma risk is minimized. First, maintain a neutral wrist posture. Second, minimize tissue compression. Third, avoid actions that repeatedly impose force on the internal structures. Fourth, minimize required wrist accelerations and motions. Fifth, consider the impact of glove use, hand size, and left-handed workers.

11.4   Analysis and Control Measures Used in the Workplace

Several analyses and control measures have been developed to evaluate and control biomechanical loading of the body during work. Since LBDs are often the objective of a biomechanical workplace analysis, most of these analyses methods have focused on spine risk. However, several of the measures also include analyses of risk to other body parts.

11.4.1   Lifting Belts

Back support belts or lifting belts have been used with increasing frequency in the workplace. However, a review of the literature related to lifting belts offers no clear biomechanical benefits of belt use. Reviews by McGill (1993) and NIOSH (1994) had concluded that there are so few well-executed studies that one can not unequivocally judge the benefits of lifting belts. However, later epidemiologic studies have indicated that there are few benefits to back belt usage for those not suffering from an LBD (Wassell et al., 2000).

Epidemiological studies have generally been limited in scope and often result in findings that were confounded by other factors such as training, the type of belt used, or the "Hawthorne Effect." Walsh and Schwartz (1990) reported a reduction in LBD injury rate with the usage of back supports (hard shell corsets) and have recommended that they would be effective at controlling the risk of LBD. However, the data from this study suggest that back supports were only effective for those workers who had previously suffered an LBD. Mitchell et al. (1994) retrospectively evaluated injury data associated with belt use over a 6-yr period at Tinker Air Force Base. Over this time period, two different types of belts were used. Leather belts were used in the first two years of the study, whereas, velcro belts were used over the last four years. No relationship between belt usage and back injury could be established, but they did find that those who wore belts suffered more costly injuries once they occurred. Reddell et al. (1992) observed that when workers stopped wearing belts the risk of injury increased. However, this study suffers from small sample size, which makes it difficult to assess the strength of the association.

Psychophysical studies (which can be used to define tolerance) have attempted to assess whether the magnitude of the weight a person was willing to lift changes when wearing a back belt. McCoy et al. (1988) found that subjects were willing to lift 19% more weight when belts were used but found no difference between belt types. Subjects reported that they preferred the elastic belt. However, this does not suggest that workers would be at lowered risk of back injury since it is not clear that spine tolerance to load would be increased with belt use.

Biomechanically based studies of lifting belts have documented their influence upon trunk motion, trunk muscle activity, and indirect indicators or predictions of trunk loading. The most consistent finding of these studies is that lateral bending and twisting trunk motion is significantly reduced with belt usage (Lantz and Schultz, 1986; McGill et al., 1994; Lavender et al., 1995). However, belt use has not resulted in a reduction of spine loading under realistic materials handling conditions (Granata et al., 1997; Marras et al., 2000a, b).

Perhaps the most important reason to be cautious of lifting belts is unrelated to biomechanical loading of the spine. There appear to be physiological reasons to be concerned with the use of lifting belts. One study has shown that lifting belts can significantly increase blood pressure (Rafacz and McGill, 1996). This could become problematic for workers who have a compromised cardiovascular system.

The brief review indicates that there is a large amount of conflicting evidence as to the benefits or liabilities associated with the use of back belts. There appears to be little biomechanical benefit to belt usage and some negative physiological consequences. Recent epidemiologic studies have not been able to find any evidence of benefit. A consistent finding among the studies is that if there is a benefit to back belts, it is probably for those who have previously experienced an LBD. The literature also suggests that belts should only be used for a limited period of time. Until more definitive studies are available it is prudent to use caution when recommending the use of back belts in a work environment. This includes a screening by an occupational physician who is familiar with the literature so that potential cardiovascular problems can be assessed.

11.4.2 1981 NIOSH Lifting Guide

The NIOSH has developed two assessment tools or guides to help determine whether a manual materials handling task is safe or risky. The lifting guide was originally developed in 1981 (NIOSH, 1981) and applies to lifting situations where the lifts are performed in the sagittal plane and to motions that are slow and smooth. Two benchmarks or limits are defined by this guide. The first limit is called the *action limit* (AL) and represents a magnitude of weight in a given lifting situation, which would impose a spine load corresponding to the beginning of LBD risk along a risk continuum. The AL is associated with the point in Figure 11.10 at which people under 40 yr of age just begin to experience a risk of vertebral endplate microfracture (3400 N of compressive load). The guide estimates the force imposed upon the spine of a worker as a result of lifting a weight and compares this spine load to the AL. If the weight of the object results in a spine load that is below the AL, the job is considered safe. If the weight lifted by the worker is larger than the AL, there is at least some level of risk associated with the task. The general form of the AL is defined according to Equation (11.1).

$$AL = k(HF)(VF)(DF)(FF), \tag{11.1}$$

where AL is the action limit in kg or lb; k is the load constant (40 kg or 90 lb), which is the greatest weight a subject could lift if all lifting conditions are optimal; HF is the horizontal factor defined as the horizontal distance from a point bisecting the ankles to the center of gravity of the load at the lift origin. Defined algebraically as $15/H$ (metric) or $6/H$ (US units); VF is the vertical factor or height of the load at lift origin. Defined algebraically as $(0.004)\,|V-75|$(metric) or $1\text{-}(0.01)|V-30|$(US units); DF is the distance factor or the vertical travel distance of the load. Defined algebraically as $0.7+7.5/D$ (metric) or $0.7+3/D$ (US units); FF is the frequency factor or lifting rate defined algebraically as $1-F/F_{\max}$ F = average frequency of lift, $F_{\max}$ is shown in Table 11.3.

The logic associated with this equation assumes that if the lifting conditions are ideal a worker could safely hold (and implies lift) the load constant, k (40 kg or 90 lb). If the lifting conditions are not ideal the allowable weight is discounted according to the four factors HF, VF, DF, and FF. These four factors are shown in monogram form in Figure 11.26 through Figure 11.29. According to the load discounting

TABLE 11.3 $F_{\max}$ Table

Period	Average Vertical Location (cm) (in)	
	Standing $V > 75$ (3)	Stooped $V \leq 75$ (3)
1 h	18	15
8 h	15	12

Source: Reprinted from NIOSH, *Work Practices Guide for Manual Lifting*, Department of Health and Human Services (DHHS) NIOSH, Cincinnati, OH, 81–122, 1981. With permission.

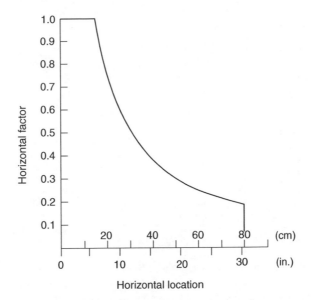

FIGURE 11.26   Horizontal factor, (*HF*) varies between the body interference limit and the limit of functional reach. (Adapted from National Institute for Occupational Safety and Health (NIOSH), Work practices guide for manual lifting, Department of Health and Human Services (DHHS), NIOSH, Cincinnati, OH, No. 81–122, 1981. With permission.)

associated with these figures, the *HF*, which is associated with the external moment has the most dramatic effect on acceptable lifting conditions. *VF* and *DF* are associated with the back muscle's length–strength relationship. *FF* attempts to account for the cumulative effects of repetitive lifting.

The second benchmark associated with this guide is the *maximum permissible limit* or MPL. The MPL represents the point at which significant risk, defined in part, as a significant risk of vertebral endplate microfracture (Figure 11.10). The MPL is associated with a compressive load on the spine of 6400 N, which corresponds to a point at which 50% of the people would be expected to suffer a vertebral endplate

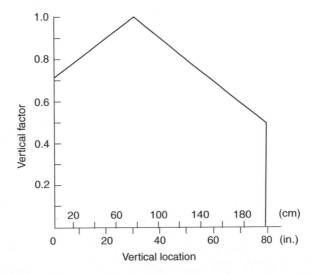

FIGURE 11.27   Vertical factor, (*VF*) varies both ways from knuckle height. (Adapted from National Institute for Occupational Safety and Health (NIOSH), Work practices guide for material lifting, Department of Health and Human Services (DHHS), NIOSH, Cincinnati, OH, 81–122, 1981. With permission.)

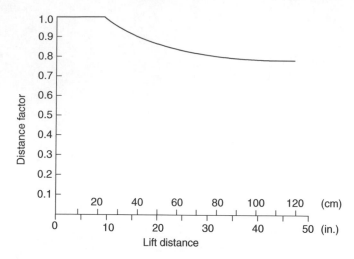

FIGURE 11.28   Distance factor, (*DF*) varies between a minimum vertical distance of 25 cm (10 in.) that was moved to a maximum distance of 200 cm (80 in.). (Adapted from National Institute for Occupational Safety and Health (NIOSH), Work practices guide for manual lifting. Department of Health and Human Services (DHHS), National Institute for Occupational Safety and Health (NIOSH), Cincinnati, OH, 81–122, 1981. With permission.)

microfracture. Equation (11.2) indicates that the MPL is a function of the AL and is defined as follows:

$$\text{MPL} = 3(\text{AL}). \tag{11.2}$$

The weight that the worker expected to lift in a work situation is compared to the AL and MPL. If the magnitude of weight falls below the AL the work is considered safe and no adjustments are necessary. If the magnitude of the weight falls above the MPL then the work is considered risky and engineering changes involving the adjustment of *HF, VF,* and/or *DF* are required to reduce the AL and MPL. If the weight falls between the AL and MPL then either engineering changes or administrative changes, defined as selecting workers who are less likely to be injured or rotating workers, are recommended.

The AL and MPL were also indexed to nonbiomechanical benchmarks. According to NIOSH (1981) these limits also correspond to strength, energy expenditure, and psychophysical acceptance points.

11.4.3   1993 Revised NIOSH Equation

The 1993 NIOSH revised lifting equation was introduced in order to address those lifting jobs that violate the sagittally symmetric lifting assumption (Waters et al., 1993). The concept of AL and MPL was replaced with a concept of a *lifting index* or *LI*. The *LI* is defined in Equation (11.3).

$$LI = \frac{L}{\text{RWL}}, \tag{11.3}$$

where *L* is the load weight or the weight of the object to be lifted; RWL is the recommended weight limit for the particular lifting situation; *LI* is the lifting index used to estimate relative magnitude of physical stress for a particular job.

If the *LI* is greater than 1.0, an increased risk for suffering a lifting-related LBD exists. The RWL is similar to the 1981 lifting guide AL equation [Equation (11.1)] in that it contains factors that discount the allowable load according to the horizontal distance, vertical location of the load, vertical travel distance, and frequency of lift. However, the form of these discounting factors was changed. Moreover, two additional discounting factors have been included. These additional factors include a lift asymmetry factor to account for asymmetric lifting conditions and a coupling factor that accounts for whether or not the load lifted has handles. The RWL is represented algebraically in Equation (11.4) (metric units) and Equation (11.5) (US units).

$$\text{RWL (kg)} = 23(25/H)[1 - (0.003|V - 75|)](0.82 + 4.5/D)(FM)[1 - (0.0032A)](CM), \quad (11.4)$$

$$\text{RWL (lb)} = 51(10/H)[1 - (0.0075|V - 30|)](0.82 + 1.8/D)(FM)[1 - (0.0032A)](CM), \quad (11.5)$$

where H is the horizontal location forward of the midpoint between the ankles at the origin of the lift. If significant control is required at the destination then H should be measured both at the origin and destination of the lift; V is the vertical location at the origin of the lift; D is the vertical travel distance between origin and destination of the lift; FM is the frequency multiplier shown in Table 11.4; A is the angle between the midpoint of the ankles and the midpoint between the hands at the origin of the lift; CM is the coupling multiplier ranked as either food, fair, or poor as described in Table 11.5.

In this revised equation the load constant has been significantly reduced compared to the 1981 equation. The adjustments for load moment, muscle length–strength relationships, and cumulative loading are still integral parts of this equation. However, these adjustments or discounting factors have been changed (compared to the 1981 Guide) to reflect the most conservative value of the biomechanical, physiological, psychophysical, or strength data upon which they are based. Recent studies report that the 1993 revised equation yields a more conservative (protective) prediction of work-related LBD risk (Marras et al., 1999).

11.4.4 Static Models

Biomechanically based spine models have been developed to help assess occupationally related manual materials handling tasks. These models assess the task based upon both spine loading criteria as well as through an evaluation of the strength required at the various major body joints in order to perform the task. One of the early static assessment models was developed by Chaffin at the University of Michigan (Chaffin, 1969). This original two-dimensional (2D) model has been expanded to a three-dimensional (3D) static model (Chaffin and Muzaffer, 1991; Chaffin et al., 1999) and has been developed to help

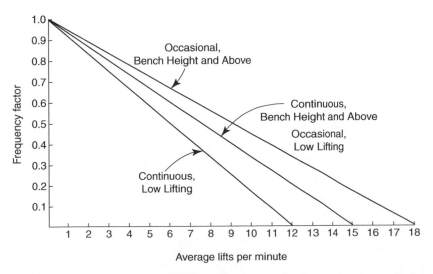

FIGURE 11.29 Frequency factor (*FF*) varies with lifts/minute and the $F_{\max}$ curve. The $F_{\max}$ depends upon lifting posture and lifting time. (Adapted from National Institute for Occupational Safety and Health (NIOSH), Work practices guide for manual lifting. Department of Health and Human Services (DHHS), National Institute for Occupational Safety and Health (NIOSH), Cincinnati, OH, 81–122, 1981. With permission.)

TABLE 11.4   Frequency Multiplier Table (*FM*)

| Frequency Lifts/min(*F*)[b] | Work Duration | | | | | |
| | ≤ 1 h | | > 1 but ≤ 2 h | | > 2 but ≤ 8 h | |
	V < 30[a]	*V* ≥ 30	*V* < 30	*V* ≥ 30	*V* < 30	*V* ≥ 30
≥0.2	1.00	1.00	0.95	0.95	0.85	0.85
0.5	0.97	0.97	0.92	0.92	0.81	0.81
1	0.94	0.94	0.88	0.88	0.75	0.75
2	0.91	0.91	0.84	0.84	0.65	0.65
3	0.88	0.88	0.79	0.79	0.55	0.55
4	0.84	0.84	0.72	0.72	0.45	0.45
5	0.80	0.80	0.60	0.60	0.35	0.35
6	0.75	0.75	0.50	0.50	0.27	0.27
7	0.70	0.70	0.42	0.42	0.22	0.22
8	0.60	0.60	0.35	0.35	0.18	0.18
9	0.52	0.52	0.30	0.30	0.00	0.15
10	0.45	0.45	0.26	0.26	0.00	0.13
11	0.41	0.41	0.00	0.23	0.00	0.00
12	0.37	0.37	0.00	0.21	0.00	0.00
13	0.00	0.34	0.00	0.00	0.00	0.00
14	0.00	0.31	0.00	0.00	0.00	0.00
15	0.00	0.28	0.00	0.00	0.00	0.00
>15	0.00	0.00	0.00	0.00	0.00	0.00

[a]Values of V are in inches.
[b]For lifting less frequently than once per 5 min, set *F* = 0.2 lifts/min.
Source: Reprinted from NIOSH, *Applications Manual for the Revised NIOSH Lifting Equation*, Cincinnati, OH, Publication No. 94–122, 1994. With permission.

assess the risk of injury during manual materials handling activities. In both models the moments imposed upon the various joints of the body due to the object lifted are evaluated assuming that a static posture is representative of the instantaneous loading of the body. These models then compare the imposed moments about each joint with the static strength capacity derived from a working population. The static strength capacity of the major articulations (assessed by this model) have been documented in a database of over 3000 workers. In this manner the proportion of the population capable of performing a particular static exertion is predicted. In addition, the joint that limits the capacity to perform the task can be identified via this method. These models assume that a single equivalent muscle (internal force) supports the external moment about each joint. By considering the contribution of the externally applied load and the internally generated single muscle equivalent, spine compression acting on the lumbar discs is predicted. The predicted compression can then be compared to the tolerance limits of the vertebral endplate (Figure 11.10). An important assumption of these models is that no significant motion occurs during the exertion since it is a static model. The implications of these assumptions are discussed further in Chapter 28. Figure 11.30 shows the output screen for this computer model where the lifting posture, lifting distances, strength predictions, and spine compression are shown.

TABLE 11.5   Coupling Multiplier

| Coupling Type | Coupling Multiplier | |
	V < 30 inches (75 cm)	*V* ≥ 30 inches (75 cm)
Good	1.00	1.00
Fair	0.95	1.00
Poor	0.90	0.90

Source: Reprinted from NIOSH, *Application Manual for Revised NIOSH Equation*, Cincinnati, OH, Publication No. 94–122, 1994. With permission.

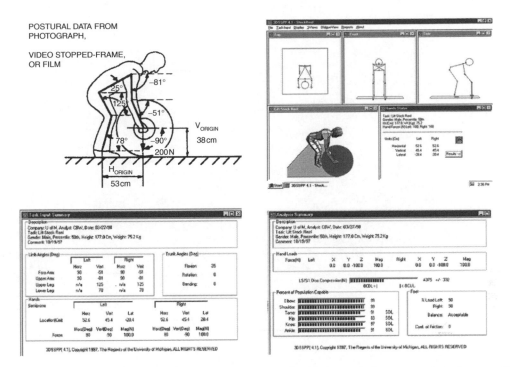

FIGURE 11.30 The 2D-static strength prediction model. (Adapted from Chaffin, D.B. and Andersson, G.B., *Occupational Biomechanics*, John Wiley & Sons, Inc., New York, 1991. With permission.)

11.4.5   Multiple Muscle System Models

One of the significant simplifying assumptions inherent in most static models is that the coactivation of the trunk musculature during a lift is negligible. The trunk is truly a multiple muscle system with many major muscle groups supporting and loading the spine (Schultz and Andersson, 1981). This can be seen in the cross-section of the trunk shown in Figure 11.31. Studies have shown that there is significant coactivation occurring in many of the major muscle groups in the trunk during realistic *dynamic* lifting (Marras and Mirka, 1993). This coactivation is important because all the trunk muscles have the ability to load the spine since antagonist muscles can oppose each other during occupational tasks and increase the total load on the spine. Thus, assumptions regarding single-equivalent muscles within the trunk can lead to erroneous conclusions about spine loading during a task. Studies have indicated that ignoring the coactivation of the trunk muscles during dynamic lifting can misrepresent spine loading by 45 to 70% (Granata and Marras, 1995a; Thelen et al., 1995). In an effort to more accurately estimate the loads on the lumbar spine especially under complex, changing (dynamic) postures multiple muscle system models of the trunk have been developed. Much of the recent research has been focused upon predicting how the multiple trunk muscles coactivate during dynamic lifting.

11.4.5.1   EMG-Assisted Multiple Muscle System Models

People recruit their muscles in various manners when moving dynamically. For example, when moving slowly the agonist muscle may dominate the muscles activities during a lift. However, when moving cautiously, asymmetrically, or rapidly there may be a great deal of antagonistic coactivation present. During occupational lifting tasks these latter dynamic conditions are typically the rule rather than the exception during lifting. As line speeds increase, highly dynamic motions are becoming more common and it is becoming more important to understand the role of muscle coactivation

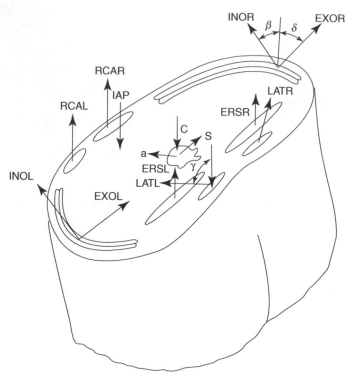

FIGURE 11.31   Cross-sectional view of the human trunk at the lumbrosacral junction. (Adapted from Schultz, A.B. and Andersson, G.B.J., *Spine*, 6, pp. 76–82, 1981. With permission.)

during work. Because of the variability in muscle recruitment patterns it has been virtually impossible to predict the instantaneous coactivation and resultant loading on the spine during dynamic trunk exertions. One of the few means to accurately account for the effect of the trunk muscle system coactivation upon spine loading is through the use of biologically assisted models. The most common of these models are electromyographic or EMG-assisted models. These models take into account the individual recruitment patterns of the muscles during a specific lift for a specific individual. By directly monitoring muscle activity the EMG-assisted model can determine individual muscle force and the subsequent spine loading. These models have been developed and tested under bending and twisting dynamic motion conditions and have been validated (McGill and Norman, 1985, 1986; Marras and Reilly, 1988; Reilly and Marras, 1989; Marras and Sommerich, 1991a, b; Granata and Marras, 1993, 1995b; Marras and Granata, 1995, 1997a, b; Marras et al., 2001). Figure 11.32 shows how such models can assess the effects of lifting dynamics upon spine loading. These models are the only ones that can predict the *multi-dimensional loads* on the lumbar spine under many 3D complex dynamic lifting conditions. The limitation of such models is that they require significant instrumentation of the worker.

11.4.5.2   Stability-Based Models

Efforts have also been attempted to use stability as criteria to govern detailed biologically assisted biomechanical models of the torso (Panjabi, 1992a, b; Cholewicki and McGill, 1996; Solomonow et al., 1999; Cholewicki et al., 2000; Granata and Marras, 2000; Granata and Orishimo, 2001; Granata and Wilson, 2001; Cholewicki and VanVliet, 2002). One potential injury pathway for LBDs suggests that the unnatural rotation of a single spine segment may create loads on passive tissues or other muscle tissues that result in spine injury (McGill, 2002a). Most of the work performed in this area to date has been

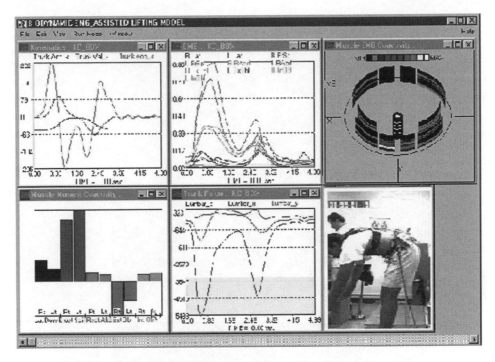

FIGURE 11.32 Windows EMG-assisted model.

directed towards static response of the trunk as well as sudden loading responses (Cholewicki et al., 2000a, b; Granata and Orishimo, 2001; Granata et al., 2001; Granata and Wilson, 2001; Cholewicki and VanVliet, 2002). While these analyses may consider muscle coactivation beneficial from a stability point of view, the point at which the stability benefits of coactivation are overcome by the increased loading remains yet to be determined.

11.4.6 Dynamic Motion and LBD

As discussed throughout this chapter it is clear that dynamic activity may significantly increase the risk of LBD, yet there are few assessment tools available to assess the biomechanical demands associated with workplace dynamics and the risk of LBD. In order to control this biomechanical situation at the worksite, one must know the type of motion that increases biomechanical load and determine "how much motion exposure is too much motion exposure" from a biomechanical standpoint. These issues were the focus of several industrial studies performed over a 6-yr period in 68 different industrial environments. Trunk motion and workplace conditions were assessed in workers exposed to high risk of LBD jobs and compared to trunk motions and work place conditions of low-risk jobs (Marras et al., 1993, 1995). A trunk goniometer (lumbar motion monitor or LMM) that has been used to document the trunk motion patterns of workers at the workplace is shown in Figure 11.33. Trunk motion and workplace conditions associated with the high-risk and low-risk environments are listed in Table 11.6. Based upon these findings, a five factor multiple logistic regression model was developed that is capable of discriminating between task exposure that indicate probability of high-risk group membership. These factors include: (1) frequency of lifting, (2) load moment (load weight multiplied by the distance of the load from the spine), (3) average twisting velocity (measured by the LMM), (4) maximum sagittal flexion angle through the job cycle (measured by the LMM), and (5) maximum lateral velocity (measured by the LMM). This LMM risk assessment model is the only model capable of assessing the *risk* of 3D trunk motion on the job. This model

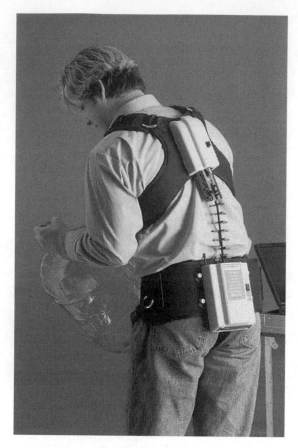

FIGURE 11.33   The lumbar motion monitor (LMM).

has been shown to have a high degree of predictability (odds ratio = 10.7) compared to previous attempts to assess work-related LBD risk. The advantage of this assessment is that the evaluation provides information about risk that would take years to derive from historical accounts of incidence rates. The model has also been validated in a prospective study (Marras et al., 2000a, b). Chapter 49 further explains the logic and validity of this tool.

11.4.7   TLVs

Threshold Limit Values or TLVs have been recently introduced for controlling biomechanical risk to the back in the workplace. These limits have been introduced through the American Conference of Governmental Industrial Hygienists (ACGIH) and provide lifting weight limits as a function of lift origin "zones" and repetitions associated with occupational tasks. The lift origin zones are defined by the lift height off the ground and lift distance from the spine associated with the lift origin. Twelve zones are defined that related to lifts within $\pm 30°$ of asymmetry from the sagittal plane. These zones are represented in three figures with each figure corresponding to different lift frequency and time exposures. Within each zone limits are specified based upon the best information available from several sources, which include: (1) EMG-assisted biomechanical models, (2) the 1993 revised lifting equation, and (3) the historical risk data associated with the LMM database. This tool is further described in Chapter 50.

TABLE 11.6 Descriptive Statistics of the Workplace and Trunk Motion Factors in Each of the Risk Groups

Factors	High Risk (N = 111)				Low Risk (N = 124)				Statistics t
	Mean	SD	Minimum	Maximum	Mean	SD	Minimum	Maximum	
WORKPLACE FACTORS									
Lift rate (lifts/h)	175.89	8.65	15.30	900.00	118.83	169.09	5.40	1500.00	2.1[a]
Vertical load location at origin (m)	1.00	0.21	0.38	1.80	1.05	0.27	0.18	2.18	1.4
Vertical load location at destination (m)	1.04	0.22	0.55	1.79	1.15	0.26	0.25	1.88	3.2[b]
Vertical distance traveled by load (m)	0.23	0.17	0.00	0.76	0.25	0.22	0.00	1.04	0.8
Average weight handled (N)	84.74	79.39	0.45	423.61	29.30	48.87	0.45	280.92	6.4[b]
Maximum weight handled (N)	104.36	88.81	0.45	423.61	37.15	60.83	0.45	325.51	6.7[b]
Average horizontal distance between load and L_5-S_1 (N)	0.66	0.12	0.30	0.99	0.61	0.14	0.33	1.12	2.5[a]
Maximum horizontal distance between load and L_5-S_1 (N)	0.76	0.17	0.38	1.24	0.67	0.19	0.33	1.17	3.7[b]
Average moment (Nm)	55.26	51.41	0.16	258.23	17.70	29.18	0.17	150.72	6.8[b]
Maximum moment (Nm)	73.65	60.65	0.19	275.90	23.64	38.62	0.17	198.21	7.4[b]
Job satisfaction	5.96	2.26	1.00	10.00	7.28	1.95	1.00	10.00	4.7[b]
TRUNK MOTION FACTORS									
Sagittal Plane									
Maximum extension position (°)	−8.30	9.10	−30.82	18.96	−10.19	10.58	−30.00	33.12	3.5[b]
Maximum flexion position (°)	17.85	16.63	−13.96	45.00	10.37	16.02	−25.23	45.00	1.5
Range of motion (°)	31.50	15.67	7.50	75.00	23.82	14.22	399.00	67.74	3.8[b]
Average velocity (°/sec)	11.74	8.14	3.27	48.88	6.55	4.28	1.40	35.73	6.0[b]
Maximum velocity (°/sec)	55.00	38.23	14.20	207.55	38.69	26.52	9.02	193.29	3.7[b]
Maximum acceleration (°/sec²)	316.73	224.57	80.61	1341.92	226.04	173.88	59.10	1120.10	4.2[b]
Maximum deceleration (°/sec²)	−92.45	63.55	−514.08	−18.45	−83.32	47.71	−227.12	−4.57	1.2
Lateral Plane									
Maximum left bend (°)	−1.47	6.02	−16.80	24.49	−2.54	5.46	−23.80	13.96	1.4
Maximum right bend (°)	15.60	7.61	3.65	42.11	13.24	6.32	0.34	34.14	2.6[a]
Range of motion (°)	24.44	9.77	7.10	47.54	21.59	10.34	5.42	62.41	2.2[a]
Average velocity (°/sec)	10.28	4.54	3.12	33.11	7.15	3.16	2.13	18.86	6.1[b]
Maximum velocity (t/sec)	46.36	19.12	13.51	115.94	35.45	12.88	11.97	76.25	4.9[b]
Maximum acceleration (°/sec²)	301.41	166.69	82.64	1030.29	229.29	90.90	66.72	495.88	4.1[b]
Maximum deceleration (°/sec²)	−103.65	60.31	−376.75	0.00	−106.20	58.27	−294.83	0.00	0.3

(Table continued)

TABLE 11.6   *Continued*

Factors	High Risk ($N = 111$)				Low Risk ($N = 124$)				Statistics t
	Mean	SD	Minimum	Maximum	Mean	SD	Minimum	Maximum	
Twisting Plane									
Maximum left twist (°)	1.21	9.08	−27.56	29.54	−1.92	5.36	−30.00	11.44	3.2[b]
Maximum right twist (°)	13.95	8.69	−13.45	30.00	10.83	6.08	−11.20	30.00	2.2[a]
Range of motion (°)	20.71	10.61	3.28	53.30	17.08	8.13	1.74	38.59	2.9[b]
Average velocity (°/sec)	8.71	6.61	1.02	34.77	5.44	3.19	0.66	17.44	3.8[b]
Maximum velocity (°/sec)	46.36	25.61	8.06	136.72	38.04	17.51	5.93	91.97	4.7[a]
Maximum acceleration (°/sec^2)	304.55	175.31	54.48	853.93	269.49	146.65	44.17	940.27	2.9[b]
Maximum deceleration (°/sec^2)	−88.52	70.30	−428.94	−5.84	−100.32	72.40	−325.93	−2.74	1.6[a]

[a]Significant at $\alpha \leq 0.05$ (two-sided).
[b]Significant at $\alpha \leq 0.01$ (two-sided).
Source: Adapted from Marras et al., *Spine* 18, pp. 617–628, 1993. With permission.

11.5 Summary

This chapter has shown that biomechanics provides one of the few means to *quantitatively* consider the implications of workplace design. Biomechanical design is important when a particular job is suspected of imposing large or repetitive forces on a particular structure of the body. It is particularly important to recognize that the internal structures of the body such as muscles are the primary loaders of the joint and tendon structures. In order to evaluate the risk of injury from a particular task, one must consider the contribution of both the external loads and internal loads upon the structure. Several quantitative models and assessment methods have been developed that systematically consider the internal loading imposed on the worker due to workplace layout and task requirements. Proper use of these models and methods involves recognizing the limitations and assumptions of each technique so that they are not applied inappropriately. When properly used, these assessments can help assess the risk of work-related injury and illness.

References

Adams, M.A., Freeman, B.J., Morrison, H.P., Nelson, I.W., and Dolan, P. Mechanical initiation of intervertebral disc degeneration, *Spine.*, 25, pp. 1625–1636, 2000.

Anderson, C.K., Chaffin, D.B., Herrin, G.D., and Matthews, L.S. A biomechanical model of the lumbosacral joint during lifting activities, *J. Biomech.*, 18, pp. 571–584, 1985.

Andersson, B.J., Ortengren, R., Nachemson, A.L., Elfstrom, G., and Broman, H. The sitting posture: an electromyographic and discometric study, *Orthop. Clin. North Am.*, 6, pp. 105–120, 1975.

Andersson, G.B. The epidemiology of spinal disorders, in *The Adult Spine: Principles and Practice*, Frymoyer, J.W., Ed., Lippincott-Raven Publishers, Philadelphia, 1997, 93–141.

Basmajian, J.V. and De Luca, C.J. *Muscles Alive: Their Functions Revealed by Electromyography*, 5th ed., Williams and Wilkins, Baltimore, MD, 1985.

Brinkmann, P., Biggermann, M., and Hilweg, D. Fatigue fracture of human lumbar vertebrac, *Clin Biomech (Bristol, Avon).*, 3, pp. S1–S23, 1988.

Cavanaugh, J.M. Neural mechanisms of lumbar pain, *Spine*, 20, pp. 1804–1809, 1995.

Cavanaugh, J.M., Ozaktay, A.C., Yamashita, T., Avramov, A., Getchell, T.V., and King, A.I. Mechanisms of low back pain: a neurophysiologic and neuroanatomic study, *Clin. Orthop.*, pp. 166–180, 1997.

Chaffin, D.B. A computerized biomechanical model: development of and use in studying gross body actions, *J. Biomech.*, 2, pp. 429–441, 1969.

Chaffin, D.B., Localized muscle fatigue — definiton and measurement, *J. Occup. Med.*, 15, pp. 346–354, 1973.

Chaffin, D.B. and Muzaffer, E. Three-dimensional biomechanical static strength predcition model sensitivity to postural and anthropometric innaccuracies, *IIE Transac.*, 23, pp. 215–227, 1991.

Chaffin, D.B., Andersson, G.B.J., and Martin, B.J. Occupational Biomechanics, New York, 1999.

Cholewicki, J. and S. McGill. Mechanical stability of the in vivo lumbar spine: Implications ofr injury and chronic low back pain, *Clin. Biomech. (Bristol, Avon)*, 11, pp. 1–15, 1996.

Cholewicki, J. and I.J. VanVliet. Relative contribution of trunk muscles to the stability of the lumbar spine during isometric exertions, *Clin. Biomech. (Bristol, Avon)*, 17, pp. 99–105, 2002.

Cholewicki, J., Polzhofer, G.K., and Radebold, A. Postural control of trunk during unstable sitting, *J Biomech.*, 33, pp. 1733–1737, 2000a.

Cholewicki, J., Simons, A.P., and Radebold, A. Effects of external trunk loads on lumbar spine stability, *J Biomech* 33, pp. 1377–1385, 2000b.

Gedalia, U., Solomonow, M., Zhou, B.H., Baratta, R.V., Lu, Y., and Harris, M. Biomechanics of increased exposure to lumbar injury caused by cyclic loading. Part 2. Recovery of reflexive muscular stability with rest, *Spine*, 24, pp. 2461–2467, 1999.

Granata, K.P. and Marras, W.S. An EMG-assisted model of loads on the lumbar spine during asymmetric trunk extensions, *J. Biomech.*, 26, pp. 1429–1438, 1993.

Granata, K.P. and Marras, W.S. The influence of trunk muscle coactivity on dynamic spinal loads, *Spine*, 20, pp. 913–919, 1995a.

Granata, K.P. and Marras, W.S. An EMG-assisted model of trunk loading during free-dynamic lifting, *J. Biomech.*, 28, pp. 1309–1317, 1995b.

Granata, K.P. and Marras, W.S. Cost-benefit of muscle cocontraction in protecting against spinal instability [In Process Citation], *Spine*, 25, pp. 1398–1404, 2000.

Granata, K.P. and Orishimo, K.F. Response of trunk muscle coactivation to changes in spinal stability, *J. Biomech.*, 34, pp. 1117–1123, 2001.

Granata, K.P. and Wilson, S.E. Trunk posture and spinal stability, *Clin. Biomech. (Bristol, Avon)*, 16, pp. 650–659, 2001.

Granata, K.P., Marras, W.S. and Davis, K.G. Biomechanical assessment of lifting dynamics, muscle activity and spinal loads while using three different styles of lifting belt, *Clin. Biomech. (Bristol, Avon)*, 12, pp. 107–115, 1997.

Granata, K.P., Orishimo, K.F., and Sanford, A.H. Trunk muscle coactivation in preparation for sudden load, *J. Electromyogr. Kinesiol.*, 11, pp. 247–254, 2001.

Grandjean, E. *Fitting the Task to the Man: An Ergonomic Approach*, Taylor & Francis, Ltd., London, 1982.

Guo, H.R. Back Pain and U.S. Workers, American Occupational Health Conference, 1993.

Guo, H.R., Tanaka, S., Halperin, W.E., and Cameron, L.L. Back pain prevalence in US industry and estimates of lost workdays, *Am. J. Public Health.*, 89, pp. 1029–1035, 1999.

Hollbrook, T.L., Grazier, K., Kelsey, J.L., and Stauffer, R.N. The Frequency of Occurrence, Impact and Cost of Selected Musculoskeletal Conditions in the United States, American Academy of Orthopaedic Surgeons, Chicago, IL, 1984.

Hoy, M.G., Zajac, F.E., and Gordon, M.E. A musculoskeletal model of the human lower extremity: the effect of muscle, tendon, and moment arm on the moment-angle relationship of musculotendon actuators at the hip, knee, and ankle, *J. Biomech.*, 23, pp. 157–169, 1990.

Jager, M., Luttmann, A. and Laurig, W. Lumbar load during one-hand bricklaying, *Int. J. Indust. Ergo.*, 8, pp. 261–277, 1991.

Khalsa, P.S. Biomechanics of musculoskeletal pain: dynamics of the neuromatrix, *J. Electromyogr. Kinesiol.*, 14, pp. 109–120, 2004.

Kim, J., Stuart-Buttle C., and Marras, W.S. The Effects of Mats on Back and Leg Fatigue, *Applied Ergonomics* 25, pp. 29–34, 1994.

Konz, S.A. *Work Design: Industrial Ergonomics*, 2nd ed., Grid Publishing, Inc., Columbus, OH, 1983. With permission.)

Kovacs, K., Splittstoesser, R., Maronitis, A., and Marras, W.S. Grip force and muscle activity differences due to glove type, *AIHA J (Fairfax, Va)* 63, pp. 269–274, 2002.

Kroemer, K.H.E. Biomechanics of the Human Body, *In* G. Salvendy ed., Handbook of Human Factors. John Wiley and Sons, Inc., New York, 1987.

Lantz, S.A. and Schultz, A.B. Lumbar spine orthosis wearing. I. Restriction of gross body motions. *Spine* 11, pp. 834–837, 1986.

Lavender, S.A., Thomas, J.S., Chang, D., and Andersson, G.B. Effect of lifting belts, foot movement, and lift asymmetry on trunk motions, *Hum Factors*, 37, pp. 844–853, 1995.

Luo, X., Pietrobon, R., X.S. S, Liu, G.G. and Hey, L. Estimates and patterns of direct health care expenditures among individuals with back pain in the United States, *Spine*, 29, pp. 79–86, 2004.

Marras, W.S. and Granata, K.P. A biomechanical assessment and model of axial twisting in the thoracolumbar spine, *Spine*, 20, pp. 1440–1451, 1995.

Marras, W.S. Spine loading during trunk lateral bending motions, *J. Biomech.*, 30, pp. 697–703, 1997a.

Marras, W.S. The Development of an EMG-Assisted Model to Assess Spine Loading during Whole-Body Free-Dynamic Lifting, *J. Electromyogr. Kinesiol.*, 7, pp. 259–268, 1997b.

Marras, W.S. and Mirka, G.A. Electromyographic studies of the lumbar trunk musculature during the generation of low-level trunk acceleration, *J. Orthop. Res.*, 11, pp. 811–817, 1993.

Marras, W.S. and Reilly, C.H. Networks of internal trunk-loading activities under controlled trunk-motion conditions, *Spine*, 13, pp. 661–667, 1988.

Marras, W.S. and Schoenmarklin, R.W. Wrist motions in industry, *Ergonomics*, 36, pp. 341–351, 1993.

Marras, W.S. and Sommerich, C.M. A three-dimensional motion model of loads on the lumbar spine: I. Model structure, *Hum Factors*, 33, pp. 123–137, 1991a.

Marras, W.S. A three-dimensional motion model of loads on the lumbar spine: II. Model validation, *Hum Factors*, 33, pp. 139–149, 1991b.

Marras, W.S., Davis, K.G., and Splittstoesser, R.E. Spine loading during whole body free dynamic lifting, The Ohio State University, Columbus, OH, Grant #:R01 OH03289, 2001.

Marras, W.S., Fine, L.J., Ferguson, S.A. and Waters, T.R. The effectiveness of commonly used lifting assessment methods to identify industrial jobs associated with elevated risk of low-back disorders, *Ergonomics*, 42, pp. 229–245, 1999.

Marras, W.S., Jorgensen, M.J. and Davis, K.G. Effect of foot movement and an elastic lumbar back support on spinal loading during free-dynamic symmetric and asymmetric lifting exertions, *Ergonomics*, 43, pp. 653–668, 2000a.

Marras, W.S., Allread, W.G., Burr, D.L., and Fathallah, F.A. Prospective validation of a low-back disorder risk model and assessment of ergonomic interventions associated with manual materials handling tasks [In Process Citation]. *Ergonomics*, 43, pp. 1866–1886, 2000b.

Marras, W.S., Lavender, S.A., Leurgans, S.E., Fathallah, F.A., Ferguson, S.A., Allread, W.G., and Rajulu, S.L. Biomechanical risk factors for occupationally related low back disorders, *Ergonomics*, 38, pp. 377–410, 1995.

Marras, W.S., Lavender, S.A., Leurgans, S.E., Rajulu, S.L., Allread, W.G., Fathallah, F.A., and Ferguson, S.A. The role of dynamic three-dimensional trunk motion in occupationally-related low back disorders. The effects of workplace factors, trunk position, and trunk motion characteristics on risk of injury, *Spine*, 18, pp. 617–628, 1993.

McCoy, M., Congleton, J., Johnson, W. and Jaing, B. The role of lifting belts in manual lifting, *Int. J. Indus. Ergon.*, 2, pp. 256–259, 1988.

McGill, S. Low back disorder, *Human Kinetics*, 2002a.

McGill, S. Low back disorders: evidence-based prevention and rehabilitation, *Human Kinetics, Champaign, IL*, 2002b.

McGill, S., Seguin, J., and Bennett, G. Passive stiffness of the lumbar torso in flexion, extension, lateral bending, and axial rotation. Effect of belt wearing and breath holding, *Spine*, 19, pp. 696–704, 1994.

McGill, S.M. Abdominal belts in industry: a position paper on their assets, liabilities and use, *Am. Ind. Hyg. Assoc. J.*, 54, pp. 752–754, 1993.

McGill, S.M. The biomechanics of low back injury: implications on current practice in industry and the clinic, *J. Biomech.*, 30, pp. 465–475, 1997.

McGill, S.M. and Norman, R.W. Dynamically and statically determined low back moments during lifting, *J. Biomech.*, 18, pp. 877–885, 1985.

McGill, S.M. and Norman, R.W. Partitioning of the L4-L5 dynamic moment into disc, ligamentous, and muscular components during lifting [see comments], *Spine*, 11, pp. 666–678, 1986.

McGill, S.M., Hughson, R.L. and Parks, K. Changes in lumbar lordosis modify the role of the extensor muscles [In Process Citation], *Clin. Biomech. (Bristol, Avon)*, 15, pp. 777–780, 2000.

Mitchell, L.V., Lawler, F.H., Bowen, D., Mote, W., Asundi, P., and Purswell, J. Effectiveness and cost-effectiveness of employer-issued back belts in areas of high risk for back injury, *J. Occup. Med.*, 36, pp. 90–94, 1994.

Nachemson, A. Towards a better understanding of low-back pain: a review of the mechanics of the lumbar disc, *Rheumatol Rehabil*, 14, pp. 129–143, 1975.

NIOSH 1981. Work Practices Guide for Manual Lifting, Department of Health and Human Services (DHHS) NIOSH, Cincinnati, OH, 81–122.

NIOSH Workplace use of back belts, Cincinnati, OH, Publication No. 94–122, 1994.

Nordin, M., and Frankel, V. Basic Biomechanics of the Musculoskeletal System. Lea and Febiger, Philadelphia, PA, 1989.

NRC Work-related musculokeletal disorders: report, workshop summary, and workshop papers. National Academy Press, Washington D. C, 1999.

NRC Musculoskeletal disorders and the workplace: low back and upper extremity. National Academy Press, Washington D. C, 2001.

Panjabi, M.M. The stabilizing system of the spine. Part II. Neutral zone and instability hypothesis, *J. Spinal. Disord.*, 5, pp. 390–396; discussion 397, 1992a.

Panjabi, M.M. The stabilizing system of the spine. Part I. Function, dysfunction, adaptation, and enhancement, *J. Spinal. Disord.*, 5, pp. 383–389; discussion 397, 1992b.

Park, K. and Chaffin, D. A biomechanical evaluation of two methods of manual load lifting, *AIIE Trans.* 6, pp. 105–113, 1974.

Pope, M.H. Muybridge Lecture, International Society of Biomechanics XIVth Congress, Paris, France, 1993.

Praemer, A., Furner, S. and Rice, D.P. Musculoskeletal Conditions in the United States, pp. 23–33. American Academy of Orthopaedic Surgeons, Park Ridge, IL, 1992.

Rafacz, W. and McGill, S.M. Wearing an abdominal belt increases diastolic blood pressure, *J. Occup. Environ. Med.*, 38, pp. 925–927, 1996.

Reddell, C., Congleton, J., Hutchingson, R. and Montgomery, J. An evaluation of weight-lifting belt and back injury prevention training class for airline baggage handlers, *Appl. Ergon.*, 22, pp. 319–329, 1992.

Reilly, C.H. and Marras, W.S. Simulift: a simulation model of human trunk motion, *Spine*, 14, pp. 5–11, 1989.

Sanders, M.S. and McCormick, E.F. *Human Factors in Engineering and Design*, McGraw-Hill Inc., New York, 1993.

Schoenmarklin, R.W., Marras, W.S., and Leurgans, S.E. Industrial wrist motions and incidence of hand/wrist cumulative trauma disorders, *Ergonomics*, 37, pp. 1449–1459, 1994.

Schultz, A.B. and Andersson, G.B. Analysis of loads on the lumbar spine, *Spine*, 6, pp. 76–82, 1981.

Silverstein, M.A., Silverstein, B.A., and Franklin, G.M. Evidence for work-related musculoskeletal disorders: a scientific counterargument, *J. Occup. Environ. Med.*, 38, pp. 477–484, 1996.

Silverstein, B.A., Stetson, D.S., Keyserling, W.M., and Fine, L.J. Work-related musculoskeletal disorders: comparison of data sources for surveillance, *Am. J. Ind. Med.*, 31, pp. 600–608, 1997.

Solomonow, M. Ligaments: a source of work-related musculoskeletal disorders, *J. Electromyogr. Kinesiol.*, 14, pp. 49–60, 2004.

Solomonow, M., Zhou, B.H., Harris, M., Lu, Y. and Baratta, R.V. The ligamento-muscular stabilizing system of the spine, *Spine*, 23, pp. 2552–2562, 1998.

Solomonow, M., Zhou, B.H., Baratta, R.V., Lu, Y. and Harris, M. Biomechanics of increased exposure to lumbar injury caused by cyclic loading: Part 1. Loss of reflexive muscular stabilization, *Spine*, 24, pp. 2426–2434, 1999.

Solomonow, M., Zhou, B., Baratta, R.V., Zhu, M. and Lu, Y. Neuromuscular disorders associated with static lumbar flexion: a feline model, *J. Electromyogr. Kinesiol.*, 12, pp. 81–90, 2002.

Solomonow, M., He Zhou, B., Baratta, R.V., Lu, Y., Zhu, M. and Harris, M. Biexponential recovery model of lumbar viscoelastic laxity and reflexive muscular activity after prolonged cyclic loading, *Clin. Biomech., (Bristol, Avon)* 15, pp. 167–175, 2000.

Stubbs, M., Harris, M., Solomonow, M., Zhou, B., Lu, Y., and Baratta, R.V. Ligamento-muscular protective reflex in the lumbar spine of the feline, *J. Electromyogr. Kinesiol.*, 8, pp. 197–204, 1998.

Thelen, D.G., Schultz, A.B. and Ashton-Miller J.A. Co-contraction of lumbar muscles during the development of time-varying triaxial moments, *J. Orthop. Res.*, 13, pp. 390–398, 1995.

Tichauer, E.R. The Biomechanical Basis of Ergonomics: Anatomy applied to the design of work situations, John Wiley & Sons, New York, 1978.

Van Dieen, J.H., Hoozemans, M.J., and Toussaint, H.M. Stoop or squat: a review of biomechanical studies on lifting technique, *Clin. Biomech., (Bristol, Avon)* 14, pp. 685–696, 1999.

Videman, T., Nurminen, M., and Troup, J.D. 1990 Volvo Award in clinical sciences. Lumbar spinal pathology in cadaveric material in relation to history of back pain, occupation, and physical loading, *Spine*, 15, pp. 728–740, 1990.

Walsh, N.E. and Schwartz, R.K. The influence of prophylactic orthoses on abdominal strength and low back injury in the workplace [see comments], *Am. J. Phys. Med. Rehabil.*, 69, pp. 245–250, 1990.

Wang, J.L., Parnianpour, M., Shirazi-Adl, A., and Engin, A.E. Viscoelastic finite-element analysis of a lumbar motion segment in combined compression and sagittal flexion. Effect of loading rate, *Spine*, 25, pp. 310–318, 2000.

Wassell, J.T., Gardner, L.I., Landsittel, D.P., Johnston, J.J., and Johnston, J.M. A Prospective Study of Back Belts for Prevention of Back Pain and Injury, *J. Am. Me. Assoc.*, 284, pp. 2727–2732, 2000.

Waters, T.R., Putz-Anderson, V., Garg, A., and Fine, L.J. Revised NIOSH equation for the design and evaluation of manual lifting tasks, *Ergonomics*, 36, pp. 749–776, 1993.

12

Fundamentals of Manual Control

Max Mulder

Marinus M.
  Van Paassen

Delft University of Technology

John M. Flach
Wright State University

Richard J. Jagacinski
The Ohio State University

12.1 Introduction

For their work, for transportation or simply for entertainment, human beings are often involved in the manual control of devices. Vehicles such as cars, bicycles, ships, and airplanes are some examples, but also video games and many work situations involve manual control. Normally, after learning the task, the human operator in such a control situation behaves like a well-designed controller. In fact, in their paper on "Quasi-linear pilot models," McRuer and Jex make the remark that data of measured pilot behavior matches very well with the *Primary Rule of Thumb for Frequency Domain Synthesis*, a design rule for automatic controllers. It is not surprising, therefore, that many of the models used in modeling manual control situations are based on various control system design techniques. Two of the most commonly applied are the frequency domain design methods, which serve as the basis for the cross-over model (COM) and variants thereof, the precision model and the simplified precision model (McRuer and Jex, 1967), and optimal control theory, which lies at the basis of the optimal control model (OCM; Kleinman et al., 1970b).

The theories on human control behavior have matured by now, and its applications, particularly human vehicle control, have been extensively studied. However, for many application areas these theories and their applications are still very relevant today. Some examples are:

- Investigation of the roles of multi-modal (visual, vestibular, tactile) feedback on human manual control behavior in virtual environments such as flight and driving simulators (Mulder et al., 2004).
- The design of haptic manipulators in applications like tele-operation or the development of force-feedback systems in vehicular control (Van Paassen, 1994).
- Investigation of vehicular control (aircraft, automobile) in general, including handling qualities research.
- Investigation and evaluation of augmented systems, that is, the study of the interim between fully manual and automated control.
- Studying human perception and action cycles in active psychophysics, for example, in the determination and identification of a human's use of visual cues in multi-cue displays (Flach, 1991; Mulder, 1999).

In this chapter we will provide a short introduction into some fundamentals of control theory. This introduction is very limited, however, and for further study the reader is referred to the many good textbooks that are available. A textbook geared towards human control is "*Control Theory for Humans*," by Jagacinski and Flach (2003). Others that are recommended for their "human engineering" perspective are "*Man-Machine Systems*" by Sheridan and Ferrell (1974), and "*Engineering Psychology and Human Performance*" by Wickens (1992). More engineering-oriented textbooks are "*Control Systems Engineering*" by Nise (1995), "*Control System Design*" by Goodwin et al. (2001) "*Modern Control Systems*" by Dorf and Bishop (2005) and "*Feedback Control of Dynamics Systems*" by Franklin et al. (2002). Note that this selection is not exhaustive, and that many more excellent textbooks are available.

Furthermore, a historic overview of the modeling of human control behavior is given, followed by a more detailed description of two of the most common and widely used approaches to describe human behavior in control-theoretical terms, the COM, based on classical control theory, and the OCM, based on optimal control theory.

12.2   Fundamentals of Systems and Control Theory

Systems and control theory is a branch of mathematics that studies dynamic processes, that is, things that evolve in time. Examples of dynamic processes that are the subject of systems and control theory are artifacts like airplanes, power plants, and cars, biological systems like the heart, chemical processes, large-scale processes like economics, etc. These can be modeled, described and simulated with dynamical systems theory, and their automated control systems are designed with control theory.

The building blocks of systems and control theory are differential equations, linear matrix algebra, complex number theory, and probability theory. In this section we will focus on studying linear, time-invariant (LTI), continuous time, single-input single-output (SISO) systems. We will briefly study the response of these systems to deterministic signals, in later sections, we will also briefly address stochastic signals.

Most of the physical processes in the real-world are nonlinear, time-varying, multi-input multi-output (MIMO), stochastic systems. A deep understanding of SISO systems, however, is an important first step for understanding more complex MIMO systems. Many of the intuitions gained from working with simple control systems will generalize to more complex systems — even though it might not ever be possible to model the more complex systems with the same confidence that we can model the simpler systems.

12.2.1   Linear, Time-Invariant Systems

A useful definition of a system in systems and control theory is given by Olsder and van der Woude (1994):

> (A system is) a part of reality which we think to be a separated unit within this reality. The reality outside the system is called the surroundings. The interaction between system and surroundings is realized via quantities, quite often functions of time, which are called input and output.

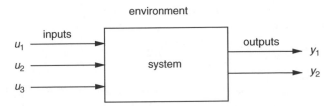

FIGURE 12.1 Representation of a system, with inputs and outputs, and a system boundary that separates the system from the environment.

This definition stresses the fact that the choice of what is considered to be the system and what belongs to the environment (surroundings) is often a subjective one. However, once the system boundaries are defined, the interaction between the system and its environment, by means of input and output signals, also becomes clear, see Figure 12.1.

A system is linear with respect to the inputs and outputs, if the output to a sum of two or more input signals, is the sum of the outputs each of these inputs would give individually. Most real-world systems are not linear, but in many control situations, where there are only small excursions around an "operating point," they can be well approximated with linear models. If also the properties of the system do not change over time, thus if we neglect processes such as wear and tear, the change of mass due to use of fuel, etc., one obtains an LTI system. The behavior of an LTI system can be described with an ordinary differential equation (ODE), with constant coefficients. For instance, consider the following ODE, describing the relationship between the *input signal* of a system, $u(t)$ and the *output signal* of that system, $y(t)$, see Figure 12.2.

$$a_0 y(t) + a_1 \frac{\mathrm{d}y(t)}{\mathrm{d}t} + a_2 \frac{\mathrm{d}^2 y(t)}{\mathrm{d}t^2} = b_0 u(t) + b_1 \frac{\mathrm{d}u(t)}{\mathrm{d}t}. \tag{12.1}$$

As in this example, the ODE generally consists of *time*-derivatives of the input and output signals, characterizing the *dynamic* response of that system to the input signals.

12.2.2 Transfer Function

The calculation of solutions for ODEs, and the combination of systems described in ODE form, is usually quite laborious. Therefore, in classical control theory, extensive use is made of the Laplace Transform, to transform the ODEs to the Laplace domain, where manipulation such as the combination of systems with other systems and signals, and the solution of the ODEs, is simpler, since all the equations become algebraic. Textbooks on control theory contain tables of Laplace transforms that show the transformation of signals to and from the Laplace domain, and Laplace transform theorems that can be used to transform system descriptions to and from the Laplace domain. One can use the Laplace real differentiation theorem, which states that:

$$L\left\{\frac{\mathrm{d}f(t)}{\mathrm{d}t}\right\} = sF(s) - f(0) \tag{12.2}$$

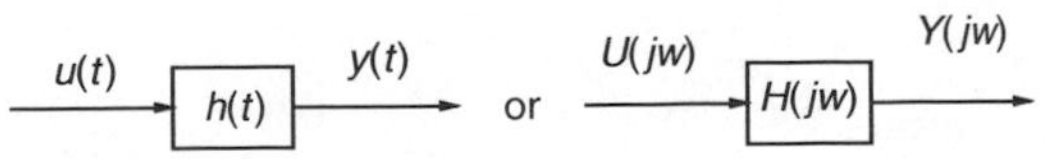

FIGURE 12.2 LTI systems in the time domain (left) and the frequency domain (right).

to transform a differential equation to the Laplace domain. Assuming that the system is at rest at $t = 0$, so $f(0) = 0$ and $df(0)/dt = 0$, one obtains:

$$a_0 Y(s) + a_1 s Y(s) + a_2 s^2 Y(s) = b_0 U(s) + b_1 s U(s) \tag{12.3}$$

This yields a constant relation between the input $U(s)$ and the output $Y(s)$:

$$H(s) = \frac{Y(s)}{U(s)} = \frac{b_0 + b_1 s}{a_0 + a_1 s + a_2 s^2}, \tag{12.4}$$

which is known as the system's *transfer function*. For an LTI system with one input and one output, the transfer function completely defines the system. The transfer function is a convenient format for manipulation of system models. For example, the transfer function for two systems placed in series, that is, the output of the first system is the input of the second, is simply the product of the two transfer functions, and, likewise, the transfer function for the total of the two systems placed in parallel is the sum of the transfer functions.

The stability of an LTI system can be determined from its transfer function by determining the *poles* of the transfer function, which are the (complex) numbers for which the denominator equals zero (these solutions are referred to as the "roots" of the denominator). Any poles with a positive real part are associated with an unstable response of the system to an input signal. This response may be oscillating for a pair of complex poles with positive real part, or it may be a-periodic when the corresponding poles are on the positive real axis. A system with such poles is unstable. Poles with negative real part produce responses with exponentially decreasing amplitude. Pole pairs on the imaginary axis produce an undamped oscillatory response of the system, a pole in the origin produces an integration. An experienced control engineer can interpret the location of the poles and zeros of a transfer function in terms of a system's dynamical behavior.

12.2.3 Frequency Response

Another method of characterizing an LTI system is by means of its *frequency response*. If a sine wave input signal is applied to an LTI system, and the system is stable, then after the initial response to the start of the sine wave has faded, the output of the system will be a sine wave with the same frequency as the input signal, but with a different amplitude, and a different phase from the input signal. Figure 12.3 shows the sine input and the output signal for a system. The ratio of the output sine to the input sine signal, as a function of frequency, is called the gain. The difference in phase between the two sine signals, is called the phase shift. Together they form the systems' frequency response. For physically implemented systems, the frequency response can be determined with a Frequency Analyzer, a device that can generate sinusoid test signals and measure the response of the system.

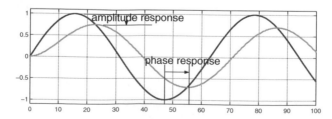

FIGURE 12.3   Input sinusoid (black) and system output (grey) as a function of time.

The frequency response can be determined from the Laplace transfer function, by substituting $j\omega$ for the Laplace variable s; proof for this is given in most control engineering textbooks. Here j is the imaginary number, and ω is the frequency of the input sine signal in radians per second. The frequency response $H(j\omega)$ is a complex function of ω. The magnitude of that function, $|H(j\omega)|$, is the *gain* of the frequency response and the angle of the complex number $H(j\omega)$ is the *phase shift*. In most cases the phase shift is negative and is called a lag. For LTI systems, the frequency response completely defines the system.

Consider an LTI system with the following transfer function:

$$H(s) = \frac{K}{(1 + \tau s)}. \tag{12.5}$$

This is known as a 'first order system' with *time constant τ*. Now consider two values of τ (0.1 and 10 sec) and look at the system response $y(t)$ for sinusoidal input signals that have various frequencies ω. The result is illustrated in Figure 12.4. The output of both LTI systems are sinusoids as well, but the amplitudes and phases of these sinusoids are different from the input sinusoids. Whereas the system with τ equal to 0.1 sec hardly changes the amplitude and phase and the input and output signals are almost the same, the system with $\tau = 10$ sec, does not (completely) "pass" the sinusoidal input signals that have a higher frequency ω. It "filters out" these higher-frequency signals, and is therefore known as a low-pass filter system. The value of τ determines which frequencies are "passed" and which frequencies are not.

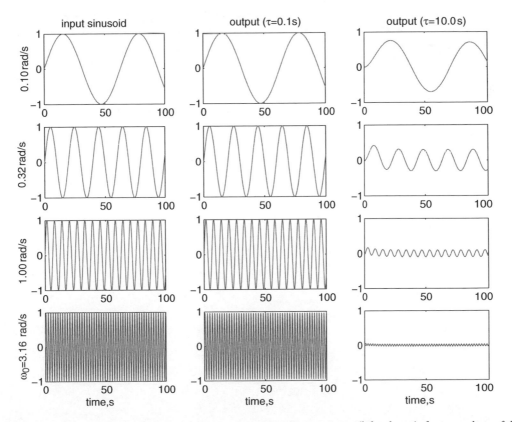

FIGURE 12.4 LTI response to sinusoidal input signal with four frequencies ω_0 (left column), for two values of the variable τ: $\tau = 0.1$ sec (middle column) and $\tau = 10$ sec (right column).

Such frequency responses are often shown as Bode plots. The magnitude and phase for different input signal frequencies are plotted separately. The magnitude is plotted on a logarithmic scale, against the frequency, which is also plotted on a logarithmic scale. At the whim of the maker of the plot, the frequency may be given in rad/sec, in Hertz, or sometimes in octaves. For the magnitude, often a decibel (dB) scale is used. The relation between a magnification M and its equivalent in dB m is:

$$m = 20 \cdot \log_{10} M \quad [dB]. \tag{12.6}$$

Figure 12.5 shows the Bode plot for this system. One can see that for low frequencies the gain of the system is 1 (0 dB). For high frequencies the gain decreases 20 dB per decade. The asymptotes for the low and high frequency behavior cross at the *corner frequency* $1/\tau$, in this case, with $\tau = 10$ sec, at 0.1 rad/sec.

12.2.4 Control

12.2.4.1 Feed-Forward and Feedback

In a control system, a controlling element (controller) provides input to a system, often called a *plant*, normally with the aim of producing outputs of the plant that are equal to given reference values. Unknown disturbances may be acting on the plant, as for example, the turbulence acting on an aircraft. The controller may be an automatic device, such as an autopilot, or it may be a human.

In feed-forward control, or open-loop control, the controller measures the disturbances on the plant, and based on the knowledge about the plant's dynamics, creates inputs that produce plant outputs as close to the reference values as possible, see Figure 12.6a.

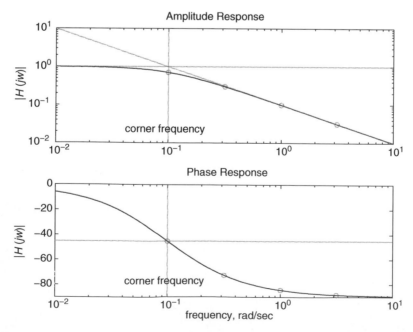

FIGURE 12.5 Bode plot for LTI system $H(j\omega) = 1/(1 + 10j\omega)$. The circles show the system response for the four frequencies shown in Figure 12.4.

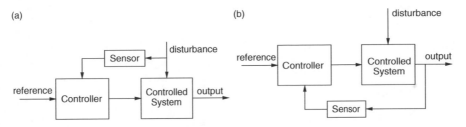

FIGURE 12.6   Open-(a) and closed-loop (b) control systems.

In feedback, or closed-loop control, the controller measures the output of the plant, and compares that to the reference values. Control input to the plant is calculated on the basis of this comparison, see Figure 12.6b. The advantage of closed-loop control over open-loop control is that in most cases "modeling errors," that is, mismatches in the plant model used to tune the controller and the real plant, have little effect on the control performance. Closed-loop control is thus said to be more *robust*, meaning that it is insensitive to variations in the controlled system and influences from the environment. Robustness is an important property of control systems and tools exist for the design and tuning of robust controllers.

Feedback is a fundamental property of many control systems. The *closed-loop* system can itself be considered as a new system, with an input (in most cases the desired output) and an output. If they are part of a larger whole, such systems are often called a servo. For example, the motor, "plant" and feedback system for moving an aircraft's elevator forms a servo system and the servo system itself is part of the automatic pilot.

The choice for a type of controller applied in a servo system is normally made on the basis of knowledge on the plant dynamics and on the desired properties of the closed loop system. Common controller types are the proportional (P) controllers that generate a control signal proportional to the error; proportional and differentiating controllers (PD) generate a control signal on the basis of a sum of error and the derivative of the error; proportional and integrating controllers (PI) generate a control signal on the basis of a sum of error and the integrated error, and the combination of the above, are the PID controllers.

After a choice for a particular type of controller has been made, its parameters need to be tuned. Several tuning methods are in use in control system design, as an example, and since it forms the basis for the commonly used COM, tuning in the frequency domain with Bode diagrams will be treated here.

Consider an LTI system, $G(s)$, Figure 12.7. The main design requirement is to design the controller $K(s)$ in such a way that the system output $z(t)$ follows the reference signal $r(t)$ as closely as possible.

Solving this problem in the Laplace domain is not too difficult, since all combinations of signals and systems can be obtained by algebraic manipulation. From Figure 12.7, one can derive the following basic equations:

$$E(s) = R(s) - Z(s), \tag{12.7}$$

$$Z(s) = G(s)C(s) = G(s)K(s)E(s), \tag{12.8}$$

$$E(s) = R(s) - G(s)K(s)E(s). \tag{12.9}$$

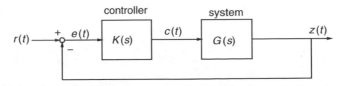

FIGURE 12.7   The feedback control of a SISO LTI system.

Solving for $E(s)$ yields:

$$(1 + G(s)K(s))E(s) = R(s), \tag{12.10}$$

so

$$E(s) = \frac{1}{1 + G(s)K(s)}R(s) \tag{12.11}$$

and

$$Z(s) = \frac{G(s)K(s)}{1 + G(s)K(s)}R(s). \tag{12.12}$$

Thus,

$$\frac{Z(s)}{R(s)} = \frac{K(s)G(s)}{1 + K(s)G(s)} \tag{12.13}$$

One can consider this solution in the frequency domain, by substituting $j\omega$ for s, which yields:

$$\frac{Z(j\omega)}{R(j\omega)} = \frac{K(j\omega)G(j\omega)}{1 + K(j\omega)G(j\omega)}. \tag{12.14}$$

The design requirement is that the system output $Z(j\omega)$ equals the reference signal $R(j\omega)$, and so $(Z(j\omega)/R(j\omega)) \approx 1$. The solution to this problem would be to achieve a high "open-loop gain" $K(j\omega)G(j\omega)$. When $K(j\omega)G(j\omega)$ is very large, one can see that $Z(j\omega) \approx R(j\omega)$ [Equation (12.14) and that $E(j\omega) \approx 0$ Equation (12.11)].

However, one should bear in mind that $K(j\omega)G(j\omega)$ is still a function of ω, with a complex-valued outcome. In general, it is not possible, and for many practical reasons not desirable to obtain a large value for $K(j\omega)G(j\omega)$ for all ω. Essentially $K(j\omega)G(j\omega)$ determines the "speed" of reaction to an error signal. When time delays are small then a faster response will yield a lower tracking error. When time delays are large, however, it is possible to respond too quickly and cause the system to become unstable. Hence, as will be discussed in more detail, in the following paras, generally there is a trade-off between response speed and accuracy.

Just as transfer functions can be considered in terms of their frequency response, that is, what (sine signal) frequencies they pass and what frequencies they block, signals can be considered in terms of their *frequency content*. A reference signal such as a block or sawtooth signal can be seen as constructed from an infinitely large sum of sine signals (see Figure 12.8), a much smoother signal has less high-frequency components.

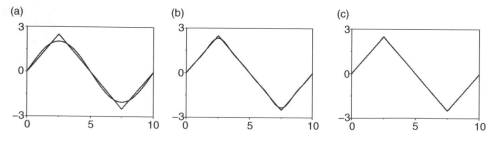

FIGURE 12.8   Triangular function approximated by sums of sine functions, base frequency 0.1 Hz, following sine components at 0.3, 0.5, 0.7 Hz, etc. (a) one sine function; (b) three sine functions; (c) ten sine functions.

In most cases, a smooth response is acceptable and even desirable from a mechanics standpoint, and thus responses to the high-frequency components in an input signal are not needed. Summarizing, at low frequencies, the gain of $K(j\omega)G(j\omega)$, called the *open-loop gain*, should be high, while at high frequencies it may be low. Note that $K(j\omega)G(j\omega)$ is a complex-valued function, and that it can have a value equal to or close to -1. To assess the behavior of the closed loop, the frequency response of the open-loop, $K(j\omega)G(j\omega)$, is studied near this point where the magnitude of the open-loop response, $|K(j\omega)G(j\omega)|$, is 1 or 0 dB. This point is the cross-over point, and the corresponding frequency is called the cross-over frequency.

The phase shift of the open-loop at the cross-over frequency determines what the gain of the closed-loop system will be at the cross-over frequency. A phase shift near 180° will cause the magnitude of the denominator of the closed-loop system in Equation 12.14 to become very small (while the numerator's magnitude is 1), and the closed-loop frequency response will have what is called an oscillatory peak. The response of the closed-loop system to, for example, a step change in the reference signal will show an oscillation with approximately the cross-over frequency. Investigation of the phase at cross-over is important for the assessment of the stability of the closed loop. The difference between the phase shift at cross-over and a phase of $-180°$ (i.e., the point -1) is called the *phase margin*. For stability of the closed-loop system, the phase margin must be positive, that is, the phase shift of the system is less negative than $-180°$. Usually, a phase margin larger than 40° is chosen.

For any feedback system, where the open-loop transfer function is not unstable, a positive phase margin is a guarantee for closed-loop stability. Stability for a system that is open-loop unstable, that is, has open-loop poles in the right-half complex plane, can be studied by means of the *Nyquist stability theorem*, which also constitutes more formal proof of stability by means of the frequency response. Proof of and explanation on the Nyquist stability theorem can be found in the engineering textbooks already mentioned.

An exemplary open-loop system, which, when used in a closed-loop feedback, is the "single in-tegrator". For a single integrator, $K(j\omega)G(j\omega) = 1/(j\omega)$. The single integrator has an infinitely high gain at $\omega = 0$, thus the output of the closed-loop system will perfectly follow the input for low frequencies. The phase margin is always 90°, whatever gain is chosen for the controller.

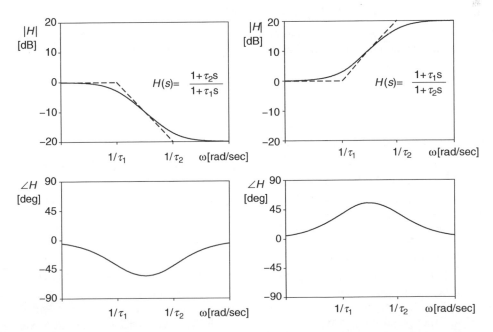

FIGURE 12.9 Bode diagram of a lag-lead compensating network (left) and of a lead-lag compensating network (right).

Of course, most systems are not single integrator systems. However, the same principles can be applied, that is, that a high gain at low frequencies is needed, and a sufficiently wide frequency range, with a frequency response locally resembling a single integrator and having an acceptable phase margin, can then be chosen for the cross-over frequency. This is summarized in the *Primary Rule of Thumb* for frequency domain design, which says that near the cross-over frequency, the following must hold:

$$Y^{OL}(j\omega) = K(j\omega)G(j\omega) = \frac{\omega_c}{j\omega}. \tag{12.15}$$

For the controlling element, when design in the frequency domain is used, a lead/lag or a lag/lead network is usually chosen. Since a logarithmic scale is used in a Bode diagram, the frequency response of the controller $K(j\omega)$ can simply be added to the frequency response of the controlled system $G(j\omega)$. Diagrams of the most common "compensating networks" are given in Figure 12.9.

12.3   Motivation and Overview of Human Manual Control Models

12.3.1   Motivation

The motivation for obtaining mathematical models of human control behavior has evolved from the need to explain the behavior of human-vehicle control systems to understanding human behavior in general. The analytical descriptions desired are in control-theoretical terms. The main purposes of the *engineering models* are:

1. To summarize behavioral data
2. To provide a basis for rationalization and understanding of human control behavior
3. To be used in conjunction with vehicle dynamics in forming predictions or in explaining the behavior of combined closed loop human-machine systems

Modeling humans using systems theory has proved to be a tremendous challenge. Humans are complex control and information processing-systems: they are time-varying, adaptive, nonlinear, and their behavior is essentially stochastic in nature (McRuer and Jex, 1967). Such systems are difficult to be characterized in mathematical terms because most of the mathematical tools are applied strictly to stationary, linear, and nonadaptive systems.

12.3.2   Quasi-Linear Function Theory

Research in the two decades after Word War II resulted in the successful application of *quasi-linear* describing function theory to the problem of modeling human control behavior in the single-axis compensatory tracking task (Krendel and McRuer, 1960). In the compensatory tracking task the human operator is controlling a dynamic system and responding to a displayed error signal in such a way that the output of that system approximates the value of a reference signal. This is essentially the same task as introduced earlier, Figure 12.7.

In the early days the model structure and model parameters obtained experimentally using the quasi-linear describing function models had predictive significance *only* in applications that were similar to the experimental conditions. No attempts were conducted to relate the model structure or model parameters to the context in which the task was conducted, the so-called task variables. This changed with the publication of McRuer et al. (1965). This landmark report provided an overview of earlier experiments, putting them in a general framework, and reported the results of new experiments that were especially conducted to show the relation between human control behavior and the task variables. The report provided convincing empirical evidence that humans systematically adapt their control behavior to the task variables.

12.3.3 The COM

This systematic adaptation of human control behavior was generalized in the COM. The COM model allows human control behavior to be *predicted*, based on knowledge about the main task variables as system dynamics and reference signal bandwidth. Attempts to parameterize the model of human control behavior resulted in the development of the so-called structural-isomorphic models such as the *precision model* and the *simplified precision model* (McRuer and Jex, 1967). The parameters of these models can be determined from the *verbal adjustment rules*, which are a set of rules that relate the parameters to the task variables.

The *descriptive models* that have become available with the COM are still widely applied in human-machine system studies. They are simple to use, require only a few parameters, and have proved to provide a good insight for many situations.

12.3.4 The OCM

The advance of "modern" control theory in the mid-1960s resulted in the concepts of optimal filtering (linear-quadratic gaussian, LQG) and optimal control (linear-quadratic regulator, LQR) (Kwakernaak and SiVan, 1972). One of the first attempts to describe human control behavior in the paradigm of optimal control and estimation theory yielded the OCM (Kleinman et al., 1970a,b). Here, the input–output relation of the human operator is compared with an *optimal* controller, instead of only a *stabilizing* controller. The algorithmic model consists of an optimal *observer*, generating an optimal state estimate of the system to be controlled and a deterministic *regulator*, which transfers the estimated state into the optimal output.

In contrast to the structural models, which *describe* what the human is doing, the OCM is a *normative* model, that is, it describes what the human *should* be doing given the inherent human limitations and the task variables at hand. The OCM was believed to be suitable for a more general application area, having a wider objective than the describing function models (Kok and Van Wijk, 1978; Wewerinke, 1989). In spite of this, however, the algorithmic model has not become as widely used as the structural models, which can be attributed inpart to the fact that the algorithmic model is over-parameterized for describing the time histories of a single tracking task (Van Wijk and Kok, 1977), complicating the model validation significantly.

12.3.5 More Recent Models

The COM can be considered as the result of applying "classical" control theory, that is, theories that emerged from the 1950s and 1960s, to model human behavior, and the OCM as the result of applying optimal control theory, that is, theories that emerged in the mid-to-late 1960s to model human behavior. It is no surprise that since then, with other control-theoretical approaches emerging such as fuzzy logic, neural networks, adaptive neuro-fuzzy models, andsoforth, that many of these new approaches have also been applied to describe human behavior. In this chapter we will concentrate on the two classic approaches, however.

12.4 The Cross-Over Model

12.4.1 The Problem with Modeling Humans

The human operator is an adaptive controller and the observed control behavior is likely to change, either consciously or unconsciously, to the environment. The ability to change and adapt is undoubtedly the most prominent and valuable human capabilities. Yet, when one attempts to describe the behavior and grasp it into a mathematical model, one encounters a nontrivial problem, as most of the available techniques only work out well in the time-invariant case.

The earliest attempts in describing human behavior with control-theoretical models failed to pay much attention to the adaptivity of human behavior. Research showed that when experiments are not done under (almost) exactly the same circumstances, the human will *adapt* and the observed control behavior will be different. The lack of a systematic approach to this problem resulted in scattered data and many different and unexpected findings, and theory progressed only slowly.

This changed in the late 1950s and early 1960s when McRuer et al. started working on this problem. Learning from past experience, they first determined and classified a list of variables that could possibly have an effect on human behavior. These included environmental variables (e.g., conducting the task in real flight or in a fixed-base simulator), procedural variables (e.g., subject instruction, practice), and operator-centered variables (e.g., subject motivation, workload).

Most important to understand the adaptation of the human operator are the *task variables*, however, which include:

- The dynamics of the *system to be controlled*
- The properties (bandwidth) of the *forcing function*, that is, the signal to be followed (in a following task) or the disturbance signal acting on the system (in a disturbance task)
- The type of *display* (e.g., a compensatory display or a pursuit display)
- The type of *manipulator*

McRuer et al. conducted a massive number of tracking task experiments. In their approach they tried to very closely control all the variables that could have an effect on human behavior and systematically varied two of the task variables, that is, the dynamics of the system and the bandwidth of the forcing function signal. As a result of this approach, human variability decreased significantly and for the first time insight was gained into *how* and *why* humans adapt to changing circumstances. In the following paragraphs the main results of this research will be elaborated on.

12.4.2   Quasi-Linear Pilot Models

Skill Acquisition   The earliest work in this field already showed that a human operator establishes, during a learning and skill-development phase, a particular control system structure (Krendel and McRuer, 1960). The feedback connections in this system are similar to those, which would be selected for the development of an automatic controller. The loop closures selected will have the following properties (McRuer and Jex, 1967):

1. To the extent possible, the feedback loops selected and adjustments made will be such as to allow wide latitude and variation in pilot characteristics
2. The loop and equalization structure selected will exhibit the highest pilot rating of all practical loop closure possibilities
3. Delays due to scanning and sampling are minimized

In short, the human will establish, in a learning process, a control system that aims at establishing a trade-off between the requirements of performance and stability, in the same fashion as a control engineer would design an automatic control system.

Compensatory Tracking Tasks   Extensive research has been conducted on the problem of modeling human control behavior in elementary single-axis compensatory tracking tasks, see Figure 12.10a. In this task the operator must minimize the difference (error E) between the output (Z) of the system to be controlled and a reference signal (R). This is the same control situation as described earlier, Figure 12.7. The double-lined block in Figure 12.10a illustrates that pilot behavior in this closed loop is essentially nonlinear.

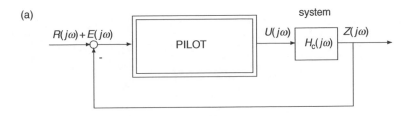

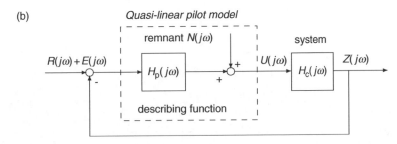

FIGURE 12.10 The quasi-linear pilot model in the elementary single-axis compensatory tracking task. In this figure, H_c depicts the dynamics of the system to be controlled, and R, E, U, and Y the reference signal, the displayed error signal, the pilot control signal, and the system output signal, respectively. The quasi-linear pilot model consists of a describing function H_p and a remnant N, (a) single-axis compensatory tracking task; (b) quasi-linear pilot model.

Experimental studies concerning the compensatory tracking task showed that, as long as the task variables remain constant, the operator control behavior remains fairly constant too. In this case, it can be described by a *deterministic* model — a linear differential equation with constant coefficients and a time delay (the describing function) — and a *remnant* model — a stationary noise process.

The result is a *quasi-linear pilot model* (Figure 12.10b): the *describing function* accounts for the portion of the human controller's output that is *linearly* related to his input and the *remnant* represents the difference between the linear model output and the experimentally measured output of the human controller.

The application of quasi-linear theory allowed human manual control behavior in single-axis compensatory control tasks to be identified. The insights gained led to the postulation of the COM theorem.

12.4.3 The COM Theorem

In a series of experiments the characteristics of two task variables, that is, the dynamics of the system to be controlled ($H_c(j\omega)$ in Figure 12.10b) and the bandwidth of the reference signal ($R(j\omega)$ in Figure 12.10b) were systematically varied.

The system dynamics were chosen to be the basic proportional (k), integrator (k/s), and double integrator (k/s^2) systems. The bandwidth of the forcing function, (ω_r), was set at 1.5, 2.5, and 4.0 rad/sec. The forcing function consisted of a sum of ten sinusoids in a shelf spectrum. These sinusoids have different frequencies and random phases and their sum results in a quasi-random signal, being quasi-random in the sense that the human operator cannot anticipate future values of this signal. The compensatory display shows only the error between the forcing function signal $R(j\omega)$ and the system output signal $Z(j\omega)$.

Figure 12.11 summarizes the main findings of McRuer's experiments. The left column shows the magnitude of the pilot frequency response function (FRF) $|\hat{H}_p(j\omega)|$ that could be estimated from the experimental data. The identification procedure allows this FRF to be determined at only the frequencies of the sinusoids in the forcing function. The center column shows the magnitude of the system FRF and because we know the system exactly we can compute this function for all frequencies. The right

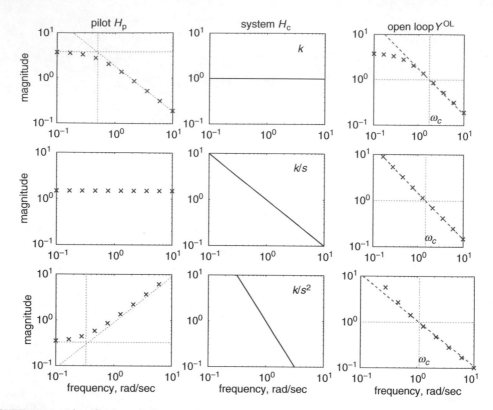

FIGURE 12.11   Identification of pilot control behavior when controlling the three basic systems k, k/s, and k/s^2. The crosses indicate the frequencies of the sinusoids in the forcing function $R(j\omega)$. The dashed lines in the right column indicate integrator-like dynamics.

column shows the magnitude of the estimated open-loop, that is, the product of the estimated pilot FRF and the system FRF: $\hat{Y}^{OL}(j\omega) = \hat{H}_p(j\omega)H_c(j\omega)$.

The dynamics of the center column show the proportional (top), integrator (center), and double integrator (bottom) properties of the system to be controlled. The left column shows that the pilot FRF is different for all three systems, a clear indication that the pilot is adapting to the dynamics of the system to be controlled. The right column, however, shows that the shape of the open-loop is *the same*. Independent of the system dynamics, the open-loop FRF resembles "integrator- like" dynamics near the frequency where the open-loop equals one, that is, near the cross-over frequency. As will be discussed later, the value of the cross-over frequency depends, among others, on the dynamics of the system and the bandwidth of the forcing function.

Apparently, human operators adapt to the system to be controlled in such a way that the open-loop, that is, the human dynamics times the system dynamics, becomes an integrator. McRuer et al. generalized this *systematic* adaptation of human control behavior with the postulation of their COM theorem (McRuer et al., 1965).

The COM theorem states that human controllers *adjust* their control behavior to the dynamics of the controlled element in such a way that the dynamic characteristics of the open-loop transfer function in the cross-over region can be described by:

$$Y^{OL}(j\omega) = H_p(j\omega)H_c(j\omega) \approx \frac{\omega_c}{j\omega}e^{-j\omega\tau_e}, \tag{12.16}$$

where ω_c is the cross-over frequency and τ_e is a time delay lumping the information-processing delays of the human operator.

The COM is a mathematical statement of the *empirical* observation that human controllers adjust their control characteristics so that the closed-loop system dynamics mimic those of a well-designed feedback system (McRuer and Jex, 1967). Then, when the dynamics of the system to be controlled are known, the COM allows a *prediction* of the human controller characteristics via:

$$H_p(j\omega) \approx \frac{\omega_c}{j\omega \, H_c(j\omega)} e^{-j\omega\tau_e}. \tag{12.17}$$

The parameters ω_c and τ_e are task-dependent and can be selected on the basis of the so-called *verbal adjustment rules* that are in turn based on an immense amount of experimental data (McRuer and Krendel, 1974).

12.4.4 Model Parametrization: Structural-Isomorphic Models

The linear describing function of the quasi-linear pilot model takes on various forms depending on the precision with which one attempts to reproduce the characteristics of the observed control behavior. In its most extensive form, the so-called *precision model*, the linear describing function can be described by (McRuer and Jex, 1967):[1]

$$H_p(j\omega) = \overbrace{K_p}^{\text{gain}} \underbrace{\overbrace{\left(\frac{1 + \tau_L j\omega}{1 + \tau_I j\omega} \right)}^{\text{lead-lag}}}_{\text{pilot equalization}} \underbrace{\overbrace{\left(\frac{\omega_n^2}{(1 + \tau_{N_1} j\omega)(\omega_n^2 + 2\zeta_n \omega_n j\omega + (j\omega)^2)} \right)}^{\text{neuromuscular system}} \overbrace{e^{-j\omega\tau}}^{\text{time delay}}}_{\text{pilot limitations}}. \tag{12.18}$$

The parameters in this model reflect the pilot adaptation characteristics as well as the limitations such as the time delay and the neuromuscular system. The equalization parameters are adjusted by the pilot in such a way that the open-loop dynamics satisfy the COM (Equation (12.16)).

The most commonly used approximation of the precision model is the *simplified precision model* (McRuer and Jex, 1967):

$$H_p(j\omega) = \underbrace{K_p}_{\text{gain}} \cdot \underbrace{\frac{1 + \tau_L j\omega}{1 + \tau_I j\omega}}_{\text{pilot equalization}} \cdot \underbrace{e^{-j\omega\tau_e}}_{\text{time delay}} \tag{12.19}$$

with:

- K_p the pilot gain
- τ_L the lead time constant (in sec)
- τ_I the lag time constant (in sec)
- τ_e the effective time delay (in sec)

The simplified precision model (SPM) is widely used and will be discussed in the following.

12.4.5 Verbal Adjustment Rules

12.4.5.1 Empirical Rules

The adjustment rules of McRuer et al. are based on the experience obtained with a large number of tracking experiments, all SISO following tasks with a compensatory display, conducted over a wide range of dynamic characteristics of controlled elements and input signal bandwidths (McRuer et al., 1965). The

---

[1]Skipping the so-called 'α'-term (McRuer and Jex, 1967).

adjustment rules enable us to *predict* the parameters used in the precision or simplified precision models, including their numerical values, for a specific combination of system dynamics and forcing function signal bandwidth.

There are six rules, only the discussion of the first and the last will follow. The reader should keep in mind that the adjustment rules have an empirical basis and should be used with care.

12.4.5.2 Rule #1: Equalization Selection and Adjustment

The first rule is the most important one, as it allows us to predict *how* humans will adapt to a particular system, that is, *what* equalization parameters of the SPM are needed in order to follow the COM.

The first rule states that the particular equalization is selected from the SPM such that the following properties occur:

1. The system can be stabilized by proper selection of gain K_p preferably over a very broad range of K_p
2. Over a wide frequency range, near the cross-over region, the magnitude ratio $|H_pH_c|$ has an "integrator-like" shape, that is, the ratio has approximately a -20 dB/decade slope
3. $|H_pH_c| \gg 1$ is obtained at low frequencies to provide good low frequency closed-loop response to system commands and suppression of disturbances

Figure 12.12 illustrates the choice of the pilot equalization parameters τ_L (lead) and τ_I (lag) such that, for all three basic systems considered, the open-loop magnitude becomes "integrator-like" near the cross-over frequency.

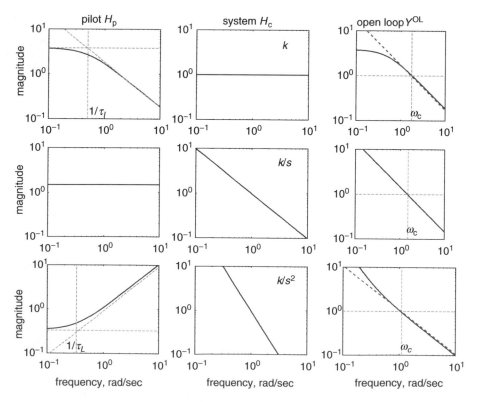

FIGURE 12.12 Pilot equalization when controlling the three basic systems k, k/s, and k/s^2.

For the velocity control system (k/s), the pilot equalization parameters can be zero or equal ($\tau_I = \tau_L$), and the pilot can act like a proportional gain with a time delay:

$$H_p = K_p e^{-j\omega\tau_e}. \tag{12.20}$$

There is no need to "fix" the system dynamics through adaptation because the system dynamics itself already mimics the integrator-like dynamics that are predicted by the COM.

For the proportional control system (k), the pilot must use the *lag* equalization (τ_I) to make the open-loop an integrator:

$$H_p = K_p \frac{1}{(1 + j\omega\tau_I)} e^{-j\omega\tau_e}. \tag{12.21}$$

The lag parameter needs to be chosen such that it creates the integrator-like characteristic *well before* the cross-over frequency, that is, $1/\tau_I \ll \omega_c$. An intuitive interpretation for the lag is that the human integrates or smoothers the input, responding to trends rather than to moment-to-moment variations. In essence, the human is responding to the "averages" of error over moving windows of observations rather than to the instantaneous error, values.

Finally, for the acceleration control system (k/s^2), the pilot needs to use the *lead* equalization (τ_L) to make the open-loop an integrator:

$$H_p = K_p(1 + j\omega\tau_L)e^{-j\omega\tau_e}. \tag{12.22}$$

The lead parameter needs to be chosen such that it creates the integrator-like characteristic *well before* the cross-over frequency, that is, $1/\tau_L \ll \omega_c$. An intuitive interpretation for the lead is that the human responds to both the position and velocity of the error signal. In essence, attending to velocity allows the human to effectively *anticipate* future position errors.

12.4.5.3 Rule # 6: Choice of Cross-Over Frequency and Effective Time Delay

Based on the empirical results of McRuer's experiments, rule # 6 yields likely values for the cross-over frequency and effective time delay for the three basic systems k, k/s, and k/s^2.

Experiments showed that τ_e decreases with ω_i with a slope, which is practically independent of the controlled element, while the basic level of τ_e depends primarily on the controlled element and is independent of ω_i:

$$\tau_e(H_c; \omega_i) = \tau_0(H_c) - \Delta\tau(\omega_i). \tag{12.23}$$

Similarly, the experiments showed that ω_c increases with ω_i with a slope that is also practically independent of the controlled element, and again the basic level of ω_c depends primarily on the controlled element and is independent of ω_i:

$$\omega_c(H_c; \omega_i) = \omega_{c_0}(H_c) - 0.18\,\omega_i. \tag{12.24}$$

The values for τ_0, ω_0 and $\Delta\tau$ are summarized in Table 12.1.

Note that these *qualitative* models are based on the empirical results of McRuer's experiments, and should be handled with care. An important phenomenon, for instance, is that for the double integrator dynamics the cross-over frequency increases with forcing function bandwith *upto a certain frequency.* For higher input signal bandwidths the cross-over frequency starts decreasing, a finding called *cross-over regression.*

TABLE 12.1   Values for τ_0, ω_0, and $\Delta\tau$ for the three basic Systems
(McRuer and Jex, 1967).

	ω_0 (rad/sec)	τ_0 (sec)	$\Delta\tau$ (sec)
k	5 to 6	0.33	0.070
k/s	4.3	0.36	0.065
k/s^2	3.3	0.50	0.065

12.4.6   Remnant

Much, if not all of the research into the applicability of quasi-linear pilot models focused on the describing function element. The reader should note that in the discussion on the COM, *no mention was made* of the remnant part in the quasi-linear model. The whole discussion focused on the adaptation that takes place in the pilot describing function, that is, the pilot frequency response function.

Numerous efforts to obtain insight into the origin and characteristics of the remnant have been conducted (Levison et al., 1969; Jex et al., 1971), and models exist for some of the more structural aspects of remnant (McRuer and Krendel, 1974; Jagacinski and Flach, 2003). In general, remnant increases when the task becomes more difficult, that is, for more difficult systems like the double integrator k/s^2, when workload increases, when more tasks have to be conducted simultaneously, and so forth.

12.4.7   Multi-Loop Pilot Models

The combination of quasi-linear pilot models and the COM theorem has become a well-established paradigm for describing and predicting human control behavior in single-axis compensatory tracking tasks. Attempts to extend the single-loop results to multi-loop or multiple loop control situations have been conducted as well (McRuer and Jex, 1967; Weir and McRuer, 1968, 1972; McRuer et al., 1977; McRuer and Krendel, 1974; McRuer and Schmidt, 1990; Hess, 1997; Mulder, 1999).

The main recipe for dealing with multi-loop pilot models is the following: start from the innermost control loop (e.g., "attitude"-control in aircraft and other vehicles), ignore all other "outer" loops and apply the COM to find the right equalization structure to stabilize the innermost loop. Compute the closed-loop system for the innermost loop and start working on the next feedback loop (e.g., "path"-control in vehicles). Repeat this procedure until all loops are closed. A useful rule-of-thumb for these multi-loop models is that the cross-over frequency reduces one-third when going from an inner loop to the next loop in the control hierarchy. In other words, the cross-over frequencies become smaller when moving towards the "slower" outer loops (Mulder, 1999).

Multi-loop tasks are more difficult to examine than single-loop tasks. Whereas the trade-offs between performance, stability, and pilot equalization efforts are relatively clear-cut in the single loop-task, the number of alternatives increase rapidly in multi-loop situations.

Furthermore, the model identification and validation efforts are difficult in these situations and involve many intricacies. First, there is the problem of determining the particular feedback loops the operator will close: the model structure itself cannot be easily identified from experimental data (Stapleford et al., 1969). Another problem that was only properly realized in the mid 1970s was the conceptual difficulty of identifying systems operating in closed loop. The rather elaborate identification techniques for these closed-loop multi-loop applications were not available until the pioneer work of Stapleford et al., (1967, 1969), and it lasted until the late 1970s until they were formalized mathematically (Van Lunteren, 1979).

12.5   The Optimal Control Model

12.5.1   Introduction

The COM and the models originating from it operate primarily in the frequency domain. This is not surprising, since the techniques involved come from the "classical" frequency-domain approach to designing feedback control systems, dating back to the early 1950s. With the advent of optimal control and estimation theory in the 1960s, a second approach to modeling human control behavior became popular, operating primarily in the time domain, resulting in the human OCM.

The basic assumption of the OCM is that "the well-trained, well-motivated human operator behaves in an optimal manner, subject to his inherent limitations and to the requirements of the control task" (Kleinman et al., 1970b). It is a normative model in the sense that parameters in the model (the observer and control gains) are set to minimize errors of observation and control based on the analytical solution of optimal algorithms. The model describes what operators *should* do given their inherent constraints limiting their behavior and the extent to which they understand the task objectives. The OCM structure is shown in Figure 12.13.

The operator task is to control a dynamic system (state $\underline{x}$) perturbed by external disturbances ($\underline{w}$). The outputs of the system (y) are presented with a display. The operator uses the information gathered about the system state to generate a control signal ($\underline{u}$), which maintains a system reference state, compensating for the effects of the disturbances. The modeling through optimal control of the stationary input–output relation of the operator has the following starting points (Kok and Van Wijk, 1978):

- The system to be controlled can be described as an LTI system
- The observed outputs are an LTI combination of the system states
- The operator minimizes a quadratic cost functional
- The limitations of the operator can be modeled with:
  - (i) A *single time delay* τ, lumping the operator information-processing lags
  - (ii) An *observation noise* $\underline{v}_y$, a Gaussian white noise signal, which represents operator uncertainties concerning the observed variables
  - (iii) A *motor noise* $\underline{v}_u$, a Gaussian white noise signal, which represents operator uncertainties in generating their control input $\underline{u}(t)$
  - (iv) A *neuromuscular lag* τ_N, representing the operator neuromuscular system
- Operators have internal models of:
  - (i) The dynamics of the system to be controlled
  - (ii) The statistics of the system disturbances
  - (iii) The relation between the observed outputs and the system state

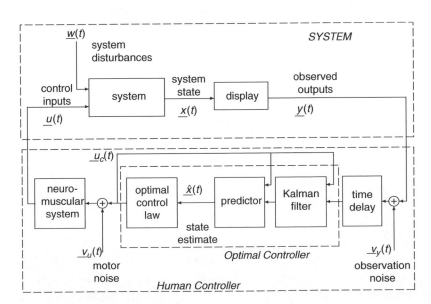

FIGURE 12.13 Structure of the OCM.

(iv) The task to be performed
(v) Their own inherent limitations

When these assumptions are satisfied, the hypothesis that the operator behaves in an optimal manner can be elaborated with modern optimal control and estimation theory (Kwakernaak and Sivan, 1972). The description of the control situation and the incorporation of the human limitations were very carefully chosen (Kleinman et al., 1970b) to *allow* a solution of the optimal control problem. The *separation theorem* states that the optimum stochastic controller is realized by cascading an optimal state estimator (LQG) with a deterministic optimal controller (LQR) (Kwakernaak and SiVan, 1972). Hence, the OCM consists of two elements (Kleinman et al., 1970b): (i) an *optimal observer*, a cascade combination of a Kalman filter and a predictor, which use the available information to obtain an optimal estimate of the system state ($\hat{x}$); and (ii) an *optimal regulator*, an optimal controller, which transfers the state estimate into an optimal control signal ($\underline{u}_c$).

12.5.2 Mathematical Formulation of the OCM Parameters

A mathematical formulation allows a derivation of the model structure and the model computation, for which the reader is referred to Kleinman and Baron (1971).

12.5.2.1 System Description

The dynamics of the system to be controlled are represented by an LTI state description:

$$\dot{\underline{x}}(t) = A\underline{x}(t) + B\underline{u}(t) + E\underline{w}(t), \tag{12.25}$$

where $\underline{x}(t)$ is the ($n \times 1$) state vector and $\underline{u}(t)$ are the ℓ inputs to the system. The ($k \times 1$) disturbance vector is defined as an independent, zero-mean, Gaussian white noise vector with auto-covariance: $E\{\underline{w}(t)\underline{w}^T(t + \tau)\} = W\delta(\tau)$. It is assumed that several system outputs are presented to the human in a continuous way via some instrument panel:

$$\underline{y}(t) = C\underline{x}(t) + D\underline{u}(t) + H\underline{w}(t), \tag{12.26}$$

where $\underline{y}(t)$ is the ($m \times 1$) observation vector representing the information set upon which the operator bases the control actions. It is assumed that if a quantity y_i is displayed, the derivative of that quantity — $\dot{y}_i$ — is also perceived; the observation vector contains *pairs* of variables explicitly displayed to, as well as those implicitly derived by the operator.

12.5.2.2 Human Limitations

The inherent psychophysical limitations of the operator are represented by: (i) a perceptual time delay τ; (ii) a neuromotor lag τ_N, and (iii) the observation and motor noise vectors $\underline{v}_y$ and $\underline{v}_u$. As illustrated in Figure 12.13, the observation noise — representing errors made in observing the displayed quantities — is placed at the input of the operator model. The ($m \times 1$) observation noise vector is defined as an independent, zero-mean, Gaussian white noise vector with auto-covariance: $E\{\underline{v}_y(t)\underline{v}_y^T(t + \tau)\} = \text{diag}(V_y)\delta(\tau)$. The operator perceives a delayed, noisy replica of the system outputs:

$$\underline{y}_p(t) = \underline{y}(t - \tau) + \underline{v}_y(t - \tau). \tag{12.27}$$

The perceived output is processed by the operator who generates a commanded control input $\underline{u}_c(t)$ that is considered *optimal* for the task at hand. A motor noise $\underline{v}_u(t)$ — representing errors in executing the intended control movements and the fact that the operator does not have perfect knowledge of the system output $\underline{u}(t)$ — is added to $\underline{u}_c(t)$:

$$u(t) = \underline{u}_c(t) + \underline{v}_u(t). \tag{12.28}$$

The ($\ell \times 1$) motor noise vector is defined as an independent, zero-mean, Gaussian white noise vector with auto-covariance: $E\{\underline{v}_u(t)\underline{v}_u^T(t + \tau)\} = \text{diag}(V_u)\delta(\tau)$.

12.5.2.3 Control Task Representation

It is assumed that the task is reflected in the operator's choice of a feedback control $\underline{u}^*(\cdot)$ which, in steady-state, minimizes the cost functional:

$$J(\underline{u}) = E\left\{\sum_{i=1}^{m} q_i y_i^2 + \sum_{i=1}^{\ell} r_i u_{c_i}^2 + \sum_{i=1}^{\ell} g_i \dot{u}_{c_i}^2\right\}, \tag{12.29}$$

conditioned on the perceived information $\underline{y}_p(\cdot)$. The cost functional weightings ($Q \geq 0$; $R \geq 0$; $G > 0$) may be either *objective* (specified by the experimenter) or *subjective* (adopted by the operator). The rate of the control input is also weighted in the cost functional. This term may also represent an *objective* (to account for human physiological limitations on the rate at which a control action can be effected) or a *subjective* (reflecting that trained operators are reluctant to make rapid control movements) weighting on control rate. The control rate weighting introduces a first-order lag in the optimal controller (Kleinman et al., 1971). Hence, the control rate weighting matrix G is used to include a first-order representation of the neuromuscular system in the model.

12.5.3 Model Parameters, Outputs, Solution, and Identification

12.5.3.1 Model Parameters

In order to apply the OCM, the following quantities must be defined: (i) the *system* parameters: the characteristics of the linear system (A, B, C, D, E, H), and the statistics of the system disturbance (W); (ii) the *task-related* parameters: the weightings of the cost functional (Q, R); and (iii) the *human response* parameters (τ, τ_N, V_y, and V_u). The OCM parameters describe the control task and the operator control behavior in terms of optimal control.

The formulation of the operator's characteristics in mathematical terms is difficult. First of all, the task-related parameters — determining the *balance* of the model, or, equivalently, the control strategy of the operator — are not easy to define beforehand. The OCM describes the operator as an optimal controller with respect to task-related quantities, which do not necessarily relate to human-centered optimization strategies. Second, the operator is assumed to have a good mental model of the system dynamics and the disturbance characteristics. Although it is reasonable to assume that this model indeed becomes well-developed through a learning process (Krendel & McRuer, 1960; Stassen et al., 1990), the OCM results can and should be considered as the *best possible* operator performance for a given quadratic cost functional.

12.5.3.2 Model Outputs

Unlike the structural pilot models of Section 4, which operate almost exclusively in the frequency domain, the OCM is essentially a time-domain model. Nonetheless, it provides results in both domains (Kleinman et al., 1970b, 1971). The *time-domain* outputs are the variances of all signals in the closed loop, that is, $E\{u_i^2\}|_{i=1,\dots,\ell}$, $E\{x_i^2\}|_{i=1,\dots,n}$, and $E\{y_i^2\}|_{i=1,\dots,m}$. Due to the incorporation of the pilot remnant through the observation and motor noises, these variances allow a direct comparison with the values obtained in the actual experiment.

The frequency-domain output is the multi-dimensional ($\ell \times m$) transfer function matrix $\underline{H}(s)$, relating the m pilot inputs to the ℓ pilot outputs. The OCM assumes a *parallel* model structure: each of the operator outputs consists of the *sum* of the individual loop closures of the displayed variables. It has been stated earlier that the observation vector contains multiple observation pairs $(y_i, \dot{y}_i)$ of a signal y_i and its derivative $\dot{y}_i$. Therefore, the *equivalent* transfer function from a displayed variable to the pilot's output always includes the effect of the implicitly perceived derivative of this variable, that is, $h_e(s) = h_{y_i}(s) + s h_{\dot{y}_i}(s)$. The equivalent frequency domain transfer function allows a comparison with the experimentally measured operator describing functions.

12.5.3.3 Model Solution

Parameter Initialization   Typical values used for the time delay τ are 0.15 to 0.35 sec. The neuromuscular lag τ_N lies in the range of 0.1 to 0.6 sec, with a value of 0.1 sec that is commonly used in compensatory tracking tasks (Kleinman et al., 1970a; Wewerinke, 1989). Because of the close relationship between τ_N and the control rate weighting matrix G the neuromuscular lag determines the nature of control behavior considerably. For an 'outer loop' type of controller — smooth control without fast and jerky control excursions — the weighting on control rate must be large yielding a large τ_N. For an 'inner loop' type of controller — a tight control mode with fast and short control responses — the weighting on control rate is small. Determining the observation and motor noise intensities, V_y and V_u, is difficult. These quantities depend on the display, the environment, and intrinsic human characteristics. The following procedure (Kleinman et al., 1970ab) has been reported for determining the observation and motor noise intensities:

$$V_{y_i} = \pi \rho_{y_i} \mathrm{E}\{y_i^2\}|_{i=1,2,\dots,m} \quad \text{and} \quad V_{u_i} = \pi \rho_{u_i} \mathrm{E}\{u_{c_i}^2\}|_{i=1,2,\dots,\ell}. \tag{12.30}$$

The observation noise *ratio* ρ_{y_i}, defined as the noise intensity V_{y_i} normalized with respect to the signal variance has a typical value of 0.01. In other words, the normalized observation noise intensity has a power density level of -20 dB. The motor noise ratio ρ_{u_i}, defined similarly as ρ_{y_i}, has a typical value of 0.003. Thus, the normalized motor noise intensity has a power density level of -25 dB. The value of the motor noise ratio has not been validated as extensively as the observation noise ratio, for which the value of -20 [dB] is substantiated in several psychophysical studies (Kleinman et al., 1971; Baron, 1976).

Iterative OCM Solution   Due to the separation theorem, the OCM can be computed in two iterations as shown in Figure 12.14 (Thompson, 1987). First, the OCM *regulator* is computed. The ℓ diagonal elements g_i of G are computed iteratively until the resulting first-order lag time constants τ_{N_i} equal those defined by the user (or, when matching experimental data, the value of τ_{N_i} as exhibited by the tracker). Second, the OCM *observer* is computed. The observation and motor noise ratios are defined, and the model iteratively changes V_{y_i} and V_{y_i} — resulting in different variances — to obtain the defined ratios.

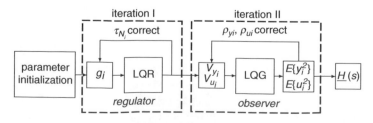

FIGURE 12.14   Computation flow of the OCM.

12.5.3.4 Model Identification

An important drawback of the OCM is that it is *over-parametrized*, that is, it contains more parameters than necessary to uniquely describe the input–output behavior of the operator (Phatak et al., 1976; Kok and Van Wijk, 1978). Due to this over-parametrization, the complete set of model parameters cannot be identified from the time histories of a single tracking task (Van Wijk and Kok, 1977). The elements of the regulator feedback matrix and the observer gain matrix can be determined only upto a similarity transformation of the model and the observation and motor noise vectors are not resolvable (Van Wijk and Kok, 1977).

Some model-matching procedures have been proposed to resolve some of these issues and identification methods have been successful for OCM structures in which a subset of the parameter vector is either neglected (Phatak et al., 1976) or fixed (Wewerinke, 1979). Therefore, the over-parameterization is only a problem when all parameters are treated as "free," and in practice many of the parameters are fixed to values that reflect the results of other psychophysical studies.

12.5.4 Allocation of Attention

12.5.4.1 Theoretical Background

A common application of the OCM is the analysis of the effect of display parameters and dynamics on manual control behavior (Kleinman and Baron, 1973; Baron and Levison, 1975; Wewerinke, 1976; Hess and Wheat, 1976; Hess, 1977, 1981; Kleinman and Korn, 1981). Generally speaking, operators allocate their mental resources among several tasks and numerous displays. The characteristics of the displayed variables, the control tasks, and the control task objectives determine the allocation of attention strategy of the operator. In Baron and Levison (1975) the effects of attention sharing are modeled by an increase in the so-called *nominal* observation noise ratio ρ_0 by:

$$\rho_{y_i} = \frac{\rho_0}{f_i} \quad (i = 1, 2, \ldots, m), \quad \text{subject to} \quad \sum_{i=1}^{i=m} f_i = 1 \tag{12.31}$$

The nominal observation noise ratio is defined as the observation noise ratio when attention is not shared and f_i represents the *fraction of attention* allocated to a display variable. Then, ρ_{y_i} is the observation noise ratio associated with the ith display variable when attention is being *shared*. In accordance with the fundamental assumption of optimality it can be hypothesized that the operator divides attention among the displayed variables in an *optimal* manner. Then, only the nominal level of attention ρ_0 needs to be defined and the model iteratively computes the *optimal* set f_i^{*} ($i = 1, \ldots, m$), yielding the observation noise ratios ρ_{y_i} for all displayed quantities (Kleinman, 1976). The optimal allocation set depends strongly on the specification of the control task. It shows the *relative* importance of the displayed quantities to the operator in performing the task and can be informative when designing and analyzing displays (Baron, 1976; Korn et al., 1982).

12.5.4.2 Inverse Optimal Allocation of Attention

The attention allocation method is often used in an "inverted" manner. First, the m observation noise ratios ρ_i are identified. When these quantities are known, a set of $(m + 1)$ equations and $(m + 1)$ unknown parameters results. The solution for the nominal level of attention is given by:

$$\rho_0 = \frac{1}{\left(\frac{1}{\rho_1} + \frac{1}{\rho_2} + \ldots + \frac{1}{\rho_m} \right)}, \tag{12.32}$$

which also allows a computation of the fractions of attention f_i.

12.6   Final Remarks

This chapter has focused on only a narrow aspect of the larger topic of manual control, specifically the application of control theory to modeling human performance in closed-loop control such as piloting an aircraft. This context was chosen to illustrate the value of a control-theoretic orientation to modeling performance in human-machine systems. Control theory provides an important language for describing any dynamical system (including human-machine systems), where perception and action are coupled to allow goal-directed behavior.

We consider the COM and the OCM described in this chapter to be two of the best examples of how control theory can help to frame questions about the performance of human-machine systems. However, we caution people that these models represent the start, not the end of our search to understand the dynamics of perception and action. It is important to appreciate the limitations of both of these models. One limitation is that the assumptions underlying these models fit a very restricted range of experimental conditions. One must be very skeptical about whether the assumptions underlying either model are met in any situations, but relatively simple tracking contexts. A second limitation is that both models have a large number of parameters. Thus, confidence about any of the parameter values depends on converging operations. Because of the large numbers of parameters, fitting the models to human performance data in any single context is still a bit of an "art" in that good scientific/engineering judgment is required.

Despite the limitations of these models, we feel strongly that familiarity with control theory and its application for modeling human-machine systems is a valuable asset for anyone interested in human-machine systems or cognitive engineering. It is unrealistic to expect that we will be able to capture the full complexity of many natural systems and situations in closed form mathematical models such as the COM and OCM. However, we believe that people who have a deep understanding of the dynamics of simple closed-loop systems will be in a better position to ask smart questions and to avoid the trap of simple, obvious, wrong answers to the complex problems associated with modern socio-technical systems. We believe that control theory is a good foundation for developing the intuition and judgment needed for smart cognitive systems engineering.

References

Baron, S., A model for human control and monitoring based on modern control theory, *J. Cyber. Inf. Sci.*, 4(1), pp. 3–18, 1976.

Baron, S. and Levison, W.H., An optimal control methodology for analyzing the effects of display parameters on performance and workload in manual flight control. *IEEE Trans. Sys., Man, Cyber.*, SMC-5(4), pp. 423–430, 1975.

Dorf, R.C. and Bishop, R.H., *Modern Control Systems*, 10th ed., Pearson Education, Upper Saddle River, NJ, 2005.

Flach, J.M., Control with an eye for perception: precursors to an active psychophysics, Proceedings of a workshop held at NASA Ames Research Center, Moffett Field (CA), in Johnson W.W. and Kaiser, M.K. eds., *Visually Guided Control of Movement*, NASA Ames Research Center, 1991, pp. 121–149, June 26 to July 14, 1989.

Franklin, G.F., Powell, J.D., and Emami-Naeini, A., *Feedback Control of Dynamic Systems*, Prentice Hall, Upper Saddle River, NJ, 2002.

Goodwin, G.C., Graebe, S.F., and Salgado, M.E., *Control System Design*, Prentice Hall, Upper Saddle River, NJ, 2001.

Hess, R.A., Analytical display design for flight tasks conducted under instrument meteorological conditions, *IEEE Trans. Sys., Man, Cyber.*, SMC-7(6), pp. 453–462, 1977.

Hess, R.A., Aircraft control-display analysis and design using the optimal control model of the human pilot, *IEEE Trans. Systems, Man, and Cybernetics*, SMC-11(7), pp. 465–480, 1981.

Hess, R.A., Feedback Control Models: Manual control and tracking, in *Handbook of Human Factors and Ergonomics* Salvendy, G., ed., John Wiley & Sons, New York, pp. 1249–1294, 1997.

Hess, R.A. and Wheat, L.W., A model-based analysis of a display for helicopter landing approach, *IEEE Trans. Sys., Man, Cyber.*, SMC-6(7), pp. 505–511, 1976.

Jagacinski, R.J. and Flach, J.M., *Control Theory for Humans — Quantitative Approaches to Modeling Performance*, Lawrence Erlbaum, Mahwah, NJ, 2003.

Jex, H.R., Allen, R.W., and Magdaleno, R.E., Effects of display format on pilot describing function and remnant, *Proc. Seventh Annual Conference on Manual Control, NASA SP-281*, pp. 155–159, 1971.

Kleinman, D.L., Solving the optimal attention allocation problem in manual control, *IEEE Trans. Automatic Control*, AC-21(6), pp. 813–821, 1976.

Kleinman, D.L. and Baron, S., *Manned-Vehicle Systems Analysis by Means of Modern Control Theory* (NASA Contractor Report No. CR-1753), Washington, D.C., 1971.

Kleinman, D.L. and Baron, S., A control theoretic model for piloted approach to landing, *Automatica*, 9, pp. 339–347, 1973.

Kleinman, D.L., Baron, S., and Levison, W.H., An optimal control model of human response. Part II: prediction of human performance in a complex task, *Automatica*, 6, pp. 371–383, 1970a.

Kleinman, D.L., Baron, S., and Levison, W.H., An optimal control model of human response. Part I: theory and validation, *Automatica*, 6, pp. 357–369, 1970b.

Kleinman, D.L., Baron, S., and Levison, W.H., A control theoretic approach to manned-vehicle systems analysis, *IEEE Trans. Automatic Control*, AC-16(6), pp. 824–832, 1971.

Kleinman, D.L. and Korn, J., Information and display requirements for aircraft terrain following, *Proc. Seventeenth Annual Conference Manual Control*, pp. 329–331, 1981.

Kok, J.J. and Van Wijk, R.A., *Evaluation of Models Describing Human Operator Control of Slowly Responding Complex Systems*. Ph.D. dissertation, Faculty of Mechanical Engineering, Delft University of Technology, 1978.

Korn, J., Gully, S.W., and Kleinman, D.L., Validation of an advanced cockpit design methodology via a workload/monitoring tradeoff analysis, *Proc. Eighteenth Annual Conference on Manual Control*, pp. 268–292, 1982.

Krendel, E.S. and McRuer, D.T., A servomechanics approach to skill development, *J. Franklin Institute*, 269(1), pp. 24–42, 1960.

Kwakernaak, H. and Sivan, R., *Linear Optimal Control Systems*, John Wiley & Sons, New York, 1972.

Levison, W.H., Baron, S., and Kleinman, D.L., A model for human controller remnant, *IEEE Trans. Man-Machine Sys.*, MMS-10(4), pp. 101–108, 1969.

McRuer, D.T., Allen, R.W., Weir, D.H., and Klein, R.H., New Results in Driver Steering Control Models. *Hum. Factors*, 19(4), pp. 381–397, 1977.

McRuer, D.T., Graham, D., Krendel, E.S., and Reisener W. Jr., *Human Pilot Dynamics in Compensatory Systems. Theory, Models, and Experiments with Controlled Element and Forcing Function Variations* (Technical Report No. AFFDL-TR-65-15), Air Force Flight Dynamics Laboratory, Wright-Patterson AFB, OH, 1965.

McRuer, D.T. and Jex, H.R., A review of quasi-linear pilot models, *IEEE Trans. Hum. Factors Electron.*, HFE-8(3), pp. 231–249, 1967.

McRuer, D.T. and Krendel, E.S., *Mathematical Models of Human Pilot Behaviour* (Agardograph No. 88), AGARD, 1974.

McRuer, D.T. and Schmidt, D.K., Pilot-vehicle analysis of multiaxis tasks, *J. Guidance, Control Dynamics*, 13(2), pp. 348–355, 1990.

Mulder, M., *Cybernetics of Tunnel-in-the-Sky Displays*, Ph.D. dissertation, Faculty of Aerospace Engineering, Delft University of Technology, 1999.

Nise, N.S., *Control Systems Engineering*, 2nd ed., Benjamin/Cummings, Redwood City, CA, 1995.

Olsder, G.J. and van der Woude, J.W., *Mathematical Systems Theory.*, Delftse Uitgevers Maatschappij, Delft, 1994.

Phatak, A.V., Weinert, H., Segall, I., and Day, C.N., Identification of a modified optimal control model for the human operator, *Automatica*, 12, pp. 31–41, 1976.

Sheridan, T.B. and Ferrell, W.R. *Man-Machine Systems*, MIT Press, Cambridge, MA, 1974.

Stapleford, R.L., Craig, S.J., and Tennant, J.A., *Measurement of Pilot Describing Functions in Single-Controller Multiloop Tasks* (NASA Contractor Report No. CR-1238), Washington, D.C., 1969.

Stapleford, R.L., McRuer, D.T., and Magdaleno, R. Pilot describing function measurements in a multiloop task. *IEEE Trans. Hum. Factors Electron.*, HFE-8(2), pp. 113–125, 1967.

Stassen, H.G., Johannsen, G., and Moray, N., Internal representation, internal model, human performance and mental workload, *Automatica*, 26(4), pp. 811–820, 1990.

Thompson, P.M. Program CC's implementation of the human optimal control model, *Proc. AIAA Guidance, Navigation and Control Conference*, Monterey, CA, August 17–19, 1987.

Van Lunteren, A., *Identification of Human Operator Describing Function Models with One or Two Inputs in Closed Loop Systems*, Ph.D. dissertation, Faculty of Mechanical Engineering, Delft University of Technology, 1979.

Van Paassen, M.M. *Biophysics in Aircraft Control. A Model of the Neuromuscular System of the Pilot's Arm*, Ph.D. dissertation, Faculty of Aerospace Engineering, Delft University of Technology, 1994.

Van Wijk, R.A. and Kok, J.J., Theoretic aspects of the identification of the parameters in the optimal control model, *Proc. Thirteenth Annual Conference on Manual Control*, pp. 27–34, 1977.

Weir, D.H. and McRuer, D.T., Models for steering control of motor vehicles, *Proc. of the Fourth Annual Conference on Manual Control*, NASA SP-192, pp. 135–169, 1968.

Weir, D.H. and McRuer, D.T., *Pilot Dynamics for Instrument Approach Tasks: Full Panel Multiloop and Flight Director Operations* (NASA Contractor Report No. CR-2019), Washington, D.C., 1972.

Wewerinke, P.H., *An Analysis of In-Flight Helicopter Pilot Control Behaviour and Work-load* (NLR Technical Report No. TR 76146 C), National Aerospace Laboratory, Amsterdam, 1976.

Wewerinke, P.H., *Visual Scene Perception — Frequency-Domain Data and Model Para-meter Estimation Procedure* (NLR Memorandum No. MP 79009 U), National Aerospace Laboratory, Amsterdam, 1979.

Wewerinke, P.H., *Models of the Human Observer and Controller of a Dynamic System*, Ph.D. dissertation, Department of Applied Mathematics, University of Twente, 1989.

Wickens, C.D., *Engineering Psychology and Human Performance*, 2nd ed., Harper-Collins, 1992.

13
Cumulative Spine Loading

Jack P. Callaghan
University of Waterloo

13.1 Introduction

Cumulative loading is an approach to assess the potential for injury or pain reporting that is garnering a great deal of attention in both industry and ergonomic communities. This chapter will examine the theoretical foundations linking cumulative exposure and injury, review the studies that have linked cumulative exposure and pain reporting, detail current barriers to the wide spread implementation of cumulative loading assessments in industrial settings and present the author's opinions on the future directions of the field.

It is important to realize that cumulative loading at its core is an extension of acute loading (Figure 13.1) analyses employed to document the time varying exposures that workers experience. It is encumbered with the same errors and limitations that biomechanical models of the low back have, with many added factors that do not occur when assessing peak task exposures. Cumulative loading assessments are utilized to capture a better representation of the factors that can contribute to injury and as such it promises a much greater ability to examine interworker variability in task performance

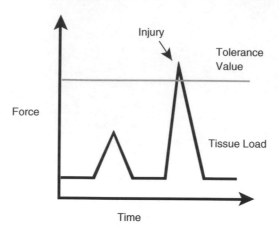

FIGURE 13.1   Acute injury scenario where a single load application exceeds the tolerance of a tissue resulting in an injury.

and injury reporting. In other words the individual methods used by different workers performing the same job would result in similar risk assessments with many acute tools that only incorporate task demands (i.e., rate of lifting, height of lift, etc.) or examine maximum demands (postures or forces). In contrast by examining the full work cycle of a task the individual differences between workers would be revealed in loading and postural demands both within a single cycle and over an extended period, typically expressed as shift exposure.

While cumulative loading is a relatively new terminology in spine biomechanics or low back ergonomics it has a relation to cumulative trauma disorders (CTD) of the upper extremity that warrants a brief examination. CTDs can be traced back in the ergonomics literature to the 1980s. It appears that Armstrong et al. (1982) were among the first to use the specific terminology in the peer reviewed literature. While the underlying rationale between CTD and cumulative loading are similar, the two terms have some very clear differences. CTD is defined as pathology to soft tissues as a result of exposure to excessively frequent use of biological tissue, compounded by excessive loads, awkward joint positions and inadequate recovery (Stramler, 1993). In contrast cumulative loading is a direct quantification of the magnitude of loading that a joint experiences over a given period of time, which is influenced by posture, repetition, duration and force. Evidently, both CTD and cumulative loading are tied to injury through common risk factors such as force, posture and rate of repetition or duration. However, the important difference between the two terms is that cumulative loading is a direct quantification of loading exposure in response to these risk factors whereas CTD is an injury definition. The origin of the terminology "cumulative loading" in the spine can be traced back to the 1990s when it first appeared in the research literature (Kumar, 1990).

The ideas of cumulative loading, repetition rate and duration of exposure, while being related, have very different meanings, which will be brought to light in this chapter. Since the publication of the phrase "cumulative loading" the idea has been quickly adopted by the industrial community, including ergonomists, health care providers, workers and management. The most frequent question asked by these groups is invariably "How much cumulative exposure is too much?" Unfortunately the question is not readily answerable with the currently available knowledge base.

This chapter will focus exclusively on cumulative loading of the lumbar spine and associated issues. Evidence will be presented in support of a cumulative injury mechanism and the hurdles that need to be addressed before cumulative loading will be usable as an injury prevention approach in industry will be discussed.

13.2 A Description of Cumulative Loading

Much of the confusion between the terminology cumulative loading and CTD could have been avoided if a more appropriate or descriptive term had been chosen in place of cumulative loading. Cumulative implies a simple addition of the loads that the individual or body joint has experienced over a fixed duration of time. In reality, the exposure would be more accurately described as integrative loading. Cumulative loading is a summation, but not of the loading magnitudes, rather of the areas under a sequence of force-time relationships representing a worker's daily exposure (Figure 13.2).

Since factors such as cycle time and magnitude of load will alter the magnitude of cumulative loading for a task, a common exposure period is used to assess the relationship between cumulative loading exposure and risk of injury. The most commonly used duration is shift exposure, typically an 8-h period. The shift exposure is determined by incorporating each of the tasks (where n = total number of tasks and i represents each task) performed over the shift (Equation [13.1]).

$$Shift\ Cumulative\ Loading = \sum_{i=1}^{n} Cumulative\ Task\ Loading\ (i) \qquad (13.1)$$

Often this is simplified by taking the cumulative loading in one cycle and extrapolating the data to represent what the equivalent exposure would be for an 8-h shift. The example presented in Figure 13.2 has a cycle time of 5 sec. To extrapolate this value to a shift dosage would require a multiplication by 5760 (12 times per minute × 60 minutes per hour × 8 hours per shift). This would yield a shift cumulative compression value of 37.5 MN s. An understanding of how cumulative loading is calculated is fundamental to appreciate the unique challenges associated with developing a method to assess the risk of injury from cumulative loading exposure. This will be explored in the following sections.

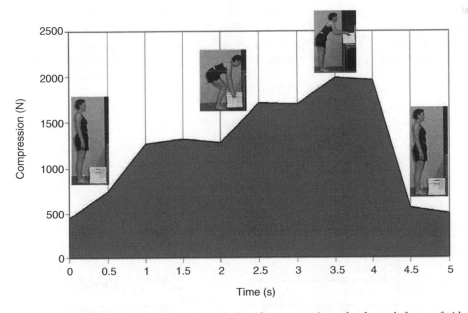

FIGURE 13.2 A biomechanical model was used to calculate the compression value for each frame of video for a sagittal lifting task. The area under the curve (solid area) is integrated to determine the amount of compression to which the L4/L5 joint is exposed over the course of the task. The cumulative compression for this lift is the value of this area (6514 N s).

13.3   Tissue-Based Evidence for Cumulative Injuries

The overriding principle behind cumulative loading is that when a biologic tissue is exposed to repeated or prolonged loading it can fail at levels below the maximal levels seen in acute exposure. The rate of loading outpaces the tissue's ability for self-repair and an injury results (Figure 13.3). This section will examine the tissue-based evidence that supports this mechanism of injury.

Evidence of injuries that occur from prolonged or chronic exposure to sub-acute loading is plentiful in published research papers. Repetitive loading with varying magnitudes of loads has been studied on hard and soft tissues both *in vitro* and *in situ*. The greatest prevalence of research has been performed on isolated bone specimens. The most direct relationship is between the magnitude of loading and the number of cycles to failure. Bone has been demonstrated to have a strong relationship between the number of cycles to failure and the magnitude of stress applied, with lower stresses having higher cycles of loading to failure (Carter and Hayes, 1976). This relationship has been described as nonlinear (Zioupos et al., 1996a). Micro damage in bone will accumulate at an increasing rate (i.e., exponential) as the number of loading cycles increases and the initial occurrence of micro-damage is influenced by the magnitude of loading (Zioupos et al., 1996b). In whole canine tarsal bones of greyhound racing dogs micro fractures were found to propagate into gross tissue damage when ongoing cyclic exposure outpaced the remodeling process (Muir et al., 1999). This type of injury pathway is also evident in humans and is the accepted model for stress fractures in runners. This fundamental research is considered as secondary level evidence of cumulative loading and injury. It does not directly measure cumulative loading or test spine specimens, but it provides strong support to the injury scenarios presented in Figure 13.3.

While the effect of repetitive loading on injury response is linked to cumulative loading, there is a large difference between the two exposures. It is entirely possible to have equal cumulative loading exposure to failure with entirely different frequencies of loading. In fact this is the case in tensile tests of bone specimens where failure is dependent on the time or duration of testing and not the repetition rate (Carter and Caler, 1983; Zioupos et al., 2001). In other words, tensile behavior of bone is governed by accumulation of creep damage based on time of exposure, which is independent of the number of cycles that the bone is exposed to. This difference between cumulative loading and repetition rate is illustrated in Figure 13.4 where the two repetition rates are dramatically different, yet represent the same cumulative exposure. Even though there is this fundamental difference between repetition rate and cumulative loading, alteration to the magnitude of loading at a fixed frequency would alter cumulative loading and is linked to changes in the rate of micro-fracture development, as described in the previous paragraph. Further, there has been very little work done that has quantified the relationship of cumulative loading and injuries directly. The insensitivity of bone in tension to frequency of loading would appear to argue

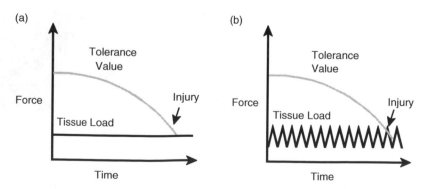

FIGURE 13.3   Cumulative loading scenarios, sustained (a) and repetitive (b) exposures that result in a tissue injury at sub-acute force values.

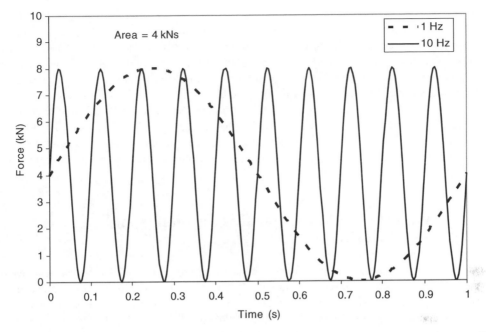

FIGURE 13.4 *In vitro* loading profiles of two different rates. The cumulative loading for the two different frequencies is equal due to the same percentage of time spent at any given loading magnitude.

that repetition is not an injury risk factor. However, these findings have only been illustrated for bone in tension. While bone in compression (Caler and Carter, 1989) and muscle tendon (Wang et al., 1995; Wren et al., 2003) both respond to different frequencies of exposure supporting repetition as a risk factor for developing CTDs.

Specific evidence of the time varying response of the spine to repetitive loading has also been documented in the research literature. At a fundamental level, research linking biological response to prolonged exposures has shown that cell death in the intervertebral disc of a rat model, tested under sustained compressive loads, was directly linked to the magnitude and duration of the loading (Lotz and Chin, 2000). The loads that were employed were comparable in relative magnitude to human spine loading of 500 to 1500 N, which are representative of the range of compressive spine loads experienced in upright standing to very light industrial lifting tasks. A porcine model of the spine exhibited morphological changes to the intervertebral disc, such as shape changes, in response to repetitive compressive loading in the elastic range of the motion segments employed (Yu et al., 2003). These changes are thought to be indicative of early stages of disc degeneration. Furthermore, an *in situ* rabbit spine model with a higher number of loading cycles in a fixed time frame (equivalent to larger cumulative exposure) revealed that annular damage was elevated with increased loading cycles (Wada et al., 1992). These studies, which provide primary evidence of sub-acute injury pathways of the spine are still secondary evidence of cumulative loading as they have not directly quantified injury due to cumulative loading magnitude.

A final property of the spine, with primary focus on the intervertebral disc, that merits consideration is its somewhat unique response to cumulative exposure and its poor ability for self-repair once an injury pathway has been initiated. Known bone healing times are in the order of several weeks or months, so it is fair to assume that even micro-fractures to trabeculae would require weeks to repair while the tissue is continually being loaded. It is hard or near impossible to isolate the spine from loading, similar to the immobilization that would be applied to an injured appendage. The avascular nature of the interior margins of the intervertebral disc is a factor that constrains its ability for self-repair. The proteoglygan turnover of the intervertebral disc has been shown to take 500 days in a canine model (Urban et al., 1978) and collagen production is thought to take even longer (Adams and Hutton, 1982; Porter et al., 1989). This is supported in an *in situ* animal study using rats that revealed that intervertebral disc annular damage was still present after a month of recovery (Lotz et al., 1998). Clearly once an injury is initiated in the spine, cumulative loading has the potential to outpace the repair mechanism. This pathway is further compounded by the lack of pain sensing fibers in the interior of the intervertebral disc. Only the outer third of the annulus is innervated with pain sensing fibers (Bogduk, 1983; Cavanaugh et al., 1995), so an injury can progress from the interior margin to the exterior boundary before there is the potential for direct pain generation. The mechanical changes in this process are not well understood and secondary pathways for pain generation could occur due to disc height change or altered mechanics.

13.4   *In Vitro* Cumulative Loading Response and Tolerance Limits

Historically the investigation into tissue tolerance has mainly focused on single-cycle destructive compressive testing of spinal motion segments *in vitro*. While it is acknowledged that compressive tolerance values can be modified by factors such as load rate, gender and age a single tolerance has been adopted for acute compressive loading (i.e., 3400N [NIOSH, 1981; Waters et al., 1993]). While there has been a relative scarcity of repetitive *in vitro* spine tissue testing, it has been examined by researchers as early as the 1950s (Hardy et al., 1958). In the 1980s it was demonstrated that *in vitro* repetitive compressive loading can generate spine injuries at sub-maximal levels (Adams and Hutton, 1983; Liu et al., 1983). These studies in conjunction with the isolated tissue-based evidence presented earlier in this chapter form a biomechanical basis for a sub-maximal spinal injury pathway.

Examination of the influence of load magnitude on cycles to failure was examined in work by Hansson et al. (1987) and Brinckmann et al. (1988) using repetitive *in vitro* compressive testing. When the data from both of these studies were examined (Figure 13.5) a nonlinear relationship between the magnitude of loading and cycles to failure was evident (Callaghan, 2002). The curve fits for this data are relatively weak ($r^2 = 0.13 - 0.27$), explaining a relatively small amount of the variance. However, they clearly indicate a nonlinear trend (linear curve fit coefficients of determination values were 0.08 to 0.12) with loading at a higher percentage of maximum strength having greater risk of injury with fewer loading cycles. The variability associated with the human spine specimens used in both these studies introduces a great deal of scatter caused by the lack of experimental control over factors such as age, prior loading exposure, activity level, etc. Both of these studies examined loads that were in the upper range of loads that workers would experience in a repetitive lifting scenario. The majority of specimens that failed within these two *in vitro* experimental paradigms were tested above 60% of maximum strength (Brinckmann et al., 1988; Hansson et al., 1987). The difficulty of testing *in vitro* specimens at lower magnitudes of compressive loading is that a longer testing period is required, making the results questionable due to the absence of biological processes and repair mechanisms that occur *in vivo*. In fact it has been hypothesized that for lumbar vertebrae the endurance limit is 30% of maximum load (Brinckmann et al., 1988). This suggests that below this value the body's ability for self-repair will be able to keep any micro-damage in check for the majority of loading scenarios.

While there is clearly a direct relationship between the number of loading cycles applied to an *in vitro* test specimen and the cumulative loading exposure it sustains, to date there has been no published work, which has studied *in vitro* cumulative loading and injury mechanisms. An examination of the data of Brinckmann et al. (1988) revealed a linear relationship between the number of loading cycles and the cumulative loading sustained to failure, see Figure 13.6. (Callaghan, 2002). While it is intuitive that an increased number of cycles will increase the cumulative loading, it highlights the fact that *in vitro* specimens failed at varying levels of cumulative compression. In other words, there was no single exposure value linking *in vitro* cumulative compression exposure and injury. The basic tissue evidence

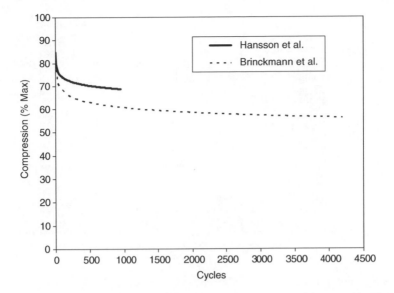

FIGURE 13.5 Logarithmic trend lines fit to the data demonstrating the nonlinear relationship between magnitude of compressive loading and cycles to failure. (From Brinckmann, P., Biggemann, M., and Hilweg, D., *Clin. Biomech.*, 3 (Suppl. 1), pp. s1–s23, 1988; Hansson, T.H., Keller, T.S., and Spengler, D.M., *J. Orthop. Res.*, 5, pp. 479–487, 1987.)

discussed earlier in the chapter supports this relationship for bone in compression (Caler and Carter, 1989) and muscle tendon (Wang et al., 1995; Wren et al., 2003). This lack of a single cumulative loading failure threshold was substantiated in work that examined the *in vitro* generation of intervertebral disc herniations from combined loading (Callaghan and McGill, 2001). Varying levels of compression were applied in conjunction with repetitive flexion/extension motions. The cumulative compression to injury was affected by the magnitude of the load applied (Figure 13.7).

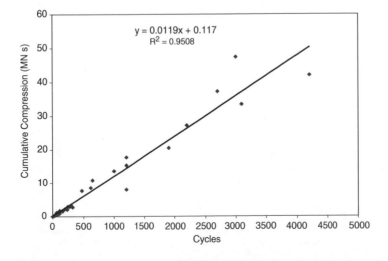

FIGURE 13.6 The linear relationship of cumulative loading sustained to failure based on the data. (From Brinckmann, P., Biggemann, M., and Hilweg, D., *Clin. Biomech.*, 3 (Suppl. 1), pp. s1–s23, 1988.)

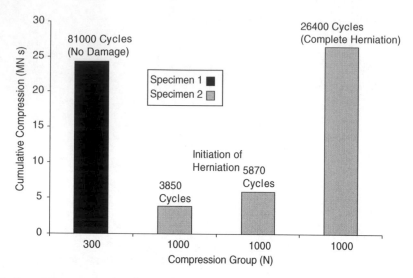

FIGURE 13.7 Cumulative compression for two *in vitro* spine specimens tested in combined flexion/extension motion with static compression. Specimen 1 (300 N) exhibited no injury. The three columns representing specimen 2 (1000 N) demonstrate the development of initiation of an injury and complete intervertebral disc herniation (based on data from Callaghan and McGill, *Clin. Biomech.*, 16, pp. 28–37, 2001. With permission.)

In summary, the idea of cumulative exposure being linked to injury is well supported by both repetitive testing on isolated tissues at sub-acute levels and existing *in vitro* spine biomechanics research. However, the clear influence of loading factors (i.e., load magnitude) as modifiers of cumulative loading to failure that the spine can sustain raises the question of whether a single tolerance value can be employed to assess a worker's risk of injury from cumulative loading exposure.

13.5 Cumulative Injury Theories

Ultimately the appeal of cumulative loading is in its ability to predict either injury or the reporting of low back pain and thereby allow for intervention to prevent these events. One misleading outcome from the presented *in vitro* work is that injury rate tends to be related to an increased magnitude of exposure (Figure 13.8A). This response holds true for *in vitro* compressed time loading paradigms where exposure is accelerated to represent weeks of loading in hours of testing. However, when biological repair processes

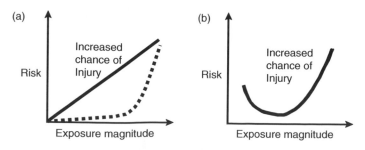

FIGURE 13.8 Injury exposure relationships. Simple models (a) that represent increasing risk with increasing exposure, either linearly or nonlinearly. The "J" or "U" nonlinear risk model (b), where there is an optimum loading that results in minimal risk of injury.

and remodeling behaviors are considered it does not appear to be a plausible model of the *in vivo* response to loading.

Increased levels of motion have been shown to be beneficial in providing nutrition to the structures of the intervertebral disc (Holm and Nachemson, 1983). This contrasts with the *in vitro* research and the linear injury models that have demonstrated that increased loading cycles resulted in an increased probability of injury. Intervertebral disc degeneration has been associated with decreased nutrition (Buckwalter, 1995). In an examination of work demands and injury risk (Videman et al., 1990), sedentary work characterized by small spinal motions and loading was shown to result in intervertebral disc injury. Additionally, workers who performed heavy work were also at increased risk of developing a spinal injury. Workers who were involved in varying or mixed work had the lowest risk of developing a spine injury. This presents the idea that too little or too much motion or load can both present an increased risk of spinal injury. An *in situ* rat intervertebral disc study (Iatridis et al., 1999) appears to support this "U"- or "J"-shaped relationship between exposure and risk of injury (Figure 13.8b). Three groups of rats were used that underwent different mechanical loading: a control group (equivalent to mixed exposure), an immobilized group (sedentary) and a chronic compression group (increased loading). The control group had the fewest biomechanical changes to the intervertebral disc with the immobilized and chronic compression group exhibiting similar degenerative responses. However, the chronic compression degenerative changes occurred earlier and were more pronounced (Iatridis et al., 1999) supporting a "J"-shaped injury risk model.

The importance of these injury models to cumulative loading is paramount in how cumulative loading will be used in the future and its potential success at preventing low back injuries. If the scenario in Figure 13.8B is the response of the spine to cumulative loading then a simple tolerance value cannot be considered without at least considering the type of work to which an individual is exposed. Further there are likely other factors that will modify the relationship between cumulative loading exposure and risk of injury that will be examined later in this chapter.

13.6 Workplace Studies Documenting Cumulative Exposure

A handful of studies have documented cumulative loading exposure and the risk of reporting back pain. These studies are the strongest foundation for the link between cumulative loading and pain reporting and are responsible for the increased attention that is being given to cumulative exposure. Invariably, all of the studies that have examined cumulative loading exposure as a risk factor have reported a positive finding (Kumar, 1990; Norman et al., 1998; Seidler et al., 2001, 2003). These studies, while providing strong evidence of the relationship between spine pain or injury and cumulative exposure, all employed different methodologies and provided different pieces of information. Similar to the point made regarding *in vitro* research previously, one of the end goals for cumulative loading research is to develop a standard that can be used to assess risk of injury for workers. To date the studies that have been performed contribute information about the magnitudes of exposures present from a cumulative perspective but invariably compare a case (back pain or spinal injury) with a control group. This information does not provide the point where risk increases, but merely shows that exposure is higher in the case group (Figure 13.9).

Two published studies from Germany (Seidler et al., 2001, 2003) using this case control approach have convincingly shown that cumulative compressive exposure is linked to spinal injuries (osteochondrosis, spondylosis, and intervertebral disc herniation). Two studies performed in Canada (Kumar, 1990; Norman et al., 1998) have linked cumulative compression and shear with the reporting of nonspecific low back pain. Two other studies examined the magnitude of cumulative loading, one in industry for four occupations (Jager et al., 2000) and the other in activities external to work (Godin et al., 2003) with neither of these studies tracking pain or injuries in the low back. Unfortunately none of these studies used common methods or outcome measures. However, this in part strengthens the evidence of the relationship between cumulative exposure and low back pain or injury. The variety of approaches

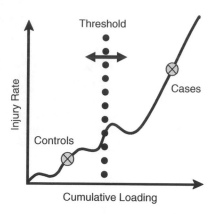

FIGURE 13.9   Representation of the unknown location where a tolerance limit value would lie when observing case control cumulative loading studies.

used in these studies will be reviewed briefly with the remainder of the chapter addressing issues that are critical to the development of a cumulative threshold limit value and measurement method.

Methods to document cumulative loading have involved questionnaires (Kumar, 1990), self-recall and report (Seidler et al., 2001, 2003) and varying forms of video documentation (Godin et al., 2003; Jager et al., 2000; Norman et al., 1998). Each of these approaches possess strengths and weaknesses and it is beyond the scope of this work to debate these inherent properties in any depth. Two studies have relied on 2D regression-based equations to yield lumbar spine compression from position of the external load and weight of the object lifted (Seidler et al., 2001, 2003). It is difficult to assess the validity of this approach as the studies do not report validation data of the modified Mainz-Dortmund dose model (MDD) used. Two studies use 2D biomechanical rigid link models with a single muscle equivalent model and examined joint compression and shear (Kumar, 1990) as well as net joint flexion/extension moment (Norman et al., 1998). The final two studies used relatively sophisticated 3D biomechanical models to yield the forces and moments on the lumbar spine (Godin et al., 2003; Jager et al., 2000). The two studies employing the regression approach examined the total cumulative compression from lifting and carrying over the individual's work life prior to spinal injury diagnosis (Seidler et al., 2001, 2003). These values are obviously extremely large, on the order of 10^6 N h or 3600 MN s. These values hold little value for ergonomists in evaluating a shift dosage in order to assess risk of injury. However, these studies do provide valuable information about the relationship of injury, diagnosed medically, and cumulative compression as stated previously. The remaining industrial studies have evaluated cumulative loading over varying periods but all included a shift exposure. These values are presented in Table 13.1.

Briefly expanding on the methodologies used in these studies, the most common approach has been to incorporate video analysis to record the working activities and then assess the exposures later. Norman et al. (1998) reduced data collection and processing time by using the peak static spinal load multiplied by the number of repeats and duration of each task to estimate the shift cumulative loading. Kumar (1990) used a similar static approach; however, it was based on subject recall of end point postures of tasks and then the generation of intervening frames to represent data collected at 5 frames/sec. The task data was then combined and similarly extrapolated out to shift exposures and longer periods. The DOLLY study (Jager et al., 2000) used a posture sampling approach and a 3D-biomechanical model to assess cumulative exposure for a large number of tasks over entire shifts of eight individuals. The posture sampling approach utilized results with a lower level of fidelity of documented postures (i.e., trunk flexion was divided into seven gradations of $15°$) and had equivalent sampling rates of 0.4 to 1.8 frames/sec. Our own approach to calculate cumulative compression also involves a posture matching approach allowing for 3D assessments from a single 2D video recording (Callaghan et al., 2003; Jackson et al., 2003). Figure 13.10 illustrates the main user interface of 3DMatch, which allows 3D

TABLE 13.1 Cumulative Loading Magnitudes Reported in the Research Literature. Values Given Represent Averages ± 1 Standard Deviation Unless Noted Otherwise.

	Cases	Controls
Compression (MN s)		
Jager et al. (2000)	n/a	10.8 to 36.0 (range) ($n = 8$)
Kumar (1990)	♂ 15.6 ± 5.0 ($n = 6$)	♂ 6.6 ± 5.5 ($n = 8$)
	♀ 13.5 ± 12.1 ($n = 52$)	♀ 9.3 ± 7.7 ($n = 95$)
Norman et al. (1998)	21.0 ± 4.72 ($n = 104$)	19.5 ± 3.84 ($n = 130$)
Shear (MN s)		
Jager et al. (2000)	n/a	n/a
Kumar (1990)	♂ 2.5 ± 0.9	♂ 1.0 ± 1.0
	♀ 2.2 ± 4.2	♀ 1.6 ± 1.5
Norman et al. (1998)	1.52 ± 0.64	1.32 ± 0.45
Moment (MN m s)		
Jager et al. (2000)	n/a	n/a
Kumar (1990)	n/a	n/a
Norman et al. (1998)	0.55 ± 0.24	0.47 ± 0.15

Note: n/a is not available.

hand forces and uses postures of the elbow (3), shoulder (10), neck (8) and trunk (18) as inputs to a 3D-biomechanical model. The video recordings with rates of 2 to 3 frames/sec are opened directly into the software and evaluated. The justification for this sample rate will be presented later in this chapter.

Alternative approaches have been used to yield cumulative loading. Electromyography (EMG) to compression relationships (Mientjes et al., 1999; Potvin et al., 1990) have been examined as one approach to provide estimates of cumulative loading. Electromagnetic tracking devices have also been used to provide valid real-time cumulative compression in 2D (Agnew et al., 2002, 2003). The relationship between heart

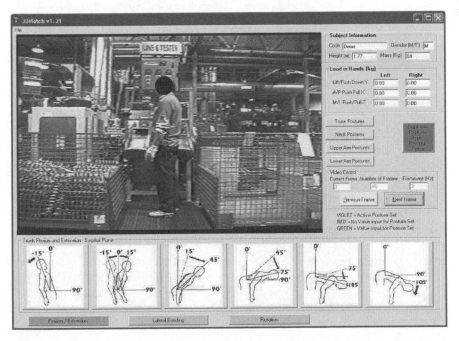

FIGURE 13.10 3DMatch user interface for using postural matching and external forces to calculate 3D spine loads using a biomechanical model.

rate determined physical activity level (HR-PAL) and cumulative low back loads has also been explored by our group (Azar, 2004). None of these approaches have been used in large-scale studies and all possess limitations. EMG to compression approaches to date have not been able to track shear and joint moment or other 3D variables. The electromagnetic approaches are very susceptible to metallic or electrical noise interference, both features that are prevalent in industry. Self-report type measures are promising for the great reduction in quantified data collection time that they provide. These have been employed to document cumulative loading in large-scale studies (Seidler et al., 2001, 2003) and have been investigated by our research group (Azar et al., 2005). Our own work revealed a strong agreement between loading calculated with a biomechanical model and the LOG method, where individuals recorded the demands of a task immediately after completing it ($r = 0.989$) or a RECALL at the end of a 2-h session ($r = 0.403$). Accuracy of recall estimates, particularly task durations are strongly affected by the length of time (Akesson et al., 2001) from the event to recall time and is a major limitation of this approach. HR-PAL accounted for over 80% of the variance in cumulative compression forces during nonoccupational tasks estimated over a period of 2 h (Azar, 2004). Relationships between HR-PAL and other cumulative loading variables were not as strong, but it was concluded that the use of physiological variables such as heart rate might provide a way to document cumulative load exposure in highly nonrepetitive activities, at a much reduced processing cost.

13.7 Towards a Cumulative Compression Tolerance Limit Value (TLV)

Cumulative compression has been the most examined cumulative loading variable in published studies. This is evident in both the study values presented in Table 13.1 and its use as the cumulative spinal exposure variable in the other two studies discussed (Seidler et al., 2001, 2003). There has been little progress or movement made to establish a dose limit for cumulative exposure. Germany has led on this front with a recommended shift dose of 19.8-MN s, see Jager et al., (2000) for a description. However, this value was not based on calculations from all tasks the workers were exposed to. In fact, this proposed standard appears to be based on only exposures of significant magnitude that occur in a shift (i.e., above 3.2 kN). If the magnitude of cumulative compression in the two studies providing shift dosages are examined (Figure 13.11) another issue quickly comes to light when trying to determine the magnitude of a cumulative loading TLV.

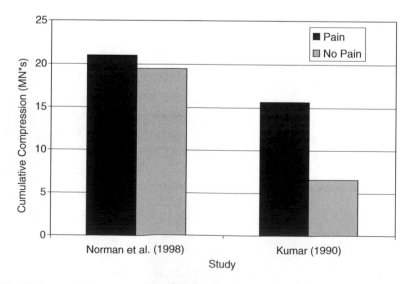

FIGURE 13.11   Shift cumulative compression values for two case control studies.

If a single TLV was to be implemented it would not isolate the pain from the no pain groups across the two studies shown in Figure 13.11. In other words the pain group in the Kumar study (1990) had less cumulative compression than the control or no pain group in the Norman et al. (1998) study. This can be attributed to many factors including different industrial samples (institutional aides versus automotive assembly workers) and the methods employed to document cumulative compression. The magnitude of peak spine compression exposure in the nursing and patient handling population tends to be quite large whereas an automotive plant with ergonomic staff and programs, works to eliminate large peak exposures. This is supported by the peak data of these two studies, where the institutional aides (Kumar, 1990) had mean acute loads of sample tasks exceeding the NIOSH MPL (6400 N) and the automotive group (Norman et al., 1998) had mean peak compression below the NIOSH AL (3400 N). This links well with the tissue-based rationale for cumulative injuries and the nonlinear relationship between magnitude of loading and cumulative loading to failure. This issue has also been raised by the research group in Germany (Jager et al., 2000) and different weighting values for the force-time exposures have been implemented in part based on the work of Brinckmann et al. (1988). The most used weighting factor appears to involve squaring of the spine forces, which would effectively weight higher forces more heavily as the time base is unaltered. Unfortunately, returning briefly to tissue evidence, weighting approaches while sound in theory are poorly supported by actual data at present. While this author strongly supports the idea and need for weighting functions, and not only for magnitude of loading, the scatter of the tissue results from Brinkmann et al. (1988) show no clear relationship between magnitude of loading and cumulative loading to failure (Figure 13.12). Regardless of the curve fitting approach employed the R^2 values were on the order of 0.1, which would not support a weighting factor being applied to compressive magnitudes prior to integrating exposures. In fact if weighting schemes such as the force squaring or quadrupling approaches as suggested by Jager et al. (2000) are applied to this data set the R^2 values approach zero. The use of weighting functions is important and clearly warranted based on the tissue mechanisms of injury.

Recent work that we have completed (Parkinson & Callaghan, 2004) has examined the association between cumulative exposure and compression injuries *in vitro*. The study modified the magnitude of the peak cyclic compression as a percentage of estimated ultimate strength. Specifically, the spine specimens were cyclically loaded at 40, 50, 70, and 90% of their estimated compressive strength up to a maximum of 21600 cycles and injuries were recorded. The magnitude of loading had a strong non-linear

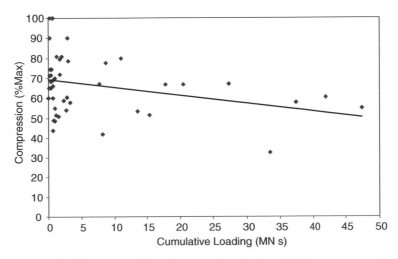

FIGURE 13.12 Cumulative loading sustained to failure calculated for different magnitudes of *in vitro* repetitive compressive testing. Only specimens were used in the reanalysis that sustained failure in the 5000-cycle test window. (From Brinckmann, P., Biggemann, M., and Hilweg, D., *Clin. Biomech.*, 3 (Suppl. 1), pp. s1−s23, 1988. With permission.)

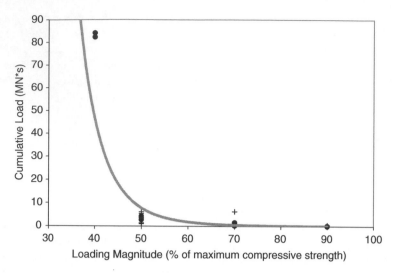

FIGURE 13.13   Scatter plot of cumulative load tolerated at failure (MN*s) vs. loading magnitude (% of maximum compressive strength), with the power curve fit to the data ($r^2 = 0.9024$). Symbols indicate the cumulative magnitude sustained until failure by specimens in the different compressive exposure groups.

relationship ($r^2 = 0.9024$) with the cumulative loading a specimen could withstand prior to failure (Figure 13.13). Based on this relationship a weighting factor was developed that created an equal cumulative exposure magnitude for a given risk of injury. The equation effectively increases the cumulative exposure for higher acute exposures to represent the same risk of injury that would result from a longer exposure at a lower force magnitude. This nonlinear weighting equation (Equation 13.2) incorporates the magnitude of exposure as a percentage of maximum strength. This can be implemented in biomechanical models by using regression approaches to predict an individual's lumbar spine compressive strength. An example of such an equation is developed by Genaidy et al. (1993) which includes age, gender, body weight and spinal level as input variables and is based on the work of Hansson et al. (1987). This work provides an initial estimate and a usable equation of the force weighting relationship when calculating cumulative exposure. However, this theoretical foundation should be assessed on a population data set of worker exposure and corresponding injury reporting to determine if it strengthens this association.

$$
\begin{aligned}
\textit{Weighting Factor} = {} & 5.4617470 \times 10^{-8} \times (\textit{loading magnitude})^5 - 1.3802063803 \times 10^{-5} \\
& \times (\textit{loading magnitude})^4 + 1.460100508173 \times 10^{-3} \times (\textit{loading magnitude})^3 \\
& - 7.8813626392557 \times 10^{-2}(\textit{loading magnitude})^2 + 2.141251917806310 \\
& \times (\textit{loding magnitude}) - 22.2341862486371
\end{aligned}
$$

(13.2)

A final issue related to setting a dose exposure limit is the length of standardized exposure period that should be adopted. Should an hourly dose be set, a shift, a day, a week or an even longer period? The most common reported exposure dose is a shift exposure. This will allow for a point exposure to be assessed and intervention decisions to be made without requiring information on long periods of exposure. However, biologically, injury pathways can progress over long periods of time and an injury is in response to the loading history prior to the injury or pain event. This extended cumulative exposure was the rationale behind the two studies linking medically documented injuries and cumulative loading exposure (Seidler et al., 2001, 2003). An additional factor is that loading exposure does not end when a worker leaves the work place. Cumulative and peak loading documented in activities external to the industrial

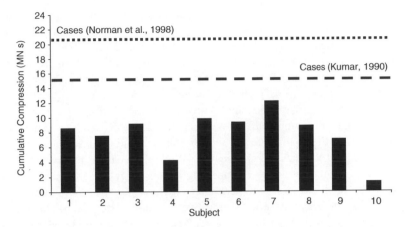

FIGURE 13.14 Cumulative spine compression forces from the work of Godin et al. (2003) for 2 h of at home activities. The horizontal bars represent the pain groups' (cases) cumulative compression shift exposure averages for industrial workers.

work environment resulted in substantial peak and cumulative loading (Godin et al., 2003; Lauder et al., 2002). In following individuals for a 2-h period around their homes, the magnitude of peak spinal compression forces exceeded the NIOSH AL in 60% of the subjects (Godin et al., 2003). The cumulative compression over this same 2-h period was also substantial (Figure 13.14). In fact if the 2 h of exposure is extrapolated out to represent a shift dose, the values would surpass the cumulative exposure reported for back pain cases in industrial workers (Kumar, 1990; Norman et al., 1998). Also noteworthy is the variability of the cumulative loading that was present in the ten subjects. This is not surprising when differences in activity levels and lifestyles are considered. However, it does raise the interesting question of whether this loading, when combined with workplace exposure, would explain any of the variance in which workers report an injury from a group experiencing homogeneous external task demands. While a shift dose limit will likely prove to be the desirable and most useful standard for workplace ergonomic assessments, a daily exposure may be more insightful into risk of developing an injury or pain given the significant magnitudes of cumulative loads documented outside of work.

13.8 Challenges of Documenting Cumulative Exposure

When examining peak loading, there is consensus on the factors that lead to the most demanding instant of a task: greatest horizontal moment arm, lowest vertical height, highest external load, etc. While there may be some small variance in selecting the instant to examine, the joint modeling approach and assumptions employed, the calculation of the mechanical loading on the spine for a single instance in time would have relatively small variance between different observers.

Unlike identifying the peak or highest instance of loading, cumulative loading presents the difficulty of documenting the variation of the spinal loads with respect to time. This is compounded by both the approach used to document exposure and the method to transform postural and external force exposures to spinal kinetics. Given the current state of technology and being unable to directly measure spine forces in the human body, the criterion approach for assessing spine forces is a biomechanical rigid link model to calculate net joint moments and reaction forces in conjunction with a suitable lumbar spine joint model (typically L4/L5 or L5/S1) to yield joint compression and shear forces. Many secondary approaches have been employed such as regression equations, heart rate and EMG to simplify this process and more importantly reduce the large processing time inherent in the biomechanical modeling approach. At the heart of addressing this issue is what cumulative loading really represents. To faithfully document cumulative loading of a single worker, even for a single shift, would require measurement, at a sufficient sampling rate — 30 frames/sec, in three dimensions for every second of the shift. This would represent almost 1 million frames of data that would need to be analyzed! Even in a controlled research environment this would be a fairly daunting task. A further consideration is the end-user of cumulative loading, the ergonomists and other safety personnel performing job assessments. It would be both impractical and infeasible to require a job assessment to track an entire shift of a job and spend weeks analyzing the data to assess risk, and then do it again after any intervention had been performed. Clearly, documenting true cumulative loading is nearly impossible with current measurement techniques. In order for ergonomists to be able to perform cumulative exposure assessments and research studies examining cumulative loading to be able to collect a sufficient sample size, more efficient means to document cumulative loading are required. While secondary approaches such as regression approaches or subject self-reports, are very appealing metrics, they are ultimately compared against quantified spine force-time histories as the criterion measure. In order to validate these and other secondary approaches, documentation of cumulative exposure using biomechanical modeling is required for comparison. To perform biomechanical modeling on a sample of any reasonable size, various strategies to reduce the quantity of data required to calculate cumulative exposure have been employed by researchers. These data reduction approaches and their validity will be addressed in the remainder of this chapter.

13.8.1   2D- vs. 3D-Biomechanical Models

A review of biomechanical models and the issue of whether 3D is required or not has been previously presented (Norman and McGill, 2002). Briefly, 2D-models can provide reliable information in a great range of activities primarily because many industrial tasks are dominated by trunk flexion/extension as the primary motion axis. However, it would be naive to believe that a 2D-model is an accurate representation of the 3D-environment that workers interact with and perform their task in. Collecting true 3D-data requires a calibrated space in a controlled environment, a situation that is difficult to achieve in an industrial setting. To date there have been no published studies on cumulative loading that have used a 3D-data collection process. Our own 3DMatch model and the model employed by Jager et al. (2000) use a user interpretation of planar video to create 3D-model postures and thereby treat the data as 3D. This allows for out-of-plane movements to be accounted for, albeit in course posture bins, and the modeling approach to calculate 3D-forces and moments on the lumbar spine. The most common approach has been to evaluate cumulative exposure with a 2D-model (Daynard et al., 2001; Kumar, 1990; Norman et al., 1998) and regression approaches based on 2D-exposures (Seidler et al., 2001, 2003). The use of 2D-models limits the model outputs that can be examined to compression, anterior/posterior shear forces and flexion/extension moment.

13.8.2   Static, Quasi-Dynamic or Quasi-Static and Dynamic Models

Static biomechanical models ignore segmental accelerations and the accelerations of any external masses that workers interact with. In other words it treats each instant in time as if the body and external objects being manipulated were not moving or in static equilibrium. Quasi-static or quasi-dynamic models ignore segmental accelerations of the body, but include the accelerations of objects being lifted or handled. Dynamic models incorporate segmental accelerations and mass accelerations external to the body. In acute or peak loading assessments McGill and Norman (1985) reported that peak L4/L5 moments from a static model were 19% lower than a dynamic model while the quasi-static model findings were 25% greater than the dynamic model. Lindbeck and Arborelius (1991) reported that static analyses underestimate the peak moment imposed during a dynamic lift by 23 to 100% depending on the type of lifting task. A quasi-static approach rendered data that agreed well with the dynamic model, with errors of less than 3% (Lindbeck and Arborelius, 1991). Additionally, the use of dynamic hand loads alone have been shown to exhibit peak hand forces of approximately 41 to 111% of the static load magnitude dependent on lifting speed and load magnitude (Danz and Ayoub, 1992). None of the published cumulative loading studies have used a dynamic modeling approach. Another data reduction method to reduce the quantity of data precludes dynamic analyses and will be discussed shortly. Several studies appear to have used quasi-dynamic approaches with representative peak measures taken of hand loads (Daynard et al., 2001; Godin et al., 2003; Norman et al., 1998) with all other studies using static models (Jager et al., 2000; Kumar, 1990) and recall of static loads (Seidler et al., 2001, 2003). The use of a static assumption in cumulative analyses does not induce the same magnitude of errors seen in peak analyses. Cumulative exposure average errors across varying tasks, masses and lifting speeds were approximately 10% compared to a dynamic model (Callaghan et al., 2005). The quasi-dynamic model faired much better with average errors well below 5% compared to a dynamic model. These lower errors in cumulative loading compared to peak loading can be attributed to the changing acceleration of hand loads when being lifted and lowered. When being lifted the hand load will exceed the actual static load of the object. But when being lowered the effective hand load will be less than the static load of the object thereby letting over- and underprediction errors cancel out when dynamic components are not included. This study used a model with sampled hand loads so that at any instant in time the hand load was known. This is impractical in an industrial application and the studies that have employed quasi-dynamic models have utilized a representative hand load for the entire lifting period or fraction of the task when the load is in contact with the body. This is effectively a static or nonchanging load and may prove to be no better than simply using the mass of the objects lifted or manipulated.

13.8.3   Extrapolation of Short-Time Periods to Represent a Shift Exposure

Determining the cumulative loading of an entire shift involves documenting the loading in all of the tasks a worker performs. To document the task cumulative exposure involves calculating the force-time or moment-time histories that are exerted on the lumbar joint at a sufficient sampling rate to capture an accurate representation of exposure. Rather than having to measure the entire shift or exposure period, would it be sufficient to document the components of a job, the number of times each was performed and then extrapolate to predict a shift exposure? This approach has not been directly validated but has been widely employed by most cumulative studies with the exception of Jager et al. (2000), who undertook the commendable task of documenting with great detail the loading exposure of eight individuals for entire shifts. Secondary evidence that this approach may be acceptable is found in research that has examined the variability of how workers perform their jobs. Granata et al. (1999) demonstrated that peak compressive loads in repetitive tasks have a fair amount of variability between trials. By taking three repeats of a task for a worker, the standard errors of the trunk kinematics can be reduced (Allread et al., 2000). This led to some of our own work examining the repeatability of trial to trial cumulative load (Keown et al., 2002). If a single cycle is used to extrapolate to shift exposure, the potential to substantially under- or overestimate actual cumulative loading is high. Keown et al. (2002) found that by averaging 3 to 6 cycles a more stable estimate of cycle loading was achieved, which would give a better representation of shift exposure when extrapolation techniques are used. Additionally, the time of day that the samples were taken at or the day that they were taken on had little influence on the average cycle cumulative loading (Keown et al., 2002).

13.8.4   Reducing the Information to Represent Task Exposure

If it is possible to reduce the documentation time down to several task repeats and then extrapolate to a shift, would it be equally feasible to reduce the information content required to determine cycle cumulative exposure? Several different approaches have been employed to reduce the data required to represent task-based exposures. Many of the different approaches to represent cycle cumulative loading were examined to determine the error that the various assumptions introduced (Callaghan et al., 2001).

A reduction in the quantity of data used to describe cumulative loading on the low back substantially increased the amount of error. The approaches that used a representative posture to quantify cumulative loading for a few portions of a cycle resulted in substantial error (square — 70%, work/ rest — 27%, component — 39%, work only — 35%) relative to the complete data set at 30 Hz (Figure 13.15).

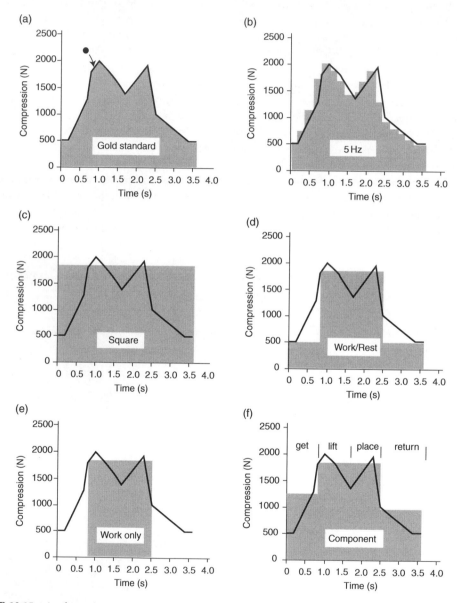

FIGURE 13.15   A schematic representation of actual data demonstrating the six approaches used to calculate the cumulative loading at L4/L5. (a) Gold standard — rectangular integration of all frames collected at 30 Hz. (b) 5 Hz - rectangular integration of data sampled at 5 Hz based on the methods of Kumar (1990). (c) Square — loading at the beginning of lift multiplied by the duration of the task based on the methods of Norman et al. (1998). (d) Work/rest — work, loading at the beginning of lift multiplied by the time the mass was in subjects hands, and rest, loading during upright standing multiplied by remaining time of the cycle. (e) Work only — the work component of work/rest alone. (f) Component — cycle divided into four components of the task: get load, lift load, place load, and return. The ● in (a) indicates the start point of the lifting phase. (From Callaghan et al. (2001) *Ergonomics*, Taylor & Francis, 44 (9), 825–837. With permission.) http://www.tandf/co.uk/journals.

With errors between different calculation methods using the same data set approaching more than 100% in some cases (Callaghan et al., 2001), it presents the problem that researchers will not be able to combine exposure data from different studies or adopt a universal exposure TLV. Arguably, the absolute error in cumulative load estimates might be of little concern when comparing relative exposure between two groups (i.e., low back pain and controls). The assumption that error would be equal in the two groups is also brought into question as error of these reduction approaches was influenced by the type of task performed. The use of a reduced sampling rate (5 Hz approach) resulted in very small errors across all tasks and subjects. This suggests that significant digitizing time might be saved without compromising the accuracy of cumulative loading estimates. The sensitivity of some of the approaches employed, in particular the square, work only and work/rest approaches, to the type of task examined further brings into question the validity of using a single point in time to represent cumulative exposure. Further reductions in sample rate were deemed to still produce accurate results (Andrews and Callaghan, 2003). In fact, representing a task with kinetics calculated from data sampled at 2 frames/sec only introduced an average of 3% error compared to kinetics at 60 frames/ sec. While this represents a great reduction in the data required to document a task, processing this volume of data is still quite labor-intensive. The reduction of sample rate to document exposure has a further impact on model usage. As alluded to earlier in the discussion of dynamic models, if data is reduced to a point where it does not sufficiently represent the frame by frame worker movement, dynamic calculations of segmental accelerations cannot be performed or they produce erroneous values. The use of 2 frames/sec is well below the sample rates required for even slow moving activities.

13.8.5   What Exposure Variable Should Be Used to Quantify Dose Exposure?

Mechanical loading of the spine has been used as a means to identify the possibility of developing a low back injury for decades. Peak compression has formed the foundation for most ergonomic tools and investigations of jobs with high rates of back injuries. Shear loading has more recently been identified (Kumar, 1990; Marras and Granata, 1997; Norman et al., 1998) as another important biomechanical factor to be considered when assessing jobs for the potential of back injuries. Both of these variables have direct physiological relevance, although all variables from a biomechanical rigid link or joint model also possess this relevance. One of the rationales for selecting a variable to be used is whether it has a relationship to the injury or pain modes that it is trying to predict. While compressive loading can definitely produce injuries, the primary injury (typically endplate failure) (Yingling et al., 1997) may only lead to pain pathways through secondary developments in response to this injury. In contrast, repetitive flexion (Callaghan and McGill, 2001) or non-neutral compression (Gunning et al., 2001) has the potential to damage both hard and soft tissues in the spine and joint moments may be a better indicator of this mode of loading. Similarly, shear can also damage structures that have a much higher density of pain sensing fibers such as the articular capsule or external margins of the intervertebral disc (Yingling and McGill, 1999a). One of the strongest justifications at present for choosing a variable would be the association between reporting of pain and the cumulative exposure that has been established in the large-scale published epidemiology studies. These studies have primarily used cumulative compression (Jager et al., 2000; Kumar, 1990; Norman et al., 1998; Seidler et al., 2001, 2003) with two reporting cumulative anterior/posterior shear (Kumar, 1990; Norman et al., 1998) and only one examining cumulative moment (Norman et al., 1998). A 2D-biomechanical model that incorporates both a rigid link model and a lumbar joint model has the potential to examine five output variables where the equivalent approach in 3D yields nine kinetic variables (Table 13.2).

Reaction forces are output from a rigid link segment biomechanical model and are representative of forces at the lumbar joint caused by body weight and the forces in the hands. Joint forces, also referred to as "bone on bone" or net joint forces are calculated by using a joint model that incorporates muscle forces and potentially passive tissue forces that are then combined with the reaction forces. Joint

TABLE 13.2   Kinetic Variables that can be Calculated Using Two and Three-Dimensional Rigid Link Segment Models Partnered with a Lumbar Joint Biomechanical Model

Model	Variable Type	Anatomical Description
Two-dimensional (2D)	Moment (Nm s)	Flexion/Extension
	Reaction forces (N s)	Compression
		Anterior/Posterior shear
	Joint forces (N s)	Compression
		Anterior/Posterior shear
Three-dimensional (3D)	Moments (Nm s)	Flexion/Extension
		Lateral bend
		Axial twist
	Reaction forces (N s)	Compression
		Anterior/Posterior shear
		Medial/Lateral shear
	Joint forces (N s)	Compression
		Anterior/Posterior shear
		Medial/Lateral shear

models partition net joint moments into tissue forces that are anatomically and biomechanically modeled.

While the number of variables has been listed as 5 and 9, for two- and three-dimensional models, respectively, if the initially stated rationale for choosing a variable is considered, namely that the variable should be linked to injury mechanics, the number of variables becomes eight for a two-dimensional

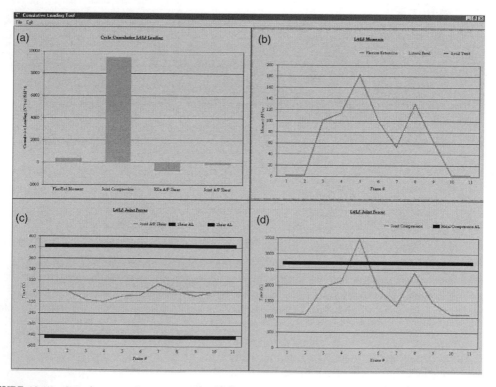

FIGURE 13.16   Sample output from 3DMatch, (Callaghan et al., 2003) which provides a frame by frame force-time history for net joint moments (b), joint shear (c) and compression (d) for a task as well as the cumulative exposure (a). The horizontal bars on the shear (c) and compression (d) outputs represent acute exposure TLVs.

model and 16 for a three-dimensional model. As an example, let us consider the anterior/posterior shear forces. When spinal motion segments are tested to failure in anterior shear, the primary site of injury is the pars interarticularis (Yingling and McGill, 1999a) whereas when tested in posterior shear the end-plate is the most common injury site (Yingling and McGill, 1999b). This would indicate that the variables with a directional component should have each direction accounted for or accumulated separately. This reasoning would also hold for the net joint moments where the musculature to generate an extension moment is completely different from the musculature for a flexor moment. Unfortunately this greatly increases the number of variables and raises the potential problem that some variables could show little or zero cumulative exposure depending on the task demands. The calculation of this increased number of variables is a relatively simple addition to any biomechanical model used for assessing cumulative loading. Modeling approaches that use sufficient data to generate force-time histories can provide outputs for both acute and cumulative exposure, an example is presented in Figure 13.16.

The real challenge with generating so many variables is interpreting their relationship with the development of pain. As mentioned previously, cumulative compression has been the most commonly used variable in published studies and is also the most used peak exposure variable in ergonomic assessments. It also has the added benefit of being unidirectional for joint compression. This is a direct result of the imposed joint compression associated with activating trunk musculature to generate movement in any dimension. These factors in combination with the epidemiological evidence and widespread use would make it the forerunner in becoming the variable of choice when setting and implementing a cumulative exposure TLV.

13.9 Summary

Taking into account all of the potential barriers and processing difficulties associated with documenting cumulative exposure, the fact that all of the studies that have examined the relationship between cumulative loading and low back pain have found a significant positive relationship (Kumar, 1990; Norman et al., 1998; Seidler et al., 2001, 2003) demonstrates the strength and potential of this approach. The positive association was independent of whether primary or secondary assessment methods between exposure and developing injury/pain were used.

This chapter has focused on spinal cumulative loading exposure. This approach is not meant to replace assessments of peak exposure. In fact, peak exposure is really just an instantaneous component of a cumulative exposure dosage. Both peak and cumulative evaluations of work provide complimentary but unique pieces of information. This is supported by the fact that in the calculation of multivariate odds ratios for the reporting of low back pain, peak and cumulative exposure were independent components in a factor analysis (Norman et al., 1998).

One of the primary barriers to the transfer of usable techniques for the assessment of cumulative exposure to practicing ergonomists is the lack of consensus in data sets. Clearly, there is a need to determine a common method, or at the least provide an estimate of the error inherent in the approach used, before any progress can be made towards developing a standard for assessing risk of injury from exposure to cumulative loading. This chapter has attempted to highlight and where possible quantify the errors in documenting cumulative loading. The knowledge presented spans from the theoretical foundations of the link between injury and cumulative loading to applied techniques such as 3DMatch for quantifying a worker's exposure. There is a strong evidence base to support the implementation of cumulative loading assessments and highlights the need for a TLV and supporting assessment techniques for practicing ergonomists.

Acknowledgments

The author's research work reported in this chapter was funded by the Natural Sciences and Engineering Research Council of Canada, and the AUTO21 Network Centers of Excellence whose funding is provided by the Canadian federal government.

The author would like to acknowledge the contribution of Dave Andrews, Ph.D. and Wayne Albert, Ph.D., who have collaborated and shared their ideas on cumulative loading. Much of the work presented here would not have been possible without a very dedicated group of graduate students: Diana Sullivan, M.Sc., Kiera Keown, M.Sc. and Jennie Jackson, M.Sc. who was instrumental in the development of 3DMatch. Jack Callaghan, Robert Parkinson, M.Sc. a Canada Research Chair in Spine Biomechanics and Injury Prevention.

References

Adams, M.A. and Hutton, W.C., Prolapsed intervertebral disc: a hyperflexion injury, *Spine*, 7, pp. 184–191, 1982.

Adams, M.A. and Hutton, W.C., The effect of fatigue on the lumbar intervertebral disc, *J. Bone Joint Surg. (B)*, 65, pp. 199–203, 1983.

Agnew, M.J., Andrews, D.M., and Callaghan, J.P., Using a magnetic tracking device as a means of measuring cumulative spine loading, in XVI International Annual Occupational Ergonomics and Safety Conference, University of Windsor, Toronto, 2002.

Agnew, M.J., Andrews, D.M., Potvin, J.R., and Callaghan, J.P., Dynamic 2-d measurements of cumulative spine loading using an electromagnetic tracking device, in 34th Annual Conference of the Association of Canadian Ergonomists, Association of Canadian Ergonomists, London, 2003.

Akesson, I., Balogh, I., and Skerfving, S., Self-reported and measured time of vibration exposure at ultrasonic scaling in dental hygienists, *Appl. Ergon.*, 32, pp. 47–51, 2001.

Allread, W.G., Marras, W.S., and Burr, D.L., Measuring trunk motions in industry: variability due to task factors, individual differences, and the amount of data collected, *Ergonomics*, 43, pp. 691–701, 2000.

Andrews, D.M. and Callaghan, J.P., Determining the minimum sampling rate needed to accurately quantify cumulative spine loading from digitized video, *Appl. Ergon.*, 34, pp. 589–595, 2003.

Armstrong, T.J., Foulke, J.A., Joseph, B.S., and Goldstein, S.A., Investigation of cumulative trauma disorders in a poultry processing plant, *Am. Ind. Hyg. Assoc. J.*, 43, pp. 103–116, 1982.

Azar, N., *The relationship between cumulative low back loads and heart rate determined physical activity level during non-occupational tasks*, M.Sc. Thesis Dissertation, University of Windsor, Windsor, ON, Canada, 2004.

Azar, N., Andrews, D.M., Callaghan, J.P., Accuracy of spine cumulative loading using self-reported duration and frequency information during non-occupational tasks, *International Journal of Industrial Ergonomics*, 43(8), pp. 459–461, 2005.

Bogduk, N., The innervation of the lumbar spine, *Spine*, 8, pp. 286–293, 1983.

Brinckmann, P., Biggemann, M., and Hilweg, D., Fatigue fracture of human lumbar vertebrae, *Clin. Biomech.*, 3 (Suppl. 1), pp. s1–s23, 1988.

Buckwalter, J.A., Spine update: aging and degeneration of the human intervertebral disc, *Spine*, 20, pp. 1307–1314, 1995.

Caler, W.E. and Carter, D.R., Bone creep-fatigue damage accumulation, *J. Biomech.*, 22, pp. 625–635, 1989.

Callaghan, J.P., Cumulative loading as an injury mechanism: tissue considerations, in *Proceedings of the IV World Congress of Biomechanics*, University of Calgary, Calgary, AB, Canada, 2002.

Callaghan, J.P., Jackson, J., Andrews, D.M., Albert, W.J., and Potvin, J.R., The design and preliminary validation of '3D-Match' — a posture matching tool for estimating three dimensional cumulative loading on the low back, in *34th Annual Conference of the Association of Canadian Ergonomists*, Association of Canadian Ergonomists, London, 2003.

Callaghan, J.P., Keown, K., and Andrews, D.M., Influence of dynamic factors on calculating cumulative low back loads, *Occupational Ergonomics*, In Press, 2005.

Callaghan, J.P. and McGill, S.M., Intervertebral disc herniation: studies on a porcine model exposed to highly repetitive flexion/extension motion with compressive force, *Clin. Biomech.*, 16, pp. 28–37, 2001.

Callaghan, J.P., Salewytsch, A.J., and Andrews, D.M., An evaluation of predictive methods for estimating cumulative spinal loading, *Ergonomics*, 44, pp. 825–837, 2001.

Carter, D.R. and Caler, W.E., Cycle-dependent and time-dependent bone fracture with repeated loading, *J. Biomech. Eng.*, 105, pp. 166–170, 1983.

Carter, D.R. and Hayes, W.C., Fatigue life of compact bone — I. Effects of stress amplitude, temperature and density, *J. Biomech.*, 9, pp. 27–34, 1976.

Cavanaugh, J.M., Kallakuri, S., and Özaktay, A.C., Innervation of the rabbit lumbar intervertebral disc and posterior longitudinal ligament, *Spine*, 20, pp. 2080–2085, 1995.

Danz, M.E. and Ayoub, M.M., The effects of speed, frequency, and load on measured hand forces for a floor to knuckle lifting task, *Ergonomics*, 35, pp. 833–843, 1992.

Daynard, D., Yassi, A., Cooper, J.E., Tate, R., Norman, R., and Wells, R., Biomechanical analysis of peak and cumulative spinal loads during simulated patient-handling activities: a substudy of a randomized controlled trial to prevent lift and transfer injury of health care workers, *App. Ergon.*, 32, pp. 199–214, 2001.

Genaidy, A.M., Waly, S.M., Khalil, T.M., and Hidalgo, J., Spinal compression tolerance limits for the design of manual material handling operations in the workplace, *Ergonomics*, 36, pp. 415–434, 1993.

Godin, C.A., Andrews, D.M., and Callaghan, J.P., Cumulative low back loads of non-occupational tasks using "3DMatch", a 3-dimensional video-based posture sampling approach, in *34th Annual Conference of the Association of Canadian Ergonomists*, Association of Canadian Ergonomists, London, 2003.

Granata, K.P., Marras, W.S., and Davis, K.G., Variation in spinal load and trunk dynamics during repeated lifting exertions, *Clin. Biomech.*, 14, pp. 367–375, 1999.

Gunning, J.L., Callaghan, J.P., and McGill, S.M., Spinal posture and prior loading history modulate compressive strength and type of failure in the spine: a biomechanical study using a porcine cervical spine model, *Clin. Biomech.*, 16, pp. 471–480, 2001.

Hansson, T.H., Keller, T.S., and Spengler, D.M., Mechanical behavior of the human lumbar spine. II. Fatigue strength during dynamic compressive loading, *J. Orthop. Res.*, 5, pp. 479–487, 1987.

Hardy, W.G., Lissner, H.R., Webster, J.E., and Gurdjian, E.S., Repeated loading tests of the lumbar spine, *Surg. Forum*, 9, pp. 690–695, 1958.

Holm, S. and Nachemson, A., Variations in the nutrition of the canine intervertebral disc induced by motion, *Spine*, 8, pp. 866–874, 1983.

Iatridis, J.C., Mente, P.L., Stokes, I.A., Aronsson, D.D., and Alini, M., Compression-induced changes in intervertebral disc properties in a rat tail model, *Spine*, 24, pp. 996–1002, 1999.

Jackson, J., Reed, B., Andrews, D.M., Albert, W.J., and Callaghan, J.P., Usability of *3D-Match* — An evaluation of the inter- and intra-observer reliability of posture matching to calculate cumulative low back loading, in *34th Annual Conference of the Association of Canadian Ergonomists*, Association of Canadian Ergonomists, London, 2003.

Jager, M., Jordan, C., Luttmann, A., and Laurig, W., Evaluation and assessment of lumbar load during total shifts for occupational manual materials handling jobs within the Dortmund Lumbar Load Study — DOLLY, *Int. J. Ind. Ergon.*, 25, pp. 553–571, 2000.

Keown, K.J., Andrews, D.M., and Callaghan, J.P., Predicting cumulative low back loading during repetitive lifting: determining the number of trials needed to represent workday conditions, in *XVI International Annual Occupational Ergonomics and Safety Conference*, University of Windsor, Toronto, ON, Canada, 2002.

Kumar, S., Cumulative load as a risk factor for back pain, *Spine*, 15, pp. 1311–1316, 1990.

Lauder, C.L., Andrews, D.M., and Callaghan, J.P., Peak and cumulative loading of the lumbar spine during simulated non-work activities, in *XVI International Annual Occupational Ergonomics and Safety Conference*, University of Windsor, Toronto, ON, Canada, 2002.

Lindbeck, L. and Arborelius, U.P., Inertial effects from single body segments in dynamic analysis of lifting, *Ergonomics*, 34, pp. 421–433, 1991.

Liu, Y.K., Njus, G., Buckwalter, J.A., and Wakano, K., Fatigue response of lumbar intervertebral joints under axial cyclic loading, *Spine*, 8, pp. 857–865, 1983.

Lotz, J.C. and Chin, J.R., Intervertebral disc cell death is dependent on the magnitude and duration of spinal loading, *Spine*, 25, pp. 1477–1483, 2000.

Lotz, J.C., Colliou, O.K., Chin, J.R., Duncan, N.A., and Liebenberg, E., Compression-induced degeneration of the intervertebral disc: an *in vivo* mouse model and finite-element study, *Spine*, 23, pp. 2493–2506, 1998.

Marras, W.S. and Granata, K.P., Changes in trunk dynamics and spine loading during repeated trunk exertions, *Spine.*, 22, pp. 2564–2570, 1997.

McGill, S.M. and Norman, R.W., Dynamically and statically determined low back moments during lifting, *J. Biomech.*, 18, pp. 877–885, 1985.

Mientjes, M.I., Norman, R.W., Wells, R.P., and McGill, S.M., Assessment of an EMG-based method for continuous estimates of low back compression during asymmetrical occupational tasks, *Ergonomics*, 42, pp. 868–879, 1999.

Muir, P., Johnson, K.A., and Ruaux-Mason, C.P., *In vivo* matrix microdamage in a naturally occurring canine fatigue fracture, *Bone*, 25, pp. 571–576, 1999.

NIOSH, *Work Practices Guide for Manual Lifting* (Rep. No. NIOSH Technical Report No. 81–122), National Institute for Occupational Safety and Health, Cincinnati, OH, 1981.

Norman, R.W. and McGill, S.M., Selection od 2-D and 3-D biomechanical spine models: Issues for consideration by the ergonomist, In *The Occupational Ergonomics Handbook* (edited by Karwowski,W. and Marras,W.S.), pp. 967–984, CRC Press, New York, 2002.

Norman, R.W., Wells, R.P., Neumann, P., Frank, J., Shannon, H.S., Kerr, M.S., and Ontario Universities Back Pain Study (OUBPS) Group, A comparison of peak vs cumulative physical work exposure risk factors for the reporting of low back pain in the automotive industry, *Clin. Biomech.*, 13, pp. 561–573, 1998.

Parkinson, R. and Callaghan, J.P., The role of load magnitude as a modifier of the cumulative load tolerance of porcine cervical spinal units: Progress towards a force weighting approach, *Theoretical Issues In Ergonomics Science*, In Press, 2005.

Porter, R.W., Adams, M.A., and Hutton, W.C., Physical activity and strength of the lumbar spine, *Spine*, 14, pp. 201–203, 1989.

Potvin, J.R., Norman, R.W., and Wells, R., A field method for continuous estimation of dynamic compressive forces on the L4/L5 disc during the performance of repetitive tasks, in *23rd Annual Conference of the Human Factors Associations of Canada Ottawa, Ontario*, Human Factors Association of Canada, Ottawa, ON, Canada, pp. 51–55, 1990.

Seidler, A., Bolm-Audorff, U., Heiskel, H., Henkel, N., Roth-Kuver, B., Kaiser, U., Bickeboller, R., Willingstorfer, W.J., Beck, W., and Elsner, G., The role of cumulative physical work load in lumbar spine disease: risk factors for lumbar osteochondrosis and spondylosis associated with chronic complaints, *Occup. Environ. Med.*, 58, pp. 735–746, 2001.

Seidler, A., Bolm-Audorff, U., Siol, T., Henkel, N., Fuchs, C., Schug, H., Leheta, F., Marquardt, G., Schmitt, E., Ulrich, P.T., Beck, W., Missalla, A., and Elsner, G., Occupational risk factors for symptomatic lumbar disc herniation; a case-control study, *Occup. Environ. Med.*, 60, pp. 821–830, 2003.

Stramler, J.H., *The Dictionary for Human Factors and Ergonomics*, CRC Press, Boca Raton, FL, 1993.

Urban, J.P.G., Holm, S., and Maroudas, A., Diffusion of small solutes into the intervertebral disc: an *in vivo* study, *Biorheology*, 15, pp. 203–223, 1978.

Videman, T., Nurminen, M., and Troup, J.D.G., Lumbar spinal pathology in cadaveric material in relation to history of back pain, occupation, and physical loading, *Spine*, 15, pp. 728–740, 1990.

Wada, E., Ebara, S., Saito, S., and Ono, K., Experimental spondylosis in the rabbit spine. Overuse could accelerate the spondylosis, *Spine*, 17, pp. S1–S6, 1992.

Wang, X.T., Ker, R.F., and Alexander, R.M., Fatigue rupture of wallaby tail tendons, *J. Exp. Biol.*, 198, pp. 847–852, 1995.

Waters, T.R., Putz-Anderson, V., Garg, A., and Fine, L.J., Revised NIOSH equation for the design and evaluation of manual lifting tasks, *Ergonomics*, 36, pp. 749–776, 1993.

Wren, T.A., Lindsey, D.P., Beaupre, G.S., and Carter, D.R., Effects of creep and cyclic loading on the mechanical properties and failure of human Achilles tendons, *Ann. Biomed. Eng.*, 31, pp. 710–717, 2003.

Yingling, V.R., Callaghan, J.P., and McGill, S.M., Dynamic loading affects the mechanical properties and failure site of porcine spines, *Clin. Biomech.*, 15, pp. 301–305, 1997.

Yingling, V.R. and McGill, S.M., Anterior shear of spinal motion segments. Kinematics, kinetics, and resultant injuries observed in a porcine model, *Spine*, 24, pp. 1882–1889, 1999a.

Yingling, V.R. and McGill, S.M., Mechanical properties and failure mechanics of the spine under posterior shear load: observations from a porcine model, *J. Spinal Disord.*, 12, pp. 501–508, 1999b.

Yu, C.Y., Tsai, K.H., Hu, W.P., Lin, R.M., Song, H.W., and Chang, G.L., Geometric and morphological changes of the intervertebral disc under fatigue testing, *Clin. Biomech.*, 18, pp. S3–S9, 2003.

Zioupos, P., Currey, J.D., and Casinos, A., Tensile fatigue in bone: are cycles-, or time to failure, or both, important? *J. Theor. Biol.*, 210, pp. 389–399, 2001.

Zioupos, P., Wang, X.T., and Currey, J.D., Experimental and theoretical quantification of the development of damage in fatigue tests of bone and antler, *J. Biomech.*, 29, pp. 989–1002, 1996a.

Zioupos, P., Wang, X.T., and Currey, J.D., The accumulation of fatigue microdamage in human cortical bone of two different ages *in vitro*, *Clin. Biomech.*, 11, pp. 365–375, 1996b.

14

Low-Level Static Exertions

Gisela Sjøgaard
National Institute of
Occupational Health

Bente Rona Jensen
University of Copenhagen

14.1 Low-Level Static Exertions in the Workplace

Low-level static exertions have been identified as a risk factor for the development of cumulative trauma disorders or repetitive strain injuries from epidemiological studies. The exposure in terms of static exertions in the workplace has been assessed for different jobs and tasks based on electromyographic recordings from specific muscle groups (Table 14.1). Jobs characterized by relatively high static levels in neck and shoulder showed health outcomes in terms of musculoskeletal disorders in these body regions (Table 14.2). In the 1970s, static contractions of 15% MVC (maximum voluntary contraction) were considered to be tolerated for an "unlimited" period of time for a muscle.[1] However, later studies showed that if a contraction is to be maintained for just 1 h, it may have to be as low as 8% MVC.[2] A permissible level of static muscle load of 2 to 5% MVC was then suggested.[3] However, it was observed that musculoskeletal disorders were frequent even in jobs with static levels of this magnitude, and it was suggested to reduce the acceptable static level, for example, by job rotation.[4] Static levels as low as 0.5 to 1% MVC may relate to troubles in the shoulder region,[5,6] and most recently, statements have been brought forward that static loads are not acceptable at all if sustained frequently or over a long period of time. Actually, "working hours as a risk factor in the development of musculoskeletal complaints" has been proposed.[7] Such continuous revision of recommendations can be foreseen if we do not understand why low-level static exertions cause disorders. The acceptable limits or interventions recommended in the workplace will only reduce cumulative trauma disorders if the true risk factors that elicit adverse health outcome are eliminated or minimized. Therefore, it is important to identify which aspect of these so-called low-level static exertions may be the risk factors. In this context, plausibility also plays an important role in risk identification, that is, possible physiological mechanisms of tissue degradation, which may

TABLE 14.1 Electromyographic Data on Static ($P = 0.1$), Mean ($P = 0.5$), and Peak ($P = 0.9$) Muscle Load in the Shoulder Region During Different Work Tasks Expressed in Percentage of Maximal Electromyographic Activity or Percentage of Maximal Voluntary Force Development (%MAX)

Job	Muscles	$P = 0.1$ (%MAX)	$P = 0.5$ (%MAX)	$P = 0.9$ (%MAX)	References
Typewriting	m. trapezius	4	7	10	8
Office work	m. trapezius	1	4	—	6,9
Computer mouse work	m. trapezius (r)	1/2	1/3	2/5	10
(young/elderly)	m. trapezius (l)	1/2	1/3	3/7	
	m. deltoideus	1/2	2/3	3/5	
	m. neck extensors	2/3	4/5	6/7	
CAD work (mouse side/ other side)	m. trapezius	2/1	5/2	9/6	11,12
Industrial sewing	m. trapezius (r)	9	14	21	13
	m. trapezius (l)	9	16	25	
	m. infraspinatus	4	9	20	
Floor cleaning	m. trapezius	10	25	54	14
Assembly plantelectronic work	m. trapezius	8	16	27	15
	m. deltoideus	7	13	28	
	m. infraspinatus	13	20	33	
Meat cutting	m. trapezius	6	10	17	16
Dental work	m. trapezius	9	13	18	17
Flight loading/unloading	m. trapezius	5	14	45	18
Letter sorting	m. trapezius	5	10	27	19
	m. deltoideus	5	14	19	
	m. infraspinatus	5	10	16	
Industrial production work	m. trapezius	1	5	15	11
Chocolate manufacturing work	m. trapezius	2	5	—	6

TABLE 14.2 One-Year Prevalence of Musculoskeletal Symptoms in Neck, Shoulder, Elbow/Forearm, and Hand/Wrist According to the Standardized Nordic Questionnaire[23]

Job	Sex	Number of Workers	Neck (%)	Shoulder (%)	Elbow/ Forearm (%)	Hand/Wrist (%)	References
Office work	F	643	48	48	12	22	24
	M	35	18	18	6	9	
Computer work	F	1745	53	42	—	30	25
	M	834	27	23	—	19	
CAD work	F	106	79	52	41	55	12
	M	43	68	49	20	30	
Sewing machine operators	F	77	55	51	7	26	26
	F	303	57	53	7	28	
Cleaners	F	737	63	63	27	46	27
Assembly plant electronic work	F + M	25	64	56	—	—	15
Meat cutting	F	16	67	54	7	47	24
	M	114	39	62	15	47	
Meat cutting	M, 8% F	2463	52	66	28	60	16
Dental work	F	43	49	40	19	40	17
	M	56	57	39	13	5	
Flight loading	M	808	30	31	14	22	18

Source: Kuorinka, I., et al. *Appl. Ergon.* 18, pp. 233–234, 1987.

be causally related to the identified risk aspect. The term "*low-level static exertions*" will be discussed, followed by a presentation of possible short- and long-term physiological responses. Based on this, preventive strategies are presented.

14.2 What Are "Static Exertions"?

Within the area of mechanics "static," in the strict sense means "no motion." In the workplace, truly static work postures are quite rare because most jobs include a number of movements to be performed often by the upper limbs. Even in supervision jobs, a number of objects have to be handled now and then.

According to the strict definition of static, one might suggest to use observation techniques to quantify how long time a certain posture is maintained without any movement. But most likely this variable would fail to show a relationship to musculoskeletal disorders. For instance, lying in bed or sitting relaxed in a well-supporting chair is hardly considered a risk, although highly static. The reason is, of course, that no muscle exertions or contractions need to be performed in these conditions. Therefore, quantifying the true variable "static" when trying to identify risk factors is not sufficient. What we are looking for is the static muscle contraction that may induce an overload on the musculoskeletal system.

A profile of the muscular load during a period of work may be obtained by analyzing the amplitude probability distribution function (APDF) of the electromyographic signal (EMG). Such measurements for analysis of static muscle activity have been widely used in workplace studies, where the static level is defined as the probability level $P = 0.1$. For instance, a static level of 5% MVC means that the contraction level of the muscle is 5% MVC or above for 90% of the time, or in other words, only for 10% of the time is the muscle contraction below 5% MVC. This implies that muscular rest may occur for 10% of the recording time or less. The interpretation of a static contraction according to the APDF analysis has caused some confusion because the static level is actually defined in the time domain. Also, this variable does not give the information that the muscle is really performing a 5% MVC throughout the recording period; indeed, larger contraction forces may occur. Finally, this variable does not control for length changes of the muscle, which means that the muscle contractions may well be dynamic. Nevertheless, redefining "static" in occupational settings has been a great "success." This is probably due to the time variable in essence being the real risk. But this was not intentional and no awareness has been paid to this fact by practitioners. Of note is that the risk factor probably is the *sustained* contraction.

14.3 What Is a "Low-Level"?

It may be surprising that low- rather than high-level contraction forces seem to imply a risk. Of course, high forces can cause ruptures as seen in accidents where bone, ligament, tendon, or muscle are exerted beyond their breaking point, and in this sense, high forces imply a risk. However, low forces constitute a corresponding risk if repeated or sustained for a prolonged time. All structures, inert materials as well as biological tissues, are able to withstand a force characteristic to their structure. At high forces, disruption will occur when the breaking point is exceeded, and lesser forces repeated over time will eventually cause fatigue fracture (Figure 14.1). Repetitive force exertions are accumulated and cause eventual disruption, possibly not of the tissue as a whole but in terms of microruptures.

First of all, when evaluating the force level, the maximum strength of the muscle must be taken into account. This relates to the muscle's cross-sectional area, age, and state of training; and different muscle groups and subjects show highly different muscle strength. Therefore, exposure assessment in terms of force recordings in absolute numbers in Newtons (N) will not give sufficient information regarding the level of exertion. The MVC force must be recorded as well, and data must be presented in percentage of MVC as mentioned earlier regarding the EMG data.

Second, endurance time for muscles plays a significant role in this context. The relationship between force level and the time for which it can be sustained is depicted in the endurance time curve (Figure 14.2). At low-force levels relative to the maximum strength, the muscle is capable of developing such force for long periods

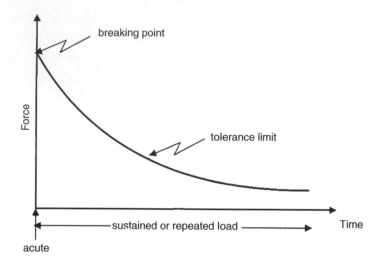

FIGURE 14.1 Tolerance limit for maximum force at breaking point depends on the type of muscle, for example, cross-sectional area, state of training, and age. Further, the contraction mode, that is, static or dynamic, is significant. For submaximal forces, the tolerance limit decreases with time, the rate of decrease depending on force magnitude and repetition frequency.

of time before being exhausted. Different muscles show highly different endurance capacity depending on muscle fiber type, anatomy, and state of training. But for every muscle there is a limit.

Third, in industry many low-level exertions include repeated static exertions or movements at quite high speed but with little displacement. When observing such tasks, often little attention is paid to the displacement, which is the cause for such exertions to be assessed as static. Also for intermittent static as well as dynamic contractions, endurance time curves exist.[2,20] It is for the dynamic contractions

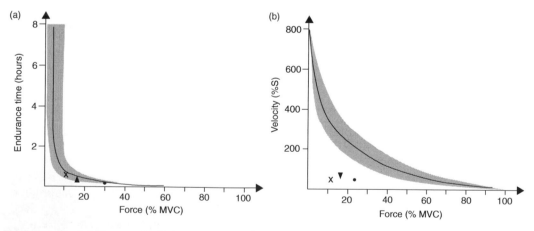

FIGURE 14.2 (a) Shows an endurance time curve for static contractions, which are defined as contractions at constant muscle length. Of note is the large range of endurance time at low forces, where the end-point of exhaustion varies significantly, for example, due to the level of motivation. Examples are given for handgrip (Δ), shoulder abduction (x), and trunk extension ($\bullet$). (b) Lower part shows a force–velocity curve for dynamic contractions, only shown for concentric contractions, that is, during muscle shortening. The maximum muscle force decreases with increasing velocity of shortening. Examples are given for shoulder movements during floor cleaning ($\bullet$), shoulder movements during forking in agriculture ($\blacktriangledown$), and wrist movements during meat cutting (x). For the latter, it is seen that although the load is only 10% MVC, it is about 20% of the dynamic strength, since the maximal dynamic strength at this velocity is approximately 50% MVC.

that the contraction level cannot be described only by the force in N or percent MVC. The force–velocity relationship must also be taken into account, the relative load being higher when a specific force is developed with increasing speed (Figure 14.2). For instance, keyboard operators may press 200,000 keys a day or 500 per minute and a piano player strikes the tangents with finger movements at very high, maybe sometimes maximum, speed. For the unloaded limb, the EMG activity increases linearly with the velocity of the movement.[21] If, in addition to the movement velocity, there is an external force to overcome during the work tasks, this could imply maximum effort at high velocities even if the force level is low. Thus, low-level cannot be assessed only in terms of percent MVC, but the mode of contractions must also be taken into account. Ideally, the maximum dynamic voluntary contraction forces should be assessed and the work task evaluated in relation to the corresponding maximum force–velocity relationship.

In short, the term "low-level" in the context of work-related static exertions refers to a working condition in which a muscle is activated at a level that can be maintained for a long period. This may be a true static contraction, sustaining a constant force and posture or varying in force within a limited range and without any movement. But even performing intermittent static or dynamic contractions (concentric/eccentric) at submaximal force velocities with small displacements and at intensities that can be maintained for a long time may be considered low-level static exertions in occupational settings.[22] Actually, when such exertions are measured by electromyography and analyzed by the previously mentioned APDF of the EMG, "static" levels of 5% MVC or more may be found. This means that a low-level static exertion is to be considered in the time domain and is characterized by workers being able to perform it for hours. The main feature is that the exertion is sufficiently low so that it can be sustained for a *prolonged* time and the duration probably implies the risk.

14.4 Which Work Requirements Induce "Low-Level Static Exertions"?

Examples of jobs in which low-level static exertions are frequent are presented in Table 14.1 and Table 14.2. Additional job titles are numerous in the literature.[28] It is important in risk assessment to identify generic work requirements that induce these exertions. At random, requirements such as precision, speed, visual demand, and mental load can be mentioned.[29] Also monotony or lack of variation is a characteristic that concerns working posture and movement as well as mental challenge. The same task is repeated over and over again most often by the hands. When operating with fast precise movements with the hands, there is a demand to stabilize the shoulder girdle. One reason is that the shoulders are the reference point for the upper limbs and if they move, the hands will be repositioned with respect to the motor control pattern for the upper limbs. Similarly, to control the position of the eyes, fixation of the neck is needed, and stable eye position is a prerequisite for most visual demands in industry. Interestingly, the fastest repositioning of the eyes can be performed when the neck and shoulder muscles are contracted upto 30% MVC.[30] When performing tasks at high speed, the stiffness of the musculoskeletal system must be increased. For this purpose, *cocontractions* are performed, which means that antagonistic muscles, that is, muscles on each side of a joint, are contracting. This is especially common for the shoulder muscles. One reason is, as mentioned earlier, that the shoulder must be the stable fix point and "take-off" for arms and hands. Also the anatomy of the shoulder is such that it has the greatest mobility of all the joints in the body. It is a joint that is highly dependent on muscle stabilization, including *cocontractions*. These *cocontractions* have been shown to increase with increasing speed and precision demands.[21,31,32]

14.5 Why Do "Low-Level Static Exertions" Imply a Risk?

As discussed earlier, it is not necessarily because exertions are static or at a low level that they imply a risk, but because such exertions are often sustained for prolonged periods of time. Additionally, often no sufficient recovery periods are allowed during such work tasks. A more informative term for the related risk factor would be *prolonged sustained* or *repeated muscle contractions*. According to the endurance curve, it is possible to sustain low-level exertions for a longer time than high-level exertions. It is likely, that this time factor is the risk. This hypothesis is supported by the physiological responses to such exertions, which constitute the plausibility. Standardized muscle contractions have been studied in combination with detailed physiological responses. In the following discussion, focus will be on mechanisms, which may induce muscle damage.

An example of a standardized setup for studying muscle contractions is shown in Figure 14.3. The test chair can be regulated for the subject to adopt any working posture and the force transducers connected to the handles allow for three-dimensional recordings. During specific work tasks, biomechanical calculations may then assess the relative load on various muscles or muscle parts/groups based on maximum contractions performed in identical postures and directions.[33]

Intramuscular Pressure and Blood Flow   With each muscle contraction, the tissue pressure (hydrostatic pressure) in the muscle increases in proportion to the force development. The absolute level in terms

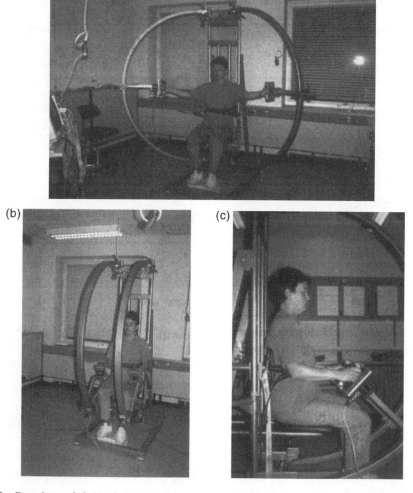

FIGURE 14.3   Experimental chair, where arm posture can be adjusted in any position of abduction (a) and flexion (b). The hands are grasping handles connected to three-dimensional force transducers (c). Professor Bjørn Quistorff, University of Copenhagen, is acknowledged for the design.

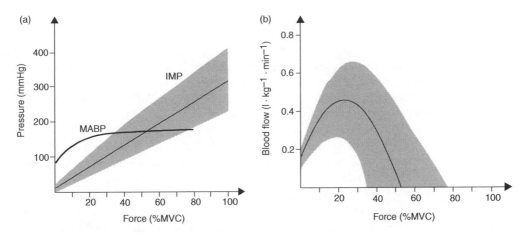

FIGURE 14.4 (a) Shows mean arterial blood pressure (MABP) and intramuscular pressure (IMP) with increasing contraction force. (b) Shows corresponding blood flow.

of mmHg varies widely between muscles and depends, among other things on the anatomy of the muscle itself as well as its surroundings. A bulky muscle attains higher pressures than a thin muscle, and a muscle with bony surroundings or tight fascia shows relatively large increases because of the low compliance of these surroundings. At high contraction forces, the intramuscular pressure may attain values far above blood pressure (Figure 14.4) and obviously cause muscle blood flow to be occluded in areas where intramuscular pressure exceeds blood pressure, the highest pressures normally occurring deep in the muscles. However, even at low-level contractions, the complex microcirculatory regulation may become impeded. First of all, at low blood flow velocities it is not the mean blood pressure but the diastolic pressure that is decisive for maintenance of blood flow.[34] Further, with prolonged contractions, the muscle water content will increase[35] and correspondingly, the thickness of the muscle has been shown to increase.[36] Such a state of edematic tissue with increased volume will *per se* increase tissue pressure in a deliminated closed muscle compartment with low compliance. At contraction levels in the order of 5 to 10% MVC, intramuscular pressures of 40 to 60 mmHg or more have been reported in muscles such as the *m. supraspinatus* in the shoulder.[37,38] Of note is that intramuscular pressure may increase to higher levels when dynamic contractions are performed at a force corresponding to the static contractions.[39] However, it may rather be the duration of increased intramuscular pressure than the absolute level that may be deleterious for the muscle and therefore static and not dynamic muscle contractions constituting a risk factor. Thus, causal relationships between prolonged moderately increased tissue pressure and pathogenic changes have been studied extensively in relation to compartment syndromes.[40] Pressures above 30 mmHg maintained for 8 h have been shown to induce necrotic changes in the muscle even if no active contraction was performed and energy demand therefore was minimal.[41] One possible mechanism is that although initially blood flow is sufficient during low-level contractions, this may not be the case when the contraction is maintained for prolonged periods. Conditions with low flow and low perfusion pressure may provoke granulocyte plugging in the capillaries, which effects microcirculation and may also facilitate formation of free radicals, which have a highly toxic effect.[42,43] Not only muscle tissue but also the peripheral nerves are sensitive to prolonged or repetitive elevated mechanical pressure. Dysfunction of the sensory perception indicating entrapment of n. medianus and n. ulnaris has been found among computer users with pain in the hand/wrist and forearm/elbow regions.[44]

Metabolism Adequate muscle blood flow is essential for muscle function because force development relies on the conversion of chemically bound energy to mechanical energy, a process also called energy turnover or metabolism. Some chemically bound energy or substrate is located in the muscle tissue (especially glycogen), but these may become depleted during prolonged activities. Therefore, the supply of substrates (including oxygen) to the muscle is crucial for such activities. A decrease in muscle oxygenation in response to a certain exertion indicates an acute imbalance between oxygen supply and oxygen demand. Oxygenation of the muscle is significantly reduced at an intramuscular pressure of 30–40 mmHg[45] and a 7% or greater reduction in muscle oxygenation relative to resting values is related to a reduced force generating capacity of the muscle.[46] The ultimate substrate in the conversion of chemical energy to mechanical energy is ATP, which is broken down in the myofibrils during the actin–myosin reaction. ATP is significant for the detachment of actin and myosin, and insufficiency of this process may cause rigor or contracture with massive pain. In normal muscle contractions, the actin–myosin reaction is initiated by the release of Ca^{2+} from the sarcoplasmic reticulum into the cytosol and has been the focus in a number of studies on muscle fatigue. However, during the last decade, attention has been drawn also to the pathogenesis of Ca^{2+}-induced damage of muscle cells.[47,48] The reuptake of Ca^{2+} into the sarcoplasmic reticulum is an ATP-dependent process, which may be insufficient during prolonged activity since it accounts for upto 30% of the energy turnover during muscle activity. Further, energy crisis may result in an influx of Ca^{2+} from the extracellular space. Consequently, the cytosolic-free Ca^{2+} is likely to be increased above normal for a prolonged time. This has serious implications for the phospholipids, including those in the muscle membrane. Ca^{2+} has a direct effect on phospholipase activity and, in addition, increases the susceptibility of the membrane lipids to free radicals, which have a highly toxic effect as mentioned earlier. Both these processes promote breakdown of the muscle membrane.[49] Finally, prolonged increased cytosolic Ca^{2+} concentration induces a Ca^{2+} load on the mitochondria and may eventually impair ATP formation, a sufficient concentration of which is a prerequisite for active force production (for more details see Ref. 50).

Motor Control   Another important aspect during low force development is that although the muscle as a whole may not be metabolically exhausted, this may well be the case for single-muscle fibers. The muscles are composed of different muscle fiber types and motor units with different recruitment thresholds. A stereotype recruitment order has been documented, which means that with increasing force, the low threshold motor units are always being recruited first.[51] Within a motor unit pool, various motor units may be alternating in activity pattern during a submaximal muscle contraction postponing fatigue to develop in each of the involved fibers.[34] However, performing highly skilled movements and accurate manipulations, it is likely that the very same motor units are being recruited continuously. This holds true for pure static as well as slow force-varying and low-velocity dynamic contractions.[52,53] Additionally, contractions may be elicited due to reflexes, causing even more stereotype recruitment than during voluntary contractions. Mental load has been demonstrated to generate nonpostural muscle tension in shoulder and forearm muscles,[54–60] and the same holds true for visual demands and neck muscles and even during breaks, for example, in computer work, muscle activity may stay above resting value.[61] Also reflexes originating in the muscle itself from chemo- as well as mechanoreceptors may play a role and recently the gamma-loop has been proposed to play a role in developing a potentially vicious circle.[62,63] The muscle fibers being continuously activated have been termed Cinderella fibers, because they are working from early to late.[64] A high energy turnover occurs in these fibers and most likely, they receive the least blood flow because tissue pressure increases in their vicinity due to the mechanical contraction impeding blood flow.[34] The pathogenic mechanisms described above regarding accumulation of Ca^{2+} and free radicals may be a concern, especially at the single-muscle fiber level. Prolonged activity of specific motor units throughout an 8-h working day may cause insufficient time for full recovery of these motor units due to a long-lasting element of fatigue,[65] which has been shown to occur in simulated occupational settings.[66] This may cause necrosis and, finally, cell destruction in these fibers. In line with this, fibers with marked degenerative characteristics have been found more frequently in muscle biopsies from patients with work-related chronic myalgia than in normal subjects in the trapezius muscle.[67] Interestingly, the degenerative fibers identified are slow twitch fibers, which connect with low-threshold motor nerves.[68]

Perception of Fatigue   When muscular work is performed over a prolonged period, fatigue develops. Fatigue may cause the work to be performed with less care or precision and an accident can result. A fatigued worker is more likely to make a wrong movement, such as a slip and fall, leading to injury. However, even if an accident does not occur, prolonged fatigue without adequate time for recovery can lead to the development of musculoskeletal disorders. From the beginning of every muscle activity, the muscle is fatiguing and muscle function is decreasing.[69] This condition is normally perceived as muscle fatigue. The perception of fatigue is a very useful mechanism for protecting the muscle against overload. Among other factors, the work-induced increase in potassium concentration in the interstitial space can help mediate the perception of fatigue to the central nervous system (CNS).[70] However, during very low-level contractions, the accumulated increase in interstitial potassium may be subliminal to the threshold of the sensory afferents mediating the information to the CNS.[71] Also, in situations of machine-controlled work or heavy work pressure, the fatigue message is depressed, when it is not possible for the employee to take a rest. In other words, fatigue is ignored — consciously or subconsciously — which in the long-term can have serious consequences.

The processes that take place in relation to fatigue are normally reversible for biological tissues, which is in contrast to inert tissues and a reason why we normally do not consider fatigue as dangerous. This means that muscles recover when resting after exertion. A rest period following muscular activity is therefore essential to enable the muscle to recover its full functional potential with regard to strength and endurance. Even an improvement or training effect of these variables may be obtained if optimal performance of activity and recovery periods is planned. There is no simple time equation for length of work and adequate length of a subsequent resting period. The process of recovery depends on the type of work that caused the muscle fatigue. For instance, the so-called low-frequency fatigue and high-frequency fatigue are caused by fundamentally different biochemical changes in the muscle.[71] If fatigue is due to relatively high loads over a short time, the necessary recovery will be quicker than if fatigue is due to prolonged working at low-load levels. Thus, if the same muscle group or group of fibers are activated continuously for a full working day of 7 or 8 h, there is a risk that the muscle will not even be fully recovered by the next day. If such conditions persist for months or even years, they can ultimately inflict irreversible or chronic changes that may result in pain and impaired function.

14.6 How to Prevent Musculoskeletal Disorders from "Low-Level Static Exertions"?

A prerequisite for effective prevention is knowledge of the cause for the disorder. The documentation so far of the time factor being essential gives the simple answer to this question: limit the time for each specific sustained muscle effort. Each single-muscle cell and corresponding motor nerve and tendon demand recovery periods sufficiently long to attain full recovery. Time for recovery is not linearly related to time for activity. Rather, it increases exponentially. For example, if exhaustion is elicited by a high contraction force for 1 min, then recovery is very fast, and after 2 to 3 min the same force can be performed again. But when a muscle is fatigued for 1 h, it may take many hours for full recovery, and if the fatiguing process has lasted for an 8-h working day, full recovery may not even have occurred the next morning when the next working day starts and the same tasks are to be managed. Interestingly, sports activities as a marathon (lasting about 2 to 5 h, depending on the state of training) are only performed a few times a year by even top athletes. Limits to prolonged activities are also seen in sport events such as the Tour de France or other endurance activities. Normally in the workplace, the worker is not totally exerted and often only part of the body is exerted. This means that somewhat shorter recovery will be acceptable, but still it is essential that the activity period is followed by a recovery period and that the duration of both is matched to the intensity and mode of contraction in the activity period (Figure 14.5).

In summary, it can be stated in concordance with earlier discussions[72,73] that human skeletal muscles are not adapted for continuous long-lasting activity. Indeed, no matter how low the exertion level is, rest periods are needed for the muscle to recover. Guidelines for low-level static exertions should therefore deal not just with the acceptable static level in percentage MVC. To recommend a reduction in exertion level from, for instance, 5 to 2% MVC will not help much physiologically; also it is not practical. Instead,

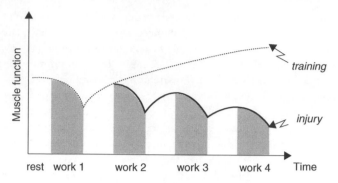

FIGURE 14.5   Muscle function in relation to work and recovery (duty circle).

we need guidelines for maximum acceptable *time limits* for prolonged sustained or repeated muscle contractions. This means that we must focus our attention on how long a *time* a muscle or group of muscles can tolerate maintaining or repeating the same low-level exertions. This is especially true when these same low-level exertions are imposed on a muscle day after day.

14.7   Recommendation for the Practitioner Regarding Job Profile and Workplace Design

1. The workplace must allow for variation in working postures. This means, for example, that table and chair are easily adjustable. For instance, shifting frequently between sitting and standing is recommended and instructions should be given to implement adjustments at frequent *time* intervals.
2. The workplace must be designed based on principles of optimization and not minimization of mechanical workload. Therefore, it is recommended that work cycles include loads ranging from complete relaxation to moderately high contraction forces and velocities. Workers should be given the possibility to optimize the phases of a duty cycle according to their capacity. It is of importance that the level or intensity of exertion changes over a wide range continuously over *time*.
3. The job profile must allow for performing a variety of different tasks. The task variation should include variation regarding mental load as well as physical (mechanical) load on the musculoskeletal system. If only specialized work tasks can be performed at each workstation, a job must include tasks at different workstations. For the use of tools, it is recommended that a variety of tools with different designs are used interchangeably. In combination, these variations should cause loading of different body regions and muscle groups regularly over *time*.

References

1. Rohmert, W. Problems of determination of rest allowances, *App. Ergon.*, 4(3), pp. 158–162, 1973.
2. Björkstén, M. and Jonsson, B. Endurance limit of force in long-term intermittent static contractions, *Scand. J. Work. Environ. Health*, 3, pp. 23–27, 1977.
3. Jonsson, B. Kinesiology. With special reference to electromyographic kinesiology, in *Contemporary Clinical Neurophysiology.*, Cobb, W.A. and Van Duijn H., Eds., Elsevier Scientific Publishing Company, Amsterdam, pp. 417–428, 1978.
4. Jonsson, B. The static load component in muscle work, *Eur. J. Appl. Physiol.*, 57, pp. 305–310, 1988.
5. Veiersted, K.B., Westgaard, R.H., and Andersen, P. Pattern of muscle activity during stereotyped work and its relation to muscle pain, *Int. Arch. Occup. Environ. Health.*, 62, pp. 31–41, 1990.

6. Jensen, C., Nilsen, K., and Hansen, K., et al., Trapezius muscle load as a risk indicator for occupational shoulder-neck complaints, *Int. Arch. Occup. Environ. Health.*, 64, pp. 415–423, 1993.

7. Wærsted, M. and Westgaard, R.H., Working hours as a risk factor in the development of musculoskeletal complaints, *Ergonomics*, 34(3), pp. 265–276, 1991.

8. Björkstén, M., Itani, T., and Jonsson, B. et al. Evaluation of muscular load in shoulder and forearm muscles among medical secretaries during occupational typing and some nonoccupational activities, in: *Biomechanics X-A*, Jonsson, B., Ed., pp. 35–39, 1987.

9. Blangsted, A.K., Hansen, K., and Jensen, C. Muscle activity during computer-based office work in relation to self-reported job demands and gender, *Eur. J. Appl. Physiol.*, 89, pp. 352–358, 2003.

10. Laursen, B., Jensen, B.R., and Ratkevicius, A. Performance and muscle activity during computer mouse tasks in young and elderly adults, *Eur. J. Appl. Physiol.*, 84, pp. 329–336, 2001.

11. Jensen, C., Finsen, L., and Hansen, K. et al. Upper trapezius muscle activity patterns during repetitive manual material handling and work with a computer mouse, *J. Electromyogr. Kinesiol.*, 9, pp. 317–325, 1999.

12. Jensen, C., Borg, V., and Finsen, L. et al. Job demands, muscle activity and musculoskeletal symptoms in relation to work with the computer mouse, *Scand. J. Work Environ. Health.*, 24, pp. 418–424, 1998.

13. Jensen, B.R., Schibye, B., and Søgaard, K. et al. Shoulder muscle load and muscle fatigue among industrial sewing-machine operators, *Eur. J. Appl. Physiol.*, 67, pp. 467–475, 1993.

14. Søgaard, K., Fallentin, N., and Nielsen, J. Work load during floor cleaning, The effect of cleaning methods and work technique, *Eur. J. Appl. Physiol.*, 73, pp. 73–81, 1996.

15. Christensen, H. Muscle activity and fatigue in the shoulder muscles of assembly-plant employees, *Scand. J. Work Environ. Health.*, 12, pp. 582–587, 1986.

16. Udbeningsarbejde i svineslagterier, 1 ed. København: Arbejdsmiljøinstituttet, 1996.

17. Finsen, L., Christensen, H., and Bakke, M. Musculoskeletal disorders among dentists and variation in the dental work, *Appl. Ergon.*, 29, pp. 119–125, 1998.

18. Jørgensen, K., Jensen, B., and Stokholm, J. Postural strain and discomfort during loading and unloading flight. An ergonomic intervention study, in *Trends in Ergonomics/Human factors IV*, Asfour, S.S., Ed. Elsevier Science Publishers B.V., North-Holland, pp. 663–673, 1987.

19. Jørgensen, K., Fallentin, N., and Sidenius, B. The strain on the shoulder and neck muscles during letter sorting, *Int J Ind Erg.*, 3, pp. 243–248, 1989.

20. Sjøgaard, G., Sejersted, O.M., and Winkel, J. et al. Exposure assessment and mechanisms of pathogenesis in work-related musculoskeletal disorders: Significant aspects in the documentation of risk factors, in *Work and health, Scientific Basis of Progress in the Working Environment*, Svane, O. and Johansen, C., Ed. European Commission, Directorate-General V, Luxembourg, 1995, pp. 75–87.

21. Carpentier, A., Duchateau, J., and Hainaut, K. Velocity-dependent muscle strategy during plantarflexion in humans, *J. Electromyogr. Kinesiol.*, 6, pp. 225–233, 1996.

22. Jørgensen, K., Fallentin, N., and Krogh-Lund, C., et al. Electromyography and fatigue during prolonged, low-level static contractions, *Eur. J. Appl. Physiol.*, 57, pp. 316–321, 1988.

23. Kuorinka, I., Jonsson, B., and Kilbom, Å. et al. Standardised Nordic questionnaires for the analysis of musculoskeletal symptoms, *Appl. Ergon.*, 18, pp. 233–237, 1987.

24. Ydreborg, B., Bryngelsson, I.-L., and Gustafsson, C. Referensdata till Örebroformulären FHV 001 D (200 D), 002 D (202 D), 003 D, 004 D och 007 D, Data från 95 yrkesgrupper insamlade åren 1984–1989. Örebro: Stiftelsen för yrkes- och miljömedicinsk forskning och utveckling i Örebro, 1989.

25. Jensen, C., Finsen, L., and Søgaard, K. et al. Musculoskeletal symptoms and duration of computer and mouse use, *Int. J. Ind. Ergon.*, 30, pp. 265–275, 2002.

26. Schibye, B., Skov, T., and Ekner, D. et al. Musculoskeletal symptoms among sewing machine operators, *Scand. J. Work Environ Health.*, 21, pp. 427–434, 1995.

27. Nielsen, J., Occupational health among cleaners. (In Danish with English summary.) Ph.D. thesis, University of Copenhagen, National Institute of Occupational Health, Copenhagen, 1995.

28. Armstrong, T.J., Buckle, P., and Fine, L.J., et al. A conceptual model for work-related neck and upper-limb musculoskeletal disorders, *Scand. J. Work Environ. Health.*, 19, pp. 73–84, 1993.

29. Sjøgaard, G., Lundberg, U., and Kadefors, R. The role of muscle activity and mental load in the development of pain and degenerative processes at the muscle cell level during computer work, *Eur. J. Appl. Physiol.*, 83, pp. 99–105, 2000.

30. Kunita, K. and Fujiwara, K. Relationship between reaction time of eye movement and activity of the neck extensors, *Eur. J. Appl. Physiol.*, 74, pp. 553–557, 1996.

31. Laursen, B., Jensen, B.R., and Sjøgaard, G. Effect of speed and precision demands on human shoulder muscle electromyography during a repetitive task, *Eur. J. Appl. Physiol.*, 78, pp. 544–548, 1998.

32. Birch, L., Juul-Kristensen, B., and Jensen, C. et al. Acute response to precision, time pressure and mental demand during simulated computer work, *Scand. J. Work Environ. Health.*, 26, pp. 299–305, 2000.

33. Laursen, B., Jensen, B.R., and Németh, G., et al. A model predicting individual shoulder muscle forces based on relationship between EMG and 3D external forces in static position, *J. Biomech.*, 31, pp. 731–739, 1998.

34. Sjøgaard, G., Kiens B., and Jørgensen K., et al. Intramuscular pressure, EMG and blood flow during low-level prolonged static contraction in man, *Acta. Physiol. Scand.*, 128, pp. 475–484, 1986.

35. Sjøgaard, G. Muscle energy metabolism and electrolyte shifts during low-level prolonged static contraction in man, *Acta. Physiol. Scand.*, 134, pp. 181–187, 1988.

36. Jensen, B.R., Jørgensen, K., and Sjøgaard, G. The effect of prolonged isometric contractions on muscle fluid balance, *Eur. J. Appl. Physiol.*, 69, pp. 439–444, 1994.

37. Jensen, B.R., Jørgensen, K., and Huijing, P.A., et al. Soft tissue architecture and intramuscular pressure in the shoulder region, *Eur. J. Morphol.*, 33, pp. 205–220, 1995.

38. Järvholm, U., Palmerud, G., and Herberts, P., et al. Intramuscular pressure and electromyography in the supraspinatus muscle at shoulder abduction, *Clin. Orthop.*, 245, pp. 102–109, 1989.

39. Sjøgaard, G., Jensen, B.R., and Hargens, A.R., et al. Intramuscular pressure and EMG relate during static contractions but dissociate with movement and fatigue, *J. Appl. Physiol.*, 96, pp. 1522–1529, 2004.

40. Pedowitz, R.A., Hargens, A.R., and Mubarak, S.J. et al. Modified criteria for the objective diagnosis of chronic compartment syndrome of the leg, *Am. J. Sports Med.*, 18(1), pp. 35–40, 1990.

41. Hargens, A.R., Schmidt, D.A., and Evans, K.L. et al. Quantitation of skeletal-muscle necrosis in a model compartment syndrome, *Bone Joint Surg. (Am).*, 63-A, pp. 631–636, 1981.

42. Schmid-Schönbein, G.W. Capillary plugging by granulocytes and the no-reflow phenomenon in the microcirculation, *Fed. Proc.*, 46(7), pp. 2397–2401, 1987.

43. Jensen, B.R., Sjøgaard, G., and Bornmyr, S., et al. Intramuscular laser-Doppler flowmetry in the supraspinatus muscle during isometric contractions, *Eur. J. Appl. Physiol.*, 71, pp. 373–378, 1995.

44. Jensen, B.R., Pilegaard, M., and Momsen, A. Vibrotactile sense and mechanical functional state of the arm and hand among computer users compared with a control group, *Int. Arch. Occup. Environ. Health.*, 75, pp. 332–340, 2002.

45. Jensen, B.R., Jørgensen, K., and Hargens, A., et al. Physiological response to submaximal isometric contractions of the paravertebral muscles, *Spine*, 24, pp. 2332–2338, 1999.

46. Murthy, G., Hargens, A.R., and Lehman, S., et al. Ischemia causes muscle fatigue, *J. Orthop. Res.*, 19, pp. 436–440, 2001.

47. Jackson, M.J., Jones, D.A., and Edwards, R.H.T. Experimental skeletal muscle damage: The nature of the calcium-activated degenerative processes, *Eur. J. Clin. Invest.*, 14, pp. 369–374, 1984.

48. Gissel, H. Ca2+ accumulation and cell damage in skeletal muscle during low frequency stimulation, *Eur. J. Appl. Physiol.*, 83, pp. 175–180, 2000.

49. Das, D.K. and Essman, W.B. Oxygen Radicals: Systemic Events and Disease Processes, Karger, 1990.

50. Sjøgaard, G. and Jensen, B.R. Muscle pathology with overuse, in *Chronic musculoskeletal injuries in the workplace*, Ranney, D., Ed., W.B. Saunders Company, Philadelphia, 1997, pp. 17–40.

51. Henneman, E. and Olson, C.B. Relations between structure and function in the design of skeletal muscles, *J. Neurophysiol.*, 28, pp. 581–598, 1965.

52. Søgaard, K., Christensen, H. and Jensen, B.R., et al. Motor control and kinetics during low level concentric and eccentric contractions in man, *Electroenceph. Clin. Neurophysiol.*, 101, pp. 453–460, 1996.

53. Christensen, H., Søgaard, K. and Jensen, B.R., et al. Intramuscular and surface EMG power spectrum from dynamic and static contractions, *J. Electromyogr. Kinesiol.*, 5, pp. 27–36, 1995.

54. Westgaard, R.H. and Bjørklund. R. Generation of muscle tension additional to postural muscle load, *Ergonomics*, 30(6), pp. 911–923, 1987.

55. Wærsted, M. and Westgaard, R.H. Attention-related muscle activity in different body regions during VDU work with minimal physical activity, *Ergonomics*, 39, pp. 661–676, 1996.

56. Wærsted, M., Eken, T. and Westgaard, R.H. Activity of single motor unit in attention-demanding tasks: firing pattern in the human trapezius muscle, *Eur. J. Appl. Physiol.*, 72, pp. 323–329, 1996.

57. Finsen, L., Jensen, C. and Søgaard, K., et al. Muscle activity and cardiovascular response during computer mouse work with and without memory demands, *Ergonomics*, 44, pp. 1312–1329, 2001.

58. Finsen, L., Søgaard, K. and Christensen, H. Influence of memory demand and contra lateral activity on muscle activity, *J. Electromyogr. Kinesiol.*, 11, pp. 373–380, 2001.

59. Laursen, B., Jensen, B.R. and Garde, A.H., et al. Effect of mental and physical demands on muscular activity during the use of a computer mouse and a keyboard, *Scand. J. Work Environ. Health.*, 28, pp. 215–221, 2002.

60. Søgaard, K., Sjøgaard, G. and Finsen, L., et al. Motor unit activity during stereotyped finger tasks and computer mouse work, *J. Electromyogr. Kinesiol.*, 11, pp. 197–206, 2001.

61. Blangsted, A.K., Søgaard, K. and Christensen, H., et al. The effect of physical and psychosocial loads on the trapezius muscle activity during computer keying tasks and rest periods, *Eur. J. Appl. Physiol.*, 91, pp. 253–258, 2004.

62. Johansson, H. and Sojka, P. Pathophysiological mechanisms involved in genesis and spread of muscular tension in occupational muscle pain and in chronic musculoskeletal pain syndromes: a hypothesis, *Med. Hypotheses.*, 35, pp. 196–203, 1991.

63. Mense, S. Considerations concerning the neurobiological basis of muscle pain, *Can. J. Physiol. Pharmacol.*, 69, pp. 610–616, 1991.

64. Hägg, G.M. Static work loads and occupational myalgia — a new explanation model, in *Electromyographical Kinesiology*, Anderson, P.A., Hobart, D.J., Danoff, J.V., Eds., Amsterdam, Elsevier Science Publishers B.V., 1991, pp. 141–144.

65. Edwards, R.H.T., Hill, D.K., and Jones, D.A., et al. Fatigue of long duration in human skeletal muscle after exercise, *J. Physiol. (Lond).*, 272, pp. 769–778, 1977.

66. Byström, S. Physiological response and acceptability of isometric intermittent handgrip contractions, *Arbete. och. Hälsa.*, 38, pp. 1–108, 1991.

67. Larsson, S.-E., Bengtsson, A., and Bodegård, L. et al. Muscle changes in work-related chronic myalgia, *Acta. Orthop. Scand.*, 59(5), pp. 552–556, 1988.

68. Henriksson, K.G. Muscle pain in neuromuscular disorders and primary fibromyalgia, *Eur. J. Appl. Physiol.*, 57, pp. 348–352, 1988.

69. Bigland-Ritchie, B., Cafarelli, E., and Vøllestad, N.K. Fatigue of submaximal static contractions, *Acta. Physiol. Scand.*, 128 (Suppl 556), pp. 137–148, 1986.

70. Sjøgaard, G. Exercise-induced muscle fatigue: The significance of potassium, *Acta. Physiol. Scand.*, 140(suppl.593), pp. 1–64, 1990.

71. Sjøgaard, G. Potassium and fatigue: the pros and cons, *Acta Physiol Scand.*, 156, pp. 257–264, 1996.

72. Sjøgaard, G. Intramuscular changes during long-term contraction, in *The Ergonomics of Working Postures: Models, Methods and Cases*, Corlett, N., Wilson, J., and Manenica, I., Eds. Taylor & Francis, London and Philadelphia, 1986, pp. 136–143.

73. Kadefors, R., Christensen, H., and Sjøgaard, G., et al. PROCID Recommendations for healthier computer work, *Work & Stress.*, 16, pp. 91–93, 2004.

15

Soft-Tissue Pathomechanics

Robert G. Cutlip
*NIOSH Health Effects
Laboratory Division*

15.1 Soft-Tissue Pathomechanics

Although epidemiological studies have been beneficial in identifying the prevalence of musculoskeletal disorders, the demographics and job types most affected, and types of injuries most sustained, they do not address the soft-tissue injury mechanisms that result in pain, injury, and impaired function. This chapter briefly reviews the relevant and related research in soft-tissue pathomechanics of muscle. Pathomechanics is defined generally as the study of the mechanisms of soft-tissue injury that result from physical loading exposures. The study of muscle pathomechanics focuses on the effects of both short- and long-term static and dynamic muscle contractions intrinsic during physical loading on the functional and cellular changes that lead to injury, pain, and loss of function. Because the internal and external forces involved in any work-related activity act on multiple structures and tissues in the body, multiple systems often are affected. Injury, pain, and loss of function may involve damage to bone, and to soft tissues in the body, such as cartilage, tendon, ligaments, muscle, nerve, or the vasculature. Because the pathomechanics of these tissues may involve different mechanisms that often are studied independently, separate treatments are warranted. This review will focus specifically on the pathomechanics of contraction-induced muscle injury.

The field of pathomechanics is not new; however, much of relevant basic and applied research has been conducted under more established but related scientific disciplines. For example, much of the research on skeletal muscle pathomechanics comes from the fields of muscle physiology, exercise physiology, and sports medicine where the study of the risk factors and physiological mechanisms associated with contraction-induced muscle injury have long been investigated. It is therefore a goal of this chapter to integrate this vast, diverse body of knowledge into a brief, yet helpful resource for applied ergonomists.

This chapter is organized into three sections. The first section introduces and illustrates the fundamental terms and concepts pertaining to basic muscle anatomy and physiology. The second section reviews the various methodological approaches that have been used with both human and nonhuman subjects. The third section examines the body of research that sheds light on the underlying molecular, cellular, and functional mechanisms associated with muscle injury caused by the major work-related risk factors such as force, repetition, posture, and vibration. The chapter concludes with a discussion of the current knowledge gaps and possible directions for further research.

15.2   Skeletal Muscle Physiology

Occupationally related musculoskeletal disorders have been associated with exposure to excessive physical loads, repetitive movements, awkward postures, and vibration.[24] A number of different tissues, including skeletal muscles, can be injured by exposure to these various factors.[24] To understand how exposure to these factors results in muscle injury, it is necessary to understand the biological and physiological mechanisms that allow skeletal muscles to generate movement, maintain posture, and support loads. The goal of this part of the chapter is to provide a basic description of skeletal muscle physiology and cellular biology.

15.2.1   Muscle Physiology and Anatomy

Individual skeletal muscles are comprised of bundles of muscle cells or myofibers (Figure 15.1a). Each myofiber is surrounded by a collagenous basement membrane (basal lamina) in addition to a cellular membrane called the sarcolemma. Myofibers are similar to other cells in the body, but they have a couple of unique features. First, myofibers contain a modified endoplasmic reticulum called a sarcoplasmic reticulum (SR). The SR functions as a protein processing and distribution organelle, and it regulates the levels of free intracellular calcium (Ca^{2+}) within the myofiber. Second, most of the intracellular space within the myofiber is comprised of the contractile elements or myofibrils (80% of a muscle's volume) (Figure 15.1b). Each myofibril is comprised of thick and thin filaments. Each thick filament (12−18 nm diameter) is composed of several hundred myosin proteins. Within the thick filaments, each myosin protein has a projection or a globular head (Figure 15.1c). These globular heads have binding sites that can interact with and form cross-bridges with the thin filaments, and an ATPase binding site. Thin filaments (5−8 nm diameter) are made of actin molecules that are organized in two strands twisted together to form a helix, that are covered by thread-like tropomyosin molecules and spherical troponin molecules (Figure 15.1d). Thin and thick filaments are organized in a specific pattern, which is repeated down the length of the muscle. It is this patterning that gives skeletal muscle its striated appearance.[237] Each repeated segment of thin and thick fibers forms a sarcomere.[237] The sarcomere is defined as the area between the Z-disk. Each sarcomere is comprised of dark areas (A bands) that contain the thick filaments, and light areas (I bands) that contain the ends of the thin filaments that do not overlap with the thick filaments. At each end of the sarcomere is a three-dimensional structure referred to as a Z-band or Z-disk (Figure 15.1e). Muscle contractions are produced when cross-bridges are formed between overlapping thin and thick fibers in the sarcomere, making the sarcomere the smallest contractile unit in the myofiber.

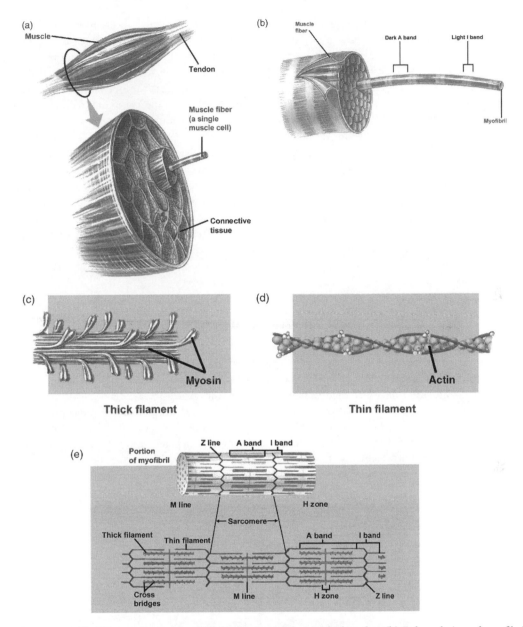

FIGURE 15.1 (a) Cross-sectional view of whole muscle and the attached tendon. (b) Enlarged view of myofibrils within a muscle fiber. (c) View of the thick filament (Myosin). (d) View of the thin filament (Actin). (e) Cytoskeletal components of a myofibril showing cross-bridge arrangement. a–e (Copyright Brooks/Cole — Thomson Learning.)

15.2.2   Cellular Processes Initiating Muscle Contractions

To initiate a contraction, myofibers must receive stimulation from motor neurons located in the ventral horn of the spinal cord. Motor neurons are considered the final common pathway where skeletal muscle activity can be governed only by input from these neurons. When activated, an action potential is propagated down the nerve axon and terminates on the neuromuscular junction (NMJ). The NMJ is a space where the action potential cannot cross from the nerve axon to the muscle fibers it innervates. Thus, a chemical messenger is used to transmit the signal from the nerve axon to the muscle fibers. As the signal is transmitted down the nerve axon, voltage-gated channels open to release Ca^{2+} into the terminal button of the NMJ. This facilitates the release of the chemical messenger Acetylcholine (ACH) that crosses the space to the motor end plate. This causes an ionic shift, which results in the propagation of the action potential down the basement membrane of the muscle fiber and then down the T-tubules of the muscle cell (Figure 15.2). The action potential activates the voltage-gated dihydro-pyridine receptors in the T-tubule. This change in the T-tubules triggers the opening of Ca^{2+} release channels (ryanodine receptors) on the SR. Ca^{2+} leaves the SR through the ryanodine receptors, enters the cytoplasm, and binds to troponin, one of the proteins on the thin filaments.

Troponin has three polypeptide units; one binds to tropomyosin, one binds to actin, and a third one binds to Ca^{2+}. Under resting conditions, tropomyosin is bound to actin and it blocks the myosin-binding site on the actin protein, preventing the formation of cross-bridges (Figure 15.3). However, when free Ca^{2+} rises in the cytoplasm of a myofiber, it binds to troponin and tropomyosin is pulled away from the myosin-binding site on actin, leaving it open for cross-bridge formation. Once cross-bridges are formed, the ATPase located on the myosin head increases its activity and hydrolyzes ATP. This causes the cross-bridge to break, and Ca^{2+} then dissociates from its binding site on tropomyosin. When Ca^{2+} is removed, tropomyosin slides back into the blocking position and the muscle relaxes. Thus, troponin and tropomyosin are referred to as regulatory proteins in muscle contraction (Figure 15.3).

15.2.3   Force Generation and Transmission in Skeletal Muscle

One of the main functions of skeletal muscle is to generate and transmit force. Force, or muscle tension, is directly related to the number of actin and myosin cross-bridges that are formed and the frequency of stimulation. A single action potential results in a single muscle contraction referred to as "twitch." As the frequency of stimulation increases, the resultant twitch tension (Figure 15.4a) increases with increasing stimulation frequency (Figure 15.4b) until a force plateau results (Figure 15.4c). Force is produced at each attached cross-bridge, so the total force development is proportional to the number of attached cross-bridges. The number of cross-bridges that can be formed depends upon the degree of overlap between the thin and thick filaments (Figure 15.5). When a sarcomere is overstretched or compressed,

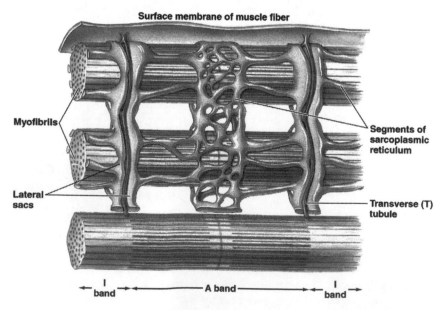

FIGURE 15.2   The T-tubules and SR in relationship to the myofibrils. (Copyright Brooks/Cole — Thomson Learning, 2001.)

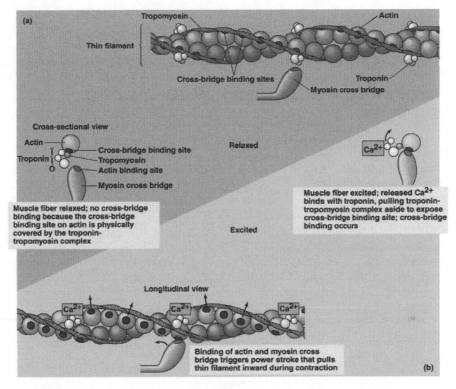

FIGURE 15.3 The role of Ca^{2+} in activating the cross-bridges. (Copyright Brooks/Cole — Thomson Learning, 2001.)

the area over which thin and thick filaments overlap is reduced, and thus there is a decrease in the number of cross-bridges that can be formed resulting in a reduction in force (Figure 15.5). Thus, maximal force is generated when sarcomeres are at a length that produces the optimal overlap between thin and thick fibers.

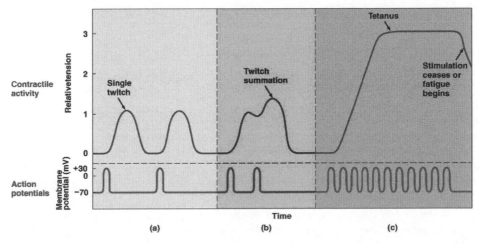

FIGURE 15.4 Muscle twitch, twitch summation, and tetanus. (a) If a muscle fiber is restimulated after complete relaxation, the second response is the same as the initial response. (b) If the muscle fiber is restimulated before complete relaxation takes place, the second twitch is added to the first twitch. (c) If the muscle fiber is stimulated rapidly such that it does not have the opportunity to relax, a maximal contraction or tetanus occurs. (Copyright Brooks/Cole — Thomson Learning, 2001.)

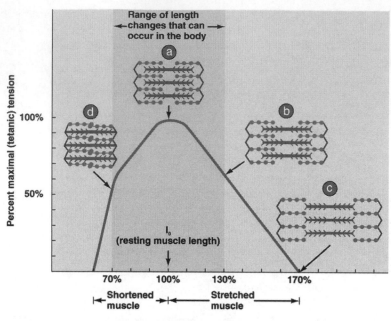

FIGURE 15.5 Length–tension relationship of muscle. At point A optimal overlap of thick and thin filaments results in maximal tension developed. This is referred to as the normal resting length in the body (l_o). As the muscle lengthens (point B), less cross-bridges are attached, which results in a decrease in tension. Further increases in length correspond with less cross-bridge attachment and further declines in tension (Point C). The response from point A to point C is usually referred to as the descending limb of the length–tension relationship. If shortening occurs at less than l_o, fewer filament-binding sites are exposed to filament cross-bridges, thus tension decreases (Point D). (Copyright Brooks/Cole — Thomson Learning, 2001.)

Force is generated at the cross-bridges, but it is transmitted longitudinally and radially along myofibrils. The longitudinal transmission of force occurs down the thick myosin filament to the Z-disk, and on to the next serial set of myofibrils. Two proteins titin and nebulin, maintain length registry of the sarcomere and aid in axial transmission of contractile forces. The actions of titin and nebulin maintain registry of the A-band with the Z-disk, which is important for sarcomere integrity. Nebulin maintains length registry of the thin filaments[49,306] by interacting with tropomyosin and troponin to form a lateral network with actin to regulate thin filament length. Titin functions as a two part spring to transmit force from the thick filaments to the corresponding Z-disk.

Radial forces are transmitted via lateral stabilization of adjacent myofibrils. The protein responsible for maintaining lateral registry of adjacent myofibrils at the Z-line is desmin.[160] The Z-disk structure is thought to be three-dimensional in nature and comprised of the proteins desmin, actin, and α-actinin. The radial enclosure of these three proteins also extends longitudinally along the myofibrils to provide both radial and longitudinal stability.[207] These proteins are thought to be anchored to the Z-disk via intermediate filament-associated proteins (IFAP). The cytoskeletal lattice extends radially from the Z-disk to the sarcolemma via the transmembrane proteins. The transmembrane proteins are thought to anchor the myofilaments to the sarcolemma via focal adhesions.[205] These adhesions or "costameres" are made up of a variety of transmembrane proteins. The basement membrane is then attached to the sarcolemma via the dystroglycan complex.[205,207] Radial transmission of forces occurs through structural proteins located inside and outside of the sarcomeric region via the intermediate filament network, and to the sarcolemma via the transmembrane proteins.[207] Capability of radial force transmission is necessary for redundancy in case of fiber injury. Thus, force can be transmitted in any direction in relation to the axis of the muscle fibers via endosarcomeric and exosarcomeric protein lattices.

15.2.4 Types of Muscle Contractions

There are three primary types of muscle contractions. These contraction types are distinguished by how the muscle length changes during the contraction.[13,48] Isometric contractions are defined as muscle activity where tension is generated without a change in length. This is also referred to as a static contraction where muscle is generating tension, but does not result in a change in length, and thus, there is no segmental (about a single joint) or whole body motion. Shortening contractions (often referred to as concentric contractions[13] are defined as the muscle generating tension while getting shorter. Concentric contractions usually generate segmental or whole body motion. Lengthening (or eccentric) muscle contractions are defined as the muscle generating tension while the muscle is lengthening. Lengthening contractions are usually used to absorb work or energy, thereby applying braking to segmental or whole body motion.

During concentric muscle actions, the tension varies as a function of shortening velocity where tension decreases as shortening velocity increases. The "force–velocity" relation is hyperbolic and also depicts that maximum shortening velocity occurs at zero load and zero velocity occurs at maximum isometric force.[295] During isometric muscle contractions, it is well understood that force varies as a function of muscle length. It been shown that muscle tension is lowest at very short and very long muscle lengths and develops higher tension in the intermediate lengths.[68,103] This is due to the degree of sarcomere overlap in the cross-bridge. Thus, the length–tension curve has an ascending and descending limb as length increases (see Figure 15.5). The ascending limb, defined as the increase in force with increase in length, is due to more actin-binding sites being available to bind with the myosin filaments. As tension plateaus, this is thought to be due to all the actin-binding sites being bound to the myosin filaments. The descending limb, defined as the decrease in muscle tension with increasing length, is due to less actin-binding sites being available as the actin filaments are pulled out of register with the myosin filaments. Thus, the length–tension relationship of muscle is due to myofilament overlap in the sarcomere (as shown in Figure 15.5).

It is now well known that muscle can generate more tension during eccentric muscle actions than during concentric or isometric contractions. This was first reported in a study involving human muscles under volitional control.[238] It is also interesting that while muscles generate more tension during eccentric muscle actions than concentric muscle actions, EMG activity is less in muscles during stretch than during shortening at the same tension. During maximal effort, the EMG signature remains constant and force varies due to the length–tension relation of that specific muscle or muscle group, however force during volitional eccentric activity never exceeded 140% of maximal shortening forces.[148] In animal studies that employ electrical stimulation to activate the muscle of interest, forces of 180% of maximum isometric force are typical.[284] High eccentric forces in humans with spastic paresis have been attained to levels similar to those seen in animal studies.[142] In addition, if muscles in humans are stimulated by external electrical stimulators, as in the case of spinal cord injured patients, the external forces generated during eccentric muscle actions are nearly 200% of the forces generated concentrically using the same electrical stimulation paradigm.[275] Thus, exogenous electrical stimulation overrides the inhibitory influences that moderate muscle output force.

Stretch-shortening cycles (SSC) are a type of muscle action that incorporates both concentric and eccentric muscle actions. In most sports-related activities, it involves a prior stretch before shortening to enhance the shortening phase of the movement. Activities that typically use SSCs are jumping, walking, running, and movement in and around obstacles. In occupational-related activities, it is mostly related to reciprocal lifting and lowering activities and repetitive lift and carry tasks. It is an excellent model to study physiological muscle function.[149] It also allows for simultaneous study of concentric and eccentric muscle function and their synergism.

15.2.5 Musculotendon Actuator

Muscle and tendon have typically been studied in isolation although they function synergistically. Their integrative function has been defined as the musculotendon actuator.[311] While physiologists have long recognized that muscle and tendon act in a synergistic fashion, they have studied those tissues in isolation to better understand the function of muscle and tendon separately. The musculotendon actuators interact with body segments to produce movement and the dynamics of movement are dependent upon the contraction dynamics of the actuator. This system also functions as a feedback loop where the dynamics of body segments affect the force output of the actuator via the length and velocity of the actuator.[311]

Tendon compliance affects the contraction dynamics of the muscle. In actuators with highly compliant tendons, a length change of the actuator would be mostly realized by the length change in the tendon, with very little concomitant length change in muscle. Compliance of an actuator is defined by the ratio of tendon slack length to muscle fiber length. Muscle length changes will be commensurate with length perturbations of the musculotendon actuator if the actuator is stiff. However in compliant actuators, muscle length changes will not follow exogenous length perturbations. This is quite relevant since changes in muscle length are rarely measured directly, indeed in human studies of muscle function the kinematics of the musculotendon are measured. The assertion that changes in musclotendon length are representative of muscle length changes may be incorrect, particularly for actuators with highly compliant tendons.[311] The active and passive force–length relation of muscles that have been published to-date may be erroneous due to the fact that muscle fibers are at different lengths in the active versus passive state even though the actuator is at the same length. This is due to differing amounts of tendon stretch, which are caused by different forces exerted by the muscle fibers. Thus, the stretch of the tendon must be accounted for to accurately represent muscle stretch.[118] The muscle–tendon interface (at the aponeurosis) exists in a state of dynamic equilibria, where force transients are equalized via stretch of the tendon and muscle activation and muscle length change. The dynamic equilibria are also governed by the response time of the tendon and muscle, which are often different. Actuator compliance varies depending on the muscle group and animal species. In humans, actuator compliance appears to be highest on the plantar flexor group and lowest on proximal groups, such as biceps and triceps.[31,128,293] In summary, one must think in terms of the musculotendon when investigating *in vivo* muscle function or reviewing scientific studies of *in vivo* function. While most studies refer specifically to muscle function, the measurements are typically made on the musculotendon group. Thus one must be cognizant of the influence of tendon mechanics on muscle function and the musculotendon unit.

15.3 Contraction-Induced Injury Models

15.3.1 Human Studies

In the past four decades, there has been substantial inquiry into the physiological effects of eccentric exercise on physiological responses, such as oxygen consumption,[29,140,143,145,147] muscle metabolites,[32,33,208] and performance via bicycle ergometry or dynamometry[28,78,115,144,146,150,151] just to name a few. However, only in the past two decades there has been an increasing amount of interest in the area of skeletal muscle injury. Formal inquiry in the area of muscle injury was initiated by Hough in 1902[126] where it was noted that delayed onset muscle soreness resulted days after the exposure and probably was due to microtears in the muscle tissue. It is interesting that there was not much scientific inquiry in the area of muscle injury until the early 1980s when Friden et al.[90] were one of the first to provide evidence of muscle fiber damage after exercise. Fortunately, the study of muscle injury has rapidly increased since then.

It has been well documented that muscle injury primarily results from lengthening contractions but not concentric or isometric contractions.[53,58,253] Thus, studies of muscle damage have employed eccentric-biased segmental exercises using either isolated muscle groups, such as the elbow flexors,[193–197,221,222,231] knee flexors,[173,174,294] hamstring,[179] pectoralis and anterior deltoids (chest press),[245] and calf and biceps.[136] Whole body movements such as downhill stepping exercise,[187,189] downhill running,[199] or cycling[175] have also been used to create muscle injuries. The amount of acute resistance for most human studies ranged from 10 to 180 repetitions for the segmental exercise studies, and as little as 20 min of downhill stepping exercise[187] to produce muscle soreness and evidence of myofiber injury. Thus, even a low number of repetitions or short exposure to lengthening contractions can result in a strain injury.

It is interesting to note that exposures that involve both whole body movements and segmental loading result in muscle damage, loss of force, and delayed onset muscle soreness. Different types of information can be attained from these two types of exposures. Whole body exercise typically involves closed kinetic chain movements where the level of exercise (treadmill or cycling speed, metabolic load, angle of inclination) is controlled but muscle forces or torques, velocity, range of motion, and number of repetitions are not typically measured. In contrast, most segmental exercise models are open kinetic chain movements involving isolated loading of the limbs, and are administered by either isokinetic or isotonic dynamometry (computer-controlled strength testing equipment that operates in either constant velocity or constant torque mode), or by isoinertial loads (using either free weight or weight attached to an apparatus). Joint torques or forces, as well as velocity, range of motion, and number of repetitions are measured.

Prolonged strength loss, as measured by maximal isometric force, is considered to be the best method to quantify the degree and time course of muscle injury and recovery after exposure to damaging lengthening contractions.[289] It is also the primary means to quantify muscle function in humans where muscle function is defined as the ability of the muscle group of interest to generate force over a prescribed range of motion, or fixed length at a given level of muscle activation.[289] Other measures, such as biochemical markers in the blood and urine, and level of histological damage as obtained from biopsy, are not well associated temporally with functional performance.[289]

Muscle injury studies with humans have been very beneficial in elucidating the type and intensity of exercise and muscle actions that produce injury. Those studies have also been beneficial for examining the resultant myofiber changes after injurious exposure and the recovery time after injury. The corresponding levels of pain perception and muscle soreness after exposure was also consonant with the degree of performance deficit.[58] However, though many studies of muscle injury with humans have been conducted in controlled laboratory settings, some experimental questions cannot be fully addressed using human subjects. Confounding factors such as lifestyle, level of psychological stress, pre-existing disease states such as diabetes and hypertension, and genetic polymorphisms are difficult to control. In addition, key issues such as the amount of fiber strain and muscle biomechanics necessary for injury, the amount and site of resultant muscle injury from longitudinal and cross-sectional tissue analysis, the effect of structural protein knockouts and inflammatory mediator blockade on muscle injury and repair, and biochemical analyses of muscle tissue, is either difficult or impossible with human subjects.

Animal models have been developed that reduce or eliminate these confounding factors as well as provide for control of the degree of motivation and muscle activation strategies. In depth physiological questions about the pathways involved in muscle injury and repair, the site and extent of injury, the role of structural proteins in muscle force transmission and injurious response, and biomechanical loading signature necessary for injury can be more easily addressed in animal models. Animal models that have been developed allow for more controlled study from the level of isolated muscle fibers to fully intact muscle groups that contain intact neural and vascular supplies.

15.3.2 Animal Models of Muscle Performance and Injury

Animal models provide a good platform to study the physiological responses to injurious and noninjurious muscle contractions and the biological aspects underlying muscle injury and adaptation. The majority of animal models used to investigate skeletal muscle performance and injury have used rodents. The majority of studies of acute injury in animals have focused on the temporal force response of muscle to a single exposure (1 to 1800 repetitions) of high force isometric (muscle generating force at a fixed length), concentric contractions, or lengthening contractions.[13,35,36,39,72,74,75,129,161,163–165]

The similarities between the micro-architecture of rodent and human skeletal muscle and the ability to precisely control the biomechanics of contractile activity in rodents through various *in vitro, in situ,* or *in vivo* methods, as well as the ability to investigate in depth physiological questions such as injury pathways, are major advantages using animal preparations. Each type of animal model has inherent advantages and disadvantages for the study of muscle injury. The models that have been used to investigate muscle pathomechanics are described here.

15.3.2.1 *In Vitro* Models

In vitro models use muscle that has been excised from an animal to study muscle function. The muscle group or muscle fibers are placed in a sealed physiological bath and activated by plate electrodes attached to an electrical stimulator. The resultant muscle tension is measured at the ends of the muscle by ergometry using strain gage force transducers. Length changes also are produced via the attached ergometer and invoked after maximal activation of the whole muscle or isolated muscle fibers. The target muscles studied were mouse soleus,[167,259,286,290,283] mouse extensor digitorum longus (EDL),[36,210] toad sartorius,[258,260] rat soleus,[171,172,284,285,309] and mouse fifth toe muscle.[226]

The main advantage of *in vitro* preparations is the ability to study a single muscle group or isolated muscle fibers of interest without the confounding effects of adjacent or antagonist muscle groups. In addition, exact length changes and the velocity of length changes of the total muscle and individual sarcomeres can be accurately measured in real time. The release of muscle proteins and enzymes into the physiological bath can also be detected that may be indicative of injury or metabolic fatigue. Measures of performance, such as isometric twitch tension, maximal isometric force, work done during stretch or shortening actions, and power absorbed during stretch or produced during shortening can be quantified. Much information regarding muscle function and injury mechanics has been attained using *in vitro* models.

There are several shortcomings of *in vitro* preparations. Because the muscle has been excised from the animal and tissue viability is time-limited, *in vitro* preparations are only suited for single exposures. Thus, the effect of repetitive exposures on muscle response, adaptation, and injury cannot be studied. In addition, the effect of changes in muscle performance on biomechanics about the joint axis cannot be examined. Also, the effect of muscle synergists on performance about the joint axis cannot be studied. Because human muscle testing is performed about the joint axis of interest with functioning muscle synergists, generalizing from *in vitro* results to *in vivo* function can be difficult. Also, because normal neural and vascular supplies have been ablated, muscle performance, response to injurious perturbations, and changes in neural recruitment patterns can be affected. For example, an intact vascular supply is instrumental in replenishing depleted energy stores and removing toxins, and bringing in appropriate cellular infiltrates for muscle repair and remodeling. *In vitro* models by nature are intrinsically ischemic due to the absence of an intact vascular supply. An intact neural supply is beneficial for proper muscle activation and to facilitate the study of neural changes in response to injury and repair. Other models that maintain intact neural and vascular supplies and connections with supporting tissues can provide data that is more physiologically representative about intact muscle function.

15.3.2.2 *In Situ* Models

In situ models address the need for muscle injury models that have the advantages of the *in vitro* models but also have normal neural and vascular supplies intact, and the ability to test whole muscles or muscle groups. Typical *in situ* models involve surgical ligation of the distal tendon of the muscle or muscle group of interest leaving the neural and vascular supplies intact. The target muscle or muscle group is stimulated via the exposed nerve or by use of percutaneous electrodes. Measurement of muscle contractile forces and control of length changes is usually via attachment of the distal tendon that was surgically ligated to a computer-controlled servomotor and load measurement transducer. This model of contraction-induced injury provides a more physiologically representative preparation than the ischemic, noninnervated *in vitro* model.

In situ models have been used to study injury mechanics in mice EDL muscles,[35,37–39,65,72,75,129,177,178,313] rat EDL,[176,276–278] rat adductor longus,[268,281] rabbit tibialis anterior (TA),[27,161,162,165,191,192,201,206] rabbit soleus muscle,[23] rabbit EDL,[219] and rabbit triceps surae.[264] The parameters used to produce injury in those studies ranged from single stretch models that consisted of stretch outside of the typical physiological range (usually greater than 130% of the resting or optimal muscle length, l_o) of the muscle[27,38,39,129,191,192,201,219] upto 1800 contractions within the normal physiological range (typically 70% l to 130% l_o).[161,165]

In situ models are well suited for acute muscle injury studies where changes in contractile forces subsequent to an injurious perturbation are of interest. Exact length changes and lengthening velocity of the muscle–tendon group can be controlled and monitored during testing. However, the invasive nature of this model precludes it from being applicable to repetitive injury models because the target muscle–tendon complex cannot be left exposed for more than a single session. Also, in most *in situ* studies, the forces or torques are not tested about the normal joint axis where effects of synergist muscles will have an effect on the resultant forces or torques about the joint axis. Also, the transmission of muscle–tendon forces through the joint axis to the target output limb via the mechanical advantage of that joint could not be studied. However, results from *in situ* studies have provided much information about the causal factors in acute muscle injury and the resultant physiological responses.

15.3.3.3 *In Vivo* Animal Models

It was apparent from *in vitro* and *in situ* findings that it would be beneficial to investigate muscle response and injury mechanics by testing about the normal joint axis of the target muscle and also be concerned with the invasiveness of the procedure. By using a noninvasive procedure, the confounding effects implicit in the required surgical procedure of *in situ* preparations are removed, and the temporal response after exposure can be examined. Also, a noninvasive preparation would be ideally suited for the study of muscle response from repeated exposures. *In vivo* models address these issues by facilitating testing about the normal joint axis in a noninvasive manner, with intact neural and vascular systems, and intact muscle–tendon systems.

Most *in vivo* models can be categorized as either volitional or nonvolitional models. Volitional models are those in which the movement tasks are performed voluntarily using different types of motivational tools. In contrast, nonvolitional models are those in which the animal typically is anesthetized and muscle contraction is initiated and controlled by an external electrical source.

15.3.3.3.1 *Nonvolitional Models*

In order to fulfill a need for more control and quantitation of *in vivo* muscle function, Wong and Booth in the late 1980s and early 1990s used electrical stimulation of the rat plantar flexors and a weighted pulley bar apparatus that would provide isotonic resistance to the plantar flexors.[302–304] This approach controlled the number of repetitions, the activation of the muscle group, the temporal arrangement of the repetitions, and loading of the plantar flexors about the joint axis. However, this model did not use a servomotor to control the range of motion or velocity and acceleration of the movement, and did not measure dynamic forces of the plantar flexors.

This approach was refined in the early 1990s to provide better control of the kinematics of the movement by use of an electrical servomotor. *In vivo* dynamometry incorporated electrical servomotors, load cells to measure forces, and potentiometers and tachometers to measure the kinematics of the movement to comprise a total testing system[63] (Figure 15.6). Dynamometry can be used to control and measure the biomechanical loading signature in real-time via control of muscle activation levels, range of motion of the muscle action, type of muscle action (isometric, shortening contractions, lengthening contractions, or stretch-shortening), velocity, acceleration, number of repetitions, duty cycle, and exposure duration.[63] The main difference between dynamometry and the Wong and Booth model is that the electrical

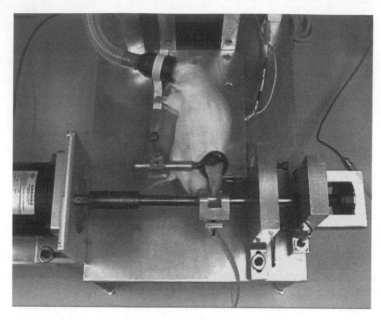

FIGURE 15.6   *In vivo* dynamometer with an anaesthetized rat on the heated $X-Y$ positioning table. The left foot is secured in the load cell fixture and the knee is secured in 90° flexion.

servomotor (in the dynamometer) controls the movement kinematics and measures the resultant muscle response during those movements. In the Wong and Booth model, the muscle forces produced the shortening-only movement about the joint axis where the kinematics was not controlled.

The first reported *in vivo* rodent dynamometer was developed by Ashton-Miller for the study of biomechanical behavior of the plantar and dorsi flexor muscles of the mouse hindlimb.[12] This approach also has been used to study rabbit dorsiflexors,[26,86,163,164,206] rat dorsiflexors,[62,97,116,152,154,156,157] mouse dorsiflexors,[132,168,283] and rat plantar flexors.[63,296–299] Typical exposures ranged from 20 repetitions of the rat plantar flexors[298] to 900 repetitions in the rabbit dorsiflexors.[86,163,164,206] Typical angular velocities were based on the animal being tested: 75 deg/sec for rabbit dorsiflexors,[163] upto 500 deg/sec for rat dorsiflexors,[97,152] and upto 2000 deg/sec for mouse dorsiflexors.[132,168,283] Ranges of angular velocities were selected based on the volitional capability of the muscle group and animal species in order to be physiologically representative.

The major benefit of nonvolitional *in vivo* models is the ability to study muscle function and injury mechanics about the joint axis of the target muscles. Thus, the normal muscle, tendon, and bone attachments are intact as well as the neural and vascular supplies. The synergistic function of muscle agonists and lateral transmission of adjacent muscle forces is also preserved.

One major limitation of nonvolitional models is the use of artificial electrical stimulation to invoke muscle contractions. Unlike voluntary contractile activity, which is submaximal and characterized by a selective recruitment of motor units, nonvolitional contractile activity is typically supramaximal because electrical stimulation involves the activation of all motor units of the target muscle. Thus, caution must be exercised when making inferences from comparisons between muscle responses from supramaximal electrical stimulation and voluntary submaximal contractions.

15.3.3.3.2 *Volitional Models*

Volitional *in vivo* models represent a more physiologically representative animal model for the study of muscle injury and adaptation. Volitional models differ from nonvolitional models in that normal muscle recruitment is employed via normal central nervous system control, and the pace of the activity is controlled by the animal, not the testing equipment or the investigator. One of the earliest reported models of volitional lifting was developed by Gordon in 1967 using weights attached to the back of rats during vertical crawling and other exercises.[104–107] This type of work was furthered by Stone et al.[261,262] and Ho et al.[120] in rats. Gonyea et al. extended this model to cats, also using weight lifting exercise.[100–102] Weights also have been added to wing muscles of chickens, roosters, and other birds to produce an overload model designed to study the skeletal muscle response to persistent overload.[3,83,166,246]

Animal treadmill models were developed in the early 1980s as a way to invoke voluntary repetitive eccentric muscle actions capable of producing muscle injury.[10,234] It was found that downhill treadmill locomotion produced an eccentric bias on the soleus muscle in the plantar flexor group that resulted in distinguishable signs of injury.[10,66,202,234] Treadmill exposure has been used to study a wide variety of physiological variables and typical exposures range from 30 to 150 min per session for a single session in rats[10,66,153,202,277] upto five sessions per week for 10 weeks duration,[16,109] and upto 9 h exposure in a single session for mice[185] to study acute injury response and adaptation and reduction of injury susceptibility after repetitive exposure. The treadmill studies are similar to those conducted in humans although the animal exposures are typically longer in duration.

In volitional treadmill and resistance training models, the exposure biomechanic, such as muscle forces or torques, or the number of muscle contractions (repetitions) are not controlled or quantified during the activity. This lack of quantitation makes it difficult to relate physiological outcomes to specific parameters of performance and loading history, which can differ widely across individual animals.

Some researchers have employed operant conditioning procedures to produce the kinds of repetitive muscle loadings that are relevant for the study of exercise-induced physiological responses. In these approaches, voluntary responses were motivated by various consequences, such as food rewards,[102,141,308] intracranial stimulation,[94] or electric shock to the tail or feet.[71,84,120,266] The species, target muscle groups, and training protocols, however, differed widely among these models. For example, Gonyea et al.[98,101,102] trained cats to grasp and move a weighted bar with the forelimb repeatedly in 30-min sessions conducted 5 days per week for upto 87 weeks. Barbe et al.[19] trained rats upto 8 weeks to repeatedly reach their forelimbs into a small tube to retrieve food pellets. Yarasheski et al.[308] trained rats over an 8-week period to climb a wire-mesh ladder with weights secured to their tails, and Klitgaard[141] trained rats over a 36-week period to enter a vertical tube and use their hindlimbs to lift a weighted ring. In other approaches, rats wore weighted jackets and were trained in sessions conducted 8 to 16 weeks to rear up on their hindlimbs to avoid an electric shock[71,120,266] or to receive brain stimulation.[94]

Many of these approaches have been developed for the study of adaptive or regenerative processes. For example, there is considerable evidence that under some conditions, voluntary repetitive exertions performed over several weeks or months can lead to muscle hypertrophy as evidenced by increases in either myofiber number or size.[71,101,266,301] Under other conditions, however, similar patterns of repetitive exertions have resulted in degenerative morphology[98] and inflammatory responses.[19] Unfortunately, many of these approaches lacked the necessary control and quantitation of the biomechanics of the movements to allow for thorough assessments of external and internal loadings that would be necessary to characterize dose–response relationships. In addition, few studies have specifically examined the effects of external loads on both internal loads and tolerances (as measured by biomechanical performance and physiological force tremor) and physiological or biochemical processes. Thus, little is known about the relation between specific parameters or changes in performance or physiological force tremor and physiological or biochemical processes.

15.4 Injury Mechanisms

15.4.1 Acute Muscle Injury

Skeletal muscle is a unique tissue in the body since it has both passive and active properties that are exhibited during muscle contraction. Because skeletal muscle can generate force during contraction for movement of the limbs and external work, as well as the absorption of work, it can produce loads on other tissues such as tendons, joints, and nerves. There have been extensive studies to date on acute contraction-induced muscle injury using both animal and human models. Studies of soft-tissue injury resulting from acute strain overload have been conducted using animal and human models. Indeed, a number of studies on contraction-induced injury have been conducted in rodents and rabbits. These studies have used *in vivo, in situ,* and *in vitro* preparations to investigate muscle injury.

15.4.2   Eccentric Muscle Actions in Acute Myofiber Injury

Eccentric contractions are known to cause a greater amount of damage in muscles. This suggests that high load tensions in fibers may be more important than physiologic considerations in the etiology of the injury process.[7,253,254] High mechanical forces produced during eccentric muscle actions have been causal in the underlying etiology of muscle strain injuries.[10,284,285] This was thought to be due to high fiber stresses in the contractile apparatus due to high forces transmitted axially to the contractile proteins. High mechanical forces produced during muscular contractions, particularly in eccentric exercise, where forces are distributed over relatively small cross-sectional areas of muscles, cause disruption of proteins in skeletal muscle fibers and connective tissues.[6,11] Eccentric contractions have been shown to result in ultrastructural damage immediately after exposure,[87] and 1–3 days after exposure.[116,178] The extent of histological damage is difficult to quantify by light or electron microscopy immediately after injurious exposure because only single sarcomeres or small groups of scattered sarcomeres are affected.[39] Damage is accompanied by a loss of contractile force with visible interfiber damage.[11] The reduction in contractile force is temporary (lasting days) and is accompanied by muscle soreness.[8] The isometric force deficit, a functional measure defined as the difference in isometric force before and after an eccentric contraction protocol, has been shown to be the best indicator of the magnitude of contraction-induced injury.[75,190,289]

Most work in eccentric contraction-induced injury has been done on small mammals *in situ*. To determine which parameters were responsible for muscle injury, some of the early studies focused on comparing muscles that were passively stretched through a range of motion to muscles that were actively stretched through the same range of motion. Muscles that were exposed to 30 min of passive stretches (TA and EDL) showed no loss in force; however, force loss was evident in the group that underwent 30 min of lengthening contractions of the same muscle group.[73,74] The role of passive stiffness was investigated in rat TA *in situ*, and stiffness increased after eccentric and isometric protocols; however, there was no correlation between injury and increased passive stiffness.[155] McCully and Faulkner[177] used mouse EDL muscles and found that the force decrement due to eccentric contractions did not recover and thus was due to mechanical insult to the tissue. Those results were corroborated by Lieber and Fridén[161] using New Zealand white rabbit TA muscle *in situ* and by Van Der Meulen,[276,277] using rat TA muscle *in situ*. Initial muscle length in conjunction with work input was also determined to be a factor in eccentric contraction-induced injury based on *in situ* single stretches to 170% L_o ("L_o" denotes the length of a particular muscle where it generates the highest force).[129] In contrast, Warren et al.,[284] observed no difference in force deficit due to initial length although L_o was less than 100%. Thus, results from *in situ* studies indicated that eccentric contraction-induced injury results from high mechanical forces in conjunction with the muscle being at long fiber lengths on the descending limb of the length–tension relationship.

The results of studies conducted using *in vivo* models of humans and animals have been consistent with *in vitro* and *in situ* results. Armstrong[10] and Ogilvie[202] used rodent treadmill testing to investigate eccentric contraction-induced injury. The purpose of these studies was to investigate the relationship between eccentrically biased treadmill exercise and skeletal muscle injury. The response of the concentrically contracting TA was compared to the eccentrically contracting soleus muscle. The control group (no exercise) and the concentrically contracting TA muscle showed no signs of injury but the eccentrically contracting soleus muscle did result in fiber injury. Newham[186] used human subjects and stimulated elbow flexors superimposed over maximum voluntary eccentric contractions to assure that muscle contractions were maximal. The results of their study showed that the loss of force was due to changes in contractile elements, not the level of muscle activation. This was the first study to suggest that the force decrement resulting from eccentric muscle actions is not the result of less muscle activation, but instead may have a mechanical etiology. Thus, the force deficit seen after eccentric contraction-induced injury in both humans and animals is due to damage of the contractile proteins and supporting structures, not central nervous system activation level.

Once eccentric contractions were identified as causing muscle injury, it was important to investigate how the injurious response could be modified by mechanical exposure factors (e.g., force, strain, strain rate, number of repetitions, and velocity). The primary factors that have been studied have some generalizability to occupational physical exposures.

15.4.3 Factors Affecting Acute Muscle Injury

15.4.3.1 The Effect of Muscle Force on Myofiber Injury

The *in situ* results of eccentric contraction-induced injury have been supported in general by *in vitro* animal models. Warren[285] used isolated rat soleus muscles to investigate mechanical factors associated with the initiation of eccentric contraction-induced muscle injury. The results indicated that a reduction in contractile force was most related to high forces during lengthening. This finding was also supported by other work.[110,178,253] In Warren's model, the primary criterion used to quantify injury was a reduction in twitch tension (P_t), which has previously been shown in muscles injured by eccentric contractions. The eccentric contraction group was compared against muscles performing isometrically using the same stimulation protocol. This study clearly demonstrated that eccentric contraction-induced muscle injury has a mechanical etiology. The predominant factor was mechanical force during lengthening with failure occurring above 113% P_o ("P_o" denotes maximum isometric force of the target muscle). In a follow-up study by Warren,[284] the focus was to investigate whether injury is the result of high tensile force after one contraction or the result of multiple contractions. The protocol consisted of 0.25 L_o excursions with a velocity of 1.5 l_o/sec at a force of 180% P_o. Muscles performing more than eight eccentric contractions resulted in injury since marked force decrements were observed after the eighth contraction. This suggests that it requires more than one repetition to result in myofiber injury within the physiological range.

15.4.3.2 The Effect of Muscle Length Changes on Myofiber Injury

To determine if muscle strain was important in susceptibility to injury, Zerba and Faulkner[314] studied mouse EDL muscles *in situ*. Isometric force was checked immediately post-test and 3 days after the protocol. Only muscles stretched to 75% L_f (length of the muscle fiber) at L_f/sec produced injury 3 days later. Muscles stretched at lower velocities and fiber strain did not exhibit any signs of injury. They hypothesized that strain and strain rate were synergistic in the etiology of muscle injury. Lieber and Fridén[162,165] also investigated whether muscle damage was a function of muscle force or muscle strain. Rabbit TA muscle was used *in situ* via securing the distal tendon to a servomotor. The TA muscle was selected due to a 30° pennation angle and negligible angular rotation during stretch. Final results indicated that muscle strain (change in length of the muscle) produced the most profound changes in muscle performance (via a force decrement) and that muscle strain was more responsible than force during lengthening in producing contraction-induced injury. Warren et al.[284] found no observed difference in force deficit when stretches were initiated from either 85% L or 90% L_o. In contrast, Hunter and Faulkner,[129] MacPherson et al.,[172] and Brooks and Faulkner[39] found that 30% strain was necessary to produce a force deficit after a single stretch and larger force deficits resulted when stretches were initiated from a longer initial length, or terminated at a longer final length. This finding was also supported in multiple repetition models.[110,305] Thus, it appears from both single- and multiple repetition models that muscle length during stretch has an impact on the resultant force deficit and myofiber injury. However, the effect of muscle length repetitive exposures of eccentric muscle actions has not been investigated to date.

15.4.3.3   The Effect of Repetitions on Muscle Injury

There is clear evidence that the number of eccentric or SSC repetitions has an effect on the amount of resultant muscle injury and force deficit.[97,116] Models that have induced single stretches in muscle within the physiological range have not resulted in muscle damage or a pronounced force deficit.[39,129] In other studies, it required more than one stretch within the physiological range to produce muscle injury.[97,110,284,297,298] Repeated stretches that varied from 225 to 900 at a final length of 110% L_o have resulted in myofiber damage and a resultant force deficit.[35,163,177,178,313] Thus, the amount of loading does have a graded effect on both changes in muscle performance and the extent of myofiber injury.[97,116] However, the effect of repetitions on repetitive exposures of either eccentric muscle actions or SSC has not been studied to date.

15.4.3.4   The Effect of Other Mechanical Factors on Muscle Injury

Dynamic muscle forces and length changes that are measured during eccentric muscle actions can be dissected into components of the dynamic signature. Components such as peak force,[110,178] average force,[39] work during the stretch,[129] and fiber length[110,129] have been found to affect the magnitude of contraction-induced injury. The force deficit resultant from an injurious exposure has been predicted by: (i) work done during the stretch when initiated from optimal length,[39,172] (ii) initial length and work during the stretch when not initiated at optimal length,[129] or by (iii) peak force and initial length.[110] Within a given level of force output, eccentric muscle actions performed at longer ranges of motion or fiber length have resulted in larger isometric force deficits in both humans[188] and animals[110,129,305] than stretches performed at a shorter range of motion. However, in these studies, the change in work (calculated by integration of the force–muscle displacement curve) during repeated stretches was not reported.[39,110,305] Work during the eccentric phase or stretch (negative work) has been shown to be well correlated with the isometric force deficit after a single eccentric contraction.[129,172] However, the length perturbation in these studies was beyond the normal physiological range of the target muscle. However, muscles stretched within the normal physiological range have required more than one repetition to produce injury.[284,297,298] Studies of repetitive eccentric muscle actions in the physiological range may have more external validity than single stretch models that have been studied outside of the normal physiological range.

15.4.3.5   The Effect of Exposure Duration and Lengthening Velocity

To determine if exposure duration and lengthening velocity affected muscle injury, McCully and Faulkner[178] used EDL muscles of mice. Although measured for 15 min exercise duration, there was no change in P_o due to muscle fatigue after the initial 5 min. A velocity of 1 L_f/sec produced a deficit in P_o after 3 days while the 0.2 L_f/sec and 0.5 L_f/sec lengthening velocities did not. A drop in P_o was mostly associated with stretch velocity. It was theorized that loss of peak force after fatigue prevents further muscle damage. Muscle injury increased with eccentric exercise duration for upto 5 min (no further force decrements were observed with subsequent eccentric muscle actions), higher velocities shortened the duration time for injury, and muscle force was a critical component in producing injury. Warren et al.'s[284] *in vitro* results in rats also indicated that higher lengthening velocities produced larger force decrements. The velocity component in muscle injury was investigated by Scifres and Martin[235] using a Kin-Com dynamometer to test human subjects for eccentric leg extension performance. One leg was tested at 30°/sec while the other leg was tested at 120°/sec. Delayed onset muscle soreness (DOMS) was more pronounced at the higher velocity, which indicated that higher velocity may produce additional muscle fiber injury. To further examine the effect of velocity on contraction-induced injury, Lynch and Faulkner[169] used single permeabilized fibers from mice EDL muscles. The severity of contraction-induced injury was not affected by the velocity of stretch. Controversy still exists as to the role of lengthening velocity in acute injury to skeletal muscle fibers. Furthermore, the role of lengthening velocity in chronic contraction-induced injury has not been investigated thus far. The effect of exposure duration (length of exposure) on repetitive injury has not been investigated to date, however it may be an important factor in repetitive injury causation.

15.4.3.6 The Effect of Loading History on Injury Susceptibility

The understanding of the injury pathophysiology is important in preventing contraction-induced injury. The predisposing factors, which mitigate injury are also important. Anecdotal observations of physical activities that result in a high incidence of injury has led to the hypothesis that if physical activities are repeated with adequate recovery time, muscles will eventually become "trained" and no injury will occur. Conversely, disuse could increase contraction-induced injury susceptibility[286] (Cutlip et al., unpublished observations). Human muscle performance studies assessed the relationship between muscle strength, soreness, and the release of intracellular proteins into the serum. Komi and Buskirk[150] used a training protocol of isometric, shortening, and eccentric muscle actions of the elbow flexors for 4 days/week for 7 weeks. Muscle soreness peaked after the first week and then disappeared. Maximum force during shortening and eccentric muscle actions increased throughout the 7-week test period. Newham et al.[186] also studied the effect of eccentric muscle actions of the elbow flexors performed on three occasions separated by 2 weeks. Muscle soreness was reduced after the second and third sessions, but was still present. Plasma levels of creatine kinase (a muscle-associated enzyme) were also greatly elevated after the first session, but were not elevated after the second and third sessions. Recovery of the maximum force was more rapid and complete after each subsequent session. A similar study indicated that a pre-training session of 24 eccentric contractions of the elbow flexors reduced muscle soreness and force deficit due to a 70 eccentric contraction protocol administered 2 weeks later.[60] Serum creatine kinase that showed an increase after the first session, diminished after the later session, which indicates a temporal relationship between creatine kinase levels and muscle soreness.

Results of studies employing small rodents (mice and rats) have been consistent with human studies. In one study, 198 eccentric contractions of the dorsi flexors were administered to anesthetized rodents *in vivo* once every 7 days for 6 weeks using a dynamometer.[76] The initial exposure to the protocol produced a 60% force deficit, which returned to 80% of the pretest value at 7 days. By the sixth week, no decline in force was observed. Muscles demonstrated an adaptive response, with increased whole muscle mass. Results from human and animal studies indicate that muscles can be trained to perform maximal eccentric muscle actions without injury; however, training must be continuous. The result of these studies raises an important issue in the study of muscle injury. Instituting training sessions containing eccentric contractions produces short-term force deficits and resultant myofiber disruption and inflammation, which later ceases and the muscle returns to normal function. The transient effects (temporary soreness and inflammation) are reduced with subsequent sessions. The preceding results pose an important issue about how the desirable effects of training can be distinguished from pathological changes that occur due to repetitive motion? Also, what characteristics of the dynamic inputs (force, strain, strain rate, number of repetitions) produce adaptive versus pathological responses in a chronic model?

15.4.3.7 The Effect of Age

Age can be an important factor in injury susceptibility and recovery. Zerba et al.[313] tested the hypothesis that muscles of old mice were more susceptible to injury than muscles of young and adult mice by attaching the distal tendon of the EDL muscle to a servomotor. The left leg was exposed to stretch while the right leg served as control. Results indicated a 27% deficit in P_o, and that old mice (43% deficit) were more susceptible to injury than young and adult mice. *In vitro* results of mice EDL muscles conducted by Faulkner et al.[72] indicated that old mice are more susceptible to injury and intrinsic differences in single permeabilized fibers account for the difference. Subsequent *in situ* and *in vitro* work by Brooks and Faulkner[37] and Faulkner et al.[72] supports earlier results that initial eccentric contraction-induced injury increases in old age. Single stretches of whole muscle and single permeabilized fibers (which rules out effects of excitation–contraction coupling, or membrane and extracellular effects) represented an effective method of focusing on factors that contribute to contraction-induced injury. Because their *in vitro* findings supported their *in situ* work, it suggested that failure has a mechanical etiology (due to work input) and increased injury susceptibility with age may be due to decreased protein synthesis, and thus an increased population of weak sarcomeres. In contrast, Willems and Stauber did not find a difference in isometric force deficit after 30 eccentric stretches in old (39%) and young rats (35%).[297] While muscle weights were similar in the young (4 months) and old (24 months) groups, peak force was lower in the old group during the eccentric stretches. Thus, injury studies using rodents should be cognizant of age as a possible contributing or confounding factor in the response to contraction-induced injury. The effect of age on the adaptive or pathological response to a chronic administration of high force eccentric muscle actions has not been studied thus far. The age of the animals selected for study is an important factor in acute injury and may affect the response to repetitive exposure.

15.4.3.8 Recovery Kinetics

Recovery kinetics also is important in the evolution of the injury process. The primary determinant of the time course required for recovery is the magnitude of the secondary injury that results after the initial mechanically induced injury. The secondary injury cascade includes an inflammatory response, free radical damage, phagocyte infiltration, and eventual phagocytosis of the cytoplasm in areas of damaged fibers.[169] Depending on the magnitude of the initial contraction-induced injury, full recovery of normal structure and function requires from 7 to 30 days.[35,75,126,127,177,178,276] Faulkner et al.[75] investigated the recovery kinetics (sampled from 1 h post-test to 30 days post-test) after an acute exposure to eccentric contractions while leaving the distal tendon intact. The extent of injury to the EDL and TA muscles of mice was determined from 1 h to 30 days after passive shortening and lengthening, and eccentric and concentric contractions. No injury was produced in the control or passively lengthened and shortened groups. The active eccentric protocol produced a 50% decrease in P_o, which did not recover by 3 days; however, total recovery was evident at 30 days thus indicating that recovery from eccentric contractions is a prolonged process.[75,178] These findings were also supported in both rats[157,276] and rabbits[191] *in situ*, and mice,[130,132,168,288] *rabbits*,[86,163] rats (Cutlip, unpublished observations), and humans[58] *in vivo*. The results of these studies clearly showed that the recovery process is not due to metabolic fatigue, but fiber injury that requires days or weeks to recover functional performance and repair fiber lesions.

15.4.3.9 The Effect of Sex on Injury Susceptibility

There have been reports of sex differences in both human and animal models of exercise-induced muscle damage.[57,58] Creatine kinase activity in rats[4,5,18] and humans[70,247,263] after muscle injury have been reported with differing results. Differences in the inflammatory response after muscle injury have also been reported in both humans[263] and rats[252] with females exhibiting less cellular infiltrates into the damaged area of the target muscle. Komulainen et al. reported that there was less myofiber damage via changes in structural proteins in female than male rats after exposure to downhill treadmill exercise.[153] In contrast, Stupka et al. reported that muscle damage was similar in males and females after exposure to injurious eccentric exercise.[263] It has been shown that males have higher baseline plasma levels of muscle proteins, such as creatine kinase, myoglobin, and skeletal troponin-I, but when expressed with respect to baseline levels, the increase after injurious exercise was not different between men and women.[247] Sex-related hormones such as estradiol may be responsible for reducing the inflammatory response after injurious exposure. Work done in female rodents and male rodents supplemented with estradiol showed a reduced amount of leukocyte invasion, less focal inflammation, and resultant myofiber necrosis.[252,271] Indeed, estradiol may have a protective effect that reduces the amount of membrane damage resulting from eccentric damage.[270] Several studies have also reported no difference in force loss after injury between males and females in rats[298] and humans[34,220,233] or in the time required for recovery. However, females did have a greater loss of range of motion after injury than males.[220] The results from animal studies indicate that females showed a blunted inflammatory response as compared to their male counterparts, but in humans this does not seem to be the case. Results from human studies tend to show a delayed recovery in females, particularly with the use of oral contraceptives.[230]

15.4.3.10 SSC-Induced Injury

A viable method to study eccentric and concentric muscle performance simultaneously in the context of muscle injury is via SSCs. SSCs (reciprocal eccentric/shortening contractions) have been studied in the context of human locomotion and athletic performance[14] and have been shown to produce muscle injury due to the eccentric component of the cycle.[97,125] Natural muscle function involves SSCs and this model provides a sound physiological foundation in which to study muscle mechanics and injury.[149] Both the change in concentric and eccentric muscle function before and after an injury protocol (that includes eccentric contractions), and the change in concentric and eccentric muscle function during the injury protocol can be investigated using SSCs. Recently, the relationship between changes in negative and positive work and the isometric force deficit after an injurious exposure has been investigated and the study showed a positive correlation between the change in negative work and the isometric force deficit resultant from injury.[62] Also, changes in real-time muscle function during SSCs (Figure 15.7) are also positively correlated with isometric force deficit and the degree of myofiber injury after an injurious exposure.[97] Specifically, the decay in peak eccentric forces and the decay in force enhancement during each stretch during an injury protocol were positively correlated with isometric force deficit and the degree of fiber injury. This study also demonstrated that changes in real-time eccentric force production during SSCs are indicative of resultant performance decrement (Figure 15.8) and myofiber damage days later (Figure 15.9). Exposure to equivalent isometric contractions did not result in myofiber injury (Figure 15.9). Increasing the number of repetitions of SSCs results in an increase in the degree of myofiber injury (Figure 15.10).

15.4.3.11 Chronic Injury

It is well understood that acute exposure to unaccustomed eccentric muscle actions can result in injury, but an initial exposure can be beneficial to reduce injury resulting from a subsequent exposure.[58,60] This protective effect from an initial exposure has been shown to reduce the effects of a subsequent exposure 1 week upto 6 months from the initial exposure (with no intervening exercise).[59,179] Thus, there is a

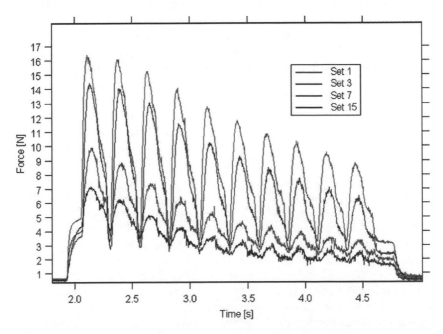

FIGURE 15.7 Fifteen sets of injurious SSCs performed on active dorsiflexor muscles at 1-min intervals. The curves are the force response (N) of the dorsi flexor muscles during sets 1, 3, 5, 7, and 15. The SSCs were conducted at a range of motion of 70–120–70° in a reciprocal fashion at 500 deg/sec.

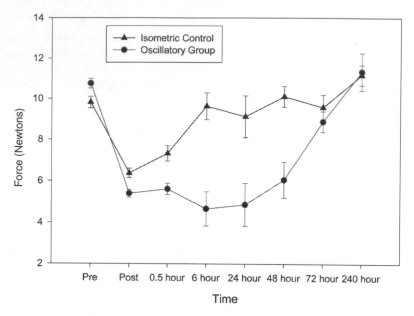

FIGURE 15.8   Time course of change in isometric force resultant from exposure to SSCs (oscillatory group) and isometric contractions (isometric control) over a 10-day period.

pronounced reduction in soreness, loss of strength, and release of enzymes, such as creatine kinase after the second exposure than the initial exposure.[17,58,60] Also, a second exposure of eccentric contractions does not appear to delay the recovery from the first exposure. Typically, recovery from eccentric contraction-induced injury takes at least 10 days for recovery of performance and remodeling of muscle fiber disruption.[96] If the second session is administered within 6 days after the first exposure, the recovery time was unaffected.[194] Thus, the normal recovery process is not affected by at least one intervening injurious exposure. Clearly there is an adaptation that takes place after one exposure to injurious eccentric contractions that ameliorates the injurious response to subsequent exposures. The mechanism by which this takes place is not clearly understood at this time. Exposure to a mild session of eccentric contractions still provides a protective effect to further sessions of more intense eccentric exposure.[40,60]

Some authors postulate that the adaptive mechanism could be at the level of the central nervous system where recruitment patterns could be adapting to recruit in a different fashion such that motor unit recruitment would be appropriately synchronized to reduce asymmetric stresses in the muscle fibers.[126,194] However, the repeated bout effect has also been found in electrically stimulated animal models, which would argue against the hypothesis that neural factors provide the adaptive effect.[179] Thus, the mechanism for adaptation must be in the muscle fibers themselves. There could be weak sarcomeres as a result of deconditioning that are more susceptible to stresses generated during eccentric contractions. The notion that stress-susceptible fibers exist was first postulated by Armstrong et al.[8] If fragile fibers were compromised after an initial session of eccentric muscle actions, the target muscle should lose muscle volume as a result of the loss of those fragile fibers. This hypothesis was supported by findings in human elbow flexors where there was a loss of muscle volume (approximately 10%) 14 days after an injurious exposure to eccentric muscle actions. In contrast, after a second session performed 8 weeks later, there was no loss of muscle volume. The authors concluded that the fragile fibers were lost after the first session and are repaired over time. As repaired fibers replace the fragile fibers, the muscle becomes more resistant to eccentric contraction-induced damage.[85] However, if muscle is not used, fragile fibers may again develop that can be injured at some juncture. There is also evidence that one session of injurious eccentric contractions can ameliorate the resultant damage from the

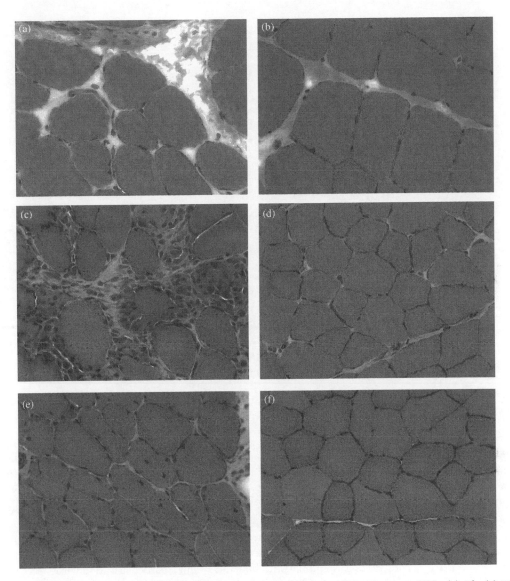

FIGURE 15.9 Cross-sections of TA muscle (H&E stain, 40× objective). After exposure to SSCs: (a) 6 h, (c) 72 h, and (e) 240 h. After isometric contractions: (b), 6 h, (d) 72 h, and (f) 240 h.

second session, even if the sessions are spaced 6 months apart.[198] This indicates that in normal subjects, the protective effect can last for some months, but the protective effect predictably dissipates with time.

Another explanation for this protective effect is the notion that muscles can add sarcomeres after an initial injurious exposure as a means to reduce the number that are at their extreme length, thus reducing injury.[58] The addition of sarcomeres after an injurious session of eccentric-biased exercise has been shown in both human[292] and animal[170] studies. Apparently, concentric muscle training reduces the number of sarcomeres and does not provide the level of protection in repeated sessions that eccentric training does.[292]

Few studies have investigated muscle response to the chronic administration of eccentric, concentric, and isometric muscle actions. The limited number of studies which have been conducted did not control the dynamics of movement, quantify the forces during the movement, or quantify the changes in performance longitudinally throughout the protocol. Investigation of *in vivo* muscle functional,

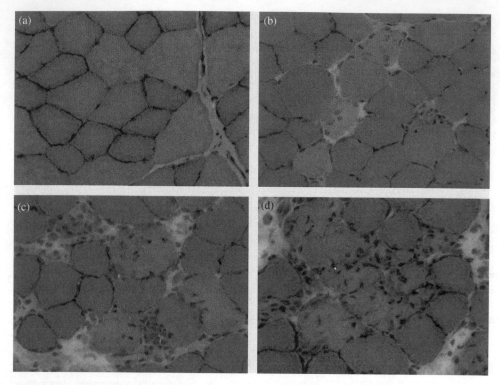

FIGURE 15.10 Histology of control muscle with no injury (a) and exposure to SSCs of 30 repetitions (b), 70 repetitions (c), and 150 repetitions (d).

histological, and biochemical responses to a chronic administration of muscle actions will further elucidate the factors, which predispose or mitigate contraction-induced injury. It will also further the understanding of the functional changes that result, characterize the inflammatory and repair kinetics longitudinally, and determine whether biochemical markers are representative of muscle performance or injury.

In summary, the study of different factors and their effect on muscle injury indicates that increasing force, task duration, number of repetitions, and range of motion exacerbate muscle injury. Older animals also have an increased susceptibility to injury. Training, particularly if involves eccentric muscle actions, can provide a protective effect from contraction-induced muscle injury. These findings should be of use to those attempting to ameliorate occupational muscle injury incidence.

15.5 Cellular Biology of Muscle Damage

Our understanding of the cellular and molecular mechanisms involved in mediating muscle damage and repair after strain, or low force, repetitive motion inducedinjuries is rapidly increasing.[20,58] Understanding how muscle tissue responds to these various stressors, and the time course of those responses, is necessary for defining approaches that can be taken to reduce the chances of obtaining an injury, or approaches to enhance the rate of recovery once an injury occurs. The cellular changes that occur as a result of acute muscle strain and repetitive motion are different. Strain injuries are associated with structural damage to myofibers, blood vessels and nerves, and cause edema and inflammation.[90,255] Repetitive motion damage can be associated with muscle ischemia, pain, and mitochondrial dysfunction.[137,243] Details regarding the cellular responses to repetitive use and strain injuries are discussed here.

15.5.1 Strain Injuries and Skeletal Muscle

Strain injuries are the result of three basic processes: (1) Initially, excessive forces coupled with lengthening result in structural damage to muscle cells (myofibers), including tearing of the cell membrane (i.e., the sarcolemma). (2) Structural damage to the sarcolemma results in an increase in intracellular Ca^{2+} levels, modifications of myofiber proteins and lipids, and the activation of intracellular pathways that regulate the injured muscle's response to damage. (3) Pro- and anti-inflammatory factors (i.e., cytokines and chemokines) are released by local tissues and infiltrating immune cells. Cytokines and chemokines cause inflammation and stimulate cellular pathways mediating muscle regeneration and repair. These three processes are involved in mediating the extent of physical damage, functional changes of the muscle, pain, and repair.

15.5.2 Physical Damage to Tissue

Strain injuries are caused by exposure of muscle and other soft tissues, including vasculature, nerves, and tendons to excessive strain or lengthening.[86,89,165] Studies performed in both humans,[88,90,91,187,190] and animals[10,161,257] have demonstrated that exposure to lengthening contractions results in physical damage to muscle tissue that can include shearing of myofibers, the loss of intermediate filaments and mitochondria, damage to the extracellular membrane, and disruptions in Z-line alignment (i.e., Z-line streaming). Immunostaining for structural proteins that maintain the integrity of the myofiber, such as desmin, titin, and fibronectin, have demonstrated that there are disruptions of the exo- and endosarcomeric membranes,[86,163,164] and of the extracellular matrix[163,257] in strain-injured muscle tissue. In lengthening contraction-induced injuries, damage within the muscle is most often seen at the myotendinous junction and at specific sarcomeres.[95,114,192,201] In fact, it has been hypothesized that there is a population of sarcomeres that are weaker, and tear more easily under lengthening conditions.[86,183,265] The stretch-induced damage to the extracellular matrix, sarcomeres, and critical cell organelles is associated with reduced isometric muscle force,[62,96,286,289] but not with pain. It has been hypothesized that these initial structural changes initiate a chain of events that may maintain an injury-induced force deficit, result in inflammation and pain, and stimulate pathways important for regulating muscle repair and regeneration.

15.5.3 Intracellular Ca^{2+} and Muscle Damage

Studies examining the effects of lengthening contractions on myofibers have demonstrated that there is an influx of Ca^{2+} into muscle cells[131,287,290] and mitochondria[66,267] in muscle exposed to damaging contractions. Calcium influx into myofibers can be increased in two ways. First, extreme stretch and strain can open the stretch-sensitive Ca^{2+} channels in the cell membrane. However, treating animals with a calcium channel blocker (i.e., verapamil) prior to injury only partially reduces the calcium influx into the cell,[9] and therefore, Ca^{2+} must be entering via an additional pathway. Injury-induced tears and damage to the sarcolemma can also allow Ca^{2+} to leak into the cell from the extracellular fluid.[290] In *in vitro* studies, membrane damage, and enzyme efflux can be prevented in muscle exposed to contractions by removing Ca^{2+} from the extracellular buffer.[135] Treating animals with Ca^{2+} chelators (molecules that preferentially bind Ca^{2+} and prevent it from entering the cell) during and after treadmill running also reduces tissue damage and the influx of Ca^{2+} in myofibers (66). Thus, this early influx of Ca^{2+} appears to play a critical role in mediating contraction-induced injuries to muscle tissue.

Increases in free Ca^{2+} within myofibers can result in the degradation of myofiber proteins and lipids, and result in the degeneration of damaged myofibers.[42] As mentioned earlier, intracellular Ca^{2+} homeostasis within myofibers is maintained by the SR. As intracellular Ca^{2+} levels rise, the SR normally increases the rate of Ca^{2+} uptake to keep free intracellular concentrations fairly stable. However, it has been demonstrated that the SR is less capable of sequestering additional Ca^{2+} after muscle has been exposed to lengthening contractions.[42] This reduction in Ca^{2+} up- take by the SR is associated with a reduction in muscle force[282] and may contribute to the rise in free intracellular Ca^{2+} after injury-inducing contractions.[42] Thus, the damage-induced loss of Ca^{2+} homeostasis in injured myofibers may mediate the acute reduction in isometric force, and it may initiate other cellular mechanisms that exacerbate myofiber damage.

Increases in intracellular Ca^{2+} can also alter myofiber structure and integrity by modifying proteins and lipids in damaged cells.[163,304] For example, calpain proteases, which are activated by increases in intracellular Ca^{2+}, are elevated in muscle after exercise.[21] Calpain can cleave signaling, cytoskeletal, and myofibrillar proteins.[228,229] This cleavage may act to target proteins for degradation by other proteases including ubiquitin.[228] *In vitro* studies have demonstrated that activated calpain degrades desmin and stimulates the release of α-actinin, thereby inducing Z-line streaming.[22] Calpain-mediated modifications of proteins can also stimulate the activity of a number of intracellular signaling proteins important for mediating cellular responses to damage. For example, the signaling molecule, protein kinase C (PKC), is activated by calpain cleavage.[228] This active form of PKC can affect myofiber function by modifying proteins that are already present in the cell or by acting upon transcriptional pathways to regulate gene expression.[232] Calpain may also act as a chemotactic signal to enhance neutrophil infiltration into the damaged tissue.[22]

A number of other intracellular signaling systems are also activated by increased intracellular Ca^{2+} levels in myofibers. For example, intracellular Ca^{2+} liberates phospholipase A2 (PLA2) from the extracellular membrane. PLA2 acts to increase arachidonic acid production and the synthesis of prostaglandins, which can stimulate inflammation and cause pain.[204,244,279,280]

Other signaling systems that may be indirectly activated in response to contraction-induced injury and increases in intracellular Ca^{2+} include the extracellular receptor kinase 1-2 (ERK1-2), p38 mitogen activating protein kinase,[51] and c-JUN NH_2-terminal kinase pathways.[47,117,200] These pathways are stimulated in response to growth factors, cellular stress, and injury. Activation of these pathways regulates transcriptional and translational activity in many cell types including muscles. Although the increase in intracellular Ca^{2+} and associated proteolysis and lipolysis have been traditionally thought to exacerbate muscle injury, Ca^{2+}-induced activation of cell signaling pathways and proteolysis might also activate cell systems necessary for initiating cellular repair and regeneration.

15.5.4 Muscle Inflammation

The physical disruptions of muscle fibers along with increases in intracellular Ca^{2+} are the initial effects of muscle injury. However, force deficits, muscle swelling, and soreness occur 1 to 7 days after the initial injury, and are associated with muscle inflammation.[10,89,97,163] Neutrophils, the first immune cells to enter damaged tissue, actually infiltrate muscle within 2 h of the initial injury.[79,269] We are only beginning to understand the complex roles that various immune cell types play in the damage and repair processes. However, it has been demonstrated that neutrophils phagocytize degenerating fibers and debris produced by injury. In addition, neutrophils can participate in the production and release of free radicals from damaged tissue, which can exacerbate the damage.[25] Neutrophils can also produce proteases and a number of cytokines, including tumor necrosis factor-α (TNF-α;[67]) and IL1-β.[272] These cytokines might increase muscle catabolism and degradation and act to attract monocytes to the site of the injury.[2,113,182]

Monocytes/macrophages, the other inflammatory cells commonly seen in injured muscle, can be found between 12 h and 14 days after the initial muscle injury.[224,251] In rats, macrophages expressing specific cell-surface molecules, including ED1 and ED2, have been identified in damaged muscle tissues.[156,157,251] ED1 expressing macrophages infiltrate damaged and necrotic tissues and remove debris. These macrophages also express pro-inflammatory cytokines including TNF-α.[64,310] Besides increasing muscle catabolism and promoting protease activity, TNF-α also activates the transcription factor, nuclear factor κB (NF-κB), to stimulate transcription of genes encoding for proteins that are part of the ubiquitin proteolytic pathway.[112,180] TNF-α may also stimulate the transcription of other pro-inflammatory cytokines including IL-1β and IL-6 and the chemokine monocyte chemoattractant protein (MCP-1[15,307]). The increased production of cytokines and chemokines by inflammatory cells in damaged tissues enhance local pathways mediating tissue inflammation and may act to exacerbate damage during the first 5 days after muscle injury.

15.5.5 Muscle Regeneration and Repair

The infiltration of immune cells along with the release of pro-inflammatory cytokines appears to enhance muscle damage. During the acute phase of the injury, both pain and force deficits appear to be reduced by treating animals with nonsteroidal anti-inflammatory drugs (NSAIDs). For example, rabbits, which were exposed to a session of repeated eccentric contractions and treated with the NSAID flurbiprofen, showed improved functional recovery during the first week after injury as compared to controls.[181] However, 4 weeks after the injury, the NSAID-treated animals demonstrated reduced force generation. The authors hypothesized that treatment with NSAIDs may have interfered with or delayed the recovery process in these animals.[181] In humans with muscle damage caused by downhill walking, treatment with over the counter doses of acetaminophen or ibuprophen results in a decrease in pain, but these anti-inflammatory agents also decrease protein synthesis, which may be needed for muscle repair.[209,273,274] These findings suggest that inhibition of the inflammatory response interferes with normal recovery of muscle after a strain-induced injury.

Inflammatory cells, particularly macrophages, may stimulate myofiber regeneration through a number of different mechanisms. ED2 expressing macrophages, also referred to as resident macrophages, are thought to play a role in muscle repair and myofiber regeneration.[43,156,157,251] These macrophages may stimulate growth and repair by releasing a number of factors that could stimulate the division, migration, and differentiation of muscle precursor cells. These factors include, fibroblast growth factor 2 (FGF 2[77,236]), insulin like growth factor 1 (IGF-1[1,119,216]) and hepatocyte growth factor (HGF[236,312]). In addition, ED2-expressing macrophages may also release the anti-inflammatory cytokines IL-6[44,93] and IL-15.[45,214,215] Thus, although inflammation causes pain and appears to exacerbate myofiber damage, the inflammatory process also appears to be necessary for complete repair of tissues and recovery of muscle function.[213]

15.5.6 Muscle Injuries Associated with Low-Force Repetitive Tasks

Low-force repetitive tasks involve movements that require little force generation by a muscle. Instead, the muscle action may need to be maintained over long periods of time, or repeated over and over again during a work cycle.[241,243] Injuries caused by these types of activities are not usually associated with inflammation or large areas of myofiber degeneration, but instead are characterized by muscle pain and/or rapid fatiguability of the muscle.[133,138,159,239,240] Low-force repetitive task injuries are prevalent in people working with computers. With the increase in computer use, both at home and in the workplace, understanding the mechanisms underlying this type of damage is crucial.

Injury due to overuse has been studied in people with trapezius myalgia. This disorder is often seen in workers whose job requires them to maintain stable upper body postures for extended periods of time, such as computer and clerical work.[133,138] Trapezius mylagia is associated with the appearance of ragged type I muscle fibers and with a decrease in muscle blood flow to the injured region. This ragged appearance of myofibers is an indicator of mitochondrial dysfunction in the cell.[137,158,159] The dysfunction and pain associated with trapezius myalgia and with other overuse injuries have been linked to changes in Ca^{2+} regulation in the damaged area, changes in the pH of the intra- and extracellular fluids, and changes in the local concentrations of specific ions involved in mediating muscle activity.[134,203,239] These biochemical alterations may have profound effects on myofiber metabolism and on the activity of sensory pathways carrying pain information.[243] Although recent studies have identified some of the factors that play a role in inducing injuries that are due to overuse, more research needs to be done to determine how these various physiological changes act together to result in pain and injury.

15.5.7 Long-Term Changes in Muscle in Response to Injuries

Most muscle injuries are acute, and recovery is complete within a couple weeks of the initial injury. However, long-term exposure to repetitive motion, load bearing or awkward postures in the workplace has been associated with an increased risk of developing more chronic musculoskeletal disorders.[24] These more chronic disorders include tendonitis, fibromyalgia, myositis, osteoarthritis, and synovitis.[19,41,111,211] Recent studies have focused on the biological mechanisms by which overuse, or acute strain injuries, may act to chronically affect muscle function and pain.

Understanding the biological underpinnings responsible for generating chronic disorders has been difficult because a number of factors, including age, general health, genetic predisposition, stress level, and length of exposure to a strenuous job, influence the development of these conditions.[24,111] Thus, what we know about the mechanisms responsible for generating chronic musculoskeletal disorders comes from studying people with chronic disorders (such as workers with trapezius myalgia), and by examining the effects of a few limited animal models of chronic damage or genetic models of chronic inflammation and muscular atrophy. Because the mechanisms involved in mediating trapezius mylagia were discussed previously, this discussion will focus on animal models used to study the etiology of chronic disorders.

Most animal models of chronic injury have used repeated injections of a toxin[52,225] or muscle overload of a single limb caused by paralyzing or binding the opposite limb, or by removing some muscle, and thereby forcing the remaining muscle to assume the additional load.[257] These models result in myofiber degeneration and the infiltration of macrophages.[255] In addition, there is an expansion of the extracellular matrix, along with an increase in collagen expression and fibrosis in muscle[92,256] and tendon[46] tissues, which leads to scarring. Muscle fibrosis results in decreased muscle strength and flexibility and is often associated with pain.[25,184,256,300] Thus, determining which mechanisms are responsible for increasing fibrosis, and interfering with these mechanisms may reduce the effects of chronic overload on muscle tissues.

Models of muscular dystrophy, such as dystrophin-deficient *mdx* mice, have also been used to try to determine how chronic damage may affect muscle tissue and function. *Mdx* mice, which are a model for Duchene's muscular dystrophy, are lacking the dystrophin protein. Dystrophin, along with dystroglycan, links the extracellular matrix to the cytoskeleton of the myofiber.[121,122] The absence of dystrophin compromises the strength of the sarcolemma, making it more susceptible to breakage.[217,218] Once the sarcolemma is damaged, Ca^{2+} can enter the cell.[99,123,124] This rise in intracellular Ca^{2+} activates the same proteases and lipases that are activated after muscle injury. Because there is persistent myofiber damage and necrosis occurring in *mdx* mice and people with Duchene's muscular dystrophy, there is also persistent inflammation.[108,249,250] In fact, it has been hypothesized that this inflammation exacerbates the muscular atrophy and loss of function. This is supported by the fact that treating Duchenne's patients with immunosuppressive drugs improves muscle strength.[139] Studies of *mdx* mice also demonstrate that these animals display a chronic inflammatory response, with the gene expression of many pro-inflammatory factors being chronically increased as compared to control animals.[212] The increased expression of these inflammatory factors is associated with fibrosis, muscle atrophy, and eventually necrosis.

The studies performed using these various animal models indicate that muscle overuse, or long-term inflammation, may result in chronic muscle damage. However, it is still unclear which occupational exposures, if any, result in long-term muscle inflammation. Future studies examining the cellular effects of repetitive loading on muscle tissue will help determine which mechanisms may underlie the development of chronic musculoskeletal disorders.

15.5.8 Biomarkers of Injury

One of the goals of occupational research is to find biomarkers that can be used to determine if a person is at risk for developing a muscle injury or for diagnostic purposes. The approaches that have been used to assess contraction-induced damage have been reviewed previously.[289] In humans, these approaches include examining force production, histology, blood levels of muscle-associated proteins, and reports of pain.

Force production has commonly been used to assess muscle function because reductions in force are often correlated with injury. Changes in force are immediate, and maintained for a number of days following the exposure that caused the injury.[10,61,87,96,291] Although muscle damage is often associated with a force deficit, force measurements alone should not be used to diagnose injury because muscle fatigue can also result in reductions in force.[80–82]

Overuse disorders can also be associated with losses in force, however a better marker for these disorders is muscle fatigue.[30,133] Although injured muscle does demonstrate signs of fatigue,[80,299] fatigue is also seen prior to the generation of injury, and it is believed that fatigue and the cellular changes associated with this physiological state, may be in part responsible for generating overuse disorders.[243] Thus, depending on when measurements are taken, fatigue can be used as a biomarker for predicting the development of a disorder, or for diagnosing a disorder.

Increases in the circulating concentrations of certain muscle-associated proteins, including creatine kinase, lactate dehydrogenase, and myoglobin are often associated with strain-induced muscle injuries.[54,195,223,227,248] Increases in these circulating proteins do not correlate well with injury-induced force deficits during the first 24 h of injury,[56,60,186] and thus, the concentrations of these proteins do not serve as good markers for the early diagnosis of muscle injury. In addition, certain proteins, such as creatine kinase, fluctuate in response to muscle activity, and not solely in response to muscle injury.[55,187,289]

Histological examination of muscle biopsies can be used to diagnose muscle injury. As mentioned previously, infiltration of immune cells, disruptions of the sarcolemma, and necrosis can be seen in muscles with strain-induced injuries.[90,163,255] However, these changes are not usually apparent until 24–48 h after the injury,[90] and thus, they cannot be used to identify injuries early during the process. With overuse disorders, ragged red fibers are found by histological examination of biopsy tissue.[159] These ragged red fibers appear to be strongly immunopositive for cytochrome C oxidase (an enzyme involved in mitochondrial function), or immunonegative for this marker.[158] Thus, the presence of ragged red fibers, and their cytochrome C oxidase phenotype, can be used as a marker of muscle disorders.

Pain or muscle soreness is one of the most common symptoms used to determine if there is muscle damage. However, with strain-induced injuries, muscle soreness is not apparent until 24–48 h after the injury occurred,[55,90] and it does not correlate well with injury-induced force deficits.[50,69,136] Therefore, pain is not a good marker to use for early diagnoses of strain injuries. Pain is also common in overuse disorders and is associated with muscle fatigue,[159,240] the presence of ragged red fibers,[158,159] and reductions in blood flow.[158] Pain has also been correlated with biochemical changes associated with fatigue.[30,240,242] Because fatigue often precedes actual damage, pain and muscle fatigue may serve as biomarkers for predicting muscle damage.

Based on our current knowledge the diagnoses of muscle strains can most quickly be done by checking for force deficits. Reports of muscle soreness, or the presence of myofiber proteins in the blood can be used to confirm that the force deficit is due to an injury. Depending on the timing of events, reports of pain and measurements of muscle fatigue can also be used to indicate that injury may occur, or to diagnose an injury. Although most biomarkers are currently used for diagnosis, research focusing on pain, fatigue, and their association with biomarkers may provide a means for determining when an individual is at risk for developing a muscle strain or disorder.

15.6 Recommendations for Future Work

15.6.1 The Need for More Refined *In Vivo* Models

The most refined *in vivo* models to date have used dynamometry and electrical stimulation to study the target muscle of interest. However, these models have generally been used to study acute muscle injury.

In vivo dynamometry is well suited for the study of chronic muscle injury and adaptation and studies of this kind would be beneficial in elucidating the physiological response to repetitive biomechanical loading. Biomechanical loading parameters, such as number of repetitions per day, the velocity and acceleration of the movement, the range of motion of the movement, muscle force or torque during each movement, and the rest interval between sets, should be controlled during experiments. Controlling and quantifying the biomechanical loading profile is essential for rigorous study of muscle injury and adaptation for both acute and chronic exposure studies. Also, it is important to have the ability to vary the biomechanical inputs to study the effects of different inputs, and how the level of those inputs, such as more or less repetitions, higher or lower velocities or acceleration, and so on, affects the physiological response after a single exposure and/or multiple exposures.

In addition to controlling the biomechanical inputs, it is also important in volitional models to not only provide an apparatus to facilitate controlled movement, but to appropriately instrument the apparatus to monitor movement dynamics in real time. This will allow for quantitation of the biomechanical loading profile within each exposure session for the study of both acute and chronic muscle response. This approach is similar to *in vivo* dynamometry in that the biomechanical loading signature is recorded during each exposure, but different from *in vivo* dynamometry in that the movement profile is not controlled by an external source like a servomotor, but controlled by the animal.

In vivo models, whether volitional or nonvolitional, can provide a wealth of information about the effects of biomechanical loading inputs on acute and chronic physiological responses. Refinement of these models to control and monitor the biomechanical loading signature has been done in nonvolitional models and is currently being accomplished in volitional models.

15.6.2   The Need for Tissue Mechanobiology Studies

There is a clear need for tissue mechanobiology studies to determine the failure or injury mechanics of soft tissues and ultimately the repair kinetics after acute or chronic injury. From a muscle perspective, dose–response models would be important to develop in both an acute and chronic framework. In the acute framework, the effect of number of repetitions, rest between sets of exposures, range of motion, movement kinematics such as velocity, acceleration, and jerk, and force on the amount of soft-tissue injury should be fully investigated. The changes in soft tissue at the cellular level commensurate with injury should also be investigated. Defining damage at the tissue level is important to develop a consensus about what is soft-tissue injury. The time course of these changes after an acute exposure and the mechanisms of repair should also be fully characterized. Factors such as level of conditioning, age, and gender, and other relevant comorbid factors on injury susceptibility and rate of repair are also important considerations.

In the chronic framework, the threshold for injury under sustained or repeated loading should be determined. This should be based on the foundation developed by acute injury studies regarding the effect of different biomechanical parameters and any intrinsic interplay. Additional parameters such as exposure duration (number of days, weeks, or months) and the duty cycle (rest period between exposures) should also be considered. Using this framework, the time frame for injury to develop, the sustainment of injury, and any repair that can take place during repeated exposures should be studied. Functional measurements as well as noninvasive biological measurements should be made to monitor the status of the animal. Patterns of rest and reuse after injury are also important to consider.

15.6.3   Summary

There has been much work in the area of exercise-induced muscle injury over the past 30 years. The relation between factors such as force, range of motion, number of repetitions, exposure duration, velocity, age, gender, and training on muscle injury have been studied and have a clear occupational relevance. Also, the cellular mechanisms responsible for the injury and repair processes have been well studied to date. The relation between muscle function, myofiber damage, pain, and molecular indicators of injury have also been studied. Clearly, more work needs to be done in this area, particularly regarding the physiological response to long-term repetitive loading. The findings to date indicate that eccentric muscle actions result in muscle damage and recovery from this injury can require upto 1 month. Increasing the biomechanical exposure such as force, number of repetitions, and range of motion can exacerbate the magnitude of injury response. Increased age can also increase injury susceptibility. The encouraging news is that training can reduce the injurious response and adaptation can take place, particularly if the appropriate rest intervals are included. There are biomarkers currently being studied that may indicate the evidence of myofiber injury that have the appropriate level of sensitivity and specificity needed for occupational monitoring. The area of soft-tissue pathomechanics can provide a wealth of information that will be of value to ergonomists and occupational health professionals in the quest to reduce the incidence of occupational musculoskeletal disorders.

Acknowledgments

The author would like to thank Kris Krajnak for her contributions in cellular biology, Oliver Wirth for his contributions in volitional models, and Steve Alway, Robert Mercer, and Aaron Schopper for their constructive comments.

References

1. Adams, G.R., Autocrine and/or paracrine insulin-like growth factor-I activity in skeletal muscle, *Clin. Orthop.*, (403 Suppl), S188–S196, 2002.
2. Alvarez, B., Quinn, L.S., Busquets, S., Lopez-Soriano, F.J., and Argiles, J.M., Direct effects of tumor necrosis factor alpha (TNF-alpha) on murine skeletal muscle cell lines. Bimodal effects on protein metabolism, *Eur. Cytokine Netw.*, 12(3), 399–410, 2001.
3. Alway, S.E., Attenuation of Ca(2+)-activated ATPase and shortening velocity in hypertrophied fast twitch skeletal muscle from aged Japanese quail, *Exp. Gerontol.*, 37(5), 665–678, 2002.
4. Amelink, G.J., Kamp, H.H., and Bar, P.R., Creatine kinase isoenzyme profiles after exercise in the rat: sex-linked differences in leakage of CK-MM, *Pflugers Arch.*, 412(4), 417–421, 1988.
5. Amelink, G.J., Koot, R.W., Erich, W.B., Van Gijn, J., and Bar, P.R., Sex-linked variation in creatine kinase release, and its dependence on oestradiol, can be demonstrated in an in-vitro rat skeletal muscle preparation, *Acta Physiol. Scand.*, 138(2), 115–124, 1990.
6. Armstrong, R.B., Mechanisms of exercise-induced delayed onset muscular soreness: a brief review, *Med. Sci. Sports Exerc.*, 16(6), 529–538, 1984.
7. Armstrong, R.B., Muscle damage and endurance events, *Sports Med.*, 3(5), 370–381, 1986.
8. Armstrong, R.B., Initial events in exercise-induced muscular injury, *Med. Sci. Sports Exerc.*, 22(4), 429–435, 1990.
9. Armstrong, R.B., Duan, C., Delp, M.D., Hayes, D.A., Glenn, G.M., and Allen, G.D., Elevations in rat soleus muscle [Ca2+] with passive stretch, *J. Appl. Physiol.*, 74(6), 2990–2997, 1993.
10. Armstrong, R.B., Ogilvie, R.W., and Schwane, J.A., Eccentric exercise-induced injury to rat skeletal muscle, *J. Appl. Physiol.*, 54(1), 80–93, 1983.
11. Armstrong, R.B., Warren, G.L., and Warren, J.A., Mechanisms of exercise-induced muscle fibre injury, *Sports Med.*, 12(3), 184–207, 1991.
12. Ashton-Miller, J.A., He, Y., Kadhiresan, V.A., McCubbrey, D.A., and Faulkner, J.A., An apparatus to measure in vivo biomechanical behavior of dorsi- and plantarflexors of mouse ankle, *J. Appl. Physiol.*, 72(3), 1205–1211, 1992.
13. Asmussen E. Positive and Negative Muscular Work. *ACTA Physiol. Scand.*, 28, 365–382, 1953.
14. Avela, J. and Komi, P.V., Reduced stretch reflex sensitivity and muscle stiffness after long-lasting stretch-shortening cycle exercise in humans, *Eur. J. Appl. Physiol. Occup. Physiol.*, 78(5), 403–410, 1998.

15. Baggiolini, M., Dewald, B., Moser B., Chemokines., In: Gallin, J.I., Snyderman, R., editors. *Inflammation: Basic Principles and Clinical Correlates*. 3rd ed., Lippincott Williams & Wilkins, Philadelphia, 1999, pp. 419–431.

16. Ballor, D.L., Tommerup, L.J., Smith, D.B., and Thomas, D.P., Body composition, muscle and fat pad changes following two levels of dietary restriction and/or exercise training in male rats, *Int. J. Obes.*, 14(8), 711–722, 1990.

17. Balnave, C.D. and Thompson, M.W., Effect of training on eccentric exercise-induced muscle damage, *J. Appl. Physiol.*, 75(4), 1545–1551, 1993.

18. Bar, P.R., Amelink, G.J., Oldenburg, B., and Blankenstein, M.A., Prevention of exercise-induced muscle membrane damage by oestradiol, *Life Sci.*, 42(26), 2677–2681, 1988.

19. Barbe, M.F., Barr, A.E., Gorzelany, I., Amin, M., Gaughan, J.P., and Safadi, F.F., Chronic repetitive reaching and grasping results in decreased motor performance and widespread tissue responses in a rat model of MSD, *J. of Orthopaed. Res.*, 21(1), 167–176, 2003.

20. Barr, A.E. and Barbe, M.F., Pathophysiological tissue changes associated with repetitive movement: a review of the evidence, *Phys. Ther.*, 82(2), 173–187, 2002.

21. Belcastro, A.N., Skeletal muscle calcium-activated neutral protease (calpain) with exercise, *J. Appl. Physiol.*, 74(3), 1381–1386, 1993.

22. Belcastro, A.N., Shewchuk, L.D., and Raj, D.A., Exercise-induced muscle injury: a calpain hypothesis, *Mol. Cell. Biochem.*, 179(1–2), 135–145, 1998.

23. Benz, R.J., Friden, J., and Lieber, R.L., Simultaneous stiffness and force measurements reveal subtle injury to rabbit soleus muscles, *Mol. Cell. Biochem.*, 179(1–2), 147–158, 1998.

24. Bernard, B.P., Putz-Anderson, V., Burt, S.E., Cole, L.L., Fairfield-Estill, C., Fine, L.J. et al., Musculoskeletal disorders and workplace factors. A critical review of epidemiologic evidence for work-related musculoskeletal disorders of the neck, upper extremity, and low back, Cincinnati: NIOSH; 1997. Report No.: DHHS (NIOSH) Pub no. 97–141.

25. Best, T.M. and Hunter, K.D., Muscle injury and repair, *Phys. Med. Rehabil. Clin. N. Am.*, 11(2), 251–266, 2000.

26. Best, T.M., McCabe, R.P., Corr, D., Vanderby, R., Jr., Evaluation of a new method to create a standardized muscle stretch injury, *Med. Sci. Sports Exerc.*, 30(2), 200–205, 1998.

27. Best, T.M., Shehadeh, S.E., Leverson, G., Michel, J.T., Corr, D.T., Aeschlimann, D., Analysis of changes in mRNA levels of myoblast- and fibroblast-derived gene products in healing skeletal muscle using quantitative reverse transcription-polymerase chain reaction, *J. Orthop. Res.*, 19(4), 565–572, 2001.

28. Bigland-Ritchie, B., Graichen, H., and Woods, J.J., A variable-speed motorized bicycle ergometer for positive and negative work exercise, *J. Appl. Physiol.*, 35(5), 739–740, 1973.

29. Bigland-Ritchie, B. and Woods, J., Proceedings: Oxygen consumption and integrated electrical activity of muscle during positive and negative work, *J. Physiol.*, 234(2), 39P–40P, 1973.

30. Blangsted, A.K., Sogaard, K., Christensen, H., and Sjogaard G., The effect of physical and psychosocial loads on the trapezius muscle activity during computer keying tasks and rest periods, *Eur. J. Appl. Physiol.*, 2003.

31. Bobbert, M.F., Huijing, P.A., and van Ingen Schenau, G.J., A model of the human triceps surae muscle-tendon complex applied to jumping, *J. Biomech.*, 19(11), 887–898, 1986.

32. Bonde-Petersen, F., Henriksson, J., and Knuttgen, H.G., Effect of training with eccentric muscle contractions on skeletal muscle metabolites, *Acta. Physiol. Scand.*, 88(4), 564–570, 1973.

33. Bonde-Petersen, F., Knuttgen, H.G., and Henriksson, J., Muscle metabolism during exercise with concentric and eccentric contractions, *J. Appl. Physiol.*, 33(6), 792–795, 1972.

34. Borsa, P.A. and Sauers, E.L., The importance of gender on myokinetic deficits before and after microinjury, *Med. Sci. Sports Exerc.*, 32(5), 891–896, 2000.

35. Brooks, S.V. and Faulkner, J.A., Contraction-induced injury: recovery of skeletal muscles in young and old mice, *Am. J. Physiol.*, 258(3 Pt 1), C436–C442, 1990.

36. Brooks, S.V. and Faulkner, J.A., Isometric, shortening, and lengthening contractions of muscle fiber segments from adult and old mice, *Am. J. Physiol.*, 267(2 Pt 1), C507–C513, 1994.

37. Brooks, S.V. and Faulkner, J.A., The magnitude of the initial injury induced by stretches of maximally activated muscle fibres of mice and rats increases in old age, *J. Physiol.*, 497(Pt 2), 573–580, 1996.

38. Brooks, S.V. and Faulkner, J.A., Severity of contraction-induced injury is affected by velocity only during stretches of large strain, *J. Appl. Physiol.*, 91(2), 661–666, 2001.

39. Brooks, S.V., Zerba, E., and Faulkner, J.A., Injury to muscle fibres after single stretches of passive and maximally stimulated muscles in mice, *J. Physiol.*, 488(Pt 2), 459–469, 1995.

40. Brown, S.J., Child, R.B., Day, S.H., and Donnelly, A.E., Exercise-induced skeletal muscle damage and adaptation following repeated bouts of eccentric muscle contractions, *J. Sports Sci.*, 15(2), 215–222, 1997.

41. Byl, N., Wilson, F., Merzenich, M., Melnick, M., Scott, P., Oakes, A. et al., Sensory dysfunction associated with repetitive strain injuries of tendinitis and focal hand dystonia: a comparative study, *J. Orthop. Sports Phys. Ther.*, 23(4), 234–244, 1996.

42. Byrd, S.K., Alterations in the sarcoplasmic reticulum: a possible link to exercise-induced muscle damage, *Med. Sci. Sports Exerc.*, 24(5), 531–536, 1992.

43. Cantini, M., Massimino, M.L., Bruson, A., Catani, C., Dalla Libera, L., and Carraro U., Macrophages regulate proliferation and differentiation of satellite cells, *Biochem. Biophys. Res. Commun.*, 202(3), 1688–1696, 1994.

44. Cantini, M., Massimino, M.L., Rapizzi, E., Rossini, K., Catani, C., Dalla Libera, L. et al., Human satellite cell proliferation in vitro is regulated by autocrine secretion of IL-6 stimulated by a soluble factor(s) released by activated monocytes, *Biochem. Biophys. Res. Commun.*, 216(1), 49–53, 1995.

45. Carbo, N., Lopez-Soriano, J., Costelli, P., Busquets, S., Alvarez, B., Baccino, F.M. et al., Interleukin-15 antagonizes muscle protein waste in tumour-bearing rats, *Br. J. Cancer*, 83(4), 526–531, 2000.

46. Carpenter, J.E., Flanagan, C.L., Thomopoulos, S., Yian, E.H., and Soslowsky, L.J., The effects of overuse combined with intrinsic or extrinsic alterations in an animal model of rotator cuff tendinosis, *Am. J. Sports Med.*, 26(6), 801–807, 1998.

47. Carrasco, M.A., Riveros, N., Rios, J., Muller, M., Torres, F., Pineda, J. et al., Depolarization-induced slow calcium transients activate early genes in skeletal muscle cells, *Am. J. Physiol. Cell Physiol.*, 284(6), C1438–C1447, 2003.

48. Cavanagh, P., On "muscle action" vs "muscle contraction", *J. Biomech.*, 21, 69, 1988.

49. Chen, M.J., Shih, C.L., and Wang, K., Nebulin as an actin zipper. A two-module nebulin fragment promotes actin nucleation and stabilizes actin filaments, *J. Biol. Chem.*, 268(27), 20327–20334, 1993.

50. Chen, T.C. and Hsieh, S.S., Effects of a 7-day eccentric training period on muscle damage and inflammation, *Med. Sci. Sports Exerc.*, 33(10), 1732–1738, 2001.

51. Childs, T.E., Spangenburg, E.E., Vyas, D.R., and Booth, F.W., Temporal alterations in protein signaling cascades during recovery from muscle atrophy, *Am. J. Physiol. Cell Physiol.*, 285(2), C391–C398, 2003.

52. Chou, S.M. and Mizuno, Y., Induction of spheroid cytoplasmic bodies in a rat muscle by local tetanus, *Muscle Nerve*, 9(5), 455–464, 1986.

53. Clarkson, P.M., Eccentric exercise and muscle damage. *International Journal of Sports Medicine*, 18(Suppl 4), S314–S317, 1997.

54. Clarkson, P.M., Apple, F.S., Byrnes, W.C., McCormick, K.M., and Triffletti, P., Creatine kinase isoforms following isometric exercise, *Muscle Nerve*, 10(1), 41–44, 1987.

55. Clarkson, P.M., Byrnes, W.C., McCormick, K.M., Turcotte, L.P., and White, J.S., Muscle soreness and serum creatine kinase activity following isometric, eccentric, and concentric exercise, *Int. J. Sports Med.*, 7(3), 152–155, 1986.

56. Clarkson, P.M. and Ebbeling, C., Investigation of serum creatine kinase variability after muscle-damaging exercise, *Clin. Sci. (Lond.)*, 75(3), 257–261, 1988.

57. Clarkson, P.M. and Hubal, M.J., Are women less susceptible to exercise-induced muscle damage?, *Curr. Opin. Clin. Nutr. Metab. Care*, 4(6), 527–531, 2001.

58. Clarkson, P.M. and Hubal, M.J., Exercise-induced muscle damage in humans. *Am. J. Phys. Med. Rehabil.*, 81(11 Suppl), S52–S69, 2002.

59. Clarkson, P.M., Nosaka, K., and Braun, B., Muscle function after exercise-induced muscle damage and rapid adaptation, *Med. Sci. Sports Exerc.*, 24(5), 512–520, 1992.

60. Clarkson, P.M. and Tremblay, I., Exercise-induced muscle damage, repair, and adaptation in humans, *J. Appl. Physiol.*, 65(1), 1–6, 1988.

61. Cutlip, R.G. The Dynamic Parameters that Affect Eccentric-Contraction Induced Injury in an In Vivo Rodent Model [Doctoral Dissertation]: West Virginia University; 1995.

62. Cutlip, R.G., Geronilla, K.B., Baker, B.A., Kashon, M.L., Miller, G.R., and Schopper, A.W., Impact of muscle length during stretch-shortening contractions on real-time and temporal muscle performance measures in rats in vivo, *J. Appl. Physiol.*, 96(2), 507–516, 2004.

63. Cutlip, R.G., Stauber, W.T., Willison, R.H., McIntosh, T.A., and Means, K.H., Dynamometer for rat plantar flexor muscles in vivo, *Med. Biol. Eng. Comput.*, 35(5), 540–543, 1997.

64. De Bleecker, J.L., Meire, V.I., Declercq, W., and Van Aken, E.H., Immunolocalization of tumor necrosis factor-alpha and its receptors in inflammatory myopathies, *Neuromuscul. Disord.*, 9(4), 239–246, 1999.

65. Devor, S.T. and Faulkner, J.A., Regeneration of new fibers in muscles of old rats reduces contraction-induced injury, *J. Appl. Physiol.*, 87(2), 750–756, 1999.

66. Duan, C., Delp, M.D., Hayes, D.A., Delp, P.D., and Armstrong, R.B., Rat skeletal muscle mitochondrial [Ca2+] and injury from downhill walking, *J. Appl. Physiol.*, 68(3), 1241–1251, 1990.

67. Dubravec, D.B., Spriggs, D.R., Mannick, J.A., and Rodrick, M.L., Circulating human peripheral blood granulocytes synthesize and secrete tumor necrosis factor alpha, *Proc. Natl. Acad. Sci. U.S.A.*, 87(17), 6758–6761, 1990.

68. Edman, K.A., The velocity of unloaded shortening and its relation to sarcomere length and isometric force in vertebrate muscle fibres, *J. Physiol.*, 291, 143–159, 1979.

69. Eston, R.G., Finney, S., Baker, S., and Baltzopoulos, V., Muscle tenderness and peak torque changes after downhill running following a prior bout of isokinetic eccentric exercise, *J. Sports Sci.*, 14(4), 291–299, 1996.

70. Eston, R.G., Lemmey, A.B., McHugh, P., Byrne, C., and Walsh, S.E., Effect of stride length on symptoms of exercise-induced muscle damage during a repeated bout of downhill running, *Scand. J. Med. Sci. Sports*, 10(4), 199–204, 2000.

71. Farrell, P.A., Fedele, M.J., Hernandez, J., Fluckey, J.D., Miller, J.L., 3rd, Lang, C.H. et al., Hypertrophy of skeletal muscle in diabetic rats in response to chronic resistance exercise, *J. Appl. Physiol.*, 87(3), 1075–1082, 1999.

72. Faulkner, J.A., Brooks, S.V., and Zerba, E., Muscle atrophy and weakness with aging: contraction-induced injury as an underlying mechanism, *J. Gerontol. A. Biol. Sci. Med. Sci.*, 50 Spec No:124–129, 1995.

73. Faulkner, J.A., Jones, D.A., and Round, J.M., An Apparatus for Unilateral Exercise of the Lower Limb Muscles of Small Mammals, *J. Physiol. Lond.*, 365, 11P, 1985.

74. Faulkner, J.A., Jones, D.A., and Round, J.M., Injury to Skeletal Muscle of Mice by Lengthening Contractions. *J. of Physiol., Lond.*, 365, 75P, 1985.

75. Faulkner, J.A., Jones, D.A., and Round, J.M., Injury to skeletal muscles of mice by forced lengthening during contractions, *Q. J. Exp. Physiol.*, 74(5), 661–670, 1989.

76. Faulkner, J.A., Opiteck, J.A., and Brooks, S.V., Injury to skeletal muscle during altitude training: induction and prevention, *Int. J. Sports Med.*, 13 Suppl 1, S160–S162, 1992.

77. Fibbi, G., D'Alessio, S., Pucci, M., Cerletti, M., and Del Rosso, M., Growth factor-dependent proliferation and invasion of muscle satellite cells require the cell-associated fibrinolytic system, *Biol. Chem.*, 383(1), 127–136, 2002.

78. Fielding, R.A., Evans, W.J., Hughes, V.A., Moldawer, L.L., and Bistrian, B.R., The effects of high intensity exercise on muscle and plasma levels of alpha-ketoisocaproic acid, *Eur. J. Appl. Physiol. Occup. Physiol.*, 55(5), 482–485, 1986.

79. Fielding, R.A., Manfredi, T.J., Ding, W., Fiatarone, M.A., Evans, W.J., and Cannon, J.G., Acute phase response in exercise. III. Neutrophil and IL-1 beta accumulation in skeletal muscle, *Am. J. Physiol.*, 265(1 Pt 2), R166–R172, 1993.

80. Fitts, R.H., Cellular mechanisms of muscle fatigue, *Physiol. Rev.*, 74(1), 49–94, 1994.

81. Fitts, R.H., Muscle fatigue: the cellular aspects, *Am. J. Sports Med.*, 24(6 Suppl), S9–S13, 1996.

82. Fitts, R.H. and Holloszy, J.O., Effects of fatigue and recovery on contractile properties of frog muscle, *J. Appl. Physiol.*, 45(6), 899–902, 1978.

83. Fluck, M., Carson, J.A., Schwartz, R.J., and Booth, F.W., SRF protein is upregulated during stretch-induced hypertrophy of rooster ALD muscle, *J. Appl. Physiol.*, 86(6), 1793–1799, 1999.

84. Fluckey, J.D., Kraemer, W.J., and Farrell, P.A., Pancreatic islet insulin secretion is increased after resistance exercise in rats, *J. Appl. Physiol.*, 79(4), 1100–1105, 1995.

85. Foley, J.M., Jayaraman, R.C., Prior, B.M., Pivarnik, J.M., and Meyer, R.A., MR measurements of muscle damage and adaptation after eccentric exercise, *J. of Appl. Physiol.*, 87(6), 2311–2318, 1999.

86. Friden, J. and Lieber, R.L., Segmental muscle fiber lesions after repetitive eccentric contractions, *Cell Tissue Res.*, 293(1), 165–171, 1998.

87. Friden, J., Lieber, R.L., and Thornell, L.E., Subtle indications of muscle damage following eccentric contractions, *Acta Physiol. Scand.*, 142(4), 523–524, 1991.

88. Friden, J., Seger, J., Sjostrom, M., and Ekblom B. Adaptive response in human skeletal muscle subjected to prolonged eccentric training, *Int. J. Sports Med.*, 4(3), 177–183, 1983.

89. Friden, J., Sfakianos, P.N., and Hargens, A.R., Muscle soreness and intramuscular fluid pressure: comparison between eccentric and concentric load, *J. Appl. Physiol.*, 61(6), 2175–2179, 1986.

90. Friden, J., Sjostrom, M., and Ekblom, B., A morphological study of delayed muscle soreness, *Experientia*, 37(5), 506–507, 1981.

91. Friden, J., Sjostrom, M., and Ekblom, B., Myofibrillar damage following intense eccentric exercise in man, *Int. J. Sports Med.*, 4(3), 170–176, 1983.

92. Fritz, V.K. and Stauber, W.T., Characterization of muscles injured by forced lengthening, II. Proteoglycans. *Med. Sci. Sports Exerc.*, 20(4), 354–361, 1988.

93. Gallucci, S., Provenzano, C., Mazzarelli, P., Scuderi, F., and Bartoccioni, E., Myoblasts produce IL-6 in response to inflammatory stimuli, *Int. Immunol.*, 10(3), 267–273, 1998.

94. Garner, R.P., Terracio, L., Borg, T.K., and Buggy, J., Intracranial self-stimulation motivates weight-lifting exercise in rats, *J. Appl. Physiol.*, 71(4), 1627–1631, 1991.

95. Garrett, W.E., Jr., Muscle strain injuries. *Am. J. Sports Med.*, 24(6 Suppl), S2–S8, 1996.

96. Geronilla, K.B., Miller, G.R., Mowrey, K., Kashon, M.L., and Cutlip, R.G., The Mechanical and Histological Response of Rat Skeletal Muscle to Oscillatory Contractions, In: American College of Sports Medicine; 2001, *Medicine and Science in Sports and Exercise, Baltimore*, MD 2001, p. 474.

97. Geronilla, K.B., Miller, G.R., Mowrey, K.F., Wu, J.Z., Kashon, M.L., Brumbaugh, K. et al., Dynamic force responses of skeletal muscle during stretch-shortening cycles, *Eur. J. Appl. Physiol.*, 90(1–2), 144–153, 2003.

98. Giddings, C.J., Neaves, W.B., and Gonyea, W.J., Muscle fiber necrosis and regeneration induced by prolonged weight-lifting exercise in the cat, *Anat. Rec.*, 211(2), 133–141, 1985.

99. Gillis, J.M., Membrane abnormalities and ca homeostasis in muscles of the mdx mouse, an animal model of the duchenne muscular dystrophy: a review, *Acta Physiol. Scand.*, 156(3), 397–406, 1996.

100. Gonyea, W. and Bonde-Petersen, F., Alterations in muscle contractile properties and fiber composition after weight-lifting exercise in cats, *Exp. Neurol.*, 59(1), 75–84, 1978.

101. Gonyea, W.J., Role of exercise in inducing increases in skeletal muscle fiber number, *J. Appl. Physiol.*, 48(3), 421–426, 1980.

102. Gonyea, W.J. and Ericson, G.C., An experimental model for the study of exercise-induced skeletal muscle hypertrophy, *J. Appl. Physiol.*, 40(4), 630–633, 1976.

103. Gordon, A.M., Huxley, A.F., and Julian, F.J., The variation in isometric tension with sarcomere length in vertebrate muscle fibres, *J. Physiol.*, 184(1), 170–192, 1966.

104. Gordon, E.E. Anatomical and biochemical adaptations of muscle to different exercises, *Jama.*, 201(10), 755–758, 1967.

105. Gordon, E.E., Kowalski, K., and Fritts M., Adaptations of muscle to various exercises, *Studies in rats. Jama.*, 199(2), 103–108, 1967.

106. Gordon, E.E., Kowalski, K., and Fritts, M., Changes in rat muscle fiber with forceful exercises, *Arch. Phys. Med. Rehabil.*, 48(11), 577–582, 1967.

107. Gordon, E.E., Kowalski, K., and Fritts, M. Protein changes in quadriceps muscle of rat with repetitive exercises, *Arch. Phys. Med. Rehabil.*, 48(6), 296–303, 1967.

108. Gorospe, J.R., Tharp, M.D., Hinckley, J., Kornegay, J.N., and Hoffman, E.P., A role for mast cells in the progression of Duchenne muscular dystrophy? Correlations in dystrophin-deficient humans, dogs, and mice, *J. Neurol. Sci.*, 122(1), 44–56, 1994.

109. Gosselin, L.E., Attenuation of force deficit after lengthening contractions in soleus muscle from trained rats, *J. Appl. Physiol.*, 88(4), 1254–1258, 2000.

110. Gosselin, L.E. and Burton H., Impact of initial muscle length on force deficit following lengthening contractions in mammalian skeletal muscle, *Muscle Nerve*, 25(6), 822–827, 2002.

111. Hales, T.R. and Bernard, B.P., Epidemiology of work-related musculoskeletal disorders, *Orthop. Clin. North Am.*, 27(4), 679–709, 1996.

112. Hasselgren, P.O., Role of the ubiquitin-proteasome pathway in sepsis-induced muscle catabolism, *Mol. Biol. Rep.*, 26(1–2), 71–76, 1999.

113. Hasselgren, P.O., Pedersen, P., Sax, H.C., Warner, B.W., and Fischer, J.E., Current concepts of protein turnover and amino acid transport in liver and skeletal muscle during sepsis, *Arch. Surg.*, 123(8), 992–999, 1988.

114. Hasselman, C.T., Best, T.M., Seaber, A.V., and Garrett, W.E., Jr. A threshold and continuum of injury during active stretch of rabbit skeletal muscle, *Am. J. Sports Med.*, 23(1), 65–73, 1995.

115. Henriksson, J., Knuttgen, H.G., and Bonde-Petersen, F., Perceived exertion during exercise with concentric and eccentric muscle contractions, *Ergonomics*, 15(5), 537–544, 1972.

116. Hesselink, M.K., Kuipers, H., Geurten, P., and Van Straaten, H., Structural muscle damage and muscle strength after incremental number of isometric and forced lengthening contractions, *J. Muscle Res. Cell. Motil.*, 17(3), 335–341, 1996.

117. Hilder, T.L., Tou, J.C., Grindeland, R.E., Wade, C.E., and Graves, L.M., Phosphorylation of insulin receptor substrate-1 serine 307 correlates with JNK activity in atrophic skeletal muscle, *FEBS Lett.*, 553(1–2), 63–67, 2003.

118. Hill, A.V., The heat of shortening and the dynamic constants of muscle, *Proc. R. Soc. London Ser. B*, 126, 1938.

119. Hill, M., Wernig, A., and Goldspink, G., Muscle satellite (stem) cell activation during local tissue injury and repair, *J. Anat.*, 203(1), 89–99, 2003.

120. Ho, K.W., Roy, R.R., Tweedle, C.D., Heusner, W.W., Van Huss, W.D., and Carrow, R.E., Skeletal muscle fiber splitting with weight-lifting exercise in rats. *Am. J. Anatomy*, 157, 433–440, 1980.

121. Hoffman, E.P., Brown, R.H., Jr., and Kunkel, L.M., Dystrophin: the protein product of the Duchenne muscular dystrophy locus, *Cell*, 51(6), 919–928, 1987.

122. Hoffman, E.P., Knudson, C.M., Campbell, K.P., and Kunkel, L.M., Subcellular fractionation of dystrophin to the triads of skeletal muscle, *Nature*, 330(6150), 754–758, 1987.

123. Hopf, F.W., Reddy, P., Hong, J., and Steinhardt, R.A., A capacitative calcium current in cultured skeletal muscle cells is mediated by the calcium-specific leak channel and inhibited by dihydropyridine compounds, *J. Biol. Chem.*, 271(37), 22358–22367, 1996.

124. Hopf, F.W., Turner, P.R., Denetclaw, W.F., Jr., Reddy, P., and Steinhardt, R.A., A critical evaluation of resting intracellular free calcium regulation in dystrophic mdx muscle. *Am. J. Physiol.*, 271(4 Pt 1), C1325–C1339, 1996.

125. Horita, T., Komi, P.V., Nicol, C., and Kyrolainen, H., Effect of exhausting stretch-shortening cycle exercise on the time course of mechanical behaviour in the drop jump: possible role of muscle damage, *Eur. J. Appl. Physiol. Occup. Physiol.*, 79(2), 160–167, 1999.

126. Hough, T., Ergographic Studies in Muscular Soreness. *Am. J. Physiol.*, 7, 76–92, 1902.

127. Howell, J.N., Chleboun, G., and Conatser R., Muscle stiffness, strength loss, swelling and soreness following exercise-induced injury in humans. *J. Physiol.*, 464, 183–196, 1993.

128. Huijing, P.A., Architecture of the human gastrocnemius muscle and some functional consequences, *Acta Anat. (Basel)*, 123(2), 101–107, 1985.

129. Hunter, K.D. and Faulkner, J.A., Pliometric contraction-induced injury of mouse skeletal muscle: effect of initial length, *J. Appl. Physiol.*, 82(1), 278–283, 1997.

130. Ingalls, C.P., Warren, G.L., and Armstrong, R.B., Dissociation of force production from MHC and actin contents in muscles injured by eccentric contractions, *J. Muscle Res. Cell Motil.*, 19(3), 215–224, 1998.

131. Ingalls, C.P., Warren, G.L., and Armstrong, R.B., Intracellular Ca2+ transients in mouse soleus muscle after hindlimb unloading and reloading, *J. Appl. Physiol.*, 87(1), 386–390, 1999.

132. Ingalls, C.P., Warren, G.L., Williams, J.H., Ward, C.W., and Armstrong, R.B., E-C coupling failure in mouse EDL muscle after in vivo eccentric contractions, *J. Appl. Physiol.*, 85(1), 58–67, 1998.

133. Jensen, B.R., Schibye, B., Sogaard, K., Simonsen, E.B., and Sjogaard, G., Shoulder muscle load and muscle fatigue among industrial sewing-machine operators, *Eur. J. Appl. Physiol. Occup. Physiol.*, 67(5), 467–475, 1993.

134. Jensen, B.R., Sjogaard, G., Bornmyr, S., Arborelius, M., and Jorgensen, K., Intramuscular laser-Doppler flowmetry in the supraspinatus muscle during isometric contractions, *Eur. J. Appl. Physiol. Occup. Physiol.*, 71(4), 373–378, 1995.

135. Jones, D.A., Jackson, M.J., McPhail, G., and Edwards, R.H., Experimental mouse muscle damage: the importance of external calcium, *Clin. Sci. (Lond)*, 66(3), 317–322, 1984.

136. Jones, D.A., Newham, D.J., Round, J.M., and Tolfree, S.E., Experimental human muscle damage: morphological changes in relation to other indices of damage, *J. Physiol.*, 375, 435–448, 1986.

137. Kadi, F., Hagg, G., Hakansson, R., Holmner, S., Butler-Browne, G.S., and Thornell, L.-E., Structural changes in male trapezius muscle with work-related myalgia, *Acta Neurophathol.*, 95, 352–360, 1998.

138. Keller, K., Corbett, J., and Nichols, D., Repetitive strain injury in computer keyboard users: Pathomechanics and treatment principles in individual and group intervention, *J. Hand Therapy*, 11, 9–26, 1998.

139. Kissel, J.T., Lynn, D.J., Rammohan, K.W., Klein, J.P., Griggs, R.C., and Moxley, R.T., 3rd, et al., Mononuclear cell analysis of muscle biopsies in prednisone- and azathioprine-treated Duchenne muscular dystrophy, *Neurology*, 43(3 Pt 1), 532–536, 1993.

140. Klausen, K. and Knuttgen, H.G., Effect of training on oxygen consumption in negative muscular work, *Acta. Physiol. Scand.*, 83(3), 319–323, 1971.

141. Klitgaard, H., A model for quantitative strength training of hindlimb muscles of the rat, *J. Appl. Physiol.*, 64(4), 1740–1745, 1988.

142. Knutsson, E., Analysis of spastic paresis. In: Proceedings of the Tenth International Congress of the World Confederation for Physical Therapy, 1987, Sydney; 1987. p. 629–633.

143. Knuttgen, H.G., Oxygen debt, lactate, pyruvate, and excess lactate after muscular work, *J. Appl. Physiol.*, 17, 639–644, 1962.

144. Knuttgen, H.G., Human performance in high-intensity exercise with concentric and eccentric muscle contractions. *Int. J. Sports Med.*, 7 Suppl 1, 6–9, 1986.

145. Knuttgen, H.G. and Klausen K., Oxygen debt in short-term exercise with concentric and eccentric muscle contractions, *J. Appl. Physiol.*, 30(5), 632–635, 1971.

146. Knuttgen, H.G., Nadel, E.R., Pandolf, K.B., and Patton, J.F., Effects of training with eccentric muscle contractions on exercise performance, energy expenditure, and body temperature. *Int. J. Sports Med.*, 3(1), 13–17, 1982.

147. Knuttgen, H.G., Petersen, F.B., and Klausen K., Oxygen uptake and heart rate responses to exercise performed with concentric and eccentric muscle contractions. *Med. Sci. Sports*, 3(1), 1–5, 1971.

148. Komi, P.V., Measurement of the force-velocity relationship in human muscle under concentric and eccentric contractions, *Medicine and Sport*, 8, 224–229, 1973.

149. Komi, P.V., Stretch-shortening cycle: a powerful model to study normal and fatigued muscle, *J. Biomech.*, 33(10), 1197–1206, 2000.

150. Komi, P.V. and Buskirk, E.R., Effect of eccentric and concentric muscle conditioning on tension and electrical activity of human muscle, *Ergonomics*, 15(4), 417–434, 1972.

151. Komi, P.V. and Rusko, H., Quantitative evaluation of mechanical and electrical changes during fatigue loading of eccentric and concentric work. *Scand. J. Rehabil. Med.*, Suppl 3, 121–126, 1974.

152. Komulainen, J., Kalliokoski, R., Koskinen, S.O., Drost, M.R., Kuipers, H., and Hesselink, M.K., Controlled lengthening or shortening contraction-induced damage is followed by fiber hypertrophy in rat skeletal muscle, *Int. J. Sports Med.*, 21(2), 107–112, 2000.

153. Komulainen, J., Koskinen, S.O., Kalliokoski, R., Takala, T.E., and Vihko, V., Gender differences in skeletal muscle fibre damage after eccentrically biased downhill running in rats, *Acta. Physiol. Scand.*, 165(1), 57–63, 1999.

154. Komulainen, J., Takala, T.E., Kuipers, H., and Hesselink, M.K., The disruption of myofibre structures in rat skeletal muscle after forced lengthening contractions, *Pflugers Arch.*, 436(5), 735–741, 1998.

155. Kuipers, H. and Van der Meulen, J.H., Increased Passive Stiffness of Rat Muscle after Eccentric and Isometric Work, *Medicine and Science in Sports and Exercise*, 24(5), S1007, 1992.

156. Lapointe, B.M., Fremont, P., and Cote, C.H., Adaptation to lengthening contractions is independent of voluntary muscle recruitment but relies on inflammation. *Am. J. Physiol. Regul. Integr. Comp. Physiol.*, 282(1), R323–R329, 2002.

157. Lapointe, B.M., Frenette, J., and Cote, C.H., Lengthening contraction-induced inflammation is linked to secondary damage but devoid of neutrophil invasion, *J. Appl. Physiol.*, 92(5), 1995–2004, 2002.

158. Larsson, B., Bjork, J., Henriksson, K.G., Gerdle, B. and Lindman, R., The prevalences of cytochrome c oxidase negative and superpositive fibres and ragged-red fibres in the trapezius muscle of female cleaners with and without myalgia and of female healthy controls, *Pain*, 84(2–3), 379–387, 2000.

159. Larsson, S.E., Bodegard, L., Henriksson, K.G., and Oberg, P.A., Chronic trapezius myalgia. Morphology and blood flow studied in 17 patients, *Acta Orthop. Scand.*, 61(5), 394–398, 1990.

160. Lazarides, E., Intermediate filaments as mechanical integrators of cellular space, *Nature*, 283(5744), 249–256, 1980.

161. Lieber, R.L. and Friden, J., Selective damage of fast glycolytic muscle fibres with eccentric contraction of the rabbit tibialis anterior, *Acta Physiol. Scand.*, 133(4), 587–588, 1988.

162. Lieber, R.L. and Friden, J., Muscle damage is not a function of muscle force but active muscle strain, *J. Appl. Physiol.*, 74(2), 520–526, 1993.

163. Lieber, R.L., Schmitz, M.C., Mishra, D.K., and Friden, J., Contractile and cellular remodeling in rabbit skeletal muscle after cyclic eccentric contractions, *J. Appl. Physiol.*, 77(4), 1926–1934, 1994.

164. Lieber, R.L., Thornell, L.E., and Friden, J., Muscle cytoskeletal disruption occurs within the first 15 min of cyclic eccentric contraction, *J. Appl. Physiol.*, 80(1), 278–284, 1996.

165. Lieber, R.L., Woodburn, T.M., and Friden, J., Muscle damage induced by eccentric contractions of 25% strain, *J. Appl. Physiol.*, 70(6), 2498–2507, 1991.

166. Lowe, D.A., and Alway, S.E., Stretch-induced myogenin, MyoD, and MRF4 expression and acute hypertrophy in quail slow-tonic muscle are not dependent upon satellite cell proliferation, *Cell Tissue Res.*, 296(3), 531–539, 1999.

167. Lowe, D.A., Warren, G.L., Hayes, D.A., Farmer, M.A., and Armstrong, R.B., Eccentric contraction-induced injury of mouse soleus muscle: effect of varying $[Ca2+]o$, *J. Appl. Physiol.*, 76(4), 1445–1453, 1994.

168. Lowe, D.A., Warren, G.L., Ingalls, C.P., Boorstein, D.B., and Armstrong, R.B., Muscle function and protein metabolism after initiation of eccentric contraction-induced injury, *Journal of Applied Physiology*, 79(4), 1260–1270, 1995.

169. Lynch, G.S. and Faulkner, J.A., Contraction-induced injury to single muscle fibers: velocity of stretch does not influence the force deficit, *Am. J. Physiol.*, 275(6 Pt 1), C1548–C1554, 1998.

170. Lynn, R. and Morgan, D.L., Decline running produces more sarcomeres in rat vastus intermedius muscle fibers than does incline running, *J. Appl. Physiol.*, 77(3), 1439–1444, 1994.

171. Macpherson, P.C., Dennis, R.G., and Faulkner, J.A., Sarcomere dynamics and contraction-induced injury to maximally activated single muscle fibres from soleus muscles of rats, *J. Physiol.*, 500 (Pt 2), 523–533, 1997.

172. Macpherson, P.C., Schork, M.A., and Faulkner, J.A., Contraction-induced injury to single fiber segments from fast and slow muscles of rats by single stretches, *Am. J. Physiol.*, 271(5 Pt 1), C1438–C1446, 1996.

173. Mair, J., Koller, A., Artner-Dworzak, E., Haid, C., Wicke, K., Judmaier, W. et al., Effects of exercise on plasma myosin heavy chain fragments and MRI of skeletal muscle, *J. Appl. Physiol.*, 72(2), 656–663, 1992.

174. Mair, J., Mayr, M., Muller, E., Koller, A., Haid, C., Artner-Dworzak, E. et al., Rapid adaptation to eccentric exercise-induced muscle damage, *Int. J. Sports Med.*, 16(6), 352–356, 1995.

175. Malm, C., Nyberg, P., Engstrom, M., Sjodin, B., Lenkei, R., Ekblom, B. et al., Immunological changes in human skeletal muscle and blood after eccentric exercise and multiple biopsies, *J. Physiol.*, 529 Pt 1, 243–262, 2000.

176. McArdle, A., van der Meulen, J.H., Catapano, M., Symons, M.C., Faulkner, J.A., and Jackson, M.J., Free radical activity following contraction-induced injury to the extensor digitorum longus muscles of rats, *Free Radic. Biol. Med.*, 26(9-10), 1085–1091, 1999.

177. McCully, K.K. and Faulkner, J.A., Injury to skeletal muscle fibers of mice following lengthening contractions, *J. Appl. Physiol.*, 59(1), 119–126, 1985.

178. McCully, K.K. and Faulkner, J.A., Characteristics of lengthening contractions associated with injury to skeletal muscle fibers, *J. Appl. Physiol.*, 61(1), 293–299, 1986.

179. McHugh, M.P., Connolly, D.A., Eston, R.G., and Gleim, G.W., Exercise-induced muscle damage and potential mechanisms for the repeated bout effect, *Sports Medicine*, 27(3), 157–170, 1999.

180. Medina, R., Wing, S.S., and Goldberg, A.L., Increase in levels of polyubiquitin and proteasome mRNA in skeletal muscle during starvation and denervation atrophy, *Biochem. J.*, 307 (Pt 3), 631–637, 1995.

181. Mishra, D.K., Friden, J., Schmitz, M.C., and Lieber, R.L., Anti-inflammatory medication after muscle injury. A treatment resulting in short-term improvement but subsequent loss of muscle function, *J. Bone Joint Surg. Am.*, 77(10), 1510–1519, 1995.

182. Moldawer, L.L., Svaninger, G., Gelin, J., and Lundholm, K.G., Interleukin 1 and tumor necrosis factor do not regulate protein balance in skeletal muscle, *Am. J. Physiol.*, 253(6 Pt 1), C766–C773, 1987.

183. Morgan, D.L., New insights into the behavior of muscle during active lengthening, *Biophys. J.*, 57(2), 209–221, 1990.

184. Mutsaers, S.E., Bishop, J.E., McGrouther, G., and Laurent, G.J., Mechanisms of tissue repair: from wound healing to fibrosis, *International Journal of Biochemistry & Cell Biology*, 29(1), 5–17, 1997.

185. Myllyla, R., Salminen, A., Peltonen, L., Takala, T.E., and Vihko, V., Collagen metabolism of mouse skeletal muscle during the repair of exercise injuries, *Pflugers Arch.*, 407(1), 64–70, 1986.

186. Newham, D.J., Jones, D.A., and Clarkson, P.M., Repeated high-force eccentric exercise: effects on muscle pain and damage, *J. Appl. Physiol.*, 63(4), 1381–1386, 1987.

187. Newham, D.J., Jones, D.A., and Edwards, R.H., Large delayed plasma creatine kinase changes after stepping exercise, *Muscle Nerve*, 6(5), 380–385, 1983.

188. Newham, D.J., Jones, D.A., Ghosh, G., and Aurora, P., Muscle fatigue and pain after eccentric contractions at long and short length, *Clin. Sci. (Lond)*, 74(5), 553–557, 1988.

189. Newham, D.J., Jones, D.A., Tolfree, S.E., and Edwards, R.H., Skeletal muscle damage: a study of isotope uptake, enzyme efflux and pain after stepping, *Eur. J. Appl. Physiol. Occup. Physiol.*, 55(1), 106–112, 1986.

190. Newham, D.J., McPhail, G., Mills, K.R., and Edwards, R.H., Ultrastructural changes after concentric and eccentric contractions of human muscle, *J. Neurol. Sci.*, 61(1), 109–122, 1983.

191. Nikolaou, P.K., Macdonald, B.L., Glisson, R.R., Seaber, A.V., and Garrett, W.E., Jr., Biomechanical and histological evaluation of muscle after controlled strain injury, *Am. J. Sports Med.*, 15(1), 9–14, 1987.

192. Noonan, T.J., Best, T.M., Seaber, A.V., and Garrett, W.E. Jr., Identification of a threshold for skeletal muscle injury, *Am. J. Sports Med.*, 22(2), 257–261, 1994.

193. Nosaka, K. and Clarkson, P.M., Effect of eccentric exercise on plasma enzyme activities previously elevated by eccentric exercise, *Eur. J. Appl. Physiol. Occup. Physiol.*, 69(6), 492–497, 1994.

194. Nosaka, K. and Clarkson, P.M., Muscle damage following repeated bouts of high force eccentric exercise, *Med. Sci. Sports Exerc.*, 27(9), 1263–1269, 1995.

195. Nosaka, K. and Clarkson, P.M., Changes in indicators of inflammation after eccentric exercise of the elbow flexors, *Medicine & Science in Sports & Exercise*, 28(8), 953–961, 1996.

196. Nosaka, K. and Clarkson, P.M., Variability in serum creatine kinase response after eccentric exercise of the elbow flexors, *International Journal of Sports Medicine*, 17(2), 120–127, 1996.

197. Nosaka, K., Clarkson, P.M., and Apple, F.S., Time course of serum protein changes after strenuous exercise of the forearm flexors, *Journal of Laboratory & Clinical Medicine*, 119(2), 183–188, 1992.

198. Nosaka, K., Clarkson, P.M., McGuiggin, M.E., and Byrne, J.M., Time course of muscle adaptation after high force eccentric exercise, *Eur. J. Appl. Physiol. Occup. Physiol.*, 63(1), 70–76, 1991.

199. Nurenberg, P., Giddings, C.J., Stray-Gundersen, J., Fleckenstein, J.L., Gonyea, W.J., and Peshock, R.M., MR imaging-guided muscle biopsy for correlation of increased signal intensity with ultra-structural change and delayed-onset muscle soreness after exercise, *Radiology*, 184(3), 865–869, 1992.

200. Oak, S.A., Zhou, Y.W., and Jarrett, H.W., Skeletal muscle signaling pathway through the dystrophin glycoprotein complex and Rac1, *J. Biol. Chem.*, 278(41), 39287–39295, 2003.

201. Obremsky, W.T., Seaber, A.V., Ribbeck, B.M., and Garrett, W.E., Jr., Biomechanical and histologic assessment of a controlled muscle strain injury treated with piroxicam, *Am. J. Sports Med.*, 22(4), 558–561, 1994.

202. Ogilvie, R.W., Armstrong, R.B., Baird, K.E., and Bottoms, C.L., Lesions in the rat soleus muscle following eccentrically biased exercise, *Am. J. Anat.*, 182(4), 335–346, 1988.

203. Ortenblad, N., Sjogaard, G., and Madsen, K., Impaired sarcoplasmic reticulum Ca(2+) release rate after fatiguing stimulation in rat skeletal muscle, *J. Appl. Physiol.*, 89(1), 210–217, 2000.

204. Ortenblad, N., Young, J.F., Oksbjerg, N., Nielsen, J.H., and Lambert, I.H., Reactive oxygen species are important mediators of taurine release from skeletal muscle cells, *Am. J. Physiol. Cell Physiol.*, 284(6), C1362–C1373, 2003.

205. Pardo, J.V., Siliciano, J.D., and Craig, S.W., A vinculin-containing cortical lattice in skeletal muscle: transverse lattice elements ("costameres") mark sites of attachment between myofibrils and sarcolemma, *Proc. Natl. Acad. Sci. USA*, 80(4), 1008–1012, 1983.

206. Patel, T.J., Cuizon, D., Mathieu-Costello, O., Friden, J., and Lieber, R.L., Increased oxidative capacity does not protect skeletal muscle fibers from eccentric contraction-induced injury, *Am. J. Physiol.*, 274(5 Pt 2), R1300–R1308, 1998.

207. Patel, T.J. and Lieber, R.L., Force transmission in skeletal muscle: from actomyosin to external tendons, *Exerc. Sport Sci. Rev.*, 25, 321–363, 1997.

208. Petersen, F.B. and Knuttgen, H.G., Effect of training with eccentric muscle contractions on human skeletal muscle metabolites, *Acta. Physiol. Scand.*, 80(4), 16A–17A, 1970.

209. Peterson, J.M., Trappe, T.A., Mylona, E., White, F., Lambert, C.P., Evans, W.J. et al., Ibuprofen and acetaminophen: effect on muscle inflammation after eccentric exercise, *Med. Sci. Sports Exerc.*, 35(6), 892–896, 2003.

210. Petrof, B.J., Shrager, J.B., Stedman, H.H., Kelly, A.M., and Sweeney, H.L., Dystrophin protects the sarcolemma from stresses developed during muscle contraction, *Proc. Natl. Acad. Sci. USA*, 90(8), 3710–3714, 1993.

211. Piligian, G., Herbert, R., Hearns, M., Dropkin, J., Landsbergis, P., and Cherniack, M., Evaluation and management of chronic work-related musculoskeletal disorders of the distal upper extremity, *American Journal of Industrial Medicine*, 37(1), 75–93, 2000.

212. Porter, J.D., Khanna, S., Kaminski, H.J., Rao, J.S., Merriam, A.P., Richmonds, C.R. et al., A chronic inflammatory response dominates the skeletal muscle molecular signature in dystrophin-deficient mdx mice, *Hum. Mol. Genet.*, 11(3), 263–272, 2002.

213. Prisk, V. and Huard, J., Muscle injuries and repair: the role of prostaglandins and inflammation, *Histol. Histopathol.*, 18(4), 1243–1256, 2003.

214. Quinn, L.S., Anderson, B.G., Drivdahl, R.H., Alvarez, B., and Argiles, J.M., Overexpression of interleukin-15 induces skeletal muscle hypertrophy in vitro: implications for treatment of muscle wasting disorders, *Exp. Cell Res.*, 280(1), 55–63, 2002.

215. Quinn, L.S., Haugk, K.L., and Grabstein, K.H., Interleukin-15: a novel anabolic cytokine for skeletal muscle, *Endocrinology*, 136(8), 3669–3672, 1995.

216. Rabinovsky, E.D., Gelir, E., Gelir, S., Lui, H., Kattash, M., DeMayo, F.J. et al., Targeted expression of IGF-1 transgene to skeletal muscle accelerates muscle and motor neuron regeneration, *Faseb J.*, 17(1), 53–55, 2003.

217. Rafael, J.A. and Brown, S.C., Dystrophin and utrophin: genetic analyses of their role in skeletal muscle, *Microsc. Res. Tech.*, 48(3–4), 155–166, 2000.

218. Rafael, J.A., Townsend, E.R., Squire, S.E., Potter, A.C., Chamberlain, J.S., and Davies, K.E., Dystrophin and utrophin influence fiber type composition and post-synaptic membrane structure, *Hum. Mol. Genet.*, 9(9), 1357–1367, 2000.

219. Reddy, A.S., Reedy, M.K., Best, T.M., Seaber, A.V., Garrett, W.E., Jr. Restriction of the injury response following an acute muscle strain, *Med. Sci. Sports Exerc.*, 25(3), 321–327, 1993.

220. Rinard, J., Clarkson, P.M., Smith, L.L., and Grossman, M., Response of males and females to high-force eccentric exercise, *J. Sports Sci.*, 18(4), 229–236, 2000.

221. Rodenburg, J.B., Bar, P.R., and De Boer, R.W., Relations between muscle soreness and biochemical and functional outcomes of eccentric exercise, *J. Appl. Physiol.*, 74(6), 2976–2983, 1993.

222. Rodenburg, J.B., Steenbeek, D., Schiereck, P., and Bar, P.R., Warm-up, stretching and massage diminish harmful effects of eccentric exercise, *Int. J. Sports Med.*, 15(7), 414–419, 1994.

223. Rogers, M.A., Stull, G.A., and Apple, F.S., Creatine kinase isoenzyme activities in men and women following a marathon race, *Medicine & Science in Sports & Exercise*, 17(6), 679–682, 1985.

224. Round, J.M., Jones, D.A., and Cambridge, G., Cellular infiltrates in human skeletal muscle: exercise induced damage as a model for inflammatory muscle disease? *J. Neurol. Sci.*, 82(1–3), 1–11, 1987.

225. Sadeh, M., Czyewski, K., and Stern, L.Z., Chronic myopathy induced by repeated bupivacaine injections, *J. Neurol. Sci.*, 67(2), 229–238, 1985.

226. Sam, M., Shah, S., Friden, J., Milner, D.J., Capetanaki, Y., and Lieber, R.L., Desmin knockout muscles generate lower stress and are less vulnerable to injury compared with wild-type muscles, *Am. J. Physiol. Cell Physiol.*, 279(4), C1116–C1122, 2000.

227. Sargeant, A.J. and Dolan, P., Human muscle function following prolonged eccentric exercise, *Eur. J. Appl. Physiol. Occup. Physiol.*, 56(6), 704–711, 1987.

228. Sato, K. and Kawashima, S., Calpain function in the modulation of signal transduction molecules, *Biol. Chem.*, 382(5), 743–751, 2001.

229. Sato, S., Yanagisawa, K., and Miyatake, T., Conversion of myelin-associated glycoprotein (MAG) to a smaller derivative by calcium activated neutral protease (CANP)-like enzyme in myelin and inhibition by E-64 analogue, *Neurochem. Res.*, 9(5), 629–635, 1984.

230. Savage, K.J. and Clarkson, P.M., Oral contraceptive use and exercise-induced muscle damage and recovery, *Contraception*, 66(1), 67–71, 2002.

231. Saxton, J.M., Clarkson, P.M., James, R., Miles, M., Westerfer, M., Clark, S., et al. Neuromuscular dysfunction following eccentric exercise, *Med. Sci. Sports Exerc.*, 27(8), 1185–1193, 1995.

232. Sayeed, M.M., Alterations in calcium signaling and cellular responses in septic injury, *New Horiz.*, 4(1), 72–86, 1996.

233. Sayers, S.P. and Clarkson, P.M., Force recovery after eccentric exercise in males and females. *Eur. J. Appl. Physiol.*, 84(1–2), 122–126, 2001.

234. Schwane, J.A. and Armstrong, R.B., Effect of training on skeletal muscle injury from downhill running in rats, *J. Appl. Physiol.*, 55(3), 969–975, 1983.

235. Scifres, J.C. and Martin, D.T., Velocity of Eccentric Contraction affects the Magnitude of Delayed Onset Muscle Soreness. *Medicine & Science in Sports & Exercise*, 24(5), S851, 1992.

236. Sheehan, S.M. and Allen, R.E., Skeletal muscle satellite cell proliferation in response to members of the fibroblast growth factor family and hepatocyte growth factor, *J. Cell Physiol.*, 181(3), 499–506, 1999.

237. Sherwood, L., Human Physiology: From cells to systems. 4th ed., Brooks–Cole, Pacific Grove, CA, 2001.

238. Singh, M. and Karpovich, P.V., Isotonic and isometric forces of forearm flexors and extensors, *J. Appl. Physiol.*, 21(4), 1435–1437, 1966.

239. Sjogaard, G., Muscle energy metabolism and electrolyte shifts during low-level prolonged static contraction in man, *Acta Physiol. Scand.*, 134(2), 181–187, 1988.

240. Sjogaard, G. and Jensen, B.R., Muscle Pathology with Overuse. In: Ranney, D., editor. Chronic Musculoskeletal Injuries in the Workplace. Philadelphia, PA: W.B. Saunders Company; 1997. p. 17–40.

241. Sjogaard, G., Lundberg, U., and Kadefors, R., The role of muscle activity and mental load in the development of pain and degenerative processes at the muscle cell level during computer work, *Eur. J. Appl. Physiol.*, 83(2–3), 99–105, 2000.

242. Sjogaard, G. and McComas, A.J., Role of interstitial potassium, *Adv. Exp. Med. Biol.*, 384, 69–80, 1995.

243. Sjogaard, G. and Sogaard, K., Muscle injury in repetitive motion disorders, *Clin. Orthop.*, (351), 21–31, 1998.

244. Smith, H.J. and Tisdale, M.J., Signal transduction pathways involved in proteolysis-inducing factor induced proteasome expression in murine myotubes, *Br. J. Cancer.*, 89(9), 1783–1788, 2003.

245. Smith, L.L., Fulmer, M.G., Holbert, D., McCammon, M.R., Houmard, J.A., Frazer, D.D. et al., The impact of a repeated bout of eccentric exercise on muscular strength, muscle soreness and creatine kinase, *Br. J. Sports Med.*, 28(4), 267–271, 1994.

246. Sola, O.M., Christensen, D.L., and Martin, A.W., Hypertrophy and hyperplasia of adult chicken anterior latissimus dorsi muscles following stretch with and without denervation, *Exp. Neurol.*, 41(1), 76–100, 1973.

247. Sorichter, S., Mair, J., Koller, A., Calzolari, C., Huonker, M., Pau, B. et al., Release of muscle proteins after downhill running in male and female subjects, *Scand. J. Med. Sci. Sports*, 11(1), 28–32, 2001.

248. Sorichter, S., Puschendorf, B., and Mair, J., Skeletal muscle injury induced by eccentric muscle action: muscle proteins as markers of muscle fiber injury, *Exercise Immunology Review*, 5, 5–21, 1999.

249. Spencer, M.J., Montecino-Rodriguez, E., Dorshkind, K., and Tidball, J.G., Helper (CD4(+)) and cytotoxic (CD8(+)) T cells promote the pathology of dystrophin-deficient muscle, *Clin. Immunol.*, 98(2), 235–243, 2001.

250. Spencer, M.J. and Tidball, J.G., Do immune cells promote the pathology of dystrophin-deficient myopathies? *Neuromuscul. Disord.*, 11(6–7), 556–564, 2001.

251. St. Pierre, B.A. and Tidball, J.G., Macrophage activation and muscle remodeling at myotendinous junctions after modifications in muscle loading, *Am. J. Pathol.*, 145(6), 1463–1471, 1994.

252. St. Pierre Schneider, B., Correia, L.A., and Cannon, J.G., Sex differences in leukocyte invasion in injured murine skeletal muscle, *Res. Nurs. Health*, 22(3), 243–250, 1999.

253. Stauber, W.T., Eccentric action of muscles: physiology, injury, and adaptation, *Exerc. Sport Sci. Rev.*, 17, 157–185, 1989.

254. Stauber, W.T., Measurement of Muscle Function in Man. In: Sports Injuries, International Perspectives in Physical Therapy: Churchhill-Livingstone; 1989.

255. Stauber, W.T., Fritz, V.K., Vogelbach, D.W., and Dahlmann, B., Characterization of muscles injured by forced lengthening I. Cellular infiltrates, *Med. Sci. Sports Exerc.*, 20(4), 345–353, 1988.

256. Stauber, W.T., Knack, K.K., Miller, G.R., and Grimmett, J.G., Fibrosis and intercellular collagen connections from four weeks of muscle strains, *Muscle Nerve*, 19(4), 423–430, 1996.

257. Stauber, W.T., and Smith, C.A., Cellular responses in exertion-induced skeletal muscle injury, *Mol. Cell Biochem.*, 179(1–2), 189–196, 1998.

258. Stevens, E.D., Relation between work and power calculated from force-velocity curves to that done during oscillatory work, *J. Muscle. Res. Cell. Motil.*, 14(5), 518–526, 1993.

259. Stevens, E.D., Effect of phase of stimulation on acute damage caused by eccentric contractions in mouse soleus muscle, *J. Appl. Physiol.*, 80(6), 1958–1962, 1996.

260. Stevens, E.D., and Syme, D.A., Effect of stimulus duty cycle and cycle frequency on power output during fatigue in rat diaphragm muscle doing oscillatory work, *Can. J. Physiol. Pharmacol.*, 71(12), 910–916, 1993.

261. Stone, M.H. and Lipner, H., Responses to intensive training and methandrostenelone administration. I. Contractile and performance variables. *Pflugers Arch.*, 375(2), 141–146, 1978.

262. Stone, M.H., Rush, M.E., and Lipner, H., Responses to intensive training and methandrostenelone administration: II. Hormonal, organ weights, muscle weights and body composition, *Pflugers Arch.*, 375(2), 147–151, 1978.

263. Stupka, N., Lowther, S., Chorneyko, K., Bourgeois, J.M., Hogben, C., and Tarnopolsky, M.A., Gender differences in muscle inflammation after eccentric exercise, *J. Appl. Physiol.*, 89(6), 2325–2332, 2000.

264. Sun, J.S., Tsuang, Y.H., Hang, Y.S., Liu, T.K., Lee, W.W., and Cheng, C.K., The deleterious effect of tetanic contraction on rabbit's triceps surae muscle during cyclic loading, *Clin. Biomech. (Bristol, Avon)*, 11(1), 46–50, 1996.

265. Talbot, J.A. and Morgan, D.L., Quantitative analysis of sarcomere non-uniformities in active muscle following a stretch, *J. Muscle Res. Cell Motil.*, 17(2), 261–268, 1996.

266. Tamaki, T., Uchiyama, S., and Nakano, S., A weight-lifting exercise model for inducing hypertrophy in the hindlimb muscles of rats, *Med. Sci. Sports Exerc.*, 24(8), 881–886, 1992.

267. Tate, C.A., Bonner, H.W., and Leslie, S.W., Calcium uptake in skeletal muscle mitochondria. I. The effects of chelating agents on the mitochondria from fatigued rats, *Eur. J. Appl. Physiol. Occup. Physiol.*, 39(2), 111–116, 1978.

268. Thompson, C.B., Choi, C., Youn, J.H., and McDonough, A.A., Temporal responses of oxidative vs. glycolytic skeletal muscles to K+ deprivation: Na+ pumps and cell cations, *Am. J. Physiol.*, 276(6 Pt 1), C1411–C1419, 1999.

269. Tidball, J.G., Inflammatory cell response to acute muscle injury, *Med. Sci. Sports Exerc.*, 27(7), 1022–1032, 1995.

270. Tiidus, P.M., Can estrogens diminish exercise induced muscle damage? *Can. J. Appl. Physiol.*, 20(1), 26–38, 1995.

271. Tiidus, P.M. and Bombardier, E., Oestrogen attenuates post-exercise myeloperoxidase activity in skeletal muscle of male rats, *Acta. Physiol. Scand.*, 166(2), 85–90, 1999.

272. Tiku, K., Tiku, M.L., Liu, S., and Skosey, J.L., Normal human neutrophils are a source of a specific interleukin 1 inhibitor, *J. Immunol.*, 136(10), 3686–3692, 1986.

273. Trappe, T.A., Fluckey, J.D., White, F., Lambert, C.P., and Evans, W.J., Skeletal muscle PGF(2)(alpha) and PGE(2) in response to eccentric resistance exercise: influence of ibuprofen acetaminophen, *J. Clin. Endocrinol. Metab.*, 86(10), 5067–5070, 2001.

274. Trappe, T.A., White, F., Lambert, C.P., Cesar, D., Hellerstein, M., and Evans, W.J., Effect of ibuprofen and acetaminophen on postexercise muscle protein synthesis, *Am. J. Physiol. Endocrinol. Metab.*, 282(3), E551–E556, 2002.

275. Triolo, R., Robinson, D., Gardner, E., and Betz, R., The eccentric strength of electrically stimulated paralyzed muscle, *IEEE Transactions on Biomedical Engineering*, 651–652, 1987.

276. van der Meulen, J.H., Exercise Induced Muscle Damage: Morphological, Biochemical, and Functional Aspects [Doctoral Dissertation]. Amsterdam: Universiteit van Amsterdam; 1991.

277. van der Meulen, J.H., Kuipers, H., and Drukker, J., Relationship between exercise-induced muscle damage and enzyme release in rats, *J. Appl. Physiol.*, 71(3), 999–1004, 1991.

278. van der Meulen, J.H., McArdle, A., Jackson, M.J., and Faulkner, J.A., Contraction-induced injury to the extensor digitorum longus muscles of rats: the role of vitamin E, *J. Appl. Physiol.*, 83(3), 817–823, 1997.

279. van der Vusse, G.J., de Groot, M.J., Willemsen, P.H., van Bilsen, M., Schrijvers, A.H., and Reneman, R.S., Degradation of phospholipids and triacylglycerol, and accumulation of fatty acids in anoxic myocardial tissue, disrupted by freeze-thawing, *Mol. Cell Biochem.*, 88(1–2), 83–90, 1989.

280. van der Vusse, G.J., van Bilsen, M., and Reneman, R.S., Is Phospholipid Degradation a Critical Event in Ischemia and Reperfusion-Induced Damage? *News in Physiological Sciences*, 49–53, 1989.

281. Vijayan, K., Thompson, J.L., Norenberg, K.M., Fitts, R.H., and Riley, D.A., Fiber-type susceptibility to eccentric contraction-induced damage of hindlimb-unloaded rat AL muscles, *J. Appl. Physiol.*, 90(3), 770–776, 2001.

282. Vollestad, N.K. and Sejersted, O.M., Biochemical correlates of fatigue. A brief review, *Eur. J. Appl. Physiol. Occup. Physiol.*, 57(3), 336–347, 1988.

283. Warren, G.L., 3rd, Williams, J.H., Ward, C.W., Matoba, H., Ingalls, C.P., Hermann, K.M., et al. Decreased contraction economy in mouse EDL muscle injured by eccentric contractions, *J. Appl. Physiol.*, 81(6), 2555–2564, 1996.

284. Warren, G.L., Hayes, D.A., Lowe, D.A., and Armstrong, R.B., Mechanical factors in the initiation of eccentric contraction-induced injury in rat soleus muscle, *J. Physiol.*, 464, 457–475, 1993.

285. Warren, G.L., Hayes, D.A., Lowe, D.A., Prior, B.M., and Armstrong, R.B., Materials fatigue initiates eccentric contraction-induced injury in rat soleus muscle, *J. Physiol.*, 464, 477–489, 1993.

286. Warren, G.L., Hayes, D.A., Lowe, D.A., Williams, J.H., and Armstrong, R.B., Eccentric contraction-induced injury in normal and hindlimb-suspended mouse soleus and EDL muscles, *J. Appl. Physiol.*, 77(3), 1421–1430, 1994.

287. Warren, G.L., Ingalls, C.P., and Armstrong, R.B., Temperature dependency of force loss and Ca(2+) homeostasis in mouse EDL muscle after eccentric contractions, *Am. J. Physiol. Regul. Integr. Comp. Physiol.*, 282(4), R1122–R1132, 2002.

288. Warren, G.L., Ingalls, C.P., Shah, S.J., and Armstrong, R.B., Uncoupling of in vivo torque production from EMG in mouse muscles injured by eccentric contractions, *J. Physiol.*, 515 (Pt 2), 609–619, 1999.

289. Warren, G.L., Lowe, D.A., and Armstrong, R.B., Measurement tools used in the study of eccentric contraction-induced injury, *Sports Medicine*, 27(1), 43–59, 1999.

290. Warren, G.L., Lowe, D.A., Hayes, D.A., Farmer, M.A., and Armstrong, R.B., Redistribution of cell membrane probes following contraction-induced injury of mouse soleus muscle, *Cell & Tissue Research*, 282(2), 311–320, 1995.

291. Warren, G.L., Lowe, D.A., Hayes, D.A., Karwoski, C.J., Prior, B.M., and Armstrong, R.B., Excitation failure in eccentric contraction-induced injury of mouse soleus muscle, *J. Physiol.*, 468, 487–499, 1993.

292. Whitehead, N.P., Allen, T.J., Morgan, D.L., and Proske, U., Damage to human muscle from eccentric exercise after training with concentric exercise, *J. Physiol.*, 512 (Pt 2), 615–620, 1998.

293. Wickiewicz, T.L., Roy, R.R., Powell, P.L., and Edgerton, V.R., Muscle architecture of the human lower limb, *Clin. Orthop.*, (179), 275–583, 1983.

294. Widrick, J.J., Costill, D.L., McConell, G.K., Anderson, D.E., Pearson, D.R., and Zachwieja, J.J., Time course of glycogen accumulation after eccentric exercise, *J. Appl. Physiol.*, 72(5), 1999–2004, 1992.

295. Wilkie, D.R., The relation between force and velocity in human muscle, *J. Physiol.*, 110, 249–280, 1950.

296. Willems, M.E. and Stauber, W.T., Isometric and concentric performance of electrically stimulated ankle plantar flexor muscles in intact rat, *Exp. Physiol.*, 84(2), 379–389, 1999.

297. Willems, M.E. and Stauber, W.T., Changes in force by repeated stretches of skeletal muscle in young and old female Sprague Dawley rats, *Aging (Milano)*, 12(6), 478–481, 2000.

298. Willems, M.E. and Stauber, W.T., Force deficits after repeated stretches of activated skeletal muscles in female and male rats, *Acta Physiol. Scand.*, 172(1), 63–67, 2001.

299. Willems, M.E. and Stauber, W.T., Fatigue and recovery at long and short muscle lengths after eccentric training, *Med. Sci. Sports Exerc.*, 34(11), 1738–43, 2002.

300. Williams, P., Kyberd, P., Simpson, H., Kenwright, J., and Goldspink, G., The morphological basis of increased stiffness of rabbit tibialis anterior muscles during surgical limb-lengthening, *J. Anat.*, 193 (Pt 1), 131–138, 1998.

301. Wirth, O., Gregory, E.W., Cutlip, R.G., and Miller, G.R., Control and quantitation of voluntary weight-lifting performance of rats, *J. Appl. Physiol.*, 95(1), 402–412, 2003.

302. Wong, T.S. and Booth, F.W., Skeletal muscle enlargement with weight-lifting exercise by rats, *J. Appl. Physiol.*, 65(2), 950–954, 1988.

303. Wong, T.S. and Booth, F.W., Protein metabolism in rat gastrocnemius muscle after stimulated chronic concentric exercise, *J. Appl. Physiol.*, 69(5), 1709–1717, 1990.

304. Wong, T.S. and Booth, F.W., Protein metabolism in rat tibialis anterior muscle after stimulated chronic eccentric exercise, *J. Appl. Physiol.*, 69(5), 1718–1724, 1990.

305. Wood, S.A., Morgan, D.L., and Proske, U., Effects of repeated eccentric contractions on structure and mechanical properties of toad sartorius muscle, *Am. J. Physiol.*, 265(3 Pt 1), C792–C800, 1993.

306. Wright, J., Huang, Q.Q., and Wang, K., Nebulin is a full-length template of actin filaments in the skeletal muscle sarcomere: an immunoelectron microscopic study of its orientation and span with site-specific monoclonal antibodies, *J. Muscle. Res. Cell Motil.*, 14(5), 476–483, 1993.

307. Wung, B.S., Cheng, J.J., Hsieh, H.J., Shyy, Y.J., and Wang, D.L., Cyclic strain-induced monocyte chemotactic protein-1 gene expression in endothelial cells involves reactive oxygen species activation of activator protein 1, *Circ. Res.*, 81(1), 1–7, 1997.

308. Yarasheski, K.E., Lemon, P.W., and Gilloteaux, J., Effect of heavy-resistance exercise training on muscle fiber composition in young rats, *J. Appl. Physiol.*, 69(2), 434–437, 1990.

309. Yeung, E.W., Bourreau, J.P., Allen, D.G., and Ballard, H.J., Effect of eccentric contraction-induced injury on force and intracellular pH in rat skeletal muscles, *J. Appl. Physiol.*, 92(1), 93–99, 2002.

310. Zador, E., Mendler, L., Takacs, V., de Bleecker, J., and Wuytack, F., Regenerating soleus and extensor digitorum longus muscles of the rat show elevated levels of TNF-alpha and its receptors, TNFR-60 and TNFR-80, *Muscle & Nerve*, 24(8), 1058–1067, 2001.

311. Zajac, F.E., Muscle and tendon: properties, models, scaling, and application to biomechanics and motor control, *Crit. Rev. Biomed. Eng.*, 17(4), 359–411, 1989.

312. Zeng, C., Pesall, J.E., Gilkerson, K.K., and McFarland, D.C., The effect of hepatocyte growth factor on turkey satellite cell proliferation and differentiation, *Poult. Sci.*, 81(8), 1191–1198, 2002.

313. Zerba, E. and Faulkner, J.A., A Single Lengthening Contraction can Induce Injury to Skeletal Muscle Fibers, *Physiologist*, 33, A122, 1990.

314. Zerba, E., Komorowski, T.E., and Faulkner, J.A., Free radical injury to skeletal muscles of young, adult, and old mice, *Am. J. Physiol.*, 258(3 Pt 1), C429–C435, 1990.

16

Mechanisms for Pain and Injury in Musculoskeletal Disorders

Beth A. Winkelstein
University of Pennsylvania

Painful musculoskeletal disorders are a common problem in today's society, affecting an estimated one-third of the population. The societal costs (including litigation, work-lost, treatment and disability) for painful musculoskeletal disorders are staggering. Financially, considering all chronic pain syndromes together, there is an annual cost of $90 billion.[2] For example, the cost of low back pain alone has been estimated between $40–50 billion annually.[20,21] Until a better understanding of the pathomechanisms in chronic pain and the injuries which cause them is defined, the effective prevention and treatment of these disorders and their symptoms will remain elusive. It is the intent of this chapter to review and highlight traditional and more recently emerging theories explaining pain sensation, signaling and transmission in the context of injury and musculoskeletal disorders.

This review offers an overview of the mechanistic pathways of persistent (chronic) pain associated with musculoskeletal disorders (MSDs) and injuries. A discussion of the neurophysiology of pain addresses concepts of injury and pain processing and describes more recent hypotheses of the central nervous system's (CNS) neuroimmunologic involvement in persistent pain. These concepts, together with the associated neurochemical nociceptive responses are addressed and discussed in the context of findings from animal models of persistent pain with behavioral hypersensitivity. Physiologic responses in the CNS are addressed as they pertain to the interplay of the electrophysiological and immune research areas, and also as they relate to MSDs and biomechanics of injury. In particular, one such injury leading to persistent pain is radiculopathy, which results from nerve root compression or impingement and is a common source of low back pain. This painful syndrome is used here as an example to provide a context for presenting immune mechanisms of chronic pain and their relationship to specific injury parameters. Measures of injury biomechanics are presented for work with these models, including behavioral sensitivity, local structural changes, and cellular and molecular changes in the CNS, as they apply to kinematics, kinetics, and injury. Incorporating effects of injury parameters on mechanisms of persistent pain, the text discusses implication of these and other factors confounding pain in MSD. Lastly, based on these findings and others, a discussion is provided highlighting areas of future work to help elucidate methods of injury prevention, diagnosis and development of therapeutic treatments.

It is important to define, at the outset, relevant distinctions in terminology. "Pain" is a complex perception that is influenced by prior experience and by the context within which the noxious stimulus occurs. Likewise, "nociception" is the physiologic response to tissue damage or prior tissue damage. Similarly, for discussion in this chapter, "hyperalgesia" is defined as enhanced pain to a noxious stimulus.[55] Strictly speaking, this is a leftward-shift of the stimulus-response function relating pain to intensity. The corresponding pain threshold is lowered and there is enhanced response to a given stimulus. Hyperalgesia is mediated by nociceptor sensitization, where "sensitization" describes a corresponding shift in the *neural* response curve for stimulation. Sensitization is characterized by a decrease in threshold, an increased response to suprathreshold stimulus and spontaneous neural activity.

For this chapter, many of the examples are drawn from painful injuries related to the spine. These include both low back and neck pain from radiculopathy. While it is recognized that these examples are by no means all-inclusive of the painful musculoskeletal injuries, they do provide an ideal context for discussing many of the pain mechanisms presented here. Certainly, syndromes such as those affecting the carpal tunnel are important injuries in their own right, yet many of the same issues apply regarding neural tissue damage, mechanical injury and inflammation. In addition, in this regard, within these examples comments are made in this text to delineate the differences between persistent and resolving (acute) pain syndromes.

16.1 Neurophysiologic Mechanisms of Pain

There are a host of physiologic mechanisms by which tissue injury leads to nociception, and ultimately pain. In persistent pain, CNS signals can result in a hypersensitivity of neurons in the CNS. In addition to alterations in electrical signaling, neuroimmune responses are initiated both locally at the site of injury and in the spinal cord due to the altered neuronal function from perceived injury. Together, these changes can further contribute to sensitization and persistent pain symptoms. Findings with regard to both neuronal signaling and neuroimmunity are reviewed here to form a basis for discussing more recent views of nociceptive mechanisms of persistent pain.

16.1.1 Neural Anatomy Relevant to Pain

In order to provide a context for presentation and discussion of pain mechanisms, it is first necessary to describe the relevant anatomical structures, connections and relationships of neural sensory and processing components. These are reviewed only briefly here and are described in much greater detail in texts specializing in neural science and pain.[30,55]

Primary afferents, which are either directly injured or which relay pain signals from injured tissues, terminate in the dorsal horn of the spinal cord (Figure 16.1). At each spinal level, dorsal nerve roots carry sensory information from the periphery into the spinal cord. Dorsal roots contain sensory neurons with cell bodies housed in the enlarged dorsal root ganglion (DRG) outside the spinal column. In contrast, the ventral root contains the axons of neurons whose cell bodies are within the ventral horn of the spinal cord and transmits efferent signals. Outside of the spinal column, the dorsal and ventral nerve roots come together distal to the DRG at each spinal level and combine to form a

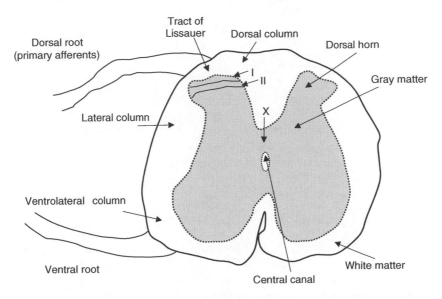

FIGURE 16.1 Schematic illustrating the spinal nerve roots, spinal cord and its regions of white and gray matter. Also indicated are columns of the neuronal tracts in the white matter. Those regions of the gray matter (laminae) of particular relevance to pain sensation and transmission are also labeled.

spinal nerve, which communicates with the peripheral nervous system. Spinal nerves further branch into smaller nerves, which innervate various tissues and different organs of the body. These peripheral nerves branch out into the periphery to innervate bones, ligaments, joints, discs, muscles, organs, and many other tissue-types (Figure 16.2).

The spinal cord is anatomically composed of two regions (Figure 16.1) based on appearance, function and cell populations. The gray matter, which has a darker appearance, contains the cell bodies of spinal neurons and comprises the central region of the spinal cord. The white matter surrounds the gray matter and contains the axons of the spinal neurons. The spinal columnar tracts are regionally specialized according to the information they carry, with the lateral column containing motor neurons, the dorsal column carrying information related to mechanoreception and the ventrolateral column housing neurons which communicate information regarding pain, temperature and motor signals (Figure 16.1). In general, the sensory system is an ascending pathway that comprises the dorsal portion of the spinal cord, while the descending pathway of the motor control system comprises the ventral aspect of the spinal cord.

Afferents of the dorsal root enter the spinal cord dorsolaterally and branch in the white matter, with collaterals terminating in the gray matter. Nerve fibers mediating pain pass through the tract of Lissauer and have branches terminating in the most superficial regions of the dorsal horn, laminae I and II (Figure 16.1). Neurons in these laminae synapse on secondary neurons in laminae IV–VI of the dorsal horn. These secondary neurons then cross the midline of the cord and ascend to the brain contralaterally in the anterolateral region of the cord. Also of note, lamina X, which is located in the gray matter region closest to the central canal, also receives pain sensory inputs. Nociceptive information is transmitted from the spinal cord to supraspinal sites, primarily in the pons, medulla, and thalamus. The anterolateral ascending system has three tracts: spinothalamic, spinoreticular, and spinomesencephalic. The spinothalamic is the most prominent of the tracts. Briefly, the spinothalamic and spinoreticular tracts mediate noxious sensations, with axons terminating on neurons in the reticular formation of the medulla and pons. From there, signals are relayed to the thalamus, and then, neurons project to the somatosensory cortex.

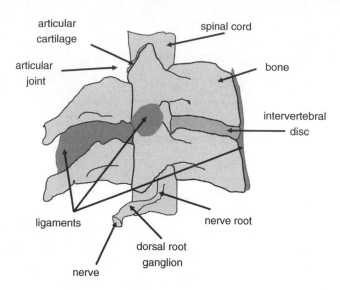

FIGURE 16.2 Schematic illustrating varied tissue components capable of sensing pain, particularly following mechanical or inflammatory injuries. While this schematic is based on the spinal column and highlights tissues relevant to its anatomy, it highlights all such tissues in the body, including neural tissue, hard and soft tissue, ligaments and bone. Muscle tissue is not shown here, but is innervated and is the tissue component particularly relevant to pain sensation.

16.1.2   Tissue Injury, Central Sensitization, and Pain

Injury to any of a variety of tissue components, including muscle, disc, ligament, and neural tissue can induce many cascades and physiological signals leading to pain perception. Most broadly, neuroplasticity and subsequent CNS sensitization include altered function of chemical, electrophysiological and pharmacological systems.[3,13,15,55,58,64] The interplay of these systems is intricate and involves complicated cross-system effects between injury and changes in both the peripheral nervous system and CNS (Figure 16.3). The integration of multiple physiologic systems that occurs as a result of a painful injury contributes to the overall challenges in preventing such syndromes since a given mechanical injury may initiate a host of different responses, which can be established and maintained remote from the actual site of injury. Nonetheless, there is a generalized series of responses, which occurs following a painful injury in the periphery.

Nociceptive afferents are specific for sensing different noxious stimuli (i.e., thermal, mechanical, and chemical). Some nociceptors are polymodal and sense all types of stimuli. Sensory nerve fibers range in diameter from <0.05 to 20 μm and can be thickly, thinly, or not myelinated. Conduction velocities of nerve fibers range from 0.5 to 120 m/sec, depending on their axon diameter and the presence of myelination. The fibers evoke sharp and pricking pain sensation, sometimes also aching pain. The largest, myelinated sensory axons, Aα, are generally classified as mechanoreceptiors. Aβ fibers are primarily proprioceptive (sensing mechanical movement in joints and muscles) and have diameters of ∼10 μm in size and slower conduction velocities than Aα fibers. The smallest myelinated fibers are Aδ fibers; they primarily mediate pin-prick, itching, and other mechanical sensations of pain. Unmyelinated, C fibers, mediate thermal sensation, in addition to mechanical pain. Stimulation of C fibers at high enough magnitude induce a burning sensation. C and some A fibers are primary high-threshold nociceptors. They have similar functions with A fibers having greater response frequencies and more communications to the spinal cord.

Pain can result from direct injury or from inflammation, which induces alterations in the local peripheral milieu. For a given insult, local nerve fibers become activated. The sensation of pain is initiated

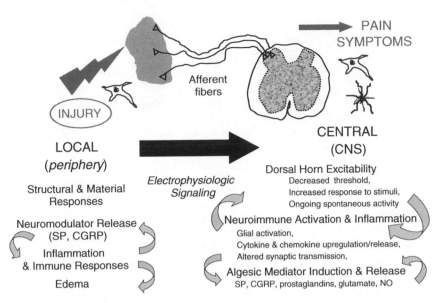

FIGURE 16.3 This schematic illustrates the collection of physiologic mechanisms following an injury in the periphery, which initiates and contributes to pain. Nociceptive responses are complicated and involve a host of changes both locally and in the CNS. While the schematic suggests a simple linear cascade (from left to right) of events following injury, which lead to pain, events are quite dynamic in nature and involve aspects of electrophysiology, immunology and an interplay between both. Some changes occur both at the site of injury and in the spinal cord and CNS. However, the degree to which these alterations occur is variable and dependent on both the nature of the injury and the type of response. Moreover, many of these individual responses affect each other (small arrows) and are themselves directly altered by confounding physiologic, anatomical and mechanical factors.

by nociceptors, the nerve endings of the $A\delta$ and C fibers. Conduction velocities of $A\delta$ fibers are approximately 10 times those of C fibers (5–30 and 0.5–2 m/sec, respectively) due to the saltatory conduction resulting from the myelin sheathing. As a consequence of this difference in conduction velocity, a sharp pain is first detected in response to a stimulus, followed by a second, longer lasting dull and burning pain mediated by the C fibers.

Pain sensation and signaling begin at the injury site where $A\delta$ and C fibers are activated by thermal, chemical, mechanical, and electrical stimuli. These nociceptors can become sensitized; this both lowers their thresholds for firing and increases their firing rates when stimulated at levels similar to those before injury.[4] Following injury, inflammation is induced in an effort to promote healing and recovery. In this process, inflammatory mediators such as prostaglandin E, serotonin, bradykinin and histamine, among others, can have altering effects on fiber responses for a given mechanical stimulation. These and other mediators activate nociceptors or further sensitize those nociceptors which are already responding to stimuli — inducing increased activity. Specifically, bradykinin, serotonin, excitatory amino acids, and hydrogen ions are all responsible for directly *activating* afferents.[4,31] Similarly, prostaglandins, serotonin, noradrenaline, adenosine, NO, and nerve growth factor *sensitize* nociceptors.

In addition to altered electrical activity, injury initiates the local synthesis and release of inflammatory mediators that induce inflammation and edema as part of the healing process (Figure 16.3). However, these same processes that provide healing also sensitize nociceptors and recruit new nociceptors that enhance pain.[17,19] Cytokines are also released in the periphery in association with tissue injury and inflammation. These small proteins, in turn, contribute to the local inflammatory response, while further affecting the electrophysiologic responses of nerve fibers in the region and altering nociception.

Additional details regarding specific biochemical mediators of pain and their mechanisms of action throughout the pain signaling pathway (i.e., injury site, spinal cord, brain) are provided in the following section specifically addressing biochemical mediators of pain.

Primary afferents communicate with spinal neurons via synaptic transmission. A variety of neurotransmitters (i.e., glutamate, NMDA, substance P; see following section for details) modulate postsynaptic responses, with further transmission to postsynaptic spinal neurons and supraspinal sites via the ascending pathways.[4] Tissue damage (injury) generates an increased neuronal excitability in the spinal cord.[64] Associated with this sensitization is also a decreased activation threshold, increased response magnitude and increased recruitment of receptive fields.[55] The continuous input from nociceptive afferents can drive the spinal circuits, leading to central sensitization, and maintaining a chronic pain state.[15] These neuroplastic changes are accompanied by other electrophysiological manifestations that cause neurons to fire with increased frequency or even spontaneously.[58] In addition, spinal processing is further directly altered by descending inhibitory and facilitory pathways that provide additional modulation of spinal interneurons.[54]

Ultimately, *persistent* pain results from the sensitization of the CNS. While the exact mechanism by which the spinal cord reaches a "hyperexcitable" state of sensitization is not fully known at present, many hypotheses have emerged. Only the highlights of these theories are provided here as an overview. More extensive and detailed discussions can be found elsewhere in the literature.[5,16,18,64] Central hyperexcitability is characterized by a "windup" response of repetitive C fiber stimulation, expanding receptive field areas and spinal neurons taking on properties of wide dynamic range neurons.[7] Low threshold Aβ afferents, which normally do not serve to transmit pain, become recruited to signal spontaneous and movement-induced pain.[16] Ultimately, Aβ fibers stimulate postsynaptic neurons to transmit pain, where these Aβ fibers previously had no effect. This plasticity of neuronal function in the spinal cord contributes to central sensitization.

16.1.3 Biochemical Mediators of Pain

A broad variety of chemical mediators are involved in pain, from the local sensory terminals, to synaptic transmitters, to afferent sensitizers in the periphery, neuronal activators, and modulators and spinal mediators. Many of these mediators are endogenous; some are inducible. Many are directly involved in pain; some are primarily inflammatory, with secondary actions on nociception. For a complete review of all of these the reader is directed to a comprehensive source of pain information.[55] However, a brief synopsis is provided here to offer context for understanding the larger picture of pain signaling.

Bradykinin is a local hormone that is heavily involved in inflammation and vascular responses, as well as contributing to pain signaling pathways. This protein works through a series of transmembrane G-protein coupled receptors, most notably bradykinin B1, which is implicated in chronic pain, and bradykinin B2, which is involved in acute inflammation and pain.[9] Both receptors are expressed on primary sensory neurons where they are positioned to participate in nociception. Bradykinin sensitizes nociceptors and thereby produces hyperalgesia. While the B2 receptor is constitutive, the B1 receptor is induced in tissue that is damaged either by infection, disease or injury. For example, 48 h after peripheral nerve injury, B2 receptor mRNA is increased, while it was not until 14 days later that B1 expression reaches similar levels.[9] B2 receptor kinetics are fast relative to B1, including the processes of ligand binding, release, and internalization. Through the B2 receptor, bradykinin causes the release of substance P and calcitonin gene-related peptide (CGRP) from primary sensory neurons. Of particular relevance to pain, this release is increased in the presence of prostaglandins, which are upregulated at the site of insult.

Prostanoids are a class of pro-nociceptive molecules such as prostaglandin E2 that are produced when arachindonic acid is metabolized by the COX enzyme, which exists in isoforms COX-1 and COX-2. Specifically the COX-2 isoform is dramatically upregulated in the spinal cord after injury in the periphery. The release of prostaglandin E2, which occurs in response to the upregulation of the enzyme causes a shift in the peripheral terminal of the nociceptor lowering its threshold and increasing its excitability.

Likewise, there are also a host of other inflammatory-related mediators, which also have direct and indirect effects on pain and nociception. Serotonin (5HT) is elevated in inflammation. It is released from platelets downstream from mast cell degranulation, as occurs during inflammatory responses. Sertotonin can cause pain by locally activating primary afferents and can further enhance sensitivity of sensory responses to bradykinin. Also, leukotriene B4 is a neutrophil attractant that can directly sensitize afferents. Certainly, each of these mediators has a complicated mechanism of action far more complicated than has been reviewed here.

Substance P is a potent pro-nociceptive neurotransmitter, which is transported to the periphery after afferent activation. It has inflammatory effects by causing vasodilation and release of prostaglandin E2 and cytokines. These protein releases further impact the local responses of inflammation as well as nociception. Substance P can cause a release of calcium from intracellular stores and in turn lead to NO production and neuronal excitability and long-term sensitization.[34,36] Substance P is also shown to have a role in pain in the CNS. Also contributing to sensitization, CGRP, a neuropeptide often colocalized with substance P in the spinal cord, regulates nociceptive responses by further promoting the release of substance P as well as glutamate from primary afferents and retarding the metabolism of substance P.[1,35] Antibodies to substance P and CGRP have been demonstrated to attenuate pain symptoms in inflammatory models of carageenan-induced hyperalgesia and painful nerve injury.[34,46] These results strongly implicate both neuropeptides in the transmission of pain. Additionally, application of antagonists to the substance P receptor, NK-1, has induced antinociception in the CNS after chronic nerve constriction,[35] as well as globally in rodent models of inflammatory arthritis.[28] However, despite research indicating the potent role of substance P and CGRP in many types of pain, little is known about the relative contributions of these neuropeptides to the onset or maintenance of pain symptoms, either in persistent or resolving pain.

Excitatory amino acids, such as glutamate, have potent roles in pain, both locally at the site of injury and in the CNS. Indeed, glutamate is produced by both non-neuronal and neuronal cells in the periphery. Both of these sources act on primary afferents by activating them when bound to any of the NMDA, kainite, AMPA or metabotropic glutamate receptors.[55] Certainly, this positive feedback mechanism of nociceptor excitation leads to peripheral sensitization and correspondingly to altered afferent signaling into the spinal cord. Glutamate receptors are expressed in dorsal root ganglion cells,[46] suggesting its direct involvement in afferent signaling to the cord. As the NMDA glutamate receptor has a key role in regulating synaptic efficacy, there is also a direct role of this receptor in the plasticity changes, which occur in the spinal cord associated with central sensitization changes of persistent pain.

Cytokines can directly or indirectly regulate cascades that can lead to the transmission and modulation of pain. The broad collection of cytokines includes both pro-inflammatory and anti inflammatory proteins, both of which are upregulated in painful injuries.[10,11,25,56,57] In particular, in models of neural injury either distal (peripheral injury) or proximal (nerve root injury) to the DRG, spinal IL-1β, IL-6, IL-10 and TNF mRNA are all significantly elevated.[61] However, not all of these factors are pro-inflammatory. In fact, IL-10 has been shown to suppress NO production in cultured astrocytes and also suppress proliferation in macrophages,[66] providing an anti-inflammatory mechanism. Additionally, the presence of cytokines often stimulates a positive feedback loop as demonstrated in the case of IL-1, which can stimulate its own production.[56] Cytokines mediate cellular processes through the production or suppression of NO. NO has an immunoregluatory role in the CNS. Its production leads directly to the hyperalgesic NMDA pathway as discussed earlier. Cytokines regulate NO by interfering with the production of NOS (nitric oxide synthase). Of the two main forms of NOS, inducible and constitutive, astrocytes can produce both types, whereas microglia are only responsible for inducible NOS.[66] TNF-α and IL-1β control the stimulation of the production of iNOS in both astrocytes and glia therefore inducing the production of NO from those cell types. Alternatively, TGF-β suppresses NO production in both astrocytes and microglia, whereas IL-10 only affects NO production in astrocytes.

16.1.4 Neuroimmunologic Responses in the CNS

While central sensitization contributes to nociceptive mechanisms of persistent pain in the CNS, recent research has demonstrated the role of spinal neuroimmune responses as contributing to persistent pain.[14] Several research groups have documented CNS immune changes associated with persistent pain due to a host of painful syndromes, including radiculopathy, neuropathy, diabetes, and HIV, among others.[10,43,51,56,57,61] This work has provided evidence supporting the role of centrally produced pro-inflammatory cytokines, glial activation and leukocyte trafficking in rodent pain models of L5 radiculopathy,[6,23,44,50,61] lending support for these immunologic changes contributing to pain. From this body of work, a cascade of events in the CNS has been proposed following injury.[13,14] Glial cells and neurons become activated and can produce and release cytokines, which lead to increased activation of these cell types. They also lead to further release of pain mediators, such as many of those described in the previous section.[10,11,56] Glial and neuronal pro-inflammatory cytokines can sensitize peripheral nociceptive fields[29] and sensitize dorsal root ganglia.[39]

Events that induce behavioral hypersensitivity also activate immune cells both centrally and in the periphery, mediating chronic pain.[10,11,56] Cytokines and growth factors have been strongly implicated in the generation of pathological pain states throughout the nervous system; in particular, pro-inflammatory cytokines, such as IL-1, IL-6, and TNF, are upregulated both locally and in the spinal cord in persistent pain.[10,65] Immune activation with cytokine production may indirectly induce the expression of many pain mediators such as glutamate, NO, and prostaglandins in the CNS, leading further to spinal sensitization. In conjunction with this neuroimmune activation, neuroinflammation occurs in which immune cells migrate from the periphery into the CNS in association with pain.[13,14,44] This infiltration may lead to further changes in the CNS and potentially to central sensitization. Infiltrating immune cells contribute to neuronal activation and algesic mediator release, further perpetuating the maintained excitability and sensitization in the CNS, which leads to behavioral sensitivity and pain. The spinal immune response of nociception has many facets, forming a complicated cascade of events leading to pain.

16.2 Implications for MSD: Pain Mechanisms and Injury Biomechanics

Electrophysiologic and neuroimmune responses of the CNS likely work together to affect pain for MSDs, with local biomechanics at the injury site modulating both such response cascades. Low back pain is an ideal representative syndrome to use as an illustrative example for discussing injury mechanisms and cellular response cascades of a chronic painful MSD in light of the previous sections on mechanisms. In this discussion, injury conditions are presented as examples of how mechanical loading modulates nociception in low back pain, with particular emphasis on nerve root injury (i.e., radiculopathy).

Many animal studies report altered electrophysiologic and cellular function for graded cauda equina compression. Compression increases endoneurial pressure locally in the rat sciatic nerve and DRG in proportion to mechanical loading.[33,45] In addition, edema patterns and intensity are modulated by the nature of the mechanical insult.[37,38,40,45] Nerve root loading produces changes in electrical impulse propagation and conduction velocity[8,22,47] and repetitive neuronal firing in the dorsal horn of the spinal cord,[22,67] which are all physiologic correlates suggestive of sensitization leading to pain. While this collective body of work suggests a mechanism of spinal cord plasticity and central sensitization for mechanical injuries, it is only inferential for understanding production and maintenance of pain.

Use of imaging techniques has quantified nerve root tissue deformation in a rodent model of painful lumbar radiculopathy[62,63] via root ligation and examined this injury parameter in the context of pain (behavioral hypersensitivity). In that work, local injury mechanics were found to modulate pain behaviors; there was a significant positive correlation in this pain model between behavioral sensitivity and the amount of tissue injury.[62] Most simply, the greater nerve root compression at injury, the more severe the clinical symptoms of behavioral sensitivity and pain. From this series of work, mechanical compressive *deformation* thresholds for pain behaviors and hypersensitivity were defined based on the amount of nerve root compression.[60] More recent work examining cervical nerve root compression magnitude and behavioral sensitivity in the forepaw also suggests a *force*-based threshold for pain symptoms, below 10gf applied to the dorsal nerve root.[26] These mechanical parameters defining painful injuries provide added utility for clinicians in diagnosing painful injuries, directly linking the injury event to the likelihood of pain symptoms. Moreover, in future, it will hopefully provide insight into predicting clinical outcomes for this class of injuries.

While defining the relationship between injury events and pain is necessary for understanding the clinical context of these pathologies, defining the relationship between injury and specific and relevant *nociceptive* responses is crucial for understanding the central mechanisms of persistent pain in MSD. Using RNase Protection Assays to detect spinal mRNA of a panel of cytokines (TNFα, IL-1α/β, IL-6, IL-10), a statistically significant correlation was found between mRNA levels at postoperative day 7 and the degree of tissue deformation at injury.[62] This suggests a modulatory effect of injury magnitude on one aspect of spinal nociception. Using immunohistochemistry, lumbar spinal expression of the pro-inflammatory cytokine IL-1β was previously found to depend on nerve root compression intensity;[23] suggesting preservation of these changes at both the message and protein levels for the spinal cytokines involved in chronic low back pain responses. Further, in comparative models of cervical nerve root injury, either from compression or transection, with different symptom presentation, spinal IL-6 protein at day 1 was elevated over sham levels, suggesting a potential relationship to hypersensitivity on the day of assay.[25] While continued research into these and other cytokine responses is needed for understanding cervical nerve root injuries and pain mechanisms, findings suggest that similar cytokine responses may be evident throughout the spine (i.e., lumbar versus cervical injuries).

Consistent with the grading of behavioral responses and spinal cytokine expression according to injury severity,[23,62,63] spinal microglial activation has been demonstrated to be more intense for greater nerve root deformation in lumbar injuries.[23,59] Yet, in these same studies, astrocytic activation did not follow injury magnitude, highlighting that biomechanics at injury in lumbar radiculopathy models may differentially modulate some neuroimmune responses and not others (Figure 16.3). Recent work from our laboratory has examined these same spinal glial responses in two different cervical spine injuries;[26,32] in these studies of nerve root compression and facet joint tension, different astrocytic responses were observed. Specifically, for the nerve root compression, spinal astrocytic expression was elevated over sham (Figure 16.4), and followed the behavioral hypersensitivity patterns. Despite this, astrocytic activation did not show a dependence on mechanical force magnitude. This is similar to the earlier findings of nerve root injury in the lumbar spine.[59] In contrast, in our mechanical facet-mediated painful injury model, spinal astrocytic activation *did* demonstrate a significant correlation with injury magnitude (Figure 16.4).[32] This finding may suggest that different spinal immune response cascades exist for mechanical injuries to different tissue types. It further highlights the need for continued integrative research to identify common and different physiologic mechanisms for injuries within the musculoskeletal system.

16.3 Confounding Factors Affecting Pain

Given the extreme complexity of understanding painful injuries and their mechanisms, it is worthwhile noting that a variety of factors can affect injury mechanisms and resulting pain. For example, many biomechanical factors have been shown to alter neuronal function: rate, loading duration, and load magnitude, among others. Given the complexity of pain mechanisms, the factors that confound the injuries themselves, can play a very large role in pain (Table 16.1). Among these, the physiologic milieu at the time of injury is also believed to modulate (i.e., potentially worsen) painful outcome. For example, it has been shown that the pain-associated behaviors resulting from a second radiculopathy injury are significantly greater than those behavioral responses of the first injury.[27] Expanding on these findings, it can be hypothesized that injury following a pre-existing "minor" tissue injury may result in a more severe nociceptive response than in the absence of any pre-existing damage. It is not unlikely that an initial

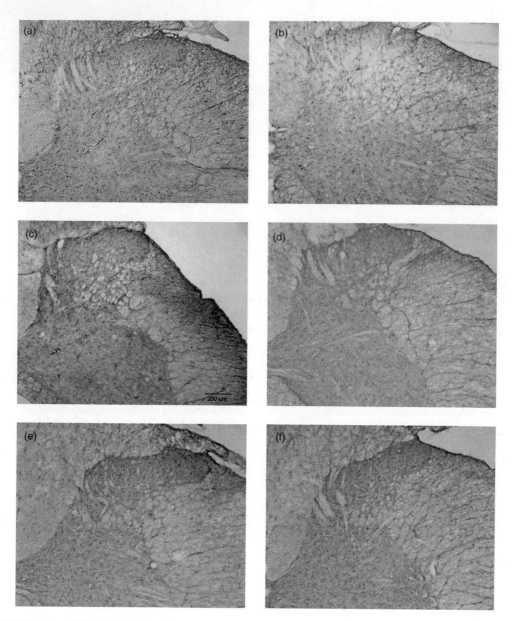

FIGURE 17.4 Glial cell activation responses (as detected by GFAP) in the ipsilateral dorsal horn of the spinal cord at the level of injury for facet joint (a, b, c) and nerve root (d, e, f) injuries. Sham levels (a, d) are the same as normal naïve responses. For the facet injury, astrocytic activation was increased for 0.89 mm of distraction (c) with significant increases in behavioral sensitivity, compared to 0.17 mm of distraction (b), which produced no symptoms of hypersensitivity. In contrast, while both 10gf (e) and 60gf (f) nerve root compression produced behavioral hypersensitivity, there was no difference in GFAP reactivity for such six-fold difference in injury force. Bar in c is 200 μm, and applies to all. Images are modified from those published in Lee, K.E., Davis, M.B., Mejilla, R.M., Winkelstein, B.A. Proceedings of 48th Stapp Car Crash Conference, Paper pp. 2004-22-0016, 48, pp. 373–393, 2004 and Hubbard and Winkelstein *Spine*, 30, pp. 1924–1932, 2005. With permission.

injury produces the local inflammatory changes discussed earlier, which in turn may lower that tissue's threshold for mechanical injury. In fact, it has been shown that the mechanical insult required to produce behavioral hypersensitivity in the presence of an inflammatory insult is nearly one-half that required in its absence, despite resulting in the same degree of sensitivity.[13]

TABLE 16.1 Factors Affecting Pain Outcomes

Mechanical Factors

Injury parameters
—Magnitude (load and deformation)
—Direction
—Rate
—Duration
Initial geometry (configuration) of anatomical structure
Loading frequency (i.e., repetitive or single application)
Responses of specific anatomical structures (i.e., individual tissue responses)

Physiological Factors

Anatomic features (i.e., geometry, stenosis, etc.)
Inflammation (pre-existing)
Degeneration (pre-existing, tissue and neural circuit)
Electrophysiological sensitivity
Electrophysiological preconditioning before injury
Immunological preconditioning before injury

Confounding Factors

Genetics
Gender
Age
Comorbidities
Psychological
Psychosocial and environmental issues

Inferences can be made in light of these findings for painful MSDs. For example, epidemiologic studies indicate that patients with pre-existing spinal degeneration at the time of injury, experience more severe and longer lasting neck pain symptoms.[41,42,49] It is possible that such degeneration can contribute to inflammatory changes in the facet joint, which may increase this joint's susceptibility to mechanical injury. Therefore, when undergoing motions or loading, which may not normally elicit nociceptive changes, the pain fibers may be sensitized and fire under mechanical conditions, which are much less severe than previously required to initiate nociception. The same may be true for degenerative changes in other tissues. In this light, it may further be possible that *repetitive* loading of tissues at normally noninjurious level, when sustained over many times, can lead to painful outcomes or lowered thresholds for stimulating painful injury.

Additional geometric and anatomic factors contributing to pain risk are gender, existing spinal degeneration, stenosis, and genetics[48] (Table 16.1). For example, for the case of whiplash, females experience increased symptom persistence when compared to males.[48] The anatomical considerations specifically related to gender, which include decreased neck muscle strength and spinal canal size, add support to the role of neck mechanics in affecting a pain mechanism. Moreover, anecdotal evidence has shown that smaller spinal canal size is associated with more symptomatic responses in whiplash.[41] Similar geometric constraints of anatomy also apply to regions of the body such as the wrist (i.e., carpal tunnel syndrome). Finally, more recent clinical research into spinal pain in general has shown that genetics may play a very key role in pain persistence for a given injury, accounting for many discrepancies observed among different patients for seemingly similar injuries.[13] Future research into neck pain mechanisms would be strengthened if it considered the role of genetics in many of the issues discussed in this chapter.

16.4 Implications for MSD: Applications and Future Research

Emerging out of this discussion, a number of areas of research focuses, which remain to be investigated are identified for painful MSD. From the broad coverage presented earlier, it can be appreciated that many aspects of injury, physiology, and cellular mechanisms contribute to chronic pain in MSDs. In this context, then, it is possible to synthesize these findings to discuss preventing these injuries and treating them. As continued biomechanical research is performed to determine conditions under which tissue injury occurs and initiates physiologic responses, it becomes clear that findings can help guide preventive strategies to protect some of these anatomic structures and tissues from undergoing kinematically and kinetically risky situations and injury. In addition, the cellular findings presented here highlight the need for defining the relationship of an injury event, its physiologic responses, and their relationship to behavioral manifestations of pain symptoms.

As the understanding of the mechanisms of persistent pain expands, increased research is being focused on development of effective treatment modalities. A broad variety of approaches exist for offering pain relief: joint blocks, TENS, manipulation, pharmacology, and many others.[13] However, the exact mechanisms of injury often remain elusive, making it extremely challenging to act at the structural site of injury for therapy. Pharmacologic treatment options offer a promising approach for manipulating those aspects of the CNS response, which contribute to chronic nociception. For example, global immunosuppressants have been shown to ameliorate pain behaviors in both neuropathic and radiculopathic rodent pain models.[61] Likewise, manipulation of specific spinal cytokines to alter sensory processing and other select agents have been effective in reducing allodynia in a variety of pain models.[12,24,51,52,61] Pharmacologic antagonists to and inhibitors of particular pro-inflammatory cytokines and other algesic mediators (IL-1, TNF, COX-2) have shown effectiveness in animal pain models for attenuating both behavioral hypersensitivity and elements of the CNS neuroimmune cascade.[12,24,51,61] Indeed, combinations of some of these agents may have promise for effectiveness in reducing pain. As continued research identifies the specific physiologic pathways (both electrophysiologic and immunologic), which are responsible for chronic pain, it will become more feasible and even more tractable to target specific sites along these pathways for selectively manipulating and modulating a persistent pain response. With continued integrative efforts, progress will be made in this area.

16.5 Summary

It is recognized that spinal injuries are by no means the only chronically painful MSDs. As such, it should be noted that many of the theories described previously may assist with developing a more broad understanding in the context of other painful MSDs, such as carpal tunnel syndrome and other repetitive motion injuries. While magnitude, rate, and duration of loading all modulate electrical signaling patterns (amplitude, frequency) and local tissue changes (edema, pressure), and the neuroimmune cascade for painful radiculopathy, their effects for other painful syndromes may be similar. Continued integration of multidisciplinary approaches applied to a broader class of MSDs will help define nociceptive responses in these disorders.

In the typical response of an acutely painful episode, the balance of injury, repair, and healing is achieved and the cascade of electrophysiologic and chemical events resolves following inflammation and injury. However, for *persistent* pain, the local, spinal, and even supraspinal, responses are undoubtedly altered from that described previously. Based on the discussion presented in the previous sections regarding persistent pain, a comprehensive picture is emerging for neural injury and CNS responses of nociception: spinal cytokine upregulation, microglial, and astrocytic activation, altered neuronal–glial interactions, cellular adhesion molecule upregulation, and immune cell infiltration into the spinal cord.[13,14,50,53,56] These aspects of neuroimmune activation induce the expression and release of pain mediators (substance P, glutamate, NO) and also lead to neuronal hypersensitivity.

In this context, it is important to consider novel methods for preventing and treating painful injuries. Clinical emphasis has largely been focused on local interventions at the injury site. However, the previous discussion points to the spinal cord physiology as having equal, if not stronger, contribution for maintenance of pain. Continued understanding of spinal and supraspinal mechanisms and mediation of central sensitization can hopefully provide valuable contributions to this understanding. It is the hope that this review has provided a summary of current thinking in pain mechanisms with a particular emphasis on how these mechanisms relate to musculoskeletal injury. Likewise, it was the intent to illuminate interesting new work within the study of pain, highlighting the complications and intricacies of its nature. Finally, through this presentation, areas of future work have been indicated. It is only through continual efforts that meaningful advances will be made in preventing and treating painful musculoskeletal disorders.

Acknowledgments

The author gratefully acknowledges the following for financial support: National Institute of Arthritis and Musculoskeletal and Skin Diseases (AR47564), the Whitaker Foundation and the Catharine Sharpe Foundation.

References

1. Allen, B., Li, J., Menning, P., Rogers, S., Ghilardi, J., Mantyh, P., and Simone, D. Primary afferent fibers that contribute to increased substance P receptor internalization in the spinal cord after injury. *J. Neurophysiol.*, 81(3), pp. 1379–1390, 1999.
2. American Chronic Pain Association, http://www.theacpa.org/.
3. Black, J., Langworthy, K., Hinson, A., Dib-Hajj, S., and Waxman, S. NGF has opposing effects on Na+ channel III and SNS gene expression in spinal sensory neurons. *Neuroreport* 8, pp. 2331–2335, 1997.
4. Cavanaugh, J.M. Neurophysiology and neuroanatomy of neck pain. In: *Frontiers in Whiplash Trauma: Clinical and Biomechanical.* Yoganandan, N., Pintar, F.A., Eds. IOS Press. Amsterdam, 2000, pp. 79–96.
5. Coderre, T.J., Katz, J., Vaccarino, A., and Melzack, R. Contribution of central neuroplasticity to pathological pain: review of clinical and experimental evidence. *Pain*, 52, pp. 259–285, 1993.
6. Colburn, R., Rickman, A., and DeLeo, J. The effect of site and type of nerve injury on spinal glial activation and neuropathic pain behavior. *Exp. Neurol.*, 157, pp. 289–304, 1999.
7. Cook, A.J., Woolf, C.J., Wall, P.D., and McMahon, S.B. Dynamic receptive field plasticity in rat spinal cord dorsal horn following C-primary afferent input, *Nature*, 325, pp. 151–153, 1987.
8. Cornefjord, M., Sato, K., Olmarker, K., Rydevik, B., and Nordborg, C. A model for chronic nerve root compression studies. Presentation of a porcine model for controlled slow-onset compression with analyses of anatomic aspects, compression onset rate, and morphologic and neurophysiologic effects. *Spine*, 22, pp. 946–957, 1997.
9. Couture, R., Harrisson, M., Vianna, R.M., and Cloutier, F. Kinin receptors in pain and inflammation. *Eur. J. Pharmacol.*, 429, pp. 161–176, 2001.
10. DeLeo, J. and Colburn, R. Low Back Pain: A Scientific and Clinical Overview. The role of cytokines in nociception and chronic pain, in: Weinstein, J., Gordon, S., Eds. Rosemont, IL: AAOS, pp. 163–185.
11. DeLeo, J., Colburn, R., Nichols, M., and Malhotra, A. Interleukin (IL)-6 mediated hyperalgesia/allodynia and increased spinal IL-6 in two distinct mononeuropathy models in the rat. *J. Interferon Cytokine Res.*, 16, pp. 695–700, 1996.
12. DeLeo, J., Hashizume, H., Rutkowski, M., and Weinstein, J. Cyclooxygenase–2 inhibitor SC-236 attenuates mechanical allodynia following nerve root injury in rats. *J. Orthop. Res.*, 18(6), pp. 977–982, 2000.
13. DeLeo, J. and Winkelstein, B. Physiology of chronic spinal pain syndromes: From animal models to biomechanics. *Spine*, 27(22), pp. 2526–2537, 2002.
14. DeLeo, J.A. and Yezierski, R.P. The role of neuroinflammation and neuroimmune activation in persistent pain. *Pain*, 91, pp. 1–6, 2001.
15. Devor, M. Neuropathic pain and injured nerve: peripheral mechanisms. *Br. Med. Bull.*, 47, pp. 619–630, 1991.

16. Devor, M. Pain arising from the nerve root and the dorsal root ganglion. in: *Low Back Pain: A Scientific and Clinical Overview*. Weinstein, J., Gordon, S., Eds. Rosemont, IL: AAOS, 1986, pp. 187–208.

17. Dray, A. and Perkins, M. Bradykinin and inflammatory pain. *Trends Neurosci.*, 16(3): pp. 99–104, 1993.

18. Dubner, R. and Basbaum, A.I. Spinal dorsal horn plasticity following tissue or nerve injury. in: *Textbook of Pain*. Wall, P.D., Melzak, R., Eds. Churchill-Livingstone: Edinburgh, 1994, pp. 225–241.

19. Dubner, R. and Hargreaves, K.M. The neurobiology of pain and its modulation. *Clin. J. Pain*, S1–S6, 1989.

20. Frymoyer, J. and Cats-Baril, W. An overview of the incidences and costs of low back pain. *Orthop. Clin. North Am.*, 22, pp. 263–271, 1991.

21. Frymoyer, J. and Durett, C. *The economics of spinal disorders*. In: *The Adult Spine: Principles and Practice*. Frymoyer, J.W., Eds. Lippincott-Raven Publishers: Philadelphia 1997.

22. Hanai, F., Matsui, N., and Hongo, N. Changes in responses of wide dynamic range neurons in the spinal dorsal horn after dorsal root or dorsal root ganglion compression. *Spine*, 21, pp. 1408–1415, 1996.

23. Hashizume, H., DeLeo, J., Colburn, R., and Weinstein, J. Spinal glial activation and cytokine expression following lumbar root injury in the rat. *Spine*, 25, pp. 1206–1217, 2000.

24. Hashizume, H., Rutkowski, M., Weinstein, J., and DeLeo, J. Central administration of methotrexate reduces mechanical allodynia in an animal model of radiculopathy/sciatica. *Pain*, 87(2), pp. 159–169, 2000.

25. Hubbard, R., Rothman, S., and Winkelstein, B. Mechanisms of persistent neck pain following nerve root compression injury: understanding behavioral hypersensitivity in the context of spinal cytokine responses and tissue biomechanics. North American Spine Society 19th Annual Meeting, #P49, Chicago, IL, October, 2004.

26. Hubbard, R.D. and Winkelstein, B.A. Transient cervical nerve root compression in the rat induces bilateral forepaw allodynia and spinal glial activation: mechanical factors in painful neck injuries. *Spine*, 30, pp. 1924–1932, 2005.

27. Hunt, J.L., Winkelstein, B.A., Rutkowski, M.D., Weinstein, J.N., and DeLeo, J.A. Repeated injury to the lumbar nerve roots produces enhanced mechanical allodynia and persistent spinal neuroinflammation. *Spine*, 26, pp. 2073–2079, 2001.

28. Hong, S., Han, J., Min, S., Hwang, J., Kim, Y., Na, H., Yoon, Y., and Han, H. Local neurokinin-1 receptor in the knee joint contributes to the induction, but not maintenance, of arthritic pain in the rat. *Neurosci. Lett.*, 322(1), pp. 21–24, 2002.

29. Junger, H. and Sorkin, L.S. Nociceptive and inflammatory effects of subcutaneous TNF alpha. *Pain*, 85(1–2), 145–151, 2000.

30. Kandel, E.R., Schwartz, J.H., and Jessell, T.M. Principles of Neural Science. 3rd edn, Elsevier: New York, 1991.

31. Kawakami, M. and Weinstein, J.N. Associated neurogenic and nonneurogenic pain mediators that probably are activated and responsible for nociceptive input. in: *Low Back Pain: A Scientific and Clinical Overview*. Weinstein, J., Gordon, S., Eds. Rosemont, IL: AAOS, pp. 265–273, 1986.

32. Lee, K.E., Davis, M.B., Mejilla, R.M., and Winkelstein, B.A. *In vivo* cervical facet capsule distraction: mechanical implications for whiplash and neck pain. Proceedings of 48th Stapp Car Crash Conference, Paper #pp. 2004-22-0016, 48, pp. 373–393, 2004.

33. Lundborg, G., Myers, R., and Powell, H. Nerve compression injury and increased endoneurial fluid pressure: a "miniature compartment syndrome." *J. Neurol. Neurosurg. Psychiatry*, 46, pp. 1119–1124, 1983.

34. Ma, W. and Eisenach, J. Intraplantar injection of a cyclooxygenase inhibitor ketorolac reduces immunoreactivities of substance P, calcitonin gene-related peptide, and dynorphin in the dorsal horn of rats with nerve injury or inflammation. *Neuroscience*, 121, pp. 681–90, 2003.

35. Meert, T., Vissers, K., Geenan, F., and Kontinen, V. Functional role of exogenous administration of substance P in chronic constriction injury model of neuropathic pain in gerbils. *Pharmacol. Biochem. Behav.*, 76(1), pp. 17–25, 2003.

36. Millan, M. The induction of pain: An integrative review. *Prog. Neurobiol.*, 57, pp. 1–164, 1999.

37. Olmarker, K., Holm, S., and Rydevik, B. Importance of compression onset rate for the degree of impairment of impulse propagation in experimental compression injury of the porcine cauda equina. *Spine*, 15, pp. 416–419, 1990.

38. Olmarker, K., Rydevik, B., and Holm, S. Edema formation in spinal nerve roots induced by experimental, graded compression. An experimental study on the pig cauda equina with special reference to differences in effects between rapid and slow onset of compression. *Spine*, 14, pp. 569–573, 1989.

39. Ozaktay, A.C., Cavanaugh, J.M., Asik, I., DeLeo, J.A., and Weinstein, J.N. Dorsal root sensitivity to interleukin-1 beta, interleukin-6 and tumor necrosis factor in rats. *Eur. Spine J.*, 11(5), pp. 467–475, 2002.

40. Pedowitz, R., Garfin, S., Massie, J., Hargens, A., Swenson, M., Myers, R., and Rydevik, B. Effects of magnitude and duration of compression on spinal nerve root conduction. *Spine*, 17, pp.194–199, 1992.

41. Pettersson, K., Karrholm, J., Toolanen, G., and Hidlingsson, C. Decreased width of the spinal canal in patients with chronic symptoms after whiplash injury. *Spine*, 20, pp. 1664–1667, 1995.

42. Radanov, B.P., Sturzenegger, M., and DiStefano, G. Long-term outcome after whiplash injury. *Medicine*, 74, pp.281–297, 1995.

43. Rutkowski, M., Pahl, J., Sweitzer, S., and DeLeo, J. Limited role of macrophages in generation of nerve injury-induced mechanical allodynia. *Physiol. Behav.*, 71, pp. 225–235, 2000.

44. Rutkowski, M.D., Winkelstein, B.A., Hickey, W.F., Pahl, J.L., and DeLeo, J.A. Lumbar nerve root injury induces CNS neuroimmune activation and neuroinflammation in the rat: Relationship to painful radiculopathy. *Spine*, 27(15): pp. 1604–1613, 2002.

45. Rydevik, B., Myers, R., and Powell, H. Pressure increase in the dorsal root ganglion following mechanical compression. Closed compartment syndrome in nerve roots. *Spine*, 14, pp. 574–576, 1989.

46. Satoh, M., Kuraishi, Y., and Kawamura, M. Effects of intrathecal antibodies to substance P, calcitonin gene-related peptide and galanin on repeated cold stress-induced hyperalgesia: comparison with carrageenan-induced hyperalgesia. *Pain*, 49(2), pp. 273–278, 1992.

47. Skouen, J., Brisby, H., Otami, K., Olmarker, K., Rosengren, L., and Rydevik, B. Protein markers in cerebrospinal fluid experimental nerve root injury. A study of slow-onset chronic compression effects or the biochemical effects of nucleus pulposus on sacral nerve roots. *Spine*, 24, pp. 2195–2200, 1999.

48. Spitzer, W.O., Skovron, M.L., Salmi, L.R., Cassidy, J.D., Duranceau, J., Suissa, S., and Zeiss, E. Scientific monograph of the Quebec Task Force on whiplash-associated disorders: redefining "whiplash" and its management. *Spine*, 20, 1S–73S, 1995.

49. Sturtzenegger, M., Radanov, B.P., and DiStefano, G. The effect of accident mechanisms and initial findings on the long-term course of whiplash injury. *J. Neurol.*, 242, pp. 443–439, 1995.

50. Sweitzer, S., Arruda, J., and DeLeo, J. The cytokine challenge: methods for the detection of central cytokines in rodent models of persistent pain. in: *Methods in Pain Research*. Kruger, L., Ed. CRC Press, 2001, pp. 109–132.

51. Sweitzer, S., Martin, D., and DeLeo, J. IL-1ra and sTNFr reduces mechanical allodynia and spinal cytokine expression in a model of neuropathic pain. *Neuroscience*, 103, pp. 529–39, 2001.

52. Sweitzer, S., Schubert, P., and DeLeo, J. Propentofylline, a glial modulating agent, exhibits antiallodynic properties in a rat model of neuropathic pain. *J. Pharmacol. Exp. Ther.*, 297(3), pp. 1210–1217, 2001.

53. Sweitzer, S., White, K.A., Dutta, C., and DeLeo, J. The differential role of spinal MHC class II and cellular adhesion molecules in peripheral inflammatory versus neuropathic pain in rodents. *J. Neuroimmunol.*, 125(1–2): pp. 82–93, 2002.

54. Vanderah, T.W., Ossipov, M.H., Lai, J., Malan, T., and Porreca, F. Mechanisms of opioid-induced pain and antinociceptive tolerance: descending facilitation and spinal dynorphin. *Pain*, 92, pp. 5–9, 2001.

55. Wall, P. and Melzack, R. Textbook of Pain. 3rd edn., Churchill Livingstone: London, 1994.

56. Watkins, L., Maier, S., and Goehler, L. Immune activation: the role of pro-inflammatory cytokines in inflammation, illness responses, and pathological pain states. *Pain*, 63, pp. 289–302, 1995.

57. Watkins, L., Wiertelak, E., Goehler, L., Smith, K., Martin, D., Maier, S. Characterization of cytokine-induced hyperalgesia. *Brain Res.*, 654, pp. 15–26, 1994.

58. Waxman, S., Dib-Hajj, S., Cummins, T., and Black, J. Sodium channels and pain. *Proc. Natl. Acad. Sci.*, U.S.A. 96, pp. 7635–7639, 1999.

59. Winkelstein, B.A. and DeLeo, J.A. Nerve root tissue injury severity differentially modulates spinal glial activation in a rat lumbar radiculopathy model: Considerations for persistent pain. *Brain Res.*, 956(2), pp. 294–301, 2002.

60. Winkelstein, B.A. and DeLeo, J.A. Mechanical thresholds for initiation and persistence of pain following nerve root injury: Mechanical and chemical contributions at injury. *J. Biomech. Eng.*, 126, pp.258–263, 2004.

61. Winkelstein, B., Rutkowski, M., Sweitzer, S., Pahl, J., and DeLeo, J. Nerve injury proximal or distal to the DRG induces florid spinal neuroimmune activation related to enhanced behavioral sensitivity. *J. Comp. Neurol.*, 439, pp.127–139, 2001.

62. Winkelstein, B., Rutkowski, M., Weinstein, J., and DeLeo, J. Quantification of neural tissue injury in a rat radiculopathy model: comparison of local deformation, behavioral outcomes, and spinal cytokine mRNA for two surgeons. *J. Neurosci. Methods*, 111, pp. 49–57, 2001.

63. Winkelstein, B., Weinstein, J., and DeLeo, J. Local biomechanical factors in lumbar radiculopathy: An *in vivo* model approach. *Spine*, 27, pp. 27–33, 2001.

64. Woolf, C.J. Evidence for a central component of post-injury pain hypersensitivity. *Nature*, 306, pp.686–688, 1983.

65. Woolf, C., Safieh-Garabedian, B., Ma, Q., Crilly, P., and Winter, J. Nerve growth factor contributes to the generation of inflammatory sensory hypersensitivity. *Neuroscience*, 62, pp. 277–331, 1994.

66. Xiao, B.G. and Link, H. Immune regulation within the central nervous system. *J. Neurol. Sci.*, 157, pp. 1–12, 1998.

67. Yoshizawa, H., Kobayashi, S., and Kubota, K. Effects of compression on intraradicular blood flow in dogs. *Spine*, 14, pp. 1220–1225, 1989.

17

Ergonomics and Aging

Tracy L. Mitzner
Georgia Institute of Technology

Christopher B. Mayhorn
North Carolina State University

Arthur D. Fisk
Wendy A. Rogers
Georgia Institute of Technology

The rate of population aging is unprecedented, pervasive, enduring, and it has profound implications for most aspects of society, including the workforce (Klinger, 2002). Early retirement pushes that occurred in the early twentieth century, leveled off in the middle twentieth century, and recently have been increasing (Burtless and Quinn, 2000), which has led to a growing number of older adults in the workforce. On a global scale, approximately 1 in 5 adults are 65 years and older and are in the workforce (Klinger, 2002). In more developed countries approximately 30% of adults aged 60 years or older are in the workforce; the percentage is more than 60% in less developed countries (Klinger, 2002). In addition, some unemployed older adults report wanting employment. For example, 5% of Americans aged 55 to 74 who are unemployed or retired report a desire to be employed (U.S. General Accounting Office, 2001). The U.S. older adult workforce is projected to rise by 37% by 2015, which means the older adult workforce could make up almost 20% of the total workforce by 2015 (U.S. General Accounting Office, 2001). Such growth trends have spawned a renewed interest in the older adult worker.

Defining the typical older adult worker, however, is not a simple task because the older adult workforce is an extremely diverse group. Currently men outnumber women in the workforce, however, the ratio of men to women is moving toward 1:1. In more developed countries, women make up 41% of the older adult workforce. In the U.S., women constituted approximately 52% of the older adult workforce in 2000 and this percentage is expected to grow to 61% by 2015 (U.S. Government Accounting Office, 2001).

TABLE 17.1   Selected Percentages of Workers Aged 55 and Older in an Occupation

Occupational Group	2000 (percent)	2008 (projected percent)
Executive, administration, and managerial	21.35	29.30
Professional specialty	15.16	20.65
Sales	21.75	28.65
Service	15.01	16.48
Precision production, craft, and repair	11.80	13.33
Operators, fabricators, laborers	11.93	13.36
Farming, forestry, and fishing	27.00	30.30

Source: U.S. General Accounting Office. *Older Workers*, GAO-02-85-95-152. Washington, D.C. GAO 2001. With permission.

The U.S. older adult workforce is also racially and ethnically diverse. About the same percentage of Caucasians (14.1%), African Americans (12.4%), Hispanics (13.6%), and Asians (12.5%) continue to work after age 65 (U.S. Bureau of Labor Statistics, 2004).

Older adults are employed in a wide variety of positions including executive, professional, sales, and service (Table 17.1), but they are more likely to be white-collar managers or professionals (U.S. General Accounting Office, 2001). Also, the proportion of older adults in certain occupations such as teaching and nursing is expected to grow in the future (U.S. General Accounting Office, 2001). Moreover, certain workers, such as farmers, are particularly more likely to remain in the workforce after the minimum retirement age (Table 17.1). One survey found that farmers were twice as likely to continue working past the age of 65 compared to their peers in other occupations (Sofranko, 2000). Hence, although older adults are employed in a wide variety of occupations, the proportion of older adults in any particular occupation may vary.

Economic, societal, cultural, and social factors have contributed to the growth of the older adult workforce. The decision to delay retirement is strongly based on financial need (68%) and the desire to build up income (64%) (Parkinson, 2002). Certain individuals are more likely to delay retirement based on their financial need, such as those who have low-paying jobs that do not offer sufficient pensions, divorced or widowed women who depend on their husband's pensions, and people who have dependent children, spouses, and/or parents (Szinovácz et al., 1992). Another financial factor that has grown in importance in the last two decades is healthcare cost. Rising healthcare costs along with increased life expectancy and an age-related increase in healthcare needs has made healthcare coverage especially important for older adults. In fact, many American workers may delay retirement to retain their health insurance coverage (Gruber et al., 1995).

Some older adults may delay retirement to retain the salary increase they have attained from reaching a level of tenure. Not surprisingly, median years of tenure tend to increase with age, such that workers aged 55 to 64 have more than three times the median tenure of workers aged 25 to 34 (Bureau of Labor Statistics, 2002). In 2002, older adult workers aged 55 and above had been at their job for more than 9 years. In these situations, individuals may be more inclined to delay their retirement to benefit from the economic rewards of tenure.

Societal trends and laws concerning aging and retirement also influence the likelihood that an older adult will remain in the workforce. Norms regarding retirement vary from country to country, with American, Japanese, and Scandinavian workers leaving the workforce later than most other industrialized countries (Burtless and Quinn, 2000). In spite of such variation, there is a strong relationship across countries between the incentives for continued work provided by social security and labor participation of older adults. That is, individuals are less likely to retire when additional work results in larger increases in "social security wealth" (Gruber et al., 1995). In the U.S., the Social Security Act mandates that 67 is the minimum age for the receipt of full benefits. It is likely that the Social Security Act along with the Age Discrimination in Employment Act, which eliminated a mandatory retirement

TABLE 17.2 Labor Force Participation Rates for Older Adults (ages 55 to 64) in 2001

Country	Percent
Sweden	70.4
Norway	68.5
Japan	65.8
New Zealand	62.9
United States	60.2
Korea	59.2
Denmark	58.9
United Kingdom	54
Portugal	52
Australia	48.6
Ireland	47.9
Spain	41.9
Germany	41.5
Netherlands	39.9
France	38.8
Austria	29
Italy	19.4

Source: Bureau of Labor Statistics (2001). With permission.

age for most occupations, have contributed substantially to the growing older adult workforce in the U.S. In contrast, under current Italian law individuals, who have worked for 35 years can retire at 57. Hence, the age at which individuals are eligible for benefits varies from country to country, which may contribute to a certain extent to differences in older adult worker participation rates (Table 17.2).

Whereas some people work out of financial necessity, others work because they enjoy their jobs (AARP, 2002). Even people who have retired from their careers may return to work because they desire to be productive. That is, they may enjoy the satisfaction they get from accomplishing tasks. Atchley (1976) noted several stages of adjustment to retirement, two of which are often accompanied by feelings that may be related to an individual's desire to return to work. For example, a "honeymoon stage" often occurs immediately following retirement, which involves feeling excited to be retired but also feeling a sense of loss about leaving a job. During the "reorientation stage" an individual re-evaluates their decision to retire after which some individuals may decide to return to work. In addition, for financial need and work enjoyment, AARP's (2002) Work and Career Study found that older adults are motivated to work past typical retirement age to have something interesting to do and to stay physically active.

In sum, the demographic characteristics of older adult workers are quite diverse. Moreover, there are a number of factors that have contributed to the growing older adult workforce. Given that many older adults need and/or desire to remain in the workplace, their demographic is important to those concerned with occupational ergonomics. Aging is associated with a variety of changes in abilities that may affect the older adult worker, including sensation and perception, cognition, and motor control. To ensure the safety and effectiveness of older workers, it is essential to understand these changes and the impact they may have on older adult workers.

17.1 Sensation and Perception

Several sensory and perceptual changes occur with age. Given that most occupations involve tasks that rely on visual and auditory performances to a certain extent, these changes could have significant effects on older workers. Age-related anatomic and physiologic changes have been well-documented. For example, pupil size decreases with age, which is associated with a condition called senile miosis (Loewenfeld, 1979). Senile miosis is most severe and disruptive in low-illumination conditions (Loewenfeld, 1979). Age-associated yellowing and opacification of the lens can also lead to vision difficulties in low-illumination conditions (Weale, 1992).

In addition to anatomic and physiologic changes, aging is associated with changes in visual functioning, the most notable of which are declines in visual acuity, which occur progressively after 50 years of age (Schieber and Baldwin, 1996). Presbyopia, a condition associated with loss of flexibility in the lens resulting in a decreased ability to accommodate (ciliary muscles are not strong enough to focus thickened lens on either near or far targets in the environment), is the major factor that affects visual acuity (Fozard and Gordon-Salant, 2001). Visual acuity declines are amplified in low-illumination and with low-contrast objects (Sturr et al., 1990). The acuity of peripheral vision is also reduced with age (Kline and Scialfa, 1996). Similarly, the useful field of view, which is the functional visual field, declines with age (Cerella, 1985). Age-related declines in visual acuity may affect the reading speed of older adults. However, Akutsu (1991) found that older adults read as fast as younger adults, unless the characters are very small (less than 0.3°) or very large (1.0°). The same range of character sizes that maximized reading speed for older adults was optimal for younger adults as well. Hence, older adults may benefit from adjusting the size of icons and text presented on computers. In addition, printed materials should be designed such that the character size is between 0.3 and 1.0°.

The ability of the eye to adapt to darkness is reduced with age, which may contribute to night vision problems, which are frequently experienced by older adults (Jackson et al., 1999). Contrast sensitivity refers to the amount of contrast that is necessary between a target and its background for the target to be perceived. Contrast sensitivity loss is associated with aging, especially at higher spatial frequencies (Owsley et al., 1983). Glare is also much more disruptive for older adults than for young adults and older adults' eyes take longer to recover from glare (Pulling, 1980). Increased sensitivity to glare may make reading from a computer screen particularly difficult for older adults (Park, 1992). Aging is also associated with loss of color discrimination (Fiorentini et al., 1996), which may make color discrimination more difficult for older adults, particularly blue/green comparisons (Charness and Bosman, 1992). Visual declines associated with aging may impact the work performance of older adults, particularly if tasks rely heavily on processing visual information or if task parameters are not changed to accommodate age-related declines in vision. Glare-reducing monitors and individually adjusted contrast levels on monitors could aid in reducing the effects of age-related contrast sensitivity loss and glare sensitivity.

Age-related hearing loss is one of the most prevalent chronic disorders reported by older adults (Willott, 1991). Physiological changes in the inner ear that cooccur with chronological age often result in presbycusis, a condition associated with decreased sensitivity to higher frequency sounds (i.e., over 6000 Hz). Accompanying presbycusis, a lifetime of previous work experience in noisy work environments may result in noise-induced hearing loss such that sensitivity to mid-range frequencies (i.e., 4000 Hz) is also reduced. Because presbycusis and noise-induced hearing are additive, older workers may demonstrate decreased sensitivity to much of the auditory spectrum. As a consequence, older adults often experience difficulty in processing auditory information such as speech and demonstrate a reduced ability to filter out background noise (Kline and Scialfa, 1997). Deficits in speech perception are quite small when older adults are presented with auditory stimuli in a quiet environment, however, these deficits become substantial as background noise increases (Helfer, 1992). Furthermore, older adults tend to miss information from multiple speakers in a noisy environment because they have to selectively attend to pertinent information received by one ear while ignoring competing information from the other (Barr and Giambra, 1990). Many of these factors associated with age-related auditory decline may interact in the work environment to reduce older adults' abilities to meet job-related demands.

17.2 Intelligence

Current theories view intelligence as having multiple aspects, rather than being a unitary ability. One conceptualization of intelligence incorporates two types of abilities: fluid intelligence, which relate to understanding new, complex relationships and making inferences and conclusions that resolve complexities (Masunaga and Horn, 2001) and crystallized intelligence, which includes breadth and depth of knowledge (Cattell, 1941, 1943). Fluid abilities are more likely to show age-related declines, whereas crystallized abilities are less likely to show age-related declines (e.g., Salthouse, 1991; Smith et al., 1989) (see Figure 17.1 for theoretical representation).

One drawback of traditional measures of intelligence is their inability to accurately gauge individual-specific intelligence, such as that obtained by work and life experience. Many researchers have argued that traditional conceptualizations of intelligence are not comprehensive because they do not incorporate real-world intelligence, sometimes called practical intelligence (Colonia-Willner, 1998; Dixon and Baltes, 1988; Wagner and Sternberg, 1988; Willis, 1987). An aspect of practical intelligence is tacit knowledge, which is knowledge about procedures that are relevant for everyday life functioning (Sternberg and Caruso, 1985). Tacit knowledge is a major contributing factor to achieving success in the workplace (Wagner and Sternberg, 1991). For example, in a study that investigated managerial practical intelligence of older bank managers, Colonia-Willner (1998) found stabilization of practical intelligence with aging, in spite of age-related declines in standard reasoning test performance. It is possible that preserved practical intelligence may serve as a compensatory mechanism that older adults use to maintain functioning (Colonia-Willner, 1998).

Another nontraditional aspect of intelligence that is related to real-world knowledge is wisdom. Wisdom has been conceptualized as "expert knowledge in the pragmatics of life" (Smith et al., 1994, p. 989), recognition and response to human limitations (Taranto, 1989), "the ability to understand human nature" (Clayton, 1982, p. 315). That is, wisdom is thought to be related to fundamental questions about the conduct, interpretation, and meaning of life (Baltes et al., 1995). Many studies of wisdom used real-world dilemmas, to which participants responded. Participants' answers were then scored with

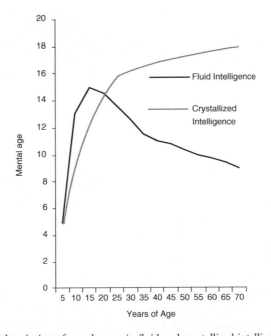

FIGURE 17.1 Theoretical description of age changes in fluid and crystallized intelligence (Cattell, 1987).

respect to criteria, such as knowledge about the conditions of life and its variations, knowledge about strategies of judgment and advice about life matters, knowledge about the contexts of life and their developmental relations, knowledge about differences in values, goals, and priorities, and knowledge about the relative indeterminism and unpredictability in life (Baltes et al., 1995; Smith et al., 1994). These studies and others found age invariance in wisdom (e.g., Smith et al., 1994; Staudinger et al., 1992). One specific characteristic of wisdom is theory of mind ability. That is, "the ability to attribute independent mental states to self and others to predict and explain behavior" (Happe et al., 1998, p. 358). Research suggests that theory of mind is also preserved with age and, moreover, that older adults may have superior theory of mind compared to younger adults. Hence, while older adults may experience declines on traditional intelligence measures, particularly those that reflect fluid abilities, they appear to have preserved practical intelligence and wisdom with which they can compensate.

17.3 Language Production and Comprehension

Communication is an integral aspect of life in general, and the ability to communicate effectively is also important for many occupations. Verbal communication is particularly important for procuring employment, as it may be one of the first skills identified by a potential employer in an interview. Some aspects of communicative ability are stable with age, some decline, and some improve. Vocabulary levels are fairly stable into adulthood (e.g., Wechsler, 1981), along with several other aspects of language related to semantic processes, which relate to the meaning of words (Burke and Peters, 1986; Light and Anderson, 1983; Light et al., 1991). Furthermore, research suggests that aging may be associated with some communicative gains in oral language production. For example, older adults' descriptions of pictures (James et al., 1998) and autobiographical narratives (James et al., 1998; Pratt and Robins, 1991) were rated as more interesting and informative than those produced by young adults.

In contrast, older adults frequently report word-finding difficulties and tip of the tongue experiences, especially for proper names (e.g., Ryan et al., 1994). Older adults produce more pronouns and ambiguous references compared to younger adults (Cooper, 1990; Heller and Dobbs, 1993; Pratt et al., 1989; Ulatowska et al., 1986), which is thought to reflect their difficulty in retrieving appropriate nouns (Burke, 1997). An age-related increase in disfluent speech, such as filled pauses, repetitions, and hesitations may also be related to older adults' word-finding difficulties (Cooper, 1990; Kemper, 1992). In addition, research has shown that older and younger adults use different strategies to compensate for doing two things at once: speaking while concurrently performing a second task. Whereas younger adults reduce the length and grammatical complexity of their sentences, older adults reduce their speech rate and increase their disfluencies (Kemper et al., 2003). These results suggest that job tasks requiring multitasking may exaggerate age-related declines in language production.

Written communication may be particularly essential for certain professions such as technical writing. Whereas older adults' writing is more diverse in terms of vocabulary (Bromley, 1991), research has demonstrated that older adults have a decrement in spelling that is, difficult-to-spell words (MacKay and Abrams, 1998). Aging is also associated with linguistic declines related to sentence complexity (Bromley, 1991; Kemper, 1990; Kemper et al., 1989). For example, an analysis of oral and written language samples revealed an age-associated decline in the mean number of clauses in each sentence, a general measure of language complexity (Kemper et al., 1989). The decline was not equal across modalities; there was a 23% decline in oral language samples, compared to a 44% decline in written samples. Older adults also used a more limited repertoire of syntactic constructions. For example, older adults produced fewer left-branching clauses in the oral and written samples, for example, "When I was on my way to work today, I stopped to get a coffee." These types of grammatical constructions are thought to be more difficult in that they place demands on working memory. Together, these results suggest that older adults produce writing that is less varied in terms of vocabulary and grammatical complexity.

Spoken language comprehension may also be a significant job-related skill depending on the occupation. Given that spoken language comprehension heavily involves hearing ability, which declines with age, one would expect significant difficulties related to listening comprehension. However, older adults rarely complain of difficulties understanding speech (Burke and Harrold, 1988). Yet, research has found some age-associated decrements with respect to listening comprehension. For example, Titone et al. (2000) found evidence that older adults do not adjust their listening behavior to meet the demands of increased difficulty during language processing. In contrast, studies have shown that older adults do adjust their language processing strategies when other text demands increase, such as perceptual difficulty (e.g., Pichora-Fuller et al., 1995). For a task identifying sentence-final words presented with varying degrees of noise where words were either predictable or unpredictable, older adults compensated for difficult listening conditions by taking advantage of supportive contextual information to a greater extent compared to younger adults (Pichora-Fuller et al., 1995). Additional research exploring the effects of listening in distracting environments, demonstrated that older adults' listening comprehension declines are primarily related to hearing decrement (Schneider et al., 2000). Hence, while older adults can use compensatory mechanisms to maintain listening comprehension, they would benefit from working in environments that do not have noise distractions.

The ability to understand written discourse, such as legal documentation, contractual materials, and training manuals, is essential in the work environment. To successfully apply for a job, potential employees must complete the application and when offered a position, they must be able to read and comprehend the terms of employment. When faced with learning to use new office equipment, workers must comprehend instructional materials. According to the situation model approach to comprehension (van Dijk and Kintsch, 1983), readers determine the meaning of text by interpreting it in terms of what they already know and drawing inferences (i.e., making conclusions that are not explicitly stated). Recent evidence suggests that older adults may be at a disadvantage when they have to draw inferences in novel situations where they cannot utilize their prior knowledge (Hancock et al., 2005).

Factors that might improve older adults' comprehension of written instructions such as those presented on office equipment include the use of large font sizes to accommodate visual deficits, simplified sentence structure, and nontechnical terminology presented at the maximum of a sixth-grade reading level (Wickens and Hollands, 1999). Other solutions that might facilitate comprehension and retention of written materials include the use of elaborative memory strategies (Qualls et al., 2001) and the use of explicit signals that highlight the main ideas and relations in the text (Meyer et al., 1998). Consistent with the work of Hancock et al. (2005), procedures should be explicitly stated so that older adults do not have to rely on inferential information.

17.4 Memory

Coupled with the ability to comprehend instructional materials, memory is another age-sensitive cognitive factor that might impact many aspects of an older worker's performance. For example, to use office technology a worker must remember the procedures involved in operating a device, but he or she may also be required to initiate use at specific times or simultaneously store and manipulate incoming information during use. As the next sections illustrate, specific types of memory decline with age whereas others are spared (Smith, 2002).

Working memory tasks require temporary storage and manipulation of information in memory (Baddeley and Hitch, 1974). The mental calculation required during a visit to the warehouse for office supplies is an example of a working memory task because a person must mentally calculate the cost of the items being purchased by constantly updating the total for the items placed on and removed from the corporate invoice. Age-related differences in working memory are well documented (Craik, 2000) and there is some evidence that working memory decrements increase with task complexity (Craik et al., 1990).

One instance where age-related differences in working memory might impact an older adult's interaction with office technology is in the use of telephone voicemail menu systems. Older workers using telephone voicemail menu systems to check on work-related responsibilities or to inquire about the status of their paycheck are required to store and process the menu options while attempting to make navigational decisions. If the structure of the menu system is very broad such that a large number of options must be considered before a choice can be made, older adults may find themselves forgetting the content of the options because their working memory capacity is being taxed. As only one option can be chosen at a time, it must be appropriate for the goals of the older adult. Thus, all options other than the desired option must be considered as unwanted information and should be ignored. Furthermore, the speed of menu option presentation is another factor to consider when designing phone menu systems because older workers may be slower to process information than younger coworkers.

System modifications used to combat the age-related decline in working memory should result in increased usability of the telephone menu system. Reducing the number of menu items that have to be considered at each level of menu hierarchy should reduce working memory demands (Reynolds et al., 2002). Yet another solution might be to present the most commonly requested menu items first, thereby reducing the need to ignore unwanted options. Slowing the speed of menu item presentation may also result in a more usable menu system for older adults because they will be given the opportunity to process all of the available menu options (Reynolds et al., 2002).

Semantic memory refers to the store of factual information that accrues through a lifetime of learning. Remembering the meaning of symbolic codes used in an account invoice, the telephone number for the employee lounge, and the route to the purchasing department are all examples of semantic memory because this is information that was acquired through experience. Age-related differences in the organization and use of semantic information are slight or nonexistent, although it may take older adults longer to learn new information. Thus, semantic memory remains intact throughout the lifespan (Light, 1992). When new information is encountered, it is often interpreted in the context of the pre-existing knowledge base.

Working memory and semantic memory are forms of retrospective memory or memory for past events. Prospective memory refers to remembering to do things in the future. For example, remembering to check a gauge at a certain time or to attend a meeting requires one to remember a task that has to be performed at some time in the future. Event-based and time-based prospective memory tasks vary by the demands of the task characteristics (Einstein and McDaniel, 1990). For an event-based task, a cue in the environment reminds one to perform a prospective task (e.g., placing a note next to the computer as a reminder to write an office memo). In this context, cues in the environment act as mnemonic or environmental support that increases the likelihood of remembering the prospective task. In contrast, time-based tasks lack environmental support because they have few external cues. Time-based tasks are largely self-initiated and require one to perform an action at a certain time or after a specified amount of time has elapsed (e.g., remembering to make a phone call at 4:30). Age differences in prospective memory are usually much greater for time-based than event-based tasks (Park et al., 1997). Hence, older workers are less likely to exhibit declines when performing event-based tasks.

17.5   Decision-Making

On a daily basis, workers encounter situations that require them to make decisions such as when to schedule a meeting or which lever to use to make a particular adjustment. While these decision contexts seem quite different, they both rely on a decision-making process that requires people to consider multiple pieces of information, interpret that information in relation to some goal, then select the best course of action from the alternatives available. Although decision-making is obviously an important cognitive factor in the workplace, relatively little is known about the decision-making capacity of older adults because it has been neglected in research on aging (Sanfey and Hastie, 2000).

Of the few studies that have been conducted, it appears that older adults consider fewer pieces of information so as not to tax working memory when determining courses of action (Johnson, 1990) and that they are susceptible to biases such as the framing effect that also influence younger adults' decisions (Mayhorn et al., 2002). While some suggest that older adults are more cautious decision-makers than younger adults (Green et al., 1994), the speed and quality of decision-making is often found to be age invariant (Dror et al., 1998).

17.6 Attention

Attention is a ubiquitous component to information processing; it involves selection, vigilance, and control (Parasuraman, 1998). In the workplace, attention ability would be most pertinent for tasks that involve finding given information among similar information, doing two tasks at the same time, selecting relevant information while inhibiting irrelevant information, switching from one task to another, and detecting rarely occurring information over a prolonged period of time. The ability to apply mental effort to something for a sustained period of time appears to be relatively preserved with age, unless the task involves perceptual discrimination (Deaton and Parasuraman, 1993). However, there is evidence that older adults are less able to select relevant information (e.g., Allen et al., 1994) and to inhibit irrelevant information compared to younger adults (e.g., Hasher and Zacks, 1988). Furthermore, some research suggests that visual selective attention, which involves selecting relevant visual information while inhibiting irrelevant visual information, is influenced by aging (e.g., Carlson et al., 1995; Connelly et al., 1991; Hasher et al., 1991; McDowd and Oseas Kreger, 1991). However, simple displays (Humphrey and Kramer, 1997), low-perceptual load (Madden and Langley, 2003) and adequate practice (Fisk and Rogers, 1991) minimize age differences.

Research has also examined selective attention to auditory information. Although some research has demonstrated an age-related decline with regard to auditory selective attention (e.g., Barr and Giambra, 1990; Panek and McGown, 1981), more recent research showed that once age-differences in auditory perception were controlled older and younger adults were equally affected by irrelevant information (Murphy et al., 1999).

Furthermore, in a recent meta-analysis of dual task performance, older adults' speed and accuracy were compromised to a greater extent compared to younger adults (Verhaeghen et al., 2003). The ramifications of these findings for the workplace are that older adults may perform attention-demanding tasks better if they are less complex and rely less on working memory. Furthermore, age-related declines in attention may be minimized when older adults are given adequate practice, when they are not distracted by irrelevant information, and if they perform complex tasks separately rather than concurrently.

17.7 Motor Control

Accompanying age-related declines in perception and cognition are declines in older adults' abilities to initiate and control motor movement. Older workers in certain professions such as factory assembly lines might be differentially affected by these declines than workers whose job-related performance is less reliant on physical activities (Panek, 1997). As people age, motor function slows such that movements require more time for completion, continuous movements become more difficult, and deficits in coordination are observed (Vercruyssen, 1997).

The relationship between movement time and accuracy is reflected by Fitts' Law (Fitts, 1954), which states that movement time is linearly related to the difficulty of the movement. Thus, a speed-accuracy trade-off results when more time is required to complete difficult movements that require increased precision. General findings from previous research that investigated older adults' motor control by observing the speed-accuracy trade-off indicated that older adults are slower than younger adults when performing the same movements because they tend to place greater importance on accuracy (for a review, see Walker et al., 1996). The issue of speed-accuracy trade-off may be important in many types of work tasks. For instance, Fitts' Law has been applied to movement tasks as diverse as using a computer mouse to click on a target icon (Card et al., 1978) to manipulating a microscope (Langolf et al., 1976). However, older adults' motor control deficits are not universal and performance among individuals is highly variable given differences in aerobic fitness and cardiovascular status (Spirduso and MacRae, 1990).

17.8   How can Employers Maximize the Productivity of Older Adults?

17.8.1   Matching KSAOs of Tasks to those of the Worker

In light of age-related changes in sensation and perception, cognition, and motor control, there are several ways that employers can maximize their older adult workers' productivity and safety. As workplace task demands vary, increased productivity should result from matching task demands to employee skill-set and ability. One method for matching job tasks with the abilities of the older adult is to compare the knowledge, skills, experience, and other factors (KSAOs) necessary for a job to those possessed by an employee (Salthouse and Mauer, 1996). Knowledge and skills refer to domain-specific aspects, as well as general aspects resulting from life experience. Abilities are capabilities of an individual, varying from physical to cognitive, that are demonstrated across situations. The "other" category includes characteristics such as motivation and personality. Of course, older adults will vary greatly in terms of their individual KSAOs depending on the nature of their past work experience. However, the relative weighting of KSAOs may be different for different ages (Salthouse and Mauer, 1996). For example, older workers may rely more heavily on their past experience and accumulated task-relevant skill-set to solve problems, whereas younger workers may be more likely to rely on reasoning and general problem-solving skills. That is, older adults may be more apt to solve problems using crystallized intelligence, whereas younger adults may be more likely to rely on fluid intelligence. It follows that older workers should excel at tasks that allow them to make use of their previous experience and capitalize on their intact capabilities. In contrast, ill-informed assignment of older adults to tasks reliant on abilities that decline with age will result in less than optimal performance by older workers.

17.8.2   Putting Experience to Work by Using Well-Learned Skills

Older and younger workers will benefit from the assignment of tasks that allow them to apply their previous knowledge or expertise. Because well-learned skills are highly practiced and familiar, older workers can apply their considerable previous experience in this domain to solve problems that arise. In contrast, older workers may perform at a lower level of proficiency compared to younger coworkers when faced with novel tasks highly reliant on fluid knowledge because of basic age-related cognitive declines in working memory. Perhaps for this reason, older adults tend to remain in jobs where they can use their experience and tacit knowledge of the familiar workspace to offset cognitively resource-demanding duties that require the acquisition of new skills (Park, 1994).

Although the most optimal match appears to pair older adult employees with tasks that utilize their well-learned skills, we do not suggest that older adults cannot acquire new skills. On the contrary, previous research indicates that older adults can acquire new skills through training (Sterns, 1986). For instance, older adults can acquire basic computer skills with practice, however, they require more time, make more errors, and need more help during training than younger adults (Kelley and Charness, 1995). Thus, older workers can learn but training programs must accommodate their special needs by considering age-related changes in perception, cognition, and motivation to learn (Mayhorn et al., 2004a).

17.8.3 Making Use of the Speed-Accuracy Trade-off

Older adults tend to take longer than younger coworkers to perform tasks that have a motor component. One behavioral explanation for this slowing is that older adults place more emphasis on being accurate than on being fast (Welford, 1981). Thus, when assigning tasks to older workers, employers should be sensitive to this generalized slowing and avoid assigning time-sensitive tasks that require immediate execution that may result in increased error production by older employees. With knowledge of the speed accuracy trade-off, employers can optimize the work quality of older employees by providing more reasonable deadlines for completion.

Having described the speed-accuracy trade-off, we do not suggest that older adults cannot complete their work in a timely fashion. Consider that much of the research that describes age-related motor declines was observed in the artificial setting of the laboratory. Real-word tasks are much different from laboratory tasks because older adults can make use of compensatory mechanisms such as skill and strategy to offset declining motor capacity. For instance, Salthouse (1984) found that older adult typists compensated for declining speed by scanning farther ahead in the document they were transcribing. Thus, tasks that are well-learned and allow older adults to apply compensatory mechanisms may not be as susceptible to motoric slowing as more novel tasks.

17.8.4 Complex Tasks and Multiple Task Environments

Declines in working memory capacity may cause difficulties for older workers in a number of contexts such as completing unfamiliar procedurally complex tasks or operating in a multiple task environment. Consider the large sequence of steps, necessary to perform computerized tasks such as text processing or accessing email. Unless this sequence of steps is well-learned and familiar, older adults may experience great difficulty because they must remember not only the correct sequence, but also the actions they must take to complete each step. While learning to acquire new skills during computer training, older adults often express difficulty with complex tasks that require completion of a long sequence of steps, which tax their working memory capacity (Mayhorn et al., 2004b; Morrell et al., 2000).

Reduced working memory capacity is also associated with deficits in the ability to perform multiple tasks at the same time. When required to perform several independent tasks simultaneously, older adults exhibit performance deficits when compared to younger adults (Sit and Fisk, 1999). Aircraft pilots represent one occupation where age-related declines in working memory often result in the reduced ability to perform multiple work-related tasks. When pilot communication with air traffic control and routine flying tasks were assessed, older pilots were less able to read back and execute air traffic instructions than younger pilots even when they were given the opportunity to practice (Morrow et al., 1993). Expertise in the occupation helped to ameliorate age-related declines, but age differences between young and older pilots remained significant (Morrow et al., 2001).

17.8.5 Using Environmental Support to Reduce Memory Load

Older adults in general might benefit from work environments that do not require multi-tasking. In addition, older workers employed in cognitively demanding positions should benefit from the use of environmental support. Environmental support refers to the provision of task characteristics that decrease the processing demands placed on memory (Craik and Byrd, 1982). Investigations with older pilots suggested that they make use of environmental support by engaging in note-taking to record air traffic control instructions thereby decreasing the amount of information they must store in memory (Morrow et al., 2003). This finding is consistent with the ergonomic principle of placing knowledge in the world, not in the head, when tasks are complex or only recently learned (Norman, 1988).

Environmental support is also useful for aiding in the execution of prospective memory tasks (i.e., remembering to do something in the future). By including environmental support in the form of a reminder cue that is specific to the task, time-based tasks can be transformed into event-based tasks

(Brandimonte et al., 1996). For instance, older workers may remind themselves to attend a late afternoon meeting by leaving their umbrella by the door or placing the appointment in a planner and leaving it in view on their desk. The use of an alarm or automated reminder is similarly beneficial. The ability to effectively use these reminder cues as memory aids should decrease the likelihood of forgetting work-related prospective tasks.

17.8.6   Experience and Decision-Making

Naturalistic decision-making emphasizes the ability of decision-makers to use their previous experience in real-world settings (Zsambok and Klein, 1997). The recognition-primed decision-making model championed by Klein (1988) postulates that decisions are formulated by examining decision contexts for familiar patterns recognized from earlier personal experience. Because older adults retain extensive knowledge in semantic memory, they may excel in making decisions in familiar environments. For instance, an older employee might remember the last time a particular problem occurred with a photo-copier. The older worker with access to this past experience might offer an immediate solution that works or readily eliminate solution options that failed in the past. In this fashion, an older adult's extensive domain knowledge and experience should be considered a resource that can benefit the older worker, coworkers, and the organization.

17.8.7   Equipment Factors for Improving Worker/Task Match

In addition to pairing older adults with appropriate tasks that make use of their abilities and minimize their limitations, employers must realize that work occurs in a physical environment that often requires the use of tools such as computers and other equipment. Employers can benefit from knowledge concerning how older adults interact with computers, the problems they experience during use, and the relatively simple adjustments that can be made to the equipment to make it more usable by older workers.

Many of the perceptual, motor, and cognitive changes that cooccur with age directly impact the ability to use a computer. Perceptual changes in vision such as presbyopia and sensitivity to glare actively inter-act with the display characteristics of a computer monitor. Presbyopia can diminish an older adult's ability to see information that appears on the monitor; however, the use of multifocal corrective lenses, particularly bifocals or trifocals, is one means of compensating for this decrease in visual percep-tion. Other methods of accommodation include altering the settings on the computer to increase font and icon size. Increased sensitivity to glare may also reduce older adults' ability to see the display. Accom-modations to reduce glare include eliminating the light source causing the glare, tilting the computer monitor, and placing a screen cover over the monitor.

Although declines in vision seem most pertinent, age-related auditory declines may also reduce the usability of computer displays. For instance, auditory feedback is useful in determining whether a word processing function was executed or whether an icon was single or double-clicked in many current computer systems. As computer systems evolve to include more complex auditory feedback such as voice interfaces, age-related declines such as presbycusis might severely impact an older worker's ability to use a computer effectively. Accommodations that might reduce the impact of presby-cusis include changing the computer settings to increase the volume of feedback, using a headset to funnel feedback directly to the ears, and reducing background noise in the office environment.

Age-related declines in motor control influence the usability of input devices such as computer mice (Smith et al., 1999; Rogers et al., 2005; Walker et al., 1997). In each of these studies, older adults displayed deficits when asked to point and click on specific objects, clicking and dragging objects on the computer screen, and single-clicking some objects while double-clicking others. Results from a recent study inter-viewing older adults who had completed a computer training course at a community center indicated that playing computer games, such as solitaire, is useful for improving motor coordination with com-puter mice (Mayhorn et al., 2004a). Other accommodations that can remedy clicking problems include altering the speed of the mouse clicks or making use of the keyboard to compensate for slow clicking. Should these accommodations with computer mice fail, older adults might also benefit from the use of other input devices such as touchscreens that allow direct input or a rotary encoder that allows precise control (Rogers et al., 2005).

Cognitive changes that influence computer usage include declines in working memory as well as retention of semantic memory. Working memory declines appear to explain why older adults have difficulty performing complex computerized tasks such as Internet searches. While searching for information within a website, older adults must remember what information they are searching for, where they have looked for it, and where they are currently located within the site (Stronge et al., 2002). As a result of declining working memory capacity, older adults have been observed to revisit pages within a website that they have already browsed and have trouble keeping track of the current page that they are visiting (Mead et al., 1997). Thus, poor working memory might result in an inaccurate mental representation of the hierarchical structure of a website or more generally, a computer system.

As semantic memory is preserved in older adults, well-learned tasks that require the use of familiar computer systems should be minimally impacted by the aging process. Problems arise when technology and systems are upgraded, thereby forcing older workers to acquire new skills through training. When new information is encountered during training, it is often interpreted in the context of the pre-existing knowledge base. For this reason, instructional materials should be presented using familiar terminology that makes reference to earlier technology. Because metaphors tap well-developed crystallized knowledge, they may be an effective means of simulating experience when teaching older adults (Bowles et al., 2002).

17.8.8 Environmental Factors for Improving Worker/Task Match

In addition to equipment design changes, environmental design changes that take into consideration age-related changes can also improve productivity and safety for older adult workers. In light of vision and hearing declines that often occur with aging, lighting optimization and noise reduction may greatly improve the working environment for older adults. Older adults require higher levels of illumination to see well compared to younger adults, although the optimal amount is a debated issue (e.g., Jaschinski, 1982). On a series of legibility tasks, Charness and Dijkstra (1999) found that the performance of older office workers improved to a greater extent than younger adults when luminance increased. Specifically, performance improved when luminance was in the 450 to 600 cd/m^2 range. In general, higher levels of illumination are needed as the size of the relevant visual details decrease and as the demand for speed and accuracy increase (Charness and Bosman, 1992). Charness and Dijkstra (1999) recommended that adults over the age of 40 should increase lighting when performing clerical tasks such as finding information in a phonebook. Furthermore, the authors suggested that both younger and older adults should increase lighting for clerical tasks such as proofreading. Given that the optimal level of illumination may vary from person to person, Fozard and Popkin (1978) recommended using individual preference to determine the level selected. While some findings suggested that individual preference for illumination can predict optimal performance (e.g., Hughes and McNelis, 1978), other results suggested that they do not (Boyce, 1981). Thus optimal lighting conditions are best determined through performance comparisons under different lighting conditions.

Reducing background noise would also aid in optimizing environmental conditions for older workers. This can be achieved by incorporating sound absorbing materials in offices and factories. One recent invention called "quiet curtains," were designed in an effort to reduce nocturnal noise in nursing homes (Ahuja, 1999). The curtains are made by placing noise reducing materials between two pieces of fabric. Ahuja's prototype reduced noise by approximately 7 (dB), and by adding a similarly made floor extension and valance, noise was reduced by 12 (dB), which indicates a reduction of sound intensity by a factor of 16. Given age-associated sensitivities to background noise, environments that make use of sound reducing materials would be particularly beneficial for older adult workers.

17.8.9   Promoting Safety on the Job with Warnings

Work-related injuries occur at a lower rate for older workers than for younger workers for almost all occupations (Root, 1981). Experience on the job is often cited to explain this result. However, another factor may be the importance that older adults place on warnings (Hancock et al., 2001; Mayhorn et al., 2004b). Warnings are communications that are meant to promote safety by alerting people to the presence of potential hazards in the environment and providing instructions to avoid injury.

The communication-human information processing (C-HIP) model proposed by Wogalter et al. (1999) described three components of the warning process: (1) the *source* or sender represents the originator of the warning, (2) the *channel* represents the medium (e.g., visual or auditory) used to deliver the warning, and (3) the *receiver* is the end-user who progresses through a series of information processing stages to determine a course of action. Cognitive and perceptual changes that cooccur with age are receiver characteristics that can influence how an older worker might interact with a warning (Mayhorn and Podany, in press; Rousseau et al., 1998).

Warnings can be transmitted through almost any sensory modality or channel but most commonly encountered warnings are presented visually (e.g., medication labels) or auditorily (e.g., fire alarm sirens). For instance, older workers in a factory environment might not notice an auditory fire alarm if there is a high level of ambient background noise. Likewise, the same older factory workers may not notice a visually displayed warning if glare from an adjacent window obscures the message. Given age-related perceptual declines in vision and audition, it is important for employers to understand any age-related changes in perception before deploying warnings in the work environment.

Receiver characteristics such as age-related cognitive changes in text comprehension and memory can also impact how an older worker might interact with a warning. Given the reduced ability of older adults to make inferences when they encounter novel warnings (Hancock et al., 2005) and well-documented working memory deficits, older workers may be at a particular disadvantage when they encounter risk communications designed to deliver a complex set of compliance instructions. Consider the "talking box" warning investigated by Conzola and Wogalter (1999), which used a miniaturized voice system to transmit numerous precautionary steps to be taken when installing the computer peripheral device stored in the box. Older workers of the future might be faced with such warnings, therefore it is imperative that employers and warning designers understand why this form of warning might be ineffective given the receiver characteristics of older adults. Instead, future warning systems might use technology to provide cognitive support by giving either carefully timed or prompted instructions initiated by the user during installation (Wogalter and Mayhorn, 2005). In this fashion, employers armed with knowledge concerning the receiver characteristics associated with older workers can provide a safe working environment for all of their employees.

17.8.10   Training

Although eluded to in previous sections of this chapter, training is an aspect of the workplace that should not be underrated. Computers and other technology have revolutionized the way almost all workers do their jobs, from the office to the factory to the farm. Particularly with this ever changing technology, comes a need for workers to be willing and able to learn new skills. Occupations that require extensive computer skills tend to hire fewer older adult workers possibly because of perceptions that older workers have difficulty learning new skills. Indeed, changing technology can be especially difficult for older adults because they were educated and trained for a different work culture. However, some research has shown that older adults are able and willing to learn new skills (for a comprehensive review, see Sterns and Doverspike, 1989). In contrast, one meta-analysis of studies related to aging and training concluded that older adults demonstrate poorer performance overall, less mastery of relevant material, and slower performance compared to younger adults (Kubeck et al., 1996). Although the evidence is not conclusive whether one type of training is more beneficial for older adults compared to younger adults

(Charness and Bosman, 1997), evidence suggests that older adults are more likely to benefit from training when they can go at their own pace and when they learn with their age peers (Gist et al., 1988; Knowles, 1987; Shea, 1991). Older adults also benefit from training when it is goal-oriented, allows exploratory learning, involves a task that is interesting to the learner, permits active participation in the learning process, involves adequate practice, and provides prompt feedback (Belbin and Belbin, 1968; Bolton, 1978; Fisk and Rogers, 1991; Hiemstra, 1972; Hollis-Sawyer and Sterns, 1999; McGehee and Thayer, 1961; Neale et al., 1968). Moreover, the match between training format and task demands is more critical to successful performance for older adults (Jamieson and Rogers, 2000; Mea and Fisk, 1998; McLaughlin et al., in press).

17.9 Conclusions and Recommendations

In this chapter, we have presented information that may aid in increasing the productivity and safety of older adult workers. There are many benefits of hiring and retaining older adults. To begin with, the workforce is aging: the median age in the workforce is rising and is projected to be over 40 years of age by 2010 (Fullerton and Toossi, 2001). Also, research suggests that there is little if any relation between age and work performance and that any negative correlation between age and work performance may be modulated by training (e.g., Sparrow and Davies, 1988; McEvoy and Cascio, 1989). In addition, older adults have many positive attributes in addition to experience-based knowledge. For example, older adults have been reported to be absent less often, have fewer accident rates, and have higher levels of job satisfaction compared to younger adults (Davies and Sparrow, 1985).

The benefit of hiring older adults has not gone unnoticed. In The Older Worker Survey (Taylor, 2000), most human resource professionals reported several advantages of hiring older adults, such as their willingness to work in different schedules and their invaluable experience. Respondents also reported that older workers had stronger work ethics and were more reliable compared to younger adults. Unfortunately, most of the respondents reported that the changing demographics of the workforce had little or no impact on the recruiting, retention, or management policies of their organizations, reflecting a lack of preparation for the rising age of the adult workforce.

Research suggests that employers would benefit from using a variety of flexible work arrangements to retain and extend the career of older workers such as rehiring retirees, reduced work schedules, phased retirement, and job sharing (U.S. General Accounting Office, 2001). Although flexible work schedules exist, they are often limited and not common (U.S. General Accounting Office, 2001). By finding solutions to policy and ergonomic issues related to aging, employers can extend and optimize the work lives of older adults. Capitalizing on the capabilities of older adults, recognizing their age-limitations, and designing work environments to accommodate such changes is crucial. The nation as a whole will benefit from these solutions by ensuring a future supply of skilled workers, enhancing economic growth, and helping secure sufficient retirement income for many workers.

Bibliography

AARP, *Staying Ahead of the Curve: The AARP Work and Career Study*, American Association of Retired Persons, Washington, D.C., 2002.

Ahuja, K., Sounds of silence: "Quiet curtains" combine audio privacy and aesthetics for nursing homes, hospitals, hotels, and offices, Georgia Institute of Technology News Release, http://gtresearch-news.gatech.edu/newsrelease/CURTAINS.html, 1999.

Akutsu, H. Psychophysics of reading: X. Effects of age-related changes in vision, *J. Gerontol*, 46(6), pp. P325–P331, 1991.

Allen, P.A., Weber, T.A. and Madden, D.J. Adult age differences in attention: filtering or selection?, *J. Gerontol.*, 49(5), pp. P213–P222, 1994.

Atchley, R.C. *The Sociology of Retirement*, Schenkman, Cambridge, MA, 1976.

Baddeley, A.D. and Hitch, G.J. Working memory, in *The Psychology of Learning and Motivation*, Bower, G.H. ed., Vol. 8, Academic Press, New York, 1974, pp. 47–90.

Baltes, P.B., Staudinger, U.M., Maercker, A., Smith, J. People nominated as wise: a comparative study of wisdom-related knowledge, *Psychol. Aging*, 10, pp. 155–166, 1995.

Barr, R.A. and Giambra, L.M. Age-related decrement in auditory selective attention, *Psychol. Aging*, 5(4), pp. 597–599, 1990.

Belbin, E. and Belbin, R.M. New careers in middle age, in *Middle Age and Aging: A Reader in Social Psychology*, Neugarten B. ed., University of Chicago Press, Chicago, 1968, pp. 341–350.

Bolton, E.B. Cognitive and noncognitive factors that affect learning in older adults and their implications for instruction, *Educ. Gerontol.*, 3, pp. 331–344, 1978

Bowles, C.T., Fisk, A.D. and Rogers, W.A. Inference and the use of similes and metaphors in warnings. *Proceedings of the Human Factors and Ergonomics Society* Santa Monica, CA: HFES, 2002, pp. 1703–1707.

Boyce, P.R. *Human factors in lighting*, Essex, England: Applied Science, 1981.

Brandimonte, M., Einstein, G.O. and McDaniel, M.A. *Prospective memory: Theory and applications*, Lawrence Erlbaum Associates Mahwah, NJ, 1996.

Bromley, D.B. Aspects of written language production over adult life. *Psychol. Aging*, 6(2), pp. 296–308, 1991.

Bureau of Labor Statistics (2002). ftp://ftp.bls.gov/pub/news.release/tenure.txt.

Burke, D. Language, aging, and inhibitory deficits: Evaluation of a theory. *J. Gerontol. Psychol. Sci.*, 52B, pp. P254–P264, 1997.

Burke, D.M. and Harrold, R.M. Automatic and effortful semantic processes in old age; Experimental and naturalistic approaches, in *Language, Memory, and Aging*, L.L. Light and D.M. Burke eds., Cambridge University Press, New York, 1988, (pp. 100–116),

Burke, D.M. and Peters, L. Word associations in old age: Evidence for consistency in semantic encoding during adulthood. *Psychol. Aging*, 1(4), pp. 283–291, 1986.

Burtless, G. and Quinn, J.F. Retirement trends and policies to encourage work among older Americans. Presented at the *Annual Conference of the National Academy of Social Insurance*, Washington, DC, 2000.

Card, S.K., English, W.K., and Burr, B.J. Evaluation of mouse, rate-controlled isometric joystick, step keys, and task keys for text selection on a CRT. *Ergonomics*, 21(8), pp. 601–613, 1978.

Carlson, M.C., Hasher, L., Connelly, S.L., and Zacks, R.T. Aging, distraction, and the benefits of predictable location. *Psychol. Aging*, 10, pp. 427–436, 1995.

Cattell, R.B. Some theoretical issues in adult intelligence testing. *Psychol. Bull.*, 38, 592, 1941.

Cattell, R.B. The measurement of adult intelligence. *Psychol. Bull.*, 40, pp. 153–193, 1943.

Cerella, J. Age-related decline in extrafoveal letter perception. *J. Gerontol.*, 40(6), pp. 727–736, 1985.

Charness, N. and Bosman, E.A. Human factors and age, in *The Handbook of Aging and Cognition*, Lawrence Erlbaum Associates, Inc, F.I.M. Craik and T.A. Salthouse, eds., Hillsdale, NJ, 1992, pp. 495–551.

Charness, N. and Dijkstra, K. Age, luminance, and print legibility in homes, offices, and public places. *Hum. Factors*, 41(2), pp. 173–193, 1999.

Clayton, V. Wisdom and intelligence: the nature and function of knowledge in the later years. *Int. J. Aging Hum. Dev.*, 15, pp. 315–321, 1982.

Colonia-Willner, R. Practical intelligence at work: Relationship between aging and cognitive efficiency among managers in a bank environment. *Psychol. Aging*, 13(1), pp. 45–57, 1998.

Connelly, S.L., Hasher, L., and Zacks, R.T. Age and reading: The impact of distraction. *Psychol. Aging*, 6, pp. 533–541, 1991.

Conzola, V.C. and Wogalter, M.S. Using voice and print directives and warnings to supplement product manual instructions. *Int. J. Ind. Ergon.*, 23, pp. 549–556, 1999.

Cooper, P.V. Discourse production and normal aging: Performance on oral picture description tasks. *J. Gerontol. Psychol. Sci.*, 45, pp. P210–P214, 1990.

Craik, F.I.M. Age-related changes in human memory. in *Cognitive Aging: A Primer* D.C. Park and N. Schwartz eds., PA: Taylor and Francis, Philadelphia, 2000, pp. 75–92.

Craik, F.I.M. and Byrd, M. Aging and cognitive deficits: The role of attentional resources. In *Aging and Cognitive Processes*, F.I.M. Craik and S.E. Trehub eds., Plenum, New York, 1982, pp. 191–211.

Craik, F.I.M., Morris, R.G., and Gick, M.L. Adult age differences in working memory. in *Neuropsychological Impairments of Short-term Memory*, G. Vallar, and T. Shallice eds., Cambridge University Press, Cambridge, England,1990, pp. 247–267.

Davies, D.R. and Sparrow, P.R. Age and work behaviour. in *Aging and Human Performance*, Charness ed., England Wiley, Chichester, 1985, pp. 293–332.

Deaton, J.E. and Parasuraman, R. Sensory and cognitive vigilance: Effects of age on performance and subjective workload. *Hum. Perform.*, 6, pp. 71–97, 1993.

Dixon, R.A. and Baltes, P.B. Toward life-span research on the functions and pragmatics of intelligence. in *Practical Intelligence: Nature and Origins of Competence in the Everyday World*, R.J. Sternberg and R.K. Wagner eds., Cambridge University Press, New York, 1988, pp. 203–235.

Dror, I.E., Katona, M., and Mungur, K. Age differences in decision-making: To take a risk or not? *Gerontology*, 44, pp. 67–71, 1998.

Einstein, G.O. and McDaniel, M.A. Normal aging and prospective memory. *J. Exp. Psychol. Mem., Learn., and Cogn.*, 16, pp. 717–726, 1990.

Fiorentini, A., Porciatti, V., Morrone, M.C., and Burr, D.C. Visual ageing: unspecific decline of the response to luminance and colour. *Vision Research*, 36, pp. 3557–3566, 1996.

Fisk, A.D. and Rogers, W.A. Toward an understanding of age-related memory and visual-search effects. *J. Exp. Psychol: General*, 120(2), pp. 131–149, 1991.

Fitts, P.M. The information capacity of the human motor system in controlling the amplitude of movement. *J. Exp. Psychol.*, 47, pp. 381–391, 1954.

Fitts, P.M. and Posner, M.I. *Human Performance*, Brooks/Cole, Belmont, CA, 1967.

Fozard, J.L. and Popkin, S.J. Optimizing adult development: Ends and means of an applied psychology. *Am. Psychol.*, 33, pp. 975–989, 1978.

Fozard, J.L. and Gordon-Salant, S. Changes in vision and hearing with aging. in *Handbook of the Psychology of Aging*. J.E. Birren and K.W. Schaie, eds., 5th ed., Academic Press, San Diego, CA, 2001, pp. 241–266.

Fullerton H.N. and Toossi, M. Labor force projections to 2010: Steady growth and changing composition. *Mon. Labor Rev.*, 11, pp. 21–38, 2001.

Gist, M., Rosen, B., and Schwoerer, C. The influence of training method and trainee age on the acquisition of computer skills. *Personnel Psychol.*, 41, pp. 255–265, 1988.

Green, L., Fry, A.F., and Myerson, J. Discounting of delayed rewards: A lifespan comparison. *Psychol. Sci.*, 5, pp. 33–36, 1994.

Gruber, J., Madrian, B.C., and National Bureau of Economic Research. *Non-employment and Health Insurance Coverage*, National Bureau of Economic Research, Cambridge, MA, 1995.

Hancock, H.E., Rogers, W.A., and Fisk, A.D. An evaluation of warning habits and beliefs across the adult lifespan. *Hum. Factors*, 43(3), pp. 343–354, 2001.

Hancock, H.E., Fisk, A.D., and Rogers, W.A. Comprehending product warning information: Age-related effects of memory, inferencing, and knowledge. *Human Factors*, 47, pp. 219–234, 2005.

Happe, F.G.E., Brownell, H., and Winner, E. The getting of wisdom: theory of mind in old age. *Dev. Psychol.*, 34(2), pp. 358–362, 1998.

Hasher, L. and Zacks, R.T. Working memory, comprehension, and aging: A review and a new view. in *The Psychology of Learning and Motivation*, G.H. Bower, ed., Vol. 22, Academic, New York, 1988, pp. 193–226.

Hasher, L., Stoltzfus, E.R., Zacks, R.T., and Rypma, B. Age and inhibition. *J. Exp. Psychol. Learn., Mem. and Cogn.*, 17, pp. 163–169, 1991.

Helfer, K.S. Aging and the binaural advantage in reverberation and noise. *J. Speech Hear. Res.*, 35, pp. 1394–1401, 1992.

Heller, R.B. and Dobbs, A.R. Age differences in word finding in discourse and nondiscourse situations. *Psychol. Aging*, 8, pp. 443–450, 1993.

Hiemstra, R.P. Continuing education for the aged: A survey of needs and interests of older people. *Adult Educ.*, 22, pp. 100–109, 1972.

Hollis-Sawyer, L.A. and Sterns, H.L. A novel goal-oriented approach for training older adult computer novices: Beyond the effects of individual-difference factors. *Educ. Gerontol.*, 25(7), pp. 661–684, 1999.

Hughes, P.C. and McNelis, J.F. Lighting, productivity, and the work environment. *Lighting Res. Des.*, 8, pp. 32–40, 1978.

Humphrey, D.G. and Kramer, A.F. Age differences in visual search for feature, conjunction, and triple-conjunction targets. *Psychol. Aging*, 12(4), pp. 704–717, 1997.

Jackson, G.R., Owsley, C., and McGwin, G. Aging and dark adaptation. *Vision Research*, 39, pp. 3975–3982, 1999.

James, L.E., Burke, D.M., Austin, A., and Hulme, E. Production and perception of "verbosity" in younger and older adults. *Psychol. Aging*, 13(3), pp. 355–367, 1998.

Jamieson, B.A. and Rogers, W.A. Age-related effects of blocked and random practice schedules on learning a new technology. *J. Gerontol.: Psychol. Sci.*, 55B, pp. P343–P353, 2000.

Jaschinski, W. Conditions of emergency lighting. *Ergonomics*, 25(5), pp. 363–372, 1982.

Johnson, M.M.S. Age differences in decision making: A process methodology for examining strategic information processing. *J. Gerontol. Psychol. Sci.*, 45(2), pp. 75–78, 1990.

Kelley, C.L. and Charness, N. Issues in training older adults to use computers. *Behav. Inf. Technol.*, 14(2), pp. 107–120, 1995.

Kemper, S. Adults' diaries: Changes made to written narratives across the life-span. *Discourse Process.*, 13, pp. 207–223, 1990.

Kemper, S. Adults' sentence fragments: Who, what, when, where, and why. *Commun. Res.*, 19, pp. 332–346, 1992.

Kemper, S., Kynette, D., Rash, S., O'Brien, K., and Sprott, R. Life-span changes to adults' language: Effects of memory and genre. *Appl. Psycholing.*, 10, pp. 49–66, 1989.

Kemper, S., Herman, R.E., and Lian, C.H.T. The costs of doing two things at once for young and older adults: Talking while walking, finger tapping, and ignoring speech of noise. *Psychol. Aging*, 18(2), pp. 181–192, 2003.

Klein, G. *Sources of Power: How People Make Decisions*. MIT Press, Cambridge, MA, 1998.

Kline, D.W. and Scialfa, C.T. Visual and auditory aging. in *Handbook of the Psychology of Aging*, J. Birren and K.W. Schaie, eds., 4th ed., Academic Press, San Diego, CA, 1996, pp. 181–203.

Kline, D.W. and Scialfa, C.T. Sensory and perceptual functioning: Basic research and human factors implications. in *Handbook of Human Factors and the Older Adult*, A.D. Fisk and W.A. Rogers, eds., Academic Press, San Diego, CA, 1997, pp. 27–54.

Klinger, A. Issue Paper: *Labour market response to population ageing and other socio-demographic change*, United Nations Economic Commission: United Nations, 2002.

Knowles, M. Adult learning. in *Training and Development Handbook*, Craig, R., ed., McGraw-Hill, New York, 1987, pp. 168–179.

Kubeck, J.E., Delp, N.D., Haslett, T.K., and McDaniel, M.A. Does job-related training performance decline with age? *Psychol. Aging*, 11(1), pp. 92–107, 1996.

Langolf, C.D., Chaffin, D.B., and Foulke, S.A. An investigation of Fitts' Law using a wide range of movement amplitudes. *J. Mot. Behav.*, 8, pp. 113–128, 1976.

Light, L.L. The organization of memory in old age. in *The Handbook of Aging and Cognition*, F.I.M. Craik and T.A. Salthouse, Eds., Lawrence Erlbaum Associates, Mahwah, NJ, 1992, pp. 111–165.

Light, L.L. and Anderson, P.A. Memory for scripts in young and older adults. *Memory*, 11, pp. 435–444, 1983.

Light, L.L., Valencia-Laver, D. and Zavis, D. Instantiation of general terms in young and old adults. *Psychol. Aging*, 6, pp. 337–351, 1991.

Loewenfeld, I.E. Pupillary changes related to age. in *Topics in Neuro-Opthalmology*, H.S. Thompson, ed., Williams and Wilkens, Baltimore, MD, 1979.

MacKay, D.G. and Abrams, L. Age-linked declines in retrieving orthographic knowledge: Empirical, practical, and theoretical implications. *Psychol. Aging*, 13(4), pp. 647–662, 1998.

Madden, D.J. and Langley, L.K. Age-related changes in selective attention and perceptual load during visual search. *Psychol. Aging*, 18(1), pp. 54–67, 2003.

Masunaga, H. and Horn, J. Expertise and age-related changes in components of intelligence. *Psychol. Aging*, 16(2), pp. 293–311, 2001.

Mayhorn, C.B., Fisk, A.D. and Whittle, J.D. Decisions, decisions: Analysis of age, cohort, and time of testing on framing of risky decision options. *Hum. Factors*, 44(4), pp. 515–521, 2002.

Mayhorn, C.B., Nichols, T.A., Rogers, W.A., and Fisk, A.D. Hazards in the home: Using older adults' perceptions to inform warning design. *Injury Control and Safety Promotion*, 11, pp. 211–218, 2004b.

Mayhorn, C.B. and Podany, K.I. Older adults and warnings. in *The Handbook of Warnings*, M.S. Wogalter, ed., (in press).

Mayhorn, C.B., Stronge, A.J., McLaughlin, A.C., and Rogers, W.R. Older adults, computer training, and the systems approach: A formula for success. *Educ. Gerontol.*, 30, pp. 185–203, 2004a.

McDowd, J.M. and Oseas Kreger, D.M. Aging, inhibitory processes, and negative priming. *J. Gerontol.*, 46(6), pp. 340–P345, 1991.

McEvoy, G.M. and Cascio, W.F. Cumulative evidence of the relationship between employee age and job performance. *J. Appl. Psychol.*, 74(1), pp. 11–17, 1989.

McGehee, W. and Thayer, P.W. *Training in Business and Industry*, John Wiley & Sons, New York, 1961.

McLaughlin, A.C., Rogers, W.A. and Fisk, A.D. (in press). Helping patients follow their doctor's instructions: Matching instructional media to task demands. in *Social and Cognitive Perspectives on Medical Adherence*, D.C. Park and L. Liu, eds., American Psychological Association, Washington, D.C., (in press).

Mead, S.E. and Fisk, A.D. Measuring skill acquisition and retention with an ATM simulator: The need for age-specific training. *Hum. Factors*, 40, pp. 516–523, 1998.

Mead, S.E., Spaulding, V.A., Sit, R.A., Meyer, B. and Walker, N. Effects of age and training on World Wide Web navigation strategies. *Proceedings of the Human Factors and Ergonomics Society* (pp. 152–156). HFES, Santa Monica, CA, 1997.

Meyer, B.J.F., Talbot, A., Stubblefield, R.A., and Poon, L.W. Interests and strategies of young and old readers differentially interact with characteristics of texts. *Educ. Gerontol.*, 24, pp. 747–771, 1998.

Morrell, R.W., Park, D.C., Mayhorn, C.B., and Kelley, C.L. The effects of age and instructional format on teaching older adults how to use ELDERCOMM: An electronic bulletin board system. *Educ. Gerontol.*, 26, pp. 221–236, 2000.

Morrow, D.G., Yesavage, J., Leirer, V. and Tinklenberg, J. Influence of age and practice on piloting tasks. *Exp. Aging Res.*, 19(1), pp. 53–70, 1993.

Morrow, D.G., Menard, W.E., Stine-Morrow, E.A.L., Teller, T. and Bryant, D. The influence of expertise and task factors on age differences in pilot communication. *Psychol. Aging*, 16(1), pp. 31–46, 2001.

Morrow, D.G., Ridolfo, H.E., Menard, W.E., Sanborn, A., Stine-Morrow, E.A.L, Magnor, C., Herman, L., Teller, T., and Bryant, D. Environmental support promotes the expertise-based mitigation of age differences on pilot communication tasks. *Psychol. Aging*, 18(2), pp. 268–284, 2003.

Murphy, D.R., McDowd, J.M., and Wilcox, K.A. Inhibition and aging: similarities between younger and older adults as revealed by the processing of unattended auditory information. *Psychol. Aging*, 14(1), pp. 44–59, 1999.

Neale, J.C., Toye, M.H., and Belbin, E. Adult training: The use of programmed instruction. *Occup. Psychol.*, 42(1), pp. 23–31, 1968.

Norman, D.A. *The Psychology of Everyday Things*, Basic Books, New York, 1988.

Owsley, C., Sekuler, R., and Siemsen, D. Contrast sensitivity throughout adulthood. *Vis. Res.*, 23(7), pp. 689–699, 1983.

Panek, P.E. The older worker. in *Handbook of Human Factors and the Older Adult*, A.D. Fisk and W.A. Rogers, eds., Academic Press, San Diego, CA, 1997, pp. 363–394.

Panek, P.E. and McGown, W.P. Risk-taking across the life-span as measured by an intrusion-omission ratio on a selective attention task. *Percep. Mot. Skills*, 52(3), pp. 733–734, 1981.

Parasuraman, R. The attentive brain: Issues and prospects. in *The Attentive Brain* R. Parasuraman, ed., MIT Press, Cambridge, MA, 1998, (pp. 3–15).

Park, D.C. Aging, cognition, and work. *Hum. Performance*, 7, pp. 181–205, 1994.

Park, D.C. Applied cognitive aging research. in *The Handbook of Aging and Cognition*, F.I.M. Craik and T.A. Salthouse, eds., Lawrence Erlbaum Associates Inc, Hillsdale, NJ, 1992, (pp. 449–493).

Park, D.C., Hertzog, C., Kidder, D.P., Morrell, R.W. and Mayhorn, C.B. The effect of age on event-based and time-based prospective memory. *Psychol. Aging*, 12(2), pp. 314–327, 1997.

Parkinson, D. *Voices of experience: mature workers in the future workforce*, (Report No. R-1319-02-RR). The Conference Board. 2002.

Pichora-Fuller, M.K., Schneider, B.A., and Daneman, M. How young and old adults listen to and remember speech heard in noise. *J. Acoust. Soc. Am.*, 97, pp. 593–608, 1995.

Pratt, M.W., Boyes, C., Robins, S.. and Manchester, J. Telling tales: Aging, working memory, and the narrative cohesion of storytellers. *Dev. Psychol.*, 25, pp. 628–635, 1989.

Pratt, M.W. and Robins, S.L. That's the way it was: Age differences in the structure and quality of adults' personal narratives. *Discourse Process.*, 14, pp. 73–85, 1991.

Pulling, N.H. Headlight glare resistance and driver age. *Hum. Factors*, 22(1), pp. 103–112, 1980.

Qualls, C.D., Harris, J.L., and Rogers, W.A. Cognitive-linguistic aging: Considerations for home health care environments. in *Human Factors Interventions for the Health Care of Older Adults*, W.A. Rogers and A.D. Fisk, eds., Lawrence Erlbaum Associates, Mahwah, NJ, 2001, pp. 47–67.

Reynolds, C., Czaja, S.J., and Sharit, J. Age and perceptions of usability on telephone menu systems. *Proceedings of the Human Factors and Ergonomics Society 46th Annual Meeting* HFES, Santa Monica, CA, 2002, pp. 175–179.

Rogers, W.A., Fisk, A.D., McLaughlin, A.C., and Pak, R. Touch a screen or turn a knob: Choosing the best device for the job. *Hum. Factors*, 47, pp. 271–288, 2005.

Root, W. Injuries at work are fewer among older employees. *Mon. Labor Rev.*, 104, pp. 30–34, 1981.

Rousseau, G.K., Lamson, N., and Rogers, W.A. Designing warnings to compensate for age-related changes in perceptual and cognitive abilities. *Psychol. Market.*, 15, pp. 643–662, 1998.

Ryan, E.B., See, S.K., Meneer, W.B., and Trovato, D. Age-based perceptions of conversational skills among younger and older adults. in *Interpersonal Communication in Older Adulthood: Interdisciplinary Theory and Research*, M.L. Hummert, J.M. Wieman and J.F. Nussbaum, eds., Sage, Thousand Oaks, CA, 1994, pp. 15–39.

Salthouse, T.A. Effects of age and skill in typing. *J. Exp. Psychol. Gen.*, 113, pp. 345–371, 1984.

Salthouse, T.A. *Theoretical Perspectives on Cognitive Aging*, L. Erlbaum Associates, Hillsdale, NJ, 1991.

Salthouse, T.A. and Mauer, Aging, job performance, and career development. In *Handbook of the Psychology of Aging*, 4th ed., J. Birren and K.W. Schaie, eds., Academic Press, San Diego, CA, 1996. pp. 353–364.

Sanfey, A.G. and Hastie, R. Judgment and decision-making across the adult life span: A tutorial review of psychological research. in *Cognitive Aging: A Primer*, D.C. Park and N. Schwartz, eds., Taylor and Francis, Philadelphia, 2000, pp. 253–273.

Schieber, F. and Baldwin, C.L. Vision, audition, and aging research. In *Perspectives on Cognitive Change in Adulthood and Aging*, F. Blanchard-Fields and T.M. Hess, eds., McGraw-Hill, New York, 1996, pp. 122–162.

Schneider, B.A., Daneman, M., Murphy, D.R., and Kwong See, S. Listening to discourse in distracting settings: The effects of aging. *Psychol. Aging*, 15(1), pp. 110–125, 2000.

Shea, G. *Managing Older Employees*, Jossey-Bass Inc, San Francisco, 1991.

Sit, R.A. and Fisk, A.D. Age-related performance in a multiple-task environment. *Hum. Factors*, 41(1), pp. 26–34, 1999.

Smith, A.D. Consideration of memory functioning in healthcare intervention with older adults. in *Human Factors Interventions for the Healthcare of Older Adults*, W.A. Rogers and A.D. Fisk, eds., Lawrence Erlbaum Associates, Mahwah, NJ, 2002, (pp. 31–46).

Smith, J. Dixon, R.A., and Baltes, P.B. Expertise in life planning: A new research approach to investigating aspects of wisdom. in *Adult development, Vol. 1: Comparisons and Applications of Developmental Models*, M.L. Commons and J.D. Sinnott, eds., England: Praeger Publishers, New York, 1989, (pp. 307–331).

Smith, J., Staudinger, U.M., and Baltes, P.B. Occupational settings facilitating wisdom-related knowledge: The sample case of clinical psychologists. *J. Consult. Clin. Psychol,,* 62(5), pp. 989–999, 1994.

Smith, M.W., Sharit, J., and Czaja, S.J. Aging, motor control, and the performance of computer mouse tasks. *Hum. Factors*, 41(3), pp. 389– 396, 1999.

Sofranko, A. "*Rocking chair remains idle for many Illinois farmers*." ACES News, July 12, 2000. College of Agricultural, Consumer and Environmental Sciences. University of Illinois at Urbana-Champaign, (2000)

Sparrow, P.R. and Davies, D.R. Effects of age, tenure, training, and job complexity on technical perform- ance. *Psychol. Aging*, 3, pp. 307–314, 1988.

Spirduso, W.W. and MacRae, P.G. Motor performance and aging. in *Handbook of the Psychology of Aging* 3rd ed., J.E. Birren and K.W. Schaie, eds., Academic Press, San Diego, CA, 1990, pp. 183–200.

Staudinger, U.M., Smith, J., and Baltes, P.B. Wisdom-related knowledge in a life review task: age differ- ences and the role of professional specialization. *Psychol. Aging*, 7, pp. 271–281, 1992.

Sternberg, R.J. and Caruso, D. Practical modes of knowing. in *Learning the Ways of Knowing* E. Eisner, ed., Chicago University Press, Chicago, 1985, pp. 133–158.

Sterns, H.L. Training and retraining adult and older adult workers. n *Age, Health, and Employment* J.E. Birren, P.K. Robinson, and J.E. Livingston, eds., Prentice-Hall, Englewood Cliffs, NJ, 1986, pp. 93–113.

Sterns, H.L. and Doverspike, D. Aging and the training and learning process. in *Training and Development in Organizations* I.L. Goldstein, ed., Jossey-Bass/Pfeiffer, San Francisco, 1989, pp. 299–332.

Stronge, A.J., Walker, N., and Rogers, W.A. Searching the World Wide Web: Can older adults get what they need? in *Human Factors Interventions for the Healthcare of Older Adults*, W.A. Rogers and A.D. Fisk, eds., Erlbaum Associates, Mahwah, NJ, 2002, (pp. 255–269).

Sturr, J.F., Kline, G.E., and Taub, H.A. Performance of young and older drivers on a static acuity test under photopic and mesopic luminance conditions. *Hum. Factors*, 32(1), pp. 1–8, 1990.

Szinovácz, M., Ekerdt, D.J., and Vinick, B.H. *Families and Retirement*, Newbury Park: Sage Publications, 1992.

Taranto, M.A. Facets of wisdom: A theoretical synthesis. *Int. Aging Hum. Dev.*, 29(1), pp. 1–21, 1989.

Taylor, H.J. Older Workers: A valuable resource for the workplace. in *Working Through Demographic Change: How OlderAamericans Can Sustain the Nation's Prosperity*, W. Zinke and S. Tattershall, eds., Human Resource Services, Inc, Boulder, CO, 2000, pp. 5–14.

Titone, D., Prentice, K.J., and Wingfield, A. Resource allocation during spoken discourse processing: Effects of age and passage difficulty as revealed by self-paced listening. *Mem. Cogn.*, 28(6), pp. 1029–1040, 2000.

Ulatowska, H.K., Hayashi, M.M., Cannito, M.P., and Fleming, S. Disruption of reference in aging. *Brain Lang.*, 28, pp. 24–41, 1986.

U.S. Bureau of Labor Statistics (2004). *Employment Status of the Civilian Noninstitutional Population by Age, Sex, and Race*. Retrieved September 18, 2005, from http://www.bls.gov/cps/cpsaat3.pdf.

U.S. General Accounting Office. *Older Workers*, GAO-02-85-95-152. Washington, D.C. GAO, 2001.

van Dijk, T. and Kintsch, W. *Strategies of Discourse Comprehension*, Academic Press, New York, NY, 1983.

Vercrucyssen, M. Movement control and speed of behavior. in *Handbook of Human Factors and the Older Adult*, A.D. Fisk and W.A. Rogers, eds., Academic Press, San Diego, CA, 1997, (pp. 55–86).

Verhaeghen, P., Steitz, D.W., Sliwinski, M.J., and Cerella, J. Aging and dual-task performance: A meta-analysis. *Psychol. Aging*, 18(3), pp. 443–460, 2003.

Wagner, R.K. and Sternberg, R.J. Tacit knowledge and intelligence in the everyday world. in *Practical Intelligence: Nature and Origins of Competence in the Everyday World*, R.J. Sternberg and R.K. Wagner, eds., Cambridge University Press, New York, 1988, (pp. 51–83).

Wagner, R.K. and Sternberg, R.J. Tacit knowledge: its uses in identifying, assessing, and developing managerial talent. in *Applying Psychology in Business: The Manager's Handbook*, J. Jones, B. Steffy, and D. Bray, eds., Human Sciences Press, New York, 1991, (pp. 333–344).

Walker, N., Philbin, D.A., and Fisk, A.D. Age-related differences in movement control: Adjusting submovement structure to optimize performance. *J. Gerontol. Psychol. Sci.*, 52B, pp. P40–P52, 1997.

Walker, N., Philbin, D.A., and Spruell, C. The use of signal detection theory on age-related differences in movement control. in *Aging and Skilled Performance: Advances in Theory and Application*, W.A. Rogers, A.D. Fisk and N. Walker, eds., Lawrence Erlbaum Associates, Mahwah, NJ, 1996, pp. 45–64.

Weale, R.A. *The Senescence of Human Vision*, Oxford University Press, Oxford, England, 1992.

Wechsler, D. *WAIS-R Manual*, Psychological Corporation, New York, 1981.

Welford, A.T. Signal noise, performance, and age. *Hum. Factors*, 23, pp. 97–109, 1981.

Wickens, C.D. and Hollands, J.G. *Engineering Psychology and Human Performance* 3rd ed.. Prentice Hall, Upper Saddle River, NJ, 1999.

Willott, J. *Aging and the Auditory System*, Singular Publishing Group, San Diego, CA, 1991.

Wogalter, M.S., Dejoy, D.M., and Laughery, K.R. *Warnings and Risk Communication*, Taylor and Francis, London, 1999.

Wogalter, M.S. and Mayhorn, C.B. Providing cognitive support with technology-based warning systems. *Ergonomics*, 48(5), pp. 522–533, 2005.

Zsambok, C.E. and Klein, G. *Naturalistic Decision Making*, Lawrence Erlbaum Associates, Mahwah, NJ, 1997.

18

Vision and Work

James Sheedy
The Ohio State University

Vision is an integral part of nearly every job. Since vision is integral to nearly every job, a corollary is that each job has a minimum level of vision or visual skills that is required for its proper performance.[1]

For some jobs the minimum level of vision is specified as a job requirement. Such vision requirements are strictly applied in situations where job performance is specifically dependent upon particular visual characteristics or skills and where the cost of nonperformance is high in terms of safety and/or clearly identifiable costs. A common example is driving, a job (or task) for which a minimum level of corrected visual acuity is required. Some states also have visual field (peripheral vision) requirements for driver licensure. Other critical occupations such as law enforcement officers, firefighters, pilots, and military among others have minimum vision requirements to obtain and/or retain employment. These vision requirements can include corrected visual acuity, uncorrected visual acuity, limits on refractive error magnitude, binocular vision, depth perception, color discrimination, visual fields, and limits on pathological conditions. Another example is requiring normal color vision for a job such as quality control inspection where good color discrimination is important. In many cases minimum visual requirements have been challenged legally on the grounds that they are discriminative. If the visual requirements can be shown to be bonafide as related to critical task performance then they are usually defensible.

Although relatively few jobs have formal vision standards, good vision is a prerequisite for optimal performance and visual comfort for most jobs. Vision is often screened at the time of application or hire as a means to determine if the applicants have the necessary visual skills to perform certain tasks and/or to identify those applicants who may be in need of better visual correction prior to hire.

Besides poor performance, inadequate visual skills and/or improper visual correction can result in eye-related symptoms such as eyestrain, headache, irritated eyes, blurred vision, and fatigue. These problems are common when performing tasks at near working distances because they place great demands upon accommodation (eye focusing mechanism) and binocular alignment. Such symptoms have been particularly common among computer workers where specific workplace ergonomic factors often increase stress upon the visual system. The costs of sub-optimally corrected vision can be quite high with respect to the costs of providing eye care (Daum et al., 2004).

---

[1]Although people without vision can also work, considerable adaptations to the job are usually required. This chapter addresses the visual aspects of work for those who are able to attain relatively normal visual acuity either with or without optical correction. Adaptation of the work environment to the blind or those with low vision is not covered in this chapter.

The objectives of this chapter are to describe and identify the parameters of vision that are important to job performance, and to provide information about those parameters that, based upon experience, will be useful to ergonomic practitioners and researchers.

18.1 Visual Acuity and Refractive Error

18.1.1 Visual Acuity

Visual acuity is the ability to resolve or discriminate small detail. It is the most common visual parameter used to assess visual ability, and, although it can be very important for assessing the ability to perform tasks, it is certainly not the only visual skill or parameter. Visual discrimination may be assessed by several different methods, but the most common method by far is to present rows of standardized letters in decreasing size to determine the smallest angular size at which the subject can identify the letters. This is called Snellen visual acuity, named after the man who developed the method.

Snellen visual acuity is designated as a fraction in which the numerator indicates the testing distance and the denominator indicates the distance at which the vertical dimension of the smallest identified line of letters subtended 5′ of arc (1/12th of a degree). "Normal" visual acuity is considered to be the identification of letters that subtend 5 min of arc at the eye. A measurement of 20/20 indicates acuity was tested with a chart at 20 ft (the numerator) and the smallest line the subject could identify contained letters that subtended 5 min of arc at 20 ft (the denominator). A measurement of 20/60 indicates that vision was tested at 20 ft and the patient was able to read the line of letters on which the letters subtended 5 min of arc at 60 ft. Another way to look at 20/60 visual acuity is that the smallest letters the patient could read at 20 ft were three times as large as the letters that correspond to a 20/20 acuity measurement.

The standard visual acuity steps are logarithmically scaled, each succeeding row is 25% larger: 20/12.5, 20/16, 20/20, 20/25, 20/32, 20/40, 20/50, 20/63, 20/80, 20/100, 20/125, 20/160, 20/200 . . . . Some acuity charts, however, display other nonstandard acuity steps. Further information about visual acuity and its measurement is found in Appendix 18.1.

18.1.2 Refractive Errors

Figure 18.1 depicts an eye that has no refractive error — a condition called "emmetropia." A comparison of the two schematics in Figure 18.1 shows that the top eye is focused for a long viewing distance (infinity) and the bottom is the same eye focused for a near viewing distance. In order to focus the near object on the retina, more refractive power needs to be added to the eye. This is accomplished by contracting the ciliary muscle inside the eye thereby increasing curvature of the crystalline lens inside the eye. This adds refractive power to the eye to enable focus on near objects as shown in Figure 18.1. This process is called "accommodation." The added refractive power enables the near by object to be focused on the retina. Diopter (D) is the unit of refractive power and is the reciprocal of the focal distance in meters. If the near target is located 40 cm from the eyes, this represents an accommodative stimulus of 2.5 D (1/0.4 M).

The refractive condition of the eye is always defined with the lens of the eye in the relaxed or unaccommodated condition. If the distant object is properly focused on the retina with accommodation fully relaxed, then the condition is referred to as emmetropia.

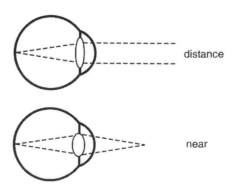

FIGURE 18.1 An eye without refractive error (emmetropia) viewing at a distance with accommodation relaxed and at near with accommodation activated.

The condition of myopia or "near-sightedness" is shown in Figure 18.2. The image of the distant object is formed in front of the retina because the optics of the eye have too much refractive power and/or because the eye is too long. Of course, the rays of light do not stop where the image is formed and are out of focus when they impinge the retina and the image of the distant object is blurred. As shown in Figure 18.2, myopia is corrected with a negative powered lens. The lower image in Figure 18.2 shows that, without correction, there is a near distance at which the image is focused on the retina. The reciprocal of the distance of this near point from the eye (in meters) is the magnitude of the myopia (in Diopters).

The condition of hyperopia or "far-sightedness" is depicted in Figure 18.3. The refractive power of the eye is too weak and/or the eye is too short. A distant object is focused behind the retina with the accommodative mechanism relaxed. Of course, the retina intercepts the light rays before the image is formed and the image that is on the retina is out of focus and blurred. A hyperope

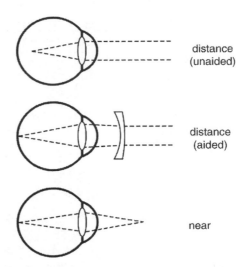

FIGURE 18.2 An eye with myopia. The top figure demonstrates improper focus because the eye is too powerful and/or too long. The middle figure shows spectacle correction with a minus lens, the bottom figure shows that a near object is properly focused without the need for accommodation.

with an adequate amount of accommodation can use accommodative effort to add refractive power to the eye and thereby see distant objects clearly (bottom image). However, if the hyperope does not have an adequate amount of accommodation (which eventually happens with age), then he or she

will be unable to see distant objects clearly. This situation is even worse when viewing near objects. If the amount of hyperopia is low and the individual is young with a large amount of accommodation, then near objects might be able to be cleared comfortably. Because the ability to accommodate decreases with age (see Section 18.2.2.1 on Accommodation), the ability to compensate for hyperopia decreases with age. The dual demands of hyperopia and a near viewing distance can exceed the accommodative abilities of the individual resulting in blur and/or discomfort. Hyperopia is corrected with a positive power lens as shown in Figure 18.3.

Astigmatism is a refractive error in which the power of the eye is different for different meridians or orientations of a plane of light that enters the eye. This condition can be visualized by imagining that the front surface of the eye is shaped like a football, in which the curvature is greater in one orientation than in the other. Astigmatism also results in a blurred image at all viewing distances. Astigmatism can coexist with myopia or with hyperopia, or it is also possible for one meridian to be hyperopic and the other myopic.

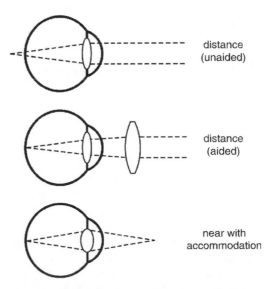

FIGURE 18.3 An eye with hyperopia. Top figure shows hyperopic eye with relaxed accommodation — distant object is unfocused because eye is too short and/or optics are too strong, middle figure shows correction with a plus lens. Bottom figure shows that (excessive) accommodation can be used to focus on the retina.

TABLE 18.1 Myopia and Uncorrected Visual Acuity

Myopia (D)	Pincus (1946)	Crawford et al. (1945)	Peters (1961)
Plano	20/20	20/20	20/20
−0.25		20/22	20/20–20/25
−0.5	20/30	20/35	20/30
−0.75		20/44	20/40–20/50
−1.00	20/50	20/52	20/50–20/60
−1.25		20/63	20/70–20/80
−1.50	20/80	20/91	20/80–20/100
−1.75			20/100–20/200
−2.00	20/100		20/200 or worse
−2.50	20/200	20/100–20/200	
−3.00	20/300	20/400	
−3.50	20/300		
−4.00	20/400		
−4.5	20/470		
−5.00	20/533		
−5.50	20/615		
−6.00	20/890		

Myopia and astigmatism will certainly reduce visual acuity. Uncorrected visual acuity for different levels of myopia is shown in Table 18.1 (Pincus, 1946; Crawford et al., 1945; Peters, 1961). Hyperopia will reduce visual acuity less predictably because younger individuals can use their accommodative mechanism (eye focusing ability) to add power to the eye in order to eliminate or reduce the blur caused by the hyperopia. Accommodative ability predictably declines with age (thus necessitating reading glasses or bifocals), therefore the visual acuity losses due to myopia and hyperopia approach one another in older age groups. Proper spectacle and/or contact lens correction will significantly improve visual acuity in most eyes with a refractive error. It was reported (Zerbe and Hofstetter, 1958) that 98% of the population between the ages of 14 and 40 can be corrected to 20/20 vision in at least one eye.

18.2 Visual Acuity, Refractive Error, and Job Performance

18.2.1 Visual Acuity

Many jobs have a formal requirement for a minimum level of visual acuity and nearly all jobs have a *de facto* visual acuity requirement; however, it is often difficult to exactly determine what the minimum required level of acuity should be. This is because job performance depends upon identifying many complex visual stimuli that are difficult to compare to acuity letters.

Reading text from a computer display at a typical viewing distance of 60 cm is a task example. The angular size of 12 pt font at this distance is approximately 17′ of arc, or would require visual acuity of approximately 20/70 to identify. However, a person cannot comfortably or optimally function by viewing text at threshold size. As a rule-of-thumb, text should be at least three times (3× rule) the threshold size (Sheedy and Shaw-McMinn, 2002), smaller sizes result in less than optimal performance and can also result in eye-related symptoms such as eyestrain and fatigue. Therefore, if the acuity demand of the text on the display is 20/70, then users should have acuity of 20/23.3 to read it comfortably. This requires near-perfect vision of computer users, which is one factor in the high number of computer users experiencing symptoms and argues for optimal visual correction of computer users. The 3× rule can be tested for a given worker by establishing the typical reading distance, then having the person view the material or computer display from 3× the typical distance — if the worker is still able to discern the words then s/he passes the test. This test is not easily performed on a person with presbyopia (see

Section 18.2.3.1 on Presbyopia) in which it is complicated with the need for optical assistance at near viewing distances.

Driving is a task for which there is a minimum visual acuity requirement. Most studies on the relationships between driving and visual measurements have shown weak relationships at best (Bailey and Sheedy, 1988). The reasons for these poor relationships are many, including the facts that licensure screening eliminates people with poor vision from the driving population and individuals with poorer vision self-limit their driving. Most (41 of 50) states require a visual acuity of 20/40 in order to obtain a regular driver's license. Most states also will consider issuing a driver's license (usually restricted) for reduced levels of visual acuity. Most states also have stricter acuity requirements for commercial driver licenses. Another aspect of visual acuity and driving is the standard to which highway signs are designed, which is 1 in. of letter height for every 50 ft of recognition distance (Mace, 1988). Mathematically this transforms to a requirement for 20/23 visual acuity.

Many occupations have specific visual acuity requirements (Mahlman, 1982). The requirements are usually for corrected visual acuity, that is, the worker can meet the requirement with spectacles or contact lenses. For example, a corrected visual acuity standard of 20/20 for police officers was justified on the basis of the need to identify city street signs at appropriate distances when driving in emergency conditions (Sheedy, 1980). Corrected visual acuity standards apply to occupations such as pilots (20/20 each eye), FBI (20/20 in one eye, 20/40 in the other) U.S. Border Patrol (20/20 each eye), firefighters (20/30 binocular), and commercial truck drivers (20/40 in each eye), among others.

Some occupations also have requirements for uncorrected visual acuity, that is, a minimum level of visual acuity is required without optical correction. Uncorrected visual acuity standards typically apply for occupations where performance is critical and especially if it is feasible that glasses or contact lenses could be dislodged during work such as for police officers. Examples of uncorrected acuity requirements include firefighters (20/100 binocular), police officers (usually between 20/40–20/200 dependent on jurisdiction), FBI (20/200), and U.S. Secret Service (20/60 each eye). In many occupations with an uncorrected visual acuity standard, job applicants are allowed to have refractive surgery to meet the standard.

The effects of visual acuity upon performance of only a few tasks have been studied. One study (Good and Augsburger, 1987) analyzed the level of visual acuity required for adequate performance with a firearm. Fifty subjects performed a "friend or foe" task under different levels of visual acuity (20 ft viewing distance, 10 cd/M2), analysis determined that 20/45 vision was required to perform at the threshold performance level. Facial recognition is a complex visual task, but one that can be related to visual acuity (Bullimore et al., 1991). With photopic (daytime) lighting (100 cd/M2) subjects could effectively identify faces at 20 ft with 20/72 vision and identify facial expressions with 20/80 vision. The equivalent acuities required at 20 ft viewing distance for lower light levels were 20/44 and 20/52.

Another means to assess visual capabilities associated with different levels of visual acuity is to utilize accepted medical or insurance categorizations. For example, best-corrected vision of 20/200 or worse is defined as legal blindness in nearly all states and allows income tax advantage. Another useful categorization of abilities related to visual acuity is provided by the International Classification of Diseases (ICD-9-CM, 2004) that is used for insurance reimbursements, shown in Table 18.2. Functionality of different

TABLE 18.2   Classification of Levels of Impairment by Visual Acuity

Best Corrected Acuity	Classification
20/10–20/25	Normal vision
20/30–20/60	Near normal vision
20/70–20/160	Moderate visual impairment or low vision
20/200–20/400	Severe visual impairment or low vision, legal blindness (U.S.)
20/500–20/1000	Profound visual impairment or low vision, moderate blindness
less than 20/1000	Near total visual impairment, severe blindness, near total blindness

Source: ICD-9-CM, American Medical Association, Ann Arbor, MI, 2004.

visual acuity levels can also be assessed with the visual efficiencies accepted by the American Medical Association (PDR, 1998) used to determine post-trauma compensation, shown in Table 18.3.

The acuity level at which individuals decide to wear glasses or contact lenses habitually is an indication of the acuity level at which people feel compromised in the performance of everyday tasks. Numerous variables can influence the acuity level at which ophthalmic correction is worn habitually: age, demands of visual tasks, illumination levels (day or night), individual tolerance to blur, and cosmetic concerns among others. In a survey (U.S., 1964) of 7710 persons, only 2% of those not wearing glasses had a visual acuity of less than 20/100. Unfortunately, they did not provide any details between 20/20 and 20/100. An analysis of 9468 screening records of West German factory workers, reported the habitual binocular visual acuities (with usual eye correction if worn) at work (Schober, 4–6 April, 1968). Only 3.62% of the workers had distance visual acuity of 20/50 or worse, and only 2.29% had near acuity of 20/50 or worse. Acuities between 20/29 and 20/40 at distance were measured in 13.8% of the workers and at near in 9.09%. These data indicate that very few factory workers have habitual acuities of 20/50 or worse — an indication that for these acuity levels most workers choose to habitually wear ophthalmic correction. It appears that most people with acuity in the range of 20/32 to 20/50 will choose to habitually wear optical correction.

18.2.2   Refractive Error

The most common reason for reduced visual acuity is an uncorrected refractive error. Refractive error is also sometimes utilized in occupational vision standards. As an example, the U.S. Air Force Academy sets refractive error limits of +2.00 D/−1.00 D in any meridian and 0.75 D astigmatism (pilot) and +3.00 D/−2.25 D in any meridian and 2.00 D astigmatism (navigator).

Uncorrected myopia and uncorrected astigmatism will reduce visual acuity. Uncorrected astigmatism in magnitudes as low as 0.50 D can create blur at all working distances and has been shown (Wiggins, 1991) to be significantly associated with visual discomfort at work. Some myopic individuals choose to perform near work without correction. Uncorrected myopia is seldom a reason for eye-related symptoms such as eyestrain. In workplace visual screenings, uncorrected myopic individuals will have reduced visual acuity at distance, but normal or much better visual acuity at near distances.

Persons with high amounts of myopia, generally −8.00 D or greater, are at increased risk for retinal detachment and other retinal disorders. It is best that these persons not be placed in jobs that place them at higher risk for head trauma and some occupations specifically exclude persons with high myopia from employment consideration.

Uncorrected hyperopia can be a significant source of visual discomfort — especially for workers with visually demanding work. This is because the hyperopic eye must use excessive accommodative effort to see clearly at near working distances resulting in visual fatigue and eyestrain. Hyperopic workers can usually exert the needed accommodation to exhibit good visual acuity at visual acuity screenings. Hyperopia, however, can be identified by re-measuring distance visual acuity through a +1.00 D lens. This lens

TABLE 18.3   Percent Visual Efficiency Corresponding to Visual Acuity Level

Visual Acuity	% Visual Efficiency	Visual Acuity	% Visual Efficiency
20/20	100	20/100	50
20/25	95	20/125	40
20/32	90	20/160	30
20/40	85	20/200	20
20/50	75	20/300	15
20/63	65	20/400	10
20/80	60	20/800	5

Source: PDR; Medical Economics Company, Inc. Montvale, NJ: 1998.

will blur distance visual acuity for most people, but a person with hyperopia will relax accommodation and hence still attain good visual acuity through the +1.00 lens. Hence, a worker fails the screening test if visual acuity is not reduced with the +1.00 D lens.

18.2.2.1 Accommodation

Accommodation is the mechanism by which the eye changes its focus to look at near objects (see Figure 18.1). The maximum amount by which the eye can change its power is termed the "amplitude of accommodation." This is measured for a corrected eye by slowly moving an object with small detail towards the eye and noting the distance at which the subject reports first noticeable blur. The distance (in meters) is measured from the spectacle plane (14 mm in front of the corneal apex) to the point of first blur. The inverse of this distance is the amplitude of accommodation in Diopters. The amplitude of accommodation quite predictably decreases with age (Donder, 1864; Duane, 1922; Turner, 1958) as shown in Table 18.4. Some workers have amplitudes that are reduced compared to age-expected values — often causing eye-related symptoms.

The near point of accommodation represents the closest distance to which a worker can bring an object and keep it focused without optical assistance. However, accommodative fatigue limits a worker from viewing at the near point of accommodation for very long. Clinical experience indicates that individuals can sustain only approximately 50% of their maximum amplitude of accommodation. The 50% accommodative effort distance is also shown in Table 18.4.

Most people under the age of 40 can comfortably use accommodation to meet typical near viewing needs. However, some workers have accommodative disorders such as reduced amplitude of accommodation for their age or accommodative infacility — a condition in which the accommodative state is slow to change. Such disorders are very common in clinical practice (Sheedy and Parsons, 1990) and clinical studies have shown that workers with these conditions have symptoms of discomfort such as eyestrain (Hennessey et al., 1984; Levine et al., 1985). Intermittent blurring of near objects is caused by inability of the accommodative mechanism to maintain steady focus on near objects. Blurring of distant objects occurs when, after extended near work, the ciliary muscle remains somewhat contracted in the near position. This effectively makes the eye myopic when looking at distance and can last several hours after extended near work, affecting after-work driving. These disorders are also associated with generalized symptoms such as general fatigue, concentration difficulties, dizziness, and headaches (Jaschinski-Kruza and Schweflinghaus, 1992). Workers with accommodative disorders can usually be treated with

TABLE 18.4 The Normal Relationships of Amplitude of Accommodation and Sustainable Near Working Distance to Age

Age	Max Amp of Accom. (D)	Near Point of Accom. (cm)	Sustainable Accom. (D)	Sustainable Near Distance (cm)	Sustainable Near Distance (in.)
10	14.0	7.1	7.0	14.3	5.6
15	12.0	8.3	6.0	16.7	6.6
20	10.0	10.0	5.0	20.0	7.9
25	8.5	11.8	4.3	23.5	9.3
30	7.0	14.3	3.5	28.6	11.2
35	5.5	18.2	2.8	36.4	14.3
40	4.5	22.2	2.3	44.4	17.5
45	3.5	28.6	1.8	57.1	22.5
50	2.5	40.0	1.3	80.0	31.5
55	1.8	57.1	0.9	114.3	45.0
60	1.0	100.0	0.5	200.0	78.7
65	0.5	200.0	0.3	400.0	157.5
70	0.3	400.0	0.1	800.0	315.0
75	0.0	∞	0	∞	∞

low power plus spectacle lenses that relieve the accommodative effort required to see clearly at near working distances.

18.2.3   Short-Working Distances

The sustainable near point in Table 18.4 represents the closest working distance at which workers with normally functioning accommodation can be expected to perform without accommodative fatigue. Jobs that require continuous work at near distances and/or at distances closer than the typical reading distance of 40 cm can be particularly fatiguing to accommodative function. Usually these are jobs in which the worker must see small details. In order to be able to resolve the details so that they are comfortably above the acuity threshold (see the 3× rule in visual acuity section), the work must be brought close to the eyes so that the visual angles are greater. Although some young workers may be able to work continuously at distances of 4–5 D (25–20 cm), most people are unable to continuously work at such close distances, especially people after age 30 and beyond. If a job requires routine working distances of 30 cm or less, loupes or magnifiers are often indicated.

18.2.3.1   Presbyopia

As shown in Table 18.4, the amplitude of accommodation gradually decreases throughout life and the sustainable near working distance recedes with age. A person has presbyopia when, as result, they can no longer comfortably see their near work. Workers typically begin to have problems with normal near working distances of 40–50 cm between the ages of 40 to 45. The primary symptoms are intermittent blur at near working distances, although the problem may also express as eyestrain.

Presbyopia is most commonly corrected with plus power lenses in the form of single-vision lenses or multifocal lenses, discussed later in this section. The lens prescribed for a presbyopic individual depends upon the amount of remaining accommodation and the working distance. Once a person reaches the age of 60, accommodation has essentially reduced to zero and the lens power is nearly totally based upon the required viewing distance.

18.2.3.2   Optical Correction of Presbyopia

There are several different spectacle options for correction of presbyopia. Out of necessity, most people acquire an optical correction mode that suits their needs and lifestyle. However, there are many cases in which the optical correction chosen by or prescribed for the worker may be well designed to meet daily viewing needs but does not provide optimal correction for visual needs at work. Improperly designed spectacle correction can cause compromised visual performance, eye-related symptoms, and awkward posture with ensuing musculoskeletal symptoms. Properly designed work-related glasses provide better worker vision and comfort (Butzon et al., 2002).

The fundamental visual limitation in the spectacle correction of presbyopia is that, because presbyopia results in a fixed focus eye, a refractive correction can only correct for a single viewing distance — although in reality there is a range of clear vision because of the depth of focus of the human eye (approximately ± 0.50D). This problem is usually resolved by wearing multifocal lenses (bifocal, trifocal, or progressive addition lenses) in which the power of the lens varies depending on the gaze angle (Figure 18.4 and Figure 18.5).

Because distant objects are generally higher in the field of view and near objects lower, optical corrections of presbyopia generally provide the distance correction in the top of the lens and the near correction in the bottom of the lens. The power in the bottom of the lens is usually that which will provide clear vision at 40 cm (16 in.) — this is a typical reading distance and the standard near testing distance used by eye doctors. If the viewing distance and/or gaze requirements of the job are different than this, then different glasses are indicated for the job. This is particularly true for workers who perform visual tasks at intermediate viewing distance (50–100 cm) and/or for those with needs to see at near or intermediate viewing distances straight ahead or overhead.

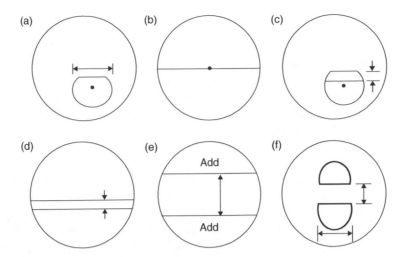

FIGURE 18.4 Lens designs for correction of presbyopia (a) flat top bifocal; (b) executive bifocal; (c) flat top trifocal; (d) executive trifocal; (e) double executive; (f) double flat top.

The most common general usage presbyopic lens designs are the flat top bifocal (Figure 18.4a) and the progressive addition lens (PAL — Figure 18.5a). As usually prescribed and fitted, the bifocal provides a wide clear field of vision at 40 cm viewing distance. The near field of vision is typically attained with 20–25° of downward gaze. The usual width of 28 mm provides a wide fixation field of approximately 50°, however, wider bifocal segments are available.

Computer workers commonly have problems with flat top bifocals because the computer is at an intermediate viewing distance (50–70 cm) and intermediate gaze angle (10–15° downward). The bifocal wearer must tilt their head backward and lean inward, resulting in awkward posture and musculoskeletal discomfort. The worker can either see the computer display clearly or have comfortable posture — but not both. Solutions include a trifocal lens (Figure 18.4c) that includes the intermediate power in the intermediate section (usually with a greater than normal height of the intermediate segment), or a bifocal design in which the intermediate power is placed on the top of the lens and the near power in the bottom. These lenses are occupational glasses and will meet vision needs at work, but they do not meet the general visual needs outside of the workplace. The occupational progressive lens (Figure 18.5b) could also be a solution, however, if the person wears a segmented bifocal for general wear it is usually more acceptable to also wear one for work. With proper selection of powers in the

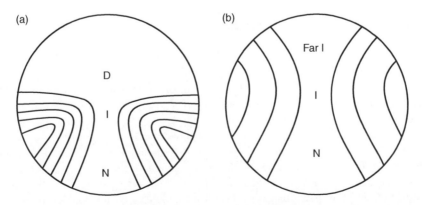

FIGURE 18.5 (a) Typical progressive addition lens and (b) typical occupational progressive lens. D, distance, I, intermediate, N, near. Contours show increasing levels of unwanted aberrations.

work lenses, the computer worker can easily navigate the workplace while wearing the computer glasses. This is important for most computer users and other workers with intermediate viewing needs — because it is inconvenient to change glasses each time they move to or from the workstation.

Progressive addition lenses (Figure 18.5a) are another common lens design used to correct presbyopia. These lenses are characterized by a smooth continuous change of power across the lens surface but with undesired aberrations in the lower quadrants that limit viewing through those areas. PALs have cosmetic advantages and also optical advantages compared to standard bifocals and are generally preferred over bifocals by most wearers. However, even though progressive addition lenses provide a region with inter-mediate power, this is the portion of the lens with the narrowest field of clear vision. If the worker requires significant viewing at intermediate distances such as at a computer, then they must continually find the small sweet spot on the lens and use their neck to move the head rather than moving their eyes. This is inefficient and results in blur and musculoskeletal stress.

Occupational progressive lenses (OPL — Figure 18.5b) can very successfully meet the needs of pres-byopic workers with extensive intermediate viewing needs such as computer users, assembly line workers, assemblers, clerks, janitorial, general office work, etc. The lens design provides a reasonably large inter-mediate viewing zone straight ahead, a wide near viewing zone in the bottom of the lens, and far-intermediate vision on the top of the lens to enable the worker to navigate the workplace. The magnitude of unwanted aberrations in an OPL is significantly less than in a PAL because the total power change is less and the power poles are farther apart, resulting in wider viewing zones compared to the PAL. OPLs are also a work pair of spectacles that do not meet general viewing needs outside of the workplace.

Workers with extensive near or intermediate viewing needs, especially at a wide workstation, benefit from an Executive bifocal or trifocal (Figure 18.4b and Figure 18.4d). This design provides very wide fields of clear, near, and intermediate vision, however, the complete near correction in the bottom of the lens can make workplace navigation difficult.

Many presbyopic workers with near or intermediate viewing needs and superior gaze angles need double segment lens designs such as in Figure 18.4e and Figure 18.4f. These jobs include pharmacists, librarians, mechanics, drapery hangers, painters, etc. The double segment design enables better job per-formance and posture.

Presbyopic workers are generally unaware of the mismatch between their visual needs at work and the vision provided by their general wear glasses. Likewise such mismatches are not frequently identified during routine eye examinations. The problems discussed above are not common among younger pres-byopic individuals who still have some remaining accommodation because the lens power prescribed for near viewing distances is relatively small and they have greater ranges of viewing distances through the distance and near portions of their lenses. The problems noted earlier become most common for workers aged 50 and above. When awkward postures are noted in presbyopic workers, the typical viewing dis-tances and gaze angles should be noted and conveyed to the eye doctor for optimal occupational prescrip-tion and lens design. In the situations discussed earlier, the spectacles are prescribed and designed to meet the specific visual tasks in the workplace and should be provided by the employer.

18.2.3.3 Eye Movements

Of course, workers do not always look straight ahead. In order to fixate a noncentral target, the eyes rotate so the image of the object falls upon the central or foveal portion of the retina. The maximum magnitudes of ocular rotation superiorly are $42°$ (range $33–56°$) and inferiorly are $50°$ (range $33–62°$), and the maximum lateral ocular rotation is approximately $55°$ (range $45–65°$) (Yamashiro, 1957). However, the eyes seldom rotate to these extremes. When a worker fixates a peripheral object, the object is initially fixated nearly entirely with eye movement (Uemura et al., 1980), but the initial eye movement is quickly followed by head rotation. The final resting state while fixating the object involves significantly more of head than eye movement (Table 18.5). Similar findings apply to vertical eye movements. Over a $39°$ range of vertical computer display location, neck angle was found to change by $25°$, ocular rotation by $8°$, and thoracic angle by $5°$ (Villanueva et al., 1997).

TABLE 18.5 Initial and Final Resting Position of the Eyes When Fixating Peripheral Stimuli at Different Angles (degress)

Lateral stimulus amount	10	20	30	40	50
Initial eye rotation	10	20	28	33	41
Final eye rotation	2	5	11	15	19

Because of the strong preference to use head rather than eye movements, workers primarily adjust body posture to a peripheral target in order that the eyes can operate comfortably. Clearly, "the eyes lead the body." If the worker must fixate a peripheral target for extended periods, the awkward posture can result in musculoskeletal symptoms. It has been shown that optimal performance and comfort are attained with a downward gaze angle of $10-15°$ and straight ahead in the lateral meridian (Sheedy and Shaw-McMinn, 2002). Ideally visual tasks should be located so the eyes are depressed by approximately $10°$ with the worker in a neutral musculoskeletal posture. For example, computer displays should be located straight in front of the worker laterally, and the height adjusted so that screen center is approximately 10 cm lower than the eyes (60 cm viewing distance) — resulting in $10°$ downward gaze.

18.2.3.4 Peripheral Vision — Visual Fields

The clinical term for peripheral vision is "visual fields." There are several methods of clinically measuring visual fields. Each method requires the eye being tested to maintain fixation on a single spot while peripheral test spots or objects are shown and the patient responds when they are seen. The normal extent of visual fields as tested with a fairly large peripheral target are listed in Table 18.6 (PDR, 1998) and displayed in Figure 18.6.

Figure 18.6 shows that the visual fields of the two eyes do not perfectly overlap, resulting in a binocular visual field that is larger than the component monocular fields. It has been shown (Good and Fogt, 1998) that the area of the monocular visual field varies from 53.5 to 78.2% of the binocular visual field depending upon the direction of fixation.

Workers missing an eye (monocular individuals) have significant loss of peripheral vision compared to normal binocular individuals. In a study (Johnson and Keltner, 1983) of 10,000 drivers, those with binocular visual field loss had significantly higher accident and conviction rates. Although monocular individuals are able to obtain passenger car driver licenses, some states require additional rearview mirrors and require two eyes for commercial licenses. One way that a monocular individual can help to compensate for loss of visual field is to move the head more. However, there is no evidence that monocular individuals make more head movements nor is it observed clinically. In tasks where peripheral awareness is a requirement (e.g., professional driving, police activities), the reduced visual field of the monocular individual can greatly decrease visual efficiency and create dangerous situations. Several occupations such as police officer, firefighter, pilot, and commercial truck driver among others require two eyes, often with a minimal required visual field in each eye. Monocular workers can be at greater risk

TABLE 18.6 Normal Extent of the Visual Field in Eight Principle Meridians

Direction from Fixation	Degrees of Visual Field
Temporal	85
Down and temporal	85
Down	65
Down and nasal	50
Nasal	60
Up and nasal	55
Up	45
Up and temporal	55

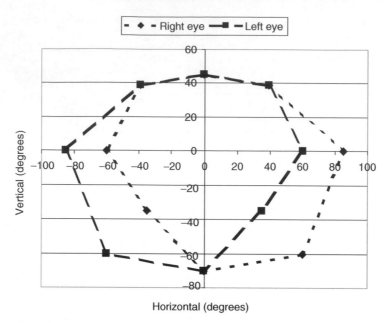

FIGURE 18.6   Peripheral vision (visual fields) with eyes fixating straight ahead. Monocular fields overlap to create a larger binocular visual field.

for injury and caution should be used in placing monocular workers. Monocular workers should always wear eye protection — they don't have a spare.

18.2.3.5   Binocular Vision

One aspect of binocular vision, that is, enhanced visual fields compared to one eye alone, had been discussed earlier. Other aspects of binocular vision will be discussed below.

18.2.3.5.1   Depth Perception

The cues we use to interpret depth in our environment are generally divided into two categories — monocular cues and binocular cues. The monocular cues are obtained with only one eye and are the same cues that exist in photographs, paintings, and motion pictures. These include cues such as size (distant objects are visually smaller), motion parallax, overlay (nearer objects visually overlay distant objects), geometric perspective, and relative height (distant objects are usually higher in the field of view). A monocular individual has full use of these cues for judging depth.

A binocular individual has an additional sensation to judge depth called "stereopsis." This binocular sensation of depth arises from the fact that the two eyes are displaced laterally in our head. Each eye views the scene from a slightly different angle and neural processing of these differences creates the sensation of depth. The mechanism is *not* like a range finder that relies upon calibration of the rotational positions of the two viewing points (in this case, the eyes), but rather it depends upon slight differences in the angles between objects as viewed by the two eyes. A rangefinder gives absolute information about the distance from observer to object, steropsis gives relative information about depth — that is, it allows accurate determination whether one object is closer or farther than another. The effectiveness of stereopsis decreases with viewing distance because the angular difference between the two eyes decreases with viewing distance. Stereopsis is the sensation of depth that is obtained in 3D movies that require special glasses.

Stereopsis is important for dentists, jewelers, assemblers, and crafts where the worker is manipulating relatively small objects at near viewing distances. Normal individuals use stereopsis in the performance of occupational-type tasks, performing depth-dependent tasks upto 30% slower with one eye occluded compared to both eyes opened (Sheedy et al., 1986). Stereopsis provides the worker with enhanced ability to properly maneuver objects with respect to one another. Stereopsis can also be important for some workers with critical intermediate distance tasks — such as forklift drivers.

Individuals without stereopsis or with significantly reduced stereopsis should probably not be placed in jobs that benefit highly from it. Monocular individuals, of course, have a total loss of stereopsis. The effects of the loss of stereopsis in everyday task performance have been described by Brady, who wrote a book based upon his traumatic loss of an eye (Brady, 1972). Likewise, individuals who have two eyes but are unable to keep them aligned (a condition named "strabismus") do not have stereopsis. Also, some individuals are able to keep their eyes aligned but have amblyopia (reduced acuity in one eye) and hence significantly reduced stereopsis. Likewise, stereopsis is reduced if a worker has reduced acuity in one eye due to uncorrected refractive error, this situation can normally be resolved by obtaining proper correction. Various vision screening devices are available to identify individuals with reduced or absent stereopsis.

18.2.3.5.2 *Binocular Alignment — Phoria and Convergence*

Even though most workers keep both eyes aligned when viewing an object, many have difficulty maintaining this ocular alignment resulting in symptoms such as fatigue, headache, blur, double vision, and general ocular discomfort. These problems are much more common among workers with demanding near and intermediate visual needs such as viewing printed text or computer displays. Binocular alignment at near viewing distances is more complex than for distance viewing because of the required ocular convergence and the interaction between ocular convergence and accommodation.

Phoria Binocular fusion, the sensory process by which the images from each eye are combined to form a single percept, requires sensory input from each of the two eyes. If the eyes are not properly aligned on the fixation object, double vision will result. The brain continuously feeds information back to the eye alignment muscles in order to maintain ocular alignment and avoid double vision. Without sensory feedback from each eye (e.g., occluding an eye as in Figure 18.7), the eyes assume their "position of rest" with respect to one another. If the position of rest is outward or diverged, the patient has exophoria. If it is inward or converged, the condition is esophoria.

Most workers have at least a small phoria. Whether a worker experiences symptoms depends upon the amount of the misalignment, the ability of the worker to overcome that misalignment, and the task demands. The symptoms associated with phoria can be eyestrain, double vision, headache, eye irritation, and general fatigue. Clinical studies have documented the relationships between the visual clinical measurements and the symptoms that are experienced (Sheedy and Saladin, 1978). Various vision screening devices enable measurement of the phoria and provide guidelines for eye care referral.

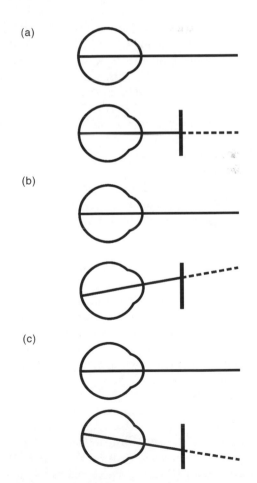

(a)

(b)

(c)

FIGURE 18.7 Binocular alignment when fusion is denied by occluding an eye. (a) Orthophoria. (b) Esophoria. (c) Exophoria.

Convergence   In order to maintain binocular fixation at near viewing distances, it is necessary to converge (or, "cross") the eyes. In the extreme, this results in noticeably crossed eyes when attempting to look at the nose. The most common method to quantify convergence ability is to determine the closest distance to which one can converge the eyes and maintain fusion. Patients with poor convergence ability have a convergence insufficiency.

Workers with convergence insufficiency commonly have symptoms of eyestrain and eye fatigue (Grisham, 1988). Convergence insufficiency is quite common in the population, affecting between 10 and 30% of adults (Grisham, 1988). Convergence insufficiency can be treated — typically with eye exercises to improve convergence ability.

18.2.3.6   Color Vision

Many jobs require the worker to distinguish color. In some jobs such as pilots, boat captains, and electricians, basic color vision is essential for safe performance. For other jobs such as color inspectors and fashion designers, fine color vision ability is important. For most jobs however, poor color vision ability is at most an inconvenience.

Three different cone systems or photoreceptor types in the retina mediate color vision. Each cone type has a different photopigment with different wavelength sensitivity. Because the various objects in our environment have different wavelength reflectance patterns, the relative stimulations of the three cone systems varies with the viewed object. The relative outputs of these three systems are neurologically assessed and the sensation of color is appreciated on this basis.

Color vision ability is best measured in terms of the ability to discriminate small differences in perceived color. Normal individuals are literally able to discriminate millions of different colors. However, there can be differences in fine color discrimination even among those with "normal" color vision. The Farnsworth Munsell 100 hue test[2] discriminates among color normals and can be used to measure for fine color discrimination in those jobs where it is important, such as some quality control inspectors or laboratory testing jobs. The test presents the worker with an array of 85 color chips that are to be arranged in proper color order.

Approximately 8% of the male population and 0.5% of the female population have congenital color vision deficiencies. These individuals are not completely "color blind," but they do have significantly reduced color vision discrimination compared to color normals. Color vision deficiencies are divided into two categories: dichromacy and anomalous trichromacy. An individual with dichromacy can be characterized as missing one of the three cone systems and is unable to distinguish many basic colors from one another (such as red from yellow from green). These dichromats have a relatively severe loss of color discrimination and represent 2% of the male population. Anomalous trichromats can be characterized as still possessing the three cone systems, however, as being "weak" in one of them. These individuals cannot distinguish as many gradations of colors as normal individuals, but they can distinguish and name all of the basic colors. Anomalous trichromacy occurs in 6% of the male population. Because congenital color vision deficiencies are sex-linked, the frequencies of the disorders in the female population are approximately the square of those in the male population. Although colored filters have been advocated to improve color discrimination (Zeltzer, 1979) in color-deficient individuals, they provide marginal improvement at best and do not significantly improve color discrimination (Sheedy and Stocker, 1984).

For jobs that require normal color vision, color vision deficiencies can be identified by testing with pseudoisochromatic plates. These are pictures composed of colored dots, in which the pattern of colored dots enables a color normal observer to identify a figure whereas the color-deficient individual cannot.

Some jobs require basic color identification although fine color discrimination may not be critical. These include jobs such as police officers (clothing and car descriptions), electricians, some assembly work, etc. Workers with anomalous trichromacy but not dichromacy have adequate color discrimination to meet these job demands. A good screening test to differentiate the two is the Farnsworth Munsell D-15

---

[2]Color vision tests are available at Richmond Products, Richmond, CA.

test, more exact testing requires examination with an anomaloscope available only in large clinics. In some cases such as airline pilots, boat captains, and train engineers, it is critical to identify the color of small lights. Job-related color vision testing can be performed with the Farnsworth Lantern test.

18.2.3.7   Vision and Work

Nearly every job has minimum *de facto* vision requirements. If a worker does not have the appropriate visual abilities to meet those requirements it can result in nonperformance, inefficiency, and/or eye-related or musculoskeletal symptoms. Through workplace screening of vision, worker education, ergonomic observation and intervention, and provision of appropriate eye care, most employees can attain the necessary visual abilities to perform most jobs.

Some jobs, however, because of their critical nature and high cost of nonperformance, require visual abilities that exclude some individuals from consideration. In these situations it is important to set visual requirements that are clearly job-related and not unnecessarily exclusionary.

Appendix 18.1   Measurement of Visual Acuity

Because visual acuity is an angular measurement, the measure of visual acuity can be used to quite reliably predict the visual resolution abilities in comparison to an individual with 20/20 acuity. An individual with 20/40 acuity will need to be at half the distance to see the same level of detail in a given target as a person with 20/20 vision. If 20/20 vision enables detection of certain visual details at 480 ft, 20/40 will be able to see it at 240 ft. If 20/20 can see it at 12 ft, then 20/40 can see it at 6 ft. For a visual acuity of 20/30, the viewing distance multiplier is 1/1.5, for 20/60 it is 1/3, for 20/80 it is 1/4, etc. This type of analysis is accurate for any distances greater than 5 or 6 ft from the individual. At distances closer than this, the visual acuity of an individual can change depending upon their refractive condition.

The standard methods for designing visual acuity charts have been specified (Bailey and Lovie, 1976) and standardized (ISO, 1994; ANSI, 1992). Most visual acuity charts have five letters per row. Typical acuity letters have a stroke width to vertical size ratio of 1:5. Although the step size between rows can vary among charts, the standard size progression to the next largest row is $10^{0.1}$ or 1.2589. This results in a logarithmic size progression, typical steps are shown in the following table. Although Snellen equations are used for most clinical and occupational designations of visual acuity, logMAR units are typically used for scientific investigational work. The minimum angle of resolution (MAR) is the stroke width of the smallest identified row of letters. For 20/20 vision, this is 1' of arc, hence logMAR = 0.0. Each incremental row is 0.1 logMAR units. For critical measurements, each letter on a five-letter row gives 0.02 credit towards final logMAR measure. Typical acuity steps in Snellen and logMAR units are shown in the following (Appendix Table 18.1).

Appendix Table 18.1 Standard Incremental Steps in Visual Acuity

20 ft Acuity Designation	6 m Acuity Designation	Letter Height (min arc)	Stroke Height (min arc)	logMAR
20/10	6/3	2.5	0.5	−0.3
20/12.5	6/3.8	3.2	0.63	−0.2
20/16	6/4.8	4.0	0.8	−0.1
20/20	6/6	5.0	1.0	0
20/25	6/7.5	6.3	1.25	0.1
20/32	6/9.5	8.0	1.6	0.2
20/40	6/12	10	2.0	0.3
20/50	6/15	12.5	2.5	0.4
20/63	6/19	16	3.2	0.5
20/80	6/24	20	4.0	0.6
20/100	6/30	25	5.0	0.7
20/125	6/38	32	6.3	0.8
20/160	6/48	40	8.0	0.9
20/200	6/60	50	10.0	1.0

The testing distance from the chart to the subjects' eyes should be maintained and specified. For each acuity measurement it is important to note the eye being tested (right or left), corrective condition of the eye (uncorrected, spectacles, contact lenses), and testing distance. Most commonly, visual acuity is measured with each eye individually (monocular visual acuity), but may also be measured with both eyes open (binocular visual acuity). The subject should be prompted to guess at letters near the size threshold, many subjects will identify 1–2 additional lines with encouragement. Subjects should be instructed to not squint the eyelids during testing and the examiner should enforce nonsquinting.

Dependent upon the refractive condition, visual acuity can be different at far distances compared to near distances. Hence acuity charts are also available for testing of near visual acuity. For near visual acuity testing, it is particularly important to maintain the correct distance of the chart to the eye.

Appendix 18.2 Refractive Error Distribution

Refractive error distribution within the population is shown in the following table (Sorsby, 1967) based upon distribution 2066 eyes of men aged 17–27 (Appendix Table 18.2).

Appendix Table 18.2 Refractive Error Distribution in the Population

Hyperopia (D)	Percentage	Myopia (D)	Percentage
0.00 to +0.90	40	−0.10 to −1.00	5.1
+1.00 to +1.90	33.4	−1.10 to −2.00	2.4
+2.00 to +2.90	6.4	−2.10 to −3.00	1.5
+3.00 to +3.90	4.0	−3.10 to −4.00	0.9
+4.00 to +4.90	1.7	−4.10 to −5.00	0.4
+5.00 to +5.90	1.2	−5.10 to −6.00	0.4
+6.00 to +6.90	0.9	−6.10 to −7.00	0.4
+7.00 to +7.90	0.4	−7.10 to −8.00	0.2
+8.00 to +8.90	0.2	−8.10 to −9.00	0.1
+9.00 to +9.90	0.1	−9.10 to −10.00	0.2
		−10.10 to −11.00	0.0
		−11.10 to −12.00	0.0
		−12.10 to −13.00	0.1

Source: Sorsby, A., *Refractive Anomalies of the Eye*, Washington, DC: U.S. Dept of HEW, 1967.

References

ANSI, American national standard for ophthalmics — instruments — general purpose clinical visual acuity charts, American National Standards Institute, 1992.

Bailey, I.L. and Lovie, J.E., New design principles for visual acuity letter charts, *Am. J. Optom. Physiol. Opt.*, 53(11), pp. 740–745, 1976.

Bailey, I.L. and Sheedy, J.E., Vision screening and driver licensure, in *Transportation in an Aging Society: Improving Mobility and Safety for Older Persons. TRB Special Report 218*, Transportation Research Board, Washington, D.C., 1988.

Brady, F.B., *A Singular View: The Art of Seeing with One Eye*, Medical Economics, Oradell, NJ, 1972.

Bullimore, M.A., Bailey, I.L., and Wacker, R.T., Face recognition in age-related maculopathy, *Invest. Ophthalmol. Visual Sci.*, 32(7), pp. 2020–2029, 1991.

Butzon, S.P., Sheedy, J.E., and Nilsen, E., The efficacy of computer glasses in reduction of computer worker symptoms, *Optometry*, 73(4), pp. 221–230, 2002.

Crawford, J.S., Shagass, C, and Pashby, T.J., Relationship between visual acuity and refractive error in myopia, *Amer. J. Ophth.*, 28, pp. 1220–1225, 1945.

Daum, K.M., Clore, K.A., Simms, S.S., Vesely, J.W., Wilczek, D.D., Spittle, B.M., and Good, G.W., Productivity associated with visual status of computer users, *Optometry*, 75(1), pp. 33–47, 2004.

Donder, F.C., *On the Anomalies of Accommodation and Refraction of the Eye*. The New Sydenham Society, London, 1864.

Duane, A., Studies in monocular and bincocular accommodation with their clinical applications, *Am. J. Ophthalmol.*, 5, p. 865, 1922.

Good, G.W. and Augsburger, A.R., Uncorrected visual acuity requirements for police applicants, *J. Police. Sci. Admin.*, 15, pp. 18–23, 1987.

Good, G.W. and Fogt, N., Dynamic visual fields of one-eyed observers, *Opto. Vis. Sci.*, 75(12S), p. 104, 1998.

Grisham, J.D., Visual therapy results for convergence insufficiency: a literature review, *Am. J. Optom. Physiol. Opt.*, 65(6), pp. 448–454, 1988.

Hennessey, D., Iosue, R.A., and Rouse, M.W., Relation of symptoms to accommodative infacility of school-aged children, *Am. J. Optom. Physiol. Opt.*, 61(3), pp. 177–183, 1984.

ICD-9-CM, *International Classification of Diseases, 9th Revision, Clinical Modification*, American Medical Association, Ann Arbor, MI, 2004.

ISO, Ophthalmic optics — visual acuity testing — standard optotype and its presentation, in *International Standards Organization*, American National Standards Institute, New York, 1994.

Jaschinski-Kruza, W. and Schweflinghaus, W., Relations between dark accommodation and psychosomatic symptoms, *Ophthal. Physiol Opt.*, 12(1), pp. 103–105, 1992.

Johnson, C.A. and Keltner, J.L., Incidence of visual field loss in 20,000 eyes and its relationship to driving performance, *Arch. Ophthalmol.*, 101(3), pp. 371–375, 1983.

Levine, S., Ciuffreda, K.J., Selenow, A. and Flax, N., Clinical assessment of accommodative facility in symptomatic and asymptomatic individuals, *J. Am. Optom. Assoc.*, 56(4), pp. 286–290, 1985.

Mace, D.J., Sign legibility and conspicuity, in *Transportation in an Aging Society: Improving Mobility and Safety for Older Persons. TRB Special Report 218*, Washington, D.C., 1988.

Mahlman, H.E., *Handbook of federal vision requirements*, Chicago: Professional Press, Inc, 1982.

PDR., *Physicians Desk Reference for Ophthalmology*, Montvale, NJ: Medical Economics Company, Inc, 1998.

Peters, H.B., The relationship between refractive error and visual acuity at three age levels, *Amer. J. Optom. Arch. Amer. Acad. Optom.*, 38, pp. 194–198, 1961.

Pincus, M.H., Unaided visual acuities correlated with refractive errors, *Am. J. Ophth.*, 29, pp. 853–858, 1946.

Schober, H.A.W., April 4-6, The role of exact eye correction in industrial hygiene. Paper read at First International Seminar Vision Science, at Bloomington, IN, 1968.

Sheedy, J.E., Police vision standards, *J. Police. Sci. Admin.*, 8(3), pp. 275–285, 1980.

Sheedy, J.E. and Parsons, S.D., The Video Display Terminal Eye Clinic: clinical report, *Optom. Vis. Sci.*, 67(8), pp. 622–626, 1990.

Sheedy, J.E. and Saladin, J.J., Association of symptoms with measures of oculomotor deficiencies, *Am. J. Optom. Physiol. Opt.*, 55(10), pp. 670–676, 1978.

Sheedy, J.E. and Shaw-McMinn, P., *Diagnosing and treating computer-related vision problems*, Woburn, MA: Butterworth-Heinemann, 2002.

Sheedy, J.E. and Stocker, E.G., Surrogate color vision by luster discrimination, *Am. J. Optom. Physiol. Opt.*, 61(8), pp. 499–505, 1984.

Sheedy, J.E., Bailey, I.L., Buri, M., and Bass, E., Binocular vs. monocular task performance, *Am. J. Optom. Physiol. Opt.*, 63(10), pp. 839–846, 1986.

Sorsby, A., The nature of spherical refractive errors, in *Refractive Anomalies of the Eye*, Washington, D.C.: U.S. Dept of HEW, 1967.

Turner, M.J., Observations of the normal subjective amplitude of accommodation, *Bri. J. Physiol. Opt.*, 15(2), pp. 70–100, 1958.

Uemura, T., Arai, Y., and Shimazaki, C., Eye-head coordination during lateral gaze in normal subjects, *Acta. Otolaryngol.*, 90(3–4), pp. 191–198, 1980.

U.S., Binocular visual acuity of adults. Washington, D.C.: U.S. Dept of Health Education and Welfare, National Center for Health Statistics, 1964.

Villanueva, M. B., Jonai, H., Sotoyama, M., Hisanaga, N., Takeuchi, Y., and Saito, S., Sitting posture and neck and shoulder muscle activities at different screen height settings of the visual display terminal, *Ind. Health.*, 35(3), pp. 330–336, 1997.

Wiggins, N. P., Visual discomfort and astigmatic refractive errors in VDT use. *J. Am. Optom. Assoc.*, 62(9), pp. 680–684, 1991.

Yamashiro, M., Objective measurement of the limit of uniocular movement, *Japanese J. Ophthalmol.*, 1, pp. 130–136, 1957.

Zeltzer, H.I., Use of modified X-Chrom for relief of light dazzlement and color blindness of a rod monochromat, *J. Am. Optom. Assoc.*, 50(7), pp. 813–818, 1979.

Zerbe, L.B. and Hofstetter, H.W., Prevalence of 20/20 with best, previous and no lens correction, *J. Am. Optom. Assoc.*, 29, pp. 772–776, 1958.

19

Individual Factors and Musculoskeletal Disorders

Donald C. Cole
Institute for Work and Health

Irina Rivilis
University of Toronto

19.1   Introduction

Musculoskeletal (MSK) disorders are common experiences in the lives of many people worldwide (Volinn, 1997). They are also an important source of disability in both Canada (Cole et al., 2001) and the U.S. The latter is well described in the chapter "Dimensions of the Problem" of a recent report by the National Research Council/Institute of Medicine (NRC/IOM) Panel on Musculoskeletal Disorders (NRC/IOM Panel, 2001). Yet not everyone has an MSK disorder and many with MSK disorders do not suffer much disability, leading the public, practitioners, and researchers to ask: what factors increase the likelihood of experiencing an MSK disorder and suffering an MSK disability?

Among the factors considered are individual factors, although the term may mean different things to different practitioners, policy-makers, and researchers. From a social epidemiology perspective (Berkman and Kawachi, 2000), individuals are nested within families, schools, neighborhoods, workplaces, and other social institutions, which are themselves located in different cities, states, provinces, and countries. Individual factors are therefore attributes of individual persons rather than the social organizations or geographic entities of which they are a part. As with most health outcomes, MSK disorders are likely multi-factorial in origin, with influences at multiple levels spinning webs of causation (Krieger, 1994). For occupational epidemiologists, such as those that participated in the U.S. National Institute of Occupational Safety and Health (NIOSH) review (Bernard, 1997), individual factors are often construed as nonworkplace factors in contrast to workplace factors that contribute to work-related MSK disorders (see appendix B in NIOSH report). In the more resent NRC/IOM Panel report, individual factors are thought to affect personal responses to workplace exposures and tend to be thought of as physiological and psychological attributes in contrast to biomechanical characteristics (NRC/IOM Panel on Musculoskeletal Disorders, 2001).

Conceptualizing the nature and meaning of individual factors therefore remains a challenge. The aim of this article is to share an approach to consideration of individual factors in MSK disorders. It draws on experience of the first author as a primary care physician, occupational medicine/community medicine specialist, and clinical/population epidemiologist. Given the multiple specific questions that might be involved, our citation of evidence is illustrative rather than comprehensive, making reference to systematic reviews when available or individual studies when useful, and pointing to complementary treatment in other articles in this volume when appropriate. We start by exploring what individual factors represent. We go on to consider the similar and different roles of individual factors as they operate in the course of an MSK disorder and the ways that their contribution can be estimated. We then note both the rationale for consideration of individual factors and the directions we might go in considering individual factors in our joint efforts to reduce the burden of MSK disability.

19.2   What Do Individual Factors Associated with MSK Disorders Represent?

People are essentially unique, resulting in distributions of most factors that we measure at the level of the individual person. Hence variation that we might attribute to individuals is the rule rather than the exception. However, the source of the variation that we might observe across individuals may be conceptualized in multiple ways (see Table 19.1). For heuristic purposes, we can group these into those that are potentially work-related, those that are best understood as concurrent exposures, and those that can be thought of as "vulnerabilities."

19.2.1   Work-Related Factors

Gender in most societies is associated with differential work roles for women and men (Messing et al., 1995). The traditional division of labor into "light" and "heavy" work is particularly apparent in the manufacturing and service sector, where women are typically assigned to jobs characterized by repetitive movements, rapid work rate, involving little force, whereas men are often found working in tasks with more extreme physical demands, but with less repetitiveness and often performed at slower speed (Messing et al., 1998). Where men have been placed in traditional women's working conditions in poultry slaughterhouses, they reported similar health symptoms as women coworkers (Mergler et al., 1987).

Social and age stratification in psychosocial exposures and associated health effects are also apparent (Brisson et al., 2001; Montreuil et al., 1996). Yet age is a difficult variable to de-construct. Is it a measure of differential exposure (lighter jobs for older workers), cumulative exposure, declining tissue tolerance, greater experience and skill, other factors, or a combination? Research in slaughterhouses found that inexperienced workers had to invest more physical effort for the same task (knife-sharpening) compared to workers with more training (Vezina et al., 1996, 2000). These differences in work procedures in turn lead to disproportionately more health problems among inexperienced workers.

Varying individual workstyles can put certain workers at differential risk for developing MSK disorders. Different methods of lifting loads and spine stabilization have been observed in laboratory, field and clinical settings, which, when altered, can reduce low back symptoms (McGill, 2001). In a group of sign language interpreters, those with a tendency to work with pain to ensure work quality were at greater risk of having upper extremity symptoms (Feuerstein et al., 1997).

TABLE 19.1 What do Individual Factors Represent?

Usual Naming of Factor Types*	Individual Factor(s)	Potential Construct(s)
Demographic	Gender	Differential labour market, different tasks, capacities and reactions to stress — all resulting in different exposures
	Age	Cumulative exposure, decreased tolerance and different skills and experience
Work	Work-style	Different biomechanical exposures
Anthropometry	Height and weight	Mismatch between equipment and person, differential tissue demands
Psychological	Personality	Differential kinematics, differential coping capacity
Lifestyle	Physical activity, hobbies, sports	Additional loads or physical exposures
	Smoking, drugs	Additional exposures
Comorbidity	Diabetes, pregnancy	Additional internal exposures
	Distress, depression	Altered biochemistry, different pain perception threshold
Past history	Episode of MSK disorder	Lower tolerance
Social	Divorced-widowed	Lower social support
	Minority race	Discrimination
	Poverty	Complex socio-health contexts

*Note different grouping in text which tries to group underlying constructs.

Physical differences between workers, such as variability in body size and height, can give rise to anthropometric mismatches between individuals and their job demands, such as when taller or shorter people are placed in "average" nonadjustable workstations (Chung et al., 1997). Equipment usability problems may be exacerbated by anthropometric differences (Botha and Bridger, 1998). Variation in muscle bulk or physical capacity may influence the biomechanical exposures from tools such as a hand-held powered nutrunner, with small women experiencing greater demands than large men (Oh and Radwin, 1998). Psychological differences can also result in different biomechanical exposures at the tissue level. Those scoring higher on different personality traits, as measured by the Meyer-Briggs Trait Inventory, showed differential trunk kinematics while carrying out standard lifting tasks (Marras et al., 2000).

19.2.2 Concommitant External/Internal Exposures

Participation in sports, vibration experienced during driving, or certain household and care-taking activities, are examples of exposures outside the work environment that add to the overall physical demands placed on individuals. Furthermore, chemical exposures such as smoking may increase risk of low back pain (Leino-Arjas, 1999). An individuals' physiological status is another important factor that affects the development of certain MSK conditions, such as circulating endogenous hormone levels during pregnancy can contribute to carpal tunnel syndrome (CTS), through alteration of fluid balance among women (Weimer et al., 2002). Some pathophysiological states likely alter chemical environments for joints as well, giving rise to condtions such as shoulder adhesive capsulitis and limited joint mobility among patients with type II diabetes mellitus (Balci et al., 1999).

19.2.3 Vulnerabilities

Closely related are underlying genetic factors with their resultant contribution to both physical structure and chemical environments in ways that make individuals more vulnerable to MSK disorders. Familial tendencies have long been recognized for different forms of arthritis (Hirsch et al., 1998) but recently, researchers investigating CTS among twin pairs, estimated that upto half of the variability in prevalence among women may be genetically determined (Hakim et al., 2002).

Acquired vulnerabilities are more commonly recognized, particularly the contribution of one MSK episode to the risk of subsequent ones. Among a cohort of soldiers, the observed risk of injury was seven times greater among previously injured individuals (Schneider et al., 2000). Earlier distress or depression, often called individual psychosocial factors in contrast to workplace or job psychosocial factors, have been shown to predict subsequent MSK pain or poorer recovery in an episode of pain in a wide range of studies, as set out in "Table 4.8 of the NRC/IOM Panel report."

Finally, indicators of social vulnerability have also been associated with MSK disorders. Divorced and widowed injured workers took considerably longer to return to work or go off with temporary total workers' compensation benefits than those who were single or married (Clarke et al., 1999). Racial discrimination may affect subordinate groups, as suggested by the higher levels of MSK pain but lower intensity of diagnostic assessment of low back pain among black patients in the U.S. (Carey and Garret, 2003). Further evidence of complex socio-economic and health relationships comes from examination of healthcare and income support utilization prior to and subsequent to workers' compensation claims in British Columbia, Canada. It appeared that a particularly vulnerable group had consistently higher utilization both before and after the accident date than the rest of the population cohort (Hertzman et al., 1999).

19.3   Where in the Course of MSK Disorders Do Individual Factors Operate?

Individual factors and the constructs they represent may contribute to or impact upon the burden of MSK disorders in a variety of ways (see Kerr, 2000). Early on, etiologic factors may "cause" an MSK disorder. Later, prognostic factors may influence the "recovery" process. When an intervention occurs in a primary prevention program either before or after an MSK disorder develops in a treatment or secondary prevention program, individual factors could influence the effectiveness of the intervention (Figure 19.1).

19.3.1   Etiology

Etiology was the main focus of the NIOSH systematic review (Bernard, 1997) and the NRC/IOM Panel report. Particularly consistent in those reviews were the role of age and gender in contributing to explanation of variations in prevalence or incidence of MSK disorders. Multiple reviews of etiology of MSK disorders have included individual factors, for example, age, personality, and work technique, were all factors, which were found to modify the burden of shoulder-neck complaints (Winkel and Westgaard, 1992).

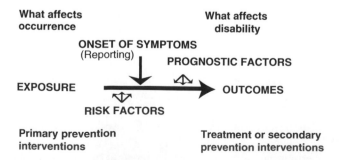

FIGURE 19.1   Where do individual factors operate in the course of a MSK disorder?

Linked to etiology is the role of individual factors in reporting and recognizing MSK disorders. Gender has been shown to influence both reporting and recognition of MSK disorders in relation to workplace exposures (Chung et al., 2000). Similarly, those of lower socio-economic status or only temporarily employed may feel anxious about reporting an MSK disorder to their workplace.

19.3.2 Prognosis

Prognostic studies examine the factors that affect recovery from (or conversely chronicity of) an MSK disorder, either in a clinical setting or using workers' compensation claim (WC) databases. Clinical studies are often quite specific as to the kind of disorder, while the WC studies may group heterogeneous soft tissue MSK conditions. Studies using WC claimants often have cost figures associated with them. The review papers by Baldwin (2004) and Gatchel (2004) comment extensively on such studies, with the latter including a number of individual factors that predict prognosis.

Across both kinds of studies, condition-related factors are among the most important. These include duration of symptoms and initial pain severity. The former is important for low back pain (Frank et al., 1996; Carey et al., 2003; van Tulder et al., 2002) and upper extremity disorders (Cole and Hudak, 1996), while the latter is also important for low back pain and for neck pain (Coté et al., 2001). Closely allied to these are specific signs such as radiation below knee for low back pain or location of pain for lateral elbow pain (Hudak et al., 1996).

The way individuals respond to their MSK disorder is a second important group of characteristic. Baseline mental health status explained variation in recovery from CTS in the main cohort health (Katz et al., 1997), while depression is a common predictor of poor recovery among low back pain patients. Patient or claimant expectations of recovery were found to explain about 15% of the variation in time on total temporary WC benefits for MSK soft-tissue disorders, with those having more positive expectations of returning to work or going off with benefits sooner (Cole et al., 2002a).

Some might regard both condition characteristics and individual responses as clinical factors rather than "individual" factors, though both would be nonworkplace and nonbiomechanical. As noted earlier, age and gender also influence prognosis for MSK disorders including a broad group of WC claimants (Hogg-Johnson and Cole, 2003) as well as more specific conditions such as whiplash (Coté et al., 2001).

Of note, is it that individual factors may make similar or different contributions, depending on where in the course of a disorder they are operating? For example, in modeling change in symptom and disability levels between two surveys of newspaper workers, 1 yr apart, we found that gender and job tenure acted in different ways for those who were not cases at one time and those who had more severe problems (Cole et al., 2002b). Caseness was defined on the basis of frequency, intensity, and duration of MSK symptoms or the extent of limitation of activities. Being a women was protective for non-cases, that is, they were less likely than men to become cases between surveys, yet if they were more severe cases, they had reduced chances of recovery, both in symptoms and disability. Similarly, those with longer tenure at the newspaper were less likely to become cases over the year (survivors), but if they already had a more severe problem, they were less likely to get better between the surveys. Hence the importance of clarifying where in the course individual factors are operating.

19.3.3 Intervention Effectiveness

The impact of combined interventions can be substantially modified by individual factors, such as the differential responsiveness to back-school programs during the subacute phase of low back pain episodes, in terms of time to return to work (Elders et al., 2000). Less researched is the differential impact of workplace interventions aiming at primary prevention, with often only the mean effect being displayed accompanied by wide standard deviations. Two studies, which did differentiate responsiveness showed that younger video display terminal operators (<40-yr-old) responded better than older workers to an office equipment and ergonomic training program, with greater reductions in MSK pain (Brisson et al., 1999) and that shorter assembly line workers experienced greater reductions in discomfort in legs and low back, when flooring was modified do be more absorbent (King, 2002).

19.4 How Can We Estimate the Contribution of Individual Factors to MSK Disorders?

Clinical researchers and practitioners have often included individual factors as part of the descriptive epidemiology of MSK disorders. Some have argued the primacy of individual factors, construed narrowly, as causes of MSK disorders. An example of such attribution occurs in the dissent appendix of the NRC/IOM Panel report (Szabo, 2001), where the contribution of a wide variety of individual factors to CTS etiology is argued to be stronger than occupational exposures. Such an assumption that if individual factors are present, they must be the primary cause, must be examined in relation to one's conception of what individual factors represent and one's understanding of the mechanisms of production of an MSK disorder.

A common practice in occupational epidemiology is to "control for" or "adjust for" individual factors while examining the role of occupational risk factors, either through stratified analyses or multivariate models, in order to avoid the bias of confounding (Rothman and Greenland, 1998). A recent systematic review on etiology of shoulder pain noted that most studies adjusted for age, with some also adjusting for sex, smoking, hobbies, or even intelligence (van der Windt, 2000). Controlling for individual factors does permit calculations of population attributable fractions due to work exposures among those exposed, as in the NRC/IOM Panel report, chapters 3 and 4 (NRC/IOM Panel, 2001). Unfortunately, adjustment means that the independent effects of individual factors are not apparent. Hence, the report also estimated population attributable fractions due to distress or depression, often called individual psychosocial factors, with attributable fractions among the exposed (i.e., those with distress or depression) ranging from 14 to 63% for MSK disorders (see Table 4.5, p. 107 of NRC/IOM Panel, 2001).

Unfortunately, adjustment of either individual factors or work exposures makes it impossible to compare the independent contributions of each to the production of MSK disorders. Fortunately, rather than treating either type of factor as less interesting and therefore needing to be controlled for to eliminate confounding, researchers are increasingly adopting a multi-causal framework and laying out the contribution of different factors. A good example can be found in the report on a case-control study of low back pain reporting among automotive assembly workers, where the variance explained by each domain of risk factors is set out (Kerr et al., 2001). The individual factors of body mass index and a prior WC claim for low back pain accounted for 4.7% of the overall variance, while workplace risk factor domains independently accounted for 11.5% (psychosocial), 11.8% (psychophysical), and 18.3% (biomechanical) of the variance, respectively.

When study populations are large enough, it becomes possible to go even further and examine the ways that factors from different domains may influence the contribution of each other. Several analytic techniques are available to assess such effect-measure modification or heterogeneity of effects of different factors (Rothman and Greenland, 1998). In a study of predictors of time on total temporary wage loss of WC benefits among WC claimants with soft-tissue disorders, we used interaction terms to take into account the associations between changes in pain intensity (from the baseline to 4-week interviews) and workplace offers of special arrangements to help injured workers return to work (workplace offers) (Hogg-Johnson and Cole, 2003). We had found that among those with worsening of their MSK pain (and hence a negative coefficient), the positive impact of a workplace offer (positive coefficient) was even greater, that is, the interaction term between change in pain and workplace offers had a negative coefficient, which when multiplied by the negative for change in pain, created a big positive impact, reducing the time on benefits to less than a third of that for those who didn't receive workplace offers. This was a much bigger effect than that for those whose pain was improving, where reductions in time on benefits with workplace offers ranged from almost the same to one-half of those who did not receive workplace offers.

A word of caution in interpreting multivariate models is warranted, however. So-called independent variables may have complex correlational relationships with other factors in the model. Although these can be detected through model diagnostic procedures, harnessing them to improve understanding may require consideration of causal pathway models, which better reflect "webs of causation" (Krieger, 1994). Path analysis extends regression models by incorporating intermediate explanatory variables, which permit examination of both direct and indirect effects of particular variables on the outcome, while structural equation modeling (SEM) adds measurement models to path structures in order to deal with uncertainties in the measurement of constructs. In one SEM application, physiological reactivity to pain was shown to fully mediate the relationship between neuroticism and subsequent pain among rheumatoid arthritis patients (Evers et al., 2001). In another application, history of systemic disease (an individual factor) was shown to have variable relationships with complex sets of work organizational, postural, force, and vibration exposures and symptoms in the neck and upper extremity (Punnett and van der Beek, 2000).

19.5   Directions

Measurement of individual factors will remain important in research on MSK disorders, particularly for the growing area of genetic factors. We face considerable challenges in understanding what individual factors represent and the mechanisms by which individual factors affect MSK disorders. We need to better link individual factors to the different periods in the course of MSK disorders, comparing etiological and prognostic roles. In keeping with notions of a web of causation (Krieger, 1994), we need to maintain the perspective of multiple versus single causes linked over time in pathways of causation. The latter would emphasize independent and combined contributions to explanation of variance and the role of heterogeneity, not only within and across populations but also across effect measures. Finally, we must incorporate our understandings into practice and policy in ethical ways, improving protection for the vulnerable through prevention and accommodation and better matching of contexts, people and interventions, to improve intervention effectiveness.

Acknowledgments

Epidemiological, clinical, biomechanical, and other colleagues for ongoing discussions of these ideas. Bill Marras, Tom Waters, and STAR-MSK participants for stimulating the development of the framework and providing a forum for discussion of these ideas. Participating workplaces, practitioners, organizations, and individuals in studies cited. Workplace Safety and Insurance Board of Ontario funding to the Institute for Work and Health. The U.S. National Institute of Occupational Safety and Health and other funders for support.

References

Balci, N., Balci, M.K., and Tuzuner, S., Shoulder adhesive capsulitis and shoulder range of motion in type II diabetes mellitus: association with diabetic complications, *J. Diabetes Complications*, 13(3), pp. 135–140, 1999.

Baldwin, M.L., Reducing the costs of work-related musculoskeletal disorders: targeting strategies to chronic disability cases, *J. Electromyog. Kinesiol.*, 14(1), pp. 33–41, 2004.

Berkman, L.F. and Kawachi, I., eds., *Social Epidemiology*, Oxford University Press, Oxford, 2000.

Bernard, B.P. ed., Musculoskeletal disorders and workplace factors. A critical review of epidemiologic evidence for work-related musculoskeletal disorders of the neck, upper extremity and low back, National Institute for Occupational Safety and Health, Cincinnati, OH, DHHS (NIOSH) Publication No. 97-141, 1997.

Bland, J.D.P., Do nerve conduction studies predict the outcome of carpal tunnel decompression?, *Muscle Nerve*, 24(7), pp. 935–940, 2001.

Bongers, P.M., Kremer, A.M., and ter Laak, J. Are psychosocial factors, risk factors for symptoms of the shoulder, elbow or hand/wrist?: a review of the epidemiological literature, *Ame. J. Ind. Med.*, 41(5), pp. 315–342, 2002.

Botha, W.E. and Bridger, R.S. Anthropometric variability, equipment usability and musculoskeletal pain in a group of nurses in the Western Cape. *Occup. Health Ind. Med.*, 39(6), 252, 1998.

Brisson, C., Montreuil, S., and Punnett, L. Effects of an ergonomic training program on workers with video display units. *Scand. J. Work Environ. Health*, 25(3), pp. 255–263, 1999.

Brisson, C., Larocque, B., and Bourbonnais, R., Les contraintes psychosociales au travail chez les Canadiennes et les Canadiens, *Canadian Journal of Public Health* 92(6), 460–467, 2001.

Buckle, P. Upper limb disorders and work: The importance of physical and psychosocial factors, *J. Psychosom. Res.*, 43(1), pp. 17–25, 1997.

Carey, T.S. and Garrett, J.M. The relation of race to outcome and the use of health care services for acute low back pain. *Spine*, 28(4), pp. 390–394, 2003.

Chung, J., Cole, D.C., Clarke, and J. Women, work and injury. Chapter 4 in: T. Sullivan ed., *Injury and the New World of Work*. University of British Columbia Press: Vancouver 2000, pp. 69–90.

Chung, M.K. and Kyungim, C. Ergonomic anlaysis of musculoskeletal discomforts among conversational V.D.T. operators, *Comput. Ind. Eng.*, 33(3–4), pp. 521–524, 1997.

Clarke, J., Chung, J., Cole, D.C., Hogg-Johnson, S., Haidar and the ECC Prognosis Working Group. Gender and benefit duration in lost time work-related soft tissue disorders: relationships with work and social factors. I.W.H. Working Paper #85. Toronto, O.N., Canada: Institute for Work & Health 1999.

Cole, D.C. and Hudak, P.L. Understanding prognosis of non-specific work related musculoskeletal disorders of the neck and upper extremity, *Am. J. Ind. Med.*, 29, pp. 657–668, 1996.

Cole, D.C., Ibrahim, S.A., Shannon, H.S., Scott, F., and Eyles, J. Work correlates of back problems and activity restriction due to musculoskeletal disorders in the Canadian National Population Health Survey (NPHS) 1994/95 Data, *Occup. Environ. Med.*, 58, pp. 728–734, 2001.

Cole, D.C., Mondloch, M.V., and Hogg-Johnson, S., ECC Prognostic Modeling Group. Listening to injured workers: how recovery expectations predict outcomes?, *Can. Med. Assoc. J.* 166, pp. 749–754, 2002a.

Cole, D.C., Manno, M., Beaton, D., and Swift, M. Transitions in self-reported musculoskeletal pain and interference with activities among newspaper workers. *J. Occup. Rehab.*, 12(3), pp. 163–174, 2002b.

Coté P, Cassidy, J.D., Carroll, L., Frank, J.W., and Bombardier, C. A systematic review of the prognosis of acute whiplash and a new conceptual framework to synthesize the literature. *Spine*, 26(19), pp. 445–458, 2001.

Elders, L.A.M., van der Beek, A.J., and Burdorf, A. Return to work after sickness absence due to back disorders-a systematic review on intervention strategies. *Int. Arch. Occup. Environ. Health*, 73(5), pp. 339–348, 2000.

Evers, A.W.M., Kraaimaat, F.W., van Riel, P.L.C.M, and Bijlsma, J.W.J. Cognitive, behavioral and physiological reactivity to pain as a predictor of long-term pain in rheumatoid arthritis patients. *Pain*, 93, pp. 139–146, 2001.

Feuerstein, M. Workstyle: Definition, empirical support and implications for prevention, evaluation and rehabilitation of occupational upper-extremity disorders. In: Moon, S.D., Sauter, S.L. eds. *Beyond Biomechanics: Psychosocial Aspects of Musculoskeletal Disorders in Office Work*. Taylor and Francis:-Bristo, PA, pp. 177–206, 1996.

Feuerstein, M., Carosella, A.M., and Burrell, L.M. Occupational upper extremity symptoms in sign language interpreters: Prevalence and correlates of pain, function, and work disability, *J. Occup. Rehab.*, 7(4), pp. 187–205, 1997.

Frank, J.W., Brooker, A.S., Demaio, S.E., Kerr, M.S., Maetzel, A., Shannon, H.S., Sullivan, T., Norman, R.W., and Wells, R.P. Disability resulting from occupational low back pain. Part, II: What do we know about secondary prevention? A review of the scientific literature on prevention after disability begins. *Spine*, 21(24), pp. 2918–2929, 1996.

Gatchel, R.J. Musculoskeletal disorders: primary and secondary interventions, *J. Electromyogr. Kinesiol.*, 14(1), pp. 161–170, 2004.

Gatchel, R.J., Polatin, P.B., and Mayer, T.G. The dominant role of psychosocial risk factors in the development of chronic low back pain disability. *Spine*, 20(24), pp. 2702–2709, 1995.

Hakim, A.J., Cherkas, L., El Zayat, S.E., Macgregor, A.J., and Spector, T.D. The genetic contribution to carpal tunnel syndrom in women: a twin study. *Arthritis Rheum.*, 47(3), pp. 275–279, 2002.

Hertzman, C., McGrail, K., and Hirtle, B. Overall pattern of health care and social welfare use by injured workers in the British Columbia cohort. *Int. J. Law Psychiatry*, 22 (5–6), pp. 581–601, 1999.

Hirsch, R., Lin, J.P., Scott, W.W. Jr, Ma, L.D., Pillemer, S.R., Kastner, D.L., Jacobsson, L.T., Bloch, D.A., Knowler, W.C., Bennett, P.H., and Bale, S.J. Rheumatoid arthritis in the Pima Indians: the intersection of epidemiologic, demographic and genealogic data. *Arthritis Rheum.*, 41(8), pp. 1464–1469, 1998.

Hogg-Johnson, S. and Cole, D.C., Early prognostic factors for duration on temporary total benefits in the first year among workers with compensated occupational soft tissue injuries, *Occupational and Environmental Medicine* 60, 244–253, 2003.

Hudak, P.L., Cole, D.C., and Haines, A.T. Understanding prognosis to improve rehabilitation: the case of lateral elbow pain. *Arch. Phys. Medi. Rehab.*, 77, pp. 586–593, 1996.

Katz, J.N., Keller, R.B., Fossel, A.H., and Punnett, L. Predictors of return to work following carpal tunnel release, *Am. J. Ind. Med.*, 31(1), pp. 85–91, 1997.

Kerr, M. The importance of psychosocial risk factors in injury. Ch 5 in: T. Sullivan ed., *Injury and the new world of work*. University of British Columbia Press:Vancouver, 2000 pp. 93–114.

Kerr, M.S., Frank, J.W., Shannon, H.S., Norman, R.W., Wells, R.P., and Neumann, W.P. Examining the overlap between measures of the psychosocial and physical demands of work: data from case-control study of risk factors for low back pain. Institute for Work & Health working paper #181. 2001.

King, P.M., A comparison of the effects of floor mats and show in-soles on standing fatigue, *Appl. Ergon.*, 33, pp. 477–484, 2002.

Krieger, N., Epidemiology and the web of causation: has anyone seen the spider? *Social Sci. Med.*, 39(7), pp. 887–903, 1994.

Leino-Arjas, P. Smoking and musculoskeletal disorders in the metal industry: A prospective study. *Occup. Health Ind. Med.*, 40(2), 99, 1999.

Marras, W.S., Davis, K.G., Heaney, C.A., Maronitis, A., and Allread, W.G. The influence of psychosocial stress, gender, and personality on mechanical loading of the lumber spine. *Spine*, 25(23), pp. 3045–3054, 2000.

McGill, S.M. Low back stability: from formal description to issues for performance and rehabilitation. *Exerc. Sports Sci. Rev.*, 29(1), pp. 26–31, 2001.

Mergler, D., Brabant, C., Vézina N., and Messing, D. The weaker sex? Men in women's working conditions report similar health symptoms. *J. Occup. Med.*, 29, pp. 417–421, 1987.

Messing, K., Chatigny, C., and Courville, J. 'Light' and 'heavy' work in the housekeeping service of a hospital. *Appl. Ergon.*, 29(6), pp. 451–459, 1998.

Messing, K., Neis, B., and Dumais, L., eds. Invisible: Issues in women's occupational health. Charlottetown, PEI: Gynergy Books, 1995.

Montreuil, S., Laflamme, L., and Tellier, C. Profile of the musculoskeletal pain suffered by textile tufting workers handling thread cones according to work, age and employment duration, *Ergonomics* 39(1), 76–91, 1996.

National Research Council and Institute of Medicine Panel on Musculoskeletal Disorders and the Workplace (NRC/IDM), Commission on Behavioral and Social Sciences and Education. Musculoskeletal disorders and the workplace: low back and upper extremities, National Academy Press: Washington, D.C., U.S.A. 2001. Available at http://www.nap.edu

Oh, S.A. and Radwin, R.G. The influence of target torque and torque build-up time on physical stress in right angle nutrunner operation. *Ergonomics*, 41(2), pp. 188–206, 1998.

Punnett, L. and van der Beek, A.J. A comparison of approaches to modelling the relationship between ergonomic exposures and upper extremity disorders, *American Journal of Industrial Medicine* 37(6), 645–655, 2000.

Rothman, K.J. and Greenland, S., eds., Modern epidemiology, 2nd Edn., Lippincott Williams & Wilkins, Philadelphia, 1998.

Schneider, G.A., Bigelow, C., and Amoroso, P.J. Evaluating risk of re-injury among 1214 army airborne soldiers using a stratified survival model. *Am. J. Prev. Med.*, 18(3 Suppl), pp. 156–163, 2000.

Spector, T.D. The menopause, oestrogens and arthritis. *Maturitas*, 27(1), 15, 1997.

Szabo, R.M. Dissent. Appendix, B in: *National Research Council and Institute of Medicine Panel on Musculoskeletal Disorders and the Workplace, Commission on Behavioral and Social Sciences and Education. Musculoskeletal Disorders and the Workplace: Low Back and Upper Extremities.* National Academy Press, Washington, D.C., 2001, pp. 439–457.

Van der Windt, D.A.W.M., Thomas, E., Pope, D.P., de Winter, A.F., Macfarlane, G.J., Bouter, L.M., and Silman, A.J. Occupational risk factors for shoulder pain: a systematic review. *Occup. Environ. Med.*, 57, pp. 433–442, 2000.

Van Tulder, M., Koes, B., and Bombardier, C. Low back pain. *Best Pract. Res. Clin. Rheumatol.*, 16(5), pp. 761–775, 2002.

Vézina, N., Prévost, J., and Lajoie, A. Élaboration d'une formation à l'affilage des couteaux dans six usines d'abattage et de transformation du porc: une étude ergonomique. Montréal, Québec: Institut de Recherche en Santé et Securité au Travail, 2000. 48 pages

Vezina, N. and Chatigny, C. Training in factories: A case study of knife-sharpening. *Saf. Sci.*, 23(2/3), 195, 1996.

Volinn, E. The epidemiology of low back pain in the rest of the world. A review of surveys in low- and middle-income countries. *Spine*, 22, pp. 1747–1754, 1997.

Weimer, L.H., Yin, J., Lovelace, R.E., and Gooch, C.L. Serial studies of carpal tunnel syndrome during and after pregnancy. *Muscle Nerve*, 25(6), pp. 914–917, 2002.

Winkel, J. and Westgard, R. Occupational and individual risk factors for shoulder-neck complaints: part II — the scientific basis (literature review) for the guide, *Int. J. Ind. Ergon.*, 10: pp. 85–104, 1992.

20

Rehabilitating Low Back Disorders

Stuart M. McGill

University of Waterloo

REDUCING THE IMPACT OF LOW BACK DISORDERS (LBD) requires the best efforts of those experts in prevention together with those in rehabilitation. Rehabilitation approaches will not work well if the patient/worker remains exposed to the cause — hence, optional rehabilitation must incorporate excellent prevention or ergonomics. By the same token, prevention involves not only primary prevention but, since a problem often pre-exists, secondary level prevention is also required. Thus the best prevention efforts must understand and involve evidence-based principles of rehabilitation. The purpose of this chapter is to illustrate this symbiosis and compliment the ergonomics concepts explained in other chapters with some notion of rehabilitation exercise for LBDs.

Injury to low back tissues leads to joint instability and very well documented changes in motor/motion patterns. Because of the strong link to instability and the effective use of specific exercise approaches to restore stability, reduce pain, and enhance function, this chapter will focus on the injury process and develop a synthesis of the scientific foundation and formalization of the notion of stability as it pertains to the lumbar spine, and then provide specific guidelines for enhancing stability to advance spine rehabilitation. While a large book could be written to describe ideal exercise programs for the entire population including chronic low back pain sufferers, adolescents to geriatrics, through to elite athletes, the focus of the exercises discussed here is more towards the beginner's program — developing the safest exercise for enhancing stability and for acquiring and maintaining low back health. For the interested reader, more extensive references together with tabulated data of specific muscle activation profiles, resultant spine loads, etc., can be found in the authors' review chapters and original papers listed at www.ahs.uwaterloo.ca/kin/kinfac/mcgill.html or in my recent textbooks. "*Low Back Disorders: Evidence Based Prevention and Rehabilitation*" and "*Ultimate Backfitness and Performance.*"[35]

In most traditional approaches to designing low back exercise, an emphasis has been placed on the immediate restoration, or enhancement, of spine range of motion and muscle strength. Generally, this approach has not been sufficiently efficacious in reducing back troubles, in fact a review of the evidence suggests only a weak link with improving back symptoms while some studies suggest a link with negative outcome in significant numbers of people.[31] It appears that the emphasis on early restoration of spine range of motion continues to be driven by legislative definitions of low back disability — namely loss of range of motion (ROM). Thus, therapeutic success is often judged on motion restored. Most recent work suggests little correlation between ROM and work versatility ratings.[39] The underlying theme of this chapter, and in fact book, reflects the developing philosophy based on mechanisms of injury and stability — that a spine must first be stable before moments and forces are produced to enhance performance but to do so in a way that spares the spine from potentially injurious load. Preliminary field evidence (although not yet definitive) suggests that the approach has promise.

20.1   The Injury Process — Tissues Damage

There is a tendency among those reporting or describing the back injury to identify a single specific event as the cause of the damage, such as lifting a box and twisting. This description of low back injury is common, particularly among the occupational/medical community who are often required to identify a single event when filling out injury reporting forms. However, relatively few low back injuries occur from a single event. Rather, the culminating injury event was preceded by a history of excessive loading, which gradually, but progressively, reduced the level of tolerance to tissue failure.[30] Thus other scenarios where sub-failure loads can result in injury are probably more important. For example, the ultimate failure of a tissue (i.e., injury) can result from accumulated trauma produced by either repeated application of load (and failure from fatigue) or of a sustained load that is applied for long duration or repetitively applied (and failure from deformation and strain). Thus, the injury process may not always be associated with loads of high magnitude. Finally, it goes without saying that loss of mechanical integrity in any load-bearing tissue of the spine will result in stiffness losses and an increased risk of unstable behavior. Thus, documenting the injury process is a necessary foundation for understanding, formulating and utilizing the concepts of spine instability and stability.

While excellent progress has been made in the laboratory documenting specific instabilities in flexion–extension, lateral bend and axial rotation modes in animal preparations,[38] understanding the injury process in humans (the cause of back troubles in real life) has perhaps been hampered by the focus on exposure to a single variable — namely acute, or single maximum exposure to, lumbar compression. A few studies have suggested that higher levels of compression exposure increased the risk of LBD[24] although the correlation was low. Further some studies show that higher rates of LBDs occur when levels of lumbar compression are reasonably low. Are there other mechanical variables that modulate the risk of LBDs?

There are many tissues in the lower back and many different modes of loading that occur when performing work and exercise. Apart from joint compression, joint shear has been shown to be very important as a metric for injury risk in the study of Norman et al.,[37] particularly cumulative shear over a work day. Shear is an interesting variable because while most studies report reaction shear (that is the action of gravity and load in the hands to shear forward the ribcage on the pelvis through the lumbar spine), this is not the form of shear load that is experienced by the lumbar joints. In a series of work, the Waterloo group[25,29,41] has shown that if the spine maintains a neutral curvature (the torso is flexed forward about the hips, neither flexing nor extending the spine itself) then the dominant low back extensors with their unique force vector direction (specifically longisimus thoracis and iliocostalis lumborum) support the shear reaction forces caused by the action of gravity on the flexed torso, resulting in a lowering of the shear load experienced by the joint. These forces would normally be borne by the disc and facet joints. However, if the individual elects to flex the spine itself when bending forward sufficiently so as to stretch the posterior ligaments with full spinal flexion, then the architecture of the interspinous ligaments cause anterior shear forces[21] to add to the shearing reaction from gravity. Furthermore, ligamentous involvement disables the lumbar muscles (specifically noted above) from supporting the reaction shear as they reorientate to a line of action more parallel to the compressive axis[33](see Figure 20.1). With full spine flexion and a modest amount of gravitational reaction shear, it is not difficult to exceed shear failure loads of the spine, which have been found to be in the neighborhood of 2000 to 2800 N in adult

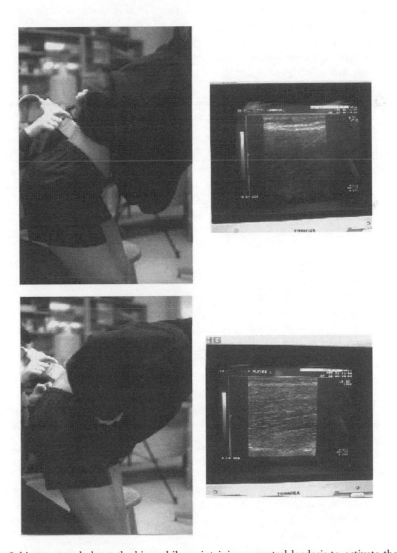

FIGURE 20.1 Subjects rotated about the hips while maintaining a neutral lordosis to activate the longissimus/ iliocostalis complex — note the oblique angle of the fibers in the ultrasound image with respect to the compressive axis suggesting these muscles produce large shear forces. This angle is lost, together with the protective shear forces, when the spine is flexed.

cadavers.[15] This paragraph suggests that personal work technique, or more specifically, spine motion can effect the risk of spine damage. Recent work by Yingling et al.[46] on pig spines has shown that load rate is not a major modulator of shear tolerance unless the load is very ballistic such as what might occur during a slip and fall. Summarizing the lumbar sagittal motion and shear issue, evidence from tissue-specific injury studies generally supports the notion of avoiding full lumbar flexion when performing loading tasks to minimize the risk of low back injury. There is no evidence to support a conscious effort to perform "pelvic tilts" (i.e., hyperlordosis or lumbar flexion) during lifting or exertion.

While twisting has been named in several studies as a risk factor for low back injury, the literature appears confused by not making the distinction between the kinematic variable of twisting and the kinetic variable of generating twisting torque. While many epidemiological surveillance studies link a higher risk of LBD with twisting, twisting with low twist moment results in a relatively low muscle activity and correspondingly low spine load.[26,27] Further, passive tissue loading is not substantial until the end of the twist range of motion.[17] However, developing twisting moment places very large compressive loads on the spine due to the enormous coactivation of the spine musculature[27] and this can occur when the spine is not twisted, but in a neutral posture where the ability to tolerate loads is higher. It would appear that either single variable (the kinematic act of twisting or generating the kinetic variable of twist torque while not twisting) is less dangerous than may be suggested by epidemiological surveys. However, it would appear that elevated risk from very high tissue loading occurs when the spine is fully twisted at the same time where there is a need to generate high twisting torque.[27]

There are many personal factors, which appear to affect spine tissue tolerance, for example, age and gender. Jager et al.[22] compiled the available literature that passed their inclusion criteria on the tolerance of lumbar motion units to bear compressive load. Their results revealed that if males and females are matched for age, females were able to sustain only approximately two-thirds of the compressive loads of males. Furthermore, Jager et al.'s data showed that within a given gender, the 60-yr-old spine was able to tolerate only about two-thirds of that tolerated by a 20-yr-old. There are other personal factors such as motor control system fitness where it appears that a motor control error can lead to a back injury during very benign tasks such as picking up a pencil from the floor. (This will be explained in a subsequent section.)

Many factors appear to modulate the risk of specific low back tissues damage other than load magnitude and loading mode. While disc herniations have been produced under controlled conditions,[19] Callaghan and McGill[9] have been able to consistently produce disc herniations by mimicking spine motion and load patterns seen in workers and in replicating the motion and loads of some lumber extension exercise machines. Specifically, it appears that only a very modest amount of spine compression force is required (only 800–1000 N) but the spine specimen must be repeatedly flexed — mimicking repeated torso-spine flexion from continual bending to a fully flexed posture. In these experiments, the progressive tracking of disc nucleus material travelling posteriorly through the annulus of the disc was documented with sequestration of the nucleus material around 18,000–25,000 cycles of flexion (fewer cycles were required for herniation with higher simultaneous compressive loads). This study included the utilization of a pig degenerative trauma model, which on one hand was an animal model but on the other, control over age, diet, physical activity provided a unique opportunity. Spines and discs obtained from humans are typically older and have lost sufficient disc hydration to match the hydration levels, and potential for herniation, seen in the age groups of workers at risk for this specific type of event (typically 30–50-yr-olds). But the important point here is that herniation appears to be more strongly linked to repeated flexion motion rather than load.

Another modulator for tissue damage appears to be the posture of the joint resulting from the curvature of the spine *in vivo*. For example, Adams et al.[1] showed that a fully flexed spine is weaker than the one that is moderately flexed. In a most recent study, Gunning and McGill[20] have shown that a fully flexed spine (using a controlled porcine spine model) is 20 to 40% weaker than if it were in a neutral posture, and that hydration levels matched to the changes seen in peoples' discs throughout the work day also modulate the tolerance. For example, the spinal discs are more easily damaged first thing in the morning upon rising from bed when they are fully hydrated. A fascinating study, reported by Snook et al.[43] demonstrated that of 85 patients randomly assigned to a group that controlled the amount of early morning lumbar flexion, had significant reduction in pain intensity, compared to a control group. Then, when the control group received the experimental treatment they responded with similar reductions.

Collectively, the evidence suggests that the risk of spine tissue damage is a function of load magnitude, directional mode of the applied load, motion repetition, spine posture, hydration level and time of day, motor control and instantaneous stability, and individual age and gender. Injury history and tissue damage is an overlaying modulator. Collectively this data supports the notion of an envelope of motion and loading for optimal tissue health. In addition it is well known that tissues adapt and remodel with load (e.g., Bone — Carter,[11] Ligament — Woo et al.,[45] Disc — Porter,[40] Vertebrae — Brinckmann et al.[8]), which is at the core of any rehabilitation program. However, biological variability prevents the identification of specific levels of loading, which either build tissue or initiate breakdown, together with the optimal rest periods and days off, which promote healthy tissue adaption for a given individual. Thus it would appear that the wisest philosophical approach for the optimal design of activity, either during the activities of daily living or during rehabilitation efforts, may be to adopt the notion that "too much of any single activity is problematic." No rehabilitation program can be fully effective if patients undo the beneficial responses of therapy with inappropriate activities of daily living.

20.2   The Injury Process — Motor Changes

Those reporting debilitating low back pain conclusively suffer simultaneous changes in their motor control systems. Recognizing these changes is important since they affect the stabilizing system and are therefore are a focal point for optimal rehabilitation. Richardson et al.[42] have produced quite a comprehensive review of this literature together with making a case for targeting specific muscle groups during rehabilitation. Specifically, their objective is to re-educate faulty motor control patterns postinjury. The challenge is to train the stabilizing system during steady-state activities together with stabilizing during rapid voluntary motions and to withstand sudden surprise loads.

In a most recent study of our own,[36] motor control changes were noted as one of the major distinguishing features of those workers who had a history of missing work due to the low back troubles. In fact they demonstrated several patterns that compromised their ability to stabilize their backs even though they were working and symptom-free at the time of testing. It is this collection of evidence that strongly supports approaches to stabilization exercises that promote patterns of muscular cocontraction observed in fit spines.

20.3   Instability as a Cause of Injury

While biomechanists have been able to successfully explain how strenuous exertions cause specific tissue damage of the lower back, explaining how injury occurs from tasks such as picking up a pencil from the floor has been more challenging. Recent evidence suggests that such injuries are real, and result from the spine "buckling" or exhibiting unstable behavior. But this buckling mechanism can occur during far more challenging exertions as well.

A number of years ago we were investigating the mechanics of powerlifter spines while they lifted extremely heavy loads using video fluoroscopy to view their vertebrae in the sagittal plane. During their lifts, even though the lifters outwardly appeared to fully flex their spines, in fact their spines were 2 to 3° per joint from full flexion, thus explaining how they could lift magnificent loads without sustaining injury — the risk of disc and ligamentous damage is greatly elevated when the spine is fully flexed (which the lifters skillfully avoided). We happened to capture one injury on the flouroscopic motion film — the first such observation that we know of. During the injury incident, just as the semi-squating lifter had lifted the load about 10 cm off the floor, only the L2/L3 joint briefly rotated to the full flexion calibrated angle and exceeded it by one-half a degree, while all other lumbar joints maintained their static positions (not fully flexed).[12] The spine buckled! Sophisticated modeling analysis revealed that buckling can occur from a motor control error where a short and temporary reduction in activation to one, or more, of the intersegmental muscles would cause rotation of just a single joint so that passive or other tissues become irritated or possibly injured.[13]

Adams and Dolan[2] have noted that passive tissues begin to damage with bending moments of 60 Nm — this occurs simply with the weight of the torso when bending over and a temporary loss of muscular support. This scenario is not an excessive task, but it is often reported to clinicians by patients as the event that caused their injury (i.e., picking up a pencil). However, reporting of such an event will not be found in the scientific literature. Medical personnel would not record this event since in many jurisdictions it would not be deemed as a compensable injury — the medical report attributes the cause elsewhere.

Other evidence linking poor motor coordination with higher risks for the lumbar spine reaching critical points of instability exists and is revealing. Cholewicki and McGill[13] have identified through a modeling analysis, the nodal points, or specific spinal joint, where buckling could occur from specific motor control errors. Such inappropriate muscle sequencing has been observed in men who are challenged by holding a load in the hands while breathing 10% CO_2 to elevate breathing. On one hand, the muscles must cocontract to ensure sufficient spine stability, but on the other, challenged breathing is often characterized by rythmic/contraction/relaxation of the abdominal wall.[30] Thus, the motor system is presented with a conflict — should the torso muscles remain active isometrically to maintain spine stability or will they rhythmically relax and contract to assist with active expiration (but sacrifice spine stability). Fit motor systems appear to meet the simultaneous breathing and spine support challenge — unfit ones may not. All of these deficient motor control mechanisms will heighten biomechanical susceptibility to injury or reinjury.[13,14]

In vitro, a ligamentous lumbar spine buckles under compressive loading at about 90 N (about 20 lb) highlighting the critical role of the musculature to stiffen the spine against buckling (the critical work and analysis of the passive tissues being performed by Crisco and Panjabi[16]). Anatomical arrangement of muscle around the spine, coupled with critically important patterns of activation, enables the spine to bear a much higher compressive load as it stiffens and becomes more resistant to buckling but in so doing, the spine bears even more load due to the "stiffening" muscle activity. As noted previously, aberrant patterns of activation can result in instantaneous spine instability[18] and acute tissue overload. But over the longer term, the Queensland group[42] have developed a tissue damage model, which suggests that chronically poor motor control (and motion patterns) initiates microtrauma in tissues, which accumulates leading to symptomatic injury. Injury leads to further deleterious change in motor patterns such that chronicity can only be broken with specific techniques to re-educate the local muscle-motor control system. Both acute and chronic instability-tissue models have been proposed. But given the wide range of individuals and physical demands, question remain as to what is the optimal balance in terms of stability, motion facilitation and moment generation — if stability is achieved through muscular cocontraction, how much is necessary and how is it best achieved?

20.4 On Stability: The Foundation

This section shall formalize the notion of stability from a spine perspective. During the 1980's, Professor Anders Bergmark of Sweden, very elegantly formalized stability in a spine model with joint stiffness and 40 muscles.[6] In this classic work he was able to formalize mathematically, the concepts of "energy wells," stiffness, stability and instability. For the most part, this seminal work went unrecognized largely because the engineers who understood the mechanics did not have the biological — clinical perspective, and the clinicians were hindered in the interpretation and implications of the engineering mechanics. This pioneering effort, together with its continued evolution by several others will be synthesized here — the current author has attempted to encapsulate the critical notions without mathematical complexity but directs the mathematically inclined reader to Reference 13 and Reference 28.

The concept of stability begins with potential energy, which for the purposes here, is of two basic forms. In the first form, objects have potential energy by virtue of their height above a datum.

$$PE = \text{mass} \times \text{gravity} \times \text{height}.$$

Critical to measuring stability are the notions of energy "wells" and minimum potential energy. If a ball is placed into a bowl it is stable; if a force is applied to the ball (or a perturbation) the ball will rise up the side of the bowl but then come to rest again in the position of least potential energy at the bottom of the bowl — or the "energy well." As noted by Bergmark, "stable equilibrium prevails when the potential energy of the system is minimum." The system is made more stable by deepening the bowl and/or by increasing the steepness of the sides of the bowl (see Figure 20.2). Thus, the notion of stability requires the specification of the unperturbed energy state of a system followed by the study of the system following perturbation — if the "joules" of work done by the perturbation is less than the "joules" of potential energy inherent to the system then the system will remain stable (i.e., the ball will not roll out of the bowl). The corollary is that the mechanical system will collapse if the applied load exceeds a critical value (determined by potential energy and stiffness).

The previous ball analogy is a two-dimensional example. This would be analogous to a hinged skeletal joint that only has the capacity for flexion/extension. Spinal joints can rotate in three planes and translate along three axes requiring a six-dimensional bowl for each joint — mathematics enables the examination of a 36-dimensional bowl (six lumbar joints with six degrees of freedom) representing the whole lumbar spine. If the height of the bowl were decreased in any one of these 36 dimensions, the ball could roll out. In clinical terms, a single muscle having an inappropriate force (and thus stiffness), or a damaged passive tissue, which has lost stiffness can cause instability that is both predictable and quantifiable.

While potential energy by virtue of height is useful for illustrating the concept, potential energy as a function of stiffness and storage of elastic energy is actually used for musculoskeletal application. Elastic

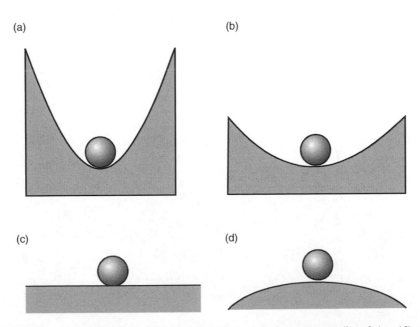

FIGURE 20.2 The continuum of stability — (a) is most stable through the continuum (b and c) to (d) which is least stable. The ball in the bowl seeks the "energy well" or position of minimum potential energy (m^*g^*h). Deepening the bowl or increasing the steepness of the sides increases the ability to survive perturbation — this increases stability.

potential energy is calculated from stiffness (k) and deformation (x) in the elastic element:

$$PE = 1/2 \times k \times x^2.$$

In other words the greater the stiffness (k) the greater the steepness of the sides of the bowl (from the previous analogy), and the more stable the structure. Thus stiffness creates stability (see Figure 20.3). Active muscle produces a stiff member and in fact the greater the activation of the muscle, the greater this stiffness — it has long been known that joint stiffness increases rapidly and nonlinearly with muscle activation such that only very modest levels of muscle activity create sufficiently stiff and stable joints. Furthermore, joints possess inherent joint stiffness as the passive capsules and ligaments contribute stiffness particularly at the end range of motion. The motor control system is able to control stability of the joints through coordinated muscle coactivation and to a lesser degree by placing joints in positions, which modulate passive stiffness contribution. However, a faulty motor control system can lead to inappropriate magnitudes of muscle force and stiffness, allowing a "valley" for the "ball to roll out" or clinically, for a joint to buckle or undergo shear translation. But mechanical systems and particularly musculoskeletal linkages, are limited to the analysis of "local stability" since the energy wells are not infinitely deep and the many anatomical components contribute force and stiffness in synchrony to create "surfaces" of potential energy where there are many local wells. Thus local minima are located from examination of the derivative of the energy surface (see Reference 13 for mathematical details). Spine stability then, is quantified by forming a matrix where the total "stiffness energy" for each degree of freedom of joint motion is represented by a number (or eigenvalue) and the magnitude of that number represents its contribution to forming the "height of the bowl" in that particular dimension. Eigenvalues less than zero indicate the potential for instability. The eigenvector (different from the eigenvalue) can then identify the mode in which the instability occurred while sensitivity analysis may reveal the possible contributors allowing unstable behavior. Gardner-Morse et al.[18] have initiated interesting investigations into eigenvectors by predicting patterns of spine deformation due to impaired muscular intersegmental control) or for clinical relevance — what muscular pattern would have prevented the instability?

Activating a group of muscle synergists and antagonists in the optimal way now becomes a critical issue. In clinical terms the full complement of the stabilizing musculature must work harmoniously to both ensure stability together with generation of the required moment and desired joint movement. But only one muscle with inappropriate activation amplitude may produce instability, or at least unstable behavior could result at lower applied loads.

How much stability is necessary — obviously insufficient stiffness renders the joint unstable but too much stiffness and coactivation imposes massive load penalties on the joints and prevents motion.

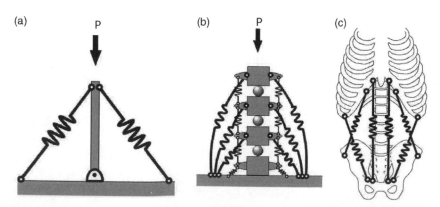

FIGURE 20.3   (a) Increasing the stiffness of the cables (muscles) increases the stability (or deepens the bowl) and increases the ability to support larger applied loads "p" without falling. (b) Spine stiffness (and stability) is achieved by a complex interaction of stiffening structures along the spine and (c) those forming the torso wall (right panel).

"Sufficient stability" is a concept that involves the determination of how much muscular stiffness is necessary for stability together with a modest amount of extra stability to form a margin of safety. Interestingly enough, given the rapid increase in joint stiffness with modest muscle force, large muscular forces are rarely required. In our recent papers, stabilization exercises were quantified and ranked for muscle activation magnitudes together with the resultant spine load.[35] (Quantification of individual tissue loads in the spine is a complex procedure and an issue outside the constraints of this article — the interested reader is directed to Reference 28.) Furthermore, Cholewicki's work[14] has demonstrated that sufficient stability of the lumbar spine is achieved, in an undeviated spine, in most people with modest levels of coactivation of the paraspinal and abdominal wall muscles. This means that people, from patients to athletes, must be able to maintain sufficient stability in all activities — with low, but continuous, muscle activation. Thus, maintaining a stability "margin of safety" when performing tasks, particularly the tasks of daily living, is not compromised by insufficient strength but rather insufficient endurance. We are now beginning to understand the mechanistic pathway of those studies showing the efficacy of endurance training for the muscles that stabilize the spine. Having strong abdominals does not necessarily provide the prophylactic effect that had been hoped for — but several works suggest that endurable muscles reduce the risk of future back troubles.[7]

20.5 A Philosophy of Low Back Exercise Prescription

Many traditional notions that exercise professionals consider to be principles for exercise prescription, particularly when dealing with the low back, may not be as well supported with data as generally thought. A review of the efficacy of traditional exercise versus stabilization programs both identified and motivated a re-examination of conventional thought (summarized in Reference 31). For example, there is a widely held view that sit-ups should be performed with bent knees, but it is becoming apparent that the resultant spinal loading (well over 3000 N of compression to a fully flexed l-spine) suggests sit-ups are not suitable for most people at all — other abdominal challenges are more effective and safer. Other examples include: contrary to the belief of many, adopting a posterior pelvic tilt when performing many types of low back exercise actually increases the risk of injury by flexing the lumbar joints and loading passive tissues; having stronger back and abdominal muscles appears to have no prophylactic value for reducing bad back episodes — however, muscle *endurance* has been shown to be protective; greater lumbar mobility leads to more back troubles — not less[7]; and in fact, lumbar ROM appears to have little correlation with work disability status.[39] It is also troubling that replicating the motion, and spine loads, that occur during the use of many low back extensor machines used for training and therapy, produces disc herniations when applied to spines in our laboratory! It is clear that some current "clinical wisdom" needs to be re-examined in the light of relatively recent scientific evidence (those interested in the literature evidence should consult my review in the *Low Back Disorders Textbook*[35]). It appears that the safest, and mechanically justifiable approach to enhancing lumbar stability through exercise entails a philosophical approach consistent with endurance — not strength; that ensures a neutral spine posture when under load; and that encourages abdominal cocontraction and bracing in a functional way. (A "neutral posture" is defined as one where the joints and surrounding passive tissues are in elastic equilibrium and thus at an angle of minimal joint load). It is also acknowledged that optimal athletic performance, which demands reaction and prehension challenges, is not synonymous with health objectives and that additional risk is accepted for extreme ranges of motion and particular motor patterns. The most recent insights provided by Cholewicki suggest that while steady-state motor patterns are important for daily activity, the health of reflexive motor patterns is critical for maintaining stability during sudden events[14] — achieving a fit and effective motor control system probably requires training in a variety of static and dynamic, expected and unexpected, stable and labile, conditions.

While many muscles have been regarded as primary spine stabilizers, confirmation of their role requires two levels of analysis. First, engineering — stability analysis must be conducted on anatomically robust spine models to document the ability of each component to stiffen and stabilize. Second, electromyographic recordings of all muscles (even deep muscles requiring intramuscular electrodes) are necessary to confirm the extent that the motor control system involves each muscle to ensure sufficient stability. For some time our limited intramuscular EMG and modeling studies, and those of others, suggested that virtually all torso muscles play a role in stabilization. (Our most recent quantification breakthroughs appears at the end of this section.) However, while multifidus, the other extensors, and the abdominal wall, have been highlighted before, the architecture of quadratus lumborum (QL) suggests that it can be a stabilizer. This notion is further strengthened by some earlier observation that the motor control system involves this muscle together with the abdominal wall when stability is required in the absence of major moment demands. The fibers of QL cross-link the vertebrae, they have a large lateral moment arm via the transverse process attachments, and traverse to the rib cage and iliac crests. Thus, the quadratus could buttress shear instability, and be effective in all loading modes, by design. Typically, the first mode of buckling is lateral — the quadratus can play a significant role in local lateral buttressing. Further, activation profiles support the notion of the stabilizing role of quadratus. It is active during a variety of flexion dominant, extensor dominant and lateral bending tasks. Specifically, Andersson et al.[4] found that the QL did not relax with the extensors during the flexion–relaxation phenomonon. The flexion–relaxation phenomonon is an interesting task since there is no substantial lateral or twisting torques and the extensor torque appears to be supported passively — suggesting some stabilizing role for QL. Other very limited data suggest (our laboratory techniques to obtain QL activation were rather imprecise at the time) that in an experiment where subjects stood upright, but held buckets in either hand, where load was incrementally added to each bucket, the QL appeared to increase its activation level (together with the obliques) as more stability was required. This task forms a special situation since only compressive loading is applied to the spine in the absence of any bending moments. The three layers of the abdominal wall are also important for stability together with muscles, which attach directly to vertebra — the multisegmented longissimus and iliocostalis and the unisegmental multifidii. Cholewicki[18] has also presented an argument for the role of the small intertransversarii in producing small but critical stabilizing forces. On the other hand, psoas activation appears to have little relationship with low back demands — the motor control system activates it when hip flexor moment is required (data is presented by Andersson et al.[3] and Juker et al.[23]).

Most recently we have completed evolution of our model to quantify the role of individual muscles to contribute to stability. Once again the conclusion is that all muscles are important and that the most important muscle at any instant or task is a transient variable — they continually change their relative contribution (see Figure 20.4 and Figure 20.5 for the ranking in a selected group of exercises). So which are the wisest ways to challenge and train these identified stabilizers?

20.6   Training QL

Given the architectural and electromyographic evidence for QL as a spine stabilizer, the optimal technique to maximize activation but minimize the spine load appears to be the side-bridge (Figure 20.6) — beginners bridge from the knees while advanced bridges are from the feet. When supported with the feet and elbow the lumbar compression is a modest 2500 N, but the quadratus closest to the floor, appears to be active upto 50% of MVC (the obliques experience similar challenge). Advanced technique to enhance the motor challenge is to roll from one elbow to the other while abdominally bracing (Figure 20.7) rather than repeated "hiking" of the hips off the floor into the bridge position. Higher levels of activation would be reached with the feet on a labile surface.[35]

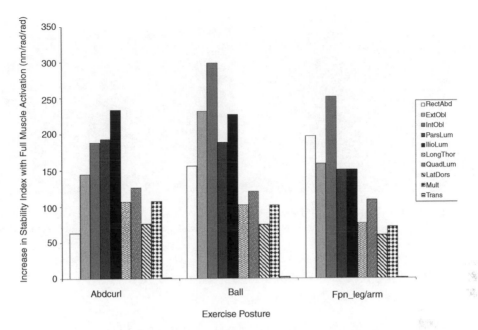

FIGURE 20.4   In an attempt to understand the contributions of each muscle pair on spine stability, the increase in the stability index is shown as a function of setting each muscle pair activation in turn to 100% MVC — in this way their relative contrive could be assessed. Note that the relative order of muscles that increase stability changes across exercises. As well, in flexion tasks, the pars lumborum (in this example) plays a larger stabilizing role over the rectus abdominis. In contrast, during the extension tasks the opposite holds true suggesting a task-dependent role reversal between moment generation and stability. The exercises were: Abcurl — curl-up on the stable floor, Ball — sitting on a gym ball, Fpn_leg/arm — four-point kneeling while extending one leg and the opposite arm.

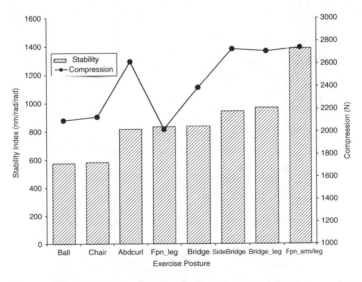

FIGURE 20.5   Stability versus L4-L5 compression for eight different stabilization exercises. All exercises were performed with the abdominal wall active and were as follows: Ball — sitting on a gym ball, Chair — sitting on a chair, Abcurl — curl-up on the floor, Fpn_leg — four-point kneeling while extending one leg at the hip, Bridge — back bridge on the floor, SideBridge — side-bridge with the elbow and feet on the floor, Bridge_leg — back-bridge but extending the knee and holding one leg against gravity, Fpn_arm/leg — four-point kneeling while extending one leg and the opposite arm. Exercises are rank ordered based on increasing lumbar spine stability.

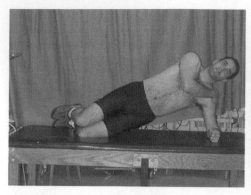
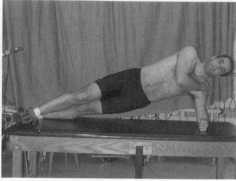

FIGURE 20.6 The horizontal isometric side-bridge. Supporting the lower body with the knees on the floor reduces the demand further for those who are more concerned with safety while supporting the body with the feet increases the muscle challenge, but also the spine load.

20.7   Training Rectus Abdominis, the Obliques and Transverse Abdominis

Given the evidence for the obliques and transverse abdominis in ensuring sufficient stability, quantitative data has confirmed that there is no single abdominal exercise that challenges all of the abdominal musculature,[5,23] requiring the prescription of more than one single exercise. Calibrated intramuscular and surface EMG evidence suggests that the various types of curl-ups challenge mainly rectus abdominis since psoas and abdominal wall (internal and external oblique, transverse abdominis) activity is relatively

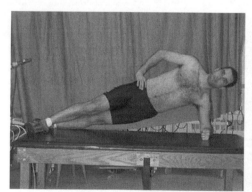
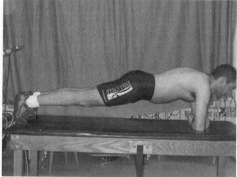
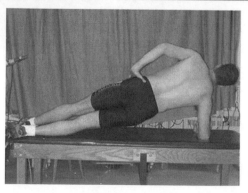

FIGURE 20.7 Advanced side-bridge — following each side-bridge hold, one "rolls" from one elbow to the other while abdominally bracing locks the pelvis and the rib cage.

low. Sit-ups (both straight-leg and bent-knee) are characterized by higher psoas activation and higher low back compressive loads that exceed NIOSH occupational guidelines, while leg raises cause even higher activation and also spine compression (the interested reader is directed to Reference 8 for actual data). It is also interesting that myoelectric evidence suggests that there is no functional distinction between an "upper" and "lower" rectus abdominis in most people but, in contrast, the obliques are regionally activated with "upper" and "lower" motor point areas together with medial and lateral components. Transverse abdominis is selectively activated by dynamically "hollowing" in the abdominal wall,[42] while an isometric abdominal brace coactivates transverse abdominis together with the external and internal obliques to ensure stability in virtually all modes of possible instability.[13]

Several relevant observations were made regarding abdominal exercises in our investigations. The challenge to psoas is lowest during curl-ups, followed by higher levels during the horizontal side-bridge, while bent-knee sit-ups were characterized by larger psoas activation than straight-leg sit-ups, through to the highest psoas activity observed during leg raises and hand-on-knee flexor isometric exertions. It is interesting to note that the often recommended "press-heels" sit-up, which has been hypothesized to activate hamstrings and neurally inhibit psoas was actually confirmed to increase psoas activation! (Original data can be found in Reference 23 — we note here that some clinicans and coaches who intentionally wish to train psoas and will find this data informative). Once again the horizontal side support appears to have merit as it challenges the lateral obliques and transverse abdominis without high lumbar compressive loading.

Clearly, curl-ups excel at activating the rectus abdominis but produce relatively low oblique activity. Curl-ups with a twisting motion is expensive in terms of lumbar compression due to the additional oblique challenge. A wise choice for abdominal exercises, in the early stages of training or rehabilitation, for simple low back health objectives, would consist of several variations of curl-ups for rectus abdominis and the side-bridge for the obliques and quadratus, the variation of which is chosen commensurate with patient/athlete status and goals.

20.8 Training the Back Extensors (and Stabilizers)

Most traditional extensor exercises are characterized with very high spine loads, which result from externally applied compressive and shear forces (either from free weights or resistance machines). From our search for methods to activate the extensors (including longissimus, iliocostalis and multifidii) with minimal spine loading, it appears that the single leg extension hold, minimizes the spine load (<2500 N) and activates one side of the lumbar extensors to approximately 18% of MVC.[10] Simultaneous leg extension with contralateral arm raise ("birddog") increases the unilateral extensor muscle challenge (approximately 27% MVC on one side of the lumbar extensors and 45% MVC on the other side of the thoracic extensors), but also increases lumbar compression to well over 3000 N. This exercise can be enhancing with abdominal bracing and deliberate "mental imaging" of activation of each level of the local extensors. The often performed exercise of laying prone on the floor and raising the upper body and legs off the floor is contraindicated for anyone at risk of low back injury — or reinjury. In this task the lumbar region pays a high compression penalty to a hyperextended spine (usually much higher than 4000 N), which transfers load to the facets and crushes the interspinous ligament.

20.9 The Beginner's Program for Stabilization

Some specific recommended low back exercises have been shown. We recommend that the program begin with the flexion–extension cycles (cat-camel; Figure 20.8) to reduce spine viscosity and floss the nerve roots as they outlet at each lumbar level, followed by hip and knee mobility exercises. Note that the cat-camel is intended as a motion exercise — not a stretch, so the emphasis is on motion rather than "pushing" at the end ranges of flexion and extension. We have found that 5 to 6 cycles is often sufficient to reduce most viscous stresses. This is followed by anterior abdominal exercises, in this case the curl-up with the hands under the lumbar spine to preserve a neutral spine posture (Figure 20.9) and one knee flexed but with

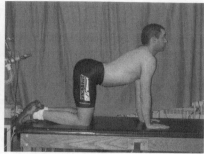

FIGURE 20.8 The flexion–extension stretch is performed by slowly cycling through full spine flexion to full extension. Spine mobility is emphasized rather than "pressing" at the end range of motion. This exercise facilitates motion for the spine with very low loading of the intervertebral joints, reduces viscous stresses for subsequent exercise and "flosses" the nerve roots through the foramenae at each spine joint (hence coordination of full cervical, thoracic and lumbar flexion–extension).

the other leg straight to lock the pelvis-lumbar spine and minimize the loss of a neutral lumbar posture. Then, lateral musculature exercises are performed — namely the side-bridge, for QL and muscles of the abdominal wall for optimal stability (Figure 20.6). Advanced variations involve placing the upper leg-foot in front of the lower leg-foot to facilitate longitudinal "rolling" of the torso to challenge both anterior and posterior portions of the wall. The extensor program consists of leg extensions and the "birddog" (Figure 20.10). In general, we recommend that these isometric holds be held no longer than 7 to 8 sec given recent evidence from near infrared spectroscopy indicating rapid loss of available oxygen in the torso muscles contracting at these levels — short relaxation of the muscle restores oxygen.[34]

Motivated by the evidence for the superiority of extensor endurance over strength as a benchmark for good back health, we have recently documented "normal" ratios of endurance times for the torso flexors relative to the extensors (e.g., it is "normal" to hold a flexor posture — see Reference 32, about 0.98 of the maximum time holding a reference extensor posture) and for the lateral musculature relative to the extensors (0.5) to assist clinicians to identify endurance deficits — both absolute values and for one muscle group relative to another. Our most recent evidence suggests that these endurance ratios (both right to left sides and flexor to extensor) are significantly "out of balance" in those who have had a history of low back troubles with work loss.[36] Finally, as patients progress with these isometric stabilization exercises, we recommend conscious simultaneous contraction of the abdominals (i.e., bracing — simply isometrically activating the abdominals for maximum stability).

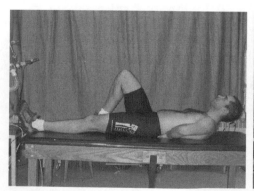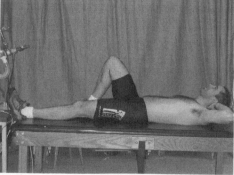

FIGURE 20.9 The curl-up, where the head and shoulders are raised off the ground with the hands under the lumbar region to help stabilize the pelvis and support the neutral spine. Only one leg is bent to assist in pelvic stabilization and preservation of a "neutral" lumbar curve. Additional challenge can be created by raising the elbows from the floor and generating an abdominal brace or cocontraction.

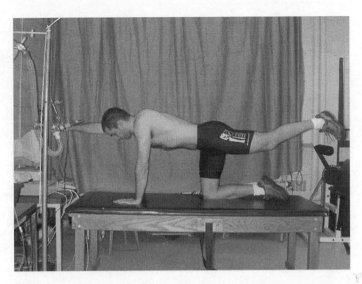

FIGURE 20.10 Single leg extension holds, while on the hands and knees, produces mild extensor activity and lower spine compression (<2500 N). Raising the contralateral arm increases extensor muscle activity but also spine compression to levels over 3000 N. Sufficient stability is ensured with mild abdominal bracing.

20.10 Advanced Techniques — A General Approach to Preparing the Occupational Athlete

The beginner's program should be sufficient for daily spine health. Preparing the occupational athlete demands higher challenges of low back training, but is achieved with much higher risk of tissue damage from overload. Furthermore, specific occupational objectives require specific training techniques — space restrictions do not permit their in-depth discussion here. However, the general approach is to first document the specific demands of the work (found in other chapters of this book), and train for these specific demands. Within this process is the need to perform provocative tests on the worker to determine not only current capabilities but also which motion/motor patterns exacerbate the pain. This expertise is outside of the scope of this chapter and the interested reader is urged to consult to consult the textbooks mentioned in Reference 35.

One example may involve the need to generate torsional moments, which are often required for maintenance work (opening a large value for example) but the question must address how to maximize stability and minimize injury during training for trunk torsion. The fact that generating torque about the twist axis imposes approximately four times the compression on the spine than for an equal torque about the flexion–extension axis cannot be dismissed. The technique we have found for producing low spine loads while challenging the torsional moment generators is to raise a hand-held weight while supporting the upper body with the other arm, and abdominally bracing (Figure 20.11) to resist the torsional torque with an isometrically contracted and neutral spine. Dynamic challenged twisting is reserved for the most robust of the athletes' backs.

Finally, it is recognized that challenges to the spine during daily activity include the maintenance of stability during stable, steady-state posture maintenance, and during unexpected loading events togther with ballistic movement that is prehensively planned. This has motivated some clinicians to utilize labile surfaces such as gym balls. Certainly, these labile surfaces challenge the motor system to meet the dynamic tasks of daily living. But is this type of training of concern for some patients? Our recent quantification of elevated spine loads and muscle coactivation when performing a curl-up on labile surfaces[44] suggests that the rehabilitation program should begin on stable surfaces. Labile surfaces should be introduced once the spine load-bearing capacity has been sufficiently restored.

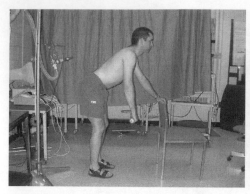 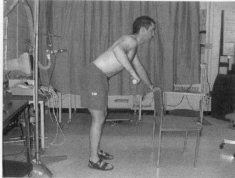

FIGURE 20.11 A challenge for the torsional components that produces low spine loads is to support the extensor moment of the flexed torso with one hand while the other raises a modest weight. The lumbar torso is braced (including all layers of the abdominal wall) in a neutral posture resisting the twisting moments generated by the weight.

In summary, progression of a patient through rehabilitation to function follows the three-step procedure:

1. Establish motion and motor patterns that avoid the injury mechanisms identified earlier in this chapter. Painful motion/motor patterns may have also been identified through provocative testing of the worker and should be avoided in this first step
2. Enhance lumbar stability
3. Progress to functional tasks while ensuring sufficient stability, but staying with the current tolerance to withstand the elevated loads placed on the spine

This chapter focussed on the second step in the process.

20.11   Looking Forward

Rehabilitation endeavors are continuing to embrace techniques that consider notions of lower torso, or "core," stability. While there is no question that first a system must be stable before presented with a physical challenge, the enhancement of low back health and the avoidance of troubles have motivated scientific inquiry into the mechanics of stability. Many groups continue to work to understand the contributions to stability of various components of the anatomy at particular joints — and the ideal ways to enhance their contribution; to understand what magnitudes of muscle activation are required to achieve sufficient stability; to identify the best methods to re-educate faulty motor control systems to both achieve sufficient stability and reduce the risk of inappropriate motor patterns occurring in the future. Clearly, it is these objectives that are needed to deal with the well-documented deficits in those workers with back troubles — namely the need to ensure sufficient spine stability.

Acknowledgments

The continual financial support from the Natural Science and Engineering Research Council, Canada is gratefully acknowledged and has made this series of work possible.

References

1. Adams, M.A., McNally, D.S., Chinn, H., and Dolan, P., Posture and the compressive strength of the lumbar spine, *Clin. Biomech.*, 9, pp. 5–14, 1994.
2. Adams, M.A. and Dolan, P., Recent advances in lumbar spine mechanics and their clinical significance, *Clin. Biomech.*, 10, pp. 3–19, 1995.

3. Andersson, E., Oddsson, L., Grundstrom, H., and Thorstensson, A., The role of the psoas and iliacus muscles for stability and movement of the lumbar spine, pelvis and hip, *Scand. J. Med. Sci. Sports*, 5, pp. 10–16, 1995.

4. Andersson, E.A., Oddsson, L.I.E., Grundström, H., and Nilsson, J., EMG activities of the quadratus lumborum and erector spinae muscles during flexion-relaxation and other motor tasks, *Clin. Biomech.*, 11(7), pp. 392–400, 1996.

5. Axler, C.T., and McGill, S.M., Low back loads over a variety of abdominal exercises: Searching for the safest abdominal challenge, *Med. Sci. Sports Exerc.*, 29, pp. 804–811, 1997.

6. Bergmark, A., Mechanical stability of the human lumbar spine, Doctoral Dissertation, Department of Solid Mechanics, Lund University, Sweden, 1987.

7. Biering-Sorensen, F., Physical measurements as risk indicators for low back trouble over a one year period, *Spine*, 9, pp. 106–119, 1984.

8. Brinckmann, P., Biggemann, M., and Hilweg, D., Prediction of the compressive strength of human lumbar vertebrae, *Clin. Biomech.*, 4, Suppl 2, pp. s1–s27, 1989.

9. Callaghan, J. and McGill, S.M., Intervertebral disc herniation: Studies on a porcine spine exposed to highly repetitive flexion/extension motion with compressive force, *Clin. Biomech.*, 16(1): 28–37.

10. Callaghan, J.P., Gunning, J.L., and McGill, S.M., Relationship between lumbar spine load and muscle activity during extensor exercises, *Phys. Ther.*, 78, pp. 8–18, 1998.

11. Carter, D.R., Biomechanics of bone, in *Biomechanics of Trauma*, Nahum, H.M., and Melvin, J., eds., Appleton Century Crofts, Norwalk, Connecticut, 1985, pp. 135–165.

12. Cholewicki, J. and McGill, S.M., Lumbar posterior ligament involvement during extremely heavy lifts estimated from flouroscopic measurements, *J. Biomech.*, 25(1), pp. 17–28, 1992.

13. Cholewicki, J. and McGill, S.M., Mechanical stability of the in vivo lumbar spine: Implications for injury and chronic low back pain, *Clin. Biomech.*, 11(1), pp. 1–15, 1996.

14. Cholewicki, J., Simons, A.P.D., Radebold, A., Effects of external trunk loads on lumbar spine stability, *J. Biomech.*, 33(11), pp. 1377–1385, 2000.

15. Cripton P., Berleman U., and Visarino H., et al., Response of the lumbar spine due to shear loading, in *Injury prevention through biomechanics*, Symposium proceedings, May, Wayne State University, 1995, pp. 4–5.

16. Crisco, J.J. and Panjabi, M.M., Euler stability of the human ligamentous lumbar spine, Part I Theory 7, pp. 19–26 and Part II, Experiment, *Clin. Biomech.*, 7, pp. 27–32, 1992.

17. Duncan, N.A. and Ahmed, A.M., The role of axial rotation in the etiology of unilateral disc prolapse: An experimental and finite-element analysis, *Spine*, 16, pp. 1089–1098, 1991.

18. Gardner-Morse, M., Stokes, IAF., and Laible, JP., Role of the muscles in lumbar spine stability in maximum extension efforts, *J. Orthop. Res.*, 13, pp. 802–808, 1995.

19. Gordon, S.J., Yang, K.H., Mayer, P.J., et al., Mechanism of disc rupture-a preliminary report, *Spine*, 1991, 16, pp. 450–456.

20. Gunning, J., Callaghan, J.P., and McGill, S.M., The role of prior loading history and spinal posture on the compressive tolerance and type of failure in the spine using a porcine trauma model, *Clin. Biomech.*, 16(6), pp. 471–480, 2001.

21. Heylings, D.J., Supraspinous and interspinous ligaments of the human lumbar spine, *J. Anat.*, 123, pp. 127–131, 1978.

22. Jäger, M. and Luttmann, A., The load on the lumbar spine during asymmetrical bi-manual materials handling, *Ergonomics*, 35, pp. 783–805, 1992.

23. Juker, D., McGill, S.M., Kropf, P., and Steffen T., Quantitative intramuscular myoelectric activity of lumbar portions of psoas and the abdominal wall during a wide variety of tasks, *Med. Sci. Sports Exerc.*, 30(2), pp. 301–310, 1998.

24. Marras, W.S. Occuaptional low back disorders causation and control, *Ergonomics*, 2000, 43, pp. 880–902.

25. McGill, S.M. and Norman, R.W., Effects of an anatomically detailed erector spinae model on L4/L5 disc compression and shear, *J. Biomech.*, 20, pp. 591–600, 1987a.

26. McGill, S.M., Kinetic potential of the lumbar trunk musculature about three orthogonal orthopaedic axes in extreme postures, *Spine*, 16, pp. 809–815, 1991a.

27. McGill, S.M., Electromyographic activity of the abdominal and low back musculature during the generation of isometric and dynamic axial trunk torque: Implications for lumbar mechanics, *J. Orthop. Res.* 9, pp. 91–103, 1991b.

28. McGill, S.M., A myoelectrically based dynamic three dimensional model to predict loads on lumbar spine tissues during lateral bending, *J. Biomech.*, 25, pp. 395–414, 1992.

29. McGill, S.M. and Kippers, V., Transfer of loads between lumbar tissues during the flexion-relaxation phenomenon, *Spine* 19, pp. 2190–2196, 1994.

30. McGill, S.M., ISB Keynote Lecture-The biomechanics of low back injury: Implications on current practice in industry and the clinic, *J. Biomech.*, 30, pp. 465–475, 1997.

31. McGill, S.M., Low Back Exercises: Evidence for improving exercise regimens, *Phys. Ther.*, 78, pp. 754–765, 1998.

32. McGill, S.M., Childs, A., and Liebenson, C., Endurance times for stabilization exercises: Clinical targets for testing and training from a normal database, *Arch. Phys. Med. Rehab.*, 80, pp. 941–944, 1999.

33. McGill, S.M., Parks, K., and Hughson, R., Changes in lumbar lordosis modify the role of the extensor muscles, *Clin. Biomech.*, 15(10), pp. 777–780, 2000.

34. McGill, S.M., Hughson, R., and Parks, K., Erector spinae oxygenation during prolonged contractions: Implications for prolonged work, *Ergonomics*, 43, pp. 486–493, 2000.

35. McGill, S.M., Low back disorders: Evidence based prevention and rehabilitation, Human Kinetics Publishers, Champaign, Illinois, 2002 and "Ultimate back fitness and performance" at www.backfitpro.com.

36. McGill, S.M., Grenier, S., Bluhm, M., Preuss, R., Brown, S., and Russell, C., Previous history of LBP with work loss is related to lingering effects in biomechanical, physiological, personal and psychosocial characteristics, *Ergonomics*, 46(7): 731–746.

37. Norman, R.W., Wells, R., Neumann, P., Frank, J., Shannon, H., and Kerr, M., A comparison of peak vs cumulative physical work exposure risk factors for the reporting of low back pain in the automotive industry, *Clin. Biomech.*, 13, pp. 561–573, 1998.

38. Oxland, T.R., Panjabi, M.M., Southern, E.P., and Duranceau, J.S., An anatomic basis for spinal instability: A porcine trauma model, *J. Orthop. Res.*, 9, pp. 452–462, 1991.

39. Parks, K.A., Crichton, K.S., Goldford, R.J., and McGill, S.M., On the validity of ratings of impairment for low back disorders, *Spine*, 28(4), pp. 380–384, 2003.

40. Porter, R.W., Is hard work good for the back? The relationship between hard work and low back pain-related disorders, *Int. J. Ind. Ergon.*, 9, pp. 157–160, 1992.

41. Potvin, J., McGill, S.M., and Norman, R.W., Trunk muscle and lumbar ligament contributions to dynamic lifts with varying degrees of trunk flexion, *Spine*, 16(9), pp. 1099–1107, 1991.

42. Richardson, C., Jull, G., Hodges, P., and Hides, J., Therapeutic exercise for spinal segmental stabilization in low back pain. Churchill-Livingston, Edinburgh, 1999.

43. Snook, S.H., Webster, B.S., McGorry, R.W., Fogleman, M.T., and McCann, K.B., The reduction of chronic nonspecific low back pain through the control of early morning lumbar flexion, *Spine*, 23, pp. 2601–2607, 1998.

44. Vera Garcia, F.J., Grenier, S.G., and McGill, S.M., Abdominal response during curl-ups on both stable and labile surfaces, *Phys. Ther.*, 80(6), pp. 564–569, 2000.

45. Woo, S.L.-Y., Gomez, M.A., and Akeson, W.H., Mechanical behaviors of soft tissues: Measurements, modifications, injuries, and treatment, in Biomechanics of trauma, H.M. Nahum and J. Melvin, eds., Appleton Century Crofts, Norwalk, Connecticut, pp. 109–133, 1985.

46. Yingling, V.R., Callaghan, J.P., and McGill, S.M., Dynamic loading affects the mechanical properties and failure site of porcine spines, *Clin. Biomech.*, 12, pp. 301–305, 1997.

21

Human Adaptation in the Workplace

Ash Genaidy
Setenay Tuncel
University of Cincinnati

21.1 Introduction

The concept of individual health, defined as the overall condition of an individual or a condition of well-being, has been broadly expanded in the past two decades from that of an individual to that of a group of individuals and general systems. The rising costs of healthcare particularly in the workplace have prompted a recent move towards the concept of "healthy workforce" as advocated by the National Institute for Occupational Safety and Health.

Very recently, Genaidy and Karwowski (2006) have coined the term "health engineering" to signify "the identification, improvement and maintenance of the well-being of general systems (e.g., individual, technologies, facilities), through the application of engineering, medicine, management, and human sciences knowledge and methodologies". This definition emphasizes the multi-dimensional nature of health and its continuum as a function of time. Furthermore, the application of different disciplines to health acknowledges the complex web of factors influencing this issue.

In light of the above, one can deduce that health is the product of an internal dynamic process that is consistently updated (if possible) to narrow the gap with reference to the state of an individual's optimum well-being. As such, systems "adapt" or "adjust" to the challenges emanating from the environment. The objectives of this chapter are to define human adaptation, describe the occupational adaptation process and workplace factors impacting adaptation and discuss the needs for future research on the subject.

21.2 Definition

According to the American Heritage Dictionary (2000) adaptation is "the act or process of adapting." In biological terms, this term indicates an alteration or adjustment in structure or behavior by which the individual improves his/her conditions in relationship to the environment. In other words, it is a dynamic process in which behavior and physiological mechanisms of an individual continually change to adjust to variations in living conditions. In general, the adaptation process involves four parts: (1) a need in the form of a strong persisting stimulus, (2) the thwarting or nonfulfillment of this need, (3) varied activity, or exploratory behavior, and (4) some response that removes or at least reduces the initiating stimulus and completes the adjustment. The adaptation effects are both short-term as well as long-term. Short-term effects consist of behavior and physiological changes. On the other hand, long-term adaptations may be developmental, that is, environmentally induced changes in anatomy, physiology or behavior. It should be emphasized that the long-term adaptations are genetic, that is, more-or-less programmed changes in anatomy, physiology or behavior.

21.3 Occupational Adaptation Process

A comprehensive search of electronic databases resulted in the identification of the seminal work of Schkade and Schultz (1992, 2003). Schkade and Schultz developed a model for the occupational adaptation process (see Figure 21.1). This process consists of a series of actions and events that unfolds as an individual is faced with an occupational challenge that occurs as the result of person–environment interactions within an occupational role. A press for mastery is created when the individual desires mastery in his/her work role, and when the environment demands at least a minimal level of performance. A mismatch between the individual and the job creates a press for adaptation if the individual wants to achieve or maintain relative mastery. The person brings individual expectations and predispositions as well as sensorimotor, cognitive and psychosocial capabilities to the occupational role expectations. Similarly,

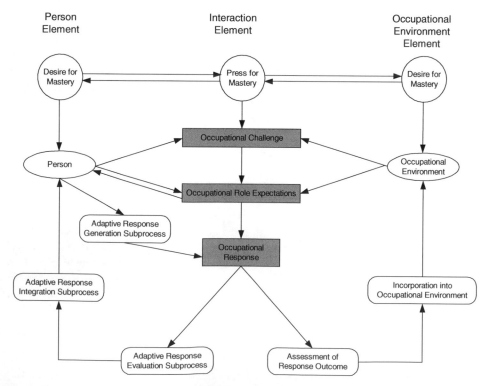

FIGURE 21.1 Occupational adaptation process (Adapted from Schkade, J.K. and Schultz, S., in *Perspectives in Human Occupation: Participation in Life*, Kramer, P., Hinojosa, J., Royeen, C.B., Eds., Lippincott Williams & Wilkins, Philadelphia, PA, 2003, pp.181–221. With permission.)

the occupational environment also contributes to those expectations through its physical, social, and cultural subsystems. The person and environment contributions to the role expectations are reflected by the arrows leading to the expectations. The arrows returning to the person indicate that the individual has a perception of the interactive internal and external expectations, which may be fairly accurate or faulty. It is this mix of expectations that prompts the person to create an occupational response, perform it, assess its efficacy, and integrate the information surrounding this occupational event into the person for subsequent use. The creation, assessment, and integration occur through the adaptive response generation subprocess, the relative response evaluation subprocess and the adaptive response integration subprocess, respectively.

21.3.1   Adaptive Response Generation Subprocess

Once the person has a perception of the interacting internal and external expectations, the two major components for the adaptive response generation subprocess come into play. First, the adaptive response mechanism produces a form of adaptation energy for use (primary response creation occurs at a high awareness level with high usage of the finite supply of adaptation energy; secondary response creation is processed at the subawareness level with a lower energy expenditure), and selects an adaptive response mode for the pattern of interaction (existing response pattern in adaptive repertoire from a previous successful one; modified changes in existing mode when existing mode fails to achieve success; new uniquely different mode developed as existing and modified mode failed to achieve), and selects an adaptive response behavior (primitive-hyperstabilized with behavior not leading to adaptation; transitional-hypermobile with behavior leading to adaptation more likely to produce response; mature-blended mobility and stability, most likely producing adaptive and masterful response to challenge).

21.3.2   Adaptive Response Evaluation Subprocess

After the occupational response occurs, the person begins to assess how adaptive and masterful was the response. Repetitive mastery, with properties of efficiency (how much of personal time, energy and available resources were used?), effectiveness (to what extent were the performance expectations met?) and satisfaction to self and society (to what extent did the experience gratification and to what extent did society regard the response as adaptive and masterful?) is the primary tool for this assessment activity. The person reaches a conclusion about the success of the overall occupational event in meeting the occupational challenge with its expectations. The decision reached is along the continuum, which ranges from dysadaptive (negative range) to adaptive (positive range), with homeostatis as midpoint (neutral).

21.3.3   Adaptive Response Integration Subprocess

Action of the adaptive response integration subprocess provides the reflective conduit through which the feedback loop to the person moves. This subprocess does the work that culminates information about the occupational event resulting in long-lasting impact on the person's adaptive capacity.

The minimal form of adaptive response integration is simply a memory of the experience. If the subprocess functions well, this memory becomes the basis for an adaptive reflection. Although less obvious, it is assumed that a negative relative mastery experience also leads to an increase in adaptive capacity.

21.4   Work Variables Impacting Individual Adaptation

Genaidy and Karwowski (2003, 2005), Abdallah et al. (2004), Salem et al. (2005) have classified the complex web of work variables into 12 categories: (1) organizational, (2) technological, (3) physical environment, (4) social/communication, (5) economic, (6) individual growth, (7) mental task content, (8) physical task content, (9) effort, (10) perceived risk/benefit, (11) performance, and (12) psychological impact. Each work variable consists of a multitude of work elements and is evaluated in terms of demands (the extent to which the work elements exert negative impact upon the person) and energizers (the extent to which the work elements induce positive effect upon the person). Genaidy and Karwowski defined work compatibility as an integration of all work demands and energizers. It is expected that the greater the incompatibility between the demands and the energizers, the greater the challenge to which the person is exposed during the "Occupational Process" and the more difficult is the human adaptation.

21.5 Human Capacity

As pointed out in Section 21.3 on Occupational Adaptation Process, adaptive capacity is a key factor in eliciting an occupational response. In fact, in addition to the work variables, human capacity may place an upper limit on the person's adaptations. It is usually conceived as the interaction of ability and motivation and may consist of the interacting entities, namely, physical, cognitive, and emotional (Shoaf et al., 2000).

21.6 Discussion

Traditionally, technological advancements have outpaced research and development in the field of human systems (Paez et al., 2004). This may be attributed to, among other things, the assumptions that humans are replaceable machines and too complex to be configured, and the narrow specializations engaged in the study of human systems. As a result, the costs of injuries and illnesses in the workplace are still prohibitive. Therefore, the National Institute for Occupational Safety and Health has initiated the concept of U.S. healthy workforce in order to adopt a holistic approach based on both health protection and promotion in order to advance research and development in the field of human systems.

It was the objective of this chapter to research the literature on occupational adaptation. In general, a significant portion of occupational injuries and illnesses is attributed to maladaptation in the response of the individual to the challenges arising from the workplace as well as maladaptation of the workplace to the needs and capabilities of human systems. In this chapter, we reviewed the state-of-the-art research on human adaptation in the workplace. Although a comprehensive framework has been established, there is a lack of quantitative methodology to address human adaptation within the context of work. This methodology should take into account both health protection and promotion. Similarly, a quantitative methodology is needed to tackle the issue of workplace adaptation to the needs and capabilities of human systems.

References

Abdallah, S., Genaidy, A., Salem, O., Karwowski, W., Shell, R., The concept of work compatibility: an integrated work design criterion for improving workplace human performance in manufacturing systems, *Hum. Factors Ergon. Manufacturing*, 14(4), pp. 1–24, 2004.

American Heritage Dictionary of the English Language, Houghton Mifflin Company, 2000.

Genaidy, A.M. and Karwowski, W., The emerging field of health engineering, *Theor. Issues Ergon. Sci.*, 2005, in press.

Genaidy, A.M. and Karwowski, W., Human performance in lean production environment: critical assessment and research framework, *Hum. Factors Ergon. Manufacturing*, 13(4), pp. 317–330, 2003.

Paez, O., Dewees, J., Genaidy, A., Tuncel, S., Karwowski, W., Zurada, J., Lean enterprise: an emerging socio-technological system integration, *Hum. Factors Ergon. Manufacturing*, 14(3), pp. 285–306, 2004.

Salem, O., Paez, O., Holley, M.B., Tuncel, S., Genaidy, A.M., Karwowski, W., Performance tracking through the work compatibility model, *Hum. Factors Ergon. Manufacturing*, 2005, in press.

Shoaf, C., Genaidy, A.M., Haartz, J., Karwowski, W., Waters, T., Hancock, P.A., Huston, R., Shell, R., An adaptive control model for assessment of work-related musculoskeletal hazards and risks, *Theor. Issues Ergon. Sci.*, 1, pp. 34–61, 2000.

Schkade, J.K. and Schultz, S., Occupational adaptation: toward a holistic approach of a contemporary practice, part I, *Am. J. Occup. Ther.*, 46, pp. 829–837, 1992.

Schkade, J.K. and Schultz, S., Occupational adaptation, in *Perspectives in Human Occupation: Participation in Life*, Kramer, P., Hinojosa, J., Royeen, C.B., Eds., Lippincott Williams & Wilkins, Philadelphia, PA, 2003, pp. 181–221.

22
Rehabilitation Ergonomics

Susan J. Isernhagen
DSI Work Solutions, Inc.

22.1 Definition

Rehabilitation Ergonomics is the practice of applying scientific and functional principles to provide a match of work and worker that prevents injury or assists in the return to work process. The practitioners are therapists and other specialists whose backgrounds include anatomy, physiology, kinesiology and pathology and ergonomics. They must analyze both the humans who perform work activities and the setting in which they work. Rehabilitation ergonomists specialize in functional evaluation, improvement of functional work performance, education of the worker and redesign of work to reduce musculoskeletal stressors.

22.2 Professional Guidelines

Guidelines for occupational health, ergonomics and the care of the injured worker have been developed by practitioners who perform the specialty services. The first groups to develop guidelines were therapists whose educational background in musculoskeletal injuries and rehabilitation formed a baseline for restoring or promoting function. Those therapists, who additionally studied ergonomics, added the ability to bring their skills into the workplace.

A. The American Physical Therapy Association developed guidelines for the practice of occupational health. The diversity of the practices parallels the diversity of the needs of those with potential or actual musculoskeletal problems related to work. Both management of the injured worker and prevention of injury are covered.

1. Occupational Health Physical Therapy Guidelines: work-related injury/illness prevention and ergonomics.[1]
2. Occupational Health Physical Therapy Guidelines: physical therapist management of the acutely injured worker.[2]
3. Occupation Health Physical Therapy Guidelines: work conditioning and work hardening programs.[3]
4. Occupational Health Physical Therapy Guidelines: evaluating functional capacity.[4]

The guidelines provide professional structure, are informational to educational programs for curricula development and are available to other professions and referral sources for delineation of the services and their goals.

B. The American Occupational Therapy Association published guidelines, "Occupational Therapy: Services for Ergonomics."[5] Within the guidelines, there are statements on ergonomics definition, ergonomic services provided by occupational therapists, goals of ergonomics, referral and payment mechanisms and occupational therapist background in relation to ergonomics. These serve as a guide for the occupational therapist and those who utilize the services.

C. In 1994, a representative group of sixteen professionals who performed rehabilitation ergonomics defined their work and standards.[6] This provided the specialty with information on training, standards, methods, and goals. The group acknowledged that the specialty of rehabilitation ergonomics was held by more than one profession. The guidelines bridged the gap of specific professional background and, instead, emphasized the commonalities of the practitioner and the practice. It has served as a guide for definition and use.

Figure 22.1 is a summary of professional and practice goals from all three groups of references.

22.3 Laws and Regulations That Effect Rehabilitation Ergonomics

A. The U.S. Occupational Safety and Health Administration (OSHA), in its ergonomic regulations, sets a goal of prevention of work-related musculoskeletal disorders. Its earliest published ergonomic guide for the meatpacking industry was published in 1993.[7] Its current guidelines, available on the internet,[8] support and enhance many of the original provisions. These guidelines call for healthcare providers to be part of the team, early intervention to be provided and job modification to be utilized as a means

---

Primary Goals of Rehabilitation Ergonomics

1. Improve the productivity of individual workers and the work group

2. Maintain the health and improve the safety of the worker in the workplace

3. Decrease lost work time due to illness or injury

4. Enhance return to work processes and minimize the likelihood of disability-induced retirement

---

FIGURE 22.1   A Summary of professional and practice goals.

to allow workers to stay at work as they improve. Other aspects of work such as education, work redesign and selection of appropriate tools are part of the scope of the rehabilitation ergonomist, although also practiced by other professionals.

B. The National Institute for Occupational Safety Health (NIOSH) discusses the roles of the healthcare provider in returning injured workers to work. In Elements of Ergonomics Programs,[9] it lists employer responsibilities that lead to involvement of rehabilitation ergonomists:

1. Provide in-house training
2. Promote early reporting and prompt evaluation by healthcare provider
3. Have healthcare provider become familiar with jobs and job tasks
4. Modify jobs or accommodate as determined by healthcare provider

C. Antidiscrimination laws such as the U.S. Americans Disabilities Act,[10] as well as Canadian and European human rights legislation, mandate accommodations for disabled persons. These allow qualified disabled persons to perform the essential functions of the job. When the disability affects physical function, ergonomic enhancements of work, work tools or work methods facilitate functional performance of the specific work.

D. Workers' compensation laws provide wage replacement while a worker is off work with an injury. To facilitate return to work, specific work-related evaluation and rehabilitation are authorized. Onsite visits to ensure that the worker can perform the job safely and provide job modifications, are part of the rehabilitation plan that leads to return to work. At the treatment and ergonomic modification level, rehabilitation ergonomists are utilized.

Following the logic of the regulations, guidelines and laws, rehabilitation ergonomists enhance the match of work and worker. This is used for primary prevention, modification of work, early return to work, prevention of disability and avoidance of discrimination in the workplace.

Figure 22.2 summarizes the structure of rehabilitation ergonomics in the return to work process. It describes the attributes of the professional specialty derived from guidelines, regulations, professional definitions, and actual practice.

22.4 Rehabilitation Ergonomics Components

Rehabilitation professionals have always treated workers with musculoskeletal injuries in their practice. In the 1980s, workers' compensation systems began to strongly emphasize reduction in work disability.

	Job Requirement		*Functional Ability*	*Match*	*Recommendation*
1.	Floor lift	50	20	No	XXX
2.	Mid Level lift	50	60	Yes	
3.	Shoulder lift	20	30	Yes	
4.	Carry	50	20	No	XXX
5.	Push/pull	60	30	No	XXX
6.	Hand grip	60	50	No	XXX
8.	Stair/ladder	10 ft	Not limited	Yes	
9.	Sit	1 h	4 h	Yes	
10.	Stand	8 h	4 h	No	XXX
11.	Hand coordination	6 h	Not limited	Yes	
	Comment:				

FIGURE 22.2 Example job and capacity match.

As a result, rehabilitation professionals developed four specialties that bridge the gap between treatment and return to work. In brief, they are:

- Functional capacity evaluation (FCE): FCE adds work relevance to testing by using functional activities such as lifting, pushing, pulling, carrying, gripping, climbing, walking, balancing, reaching, sitting, and standing.
- Work rehabilitation: Work-related rehabilitation provides a structured regime that allows the injured worker to increase function and regain work capabilities.
- Job modification: At the worksite, modifications match the work to the capacity of the worker to promote return to work and prevention of reinjury.
- Early intervention: Immediate intervention when a work injury or illness threatens work ability, reduces the lost time for the worker and increases healing and functional work capability.

22.4.1   Functional Capacity Testing

Functional capacity evaluations (FCE) were developed to evaluate the physical work-related abilities of an injured worker.[11-13] The impetus came from workers' compensation administrators who determined that physicians restrictions alone did not provide adequate specific information for an employer to bring a worker back to work. Specific work functions were listed and physicians were asked to rate the worker on each category. In turn, therapists were called upon to develop an objective means to measure work function that could be used as an adjunct for the medical release to work.

Functional evaluation is an objective measure of the ability of a worker to perform actual work tasks. There are two types, FCE and work-related tests.

FCEs utilize the listing of job tasks developed by the U.S. Department of Labor and Industry.[14] Tests are standardized for each physical task and the results are given in each category. By designing functional capacity tests for these work items, it was possible to match a worker's functional capacities with the requirements of the job. This allowed appropriate job or jobs to be selected and to define what modifications would be necessary to allow safe work return.

Figure 22.3 is an example of the matching of a job description with functional abilities of a worker. The XXX denotes areas where specific comments would be made.

As functional capacity technology grew, so did the numbers and types of tests. To ensure confidence in an FCE, studies determined which FCEs are scientifically reliable.[15-22] This provides assurance to the referrer, end user and evacuee that the test is reproducible. Validity studies demonstrate that certain FCEs can have impact outside of the clinical setting.[23-25]

Specific work-related tests differ from standardized FCEs. They are based on job descriptions. Reliability and reproducibility are more difficult, as each job description requires a different test.[26]

---

1. Rehabilitation ergonomist: rehabilitation professional educated in ergonomics and workplace issues

2. Functional evaluation that identifies the safe capacities of a worker combined with a functional job description to provide the match of the worker and work, defining what matches and what needs modification

3. Job modifications used immediately at the worksite, to keep the worker at work safely or allow an early return to work

4. Enhancement of the work abilities as the workers healing or progress toward return to work

5. Relationship with the team of medical manager, case manager, human resource professional, safety manager, and worksite supervisors

---

FIGURE 22.3   Structure of rehabilitation ergonomics in return to work.

However, validity and real-world usefulness can be enhanced by insuring that the job description is accurate and designing the test to match the functional job description.[27] Occupations, which lend themselves to work-specific tests have unique physical demands. Examples are nurse and nurse assistants when transfers are required, specific repetitive assembly processes, lineman who climbs power poles, etc.

Choosing a standardized FCE versus a work-related functional test depends on the questions that are being asked. If there is no specific job description available for the worker, or if the worker could be considered for several jobs, then a standardized FCE is most effective. It can be compared to several job descriptions in order to prioritize which jobs match the worker more closely. If only one job is being considered, however, a specific functional test for that job can be developed from the functional job description. This provides greater opportunities for work-specific tests, but the evaluators must be careful to determine that the job description is valid and that the functional tests replicate the job demands.

While functional testing is a specialty for therapists, the resultant findings and recommendations are stronger when the therapist is also a rehabilitation ergonomist. The knowledge of the worksite, the jobs and the job modification opportunities provide information for a stronger resolution of the return to work objective.

22.4.2 Work Rehabilitation

For those workers whose physical limitations prevent return to work at their previous job or at full duty, rehabilitation is indicated. In addition to traditional therapeutic exercise, work rehabilitation includes actual or simulated work task and work behavior management.[28]

In the 1980s Matheson defined and described work hardening.[29,30] Its use of work simulation ensured that work behaviors would be addressed and that return to actual work would be the goal. The behavioral aspects of returning a chronically injured or disabled person to work were emphasized. The medical and physical models were joined with psycho-social and vocational models.

A rehabilitation model is utilized and differs from a treatment model. The essence of rehabilitation is brought into the work simulation setting:

- An atmosphere is created in which the clients are responsible for their own progress. The therapist is a guide and assists, but the worker does the work to accomplish the goals.
- The focus is on function, not perfection. For those with injury or illnesses that are beyond the acute phase, improving the residual function is the focus of the program. While increase in motion and strength along with decrease of discomfort are important, they only build a foundation for which function can be obtained.
- The goals are related to ability to function outside the clinic. In work rehabilitation, the return to work is always the target. The program is designed for sequential improvement until full duty or the highest level of safe work is gained.

The Commission on Accreditation of Rehabilitation Facilities (CARF),[31] developed the first work hardening standards. It defined work hardening as a multi-disciplinary program including physical, psychological and vocational components.

In response to CARF, the American Physical Therapy Association determined that there were two types of work rehabilitation programs. Work hardening was defined in similar terms as CARF. A more direct program was additionally defined and described. Work conditioning emphasizes physical and functional strengthening for return to work. It is often performed by a single discipline rather than being multi-disciplinary. The psychological and vocational aspects are not part of the program, although they may be contracted separately.

Both types of work rehabilitation require full participation from the worker in a setting that emphasizes safe work behaviors, use work-related tasks for rehabilitation and have return to work as the goal.

22.4.3 Functional Restoration

Mayer et al.[32] designed a multi-disciplinary program to return chronically injured workers to the workplace. It used a strong medical model with the core physician's involvement. In addition, objective measurements were emphasized, including those from isokinetic exercise technology. Its highly structured program gained recognition for its ability to be replicated. In the initial functional restoration and work rehabilitation programs, the population served was primarily those chronically off work with injuries or illnesses.

22.4.4 Job Modification

Both FCE and work rehabilitation focus on the functional ability of the worker. If a worker has demonstrated capacities to do the essential functions of his/her job, then the rehabilitation ergonomist provides assurance, answers questions and facilitates communication with the supervisor and coworkers.

If there is not a match, however, modifications of job tasks are necessary to protect the worker and yet allow essential functions to be performed productively.

Clarity on the permanency of restrictions is necessary for the worker, the medical team and the employer. Both the worker and supervisor must understand if there is to be progression and how this progression will be handled. This is described in Figure 22.4. Many modifications are time-limited, as the worker's function should improve with physical work and continuation of the healing process. If a condition is permanent (e.g., spinal fusion, neurological damage), the modification may also be permanent.

Figure 22.5 demonstrates how specific essential functions can be modified in the return to work process. In some cases, merely revising the work method can be a modification. There is low or no cost to this solution. In some cases, the work task can be changed through equipment, in order to accomplish the function with less effort or stress. In the third case, the employer can decide to modify the essential function by lowering the requirement, or eliminate the function. The decision requires adequate information and dialog on the choices, the ability of the employer and supervisor to modify a job and the funds that can be allocated. Each situation must be evaluated individually for the best option, as there are often more than one potential way in which modifications can be designed.

22.4.5 Early Intervention

As evaluation of the success of work programs was studied, it became clear that the earlier the referral to the program, the better the result. While some of the late referral issues revolved around severity of the injury or illness, most of it was a result of slow or inadequate case management. Thus, innovative

---

Job Modification Categories

Permanence

A. Temporary
- Time limitation specific (e.g., 4 weeks)
- Utilized when functional performance is expected to improve through healing, therapy or actual work performance
- Can be gradually upgraded as function improves

B. Permanent
- Utilized when a worker's condition is permanent
- Utilized for a worker with a disability
- Utilized when the ergonomic job modification is desired as a permanently improved work design

---

FIGURE 22.4 The description of both the worker and supervisor how to handle the progression.

Essential Functions (EF) and Job Modifications Examples

Item	Keep EF	Change Method of EF	Eliminate EF
Lifting	Rotate lift with nonlift tasks to provide a break, yet perform complete job	Use vacuhoist for all lifts over 10#	Do all nonlifting components of the job
Repetitive hand work	Rotate unlike tasks to provide musculoskeletal relief, yet perform full job	Use hand tools, assistive devices to reduce stressors	Limit EF to a portion of worktime and add EF from a second job to complete shift
Sit	Use ergonomic chair	Change work station to sit — stand but maintain all other EF	Change position of work to standing

FIGURE 22.5 Modification of specific essential functions in the return to work process.

programs were designed to bring an injured worker very early for evaluation and treatment of their musculoskeletal injuries.[33-35]

In many cases, the early intervention, in conjunction with modified work, keeps the injured person in the workplace without lost days. For those who do have lost time, treatment is underway at once and an early return to work is more likely. In either case, the workers retain their self-image for work and avoid the "patient" mindset. The coworkers see that the worker is retained and supervisors continue to see the worker as productive. The intervention is directed at healing, protecting the injured part while healing is taking place, using the uninjured portions of the worker's body in work tasks and focusing on function rather than pain.

In order to develop early intervention, measures must be in place before an injury is reported. Industry management must approve the constructs of the program and facilitate education and responsibility of the injury team. These include the claims manager, human resource coordinator, safety officer and any other onsite medical personnel. In addition, all employees must understand that early reporting is desired. It will be met with positive response, not negative. At times it takes many reminders, as workers often fear that a report of a problem will result in negative action toward them. This is especially true when reward systems for "no OSHA recordables" is in place. Also, if the culture is to disdain a worker with an injury, or the "light" duty that might be a result, employees will be reluctant to report until the problem is severe.

Once the early intervention process is in place, education for workers and supervisors is necessary for early symptoms of musculoskeletal disorders to be recognized. Workers must be aware of the early stages of carpal tunnel syndrome, tendonitis, strains, and others, in order for the system to work.

When injury, illness or early symptoms are reported, the onsite rehabilitation ergonomist works with the employer and the medical team. Intervention includes evaluation of the condition, assessment of current functional capacity to determine if the worker can continue to work, institution of functional treatment and modification of the job when early return to work can be accomplished. Early intervention is best done onsite or in a clinic that is close to the worksite.

Outcome measurement identifies the effectiveness of early intervention compared to previous traditional treatment. Analysis can also identify jobs or job tasks in the workplace where problems occur most frequently. The safety department, with the rehabilitation ergonomist as part of the team, can then institute prevention measures. These may be ergonomic redesign, new tools, education, improved job training, stretching, ergonomic postures and problem solving. Early intervention is a bridge between injury prevention and injury management.

22.5   Rehabilitation Ergonomics as Part of a Medical Continuum

In the 1990s, the focus on return to work outcomes became an important area of research. Pransky's comprehensive review of the literature categorized the variables that affect return to work.[35] He identified that work absences were related to a higher preinjury ergonomic risk, a dissatisfaction of the worker with return to work accommodations and a negative relationship with the worker's compensation insurer. He also identified an important, but lightly studied variable, that of reduction of reinjury. He concluded, that the reinjury rates are increased in women with jobs that have both high preinjury ergonomic risk and high postinjury ergonomic risk, dissatisfaction with work accommodation, negative employer reactions, dissatisfaction with the medical services and dissatisfaction with low back statistics. Two of these items relate to the ergonomic risk being either high or a negative perception of work capacity in returning workers. If workers had high postinjury ergonomic risk, it tended to decrease their perception of capacity.

Feuerstein evaluated clinical and workplace factors associated with the return to modified duty in upper extremity disorders.[36] The model could predict or classify those not working and those on modified duty. Increased ergonomic stressors was one of the four primary predictors for those not working. Return to work should not just be limited to medical and clinical signs only.

Staal performed a descriptive review of return to work interventions for low back pain.[37] Out of the seventeen studies that he evaluated, only three had ergonomic interventions and none of those had randomized controlled studies. He noted that multi-model treatment consisting of exercise, education, behavioral training and ergonomics would be the most promising.

Matheson looked at the predictability of functional capacity to identify whether return to work would take place and at what level.[21] Data on return to work and, specifically, return to work at the original job/original employer, were collected. Functional capacities' items were evaluated for their relationship to those outcomes. FCE lifts were linked with both return to work and the level of return to work.

In the mid 1990s, Loisel et al. developed the Sherbrooke Model,[38] which postulated that ergonomic interventions should be used with clinical interventions in return to work. For subjects that had been off work 6 weeks, this model went into effect. Early and active treatment was part of the regime but, additionally, ergonomic evaluations and interventions were utilized. In 1997, his randomized clinical trial indicated that the occupational intervention, which included ergonomics on the job, was an important component of full case management.[39] The best results were in returning to work and were accomplished by combining clinical interventions with occupational interventions.

A 6-yr follow-up study in 2002[40] demonstrated that the occupational interventions, combined with clinical interventions, saved days on benefits and saved costs. In 2003, the group defined a PREVICAP model, which had three dimensions.[41] It revolved around the worker, the work environment and the interaction between the work and work environment.

Anema et al. looked at participatory ergonomics as a return to work intervention.[42] Those with low back pain were studied. Anema acknowledged the use of ergonomics for prevention and added a study for disability management. When ergonomic suggestions and interventions were developed for low back pain patients the results were positive. Over half of the ergonomic interventions were implemented and workers were satisfied with the solutions and reported that they had a stimulating effect.

Lemstra's (2003) study evaluated one industry in Canada and demonstrated the effectiveness of occupational management.[43] This included a physical therapist onsite using ergonomic reassurance and encouragement to assist injured workers to be on the job safely. The work was based on the physical and functional information from the physical therapist and the medical information from the family physician. This blending of prevention and return to work ergonomics allowed the intervention to take place sooner, onsite and with professionals known to the workers. Upper extremity and back injury claims with the new model demonstrated decrease in days lost upto 91%. This was superior to the traditional medical model of standard care or a regime of clinical physical therapy service.

As rehabilitation professionals turn to rehabilitation ergonomics as a specialty, they are bolstered by the scientific ongoing studies that describe the specifics of programs that have demonstrated effectiveness. Costs and days lost are decreased. Workers appreciate the ergonomic modifications, assistance with understanding their capacities and assistance with return to work at the worksite. The work is positively received by employers as well. With reinjury rate now being measured, the second positive effect of rehabilitation ergonomics is being identified.

22.6 Three Models of Rehabilitation Ergonomics for Return to Work

22.6.1 Overview

Three models of rehabilitation ergonomics provide return to work:

1. The clinical model utilizes an occupational rehabilitation specialist in the clinic and a rehabilitation ergonomist onsite for the re-entry to work phase. The clinical model is utilized in three situations. First, if a worker has an illness or injury that requires strict safety standards, the adherence to medical safety in the clinic provides security. This can include those with heart rate or blood pressure problems, as well as those in the healing stage of a severe sprain, fracture, etc. Second, if a worker is at a level of job readiness that is too low for even modified work, conditioning will need to be provided to raise the worker to a level that would sustain half or full day of work. Third, some occupations do not provide for anything but full duty. This could include trucking, construction work, heavy manufacturing or heavy patient handling.
2. The mixed clinical and work model combines the benefit of safe structured rehabilitation with worksite experience that prevents the worker from becoming alienated from the work and coworkers.
3. The onsite work model is utilized either when there is a clinic at the worksite or the employer approves using actual work for rehabilitation.

22.6.2 Case Study I: Clinical Model

John Jones has had two lumbar surgeries. The first was a laminectomy at the L4-5 level and the second was a laminectomy at the L3-4 level with removal of scar tissue at the site of the first surgery. John's previous job, maintenance person for the transit service, is available. The functional job description identifies standing, walking, working in awkward positions, lifting upto 40 lb, gross and fine hand coordination and performing elevated work.

The physician has referred John for initial rehabilitation for overall strengthening, improvement of aerobic capacity and spinal stability exercises as he has been off work for 6 of the past 7 months. John is deconditioned as well as in need of rehabilitation for this spine.

Goals includes improved strength and endurance with overhead work, kneeling, crouching and aerobic activity. In addition, a low back stabilization program improves musculature in the lumbar area to sustain a stabilized position. This is necessary, as John needs to reach, get into various positions to do his maintenance work.

As he heals and improves, work activities including lifting, carrying, pushing and pulling are added. With close clinical supervision and safety parameters, he can work toward the physical requirements in his job description with confidence. Monitoring of capacities takes place until John can perform seven out of ten essential functions of his job. Three heavier aspects of the job are limited but improving.

A rehabilitation ergonomist joins the team to begin return to work planning. He confers with John and his therapist. It is necessary to understand John's functional abilities and limitations, meet with the supervisors at work and design modifications for the three items he cannot yet perform.

Once John returns to work, he is monitored on a regular basis to insure safety and provided with reassurance and assistance by his supervisor and coworkers. Sequential upgrading of activity takes place until full duty is reached. A satisfactory final evaluation by the therapist and physician formally releases John to full duty.

Two carry-over changes will impact John:

1. John is interested in maintaining his new fitness level. He expressed concern that he had never recovered his preoperative work state until he participated in the work rehabilitation. His employer pays for a local health club membership. John's program will be transferred to this facility for maintenance and improvement in fitness and function.
2. The employer and medical providers became aware that workers who have been out for prolonged periods, with any medical condition, would benefit from functional testing prior to returning to heavy work. If there is a deficiency, a suitable rehabilitation program will be provided and the rehabilitation ergonomist will be able to serve such workers at the worksite. The employer has identified a high reinjury rate for returning employees and has targeted this for reduction with these policies and programs.

This combines the best of clinical rehabilitation with the use of a rehabilitation ergonomist in return to work.

22.6.3   Case Study II: Combined Clinical and Worksite Model

Shirley Walters works in the housekeeping department of a hospital. She strained both shoulders and low back and has a history of myofascial neck and upper extremity pain. Shirley previously worked at her job productively and had her myofascial symptoms under control before she suffered the shoulder and back strain moving heavy equipment. She is afraid that her strains and exacerbated myofacial pain will prevent her from returning to work.

Shirley's rehabilitation program begins with a partial functional evaluation, recognizing that she is in the healing phase. The initial evaluation shows that Shirley has the motions and capability of doing 60% of her current job requirements but her endurance continues to be low and myofacial symptoms increase after approximately 4 h of work-like activities. She will be working 8-h shifts at full time.

In order to return her to work and to alleviate her fears, she is put on a schedule of both clinical and return to work. Her employer and her case manager are willing to begin with 4 h of activity, increasing to 6 h, and ultimately progress to full-time work. The rehabilitation ergonomist, a physical therapist, works with Shirley on functional activities and strengthening related directly to her job description essential functions.

In the first week, Shirley works in the clinic performing 4 h of simulated housekeeping activity, building her endurance and her confidence. At the end of the 4 h, symptoms are high, but she learns stretches and positioning to decrease them. They have generally subsided by the next day.

On the second week, she upgrades to 6 h of clinical activity. The third week begins the return to work process with 3 h of work simulation at the clinic and 3 h at work, doing the essential functions for which she has been tested out safely.

The fourth week, she is completing 6 h of modified work, all at the worksite. The rehabilitation ergonomist oversees her for half an hour per day, but she is at normal work pace and under her normal supervision processes for the shift. The rehabilitation and training have decreased symptoms in the back and neck. The myofascial symptoms continue at the level before she suffered the strain but are able to be controlled by her.

On the fifth week, she returns to full duty, which includes all essential functions of her job. The gradual process of becoming stronger, learning new techniques of work and knowing that her supervisor is supportive have given Shirley confidence.

At discharge, she does have considerations that were not in place previously. They are contained in a new agreement with her supervisor and employer:

1. When faced with pushing/lifting that is beyond her specified job requirements, she will seek help. This will prevent the sprain/strain she suffered previously.
2. If myofascial symptoms are high, she will be allowed short breaks for her stretching. She and her supervisor will work out what is acceptable. This will prevent her suffering in silence and being absent because of another physical problem.
3. The equipment she has to move has been ergonomically redesigned with larger and more movable wheels. Force required for pushing and pulling has been reduced from 100# previously to 30# with the new wheels. If other equipment or material creates high stresses, she is to report them for analysing an ergonomic modification. To this point, the employer now has a relationship with the rehabilitation ergonomist for more work to reduce stressors in the entire facility.
4. Interest in stretches, ergonomic education and fitness increased after there was interaction with the rehabilitation ergonomist. Preventive programs were discussed and are being put into place for all employees.

22.6.4   Case Study III: Worksite Model

Makai Brown assembles furniture from parts shipped from a supplier. His employer sells the assembled models, which are fitted with customized additions. Makai has begun to have right thumb and finger numbness at night and discomfort in the lateral epicondyle area of his right arm during activity. His physician has diagnosed early carpal tunnel syndrome and lateral epicondylitis on the right.

The company's occupational health department contracts for a rehabilitation ergonomist to work at the company 2 days a week. The role is to work with the team in prevention ergonomics, to provide ergonomic education to the workers and to work with those with early musculoskeletal symptoms, illness or injury.

The physician prescribes an anti-inflammatory medicine. The rehabilitation ergonomist institutes therapeutic measures such as night splints, an epicondylitis band to support the elbow extensor muscles and a stretching and flexibility exercise program for the neck and upper extremities. In addition, modified functional testing is performed and matched with his job description. His current job requires moderate to heavy repetitive upper extremity activity and use of an impact wrench. All gripping activities increase his symptoms. He is able to do nonrepetitive work with his hands and arms and he has no limitation in walking, standing or sitting.

It is desirable from his and the employer's point of view that he remains at work. He is temporarily given the job tasks of quality inspection, inventory and tool maintenance. He is treated onsite with physical therapy for his epicondylitis. His functional statistics such as grip and pinch are taken for a baseline, his carpal tunnel symptoms are documented and monitored, and he is given a home exercise and positioning program. He maintains his restrictions on gripping and repetitive work for home and recreational activities. He also takes the ergonomic training course, which educates him on stressors and modifications that he can do himself. He works for 2 weeks at the modified job.

At 2 weeks, his sensory symptoms in fingers and hands have been reduced, the lateral elbow pain is gone, but the grip strength is only 75% of what his left hand shows. He states it feels weaker than before his problems. He continues all treatments, as he gradually resumes some of his normal activities. Four hours a day he resumes light furniture assembly. He is given a pneumatic impact wrench to replace the mechanically driven one. The new one is half the weight and much easier to grip than the former tool. The other 4 h continues to be the lighter duty tasks. He is aware of early symptoms and is to stop, stretch and resume activity slowly if there is a problem.

At 4 weeks, both carpal tunnel and epicondylitis symptoms have subsided. His active treatment stops, but self-exercise continues as needed. He resumes full duty with self-monitoring of symptoms and regular re-evaluation by the rehabilitation ergonomist. Makai and the supervisor can call the ergonomist with questions or problems and the ergonomist will consult to make adjustments as necessary.

His resumption of full duty is accompanied by removal of restrictions on home activity. The following ergonomic home and work guidelines remain:

1. He is to continue to use the pneumatic impact wrench at work to reduce work stressors permanently. He will make similar accommodations at home. He now realizes tool use and heavy repetitive gripping must be monitored in his recreation and home activities. New sensibilities are present in his home carpentry, sports and chores. They particularly revolve around ergonomic positions, proper tools, reduction of stressors and implementation of stretching as appropriate.
2. If he has symptoms again, he is to report these at work as soon as possible. The health department at work has determined that causation (work-related or not work-related) will not stop the process of working with him to avoid a reinjury. The rehabilitation ergonomist also can be brought in any time that there is an issue.
3. He participates in safety meetings to work on early intervention and ergonomic guidelines for all employees.

Note: This case was resolved with no lost time and 4 weeks of modified duty. This result was superior to former cases of workers with similar problems. In addition to the good metrics (decreased lost time and medical costs) both the worker and supervisor rated their satisfaction as high.

22.7   The Aging Worker: Special Considerations

The aging process affects the neuromusculoskeletal system and is translated into functional changes. The loss of functional work ability is noted in middle-aged and older workers. For those with chronic injuries and illnesses, aging changes add to the decrement in functional abilities.[44]

For employers and workers, the age-related decrease in functional ability can translate into decreased productivity, increased injuries and increased disability. Perceived disability for older workers is also heightened, so that when an older worker is hurt, there may be the perception that they will not be able to return to the workforce without reinjury.

To study the effect of education, training and ergonomics on workers, five studies were performed and reported by Finnish researchers.[45] A physiotherapist participated in the training for material handling and in planning of workplace health promotion. Results of the five studies demonstrated that both the health and work ability of aging workers can be promoted with ergonomic measures at the workplace. Team spirit, a result of the participative approach, was also enhanced. Education and training had a positive effect, especially on aging workers.

These studies show that there will be a functional decline in workers, but that ergonomics, education, training and participation can be used to increase work ability. They point to an opportunity to decrease the rate of injury and disability in the aging population. Exercise is a deterrent to changes in musculoskeletal ability, and when added to the suggested regime, could slow the process of declining function.

There is a need for ongoing research to specifically address the multiple interventions that can be successfully utilized in reducing work injuries, lowered productivity and diminished perceptions of work ability in the aging worker.

Regarding use of exercise, the rehabilitation ergonomist must balance the presence of slow-onset repetitive injuries in physical work (known in industry as "wear out"), and the exercises used to maintain fitness. Program design must take into consideration the potential interaction of concurrent overuse syndrome while promoting restorative fitness regimes.

With the recognition that increasing numbers of workers are reaching middle and older age, and that older workers are valuable due to experience and loyalty, employers are seeking methods to maintain health and productivity of older employees. Rehabilitation ergonomics for both return to work and for prevention of injury can make a significant contribution.

22.8 Rehabilitation Ergonomics in Prevention

While rehabilitation ergonomists first became involved with work injury in the return to work arena, there has been a cross-over into prevention of musculoskeletal injuries. This derives from cases where employers formed a positive relationship with a rehabilitation ergonomist who has learned the culture, jobs and methods of a worksite through bringing employees back to work. They work in prevention because they have been requested to do so by satisfied employers and employee groups.

In addition, the ergonomic resolutions created by modifying work for a returning worker often are appropriate permanent modifications for the entire work group. The process of reducing stressors for an individual worker is similar to reducing stressors for a group of workers.

Rehabilitation ergonomists determined that there is a dual role of return to work and prevention.[1-6] Government guidelines or injury prevention also acknowledge the relationship with return to work and describe roles for healthcare professionals who can work on both sides.[7-10] Research into return to work outcomes and ergonomic practice describe the role of the rehabilitation professional as working in the prevention mode in industry.[34,41,43,45]

Rehabilitation ergonomists specialize in the prevention of musculoskeletal injuries by utilizing ergonomic principles of worksite redesign, tool selection/modification,[46] work method design, ergonomic education, fitness and early intervention. They perform as part of a larger prevention team, often including engineer ergonomists, safety departments, production managers and other medical professional involved in prevention.

Ergonomic training and education are a foundation for the prevention practice of the rehabilitation ergonomist. By referencing other chapters in this text, one can find principles of prevention used universally.

22.9 Challenges for the Future

Rehabilitation ergonomics is in its early stages. It has been developed to meet specific needs. Rehabilitation professionals were invited into the return to work process by insurance systems and employers looking for better methods to return injured workers to work. Early intervention bridged the gap between return to work and prevention of injury. Job modifications began with individual patients and advanced into wider use in musculoskeletal injury prevention because of utility.

Scientific studies use rehabilitation ergonomists as part of the team, but refer to them by their profession (e.g., physical therapists, occupational therapists), rather than their specialty. The studies show positive results of team effort working in several related strategies. It is difficult to point to any member of the team (e.g., safety officer, engineer, physician, nurse, case manager) as the reason for success. Thus, the role for rehabilitation ergonomists is as part of a larger team.

The future challenge for rehabilitation ergonomists is similar to that of all ergonomists and specialists in work injury management and prevention. It is to define its role and set goals in the following areas:

- Create educational opportunities for rehabilitation professionals that create knowledge and practice parameters in ergonomics.
- Create educational opportunities in concert with related professions for effective use of specialties, which have the same goals.
- Delineate its role in the return to work process and design accompanying studies to validate efficacy, cost effectiveness, worker satisfaction and employer satisfaction.
- Delineate its role in the prevention of musculoskeletal injuries, specifically combining ergonomics, education and fitness and design accompanying studies to validate efficacy, cost effectiveness, worker satisfaction and employer satisfaction.
- Create models for its role in teams of professionals, both in prevention and return to work.
- Design and implement research that evaluates and categorizes the most effective interventions and those which need improvement.

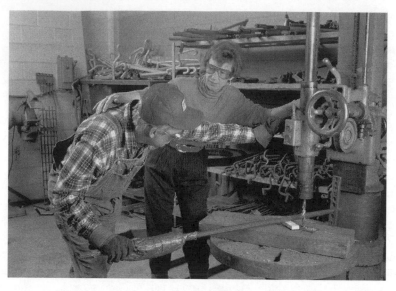

FIGURE 22.6　Working onsite, a rehabilitation ergonomist analyzes how the worker's neck and upper extremities function during work activity. This forms a base for tasks described in a job function description.

- Assist in creation of a model that defines work health as a continuum, using the team approach to problem solving for all workers whether currently healthy or temporarily unable to perform work. Minimize the need for separate systems for injured and noninjured workers

FIGURE 22.7　In the return to work process, the rehabilitation ergonomist blends knowledge of the worker's functional capacity with the demands of the job to ensure productivity and prevent reinjury.

Rehabilitation ergonomics is one of several models of broader ergonomics. With continued dialog, research and interaction across professions, it will continue to provide value for the workers and employers it serves.

References

1. Occupational Health Guidelines: evaluating functional capacity, American Physical Therapy Association, Alexandria VA, 1999.
2. Occupational Health Guidelines: physical therapist management of the acutely injured worker, American Physical Therapy Association, Alexandria VA, 2000.
3. Occupational Health Guidelines: work conditioning and work hardening programs, American Physical Therapy Association, Alexandria VA, 1997.
4. Occupational Health Guidelines: work related injury/illness prevention and ergonomics, American Physical Therapy Association, Alexandria VA, 2001.
5. Occupational therapy services in ergonomics, American Occupational Therapy Association, Bethesda MD, 1998.
6. Isernhagen, S.J., Hart, D.L., and Matheson, L.N., Rehabilitation ergonomists: standards for development, *Work*, 10, 199, 1998.
7. Ergonomic program management guidelines for meatpacking plants, U.S. Department of Labor, Occupational Safety and Health Administration, OSHA 3123,1993.
8. Effective Ergonomics: strategy for success, www.OSHA.gov/ergonomics, current.
9. Elements of Ergonomics Programs. NIOSH, DHHS Publications 97, 1997.
10. A technical assistance manual on the employment provision of the Americans with Disabilities Act, Equal Employment Opportunity Commission, U.S. Government Printing Office, Washington, DC, 1992.
11. Isernhagen, S., Functional capacity evaluation, in *Work Injury Management and Prevention*, Isernhagen, S., Ed., Aspen, Gaithersburg MD, 1988, Chap. 14.
12. Hart, D.L., Isernhagen, S.J., and Matheson, L.N., Guidelines for functional capacity evaluation of people with medical conditions, *J. Ortho. Sports Phys. Ther.*, 18, 682, 1993.
13. Key, G., Work capacity analysis, in *Physical Therapy*, Scully, R. and Barnes, M., Eds., J.B. Lippincott, Philadelphia PA, 304, 1989.
14. Selected characteristics of occupations as defined in the revised dictionary of occupational titles, U.S. Department of Labor, National Technical Information Service, Washington, DC, 94, 1993.
15. Reneman, M.F. et al., Test-retest reliability of lifting and carrying in a 2 day functional capacity evaluation, *J. Occup. Rehab.*, 12, 269, 2002.
16. Reneman, M.F. et al., The reliability of determining effort level of lifting and carrying in a functional capacity evaluation, *J. Occup. Rehab.*, 18, 23, 2002.
17. Brouwer, S. et al., Test-retest reliability of a modified Isernhagen Work Systems functional capacity evaluation in patients with chronic low back pain, *J. Occup. Rehab.*, 13, 207, 2003.
18. Gross, D.P. and Battie, M.C., Reliability of safe maximum lifting determinations of a functional capacity, *Phys. Ther.*, 4, 364, 2002.
19. Isernhagen, S., Hart, D.L., and Matheson, L.N., Reliability of independent observer judgments of level of lift effort in a kinesiophysical functional capacity evaluation, *Work*, 12, 145, 1999.
20. Lechner, D.E. et al., Reliability and validity of a newly developed test of physical work performance, *J. Occup. Med.*, 38, 997, 1994.
21. Matheson, L.N. et al., A test to measure lift capacity of physically impaired adults, Part 1, development and reliability testing, *Spine*, 20, 2119, 1995.
22. Smith, R.L., Therapists' ability to identify safe maximum lifting in low back pain patients during functional capacity evaluation, *J. Ortho. Sports Phys. Ther.*, 19, 277, 1994.
23. Gross, D.P. and Battie, M.C., The construct validity of a kinesiophysical functional capacity evaluation administered within a worker's compensation environment. *J. Occup. Rehab.*, 13, 287, 2003.

24. Matheson, L.N., Isernhagen, S.J., and Hart, L., Relationships among lifting ability, grip force and return to work, *Phys. Ther.*, 82, 249, 2002.
25. Reneman, M.F. et al., Concurrent validity of questionnaire and performance-based disability measurements in patients with chronic nonspecific low back pain, *J. Occup. Rehab.*, 12, 119, 2002.
26. Innes, E. and Straker, L., Reliability of work-related assessments, *Work*, 13, 107, 1999.
27. Innes, E. and Straker, L., Validity of work-related assessments, *Work*, 13, 125, 1999.
28. Lett, C.F., Work hardening in *Work Injury Management and Prevention*, Isernhagen, S., Ed., Aspen, Gaithersburg MD, 1988, Chap. 15.
29. Matheson, L.N. et al., Work hardening: occupational therapy in industrial rehabilitation, *Am. J. Occup. Ther.*, 39, 314, 1985.
30. Matheson, L., Work Capacity Evaluation Manual, Employment and Rehabilitation Institute of California, 1987.
31. Work Hardening Standards, Commission on Accreditation of Rehabilitation Facilities, Tucson AZ, 1992.
32. Mayer, T. et al., Objective assessment of spine function following industrial injury: a prospective study with comparison group and one year followup, *Spine*, 10, 482, 1985.
33. Isernhagen, S., Primary and secondary therapy for acute musculoskeletal disorders, in *Occupational Musculoskeletal Disorders*, Mayer, T.G., Gatchel, R.J., and Polatin, P.B., Lippincott, Williams and Wilkins, Philadelphia, 2000, Chap. 19.
34. Vance, S.R., Brown, A.M., On-site medical care and physical therapy impact, in *Comprehensive Guide to Work Injury Management*, Isernhagen, S., Ed., Aspen, Gaithersburg MD, 1995, Chap. 13.
35. Pransky, G. et al., Work-related outcomes in occupational low back pain, *Spine*, 27, 864, 2002.
36. Feuerstein, M. et al., Clinical and workplace factors associated with a return to modified duty in work-related upper extremity disorders, *Pain*, 102, 51, 2003.
37. Staal, J.B., Return to work interventions for low back pain: a descriptive review of contents and concepts of working mechanisms, *Sports Med.*, 32, 251, 2002.
38. Loisel, P. et al., Management of occupational back pain: the Sherbrooke model: results of a pilot and feasibility study, *Occup. Environ. Med.*, 51, 597, 1994.
39. Loisel, P. et al., A population-based, randomized clinical trial on back pain management, *Spine*, 22, 2911, 1997.
40. Loisel, P. et al., Cost-benefit and cost-effectiveness of a disability prevention model for back pain management: six year follow up study, *Occup. Environ. Med.*, 59, 807, 2002.
41. Durand, M.J. et al., Constructing the program impact theory for an evidence-based work rehabilitation program for workers with low back pain, *Work*, 21, 233, 2003.
42. Anema, J.R. et al., Participatory ergonomics as a return to work intervention: a future challenge? *Am. J. Ind. Med.*, 44, 273, 2003.
43. Lemstra, M. and Olszynski, W.P., The effectiveness of standard care, early intervention, and occupational management in workers compensation claims, *Spine*, 25, 299, 2003.
44. Isernhagen, S.J., Functional capacities assessment after rehabilitation, in *Ergonomics*, Bullock, M., Ed., Churchill Livingstone, London, 1990, Chap. 11.
45. Nygard, C.H., Pikkanen, M., and Arola, Hl, Promotion of health and work ability through ergonomics among aging workers, Proceedings of the 34th Congress of the Nordic Ergonomics Society, *Humans in a Complex Environment Vol II*, Kalmarden, Sweden, 611, 2002.
46. Job Accommodation Network Website, www.jan

23

Visual, Tactile, and Multimodal Information Processing

Nadine Sarter
University of Michigan

23.1 Introduction

This chapter will discuss the most, and one of the least, frequently employed sensory channels in current interface design: vision and touch. Foveal vision[1] in particular continues to be relied on heavily in display design, mostly because it affords a higher rate of information transfer than other sensory channels (Sorkin, 1987). Visual representations appear to be well-suited for conveying large amounts of complex detailed information, especially in the spatial domain. They also allow for permanent presentation, which affords delayed and prolonged attending. However, in many environments, overreliance on, and the inappropriate design of, foveal visual displays has resulted in data overload and related breakdowns in attention management and human–machine interaction.

One promising way to overcome these problems was suggested by an early version of Multiple Resource Theory (MRT; Wickens, 1984), which assumed that different modalities draw from separate pools of attentional resources and that therefore more information could be processed effectively if it was distributed across sensory channels. This assumption seems to be supported by experiences in many real-world domains where a combination of visual and auditory information presentation is

---

[1]In contrast to peripheral vision, foveal vision involves the perception of cues that are presented in the central 2° of visual angle.

being employed rather successfully. In some cases, and under certain circumstances, however, even the use of these two channels is no longer sufficient nor appropriate to handle the ever increasing amounts and complexity of available data. This has recently sparked considerable interest in haptic interfaces as an alternate or additional channel for conveying information (Sarter, 2002).

Haptic sensory information can take various forms, including proprioceptive, kinesthetic, and tactile cues. Tactile feedback — the focus of this chapter — is presented to the skin in the form of force, texture, vibration, and thermal sensations. One advantage of cues presented in the tactile modality is that they do not require a particular body or head orientation in order to be perceived. They are transient in nature yet rather difficult to miss, which makes them appropriate for indicating unexpected events (but not necessarily critical events that warrant warnings and alerts for which the more intrusive auditory channel tends to be reserved). Also, tactile cues are well-suited for providing spatial guidance.

First, we will present separate brief overviews of the anatomy and physiology of the visual and tactile systems, which focus primarily on the physical basis for the registration and early processing of sensory stimuli, rather than the phenomenal experience of perception. The chapter will also consider different approaches to perception, such as Gestalt theory, which are concerned with the question how we succeed in disambiguating retinal images and thus "why we see what we see." These perspectives reject the idea that physiology can ultimately explain perception. Instead, strong top-down influences are considered necessary to be able to "make sense" of the world around us. Perceptual phenomena and limitations that are associated with the two modalities and that are of particular interest to human factors professionals, will be described, and affordances associated with both modalities will be reviewed. The last part of the chapter is concerned with the benefits and limitations of combining vision, audition, and touch in multimodal interfaces. In this context, recently identified crossmodal constraints on attention will be discussed.

Because this chapter will provide an overview of both visual, tactile, and multimodal information processing, the level of detail in which each area can be covered is necessarily limited. The interested reader is referred to Soderquist (2002), Wade and Swanson (2001), Purves and Lotto (2003), and Gordon (1989) for more indepth accounts of the neurophysiological basis for, as well as empiricist accounts and theories of vision. More detailed descriptions of various aspects of touch can be found, for example, in Cholewiak and Collins (1991), Loomis and Lederman (1986), and Kruger (1996). Finally, multimodal information presentation and processing is examined in more detail in Stein and Meredith (1993), Spence and Driver (1997), Oviatt (2002), and Sarter (2002).

23.2   Vision

We will begin this account of vision by ignoring, for now, the process by which objects in our surroundings are selected for in-depth processing. Instead, we will assume that the process of visual perception simply starts with light passing through the cornea, a protective surface that surrounds the eyeball. The cornea covers the pigmented iris, which adjusts in diameter to control the amount of light entering the pupil, the round opening at the front of the eye. Next, the light passes through the lens, a transparent oval structure that is located directly behind the pupil and iris. Ciliary muscles adjust the shape of the lens to bring the object in focus on the retina at the back of the eye — a process called *accommodation*. When the eyes are at rest, the shape of the lens is relatively flat and allows us to view distant objects. To view nearer objects, the shape of the lens becomes more spherical and therefore has greater refractive power (e.g., Soderquist, 2002).

The process and limits of accomodation are of interest to human factors practitioners for various reasons. First, accomodation contributes to depth perception (as will be discussed later), which is important for a variety of real-world tasks. Also, accomodation is affected by aging, which often leads to reduced elasticity of the lens and thus a limited range over which objects can be brought into focus. The result can be myopia (nearsightedness where distant objects cannot be

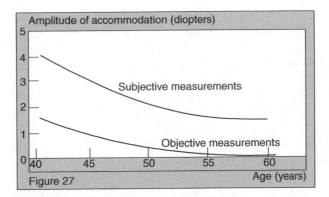

FIGURE 23.1 Average magnitude of available accommodation by age, measured subjectively and objectively. (Adapted from Hamasaki, D., Ong, J. and Marg, E., *Am. J. Optom. Physiol. Opt.*, 33, pp. 3–14, 1956.)

focused) or presbyopia (farsightedness where near stimuli cannot be brought in focus) (e.g., Soderquist, 2002).

For example, a 15-yr-old healthy user normally has approximately 10 diopters of accommodation (or is able to focus at 1/10 of a meter). However, that same person at age 40 may show less than one-third of that accommodation. Figure 23.1 illustrates how the ability to accommodate deteriorates even further for people over the age of 40 yr, both objectively and according to subjective ratings. This reduced accommodation ability can be compensated for, to a limited extent, by applying more force to the muscles that control accommodation. Still, it cannot be overcome completely and therefore needs to be considered in the design of interfaces that are intended to be used by a range of diverse users.

After passing through the lens, visual stimuli reach the retina at the back of the eye. The retina consists of layers of cells that contain two major types of photoreceptors: *rods* and *cones*. These two types of receptors contain different light-processing chemicals, called photopigments. *Rods*, which contain the photopigment rhodopsin, are much more numerous than cones. There are approximately 100 to 120 million rods in the retina, compared to only 6 to 7 million cones.

Rods and cones differ considerably with respect to the perceptual functions they support. Rods are highly sensitive to light but they are not sensitive to color nor do they support high-acuity vision. Thus, rods are particularly important for night vision and mostly ineffective during daylight because of saturation. Rods also support contrast sensitivity, a perceptual phenomenon that is of critical importance to human factors professionals. Contrast sensitivity has been defined as "the reciprocal of minimum contrast between lighter and darker spatial areas that can just be detected" (Wickens et al., 1998). Contrast sensitivity thus is a prerequisite for detecting and recognizing shapes. Several factors influence contrast sensitivity, including the illumination of an object where lower illumination reduces sensitivity, as illustrated by our difficulties with reading under low lighting conditions or detecting objects at night. Contrast sensitivity is also reduced when the object of interest is moving and with increasing age due to factors such as cataracts (increased clouding of the lens).

When stimulated, rods rapidly lose their sensitivity to light and require a long time to regain it. For example, when entering a dark room, we cannot distinguish any objects around us at first. Over several minutes, the visual system adapts to the ambient light and objects become increasingly visible. This *dark adaptation* can take as long as 20 to 30 min if a transition from photopic (cone-based vision) to scotopic (rod-based vision) conditions is required. Note that during the first 7 min in darkness, the cones require less light to perceive a visual stimulus. After that time period, the cones are more sensitive. This point is referred to as the rod-cone break. The reverse process, *light adaptation* from darkness to bright light, occurs significantly faster than dark adaptation. It requires only about 2 to 3 min. Dark and light adaptation need to be considered in the design of workspaces that involve rapid changes in illumination (see Purves and Lotto, 2003).

One way to overcome difficulties with dark adaptation through design is being suggested by the fact that, while rods are highly sensitive to light, their sensitivity to color is very limited. In particular, they are insensitive to long wavelengths (which, as we will discuss later, lead to the perception of a red hue). We can turn this apparent limitation into an advantage by illuminating objects in red light and thus minimize the stimulation of rods. By doing so, the need for dark adaptation is minimized, which can be very useful, for example, when briefing pilots before a night mission.

Rods do not only exhibit high sensitivity to light; they are also highly effective for perceiving orientation and motion (e.g., Leibowitz, 1988). One often cited illustration of how to utilize this affordance of photopic vision is the so-called Malcolm Horizon that helps pilots notice changes in an airplane's roll and pitch without having to look directly at a foveal visual display (Stokes et al., 1990). The Malcolm Horizon is a line of red laser light that is projected onto the instrument panel to the left and right of the traditional attitude indicator. The Malcolm Horizon moves along with the artificial horizon relative to the earth's surface as the plane moves through space. Unlike the attitude indicator, the Malcolm Horizon utilizes the peripheral visual system, which is represented almost exclusively by rods and, which is highly effective for processing motion and orientation cues efficiently and effortlessly.

In contrast to rods, which are highly sensitive to light and highly effective for perceiving orientation and motion, the 4 to 6 million cones, which are concentrated in the fovea (a 0.3 mm diameter area in the retina that covers approximately 2° of visual angle), support high acuity and *color vision.* The human eye can distinguish three properties of color: hue, saturation, and brightness (Figure 23.2).

If we think of visual perception as the transformation of light, that is, waves of electromagnetic energy into electrochemical neural energy, then the amplitude of the electromagnetic energy translates into the perceived brightness of a stimulus. Brightness can be further subdivided into illuminance, that is, the amount of light falling on an object, and luminance, that is, the amount of light that is reflected from a surface. Illuminance is measured in lumens per square meter (lm/m^2) or lux whereas luminance is expressed in candela per meter squared (cd/m^2).

Saturation — the second distinguishable attribute of color — refers to the extent to which a stimulus is considered "diluted" or "pure," that is, the extent to which a stimulus represents a mixture of various wavelengths. Finally, the wavelength of light determines the perceived hue. To most humans, wavelengths from 400 nm (perceived as blue-violet) to 700 nm (perceived as red) are visible. Color perception is accomplished by three types of cones that

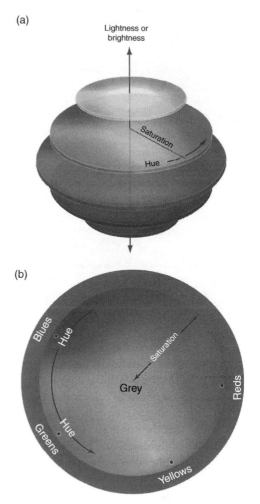

FIGURE 23.2 The human color space with its dimensions of hue, saturation, and brightness. (a) and (b) are different views of the overall organization of subjective color experience ellicited by spectral distribution of light. (Adapted from Purves, D. and Lotto, R.B., *Why We See What we Do: An Empirical Theory of Vision,* Sunderland, MA, Sinauer Associates Inc., 2003.)

differ in terms of the type of light absorbing pigment they contain. These cones are "tuned" to different portions of the visible spectrum (e.g., Soderquist, 2002; Purves and Lotto, 2003):

- Cones that absorb best at the relatively long wavelengths peaking at 575 nm (leading to the perception of red)
- Cones with a peak absorption at 535 nm (leading to the perception of green)
- Cones with a peak absorption at 445 nm (leading to the perception of blue)

A number of predictable contrast and fatigue effects are associated with color vision. These effects can be demonstrated, for example, by fixating a red square that is placed on a green surrounding. After a few moments, we begin to see a greenish tinge surrounding the red. Also, an intense green light induces a reddish afterimage. The same effect can be observed also for a blue square on a yellow background, or, more generally, for any complementary hues, that is, hues that, when mixed, form neutral grays.

While the fovea contains almost exclusively cones, the periphery of the retina is inhabited by both rods and cones, with the number of cones declining rapidly with increasing eccentricity (Figure 23.3). As a result, color discrimination is degraded at eccentricities beyond 20 to 30°. For example, the relative brightness and perceived hues change, causing red and green to appear yellow. Also, colors that are perceived in the periphery tend to be less saturated. This gradual reduction in the ability to discriminate colors in the periphery can be compensated for, to some extent, by increasing the size, luminance, or saturation of the stimulus. However, complete color blindness starts at around 40 to 50° of visual angle.

Another factor that can limit color vision is color blindness or color deficiency. Complete color blindness in people is extremely rare. Only about 0.005% of the population is truly color blind, that is, they completely lack at least one of the photopigments used to transmit color information (Cornsweet, 1970). However, for about 8% of males and 0.5% of females, some color distinctions are absent or at least not as pronounced. In these cases, all photopigments are present but their responses are slightly altered. People with color deficiencies most often have trouble distinguishing between red and green, a form of color deficiency that is referred to as "protanopia". They discriminate between the two hues based on perceived brightness instead. For example, red colors appear darker to a person with a deficiency of "red" photopigment cones (Murch, 1984). The need to account for users with color blindness or deficiency can be considered one of the reasons for the general design recommendation to create monochrome displays first and add color only at a later stage (Shneiderman, 1998).

As mentioned earlier, cones are not only responsible for color vision; they also support high acuity vision. Their ability to resolve detail is much greater than for rods which, in turn, display greater

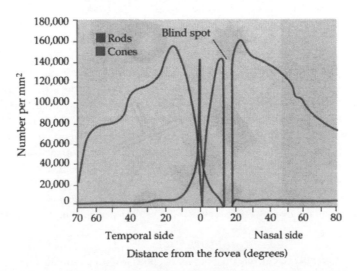

FIGURE 23.3 Distribution of rods and cones in the retina. (Adapted from Rosenzweig, M.R., Leiman, A.L., and Breedlove, S.M., *Biological Psychology*, Sunderland, MA, Sinauer Associates Inc., 1996.)

sensitivity to light. To understand this trade-off, we need to examine the structure of the retina in more detail. As shown in Figure 23.4, the retina consists not only of rods and cones but also includes amacrine, bipolar, horizontal, and ganglion cells.

A detailed treatment of the functions of these cells is beyond the scope of this chapter. Basically, both rods and cones connect to bipolar cells which, in turn, transfer impulses to a second set of neurons, called ganglion cells. Horizontal cells synapse with both bipolar cells and receptor cells as well as with other horizontal cells. And amacrine cells synapse with both bipolar cells and ganglion cells as well as with other amacrine cells. The horizontal and amacrine cells, thus, help transfer information laterally among different elements of the retina. In contrast, bipolar and ganglion cells play a role in determining the sensitivity and the acuity of the photopic (cone-based) and scotopic (rod-based) visual systems.

To understand this phenomenon, remember that there are approximately 100 to 120 million rods and 4 to 6 million cones in the retina but only approximately 1 million ganglion cells. This implies that a considerable amount of convergence and compression of information must occur as information is passed from the receptor cells via the bipolar cells to the ganglion cells. Convergence, that is, the summation of input from several receptors to a ganglion cell, supports increased sensitivity because a weak stimulus can activate several receptors to a limited extent and, once their input is combined and sent off to the ganglion cell, it may exceed the threshold for neural activity, that is, for an action potential to occur. As many as 1000 rods may pass information via their bipolar cells to a single ganglion cell, thus exhibiting a high degree of convergence.

In contrast, cones show very little convergence. There is typically a 1:1 relationship and ratio between cones in the fovea and the corresponding ganglion cells. For cones outside the fovea, the ratio is somewhat larger but never reaches that of rods. This explains the high acuity of cones, which have smaller receptive fields than rods. A receptive field can be defined as "a circumscribed area on the retina that provides the input to a ganglion cell" (Soderquist, 2002).

Table 23.1 summarizes the main differences between rods and cones that have been discussed so far.

So far, we have focused on monocular vision, that is, the structures and perceptual processes associated with the individual eye. To understand other important visual functions, such as depth perception, we need to consider binocular vision and its affordances. For example, binocular vision grants us a larger field of view, it reduces the risk of becoming disabled following damage to one eye, and it supports stereoscopic vision and thus depth perception.

Depth perception is important for a variety of tasks (e.g., flying an aircraft, driving a car), where a person needs to be able to judge distance from and between objects in the environment. It is supported by three main classes of depth cues: (a) oculomotor cues, (b) visual binocular cues, and (c) visual monocular cues. Oculomotor cues include accomodation (discussed earlier), which provides depth information by informing higher-level brain regions about the extent to which the ciliary muscles had to change the lens shape in order to bring the object of interest in focus. This information indirectly indicates the distance of an object from the observer. Convergence, that is, the amount to which inward rotation of the two eyeballs is necessary to bring an image to rest on corresponding areas of the retina of both eyes, is an example of a binocular depth cue. If an object is 6 m or more away from the observer, the line of sight is parallel. If the object moves closer, the eyeballs begin to turn inward progressively. Binocular disparity, that is, the disparity between the views obtained by each eyeball, is another example of such cues and also provides information on distance. These three mechanisms relate, for the most part, to depth perception for objects that are close to the observer (within a few meters).

Depth perception for more distant objects requires so-called "pictorial" cues, which are based on past experience and thus represent a top-down influence on perception. They include, but are not limited to:

- Linear perspective, that is, convergence of parallel lines toward a more distant point (Figure 23.5)
- Relative size, that is, if two objects that are known to be of the same size occupy different visual angles, then the one occupying a smaller angle is perceived to be farther away

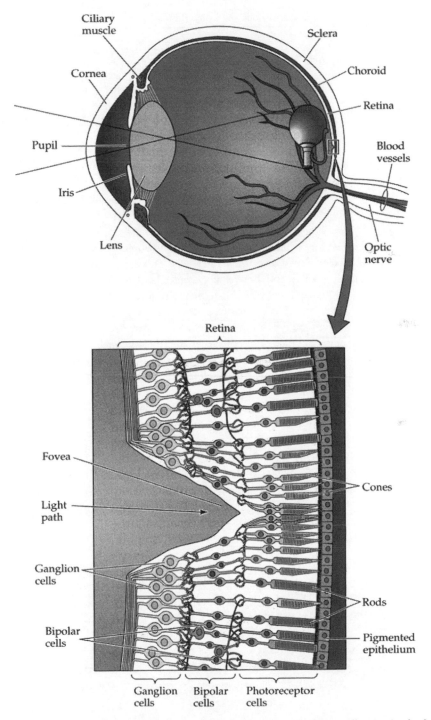

FIGURE 23.4 Basic structures of the human eye and configuration of various cell types in the human retina. (Adapted from Rosenzweig, M.R., Leiman, A.L., and Breedlove, S.M., *Biological Psychology*, Sunderland, MA, Sinauer Associates Inc., 1996.)

TABLE 23.1   Comparison of Rods and Cones Properties and Functions

Rods	Cones
High sensitivity to light	Lower absolute sensitivity
Low acuity	High acuity
Achromatic: one type of pigment	Chromatic: three types of pigments
More numerous (100 to 120 million)	Less numerous (6 to 7 million)
Highly convergent retinal pathways	Less convergent retinal pathways

- Parallax or relative motion, that is, more distant objects show smaller movement across visual field as the observer moves
- Occlusion, that is, if one object is in front of another (with respect to the viewer), it "over-writes" the other object in the image (Figure 23.6)

It is important to note that depth perception is affected by illumination and texture. For example, objects showing less texture tend to be perceived as being further away. This explained the difficulties that pilots experience when they are flying at night and/or over an extended body of snow or calm water. The lack of surface texture creates the risk that pilots overestimate their altitude.

Visual perception beyond the retinal level is carried by the axons of the ganglion cells, which form the optic nerve. The optic nerve leaves the eye through the retina in the area of the optic disk (better known as the "blind spot" because of the absence of receptor cells, and thus the inability to perceive stimuli, in this area). Part of each branch of the optic nerve crosses over in an area in front of the pituitary gland at the optic chiasm and then reaches the lateral geniculate nucleus (LGN), a walnut-sized nuclear complex in the thalamus.

The LGN comprises several layers, including two magnocellular layers and four parvocellular layers. The magnocellular layers receive input from larger retinal ganglion cells. They are characterized by high sensitivity, low spatial resolution, high temporal resolution, and little or no color selectivity. These cells and layers are concerned primarily with the perception of motion. The parvocellular layers are innervated by smaller ganglion cells and exhibit low sensitivity, high spatial resolution, low temporal resolution, and color selectivity. Thus, their role is to transmit spatially detailed visual information such as form and color sensations.

FIGURE 23.5   Linear perspective.

FIGURE 23.6   Occlusion.

The neurons in both the magno- and parvocellular layers are innervated by axons descending from the cortex, which support the top-down modulation of perception through factors such as expectations and the perceived importance of a signal. Also, the retinal ganglion cells project to other brain regions, such as brainstem cells that control pupil diameter as a function of light intensity, the superior colliculus (mediates the organization of eye movements to keep objects in focus), and the hypothalamus (involved in organizing circadian rhythm based on normal cycles of light and dark).

From the LGN, projections lead to the primary visual cortex where objects in the left visual field are represented in the right hemisphere and vice versa. At this level, only simple visual sensations are available, which maintain a topographical representation of the pattern of retinal activity. This representation is abandoned once the signals are sent to higher-order visual processing areas in the occipital, parietal, and temporal lobes of the brain (Figure 23.7).

One major difference between neurons in the primary and higher-order visual areas is the size of their receptive fields. Receptive fields of neurons in the primary cortex serve foveal vision and cover less than 1° of visual angle. In contrast, receptive fields of neurons serving peripheral vision are a few degrees of visual angle across. The overall visual field extends 180° horizontally and 130° vertically.

But why do we see what we see?

"Whilst part of what we perceive comes through our senses from the object before us, another part (and it may be the larger part) always comes out of our head."

(William James, 1890)

The previous brief overview of the morphology and physiology of the visual system and of selected perceptual phenomena does not answer two very important questions about visual perception: (a) "What will the user look at?" and (b) "What will things look like to the user?". The first question is concerned with the selection of visual objects for further processing, which, as mentioned earlier, is affected both by top-down influences (such as operator expectations) and bottom-up factors (such as the salience of objects). It has been suggested that an initial organization of the visual field that supports figure-ground perception occurs at a preattentive level. In other words, the entire visual field is processed automatically with the goal to detect basic features of objects such as colors, contrast, or size (Treisman, 1986) and determine which objects should undergo further processing (Broadbent, 1958; Neisser, 1976).

Interface design can capitalize on this bottom-up process for the purpose of attention guidance through the presentation of highly salient or conspicuous objects, that is, objects that are large, bright, colorful, or flashing (Wickens, 1984) and thus are likely to capture attention.

It is important to note that these tendencies are easily overridden by top-down influences on the selection of visual objects for in-depth processing. For example, people tend to start scanning in

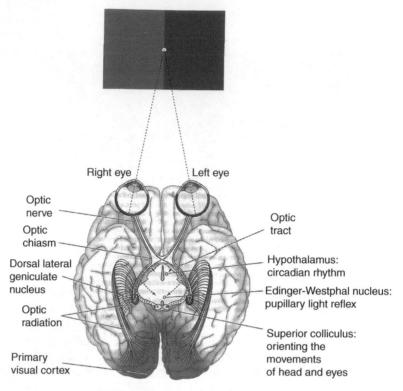

FIGURE 23.7 The pathways of visual information throughout the visual system. (Adapted from Purves, D. and Lotto, R.B., *Why We See What We Do: An Empirical Theory of Vision*, Sunderland, MA, Sinauer Associates Inc., 2003. With permission.)

the upper left corner of a display and focus on the center regions of the interface while avoiding its edges (Wickens, 1984). Another top-down influence on visual selection is expectations and schemata. The interplay between top-down and bottom-up influences on the selection of visual stimuli is captured by Neisser's (1976) perceptual cycle (Figure 23.8).

One starting point for perception in this cycle is internal schemata (mental models or expectations) that direct sensory exploration which, in turn, samples the environment for relevant information. The result of this exploration is either confirmation or modification of the existing schemata, which leads to the beginning of a new cycle, possibly involving redirection of attention. Alternatively, as mentioned earlier, highly salient stimuli in the environment can serve as the starting point for a perceptual cycle where the person's attention is captured externally. The signal that attracted attention may lead to changes of expectations or schemata and result in a subsequent search for additional related information.

The second important question that remains unanswered by any account of the neurophysiological basis of vision is "Why do things look as they do?" One critical question in this context is how the visual system deals with the inherent ambiguity of retinal stimuli. Answering this question was one of the driving forces behind the Gestaltist movement, which proposed that "the whole is greater than the sum of its parts." In other words, Gestalt theorists insisted that what matters is not isolated stimuli but entire patterns and configurations. Gestaltists also emphasize that perception is not passive. Instead, humans impose structure on observed visual stimuli and scenes. Without this ability to organize and interpret sensations, we would feel surrounded by a meaningless mishmash of colors and shapes.

One of the main questions that Gestaltists (including Wertheimer, 1880–1943, Koehler; 1887–1964; and Koffka, 1886–1941) studied was how we organize our percepts to be able to distinguish between figures and the background against which they are seen. Some of the general organizational tendencies

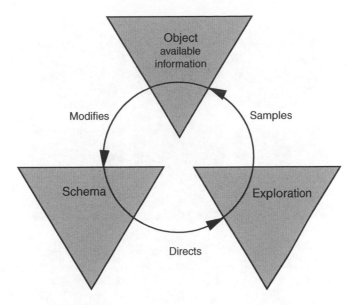

FIGURE 23.8   Neisser's perceptual cycle.

that have been observed by Gestaltists in this context are: (a) that symmetric patterns tend to be seen as a figure, (b) a region that is completely surrounded by another is perceived as a figure, (c) the smaller of two regions tends to be seen as a figure, and (d) vertically or horizontally organized region tends to be seen as figure. A number of Gestalt laws were formulated, including the laws of proximity, similarity, closure (gaps in figures are filled in), and common fate (elements with common motion or orientation are perceived as belonging together).

The law of proximity predicts that elements that are closer together will be perceived as a coherent object. This tendency is illustrated in Figure 23.9. On the left (a), the gauges in each column are likely to be perceived as belonging together whereas on the right (b), the gauges in each row are spaced more closely and therefore seem to form groups. Consideration of the law of proximity is important in interface design to achieve an appropriate system image, that is, to create a physical structure of a system (in this case, the arrangement of gauges) that suggests the correct relationship between elements.

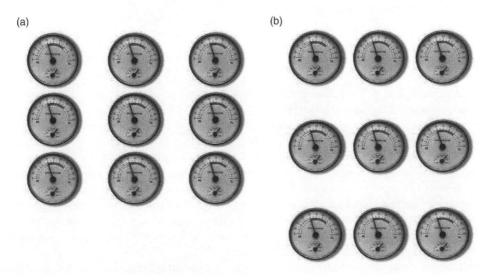

FIGURE 23.9   Illustration of the Gestalt Law of proximity.

FIGURE 23.10   Illustration of the Gestalt Law of similarity.

Another important Gestalt Law is the law of similarity, which states that elements that look similar will be perceived as part of the same form. In Figure 23.10, the gauges that are arranged in rows are likely to be perceived as belonging to one group because of their identical appearance.

The law of closure states that gaps in figures tend to be filled. Figure 23.11 illustrates this tendency, which leads to the perception of three filled circles and two triangles.

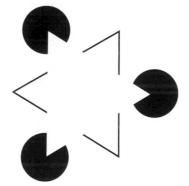

FIGURE 23.11   Illustration of the Gestalt Law of closure.

23.3 Touch

Compared to the visual (and the auditory) modality, our understanding of touch is still very limited. This may, in part, explain why touch is also still one of the least used modalities in current interface design. Other reasons likely include the relatively low bandwidth of this channel and the fact that the development of small unintrusive tactile interfaces has made this approach feasible in most real-world domains only in recent years (Zlotnik, 1988). Since then, interest in the tactile modality has increased significantly and has led to attempts to include this channel in the design of multimodal interfaces to support both input to, and output from, a system. In particular, recent developments in the field of virtual reality (where touch serves to create a greater sense of immersion) and the need to address problems with (visual) data overload in many domains have led to a growing interest in utilizing touch as yet another information channel.

Touch refers to the sensations that result from mechanical, thermal, chemical, or electrical stimulation of the skin (Cholewiak and Collins, 1991). It represents one component of the somatosensory system, which also includes proprioception and kinesthesis (sensory input from joint, muscles, and internal organs) and pain (tissue damaging high intensity stimuli). Touch is, for the most part, a proximal sense. In other words, we tend to feel stimuli that are in contact with, or at least in close proximity to, our body (Cholewiak and Collins, 1991). Touch is also our only bi-directional sense, that is, it supports both perception and acting on the environment.

We will start our overview of touch by examining its medium, the skin, which is a multi-layered sheet of 1.8 m² in area and approximately 4 kg in weight in an average adult. There are three types of skin: (a) glabrous skin, that is, hairless skin such as the skin of our palms, (b) hairy skin, and (c) mucocutaneous skin, that is, skin that borders the entrances to the body's interior (Greenspan and Bolanowski, 1996). The most active role in tactual perception is played by the glabrous skin, especially in the palmar and fingertip regions of the hand. The ridges and valleys of the skin in this area have been implied in the perception of texture and in the tactile identification of objects. Most studies on tactual perception have focused on these regions, and they are most often used to present tactile stimuli in current interfaces (e.g., CyberTouch, Tactools, and Touchmaster). In general, the skin is composed of the epidermis (its outer layer) and the dermis (the inner layer), both of which contain several types of receptors (see Figure 23.12).

In our overview of touch, we will focus on tactile sensations resulting from mechanical stimulation of the skin, which forms the basis of most current tactile displays. Mechanoreceptors can be divided into four major types (e.g., Burdea, 1996):

1. *Meissner corpuscles*: these receptors represent approximately 43% of all tactile receptors in the hand. They are found only in glabrous skin and are sensitive to stimuli such as velocity and skin curvature.
2. *Merkel's disks*: merkel's disks represent 25% of all mechanoreceptors in the hand and sense gentle localized pressure and vibration information.
3. *Pacinian corpuscles*: these receptors are located deeper in both hairy and glabrous skin. Approximately 13% of all mechanoreceptors are Pacinian corpuscles. They sense rapid variations of deformation, acceleration, and vibration.
4. *Ruffini corpuscles*: 19% of all mechanoreceptors in the hand are Ruffini corpuscles. They are located deep under the skin and are sensitive to vibrations, stretching of the skin, and thermal changes.

The distribution of these receptors varies considerably across different body regions. For example, there is a total of approximately 17,000 mechanoreceptors in the human hand (Johansson and Vallbo, 1983), which exceeds by far the number of these receptors in other body regions. Also, certain types of receptors are not represented in some body regions. For example, there are no Pacinian corpuscles in the skin of the cheek (Cholewiak and Collins, 1991).

The four types of mechanoreceptors can be classified according to the following two criteria (Kontaniris and Howe, 1995; Johansson and Vallbo, 1983):

1. The receptor's active area: small well-defined receptive fields (Type I units) and larger receptive fields with obscure borders (Type II).
2. The receptor's response to static stimuli: approximately 45% of all mechanoreceptors respond to static stimuli with a sustained discharge and are called slowly adapting (SA). The remaining receptors respond to the onset and offset of stimuli with bursts of impulses and are called fast or rapidly adapting (FA or RA).

Table 23.2 summarizes important characteristics of the four receptor types.

For the purpose of this chapter, it is not necessary to examine in detail the process of transformation of mechanical stimuli into neural events. This process is still poorly understood and, more importantly, it is of limited relevance for human factors practitioner. Ultimately, all tactile information is relayed to the somatosensory cortex, which is laid out in the form of a homunculus representing the opposite side of the body. In this representation, areas of greater sensitivity occupy larger cortex areas. This suggests

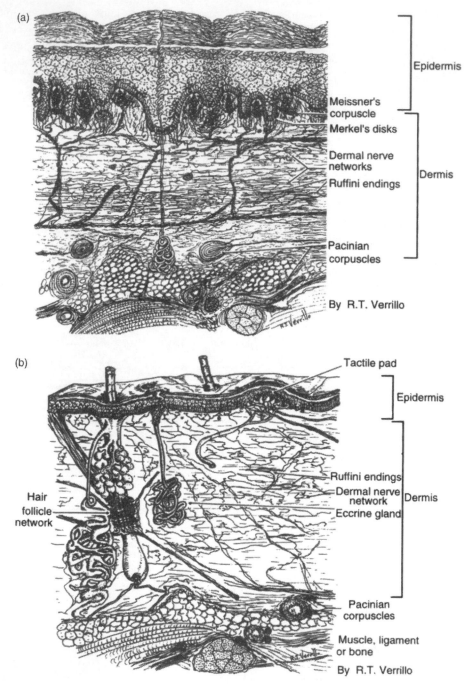

FIGURE 23.12   Cross-section of the human glabrous (a) and hairy (b) skin showing various haptic receptors. (Adapted from Kruger, L. eds., *Handbook of Perception and Cognition: Pain and Touch*, 2nd edn., Academic Press, New York, 1996. With permission.)

the most promising areas for applying tactile feedback such as the fingertips or lips. However, since it is often not feasible to use these regions for tactile displays in real-world environments where tactile devices could interfere with the performance of communication or manipulation tasks, designers need to make trade-off decisions.

TABLE 23.2 Functional Features of Cutaneous Mechanoreceptors

Characteristic	Meissner Corpuscles	Pacinian Corpuscles	Merkel's Disks	Ruffini Endings
Receptor type	FAI	FAII	SAI	SAII
Rate of adaptation	Rapid	Rapid	Slow	Slow
Location	Shallow; superficial dermis	Deep; dermis and subcutaneous	Shallow; superficial dermis	Deep; dermis and subcutaneous
Mean size of receptive field	13 mm^2	101 mm^2	11 mm^2	59 mm^2
Spatial resolution	Poor	Very poor	Good	Fair
Percent of all mechanoreceptors	43%	13%	25%	19%
Response frequency range	10 to 200 Hz	70 to 1000 Hz	0.4 to 100 Hz	0.4 to 100 Hz
Spatial summation	Yes	No	No	Unknown
Temporal summation	Yes	No	No	Yes

Source: Adopted from Shimoga, K. A survey of perceptual feedback issues in dextrous telemanipulation: Part II. Finger touch feedback. Proceeding of IEEE Virtual Reality Annual International Symposium, pp. 271–279. New York: IEEE, 1993. With permission.

The absolute threshold for perceiving tactile stimuli varies not only between different locations on a person's body but also depends on a variety of other factors (Cholewiak and Collins, 1991), including:

1. Frequency of the stimulus: for example, sensitivity for tactile stimuli reaches a maximum at 200 to 300 Hz for Pacinian corpuscles, which respond to vibratory stimulation.
2. Temperature: temperature decreases have been shown to affect the sensitivity of Pacinian corpuscles but not other skin receptors. For example, the thresholds for 100, 256, and 900 Hz vibrations were shown to vary with skin temperature, with a maximum sensitivity at 37°C (Weitz, 1941). However, for vibrations in the 30 Hz range, the effect of skin temperature was negligible (Green, 1977). When measured on hairless skin, cooling impairs sensitivity to high-frequency vibration. Warming the skin improves, to a limited extent, the sensitivity of all receptor types and a wide range of frequencies (Green, 1977).
3. Area of contact: the larger the contact area of a tactile stimulus is, the lower is the threshold for perceiving the stimulus due to spatial summation. However, this rule does not seem to apply at frequencies below 40 Hz.
4. Duration of the stimulus: temporal summation of tactile stimuli has been observed for upto 300 to 500 m sec. If the stimulus is applied for an even longer time, then adaptation may occur.
5. Age of the observer: sensitivity for vibrotactile signals deteriorates throughout aging (Verrillo, 1980). However, some studies suggest that this effect does not occur with respect to the fingertips (Stuart et al., 1999).

Variations in hormone levels can also affect sensitivity for tactile cues. In general, females appear to be slightly more sensitive to tactile stimuli although the overall pattern of sensitivity across body locations is similar to that of males (Weinstein, 1968).

In terms of the temporal resolution for tactile stimuli, it has been shown that people can distinguish between two tactile stimuli that are presented at least 5 m sec apart. Note that this represents a much higher temporal resolution than for vision where stimuli have to be presented 25 m sec apart to be perceived as two separate signals. At the other extreme, two sounds can be distinguished even if they are separated by as little as 0.01 m sec. The sequence in which signals were presented to the skin can be determined reliably for stimuli that are presented 20 or more meter seconds apart.

If a tactile stimulus is presented for a long time, adaptation, that is, an increase in threshold or a reduction in the perceived intensity of the stimulus, can occur. The time course of adaptation to tactile stimuli is not well known. In general, it appears that the adaptation period increases with increasing stimulus intensity, decreases with increasing stimulus area, and varies inversely with the sensitivity of different body regions (Greenspan and Bolanowski, 1996). Recovery from adaptation occurs quickly once the particular stimulus is removed. Note that adaptation to one type of tactile stimulus does not lead to increased thresholds for other types of tactile stimuli.

This independence of tactile channels is also supported by the observation that masking only occurs for stimuli that are presented in the same channel. Masking refers to an increased threshold or reduced perceived intensity of a stimulus as a result of a second stimulus being presented in close spatial or temporal proximity to another stimulus. For example, a 250-Hz vibratory stimulus will not mask a 20-Hz stimulus, except under conditions of very high stimulus intensity (Gescheider et al., 1983). The potential for masking creates a problem for designers of tactile interfaces who would like to achieve as high information transfer rates as possible through fast patterns and small spatial separation — the same thing that is likely to result in masking. This is an important consideration in the design of multicontactor dynamic devices that may be used to convey texture or motion.

Overall, the spatial resolution for tactile stimuli is determined by the size of the receptive field of the respective receptor. For example, the receptive fields of mechanoreceptors vary from 1 to 2 mm² upto 45 cm². The larger the receptive field of a receptor is, the lower is its spatial resolution. For example, on average, the skin on the thigh cannot discriminate two points that are closer than 67 mm whereas the palm can distinguish between two points that are as little as 11 mm apart. As indicated earlier, the fingertip region has the highest resolution with a two-point discrimination of 2.5 mm (Shimoga, 1993). The fingertips thus play a similar role in touch as the fovea of the retina does in visual perception (Krueger, 1982).

One important distinction that has been made with respect to tactile perception is between active and passive touch (Gibson, 1962). Passive touch refers to situations where a tactile stimulus is applied to the passive body. In contrast, active touch involves an individual moving their hands to manipulate and/or explore an object. Gibson emphasized that, similar to vision, some important phenomena are experienced only when a person actively explores an object by touch, thus adding kinesthetic information. According to Gibson, active touch enables a person to integrate over time and space the invariances in the stimulation that characterize an object; in that sense, "active touch is an exploratory rather than a merely receptive sense" (Gibson, 1962). This is of importance, for example, when trying to identify a relatively large object. It appears, however, that for other purposes, such as the perception of texture, active touch is not necessarily more useful than passive touch (see Kruger, 1996).

Clearly, there are connections between vision and touch, the two senses that we have discussed so far. Tactile sensing tends to be combined and aided by vision and vice versa. Vision, touch, and proprioception operate together to explore the environment. In some cases, we use touch to substitute for vision if it has become unavailable or lost completely. There is considerable empirical evidence for interactions between these two (and other) senses at the neuronal level. For example, spatially coincident stimuli in the two modalities tend to produce response enhancement. In contrast, disparate stimuli often lead to response depression (Stein and Meredith, 1993). The following sections will discuss in some detail multimodal information processing and presentation, both in terms of their benefits, limitations, and possible implications for interface design.

23.4 Combining Vision, Touch, and Other Modalities: The Benefits and Limitations of Multimodal Information Presentation

So far, we have considered vision and touch as separate information channels. However, in real-world perception, and in modern interface design, these senses tend to be combined with one another and with auditory information. One of the driving forces behind the design of multimodal interfaces is the perception that, in many environments, a state of visual, and even auditory, overload has been reached. An early version of multiple resource theory (Wickens, 1984) suggested one possible solution to this problem. It assumed that different sensory channels draw from separate pools of attentional resources, and that therefore the distribution of information across sensory channels should reduce resource competition, lead to increased bandwidth, and improve overall performance on multiple concurrent tasks. In addition to an increase in bandwidth, the distribution of information across sensory channels can support a number of other important functions.

First, sensory channels can be combined to achieve richer representations through synergy, that is, the merging of information that refers to various aspects of the same event or process but is sent via different modalities. Multiple modalities can also be used to support redundancy, that is, the processing of the exact same information in various channels. Another reason for presenting information via more than one modality is to support substitution in environments where a particular modality may become permanently or temporarily unavailable. And finally, attempts have been made to match modalities to particular functions or types of information for which they appear particularly well-suited. The following section will describe some of the affordances of vision, audition, and touch that can inform such modality assignments. It is important to realize, however, that fixed assignments cannot be made in the design of a multimodal interface where context determines, to a considerable extent, the feasibility and effectiveness of using a particular channel.

23.4.1 Modality Affordances

Visual information presentation is heavily emphasized in current interface designs. In particular, interface designers rely on foveal vision (as opposed to peripheral vision), which seems appropriate for the presentation of complex graphics and, in general, for conveying large amounts of detailed information, especially in the spatial domain. Visual information displays also allow for permanent presentation, which affords delayed and prolonged attending.

Current interfaces make use of peripheral vision to a much lesser extent. While the two channels — foveal and peripheral vision — do not, in the strict sense, represent separate modalities, they are associated with different options and constraints. Peripheral vision is well-suited for detecting motion, luminance changes, and the appearance of new objects. However, in contrast to foveal vision, it does not support the recognition of objects or details. Peripheral vision represents an early orientation mechanism (McConkie, 1983) that can be utilized by designers to help operators attend to a relevant location or critical information at the right time. One potential problem with peripheral visual feedback is that the visual field changes dynamically in response to contextual factors. With increasing foveal taskloading, for example, visual attention begins to focus on information in the center of a display at the expense of information presented in peripheral vision — a phenomenon called "attentional narrowing."

The auditory channel differs from vision along several dimensions. First, it is omnidirectional, thus allowing for information to be picked up from any direction and, to some extent, in parallel with information presented via other channels. Secondly, auditory information presentation is transient. This potential limitation is compensated for by a longer short-term storage of auditory (as opposed to visual) information so that it can be processed with some delay. Finally, since it is impossible for us to "close our ears," auditory displays tend to be intrusive and are therefore reserved for alerting functions.

The auditory channel shares a number of characteristics with haptic sensory information, which is currently underutilized in interface design. Most importantly, cues presented via these two modalities are transient in nature. Also, like vision and hearing, touch allows for the concurrent presentation and extraction of several dimensions such as frequency and amplitude in the case of vibrotactile cues. Touch differs from vision and hearing in that it is capable of both sensing and acting on the environment. In general, touch can serve a variety of purposes, including: (a) grasping and manipulating tools, (b) object identification, (c) exploring the spatial layout of objects, (d) assessing texture, temperature, weight, and other attributes of objects, (e) sensing vibrations, and (f) exploring spaces that are not accessible to vision (Lederman and Browse, 1988).

23.4.2 Multimodal Interface Design to Date

Research in the area of multimodal interfaces has expanded rapidly during the past decade and has led to the emergence of two groups of multimodal interfaces. The first group includes systems that support two or more combined user *input* modes such as speech, pen, touch, manual gestures, gaze, and head and body movements (e.g., Benoit and Le Goff, 1998; Cohen et al., 1997; Pentland, 1996; Stork and Hennecke, 1995; Turk and Robertson, 2000; Vo and Wood, 1996; Wang, 1995; Zhai et al., 1999). These systems have been developed to support functions such as increased system accessibility for diverse users, improved performance of recognition-based systems, and increased expressive power (see Oviatt and Cohen, 2000). Applications include map-based navigation systems, medical systems for mobile use in noisy environments, person identification systems for security purposes, and web-based transaction systems (for an overview see Oviatt, 2002).

The second group of interfaces presents users with multimodal system *output* to enhance awareness of their overall workspace and surroundings (so-called ambient displays — for example, MacIntyre et al., 2001) and to support time-sharing and attention management in the context of human–human and human–machine interactions (e.g., Ho et al., 2001; Nikolic and Sarter, 2001; Sklar and Sarter, 1999). Multimodal output systems have been designed primarily for virtual-reality applications and for use in a variety of high-risk data-rich event-driven domains such as future car cockpits (e.g., Means et al.,

1993) and modern flight decks (e.g., Brickman et al., 2000; Latorella, 1999; Sklar and Sarter, 1999). Given the increasing information demands on operators in these environments, the trend towards multimodal output systems is likely to continue.

As mentioned earlier, the design of multimodal interfaces tends to be based (albeit implicitly) on the assumption of an early version of MRT that modalities represent separate attentional resources and that the distribution of information therefore enhances our ability to share time. More recently, Wickens and colleagues have added some qualifications to the modality-specific aspects of MRT. First, they emphasized that MRT was intended to apply to the performance of continuous tasks only, and that information distribution across sensory channels may not matter in the context of discrete tasks where a person can switch rather than share time (Wickens, 1991). Also, Wickens and Liu (1988) pointed out that benefits that have been observed for crossmodal time sharing may not result from using different pools of attentional resources but may instead be related to peripheral factors such as visual scanning costs. More recently, additional behavioral and neurophysiological evidence has emerged that suggests the existence of crossmodal constraints on performance.

23.4.3 Crossmodal Constraints on Multimodal Information Processing

Given that operators in real-world environments are presented with numerous simultaneous and sequential cues in various modalities, it is critical to examine how the various senses interact. The following sections discuss crossmodal phenomena, such as crossmodal interference, modality expectations, the modality switching effect (MSE), and crossmodal spatial links and present behavioral and neurophysiological evidence for these phenomena.

23.4.3.1 Modality Expectations and the MSE

Modality expectations represent an attentional control setting (Bundesen, 1990; Folk et al., 1992; Pratt and Hommel, 2003; Wu and Remington, 2003) that leads to preparatory attentional modulation before stimulus onset (Driver and Frith, 2000). Modality expectations are formed based on the observed frequency or the perceived importance of a cue in a particular modality. Early laboratory studies of multimodal information processing failed to consider this top-down influence on attention allocation because they employed "neutral" stimuli and assumed a neutral attentional state of their subjects. However, more recent research has confirmed that expecting a cue to appear in a certain modality increases the detection rate and reduces the response time to that stimulus (e.g., Hohnsbein et al., 1991; Posner, 1978; Post and Chapman, 1991; Spence and Driver, 1997). These findings were obtained independent of whether a study employed a highly dynamic trial-by-trial cuing method (where attention is directed to a particular modality at the beginning of each trial) or a static blocked-cuing method (where attention is directed to one particular modality across a block of trials). For example, Spence and Driver (1997) and Spence (2002) have shown that subjects respond more rapidly to targets in the visual, auditory, and tactile modality when their attention is directed to the corresponding modality in advance than when it is divided between modalities or directed to another modality. Posner (1978) found that tactile and visual discrimination tasks were performed faster, while being comparably accurate, when the target appeared in an expected modality. In contrast, response times increase to cues in an unexpected modality and with modality uncertainty. The effects of modality uncertainty appear to be somewhat less pronounced for tactile than for visual and auditory cues (Boulter, 1977).

Findings from neuroimaging and neurophysiological studies confirm the influence of modality expectations on crossmodal information processing. For example, Eimer (2001) has shown that, when vision is designated the modality of primary importance, modality-specific ERP[2] components are enhanced by a visual stimulus. Similarly, Kawashima et al. (1995) have shown that selective direction of attention to a particular modality leads to enhancement of neural activity within the modality-specific cortex areas and

---

[2]ERP = Evoked Response Potential = electroencephalographic signals reflecting the operations of neuronal systems when transmitting and processing responses to sensory stimuli.

to a decrease of activity related to processing of stimuli in unattended modalities. Collectively, findings from these and similar studies suggest that modality expectations lead to a top-down facilitation of perceptual attentional processes, that is, to an enhanced readiness to detect and discriminate information in specific sensory channels (e.g., Posner and Petersen, 1990).

It is important to note that, in the above laboratory studies, the absolute changes in subjects' response times due to modality expectations were very small. For example, Spence and Driver (1997) found average costs and benefits ranging from a 10 m sec decrease to a 76 m sec increase in response time. However, when viewed as a percentage of the average response times, as much as a 2% decrease and a 15% increase in reaction times to stimuli were found, depending on whether or not the stimuli appeared in the expected modality. If these effects scale upto more complex environments, they are likely to result in operationally significant performance costs and benefits and therefore need to be considered in the design of future multimodal output systems.

23.4.3.2 MSE

MSE (e.g., Ferstl et al., 1994; Zubin, 1975) is another example of a crossmodal constraint on attention. It describes the strong tendency of people to "respond more slowly to a target in one modality if the preceding target was presented in a different modality than if the preceding target was presented in the same modality" (Spence and Driver, 1997). This implies that, in general, responses will be slower to targets in less frequent modalities. It appears to be particularly difficult (time-consuming) to shift attention to the visual or auditory channel away from rare events that are presented in the tactile modality (Spence et al., 2000).

MSE likely plays a role in real-world domains where signals tend to be assigned to modalities based on appropriateness (rather than experimental control) considerations. For example, auditory signals are often reserved for warning and alerts; given the rare occurrence of these events, the frequency of auditory cues is likely lower than that of visual indications, which are considered appropriate for presenting a wide range of diverse and detailed information.

23.4.3.3　Crossmodal Spatial and Temporal Links

Deliberate shifts of attention to a particular location in either the visual or auditory modality have been shown to lead to concurrent shifts in the other modality. Even though spatial attention in the secondary modality appears to be somewhat less focused than in the primary modality, this spatial linkage between modalities still results in performance benefits such as faster response times to cues in the same location (e.g., Driver and Grossenbacher, 1996; LaBerge, 1995; Quinlan and Bailey, 1995). Physiological data confirm that the processing of visual and auditory stimuli at attended spatial locations is facilitated at an early sensory level. In contrast, spatially disparate stimuli produce either response depression or no interaction (Calvert et al., 1999; Kennett et al., 2001; Macaluso et al., 2002; McDonald et al., 2000).

While visual–auditory spatial links have been studied extensively, much less is known about the interaction between vision or hearing and the sense of touch. Findings from a small number of studies suggest that auditory–visual interactions may be qualitatively different from visual-haptic links. For example, Gray et al. (2002) have shown that, while visual and auditory stimuli have to be presented in close spatial proximity to produce performance benefits, proximal haptic cues can be used to reorient visual attention to areas in distal space (e.g., to elements of a visual display in front of the person). In general, tactile cues appear to be less affected by crossmodal spatial links and thus can be decoupled from other attentional processes more easily and effectively (Eimer, 1999).

Temporal proximity between cues also affects crossmodal information processing. Approximate temporal synchrony of cues in different modalities is a requirement for multisensory integration (e.g., the ventriloquism[3] effect) and synergy. In other circumstances, however, an unintended close temporal proximity of cues can lead to a reduced ability to process the second unrelated cue. For example,

---

[3]The production of voice in such a way that the sound seems to come from a source other than the vocal organs of the speaker (often a puppet).

Soto-Faraco et al. (2002) and Arnell and Jolicoeur (1999) have found that, if a stimulus is presented in one modality, then the ability to process stimuli in a different modality is limited if the latter appear within 50 m sec of the first signal. This phenomenon has been referred to as the "crossmodal attentional blink." At the neurophysiological level, it has been confirmed that stimuli that occur in close temporal proximity affect one another, while those separated by long-time intervals are processed separately by neurons in the superior colliculus (Stein and Meredith, 1993).

23.4.3.4 Cross- and Intramodal Interference

There is also considerable empirical evidence that the concurrent use of modalities can lead to interference within and between those channels and thus to reduced detection and processing performance (e.g., Wickens, 1984; Wickens and Liu, 1988). For example, the phenomenon of visual dominance opposes the instinctive tendency of humans to switch attention to stimuli in the auditory and tactile modalities. However, most work on crossmodal attention to date has examined interference between two channels only (for the most part, between vision and hearing; for two of few exceptions, see Spence and Driver, 1997; Spence et al., 1998).

23.5 Concluding Remarks

This chapter has focused on vision and touch, the most and one of the least used sensory channels in current interface design, and on the benefits and potential limitations of combining sensory channels in the design of multimodal interfaces. Brief overviews of the structure and function of each modality in isolation included discussions of some perceptual phenomena that are of particular interest to the human factors community. While the above topics have been covered in even more detail in earlier accounts, the more important contribution of this chapter may be the review of benefits and limitations of multimodal information presentation and processing. The joint consideration and the combination of sensory channels in modern interfaces is critical for meeting the challenges created by increasingly complex and data-rich environments. And thinking of our various senses as a joint perceptual system, rather than treating them in isolation, is considered by some a prerequisite for truly understanding human perception and for using this understanding to inform design.

References

Arnell, K.M. and Jolicoeur, P., The attentional blink across stimulus modalities: evidence for a central processing limitation, *J. Exp. Psychol. Hum. Percept. Perform.*, 25, pp. 630–648, 1999.

Benoit, C. and Le Goff, B., Audio-visual speech synthesis from French text: eight years of models, designs and evaluation at the ICP, *Speech Commun.*, 26, pp. 117–129, 1998.

Boulter, L.R., Attention and reaction times to signals of uncertain modality, *J. Exp. Psychol. Hum. Percept. Perform.*, 3, pp. 379–388, 1977.

Broadbent, D.E., *Perception and Communication*, Pergamon, Oxford, England, 1958.

Brickman, B.J., Hettinger, L.J. and Haas, M.W., Multisensory interface design for complex task domains: replacing information overload with meaning in tactical crew stations, *Int. J. Aviat. Psychol.*, 10(3), pp. 273–290, 2000.

Bundesen, C., A theory of visual attention, *Psychol. Rev.*, 97(4), pp. 523–547, 1990.

Burdea, G.C., *Force and Touch Feedback for Virtual Reality*, John Wiley & Sons, New York, 1996.

Calvert, G.A., Brammer, M.J., Bullmore, E.T., Campbell, R., Iversen, S.D. and David, A.S., Response amplification in sensory-specific cortices during crossmodal binding, *Neuroreport*, 10, pp. 2619–2623, 1999.

Cholewiak, R. and Collins, A., Sensory and physiological basis of touch, in *The Psychology of Touch* Heller, M. and Schiff, W., eds., LEA, Mahwah, NJ, 1991, pp. 23–60.

Cohen, P.R., Johnston, M., McGee, S., Oviatt, S. and Pittman, J., *Quickset: Multimodal Interaction for Simulation Set-up and Control.* Proceedings of the 5th Applied Natural Language Processing Meeting, Washington, DC, 1997.

Cornsweet, T.N., *Visual Perception*, Academic Press, New York, 1970.

Driver, J. and Frith, C., Shifting baselines in attention research, *Nat. Rev. Neurosci.*, 1, pp. 147–148, 2000.

Driver, J. and Grossenbacher, P.G., Multimodal spatial constraints on tactile selective attention, in *Attention and Performance, Vol. 16, Information Integration in Perception and Communication*, Innui, T. and McClelland, J.L. eds., MIT Press, Cambridge, MA, 1996, pp. 209–235.

Eimer, M., Crossmodal links in spatial attention between vision, audition, and touch: evidence from event-related brain potentials, *Neuropsychologia*, 39, pp. 1292–1303, 2001.

Eimer, M., Can attention be directed to opposite locations in different modalities? An ERP study, *Clin. Neurophysiol.*, 110, pp. 1252–1259, 1999.

Ferstl, R., Hanewinkel, R. and Krag, P., Is the modality-shift effect specific for schizophrenic patients? *Schizophr. Bull.*, 2, pp. 367–373, 1994.

Folk, C.L., Remington, R.W. and Johnston, J.C., Involuntary covert orienting is contingent on attentional control settings, *J. Exp. Psychol. Hum. Percept. Perform.*, 18, pp. 1030–1044, 1992.

Gescheider, G.A., O'Malley, M.J. and Verrillo, R.T., Vibrotactile forward masking: Evidence for channel independence, *J. Acoust. Soc. Am.*, 74(2), pp. 474–485, 1983.

Gibson, J.J., Observations on active touch, *Psychol. Rev.*, 69, pp. 477–490, 1962.

Gordon, I.E., *Theories of Visual Perception*, John Wiley & Sons, New York, 1989.

Gray, R., Tan, H.Z. and Young, J.J., *Do Multimodal Signals Need to Come from the Same Place? Crossmodal Attentional Links between Proximal and Distal Surfaces*, Proceedings of the Fourth IEEE International Conference on Multimodal Interfaces (ICMI '02). IEEE, 2002.

Green, B.G., The effect of skin temperature on vibrotactile sensitivity, *Percept. Psychophys.*, 21(3), pp. 243–248, 1977.

Greenspan, J.D. and Bolanowski, S.J., The psychophysics of tactile perception and its peripheral physiological basis, in *Pain and Touch*, Kruger, L., ed., Academic Press, San Diego, 1996, pp. 25–104.

Hamasaki, D., Ong, J. and Marg. E., The amplitude of accommodation in presbyopia, *Am. J. Optom. Physiol. Opt.*, 33, pp. 3–14, 1956.

Hohnsbein, J., Falkenstein, M., Hoormann, J. and Blanke, L., Effects of crossmodal divided attention on late ERP components: I. Simple and choice reaction tasks, *Electroencephalogr. Clin. Neurophysiol.*, 78, pp. 438–446, 1991.

Ho, C-Y., Nikolic, M.I. and Sarter, N.B., *Supporting Timesharing and Interruption Management Through Multimodal Information Presentation*, Proceedings of the 45th Annual Meeting of the Human Factors and Ergonomics Society, Minneapolis, MN, October, 2001.

James, W., *Principles of Psychology* (Vol. 1), Holt, New York, 1890.

Johansson, R.S. and Vallbo, A.B., Tactile sensory coding in the glabrous skin of the human hand, *Trends Neurosci.*, 6(1), pp. 27–32, 1983.

Kawashima, R., O'Sullivan, B.T. and Roland, P.E., *Positron Emission Tomography Studies of Cross-Modality Inhibition in Selective Attentional Tasks: Closing of the "mind's eye"*, Proceedings of the National Academy of Sciences USA, 92, pp. 5969–5972, 1995.

Kennett, S., Eimer, M., Spence, C. and Driver, J., Tactile-visual links in exogenous spatial attention under different postures: convergent evidence from psychophysics and ERPs, *J. Cogn. Neurosci.*, 13(4), pp. 462–478, 2001.

Kontaniris, D.A. and Howe, R.D., Tactile display of vibratory information in teleoperation and virtual environments, *Presence*, 4(4), pp. 387–402, 1995.

Kruger, L. eds., *Handbook of Perception and Cognition: Pain and Touch*, 2nd edn. Academic Press, New York, 1996.

Krueger, L.E., Historical perspective, in Schiff, W. and Foulke, E. eds., *Tactual Perception: A Sourcebook*, pp. 1–54, Cambridge University Press, Cambridge, 1982.

LaBerge, D., *Attentional Processing: The Brain's Art of Mindfulness*, Harvard University Press, Cambridge, MA, 1995.

Latorella, K.A., Investigating *Interruptions: Implications for Flightdeck Performance*, NASA/Technical Memorandum (TM)-1999-209707, Virginia, Hampton, 1999.

Lederman, S.J. and Browse, R.A., The physiology and psychophysics of touch, in Dario, P. eds., *Sensors and Sensory Systems for Advanced Robots*, NATO ASI Series F43, Springer Verlag, Berlin, 1988.

Leibowitz, H., The human senses in flight, in Wiener, E. and Nagel, D. eds., *Human Factors in Aviation*, Academic Press, San Diego, CA, 1988.

Loomis, J.M. and Lederman, S.J., Tactual perception, in Boff, K.R., Kaufman, L., Thomas, J.P. eds., *Handbook of Perception and Human Performance: Cognitive Processes and Performance*, 2(31), pp. 1–41, Wiley & Sons, New York, 1986.

Macaluso, E., Frith, C.D. and Driver, J., Directing attention to locations and to sensory modalities: Multiple levels of selective processing revealed with PET. *Cereb. Cortex*, 12, pp. 357–368, 2002.

MacIntyre, B., Mynatt, E., Voida, S., Hansen, K., Tullio, J., Corso, G., *Support For Multitasking and Background Awareness Using Interactive Peripheral Displays*, Proceedings of ACM User Interface Software and Technology (UIST'01). Orlando, FL, November, 2001.

McConkie, G.W., Eye movements and perception during reading, in Raynor K. ed., *Eye Movements in Reading*, Academic Press, New York, 1983.

McDonald, J.J., Teder-Saelejaervi, W.A. and Hillyard, S.A., Involuntary orienting to sound improves visual perception. *Nature*, 407, pp. 906–908, 2000.

Means, L.G., Fleischman, R.N., Carpenter, J.T., Szczublewski, F.E., Dingus, T.A. and Krage, M.K., *Design of TravTek Auditory Interface*, In NRC-TRB: Driver performance: measurement and modeling, National Academy Press, Washington, D.C., 1993.

Murch, G.M., Physiological principles for the effective use of color, *IEEE Comput. Graph. Appl.*, November, pp. 49–54, 1984.

Neisser, U. *Cognition and Reality*, W.H. Freeman, San Francisco, 1976.

Nikolic, M.I. and Sarter, N.B., Peripheral visual feedback: A powerful means of supporting attention allocation and human-automation coordination in highly dynamic data-rich environments, *Hum. Factors*, 43(1), pp. 30–38, 2001.

Oviatt, S., Multimodal interfaces, in Jacko, J. and Sears, A. eds., *Handbook of Human-Computer Interaction*, LEA: Hillsdale, NJ, 2002.

Oviatt, S.L. and Cohen, P.R., Multimodal systems that process what comes naturally, *Commun. ACM*, 43(3), pp. 45–53, ACM Press, New York, March, 2000.

Pentland, A., Smart room, *Sci. Am.*, pp. 54–62, April, 1996.

Posner, M.I., *Chronometric Explorations of Mind*, LEA, Hillsdale, NJ, 1978.

Posner, M.I. and Petersen, S.E., The attention system of the human brain, *Ann. Rev. Neurosci.*, 13, pp. 25–42, 1990.

Post, L.J. and Chapman, C.E., The effects of cross-modal manipulations of attention on the detection of vibrotactile stimuli in humans, *Somatosen. Mot. Res.*, 8, pp. 149–157, 1991.

Pratt, J. and Hommel, B., Symbolic control of visual attention: The role of working memory and attentional control settings, *J. Exp. Psychol.: Hum. Percept. Perform.*, 29(5), pp. 835–845, 2003.

Purves, D. and Lotto, R.B., *Why We See What We Do: An Empirical Theory of Vision*, Sinauer Associates Inc., Sunderland, MA, 2003.

Quinlan, P.T. and Bailey, P.J., An examination of attentional control in the auditory modality – Further evidence for auditory orienting. *Percept. Psychophys.*, 57(5), pp. 614–628, 1995.

Sarter, N.B., Multimodal information presentation in support of human-automation communication and coordination, in Eduardo, S. eds., *Advances in Human Performance and Cognitive Engineering Research*, pp. 13–36, JAI Press, New York, 2002.

Shimoga, K., *A Survey of Perceptual Feedback Issues in Dextrous Telemanipulation: Part II. Finger Touch Feedback*, Proceeding of IEEE Virtual Reality Annual International Symposium, pp. 271–279, IEEE, New York, 1993.

Shneiderman, B., *Designing the User Interface: Strategies for Effective Human-Computer Interaction*, Addison-Wesley, Reading, MA, 1998.

Sklar, A.E. and Sarter, N.B., "Good vibrations": The use of tactile feedback in support of mode awareness on advanced technology aircraft. *Hum. Factors.*, 41(4), pp. 543–552, 1999.

Soderquist, D.R., *Sensory processes*, London, U.K.: Sage Publications, (2002).

Sorkin, R.D. Design of auditory and tactile displays, in Salvendy, G. eds., *Handbook of Human Factors* John Wiley & Sons, New York, pp. 549–576, 1987.

Soto-Faraco, S., Spence, C., Fairbank, K., Kingstone, A., Hillstrom, A.P. and Shapiro, K., A crossmodal attentional blink between vision and touch, *Psychon. Bull. Rev.*, 9(4), pp. 731–738, 2002.

Spence, C., Multisensory attention and tactile information processing. *Behav. Brain Res.*, 135, pp. 57–64, 2002.

Spence, C. and Driver, J., Crossmodal Links in Attention Between Audition, Vision, and Touch: Implications for Interface Design, *Int. J. Cogn. Ergon.*, 1(4), pp. 351–373, 1997.

Spence, C., Pavani, F. and Driver, J., Crossmodal links between vision and touch in covert endogenous spatial attention, *J. Exp. Psychol. Hum. Percept. Perform.*, 26, pp. 1298–1319, 2000.

Spence, C., Nicholls, M.E.R., Gillespie, N. and Driver, J., Cross-modal links in exogenous covert spatial orienting between touch, audition, and vision, *Percept. Psychophy.*, 60(4), pp. 544–557, 1998.

Stein, B.E. and Meredith, M.A., *The Merging of the Senses*, MIT Press, Cambridge, MA, 1993.

Stork, D.G. and Hennecke, M.E. eds., *Speechreading by Humans and Machines*, Springer Verlag, New York, 1995.

Stokes, A.F., Wickens, C.D. and Kite, K., *Display Technology: Human Factors Concepts*, Society of Automotive Engineers, Inc., Warrendale, PA, 1990.

Stuart, M., Shaw, J.A., Walsh, N., Nguyen, V. and Turman, A.B. 1999. Vibration detection thersholds at several body sites: influence of gender, preferred side and assessment method. Proc. Aust. Neuroscience Soc, 10, 173ff.

Treisman, A. Properties, parts, and objects, in Boff, K.R., Kaufman, L. and Thomas, J.P. eds., *Handbook of Perception and Human Performance*, John Wiley & Sons, New York, 1986.

Turk, M. and Robertson, G. eds., Perceptual user Interfaces, *Commun. ACM [Special Issue]*, 43(3), pp. 32–70, 2000.

Verrillo, R.T., Age related changes in the sensitivity to vibration, *J. Gerontol.*, 35, pp. 185–193, 1980.

Vo, M.T. and Wood, C., *Building an Application Framework for Speech and Pen Input Integration In Multimodal Learning Interfaces*, Proceedings of the International Conference on Acoustics, Speech, and Signal Processing., pp. 3545–3548, 1996.

Wade, N. and Swanson, M., *Visual Perception: An Introduction* 2nd edition, Psychology Press, 2001.

Wang, J., *Integration of Eye-gaze, Voice and Manual Response in Multimodal user Interface*, Proceedings of the IEEE International Conference on Systems, Man, and Cybernetics., pp. 3938–3942, 1995.

Weinstein, S., Intensive and extensive aspects of tactile sensitivity as a function of body part, sex, and laterality, in Kenshalo D.R. eds., *The Skin Senses*, Thomas Springfield, IL, pp. 195–222, 1968.

Weitz, J., Vibratory sensitivity as a function of skin temperature. *J. Exp. Psychol.*, 28, pp. 21–36, 1941.

Wickens, C.D., Processing resources in attention, in Parasuraman, R. and Davies, D.R. eds., *Varieties of Attention*, Academic Press, pp. 63–102, 1984.

Wickens, C.D., Processing resources and attention, in Damos D.L. eds., *Multiple Task Performance*, Taylor&Francis, London, pp. 3–34, 1991.

Wickens, C.D., Gordon, S.E. and Liu, Y., *An Introduction to Human Factors Engineering*, Addison Wesley Longman, New York, 1998.

Wickens, C.D. and Liu, Y., Codes and modalities in multiple resources: A success and qualification, *Hum. Factors*, 30(5), pp. 599–616, 1988.

Wu, S-C. and Remington, R., Characteristics of covert and overt visual orienting: Evidence from attentional and oculomotor capture, *J. Exp. Psychol.: Hum. Percept. Perform.*, 29(5), pp. 1050–1067, 2003.

Zhai, S., Morimoto, C. and Ihde, S., *Manual and Gaze input Cascaded (MAGIC) Pointing*, Proceedings of the Conference on Human Factors in Computing Systems (CHI'99), pp. 246–253, ACM Press, New York, 1999.

Zlotnik, M. A., *Applying Electro-Tactile Display Technology to Fighter Aircraft — Flying with Feeling Again*, In Proceedings of the IEEE 1988 National Aerospace and Electronics Conference NAECON 1988, pp. 191–197, IEEE Aerospace and Electronics Systems Society, New York, 1988.

Zubin, J., Problems of attention in schizophrenia, in Kietzman, M.L., Sutton, S. and Zubin, J. eds., *Experimental Approaches to Psychopathology*, pp. 139–166, Academic Press, New York, 1975.

24

Applying Cognitive Psychology to System Development

Philip J. Smith
R. Brian Stone
The Ohio State University

Amy L. Spencer
*Cognitive Systems Engineering,
    Inc.*

24.1   Introduction

Product designers need to wear many hats during the development process. These different perspectives range from that of the product designer who is trying to invent and integrate new concepts within a product, to the hardware/software engineer who is trying to create the physical realization of these design concepts, to the manufacturing engineer who is trying to manufacture the product, to the marketing specialist who is trying to sell the product (Bralla, 1998).

The successful designer needs to wear each of these hats in order to make sure that critical questions are asked during the design process, or make sure that a broader design team has been assembled that coordinates to deal with the implications and interactions of each of these perspectives regarding design decisions. As an example, it is not enough to design from a user's perspective. It is equally important to consider the viewpoint of the manufacturing engineer who must cope with the constraints imposed by the design when developing a manufacturing process that will be effective in terms of its impact on productivity, quality, and ergonomic/health costs.

The broad theme of this chapter is that a multidisciplinary perspective (Carroll, 2003) is needed to identify and answer all of the questions relevant to developing a cost-effective product that is useful, usable, and actually used. A failure to consider any one of these perspectives can increase costs, decrease usefulness or usability, or create barriers to the actual marketing and use of the product.

This broad theme will be illustrated by considering one of these perspectives in detail, that of the psychologist who must predict how alternative design concepts and features will influence the performances of users. This is not to say that every design team needs to include a psychologist. Rather, the point is that every design team needs to consider the psychology of the user (Norman, 2002), looking at design as a prediction task.

24.2   Defining the Design Problem: Initial User Studies

Early in the design process, there are a number of questions that need to be asked, including:

- What needs am I trying to serve? From a problem-driven design perspective, what are the short-comings of the existing products and methods that I am trying to improve upon? What benchmark tasks should my product support? See Witkin and Altshuld (1995) for a discussion of methods for conducting such needs assessments
- Who are the potential users I am trying to serve? What are the important defining dimensions that identify the different populations, such as:
  - expected frequency of use
  - knowledge of the task domain
  - familiarity with other products used for the same or similar tasks
  - relevant physical, perceptual, psychomotor, or cognitive abilities
  - the importance/value of the needs to be served by the product
  - the resources the user may be willing to expend to meet these needs (time, money, etc.)
  - personal preferences and beliefs that influence the likelihood of purchasing and using the product
  - individual differences that exist along these dimensions?
- For those needs or tasks that are met by some existing product, how is the task currently performed? See Rubenstein and Hersh (1984) and Preece (1994) for a discussion of how to complete a *descriptive* task analysis or cognitive task analysis.
- What are the broader physical, organizational, social, and legal contexts in which the benchmark tasks will be performed (Flach, 1998)?
- What different combinations of users, needs/tasks, and contexts actually exist? How can this representative set of use cases or scenarios be used to avoid cognitive narrowing (Smith and Geddes, 2003) when making design decisions, making sure that the design accommodates all of the use cases satisfactorily, rather than focusing on satisfying only a subset? See Carroll (1995) and Rosson and Carroll (2001) for a discussion of scenario-based design.
- What key constraints should be considered during the design process, including acceptable levels for
  - learning rates
  - error rates
  - productivity
  - development time
  - manufacturing costs
  - marketing constraints
  - development and operational costs?
- In terms of design for manufacturability (Helander and Nagamachi, 1992; Bralla, 1998) what constraints does the design place on the production process?
- As I consider design alternatives, what are the key discriminators that could differentiate me from my competitors, including possible differences in:
  - usefulness
  - usability
  - aesthetic appeal
  - cost
  - quality
  - durability
  - service/maintenance
  - marketing strategy (including market entry time)
  - product evolution plan?
- What are the alternative business plans for sustaining this product in the market? What implications does this have for the design?

Although answering these conceptual questions about the nature of the design problem is always an iterative process, using a top-down approach in which they are addressed early in the development process can help to ensure that the real needs of the potential users are being addressed in an effective fashion. Once the design problem is well understood, then alternative conceptual solutions can be generated and evaluated, ultimately leading to a specific design.

Clearly, in order to answer these questions during the initial user studies for a design, an understanding of cognitive, organizational, and social psychology is very useful, both in terms of the research methods developed by these fields and in terms of the models of human performance (Wickens et al., 2004) and group dynamics (Brehm et al., 1999) that they provide.

To illustrate applications of psychology to design in more detail, below we discuss a second component of the design process where psychology is involved, the evaluation of a specific product design proposal.

24.3 Consideration of the User During the Evaluation of a Proposed Design

As suggested earlier in this chapter, one of the hats that the designer needs to wear is that of a psychologist, attempting to predict how a given design will influence user performance. There are two general approaches to evaluating a proposed design from this perspective:

- Analytical evaluations
- Empirical evaluations

This chapter will focus on techniques for conducting analytical evaluations. It should be kept in mind, however, that empirical evaluations (Rubin, 1994) are an important complement to analytical studies, as they are likely to identify design concerns that are missed in the analytical evaluations.

24.3.1 Analytical Evaluations

Two of the most popular approaches to the analytical evaluation of a product's design in terms of its usefulness and usability are the use of *cognitive walkthroughs* and *heuristic analyses*. These two complementary approaches both provide structure to the evaluation process, on the assumption that such structured approaches are likely to result in more complete and accurate predictions than simple casual considerations of a design might yield.

24.3.1.1 Cognitive Walkthroughs

There are a number of variations on how to conduct a cognitive walkthough or to complete a *predictive cognitive task analysis* (Kirwan and Ainsworth, 1992; Gordon and Gill, 1997; Jonasson, et al., 1999; Vincente, 1999; Annett and Stanton, 2000; Klein, 2000; Schraagen et al., 2000; Shepherd, 2000; Hollnagel, 2003; Diaper and Stanton, 2004). The method that we will build upon in this chapter was developed by Lewis and Wharton (1997).

Step 1. Select a use case for evaluation. As with all cognitive walkthroughs, Lewis and Wharton begin by emphasizing the value of making context-specific predictions when evaluating a product. Such context can be provided by the use cases or scenarios developed during the initial stages of the design. Each such use case specifies three primary considerations:

- What are the characteristics of the user of a product that might affect its use? (the persona in the use case)
- What is the broader physical, organizational, social and legal context in which the product will be used?
- What is the high-level task or goal for which the product is being used?

Note that such use cases are solution independent. They essentially define the design problem of finding a design that supports performance for all of the use cases.

Step 2. Specify the normative (correct) paths for this use case (represented as a goal hierarchy), thus indicating the alternative sequences of steps that the user could take to *successfully* achieve the specified goal. Note that there could be more than one correct path for completing a given task.

Step 3. Identify the state of the product and the associated "world" at each node in the goal hierarchy. In the case of a software product, the state of the product would be the current appearance of the interface and any associated internal states (such as the queue of recently completed actions that would be used should an undo function be applied). The state of the "world" applies if the product or the user actually changes something in the world as part of an action, such as changing the temperature of a glass manufacturing system when using the interface to a process control system.

Step 4. Generate predictions. For each correct action (node in hierarchy):

- Predict all the relevant success stories.
- Predict all the relevant failure stories.
- Record the reasons and assumptions made in generating these stories.
- Identify potential fixes to avoid or assist in recovery from failure stories. (Keep in mind that these fixes could be local patches or bandaids or they could involve proposing a major change in the design concept.)

Note that, in developing a specific set of predictions, it may be more efficient to consider the use cases for the full range of users in parallel while stepping through the normative goal hierarchy. For instance, predictions might be generated for both a first-time user and a frequent user of the product while looking at each node in the goal hierarchy.

To make this discussion clearer, a sample cognitive walkthrough is provided next. For this illustration, we will consider the design of a specific online library search system.

Example of Step 1. Select a use case for evaluation. In completing this sample walkthrough, assume as the use case that we are dealing with a college student in the United States who is a first-time user who has used other online library search systems before (but is not an expert at library searches), and who also regularly searches for material on the Web. Abstractly, this means that the user:

- Is reasonably literate in the English language
- Is familiar with the use of a keyword entry box and the use of hot links on web pages
- Knows that menus are often shown along the top of the page as short phrases and that a given menu item can be selected by clicking on it
- Knows that library searches are often structured in terms of author, title, and subject searches
- Is not familiar with the layout and navigation of this specific library system

Assume the user is looking for wedding songs to be performed at her own wedding, that she is conducting the search at home on her own computer over a high-speed connection, and that help from a librarian is available only by phone. Assume further that she wants CDs to listen to (she only has a CD player), rather than sheet music, tapes, etc.

Example of Step 2. Specify the normative (correct) paths for this use case. Such a normative model is generally best represented as a goal/subgoal hierarchy (Preece, 1994). The high-level goal or task is represented as the top node in the hierarchy, and each level below a node represents the subgoals that will achieve the goal represented by that higher level node. Relationships among subgoals can be indicated by OR (completion of either subgoal alone is sufficient to achieve the higher level goal), AND (both subgoals must be completed to achieve the higher level goal, but in any order) or sequence (both subgoals must be completed to achieve the higher level goal, and they must be completed in a specified order) operators.

In this example, the initial screen that this student will access is shown in Figure 24.1 and Figure 24.2. For this user the top section of the normative model or goal hierarchy is shown in Figure 24.3. Note that:

- The highest node is the student's goal or task in this use case (finding wedding songs on CDs)
- The nodes below this highest node represent four different ways the student could successfully begin (and are therefore marked with OR)

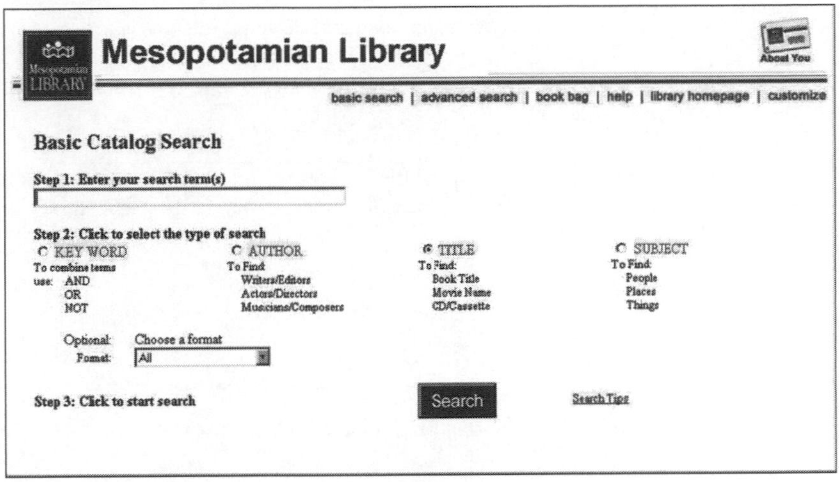

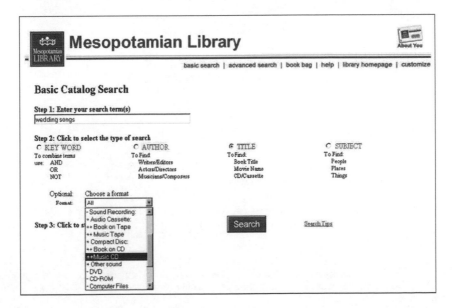

FIGURE 24.1 Initial screen seen when accessing this library search system.

    – Ask a librarian (over the phone)
    – Click on the help link to learn more about how to conduct a search
    – Click on the Search Tips link to learn more about how to conduct a search
    – Start running a search without accessing any form of assistance first
  • The subgoals to start the actual online search (Node A in Figure 24.3) are shown in Figure 24.4

Figure 24.4 shows the alternative ways to actually run this search. (For this illustration, we are restricting consideration to use of the Basic Catalog Search function.) In order to run the search the user must complete the subgoals "Enter term," "Select type", and "Choose format," but the order for completing them does not matter. The branches to these subgoals are therefore marked with AND. The user also must complete the subgoals "Enter term", "Select type," and "Choose format," before clicking on the Search button, so this is indicated with an arrow specifying the necessary sequence. Note that, in Figure 24.4, we chose not to include another layer of detail, indicating how a given step such as

FIGURE 24.2 Choices of format.

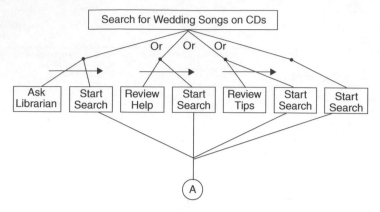

FIGURE 24.3   High-level nodes (goals) in the normative goal/subgoal hierarchy.

"Enter term," "Select type," or "Choose format" was accessed (using either the mouse or the tab key). We decided that such lower level details would not add any important insights into our predictions for this type of Web literate user. (If we were concerned with the time required to complete the task, we might want to include such details.)

Note also that we did not include all possible paths (such as "click on AUTHOR"), but only included those paths that could lead to a successful search. Since, when dealing with information retrieval, "success" is a relative concept that must be defined in terms of a tradeoff between recall and precision (Baeza-Yates and Ribeiro-Neto, 1999), we chose to define "success" as any path that leads to at least some useful retrievals. (For this example, a title search using the search term "Wedding songs" does not retrieve any CDs. Because such a search therefore represents a failure story, it is not represented in the normative/correct model.)

We could also go on to add additional nodes to this goal hierarchy to indicate iterations in the search process (e.g., capturing how the user might also try a Keyword search after trying a Subject search). However, for the purposes of this illustration, we will assume that, for each of the alternative paths, the user stops after following that one search path.

Note that, in terms of this general method of conducting a cognitive walkthrough, three points have been illustrated earlier:

- Branches to subgoals under a given node can be marked with AND, OR, or a sequence arrow to indicate the relationships among these subgoals
- To make the analysis process more efficient, only the normative paths (those paths that lead to a successful completion of the task) are shown in the goal hierarchy

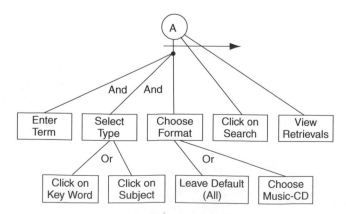

FIGURE 24.4   Subgoals showing alternative ways to successfully complete the desired search.

- Also for efficiency, decisions can be made about how deep to go in the hierarchy. In some cases, lower levels can be left out if they are judged to be unimportant in terms of predicting critical behaviors

Two final ways to increase efficiency in completing such a cognitive walkthough are to:

- Only represent those portions of the goal hierarchy that are judged to be critical by the analyst
- Limit the number of use cases to be considered based on some judgment of importance, for instance, considering only those use cases that are likely to occur frequently, that could result in some highly undesirable outcome, or that are relevant to some design decision about which the developers are uncertain (Mitta et al., 1995)

Example of Step 3. Identify the state of the product and the associated "world" at each node. Previously, we have illustrated the completion of Step 1 (Select a use case for evaluation) and Step 2 (Specify the normative paths for this use case). To complete the third step, we need to identify the screen displays associated with each node in the hierarchy represented in Figure 24.3 and Figure 24.4. (Note that the rest of the context — the associated "world" — is assumed to stay constant in this illustration, as the student is accessing the library system at home on her own computer over a high-speed connection, and that help from a librarian is available only by phone.)

Analysis of a Sample Path. Suppose the student enters "wedding songs" and changes the search type from the default (Title) to Subject, but uses the default format (All). Figure 24.5 shows the appearance of the Basic Catalog Search screen for this path, and Figure 24.6 shows the appearance of the search results displayed for this path.

To complete Step 3, we would normally identify the state of the software interface associated with *all* possible paths leading to success in the normative goal hierarchy. For the purposes of this example, however, we will limit our focus to this one sample path in which the information seeker has started by entering the term "wedding songs" in a subject search for all possible formats.

Example of Step 4. Generate predictions. The final step is to walk through the hierarchy along each path that would lead to success, and to play psychologist, generating predicted behaviors at each node (for

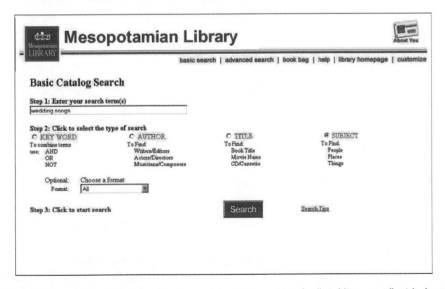

FIGURE 24.5 Basic catalog search screen for the sample path: searching for "Wedding songs" with the search type Subject and the default format (all).

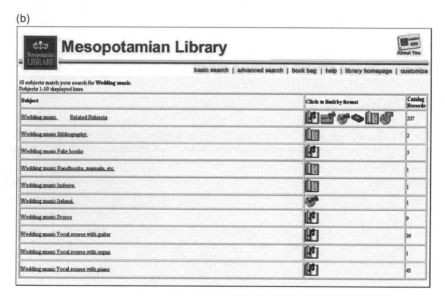

FIGURE 24.6 (a) Initial search results displayed for sample path: searching for "Wedding songs" with subject as the search type and all as the default format. (b) Next level of detail when the user clicks on the "Wedding music" link shown in (a).

each associated screen display). Lewis and Wharton (1997) suggest asking four questions to help guide this prediction process:

- "Will the user be trying to achieve the right effect?"
- "Will the user notice that the correct action is available?"
- "Will the user associate the correct action with the desired effect?"
- "If the correct action is performed, will the user see that progress is being made?"

More detailed questions can be identified by considering how different cognitive processes could influence performance (Card et al., 1983; Eysenck and Keane, 1990; Wickens and Hollands, 1999; Ashcraft, 2002; Wickens et al., 2004), such as:

- *Selective attention*: What are the determinants of attention? What is most salient in the display? Where will the user's focus of attention be drawn? (See Johnston and Dark, 1986; Pashler et al., 2001.)
- *Perception*: How will perceptual processes influence the user's interpretation? How, for instance, will the proximity of various items on the screen influence judgments of "relatedness" as predicted

by the Gestalt Law of Proximity? (See Tufte, 1983, 1990, 1997; Gibson, 1986; Bennett and Flach, 1992; Goldstein, 1996; Card et al., 1999; Vincente, 2002; Watzman, 2003.)

- *Memory*: How will the user's prior knowledge influence selective attention and interpretation? Does the knowledge necessary to perform tasks reside in the world or in the user's mind? (See Bransford and Johnson, 1972; Hutchins, 1995; Baddeley, 1998.)
- *Information Processing, mental models, and situation awareness*: What inferences/assumptions will the user make? (See Johnson et al., 1981; Gentner and Stevens, 1983; Plous, 1993; Endsley, 2003.)
- *Design-induced error*: How could the product design and the context of use influence performance and induce errors? (See Bainbridge, 1983; Roth et al., 1987; Reason, 1991, 1997; Parasuraman and Riley, 1997; Smith et al., 1997; Skirka et al., 1999; Sheridan, 2002; Smith and Geddes, 2003.)
- *Motor performance*: Can the controls be used efficiently and without error? (See Jagacinski and Flach, 2003.)
- *Group dynamics*: How will the system influence patterns of interaction among people? (See Horton and Lewis, 1991; Baecker et al., 1995; Olson and Olson, 2003.)

These predictions could be generated by members of the design team, independent experts in interaction design, and experts in the relevant domain (such as expert librarians). Because of their different backgrounds relevant to the design and the domain of interest, individuals from each group are likely to generate somewhat different predictions.

For the sample path discussed in the example for Step 3, some sample predictions include the following.

Success story. The user enters the term "Wedding songs." The user also recognizes that it is important to change the search type from the default (Title) to either Keyword or Subject, and selects Subject. The user either does not notice or does not choose to change the format, but by leaving it as the default (All), she completes a search with high recall but low precision in terms of the desired format (music CD). Because "wedding songs" happens to be a synonym that triggers search results for the controlled vocabulary term "Wedding music," the user completes a search that, in terms of her intended semantic concept, "Wedding songs," leads to 337 retrievals (Figure 24.6a) with high precision and high recall for that semantic category (plus one other relevant retrieval that is a CD under the controlled vocabulary term "Wedding songs, English Ireland").

Assuming she understands the need to click on the "SEE Wedding music" link (Figure 24.6a), the searcher will then access a page that provides access to all 337 items (Figure 24.6b). If she further recognizes that she can still limit the retrievals by format at this point by clicking on the CD icon in the column labeled "Click to limit by format," she will further improve the precision of her search by limiting the retrievals to music CDs. After viewing those retrievals, she could then back up and look at the relevant retrieval indexed under "Wedding songs, English Ireland."

Failure story 1. A very plausible failure story would be for the searcher to simply enter "Wedding songs" as the search term and hit return or click on the Search button (Figure 24.7a). Either action would result in a search with only 13 document records, none of which are music CDs (Figure 24.7b).

This failure is very plausible for two reasons. Based on their past experience with other websites, many experienced Web users are in the habit of just entering a keyword and hitting the Enter key or clicking on a "search" or "go" button in order to run a search. This prior knowledge, combined with the salience of the Search Term entry box, makes it quite likely that a number of users will make this mistake.

Note further that, because the default for this search system is a Title search, and because that search does generate retrievals, it is also quite possible that the user will not recognize that she could have run a different type of search that would have generated more retrievals. The page of retrievals for this search, shown in Figure 24.7b, does not provide any feedback on the type of search run. It merely states that "9 titles match your search for wedding songs." This is probably the worst kind of flaw in the design of an information retrieval system, as the user does not even realize that she should try a different search.

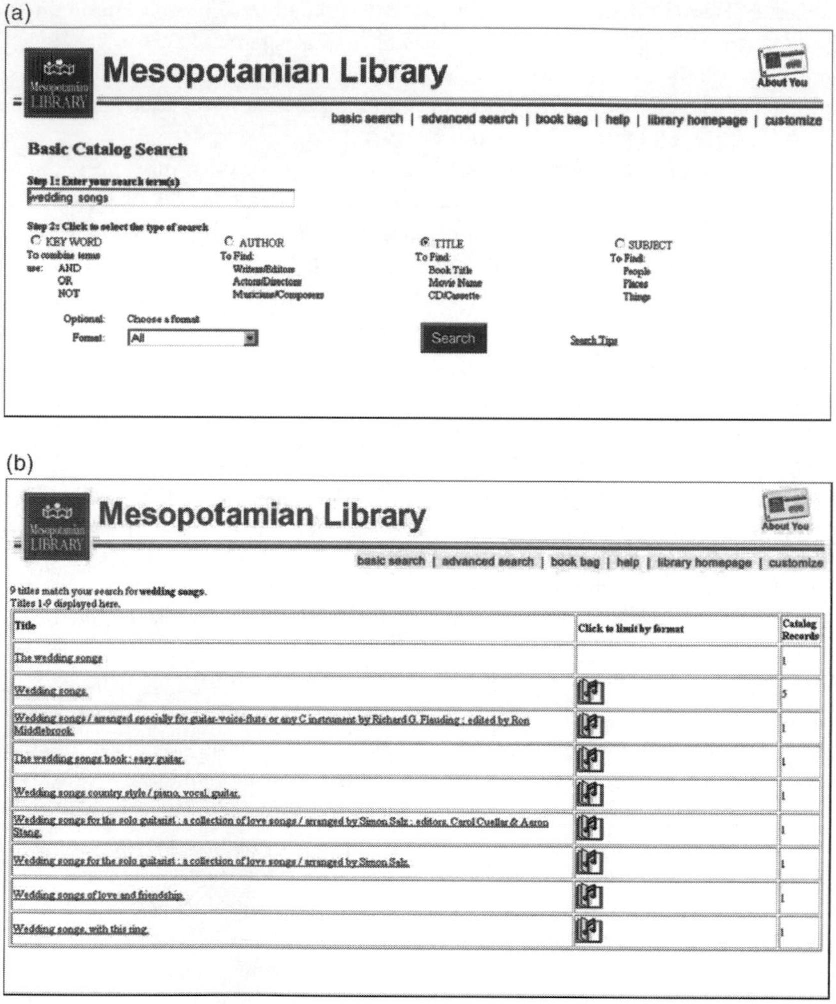

FIGURE 24.7   (a) Search definition display for failure story 1: searching for "Wedding songs" using the defaults for search type (title) and format (all). (b) Search results display for failure story 1: searching for "Wedding songs" using the defaults for search type (title) and format (all).

To avoid this problem, it would be better to pick a default for search type that is more encompassing (erring on the side of high recall at the expense of low precision). One way to accomplish this would be to use a Keyword search as the default. The designer would have to further decide whether to require the use of logical operators (AND; OR; NOT), and whether to allow phrases to be entered without separating words with logical operators.

It would probably be better to not require the use of logical operators, given this is not typical in most websites with keyword searches, and given people are known to be poor at correctly using such operators. Instead, the search engine could:

- Run the search using the broader query Wedding AND Songs.
- Sort the retrievals using some type of relevance ranking (e.g., ranking highest those retrievals indexed with the controlled vocabulary term "Wedding Music" — which has "Wedding songs" as a synonym, followed by those retrievals containing "Wedding songs" as a phrase somewhere in the document representation (but not indexed as "Wedding music"), followed by those

containing the words "Wedding" AND "Songs" but not containing the phrase "Wedding Songs" and not indexed as "Wedding Music")

- The search engine should also automatically deal with alternative spellings, cognates, etc. so that "Wedding songs" and "Wedding song" would generate the same set of retrievals

If this list is presented ungrouped, then it should probably be designed so the user can sort based on format in order to support more rapid scanning of the results of most interest in use cases such as this one.

Failure story 2. In the success story mentioned earlier, the searcher failed to use the format menu to limit the retrievals to Music CDs. This is very plausible, as it is last in the list of steps, is labeled optional, and has a label that may not be very suggestive of its function for some users. If the user also fails to recognize that icons in the column labeled "Click to limit by format" indicate the formats of the items (book vs. sheet music vs. music CD, etc.), then she might wind up having to look through the full set of 338 retrievals instead of only the subset which consists of music CDs.

Since there is a lot of unused space on the search window (Figure 24.6a), the different formats could be presented directly instead of hidden within a pull-down window. Alternatively, the ability to limit the retrievals to specific formats could be made more salient and understandable on the retrieval page (Figure 24.6b). A third alternative would to be automatically sort retrievals by format as the default display when the user clicks on "Wedding Music," showing them organized in a hierarchy such as:

Wedding Music

Books
Music CDs
Sheet Music
Music Tapes

Failure story 3. Most users accessing the display shown in Figure 24.6a would probably click on the link for "Wedding music." However, a few users might look at the column labeled "Catalog Records" and see the indication that there is 1 item for "Wedding songs, English Ireland" and blank items for "Wedding music." This could be misinterpreted as indicating that there are no items available on "Wedding music."

Failure story 4. In this example, the information seeker was shown entering a term that triggered the appropriate controlled vocabulary term. It is well known, however, that people are remarkably varied in the labels that they choose to express a concept of interest, and that they often do not understand the need to explore the use of alternative synonyms and related terms in order to conduct a thorough search.

To avoid such a failure, the thesaurus used for triggering the controlled vocabulary terms therefore needs to be carefully constructed. In addition, the search system could suggest a search strategy (entering synonyms, broader terms or related terms) when the user completes a search the doesn't generate any retrievals (Shute and Smith, 1992).

Example of a Cognitive Walkthrough — Summary. The sample analysis above serves to illustrate how a cognitive walkthrough can be conducted. There are several important considerations surrounding such an analysis:

- A person's performance is strongly influenced by his or her immediate goal context. Developing the normative goal hierarchies for the different use cases of interest, and identifying the product state (and the associated context) corresponding to each node in a goal hierarchy, helps to focus the *analyst's* attention on that immediate goal context when making predictions about success and failure stories
- The analyst is more likely to generate a thorough set of predictions by explicitly considering a list of questions like those suggested previously:
  - What is most salient in the display?
  - Where will the user's focus of attention be drawn?

- How will the user's prior knowledge influence selective attention and interpretation?
- Will the user be trying to achieve the right effect?
- To keep the time required for the analysis manageable
  - Carefully select a subset of the overall use cases for analysis, based on a judgment of which ones are most likely to produce insights of value
  - Develop the goal hierarchy for the normative paths rather than for all possible (correct and incorrect) paths
  - Make decisions regarding how deep to go in building the goal hierarchy based on judgments about how useful the inclusion of additional levels of detail will be
  - Build goal hierarchies for subtasks rather than covering a complete use case when there are only specific subtasks that have raised questions or uncertainties about the current design

In addition, it is not always necessary to formally draw the goal hierarchies. The goal is to identify sequences of steps involved in using the product for a given scenario, and to predict potential errors at each step. Unless the results need to be communicated formally to other people, it is often sufficient to just make rough sketches of the hierarchy in order to help the analyst think through the sequence of states for the product for each step (such as a sequence of screen displays for a software product). However, there are also times when it is useful to formally develop the hierarchies, either to help ensure that an analysis is sufficiently accurate and complete, or to help communicate the methods used and the findings of an analysis to other people.

24.3.1.2 Heuristic Analyses

Heuristic reviews provide a second approach to completing an analytical evaluation of some product. As a form of usability inspection, a heuristics review is a process in which experts evaluate the usefulness and usability a given product or product interface based on its consistency with certain established and agreed upon principles know as heuristics (Nielsen and Mack, 1994).

Heuristic reviews (Nielsen, 1994, 1997, 2000; Spool et al., 1998; Krug, 2000; Badre, 2002; Rosenfeld and Morville, 2002; Van Duyne et al., 2003; Silver, 2005) have become popular as an expeditious and inexpensive method to complement the results of empirical usability testing. Originally developed for computer applications, lists of heuristics have developed over the years to accommodate product interface and website development.

There are a number of variations on how to conduct a heuristic review. The "use case" method for heuristic reviews is generally considered to be most effective. If such a heuristic review based on use cases is structured such that it runs through the normative paths for each use case in a cognitive walkthrough, then the heuristic review essentially provides an additional set of questions to ask as each system/product state (corresponding to a node in the normative goal hierarchy) is reviewed during the cognitive walkthrough, as described earlier in this chapter.

Thus, the preparation for a heuristic review based on the "use case" method can be completed by following Steps 1–3 of a cognitive walkthrough as described earlier:

- *Step 1.* Select a use case for evaluation
- *Step 2.* Specify the normative (correct) paths for this use case (represented as a goal hierarchy)
- *Step 3.* Identify the state of the product and the associated "world" at each node in the goal hierarchy

Then the analyst continues with Step 4. Generate predictions. However, in addition to the broader questions outlined in our description of a cognitive walkthrough (How will the user's prior knowledge influence selective attention and interpretation? Will the user be trying to achieve the right effect? etc.), a specific set of additional questions or heuristics are addressed. These heuristics can deal with aesthetic concerns as well as weaknesses regarding usefulness or usability.

These questions or heuristics serve to further structure the analysis task, helping to make sure that key issues are not overlooked. In addition, a severity rating can be associated with each failure story. Severity ratings may be based on consideration of the probability that the defect will be encountered along with its expected impact.

Sample Heuristics. The following set of heuristics, although not exhaustive, is useful in its versatility of application:

Does the design support the users' task flows?

Look at the sequence of steps necessary to complete a given task with the new product and determine whether this sequence is efficient, effective, and clear, and whether it fits into the user's broader task flow.

Are metaphors appropriately used?

There are situations where an interface can be more easily understood by the use of a metaphor. By resembling a commonplace system such as the controls on a tape deck, an interface's functionality may be quickly ascertained (if the user is familiar with and recognizes the metaphor). Any metaphors that have been used should be consistently and clearly suggested through the product's interface. Areas where the metaphors break down should be carefully assessed to determine how to deal with such concerns.

What is the hierarchical focus of attention in the display? Does it support completion of the alternative goals that users may have when looking at that display?

Our eyes are drawn to animated areas of an interface or product display more readily than static areas. Determine whether the use of motion or a flashing cursor helps users notice important state changes, or whether it becomes a distraction to what is important. Size, font style, display format/layout, and color also have a big impact on the focus of attention.

Are the product's key functions and features salient?

Determine whether important features and functions are visibly presented. Many users prefer to see "up-front" the set of functions that are available to them. Important features and functions buried deep in an interface structure may go unrecognized. Users should be able to understand the functions afforded by the product interface through a quick visual scan of the interface.

Do the labels and icons for buttons, menu items, etc. clearly indicate their meanings, telling the user what action will be performed or what information will be accessed if that item is selected?

Whether a graphical or textual label, it is important that this label clearly indicate its meaning. In many cases, if a graphic is used, it helps to include a text label along with the graphic or icon.

Are external memory aids provided to help the user remember what actions to take or to remember what steps in some process have already been completed?

Knowledge and information can be stored in the head or in the world (Norman, 2002). While access to knowledge in the head can be very efficient, that assumes that the person has committed it to memory, and will be able to retrieve it at the right time. Often, by showing critical information within some display, this memory load can be reduced, making it easier for the person to remember how to perform various functions with the product.

When information is presented as text, has it been organized using subheadings as advance organizers, thus structuring the content to make it easy for the user to scan for specific information?

The use of subheadings and bulleted lists both make it easier for the user to scan text and to identify those portions that are relevant to his or her interests. Subheadings further serve to inform the user about the topic of the associated content, making it easier for the reader to quickly and accurately interpret that content as the details are read (Bransford and Johnson, 1972).

Are the relationships among associated controls and displays indicated through some form of functional grouping?

There are often relationships among information displays and controls based on specific tasks or subtasks that users need to perform. When tasks create such functional relationships, they should be supported by grouping or integrating the displays (Wickens et al., 2004), clearly indicating this relationship. This can be done by integrating related information into a single graphical display (Tufte, 1983, 1990, 1997), by placing the related displays and controls in close proximity to one another, or by indicating relationships among controls and display elements using techniques like color coding. It can also be done in a virtual space when all of the related information cannot be displayed at the same time, by providing links to all of the related information in the appropriate location so that the user has a reminder of its relevance and availability, and so that the information can be quickly accessed by pressing a button or clicking on a link (hot spot).

Does the product look and behave consistently?

Displays and controls should look and behave in a consistent and coherent manner. These behaviors should make sense to the user from one part of the product's interface to the next.

Does the product design take into consideration users' past experiences with related products or tasks, either to reduce resistance to change (Brehm et al., 1999), to reduce the time required to learn this new product, or to differentiate it from its competitors?

Just as consistency within a product is important, consistency with the users' past experiences can be very important, as users develop expectancies for certain design features. If there are actual or implicit standards for the design of a given type of product, or if users are likely to have had significant experience with similar products, the resultant expectancies that they have developed from these past experiences need to be considered.

Is the navigation robust enough to support the easy completion of alternative tasks, while still clear enough to help the user navigate along the correct paths without getting lost? Are landmarks provided to help the user remember where he or she is within the system (relative to the overall navigational structure), and to understand where he or she can go next to complete different tasks?

Most products are meant to support more than one goal or task. Thus, the interface needs to support navigation along a variety of different paths to support these different goals. This represents a fundamental design challenge, as the flexibility offered by multiple paths to support different user goals makes it more difficult to develop a design that ensures that the user will know which path to select and will be able to navigate along this path successfully.

Is the navigation flexible enough to accommodate novice users (for whom quickly learning to use the system for limited purposes may be most important), while also meeting the needs of frequent, experienced users, for whom shortcuts to ensure speed of use may be most important?

One of the challenges in designing a product is the accommodation of users with a variety of needs, and with different experience levels. If possible it is desirable to embed training in the interface to help novices learn shortcuts as they use the system. A good example of this is often provided in software by including a description of the shortcut (Control-P) along with the label for a specific function such as Print in the pull-down menu used to select the print function.

Does the system support different mechanisms for searching for relevant information in order to accommodate users with levels of knowledge about their topics and with different preferences regarding how to conduct a search?

Hierarchical menus are useful to support browsing when searching for information. However, in some cases, the information seeker may not recognize the correct path to follow through the menu and may therefore need to have a more direct keyword search function as an alternative to find the desired information. Keyword searches provide faster, more direct access to information if the information seeker knows the "correct" terminology. However, since different information seekers may enter a variety of different terms when searching for the same content, it is useful for the search engine to map synonyms to the same underlying information content.

Users may also have difficulty selecting the right level of detail to use when searching for a topic. They may enter a keyword or keyword phrase that is too narrow or too broad (Shute and Smith, 1992). Embedding a thesaurus function in the search engine can be very helpful for both of these cases.

While trained searchers often like the specificity enabled by search systems that allow the entry of logical operators (AND, OR, NOT, ADJACENT, etc.), many people have difficulty understanding the meanings of such operators. A well-designed relevance ranking algorithm is often an effective way to avoid requiring the use of logical operators when entering a keyword phrase, helping to direct the information seeker's attention to those retrievals that are likely to be most focused on the topic of interest by putting them at the top of the list of retrievals, while still providing access to the broader set of retrievals lower in the list in case the search engine's relevance judgments do not fully reflect the interests of that user.

Finally, structured keyword entry systems can be an effective compromise between browsing and a simple keyword search. If the information available in a system can be categorized along several dimensions, (such as the model, year, color, and cost of used cars in a database), it may be helpful to present the user with these categories so he or she can pick from the menus of items available for each category. In some cases, such menus can even be made context specific, so that as the user fills in one category, the remaining choices in the other categories are limited.

Are "cause and effect" actions represented in the interface?

Each input by the user or change in the behavior of the program should be accompanied by a corresponding change of its representation within the interface. The "current selection" (active field or object) should be indicated with some visual contrast. The current state of the product should be displayed in a consistent, clear, and unambiguous manner. If the product acts directly upon the "world" (such as with a process control system), the current state of the "world" as known to the product should also be displayed clearly.

Will users be able to detect and recover from errors?

It is inevitable that someone will hit the wrong button due to a slip or mistake. Potential errors need to be predicted, and measures to assure detection and recovery need to be identified if design changes cannot be made to prevent these errors. Minimally, it should be possible to undo or cancel actions. Actions that are irreversible like "saving over an existing file" should require a user to make an explicit commitment. These safety nets will give novice users a sense of comfort, as well as supporting more effective use of the product.

Is the interface esthetically pleasing?

Keeping in mind that beauty is in the mind of the beholder, the interface should have a sense of beauty, or at the very least not be perceived by the intended users as crude or awkward. Ask questions like:

- Are the colors harmonious?
- Is the type legible?
- Are the graphics consistent?

Also note whether users have the ability to customize the display of the interface, that is, change the colors or type font to suit their own esthetic judgments. Factors that influence the esthetics of a product also often have an effect on usability because of their impact on the salience of key information.

24.4 Conclusions

As suggested at the beginning of this chapter, one of the hats that the designer needs to wear is that of a psychologist, conducting initial user studies and needs assessments to define the design problem, and attempting to determine how a given design will influence user performance as part of its evaluation.

In terms of the evaluation of a specific design, both analytical and empirical evaluations are valuable. Both of these forms of evaluation call for expertise in cognitive and social psychology, as patterns of performance need to be predicted a priori for analytical evaluations, and need to be detected in empirical evaluations.

To improve analytical predictions, two general approaches have been developed:

- Cognitive walkthroughs
- Heuristic evaluations

Both types of reviews can be conducted using paper or functional prototypes, thus making it possible to complete them at different times during the design process. In addition, designers, experts in human-centered design and subject matter experts can all participate in such evaluations, each providing their particular insights into potential user behaviors.

The strength of these two complementary approaches is that they guide evaluators to focus on context-specific predictions of user behaviors, and they make explicit important questions to ask as part of the prediction process. By doing so, they provide structure to the evaluation process, helping to ensure that the users' underlying cognitive processes are considered and that important principles for effective design are followed. Thus, the theme of this chapter is that one important perspective is to view *design as a prediction task*, using an understanding of the psychology of the user to guide the identification of potential weaknesses in a design, so that they can be corrected during the design and implementation process.

References

Annett, J. and Stanton, N. (2000). *Task Analysis*. London: Taylor & Francis.

Ashcraft, M.H. (2002). *Cognition*, 3rd ed. Upper Saddle River, NJ: Prentice Hall.

Baddeley, A. (1998). *Human Memory: Theory and Practice*. Boston: Allyn and Bacon.

Badre, A. N. (2002). *Shaping Web Usability, Interaction Design in Context*. Boston, MA: Pearson Education, Inc.

Baecker, R., Grudin, J. Buxton, W., and Greenberg, S. (1995). *Readings in Groupware and Computer-Supported Cooperative Work: Assisting Human–Human Collaboration*. San Francisco, CA: Morgan Kaufmann.

Baeza-Yates, R. and Ribeiro-Neto, B. (eds.) (1999). *Modern Information Retrieval*. Harlow, UK: Addison-Wesley.

Bainbridge, L. (1983). Ironies of automation. *Automatica*, 19, 775–779.

Bennett, K.B. and Flach, J.M. (1992). Graphical displays: implications for divided attention, focused attention, and problem solving. *Human Factors*, 34(5), 513–533.

Bralla, J. (1998). *Design for Manufacturability Handbook*, 2nd ed. New York: McGraw Hill.

Bransford, J.D. and Johnson, M.K. (1972). Contextual prerequisites for understanding: Some investigations of comprehension and recall. *Journal of Verbal Learning and Verbal Behavior*, 11, 717–726.

Brehm, S., Kassin, S. M., and Fein, S. (1999). *Social Psychology*, (4th ed) Boston, MA: Houghton Mifflin Company.

Card, S., Moran, T., and Newell, A. (1983). *The Psychology of Human–Computer Interaction*. San Francisco, CA: Morgan Kaufmann.

Card, S., Mackinlay, J. and Shneiderman, B. (1999). *Readings in Information Visualization: Using Vision to Think*. San Francisco, CA: Morgan Kaufmann.

Carroll, J.M. (1995). *Scenario-Based Design: Envisioning Work and Technology in System Development*. New York: John Wiley & Sons.

Carroll, J.M. (2003). *HCI Models, Theories and Frameworks: Toward a Multidisciplinary Science*. San Francisco, CA: Morgan Kaufmann.

Diaper, D. and Stanton, N. (eds.) (2004). *The Handbook of Task Analysis for Human–Computer Interaction*. Mahwah, NJ: Lawrence Erlbaum.

Endsley, M. (2003). *Designing for Situation Awareness*. London: Taylor and Francis.

Eysenck, M. and Keane, M. (1990). *Cognitive Psychology: A Student's Handbook*. Hillsdale, NJ: Lawrence Erlbaum.

Flach, J.M. (1998). Cognitive systems engineering: putting things in context. *Ergonomics*, 41(2), 163–167.

Gentner, D. and Stevens, A. (1983). *Mental Models*. Hillsdale, NJ: Lawrence Erlbaum.

Gibson, J.J. (1986). *The Ecological Approach to Visual Perception*. Hillsdale, NJ: Lawrence Erlbaum.

Goldstein, B.E. (1996). *Sensation and Perception*. Pacific Grove, CA: Brooks/Cole.

Gordon, S.E. and Gill, R.T. (1997). Cognitive task analysis. In: C. Zsambok and G. Klein (eds.), *Naturalistic Decision Making*. Hillsdale, NJ: Lawrence Erlbaum, 131–140.

Helander, M. and Nagamachi, M. (1992). *Design for Manufacturability: A Systems Approach to Concurrent Engineering and Ergonomics*. London: Taylor and Francis.

Hollnagel, E. (2003). *Handbook of Cognitive Task Design*. Mahwah, NJ: Lawrence Erlbaum.

Horton, F. and Lewis, D. (eds.) (1991). *Great Information Disasters*. London: Association for Information Management.

Hutchins, E. (1995). *Cognition in the Wild*. Cambridge, MA: MIT Press.

Jagacinski, R.J. and Flach, J.M. (2003). *Control Theory for Humans: Quantitative Approaches to Modelling Performance*. Hillsdale, NJ: Lawrence Erlbaum.

Johnson, P., Duran, A., Hassebrock, F., Moller, J. Prietulla, M., Feltovich, P., and Swanson, D. (1981). Expertise and error in diagnostic reasoning. *Cognitive Science*, 5, 235–283.

Johnston, W.A. and Dark, V.J. (1986). Selective attention. *Annual Review of Psychology*, 37, 43–75.

Jonassen, D., Tessmer, M., and Hannum, W. (1999). *Task Analysis Methods for Instructional Design*. Mahwah, NJ: Lawrence Erlbaum.

Kirwan, B. and Ainsworth, L. (1992). *A Guide to Task Analysis*. London: Taylor and Francis.

Klein, G. (2000). Cognitive task analysis of teams. In: J.M. Schraagen, S. Chipman, and V. Shalin (eds.), *Cognitive Task Analysis*. Mahwah, NJ: Lawrence Erlbaum, 417–430.

Krug, S. (2000). *Don't Make Me Think — A Common Sense Approach to Web Usability*. Indianapolis, IN: New Riders.

Lewis, C. and Wharton, C. (1997). Cognitive walkthroughs. In: M. Helander, T. Landauer, and P. Prabhu (eds.), *Handbook of Human–Computer Interaction*, 2nd edn. Amsterdam: Elsevier, 717–731.

Mayhew, D. (1999). *The Usability Engineering Lifestyle: A Practitioner's Handbook for User Interface Design*. San Francisco, CA: Morgan Kaufmann.

Mirel, B. (2003). *Interaction Design for Complex Problem Solving: Developing Useful and Usable Software*. San Francisco, CA: Morgan Kaufmann.

Mitta, D., Delk, C., and Lively, W. (1995). Selecting system functionalities for interface evaluation. *Human Factors*, 37(4), 817–834.

Nielsen, J. (1994). *Usability Engineering*. San Francisco, CA: Morgan Kaufmann.

Nielsen, J. (1997). Usability testing. In: G. Salvendy (ed.), *Handbook of Human Factors and Ergonomics*. New York: John Wiley & Sons, 1543–1567.

Neilsen, J. (2000). *Designing Web Usability: The Practice of Simplicity*. Indianapolis, IN: New Riders.

Nielsen, J. and Mack, R. (1994). *Usability Inspection Methods*. New York: John Wiley & Sons.

Norman, D. (2002). *The Design of Everyday Things*. New York: Doubleday.

Olson, G.M. and Olson, J.S. (2003). Human–computer interaction: psychological aspects of the human use of computing. *Annual Review of Psychology*, 54, 491–516.

Parasuraman, R., and Riley, V. (1997). Humans and automation: use, misuse, disuse, abuse. *Human Factors*, 39(2), 230–253.

Pashler, H., Johnston, J.C. and Ruthruff, E. (2001). Attention and performance. *Annual Review of Psychology*, 52, 629–651.

Plous, S. (1993). *The Psychology of Judgment and Decision Making*. New York: McGraw-Hill.

Preece, J. (1994). *Human–Computer Interaction*. Harlow, UK: Addison-Wesley.

Reason, J. (1991). *Human Error*. Cambridge, UK: Cambridge University Press.

Reason, J. (1997). *Managing the Risks of Organizational Accidents*. Aldershot, Hampshire, UK: Ashgate.

Rosenfeld, L. and Morville, P. (2002). *Information Architecture for the World Wide Web*, 2nd ed. Sebastopol, CA: O'Reilly.

Rosson, M.B. and Carroll, J.M. (2001). *Usability Engineering: Scenario-Based Development of Human–Computer Interaction*. San Francisco, CA: Morgan Kaufmann.

Roth, E.M., Bennett, K., and Woods, D.D. (1987). Human interaction with an "intelligent" machine. *International Journal of Man–Machine Studies*, 27, 479–525.

Rubenstein, R. amd Hersh, H. (1984). *The Human Factor*. Boston, MA: Digital Press.

Rubin, T. (1994). *Handbook of Usability Testing: How to Plan, Design and Conduct Effective Tests*. New York: John Wiley & Sons.

Schraagen, J. M., Chipman, S. and Shalin, V. (eds.) (2000). *Cognitive Task Analysis.* Mahwah, NJ: Lawrence Erlbaum.

Sheridan, T. (2002). *Humans and Automation: System Design and Research Issues.* Chichester, UK: Wiley.

Shepherd, A. (2000). *Hierarchical Task Analysis.* London: Taylor & Francis.

Shute, S., and Smith, P. J. (1992). Knowledge-based search tactics. *Information Processing and Management,* 29, 29–45.

Silver, M. (2005). Exploring Interface Design. Clifton Park, NY: Delmar Learning.

Skirka, L., Mosier, K., and Burdick, M. (1999). Does automation bias decision making? *International Journal of Human–Computer Systems,* 51, 991–1006.

Smith, P.J. and Geddes, N. (2003). A cognitive systems engineering approach to the design of decision support systems. In: J. Jacko and A. Sears (eds.), *The Human–Computer Interaction Handbook: Fundamentals, Evolving Technologies and Emerging Applications.* Mahwah, NJ: Lawrence Erlbaum Associates, 656–675.

Smith, P.J., McCoy, E., and Layton, C. (1997). Brittleness in the design of cooperative problem-solving systems: the effects on user performance. *IEEE Transactions on Systems, Man, and Cybernetics,* 27(3), 360–370.

Spool, J., Scanlon, T., Snyder, C., DeAngelo, T. and Schroeder, W. (1998). *Web Site Usability: A Designer's Guide.* San Francisco, CA: Morgan Kaufmann.

Tufte, E.R. (1983). *The Visual Display of Quantitative Information.* Chesire, CT: Graphics Press.

Tufte, E.R. (1990). *Envisioning Information.* Chesire, CT: Graphics Press.

Tufte, E.R. (1997). *Visual Explanations.* Cheshire, CT: Graphics Press.

Van Duyne, D., Landay, J., and Hong, J. (2003). *The Design of Sites.* Boston, MA: Addison-Wesley.

Vincente, K. (1999). Computer Work Analysis: Toward Safe, Productive and Healthy Computer-Based Work. Mahwah, NJ: Lawrence Erlbaum.

Vicente, K.J. (2002). Ecological interface design: progress and challenges. *Human Factors,* 44(1), 62–78.

Watzman, S. (2003). Visual design principles for usable interfaces. In: J. Jacko and A. Sears (eds.), *The Human–Computer Interaction Handbook: Fundamentals, Evolving Technologies and Emerging Applications.* Mahwah, NJ: Lawrence Erlbaum Associates, 263–285.

Wickens, C.D. and Hollands, J. (1999). *Engineering Psychology and Human Performance,* 3rd edition. Upper Saddle River, NJ: Prentice Hall.

Wickens, C.D., Lee, J., Liu, Y. and Gordon-Becker, S. (2004). *An Introduction to Human Factors Engineering,* 2nd edition. Upper Saddle River NJ: Pearson/Prentice Hall.

Witkin, B.R. and Altshuld, J.S. (1995). *Planning and Conducting Needs Assessments.* Thousand Oaks, CA: Sage Publications.

Yantis, S. (1998). Control of visual attention. In: H. Pashler (ed.), *Attention.* Hillsdale, NJ: Lawrence Erlbaum, 223–256.

25

The Role Personality in Ergonomics

W. Gary Allread
The Ohio State University

25.1 Background

It is not surprising that the underlying causes of musculoskeletal disorders (MSDs) are complex, multifactorial, and difficult to determine. Despite decades of research, the numbers of individuals developing MSDs are still quite high. Yelin et al. (1999) reported that 90% of disabled older workers had MSDs. Further, it was estimated that 18.4% of the U.S. population (nearly 60 million individuals) will suffer from one or more chronic MSDs by the year 2020 (Lawrence et al., 1998). Treatment of these disorders is in the tens of billions of dollars (Praemer et al., 1999). These injury statistics suggest that the current methods for identifying and reducing or eliminating MSDs in today's industries have not been completely effective.

Research on MSD causation typically has focused on either of two primary groups of possible factors — biomechanical demands in the workplace or psychosocial influences. Biomechanical demands most often arise from performing physical work, which impact the body's tissues. These types of factors include, but are not limited to, loads handled, work repetition, the duration of the activity, joint kinematics, and contact with sources of vibration. Psychosocial influences are less clearly defined, but they typically involve social aspects of the work environment, how a job is organized, and the content of the job tasks performed (Sauter et al., 1990), in addition to the environment outside of work and individual traits (Bernard, 1997).

Each of these group's contributions to injury risk has not yet been established, and estimates have varied considerably. Regardless, neither research area alone appears capable of explaining MSD causation. Davis et al.'s (2000) review of the literature concluded that neither biomechanical nor psychosocial factors could individually explain all the variabilities in MSD reporting.

Further, research on MSD causation has not yet provided a clear picture of how physical workplace and psychosocial factors *together* contribute to MSDs. Leaders in the field, including ergonomists, physicians, and epidemiologists, generated a report suggesting that several types of factors likely are involved and interact with one another to produce an MSD (National Research Council and Institute of Medicine, 2001). Here, these experts reviewed hundreds of research studies and determined that a variety of factors must be considered to understand injury mechanisms. They developed a conceptual model of factors that may affect MSD risk in the workplace, which is shown in Figure 25.1. This model approaches MSD risk from a "whole person" perspective, meaning that it includes aspects both of the workplace (e.g., loads imposed on the body, organizational factors, and social aspects of jobs) and the individual (e.g., physiological responses to external loading, physical tissue tolerances, health outcomes, and personal factors). The model also underscores the complex nature of MSD causation. These researchers concluded that, "Because workplace disorders and individual risk and outcomes are inextricably bound, musculoskeletal disorders should be approached in the context of the whole person rather than focusing on body regions in isolation." Thus, it would appear that only by studying the entire work system (i.e., the interaction of physical, psychosocial, and individual factors) can we derive the root causes of MSDs.

Clearly, the range of factors considered psychosocial in nature is large and diverse. They include personal aspects of individuals as well as their perceptions of the work environment. Undoubtedly, different individuals can view the same environment and work situation in a number of ways. Some may view these environments as suitable to their preferred method of operating, or they may find the work organization too stressful. Thus, the psychosocial nature of an environment should perhaps be viewed from the perspective of those evaluating and working within it. An understanding of personality and personality theory may aid in this comprehension.

One's personality preferences can be included in the group of individual factors interacting with others to affect MSD risk. There is general consensus that one's personality is, at least, in part, biologically based. Jung (1923) proposed that human behavior was predictable, not random, that one's preferences emerge early in life, and that these preferences form the basis for our attractions to and aversions from people,

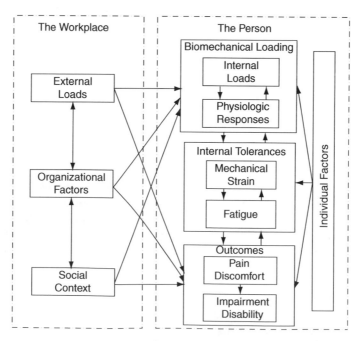

FIGURE 25.1 Conceptual model of likely groups of musculoskeletal disorder risk factors and their interactions. (From National Research Council and Institute of Medicine, 2001, *Musculoskeletal Disorders and the Workplace: Low Back and Upper Extremities* (Washington, DC: National Academy Press). With permission.)

tasks, events, etc. throughout our lifetimes. Recent research has suggested that several traits important in understanding personality have an inherited or genetic component (Bouchard et al., 1990; Plomin and Caspi, 1999; Krueger, 2000). Loehlin (1992) estimated that as much 40% of the variability in personality could be attributed to inherited traits. As further evidence that personality preferences are at least partly "hard wired," Canli et al. (2001) found clear relationships between personality and reactivity in the brain to emotional stimuli, particularly with regard to extraversion and neuroticism. These results suggest that personality is a stable trait that can be measured or observed as part of the environment that may influence how work is perceived and reacted to.

By understanding one's personality, we may be better able to understand the influence that psychosocial factors have on individuals and, as a result, on the potential for MSD development. Our personalities dictate our preferences for many life situations, including work. Theorist Allport (1937) wrote that personality is "... the dynamic organization within the individual of those psychophysical systems that determine his unique adjustments to his environment." This implies that personality is a system, that it integrates both the physical and mental aspects of an individual, and that people are active in adapting to their surroundings.

Personality theory has been applied extensively to occupational work. Several researchers have studied individuals' personalities and how these interact with one's jobs. In situations where an individual's personality preferences were less matched with the nature of their work, higher rates of turnover (Myers and Myers, 1980) and job dissatisfaction (Smart et al., 1986; Smith, 1988; Karras, 1990; Holland, 1996) were found. Such incongruence also has created communication problems, inefficiency, and, at times, health concerns (Shelton, 1996). Thus, there appears to be a link between the job environment and individuals' reactions to it, depending on their personality preferences.

Understanding and determining one's personality is not an exact science. There is a wide range of competing personality theories. The aim of this chapter is not to compare and contrast these theories or to detail the specifics of all types of personality assessments that have been used historically. The goal, however, is twofold. The first is to identify those personality theories that have a solid research foundation. The second goal is to detail those personality assessment techniques that may best be suited for use in studying occupational risk factors for the development of MSDs. Those listed have established acceptable levels of reliability and validity, and could be used to address the interaction of individual factors with risk factors for the development of MSDs.

25.2 Personality Assessments for Use in Occupational Settings

There are many different methods designed to measure personality. These differ dramatically in terms of their theoretical bases and applicability, especially with regard to use in occupation settings. However, they can be grouped based on their assessment strategies. Block (1993) categorized four such techniques. *L-data* comprise information pertaining to one's life history or life record. These have been used, for example, to assess aspects of criminal behavior. *O-data* consist of ratings by observers, such as parents, peers, or spouses and have measured hyperactivity and ego. *T-data* are the result of standardized tests or the use of experimentation. They are often used to study personality in relation to such functions as conflict, arousal, and achievement motivations. *S-data* involve self-reporting, such as to determine the various personality traits, for Type A behavior, and to study coping mechanisms. Each of these methods has its strengths and weaknesses; however, a majority of assessment techniques are of the self-reporting (S-data) type.

Five personality assessments and behaviors have been included in this review: The Minnesota Multiphasic Personality Inventory or MMPI; the Myers–Briggs Type Indicator; the Eysenck Personality Questionnaire; an assessment of Type A behavior, using the Jenkins Activity Survey; and the Five Factor Model. Descriptions of each of these are described next. This is followed by a summary of how the personality traits assessed from these tools can be used to study individuals' interactions with their work environments.

25.2.1 Minnesota Multiphasic Personality Inventory

This assessment, known more commonly as the MMPI, is designed to assess major symptoms and indicators of social and personal maladjustment, which are believed to be indicative of disabling psychological dysfunction. The MMPI has been applied across a wide variety of settings. Clinicians often use the MMPI to select the proper treatment modality for their patients. Employers in public safety have found it can assist in the selection of individuals for jobs that involve high risk. The MMPI also has been applied to the areas of career counseling, marital relations, and family therapy.

The MMPI and its most recent revision, the MMPI-2 (Butcher et al., 1989), are among the most widely used and researched of the various personality inventories (Watkins et al., 1995), which is the primary reason it is referenced here. The literature contains over 4000 references to the MMPI-2 published between the mid-1970s and the mid-1990s (Butcher and Rouse, 1996). Greene (2000) provides an extensive review of the MMPI-2's validity and notes that the interpretation of one's scores from this inventory may be affected by age, gender, educational level, or ethnicity.

The MMPI-2 is composed of 567 statements, which are written at the sixth-grade level. Respondents answer each using the choices "true," "false," or "cannot say." The entire questionnaire typically can be completed in 60 to 90 min. Scoring of the MMPI-2 not only determine one's truthfulness and test-taking motivation but also evaluates the individual on ten clinical scales. Each of these scales, along with their abbreviation and a brief description, is listed in Table 25.1. As this table shows, the ten scales of the MMPI-2 assess major categories of psychopathology. Scale 1 (hypochondriasis) distinguishes people having legitimate physical ailments from those with nonspecific complaints regarding body functioning. Scale 2 can assess individuals along the continuum from low morale to clinical depression. Hysteria (scale 3) is used to determine reactions to stressful situations. The psychopathic deviate scale (4) identifies one's tendency to accept authority or become more rebellious. Scale 5 (masculinity–femininity) relates to sexual concerns and problems. Paranoia scores (scale 6) reflect an individual's level of rigidity in opinions

TABLE 25.1   Scales of the MMPI-2

Scale Name, Number, and Description	Abbreviation
Hypochondriasis (1) An excessive preoccupation with one's health and bodily functioning	Hs
Depression (2) A condition of emotional withdrawal, dejection, hopelessness, and general dissatisfaction	D
Hysteria (3) An emotional outburst often associated with laughter, weeping, or irrational behavior	Hy
Psychopathic deviate (4) A feeling of social maladjustment, combined with an absence of pleasant experiences	Pd
Masculinity–Femininity (5) Reflects one's interests in hobbies, vocations, and esthetics, and personal sensitivity	Mf
Paranoia (6) Excessive mistrust of others and interpersonal sensitivity	Pa
Psychasthenia (obsessive–compulsive disorder) (7) The inability to resist specific actions or thoughts regardless of their nature	Pt
Schizophrenia (8) A pattern of disorganized behavior and speech, as well as bizarre thought processes such as delusions and hallucinations	Sc
Hypomania (9) Characterized by overactive behavior and an elated but unstable mood	Ma
Social introversion (10) One's level of comfort with social situations and interactions	Si

and attitudes. Among the objectives of scale 7 (psychasthenia or obsessive–compulsive disorder) is to determine short- and long-term anxieties. Items on the schizophrenia scale (8) relate to one's perceptions and thought processes. In addition to mood and activity levels, scale 9 (hypomania) assesses family relationships and moral values. Finally, social introversion (scale 10) relates to one's comfort level is group situations, from shy and uncomfortable (introverted) to friendly and talkative (extraverted).

25.2.2 Myers–Briggs Type Indicator

The Myers–Briggs Type Indicator, or MBTI (Consulting Psychologists Press, Inc., 1998), was designed as a forced-choice, self-reporting inventory of individuals' basic preferences toward life, particularly in the perception and judgment of one's surroundings. It categorizes personality across four independent scales:

- *Extraversion–introversion*. Denotes a preference from where one gains energy. For introversion, this would be from the inner self; in extraversion, it is from the outer world.
- *Sensing–intuition*. Reflects the preference for how one perceives, takes in information, or looks at life. Sensors tend to do this through facts, while intuitors do so through possibilities.
- *Thinking–feeling*. Indicates how one prefers to make judgments or decisions. Thinkers are more inclined to use logic, while feelers often rely more on personal values.
- *Judging–perceiving*. Indicates one's preference for an external lifestyle. Judgers favor more orderliness, while perceivers tend to be more flexible.

Occupationally related traits of these preferences (Myers and Myers, 1980; Kummerow et al., 1997) are listed in Table 25.2. This table lists work methods that individuals with those preferences may find either more or less suitable. Also, it suggests that what is an ideal work situation to one person (e.g., task variety, learning new skills, interaction with others) may be undesirable to another.

The MBTI has been used in numerous settings (Bayne, 1995). Research based on studies using the MBTI has been published in hundreds of publications (Carlson, 1985) and referenced in a wide variety of peer-reviewed business, education, medicine, psychology, and science journals (Willis, 1984). The MBTI is commonly administered in occupational environments to enhance personal development, increase understanding of colleagues, and improve upon managing conflict.

The score reliability of the MBTI has been validated by a number of researchers (e.g., Johnson, 1992; Myers et al., 1998), though its use is not accepted for all applications (Druckman and Bjork, 1991). Capraro and Capraro's (2002) reliability generalization analysis, a review of data from more than 200 recent research articles using the MBTI, found that, overall, the MBTI produced acceptable score reliabilities across these studies.

The MBTI consists of 93 forced-choice questions and word pairs. Users are asked to either choose an answer that most closely describes their feelings or actions, or to select one word in each pair that best appeals to them. The MBTI is written at the seventh-grade reading level and takes 15 to 25 min to complete.

25.2.3 Eysenck Personality Questionnaire

Psychologist Hans Eysenck believed that personality trait differences were the result of one's biological functioning; in other words, he theorized that an individual's personality was genetically based. From his research, he developed the Eysenck Personality Questionnaire or EPQ (Eysenck and Eysenck, 1975) and the Eysenck Personality Questionnaire-Revised or EPQ-R (Eysenck and Eysenck, 1994).

The Eysenck questionnaires assess three basic types of personality dimensions — psychoticism, extraversion–introversion, and neuroticism. The hierarchical structure (i.e., traits) of each of these dimensions (Eysenck, 1990) is shown in Figure 25.2. Persons scoring high on the psychoticism scale tend to be socially indifferent, aggressive, intolerant, and lacking in empathy. Individuals with high extraversion scores typically are outgoing and impulsive, and they prefer to take part in group activities. A high neuroticism score on the EPQ indicates overactivity and a propensity for worry and anxiety.

TABLE 25.2   Descriptions and Occupationally Related Traits of the Eight MBTI Personality Preferences

Personality Types	

Extraversion (E) is a preference for active involvement and quick action. Extraverts:
   Like variety, action
   Are impatient with long, slow jobs
   Like to have people around
	Introversion (I) is a preference for inner reflection of thoughts and ideas. Introverts:
	   Like quiet and periods of uninterrupted work
	   Are content working alone
	   Have some problems communicating

Sensing (S) reflects a preference for dealing with physical reality and facts. Sensors:
   Like established ways of doing things
   Enjoy using current skills than learning new ones
   Work more steadily
   Are patient with routine details
   Are good at precision work
	Intuition (N) is a preference for deriving meaning from personal insight. Intuitors:
	   Dislike doing the same thing repeatedly
	   Enjoy learning new skills than using them
	   Work in bursts of energy, with slack periods in between
	   Dislike taking time with precision

Thinking (T) is a preference for making decisions based on logic. Thinkers:
   Are uncomfortable dealing with people's feelings
   Like putting things into a logical order
   Are sometimes impersonal
   Can reprimand and fire people when necessary
	Feeling (F) is a preference for decision-making based on personal values. Feelers:
	   Are people-oriented and aware of others' feelings
	   Enjoy pleasing people
	   Like harmony
	   Need praise

Judging (J) is a preference to lead an orderly, planned life. Judgers:
   Like to keep and follow a plan of work
   Prefer to settle and finish things
   Dislike interruptions, even for more urgent matters
	Perceiving (P) is a preference to lead a more spontaneous, flexible life. Perceivers:
	   Adapt well to changing situations
	   May have trouble making decisions
	   Want to know all about a new job
	   Are curious about new work situation

Research generally supports the validity of Eysenck's personality questionnaires. Early studies provided evidence of construct validity of the Eysenck Personality Inventory (White et al., 1968; Platt et al., 1971), a precursor to the EPQ. Comparisons with other personality measures found the EPQ to be valid and reliable on the extraversion–introversion and neuroticism scales but less so on psychoticism (Wakefield et al., 1976; Goh et al., 1982; Caruso et al., 2001). Cronbach's alpha coefficients for these scales range from 0.66 to 0.86 (Eysenck and Eysenck, 1994).

Eysenck's questionnaires were designed to be completed quickly. The revised version of the EPQ contains 73 questions and takes about 10 to 15 min to complete. A short form contains only 57 items in which respondents answer "yes" or "no." It can be completed in 3 to 5 min.

The extraversion–introversion and neuroticism traits, in particular, have been linked to physical response outcomes. As summarized by Pervin (2003), introverts tend to be more sensitive to pain and are more easily fatigued. Those scoring high on the neuroticism scale are more likely to be anxious and to experience body aches. These characteristics suggest the possible interaction between job design and physical outcomes due to one's personality.

25.2.4   Type A Personality Behavior

Theories regarding patterns of Type A behavior were put forth most notably by Friedman and Rosenman (1974). Individuals exhibiting this trait typically have characteristics such as those listed in Table 25.3. This table implies that Type A traits can manifest themselves in one's physical conduct, personal attitudes, and cognitive behavior. Type A behavior is not considered a rigid personality trait by some (Matthews and Haynes, 1986) but rather the outcome from an interaction between one's predispositions and particular situations in which they find themselves. Type A behavior is included here because of its suspected link with a number of physical responses.

Several studies have linked Type A behavior with coronary heart disease (CHD). This arose following findings that the primary risk factors for CHD (i.e., smoking, hypertension, elevated serum cholesterol

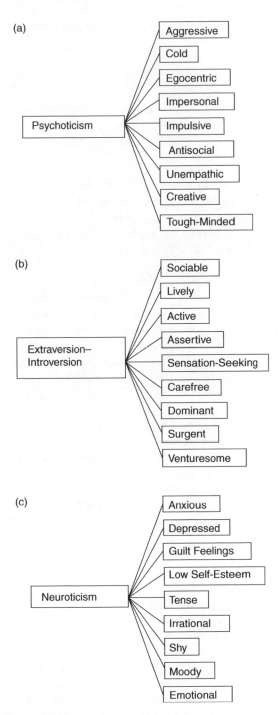

(a)

Psychoticism
- Aggressive
- Cold
- Egocentric
- Impersonal
- Impulsive
- Antisocial
- Unempathic
- Creative
- Tough-Minded

(b)

Extraversion–Introversion
- Sociable
- Lively
- Active
- Assertive
- Sensation-Seeking
- Carefree
- Dominant
- Surgent
- Venturesome

(c)

Neuroticism
- Anxious
- Depressed
- Guilt Feelings
- Low Self-Esteem
- Tense
- Irrational
- Shy
- Moody
- Emotional

FIGURE 25.2 Eysenck's three basic dimensions of personality — psychoticism (a), extraversion–introversion (b), and neuroticism (c). (Adapted from Eysenck, H.J., 1990, Biological dimensions of personality, in Pervin, L.A. (ed.), *Handbook of Personality: Theory on Research* [New York: Guilford]. With permission.)

TABLE 25.3   Characteristics Typical of those Exhibiting Type A Behavior

Characteristic Type	Description
Physical	Speaking quickly or with a loud voice; exhibiting tension in the facial muscles
Attitudinal and emotional	Becoming hostile, impatient, angry, or aggressive
Motivational	Seeking achievement, motivated, competitive, ambitious
Behavioral	Alert, hyperactive
Cognitive	Tendency toward controlling situations

levels) alone did not account for the increases seen in this disease (Friedman and Rosenman, 1957). Results from the Framingham Heart Study data (Haynes et al., 1982) and other research (e.g., Rosenman et al., 1975) established a significant link between Type A behavior and increased incidence of CHD. However, this positive association has not been found in some follow-up studies, such as the Aspirin Myocardian Infarction Study (Shekelle et al., 1985a) and the Multiple Risk Factor Intervention Trial (Shekelle et al., 1985b).

A limited number of studies have focused on links between Type A personality and MSDs. Flodmark and Aase's (1992) survey of blue-collar workers found that those reporting symptoms of MSDs had more pronounced Type A behaviors than those who did not report any symptoms. Further, tenderness in the neck and shoulder regions was found by Salminen et al. (1991) to be greater in those exhibiting Type A behavior, leading these researchers to postulate that this was due to higher levels of muscle activity exhibited by this group. Finally, Glasscock et al. (2003) found significantly higher levels of elbow flexor muscle activity in those with Type A behavior, though the effect was a function of gender. Collectively, these studies suggest that a link may exist between behavior due to personality preferences and muscle functioning.

The Jenkins Activity Survey, or JAS, was developed to measure Type A behavior (Jenkins et al., 1979). This assessment resulted from a study in which a group of males having had experienced a heart attack scored significantly higher on the survey than did the control group (Jenkins et al., 1971). The JAS is a self-administered, multiple-choice questionnaire that can be completed in 20 to 30 min. In addition to a rating on the Type A scale, it yields scores on three subscales: speed and impatience, job involvement, and being hard-driving and competitive. Estimates of test–retest reliability for the JAS, for intervals from 6 months to 4 yr, ranged from between 0.60 and 0.70.

25.2.5   Five Factor Model

The Five Factor Model (FFM), also known as the Big Five model, has recently been achieving widespread support, because it includes the basic personality traits generally agreed upon by researchers in the field. The FFM operates on the belief that most human personality traits can be grouped into five broad dimensions of personality. This model has been studied extensively, most notably by Goldberg (1993) and McCrae and Costa (1999).

The constructs of the FFM are detailed in Table 25.4. The extraversion factor relates to one's need for stimulation and interpersonal interaction. Agreeableness is the factor assessing one's preference in how they think, feel, and interact with others. Conscientiousness refers to an individuals' level of organization, motivation, and persistence. The neuroticism factor relates to one's degree of emotional stability and how they are affected by stress. Finally, openness to experience is meant to reflect one's preference for exploring the unfamiliar. The characteristics included in the FFM are believed to be innate and stable over one's lifetime. Several of these factors, such as extraversion/introversion and neuroticism, are incorporated into other personality assessments.

The NEO-PI Five-Factor Inventory (Costa and McCrae, 1992) is the questionnaire used to assess personality according to the FFM. It comprises 300 items, in which respondents answer statements reflecting their views on a five-point scale, from "strongly agree" to "strongly disagree." Shorter forms of this

TABLE 25.4   Categories of the Five Factor Model and their Descriptions

Factor	Description
Extraversion	Relates to the level of comfort one has with sensory stimulation. Those with high levels of extraversion tend to prefer larger numbers of relationships and more social interactions, while low levels (i.e., introversion) reflect a preference for being more independent and spending time alone
Agreeableness	This trait involves the quality of one's interpersonal communications. Those at one end of the spectrum on this dimension (sometimes referred to as adapters) select their norms for behavior from a number of sources. "Challengers" typically are on the other end of this continuum, as they base their behavior on their own personal beliefs
Conscientiousness	This construct refers to the focus one has on goals. On one end of this spectrum are individuals who are more self-disciplined and concentrate on fewer goals, while on the other end are individuals who pursue many goals but are more easily distracted
Neuroticism	Neuroticism is akin to anxiety. Those considered to be more reactive are bothered by a greater variety of stimuli that may be of lower strength, while more resilient individuals are affected by fewer types of stimuli that are at more intense levels.
Openness to experience	This factor refers to one's interest level in the unfamiliar. Those with "high" openness tend to have more interests but less depth of knowledge about them, while those with "low" openness prefer to focus more in-depth on fewer interests

assessment have been constructed, included the Big Five Marker Scales (taking about 10 min to complete) and the Traits Personality Questionnaire (25 min). The Hogan Personality Inventory, or HPI (Hogan and Hogan, 1995), was constructed using data from adult workers and is geared toward use in occupational environments. A complete list and descriptions of inventories, questionnaires, adjective scales, and other related instruments using FFM methodology has been provided by De Raad and Perugini (2002).

Many researchers have provided evidence of this model's validity. McCrae and Costa (1990) found considerable agreement between self-ratings using the FFM and those ratings put forth by colleagues and significant others. Significant associations between self-ratings and observer ratings using the FFM also were reported by Funder et al. (1995), Riemann et al. (1997), and Watson et al. (2000). De Raad and Perugini (2002) provided reviews of validity studies for several instruments using the FFM.

25.3   Personality Traits and Physical Outcomes

As reported in the preceding section, a number of personality theories have been put forth in the literature, and assessment techniques used to measure their associated factors have accepted levels of validity. Thus, these personality evaluations have been used to assess a variety of human factors, particularly in medical, psychological, and occupational settings. Their outcomes have allowed clinicians and mental health professionals to more effectively diagnose and treat individuals. They also have been used to increase understanding of one's own actions and improve communication among colleagues in occupational environments.

The focus of most efforts regarding personality preferences has been on behavioral outcomes, but recent research has suggested a link between personality traits and physical effects. These results have implications for those interested in studying the interaction between people and their work from an injury perspective. There is evidence to suggest that an individual's personality preferences may relate to the likelihood of experiencing physical discomfort of the body's soft tissues or to actual MSDs.

There is not an overabundance of research addressing the interactions between personality preferences and their physical effects in the context of occupational work. However, the results from several studies suggest that such a link exists.

25.3.1   Personality, Low Back Pain, and Spinal Loading

Of the few studies addressing the relationship between personality preferences and physical outcomes, several involve low back pain reporting. Gatchel et al. (1995) scored subjects' responses on the hypochondriasis, depression, and hysteria scales of the MMPI. They found that those in the group having a chronic disability due to low back pain had significantly higher mean scores on the hysteria scale than did those persons who were not disabled. A similar finding was also reported in a large-scale study of aircraft assembly employees (Bigos et al., 1991). Here, one's hysteria score was found to be a predictor variable for the development of low back pain in workers who previously were uninjured. Bongers et al. (1993) conducted an extensive literature review in an effort to evaluate the relationships between psychosocial measures and the occurrence of MSDs. Among those factors positively correlated with the reporting of back pain was Type A behavior and anxiety (which is analogous to neuroticism in the FFM).

One study specifically associated personality with loads on the lumbar spine. Marras et al. (2000) studied the muscle activation patterns and spinal loads of subjects as they handled objects under both "stressed" and "unstressed" lifting conditions. They also evaluated subjects' personalities using the MBTI and found that increases in spinal loading during the stressed condition varied depending on their personality types. Specifically, those with preferences toward introversion or intuition produced significantly higher levels of spinal compression and lateral shear. This study suggests a possible pathway between psychosocial stress and loading on the spine that may, in part, be influenced by one's personality preferences.

25.3.2   Personality and Loading on the Upper Extremities

There are a few indications that personality traits may be related to differences in physical outcomes of the upper extremities as well. Neck and shoulder muscle tension was reported by Salminen et al. (1991) to be due to higher levels of muscle activity in those exhibiting Type A behavior. Further, Glasscock et al. (2003), who also assessed Type A behavior, reported greater elbow flexor muscle activities among those who were found to be Type A using the JAS. Together with the aforementioned low back studies, this literature suggests that personality factors may play a role in determining the likelihood that one develops musculoskeletal discomfort.

The research that does exist on a personality–physical outcome link is by no means all in agreement or conclusive. A case in point relates to studies involving the extraversion–introversion traits. Scheier and Carver (1987) theorized that the extraversion–introversion trait may be linked to a number of health-related behaviors, but Bongers et al. (1993) concluded that there was not enough evidence in the literature to show a relationship between extraversion and back pain. The reason for this may be because different assessment methods were used for measuring extraversion and introversion. For example, the MMPI, the MBTI, the EPQ, and the FFM all include extraversion–introversion scales in their assessments.

25.4   Conclusions

Researchers still are learning a great deal about factors related to musculoskeletal discomfort and the causes of MSDs. However, the evidence is increasingly clear that the origins of MSDs are multifactorial in nature. As conceptualized in Figure 25.1, important risk factors relate to both the physical and psychosocial aspects of an individual's working environment, as well as to aspects of the individuals themselves. One of the many characteristics of individuals that can influence their (real or perceived) reactions to the working environment is personality.

The study of human personality has created a multitude of theories aimed at describing one's behavior. From these models arose a variety of inventories developed to assess personality factors relevant to that theory. Some of the more commonly used models were described here. These have produced generally acceptable levels of validity and have been used by researchers and practitioners to study human behavior in a wide range of settings.

This chapter summarized personality assessments that may be useful in understanding how individual influences interact with workplace factors in contributing to discomfort and MSD risk. The study of personality in occupational settings is not new, but the focus has traditionally been on improving communication among coworkers or assisting individuals to better understand themselves and their actions. Some recent research highlighted here has studied personality in terms of the possibility that a "mind–body" injury pathway exists. In other words, it may be that individuals' personalities influence how they respond to the physical and psychosocial demands of their workplace. This may translate to the creation of physical symptoms (e.g., increased muscle activity, more joint loading) that, cumulatively, lead to pain or injury.

This chapter has also made clear the fact that much work still remains in understanding the complexities of MSD causation. Finally, it suggests that new discoveries may be made by integrating knowledge of human behavior and individuals' *perceptions* of their workplace with the actual, measurable workplace factors that have been more commonly studied.

References

Allport, G. W., 1937, *Personality: A Psychological Interpretation* (New York: Holt, Rinehart and Winston).

Bayne, R., 1995, *The Myers–Briggs Type Indicator — A Critical Review and Practical Guide* (London: Chapman and Hall).

Bernard, B. P. (ed.), 1997, *Musculoskeletal Disorders and Workplace Factors: A Critical Review of Epidemiological Evidence for Work-Related Musculoskeletal Disorders of the Neck, Upper Extremity, and Low Back* (U.S. Department of Health and Human Services, Public Health Service, Centers for Disease Control, National Institute for Occupational Safety and Health).

Bigos, S. J., Battie, M. C., Spengler, D. M., Fisher, L. D., Fordyce, W. E., Hansson, T. H., Nachemson, A. L., and Wortley, M. D., 1991, A prospective study of work perceptions and psychosocial factors affecting the report of back injury, *Spine*, 16(1), 1–6.

Block, J., 1993, Studying personality the long way, in Funder, D. C., Parke, R. D., Tomlinson-Keasey, C., and Widaman, K. (eds.), *Studying Lives through Time* (Washington DC: American Psychological Association), pp. 9–41.

Bongers, P. M., De Winter, C., Kompier, M. A. J., and Hildebrandt, V. H., 1993, Psychosocial factors at work and musculoskeletal disease, *Scandinavian Journal of Work, Environment & Health*, 19(5), 297–312.

Bouchard, T. J., Jr., Lykken, D. T., McGue, M. Segal, N. L., and Tellegen, A., 1990, Sources of human psychological differences: The Minnesota study of twins reared apart, *Science*, 250, 223–250.

Butcher, J. N., and Rouse, S. V., 1996, Personality: individual differences and clinical assessment, *Annual Review of Psychology*, 47, 87–111.

Butcher, J. N., Dahlstrom, W. G., Graham, J. R., Tellegen, A., and Kaemmer, B., 1989, *MMPI-2: Manual for Administration and Scoring* (Minneapolis: University of Minnesota Press).

Canli, T., Zhao, Z., Desmond, J. E., Kang, E., Gross, J., and Gabrieli, J. D., E., 2001, An fMRI study of personality influences on brain reactivity to emotional stimuli, *Behavioral Neuroscience*, 115, 33–42.

Capraro, R. M. and Capraro, M. M., 2002, Myers-Briggs Type Indicator score reliability across studies: a meta-analytic reliability generalization study, *Educational and Psychological Measurement*, 62(4), 590–602.

Carlson, J. G., 1985, Recent assessments of the MBTI, *Journal of Personality Assessment*, 49, 356–365.

Caruso, J. C., Witkiewitz, K., Belcourt-Dittloff, A., and Gottlieb, J., 2001, Reliability of scores from the Eysenck Personality Questionnaire: a reliability generalization (RG) study, *Educational and Psychological Measurement*, 61, 675–682.

Consulting Psychologists Press, Inc., 1998, *Myers–Briggs Type Indicator, Form M* (3803 E. Bayshore Road, Palo Alto, CA, 94303).

Costa, P. T., Jr., and McCrae, R. R., 1992, *NEO PI-R: Professional Manual* (Odessa, FL: Psychological Assessment Resources).

Davis, K. G., Marras, W. S., and Heaney, C. A., 2000, The relationship between psychosocial work characteristics and low back pain: underlying methodological issues, *Clinical Biomechanics*, 15, 389–406.

De Raad, B., and Perugini, M. (eds.), 2002, *Big Five Assessment* (Goettingen, Germany: Hogrefe & Huber).

Druckman, D., and Bjork, R. A. (eds.), 1991, *In the Mind's Eye: Enhancing Human Performance* (Washington, DC: National Academy Press).

Eysenck, H. J., 1990, Biological dimensions of personality, in Pervin, L. A. (ed.), *Handbook of Personality: Theory and Research* (New York: Guilford), p. 246.

Eysenck, H. J., and Eysenck, S. B. G., 1975, *Manual of the Eysenck Personality Questionnaire* (London: Hodder & Stoughton).

Eysenck, H. J., and Eysenck, S. B. G., 1994, *Manual of the Eysenck Personality Scales – Revised* (San Diego, CA: Educational and Industrial Testing Service).

Flodmark, B. T., and Aase, G., 1992, Musculoskeletal symptoms and type A behavior in blue collar workers, *British Journal of Industrial Medicine*, 49(10), 683–687.

Friedman, M., and Rosenman, R., 1974, *Type A Behavior and Your Heart* (New York: Knopf).

Friedman, M., and Rosenman, R. H., 1957, Comparison of fat intake of American men and women: possible relationship to incidence of clinical coronary artery disease, *Circulation*, 16, 339–347.

Funder, D. C., Kolar, D. C., and Blackman, M. C., 1995, Agreement among judges of personality: interpersonal relations, similarity, and acquaintanceship, *Journal of Personality and Social Psychology*, 69, 656–672.

Gatchel, R. J., Polatin, P. B., and Kinney, R. K., 1995, Predicting outcome of chronic back pain using clinical predictors of psychopathology: a prospective analysis, *Health Psychology*, 14(5), 415–420.

Glasscock, N. F., Mirka, G. A., Sommerich, C. M., and Klein, K. W., 2003, The effects of personality type and stress on muscle activity during simulated work tasks, Proceedings of the Human Factors and Ergonomics Society 47th Annual Meeting, pp. 1159–1163.

Goh, D. S., King, D. W., and King, L. A., 1982, Psychometric evaluation of the Eysenck Personality Questionnaire, *Educational and Psychological Measurement*, 42(1), 297–309.

Goldberg, L. R., 1993, The structure of phenotypic personality traits, *American Psychologist*, 48(1), 26–34.

Greene, R. L., 2000, *The MMPI-2: An Interpretive Manual*, 2nd edition (Boston: Allyn and Bacon).

Haynes, S. G., Feinleib, M., and Eaker, E. D., 1982, Type A behavior and the ten year incidence of coronary heart disease in the framingham Heart Study, *Activitas Nervosa Superior*, Suppl. 3(1), 57–77.

Hogan, R., and Hogan, J., 1995, *The Hogan Personality Inventory Manual*, 2nd edition (Tulsa, OK: Hogan Assessment Systems).

Holland, J. L., 1996, Exploring careers with a typology: what we have learned and some new directions, *American Psychologist*, 51, 397–406.

Jenkins, C. D., Zyzanski, S. J., and Rosenman, R. H., 1971, Progress toward validation of a computer-scored test for the type A coronary-prone behavior pattern, *Psychosomatic Medicine*, 33(3), 193–202.

Jenkins, C. D., Zyzanski, S. J., and Rosenman, R. H., 1979, *Jenkins Activity Survey: JAS Manual* (New York: The Psychological Corporation).

Johnson, D. A., 1992, Test–retest reliabilities of the Myers–Briggs Type Indicator and the type differentiation indicator over a 30-month period, *Journal of Psychological Type*, 24, 54–58.

Jung, C. G., 1923, *Psychological Types* (New York: Harcourt & Brace).

Karras, J. E., 1990, Psychological and environmental types: implications of degrees of congruence in work, Unpublished doctoral dissertation, The Ohio State University, Columbus, OH.

Krueger, R. F., 2000, Phenotypic, genetic, and nonshared environmental parallels in the structure of personality: a view from the Multidimensional Personality Questionnaire, *Journal of Personality and Social Psychology*, 79, 1057–1067.

Kummerow, J. M., Barger, N. J., and Kirby, L. K., 1997, *Work Types* (New York: Warner Books).

Lawrence, R. C., Helmick, C. G., Arnett, F. C., Deyo, R. A., Felson, D. T., Giannini, E. H., Heyse, S. P., Hirsch, R., Hochberg, M. C., Hunder, G. G., Liang, M. H., Pillemer, S. R., Steen, V. D., and Wolfe, F., 1998, Estimates of the prevalence of arthritis and selected musculoskeletal disorders in the United States. *Arthritis and Rheumatism*, 41(5), 778–799.

Loehlin, J. C., 1992, *Genes and Environment in Personality Development* (Newbury Park, CA: Sage).

Marras, W. S., Davis, K. G., Heaney, C. A., Maronitis, A. B., and Allread, W.G., 2000, The influence of psychosocial stress, gender, and personality on mechanical loading of the lumbar spine, *Spine*, 25(23), 3045–3054.

Matthews, K. A., and Haynes, S. G., 1986, Type A behavior pattern and coronary disease risk: Update and critical evaluation, *American Journal of Epidemiology*, 123, 923–960.

McCrae, R. R., and Costa, P. T., Jr., 1990, *Personality Adulthood* (New York: Guilford).

McCrae, R. R., and Costa, P. T., Jr., 1999, A five-factor theory of personality, in Pervin, L. A., and John, O. P. (eds.), *Handbook of Personality: Theory and Research* (New York: Guilford), pp. 139–153

Myers, I. B. and Myers, P., 1980, *Gifts Differing* (Palo Alto: Consulting Psychologists Press, Inc.).

Myers, I. B., McCaulley, M. H., Quenk, N. L., and Hammer, A. L., 1998, *MBTI Manual: A Guide to the Development and Use of the Myers–Briggs Type Indicator*, 3rd edn (Palo Alto: Consulting Psychologists Press, Inc.).

National Research Council, Institute of Medicine, 2001, *Musculoskeletal Disorders and the Workplace: Low Back and Upper Extremities* (Washington, DC: National Academy Press).

Pervin, L. A., 2003, *The Science of Personality* (New York: Oxford University Press).

Platt, J. J., Pomeranz, D., and Eisenman, R., 1971, Validation of the Eysenck Personality Inventory by the MMPI and Internal–External Control Scale, *Journal of Clinical Psychology*, 27(1), 104–105.

Plomin, R., and Caspi, A., 1999, Behavioral genetics and personality, in Pervin, L. A., and John, O. P. (eds.), *Handbook of Personality: Theory and Research* (New York: Guilford), pp. 251–276.

Praemer, A., Furner, S., and Rice, D. P., 1999, *Musculoskeletal Conditions in the United States* (Rosemont, IL: American Academy of Orthopaedic Surgeons).

Riemann, R., Angleitner, A., and Strelau, J., 1997, Genetic and environmental influences on personality: a study of twins reared together using the self- and peer report NEO-FFI scales, *Journal of Personality*, 65, 449–476.

Rosenman, R. H., Brand, J. H., Jenkins, C. D., Friedman, M., Straus, R., and Wurm, M., 1975, Coronary heart disease in the Western Collaborative Group Study: final follow-up experience of 8.5 years, *Journal of the American Medical Association*, 233, 872–877.

Salminen, J. J., Pentti, J., and Wickstrom, G., 1991, Tenderness and pain in neck and shoulders in relation to type A behaviour, *Scandinavian Journal of Rheumatology*, 20(5), 344–350.

Sauter, S. L., Murphy, L. R., and Hurrell, J. J., Jr., 1990, Prevention of work-related psychological disorders: a national strategy proposed by the National Institute for Occupational Safety and Health, *American Psychologist*, 45, 1146–1158.

Scheier, M. F., and Carver, C. S., 1987, Dispositional optimism and physical well-being: the influence of generalized outcome expectancies on health, *Journal of Personality*, 55(2), 169–210.

Shelton, J., 1996, Health, stress, and coping, in Hammer, A. L. (ed.), *MBTI Applications* (Palo Alto, CA: Consulting Psychologists Press, Inc).

Shekelle, R. B., Gale, M., and NORUSIS, M., 1985a, Type A score (JAS) and risk of recurrent coronary heart disease in the Aspirin Myocardial Infarction Study, *American Journal of Cardiology*, 56, 221–225.

Shekelle, R. B., Hulley, S. B., Neaton, J. D., Billings, J. H., Borhani, N. O., Gerace, T. A., Jacobs, D. R., Lasser, N. L., Mittlemark, M. B., and Stamler, J., 1985b, The MRFIT behavior pattern study II: Type A behavior and incidence of coronary heart disease, *American Journal of Epidemiology*, 122, 559–570.

Smart, J. C, Elton, C. F., and McLaughlin, G. W., 1986, Person–environment congruence and job satisfaction, *Journal of Vocational Behavior*, 29, 216–225.

Smith, C. E., 1988, A comparative study of the Myers-Briggs Type Indicator and the Minnesota Importance Questionnaire in the prediction of job satisfaction. Unpublished doctoral thesis, Ball State University, Muncie, IN.

Wakefield, J. A., Sasek, J., Brubaker, M. L., and Friedman, A. F., 1976, Validity study of the Eysenck Personality Questionnaire, *Psychological Reports*, 39(1), 115–120.

Watkins, C. E., Jr., Campbell, V. L., Nieberding, R., and Hallmark, R., 1995, Contemporary practice of psychological assessment by clinical psychologists, *Professional Psychology: Research and Practice*, 26, 54–60.

Watson, D., Hubbard, B., and Wiese, D., 2000, Self-other agreement in personality and affectivity: the role of acquaintanceship, trait visibility, and assumed similarity, *Journal of Personality and Social Psychology*, 78, 546–558.

White, J. H., Stephenson, G. M., and Child, S. E., 1968, Validation studies of the Eysenck Personality Inventory, *British Journal of Psychiatry*, 114(506), 63–68.

Willis, C. G., 1984, Review of Myers-Briggs Type Indicator, in Keyser, D. J., and Sweetland, R.C. (eds.), *Test Critiques*, Vol. 1 (Austin, TX: Pro-Ed).

Yelin, E. H., Trupin, L. S., and Sebesta, D. S., 1999, Transitions in employment, morbidity, and disability among persons aged 51 to 61 with musculoskeletal and non-musculoskeletal conditions in the U. S., 1992–1994, *Arthritis and Rheumatism*, 42, 769–779.

26

Psychosocial Work Factors

Pascale Carayon
*Ecole des Mines Nancy
France*

Soo-Yee Lim
NIOSH

26.1   Introduction

This chapter examines the concept of psychosocial work factors and its relationship to occupational ergonomics. First, we provide a brief historical perspective of the development of theories and models of work organization and psychosocial work factors. Definitions and examples are then presented. Several explanations are given for the importance of psychosocial work factors in occupational ergonomics. Finally, measurement issues and methods for controlling and managing psychosocial work factors are discussed.

The role of "psychosocial work factors" in influencing individual and organizational health can be traced back to the early days of work mechanization and specialization, and the emergence of the concept of division of labor. Taylor (1911) expanded the principle of division of labor by designing efficient work systems accounting for proper job design, providing the right tools, motivating the individuals, and sharing of responsibilities between management and labor, and sharing of profits. This is known as the era of scientific management in which scientific methods are used to objectively measure work with the aim of improving its efficiency. These scientific methods involved breaking the tasks into small components or units, thus making work requirements and performance evaluations easy to define and monitor. Under these methods, work is simplified and standardized, therefore having a great impact on job and work processes. An analysis of psychosocial work factors in a job in this system would reveal that skill variety is minimal, workers have no control of the work processes, and the job is highly repetitive and monotonous. Such work system design can still be found in numerous workplaces.

As the workforce became more educated, individuals became more aware of their working conditions and environment, and began to seek avenues for improving their quality of working life. This is when the human relations movement emerged (Mayo, 1945), which raised the issue of the potential influence of the work environment on an individual's motivation, productivity, and well-being. Individual needs and wants were emphasized (Maslow, 1970). Thus, job design theorists incorporated worker behavior and work factors in their theories. The two theories of job enlargement and job enrichment formed the basis for many job design theories thereafter. These theories conceptualize the role of worker behavior and perception of the work environment in influencing personal and organizational outcomes. Job enlargement theory emphasized giving a larger variety of tasks or activities to the worker. While this was an improvement from the era of scientific management, the additional tasks or activities could be of a similar skill level and content: workers were performing multiple tasks of the same "kind." This has been called "horizontal loading" of the job, and is the opposite of job enrichment, which focused on the "vertical loading" of the job. Job enrichment aims at expanding the skills used by workers, while at the same time increasing their responsibility. Herzberg (1966), the father of the job enrichment theory, defined intrinsic and extrinsic factors (or motivation versus hygiene factors) that are important to worker motivation, thus leading to satisfaction or dissatisfaction, and psychological well-being. Intrinsic factors are related to the work (or job) conditions, such as having additional control over work schedules or resources, feedback, client relationships, skill use and development, better work content, direct communications, and personal accountability (Herzberg, 1974). Extrinsic factors are related to aspects of financial rewards and benefits and also to the physical environment. Herzberg indicated that extrinsic factors could lead to dissatisfaction with work, but not to satisfaction, while intrinsic factors could increase satisfaction with work. Herzberg's work demonstrated the complex relationships of job conditions, the individual's motivation, satisfaction, dissatisfaction, and psychological well-being. In a way similar to Herzberg's job enrichment theory, the Job Characteristics Theory (Hackman and Oldham, 1976) focused on the idea that specific characteristics of the job (i.e., skill variety, task identity, task significance, autonomy, and feedback) in combination with individual characteristics (growth need strength) would determine personal and work outcomes.

The Sociotechnical Systems Theory recognized two inter-related systems in an organization: the social system and the technical system. The main principle of the Sociotechnical Systems Theory is that the social and technical systems interact with each other, and that the joint optimization of both systems can lead to increased satisfaction and performance. The social system focused on the workers' perception of the work environment (i.e., job design factors) and the technical system emphasized the technology and the work processes used in the work (for example, automation, paced systems, and monitoring systems). In a study of coal mining (Trist and Bamforth, 1951), it was demonstrated that the technical system could impact the social system. In this study where semi-autonomous work groups were set up, workers were given opportunities to make decisions related to their work, and experienced better interactions with workers in their group, as well as task significance and completeness (see also Trist, 1981). Work by Trist and his colleagues showed that technological factors could influence both organizational and job factors. However, it was Davis (Davis 1980) who provided a conceptual framework and a set of principles that formulated the Sociotechnical theory. His framework called for a flattened management structure that would promote participation, interaction between and across groups of workers, enriched jobs, and most important, meeting individual needs. The Sociotechnical Systems Theory laid down the groundwork for the current understanding of how psychosocial work factors can be related to ergonomic factors by examining the interplay between the social and technical systems in organizations. Other recent theories and models of psychosocial work factors will be discussed later.

This rapid overview of the development of job design theories in the 20th century demonstrates the increasing role of psychological, social, and organizational factors in the design of work.

26.2　Definitions

Within the last decade, the role of psychosocial work factors on worker health has gained much popularity. However, the term of "psychosocial work factors" has been used loosely to define and represent many factors that are a part of, attached to or associated with the individuals. Some would consider what has been traditionally termed socioeconomic factors such as income, education level, and demographic or individual factors (e.g., age and marital status) as part of the psychosocial factors (Hogstedt,

Vingard et al., 1995; Ong, Jeyaratnam et al., 1995). In order to understand psychosocial factors in the workplace, one needs to take into account the ability of an individual to make a psychological connection to his or her job, thus formulating the relationship between the person and the job. For instance, the International Labour Office (ILO, 1986) defines psychosocial work factors as "interactions between and among work environment, job content, organizational conditions and workers' capacities, needs, culture, personal extra-job considerations that may, through perceptions and experience, influence health, work performance, and job satisfaction." Thus, the underlying premise in defining psychosocial work factors is the inclusion of the behavioral and psychological components of job factors. In the rest of the chapter, we will use the definitions proposed by Hagberg and his colleagues (Hagberg et al., 1995) because they are most highly relevant for occupational ergonomics.

Work organization is defined as the way work is structured, distributed, processed, and supervised (Hagberg et al., 1995). It is an "objective" characteristic of the work environment, and depends on many factors, including management style, type of product or service, characteristics of the workforce, level and type of technology, and market conditions. Psychosocial work factors are "perceived" characteristics of the work environment that have an emotional connotation for workers and managers, and that can result in stress and strain (Hagberg et al., 1995). Examples of psychosocial work factors include overload, lack of control, social support, and job future ambiguity. Other examples are described in the following section.

The concept of psychosocial work factors raises the issue of objectivity–subjectivity. Objectivity has multiple meanings and levels in the literature. According to Kasl (1978), objective data is not supplied by the self-same respondent who is also describing his distress, strain, or discomfort. At another level, Kasl (1987) feels that "psychosocial factor perception" can be less subjective when the main source of information is the employee but that this self-reported exposure is devoid of evaluation and reaction. Similarly, Frese and Zapf (1988) conceptualize and operationalize "objective stressors" (i.e., work organization) as not being influenced by an individual's cognitive and emotional processing. Based on this, it is more appropriate to conceptualize a continuum of objectivity and subjectivity. Work organization can be placed at one extreme of the continuum (that is the objective nature of work) whereas psychosocial work factors have some degree of subjectivity (see definitions above).

Psychosocial work factors result from the interplay between the work organization and the individual. Given our definitions, psychosocial work factors have a *subjective*, perceptual dimension, which is related to the *objective* dimension of work organization. Different work organizations will 'produce' different psychosocial work factors. The work organization determines to a large extent the type and degree of psychosocial work factors experienced by workers. For instance, electronic performance monitoring, or the on-line, continuous computer recording of employee performance-related activities, is a type of work organization that has been related to a range of negative psychosocial work factors, including lack of control, high work pressure, and low social support (Smith et al., 1992). In a study of office workers, information on psychosocial work factors was related to objective information on job title (Sainfort, 1990). Therefore, psychosocial work factors are very much anchored in the objective work situation, and are related to the work organization.

26.3 Examples of Psychosocial Work Factors

Psychosocial work factors are multiple and various, and are produced by different, interacting aspects of work. The Balance Theory of Job Design (Smith and Carayon-Sainfort, 1989) proposed a conceptualization of the work system with five elements interacting to produce a "stress load." The five elements of the work system are: (1) the individual, (2) tasks, (3) technology and tools, (4) environment, and (5) organizational factors. The interplay and interactions between these different factors can produce various stressors on the individual which then produce a "stress load" which has both physical and psychological components. The stress load, if sustained over time and depending on the individual resources, can produce adverse effects, such as health problems and lack of performance. The models and theories of job design reviewed at the beginning of the chapter tended to emphasize a small set of psychosocial work factors. For instance, the human relations movement (Mayo, 1945) focused on the social aspects of work, whereas the job characteristics theory (Hackman and Oldham, 1976) lists five job characteristics, i.e., skill variety, task identity, task significance, autonomy, and feedback. However, research and practice in the field of work organization has demonstrated that considering only a small number of work factors can be misleading and inefficient in solving job design problems. The balance theory proposes a systematic, global approach to the diagnosis and design or redesign of work systems that does not emphasize any one aspect of work. According to the balance theory, psychosocial work factors are multiple and of diverse nature.

Table 26.1 lists eight categories of psychosocial work factors and specific facets in each category. This list cannot be considered as exhaustive, but is representative of the most often studied psychosocial work factors.

The study of psychosocial work factors needs to be tuned in to the changes in society. Changes in the economic, social, technological, legal, and physical environment can produce new psychosocial work factors. For instance, in the context of office automation, four emerging issues are appearing (Carayon and Lim, 1994): (1) electronic monitoring of worker performance, (2) computer-supported work groups, (3) links between the physical and psychosocial aspects of work in automated offices, and (4) technological changes. The issue of technological changes applies nowadays to a large segment of the work population. Employees are asked to learn new technologies on a frequent, sometimes continuous, basis. Other trends in work organization include the development of teamwork and other work arrangements, such as telecommuting. These new trends may produce new psychosocial work factors, such as high dependency on technology, lack of socialization on the job and identity with the organization, and pressures from teamwork. Two APA publications review psychosocial stress issues related to changes in the workforce in terms of gender, diversity, and family issues (Keita and Hurrell, 1994), and some of the emergent psychosocial risk factors and selected occupations at risk of psychosocial stress (Sauter and Murphy, 1995).

TABLE 26.1   Selected Psychosocial Work Factors and their Facets

1.	Job demands	Quantitative workload
		Variance in workload
		Work pressure
		Cognitive demands
2.	Job content	Repetitiveness
		Challenge
		Utilization and development of skills
3.	Job control	Task/instrumental control
		Decision/organizational control
		Control over physical environment
		Resource control
		Control over work pace: machine-pacing
4.	Social interactions	Social support from supervisor and colleagues
		Supervisor complaint, praise, monitoring
		Dealing with (difficult) clients/customers
5.	Role factors	Role ambiguity
		Role conflict
6.	Job future and career issues	Job future ambiguity
		Fear of job loss
7.	Technology issues	Computer-related problems
		Electronic performance monitoring
8.	Organizational and management issues	Participation
		Management style

26.4   Occupational Ergonomics and Psychosocial Work Factors

The emergence of macroergonomics has strongly contributed to the increasing interest in psychosocial work factors in the occupational ergonomics field (Hendrick, 1991; Hendrick, 1996). As shown above, the work factors can be categorized into the individual, task, tools and technologies, physical environment, and the organization (Smith and Carayon-Sainfort, 1989). They can also be described as either physical or psychosocial (Cox and Ferguson, 1994). Cox and Ferguson (1994) developed a model of the effects of physical and psychosocial factors on health. According to this model, the effects of work factors on health are mediated by two pathways: (1) a direct physicochemical pathway, and (2) an indirect psychophysiological pathway. These pathways are present at the same time, and interact in different ways to affect health. Physical work factors can have direct effects on health via the physicochemical pathway, and indirect effects on health via the psychophysiological pathway, but can also moderate the effect of psychosocial work factors on health via the psychophysiological pathway. This model demonstrates the close relationship between physical and psychosocial work factors in their influence on health and wellbeing.

The importance of psychosocial work factors in the field of occupational ergonomics emerges from several considerations.

1. Physical and psychosocial ergonomics are interested in the same job factors
2. Physical and psychosocial work factors are related to each other
3. Psychosocial work factors play an important role in physical ergonomics interventions
4. Physical and psychosocial work factors are related to the same outcome, for instance, work-related musculoskeletal disorders

First, some of the concepts examined in the physical ergonomics literature are similar to concepts examined in the psychosocial ergonomics literature. For instance, the degree of repetitiveness of a task is very important from both physical and psychosocial points of view. Physical ergonomists are more interested in the effect of the task repetitiveness on motions and force exerted on certain body parts, such as hands; whereas psychosocial ergonomists are concerned about the effect of task repetitiveness on monotony, boredom, and dissatisfaction with one's work (Cox 1985). In the physical ergonomics literature, an important job redesign strategy for dealing with repetitiveness is job rotation: workers are rotated between tasks which require effort from different body parts and muscles, therefore reducing the negative effects of repetitiveness of motions in a single task. From a psychosocial point of view, job rotation is one form of job enlargement (see above for a discussion of job enlargement). However, as discussed earlier, the psychosocial benefits of job rotation are limited because workers may be simply performing a range of similar, nonchallenging tasks. From a physical ergonomics point of view, job rotation is effective only if the physical variety of the tasks is increased; whereas from a psychosocial ergonomics point of view, job rotation is effective only to the extent of the content and meaningfulness of the tasks.

Second, physical and psychosocial work factors can be related to each other. For instance, the model proposed by Lim (1994) states that the psychosocial factor of work pressure can influence the physical factors of force and speed of motions. According to this model, workers may change their behaviors under the influence of work pressure, and therefore, tend to exert more force or to speed up their work. Empirical evidence tends to confirm this relationship between work pressure (i.e., a psychosocial work factor) and physical work factors (Lim 1994). Another form of relationship between physical and psychosocial work factors is evident in the literature on control over one's physical environment. In this case, the psychosocial work factor of control is applied to one particular facet of the work, that is the physical environment. Control over one's physical environment can, therefore, have benefits from a physical point of view (i.e., being able to adapt one's physical environment to one's physical characteristics and task requirements), but also from a psychosocial point of view (i.e., having control is known to have many psychosocial benefits [Sauter et al., 1989]).

Third, psychosocial work factors are a crucial component of physical ergonomics interventions. In particular, the concept of participatory ergonomics uses the benefits of one psychosocial work factor, that is participation, in the process of implementing physical ergonomics changes (Noro and Imada, 1991). From a psychosocial point of view, using participation is important to improve the process and outcomes of ergonomic interventions. In addition, any type of organizational interventions, including ergonomic interventions, can be stressful because of the emergence of negative psychosocial work factors, such as uncertainty and increased workload (i.e., having more work during the intervention or the transitory period). Therefore, in any physical ergonomics intervention, attention should be paid to psychosocial work factors in order to improve the effectiveness of the intervention and to reduce or minimize its negative effects on workers.

Fourth, physical and psychosocial work factors can be related to the same outcome. One of these outcomes is work-related musculoskeletal disorders (WMSDs). There is increasing theoretical and empirical evidence that both physical and psychosocial work factors play a role in the experience and development of WMSDs (Hagberg et al. 1995; Moon and Sauter, 1996). Several mechanisms for the joint influence of physical and psychosocial work factors on WMSDs have been presented (Smith and Carayon, 1996). Therefore, in order to fully prevent or reduce WMSDs, both physical and psychosocial work factors need to be considered.

26.5   Measurement of Psychosocial Work Factors

From the occupational ergonomics point of view, the purpose of examining psychosocial work factors is to investigate their influence on and role in worker health and well-being. Thus, psychosocial work factors can be considered as predictors (i.e., independent variables), while worker health and well-being serve as the dependent variables or outcomes. The measurement or assessment of well-being can be classified into two levels of measures in terms of "context-free" (that is, life in general or general satisfaction) and "context-specific" (for example, job-related well-being) (Warr, 1994). It is the latter level of measure, "context-specific" that is relevant to the assessment of psychosocial work factors in the work-place. Table 26.1 shows a selected sample of the many different dimensions of jobs (for example, job demands, control, social support) that have been studied extensively. Furthermore, each dimension is made up of different facets that define and operationalize that particular dimension. For example, as shown in Table 26.1, the dimension of job demands consists of various facets, such as quantitative workload, variance in workload, work pressure, and cognitive demands; the dimension of job content includes repetitiveness, challenge on the job, and utilization of skills. It should be noted that Table 26.1 is not an exhaustive list of psychosocial work factors.

The most often used method for measuring psychosocial work factors in applied settings is the questionnaire survey. Difficulties with questionnaire data on psychosocial work factors are often due to the lack of clarity of the definitions of the measured factors or poorly designed questionnaire items that measure "overlapping" conceptual dimensions of the psychosocial work factor of interest. Measures of any one facet typically include several items that can be grouped in a "scale." Reliability of the scale is often being assessed by the Cronbach-alpha score method in which the intercorrelations among the scale items are examined for internal consistency. In general, it is recommended that existing, well-established scales be used in order to ensure the "quality" of the data (i.e., reliability and validity) and to be able to compare the newly collected data with other groups for which data has been collected with the same instrument (benchmarking).

The level of objectivity/subjectivity of the measures of psychosocial work factors will depend on the degree of influence of cognitive and emotional processing. For example, ratings of work factors by an observer cannot be considered as purely objective because of the potential influence of the observer's cognitive and emotional processing. However, ratings of work factors by an outside observer can be considered as more objective than an evaluative question answered by an employee about his/her work environment (e.g., "How stressful is your work environment?"). However, self-reported measures of psychosocial work factors can be more objective when devoid of evaluation and reaction (Kasl and Cooper, 1987). As discussed earlier, any kind of data can be placed somewhere on this objectivity/subjectivity continuum from "low in dependency on cognitive and emotional processing" (e.g., objective) to "high in dependency on cognitive and emotional processing" (e.g., subjective).

We discuss three different questionnaires which include numerous scales of psychosocial work factors. In addition, validity and reliability analyses have been performed on all three questionnaires. Two of these questionnaires have been developed and used to measure psychosocial work factors in various groups of workers or large samples of workers: (1) the NIOSH Job Stress questionnaire (Hurrell and McLaney 1988), and (2) the Job Content Questionnaire (JCQ) (Karasek, 1979). The NIOSH Job Stress questionnaire is often used in the Health Hazard Evaluations performed by NIOSH. Translations of Karasek's JCQ exist in many different languages, including Dutch and French. The University of Wisconsin Office Worker Survey (OWS) is a questionnaire developed to measure psychosocial work factors in office/computer work (Carayon, 1991). This questionnaire covers a wide range of psychosocial work factors of importance in office and computer work. In addition to many of the psychosocial work factors measured by the NIOSH Job Stress Questionnaire or Karasek's JCQ, the OWS measures psychosocial work factors related to computer technology, such as computer-related problems (Carayon-Sainfort, 1992). The OWS questionnaire has been translated into Finnish, Swedish, and German. For all three questionnaires, data exist for various groups of workers in numerous organizations of multiple countries. This data can serve as a comparison to newly collected data and for benchmarking. Numerous other questionnaires for measuring psychosocial work factors exist, such as the Occupational Stress Questionnaire in Finland (Elo et al., 1994) and the Occupational Stress Indicator in England (Cooper et al., 1988). Other questionnaires are listed in Cook et al. (1981).

26.6 Managing and Controlling Psychosocial Work Factors

It is clear from the job design and occupational stress literature that jobs with negative psychosocial work factors, such as repetitiveness, no opportunity to develop skills, and low control, can have adverse effects on job performance and mental and physical health. Various approaches have been proposed to improve the design of jobs, such as job rotation and other forms of job enlargement, and job enrichment (see above). These strategies can be efficient to increase the variety in a job, to reduce the dependence on a particular technology or tool, and to increase worker control and responsibility. In particular, lack of job control is seen as a critical psychosocial work factor (Sauter et al., 1989). Providing a greater amount of control can be achieved by, for instance, allowing workers to determine their work schedules in accordance with organizational policies and production requirements, by allowing workers to give input into decisions that affect their jobs, by letting workers choose the best work procedures and task order, and by increasing worker participation in the production process. An experimental field study of a participation program showed the positive effects of participation on emotional distress and turnover (Jackson, 1983). According to the Sociotechnical Systems theory, autonomous work groups can be an effective strategy for increasing worker control and enriching jobs. Beyond increased control and improved job content, some forms of teamwork can have other positive psychosocial benefits, such as increased opportunity for socialization and learning.

Achieving the perfect job without any negative psychosocial work factors may not be feasible or realistic, given individual, organizational, or technological constraints and requirements. The balance theory (Smith and Carayon-Sainfort, 1989) proposes a job redesign strategy that aims to achieve an optimal job design. In this process, negative psychosocial work factors need to be eliminated or reduced as much as possible. However, when this is not possible, positive psychosocial work factors can be used to reduce the impact of negative psychosocial work factors. This balancing, or compensating, effect is based on the concept of the work system of the balance theory. The five elements of the work system (the individual, tasks, technology and tools, environment, and organizational factors) are interrelated: they can influence each other, and they can also influence the impact or effect of each other or their interactions. In this systems approach, negative psychosocial work factors can be balanced out or compensated by positive work factors.

Some trends in the field of organizational design and management may have positive characteristics from a psychosocial point of view. For instance, under certain conditions, the use of quality engineering and management methods can positively affect the psychosocial work environment, such as increased opportunity for participation, and learning and development of quality-related skills (Smith et al., 1989). However, other trends in the business world can have negative effects on the psychosocial work environment. For instance, downsizing and other organizational restructuring and reengineering may create highly stressful situations of uncertainty and loss of control (DOL 1995).

26.7   Conclusion

This chapter has demonstrated the importance of psychosocial work factors in the research and practice of occupational ergonomics. In order to clarify the issue at hand, we presented definitions of work organization and psychosocial work factors. It is important to understand the long research tradition on psychosocial work factors that has produced numerous models and theories, but also valid and reliable methods for measuring psychosocial work factors. At the end of the chapter, we presented examples of methods for managing and controlling psychosocial work factors.

Psychosocial work factors need to be taken into account in the research on and practice of occupational ergonomics. We have discussed the important role of psychosocial work factors with regard to physical ergonomics. In addition, given the constantly changing world of work and organizations, we need to pay even more attention to the multiple aspects of people at work, including psychosocial work factors.

References

Carayon, P. (1991). *The Office Worker Survey.* Madison, WI, Department of Industrial Engineering, University of Wisconsin-Madison.

Carayon, P. and Lim, S.-Y. (1994). Stress in automated offices. *The Encyclopedia of Library and Information Science.* A. Kent. New York, Marcel Dekker. Vol. 53, Supplement 16: 314–354.

Carayon-Sainfort, P. (1992). The use of computers in offices: impact on task characteristics and worker stress. *International Journal of Human Computer Interaction* 4(3): 245–261.

Cook, J. D., Hepworth, S. J. et al. (1981). *The Experience of Work.* London, Academic Press.

Cooper, C. L., Sloan, S. J. et al. (1988). *Occupational Stress Indicator.* Windsor, England, NFER-Nelson.

Cox, T. (1985). Repetitive work: Occupational stress and health. *Job Stress and Blue-Collar Work.* C. L. Cooper and M. J. Smith. New York, John Wiley & Sons: 85–112.

Cox, T. and Ferguson, E. (1994). Measurement of the subjective work environment. *Work and Stress* 8(2): 98–109.

Davis, L. E. (1980). Individuals and the organization. *California Management Review* 22(2): 5–14.

DOL (1995). *Guide to responsible restructuring.* Washington, D.C. 20210, U.S. Department of Labor, Office of the American Workplace.

Elo, A.-L., Leppanen, A., et al. (1994). The Occupational Stress Questionnaire. *Occupational Medicine.* C. Zenz, O. B. Dickerson and E. P. Horvarth. St. Louis, Mosby: 1234–1237.

Frese, M. and Zapf, D. (1988). Methodological issues in the study of work stress. *Causes, Coping and Consequences of Stress at Work.* C. L. Cooper and R. Payne. Chichester, John Wiley & Sons.

Hackman, J. R. and Oldham, G. R. (1976). Motivation through the design of work: test of a theory. *Organizational Behavior and Human Performance* 16: 250–279.

Hagberg, M., Silverstein, B., et al. (1995). *Work-Related Musculoskeletal Disorders (WMSDs): A Reference Book for Prevention.* London, Taylor & Francis.

Hendrick, H. W. (1991). Human Factors in organizational design and management. *Ergonomics* 34: 743–756.

Hendrick, H. W. (1996). Human factors in ODAM: an historical perspective. *Human Factors in Organizational Design and Management -V.* O. J. Brown and H. W. Hendrick. Amsterdam, The Netherlands, Elsevier Science Publishers: 429–434.

Herzberg, F. (1966). *Work and the Nature of Man.* New York, Thomas Y. Crowell Company.

Herzberg, F. (1974). The wise old turk. *Harvard Business Review* (September/October): 70–80.

Hogstedt, C., E. Vingard, et al. (1995). *The Norrtalje-MUSIC Study — An ongoing epidemiological study on risk and health factors for low back and neck-shoulder disorders.* PREMUS'95-Second International Scientific Conference and Prevention of Work-Related Musculoskeletal Disorders, Montreal, Canada.

Hurrell, J. J. J. and M. A. McLaney (1988). "Exposure to job stress — A new psychometric instrument." *Scandinavian Journal of Work Environment and Health* 14(Suppl. 1): 27–28.

ILO (1986). *Psychosocial Factors at Work: Recognition and Control.* Geneva, Switzerland, International Labour Office.

Jackson, W. E. (1983). Participation in decision-making as a strategy for reducing job-related strain. *Journal of Applied Psychology* 68: 3–19.

Karasek, R. A. (1979). Job demands, job decision latitude, and mental strain: implications for job redesign. *Administrative Science Quarterly* 24: 285–308.

Kasl, S. V. (1987). Methodologies in stress and health: past difficulties, present dilemmas, future directions. *Stress and Health: Issues in Research and Methodology.* S. V. Kasl and C. L. Cooper. Chichester, John Wiley & Sons: 307–318.

Kasl, S. V. and C. L. Cooper, Eds. (1987). *Stress and Health: Issues in Research and Methodology.* Chichester, John Wiley & Sons.

Keita, G. P. and Hurrell, J. J. J. (1994). *Job Stress in a Changing Workforce — Investigating Gender, Diversity, and Family Issues.* Washington, D.C., APA.

Lim, S. (1994). An integrated approach to cumulative trauma disorders in computerized offices: the role of psychosocial work factors, psychological stress and ergonomic risk factors. *IE.* Madison, WI, University of Wisconsin-Madison.

Maslow, A. H. (1970). *Motivation and Personality.* New York, Harper and Row.

Mayo, E. (1945). *The Social Problems of an Industrial Civilization.* Andover, MA, The Andover Press.

Moon, S. D. and Sauter, S. L. Eds. (1996). *Beyond Biomechanics — Psychosocial Aspects of Musculoskeletal Disorders in Office Work.* London, Taylor & Francis.

Noro, K. and Imada, A. (1991). *Participatory Ergonomics.* London, Taylor & Francis.

Ong, C. N., Jeyaratnam, J., et al. (1995). "Musculoskeletal disorders among operators of video display terminals." *Scandinavian Journal of Work Environment and Health* 21(1): 60–64.

Sainfort, P. C. (1990). *Perceptions of Work Environment and Psychological Strain Across Categories of Office Jobs.* The Human Factors Society 34th Annual Meeting.

Sauter, S. L., Hurrell, J. J. Jr., et al., Eds. (1989). *Job Control and Worker Health.* Chichester, John Wiley & Sons.

Sauter, S. L. and Murphy, L. R. (1995). *Organizational Risk Factors for Job Stress.* Washington, D.C., APA.

Smith, M. J. and Carayon, P. (1996). Work organization, stress, and cumulative trauma disorders. *Beyond Biomechanics — Psychosocial Aspects of Musculoskeletal Disorders in Office Work.* S. D. Moon and S. L. Sauter. London, Taylor & Francis: 23–41.

Smith, M. J., Carayon, P., et al. (1992). Employee stress and health complaints in jobs with and without electronic performance monitoring. *Applied Ergonomics* 23(1): 17–27.

Smith, M. J. and Carayon-Sainfort, P. (1989). A balance theory of job design for stress reduction. *International Journal of Industrial Ergonomics* 4: 67–79.

Smith, M. J., Sainfort, F., et al., Eds. (1989). *Efforts to Solve Quality Problems,* Secretary's Commission on Workforce Quality and Labor Market Efficiency, U.S. Department of Labor, Washington, D.C.

Taylor, F. (1911). *The Principles of Scientific Management.* New York, Norton and Company.

Trist, E. (1981). *The Evaluation of Sociotechnical Systems.* Toronto, Quality of Working Life Center.

Trist, E. L. and Bamforth, K. (1951). Some social and psychological consequences of the long-wall method of coal getting.

27

Biomechanical Modeling of the Shoulder

Krystyna
Gielo-Perczak
*Liberty Mutual Research
Institute for Safety*

27.1 Introduction

At present, it is difficult to establish a unifying integrative approach in the study of glenohumeral stability and upper extremity strength. The glenohumeral joint is the most mobile articulation of the human body, involving interacting and interrelated geometric variables relating bones, muscles, and ligaments, that needs further exploration and understanding. The purpose of the current study was to propose a method of geometrical description of the glenoid fossa in order to deduce an unrecognized inter-relationship between glenoid concavity and the deltoid attachment on the glenohumeral head position. It supports the concept of a new biomechanical parameter, the dynamic glenohumeral stability index proposed by Lee and An,[16] which considers not only the force vectors generated by the deltoid muscle but also the concavity compression mechanism. Recent analyses of the glenohumeral joint have not focused on the contribution of geometric parameters like the shape of the glenoid fossa, radius of the humerus, the attachment of the deltoideus lateral part in relation to the glenoid fossa, and the angular measure of the articular surface of the humerus head.

The concavity of the glenoid fossa plays a significant role in the stability of the glenohumeral joint as was revealed in recent experiments by Lee and An.[16] A labrum may not contribute to glenohumeral stability as much as was previously assumed, as also was found by Halder et al.[8] It has been shown that the glenohumeral joint exhibits ball-and-socket kinematics; however, its motion is coupled with translation of the humeral head on the glenoid which requires a certain degree of mismatch of the articulating surfaces and leads to variations in the joint-contact area.[8,9,11,17]

Gagey and Hue[5,6] recently analyzed the mechanics of the deltoid muscle and suggested that the deltoid acts on the humeral head like a cable on a pulley. They pointed out "the downward-oriented force applied by the deltoid to the head depends on the angle of reflection around the humerus." These observations suggest that there exist more geometric details which never were analyzed.

Saha[18] pointed out the variance in the congruity of the glenoid and humeral head. Gielo-Perczak[7] confirmed in theoretical studies the influence of the variance in the congruity on physical capabilities. Additionally, Lee and An[16] reported the strong relationship between glenoid depth and stability of a joint. The study provided by Halder et al.[8] revealed "the degree of stability of the glenohumeral joint depends on the perpendicular component of the rotator-cuff muscle forces as well as on the radius of the articular surface." Halder et al.[8] hypothesized that "the degree of stability through concavity-compression is position-dependent." Generally, the translations in the inferior–superior direction (more than 4 mm) are larger than in the anterior–posterior direction.[3] These translations are a function of glenoid concavity, which in all previous studies was described as a part of a sphere or ellipse. Are these adequate descriptions to explain the variety of the glenohumeral head positions and the strength of the joint during abduction of the arm?

The glenoid concavity is irregular and less marked than the convexity of the humeral head,[14] thus it needs adequate radiographic data and analysis. There are two major controversies relating to the shape of the glenoid which is considered with or without the cartilage surfaces of the glenohumeral joint. One represents the concept that the glenoid is flatter[18] or apparently shallow[3] if taking into account the bone surfaces beneath the articulating surfaces as determined from radiograph. The second represents that the glenoid can be described as a part of a sphere. Bigliani and his coworkers[2] observed that out of nine fresh frozen human cadaveric shoulders (average age 50 yr; range 42–59 yr), the glenoid bone surfaces for eight had a larger radius of curvature than the matching humeral head bone surfaces (difference was less than 2.5 mm), whereas there was only one joint in which the humeral head had a larger radius of curvature than the glenoid (difference was 1.2 mm). Iannotti and his coworkers[13] (1992) shared similar observations. Are these inconsistencies and lack of consensus in the descriptions of the glenohumeral joint the result of conceptually omitting the geometric parameters in the modeling of its structure and, as a next step, disregarding them in the experiments?

A few studies have focused on measuring the glenoid geometry by:

- The lateral humeral displacement during translation across the glenoid[8,15]
- The glenoid articular surface angles in the superior–inferior and anterior–posterior planes[1]
- The depths in the superoinferior and anteroposterior directions[10]

However, none of these terms adequately provides a description of the glenoid fossa concavity which can be used in mechanical studies.

The purpose of the study was twofold. (1) Theoretically explain the influence of the glenoid concavity (without the labrum) and the middle deltoid attachment on humeral head positions in the superior–inferior direction. (2) By applying the theoretical results, propose a method of geometrical description of the glenoid and formulate a necessary number of segments tangent to the glenoid, which can be used to determine the planar shape of the glenoid fossa.

27.2 Methods

27.2.1 Theoretical Consideration

27.2.1.1 Effects of Glenoid Inclination and the Attachment of the Deltoid Lateral on Humeral Head Position

A planar model was developed to determine the influence of tangent inclination to the glenoid surface, and the distance of the deltoideus lateral part attachment on humeral head positions.

The model (Figure 27.1) was constructed based on the assumption that: (1) the analysis was provided during abduction ranging from 0° to 170° and the forces due to gravity were applied at the same point at which the external force was applied, (2) the capsuloligamentous structures were removed, (3) the

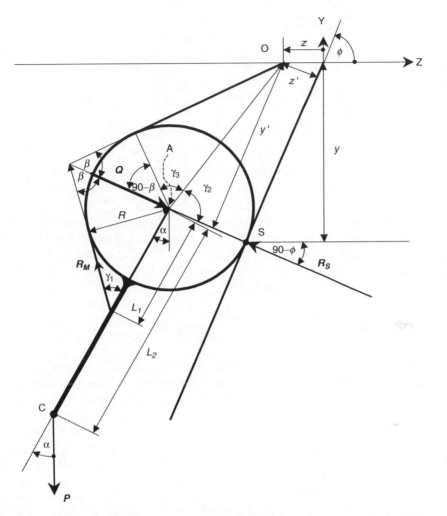

FIGURE 27.1 A planar model of the glenohumeral joint with the glenoid inclination and the distance of the deltoideus lateral part attachment.

glenohumeral joint was restrained only by the deltoid lateral part, (4) the glenohumeral joint reaction force acts perpendicular to the articular surface, (5) friction at the joint was omitted. Reference axes Y–Z were fixed in the scapula at the center of the lateral margin of the acromion, where the Y-axis was parallel to the longitudinal axis of the thorax.

The main feature of the approach was combining the possible mutual geometric relations, including the position of the humerus, the inclination of the tangents to the glenoid surface, the radius of the humerus, the angle between the middle deltoid attachment, and the muscle contact with the humerus, with the conditions for equilibrium of the structure during static loading.

These relations are:

$$\gamma_1 = \sin^{-1}\left(\frac{R}{L_1}\right) \tag{27.1}$$

$$\gamma_2 = \tan^{-1}\left(\frac{\frac{y}{\sin\phi} - z\cos\phi}{R - z\sin\phi}\right) \tag{27.2}$$

$$\gamma_3 = \cos^{-1}\left(\frac{R}{\sqrt{(\frac{y}{\sin}\phi - z\cos\phi)^2 + (R - z\sin\phi)^2}}\right) \tag{27.3}$$

$$\beta = 0.5(\gamma_2 + \gamma_3 - \gamma_1 + \phi + \alpha - 90) \tag{27.4}$$

$$Q = 2R_M \cos\beta \tag{27.5}$$

$$\left.\begin{array}{l} \sum M_A = P\ L_2 \sin\alpha - R_M\ R = 0 \\ \sum F_Z = -R_S \sin\phi - R_M \sin(\gamma_1 - \alpha) - Q\ \cos(\beta - \gamma_2 - \gamma_3 - \phi) = 0 \\ \sum F_Y = -P + R_S \cos\phi + R_M \cos(\gamma_1 - \alpha) + Q \sin(\beta - \gamma_2 - \gamma_3 - \phi) = 0 \end{array}\right\} \tag{27.6}$$

The known variables are: α — glenohumeral joint angle defined as the angle between the axis of the humerus and the Y-axis of body; ϕ — an angle at which a tangent to a given curve of the glenoid fossa at the point S in the frontal plane crosses the Z-axis; z — distance measured along the Z-axis between the middle deltoid attachment and a point at which a tangent to the glenoid fossa crosses the Z-axis; γ_1 — angle between the distal tendon of the deltoid fiber and the axis of the humerus; P — external load and reduced weight of upper limb applied at point C; L_1 — distance between the distal insertion of the middle deltoid and center of the humerus head considered as a sphere; L_2 — distance between the center of the humerus head and point of hand where an external load and reduced weight of upper limb were applied; R — radius of the humerus head.

The unknown variables are: R_S — the glenohumeral reaction force between the articular surface of the glenoid fossa and the head of the humerus which is perpendicular to the glenoid articular surface described by a tangent at its midpoint; R_M — the force in a single muscle fiber of the deltoideus lateral par, applied on the humeral insertion of the deltoideus; Q — the reaction force of the deltoideus on the curved humerus area, which is a function of R_M; y — distance along the Y-axis at which a humerus is balanced; β — angle of "glenohumeral pulley" defined as an angle of the curved contact area of muscle fiber with the head of the humerus; γ_2 — angle at the center of the humerus formed by the perpendicular line of reaction force and line connected with the distal insertion of the middle deltoid fiber muscle; γ_3 — angle at the center of the humerus formed by the perpendicular line to the muscle fiber at the curved contact area and line connected with the distal insertion of muscle. This geometric model should be applied to each fiber of the lateral part of the deltoideus.

The variable y was calculated as a distance along the Y-axis at which a humerus was in balance. The detailed trigonometric analysis revealed the relations into analytic form and the geometric equations were:

$$z' = z\sin\phi$$

$$E_z = (R - z')^2$$

$$B_z = (R - z')^2 - R^2$$

$$C_z = (R - z')R$$

$$G_z = \frac{R\sin\phi}{L_2\sin\alpha} \tag{27.7}$$

$$D_z = G_z E_z + C_z$$

$$I_z = \sqrt{4D_z^2 - 4B_z G_z D_z + B_z^2}$$

$$J_z = \frac{B_z - 2G_z D_z - I_z}{2G_z^2 - 1}$$

$$y = \sqrt{J_z}\sin\phi + z\sin\phi\cos\phi$$

Parametric analyses have been provided for changing values ϕ, z, and function of α. The positions of the humerus head were calculated for each angle of elevation/abduction $\alpha = 0°–170°$, and for the four parametric conditions: (1) $\phi = 90°$, $z = 0$ mm (Figure 27.2a); (2) $\phi = 90°$, $z = 0–32$ mm (Figure 27.2b); (3) $\phi = 10°–170°$, $z = 0$ mm (Figure 27.2c and e); (4) $\phi = 10°–170°$, $z = 0–32$ mm (Figure 27.2d and f). In the four geometrical structures related to joint articulations, the concavity of the glenoid fossa was reduced to a tangent to the glenoid surface. The deltoid lateral part was attached directly to the glenoid fossa or at the certain distance (Figure 27.2).

For the first conditions (Figure 27.2a) where $\phi = 90°$ (a tangent to the glenoid fossa was a vertical line) and $z = 0$ mm (the deltoid lateral part was attached directly to the glenoid fossa), the distance along the Y-axis was:

$$y = R\sqrt{\frac{L_2 \sin \alpha + R}{L_2 \sin \alpha - R}} \tag{27.8}$$

For the second condition (Figure 27.2b) where $\phi = 90°$ (a tangent to the glenoid fossa was a vertical line) and z ranged from 0 to 32 mm (the deltoid lateral part was attached at a certain distance from the

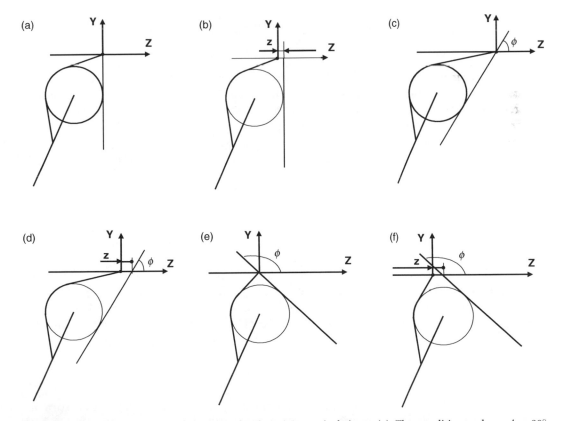

FIGURE 27.2 The geometrical structures related to joint articulations. (a) The conditions where $\phi = 90°$, $z = 0$ mm. (b) The conditions where $\phi = 90°$, $z = 0–32$ mm. (c) The conditions where $\phi = 0°–90°$, $z = 0$ mm. (d) The conditions where $\phi = 0°–90°$, $z = 0–32$ mm. (e) The conditions where $\phi = 90°–180°$, $z = 0$ mm. (f) The conditions where $\phi = 90°–180°$, $z = 0–32$ mm.

glenoid fossa), the geometric relations were:

$$E = (R - z)^2$$
$$B = (R - z)^2 - R^2$$
$$C = (R - z)R$$
$$G = \frac{R}{L_2 \sin \alpha}$$
$$D = GE + C \tag{27.9}$$
$$I = \sqrt{4D^2 - 4BGD + B^2}$$
$$J = \frac{B - 2GD - I}{2G^2 - 1}$$
$$y = \sqrt{J}$$

For the third conditions (Figure 27.2c and e) where $\phi = 10°$ to $170°$ (a tangent to the glenoid fossa was a line crossing the Z-axis in the range from $10°$ to $170°$) and $z = 0$ mm (the deltoid lateral part was attached to the glenoid fossa which was an inclined articular surface), the conditions were represented by both an acute angle (Figure 27.2c) and an obtuse angle (Figure 27.2e). The geometric relations were:

$$y = R \sin \phi \sqrt{\frac{L_2 \sin \alpha + R \sin \phi}{L_2 \sin \alpha - R \sin \phi}} \tag{27.10}$$

For the fourth conditions (Figure 27.2d and f) where $\phi = 10°$ to $170°$ (a tangent to the glenoid fossa was a line crossing the Z-axis in the range from $10°$ to $170°$) and z ranged from 0 to 32 mm (the deltoid lateral part was attached at a certain distance from the glenoid fossa), the conditions are represented by both an acute angle (Figure 27.2d) and obtuse angle (Figure 27.2f). The geometric relations were as presented in Equation (27.7).

27.2.2   Method of Geometrical Description

27.2.2.1   Concavity of the Glenoid Fossa

The problem of defining the smallest number of tangents which can be used to determine the planar shape of the glenoid fossa was resolved. Assuming that a glenoid fossa can be described as a convex function and defined as a polyhedral model, the mathematical task was formulated.[4] Reference axes Y–Z were fixed in the scapula at the center of the lateral margin of the acromion (the middle deltoid attachment), where the Y-axis was parallel to the longitudinal axis of the thorax. The concavity of the glenoid fossa, as a convex function, was considered as a polyhedral model of f based on linearizations f_i, $i = 1, \ldots, k$ with an error of

$$\varepsilon_i(z) = f(z) - f_i(z) = f(z) - \langle g_i, z - z_i \rangle - f(z_i) \tag{27.11}$$

where g_i is a subgradient of f in the point z_i (if f is differentiable). The nodes $z_i, \ldots, z_k$ on the curve of the glenoid in the frontal plane were determined when the maximal linearization error $\varepsilon(z)$ was minimal. The tangents to a given glenoid fossa in the frontal plane $f(z_i)$ are the straight lines at the points $(z_i, f(z_i))$ which cross the Z-axis at angles $\phi_1, \ldots, \phi_i$. Applying the smallest error of the polyhedral model of a convex function,[2] it was found that the minimum number of tangents necessary to approximate a glenoid shape is

$$k = \frac{c}{2\sqrt{2}e} \tag{27.12}$$

where c is the height of glenoid fossa measured along the Y-axis and e is the tolerance of measurement. Both terms are expressed in millimeters. For example, if the height of glenoid fossa c was 39.1 mm and the humerus magnetic resonance imaging (MRI) slices were 2 mm in thickness, then the number of tangents was seven (Figure 27.3). For thinner MRI slices, the minimum number of tangents needed to fit the glenoid fossa geometric surface would be larger.

27.3 Data Collection

The method of geometrical description of the glenoid fossa concavity in a frontal plane, for a cross section through the acromion, has been applied for 12 subjects participating in an experiment to determine their glenoid geometries. The participants' mean age (and standard deviation) was 40.5 ± 8 yr, with an average height of 178 ± 7.09 cm and body weight of 81.54 ± 15.60 kg. Proper informed consent forms were obtained from the participants, as approved by the Institutional Review Committee (IRC) at the Liberty Mutual Research Institute for Safety. The geometric data of the glenoid concavity means the tangents' inclination ϕ_i at the seven nodes ($i = 7$) on the curve of the glenoid and the distances z_i between the middle deltoid attachment and a point at which each tangent to the glenoid fossa crosses the Z-axis (Figure 27.3) were collected. Magnetic resonance images were taken at the same arm positions

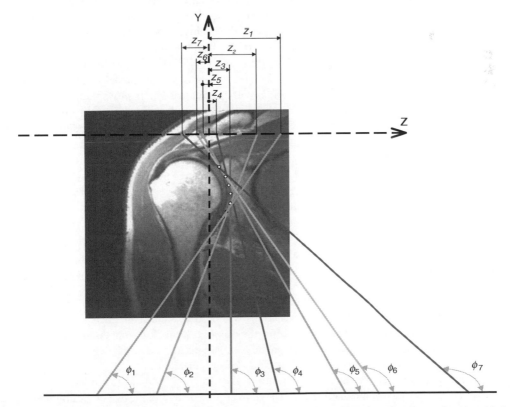

FIGURE 27.3 The group of seven tangents approximated the bone surface data that fit to the glenoid fossa geometric surface from MRI slice of 2 mm thickness.

for all 12 male participants. Data were recorded by using the 1,5-T Sigma system (GE Medical Systems, Milwaukee, WI). From 120 images taken in the frontal plane, a total of 12 characteristic images were identified. The humerus MRI slices were 2 mm in thickness. Data were digitized by using AutoCAD R14 and Alice v.4.4.9 software.

27.4   Results and Discussion

These theoretical studies revealed that a migration of the humeral head in the superior–inferior direction was related to the glenoid fossa inclination and the attachment of the lateral part of the deltoideus muscle between the glenoid concavity and the middle deltoid attachment, during abduction ranging from 0° to 170°. Generally, the largest translations of the humerus were during abduction ranging from 0° to 40° and from 140° to 170°. The translations increased when tangents ϕ increased from 10° to 90° for the deltoid lateral part attached directly to the glenoid fossa ($z = 0$). The translations were reduced when tangents ϕ decreased from 90° to 170° for the deltoid lateral part attached directly to the glenoid fossa. The increasing distance of muscle attachment, measured along the Z-axis, between the lateral margin of the acromion and a point at which a tangent to the glenoid fossa crosses the Z-axis, increased the translations of the humerus when the angle of tangent ϕ increased from 10° to 90°. However, the increasing distance of muscle attachment reduced the translations of the humerus when the angle of tangent ϕ increased from 90° to 170°.

Parametric analyses have been provided for changing values ϕ, z, and α. The positions of the humeral head were calculated for the four parametric conditions: (1) $\phi = 90°$, $z = 0$ mm (Figure 27.2a); (2) $\phi = 90°$, $z = 0$–32 mm (Figure 27.2b); (3) $\phi = 10°$–170°, $z = 0$ mm (Figure 27.2c and e); (4) $\phi = 10°$–170°, $z = 0$–32 mm (Figure 27.2d and f), during arm elevation/abduction ranging from $\alpha = 0°$ to 170°. The calculations were performed during arm abduction in 10° increments. The humeral head positions were related to the interactions among the distinguished geometric parameters.

For the first parametric condition, an inclination of 90° with the muscle attachment in line with the glenoid fossa ($\phi = 90°$, $z = 0$ mm) during abduction from 0° to 170°, the greatest inferior translation was observed. For a tangent to the shape of the glenoid fossa $\phi = 90°$ and for increased angle of abduction (α) to 90°, the humeral head positions were decreased. For increased angle of abduction above 90°, the humeral head positions were increased. When the muscle attachment was placed further from the glenoid fossa ($\phi = 90°$, $z = 0$–32 mm) for the second parametric condition, the translations were smaller. The increasing distance of the muscle attachment reduced the translations of the humerus for the whole range of arm abduction (Figure 27.4). The results revealed that for an increasing distance of the muscle attachment to $z = 24$ mm the translations of the humerus head decreased up to 90° abduction and increased above 90°. However, for the distance $z = 32$ mm the trend was the opposite with the translations of the humerus head increased to 90° and decreased above 90°.

For the third parametric condition, where the angle of inclination ϕ ranged from 10° to 170° with the muscle attachment at the glenoid fossa ($\phi = 10°$–170°, $z = 0$ mm), all observed translations of the humerus head were smaller than for conditions $\phi = 90°$ and $z = 0$ mm. For all angles of tangent inclination and increased angle of abduction to 90°, the humeral head translations decreased, but increased during abduction from 90° to 170°.

For the conditions (1) $\phi = 90°$, $z = 0$ mm; (2) $\phi = 90°$, $z = 0$–32 mm; and (3) $\phi = 10°$–170°, $z = 0$ mm, a symmetry of humeral head translations was observed along the angle of arm abduction at 90°. The results revealed that the translations of the humerus head along any tangent to the glenoid fossa were the same as for the supplementary angle to this tangent inclination. The translations were the smallest for the supplementary angles of 10° and 170° (Figure 27.5).

The results of calculations for the fourth condition ($\phi = 10°$–170°, $z = 0$–32 mm) revealed another symmetry. When the angle of inclination ϕ increased from 10° to 70° and the distance of muscle attachment of the lateral part of the deltoideus increased, the humeral head translations also increased; however, when the angle of inclination increased from 110° to 170° and the distance of the muscle

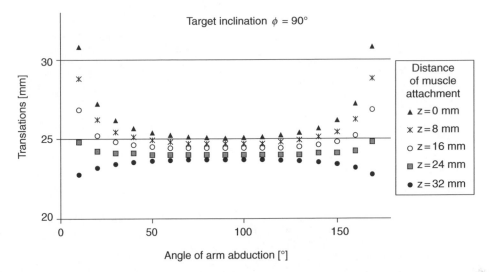

FIGURE 27.4 The translation of the humeral head center during arm abduction $\alpha = 0° - 170°$ and the conditions $\phi = 90°$ and $z = 0 - 32$ mm.

attachment of the lateral part of the deltoideus increased, the humeral head translations decreased. Generally, the increased distance of muscle attachment from the glenoid fossa and the increased inclination from 110° to 170° shortened the translations. However, these translations were all smaller than for the third parametric condition. For example, during abduction of 10° with muscle attachment of 24 and 28 mm, and the angle of tangent inclination is 150°, the translations were 2.4 and 0.6 mm, respectively (Figure 27.6a).

The smallest translations for any position of an arm during abduction were for tangent $\phi = 170°$ and muscle attachment z ranging from 16 to 28 mm. Additional analysis has been provided for the angles of abduction $\alpha = 110°$, 140°, 160°, and 170°. The calculations revealed the same results as for the supplementary angles $\alpha = 70°$, 40°, 20°, and 10°, respectively.

It was observed from MRI measurements of the glenoid fossa concavity for the 12 participants that the tangents crossed the Z-axis at angles in a range from $\phi_1 = 67.2 \pm 12.1°$ to $\phi_7 = 116.1 \pm 14.1°$ with a maximum of 136.0° and a minimum of 50.2°. The distances at which the tangents crossed the

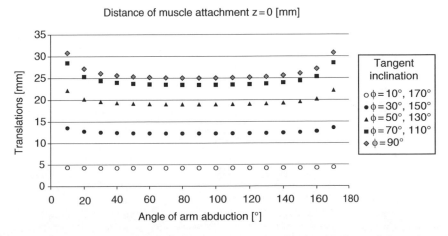

FIGURE 27.5 The translation of the humeral head contact point with the glenohumeral fossa along the Y-axis during arm abduction $\alpha = 0° - 170°$ and the conditions $\phi = 10° - 170°$ and $z = 0$ mm.

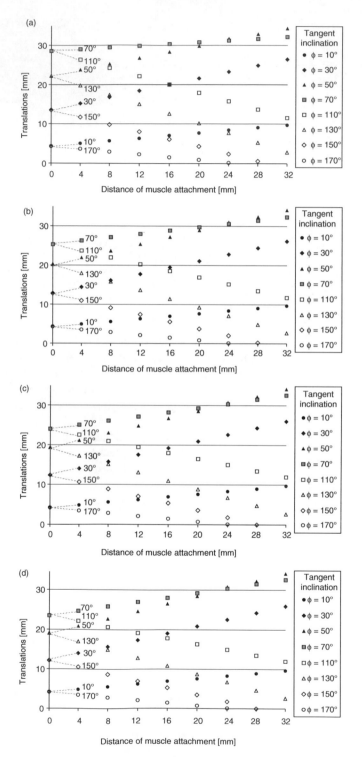

FIGURE 27.6 The translation of the humeral head contact point with the glenohumeral fossa along the Y-axis during (a) arm abduction 10° and 170° and the conditions $\phi = 10°-170°$ and $z = 0-32$ mm; (b) arm abduction 20° and 160° and the conditions $\phi = 10°-170°$ and $z = 0-32$ mm; (c) arm abduction 40° and 140° and the conditions $\phi = 10°-170°$ and $z = 0-32$ mm; and (d) arm abduction 70° and 110° and the conditions $\phi = 10°-170°$ and $z = 0-32$ mm.

TABLE 27.1 Geometric Parameters of the Glenoid Fossa Shapes at the Seven Nodes ($i = 7$) for 12 Subjects

		Subjects											
		1	2	3	4	5	6	7	8	9	10	11	12
z_1	mm	67.2	14.2	30.8	28.6	27.8	34.3	10.6	37.9	49.4	8.2	67.1	20.8
z_2	mm	56.4	11.8	24.1	30.1	33.1	23.7	16	23.7	38.3	26.5	37.3	13.4
z_3	mm	38.2	8	18.9	14.4	30.5	14.2	7.7	10.1	30.4	15.1	17.6	12.3
z_4	mm	30.3	1.1	8.9	9.4	16.5	4.7	4.6	8.3	22.4	34	11.5	10.1
z_5	mm	16.3	−1.2	2.6	5.6	8.8	−1.2	−0.7	1.7	15.1	0.9	5.3	8.8
z_6	mm	8.4	2.2	−3.4	−4	0.7	−5.6	−2.8	−5.8	5.4	−1.9	−7.2	3
z_7	mm	5.7	6	−8	−3.7	1.7	−14.6	−5.5	−12.5	−11.4	2.8	−20	−6.5
ϕ_1	°	50.2	78.5	70	72.8	69.1	63.3	75	66	58	93.3	53	57
ϕ_2	°	56.4	81.7	76	71.5	64.7	73	64	78	66	74	73	72
ϕ_3	°	70.7	87.3	81	87.1	67	83.5	84	95	73	87.5	92	74
ϕ_4	°	78.7	98.5	93	92.9	82.4	95.7	93	97	81	103.6	98	81
ϕ_5	°	95.4	102.7	101	97.8	92.2	103.9	109	106	90	107.5	105	85
ϕ_6	°	105.8	95.9	109	110.9	104.4	110.5	116	116	103	112.1	120	109
ϕ_7	°	109.6	87.3	116	110.4	102.5	123.6	125	124	124	102.4	132	136

Note: z_i is the distance between the middle deltoid attachment and a point at which a tangent to the glenoid foosa crosses the Z-axis and ϕ_i is the tangent inclination.

Z-axis from the lateral margin of the acromion, the lateral deltoid muscle attachment, were respectively from $z_1 = 33.1 \pm 18.9$ mm to $z_7 = -5.5 \pm 8.0$ mm with a maximum of 67.2 mm and a minimum of −20.0 mm. The results of the measurements are summarized in Table 27.1.

27.5   Conclusions

1. The study responds to the recent studies[12] confirming an emerging need to include glenoid inclination, mechanics of the deltoid muscle, and glenoid labrum concavity into stability analyses of the glenohumeral joint
2. The study implies the practicability of a geometric description of the glenoid fossa which should no longer be considered as a straight line but as a series of tangents
3. The current study proposed considering the whole planar shape of the glenoid fossa as a meaningful geometric component of the glenohumeral joint
4. Migration of the humeral head along the glenoid fossa is related to the glenoid fossa inclination and the distance from the glenoid fossa of the deltoid muscle attachment
5. The results for the tangents with supplementary angular values (i.e., in Figure 27.6a $\phi = 30°$ and 150°; $\phi = 50°$ and 130°; $\phi = 70°$ and 110°) demonstrate symmetry because the translations of the glenoid head along the Y-axis oscillate around the same balance position
6. A tangent to the glenoid fossa, formed by a larger than 90° angle to the z-axis improves the glenohumeral stability significantly
7. The MRI results for 12 subjects confirm variability in tangent inclinations and distances of tangents from the lateral deltoid attachment
8. Considering and measuring the angles of tangents as the parameters of the glenoid fossa will provide new information on the interface shape sensitivity of glenohumeral joint stability
9. This method may help in early diagnosis of mechanical instability, which can be considered as a condition predisposing the glenohumeral joint to degenerative change especially for a subject returning to work after work absence resulting from a work-related injury
10. These analytical mechanical considerations may provide a better biomechanical basis for joint modeling and lead to consideration of articular geometry during designing of workplaces

11. The geometrical parameters distinguished in mechanical glenohumeral joint analysis will help to modify the physical strength standards which are still too high as reflected in the high frequency of recorded shoulder injuries

12. Additional research is necessary combining geometries of the glenoid fossa with strength tests performed by the participants in order to find a functional relationship which would be applied in work environments

References

1. F.T. Ballmer, S.B. Lippitt, A.A. Romeo, and F.A. Matsen 3rd, Total shoulder arthroplasty: Some considerations related to glenoid surface contact, *Journal of Shoulder and Elbow Surgery* 3 (1994), 299–306.

2. D.P. Bertsekas, *Nonlinear Programming*, Athena Scientific, Belmont, MA, 1995.

3. L.U. Bigliani, R. Kelkar, E.L. Flatow, R.G. Pollock, and V.C. Mow, Glenohumeral stability: biomechanical properties of passive and active stabilizers, *Clinical Orthopaedics and Related Research* 330 (1996), 13–30.

4. R. Fletcher, *Practical Methods of Optimization*, John Wiley & Sons, Chichester, 1987.

5. O. Gagey and E. Hue, Mechanics of the deltoid muscle. A new approach, *Clinical Orthopaedics and Related Research* 375 (2000), 250–257.

6. O. Gagey and E. Hue, Reply to Blasier *et al.* comments on: *Clin. Orthop. 2000* Jun 375:250–257, *Clinical Orthopaedics and Related Research* 388 (2001), 259.

7. K. Gielo-Perczak, Analysis and modeling of the glenohumeral joint mechanism structure, *SOMA Engineering for the Human Body* 3 (1989), 35–46.

8. A.M. Halder, S.G. Kuhl, M.E. Zobitz, D. Larson, and K.N. An, Effects of the glenoid labrum and glenohumeral abduction on stability of the shoulder joint through concavity-compression: an in vitro study, *The Journal of Bone and Joint Surgery* 83-A (2001), 1062–1069.

9. D.T. Harryman 2nd, J.A. Sidles, J.M. Clark, K.J. Mcquade, T.D. Gibb, and F.A. Matsen 3rd, Translation of the humeral head on the glenoid with passive glenohumeral motion, *The Journal of Bone and Joint Surgery* 72 (1990), 1334–1343.

10. S.M. Howell and B.J. Galinat, The glenoid-labral socket: a constrained articular surface, *Clinical Orthopaedics and Related Research* 243 (1989), 122–125.

11. S.M. Howell, B.J. Galinat, A.J. Renzi and P.J. Marone, Normal and abnormal mechanics of the glenohumeral joint in the horizontal plane, *The Journal of Bone and Joint Surgery* 70 (1988), 227–232.

12. R.E. Hughes, C.R. Bryant, J.M. Hall, J. Wening, L.J. Huston, J.E. Kuhn, J.E. Carpenter, and R.B. Blasier, Glenoid inclination is associated with full-thickness rotator cuff tears, *Clinical Orthopaedics and Related Research* 407 (2003), 86–91.

13. Iannotti, JP, Gabriel JP, Schneck SL, Evans BG, Misra S. The normal glenohumeral relationships. An anatomical study of one hundred and forty shoulders. The Journal of Bone and Joint Surgery 1992; 74:491–500.

14. I.A. Kapandji, *The Physiology of the Joints*, 5th ed., vol. 1 Upper Limb, Churchill Livingstone, Edinburgh, 1982.

15. M.D. Lazarus, J.A. Sidles, D.T. Harryman 2nd, and F.A. Matsen 3rd, Effect of a chondral-labral defect on glenoid concavity and glenohumeral stability. A cadaveric model, *The Journal of Bone and Joint Surgery* 78 (1996), 94–102.

16. S.-B. Lee and K.-N. An, Dynamic glenohumeral stability provided by three heads of the deltoid muscle, *Clinical Orthopaedics and Related Research* 400 (2002), 40–47.

17. N.K. Poppen and P.S. Walker, Normal and abnormal motion of the shoulder, *The Journal of Bone and Joint Surgery* 58-A (1976), 195–201.

18. A.K. Saha, *Theory of Shoulder Mechanism: Descriptive and Applied*, Charles C Thomas Publishing Company, Springfield, IL, 1961.

28

Application of Ergonomics to the Low Back

Kermit G. Davis, III
University of Cincinnati

28.1 Overview

Before an individual can accurately assess the workplace, there must be a basic understanding of the principles of ergonomics, particularly with respect to how they relate to risk of injury. When evaluating the risk of low back injury, it is important to start with a load–tolerance relationship, whether that is a physical, social, or psychological load. In other words, exposures in the workplace exert loads on the body, which the body must respond to, and, at some point, a tolerance is exceeded causing the injury and resulting in an injury (McGill, 1997). The most likely tolerance relating to low back injuries is structural tolerance — loading on the spine. Another potential tolerance that may also contribute to risk assessment is pain sensation, particularly with respect to psychosocial work factors (this will be discussed further later). Based on this load–tolerance viewpoint, one must understand what factors contribute to the loading on the spine structures and their corresponding tolerances. Once these factors are identified, the proper assessment tools can be identified and developed to measure the levels of the specific risk factors that would exceed the tolerances of the spine structures. Thus, assessment tools can only be effective if they can accurately identify when tolerances are exceed either cumulatively or acutely and, often times, must be multifactorial in nature.

Many factors in the workplace may contribute to the loading on the lower back. Some of these factors are the weight of the object being lifted, height of origin and destination, horizontal distance (moment arm) at origin and destination, task asymmetry at origin and destination, repetition (lift rate), mode of exertion (lifting, squat lifting, stoop lifting, lowering, pushing, pulling, carrying, etc.), feet position, coupling (handles), and box size (Kelsey et al., 1984; Kelsey and Golden, 1988; Marras et al., 1993, 1995; Hales and Bernard, 1996; Burdorf and Sorock, 1997; Ferguson and Marras, 1997; National Academy of Science, 2001; Neumann et al., 2001). Muscles generate internal forces that must offset the loads placed on low back by these workplace (external) factors. Several factors relating to the internal response (muscle force) impact the ability to counteract the forces and moments such as strength, the ability to exert sufficient force; posture, the length–strength relationship; motion, the force–velocity relationship; and endurance, the ability to resist fatigue. Together, the external demands and the internal muscle response produce loads on the spine structures that ultimately produce the injury. Thus, a successful low back risk assessment tool must account for many of these factors either by direct assessment or using surrogate measures. The next seven sections of this chapter will discuss in detail the relationship of various factors and the resulting loads on the spine and how these specific factors can be evaluated. Finally, the role of psychosocial work characteristics and psychological responses will also be discussed with respect to injury mechanism, assessment, and interaction with other factors.

28.2   Assessment of Weight Lifted

The simplest assessment with respect to all potential risk factors is measuring the weight of the object lifted. In general, there is a direct relationship between increased weight of the object and loading on the spine (Chaffin and Park, 1973; Marras et al., 1997, 1999; Kingma et al., 1998; Dolan et al., 1999; Granata et al., 1999). Heavier load weight, particularly above 25 lb (11.4 kg), has also been found to be associated with low back pain in many epidemiological studies (Kelsey et al., 1984; Kelsey and Golden, 1988; Marras et al., 1993, 1995; Hales and Bernard, 1996; Neumann et al., 2001). As a general rule, there appears to be a tolerance level at 25 lb where workers may be at increased risk of injury. However, there is evidence (Davis and Marras, 2000a) that an increase in weight does not necessarily correspond to increased spinal loads and thus increased risk of injury. These authors found small changes in weight did not necessarily correspond to changes in loading due to kinematic compensations such as faster motions and more awkward postures. There maybe other factors that play a role in how effective reducing weight is in controlling the risk of injury.

28.3   Assessment of Trunk Moment

The role of weight may also be impacted by the individual's relative position to the box (or object) being lifted. One specific factor that relates to position is the horizontal distance between the object and the back (called the horizontal moment arm). Traditionally, the horizontal moment arm has been measured by a tape measure by locating the midpoint of the hands on the object and center of the back or spine (usually approximated by the midpoint between the heels of the feet) (NIOSH 1981, 1993; Marras et al., 1993, 1995) (see Figure 28.1 for illustration of the measurement of moment arms).

To illustrate the importance of horizontal moment arm, Figure 28.2 shows "safe" regions as it relate this to the loads on the spine for a 25 lb (or 11.4 kg) load. As one can see, the compressive loads increased significantly as the horizontal shelf distance increased from 12 in. (30.5 cm) to 24 in. (61.0 cm) (at either waist or knee height). The anterior–posterior shear forces were also impacted by the horizontal moment arm but only for the knee shelf height.

The combination of weight and horizontal moment arm is often referred to as the trunk moment (e.g., trunk moment equals horizontal moment arm multiplied by object weight). The measurement of the two factors that make up trunk moment is relatively easy since both object weight and horizontal moment arm can be measured with inexpensive and simple measures (e.g., scale and measuring tape). Similar to weight, trunk moment has been found to increase the loads on the spine (Schultz et al., 1982a,b; Marras and Sommerich, 1991) and has also been found to relate to low back pain (Marras et al., 1993, 1995; Norman et al., 1998). As a single risk factor, trunk moment (combination of object weight and horizontal moment arm) may be the most important risk factor with respect to low back injuries (Chaffin and Park, 1973; Marras et al., 1993, 1995; Norman et al., 1998). Several of the most widely used assessment tools account for moment by either an individual factor (e.g., already multiplied together) (Marras et al., 1993, 1995) or as a combination factor (e.g., two individual factors — weight and horizontal moment arm — accounted for independently) (NIOSH, 1981, 1993; Washington State Ergonomics Standard 296-62-051). Obviously, the latter assessment tools may only indirectly account for trunk moment. With the relative importance and ease of measurement, trunk moment appears to be a significant risk factor that should be a cornerstone of any low back assessment tool.

28.4 Assessment of Trunk Posture

The location of the box (or object) being moved is also a major contributor to the stresses on the lower back and the potential risk of injury. Not only does the position impact the moment arm, it will also influence the posture adopted by the worker. Trunk flexion is a function of the starting height (origin) of the lift, body stature, and reach distance. As the position of the object gets lower (e.g., height of origin) or farther away (e.g., larger reach distances) from the body, a typical response will be to flex the trunk forward (Marras et al., 1997). It is also logical that taller individuals will flex the trunk more when lifting

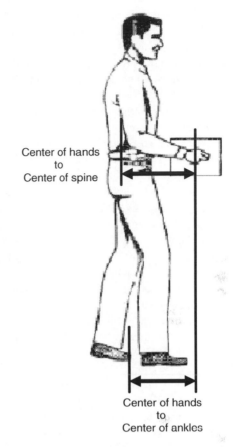

FIGURE 28.1 Measurement of horizontal moment arms (Adapted from National Institute for Occupational Safety and Health, 1993, NIOSH Technical Report 94–110. U.S. Department of Health and Human Services, NIOSH, Cincinnati, OH. With permission.)

below the knee. As a result, greater flexion produces larger trunk moments due to the upper torso (head, chest, and arms) being supported by the back, which ultimately results in larger forces on the spine (Drury et al., 1989; Marras et al., 1997). Trunk flexion has been found to be strongly associated

Compression force		Anterior–posterior shear force	
Shoulder at 12 in. 1108 N		Shoulder at 12 in. 514 N	
Waist at 12 in. 1239 N	Waist at 24 in. 2325 N	Waist at 12 in. 471 N	Waist at 24 in. 479 N
Knee at 12 in. 2929 N	Knee at 24 in. 4594 N	Knee at 12 in. 458 N	Knee at 24 in. 817 N
Floor at 12 in. 4425 N		Floor at 12 in. 1125 N	

FIGURE 28.2 Compression and anterior–posterior shear force is impacted by moment arm (horizontal distance) and height of origin. (Marras et al. 2001, *Spine*.)

with low back pain in industry (Punnet et al., 1991; Marras et al., 1993, 1995; Ono et al., 1997; Brulin et al., 1998; Josephson et al., 1998; Norman et al., 1998; Wickstrom and Pentti, 1998; Ozguler et al., 2000). Further, Figure 28.2 also provides indirect evidence of the importance of trunk flexion by having greater compression and anterior–posterior shear forces as the lift origin becomes lower (e.g., knee or floor height). It would be expected that these lower regions would have more trunk flexion (based on previous literature).

Awkward postures (e.g., lateral flexion, twisting, and combination of flexion with twisting) also pose a risk to workers with respect to low back pain (Andersson, 1981; Kelsey and Golden, 1988; Hales and Bernard, 1996; Ferguson and Marras, 1997; Hoogendoorn et al., 1999). One factor that influences the awkward postures is task asymmetry — producing greater lateral flexion and twisting (Ferguson et al., 1992; Plamondon et al., 1995; Allread et al., 1996; Kingma et al., 1998; Marras and Davis, 1998; Granata et al., 1999). Task asymmetry is the location of the object (or box) being lifted relative to the midsagittal plane of the individual at the beginning or at the end of the lift (NIOSH, 1993). Increases in nonsagittal plane motion (lateral and twist) will typically accompany more asymmetric tasks since the box is located away from the midline of the sagittal plane, causing the individual to twist and bend sideways. However, the magnitude of off-plane motion may be minimized by lower extremity compensation (e.g., moving of feet or twisting of hips).

The posture of the worker can be quantified by several methods with varying levels of accuracy and resource requirements (cost, time, effort, and knowledge) (Burdorf et al., 1997). The simplest method is using a checklist or questionnaire that identifies designated regions (see Figure 28.3 for an example). The basic checklist method uses diagrams as a reference to identify how far an individual flexes forward, flexes laterally, or twists. One useful aid is to videotape the worker performing the job so that repeated evaluations of the postures may be made. The drawback of this method is the subjectivity or the ability to distinguish the actual posture of the individual, particularly twist postures. These methods are effective for more gross assessment of posture.

For more accurate quantification of the posture, goniometers (also referred to as electrogoniometers or potentiometers) measure joint position; in this case, the lower back. There are several forms of low back goniometric systems (e.g., lumbar motion monitor, Isotechnologies Back Tracker) that are commercially available, but each requires substantial monetary resources and expertise in application. One system, the lumbar motion monitor (LMM), has been validated with a large industrial database that

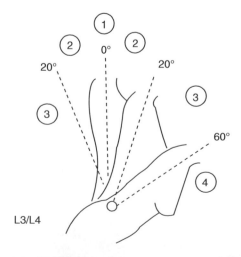

FIGURE 28.3 One example of checklist diagrams for posture: Trunk flexion diagram from REBA checklist. (Adapted from Hignett and McAtamney, 2000. With permission.)

is linked to actual injuries (Marras et al., 1993, 1995, 2000a). The LMM has also been validated with respect to posture in three dimensions (Marras et al., 1992). Thus, the LMM is a potentially powerful tool with respect to measuring the three-dimensional posture of the trunk but requires significant expertise and monetary resources.

28.5 Assessment of Trunk Motion

Another important risk factor that is related to trunk posture is trunk motion — the rate of change in trunk position or velocity. Three-dimensional trunk velocities may be important risk factors in highly dynamic work conditions, even more than just posture itself. Two major research initiatives have found trunk velocities to be significant risk factors of low back injuries and pain (Marras et al., 1993, 1995; Norman et al., 1998). The general findings have been that complex motion (e.g., motion in multiple planes) is more detrimental to the low back (Fathallah et al., 1998). The problem with complex motions is that they increase loads on the spine (Freivalds et al., 1983; Jager and Luttmann, 1989; Marras and Sommerich, 1991; Granata and Marras, 1995; Davis et al., 1998; Davis and Marras, 2000b).

Generally, the two most viable assessment techniques to evaluate trunk motion are goniometers and video motion analyses. However, both of these methods require significant resources and expertise. One potential low-level assessment of trunk velocity might be estimating the total flexion angle by questionnaire or checklist combined with the use of a stopwatch to measure the time it takes to complete the motion. While by no means would this assessment accurately measure the velocity to the same accuracy of goniometers or video assessment, this quick and easy method may provide a "ballpark" estimate that takes into account the dynamics of the task. A major drawback to these less-quantitative methods is that they have not been adequately developed or validated.

Many exposure measures use a surrogate variable — lift rate or frequency — to account for the dynamics of the lift. While this factor potentially accounts for trunk motion (in an indirect way), it fails to account for the influence of other factors such as weight, lift distance, task asymmetry, and mental demands. As the level of load (weight) being lifted is increased, trunk motions have been found to decrease (Buseck et al., 1988; Ferguson et al., 1992; Davis and Marras, 2000a,b). Lift rate also accounts for the number of actual items being lifted, which would be related to the amount of effort required over a specific time. In other words, lift rate is a composite variable that has limited usefulness in accounting for trunk motions. Thus, assessments of trunk motion should rely on more sound and objective techniques such as video and goniometeric systems.

28.6 Assessment of Other Manual Material Handling Modes

While most low back assessments focus on lifting and lowering, other modes of moving objects may also be risk factors for low back injuries. Some of these tasks would be pushing, pulling, carrying, and prolonged standing. Pushing and pulling have been identified as risk factors in many recent epidemiological studies (Snook et al., 1978; Garg and Moore, 1992; Hoozemans et al., 1998). Examples of tasks that typically require pushing or pulling are: moving a cart; transporting patients on a stretcher; using a broom, mop, rake, or hoe; and moving lift assist devices (e.g., hydraulic lift to move boxes). One potential reason for increased risk for these two modes of handling may be the nature of the loads on the spine — not as much compressive force but more shear loading. The actual nature of the force will depend upon the height of the handle and whether pushing or pulling is being performed (Gagnon et al., 1992; De Looze et al., 1995; Resnick and Chaffin, 1996).

The typical assessment method for push or pull forces is using a strain gauge. There are usually two forces that are of interest during these assessments: (1) initial force — force required to start (or accelerate) the object in motion; (2) sustained force — force required to keep the object moving (Hoozemans et al., 1998). Currently, the only method of evaluating these measured forces is to compare it to maximum acceptable forces that have been established by Snook and associates (Ciriello et al., 1990;

Snook and Ciriello, 1991). The Snook tables provide benchmarks for both sustained and initial push and pull forces as a function of height of applied force (handle height), distance of object movement, and frequency of push/pull.

The impact of carrying will depend upon the weight lifted and the length of time carrying (or distance). While it would be expected that the impact of carrying would be less than lifting, the risk may increase with large amounts of weight or long distances. Again, the only current method to evaluate carrying is the Snook tables (Snook and Ciriello, 1991).

One factor that may play an increasing role in the development of low back pain may be prolonged standing. Since major lifting has been designed out of many workplaces, more static and upright tasks have become prevalent. These types of tasks will include standing with limited walking or in a restricted pattern causing static postures in the low back. One example of a prolonged standing is working on a drill press or other machinery for long periods of time throughout the shift while an example of walking in a restricted pattern is when a worker performs a task while walking along with the moving assembly line. In both cases, the trunk oftentimes remains in a slightly flexed posture (static) for long durations. As a result, muscles remain on for long durations, potentially eliciting pain from fatigue failure. Since prolonged standing has been typically neglected as a potential risk factor, there are few methods to assess this risk factor. One method is to use an ergonomic dosimeter (one device is the ActivPAL — PAL Technologies Ltd, Glasgow, U.K.) that measures the amount of time standing without moving, time walking, and time sitting.

28.7   Assessment of Whole-Body Vibration

Whole-body vibration is another potential risk factor for the development of low back pain (Bovenzi and Zadini, 1992; Burdorf and Sorock, 1997; Bovenzi and Hulshof, 1999; Lings and Leboeuf-Yde, 2000). Whole-body vibration results from driving trucks, forklifts, cranes, lifters, and other mechanical machines as well as being in direct contact with the surface that vibrates from large machinery or presses. As the whole body vibrates at the vehicle resonance frequency, the vibration energy is absorbed into the structures of the low back, specifically the lumbar disc, and over the long term results in disc degeneration.

Whole-body vibration is usually measured by two methods: (1) questionnaire (subjective) and (2) lightweight accelerometers (objective). Most questionnaires assessing whole-body vibration ask questions about the amount of time spent riding vehicles such as forklifts or trucks. This subjective measure is a very crude assessment since different vehicles have different frequencies that result in different levels of health impact. The lightweight accelerometers measure the acceleration in the vertical direction (z) or the vector sum of the frequency-weighted root mean square acceleration (Bovenzi and Hulshof, 1999). These values can then be compared against standards that have been established in literature such as those developed by International Standards Organization (ISO, 1985) and American Conference of Governmental Industrial Hygienists (ACGIH, 1996).

28.8   Assessment of Psychosocial Work Characteristics and Responses

Another facet of the workplace that must be accounted for is psychosocial factors such as mental concentration, lack of variety and skill utilization (e.g., job monotony), job responsibility, social relationships with coworkers and supervisors, job control, mental stress, and job satisfaction. Many of these factors have been found to be associated with increased reporting of low back pain (Bongers et al., 1993; Burdorf and Sorock, 1997; Ferguson and Marras, 1997; Davis and Heaney, 2000). There have been several potential hypothesized mechanisms for psychosocial risk factors leading to low back pain. First, psychosocial factors influence the loading on the spine through changes in trunk mechanics and muscle activity (Flor et al., 1985; Bongers et al., 1993; Davis and Heaney, 2000); two recent studies have actually documented such changes in response to mental stress (Marras et al., 2000b) and mental concentration (Davis et al., 2002). A second potential mechanism is that the psychosocial factor changes chemical reactions such as accumulation of metabolites in the muscles (Backus and Dudley, 1974) or increased cortisol levels that make the muscle vulnerable to mechanical load (Theorell et al., 1993). Finally, psychosocial factors may alter the reporting of low back pain in two ways: (1) by altering the pain tolerance or threshold — increasing the likelihood of reporting (Theorell et al., 1991, 1993); (2) by avoiding the work situation — report an injury when low back pain is not even severe or call in sick (Frank et al., 1995). No matter what the reason for reporting, these variables must be accounted for in any comprehensive evaluation.

There are many well-known questionnaires in literature that assess the various components of the psychosocial environment and can be easily administrated to workers. Some examples, although not exhaustive, are: work APGAR by Bigos et al. (1991), Generic Job Stress Questionnaire (Hurrell and McLaney, 1988), and Job Content Questionnaire designed by Karasek and Theorell (Karasek et al., 1998). With any questionnaire, the reliability and validity of the questionnaire is dependent upon the questions used (e.g., degree of subjectivity, ability to delineate the levels of exposure), the number of questions assessing specific item (e.g., one question less reliable than multiple questions), and complexity of the questions (e.g., language issues). Oftentimes, the effectiveness of the questionnaire is related to the balance between the length and extensiveness and the time it takes to complete it. In most cases of job assessment, time required away from the work must be minimized so the selection of which factors, how many factors, and which questionnaire will be important since different questionnaires require different amounts of time to complete.

28.9 Summary of Assessment Tools

As the workplace has become more complex and technologically demanding, risk assessment tools must account for a wide variety of exposure factors ranging from biomechanical (physical) factors to psychosocial work characteristics. At this time, the assessment tools are typically one dimensional; that is, they concentrate on a few aspects of the workplace. Therefore, to perform a complete and comprehensive evaluation of the job, one will have to identify several potential assessment tools. The identification of which tools are most appropriate will depend on the resources available, expertise of the evaluator, availability (or lack of) of the equipment, time frame (e.g., time available for assessment), and risk factors expected to be present. The most crucial of these options is obviously determining what risk factors need to be identified since that will influence many of the other issues with the most important risk factor being trunk moment (weight multiplied by horizontal moment arm). No matter what risk factor is of interest, more objective (quantitative) measures will provide more accurate evaluations than techniques that rely on the subjective assessment from the worker or evaluator.

Although not discussed in this chapter, there are several potential environmental factors that may contribute to the development of low back disorders such as temperature (e.g., cold and hot environments), humidity, floor surfaces (e.g., slippery surfaces, uneven surfaces), and noise. These factors may impact the low back through more indirect or modifying relationships of other known risk factors (e.g., slippery floors may be more detrimental with heavier weights). Further, the evidence linking these types of factors to low back injuries is weak to moderate at best with current assessment tools oftentimes failing to even address these factors.

The important message of this chapter is that assessment of low back risk is complex and will most likely need to account for many aspects of the workplace to truly understand the potential of having an injury occurring. Furthermore, the better assessment tools rely on sound scientific approaches that quantitatively account for the stress placed on the spine structures. One must remember that one assessment tool may not provide all the answers regarding risk and more comprehensive evaluations may be needed (e.g., using multiple assessment techniques).

References

ACGIH, 1996, *Threshold Limit Values for Chemical and Physical Agents and Biological Exposure Indices: Whole-Body Vibration.* Cincinnati, OH: American Conference of Governmental Industrial Hygienists, pp. 123–131.

Allread, W.G., Marras, W.S., Parnianpour, M., 1996, Trunk kinematics of one-handed lifting, and the effects of asymmetry and load weight. *Ergonomics*, 39(2), 322–334.

Andersson, G.B.J., 1981, Epidemiologic aspects on low-back pain in industry. *Spine*, 6(1), 53–60.

Backus, F.I., Dudley, D.L., 1974, Observations of psychosocial factors and their relationship to organic disease. *International Journal of Psychiatry Medicine*, 5, 499–515.

Bigos, S.J., Battie, M.C., Spengler, D.M., Fisher, L.D., Fordyce, W.E., Hansson, T.H., Nachemson, A.L., Wortley, M.D., 1991, A prospective study of work perceptions and psychosocial factors affecting the report of back injury. *Spine*, 16, 1–6.

Bongers, P.M., de Winter, C.R., Kompier, M.A.J., Wildebrandt, V.H., 1993, Psychosocial factors at work and musculoskeletal disease. *Scandinavian Journal of Work, Environment and Health*, 19, 297–312.

Bovenzi, M. and Holshof C.J. 1999, An updated review of epidemiologic studies on the relationship between exposure to whole-body vibration and low back pain (1986–1997). *International Archives of Occupational and Environmental Health*, 72(6), 351–365.

Bovenzi, M., Zadini, A., 1992, Self-reported low-back symptoms in urban bus drivers exposed to whole-body vibration. *Spine*, 17, 1048–1059.

Brown, R., Li, G., 2003, The development of action levels for the "Quick Exposure Check" (QEC) system. In: *Contemporary Ergonomics*, McCabe P.T., Ed., London: Taylor & Francis, pp. 41–46.

Brulin, C., Gerdle, B., Granlund B., Hoog, J., Knutson, A., Sundelin, G., 1998, Physical and psychosocial work-related risk factors associated with musculoskeletal symptoms among home care personnel. *Scandinavian Journal of Caring Sciences*, 12(2), 104–110.

Burdorf, A., Sorock, G., 1997, Positive and negative evidence of risk factors for back disorders. *Scandinavian Journal of Work, Environment and Health*, 23, 243–256.

Burdorf, A., Rossignol, M., Fathallah, F.A., Snook, S.H., Herrick, R.F., 1997, Challenges in assessing risk factors in epidemiologic studies on back disorders. *American Journal of Industrial Medicine*, 32, 142–152.

Buseck, M., Schipplein, O.D., Andersson, G.B.J., Andriachi, T.P., 1988, Influence of dynamic factors and external loads on the moment at the lumbar spine in lifting. *Spine*, 13(8), 918–921.

Chaffin, D.B., Park, K.S., 1973, A longitudinal study of low-back pain as associated with occupational weight lifting factors. *American Industrial Hygiene Association Journal*, 34, 513–525.

Ciriello, V.M., Snook, S.H., Blick, A.C., Wilkenson, P.L., 1990, The effects of task duration on psychophysically-determined maximum acceptable weights and forces. *Ergonomics*, 33(2), 187–200.

Davis, K.G., Heaney, C.A., 2000, The relationship between psychosocial work characteristics and low back pain: underlying methodological issues. *Clinical Biomechanics*, 15(6), 389–406.

Davis, K.G., Marras, W.S., 2000a, Assessment of the relationship between box weight and trunk kinematics: does a reduction in box weight necessarily correspond to a decrease in spinal loading? *Human Factors*, 42(2), 195–208.

Davis, K.G., Marras, W.S., 2000b, The effects of motion on trunk biomechanics. *Clinical Biomechanics*, 15, 703–717.

Davis, K.G., Marras, W.S., Waters, T.R., 1998, Evaluation of the spinal loading during lowering and lifting, *Clinical Biomechanics*, 13(3), 141–152.

Davis, K.G., Marras, W.S., Heaney, C.A., Waters, T.R., Gupta, P., 2002, The impact of mental processing and pacing on spine loading, *Spine*, 27(23), 2645–2653.

de Looze, M.P., Stassen, A.R.A., Markslag, A.M.T., Borst, M.J., Wooning, M.M., Toussaint, H.M., 1995, Mechanical loading on the low-back in 3 methods of refuse collecting. *Ergonomics*, 38, 1993–2006.

Dolan, P., Kingma, I., van Dieen, J., de Looze, M.P., Toussaint, H.M., Baten, C.T.M., Adams, M.A., 1999, Dynamic forces acting on the lumbar spine during manual handling — can they be estimated using electromyographic techniques alone? *Spine*, 24(7), 698–703.

Drury, C.G., Deeb, J.M., Hartman, B., Wooley, S., Drury, C.E., Gallagher, S., 1989, Symmetric and asymmetric manual materials handling. Part 2: biomechanics. *Ergonomics*, 32(6), 565–583.

Fathallah, F.A., Marras, W.S., Parnianpour, M., 1998, The role of complex, simultaneous trunk motions in the risk of occupational-related low back disorders. *Spine*, 23(9), 1035–1042.

Ferguson, S.A., Marras, W.S., 1997, A literature review of low back disorder surveillance measures and risk factors. *Clinical Biomechanics*, 12(4), 211–226.

Ferguson, S.A., Marras, W.S., Waters, T.R., 1992, Quantification of back motion during asymmetric lifting. *Ergonomics*, 35(7/8), 845–859.

Flor, H., Turk, D.C., Birbaumer, N., 1985, Assessment of stress-related psychophysiological reactions in chronic back pain patients. *Journal of Consulting and Clinical Psychology*, 53, 354–364.

Frank, J.W., Pilcins, I.R., Kerr, M.S., Shannon, H.S., Stansfeld, S.A., 1995, Occupational back pain — an unhelpful polemic. *Scandinavian Journal of Work, Environment and Health*, 21, 3–14.

Freivalds, A., Chaffin, D.B., Garg, A., Lee, K.S., 1983, A dynamic biomechanical evaluation of lifting maximum acceptable loads. *Journal of Biomechanics*, 18(8), 571–584.

Gagnon, M., Beaugrand, S., Authier, M., 1992, The dynamics of pushing loads onto shelves of different heights, *International Journal of Industrial Ergonomics*, 9, 1–13.

Garg, A., Moore, J.S., 1992, Epidemiology of low-back pain in industry. *Occupational Medicine: State of the Art Review*, 7, 593–608.

Granata, K.P., Marras, W.S., 1995, An EMG-assisted model of trunk loading during free-dynamic lifting. *Journal of Biomechanics*, 28(11), 1309–1317.

Granata, K.P., Marras, W.S., Davis, K.G., 1999, Variation in spinal load and trunk dynamics during repeated lifting exertions. *Clinical Biomechanics*, 14(6), 367–375.

Hales, T.R., Bernard, B.P., 1996, Epidemiology of work-related musculoskeletal disorders. *Journal of Orthopaedic Clinics of North of America*, 27(4), 679–709.

Hoogendoorn, W.E., van Poppel, M.N.M., Bongers, P.M., Koes, B.W., Bouter, L.M., 1999, Physical load during work and leisure time as risk factors for back pain. *Scandinavian Journal of Work, Environment and Health*, 25(5), 387–403.

Hoozemans, M.J.M., Van der Beek, A.J., Frings-Dresen, M.H.W., Van Dijk, F.J.H., Van der Woude, L.H.V., 1998. Pushing and pulling in relation to musculoskeletal disorders: a review of risk factors. *Ergonomics*, 41, 757–781.

Hurrell, J.J., McLaney, M.A., 1988, Exposure to job stress — a new psychometric instrument. *Scandinavian Journal of Work, Environment and Health*, 14, 27–28.

International Standards Organization, 1985, *Evaluation of Human Exposure to Whole-Body Vibration*. Geneva, Switzerland: ISO. Report No. ISO-2631.

Jager, M., Luttmann, A., 1989, Biomechanical analysis and assessment of lumbar stress during load lifting using a dynamic 19-segment human model. *Ergonomics*, 32(1), 93–112.

Josephson, M., Vingard, E., MUSIC-Norrtalje Study Group, 1998, Workplace factors and care seeking for low-back pain among female nursing personnel. *Scandinavian Journal of Work, Environment and Health*, 24(6), 465–472.

Karasek, R.A., Brisson, C., Kawakami, N., Houtman, I., Bongers, P., Amick, B., 1998, The Job Content Questionnaire (JCQ): an instrument for internationally comparative assessments of psychosocial job characteristics. *Journal of Occupational Health Psychology*, 3(4), 322–354.

Kelsey, J.L., Golden, A.L., 1988, Occupational and workplace factors associated with low back pain. *Journal of Occupational Medicine*, 3(1), 7–16.

Kelsey, J.L., Githens, P.B., White, A.A., Holford, T.R., Walter, S.D., O'Conner, T., Ostfeld, A.M., Weil, U., Southwick, W.O., Calogero, J.A., 1984, An epidemiological study of lifting and twisting on the job and risk for acute prolapsed lumbar intervertebral disc. *Journal of Orthopaedic Research*, 2, 61–66.

Kingma, I., van Dieen, J.H., de Looze, M., Toussaint, H.M., Dolan, P., Baten, C.T.M., 1998, Asymmetric low back loading in asymmetric lifting movements is not prevented by pelvic twist. *Journal of Biomechanics*, 31, 527–534.

Lings, S., Leboeuf-Yde, C., 2000, Whole-body vibration and low back pain: a systematic critical review of the epidemiological literature 1992–1999. *International Archives of Occupational and Environmental Health*, 73, 290–297.

Marras, W.S., Davis, K.G., 1998, Spine loading during asymmetric lifting using one vs. two hands. *Ergonomics*, 41(6), 817–834.

Marras, W.S., Sommerich, C.M., 1991, A three dimensional motion model of loads on the lumbar spine: II. Model validation. *Human Factors*, 33(2), 139–149.

Marras, W.S., Fathallah, F.A., Miller, R.J., Davis, S.W., and Mirka, G.A., 1992, Accuracy of a three dimensional lumbar motion monitor for recording dynamic trunk motion characteristics. *International Journal of Industrial Ergonomics*, 9, 75–87.

Marras, W.S., Lavender, S.A., Leurgans, S.E., Rajulu, S.L., Allread, W.G., Fathallah, F.A., Ferguson, S.A., 1993, The role of dynamic three-dimensional motion in occupationally-related low back disorders. The effects of workplace factors, trunk position, and trunk motion characteristics on risk of injury. *Spine*, 18(5), 617–628.

Marras, W.S., Lavender, S.A., Leurgans, S.E., Fathallah, F.A., Ferguson, S.A., Allread, W.G., Rajulu, S.L., 1995, Biomechanical risk factors for occupationally related low back disorders. *Ergonomics*, 38(2), 377–410.

Marras, W.S., Granata, K.P., Davis, K.G., Allread, W.G., Jorgensen, M.J., 1997, Spine loading and probability of low back disorder risk as a function of box location on a pallet. *Human Factors in Manufacturing*, 7(4), 323–336.

Marras, W.S., Granata, K.P., Davis, K.G., Allread, W.G., Jorgensen, M.J., 1999b, The effects of box features on spinal loading during warehouse order selecting. Ergonomics 42(7), 980–996.

Marras, W.S., Allread, W.G., Burr, D.L., Fathallah, F.A., 2000a, Prospective validation of a low-back disorder risk model and assessment of ergonomic interventions associated with manual materials handling tasks. *Ergonomics*, 43(11), 1866–1886.

Marras, W.S., Davis, K.G., Heaney, C.A., Maronitis, A.B., Allread, W.G., 2000b, The influence of psychosocial stress, gender, and personality on mechanical loading of the lumbar spine, *Spine*, 25(23), 3045–3054.

Marras, W.S., Davis, K.G., Ferguson, S.A., Lucas, B.R., Gupta, P., 2001, Spine loading characteristics of low back pain patients compared to asymptomatic individuals, *Spine*, 26(23), 2566–2574.

McGill, S.M., 1997, The biomechanics of low back injury: implications on current practice in industry and the clinic, *Journal of Biomechanics*, 30(5), 465–475.

National Academy of Science, 2001, *Musculoskeletal Disorders and the Workplace: Low Back and Upper Extremities*. Washington, DC: National Academy Press.

National Institute for Occupational Safety and Health, 1981, *Work Practices Guide for Manual Lifting*. NIOSH Technical Report 81–122. [Cincinnati, OH:] U.S. Department of Health and Human Services, National Institute for Occupational Safety and Health.

National Institute for Occupational Safety and Health, 1993, *Application Manual for the Revised Lifting Equation*. NIOSH Technical Report 94–110. [Cincinnati, OH:] U.S. Department of Health and Human Services, National Institute for Occupational Safety and Health.

Neumann, W.P., Wells, R.P., Norman, R.W., Frank, J., Shannon, H., Kerr, M.S., OUBPS Working Group, 2001, A posture load sampling approach to determining low-back pain risk in occupational settings. *International Journal of Industrial Ergonomics*, 27, 65–77.

Norman, R., Wells, R., Neumann, P., Frank, J., Shannon, H., Kerr, M., The Ontario Universities Back Pain Study (OUBPS) Group, 1998, A comparison of peak vs cumulative physical work exposure risk factors for the reporting of low back pain in the automotive industry, *Clinical Biomechanics*, 13, 561–573.

Ono, Y., Shimaoka, M., Hiruta, S., Takeuchi, Y., 1997, Low back pain among cooks in nursery schools. *Industrial Health*, 35(2), 194–201.

Ozguler, A., Leclerc, A., Landre, M.-F., Pietri-Taleb, F., Niedhammer, I., 2000, Individual and occupational determinants of low back pain according to various definitions of low back pain. *Epidemiology and Community Health*, 54, 215–220.

Plamondon, A., Gagnon, A., Gravel, D., 1995, Moments at the L5/S1 joint during asymmetrical lifting: effects of different load trajectories and initial positions. *Clinical Biomechanics*, 10, 128–136.

Punnet, L., Fine, L.J., Keyserling, W.M., Herrin, G.D., Chaffin, D.B., 1991, Back disorders and non-neutral trunk postures of automobile assembly workers. *Scandinavian Journal of Work, Environment and Health*, 17, 337–346.

Resnick, M.L., Chaffin, D.B., 1996, Kinematics, kinetics, and psychophysical perceptions in symmetric and twisting pushing and pulling tasks, *Human Factors*, 38, 114–129.

Schultz, A.B., Andersson, G.B.J., Ortengren, R., Haderspeck, K.K., Nachemson, A.L., 1982a, Loads on the lumbar spine — validation of a biomechanical analysis by measurements of intradiscal pressures and myoelectric signals. *Journal of Bone and Joint Surgery — American Volume*, 64(5), 713–720.

Schultz, A.B., Andersson, G.B.J., Ortengren, R., Bjork, R., Nordin, M., 1982b, Analysis and quantitative myoelectric measurements of loads on the lumbar spine when holding weights in standing postures. *Spine* 7(4), 390–397.

Snook, S.H., Ciriello, V.M., 1991, The design of manual handling tasks: revised tables of maximum acceptable weights and forces. *Ergonomics*, 34(9), 1197–1213.

Snook, S.H., Campanelli, R.A., Hart, J.W., 1978. A study of three preventive approaches to low back injury. *Journal of Occupational Medicine*, 20, 478–481.

Theorell, T., Harms-Ringdahl, K., Ahiberg-Hulten, G., Westin, B., 1991, Psychosocial job factors and symptoms from the locomotor system — a multicausal analysis. *Scandinavian Journal of Rehabilitation Medicine*, 23, 165–173.

Theorell, T., Nordemar, R., Michelsen, H., Stockholm Music I Study Group, 1993, Pain thresholds during standardized psychological stress in relation to perceived psychosocial work situation. *Journal of Psychosomatic Research*, 37(3), 299–305.

Washington State Ergonomics Standard 296-62-051, http://www.lni.wa.gov/wisha/Rules/generaloccupationalhealth/HTML/ergowac.htm

Wickstrom, G.J., Pentti, J., 1998, Occupational factors affecting sick leave attributed to low-back pain. *Scandinavian Journal of Work, Environment and Health*, 24(3), 145–152.

29

Application of Ergonomics to the Legs

Steven A. Lavender
The Ohio State University

29.1 Introduction

Most of the ergonomics literature dealing with the prevention and control of musculoskeletal disorders in the workplace has focused on the upper extremity and the back. Comparatively, little attention has been given to lower extremity musculoskeletal disorders that occur in the workplace. One could argue that since lower extremity problems are not well documented in ergonomics journals, the problems may not be of much practical significance. The first objective of this chapter is to review the current literature regarding occupational musculoskeletal disorders affecting the lower extremities and to demonstrate the significance of the problem. The second objective is to describe what types of intervention strategies are available to minimize the likelihood of future or recurrent injuries to the feet, angles, knees, and hips.

29.2 Lower Extremity Injuries: Is There an Occupational Problem?

The sports medicine literature is full of lower extremity overuse injuries in athletes. All too often we have seen athletes relegated to the sidelines following some sort of soft tissue injury that is likely to be the effect of not just a single incidence, but rather a cumulative loading pattern during practice and competition. Luckily, in most occupational environments, the intensity of the exercise is greatly diminished; however, the cumulative exposure problem still persists. Studies have begun to report the relationship between occupational factors and knee, hip, and foot trauma.

29.2.1 Hip Disorders

Lindberg and Axmacher (1988) reported the prevalence of coxarthosis in the hip to be greater in male farmers than in an age-matched group of urban dwellers. Vingard et al. (1991) classified blue-collar occupations as to whether static or dynamic forces could be expected to act on the lower extremity. The authors found that those employed in occupations that experienced greater loads on the lower extremity, namely farmers, construction workers, firefighters, grain mill workers, butchers, and meat preparation workers, had an increased risk of osteoarthritis (OA) of the hip. Similarly, Vingard et al. (1992) found that disability pensions for hip OA were significantly more likely to be received by males employed as farmers, forest workers, and construction workers. Given the occupations at risk, some have postulated that the increased incidence of hip OA may be due to driving vehicles with high levels of whole body vibration. Jarvholm et al. (2004), in comparing those having undergone hip replacement surgery with appropriate reference groups, found that drivers who were exposed to higher levels of whole body vibration were not at increased risk of hip OA. Yoshimura et al. (2000) reported a significant association between occupational lifting and the incidence of hip OA. Specifically, they found that those whose first job entailed regular lifting of 25 kg or more had an increased risk, as did individuals who regularly lifted more than 50 kg in their primary job. Conversely, this same study found that those who spent greater than 2 h sitting in their first job were at a reduced risk of developing the disorder.

29.2.2 Knee Disorders

Lindberg and Montgomery (1987) reported that knee OA, as defined by a "narrowing of the joint space with a loss of distance between the tibia and the femur in one compartment, of one-half or more of the distance in the other compartment of the same knee joint or the same compartment of the other knee, or less than 3 mm," was more common in those who had performed jobs that required heavy physical labor for a long time. Kohatsu and Schurman (1990) found that, relative to controls, the individuals with severe OA were two to three times more likely to have worked in occupations requiring moderate to heavy physical work. Prolonged exposure (11 to 30 yr) to building and construction occupations increased the risk of knee OA by 3.7 times (Holmberg et al., 2004). These same authors report that farm, forestry, letter-carrying, cleaning, and health-care work were not associated with increased risk of knee OA. Manninen et al. (2002), after adjusting for body mass index (BMI), prior knee injury, and leisure activities, found that knee OA increased with in individuals with a history of high physical workload, although the results were more consistent for women than for men.

Occupations with an increased frequency of knee bending and moderate physical demands have been associated with increased knee OA in the older working population after adjusting for age, body mass, knee injury history, smoking, and education level (Anderson and Felson, 1988; Felson et al., 1991). Moreover, it has been shown that the strength demands of the job were predictive of knee OA in the women from this older age group (Anderson and Felson, 1988). The authors suggest that the increased OA in those with long exposure to occupational tasks is indicative of the role of repetitive occupational exposure. Further supporting the link between material handling jobs and knee problems is the finding by McGlothlin (1996), who recently reported that beverage delivery personnel were experiencing discomfort in the knees, in addition to the anticipated discomfort in the back and shoulders. The work performed by these delivery personnel requires heavy lifting, kneeling, squatting, and, of course, driving. Risk of knee OA has been reported to increase for those who reported prolonged kneeling, walking more than two miles per day, and lift 25 kg or more on a regular basis (Coggon et al., 2000). Those whose occupations required kneeling and squatting were more likely to experience knee OA (Manninen et al., 2002) and be referred for meniscectomy procedures (Baker et al., 2003). Climbing was also found to increase knee OA in men (Manninen et al., 2002). Similarly, the number of knee flexions ($>45°$) has been related to the onset of occupational illness/injury by Craig et al. (2003). Chen et al. (2004) found that taxi drivers were at increased risk of knee pain if their work duration exceeded 6 h per day. Further, it should be recognized that personal risk factors for OA of the knee include heredity, obesity, and significant knee injury (Kohatsu and Schurman, 1990; Holmberg et al., 2004). Sahlstrom and Montgomery (1997) report that knee OA was weakly associated with weight-bearing knee bending, which increases the dynamic load on the knee when bending. In fact, when corrected for confounders, weight-bearing knee bending was no longer significant. Being overweight was a significant risk factor, however. Likewise, Coggon et al. (2000) found that obesity, defined as a BMI greater than or equal to 30 kg/m^2, and prolonged kneeling or squatting on the job combined to substantially increase the risk of knee OA. Kohatsu and Schurman (1990) found no relationship between leisure time activities and knee OA. Baker et al. (2003) report that while occupational activities contribute to the hospital referrals for knee symptoms, participation in soccer substantially increases the risk of knee cartilage injury.

Torner et al. (1990) reported that chronic prepatellar bursitis was the predominant knee disorder in 120 fishermen who underwent an orthopedic physical examination. Forty-eight percent of the men examined showed this disorder. Interestingly, the finding was as common among the younger men as in the older men. The authors believe that this disorder is a secondary effect of the boat's motion. The knees are used to stabilize the body by pressing against gunwales or machinery as tasks are performed with the upper extremities. Furthermore, just standing in mild sea conditions (maximum roll angles of 8°) has been shown to considerably elevate the moments at the knees the motion in the lower extremities and the trunk are the primary means for counteracting a ship's motions (Torner et al., 1991). Kivimaki (1992a) reported an increased thickness in the prepatellar or infrapatellar bursa was much more common in carpet and floor layers than in a reference group of house painters. Carpet and floor layers also experienced greater laxity in the knee joint (Kivimaki et al., 1994a), had more osteophytes of the patella (Kivimaki et al., 1992b), and more frequently reported prior knee conditions than house painters (Kivimaki, 1994b).

Several musculoskeletal colloquialisms have been used to describe occupationally related knee conditions including "beat knee," "carpet-layer knee," "preacher knee," and "housemaid knee" (Lee et al., 2004). Housemaid knee is an inflammation of the prepatellar bursa whereas preacher knee is an infrapatellar bursitis that is associated with excessive kneeling (Lee et al., 2004). The etiology of "beat knee" was described by Sharrard (1963), who reported on the examination of 579 coal miners. Forty percent of those examined were symptomatic or had previously experienced symptoms. Most of the injuries could be characterized as acute simple bursitis or chronic simple bursitis. The majority of the affected miners were colliers whose job requires constant kneeling at the mine face. There was a strong relationship between the coal seam height (directly related to roof height in a mine) and the incidence of beat knee. The incidence rates were much higher in mines with a roof height under 4 ft as compared with those with greater roof heights. Obviously, this factor greatly affects the work posture of the miners. With higher roof heights miners can alternate between stooped and kneeling postures but when seams are 1 m or less, the stooped posture is no longer an alternative. Gallagher and Unger (1990), for example, present recommendations for weight limits of handled materials in underground mines. Below 1.02 m these are based on miners in kneeling postures. Sharrard (1963) also speculated on the individual factors attributable to the disorder and found a higher incidence among younger men. However, this may be due to the "healthy worker effect" (Andersson, 1991) in which older miners with severe "beat knee" have left the mining occupation.

Tanaka et al. (1982) reported that the occupational morbidity ratios for workers' compensation claims of knee-joint inflammation among carpet installers was twice that found in tile setters and floor layers, and was over 13 times greater than that of carpenters, sheet metal workers, and tinsmiths. Others have shown the knees of those involved with carpet and flooring installation were more likely to have fluid collections in the superficial infrapatellar bursa, have a subcutaneous thickening in the anterior wall of the superficial infrapatellar bursa, and have an increased thickness in the subcutaneous prepatellar region (Myllymaki et al., 1993).

Thun et al. (1987) determined the incidence of repetitive knee trauma in the flooring installation professions. While all flooring installers spend a large amount of time kneeling, the authors divided the 154 survey respondents into two groups, "tilesetters" and "floor layers," based on their use of a "knee kicker." This device is used to stretch the carpet during the installation process. These respondents were compared with a group of millwrights and brick layers whose jobs did not require extended kneeling or the use of a knee kicker. Of the 112 floor layers (those who used the knee kicker), the prevalence rate of bursitis was approximately twice that found in the 42 tilesetters, and over three times that found in the 243 millwrights and brick layers. However, the prevalence in both groups of flooring workers of having required needle aspiration of the knee was almost five times that of millwrights and bricklayers. These results suggest that long durations of occupational kneeling is related to fluid accumulation, yet the bursitis is due to the repetitive trauma endured by the floor layers using the knee kicker. Similar findings were obtained by Jensen et al. (2000), who reported that the percentage of time performing "knee straining work" for floor layers, carpenters, and compositors (56, 26, and 0%, respectively) was positively correlated with knee complaints. It is also important to note that these authors found age, seniority, weight, BMI, smoking, and knee-straining sports were not significant covariates in the their analyses.

Village et al. (1991, 1993) found that the peak impulse forces generated in the knees of carpetlayers when using the "knee kicker" were on the order of 3000 N. The opposite knee that was supporting the body during this action had an average peak force of 893 N. Bhattacharya et al. (1985) reported knee impact forces of 2469 N (about three times body weight) for a light kick and 3019 N (or about four times body weight) for a hard kick. These light and hard kicks resulted in impact decelerations of 12.3 and 20 g, respectively. The authors observed that the knee kicking action during flooring installation occurred at a rate of 141 kicks per hour. However, putting the knee injuries in perspective, pain was reported by 22% of questionnaire respondents in the tufting job in a carpet manufacturer. However, knees were only listed in 2.4% of the accident records. Thus, the knee is frequently the site of discomfort, although there may be few lost days associated with knee pain (Tellier and Montreuil, 1991).

29.2.3 Stress Fractures

Cumulative trauma injuries can take the form of stress or fatigue fractures. While there have been many studies investigating stress fractures in the lower extremity, the occupational concerns have primarily been focused on military recruits (see the review by Jones et al., 2002). Linenger and Shwayhat (1992) reported training-related injuries to the foot occurred in military personnel undergoing basic training at a rate of three new injuries per 1000 recruit days. These authors found that stress fractures to the foot, ankle sprains, and Achilles tendonitis accounted for the bulk of the injuries. Anderson (1990) found the stress fractures to be most common in the distal second and third metatarsal bones but could occur in any of the bones in the foot. Greaney et al. (1983) reported that 73% of the stress fractures occurred in the tibia, with the most common site being the calcaneal tuberosity. Similarly, Giladi et al.'s (1985) findings indicated that 71% of the stress fractures in their sample of military recruits occurred in the tibia and 25% in the femoral shaft. Moreover, they found the fractures to occur later in the training process than reported by others. Jordaan and Schwellnus (1994) reported that overuse injuries, when normalized according to training hours per week, decreased from week 1 to week 4, showed a resurgence in week 5, and a large peak in the final week of training. The injury rates corresponded to the weeks in which there was increased marching and less field training. In a sample of 21 elite military recruits, Kiuru et al. (2005) reported 75 bone stress injuries were detectable using magnetic resonance imaging; however, only 40% of the injuries were symptomatic. These findings suggest that stress factures, many of which are resolved without symptoms, may be much more common than originally thought when there is intense increase in physical activity.

In an attempt to predict overuse injuries found during training of military recruits, some investigators have looked into aspects of lower limb morphology as indicators of individual susceptibility. Giladi et al. (1991) reported the influence of individual factors on the incidence of fatigue fractures; specifically, they found that individuals with narrow tibia or a greater external rotation of the hip were more likely to experience fatigue fractures. Cowan et al. (1996) reported the relative risk of "overuse" injuries was significantly higher in military recruits with the most valgus knees. In addition, these authors showed that the "Q" angle, which defines the degree of deviation in the patellar tendon from the line of pull on the patella by the quadriceps muscles, was predictive of stress fractures.

In summary, several occupational risk factors have been identified which place an employee at increased risk for disorders in the lower extremity. The literature has shown that heavy physical labor and frequent knee bending are factors, especially in the older component of the workforce, thereby suggesting an interaction between the age degenerative processes and cumulative work experience. And clearly, the role of direct cumulative trauma in those employees who must maintain kneeling postures and use their knees to strike objects (knee kicker) cannot be overlooked when considering preventive measures.

29.3 Preventing Injury: Types of Ergonomics Controls

Several types of control mechanisms to prevent or accommodate lower extremity disorders are available. This section will focus on the techniques whereby the interface between the lower extremity and the environment can be improved. This includes interventions to prevent stress fractures, interventions aimed at improving circulation and comfort in the lower extremities for those who remain in relatively static work postures throughout the day, and interventions for those who must work in kneeling postures.

29.3.1 Floor Mats

Floor mats are often used for local slip protection. While inexpensive, they create a possible trip hazard, interfere with operations or cleanliness, and wear excessively (Andres et al., 1992). Several investigators have looked into the use of floor mats to reduce the fatigue effects observed in jobs that require prolonged standing. The subjects tested by Kuorinka et al. (1978) indicated through subjective ratings that they preferred to work on softer surfaces as opposed to harder surfaces. A foam plastic surface was rated the best and concrete the worst. These authors reported a moderate correlation between the subjective comfort ratings of the five surfaces tested and the order of surface hardness. However, integrated electromyographic (EMG) signals, median frequency of the EMG, measures of postural sway, and measures of calf circumference did not show any significant difference due to the floor covering. Hinnen and Konz (1994) asked employees in a distribution center to stand for two 8-h shifts on each of five mats tested in the study. Approximately every hour the employees rated their comfort in several body regions including the upper leg, lower leg, ankle, and back. A scale of 0 (no discomfort at all) to 10 (extreme discomfort) was used. While these worker experienced relatively little discomfort, the mats with compressibility between 3 and 4% did best in the upper leg discomfort rankings as well as subject preference rankings. Marginally significant changes in the discomfort ratings were reported for the ankle. The ratings of lower leg and back discomfort showed no significant differences.

Rys and Konz (1989, 1990) reported on several anthropometric and physiological measures including changes in foot size and skin temperature at the instep and the calf. In general, the mats included in this study were significantly different from concrete in that there was greater skin temperature at both measured locations and greater comfort ratings. These authors report that the comfort was inversely related to mat compressibility.

Cook et al. (1993) used surface EMG to study the recruitment of the anterior tibialis and paraspinal muscles when standing on linoleum-covered concrete versus a expanded vinyl 9.5-mm-thick surgical mat. After subjects stood for two sessions, of 2 h duration, on the mat and on the linoleum, it was concluded that there were no significant changes in the mean of the rectified EMG signals in either muscle due to the mats. As in the studies discussed earlier, subjective data support the use of the mat. Madeleine et al. (1997) did find changes in their objective measures, however. They reported that, when compared with standing on a hard aluminum casting, subjects standing on a soft polyurethane mat showed less swelling of the shank and less EMG activity in the soleus muscle, and reduced postural activity as detected by center of pressure displacement measures over the 2-h trial.

Kim et al. (1994) tested two types of floor mats and a control condition in which subjects stood on concrete. While these authors observed muscular fatigue, as determined by a shift in the EMG median frequencies in the gastrocnemius and anterior tibialis muscle over the trial period, the EMG median frequencies in these muscles were not affected by the use of floor mats. The median frequency shift in the erector spinae was reduced when subjects stood on the thinner and more compressible mat. The authors hypothesized that greater compressibility would have made for a less stable base of support, thereby requiring more frequent postural changes in the trunk to overcome the destabilization associated with postural sway. Thus, the dynamic use of erector spinae muscles to correct for postural sway would facilitate the oxygen delivery and the removal of contractile byproducts through increased blood flow. A further test of this hypothesis would evaluate whether this motion occurred only in the trunk, or if it occurred in the lower extremities that did not show the spectral shift due to the floor condition.

Redfern and Cham's (2000) review of the data on antifatigue floor mats suggests that there is a consistent reduction of discomfort with matting when compared with hard floor conditions. Unfortunately, the objective measures are less conclusive across studies. In part this may be due to variations in the physical properties of the mats tested. Cham and Redfern (2001) report that fatigue and discomfort were reduced when floor mats had increased elasticity, decreased energy absorption, and increased stiffness; however, the changes in these subjective responses were only detectable after 3 h of standing.

29.3.2 Shoe Insoles

The critical role that shoe design plays in the development of overuse syndromes in runners is widely recognized (Lehman, 1984; Pinshaw et al., 1984; McKenzie et al., 1985). Moreover, the role of the shoe in controlling lower extremity kinematics has been reviewed by Frederick (1986) and discussed by McKenzie et al. (1985). Similarly, the use of wedged insoles has been shown to alter the static posture of the lower extremity (Yasuda and Sasaki, 1987). Sasaki and Yasuda (1987) have shown the use of wedged insoles to be a good conservative treatment for medial OA of the knee in the early stages. These authors reported that patients with early radiographic stages of OA and who were provided a wedged insole had reduced pain and improved walking ability relative to controls without the insole. Crenshaw et al. (2000) showed that lateral-wedged insoles significantly reduced the varus moment and the medial compartment load during gait.

Clearly, the lower extremity disorders reported by runners represent extreme overuse; however, the treatment and prevention mechanisms may be applicable to occupational settings where employees must stand, walk, run, or even jump during their normal work activities. Insoles have been investigated as an alternative to floor mats for those performing jobs that require extended standing postures and greater mobility. Orlando and King (2004) investigated a small sample ($n = 11$) of assembly line workers' perception of fatigue and discomfort when standing on a woodblock floor, a floor mat, and when wearing insoles. While the findings indicated that both the floor mat and the insoles were superior to the woodblock floor with respect to general fatigue, leg fatigue, and discomfort ratings, there were no significant differences between these two intervention conditions (King, 2002). However, in the older participants and the participants with more seniority, discomfort was minimized with the insoles. Padded insoles have been investigated for the shock abating effects on the skeletal system. Loy and Voloshin (1991) used light-weight accelerometers for measuring the shock waves as subjects walked, ascended and descended stairs, and jumped off platforms of a fixed height. The peak magnitude of the shock waves during jumping activities were approximately eight times that seen during normal walking. The results indicated that the insoles reduced the amplitude of the shock wave by between 9 and 41% depending upon the activity performed. The insoles were most effective at reducing heel strike impacts and had the largest effect with the jumping activities.

Milgrom et al. (1992) tested the effects of shock attenuation on the incidence of overuse injuries in infantry recruits. Earlier studies conducted by fixing accelerometers to the tibial tubercle showed that soldiers wearing modified basketball shoes had mean accelerations that were 19% less than soldiers wearing lightweight infantry boots. These authors also found that over the 14 weeks of basic training the modified basketball shoes reduced the metatarsal stress fractures; however, the tibial and femoral stress fractures were not affected by the shoes worn. Gardner et al. (1988) compared an viscoelastic polymer insole and a standard mesh insole that were issued by platoon to over 3000 marine recruits. While the polymer insole had good shock absorbing properties, the incidence rate of lower extremity stress injuries over the 12 week basic training program were unaffected by the insole used. Schwellnus et al. (1990) found that the mean weekly incidence of total overuse injuries and tibial stress syndrome decreased significantly in the 237 military recruits provided with neoprene insoles as compared to 1151 controls. These authors also found a trend ($p < 0.10$) suggesting that the incidence of stress fractures was reduced with the insoles. However, Jones et al. (2002), in their review of the military intervention studies investigating insoles, do not find convincing evidence that shock absorbent insoles prevent stress fractures, thus suggesting they are of questionable value if the intended purpose is solely stress fracture prevention. Rome et al. (2005), in their review of the research examining the potential for shock absorbing shoe insoles to prevent injuries in military recruits, conclude that there is evidence to support their use; however, the strongest supporting trial included in their review had some methodological issues.

Several studies have been conducted to evaluate variations in insole materials. Leber and Evanski (1986) describe the characteristics of the following seven insole materials: Plastazote, Latex foam, Dynafoam, Ortho felt, Spenco, Molo, and PPT. These authors measured the plantar pressures in 26 patients with forefoot pain. All insole materials reduced the plantar pressure by between 28 and 53% relative to a control condition; however, PPT, Plastazote, and Spenco were the superior products. Viscolas and Poron were found to have the best shock absorbency of the five insole materials tested by Pratt et al. (1986). Maximum plantar pressures were found to be significantly reduced in the forefoot region with PPT, Spenco, and Viscolas, although the three materials were not significantly different (McPoil and Cornwall, 1992). In the rear foot region, however, McPoil and Cornwall (1992) report that only the PPT and the Spenco insoles reduced the maximum plantar pressure relative to the barefoot condition. The plantar pressure in the rearfoot region was not significantly reduced with the Viscolas. Interestingly, based on the shock absorbency data from Pratt's (1988) 30-day durability test, the resilience of Viscolas, PPT, and Plastazote could be described as excellent, good, and poor, respectively. Sanfilippo et al. (1992) also reported the change in foot to ground contact area as a function of insole material. Plastazote, Spenco, and PPT led to a significantly greater contact area than the other materials tested.

In summary, insoles appear to be effective at modifying the lower extremity kinematics and reducing the peak plantar pressures, although their effectiveness is dependent upon the material used. Additional research is needed to clarify the effectiveness of insoles in controlling lower extremity stress injuries. Based on the previous discussion it should be clear that the effectiveness of this control strategy will be dependent upon shock absorbing capacity, the pressure dispersion, and the durability properties of the insole materials selected.

29.3.3 Help for Those in Kneeling Postures

Sharrard (1963) reported that there was no relationship between the type of knee pads used and the incidence of beat knee in miners. This author recorded peak pressures on the order 35.7 kg/cm^2 as simulated mining tasks were performed. These compression forces were shown to vary widely throughout the 2.5-sec cycle time for a shoveling task. Unfortunately, the author had no instrumentation capable of determining the shear forces and the torsional moments placed on the knee during the simulated tasks. At the time of Sharrard's paper a "bursa pad" had been designed that allowed perspiration to escape, pushed coal particles away from the skin, and provided satisfactory cushioning. Although no control group was used, the author reported that of the 24 previously affected men selected to test the pad under working conditions only two reported a recurrence of beat knee after a 12-month period.

In general, while many types of knee pads are currently commercially available, and frequently recommended, there has been little research published on knee pad design. Clearly, there are issues with the construction of knee pads and the resulting subjective comfort/discomfort while performing kneeling tasks, how the comfort compares with the the actual force distribution within the pad, and how the pad construction interacts with ground conditions to affect a worker's postural stability. Some have proposed using imaging methods to derive knee anthropometric features that could be used to optimize knee pad design (Pellmann and Thumler, 1992). Moreover, the pads also have to be evaluated for their impact on comfort and performance of nonkneeling tasks; for example, tasks that require standing and walking. This is important for individuals who work in a variety of low extremity postures throughout the day. If the knee pads do not accommodate walking and standing, for example, they will likely not be used where they should be.

Ringen et al. (1995) report of a new tool to reduce the knee and back trauma in those who tie rebar rods together in preparation for pouring concrete. No longer will concrete workers need to kneel or stoop for extended periods to interconnect the iron rods as this tool allows the operator to work in a standing posture.

Powered carpet stretching tools are available to remove the repeated trauma experienced by carpet layers. Village et al. (1993) have provided design guidelines for improved carpet stretching devices. However, the widespread implementation of improved devices depends upon educating flooring workers on the trade-offs between the additional time necessary to operate the tool and the knee disorders associated with the conventional technique.

29.4   Summary

Ergonomics texts historically have focused relatively little attention on the prevention of lower extremity disorders or the accommodation of individuals returning to work whom have experienced a lower extremity disorder. In part this may be due to lesser appreciation of the frequency and severity of occupational lower extremity disorders. Unlike many back or upper extremity disorders, which have their origins in the repeated stresses placed on muscular, tendinous, and ligamentous tissues, many of the occupational lower extremity disorders occur through direct compression of the body tissues by a surface in the environment. As a result the occupational lower extremity disorders often involve cartilaginous tissue and bone. Therefore, accommodation and prevention of these disorders occur primarily through the optimization of the body's contact with surfaces in the environment. This chapter, in addition to highlighting some of the epidemiological findings relevant to occupational lower extremity disorders, has reviewed some of the more common intervention pathways available.

References

Anderson, E.G. (1990). Fatigue fractures of the foot. *Injury*, 21, 275–279.

Anderson, J.J., Felson, D.T. (1988). Factors associated with osteoarthritis of the knee in the first national health and nutrition examination survey (HANES I). *American Journal of Epidemiology*, 128, 179–189.

Andersson, G.B.J. (1991). The epidemiology of spinal disorders. In: J.W. Frymoyer (ed.) *The Adult Spine: Principles and Practice*. New York: Raven, pp. 107–146.

Andres, R.O., O'Conner, D., Eng, T. (1992). A practical synthesis of biomechanical results to prevent slips and falls in the workplace. In: S. Kumar (ed.) *Advances in Industrial Ergonomics and Safety IV*. London: Taylor & Francis, pp. 1001–1006.

Baker, P., Reading, I., Cooper, C., Coggon, D. (2003). Knee disorders in the general population and their relation to occupation. *Occupational and Environmental Medicine*, 60, 794–797.

Bhattacharya, A., Mueller, M., Putz-Anderson, V. (1985). Traumatogenic factors affecting the knees of carpet installers. *Applied Ergonomics*, 16, 243–250.

Cham, R., Redfern, M.S. (2001). Effects of flooring on standing comfort and fatigue. *Human Factors*, 43, 381–391.

Chen, J.C., Dennerlein, J.T., Shih, T.S., Chen, C.J., Cheng, Y., Chang, W.P., Ryan, L.M., Christiani, D.C. (2004). Knee pain and driving duration: a secondary analysis of the taxi drivers' health study. *American Journal of Public Health*, 94, 575–581.

Coggon, D., Croft, P., Kellingray, S., Barrett, D., McLaren, M., Cooper, C. (2000). Occupational physical activities and osteoarthritis of the knee. *Arthritis & Rheumatism*, 43, 1443–1449.

Cook, J., Branch, T.P., Baranowski, T.J., Hutton,W.C. (1993). The effect of surgical floor mats in prolonged standing: an EMG study of the lumbar paraspinal and anterior tibialis muscles. *Journal of Biomedical Engineering*, 15, 247–250.

Cowan, D.N., Jones, B.H., Frykman, P.N., Polly, D.W., Harman, E.A., Rosenstein, R.M., Rosenstein, M.T. (1996). Lower limb morphology and risk of overuse injury among male infantry trainees. *Medicine and Science in Sports and Exercise*, 28, 945–952.

Craig, B.N., Congleton, J.J., Kerk, C.J., Amendola, A.A., Gaines, W.G., Jenkins, O.C. (2003). A prospective field study of the relationship of potential occupational risk factors with occupational injury/illness. *AIHA Journal*, 64, 376–387.

Crenshaw, S., Fabian, E., Calton, E. (2000). Effects of lateral-wedged insoles on kinetics at the knee. *Clincial Orthopaedics and Related Research*, 375, 185–192.

Felson, D.T., Nannan, M.T., Naimark, A., Berkeley, J., Gordon, G., Wilson, P.W., Andersson, J. (1991). Occupational physical demands, knee bending, and knee osteoarthritis: results from the Framingham study. *Journal of Rheumatology*, 18, 1587–1592.

Frederick, E.C. (1986). Kinematically mediated effects of sport shoe design: a review. *Journal of Sports Sciences*, 4, 169–184.

Gallagher, S., Unger, R.L. (1990). Lifting in four restricted lifting conditions. *Applied Ergonomics*, 21, 237–245.

Gardner, L.I., Dziados, J.F., Jones, B.H., Brundage, J.F. (1988). Prevention of lower extremity stress fractures: a controlled trial of a shock absorbent insole. *American Journal of Public Health*, 78, 1663–1567.

Giladi, M., Ahronson, Z., Stein, M., Danon, Y.L., Milgrom, C. (1985). Unusual distribution and onset of stress fractures in soldiers. *Clinical Orthopaedics and Related Research*, 192, 142–146.

Giladi, M., Milgrom, C., Simkin, A., Danon, Y. (1991). Stress fractures. Identifiable risk factors. *American Journal of Sports Medicine*, 19, 647–652.

Greaney, R.B., Gerber, F.H., Laughlin, R.L., Kmet, J.P., Metz, C.D., Kilsheski, T.S., Rao, B.R., Silverman, E.D. (1983). Distribution and natural history of stress fractures in U.S. Marine recruits. *Radiology*, 146, 339–346.

Hinnen, P., Konz, S. (1994). Fatigue mats. In: F. Aghazadeh (ed.) *Advances in Industrial Ergonomics and Safety VI*. London: Taylor & Francis, pp. 323–327.

Holmberg, S., Thelin, A., Thelin, N. (2004). Is there an increased risk of knee osteoarthritis among farmers? A population-based case–control study. *International Archives of Occupational and Environmental Health*, 77, 345–350.

Jarvholm, B., Lundstrom, R., Malchu, H., Rehn, B., Vingard, E. (2004). Osteoarthritis in the hip and whole-body vibration in heavy vehicles. *International Archives of Occupational and Environmental Health*, 77, 424–426.

Jensen, L.K., Mikkelsen, S., Loft, I.P., Eenberg, W. (2000). Work-related knee disorders in floor layers and carpenters. *Journal of Occupational and Environmental Medicine*, 42, 835–842.

Jones, B.H., Thacker, S.B., Gilchrist, J., Kimsey, C.D. Jr., Sosin, D.M. (2002). Prevention of lower extremity stress fractures in athletes and soldiers: a systematic review. *Epidemiologic Reviews*, 24, 228–247.

Jordaan, G., Schwellnus, M.P. (1994). The incidence of overuse injuries in military recruits during basic military training. *Military Medicine*, 159, 421–426.

Kim, J.Y., Stuart-Buttle C., Marras, W.S. (1994). The effects of mats on back and leg fatigue. *Applied Ergonomics*, 25, 29–34.

King, P.M. (2002). A comparison of the effects of floor mats and shoe in-soles on standing fatigue. *Applied Ergonomics*, 33, 477–484.

Kiuru, M.J., Niva, M., Reponen, A., Pihlajamaki, H.K. (2005). Bone stress injuries in asymtomatic elite recruits: a clinical and magnetic resonance imaging study. *The American Journal of Sports Medicine*, 33, 272–276.

Kivimaki, J. (1992a). Occupationally related ultrasonic findings in carpet and floor layers' knees. *Scandinavian Journal of Work, Environment and Health*, 18, 400–402.

Kivimaki, J. (1992b). Knee disorders in carpet and floor layers and painters. *Scandinavian Journal of Work, Environment and Health*, 18, 310–316.

Kivimaki, J., Hanninen, K., Kujala, U.M., Osterman, K., Riihimaki, H. (1994a). Knee laxity in carpet and floor layers and painters. *Annales Chirurgiae et Gynaecologie*, 83, 229–233.

Kivimaki, J., Riihimaki, H., Hanninen, K. (1994b). Knee disorders in carpet and floor layers and painters. Part II: knee symptoms and patellofemoral indicies. *Scandinavian Journal of Work, Environment and Health*, 26, 97–101.

Kohatsu, N.D., Schurman, D.J. (1990). Risk factors for the development of osteoarthrosis of the knee. *Clinical Orthopaedics and Related Research*, 261, 242–246.

Kuorinka, I., Hakkanen, S., Nieminen, K., Saari, J. (1978). Comparison of floor surfaces for standing work. In: E. Asmussen and K. Jorgensen (eds.) *Biomechanics VI-B, Proceedings of the Sixth International Congress of Biomechanics*. Baltimore: University Park Press, pp. 207–211.

Leber, C., Evanski, P.M. (1986). A comparison of shoe insole materials in plantar pressure relief. *Prosthetics and Orthotics International*, 10, 135–138.

Lee, P., Hunter, T.B., Taljanovic, M. (2004). Musculoskeletal Colloquialisms: How did we come up with these names? *RadioGraphics*, 23, 1009–1027.

Lehman, W.L. (1984). Overuse syndromes in runners. *American Family Physician*, 29, 157–161.

Lindberg, H., Axmacher, B. (1988). Coxarthrosis in farmers. *Acta Orthopaedica Scandinauica*, 59, 607.

Lindberg H., Montgomery, F. (1987). Heavy labor and the occurrence of gonarthrosis. *Clinical Orthopaedics and Related Research*, 214, 235–236.

Linenger, J.M., Shwayhat, A.F. (1992). Epidemiology of podiatric injuries in US Marine recruits undergoing basic training. *Journal of the American Podiatric Medical Association*, 82, 269–271.

Loy, D.J., Voloshin, A.S. (1991). Biomechanics of stair walking and jumping. *Journal of Sports Sciences*, 9, 136–149.

Madeleine, P., Voigt, M., Arendt-Nielsen, L. (1997). Subjective, physiological, and biomechanical response to prolonged manual work performed standing on hard and soft surfaces. *European Journal of Applied Physiology and Occupational Physiology*, 77, 1–9.

Manninen, P., Heliovaara, M., Riihimaki, H., Suoma-Iainen, O. (2002). Physical workload and the risk factors of severe knee osteoarthritis. *Scandinavian Journal Work, Environment and Health*, 28, 25–32.

McGlothlin, J.D. (1996). *Ergonomic Interventions for the Soft Drink Beverage Delivery Industry*. U.S. Department of Health and Human Services (NIOSH) Publication No. 96–109.

McKenzie, D.C., Clement, D.B., Taunton, J.E. (1985). Running shoes, orthotics, and injuries. *Sports Medicine*, 2, 334–347.

McPoil, T.G., Cornwall, M.W. (1992). Effect of insole material on force and plantar pressures during walking. *Journal of the American Podiatric Medical Association*, 82, 412–416.

Milgrom, C., Finestone, A., Shlamkovitch, N., Wosk, J., Laor, A., Voloshin, A., Eldad, A. (1992). Prevention of overuse injuries of the foot by improved shoe shock attenuation. A randomized prospective study. *Clinical Orthopaedics and Related Research*, 281, 189–192.

Myllymaki, T., Tikkakoski, T., Typpo, T., Kivimaki, J., Suramo, I. (1993). Carpet-layer's knee. An ultrasonographic study. *Acta Radiologica*, 34, 496–499.

Orlando, A.R., King, M. (2004). Relationship of demographic variables on perception of fatigue and discomfort following prolonged standing under various flooring conditions. *Journal of Occupational Rehabilitation*, 14, 63–76.

Pellmann, P., Thumler, P. (1992). Ergonomic design of knee pads. Wirtschaftsverlag NW, Postfach 10 11 10, Am Alten Hafen 113–115, D-W-2850 Bremerhaven 1, Germany.

Pinshaw, R., Atlas, V., Noakes, T.D. (1984). The nature and response to therapy of 196 consecutive injuries seen at a runners' clinic. *Sourth African Medical Journal*, 65, 291–298.

Pratt, D.J. (1988). Medium term comparison of shock attenuating insoles using a spectral analysis technique. *Journal of Biomedical Engineering*, 10, 426–429.

Pratt, D.J., Rees, P.H., Rodgers, C. (1986). Assessment of some shock absorbing insoles. *Prosthetics and Orthotics International*, 19, 43–45.

Redfern, M.S., Cham, R. (2000). The influence of flooring on standing comfort and fatigue. *AIHA Journal*, 61, 700–708.

Ringen, K., Englund, A., Seegal, J. (1995). Construction workers. In: B.S. Levey and D.H. Wegman (eds.) *Occupational Health: Recognizing and Preventing Work-Related Disease*, Boston: Little, Brown, and Company, pp. 685–701.

Rome, K, Handoll, H.H.G., Ashford, R. (2005). Interventions for preventing and treating stress fractures and stress reactions of bone of the lower limbs in young adults. *The Cochrane Database of Systematic Reviews*, Issue 2. Art. No.: CD000450.pub2. DOI: 10.1002/14651858.CD000450.pub2.

Rys, M., Konz, S. (1989). An evaluation of floor surfaces. In: *The Proceedings of the Human Factors Society 33th Annual Meeting*, Vol. 1, pp. 517–520.

Rys, M., Konz, S. (1990). Floor mats. In: *The Proceedings of the Human Factors Society 34th Annual Meeting*, Vol. 1, pp. 575–579.

Sahlstrom, A., Montgomery, F. (1997). Risk analysis of occupational factors influencing the development of arthrosis of the knee. *European Journal of Epidemiology*, 13, 675–679.

Sanfilippo, P.B., Stess, R.M., Moss, K.M. (1992). Dynamic plantar pressure analysis. Comparing common insole materials. *Journal of the American Podiatric Medical Association*, 82, 502–513.

Sasaki. T., Yasuda, K. (1987). Clinical evaluation of the treatment of osteoarthritic knees using a newly designed wedged insole. *Clinical Orthopaedics and Related Research*, 221, 181–187.

Schwellnus, M.P., Jordaan, G., Noakes, T.D. (1990). Prevention of common overuse injuries by the use of shock absorbing insoles. A prospective study. *The American Journal of Sports Medicine*, 18, 636–641.

Sharrard, W.J.W. (1963). Aetiology and pathology of beat knee. *British Journal of Industrial Medicine*, 20, 24–31.

Tanaka, S., Smith, A.B., Halperin, W., Jensen, R. (1982). Carpet-layers knee. *The New England Journal of Medicine*, 307, 1276–1277.

Tellier, C., Montreuil, S. (1991) Pain felt by workers and musculoskeletal injuries: assessment relating to tufting shops in the carpet industry. In: Y. Queinnec and F. Daniellou (eds.) *Designing for Everyone*, Taylor and Francis, Vol. 1, pp. 287–289.

Thun, M., Tanaka, S., Smith, A.B., Haperin, W.E., Lee, S.T., Luggen, M.E., Hess, E.V. (1987). Morbidity from repetitive knee trauma in carpet and floor layers. *British Journal of Medicine*, 44, 611–620.

Torner, M., Zetterberg, C., Hansson, T., Lindell, V. (1990). Musculo-skeletal symptoms and signs and isometric strength among fishermen. *Ergonomics*, 33, 1155–1170.

Torner, M., Almstrom, C., Karlsson, R., Kadefors, R. (1991). Biomechanical calculations of musculo-skeletal load caused by ship motions, in combination with work, on board a fishing vessel. In: Y. Queinnec and F. Daniellou (eds.) *Designing for Everyone*, Taylor & Francis, London, Vol. 1, pp. 293–295.

Village, J., Morrison, J.B., Leyland, A. (1991). Carpetlayers and typesetters ergonomic analysis of work procedures and equipment. In: Y. Queinnec and F. Daniellou (eds.) *Designing for Everyone*, Taylor & Francis, London, Vol. 1, pp. 320–322.

Village, J., Morrison, J.B., Leyland, A. (1993). Biomechanical comparison of carpet stretching devices. *Ergonomics*, 36, 899–909.

Vingard, E., Alfredsson, L., Goldie, I., Hogstedt, C. (1991). Occupation and osteoarthrosis of the hip and knee: a register-based cohort study. *International Journal of Epidemiology*, 20, 1025–1031.

Vingard, E., Alfredsson, L., Fellenius, E., Christer, H. (1992). Disability pensions due to musculo-skeletal disorders among men in heavy occupations. *Scandinavian Journal of Social Medicine*, 20, 31–36.

Yasuda, K., Sasaki, T. (1987). The mechanics of treatment of the osteoarthritic knee with a wedged insole. *Clinical Orthopaedics and Related Research*, 215, 162–172.

Yooshimura, N., Sasaki, S., Iwasaki, K., Danjoh, S., Kinoshita, H., Yasuda, T., Tamaki, T., Hashimoto, T., Kellingray, S., Croft, P., Coggon, D., Cooper, C. (2000). Occupational lifting is associated with hip osteoarthritis: a Japanese case–control study. *Journal of Rheumatology*, 27, 434–440.

30

Application of Ergonomics of the Foot*

Stephan Konz
Kansas State University

30.1 Foot/Leg

30.1.1 Anatomy

Figure 30.1 shows the bones of the foot and ankle. The toes (foot fingers) are divided into *metatarsals* and three *phalanges* (except for the big toe, which only has two phalanges). In supporting the body, the *calcaneus* (heel) supports 50% of the weight, the first and second metatarsals 25%, and the third, fourth, and fifth metatarsals 25%. In between are two arches: (1) the medial arch (calcaneus, the talus, the navicular, the cuneiform bones, and the first, second, and third metatarsals) and (2) the lateral arch (calcaneus, talus, cuboid, and the fourth and fifth metatarsals). The plantar facia is a fibrous tissue that forms the arch underneath your foot from the heel to the toes. If it weakens, the facia can cause pain to either the heel end or the toe end.

Under the heel (calcaneus) is a very important shock absorber, the heel pad (about 1.8 cm thick). The bottom of the calcaneus is not spherical but has two small "mountains"; the pad reduces the pressure on these mountains, and thus on the ankle, knee, and back.

The foot is connected to the ankle with a *mortise and tenon* joint. The vertical leg of the mortise is short on the outside (*lateral side*); in addition, the ligaments holding the bottom of the fibula (lateral malleolus) to the talus and calcaneus are relatively weak. In contrast, the vertical leg of the inside (*medial*) mortise is longer, and the ligaments holding the bottom of the tibia (medial malleolus) to the talus are relatively strong.

---

*This chapter is a concise version of material in Konz and Johnson (2004).

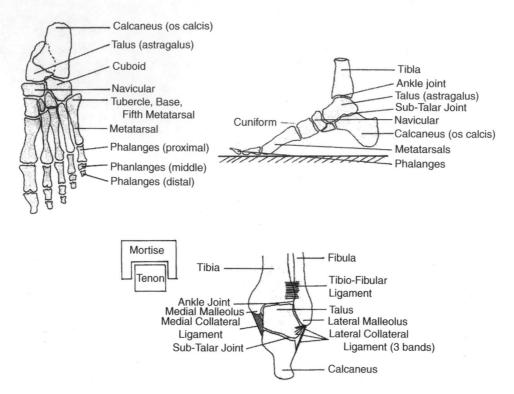

FIGURE 30.1   The foot and ankle. The right foot is viewed from below (top left) and the outside (top right); the left ankle (bottom) is viewed from the front.

Inward rotation (*inversion*) of the foot tends to pull the ligaments from the bone. With proper treatment, healing is usually complete in about 3 weeks. There is a danger that the injured person may not seek medical advice — even with a complete tear of the ligaments (connecting either the malleolus and the talus or the tibia). Then surgical repair and rigid fixation in a cast is needed for 2–3 months.

External rotation (*eversion*) of the foot tends to break one of the malleoli bones (vertical part of the mortise). The person tends to recognize this serious injury and go to a physician.

Approximately 80% of all foot fractures involve the toes; almost all of them could be prevented by safety shoes since they lie within the area protected by the metal toe cap.

Three venous systems drain the lower limbs: (1) a deep central system drains the muscles, (2) a superficial system drains the foot and the skin of the leg, and (3) a perforating system connects the deep and superficial systems.

30.1.2   Physiology

The veins store the body's blood. If the legs do not move, the blood from the heart tends to go down to the legs and stay there (*venous pooling*). This causes more work for the heart, as, for a constant supply of blood, when the blood/beat is lower, then there must be more beats. Venous pooling causes swelling of the legs (*edema*) and varicose veins. The foot swelling during stationary seated work can be overcome by modest leg activity (such as by rolling the chair about the workstation) (Winkel and Jorgensen, 1986).

Venous pressure in the ankle of sedentary people is approximately equal to hydrostatic pressure from the right auricle. Pollack and Wood (1949) gave a mean ankle venous pressure of 56 mmHg for sitting and 87 mmHg for standing. Nodeland et al. (1983) gave 48 mmHg for sitting and 80 mmHg for standing. Pollack and Wood reported walking drops ankle venous pressure to about 23 mmHg (Nodeland et al. reported 21 mmHg) in about 10 steps.

TABLE 30.1 Selected Dimensions (cm) of Nude U.S. Adult Civilians

	Mean		Std. Deviation	
	Females	Males	Females	Males
Stature	162.9 (100%)	175.6 (100%)	6.4	6.7
Crotch height	74.1 (45%)	83.7 (48%)	4.4	4.6
Knee height	51.5 (32%)	55.9 (32%)	2.6	2.8
Foot length	24.4 (15%)	27.0 (15%)	1.2	1.3
Foot breadth	9.0 (6%)	10.1 (6%)	0.5	0.5

Note: The percentages show the dimension as a percent of stature height. If only male anthropometric data for a population are available, the female dimension can be estimated as 93% of the corresponding male dimension. Shoes add 25 mm height for males and 15 mm for females. Shoes add 0.9 kg to body weight.

Source: From Konz, S. and Johnson, S. 2004. *Work Design: Occupational Ergonomics*, 6th ed. Holcomb-Hathaway, Scottsdale, AZ. With permission.

The fall occurs as the calf muscles contract in taking the next step before venous filling is complete. Thus additional blood is pumped out of the leg, causing a further drop in pressure when the calf muscles relax. The drop stabilizes in about 10 steps when the incoming flow to the vein from the capillaries equals the flow out of the leg. Thus, walking can partially compensate for posture. For example, Nodeland et al. reported standing bench work (i.e., with occasional steps around the area) had ankle pressure approximately equal to sitting at a desk (48 mmHg). Walking aids blood circulation through the "milking action" of the leg muscles; this reduces the work of the heart. Active standing (walking 2–4 min every 15 min) results in less discomfort than standing without walking.

Because of vasoconstriction, foot skin temperature (without shoes) usually is the lowest body skin temperature. Normal skin foot temperature = 33.3°C for males but 30.2° for females (Oleson and Fanger, 1973).

30.1.3 Dimensions

Table 30.1 gives some dimensions for U.S. adults. Also see Chapter 9 — Engineering Anthropometry. The torso is relatively constant in height; most of the variation in standing height is due to differences in leg length. Figure 30.2 shows the mean difference, when standing, between the inside of the two feet is about

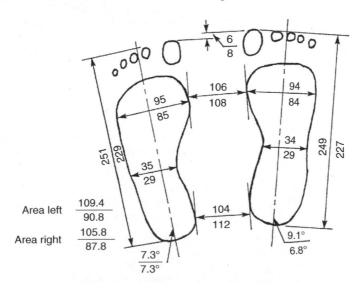

FIGURE 30.2 Footprint dimensions in mm (males above the line; females below line); areas in mm^2; angles in degrees. Toe area is 10% of contact area. They stood with the right foot slightly (6–8 mm) ahead of the left foot. The left foot (for males) averaged 7° to the left of the medial plane; the right foot averaged 9° to the right. (From Rys, M. and Konz, S. 1994. *Ergonomics*, 37(4):677–687. With permission.)

107 mm. The distance between foot centerlines is about $107 + 90 = 197$ mm (200 mm in round numbers). The distance between outside edges is $107 + 90 + 89 = 286$ mm (300 mm in round numbers). Yet mean height for males is 1756 mm! Thus there is only a base of 200–300 mm for a structure of 1756 mm.

There is no significant average difference between the left and right feet. However, for specific individuals, that there often is considerable difference between the left and right foot.

The technical name for differences in leg length in the same person is *leg length discrepancy* (LLD). Contreras et al. (1993), summarizing studies with $N = 2377$, reported that 40% of people had LLD ≤ 5 mm, 30% had LLD ≤ 11 mm, and 10% had LLD ≤ 14 mm.

Weight of leg segments, as a percent of body weight, are 1.47 for foot, 4.35 for calf, and 10.27 for thigh; a total leg is 16.1 and both legs are 32.2. Thus the leg weight of a 70 kg person would be 0.322 (70) = 22.5 kg.

When people stand at a work surface, there needs to be an indentation (150 mm deep, 150 mm high, and 500 mm wide) so they can stand close to the work surface.

30.2 Activities of the Foot

30.2.1 Standing

During standing, the legs will generally move occasionally. Figure 30.3 shows how to support the standing body. The changed height of the foot rotates the thigh bone forward, enabling the hip on that side to be fixed; it also flattens the lumbar curve and relaxes the iliopsoas muscle. When a leg is on the rail/footrest, the large back muscles on that side of the body are relaxed and receive a flow of blood (bringing nutrients and removing waste products).

30.2.2 Walking

When walking, the activity of one leg has a shorter swing phase (when the foot is being passed forward) and the longer support (stepping, contact) phase (when the foot is on the ground). The support phase starts at heel strike and ends at toe-off; it has an early, passive section and a later active (propulsion) section.

FIGURE 30.3 Bar footrests along a work station or conveyor can reduce standing stress. (From Rodgers, S. 1984. *Working with Backache*. Perinton Press, Fairport, NY. With permission.) A small platform is another alternative. (From Konz, S. and Johnson, S. 2004. *Work Design: Occupational Ergonomics*, 6th ed. Holcomb-Hathaway, Scottsdale, AZ. With permission.)

At heel strike, the forward-moving heel hits the ground (causing deceleration). Continued forward motion of the body results in the forefoot contacting the ground; propulsion (acceleration) begins. The heel rises and the foot is pushed backward under the body. This tendency is resisted by friction under the sole; the body is propelled forward. The foot is everted, increasing forefoot contact area on the inner side, until only the skin around the big toe is in ground contact. Finally, contact ceases and the cycle repeats.

Because the swing phase is shorter than the support phase, heel strike on the opposite limb occurs during the propulsion part of the support phase. At heel strike, horizontal velocity decreases from about 450 to 20 cm/sec; heel angle to the floor changes from about 20° prior to heel contact to 0° at 100 m/sec after contact (Redfern and Rhoades, 1996). During a slip, instead of stopping, the heel continues to move and the leading foot moves out in front of the body.

30.2.3   Running

Walking changes to running, for normal size adults, at about 2.5 m/sec (6 miles/h), since it uses less energy (for the same speed). Running differs from walking in that both feet are off the ground for part of the stride. In addition, the heel strike should be renamed the "foot strike," since the initial contact probably will be forward of the heel. After foot strike (usually on the outside edge of the foot), the foot rolls inward and flattens out (pronation). Then the foot rolls through the ball and rotates outward (supination).

Peak force is about three times body weight at about 0.1 sec after contact. For running, the average contact duration is 0.29 sec. In contrast, it is 0.48 sec for walking (Scanton and McMaster, 1976).

30.2.4   Stepping

Descending stairs demands a gait quite different from ascent (Templer, 1992).

For descent, the leading foot swings forward over the nosing edge and stops its forward motion when it is directly over the tread below; the toe is pointed downward. Meanwhile, the heel of the rear foot begins to rise, starting a controlled fall downward toward the tread. The heel of the forward foot then is lowered and the weight is transferred to the forward foot. The rear foot then begins to swing forward.

We tend to hold our center of gravity as far back as possible by leaning backward. Problems are over-stepping the nosing with the forward foot, catching the toe of the forward foot, and snagging the heel of the rear foot on the nosing as it swings past. Falls tend to be down the stairs.

For ascent, the leading foot has a toe-off, swing, and first contact with the upper step. The foot is approximately horizontal. The ball of the foot is well forward of the tread; the heel may or may not be on the tread. The rear foot then rises on tiptoes, pushing down and back. The rear foot then begins the swing phase. The primary problem is catching the toe, foot or heel of either foot on the stair nosing. Another problem is the rear foot slipping when it pushes backward. Falls tend to be upward.

30.3   Accidents

30.3.1   Problem of Falls

Courtney et al. (2001) provide the following statistics to indicate the magnitude of the slip, trip and fall problem.

- Slips and falls are second largest source of unintentional injury deaths
- Slips and falls are the leading reason for unintentional emergency department visits (21% of visits or about 550,000/year)
- Slips and falls have an annual direct cost/capita of occupational injuries of $50−400 (depending on the industry)

In industrial fatalities, falls account for about 12%, which is greater than the total for electrical current, fires, burns, drowning, and poisons.

Falls that occur when a person is carrying something are especially dangerous. The object carried decreases stability as a function of the torque above the ankle (weight times object height above ankle). Other problems are that the arms cannot be used for balance (to prevent a fall), to grab a railing, or to break the fall.

Not all underfoot accidents result in falls, and not all falls result in a lost-time injury. Some falls result in no lost time, some result in strains and sprains, some in broken bones, and some in death. In addition, falls often are not recorded by accident-reporting procedures. Thus, the accident reports tend to drastically underestimate the number of falls.

30.3.2 Causes/Solutions of Falls

Falls can occur from slips (unexpected horizontal foot movement), trips (restriction of foot movement) and stepping-on-air (unexpected vertical foot movement). In addition, people can fall as a result of alcohol or drug use, fainting, and so on.

30.3.2.1 Slips

Slips occur primarily during foot push off and heel strike. During pushoff, the person falls forward (less common and less dangerous); in addition, during pushoff, most of the weight has already been transferred to the other foot.

During heel strike, the person falls backward.

Slips also can occur when the "ground" slopes (front to back or side to side). Examples are ramps and ladder rungs. Slips also can occur, with stationary feet, during pushing and pulling, but there does not tend to be a fall and injury.

During a slip, a lubricant (water, oil, grease, dust, ice, snow) usually is present, either on the surface or on the shoe heel. Table 30.2 has some coefficients of friction. In the special outdoors circumstances of snow and ice, slipping can be common. The most danger occurs when the ice is "wet" (i.e., close to the freezing point). Chang et al. (2001) give 0.67 for the static coefficient of friction for shoes on ice at $-40°C$, but 0.01 as the dynamic coefficient at $-1°C$ (when there is a thin film of water).

Table 30.3 summarizes how to reduce slips.

30.3.2.2 Trips

Trips occur during foot swing. As the foot swings forward, it hits an obstacle and the person falls forward; the problem is lack of leg movement. Usually there is a visual problem. Indoor trips tend to be from obstacles on the floor. Outdoor trips often occur from uneven surfaces (e.g., walkways, parking lots) that the person expects to be even.

30.3.2.3 Stepping-on-air

Unexpected vertical movement can occur on stair steps when the distance between steps is not equal; when there is a hole in the ground; or when there is no ground (e.g., "cliff," edge of scaffold, unexpected step, step on spiral stairs, unexpected curb or ramp). "Single steps" (small changes in elevation such as curbs or one-step changes in floor level) are dangerous. Steps descending from large trucks and off-road vehicles can present problems; for such vehicles use the "three-contact rule" (at least three limbs should be in contact with steps or handles at all times). On steps, the fall usually occurs when descending; the fall can be for a considerable distance.

TABLE 30.2 Coefficients of Friction of Floors and Shoes.

Coefficient of Friction	Floors	Floor Clean	Soiled	Shoe: Soles
1.0	Soft rubber pad	0.8	0.6	Rubber-cork
0.8	End grain wood	0.75	0.55	U.S. Army/ Air Force standard
0.7	Concrete, rough finish	0.7	0.5	Rubber-crepe
0.65	Working decorative, dry	0.6	0.4	Neoprene
0.5	Working decorative, soiled	0.5	0.3	Leather
0.4	Steel			
				Shoes: Heels
		0.7	0.5	Neoprene
		0.65	0.55	Nylon

Source: From Kroemer, K. 1974. *Applied Ergonomics*, 5(2):94–102. With permission.

TABLE 30.3   Minimization of Slips and Slip Effects

1. Eliminate lubricants
   Avoid presence/spilling of lubricant
   Clean up lubricants quickly. Note that the lubricant (water, mud) could be on the shoe. Lubricants also can be solid
      objects such as coins, paper clips, hairpins, screws, and metal chips
   Avoid adding lubricants during cleaning (e.g., do not use an oil mop on a waxed floor).
   Improve drainage of rain/snow on exterior walks and stairs

2. Choose good flooring
   Carpet is best (high-friction, low effect of lubricants)
   For hard surfaces, steel and ceramic are the worst as they are the smoothest. Grooved or porous floors reduce lubricant
      problems, but are hard to clean
   Use mats, rugs, and duckboards (elevated slatted flooring) for local areas where wet floors are common. Building
      entrances often are wet due to tracked-in water or snow. Machines using oil or coolant can be a problem. Mats should
      have beveled edges to reduce tripping, and holes to encourage drainage

3. Have good visibility (especially for change-in-level, such as stairs)
   Quantity (amount) of light is important. Too much light causes glare; too little causes poor visibility
   Quality (direction, depth perception) of light should avoid camouflage of changes in level. Change surface colors at the
      different levels

4. Choose good shoes. The heel is critical
   Bevel the rear of the heel (reduce the contact angle during heel strike to $0°$ from $10°$–$15°$)
   Penetrate (squeeze out) the lubricant by tread or high sole roughness
   Have soft sole material to increase contact area and, thus, grip. Slip resistance of shoes increases after about 5   km of
      walking, so tests on new shoes underestimate the slip resistance

5. Realize a slip may occur (be alert)
   Walk carefully (slow, short steps)
   Keep the body center of gravity within the stride
   Reduce heel angle at heel strike (i.e., shuffle)

6. If there is a slip, eliminate the fall (e.g., handrails, harnesses). This is akin to auto seat belts

7. If there is a fall, decrease the fall consequences. This is akin to auto airbags
   Reduce distance to fall (e.g., have stair landings)
   Lower impact force/pressure (use soft surfaces such as carpet; minimize sharp objects)

Source: From Hanna and Konz, 2004. With permission.

In some cases, the surface is there initially but breaks or moves (e.g., step breaks, a chair used as a step-stool moves, ladder feet move, a roof gives way under the foot).

30.4   Fatigue/Comfort

The discussion is divided into walking and standing.

30.4.1   Walking/Running

The primary problem is the shock of heel strike being transmitted up the foot, leg, and back. For shoe solutions, see Figure 30.4.

The energy cost of walking depends on the terrain, with a hard surface giving the minimum cost (Pandolf et al., 1976):

$$\text{WLKMET} = C\,(2.7 + 3.2(v - 0.7)^{1.65})$$

where:

$$\text{WLKMET} = \text{walking metabolism, W/kg of body weight}$$

If the heel is "high" (some women's shoes, cowboy boots), the center of gravity is moved forward, causing a variety of biomechanical stresses while standing or walking.

The general goal is to have the hips parallel to the floor (i.e., stand with weight equally on both feet). However, unequal leg lengths (see Section 30.1.3 of this chapter) or unequal shoe wear can be a problem.

Feet often swell so, for fit, buy shoes when your feet are swollen (late in the day). In addition, your left and right foot might vary slightly. Thus buy shoes with at least four pairs of eyelets as they increase the adjustment possibilities. The shoe sole should be resistant to slipping (see Table 30.2) through either or both material or shape.

People with low arches (footprint has a broad connection of the two areas) will be more comfortable with shoes with a straighter "inner line" (difficult to distinguish left from right shoe).

Boots support the ankle and calf as well as increasing insulation; they are especially useful for side support, such as when walking outdoors. Boots also protect against chemicals and animal products (such as fats and oils); some boot materials have a better life than others.

For impact protection, use a steel-toe shoe; in some industries (such as mining), metatarsal guards are also used.

FIGURE 30.4   Shoes.

$$
\begin{aligned}
C = {}& \text{terrain coefficient} \\
= {}& 1.0 \text{ for treadmill, blacktop road} \\
= {}& 1.1 \text{ for dirt road} \\
= {}& 1.2 \text{ for light brush} \\
= {}& 1.3 \text{ for hard packed snow; } C = 1.3 + 0.082 \text{ (foot depression, cm)} \\
= {}& 1.5 \text{ for heavy brush} \\
= {}& 1.8 \text{ for swamp} \\
= {}& 2.1 \text{ for sand} \\
v = {}& \text{velocity, m/sec [for } v > 0.7 \text{ m/sec (2.5 km/h)]}
\end{aligned}
$$

The metabolic cost of carrying (walking with a load) depends on the load location. Soule and Goldman (1980) reported that loads on the head used 1.2 times the energy of carrying a 1 kg of your own body weight; in the hands, loads required 1.4–1.9 times as much; on the feet, loads required 4.2–6.3 times as much.

The energy cost of running is:

$$
\text{RUNMET} = -142/WT + 11 + 0.04 V^2
$$

where RUNMET is the running metabolism (total), W/kg, V is the velocity (km/h), and WT is the weight (body), in kg.

30.4.2   Standing

Although there are some problems with static electricity and with floor temperature, the primary problem is lack of circulation in the leg.

Static electricity. Static electricity solutions include:

- Raise humidity above 40% (Not only permits voltage to flow to ground but also moisture is a lubricator, reducing friction between moving parts.)
- Make carpets conductive (Carbon fibers added to carpets.)
- Put an antistatic floor mat under the operator's chair. Also can ground operator with a static-bleed wrist strap
- Encourage cotton clothing; discourage nylons, slips, and polyester clothing. Use shoes with static-dissipating soles
- Remove dust by blowing deionized air from an air gun

Floor temperature. When wearing normal shoes, ASHRAE recommends 23°C for standing and walking but 25°C for sedentary standing. Heavy carpet will save about 1% of the total energy used to heat the building (Hager, 1977).

Floor comfort. Teitelman et al. (1990) reported preterm births occurred more often (7.7%) when women had jobs with prolonged standing; the rate for sitting jobs was 4.2% and for active jobs was 2.8%. There are four alternatives: (1) replace standing with sitting; (2) supplement static standing with occasional walking; (3) shift posture while standing; and (4) cushion the floor. See Figure 30.5.

Replace standing with sitting. In general, heart rates are lower when sitting than standing. However, in an experiment of checkout workstations, Lehman et al. (2001) reported workstations with a standing 0operator had lower EMG than sitting. Sitting does restrict torso movement and thus reach distances.

A possible alternative is a sit–stand stool, where the person's legs are almost vertical, supporting about 2/3 of body weight. The adjustable height seat (65–85 cm) should tilt forward 15°–30°. Chester et al. (2002) recommend a footrest resulting in a 90° footrest–calf angle.

Supplement static standing with occasional walking. As discussed in Section 30.1.2 at the start of this chapter, there are cardiovascular benefits of occasional walking.

Shift posture while standing. As discussed in Section 30.2.1 and Figure 30.3, there are benefits of shifting standing posture using standing aids such as bar rails and foot rests.

Cushion the floor. See Figure 30.5.

Reduce standing stress by (1) softer floors (i.e., mats), (2) better shoes, (3) foot rests, (4) walking, and (5) sit/stand chairs.

Softer floors. Concrete is worst, plastic or cork tile is slightly better, wood is better yet, and carpet is best of all. The entire floor can be cushioned (carpeted) or the floor can be cushioned locally with a mat. A mat should have the following features (Krumweide et al., 1998):

- It should compress 3–7% under adult weight
- It should have beveled edges to reduce tripping
- The mat should not move on the floor. Achieve this with a frictional underside or taping the mat to the floor
- The mat may need to be cleaned (sanitized) periodically. An alternative is a viscoelastic material poured into a floor insert (surface top level with the floor)

Better shoes. See Box 30.1

Foot rests. See Section 30.2.1. Satzler et al. (1993) (study described in Rys and Konz, 1994) had 16 subjects spend 2 h in each of 4 footrest alternatives while standing: (1) both feet flat on the floor (control), (2) one foot on a 100 mm high flat platform, (3) one foot on a 100 mm high platform, angled 15°, and (4) one foot on a 100 mm high, 50 mm diameter bar. All three standing aids were preferred over the control; the two platforms were preferred over the bar. On average, subjects switched their foot from the floor to an aid every 90 sec.

Walking. Benefits of occasional walking are described in Section 30.1.2. Walking could be from tending multiple machines and from restocking supplies at the workstation.

Sit–stand chairs. Sit–stand chairs should have a seat with a angle from the horizontal of about 12°; the seat should be padded with a fabric (i.e., slip resistant material). The feet should rest on a floor support. A 30° footrest (i.e., a 90° foot-calf angle) is probably better than a 15° footrest (Chester et al., 2002). Occupational use of a sit–stand chair should combine use of the chair with walking about the workstation and so reduce blood pooling in the legs. Sit–stand chairs should be considered as an occasional aid to standing, not a replacement for a chair.

FIGURE 30.5 Standing Aids. (From: Adapted from Konz, S. and Rys, M. 2003. *Occupational Ergonomics*, 3:165–172. With permission.)

30.5 Foot Controls

Although most controls are operated by the hands, there also are foot controls. The foot does not have the dexterity of the hand, but since it is connected to the leg instead of the arm, it can exert more force. A leg has approximately three times the strength of an arm. A foot control also reduces use of the hand/arm.

Foot controls will be divided into pedals and switches.

30.5.1 Pedals

Pedals can be used for power and control. Power generation can be continuous (bicycle) or discrete (non-powered automobile brake pedal). For information on continuous power, see Whitt and Wilson (1982) and Brooks et al. (1986).

Discrete power generally is applied by one leg; there does not seem to be any advantage to using the left or right leg. Force using both feet is about 10% higher than using just one foot.

A control example is an auto accelerator pedal. Bend the ankle (80°–115°) by depressing the toe rather than depressing the heel or moving the entire foot.

30.5.2 Switches

A foot switch can actuate a machine (such as a punch press). Generally the foot remains on the switch so the time and effort of moving the foot/leg is not important. On–off controls (such as faucets, clamping fixtures) can be actuated by lateral motion of the knee as well as vertical motion of the foot. The knee should not have to move more than 75–100 mm; force requirements should be light. Hospitals use knee switches to actuate faucets to improve germ control of the hands.

Avoid foot pedals/switches which are operated while standing as they tend to distort posture and cause back problems.

30.6 Defining Terms

Calcaneus:	The heel bone; see Figure 30.1
Edema:	Swelling of the legs due to fluid retention
Eversion:	External rotation of the foot
Inversion:	Inward rotation of the foot
Metatarsals:	Bones in the foot; see Figure 30.1
Mortise and tenon joint:	A type of joint; see Figure 30.1
Lateral:	The outside (side farthest from the centerline)
Leg length discrepancy:	Difference in leg length (in the same person)
Medial:	The inside (side closest to the centerline)
Phalanges:	Bones in the foot; see Figure 30.1
Pronation:	Rolling inward (toward the centerline) of the foot
Supination:	Rolling outward (away from the centerline) of the foot
Three-contact rule:	Rule used on ladders and steps. At least three limbs should be in contact with steps or handles at all times
Venous pooling:	Pooling of blood in the veins of the legs

References

Brooks, A., Abbott, A., and Wilson, D. 1986. Human-powered watercraft. *Scientific American*, 256(12):120–130.

Chang, W-R, Gronqvist, R., Leclercq, S., Myung, R., Makkonen, L., Strandberg, L. Brungraber, R., Mattke, U., and Thorpe, S. 2001. The role of friction in the measurement of slipperiness. *Ergonomics*, 44(13):1217–1232.

Chester, M., Rys, M., and Konz, S. 2002. Leg swelling, comfort, and fatigue when sit/standing using a 0°, 15° and 30° footrest. *International Journal of Industrial Ergonomics*, 29, 289–296.

Contreras, R., Rys, M., and Konz, S. 1993. Leg length discrepancy. In *The Ergonomics of Manual Work*, eds. W. Marras, W. Karwowski, and L. Pacholski, 199–202, Taylor & Francis, London.

Courtney, T., Sorock, G., Manning, D., Collins, J., and Holbein-Jenny, M. 2001. Occupational slip, trip, and fall-related injuries — can the contribution of slipperiness be isolated? *Ergonomics*, 44(13):1118–1137.

Hager, N. 1977. Energy conservation and floor covering materials. *ASHRAE Journal*, 34–39.

Hanna, S. and Konz, S., 2004, *Facility Design and Engineering*, 3rd edn. Holcomb Hathaway, Scottsdale, AZ.

Konz, S. and Johnson, S. 2004. *Work Design: Occupational Ergonomics, 6th ed.* Holcomb-Hathaway, Scottsdale, AZ.

Konz, S. and Rys, M. 2003. An ergonomic approach to standing aids. *Occupational Ergonomics*, 3:165–172.

Kroemer, K. 1974. Horizontal push and pull forces. *Applied Ergonomics*, 5(2):94–102.

Krumweide, D., Konz, S., and Hinnen, P. 1998. Floor mat comfort. In *Advances in Occupational Ergonomics and Safety*, ed. S. Kumar, 159–162, IOS Press.

Lehman, K., Psihogios, J., and Meulenbroek, R., 2001, Effects of sitting versus standing and scanner type on cashiers. *Ergonomics*, 44(7): 719–38.

Nodeland, H., Ingemansen, R., Reed, R., and Aukland, K., 1983. A telemetric technique for studies of venous pressure in the human leg during different positions and activities. *Clinical Physiology*, 3:573–576.

Oleson, B. and Fanger, P. 1973. The skin temperature distribution for resting man in comfort. *Archives des Sciences Physiologigues*, 27(4):A385–A393.

Pandolf, K., Haisman, M., and Goldman, R. 1976. Metabolic energy expenditure and terrain coefficients for walking on snow. *Ergonomics*, 19:683–690.

Pollack, A. and Wood, E. 1949. Venous pressure in the saphenous vein at the ankle in man during exercise and change of posture. *Journal of Applied Physiology*, 1:649–662.

Redfern, M. and Rhoades, T., 1996, Fall prevention in industry using slip resistance testing. In *Occupational Ergonomics*, Bhattacharya, A. and McGloughlin, J. (eds.), 463–476, M. Decker, New York.

Rodgers, S. 1984. *Working with Backache*. Perinton Press, Fairport, NY.

Rys, M. and Konz, S. 1994. Standing. *Ergonomics*, (37)4:677–687.

Scanton, P. and McMaster, J. 1976. Momentary distribution of forces under the foot. *Journal of Biomechanics*, 9:45–48.

Satzler, L., Satzler, C., and Konz, S. 1993. Standing aids. *Proceedings of Ayoub Symposium*, 29–31, Texas Tech University, Lubbock, TX.

Soule, R. and Goldman, R. 1980. Energy cost of loads carried on the head, hands, or feet. *Journal of Applied Physiology*, 27:687–690.

Teitelman, A., Welch, L., Hellenbrand, K., and Bracken, M. 1990. Effect of maternal work activity on preterm birth and low birth weight. *American Journal of Epidemiology*, 131:104–113.

Templer, J. 1992. *The Staircase: Studies of Hazards, Falls, and Safer Design*. MIT Press, Cambridge, MA.

Whitt, F. and Wilson, D. 1982. *Bicycling Science*. MIT Press, Cambridge, MA.

Winkel, J. and Jorgensen, K., 1986, Evaluation of foot swelling and lower-limb temperature in relation to leg activity during long-term seated office work. *Ergonomics*, 29(2):313–328.

Further Reading

Konz, S. and Johnson, S. 2004. *Work Design: Occupational Ergonomics*, 6th ed. Holcomb Hathaway, Scottsdale, AZ. This popular textbook concisely summarizes many aspects of job design and gives detailed design guidelines.

Ergonomics. This journal publishes articles on ergonomics from authors around the world.

International Journal of Industrial Ergonomics. This journal publishes articles on ergonomics from authors around the world.

31

Noise in Industry

John G. Casali
Gary S. Robinson
Virginia Polytechnic Institute and State University

31.1 Introduction

The din of noise emanating from industrial processes pervades many occupational settings, and its effects on workers range from minor annoyance to major risk of hearing damage. Unfortunately, at least within the current limits of technology, noise is a by-product of many industries, such as manufacturing, especially those which use high-energy or impact processes such as metal cutting and mineral refinement, and service-related industries, such as air transport, construction, and farming. Workers complain about the negative effects of noise on their abilities to communicate, hear warning and other signals, and concentrate on tasks at hand. However, the effect that has been of most concern to industry has been permanent *noise-induced hearing loss*, or NIHL.

The primary intent of this chapter is to provide an introductory overview of the basic properties, measurement, effects on hearing, government regulations, and abatement of industrial noise, with a particular focus on reducing the physiological damage potential of noise as it impacts the human hearing organ. While the effects of noise exposure are serious and must be reckoned with by the hearing conservationist or safety professional, one fact is encouraging: process/machine-produced noise is a physical stimulus that can be avoided, reduced, or eliminated; therefore, occupationally related NIHL in workers is completely preventable with effective abatement and protection strategies. Total elimination of NIHL should thus be the only acceptable goal.

31.2 Sound and Noise

Because almost all aspects of hearing conservation and noise abatement in industry rely upon accurate quantification and evaluation of the noise itself, a basic understanding of sound parameters and sound measurement is needed before delving into other noise issues.

31.2.1 Basic Parameters

Sound is a disturbance in a medium (in industry, most commonly air or a conductive structure such as a plant floor) that has mass and elasticity. For example, an industrial metal-forming process wherein a hydraulic ram impacts a plate of sheet metal with great force causes the plate to oscillate or vibrate. Because the plate is coupled to the air medium, it produces a pressure wave that consists of alternating compressions (above ambient air pressure) and rarefactions (below ambient pressure) of air molecules, the *frequency* (f) of which is the number of above/below ambient pressure cycles per second, or *hertz* (Hz). The reciprocal of frequency, $1/f$, is the *period* of the waveform. The waveform propagates outward from the plate as long as it continues to vibrate, and the disturbance in air pressure that occurs in relation to ambient air pressure is heard as sound. The linear distance traversed by the sound wave in one complete cycle of vibration is the *wavelength*. As shown in Equation (31.1), wavelength (λ in m or ft) depends on the sound frequency (f in Hz) and velocity (c in m/sec or ft/sec; in air at 68°F and pressure of 1 atm, 344 m/sec or 1127 ft/sec) in the medium. The speed of sound increases about 1.1 ft/sec for each increase of 1°F.

$$\lambda = c/f \tag{31.1}$$

Noise can be loosely defined as a subset of sound; that is, noise is sound that is undesirable or offensive in some aspect. However, the distinction is largely situation-and listener-specific, as perhaps best stated in the old adage "one person's music is another's noise."

Unlike some common ergonomics-related stressors such as repetitive motions or awkward lifting maneuvers, noise is a physical stimulus that is readily measurable and quantifiable using transducers (microphones) and instrumentation (sound level meters) that are commonly available. Aural exposure to noise, and the damage potential therefrom, is a function of the *total energy* transmitted to the ear. In other words, the energy is equivalent to the product of the noise intensity and duration of the exposure. Several metrics that relate to the energy of the noise exposure have been developed, most with an eye toward expressing the exposures that occur in industrial or community settings. These metrics are covered later in this chapter. But first, the most basic unit of measurement must be understood, namely the *decibel*.

31.2.2 Physical Quantification: Sound Levels and the Decibel Scale

The unit of *decibel*, or 1/10 of a *bel*, is the most common metric applied to the quantification of noise amplitude. The decibel, hereafter abbreviated as dB, is a measure of *level*, defined as the logarithm of the ratio of a quantity to a reference quantity of the same type. In acoustics, it is applied to sound level, of which there are three types.

Sound power level is the most basic quantity, is typically expressed in dB, and is defined as

$$\text{Sound power level in dB} = 10 \log_{10} P_{w_1}/P_{w_r} \tag{31.2}$$

where P_{w_1} is the acoustic power of the sound in Watts, or other power unit, and P_{w_r} is the acoustic power of a reference sound in Watts, usually taken to be the acoustic power at hearing threshold for a young, healthy ear at the frequency of maximum sensitivity, or the quantity 10^{-12} W.

Sound intensity level, following from power level, is typically expressed in dB, and is defined as

$$\text{Sound intensity level in dB} = 10 \log_{10} I_1/I_r \qquad (31.3)$$

where I_1 is the acoustic intensity of the sound in W/m^2, or other intensity unit, and I_r is the acoustic intensity of a reference sound in W/m^2, usually taken to be the acoustic intensity at hearing threshold, or the quantity $10^{-12} W/m^2$.

Within the last decade, sound measurement instruments to measure sound intensity level have become commonplace, albeit expensive and relatively complex. Sound power level, on the other hand, is not directly measurable but can be computed from empirical measures of sound intensity level or sound pressure level.

Sound pressure level (SPL) is also typically expressed in dB. Since power is directly proportional to the square of the pressure, SPL is defined as

$$\text{SPL in dB} = 10 \log_{10} P_1^2/P_r^2 = 20 \log_{10} P_1/P_r \qquad (31.4)$$

where P_1 is the pressure level of the sound in μPa, or other pressure unit, and P_r is the pressure level of a reference sound in mPa, usually taken to be the pressure at hearing threshold, or the quantity 20 μPa, or 0.00002 Pa. Other equivalent reference quantities are: $0.0002 \, dyn/cm^2$, $20 \, \mu N/m^2$, and 20 μbar.

The application of the decibel scale to acoustical measurements yields a convenient means of collapsing the vast range of sound pressures that would be required to accommodate sounds that can be encountered into a more manageable, compact range. As shown in Figure 31.1, using the logarithmic compression produced by the decibel scale, the range of typical sounds is 120 dB, while the same

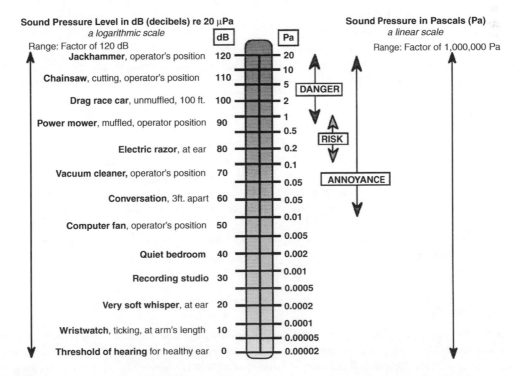

FIGURE 31.1 For typical sounds, sound pressure level values in decibels and sound pressure values in pascals.

range measured in pressure units (Pa) would be 1,000,000. Of course, sounds do occur that are higher than 120 dB (for instance, artillery fire) or lower than 0 dB (below the normal threshold on an audiometer). A comparison of decibel values of example sounds to their pressure values (in Pa) is also depicted in Figure 31.1.

In considering changes in sound level measured in decibels, a few numerical relationships emanating from the decibel formulae given earlier are often helpful in practice. An increase (decrease) in SPL by 6 dB is equivalent to a doubling (halving) of the sound pressure. Similarly, on power or intensity scales, an increase (decrease) of 3 dB is equivalent to a doubling (halving) of the sound power or intensity. This latter relationship gives rise to what is known as the "equal energy rule or trading relationship." Because sound represents energy that is itself a product of intensity and duration, an original sound that increases (decreases) by 3 dB is equivalent in total energy to the same original sound that does not change in decibels but decreases (increases) in its duration by half (twice).

31.2.3 Psychophysical Quantification: Loudness Scales

While the decibel is useful for quantifying the amplitude of a sound on a physical scale, it does not yield an absolute or relative basis for quantifying the human perception of sound amplitude, commonly called loudness. However, there are several psychophysical scales that are useful for measuring loudness, the two most prominent being *phons* and *sones*.

31.2.3.1 Phons

The decibel level of a 1000-Hz tone that is judged by human listeners to be equally loud to a sound in question is the phon level of the sound. The phon levels of sounds of different intensities are shown in the top panel of Figure 31.2; this family of curves is referred to as the *equal loudness contours*. On any given curve, the combinations of sound level and frequency along the curve produce sound experiences of equal loudness to the normal-hearing listener. Note that at 1000 Hz on each curve the phon level is equal to the decibel level. The threshold of hearing for a young, healthy ear is represented by the 0 phon level curve. The young, healthy ear is sensitive to sounds between about 20 and 20,000 Hz, although, as shown by the curve, it is not equally sensitive to all frequencies. At low- and mid-level sound intensities, low-frequency sounds and to a lesser extent high-frequency sounds are perceived as less intense than sounds in the 1000 to 4000 Hz range, where the undamaged ear is most sensitive. But as phon levels move to higher values, the ear becomes more linear in its loudness perception for sounds of different frequencies.

Because the ear exhibits this nonlinear behavior, several frequency weighting functions have been standardized for use with sound level meters. The most common curves are the A, B, and C curves, with the corresponding decibel measurement denoted as dBA, dBB, and dBC, respectively. If no weighting function is selected on the meter, the notation dB or dB(linear) is used, and all frequencies are processed without weighting factors. The actual weighting functions for the three suffix notations A, B, and C are superimposed on the phon contours of the top panel of Figure 31.2, and also depicted as actual frequency weighting functions in the bottom panel. In nearly all U.S. measurements of industrial noise made for assessment of exposure risk to workers, the dBA scale is used, and the meter is set on the "slow" dynamic response setting, which produces slow exponential averaging of a 1-sec window. For determination of the adequacy of hearing protection for a particular noise, and for application of noise control measures, the C-weighted level is often taken in addition to the A-weighted level.

31.2.3.2 Sones

While the phon scale provides the ability to equate the loudness of sounds of different frequencies, it does not afford an ability to describe how much louder one sound is compared with another. For this, the *sone* scale is needed.[1] One sone is defined as the loudness of a 1000 Hz tone of 40 dB SPL. In relation to one sone, two sones are twice as loud, three sones are three times as loud, one-half sone is half as

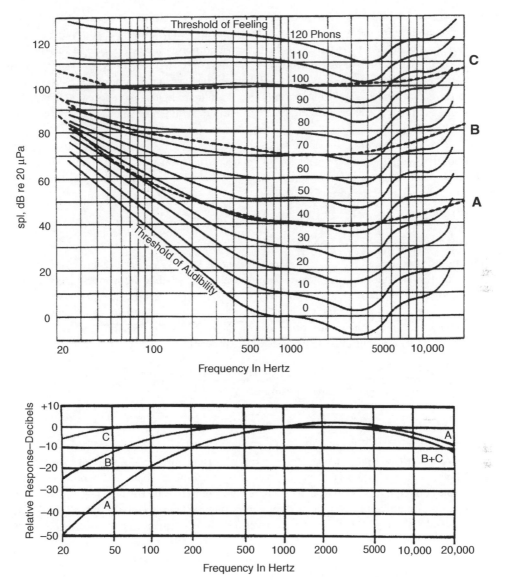

FIGURE 31.2 Top: Equal loudness contours based on the psychophysical phon scale, with sound level meter frequency weighting curves superimposed. Bottom: Decibel versus frequency values of A, B, and C sound level meter weighting curves. (Adapted from Earshen, J. J. (1986). Sound measurement: instrumentation and noise descriptors. In Berger, E. H., Ward, W. D., Morrill, J. C., and Royster, L. H. (Eds.), *Noise and Hearing Conservation Manual* (pp. 38–95). Akron, OH: American Industrial Hygiene Association. With permission.)

loud, and so on. Phon level (L_P) and sones are related by the following formula for sounds at or above a 40-phon level:

$$\text{Loudness in sones} = 2^{(L_P-40)/10} \tag{31.5}$$

According to formula 31.5, 1 sone equals 40 phons and the number of sones doubles with each 10-phon increase; therefore, it is straightforward to conduct a comparative estimate of loudness levels of sounds with different decibel levels. The "rule-of-thumb" is that each 10-dB increase in a sound (i.e., one that is above 40 dB to begin with) will result in a doubling of its loudness. For instance, a conference room that is at 45 dBA may currently be comfortable for communication. However, if a new ventilation system increases the noise level in the room by 10 dBA, the occupants will experience a doubling of loudness and will likely complain about the effects of the background noise on conversation in the room. Once again, the compression effect of the decibel scale yields a measure that does not reflect the much larger influence that an increase in sound level will have on the human perception of loudness. Although the sone scale is not widely used (one exception is that household ventilation fans typically have voluntary sone ratings), it is a very useful scale for comparing different sounds as to their perceived loudness.

It should be evident that phon levels can be calculated directly from psychological measurements in sones, but not from physical measurements of SPL in decibels. This is because the phon-based loudness and SPL relationship changes as a function of the sound frequency and the magnitude of this change depends on the intensity of the sound.

31.2.3.3 Modifications of the Sone

A modification of the sone scale *(Mark VI and subsequently, Mark VII sones)* was proposed by Stevens[2] to account for the fact that most real sounds are more complex than pure tones. Utilizing general formula (31.6), Steven's method utilizes octave band, 1/2 octave band, or 1/3 octave band noise measurements, and adds to the sone value of the most intense frequency band a fractional portion of the sum of the sone values of the other bands ($\sum S$). S_m is the maximum sone value in any band, and k is a fractional multiplier that varies with bandwidth (octave, $k = 0.3$; 1/2 octave, $k = 0.2$; 1/3 octave, $k = 0.15$):

$$\text{Loudness in sones} = S_m + k(\Sigma S - S_m) \tag{31.6}$$

31.2.3.4 Zwicker's Method

The concept of the *critical band for loudness* formed the basis for Zwicker's method of loudness quantification.[3] The *critical band* is the frequency band within which the loudness of a band of continuously distributed sound of equal SPL is independent of the width of the band. The critical bands widen as frequency increases. A graphical method is used for computing the loudness of a complex sound based on critical band results obtained and graphed by Zwicker. The noise spectrum is plotted and lines are drawn to depict the spread of masking effect (defined later in this chapter). The result is a bounded area on the graph that is proportional to total loudness. The method is relatively complex and the reader may wish to consult Zwicker[3] for computational details.

31.2.3.5 Noisiness Units

Loudness and noisiness are related but not synonymous. Noisiness can be defined as the "subjective unwantedness" of a sound. Perceived noisiness may be influenced by a sound's loudness, tonality, duration, impulsiveness, and variability.[4] Whereas a low level of loudness might be perceived as enjoyable or pleasing, a low level of unwantedness, that is, noisiness, is by definition undesirable. Equal noisiness contours, analogous to equal loudness contours, have been developed based on a unit (analogous to the phon) called the *perceived noise level* (PN_{dB}), which is the SPL in decibels of a 1/3 octave band of random noise centered at 1000 Hz, which sounds equally noisy to the sound in question. Also, an *N-* (*later D-*) sound level meter weighting curve was developed for measuring the perceived noise level of a sound. A subjective noisiness unit analogous to the sone, the *noy*, is used for comparing sounds as to their relative noisiness. One noy is equal to 40 PN_{dB}, two noys are twice as noisy as one, five noys are five times as noisy, and so on. Similar to the behavior of sone values for loudness, an increase of about 10 PN_{dB} is equivalent to a doubling of the perceived noisiness of a sound.

31.3 Effects of Noise in Occupational Settings

31.3.1 Nonauditory Effects

Noise exposure in industry has been linked to several deleterious effects, some of which are nonauditory and thus beyond the scope of this chapter. However, it is at least important to recognize that noise can degrade operator task performance. Research studies concerning the effects of noise on performance are primarily laboratory based and task/noise specific; therefore, extrapolation of the results to actual industrial settings is somewhat risky.[5] Nonetheless, on the negative side, noise is known to mask task-related acoustic cues, as well as cause distraction and disruption of "inner speech," while on the positive side, noise may at least initially heighten operator arousal and thereby improve performance on tasks that do not require substantial cognitive processing.[6] To obtain reliable effects of noise on performance, except on tasks that rely heavily on short-term memory, the level of noise must be fairly high, usually 95 dBA or greater. Tasks that are simple and repetitive often show no deleterious performance effects (and sometimes improvements) in the presence of noise, while difficult tasks that rely on perception and information processing on the part of the operator will often exhibit performance degradation.[5] It is generally accepted that unexpected or aperiodic noise causes greater degradation than predictable, periodic, or continuous noise, and that the startle response created by sudden noise can be disruptive.

Furthermore, noise has been linked to physiological problems other than those of the hearing organ, including hypertension, heart irregularities, extreme fatigue, and digestive disorders. Most physiological responses of this nature are symptomatic of stress-related disorders. Because the presence of high noise levels often induces other stressful feelings (such as sleep disturbance and interference with conversing in the home, and fear of missing oncoming vehicles or warning signals on the job), there are second-order effects of noise on physiological functioning that are difficult to predict.

31.3.2 Signal Detection and Communications Effects

31.3.2.1 Interference and the Signal-to-Noise Ratio

One of the most noticeable effects of noise is its interference with speech communications and the hearing of nonverbal signals. Workers often complain that they must shout to be heard and that they cannot hear others trying to communicate with them. Likewise, noise interferes with the detection of workplace signals, such as alarms for general area evacuation and warnings, annunciators, on-equipment alarms, and machine-related sounds, which are relied upon for feedback. The ratio (actually the algebraic difference) of the speech or signal level to the noise level, termed the *signal-to-noise ratio* (S/N) is the most critical parameter in determining whether speech or signals will be heard in noise. An S/N of 5 dB means that the signal is 5 dB greater than the noise, while a S/N of −5 dB means that the signal is 5 dB lower than the noise. Hearing protection is often blamed for exacerbating the effects of noise on the audibility of speech and signals, although, at least for individuals with normal hearing, protectors may actually facilitate hearing in some noisy situations, particularly those above about 90 dBA.

31.3.2.2 Masking

Masking is technically defined as the tendency for the threshold of a desired signal or speech (*the masked sound*) to be raised in the presence of an interfering sound (*the masker*). As an example, in a noisy airport waiting area, a pay telephone's earphone volume must often be increased to enable the listener to hear the party on the line, whereas a lower volume will be more comfortable while affording audibility when there is no crowd or public address system noise present. The *masked threshold* is defined as the SPL required for 75% correct detection of a signal when that signal is presented in a two-interval task wherein, on a random basis, one of the two intervals of each task trial contains the signal and the noise and the other contains only noise. In a controlled laboratory test scenario, a signal that is about 6 dB above the masked threshold will result in near perfect detection performance.[7] Analytical prediction (as opposed to actual experimentation with human subjects) of the interfering effects of noise on speech communications may be conducted using the Speech Intelligibility Index (SII) technique defined in ANSI S3.5-1997 (R2002).[8] Essentially, this relatively complex technique utilizes a weighted sum of the speech-to-noise ratios in specified frequency bands to compute an SII score ranging between 0.0 and 1.0, with higher scores indicative of greater predicted speech intelligibility. (The SII actually represents the proportion of the speech cues that would be available to the listener for "average speech" under the noise/speech conditions for which the calculations were performed. Hence, intelligibility would be greatest when the SII = 1.0, indicating that all of the speech cues are reaching the listener, and poorest when the SII = 0.0, indicating that none of the speech cues are reaching the listener.) This method is extremely flexible and can account for factors such as differences in speaker vocal effort, room reverberation, monaural and binaural listening, hearing loss, varying message content, and insertion loss (HPD use) or gain (amplified communications systems), as well as the existence of external masking noise. Nonverbal signal detectability predictions can also be made analytically, with the most comprehensive computational technique, based on a spectral analysis of the noise, appearing in ISO 7731–2003.[9] While a full discussion of these analytical procedures is beyond the scope of this chapter, the reader is referred to the individual ANSI and ISO standards for detail. The SII and masked threshold computational techniques provide better resolution and accuracy for speech intelligibility and signal detectability predictions than a simple evaluation of broadband S/N ratios because the techniques incorporate the frequency-specific information that simple S/N ratios do not reflect. However, the following general principles regarding masking effects on nonverbal signals and speech can be used for general guidance:

1. The greatest increase in masked threshold occurs for nonverbal signal frequencies that are equal or near to the predominant frequencies of the masking noise; this is called *direct masking*. Therefore, warning signals should not utilize tonal frequencies equivalent to those of the masker. Preferably, the signal should be in the most sensitive range of human hearing, approximately 1000 to 4000 Hz, unless the noise energy is intense at these frequencies.
2. If the signal and masker are tonal in nature, the primary masking effect is at the fundamental frequency of the masker and at its harmonics. For instance, if a masking noise has primary frequency content at 1000 Hz, this frequency and its harmonics (2000, 3000, 4000, etc.) should be avoided as signal frequencies.
3. The greater the SPL of the masker, the more the increase in masked threshold of the signal. A general rule-of-thumb is that the S/N ratio at the listener's ear should at a minimum be about 15 dB above the masked threshold for reliable signal detection. However, in noise levels above about 80 dBA, the signal levels required to maintain an S/N ratio of 15 dB above the masked threshold may increase the hearing exposure risk, especially if signal presentation occurs frequently. Therefore, if lower S/Ns become necessary, it is best to construct signals that are unlike the masker in frequency and have modulated or alternating frequencies to grab attention.
4. Warning signals should not exceed the masked threshold by more than 30 dB to avoid verbal communications interference and operator annoyance.[7]
5. As the SPL of the masker increases, the primary change in the masking effect is that it spreads upward in frequency, often causing signal frequencies that are higher than the masker to be missed. This is termed *upward masking*. Since most warning signal guidelines recommend that the midrange and high-frequency signals (about 1000 to 4000 Hz) be used for detectability, it is important to consider that the masking effects of industrial noise of lower frequencies can spread upward and cause interference in this range. Therefore, if the noise has its most significant energy in this range, a lower frequency signal, say 500 Hz, may be necessary. However, it must be kept in mind that the ear is not as sensitive to low frequencies, so the signal level must be carefully set to ensure reliable audibility.
6. Masking effects can also spread downward in frequency, causing signal frequencies below those of the masker to be raised in threshold. This is called *remote masking* and the effect is most prominent at signal frequencies that are subharmonics of the masker. With typical industrial noise sources, remote masking is generally less of a problem than direct or upward masking.

7. In extremely loud environments of about 110 dB and above, nonauditory signal channels such as visual and vibrotactile should be considered as alternatives to auditory displays.

8. Speech intelligibility in noise depends on a combination of complex factors and, as such, predictions based on simple S/N ratios should not be relied upon. However, in very general terms, S/N ratios of 15 dB or higher should result in intelligibility performance above about 80% words correct for normal-hearing individuals in broadband noise.[10] Above speech levels of about 85 dBA, there is some decline in intelligibility even if the S/N ratio is held constant.[11] In very high noise levels, it is impractical and may pose additional hearing hazard risk to amplify the voice to maintain the high S/N ratios necessary for good intelligibility performance. The S/N ratio required for reliable intelligibility may be reduced via the use of certain techniques such as reduction of speaker-to-listener distances, use of smaller vocabularies, provision of contextual cues in the message, use of the phonetic alphabet, and use of noise-attenuating headphones and noise-canceling microphones in electronic systems.

9. Electronic speech communications systems should reproduce speech frequencies in the range of 500 to 5000 Hz, which encompasses the most sensitive range of hearing and includes the speech sounds important for message understandability. More specifically, because much of the information required for word discrimination lies in the consonants, which are in the higher end of the frequency range and of low power (while the power of the vowels is in the peaks of the speech waveform), the use of electronic peak-clipping and reamplification of the waveform may improve intelligibility because the power of the consonants is thereby boosted relative to the vowels. Furthermore, it is critical that frequencies in the region of 1000 to 4000 Hz be faithfully reproduced in electronic communication systems to maintain intelligibility. Filtering out of frequencies outside this range will not appreciably affect word intelligibility, but will influence the quality of the speech.

10. Actual human speech results in higher intelligibility in noise than computer-generated speech; therefore, especially for critical message displays and annunciators, live, recorded, or digitized human speech is preferable over synthesized speech.[12]

31.3.3 Noise-Induced Hearing Loss

31.3.3.1 Scope of Hearing Loss in the United States

NIHL is one of the most widespread occupational maladies in the United States, if not the world. In the early 1980s, it was estimated that over nine million workers are exposed to noise levels averaging over 85 dBA for an 8 h workday.[13] Today, this number is likely to be higher because the control of noise sources, both in type and number, has not kept pace with the proliferation of industrial and service sector development. Due in part to the fact that before 1971 there were no U.S. federal regulations governing noise exposure in general industry, many workers over 50 yr of age now exhibit hearing loss that results from the effects of occupational noise. Of course, the total noise exposure from both occupational and nonoccupational sources determines the NIHL that a victim experiences. Of the estimated 28 million Americans who exhibit significant hearing loss due to a variety of etiologies, such as pathology of the ear, ototoxic drugs, and hereditary tendencies, over 10 million have losses that are directly attributable to noise exposure.[14] Therefore, the noise-related losses are preventable in nearly all cases. The majority of losses are due to on-the-job exposures, but leisure noise sources do contribute a significant amount of energy to the total noise exposure of some individuals.

31.3.3.2 Types and Etiologies of Noise-Induced Hearing Loss

Although the major concern of the industrial hearing conservationist is to prevent employee hearing loss that stems from occupational noise exposure, it is important to recognize that hearing loss may also emanate from a number of sources other than noise, including: infections and diseases specific to the ear, most frequently originating in the middle or conductive portion; other bodily diseases, such as multiple sclerosis, which injures the neural part of the ear; ototoxic drugs, of which the mycin family is a prominent member; exposure to certain chemicals and industrial solvents; hereditary factors; head trauma; sudden hyperbaric- or altitude-induced pressure changes; and aging of the ear (presbycusis). Furthermore, not all noise exposure occurs on the job. Many workers are exposed to hazardous levels during leisure activities, from such sources as automobile/motorcycle racing, personal stereo headsets and car stereos, firearms, and power tools. The effects of noise on hearing are generally subdivided into the following three categories.[15]

31.3.3.2.1　Acoustic Trauma

Immediate organic damage to the ear from an extremely intense acoustic event such as an explosion is known as *acoustic trauma*. The victim will notice the loss immediately and it often constitutes a permanent injury. The damage may be to the conductive chain of the ear, including rupture of the eardrum or dislodging of the ossicles (small bones) of the middle ear. Conductive losses can, in many cases, be compensated for with a hearing aid or surgically corrected. Neural damage may also occur, involving a dislodging of the hair cells or breakdown of the neural organ (Organ of Corti) itself. Unfortunately, neural loss is irrecoverable and not typically compensable with a hearing aid. Acoustic trauma represents a severe injury, but fortunately its occurrence is uncommon, including in the industrial setting.

31.3.3.2.2　Noise-Induced Threshold Shift

A *threshold shift* is defined as an elevation of hearing level from the individual's baseline hearing level and it constitutes a loss of hearing sensitivity. *Noise-induced temporary threshold shift* (NITTS), sometimes referred to as "auditory fatigue," is by definition recoverable with time away from the noise. The elevation of threshold is temporary, and usually can be traced to an overstimulation of the neural hair cells (actually, the stereocilia) in the Organ of Corti. Although the individual may not notice the temporary loss of sensitivity, NITTS is a cardinal sign of overexposure to noise. It may occur over the course of a full workday in noise or even after a few minutes of exposure to very intense noise. Although the relationships are somewhat complex and individual differences are rather large, NITTS does depend on the level, duration, and spectrum of the noise, as well as the audiometric test frequency in question.[15]

Prevention of *noise-induced permanent threshold shift* (NIPTS), for which there is no possibility of recovery, is the primary target of the industrial hearing conservationist. NIPTS can manifest suddenly as a result of acoustic trauma; however, industrial noise problems that cause NIPTS most typically constitute exposures that are repeated over a long period of time and have a cumulative effect on hearing sensitivity. In fact, the losses are often quite insidious in that they occur in small steps over a number of years of overexposure and the worker is not aware of the problem until it is too late. This type of exposure produces permanent neural damage, and although there are some individual differences as to the magnitude of loss and audiometric frequencies affected, the typical pattern for NIPTS is a prominent elevation of threshold at the 4000 Hz audiometric frequency (sometimes called the 4-kHz notch), followed by a spreading of loss to adjacent frequencies of 3000 and 6000 Hz. From a classic study on workers in the jute weaver industry, Figure 31.3 depicts the temporal profile of NIPTS as the family of audiometric threshold shift curves, with each curve representing a different number of years of exposure.[16] As noise exposure continues over time, the hearing loss will spread over a wider frequency bandwidth inclusive of midrange and high frequencies, and encompassing the range of most auditory warning signals. In some cases, the hearing loss renders it unsafe or unproductive for the victim to work in certain occupational settings where the hearing of certain signals are requisite to the job. Unfortunately, the power of the consonants of speech sounds, which heavily influence the intelligibility of human speech, also lie in the frequency range that is typically affected by NIPTS, compromising the victim's ability to understanding speech. This is the tragedy of NIPTS in that the worker's ability to communicate is hampered, often severely and always irrecoverably. Furthermore, unlike blindness or many physical disabilities, hearing loss is not overt and therefore often goes unrecognized by others. Thus, it is a particularly isolating disability because the victim is unintentionally excluded from conversations and may miss important auditory signals because others either are unaware of the loss or simply forget about the need to compensate for it.

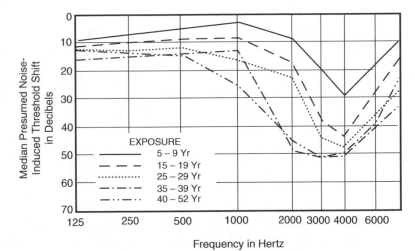

FIGURE 31.3 Cumulative auditory effects of years of noise exposure in a jute weaving industry. (Adapted from Taylor, W., Pearson, J., Mair, A., and Burns, W., (1964). *Journal of the Acoustical Society of America, 38*, 113–120. With permission.)

31.3.3.2.3 *Concomitant Auditory Maladies*

Following exposure to high-intensity noise, some individuals will notice that ordinary sounds are perceived as "muffled," and in some cases, they may experience a ringing or whistling sound in the ears, known as *tinnitus*. These manifestations should be taken as serious indications that overexposure has occurred, and that protective action should be taken if similar exposures are encountered in the future. Tinnitus may also occur by itself or in conjunction with NIPTS, but in any case it is thought to be the result of *otoacoustic emissions*, which are essentially acoustic outputs from the inner ear that are audible to the victim, apparently resulting from mechanical activity or *microphonics* of the neural cells. Some individuals report that tinnitus is always present, pervading their lives. It thus has the potential to be quite disruptive and, in severe cases, debilitating.

More rare than tinnitus, but typically quite debilitating is the malady known as *hyperacusis*, which refers to hearing that is extremely sensitive to sound. Hyperacusis can manifest in many ways, but a number of victims report that their hearing became painfully sensitive to sounds of even normal levels after exposure to a particular noise event. Therefore, at least for some, hyperacusis can be directly traced to noise exposure. Sufferers typically must use hearing protectors when performing normal activities, such as walking on city streets, visiting movie theaters, or washing dishes in a sink, because such activities produce sounds that are painfully loud to them. It should be noted that hyperacusis sufferers often exhibit normal audiograms; that is, their thresholds are not typically better than those of "normal hearers," even though their reaction to sound is one of hypersensitivity.

It is important that the industrial hearing conservationist be aware of these hearing-related maladies that may or may not arise as a result of on-the-job noise exposure, but which may influence the worker's ability to perform certain jobs or work in certain environments.

31.4 Measurement and Quantification of Noise Exposures

31.4.1 Basic Instrumentation

Measurement and quantification of sound exposure levels provide the fundamental data for assessing hearing exposure risk, speech and signal masking effects, hearing conservation program needs, and engineering noise control strategies. A vast array of instrumentation is available for sound measurement; however, for monitoring and assessment of most noise exposure situations, a basic understanding of three primary instruments (sound level meters, dosimeters, and real-time spectrum analyzers) and their data output will suffice. In instances where noise is highly impulsive in nature or selection and development of situation-specific engineering noise control solutions is anticipated, more specialized instruments may be necessary.

Because sound is propagated as pressure waves that vary over space and in time, a complete quantification would require simultaneous measurements over the continuous time periods (representing complete operator exposure durations) at all points of an occupational sound field to exhaustively document the noise level in the space. Clearly, this is typically cost- and time-prohibitive, so one must resort to sampling strategies for establishing the observation points and intervals. The hearing conservationist must also decide whether detailed, discrete time histories are needed (such as with a noise-logging *dosimeter*, discussed later), if averaging over time and space with long data records is required (with an averaging/integrating dosimeter), whether discrete samples taken with a short-duration moving time average (with a basic sound level meter) will suffice, or if frequency-band-specific SPLs are needed for selecting noise abatement materials (with a spectrum analyzer). Following is a brief discussion of the three primary types of sound measurement instruments and the noise descriptors that can be obtained therefrom.

31.4.1.1 Sound Level Meter

Most sound measurement instruments derive from the basic sound level meter (SLM), a device for which four grades and associated performance tolerances that become more stringent as the grade number increases are described in ANSI S1.4-1983 (R2001).[17] Type 0 instruments have the most stringent tolerances and are for laboratory use only. Other grades include Type 1, intended for precision measurement in the field or laboratory, Type 2, intended for general field use, especially where frequencies above 10,000 Hz are not prevalent, and Type S, a special purpose meter that may perform at grades 1 to 3, but may not include all of the operational functions of the grade. A grade of Type 2 or better is needed for occupational exposure measurements.

31.4.1.1.1 Components of a Sound Level Meter

A block diagram of the functional components of a generic SLM is given in Figure 31.4. At the top, a microphone/preamplifier senses the pressure changes caused by an airborne sound wave and converts the pressure signal into a voltage signal. Because the pressure fluctuations of a sound wave are small in magnitude, the corresponding voltage signal must be preamplified and then input to an amplifier that boosts the signal before it is processed further. The passband, or range of frequencies that are passed through and processed, of a high-quality SLM contains frequencies from about 10 to 20,000 Hz, but depending on the frequency weighting used, not all frequencies are treated the same. A selectable frequency weighting network, or filter, is then applied to the signal. These networks most commonly include the A-, B-, and C-weighting functions shown in the bottom panel of Figure 31.2. For OSHA noise monitoring measurements, the A-scale, which de-emphasizes the low frequencies and to a smaller extent the high frequencies, is used. In addition to the common A scale (which approximates the 40-phon level of hearing) and C scale (100-phon level), other scales, including dB(linear), may be included in the meter.

Next (not shown), the signal is squared to reflect the fact that the sound pressure level in decibels is a function of the square of the sound pressure. The signal is then applied to an exponential averaging network, which defines the meter's dynamic response characteristics. In effect, this response creates a moving-window, short-time average display of the sound waveform. The two most common settings are defined as FAST, which has a time constant of 0.125 sec, and SLOW, which has a time constant of 1.0 sec. These time constants were established decades ago to give analog needle indicators a rather sluggish response so that they could be read by the human eye even when highly fluctuating sound pressures were measured. Under the FAST or SLOW dynamics, the meter indicator rises exponentially toward the decibel value of an applied constant SPL. In theory, when driven by an exponential process, the indicator would reach the actual value at infinite time; however, the time constant defines the time period within which the indicator reaches 63% of the maximum value in response to a constant input. For OSHA measurements, the SLOW setting is used, and this setting is best when the average value or average as it is changing over time is desired. The FAST setting is more appropriate when the variability or range of fluctuations of a time-varying sound is desired. On certain SLMs, a third time constant, IMPULSE, may also be included for measurement of sounds that have sharp transient characteristics over time and are generally less than 1 sec in duration, exemplified by gun shots or impact machinery such as drop forges and embossing processes. The IMPULSE setting has an exponential rise time constant of 35 msec and a decay time of 1.5 sec. It is useful to afford the observer the time to view the maximum value of a burst of sound before it decays, and is more commonly applied in community and business machine noise measurements than in industrial settings.

31.4.1.1.2 *Microphone Considerations*

Most SLMs have interchangeable microphones that offer varying frequency response, sensitivity, and directivity characteristics.[18] The *response* of the microphone is the ratio of electrical output (in volts) to the sound pressure at the diaphragm of the microphone. Sound pressure is commonly expressed in pascals for free-field conditions

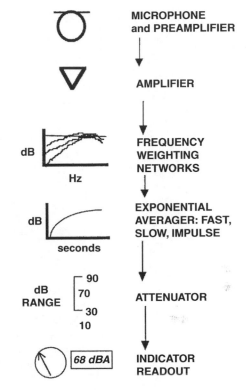

FIGURE 31.4 Block diagram of the basic functional components of a sound level meter.

(where there are no sound reflections resulting in reverberation), and the free-field voltage response of the microphone is given as millivolt per pascal. When specifications for *sensitivity* or *output level* are given, the response is based on a pure tone sound wave input. Typically, the output level is provided in dB re 1 V at the microphone electrical terminals and the reference sensitivity is 1 V/Pa.

Most microphones that are intended for industrial noise measurements are essentially *omnidirectional* (i.e., nondirectional) in their response for frequencies below about 1000 Hz. When the physical diameter of the microphone is comparable in length to the wavelength of the sound frequency (as occurs at higher frequencies), the microphone, even an omnidirectional one, will exhibit some directionality. This means that depending on the angle of the microphone's diaphragm in relation to the noise, the measurement readout can be less than or even greater than the true value. The 360° response pattern of a microphone is called its *polar response*, and the pattern is generally symmetrical about the axis perpendicular to the diaphragm. Some microphones are designed to be highly directional; one example is the cardioid design that has a heart-shaped polar response wherein the maximum sensitivity is for sounds whose direction of travel causes them to enter the microphone at 0° (or the *perpendicular incidence response*), and minimum sensitivity is for sounds entering at 180° behind the microphone. The response at 90°, where sound waves travel and enter parallel to the diaphragm, is known as the *grazing incidence response*. Another response pattern, called the *random incidence response*, represents the mean response of the microphone for sound waves that strike the diaphragm from all angles with equal probability. This response characteristic is the most versatile, and thus it is the response pattern most often applied in the United States. Hypothetical response characteristics for different sound wave incidences are shown in Figure 31.5.

Because most U.S. SLM microphones are omnidirectional and utilize the random-incidence response, it is best for an observer to point the microphone at the primary noise source and hold it at an angle of

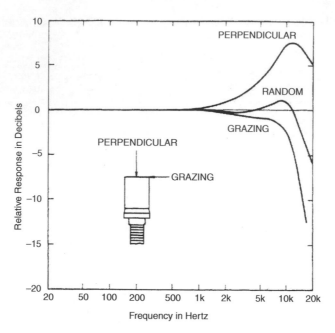

FIGURE 31.5   Frequency response of an hypothetical microphone for three angles of incidence. (Adapted from Peterson, A.P.G. (1979). Noise measurements: Instruments. In Harris, C. M. (Ed.), *Handbook of Noise Control* [pp. 5-1–5-19]. New York: McGraw-Hill. With permission.)

incidence of approximately 70°. This will produce a measurement most closely corresponding to the random-incidence response. Care must be taken to avoid shielding the microphone with the body or other structures. The response of microphones can also vary with temperature, atmospheric pressure, and humidity, with temperature being the most critical factor. Correction factors for variations in decibel readout due to temperature effects are supplied by most microphone manufacturers. Atmospheric effects are generally only significant when measurements are made in aircraft or at very high altitudes, and humidity has a negligible effect except at very high levels. In any case, microphones must not be exposed to moisture or large magnetic fields, such as those produced by transformers. When used in windy conditions, a foam windscreen should be placed over the microphone. This will reduce the contaminating effects of wind noise, while only slightly influencing the frequency response of the microphone at primarily high frequencies. In an industrial setting, the windscreen offers the additional benefit of protection of the microphone from damage due to striking and/or airborne foreign matter.

31.4.1.1.3   Root Mean Square

Because sound consists of pressure fluctuations above and below ambient air pressure for which the arithmetic average is zero, a *root mean square* (rms) averaging procedure is applied within the SLM when FAST, SLOW, or IMPULSE measurements are taken. In effect, each pressure (or converted voltage) value is squared, the arithmetic sum of all squared values is then obtained, and finally the square root of the sum is computed to provide the rms value. The rms value is what appears on the meter's display.

31.4.1.1.4   TRUE PEAK SLM

Some SLMs include an unweighted *TRUE PEAK* setting that does not utilize the rms measurement averaging technique, but instead provides an indication of the actual peak SPL reached during a pressure impulse. This measurement mode is necessary for determining if the OSHA limit of 140 dB for impulsive exposure is exceeded. A Type 1 or 2 meter must be capable of measuring a 50-μsec pulse. It is important to note that the rms-based IMPULSE dynamics setting is unsuitable for measurement of TRUE PEAK SPLs.

31.4.1.1.5 Analog vs Digital Readouts

With regard to the final component of an SLM shown in Figure 31.4, the indicator display or readout, much debate has existed over whether an analog (needle pointer or bar "thermometer-type" linear display) or digital (numeric) display is best. Ergonomics research indicates that while the digital readout affords higher precision of information to be presented in a smaller space, its Achilles heel is that the digits (particularly the least significant position) become impossible to read when the sound level is fluctuating rapidly. Also, it is more difficult for the observer to capture the maximum and minimum values of a sound, as is often desirable using the FAST response, or the maximum impulse peak attained, with a digital readout. On the other hand, if very precise measurements down to a fraction of a decibel are needed, the digital indicator is preferable as long as the meter incorporates an appropriate time integrating or averaging feature or "hold" setting so that the data values can be captured by the human eye. Because of the advantages and disadvantages of each type of display, some contemporary SLMs include both analog and digital readouts.

31.4.1.1.6 Sound Level Meter Applications

It is important to note that the standard SLM is intended to measure sound levels at a given moment in time, although certain specialized devices can perform integration or averaging of levels over an extended period of time to provide a long-term descriptor of the noise. When the nonintegrating/averaging SLM is used for noise exposure measurements in the workplace, it is necessary to sample and make multiple manual data entries on a record to characterize the exposure. This technique is usually best limited to area sampling, not individual employee measurements, because it is difficult for the observer to hold the microphone near the employee's ear and to closely shadow the employee as he or she moves about the workplace. Furthermore, the sampling process becomes more difficult as the fluctuations in a noise become more rapid and/or random in nature.

31.4.1.2 Dosimeter

The *"audio-dosimeter"* or more simply, *"dosimeter,"* is a battery-powered, highly portable device that is derived directly from an SLM but also features the ability to obtain special measures of noise exposure (discussed later) that relate to regulatory compliance and hearing hazard risk. Dosimeters are very compact and are generally worn on the belt or in the pocket of an employee, with the microphone generally clipped to the lapel or shoulder of a shirt or blouse. The intent is to obtain a noise exposure log or record over the course of a full or partial workshift, and to obtain, at a minimum, a readout of the time-weighted average (TWA) exposure and noise dose for the period measured. Depending upon the features, the dosimeter can log the time history of exposure, providing a running histogram of noise levels on a short time interval (such as 1 min) basis, compute statistical distributions of the noise exposures for the period, flag and record exposures that exceed OSHA maxima of 115 dBA continuous or 140 dB TRUE PEAK, and compute average metrics using 3 dB, 5 dB, or even other time-versus-level exchange rates. The dosimeter eliminates the need for the observer to set up a discrete sampling scheme or follow the worker, both of which are necessary with a conventional SLM. However, it is important that the observer establish rapport and gain the confidence of the worker wearing the dosimeter, and convey at least the following information: (1) to behave normally as to the work activity, (2) to not tamper with the dosimeter or microphone, (3) to return the device when visiting restrooms or entering damp areas, (4) to return the device if there is a need to approach large transformers or other magnetic fields, and (5) to understand the purpose of the dosimetry. Since they are designed to be worn on the noise-exposed employee, dosimeters are typically thought of as devices for personal measurements, but they may also be tripod-mounted or held by an observer for area or survey measurements and are very useful for obtaining community noise measurements as well.

31.4.1.3   Spectrum Analyzer

The *spectrum analyzer* is an advanced SLM that incorporates selective frequency-filtering capabilities to provide an analysis of the noise level as a function of frequency. In other words, the noise is broken down into its frequency components and a distribution of the noise energy in all measured frequency bands is available. Bands are delineated by upper and lower edge or cutoff frequencies and a center frequency. Different widths and types of filters are available, with the most common width being the *octave filter*, wherein the center frequencies of the filters are related by multiples of two (i.e., 31.5, 63, 125, 250, ..., 4000, 8000, and 16,000 Hz), and the most common type being the center frequency proportional, wherein the width of the filter depends on the center frequency (as in an octave filter set, in which the passband width equals the center frequency divided by $2^{1/2}$). The *octave band*, commonly called $1/1$ octave filter, has a center frequency, cf, which is equal to the geometric mean of the upper (f_u) and lower (f_l) cutoff frequencies. The formulae to compute the center frequency for the octave filter, as well as the band edge frequencies, are:

$$\text{Center frequency, } cf = (f_u * f_l)^{1/2}; \quad \text{upper cutoff, } \quad f_u = (cf)2^{1/2};$$

$$\text{lower cutoff, } f_l = (cf)/2^{1/2} \tag{31.7}$$

More precise spectral resolution can be obtained with other center frequency proportional filter sets with narrower bandwidths, the most common being the $1/3$ octave, and with constant percentage bandwidth filter sets, such as 1 or 2% filters. Note that in both types the filter bandwidth increases as the center frequency increases. Still other analyzers have constant bandwidth filters, such as 20-Hz-wide bandwidths that are of constant width regardless of center frequency. While in the past most spectrum analyzer filters have been analog devices with "skirts" or overshoots extending slightly beyond the cutoff frequencies, digital computer-based analyzers are now very common. These "computational" filters use fast Fourier transform (FFT) algorithms to compute sound level in a prespecified band of fixed resolution. FFT devices can be used to obtain very high resolutions of noise spectral characteristics using bandwidths as low as 1 Hz. However, in most industrial noise applications, a $1/1$- or $1/3$-octave analyzer will suffice unless the noise has considerable power in near-tonal components that must be isolated. One caution is in order: if a noise fluctuates in time or frequency, an integrating/averaging analyzer should be used to achieve good accuracy of measurements. It is important that the averaging period be long in comparison to the variability of the noise being sampled.

Inexpensive spectrum analyzers sometimes have filter sets that must be addressed individually in obtaining a measurement. Such devices are called *sequential analyzers* and the operator must manually (or via computer control) step through each filter separately and then read the result. Obviously, sequential filters are problematic when applied to the measurement of a fluctuating noise. On the other hand, *real-time analyzers* incorporate parallel banks of filters that can process all frequency bands simultaneously, and the signal output may be controlled by a SLOW, FAST, or other time constant setting, or it may be integrated or averaged over a fixed time period to provide L_{OSHA}, L_{eq}, or other average-type data.

While occupational noise is monitored with a dosimeter or SLM for the purpose of noise exposure compliance (using A-weighted broadband measurement), or assessment of hearing protection adequacy (using C-weighted broadband measurement), both of these applications can also be addressed (in some cases more accurately) with the use of spectral measurements of the noise level. For instance, the OSHA occupational noise exposure standard.[19] allows the use of octave band measurements reduced to broadband dBA values to determine if noise exposures exceed dBA limits defined in Table G-9 of the standard. Furthermore, Appendix B of the standard concerns hearing protector adequacy and allows the use of an octave band method for determining, on a spectral rather than a broadband basis, whether a hearing protector is adequate for a particular noise spectrum. It is also noteworthy that spectral analysis can help the hearing conservationist discriminate noises as to their hazard potential even though they may have similar A-weighted SPLs. This is illustrated in Figure 31.6, where both noises would be considered to

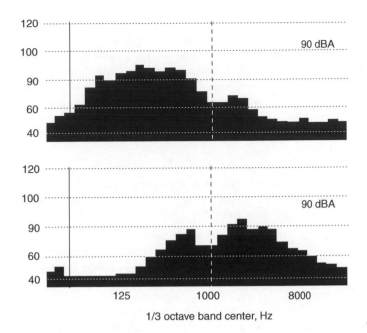

FIGURE 31.6 Spectral differences for two different noises that have the same dBA value.

be of equal hazard by the OSHA-required dBA measurements (since they both are 90 dBA), but the 1/3-octave analysis demonstrates that the lowermost noise is more hazardous as evidenced by the heavy concentration of energy in the midrange and high frequencies.

Perhaps the most important application of the spectrum analyzer is to obtain data that will provide the basis for engineering noise control solutions. For instance, in order to select an absorption material for lining interior surfaces of a workplace, the spectral content of the noise must be known so that the appropriate density and thickness of material may be identified. If the noise is found to be primarily of low frequency, the absorption techniques may not provide adequate reduction because low frequencies are more difficult to absorb than high frequencies.

Lacking a spectrum analyzer, the hearing conservationist can obtain a very rough indication of the dominant spectral content of a noise by using an SLM and taking measurements in both dBA and dBC for the same noise. If the (dBC − dBA) value is large, that is, about 5 dB or more, then it can be concluded that the noise has considerable low frequency content. If, on the other hand, the (dBC − dBA) value is negative, then the noise clearly has strong midrange components, since the A-weighting curve exhibits slight amplification in the 2000 to 4000 Hz range. Such rules-of-thumb rely on the differences in the C- and A-weighting curves shown in Figure 31.2. However, they should not be relied upon in lieu of a spectrum analysis if the noise is believed to have high frequency or narrow band components that need noise control attention.

31.4.1.4 Acoustical Calibrator and Microphone Calibration

Each of the instruments described previously contains a microphone that transduces the changes in pressure and inputs this signal into the electronics. While modern sound measurement equipment is generally stable and reliable, calibration is necessary to match the microphone to the instrument so that the accuracy of the measurement is assured. Because of its susceptibility to varying environmental conditions and damage due to rough handling, moisture, and magnetic fields, the microphone is the weakest link in the measurement equipment chain. Therefore, an acoustical calibrator should be applied before and after each measurement with an SLM. The pretest calibration insures that the instrument is indicating the correct SPL for a standard reference calibrator output at a specified SPL and frequency (most often 94 dB at 1000 Hz). The posttest calibration is done to determine if the instrumentation, including the microphone, has drifted during the measurement and if so, if the drift is large enough to invalidate the data obtained. Calibrators may be electronic transducer-type devices with loudspeaker outputs from an internal oscillator, or "pistonphones," which use a reciprocating piston in a closed cavity to produce sinusoidal pressure variations as the cylinder volume changes. Both types include adapters that allow the device to be mated to microphones of different diameters. Calibrators should be sent to the factory for annual calibration. SLMs and dosimeters used for occupational noise measurements should also be factory calibrated on an annual basis.

There are many other issues that bear on the proper application of sound level measurement equipment, such as microphone selection and placement, averaging time and sampling schemes, and statistical data reduction techniques, all of which are beyond the scope of this chapter. Further coverage of measurement and instrumentation appears in Harris.[20]

31.4.2 Measures for Quantifying Occupational Noise Levels

31.4.2.1 Exchange Rates

As alluded to earlier in the discussion on the OSHA occupational noise exposure standard, most noise regulations stipulate that a worker's exposure may not exceed a maximum daily accumulation of noise energy, and that the total energy is defined by the combination of exposure duration and intensity of the noise. In other words, in OSHA terms the product of duration and intensity must remain under the regulatory cap or *permissible exposure limit* (PEL) of 90 dBA TWA for an 8-h work period, which is equivalent to a 100% noise dose. Because both noise amplitude and noise duration determine the energy in the exposure, average-type measures of exposure are based on simple algorithms or "exchange rates" that trade amplitude for time and vice versa. Much debate has occurred over the past several decades about which exchange rate is most appropriate for prediction of hearing damage risk, and most countries currently use either a 3- or 5-dB relationship. The OSHA exchange rate is 5 dB, which means that an increase (decrease) in decibel exposure by 5 dB is equivalent to a doubling (halving) of exposure time. For instance, using the OSHA PEL of 90 dBA for 8 h, if a noise is at 95 dBA, the allowable exposure per workday is half of 8 h, or 4 h. If a noise is at 85 dBA, the allowable exposure time is twice 8 h, or 16 h. These allowable reference exposure durations (T values) are provided in Table A-1 of the OSHA[19] regulation, or they may be computed using the formula for T, given in Equation (31.14). The 5-dB exchange rate is predicated on the theory that intermittent noise is less damaging than continuous noise because some recovery from temporary hearing loss occurs during quiet periods. Arguments against it include the fact that an exchange of 5 dB for a factor of two in time duration has no real physical basis in terms of energy equivalence. Furthermore, there is some evidence that the quiet periods of intermittent noise exposures are insufficient in length to allow for recovery to occur. The 5-dB exchange rate is used for all measures associated with OSHA regulations, including the most general average measure of L_{OSHA}, the TWA referenced to an 8-h duration, and noise dose in percent.

Most European countries use a 3-dB exchange rate, also known as the "equal energy rule." In this instance, a doubling (halving) of sound intensity, which corresponds to a 3-dB increase (decrease), equates to a doubling (halving) of exposure duration. The equal energy concept stems from the fact that if noise energy is doubled or halved, the equivalent decibel change is 3 dB. An exposure to 90 dBA for 8 h using a 3-dB exchange rate is equivalent to a 120-dBA exposure of only 0.48 min. Because each increase in decibels by 10 corresponds to a tenfold increase in intensity, the 30-dB increase from 90 to 120 dBA represents a 1000-fold (10^3) increase in sound intensity, from 0.001 to 1 W/m^2. The 90-dBA exposure period is 8 h or 480 min, and this must be reduced by the same factor as the SPL increase, so 480/1000 equals 0.48 min or 29 sec. The 3-dB exchange rate is used for all measures associated with the equivalent continuous sound level, or L_{eq}.

31.4.2.2 Average and Integrated SPLs

As discussed earlier, conventional SLMs provide "momentary" decibel measurements that are based on very short moving-window exponential averages using FAST, SLOW, or IMPULSE time constants. However, since the majority of noises fluctuate over time, one of several types of average measurements is usually most appropriate as a descriptor of the central tendency of the noise. Averages may be obtained in one of two ways: (1) by observing and recording conventional SLM readouts using a short time interval sampling scheme, and then manually computing the average value from the discrete values, or (2) by using an SLM or dosimeter that automatically calculates a running average value using microprocessor circuitry that provides either a true continuous integration of the area under the sound pressure curve or obtains discrete samples of the sound at a very fast rate and computes the average. Generally, average measures obtained by method 2 yield more representative values because they are based on continuous or near-continuous sampling of the waveform, which the human observer cannot perform. For sounds that are constant or slowly fluctuating in level, either method should provide representative values, although method 1 necessitates continuous vigilance by an observer.

The average metrics discussed next are generally considered as the most useful for evaluating noise hazards and annoyance potential. In most cases for industrial hearing conservation as well as community noise annoyance purposes, the metrics utilize the A-weighting scale. The equations are all in a form where the data values are considered to be discrete sound levels. Thus, they can be applied to data from conventional SLMs or dosimeters. For continuous sound levels (or when the equations are used to describe true integrating meter functioning), the Σ sign in the equations would be replaced by the integral sign, $\int_0^T$, and the t_i replaced by dt.

Variables used in the equations are as follows: L_i is the dB level in measurement interval i, N is the number of intervals, t_i is the length of measurement interval i, T is the total measurement time period, Q is the exchange rate in dB, and

$$q = Q/\log_{10}(2) \begin{cases} \text{for 3-dB exchange, } q = 10.0 \\ \text{for 4-dB exchange, } q = 13.3 \\ \text{for 5-dB exchange, } q = 16.6 \end{cases}$$

The general form equation for *average SPL*, or L_{average}, L_{av} is

$$L_{\text{av}}(Q) = q \log_{10}\left[\frac{1}{T}\sum_{i=1}^{N}\left(10^{(L_i/q)} * t_i\right)\right] \tag{31.8}$$

The *equivalent continuous sound level*, or L_{eq}, equals the continuous sound level, which, when integrated or averaged over a specific time, would result in the same energy as a variable sound level over the same time period. The equation for L_{eq}, which uses a 3-dB exchange rate, is

$$L_{\text{eq}} = L_{\text{av}}(3) = 10 \log_{10}\left[\frac{1}{T}\sum_{i=1}^{N}\left(10^{(L_i/10)} * t_i\right)\right] \tag{31.9}$$

In applying the L_{eq}, usually the individual L_i values are in dBA. Equation (31.9) may also be used to compute the overall equivalent continuous sound level (for a single site or worker) from individual L_{eq}'s that are obtained over contiguous time intervals by substituting the L_{eq} values in the L_i variable. L_{eq} values are often expressed with the time period over which the average is obtained, for instance, L_{eq} (24) is an equivalent continuous level measured over a 24-h period. Another average measure that is derived from the L_{eq} and often used for community noise quantification is the L_{dn}. The L_{dn} is simply a 24-h L_{eq} measurement with a 10-dB penalty added to all nighttime noise levels from 10 p.m.

to 7 a.m. The rationale for the penalty is that humans are more disturbed by noise, especially due to sleep arousal, during nighttime periods.

The equation for the *OSHA average noise level*, or L_{OSHA}, which uses a 5-dB exchange rate, is

$$L_{OSHA}(5) = 16.61 \log_{10}\left[\frac{1}{T}\sum_{i=1}^{N}\left(10^{(L_{iA}/16.61)} * t_i\right)\right]$$

(where L_{iA} is in dBA, slow response)   (31.10)

OSHA's TWA is a special case of L_{OSHA}, which requires that the total time period always be 8 h, that time is expressed in hours, and that sound levels below 80 dBA, termed the *threshold level*, are not included in the measurement:

$$\text{TWA} = 16.61 \log_{10}\left[\frac{1}{8}\sum_{i=1}^{N}\left(10^{(L_{iA}/16.61)} * t_i\right)\right]$$

(where L_{iA} is in dBA, slow response; T is always 8 h;

only $L_{iA} \geq 80$ dBA is included)   (31.11)

OSHA's *noise dose* is a percentage representation of the noise exposure, where 100% is the maximum allowable dose, corresponding to a 90-dBA TWA referenced to 8 h. Dose utilizes a *criterion sound level*, which is presently 90 dBA, and a *criterion exposure period*, which is presently 8 h. A noise dose of 50% corresponds to a TWA of 85 dBA, and this is known as the OSHA *action level*. Calculation of dose, D, is as follows:

$$D = \frac{100}{T_c}\sum_{i=1}^{N}\left(10^{((L_{iA}-L_c)/q)} * t_i\right)$$

(where L_{iA} is in dBA, slow response; L_c is the criterion sound level;

T_c is the criterion exposure duration; only $L_{iA} \geq 80$ dBA is included)   (31.12)

Noise dose, D, can also be expressed as follows, for a constant sound level over the workday:

$$D = 100 * \left(\frac{C_1}{T_1} + \frac{C_2}{T_2} + \cdots + \frac{C_n}{T_n}\right)$$

(where C_i is the total time (hours) of actual exposure at

L_i; T_i is total time (hours) of reference allowed exposure at

L_i, from TableG-16a of OSHA[19];

C_i/T_i represents a partial dose at sound level i)   (31.13)

T, the *reference allowable* exposure for a given sound level, can also, in lieu of consulting Table G-16a in OSHA,[19] be computed as

$$T = \frac{8}{2^{(L-90)/5}} \quad \text{(where } L \text{ is the measured dBA level)}$$ (31.14)

Two other useful equations to compute dose, D, from TWA and vice versa are

$$D = 100 * 10^{((\text{TWA}-90)/16.61)} \qquad (31.15)$$

$$\text{TWA} = \left[16.61 \log_{10}\left(\frac{D}{100}\right)\right] + 90 \ \text{(where } L \text{ is the measured dBA level)} \qquad (31.16)$$

TWA can also be found for each value of dose, D, in Table A-1 of OSHA.[19]

A final measure that is particularly useful for quantifying the exposure due to single or multiple occurrences of an acoustical event (such as a complete operating cycle of a machine, a vehicle drive-by, or aircraft flyover), is the *sound exposure level*, or SEL. It has also been suggested for use in exposure regulations for industry, but to date has not been incorporated into OSHA requirements. The SEL represents a sound of one second length that imparts the same acoustical energy as a varying or constant sound that is integrated over a specified time interval, t_i, in seconds. Over t_i, an L_{eq} is obtained, which indicates that SEL is used only with a 3-dB exchange rate. A reference duration of 1 sec is applied for t_0 in the following equation for SEL:

$$\text{SEL} = L_{eq} + 10 \log_{10}\left(\frac{t_i}{t_0}\right) \quad \text{(where } L_{eq} \text{ is the equivalent sound pressure}$$

$$\text{level measured over time period } t_i) \qquad (31.17)$$

31.4.2.1.1 *Example Computational Problems*
Because the majority of industrial noise exposure problems in the United States involve measurements to determine OSHA compliance and hearing conservation program needs, the most common measurements from those discussed earlier entail calculation of the OSHA dose and TWA. Therefore, example computational problems using these measures follow.

Example 1. *Workshift less than 8-h, reading from SLM*
Exposures comprising a 7-h workday consist of 1 h at 95 dBA, 2 h at 90 dBA, and 4 h at 85 dBA, with measurements taken from an SLM. What is the dose and TWA?

Use Equation (31.14) (or OSHA code, Table G-16a[19]) to determine that 95 dBA is allowed for 4 h, 90 dBA for 8 h, and 85 dBA for 16 h. Then use Equation (31.13) to determine the partial doses associated with each exposure and the total dose for the workday:

$$D = 100 * \left(\frac{1}{4} + \frac{2}{8} + \frac{4}{16}\right) = 75\%$$

Since 50% action level is exceeded, a HCP program is needed

Equation (31.16) (or OSHA, Table A-1[19]) is then used to compute the TWA:

$$\text{TWA} = \left[16.61 \log_{10}\left(\frac{75}{100}\right)\right] + 90 = 87.9 \ \text{dBA per 8-h day}$$

Note: As shown in this example, regardless of the total workday, the OSHA method references everything to an 8-h criterion, with PEL of 90 dBA TWA. This problem could also have been solved by application of Equations (31.11) and (31.12).

Example 2. *Workshift greater than 8-h, reading directly from dosimeter*
A dosimeter is set up to run for a 12-h shift, and the readout at the end of the period is $D = 300\%$. If the dosimeter is programmed for an 8-h criterion exposure duration, a 90-dBA criterion sound level,

an 80 dBA threshold sound level, and a 5-dB exchange rate, then the OSHA dose may be read directly from the meter regardless of the fact that the total measurement period is 12 h. The TWA can be then be computed using Equation (31.16) (or OSHA code, Table A-1[19]):

$$TWA = \left[16.61 \log_{10}\left(\frac{300}{100}\right) \right] + 90 = 97.9 \text{ dBA}$$

Example 3. *Workshift greater than 8-h, reading from SLM*

A sound level meter is used to measure exposures in a 12-h workshift and the average levels obtained over the four time periods sampled are 3 h at 92 dBA, 2 h at 98 dBA, 6 h at 96 dBA, and 1 h meal time at 75 dBA.

First, the 75-dBA period is deleted in the TWA computation (but not in L_{OSHA} if it is to be calculated) since it is less than the OSHA 80-dBA threshold. Then, Equation (31.11) is used to compute the TWA, which is based on an 8-h criterion (therefore, $T = 8$):

$$TWA = 16.6 \log_{10}\left[\frac{1}{8}(3 * 10^{92/16.61} + 2 * 10^{98/16.61} + 6 * 10^{96/16.61}) \right]$$

$$= 16.6 \log_{10}\left[\frac{1}{8}(1037416.8 + 1588876.7 + 3612451.8) \right]$$

$$= 16.61 \log_{10} 779843.2$$

$$= 97.9 \text{ dBA}$$

Next, Equation (31.15) is used to compute the dose from the TWA:

$$D = 100 * 10^{((97.9-90)/16.61)}$$

$$D = 299\%$$

Example 4. *Workshift greater than 8-h, dosimeter measurement for only partial workshift*

A dosimeter is worn by an employee for 7 h of a 12-h workshift. It was not possible to apply the dosimeter for the full shift, but it has been determined, based on discussion with employees and direct observation that the entire workshift is consistent in regard to work activity. The dose measured for the 7-h period is 115%. Note that this dose is based on only 7 h of data and that the OSHA criterion exposure period of 8 h is reflected in the dose calculation from the meter. Since only 7 h of data are included, the dose is lower than that which would occur during a full 12-h shift.

Because the entire workshift is consistent with respect to noise-producing work activity, it is reasonable to assume that the same rate of dose per hour would continue through the complete shift.

The 7-h sampling period included: (1) one 15-min rest break and (2) one 30-min meal break. The remaining 5-h period that was not sampled does include one 15-min break

7 h sampled less the total of meal/breaks of 45 min = 375 min in noise

Total 12-h shift = (12 × 60) − 60 min of meal/breaks = 660 min in noise

The 12-h shift dose can be computed via either of the following methods:

1. Set up a proportional relationship as follows:

$$\frac{115\% \text{ dose}}{375 \text{ min}} = \frac{D\% \text{ dose}}{660 \text{ min}}$$

$$375D = 75,900$$

$$D = 202.4\%$$

Applying Equation (31.16) (or OSHA code, Table A-1[19]):

$$TWA = 95.1 \, dBA$$

2. Calculate a rate of dose per minute:

$$\frac{115\% \text{ dose}}{375 \text{ minutes}} = 0.3067\% \text{ dose per minute}$$

$$D = 660 \text{ minutes} * 0.3067\% \text{ dose/minute} = 202.4\%$$

Applying formula (31.16) (or OSHA code, Table A-1[19]):

$$TWA = 95.1 \, dBA$$

31.5 Industrial Noise Regulation and Abatement

31.5.1 Indicators of the Need for Attention to Noise

The need for management, or perhaps more appropriately, abatement of industrial noise is indicated when: (1) noise creates sufficient intrusion and operator distraction such that job performance and even job satisfaction are compromised; (2) noise creates interference with important communications and signals, such as interoperator communications, machine- or process-related aural cues, or alerting/emergency signals; and (3) noise exposures constitute a hazard for NIHL in workers. While this chapter primarily targets problem 3, which is governed by OSHA federal regulations in general industry[19] and MSHA (Mine and Safety Health Administration) regulations in mining, the principles of noise measurement, management, and abatement discussed herein may also be applied in mitigating problems 1 and 2.

31.5.1.1 OSHA Noise Exposure Limits

With regard to combating the hearing loss problem, in OSHA terms if the noise dose exceeds the OSHA action level of 50%, which corresponds to an 85-dBA TWA, the employer must institute a *hearing conservation program* (HCP) that consists of several facets, to be discussed later.[19] (It is noteworthy that the OSHA regulation specifically exempts employers in oil and gas well drilling and servicing from the HCP requirements, although they are subject to the 100% dose criterion.) If the criterion level of 100% dose is exceeded (which corresponds to the PEL of 90-dBA TWA for an 8-h day), the regulations specifically state that steps must be taken to reduce the employee's exposure to the PEL or below via administrative work scheduling or the use of engineering controls. It is specifically stated that hearing protection devices (HPDs) shall be provided if administrative and engineering controls fail to reduce the noise to the PEL. Therefore, in applying the letter of the law, HPDs are only intended to be relied upon when administrative and engineering controls are infeasible or ineffective. The final OSHA noise level requirement pertains to impulsive or impact noise, which is not to exceed a TRUE PEAK SPL limit of 140 dB.

31.5.2 Hearing Conservation Programs and the Systems Approach

31.5.2.1 Shared Responsibility between Management, Workers, and Government

A successful HCP, which includes many facets relating to the measurement, management, and control of noise, depends upon the shared commitment of management and labor, as well as the quality of services and products provided by external noise control consultants, audiology or medical personnel who conduct the hearing measurement program, and vendors (e.g., hearing protection suppliers). Furthermore, government regulatory agencies, such as OSHA and MSHA, have a responsibility to maintain and disseminate up-to-date noise exposure regulations and HCP guidance, to conduct regular in-plant monitoring of noise exposure and quality of HCPs, and to provide strict enforcement where inadequate noise control and hearing protection exists. And finally, the "end user" of the HCP, that is, the worker himself/herself, must be an informed and motivated participant. For instance, if a fundamental component of the HCP is the personal use of HPDs, the effectiveness of the program in preventing NIHL will depend most heavily on the worker's commitment to properly and consistently wear the HPD. Failure by any of these groups to carry out their responsibilities can result in HCP failure and worker hearing loss.

31.5.2.2 Hearing Conservation Program Structure and Components

Hearing conservation in industry should be thought of as a strategic, programmed effort that is initiated, organized, implemented, and maintained by the employer, with cooperation from other parties as indicated earlier. A well-accepted approach is to address the noise exposure problem from a *systems* perspective, wherein empirical noise measurements provide data input that drives the implementation of countermeasures against the noise (including engineering controls, administrative strategies, and personal hearing protection). Subsequently, noise and audiometric data, which reflect the effectiveness of those countermeasures, serve as feedback for program adjustments and improvements. Figure 31.7 illustrates the human and other system components that are typically included in an HCP, along with the links between components. Not all programs will include all of these components; for instance, personal hearing protection may be unnecessary if engineering controls provide sufficient noise reduction. A brief discussion of the major elements of an HCP, as dictated by OSHA,[19] follows.

31.5.2.2.1 Monitoring

Noise exposure monitoring is intended to identify employees for inclusion in the HCP and to provide data for the selection of HPDs. The data are also useful for identifying areas where engineering noise control solutions or administrative work scheduling may be necessary. All OSHA-related measurements, with the exception of the TRUE PEAK SPL limit, are to be made using an SLM or dosimeter set on the dBA scale, SLOW response, using a 5-dB exchange rate, and incorporating all sounds whose levels are from 80 to 130 dBA. It is unspecified, but must be assumed that sounds above 130 dBA should also be monitored. (Of course, such noise levels represent OSHA noncompliance since the maximum allowable continuous sound level is 115 dBA.) The measurement instrument should be ANSI Type 2[17] or better and calibrated to a known standard level before and after noise measurement. Monitoring strategies must take into account the effects of worker movement and noise level variation over time. Although no specific time interval between consecutive monitoring samples is specified, new samples should be taken whenever alterations in equipment or production produce changes in noise exposure. Appendix G of the OSHA regulation suggests that monitoring be conducted at least once every 1 or 2 yr.

Relating to the noise monitoring requirement is that of notification. Employees must be given the opportunity to observe the noise monitoring process and they must be notified when their exposures exceed the 50% dose (85-dBA TWA) level.

31.5.2.2.2 Audiometric Testing Program

All employees whose noise exposures are at the 50% dose level or above must be included in a pure-tone audiometric testing program wherein a baseline audiogram is completed within 6 months of the first exposure, and subsequent tests are done on an annual basis. Prior to the baseline audiogram, the worker must avoid workplace noise exposure for 14 h, or alternatively, use HPDs. Annual audiograms are compared against the baseline to determine if the worker has experienced a *standard threshold shift* (STS), which is defined as an increase in hearing threshold level relative to the baseline of an average of 10 dB at 2000, 3000, and 4000 Hz in either ear. The annual audiogram may be adjusted for age-induced hearing loss (presbycusis) using gender-specific correction data found in Appendix F of the regulation.

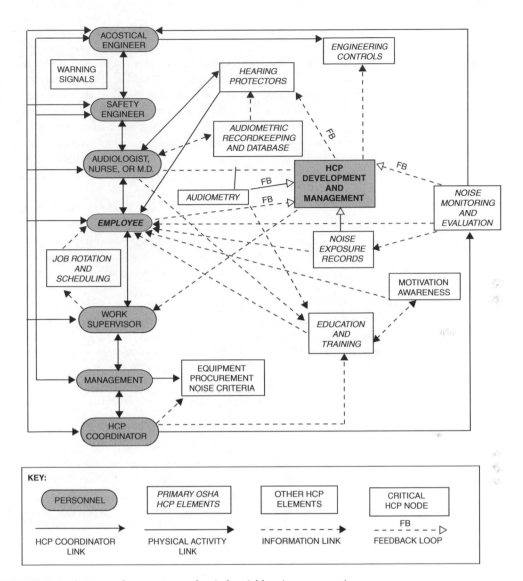

FIGURE 31.7   System and components of an industrial hearing conservation program.

All OSHA-related audiograms must include 500, 1000, 2000, 3000, 4000, and 6000 Hz, in comparison to most clinical audiograms that extend from 125 to 8000 Hz. If an STS is revealed, a licensed physician or audiologist must review the audiogram and determine the need for further audiological or otological evaluation, the employee must be notified of the STS, and the selection and proper use of HPDs must be revisited. An annual audiogram is substituted for the original baseline when the STS is determined to be persistent or when the annual audiogram indicates significant improvement over the baseline.

31.5.2.2.3   *Hearing Protection Devices*

A selection of HPDs that are suitable for the noise and work situation must be made available to all employees whose TWA exposures meet or exceed 85 dBA. *Earplugs* consist of vinyl, silicone, spun fiber-glass, cotton/wax combinations, and closed-cell foam products that are inserted into the ear canal to form a noise-blocking seal. Proper fit to the user's ears and training in insertion procedures are critical to the success of earplugs. A related device is the *semi-insert* or *ear canal cap*, which consists of earplug-like pods that are positioned at the rim of the ear canal and held in place by a lightweight headband. The headband is useful for storing the device around the neck when the user moves out of the noise. *Earmuffs* consist of earcups, usually of a rigid plastic material with an absorptive liner, that completely enclose the outer ear and seal around it with foam- or fluid-filled cushions. A headband connects the earcups, and on some models this band is adjustable so that it can be worn over the head, behind the neck, or under the chin, depending upon the presence of other headgear, such as a welder's mask. In general terms, as a group, earplugs provide better attenuation than earmuffs below about 500 Hz and equivalent or greater protection above 2000 Hz. At intermediate frequencies, earmuffs typically have the advantage in attenuation. Earmuffs are generally more easily fit by the user than earplugs or canal caps, and depending on the temperature and humidity of the environment, the earmuff can be uncomfortable (in hot or high humidity environments) or a welcome ear insulator (in a cold environment). Semi-inserts generally offer less attenuation and comfort than earplugs or earmuffs, but because they are readily storable around the neck, they are convenient for those workers who frequently move in and out of noise. A thorough review of HPDs and their application may be found in Berger and Casali.[21] Recent new technologies in hearing protection have emerged, including electronic devices offering active noise cancellation, communications capabilities, and noise-level-dependent attenuation, as well as passive, mechanical HPDs that offer level-dependent attenuation and near-flat or uniform attenuation spectra; these devices are reviewed in Casali and Berger.[22]

Regardless of its general type, HPD effectiveness depends heavily on the proper fitting and use of the devices.[23] Therefore, the employer is required to provide training in the fitting, care, and use of HPDs to all affected employees.[19] Hearing protector use becomes mandatory when the worker has not undergone the baseline audiogram, has experienced an STS, or has a TWA exposure that meets or exceeds 90 dBA. In the case of the worker with an STS, the HPD must attenuate the noise to 85 dBA TWA or below. Otherwise, the HPD must reduce the noise to at least 90 dBA TWA.

The protective effectiveness or adequacy of an HPD for a given noise exposure must be determined by applying the attenuation data required by the EPA[24] to be included on protector packaging. These data are obtained from psychophysical threshold tests at nine 1/3-octave bands with centers from 125 to 8000 Hz that are performed on human subjects, and the difference between the thresholds with the HPD on and without it constitutes the attenuation at a given frequency. Spectral attenuation statistics (means and standard deviations) and the single number noise reduction rating (NRR), which is computed therefrom, are provided. The ratings are the primary means by which end users compare different HPDs on a common basis and make determinations of whether adequate protection and OSHA compliance will be attained for a given noise environment.

The most accurate method of determining HPD adequacy is to use octave band measurements of the noise and the spectral mean and standard deviation attenuation data to determine the '*protected exposure level*' under the HPD. This is called the '*NIOSH long method*' or the '*octave band*' method. Computational procedures appear in NIOSH.[25] Because this method requires octave band measurements of the noise, preferably with each noise band's data in TWA form, the data requirements are large and the method is not widely applied in industry. However, because the noise spectrum is compared against the attenuation spectrum of the HPD, a "matching" of exposure to protector can be obtained; therefore, the method is considered to be the most accurate available.

The NRR represents a means of collapsing the spectral attenuation data into one broadband attenuation estimate that can easily be applied against broadband dBC or dBA TWA noise exposure measurements. In the calculation of the NRR, the mean attenuation is reduced by two standard deviations; this translates into an estimate of protection theoretically achievable by 98% of the population.[24] The NRR is primarily intended to be subtracted from the dBC exposure TWA to estimate the protected exposure level in dBA, as via the following equation:

$$\text{Workplace TWA in dBC} - \text{NRR} = \text{protected TWA in dBA} \tag{31.18}$$

Unfortunately, because OSHA regulations require that noise exposure monitoring be performed in dBA, the dBC values may not be readily available to the hearing conservationist. In the case where the TWA values are in dBA, the NRR can still be applied, albeit with some loss of accuracy. With dBA data, a 7-dB "safety" correction is applied to the NRR to account for the largest typical differences between C- and A-weighted measurements of industrial noise, and the equation is as follows:

$$\text{Workplace TWA in dBA} - (\text{NRR} - 7) = \text{protected TWA in dBA} \qquad (31.19)$$

While these methods are promulgated by OSHA[19] for determining HPD adequacy for a given noise situation, a word of caution is needed. The data appearing on HPD packaging are obtained under optimal laboratory conditions with properly fitted protectors and trained human subjects. In no way does the "experimenter-fit" protocol and other aspects of the current test procedure (ANSI S3.19-1974)[26] represent the conditions under which HPDs are selected, fit, and used in the workplace.[23] Therefore, the attenuation data used in the octave band or NRR formulae are highly inflated and cannot be assumed as representative of the protection that will be achieved in the field. The results of a review of research studies in which manufacturers' on-package NRRs were compared against NRRs computed from actual subjects taken with their HPDs from field settings are shown in Figure 31.8.[27] Clearly, the differences between laboratory and field estimates of HPD attenuation are large and the hearing conservationist must take this into account when selecting protectors. Efforts by ANSI Working Group S12/WG11 focused on the development of a new testing standard, ANSI S12.6-1997(R2002),[28] which utilizes subject (not experimenter) fitting of the HPD and relatively naive (not trained) subjects to yield attenuation data that are more representative of those achievable under workplace conditions wherein an HCP is operated (described in Royster, Ref. 29). However, at the time of working this new standard had not been adopted into law promulgating its use in producing the data to be utilized in labeling HPD performance (although it is likely to happen in the future).

If the currently available HPD attenuation data are inaccurate, what steps should be taken to gain a more accurate estimate of the NRR for use in determining protected exposure levels? The OSHA[30] Field Operations Manual of the Office of General Industry Compliance Assistance indicates: "Citations for violations of 29CFR 1910.95(b)(1) shall be issued when engineering and/or administrative controls are feasible, both technically and economically; and (1) Employee exposure levels are so high that hearing

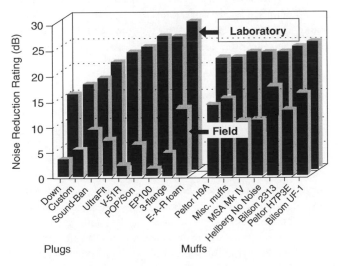

FIGURE 31.8 Comparison of hearing protection device NRRs by device type: manufacturers' laboratory data versus real-world "field" data. (Adapted from Berger, E. H., Franks, J. R., and Lindgren, F. (1996). International review of field studies of hearing protector attenuation. In Axelsson, A., Borchgrevink, H., Hamernik, R. P., Hellstrom, P., Henderson, D., and Salvi, R. J. (Eds.), *Scientific Basis of Noice-Induced Hearing Loss* (pp. 361–377). New York: Thiesse Medical Publishers, Inc. With permission.)

protectors alone may not reliably reduce noise levels received by the employee's ear to the levels specified in Tables G-16 or G-16a of the Standard. Given the present state of the art, hearing protectors which offer the greatest attenuation may not reliably be used when employees exposure levels border on 100 dBA." This guideline alludes to the importance of engineering controls as a primary countermeasure against high noise levels. The OSHA[31] Technical Manual of the Directorate of Technical Support states: "OSHA experience and the published scientific literature indicate that laboratory-obtained real-ear attenuation data for hearing protectors are seldom achieved in the workplace." ... Under "Field Attenuation of Hearing Protection": "When analyzing the attenuation a personal hearing protector may afford a noise-exposed employee in an actual work environment, the hearing protector shall be evaluated as follows: ... 2) To adjust for the lack of attainment of the laboratory-based noise reduction calculated according to Appendix B [laboratory ratings] estimating techniques, apply a safety factor of 50 percent; that is, divide the calculated laboratory-based attenuation by two. 3) For dual protection (i.e., earplugs and muffs) add 5 dB to the NRR of the higher-rated protector." For case (2), the derating factor may appear to be a reasonable strategy; however, these authors and others have argued that a constant derating factor is not appropriate because certain protectors (e.g., earmuffs) are easier to fit properly than others (e.g., user-formed earplugs), and thus the differences between laboratory and actual in-workplace performance will not be the same for all devices. In perusing Figure 31.8, this becomes quite apparent in that the laboratory NRRs for earplugs overestimate the field NRRs by an average of about 75%, while the laboratory NRRs for earmuffs overestimate the field NRRs by an average of only about 40%. These data would argue for the use of derating factors that differ by device type, not a constant derating such as the 50% OSHA recommendation. But in any case, the use of derating factors or other modifications of the NRR to adjust it for field applications is tenuous at best and should not be expected of the end user. The best solution is to establish a testing standard (and attenuation rating therefrom) that accurately predicts workplace protection achieved by HPDs, and this is the ANSI standard work described in Royster et al.[29]

31.5.2.2.4 Training Program and Access to Information and Materials

An oft-overlooked, but essential component of an industrial HCP is an annual training program for all workers included in the HCP. The required training elements to be covered are: (1) the effects of noise on hearing; (2) purpose, selection, and use of HPDs; and (3) purpose and procedures of audiometric testing. It is essential to the success of an HCP that workers become acutely aware of the need for hearing conservation, understand and believe in the merits of the program, and develop a commitment to and the motivation for protecting their hearing. Employers must make the OSHA regulations available to affected employees and, upon request, make all training materials available to OSHA representatives.

31.5.2.2.5 Recordkeeping and Intraprogram Feedback

Accurate records must be kept of all noise exposure measurements, at least from the last 2 yr, and audiometric test results for the duration of the worker's employment. It is important, but not required by OSHA, that noise and audiometric data be used as feedback for improving the program as shown in the feedback loops of Figure 31.7. Because the primary goal of the HCP is to prevent NIHL for employees, the program's effectiveness can be evaluated via *audiometric database analysis* (ADBA) for employees as a group, as opposed to individuals. By using population statistics from and inferential analysis of the database for exposed employees, problems can be identified early and corrective actions taken before significant threshold shifts appear in a number of individuals.[32] ADBA, however, is not a substitute for annual individual audiogram review and comparison against baseline. As discussed previously, this type of intra-worker analysis is essential for identifying threshold shifts and implementing preventative measures that are specific to the individual worker and job environment. Also, discussion of individual audiogram data with employees can aid in motivating them to exercise care in their daily hearing conservation practices, and audiometric feedback, sometimes posted anonymously by code number but including each individual's HPD use information, has been experimentally demonstrated to be an effective means of establishing higher HPD usage rates.[33]

Noise exposure records may be used as feedback to identify machines that need maintenance attention, to assist in the relocation of noisy equipment during plant layout efforts, to provide information for future equipment procurement decisions, and to target plant areas that are in need of noise control intervention. Some employers plot noise levels on a "contour map," delineating floor areas by their decibel levels. When monitoring indicates that the noise level in a particular contour has changed, it is taken as a sign that the machinery or work process has changed in the area and that further evaluation may be needed.

31.5.2.2.6 *Engineering Noise Control*

While OSHA does not stipulate the level of effort to be devoted to engineering noise controls or the types of controls that should be applied, the physical reduction of the noise energy, at its source, in its path, or at the worker, should be a major focus of noise management programs. Hearing protection or administrative controls should not supplant noise control engineering; the best solution, because it does not rely on employee behavior, is to reduce the noise itself, preferably at the emission source. However, in many cases where noise control is ineffective, infeasible (as on an airport taxi area), or prohibitively expensive, HPDs become the primary countermeasure.

There are many techniques used in noise control, and the specific approach must be tailored to the noise problem at hand. A noise control engineer is typically consulted to assist in the measurements, usually taken from spectrum analyzers, and in the selection of control strategies. Example noise control strategies include: (1) *isolation of the source* via relocation, enclosure, or vibration-damping using metal or air springs (below about 30 Hz) or elastomer (above 30 Hz) supports; (2) *reduction at the source or in the path* using mufflers or silencers on exhausts, reducing cutting, fan, or impact speeds, dynamically balancing rotating components, reducing fluid flow speeds and turbulence, absorptive foam or fiberglass on reflective surfaces to reduce reverberation, shields to reflect and redirect noise (especially high frequencies), and lining or wrapping of pipes and ducts; (3) *replacement or alteration of machinery*, examples including belt drives as opposed to noisier gears, electrical rather than pneumatic tools, and shifting frequency outputs such as by using centrifugal fans (low frequencies) rather than propeller or axial fans (high frequencies), keeping in mind that low frequencies propagate further than high frequencies, but high frequencies are more hazardous to hearing; and (4) *application of quieter materials*, such as rubber liners in parts bins, conveyors, and vibrators, resilient hammer faces and bumpers on materials handling equipment, nylon slides or rubber tires rather than metal rollers, and fiber rather than metal gears. Further discussion of these techniques may be found in Bruce and Toothman,[34] and an illustration of implementation possibilities in an industrial plant appears in Figure 31.9. A final approach that has just recently become available to industry is *active noise reduction (ANR)* in which an electronic system is used to transduce an offensive noise in a sound field and then process and reintroduce the noise into the same sound field such that it is exactly 180° out-of-phase with, but of equal amplitude to the original noise.[22] The superposition of the out-of-phase "antinoise" with the original noise causes physical cancellation of the noise in a target zone of the workplace. For highly repetitive, predictable noises, synthesis of the antinoise, as opposed to transduction and reintroduction, may also be used. At frequencies below about 1000 Hz, the ANR technique is most effective, which is fortuitous since the passive noise control materials to combat low-frequency noise, such as absorptive liners and barriers, are typically heavy, bulky, and expensive. At higher frequencies and their corresponding shorter wavelengths, the processing and phase relationships become more difficult and cancellation is less successful, although the technology is rapidly improving.

In designing and implementing noise control hardware, it is important that ergonomics be taken into account. For instance, in a sound-treated booth to house an operator, the ventilation system, lighting, visibility outward to the surrounding work area, and other considerations relating to operator comfort and performance must be considered. With regard to noise-isolating machine enclosures, access provisions should be designed so as to not compromise the operator/machine interface. In this regard, it is important that production and maintenance needs be met. If noise control hardware

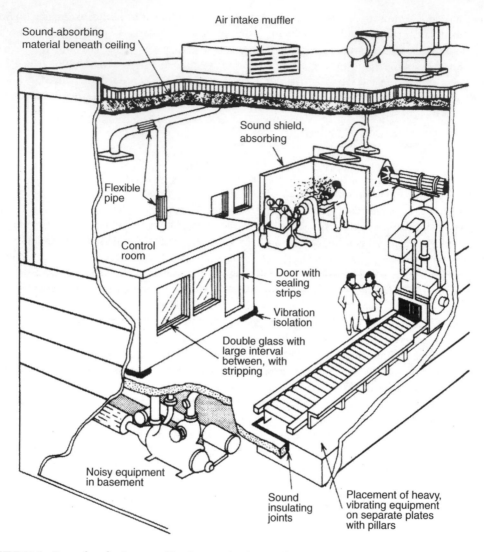

FIGURE 31.9   Examples of noise control implementation in an industrial plant. (Adapted from OSHA (1980) *Noise Control, A Guide for Workers and Employers.* Occupational Safety and Health Administration Report No. 3048. Washington, DC: U.S. Department of Labor. With permission.)

creates difficulties for the operators in carrying out their jobs, they may tend to modify or remove it, rendering it ineffective.

31.5.2.2.7   Personnel

As shown in Figure 31.7, multiple individuals play important roles in an industrial HCP, and the program should filter down from management personnel who must demonstrably support it. The key individual is the HCP coordinator (at the lower left in Figure 31.7), typically a permanent employee of the company but sometimes an outside consultant, who serves as the responsible individual and overseer for the program as well as its internal "champion." This individual, if properly qualified, may also be responsible for implementation of certain aspects of the program, including noise monitoring, audiometry on employees, selection and purchase of hearing protection devices (HPDs), and other functions. The HCP coordinator often heads a hearing conservation committee with representatives from labor, management, plant engineering, and safety. The coordinator also serves as a link between management and the workforce, and generally participates in management decisions that impact the noise environment or the HCP itself. For instance, one means of noise control is to establish a procurement policy that limits the decibel output of new equipment to a prespecified level; the HCP coordinator should be involved in such purchase decisions and in ensuring that criteria for noise emissions are met.

An audiologist, nurse, otolaryngologist, or other physician may conduct audiometric tests on employees and maintain a database for the test records. Industrial audiometry for OSHA purposes may also be conducted by a technician who is certified by the Council of Accreditation in Occupational Hearing Conservation (CAOHC), but this individual must ultimately be responsible to a professional audiologist or physician. The person who performs the audiometric test function may also be involved in helping the worker select an appropriate HPD (with input from the noise exposure records) and in educating and training the worker about the hazards of noise and the proper use of protection.

The work supervisor or foreman may also provide input to the HCP. For instance, in cases where workers are rotated on and off noisy machines to limit their exposures (a type of administrative countermeasure), the supervisor should be consulted to determine feasible rotation schemes. Furthermore, it is imperative that the supervisor exhibits good hearing conservation practice himself/herself and provide specific feedback to the HCP coordinator about occurrences that impact the success of the HCP, such as a machine that has become noisy due to lack of maintenance or a worker who is uncomfortable with his/her assigned HPD and therefore repeatedly takes it on and off. Because of his/her close relationship and proximity to production employees, the foreman or supervisor can serve as a key individual in helping to motivate the workers to exercise good hearing conservation practice, both by serving as a role model and an information resource.

Some large companies have an acoustical engineer on staff while others may need to hire such an individual when engineering noise control becomes necessary. The acoustical engineer can perform in-depth spectral analyses of specific noise sources and design noise control solutions. Furthermore, acoustical engineers can be helpful in the overall design of the HCP, in that the specialized knowledge they possess will be useful in considering tradeoffs in dollar cost-to-decibel reduction benefits when comparing various countermeasure strategies.

If the company has a safety engineer on staff, this individual should serve on the hearing conservation committee and participate in noise-related decisions that impact safety in other ways. For instance, if noise levels increase in an area where acoustic alarms signal the approach of an automated material transport vehicle, the safety engineer will need to work to increase the alarm's output to maintain detectability or use an alternate warning system, such as a flashing strobe, to maintain vehicle conspicuity. The safety engineer may also work with the HCP coordinator in selecting appropriate hearing protection for employees who must maintain communications in hazardous areas. In some small companies, the safety engineer may, in fact, have responsibility for the HCP itself.

Involvement and commitment of the proper hearing conservation and safety personnel, support of company management, and a trained and motivated workforce are all important to the success of a properly designed and implemented industrial hearing conservation program. Such a program can markedly reduce noise-induced distractions and interference on the job, and, above all, prevent the tragic and irrecoverable occurrence of occupational hearing loss in workers.

References

1. Stevens, S. S. (1936). A scale for the measurement of a psychological magnitude: loudness. *Psychological Review*, *43*, 405–416.
2. Stevens, S. S. (1972). Perceived level of noise by Mark VII and decibels (E). *Journal of the Acoustical Society of America*, *51*(2, pt. 2), 575–601.
3. Zwicker, E. (1960). En verfahren zur berechnung der lautstarke. *Acustica*, *10*, 304–308.
4. Kryter, K. D. (1994). *The Handbook of Hearing and the Effects of Noise.* New York: Academic Press.
5. Sanders, M. S. and McCormick, E. J. (1993). *Human Factors in Engineering and Design*, 7th edition. New York: McGraw-Hill.

6. Poulton, E. (1978). A new look at the effects of noise: a rejoinder. *Psychological Bulletin*, 85, 1068–1079.

7. Sorkin, R. D. (1987). Design of auditory and tactile displays. In Salvendy, G. (Ed.), *Handbook of Human Factors* (pp. 549–576). New York: McGraw-Hill.

8. ANSI S3.5–1997 (R2002). *Methods for the Calculation of the Speech Intelligibility Index*. New York: American National Standards Institute, Inc.

9. ISO 7731:2003 (2003). *Ergonomics — Danger Signals for Public and Work Places — Auditory Danger Signals*. Geneva, Switzerland: International Organization for Standardization.

10. Acton, W. I. (1970). Speech intelligibility in a background noise and noise-induced hearing loss. *Ergonomics*, 13(5), 546–554.

11. Pollack, I. (1958). Speech intelligibility at high noise levels: effects of short-term exposure. *Journal of the Acoustical Society of America*, 30, 282–285.

12. Morrison, H. B. and Casali, J. G. (1994). Intelligibility of synthesized voice messages in commercial truck cab noise for normal-hearing and hearing-impaired listeners. *Proceedings of the 1994 Human Factors and Ergonomics Society 38th Annual Conference*, Nashville, Tennessee, October 24–28, pp. 801–805.

13. EPA (1981). *Noise in America: the Extent of the Noise Problem*. Environmental Protection Agency Report No. 550/9-81-101. Washington, DC: EPA.

14. National Institutes of Health (NIH) Consensus Development Panel (1990). Noise and hearing loss. *Journal of the American Medical Association*, 263(23), 3185–3190.

15. Melnick, W. (1991). Hearing loss from noise exposure. In Harris, C. M. (Ed.), *Handbook of Acoustical Measurements and Noise Control* (pp. 18.1–18.19). New York: McGraw-Hill.

16. Taylor, W., Pearson, J., Mair, A., and Burns, W. (1964). Study of noise and hearing in jute weavers. *Journal of the Acoustical Society of America*, 38, 113–120.

17. ANSI S1.4–1983 (R2001). *Specification for Sound Level Meters*. New York: American National Standards Institute, Inc.

18. Peterson, A. P. G. (1979). Noise measurements: instruments. In Harris, C. M. (Ed.), *Handbook of Noise Control* (pp. 5-1–5-19). New York: McGraw-Hill.

19. OSHA (1983). 29CFR1910.95. *Occupational Noise Exposure; Hearing Conservation Amendment; Final Rule*. Occupational Safety and Health Administration. *Code of Federal Regulations*, Title 29, Chapter XVII, Part 1910, Subpart G, 48 FR 9776–9785. Washington, DC: Federal Register.

20. Harris, C. M. (1991), *Handbook of Acoustical Measurements and Noise Control*. New York: McGraw-Hill.

21. Berger, E. H. and Casali, J. G. (1997). Hearing protection devices. In Crocker, M. J. (Ed.), *Encyclopedia of Acoustics*. New York: John Wiley & Sons.

22. Casali, J. G. and Berger, E. H. (1996). Technology advancements in hearing protection: active noise reduction, frequency/amplitude-sensitivity, and uniform attenuation. *American Industrial Hygiene Association Journal*, 57, 175–185.

23. Park, M. Y. and Casali, J. G. (1991). A controlled investigation of in-field attenuation performance of selected insert, earmuff, and canal cap hearing protectors. *Human Factors*, 33(6), 693–714.

24. EPA (1979). 40CFR211, Noise labeling requirements for hearing protectors. Environmental Protection Agency, *Federal Register*, 44(190), 56130–56147.

25. NIOSH (1975). *List of Personal Hearing Protectors and Attenuation Data*. National Institute for Occupational Safety and Health-HEW Publication No. 76-120, pp. 21–37. Washington, DC: NIOSH.

26. ANSI S3.19-1974 (1974). *Method for the Measurement of Real-Ear Protection of Hearing Protectors and Physical Attenuation of Earmuffs*. New York: American National Standards Institute, Inc.

27. Berger, E. H., Franks, J. R., and Lindgren, F. (1996). International review of field studies of hearing protector attenuation. In Axelsson, A., Borchgrevink, H., Hamernik, R. P., Hellstrom, P., Henderson, D., and Salvi, R. J. (Eds.), *Scientific Basis of Noise-Induced Hearing Loss* (pp. 361–377). New York: Thieme Medical Publishers, Inc.

28. ANSI S12.6-1997 (R2002) *Methods for Measuring the Real-Ear Attenuation of Hearing Protectors*. New York: American National Standards Institute, Inc.

29. Royster, J. D., Berger, E. H., Merry, C. J., Nixon, C. W, Franks, J. R., Behar, A., Casali, J. G., Dixon-Ernst, C., Kieper, R. W., Mozo, B. T., Ohlin, D., and Royster, L. H. (1996). Development of a new standard laboratory protocol for estimating the field attenuation of hearing protection devices. Part I. Research of Working Group 11, Accredited Standards Committee S12, Noise. *Journal of the Acoustical Society of America*, *99*(3), 1506–1526.

30. OSHA (1989). OSHA Instruction CPL 2.45B, June 15. *Field Operations Manual* (pp. IV-33–IV-35). Rockville, MD: Government Institutes, Inc.

31. OSHA (1990). OSHA Instruction CPL 2-2.20B, February 5. *Field Technical Manual* (pp. 4-1–4-15). Rockville, MD: Government Institutes, Inc.

32. Royster, J. D. and Royster, L. H. (1986). Audiometric data base analysis. In Berger, E. H., Ward, W. D., Morrill, J. C., and Royster, L. H. (Eds.), *Noise and Hearing Conservation Manual* (pp. 293–317). Akron, OH: American Industrial Hygiene Association.

33. Zohar, D., Cohen, A., and Azar, N. (1980). Promoting increased use of ear protectors in noise through information feedback. *Human Factors*, *22*(1), 69–79.

34. Bruce, R. D. and Toothman, E. H. (1986). Engineering controls. In Berger, E. H., Ward, W. D., Morrill, J. C., and Royster, L. H. (Eds.), *Noise and Hearing Conservation Manual* (pp. 417–521). Akron, OH: American Industrial Hygiene Association.

35. Earshen, J. J. (1986). Sound measurement: instrumentation and noise descriptors. In Berger, E. H., Ward, W. D., Morrill, J. C., and Royster, L. H. (Eds.), *Noise and Hearing Conservation Manual* (pp. 38–95). Akron, OH: American Industrial Hygiene Association.

36. OSHA (1980). *Noise Control, A Guide for Workers and Employers*. Occupational Safety and Health Administration Report No. 3048. Washington, DC: U.S. Department of Labor.

32

Shiftwork

Timothy H. Monk
University of Pittsburgh Medical Center

32.1 Introduction

The ergonomic problem with regard to shiftwork is that of enabling the individual to work at abnormal hours, an activity which runs both counter to his or her own biology (*Homo sapiens* is a diurnal species) and counter to the surrounding society which is structured to protect the sleep of day workers, but not that of night workers, and expects evenings and weekends to be free for social, religious, athletic, and cultural events. Moreover, shiftwork is *not* simply restricted to a very small group of people who can be carefully selected or self-selected to experience minimal problems. Neither is it restricted to the youngest and fittest of workers who can bid their way out of abnormal hours when they advance into their middle age. Employment trends, particularly in the manufacturing sector, now dictate that the "bidding out of shiftwork by seniority" option often no longer applies. For many middle-aged and late middle-aged workers the only option is between shiftwork and no work. Thus, approximately one fifth of all employees are engaged in some form of work that requires their presence outside of the "standard" 7 am to 6 pm working day on a regular basis, and can thus be regarded as "shiftworkers."

The proportion of the working population engaged in shiftwork can be expected to rise as second jobbing and mandatory overtime increase.[1] The fastest growing sector of most Western economies is the service sector, and people are increasingly demanding and receiving around the clock availability of such services. Even in the production sector, plant machinery has become so expensive and so quickly obsolete that it has to be run 24 h per day, 7 days per week, in order for it to be profitable. Also, many nations have adopted taxation and business evaluation strategies (e.g., in assessing profitability and providing health insurance) that encourage employers to squeeze as many work hours per year as possible from their existing employees, rather than hiring new ones, as the volume of business increases. This may lead to extended work weeks and fewer different work teams covering each 24-h day.

Some people cope well with shiftwork, others poorly. Moore-Ede[2] and others have referred to a shiftwork maladaptation syndrome in those failing to cope. As noted above, shiftwork intolerance stems primarily from the fact that we are a diurnal species, designed to be asleep at night and alert and active during the day, and that we have constructed a society which is built around this biological reality.

However, shiftwork intolerance is a problem that should not be regarded as *solely* a circadian rhythms ("biological clock") issue, or, indeed, solely a sleep disorders issue, or solely a social and domestic issue.[3] Rather, it is a complex interaction of the three factors, with each factor influencing both of the other factors and the final outcome of shiftwork tolerance.[4] The long-term health consequences of shift-work have been reviewed extensively elsewhere.[5-7] In addition to sleep disorders, gastrointestinal dys-function, cancer, and cardiovascular disease have been the major complaints implicated, and shiftworkers should be encouraged to abstain from behaviors (such as unwise dietary choices) that might further exacerbate such risks. There may also be psychiatric disorders such as depression and sub-stance abuse resulting from prolonged exposure to shiftwork. The aim of this chapter is to introduce a general conceptual framework within which shiftwork coping ability may be considered, so that the ergo-nomist can better understand the various factors that are involved.

As noted earlier, shiftwork coping ability can be considered to be the product of a mutually interactive triad of factors: (1) circadian, (2) sleep, and (3) social/domestic. Circadian factors stem from the indi-vidual's biological clock, which has been shown to be endogenous and self-sustaining under conditions of temporal isolation.[8] Sleep factors are, of course, intimately bound up with the circadian ones, but have a greater significance for the shiftworkers themselves and are thus more likely to appear in shiftworker complaints.[9] Domestic factors (including social and community aspects) are often neglected in terms of empirical research[10] but can be equally important as determinants of shiftwork coping ability, and cer-tainly influence the behavior of the shiftworker in relation to the other two factors.[11] The three factors are discussed in the following sections, with emphasis placed on interactions and interrelationships.

32.2   Circadian Factors

One could argue that circadian factors constitute the essential basic determinant of shiftwork coping ability. Without an endogenous circadian system, sleep could simply be taken "at will," and society would probably be structured in a much less day-oriented fashion. Unfortunately, as noted above, it is quite clear that, like it or not, *Homo sapiens* is a diurnal species, with a circadian clock system which is biologically hard-wired to be active during the day and sleepy at night. Working at night must therefore be regarded as an inherently unnatural act. Thus, much as the deep sea diver is working in an unnatural *physical* environment, the shiftworker may be working in an unnatural *temporal* environment. In both cases, there needs to be an understanding (by both employer and employee) of the basic physiological principles involved, so that adverse health and safety consequences can be avoided. Thus, one might argue that educational initiatives and regulatory protections available to deep sea divers in their domain should also be available to shiftworkers in their domain.[12]

The prime negative influence of the circadian system stems from its inability to adjust instantaneously to the changes in routine that shiftwork schedules require.[13] For the worker adjusting to a run of night duty, a delay in the timing of the circadian system to a phase position about 9 h later than before is required, because night workers usually sleep straight after work (with a bedtime, e.g., at 8 am, rather than 11 pm), saving recreation for the evening before they go to work. The process of circadian phase adjustment is a slow one, with about a week elapsing before complete realignment occurs. This is very similar to the jet lag one might experience in flying from Berlin to San Diego, for example. Using subjects working on a socially isolated oil rig, Barnes and colleagues[14] have shown that the rate of phase adjust-ment of the circadian system to night work (as measured by the urinary melatonin sulfate rhythm) was about 90 min of phase delay per day. Thus, 5 or 6 days were needed before the melatonin onset achieved its desired timing, just prior to the (day) sleep episode. In most work situations, however, this would bring the individual to an off-duty break when a reversion to a diurnal pattern would likely ensue — see following text.

One reason that the circadian realignment of shiftworkers can take longer than that associated with jet lag is the difference between the two situations in zeitgeber (time cue giver) influence. In jet lag, both physical (daylight–darkness) and social (e.g., mealtimes and traffic noise) zeitgebers are *encouraging* the realignment of the circadian system. For the shiftworker, however, the physical zeitgebers are resolutely *opposed* to a nocturnal alignment, as are most of the social zeitgebers stemming from a day-oriented society. Much research has thus focused on enhancing the zeitgebers that may encourage a nocturnal circadian orientation in night workers. The discovery of the strong zeitgeber effects of bright lights led to a series of studies using very bright artificial light to assist in changing the phase of the circadian system. Typically, the bright light exposure regimens required at least 3 h of >3000 Lux exposure from a bank of florescent tubes in a light box. Eastman[15] conducted a careful series of experiments using both student volunteers and real shiftworkers to assess the utility of bright artificial light in helping shiftworkers to cope. As with other investigations, however, (e.g., Czeisler and colleagues[16]), these studies indicated that darkness during the sleep period was almost as important as the light at work. Thus, complete bedroom light proofing was required, and Eastman and colleagues[17] also showed that dark sunglasses or welder's goggles usually needed to be worn during the morning commute home from night work for the required circadian system phase delay to be accomplished.

The case can still be made quite strongly, however, for nighttime workplace lighting to be increased in brightness by whatever amount is financially and operationally possible. Several studies have shown that bright light on the night shift definitely increases alertness *even when the light is of insufficient intensity to induce a strong resetting of the circadian system.* Thus, the improvement in nighttime alertness rendered by the light appears to be mediated by some other alerting mechanism than an enhanced phase resetting of the circadian system. Having said that, it should be noted that several authors (e.g., Boivin and collegues,[18] Martin and Eastman[19]) have shown that even moderate levels of night shift illumination can phase shift the circadian system, albeit less strongly than is achieved by very bright light.

Another way of enhancing circadian adjustment is by taking melatonin pills — a strategy used by many night workers in the United States following the attention given to that hormone in the popular press, and its availability there without a prescription. While there is some laboratory evidence for the effectiveness of melatonin as a chronobiotic,[20] its effects are comparatively weak compared to those of daylight and are likely to be washed out for many shiftworkers. Slightly different is the concept of using melatonin pills to facilitate daytime sleep (without necessarily changing the timing of the circadian pacemaker). However, although there is good laboratory evidence for such facilitation, double-blind studies of melatonin effects in actual shiftworkers have resulted in few definitive improvements in the quality and duration of daytime sleep.[21,22] Concerns also remain regarding chronic use of melatonin pills whose safety and purity cannot always be guaranteed since their production and sale in the United States is unregulated.

In all but the most socially isolated shiftworkers, attention must be paid to behavior during off-duty ("weekend type") breaks. The process of circadian realignment for the night worker can be likened to a salmon leaping up a waterfall; it is difficult to achieve a nocturnal orientation (i.e., reach the top of the waterfall), but easy indeed to fall back down to a diurnal orientation, since that is the natural state for the human organism. That asymmetry becomes vitally important when social and domestic influences during days off lead to daytime activity, particularly when it is outdoors, resulting in daylight exposure. Few parents would forgo attending their child's Saturday morning soccer game simply to preserve their nocturnal orientation. Thus, although a worker may be a permanent night worker as far as the company is concerned, in reality the individual may be alternating between nocturnal and diurnal orientations, simply because of the social and domestic concerns. On the first night after a weekend break, even permanent night workers may have a totally diurnal circadian orientation in their temperature and subjective alertness rhythms.[23]

During the process of circadian realignment, there are three mechanisms by which mood, well-being, and performance efficiency can be adversely affected. First, sleep will be disrupted, and the individual will be in a state of partial sleep deprivation.[24] Second, the new time of wakefulness is likely to tap into the "down phases" of various psychological functions that are normally coincident with sleep in the day-oriented individual.[7,25] Third, the various individual components of the circadian system will be in a state of disarray, with the normal harmony of appropriate phase relationships destroyed.[26] An analogy of the circadian system under these conditions is that it is like a symphony orchestra, with a conductor on the rostrum making sure that the various instruments are brought in at the right time. For the night worker, it is as if a second conductor appears on the rostrum, beating at a different time. The rate at which the different instruments switch to the new conductor varies, and until they all do, there is a cacophony, with all harmony lost. In circadian terms, we speak of this cacophony as "desynchronosis" or "internal dissociation" because the component circadian rhythms no longer have appropriate phase relationships to each other. In addition to poor sleep, the symptoms of desynchronosis include malaise, gastrointestinal dysfunction, and performance decrements.

Individual differences in circadian system characteristics may also have a role in determining shiftwork coping ability. Individuals who are "night owls," or "late phasers," in their circadian system often find shiftwork considerably easier to cope with than do "morning larks," or "early phasers."[27] Phase differences may also explain why late-middle-aged people often find shiftwork difficult. A typical case is that of a 50-yr-old patient who has hitherto been fairly happy with shiftwork but now finds it increasingly difficult to cope with. In some ways, this is paradoxical, given that he has had many decades of learning shiftwork coping strategies and that he probably has a quieter house now that his children have grown up and he can afford better housing. The reason for the problem may be that he has become more of a "morning lark" in circadian phase orientation. Carrier and colleagues[28] have shown that many of the sleep decrements seen in the progression through the middle years of life (even in day workers) can be attributed to age-related changes in morningness–eveningness which can occur through a person's forties and fifties. Also, Campbell[29] has shown that circadian manipulations designed to improve night work tolerance may work much better for young adults than for those in middle age.

Before the discussion of circadian factors is concluded, the question must be addressed whether circadian realignment is actually desirable, given all the caveats regarding the weekend regression to a diurnal orientation mentioned before.[12] In Europe, many companies use "rapidly rotating" systems in which only one or two shifts are worked at a time, before a different one is worked.[30] Thus, for example, on the "continental" rotation, employees work two morning shifts, two evening shifts, and two night shifts, followed by two days off. Most European experts favor such systems because they allow the circadian system to retain its diurnal orientation, thus eliminating problems of desynchronosis. Because only one or two night shifts are worked before time off is given, sleep loss and fatigue are minimized. The drawbacks of rapid rotation are the circadian-related fatigue experienced during the night shifts, which, for some tasks, may render the approach undesirable, and the workers' difficulties in predicting when they will be at work. However, there are undoubtedly many situations in which rapid rotation is worthy of consideration.

32.3   Sleep Factors

Sleep is the major preoccupation of most shiftworkers. In both Europe [31] and the United States[32], surveys have indicated that night workers get about 10 h less sleep per week than their day-working counterparts. Thus, individuals who happen to need 9 h of sleep per 24 h in order to feel well rested very often find shiftwork extremely difficult to cope with. In his survey of field and laboratory shiftwork sleep studies, Akerstedt[33,34] concluded that the shortening in a night worker's day sleep comes primarily from a reduction in stage 2 and rapid eye movement (REM) sleep, with slow-wave sleep relatively unaffected. Not surprisingly, given the prolonged levels of partial sleep deprivation involved, sleep latency can be somewhat reduced in night workers, and some studies have found shorter REM latencies to occur. Essentially, the problem is usually one of sleep maintenance insomnia, rather than sleep onset insomnia. Although there are many social and domestic negatives to *evening* shiftwork, from a sleep point of view such shifts are much preferable to night shifts, and even preferable to daylight shifts, particularly when the daylight shifts have early start times.

A shiftworker's sleep loss is sometimes partially recouped on days off and by the taking of naps but does represent a chronic state of partial sleep deprivation which undoubtedly affects the mood and performance abilities of the worker. There are now several well-controlled studies which document the pathological sleepiness levels exhibited by many shiftworkers, both at work[35] and on the drive home after work.[36] Indeed, one could argue that the latter represents the most dangerous activity that most shiftworkers ever engage in, and one which, in aggregate, represents a major public safety concern involving significant loss of life.[37] Prophylactic naps have shown to be beneficial in reducing sleepiness in night workers before starting a run of night duty (a 2-h nap after lunch is recommended), and some experts favor short naps during the night shift itself, although controls have to be in place for the worker to recover from the grogginess of sleep inertia, before operating dangerous machinery or monitoring equipment.

Many shiftworkers assert that if only they could solve their sleep problem, then everything else would be quite tolerable. However, because of the impact of the circadian system on sleep, disrupted sleep may be as much a *symptom* of shiftwork maladjustment as a *cause* of it. This idea is demonstrated clearly in a study by Walsh and colleagues[38] who brought actual shiftworkers into a sound-attenuated, electrically shielded bedroom for their sleep periods, with the subjects commuting to their work from the laboratory rather than from home. Even in this closely protected environment, there was a highly significant difference in duration between the day sleep of night workers and the night sleep of day workers (306 vs. 401 min). In addition, there were reliable differences between the polysomnographic characteristics of the sleep, with a smaller amount of REM sleep and a greater proportion of slow-wave sleep for the night workers. Thus, even if it were economically feasible, the complete soundproofing and lightproofing of all shiftworkers' bedrooms would not eradicate the problem of sleep for shiftworkers.

Circadian factors are not the only ones having an impact on a shiftworker's sleep, however. Domestic and social factors (see Section 33.4) are also crucial in determining the patient's sleep quality and duration. First, the sleep of the shiftworker is not as protected by society's taboos as that of a day worker; for example, no one would think of phoning a day worker at 2 am, but few would have qualms about phoning a night worker at 2 pm. Similarly, unless the shiftworker is in a well-adjusted household, his (and more especially her) sleep is liable to be truncated by the demands of child care, shopping, and household management. In viewing the sleep of shiftworkers, one must therefore consider both endogenous and exogenous factors that are going to limit sleep time.

Sleep demands may also be as much of an *influence* on the other two factors in the triad as a *product* of their influence. Much domestic disharmony can be attributed to the shiftworker's need for sleep at a time when households are usually rather noisy, and impaired mood is a classic symptom of partial sleep loss.[39] Prescribed circadian rhythm coping strategies may not work because the weary shiftworker may be asleep when he or she would ideally be experiencing bright light and activity.

Finally, in discussing the sleep of shiftworkers, one must address the issue of caffeine to promote alertness, and hypnotics (sleeping pills) to promote sleep. In a study of rotating shiftworkers, Walsh and colleagues[40] found that triazolam 0.5 mg could improve the quality and duration of day sleeps. However, the study was also important in demonstrating that the drug had no significant "phase-resetting" effects. Thus, on the third- and fourth-day sleeps in a run of night duty, for which no medication was given, there were no significant differences between those who had been given triazolam on day sleeps 1 and 2 and those who had been given placebo. Moreover, when drug and placebo groups were compared in terms of nighttime alertness and performance, no reliable differences emerged, even on the days in which medication was given.[41] One must therefore recognize that hypnotics will probably ameliorate only the *sleep* factor of the triad. As a general rule, the use of hypnotics is thus inadvisable for shiftworkers because problems of tolerance and dependence are likely to occur. It is noteworthy that most hypnotics are intended only for "occasional transitory insomnia," and are thus not intended to be taken daily for months or years at a stretch. The recently available short-acting hypnotics such as zaleplon are unlikely to be helpful to the shiftworker who is usually suffering from a problem of remaining asleep, rather than falling asleep. One situation in which hypnotics might be sometimes appropriate is in rapidly rotating shift systems, in which the occasional day sleep may be improved by hypnotics, and no phase resetting is required.

Caffeine is widely used by shiftworkers in order to stay awake at work, and there is empirical evidence from the laboratory that it is effective in improving alertness. However, it is important that shiftworkers recognize that caffeine has a half-life of between 3 and 5 h. Because of the fragility of day sleeps discussed earlier, it is important that night workers do not further worsen their day sleep by having caffeine still in their bloodstream when attempting a day sleep. Thus, night workers should not consume any caffeine after 4 am.[12]

32.4 Domestic Factors

Human beings are essentially social creatures, and one could argue, as Walker[42] and others[43] have done, that the social and domestic factors are at least as important in shiftwork as the biological ones. Certainly, if a shiftworker's domestic and social life is unsatisfactory, then the individual will not be coping satisfactorily, however well adjusted the sleep and circadian rhythm factors may be. In a Connecticut shiftwork study conducted by the author, the intervention included elaborate specification of sleep and wake times and education in circadian and sleep hygiene principles. The importance of social and domestic factors was made clear when one participant pointed out that "Your advice is all well and good, but right now things for me are so bad at home with my wife that I am sleeping out in the car!" Thus, poor domestic adjustment can seriously affect the other two factors of the triad. A common example concerns the childcare and household management tasks that can be expected of a female shiftworker. Unlike her male colleagues, she is often expected by her spouse to continue to run the household and can thus find herself completely unable to comply with the routine that good sleep hygiene and circadian adjustment might require.

Another aspect of domestic disruption concerns the role of the male shiftworker as husband and parent. With regard to the former, there are three major spouse roles that are affected: sexual partner, social companion, and protector-caregiver. All three roles are compromised. Perhaps as a consequence, shiftwork has been shown in a longitudinal follow-up study to increase the risk of divorce by 57%.[44] The evening shift, which has minimal impact on the sleep and circadian factors, can have a crushing impact on the role of the shiftworker as social companion. With regard to the family role of parent, the evening shift is again the most disruptive. Often during the school week the shiftworker may only get to see his or her children when they are asleep in bed. In addition, both spouse and parent roles are heavily disrupted when the shiftworker is required to work on weekends.

In addition to disrupted family roles, the shiftworker often suffers from social isolation from day-working friends and from religious, sporting, and community organizations that work under the expectation that evenings or weekends will be free for meetings and activities. One might advance the view that perhaps a shiftworker who is denied access to community meetings and sports, social, and political associations is as much disadvantaged as a handicapped person who is denied wheelchair access to a museum.

32.5 Solutions

The best approach to the challenge of shiftwork is one that involves both management and the work force.[3] Corporate safety officers and medical officers can be extremely helpful in this regard because they are skilled at bridging the gap between workers and managers, and at developing long-running education and awareness programs. Management should realize that it has not only a *moral* but also a *financial* obligation to be sensitive to issues of shiftwork tolerance in the training of its employees and in the selection of shift schedules. Increasing medical, recruiting, and retraining costs dictate that poor employee morale, higher job turnover, and increased accident, ill-health and absenteeism rates resulting from shiftwork intolerance can become a financial burden to the company or organization.

Employee education programs should emphasize the way in which circadian, sleep, and domestic factors can influence shiftwork coping ability. Workers should be taught good sleep hygiene practice and advised how they can manipulate zeitgebers to their advantage, enhancing those that are acting in their favor and attenuating those acting against them. They should also be taught the benefit of prophylactic naps and caveats about the use of caffeine. In some cases, family counseling may be indicated to discuss solutions to some of the social and domestic problems. The creation of self-help networks can often be of benefit, lessening some of the social and community isolation that many shiftworkers feel. When educational strategies fail, and the shift schedule cannot be changed, the patient may require a change to a day-working job.

The main task with regard to management education is that of first convincing managers that there *is* a problem and that shiftwork concerns cannot simply be swept under the carpet or dismissed as a problem confined to sick or disgruntled employees who are simply not trying hard enough. Second, management must be informed of the wide range of different shift systems that are available, including the rapidly rotating systems so popular in Europe. Third, managers must be taught to recognize the factors (e.g., type of job, nature of work force, average commuting time, male–female ratio, and preponderance of moonlighting) that should influence the selection of the optimal schedule for that work group in that situation. For management, the "carrot" is a happy, healthy, and productive work force; the "stick" is the specter of human error failures, such as that at the Three Mile Island nuclear power plant, and of litigation from a work force that might consider inappropriately selected work schedules to have adversely affected their health or their safety.

A recent tool that may help management in creating a more "shiftworker tolerant" environment is the mathematic model. Several authors (e.g., Folkard and Akerstedt[45]) have developed models incorporating both circadian and sleep loss effects as determinants of "on shift" alertness and performance. Such models are currently in the early stages of development and need considerable refinement. Eventually, though, they might allow for the effects of different shift schedule choices to be evaluated in computer simulations, before they are actually imposed on the hapless shiftworker.

32.6 Conclusions

Although some people cope well with shiftwork, many others have significant problems that can adversely affect their health and well-being. These problems can become a "shiftwork sleep disorder," which may be quite debilitating to the individual. Shiftwork problems can be usefully understood using a multifaceted approach that recognizes the interaction of circadian rhythms, sleep, and social and domestic factors in determining shiftwork coping ability.

References

1. *U.S. Congress*, Office of Technology Assessment, (OTA). *Biological Rhythms: Implications for the Worker (OTA-BA-463)*. 1991. Washington, D.C.: U.S. Government Printing Office.
2. Moore-Ede MC. Jet lag, shift work, and maladaption. *News in Physiological Sciences*. 1986. 1:156–160.
3. Knauth P, Hornberger S. Preventive and compensatory measures for shift workers. *Occupational Medicine (London)*. 2003. 53:109–116.
4. Monk TH. Coping with the stress of shift work. *Work & Stress*. 1988. 2:169–172.
5. Rutenfranz J, Colquhoun WP, Knauth P, et al. Biomedical and psychosocial aspects of shift work: a review. *Scandinavian Journal of Work, Environment and Health*. 1977. 3:165–182.
6. Scott AJ, LaDou J. Shiftwork. Effects on sleep and health with recommendations for medical surveillance and screening. *Occupational Medicine*. 1990. 5:273–299.
7. Monk TH, Carrier J. Shift worker performance. *Occupational and Environmental Medicine*. 2003. 3:209–229.
8. Moore RY. The suprachiasmatic nucleus and the organization of a circadian system. *Trends in Neurosciences*. 1982. 5(11):404–407.
9. Tepas DI, Carvalhais AB. Sleep patterns of shiftworkers. *Occupational Medicine*. 1990.5:199–208.

10. Akerstedt T, Gillberg M. *Night and Shift Work: Biological and Social Aspects.* 1990. Oxford: Pergamon Press.

11. Tepas DI. Shift worker sleep strategies. *Journal of Human Ergology.* 1982. 11(Suppl):325–326.

12. Monk TH. What can the chronobiologist do to help the shift worker? *Journal of Biological Rhythms.* 2000. 15:86–94.

13. Aschoff J, Hoffman K, Pohl H, et al. Re-entrainment of circadian rhythms after phase-shifts of the zeitgeber. *Chronobiologia.* 1975. 2:23–78.

14. Barnes RG, Deacon SJ, Forbes MJ, et al. Adaptation of the 6-sulphatoxymelatonin rhythm in shift-workers on offshore oil installations during a 2-week 12-h night shift. *Neuroscience Letters.* 1998. 241:9–12.

15. Eastman CI. Squashing versus nudging circadian rhythms with artificial bright light: solutions for shift work? *Perspectives in Biology and Medicine.* 1991. 34,2:181–195.

16. Czeisler CA, Johnson MP, Duffy JF, et al. Exposure to bright light and darkness to treat physiologic maladaptation to night work. *New England Journal of Medicine.* 1990. 322(18):1253–1259.

17. Eastman CI, Stewart KT, Mahoney MP, et al. Dark goggles and bright light improve circadian rhythm adaptation to night-shift work. *Sleep.* 1994. 17:535–543.

18. Boivin DB, Duffy JF, Kronauer RE, et al. Dose–response relationships for resetting of human circadian clock by light. *Nature.* 1996. 379:540–542.

19. Martin SK, Eastman CI. Medium-intensity light produces circadian rhythm adaptation to simulated night-shift work. *Sleep.* 1998. 21:154–165.

20. Sack RL, Lewy AJ. Melatonin as a chronobiotic: treatment of circadian desynchrony in night workers and the blind. *Journal of Biological Rhythms* 1997. 12:595–603.

21. Jorgensen KM, Witting MD. Does exogenous melatonin improve day sleep or night alertness in emergency physicians working night shifts? *Annals of Emergency Medicine.* 1998. 31:699–704.

22. James M, Tremea MO, Jones JS, et al. Can melatonin improve adaptation to night shift? *American Journal of Emergency Medicine.* 1998. 16:367–370.

23. Monk TH. Advantages and disadvantages of rapidly rotating shift schedules — a circadian viewpoint. *Human Factors.* 1986. 28:553–557.

24. Weitzman ED, Kripke DF, Goldmacher D, et al. Acute reversal of the sleep-waking cycle in man. *Archives of Neurology.* 1970. 22:483–489.

25. Folkard S, Monk TH. Shiftwork and performance. *Human Factors.* 1979. 21:483–492.

26. Wever RA. *The Circadian System of Man: Results of Experiments under Temporal Isolation.* 1979. New York: Springer-Verlag.

27. Monk TH, Folkard S. Individual differences in shiftwork adjustment. In: Folkard S, Monk TH, eds. *Hours of Work — Temporal Factors in Work Scheduling.* 1985. New York: John Wiley & Sons.

28. Carrier J, Monk TH, Buysse DJ, et al. Sleep and morningness-eveningness in the "middle" years of life (20y–59y). *Journal of Sleep Research.* 1997. 6:230–237.

29. Campbell SS. Effects of timed bright-light exposure on shift-work adaptation in middle-aged subjects. *Sleep.* 1995. 18:408–416.

30. Knauth P, Rutenfranz J, Schulz H, et al. Experimental shift work studies of permanent night, and rapidly rotating, shift systems. II. Behaviour of various characteristics of sleep. *Internal Journal of Occupational and Environmental Health.* 1980. 46:111–125.

31. Knauth P, Landau K, Droge C, et al. Duration of sleep depending on the type of shift work. *International Journal of Occupational and Environmental Health* 1980. 46:167–177.

32. Tasto DL, Colligan MJ. *Health Consequences of Shift Work (Project UR11-4426).* 1978. Menlo Park,CA.: Stanford Research Institute.

33. Akerstedt T. Adjustment of physiological circadian rhythms and the sleep-wake cycle to shiftwork. In: Folkard S, Monk TH, eds. *Hours of Work: Temporal Factors in Work Scheduling.* 1985. New York: John Wiley & Sons.

34. Akerstedt T. Shift work and disturbed sleep/wakefulness. *Occupational Medicine (London).* 2003. 53:89–94.

35. Akerstedt T, Torsvall L, Gillberg M. Sleepiness and shift work: field studies. *Sleep*. 1982. 5(Suppl 2):S95–S106.

36. Richardson GS, Miner JD, Czeisler CA. Impaired driving performance in shiftworkers: the role of the circadian system in a multifactorial model. *Alcohol, Drugs and Driving*. 1990;5(4) and 6(1):265–273.

37. Pack AI, Pack AM, Rodgman E, et al. Characteristics of crashes attributed to the driver having fallen asleep. *Accidental Analysis and Prevention* 1995. 27:769–775.

38. Walsh JK, Tepas DI, Moss PD. The EEG sleep of night and rotating shift workers. In: Johnson LC, Tepas DI, Colquhoun WP et al., eds. *The Twenty-four Hour Workday: Proceedings of a Symposium on Variations in Work–Sleep Schedules*. 1981. Cincinnati, OH: Department of Health and Human Services (NIOSH).

39. Horne J. *Why We Sleep: The Functions of Sleep in Humans and Other Mammals*. 1988. Oxford, England: Oxford University Press.

40. Walsh JK, Muehlbach MJ, Scweitzer PK. Acute administration of triazolam for the daytime sleep of rotating shift workers. *Sleep*. 1984. 7:223–229.

41. Walsh JK, Schweitzer PK, Anch AM, et al. Sleepiness/alertness on a simulated night shift following sleep at home with triazolam. *Sleep*. 1991. 14 (2):140–146.

42. Walker JM. Social problems of shift work. In: Folkard S, Monk TH, eds. *Hours of Work — Temporal Factors in Work Scheduling*. 1985. New York: John Wiley & Sons.

43. Colligan MJ, Rosa RR. Shiftwork effects on social and family life. *Occupational Medicine*. 1990. 5:315–322.

44. White L, Keith B. The effect of shift work on the quality and stability of marital relations. *Journal of Marriage & Family*. 1990. 52(May):453–462.

45. Folkard S, Akerstedt T. Trends in the risk of accidents and injuries and their implications for models of fatigue and performance. *Aviation, Space, and Environmental Medicine*. 2004. 75:A161–A167.

33

Vibrometry

Donald E. Wasserman
D.E. Wasserman, Inc.

David G. Wilder
University of Iowa

33.1   Introduction

There are some 8 million workers[1] in the U.S. exposed to occupational whole-body vibration (WBV) or hand-arm vibration (HAV) with resulting severe medical consequences of WBV or HAV exposures (see text Chapter: Occupational Vibration and Cumulative Trauma Disorders). The ability to measure, quantify, and evaluate the vibration impinging on the human body and relating these results to the disease processes it produces is essential to understanding both dose–response relationships and methods for controlling human vibration exposure. The purpose of this chapter is thus threefold: (1) To provide an introduction to the occupational vibration measurement process; (2) to provide a basic understanding of the occupational WBV and HAV health and safety standards/guides currently in use in the U.S.; and (3) to demonstrate the interrelationships between these measurements and their respective WBV and HAV standards/guides.

33.2   Vibration Basics[2,3]

Vibration is a description of motion. As such, this motion is characterized by its direction *and* a corresponding magnitude; thus, by definition vibration is a *vector quantity*. A total of six vectors are needed to describe vibrating motion measured at any one point; three of these vectors portray "linear motion" and are situated mutually perpendicular to each other; the remaining three vectors portray the rotational motion around each of these linear vectors and are called pitch, yaw, and roll. Currently pitch, yaw, and roll are not measured; only the three perpendicular linear vectors are measured and evaluated in human vibration work. Figure 33.1 shows the mutually perpendicular measure coordinate system used for WBV measurements. Similarly, Figure 33.2 shows the two coordinate systems used to measure HAV. We define the directions of motion as follows: The "Z axis" motion is in the long (head-to-toe) WBV direction, and for HAV measurements the motion is a direction parallel to the hand/arm long bones. Similarly, the "Y axis" motion is in the direction across the shoulders for WBV measurements, and for HAV measurements the motion is across the knuckles of the hand. Finally, the

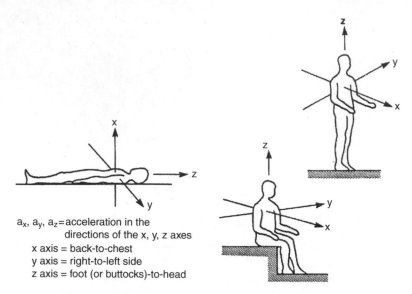

a_x, a_y, a_z = acceleration in the
              directions of the x, y, z axes
x axis = back-to-chest
y axis = right-to-left side
z axis = foot (or buttocks)-to-head

FIGURE 33.1   Whole-body vibration measurements coordinate system. (ANSI S3.18, ISO 2631, ACGIH-TLV (WBV), EU).

"X axis" motion is in the front-to-back direction (through the sternum) for WBV, and for HAV measurements the motion is through the palm of the hand.

Having defined the directions of motion, the vibration magnitude or intensity parameter(s) must be specified. We can choose between three mathematically interrelated quantities: displacement, velocity, or acceleration. Displacement is merely the distance moved away from some reference position. Velocity (or speed) is the time-rate-of-change of displacement. *Acceleration* is the time-rate-of-change of velocity.

Acceleration is usually the magnitude/intensity parameter of choice for several reasons which include ease of measurement and the belief that acceleration is both a hard and soft tissue stressor. Acceleration is expressed in units of meters/sec/sec or in terms of gravitational *g* units, where $1g = 9.81$ m/sec/sec. The "peak" acceleration or maximum values are not usually evaluated, rather an average acceleration parameter called *root-mean-squared* or *rms* acceleration is measured and evaluated and is relatable directly to

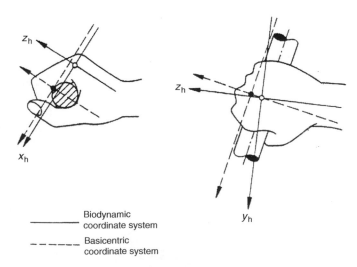

————— Biodynamic
          coordinate system
– – – – – Basicentric
          coordinate system

FIGURE 33.2   Hand-arm vibration measurements coordinate systems. (ANSI S3.34, ACGIH-TLV (HAV), EU, NIOSH #89-106).

the human vibration standards. In the rms process the measured values of acceleration are squared and subsequently averaged to get its mean value. Finally, the square root is determined resulting in an acceleration value proportional to the vibration signal's energy content (see equation 33.1).

$$a_{rms} = \sqrt{\frac{1}{T} \int_0^T a^2(t) \, dt} \tag{33.1}$$

where a = acceleration (in g's or m/sec^2); t = time (in seconds).

Vibration motion can repeat itself. This is called periodic motion. Motion need not repeat itself; it can be random or nonperiodic. The most basic form of motion is sinusoidal, or a pure tone if it were audible, which usually repeats itself; the rms value of a sinusoid is about 70% of its peak or maximum value. If a sinusoid completes its cycle in one second, before it begins to repeat itself, its vibration *frequency* is 1 Hertz (Hz); 10 Hertz simply means that 10 complete cycles have occurred in one second; two kilohertz (2 kHz) means that two thousand complete cycles have occurred in one second, etc. For WBV we are interested in a vibration frequency range (or bandwidth) of 1 Hz to 80 Hz; for HAV the bandwidth of interest is from about 6 Hz to 1400 Hz or 1.4 kHz; sometimes even higher extending up to 5000 Hz or 5 kHz. Since most motion appearing in the industrial environment is a compound mixture of many vibration frequencies at various acceleration levels, it is necessary to mathematically sort these frequencies into their individual sinusoidal frequencies and their corresponding magnitudes (see equation 2). This computer process is called *Fourier Spectrum Analysis* and is required by most human vibration standards before they can be used.

$$\begin{aligned} F(t) = a_0 &+ a_1 \sin wt \, a_2 \sin 2 \, wt + a_3 \sin 3 \, wt + \cdots + a_n \sin(n) \, wt \\ &+ b_1 \cos wt + b_2 \cos 2 \, wt + b_3 \cos 3 \, wt + \cdots b_n \sin(n) wt \end{aligned} \tag{33.2}$$

where a and b = amplitude values of each sinusoid at specific frequencies composing the spectrum; a_0 = dc term or zero Hertz value.

Specialized computers called Fast Fourier Transform (FFT) analyzers or real time analyzers (RTA) are used to transform the vibration mix into its discrete frequencies.[4] Each such frequency is graphically displayed as a series of vertical lines or spectra; the position of each line identifies its vibration frequency in Hz and thus its place in the spectrum; the height of each line is a measure of its individual vibration acceleration intensity in g units or meters/sec/sec. The entire spectrum is the sum of all these lines. Vibrating tool spectra are quite unique, depending on the tool. Vehicle spectra from trucks, buses, trains, etc., are also unique.

The final concept to be discussed is called *resonance* (or natural frequency) which is an unwelcome situation where the conditions for transferring vibration from its source (i.e., tools, vehicles, etc.) to the human receiver are optimal. Thus, a very small magnitude of vibration impinging on a human or a structure (such as a bridge) causes an uncontrollably amplified response by the human or structure. This is the reason why bridges collapse if soldiers march in cadence across them. Unfortunately, we humans have resonances too, namely, in WBV 4 to 8 Hz for Z axis vertical vibration, 1 to 2 Hz in both the X and Y axes. Spinal resonance is 4.5 to 5.5 Hz.[5] The hand-arm system seems to resonate in the 150 to 250 Hz range.[3] In general the larger the mass or weight of a structure, the lower the resonant frequency. Equation II is the resonance equation for a simple single-degree of freedom system consisting of motion in one direction consisting of a mass, spring, and damping element.

$$W = \sqrt{\frac{k}{m}} \tag{33.3}$$

where W = $2\pi f$; k = spring constant; m = mass.

Resonance thus represents the Achilles heel of human response to vibration. As is the auditory system to sound, human response to vibration is therefore frequency dependent and nonlinear because at resonance the impinging vibration finds its easiest pathway to the person; at other vibration frequencies, the vibration pathway is not as easy and thus it requires more acceleration at a nonresonant frequency to produce the same level of human response.

33.3   Vibration Measurements Basics[2-4]

The major reason for performing occupational vibration measurements is to evaluate the vibration impinging on persons. Evaluations are performed using the various WBV and HAV standards/guides. These standards/guides are the critical link between the various health and safety effects of WBV or HAV and the vibration hazard levels experienced by workers. It is important to note that there are many esoteric types of vibration measurements (i.e., mechanical: impedance, mobility, stiffness, compliance, etc.) and other methods of data analysis, such as modal analysis, but intentionally in this chapter we briefly describe only performing acceleration measurements as required by the applicable health and safety standards/guides in order to use them. Finally, note that since displacement, velocity, and acceleration are all mathematically linked, then from a measurement of acceleration, the velocity function can be derived by electronic integration; repeating the integration next on the velocity function yields the displacement function. Thus, if desired, an acceleration measurement can yield additional data.

Figure 33.3 shows a basic vibration acceleration measurement setup. Since we must *simultaneously* but separately measure in all three X, Y, Z axes acceleration data from a vibrating tool, for example, or from a driver's truck seat, three separate data channels are needed. Three perpendicularly mounted lightweight accelerometers are used to measure each axis acceleration, followed by three appropriate preamplifiers to amplify and electronically condition the tiny millivolt signals coming from each accelerometer. The outputs of each of these three X, Y, Z preamplifiers are then individually recorded and stored on a multi-track tape system known as a *digital audio tape* (DAT) for later Fourier spectrum analysis. It is also desirable to have: (1) a microphone/voice track on the DAT to note the chronology of events being recorded, and (2) an oscilloscope or similar device monitoring the X, Y, Z acceleration axes for possible signal overload conditions leading to distortion of the recorded signal(s) and resulting in erroneous data processing results. After the three channel or "triaxial" data have been measured, stored, and recorded, then a

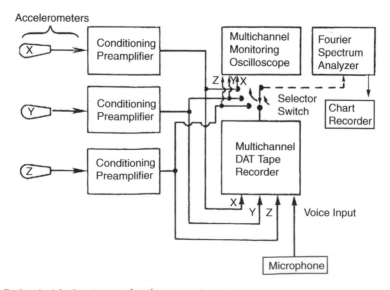

FIGURE 33.3   Basic triaxial vibration acceleration measuring system.

Fourier spectrum analysis can be performed separately on each data channel. Each spectrum is next separately evaluated using the appropriate standard(s). A key element in these measurements is the triaxial accelerometer.

Accelerometers are devices which convert mechanical motion into a corresponding electrical signal. Two distinctly different devices are used for occupational vibration measurements. For WBV measurements piezoresistive accelerometers are used; for HAV measurements piezoelectric or crystal accelerometers are used.

The former device works on the principle of an electronic four arm electronic balanced bridge; P or N semiconductors form each of the bridge arms. All of these arms are bonded to one end of a tiny metal beam. The other end of this beam is bonded to a tiny metal mass or weight to which the force of vibration is applied. With no vibration present, the bridge is "balanced," yielding zero output voltage. When vibration is applied to this tiny mass/beam combination, we obtain the acceleration of this mass against the beam because of the bending motion of the beam compressing some of the beam arms, thereby unbalancing the bridge, resulting in an electrical voltage proportional to acceleration; in effect we are calculating Newton's second law, $F = ma$, with a vibration force and a known mass bonded to the metal beam. The acceleration signal is very small (millivolts) and needs to be amplified using a so-called difference or differential amplifier which measures and amplifies the voltage or potential difference across the arms of the Wheatstone bridge. The amplifier's output voltage is next recorded by the DAT.

HAV measurements require a different type of accelerometer called a crystal, which is commonly found in nature. The crystal has a phenomenon called the piezoelectric effect; if a moving force is applied to the crystal, the crystal responds by generating a small electrical voltage across its face. The more intense the force, the larger the voltage generated. A crystal is a force-measuring device found in nature, not an accelerometer *per se*. If, however, a small weight or mass is bonded to the motion-sensitive surface of the crystal and the force of vibration is applied to this tiny mass, we have once again created an accelerometer, since, as before, it is the vibration force accelerating this mass against the crystal face which results in a corresponding voltage and charge proportional to acceleration. The acceleration signal is very small and needs to be amplified by a special charge amplifier whose output can then be recorded on a DAT recorder.

In either WBV or HAV measurements the accelerometers must be of very light weight (less than 15 grams) and small; if not, measurement errors called mass loading result, the rule is that the total weight of the triaxial accelerometers, mounting fixture, cables, etc., must collectively be less than 10% of the weight of the object whose vibration is to be measured — a tool handle, for example. In all cases great care must be taken to avoid cable entanglement and/or breakage; the shortest possible cable length and integrity must be maintained especially from accelerometers to preamplifiers to ensure high quality and low electrical noise signals. In the case of HAV measurements, some crystal accelerometers can be purchased with built-in charge amplifiers to avoid some of these problems.

WBV measurements are usually made using an instrumented hard rubber disc about the size of a pie plate (Figure 33.4). The disc is placed between the top of the driver's seat cushion and the buttocks. The center of the disc is hollow and contains three tiny accelerometers, mounted mutually perpendicular to each other to a small metal cube. The three piezoresistive accelerometer cables lead from the disc to preamplifiers and on to a DAT.

HAV measurements use three small lightweight crystal accelerometers mounted to a small metal cube, which in turn is usually welded to an inexpensive automotive hose clamp as shown in Figure 33.5. This hose clamp/accelerometer assembly is next clamped around the vibrating tool handle with the accelerometers placed very close to where the operator grasps the tool. Once again great care must be taken in arranging the three accelerometer cables such that the tool operator is free to perform the job safely during the vibration measurements.

To summarize, piezoresistive accelerometers are best suited to performing WBV measurements which are inherently very low frequency, low acceleration level measurements. Piezoelectric or crystal accelerometers are best suited to performing HAV measurements which require a wide bandwidth from a low of

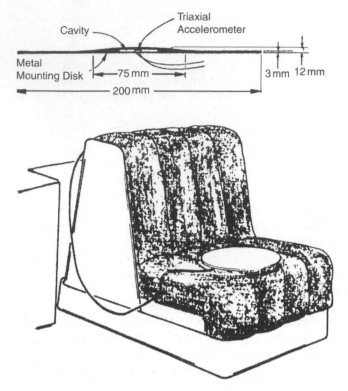

FIGURE 33.4 Whole-body vibration vehicle measurements using an instrumented seat disc. (Adapted from Wasserman, D. 1987. *Human Aspects of Occupational Vibration*, Elsevier Publishers, Amsterdam, The Netherlands. With permission.)

about 6 Hz to as high as 5,000 Hz; tool acceleration levels can be very high (several hundred *g*'s or more) and these devices must be able to measure these high *g* levels; these devices must be rugged too. In all cases care with the accelerometer cabling must be taken. It is certainly not advisable to drop any of these devices on the ground or else they can be severely damaged and/or lose their calibration. Generally the manufacturer will supply a calibration sheet with a newly purchased accelerometer. It is advisable to

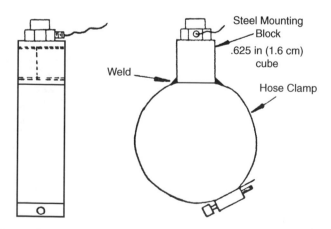

FIGURE 33.5 Hand-arm vibration tool measurements using an instrumentated automotive hose clamp. (Adapted from Wasserman, D. 1987. *Human Aspects of Occupational Vibration*, Elsevier Publishers, Amsterdam, The Netherlands. With permission.)

use a portable calibrator as an added calibration safety measure just in case the accelerometer has been unknowingly damaged.

Obtaining vibration measurements requires careful planning and first performing a walk-through tour of the worksite to be measured, or the course a vehicle takes if its vibration is to be measured. Many times, first using a video camcorder is a good way to record the details of how workers function on the job. Using the camcorder in real time while vibration measurements are obtained is also very useful for recalling the chronology of events of the test day. Finally, the minimum test time that vibration data are gathered and recorded is usually specified by the standard(s) which will be used. For example, most HAV standards require that a minimum time of one minute of continuous triaxial vibration acceleration data be collected and tape recorded per tool tested. The differences in WBV work situations and the so-called duty cycle to a large extent determine the minimum vibration measurement time. For example, the length of a complete work cycle for a delivery truck, or the duty cycle of a large vibrating metal stamping machine in a plant are quite different and should be considered individually.

Finally, a word about handheld portable human vibration meters. We have briefly described measurement methods which will yield maximum usable information for the time and expense spent in gathering, recording, and analyzing vibration data and then applying it to the human vibration standards (to be discussed next). These methods provide: (1) a permanent tape recording of the vibration data; (2) a computer spectrum analysis which provides a graphical picture of the vibration frequencies which comprise the spectrum; (3) the interaction and comparison of these spectra with these standards; and (4) numerical results indicating the total rms acceleration of the spectra. However, if only a single number total rms acceleration value for each axis is required, then there are handheld instruments available from two commercial manufacturers at this writing. The problem is that some of these instruments measure only one acceleration axis and the testing is stopped. The one accelerometer is reoriented in another axis and the testing is resumed. This is repeated until all three axes are recorded. This is *not* desirable since vibration virtually always moves *simultaneously* in all three axes; thus data can be lost as the one accelerometer is reoriented over and over again. One of the available commercial instruments has in a single handheld meter the desirable three accelerometers for simultaneous measurements of either WBV or HAV and also has triple output jacks for DAT recording and later spectrum analysis of the data. Thus, the reader should be very careful in the selection of a handheld vibration meter. Further, be aware that with the advent of miniaturized, high-density/high-speed, surface-mounted electronics technology, many of the above-mentioned functions (i.e., triaxial accelerometer and signal conditioning, data collection, analog-to-digital conversion, data storage, initial unweighted spectrum analysis/display, radio frequency remote control of functions) can all be performed onsite using rugged, battery operated/stackable, miniature solid-state modules.

33.4 Occupational Vibration Standards/Guides*

There are four whole-body vibration standards/guides and four hand-arm vibration standards/guides now in use in the U.S.:

1. International Standards Organization (Geneva, Switzerland), ISO 2631: *Guide to the Evaluation of Human Exposure to Whole-Body Vibration*, 1972–85
2. American National Standards Institute (New York, NY), ANSI S3.18: *Guide to the Evaluation of Human Exposure to Whole-Body Vibration*, 1979

---

*Because of the differences and complexity of each occupational vibration standard/guide, the reader is encouraged to obtain, read, and understand the standard(s) which are to be used *before* collecting vibration data. Herein we can only discuss some of the major elements contained in these standards.

3. American Conference of Government Industrial Hygienists (Cincinnati, Ohio), *ACGIH-Threshold Limit Values for: Whole-Body Vibration*, 1995–96
4. European Union (EU, Luxembourg, Belgium), #89/392, 91/368EEC: (Whole-Body) *Vibration Standard*, 1989–91
5. American National Standards Institute (New York, NY), ANSI S3.34: *Guide for the Measurement and Evaluation of Human Exposure to Vibration Transmitted to the Hand*, 1986
6. American Conference of Government Industrial Hygienists (Cincinnati, Ohio), *ACGIH-Threshold Limit Values for: Hand-Arm Vibration*, 1984
7. National Institute for Occupational Safety and Health (Cincinnati, Ohio), NIOSH Document #89–106: *Criteria for a Recommended Standard for Hand-Arm Vibration*, 1989
8. European Union (EU, Luxembourg, Belgium), #89/392, 91/368EEC: (Hand-Arm) *Vibration Standard*, 1989–91

33.4.1   Whole-Body Vibration Standards/Guides Used in the U.S.

ISO 2631 is the oldest standard, initially introduced in 1972. There have been several revisions over the years, but the basic evaluation criteria remain the same. In 1979, ANSI S3.18 was introduced; this document is virtually identical to the revised 1978 version of ISO 2631. In 1989–91 the EU essentially agreed with ISO 2631 and adopted a (weighted) vector sum triaxial acceleration level of 0.5 meters/sec/sec as an action level for an 8 hr/day workplace WBV exposure level. In 1995–96, the ACGIH–TLV for WBV was introduced; it too uses the basic ISO 2631 curves, but the focus is mainly on occupation health and safety criteria and calculations, while ignoring the so-called comfort criteria. Since all of these standards use the same shape weighting curves, we begin there (refer to Figure 33.1, Figure 33.6, and Figure 33.7. All WBV acceleration measurements used in all these standards use the biodynamic coordinate system defined in Figure 33.1. Figure 33.6 is used to evaluate the Z axis (vertical) rms acceleration data. Figure 33.7 is used to separately evaluate: (1) the X axis rms acceleration data, and then (2) the Y axis rms acceleration data. In order to use these WBV standards, each axis of vibration acceleration data must be converted from the time domain in which it was collected to the frequency domain, which says a Fourier spectrum analysis must be performed separately for the X,Y,Z axes. Each spectrum is then formatted into 1/3 octave bands before it can be applied to the WBV standards. Once the foregoing has taken place the data are then compared to (i.e., overlayed) the "weighted family of curves" shown in Figures 33.6 and Figure 33.7. The abscissa in each of these figures is 1/3 octave band "vibration frequency" from 1 to 80 Hz. The ordinate in each of these figures is "vibration intensity" in rms acceleration in both meters/sec/sec or g's, where 1 $g = 9.81$ m/sec/sec. Within each graph in Figure 33.6 and 33.7 are families of "weighted" parallel-time-dependent daily exposure curves which are called FDP or fatigue decreased proficiency curves. There are three levels of comparing these spectra to the ISO and ANSI standards, the acceleration data can be separately compared to: (1) the FDP curves as shown; (2) if we divide each of the FDP acceleration values by 10 dB (3.15) we generate another family of similar curves called RC or reduced comfort; (3) if we double each value of the original FDP curves, we generate another family of similar curves called EL or exposure limits. These standards tell us that the RC curves should be used for WBV *comfort* criteria, such as in a vehicle ride situation; the FDP curves are operator fatigue level curves where *safety* may well be the issue; and finally the EL curves are concerned with WBV *health effects*. The U-shape of the family of the Z axis curves in Figure 33.6 emphasizes that resonance occurs in the 4 to 8 Hz band as shown by the trough of the curves for each of the daily exposure times. Higher daily rms acceleration levels are allowed at frequencies lower than 4 Hz and greater than 8 Hz since at resonance smaller input acceleration levels produce larger responses than would occur at other frequencies. Similarly, the format for Figure 33.7 shows elbow-shaped curves where the resonant frequency range in either the X or Y axes occurs at 1 to 2 Hz; higher acceleration levels are allowed for frequencies greater than 2 Hz. In actual use, vibration spectra are overlayed on, say, the FDP curves separately for the Z axis, Y axis, and X axis. If, for example, all of the Z spectra fall below the Figure 33.6 FDP curves, then the standard has not been exceeded for that axis; if one or more spectral

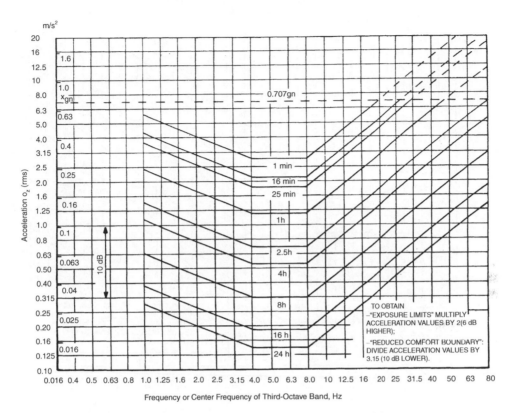

FIGURE 33.6 (FDP) Whole-body vibration curves for Z axis rms acceleration evaluation. (ANSI S3.18, ISO 2631, EU.)

peaks touches and/or exceeds an FDP weighted curve, then the standard has been exceeded for that daily exposure time. The most severe axis is defined by the highest spectral peak(s) which intersect the FDP curves. As a matter of practice, FDP curves are mostly used for *both* health and safety, and the EL curves are *not* used because researchers believe these curves are not protective enough.[2,6] Thus, the ACGIH-TLV for WBV uses only the FDP curves for health and safety, and they totally eliminate both the EL and RC curves. Further, the ACGIHTLV for WBV then recommends using a weighted vector-sum calculation for all three axes to obtain a single number which is then compared to the 0.5 m/sec./sec. action level established by the EU. In all of the cited WBV standards if in any of the three axes, the vibration crest factor (defined as the peak acceleration divided by the rms acceleration in the same direction) is less than or equal to six, the standard can be used; values greater than six cause the standard to *underestimate* the true severity of the vibration hazard. This is particularly troublesome when a vehicle, for example, goes off road and traverses numerous very steep bumps at fast speeds.

There are other methods for evaluating WBV exposure. For example, there are those who believe that the WBV severity is best described by equations raised to the fourth power of acceleration;[7] actual data support the notion that mostly subjective discomfort is best described by this fourth power concept since there is little hard epidemiological evidence at this writing to show that this concept applies to worker health. Finally, the reader should be aware that there are various proposals to revise ISO 2631, which may occur in the future.

33.4.2 Hand-Arm Vibration Standards/Guides Used in the U.S.

Figure 33.8 shows the HAV weighting curves used in ANSI S3.33 where each of the three X,Y,Z axes are evaluated using the same graph by overlaying each spectrum separately over Figure 33.4; as before, this

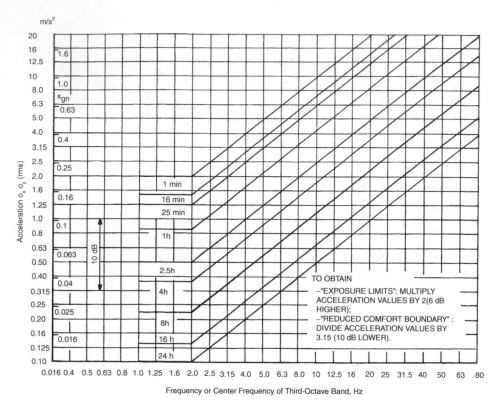

FIGURE 33.7 (FDP) Whole-body vibration curves separately used for X, Y axes rms acceleration evaluations. (ANSI S3.18, ISO 2631, EU)

standard is exceeded if one or more spectral peaks in any of the axes touches or exceeds one or more of the exposure time dependent curves; the ACGIH-TLV for HAV uses the same "shape" weighted curve but requires that each axis yield a numerical weighted sum, each of which is next compared to the acceptable values of HAV daily exposure given in Table 33.1. The EU standard also requires that each of these numerical weighted values or their weighted sum be compared to the 2.5 meters/sec./sec. "action level." Notice that the format of Figure 33.8 is similar to the formats of Figure 33.6 and Figure 33.7, where the abscissa is vibration frequency, in 1/3 octave bands, from 5.6 to 1250 Hz and the ordinate is vibration intensity in acceleration. All standards use the HAV measurement coordinate system previously shown in Figure 33.2 with the "basicentric system" the method of choice. Except for NIOSH #89-106, all of the above standards use this same "elbow shaped" weighting given in Figure 33.8.

The NIOSH standard is an interim standard without stating any acceptable acceleration level limit at any frequency; this standard asks for each axis that: (1) weighted HAV acceleration values from 5.6 to 1250 Hz be calculated, (2) unweighted acceleration values from 5.6 to 5000 Hz be calculated, and (3) the weighted and unweighted be compared in view of the severity of the prevalence of the hand-arm vibration syndrome (HAVS) determined by using the tool(s) from whence these acceleration measurements were made. NIOSH has chosen to issue this interim standard because there is an anomaly in the other HAV standards, namely that the HAV weighting network shown in Figure 34.4 was originally developed using older vibrating tool types commonly found in the workplace. Over the last few years, some very high speed vibrating hand tools have been introduced, some of which have spectral peaks extending to 5000 Hz and above. Current standards end at 1250 Hz, and hence in a few instances the current standards would rule these very high-speed tools as acceptable when that may not be the case. NIOSH has chosen to keep their interim standard, until this anomaly is resolved.

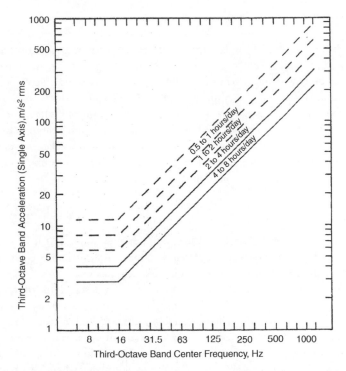

FIGURE 33.8 Hand-arm vibration curves for the separate evaluation of X,Y,Z axes rms accelerations (see text). (ANSI S3.34.)

The architects of the other HAV standards are carefully making adjustments to their standards with regard to the special case of these very high-speed tools as the vibration data and corresponding HAVS health data become available. We recommend the use of ANSI S3.34, ACGIH-TLV for HAV and the EU criteria since all provide good overall guidance, and in the special case where very high-speed tools are to be tested *caution* is advised when evaluating the triaxial vibration test data.

TABLE 33.1 ACGIH Threshold Limit Values for Exposure of the Hand to Vibration in X, Y, Z Directions

	Values of the Dominant,[b] Frequency-Weighted, rms, Component Acceleration Which Shall Not Be Exceeded a_K, $(a_{K_{eq}})$	
Total Daily Exposure Duration[a]	m/s^2	g
4 hours and less than 8	4	0.40
2 hours and less than 4	6	0.61
1 hour and less than 2	8	0.81
Less than 1 hour	12	1.22

Note: 1 $g = 9.81$ m/sec^2.

[a]The total time vibration enters the hand per day, whether continuously or intermittently.

[b]Usually one axis of vibration is dominant over the remaining two axis.

If one or more vibration axis exceeds the Total Daily Exposure then the TLV has been exceeded.

Courtesy of ACGIH.

33.5 Summary

In this chapter the basic concepts of displacement, velocity, acceleration, resonance, coordinate systems for measurements, and spectrum analysis are presented and integrated for an understanding of their application to whole-body and hand-arm occupational vibration. Generic acceleration measurement systems and methods are discussed, which included piezoelectric and piezoresistive accelerometers, conditioning preamplifiers, data recording systems, and Fourier spectrum computers. The chapter concludes with a discussion of the various occupational whole-body and hand-arm vibration standards/guides currently used in the U.S. and their application to the evaluation of triaxial acceleration data from the workplace; because of the complexity of each standard, users are encouraged to obtain copies of standards which are to be used before obtaining vibration data.

Defining Terms

Acceleration: The time rate of change of velocity of a moving object.

Accelerometer: A device designed to convert mechanical motion into a corresponding electrical analog voltage, charge, or current proportional to acceleration.

Conditioning Preamplifier: An electronic solid state amplifier designed to faithfully amplify, both in amplitude and frequency bandwidth, the minute electrical signals emanating from an accelerometer. Some preamplifiers are called "charge amplifiers" and convert the voltage generated across the face of a crystal (piezoelectric) accelerometer into corresponding charge, thereby allowing long cables to be used for measurements without loss of signal. Other preamplifiers are called "differential amplifiers," which act as an amplifying voltmeter when used with "piezoresistive" type accelerometers.

DAT or **Digital Audio Tape:** A new type of instrumentation tape system with large dynamic input range and wide frequency bandwidth, whereby an analog input signal is converted and stored on a cassette tape in digital format. The original signal so stored can be retrieved either in digital format or reconverted again into its original analog version.

Displacement: Movement traversed away from a reference position.

Fourier Spectrum Analysis: The analysis of vibration data by mathematically converting time domain information into its corresponding frequency domain; the underlying assumptions are that the data are linear and that time domain information can be dissected and represented as a mathematical series of elemental sines and cosines. Computers which perform this function are called Fast Fourier Transform (FFT) analyzers or Real Time Analyzers (RTA).

Resonance: The tendency of an object to (1) move in concert with an external vibrating source and (2) to internally amplify the impinging vibration from that source; resonance is the optimum energy transfer condition between the source and the receiver.

Vector Coordinate System: A mutually perpendicular set of vectors, originating at the same motion point, which define the vector motion of that point. Typically, there are three linear and three rotational vectors which comprise motion at a point.

Vector: A mathematical quantity defined by both its magnitude and direction.

Velocity: The time rate of change of displacement of a moving object. Also called speed.

Vibration: At any one point, vibration is motion defined by six vectors, three mutually perpendicular linear vectors and three rotational vectors moving around these linear vectors (pitch, yaw, roll).

References

1. Wasserman, D., Badger, D., Doyle, T., and Margolies, L. 1974. Industrial vibration — An overview. *J. Am. Soc. Safety Engrs.* 19(6):38–43.

2. Wasserman, D. 1987. *Human Aspects of Occupational Vibration*, Elsevier Publishers, Amsterdam, The Netherlands.

3. Pelmear, P. and Wasserman, D. 1998. *Hand-Arm Vibration: A Comprehensive Guide for Occupational Health Professionals*, Second edition, OEM Press, Beverly Farms, MA.

4. Thalheimer, E. 1996. *Practical Approach To Measurement and Evaluation of Exposure to Whole-Body Vibration in the Workplace.* Conference Proceedings: Seminars in Perinatology, International Conference on Pregnant Women in the Workplace Sound and Vibration Exposure, Univ. of Florida, Gainesville, FL.

5. Wilder, D., Wasserman, D., Pope, M., Pelmear, P., and Taylor, W. 1994. Chapter 4: Vibration, in *Physical and Biological Hazards of the Workplace*, Eds. P. Wald and G. Stave, p. 64–83. Van Nostrand Reinhold, New York, NY.

6. Bovenzi, M. and Betta, A. 1994. Low-back disorders in agricultural tractor drivers exposed to whole-body vibration and postural stress. *Appl. Ergonomics* 25(4): 231–241.

7. Griffin, M. 1990. *Handbook of Human Vibration.* Academic Press, London, U.K.

For Further Information

Some of the cited references are comprehensive and excellent sources of information. Reference 2 is principally a WBV and HAV measurements, evaluation, and control textbook. Reference 3 is principally a medical textbook on HAV with some measurements and control information; reference 5 is a basic and comprehensive book chapter for both WBV and HAV; reference 7 is a very complete tome on WBV and HAV, but definitely not for the beginner.

The WBV or HAV standards cited in this chapter can be obtained from the following: TLVs from: American Conference of Government Industrial Hygienists, 1330 Kemper Meadow Drive, Cincinnati, OH 45240 (Telephone: 513-742-2020).

ISO and ANSI Standards from: Standards Secretariat, Acoustical Society of America, 120 Wall St., 32nd. Floor, New York, NY 10005.

(Telephone 212-248-0373).

NIOSH HAV Standard #89-106 write to: NIOSH Publications Dept. Taft Labs. 4676 Columbia Parkway, Cincinnati, OH 45226.

European Union Standards write to: Commission des Communautes Europeennes, Direction generale emploi, relations industrielles et affaires sociates: Batiment Jean Monnet-L-2920 Luxembourg, Belgium.

II

Assessment Tools

34

Overview of Ergonomic Assessment

Chris Hamrick
Ohio Bureau of Workers'
  Compensation

34.1 Introduction

The statistician George P. E. Box has been quoted as saying, "All models are wrong; some models are useful." Such is the case with the models that ergonomics practitioners use in the field to perform ergonomic assessments. No single assessment tool is perfect. However, some assessment tools can be of value to the ergonomist in the field.

The purpose of this chapter is to provide an overview of some of the practical assessment tools that are readily available to ergonomics practitioners. The tools that will be discussed can be used in the field to assess the relative risk of musculoskeletal disorders (MSDs) resulting from the job or task being evaluated. Each of the models presented here have drawbacks, but if used properly each can play a key role as part of a comprehensive ergonomics process.

Organizations have great incentive to identify workplace factors contributing to MSDs; work-related MSDs are prevalent and costly. In the United States in 2001, there were 216,400 "repeated trauma" cases reported by private industry, which translates to a rate of 23.8 cases per 10,000 worker hours of exposure (United States Department of Labor, 2003). Data from the Ohio Bureau of Workers' Compensation (Hamrick, 2000) showed that in Ohio in 1996, MSDs accounted for 15.5% of all claims in the workplace, and 48.5% of all workers' compensation dollars paid went toward MSDs. Those data also indicated that approximately 40% of these claim costs resulted from back pain claims, and 9% resulted from upper extremity claims.

The ideal field assessment tool would possess the following attributes:

- *Predictive* — the tool would be able to provide a valid predictive measure of the risk of musculo-skeletal injury that would occur to a population performing the task being assessed

- *Robust* — can be used in any work situation
- *Inexpensive* — the tool would be available at minimal monetary cost
- *Noninvasive* — the assessment would neither affect the way the worker performs his or her job, nor would it affect the process workflow or quality of work
- *Quick* — the assessment and analysis could be performed quickly
- *Easy to use* — the tool could be used with minimal training

Of course, given technological limitations, no current assessment tool is perfect in all of those attributes. Consequently, assessment tool developers must often make compromises. For example, compromises are often made between robustness and ease of use. Generally speaking, the more robust a model, the more complex it becomes and, hence, the more difficult the model is to use. Conversely, the easier the model is to use, the more limitations that are placed on how the practitioner can use and interpret the results.

34.2   Use of Assessment Tools by Practitioners

Practitioners generally use assessment tools for any of four main purposes. Probably the most valuable information that can be gained from an assessment is a measure of risk of injury. However, given the complexity of determining the exact likelihood that an MSD that could result from a task, very few assessment tools can provide an assessment of absolute risk. The tool that perhaps comes closest to provide an assessment of risk is the lumbar motion monitor (LMM) system, as described by Marras et al. (1993) and later validated in the field by Marras et al. (2000). The device and the accompanying software output a probability that the task being analyzed is in a high-risk group (low back pain incidence rate >12) given trunk kinematic data and physical workplace measures.

Another way that assessment tools can be incorporated into an ergonomics process is by using them to prioritize job redesign efforts. The reality of business is that we live in a world with finite resources, and the two most valuable resources are time and money. So, ergonomists and those involved in making workplace changes must direct resources to where they can have the biggest impact. Results from analysis tools can then be used to determine where intervention efforts, which involve both time and money, can be focused. Comparisons of results from different redesign candidates can be coupled with other information, such as cost and potential impact of interventions, to allocate resources.

Assessment tools can also help when developing the appropriate intervention strategy for a task. If a tool has sufficient resolution, then the most hazardous components of a task can be determined and job redesign can be used to lessen worker exposure to that particular component. For example, an assessment tool may identify trunk flexion as a component that can lead to low back pain. An appropriate intervention strategy could then focus on eliminating the need for the worker to bend the trunk, thus reducing the risk of injury.

Finally, assessment tools can be used to evaluate the effectiveness of intervention measures. Quantitative measures can be taken before an ergonomic intervention is implemented, and then the same quantitative measure can be taken after the intervention is put into place. The results of the two assessments can then be compared to ensure that the intervention had the desired effect of reducing worker exposure to problem risk factors, and to also ensure that no additional risk factors have been introduced. Additionally, the new results can be used as part of a continuous improvement process so that future modifications can be made, if appropriate.

34.3   Health and Medical Indicators

An initial step that can be taken to determine where MSDS problems are occurring in a facility is to review existing records. Monitoring existing records is referred to as "passive surveillance" (Tanaka, 1996). One set of records that is commonly used in the United States is the OSHA 300 log, which the Occupational Safety and Health Administration (OSHA) requires most employers to keep. Information on these logs includes the type of injury or illness, the number of days away from work, and the number of days of job transfer or restriction as a result of the injury/illness.

When using this information, it is imperative to normalize any loss data to worker exposure hours, so that the injuries/illnesses can be expressed in terms of an incidence rate. OSHA recommends the following formula for calculating incidence rates (United States Department of Labor, 2004):

$$\text{Total recordable case rate} = (\text{Total number of new injuries/illnesses} * 200{,}000)/$$
$$\text{Number of hours worked by all exposed employees.}$$

Expressing injuries and illnesses in terms of a rate allows comparisons to be made between exposure groups (e.g., job titles, departments, and facilities) and for different time periods, even if those time periods are of different duration. Additionally, the number of days away from work and the number of job transfer or job restriction days can be similarly normalized to get an indication of severity. When looking at these results, we must realize that transitional work programs can have an effect on these rates.

Cost of injuries can also provide a measure of impact of MSDs. Workers' compensation records can be used to determine medical and indemnity costs. It should be noted that most workers' compensation systems were not created for the purpose of surveillance; rather, they were created to track payments to those injured at work. So, information obtained from this source is often limited. Further investigation may be required to determine the nature and cause of the injury. Additionally, workers' compensation laws vary considerably by state, so comparing results between workers' compensation systems is not advisable. Other passive surveillance sources can include employee medical visits and employee turnover rates.

Passive surveillance data sources are generally inexpensive to obtain — they are usually already being kept so no additional cost is involved. The information obtained is also valuable at pointing out real problems and their costs. Hence, any ergonomics effort should, at a minimum, incorporate passive surveillance. However, a major drawback to relying solely on passive surveillance is that although they may provide an indication of future injuries, the measurements are reactive; they occur after problems are reported. A proactive approach is preferred.

A more proactive type of surveillance method is referred to as "active surveillance" (Saldaña, 1996). These types of information gathering involve actively seeking information from the workforce on MSDs before they are reported on OSHA logs or become a workers' compensation claim. Examples of these types of methods can include questionnaires, symptoms surveys, and "discomfort surveys." To gather information, worker input is solicited concerning their work environment, health problems they are experiencing, and ways to eliminate the health problems. Many practitioners use the survey, or a variation thereof, developed by Corlett and Bishop (1976).

An advantage to using active surveillance is that the input of those closest to the task, the worker, is obtained. Worker involvement is a critical part of any comprehensive ergonomics program (NIOSH, 1997). Advantages of soliciting worker input include more feasible recommendations and better worker acceptance. Also, the tools provide a picture of the current situation. Health indicators are measured before they are recorded as workers' compensation claims or recordable injuries/illnesses.

Care must be taken, however, when developing and administering these tools. Design of questionnaires and surveys requires skill; a more detailed description of what is involved in administering questionnaires and surveys can be found in Sinclair (1995). The author cautions, "unless the information required is strictly factual and fairly easily checked, their reliability and validity can be quite low, so considerable care must be taken." Design and administration of active surveillance tools can be costly; skill and time are required for development, and workers must take time to fill out the instrument. Furthermore, it is advisable that the use of active surveillance technique not be undertaken by an organization unless management is committed to acting upon the results. Soliciting worker input and then failing to act can be counterproductive to ergonomic efforts.

It should also be noted that very few questionnaires have been demonstrated to be valid, unbiased estimators of true exposure to risk factors. Burdorf (1992) stated, "several studies have cast doubt on the determination of exposure to risk factors through questionnaire assessments. Comparisons of questionnaire assessments with observational data have shown that reports on the time spent in specific activities like walking, standing, and kneeling are not very reliable." So, questionnaires may have an appropriate use as a screening tool to direct the practitioner for appropriate follow-up analyses, but the specific results should be interpreted with caution.

34.4   Assessment Tools

Given that the surveillance methods mentioned earlier provide a measure of past and present MSD health indicators, it is also desirable to use tools that are able to predict the risk of future MSDs so that the monetary costs and human suffering can be averted through remediation efforts before they are incurred. Assessment tools that focus on identifying workplace MSD risk factors can help ergonomics practitioners be proactive by identifying problems and implementing interventions before they result in costly injuries.

34.4.1   Material Handling Assessment Tools

Manual materials handling (MMH) has long been recognized as a contributor to back pain. In a review of literature examining the effects of workplace factors on MSDs, the National Institute for Occupational Safety and Health (NIOSH) found evidence for a causal relationship between "heavy physical work" and back pain and "strong evidence" for a causal relationship between "lifting and forceful movement" and back pain (Bernard, 1997). Hence, assessment tools to quantify material handling requirements and their relationship to human limitations have been employed by ergonomists to minimize the impacts of back pain in the workplace.

34.4.1.1   Psychophysical Data Tables

One approach that has been used to develop lifting guidelines for MMH tasks is the psychophysical method. Psychophysics is a branch of psychology that studies the effects of external stimuli on the mental processes of organisms — in our case, humans. In ergonomics, the field has been used to determine safe lifting weight limits based on lifting task parameters such as lifting frequency, height of lift, and load parameters. It has also been used to determine maximum acceptable initial and sustained pushing and pulling force limits based upon frequency of the task and the handle heights.

The data are usually gathered in a laboratory, or other controlled setting, where MMH tasks can be simulated. Subjects usually perform a specific MMH task, and they are typically asked to adjust the load weight until they feel it is a load that can be safely and comfortably lifted for a specified time period. The subject monitors his or her feelings of exertion and fatigue, and adjusts the load weight accordingly. Checks such as repeating trials and ensuring the load weight remains unknown to the subject are put into place to ensure the reliability of the data.

The most notable data set was based on an industrial population and was gathered by Liberty Mutual Insurance Company, as reported by Snook (1978) and Snook and Ciriello (1991). The papers contain data tables which provide: (a) maximum acceptable lift weights, (b) maximum acceptable weights of lower, (c) maximum acceptable push forces, (d) maximum acceptable forces of pull, and (e) maximum acceptable weights of carry. Data are presented for both male and female populations. The reader is referred to the aforementioned references for the actual data tables.

One benefit of using the psychophysical approach is that it allows for a realistic simulation of work. The Liberty Mutual tables are also one of the few tools that allow an assessment of pushing and pulling tasks, which are frequently seen in the work environment. Another advantage of using the Liberty Mutual tables for a practitioner in the field is the ease of use. The user measures the MMH task characteristics, and then looks up the maximum value in the appropriate table. The user then compares the actual value to the determined for the percentage of the population that they wish to accommodate. If, however, the conditions of the task are outside of the range of those in the tables, then it may be necessary to use another assessment tool. Alternatively, the user could conduct their own psychophysical study for their specific MMH conditions, but doing so would be costly, time consuming, and require expertise, so conducting a study is usually not feasible by practitioners.

When using the tables, some limitations of the approach must be considered. During the testing, the load limits were based on subjective assessments by the test subject. So, although the limits may provide a realistic assessment of what a worker feels he or she can tolerate, it is questionable whether a subject can anticipate how much can be tolerated over the long term without incurring an injury. Marras et al. (1999), in an investigation of measurement effectiveness of different MMH assessment tools, found that in nearly two thirds of jobs that could be classified as high and medium risk the psychophysical criteria would indicate that those jobs are acceptable. So, a psychophysical approach appears to underestimate the actual level of risk.

The user should also keep in mind that the model also assumes good coupling (between the load and the hands and between the feet and the floor), two-handed symmetrical material handling, moderate load lifts, unrestricted working postures, and a favorable physical environment.

34.4.1.2 NIOSH Lifting Equation

One of the tools used most often by the practitioner, and also one of the most studied, is the Revised NIOSH Lifting Equation developed in 1993 and described by Waters et al. (1993). The equation is used to determine a lifting index (LI), an index of relative physical stress associated with MMH tasks. Inputs to the model are: (1) horizontal distance of the load from the spine, (2) vertical location of the lift, (3) vertical travel distance of the lift, (4) frequency of the lift, (5) angle of asymmetry of the lift, (6) lifting frequency, (7) quality of the coupling, and (8) weight of the actual load being lifted. A comprehensive guide to applying the equation in industry can be found in NIOSH (1994).

The revised equation built upon an equation originally released in 1981 (NIOSH, 1981). The 1993 equation was developed to reflect new findings, allow for asymmetric lifting tasks, and to account for objects with suboptimal hand–container couplings. Psychophysical, biomechanical, and physiologic criteria were used to determine the parameters of the equation, thus helping the tool to be more robust and usable in a variety of situations.

Several studies have been conducted to determine the validity of the model in identifying tasks associated with a higher risk of work-related low back pain. In the investigation by Marras et al. (1999) of the effectiveness of several methods of MMH assessment, it was found that both the 1981 and the 1991 equations had predictive power to identify jobs associated with high-risk of work-related low back disorders. Although both equations were about equally predictive, the 1981 equation achieved high specificity, meaning that it "was liberal in the assessment of risk by misidentifying most jobs as safe," while the 1993 equation, conversely, achieved high sensitivity "because it is more conservative and identified most jobs as being risky."

Waters et al. (1999) also evaluated the correlation between the LI and epidemiologic data; the data set included 50 jobs from four industrial sites. Based upon the findings that workers who perform jobs with an LI greater than 2.0 are at a significantly risk of having low back pain, the authors concluded, "The LI may be a useful indicator of risk of LBP caused by manual lifting." Interestingly, the data also showed that the risk of those who are exposed to an LI greater that 3.0, the highest exposure group, experienced *lower* risk than those in the group performing tasks where the LI was between 2.0 and 3.0. The authors theorize that this effect is due to a combination of worker selection and the survivor effect. That is, workers who are not able to perform MMH tasks where the LI is greater than 3.0 are injured or leave the jobs, thus leaving only those individuals who may have a high tolerance for manual lifting.

One attribute of using the NIOSH equation that is particularly pertinent to the practitioner concerns accuracy of the measures of the variables used as model inputs. Waters et al. (1998) investigated the accuracy of input measurements as recorded by highly educated nonergonomists with experience in data measuring and who participated in a 1-day training class. They found that in general, the students could accurately make the required measurements. However, the students did have trouble with three of the required measurements: (1) measurement of asymmetry angles, (2) the coupling factor, and (3) locating the reference point that is midpoint between the angles. The measurement of asymmetry angles was generally misestimated by 4.5° less than the reference value at the origin of lift and by 7.9° greater than the reference value at the destination. Also, the reference point location is a critical measurement because it has a large impact on the determination of the horizontal distance factor, which, in turn, carries great mathematical weight in the LI equation. It is apparent, then, that the NIOSH guide should really only be used by experts or by those given training on proper measurement and calculation of the LI.

Another investigation into use of the 1991 equation asymmetry multiplier in industry was reported by Dempsey and Fathallah (1999). The authors stated that problems with measuring asymmetry in practice occur because: (1) the definitions provided by NIOSH do not provide a consistent, objective method, (2) the analyst often must measure while the worker is in motion, and (3) there are frequently obstructions in the workplace which compound the problem. They then concluded that, in practice, field measurements of the asymmetry angle are essentially qualitative.

So, in practice, the NIOSH Lifting Equation can provide a valid assessment of the risk associated with MMH activities. However, only those who have received training on the tool should take the measurements use the equation, and interpret the results. Those using the equation should be aware of the limitations of the model. Despite the limitations, the NIOSH Lifting Equation can provide the ergonomics practitioner with a to that can be used set priorities for job design. The ergonomist can also determine which attributes of the job are good candidates for redesign by looking at the specific components of the equation. For example, if the horizontal multiplier is the lowest of the multipliers in the equation, the ergonomist would want to employ design strategies to reduce the horizontal distance of the lift. In summary, the NIOSH Lifting Equation appears to be predictive, fairly robust (when used within the limitations explicitly stated in the applications manual), inexpensive (the tool is available for free, but its use takes expert time), and relatively noninvasive. However, the tool is not necessarily easy to use and should be reserved for use by those who have been adequately trained.

34.4.1.3 ACGIH Lifting TLV

Several tools have been developed in an attempt to provide lifting guidelines that are quicker and easier for the practitioner to use and that are also scientifically valid. The American Conference of Governmental Industrial Hygienists (ACGIH) has developed a threshold limit value (TLV) for lifting (ACGIH, 2003). (At the time of this writing, the lifting guidelines are published as a "notice of an intent to establish" a TLV.) A TLV is defined by the ACGIH as "workplace lifting conditions under which it is believed nearly all workers may be repeatedly exposed, day after day, without developing work-related low back and shoulder disorders associated with repetitive lifting tasks."

The tool consists of three charts; the chart used to determine the TLV is a function of the lifting duration and lifting frequency. Then, based upon the one of four categories for height of the lift (floor to mid-shin height, mid-shin to knuckle height, knuckle height to below shoulder, and below shoulder to 30 cm above shoulder) and one of three categories of horizontal lift distance (close, intermediate, and extended), a TLV for weight is given. In some cases, the tables indicate that there is no known safe limit for repetitive lifting under those conditions.

The TLV is limited to two-handed mono-lifting tasks performed within 30° of the sagittal plane, so tasks requiring a large amount of trunk twisting should not be analyzed with this tool. Additionally, the ACGIH states that the guide does not apply to tasks where there is lifting >360 lifts per hour, lifting more than 8 h per day, constrained body posture, high heat or humidity, lifting unstable objects, poor hand couplings, or unstable footing, and professional judgment should be used to reduce weight limits under those conditions.

The TLV does, however, provide a very useful tool to quickly assess lifting tasks. It is quick, easy to use, and easy to interpret. The results can also direct the user to job redesign strategies. For example, if the lifting conditions exceed the TLV, the user may then find cells in the table where the weight would not exceed the TLV, and then redesign the job accordingly. Since the results are presented in a straightforward, intuitive format, the TLV also can be useful when requesting support from management for resources to institute ergonomic interventions.

34.4.1.4 OSU/BWC Lifting Guidelines

The Ohio State University (OSU) and the Ohio Bureau of Workers' Compensation (BWC) also developed lifting guidelines that are easy for the practitioner to use. An important feature of these guidelines is the incorporation of data for individuals currently experiencing low back pain. The data for the guidelines are based on research presented by Marras et al. (2001), which showed that the trunk loading of individuals with a low back disorder (LBD) is higher than for asymptomatic individuals under the same lifting conditions. The increased loading can be attributed to trunk muscle coactivation in individuals with low back pain.

The inputs to the lifting guide are: health status of worker (healthy or LBD), category of asymmetry angle (less than 30°, between 30° and 60°, and between 60° and 90°), horizontal reach distance (less than 12 in. or between 12 and 24 in.), and vertical lift origin (floor, knee, waist, or shoulder). The output of the model indicates the level of risk (low, medium, or high), which is given in weight ranges for the particular lifting conditions.

The OSU/BWC Lifting Guideline charts are presented in Figure 34.1, Figure 34.2, and Figure 34.3; instructions for using the charts are as follows:

1. Determine the angle of asymmetry associated with a specific lifting task. Then, choose the chart that corresponds. There are three separate charts: ±30° (Figure 34.1), between 30 and 60° (Figure 34.2), and between 60 and 90° (Figure 34.3).
2. Choose a column indicating whether the individual has an LBD or is healthy.
3. Determine the maximum horizontal reach distance from spine (measured from spine to hands) using the guide at the top of the columns. The guide is broken down into distances of 12 and 24 in.
4. Choose the vertical lift origin for the specific task from the right side of the chart. Vertical lift origin is the level from which lifts will be made. This can be from the floor, from the knees, from the waist or shoulder level. Also, choose the weight of the lift in pounds.
5. Determine if your specified lifting conditions fall within the green zone (low risk), yellow zone (moderate risk), or red zone (high risk).
6. To minimize the risk of injury, change any lifting criteria in the high or moderate risk zone so that the lifting task is in a green area of the chart.

The guidelines apply to two-handed, low-frequency lifts performed in unrestricted postures with good coupling. Ergonomics practitioners can use the results of the guidelines in a similar manner to the

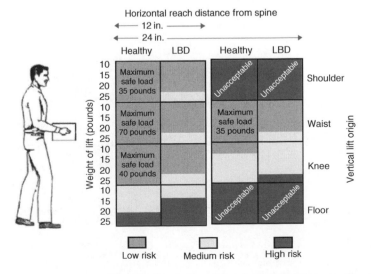

FIGURE 34.1 Guidelines for lifts involving trunk twisting angle of asymmetry between 0 and 30° (The angle between the front of the body, when facing forward, and the load being lifted).

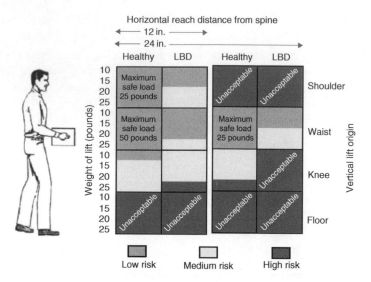

FIGURE 34.2　Guidelines for lifts involving trunk twisting angle of asymmetry between 30 and 60° (The angle between the front of the body, when facing forward, and the load being lifted).

ACGIH TLV. In addition, the guidelines can be used by medical professionals to develop realistic recommendations for injured workers' capabilities, since capabilities can be related to characteristics in the workplace. Also, transitional work providers can use the guidelines to place employees recovering from LBDs in the right tasks at the right time. The tool can help return an injured worker to work as soon as possible while minimizing the risk of aggravating an existing LBD.

34.4.2　Upper Extremity Assessment Tools

34.4.2.1　Strain Index

Moore and Garg (1995) developed a tool called the Strain Index to assess the risk of upper extremity MSDs based upon physiological, biomechanical, and epidemiological literature. The inputs to the tool

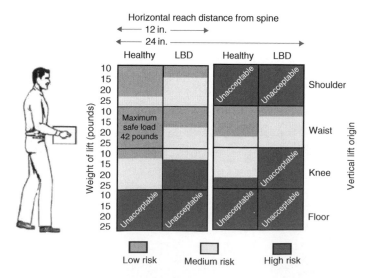

FIGURE 34.3　Guidelines for lifts involving trunk twisting angle of asymmetry between 60 and 90° (The angle between the front of the body, when facing forward, and the load being lifted).

are intensity of exertion, duration of exertion, efforts per minute, wrist posture, speed of exertion, and duration of task per day. After data collection, the analyst uses a table to assign rating values for each of the measures. These ratings values are then converted to multipliers by the analyst using the provided table. When the multipliers are multiplied together, a final index is obtained. The authors stress that this assessment tool is designed to assess jobs, not individual workers.

A validation study of the strain index was reported by Rucker and Moore (2002), who measured 28 jobs in two manufacturing facilities. The authors suggested using a strain index score of 9 as a criterion for classifying jobs as "hazardous" (>9) or "safe" (<9). They found that the sensitivity of the index was 0.92, meaning that the tool correctly identified 92% of the jobs associated with morbidity data. The sensitivity reported in the study was 1.00, meaning that the tool correctly classified all jobs not associated with morbidity. These numbers suggest that the strain index is effective at identifying jobs that do and do not expose workers to an increased risk of developing upper extremity MSDs.

The developers also point out some limitations of the index that users should be aware of as they collect data and interpret results. Notably, the index only applies to the distal upper extremity; it predicts a spectrum of upper extremity MSDs, but not specific disorders; and the training and experience needed to use the tool effectively has not been determined.

From a practitioner's standpoint, an advantage of the strain index is the flexibility in the types of measurement that can be used to obtain some of the task variables. For example, the hand/wrist posture rating value can be obtained by actually measuring wrist angles in flexion, extension, and ulnar deviation. An ergonomist could measure the posture with a goniometer. On the other hand, if an ergonomics team is using the tool and equipment is not available, then they can use the anchors provided for subjective ratings of "perceived posture." Of course, it is preferable to use the quantitative measures, but it is not always feasible in practice due to time and money constraints.

It should also be noted that the tool requires qualitative ratings for three of the input variables, and no quantitative alternative is given. Consequently, it was recommended by the developers that any subjective ratings be reached by consensus to minimize interrater reliability. To date, test–retest reliability has not yet been formally measured for this tool.

In all, however, the Strain Index can be a valuable tool for ergonomics practitioners and ergonomics teams to prioritize intervention efforts, to determine which attributes of the task to address in order to reduce worker exposure to MSD risk factors, and as a tool for follow-up analysis to ensure that any intervention efforts resulted in the desired effect.

34.4.2.2 ACGIH Hand Activity Level (HAL) TLV

In addition to developing a tool to assess lifting tasks, the ACGIH has developed a threshold limit value (TLV) and an action limit (AL) for hand–arm activity levels (ACGIH, 2003). The TLV and the AL are based on the peak hand force and the hand activity level for a given task. The peak force can be determined quantitatively (e.g., electromyography, strain gauges, or other biomechanical methods), or it can be measured qualitatively (e.g., ratings by a trained observer, or rated by the worker). Peak hand force is then normalized on a scale of 0–10, where 10 represents the strength capability of 100% of the work population. The HAL can also be determined either quantitatively, using a combination of measures of exertion frequency and percentage of time the exertion occurs in the duty cycle, or it can be determined using qualitative guidelines presented in the standard.

The HAL and the normalized peak force are then combined on a graph where they can be compared to the TLV and the AL. The ACGIH states that the TLV "represents conditions to which it is believed that nearly all workers may be repeatedly exposed without adverse health effects." The AL, which is lower than the TLV, is defined as the limit "for which general controls are recommended."

One major advantage to the HAL TLV is that it allows for a fairly wide range of measurement techniques by the user. For example, the expert ergonomist can incorporate biomechanical measures, while an ergonomics team could use measures that are more objective. Thus, those with various levels of expertise can use the tool.

The tool does not account for some workplace factors that may increase the likelihood of MSDs, namely sustained non-neutral postures, contact stresses, low temperatures, and vibration exposure. If these factors are present, then professional judgment should be used to determine if exposure is above the TLV. Also, the TLV is not meant to apply to work conditions that include multiple tasks. The TLV does, however, provide a useful tool to quickly assess mono-tasks. It can be quick, easy to use, and easy to interpret. It also allows the user flexibility to be more accurate with an additional investment of time and expertise. Appropriate uses of the TLV include identifying potential jobs as candidates for ergonomic interventions and for setting priorities for resource allocation. The TLV can also be used as a metric to determine if interventions have had the desired effect by performing an assessment before and after implementation and comparing the results.

34.4.3 Entire Body Assessment Tools

34.4.3.1 Rapid Upper Limb Assessment

McAtamney and Corlett (1993) developed a survey method to quickly screen a working population for the risk of work-related upper extremity MSDs. The method, called Rapid Upper Limb Assessment (RULA) was developed so that the screening could be performed without the need for specialized equipment. Despite its name, RULA also accounts for posture, force, and repetition of the neck, trunk, and legs.

Some MSD risk factors that can be assessed with the RULA are numbers of movements, static muscle work, force, work postures determined by equipment and furniture, and time worked without a break. To determine the final RULA score, the analyst categorizes the posture of different body parts into predetermined categories. (Up until the final score is calculated, the wrist and arm values and the neck, trunk, and leg values are calculated separately.) Charts are then used to combine the postures to obtain the general posture scores. These general posture scores are then combined with muscle use scores and force/load scores. Finally, the wrist and arm score and the neck, trunk, and leg score are combined with a table to determine the final score.

The authors reported an assessment of the validity of RULA based upon 16 computer operators, and they found significant relationships between individual RULA body part scores and the development of pain or discomfort for the neck and lower arm, but not for the trunk, upper arm or risk. A reliability test was performed by having over 120 professionals who had been trained on the use of RULA score videotaped examples of work tasks. The authors state that there was a "high consistency of scoring amongst subjects," but there are no statistics to support this claim. So, further validity and reliability studies of this assessment tool are warranted.

The main benefits of RULA for the practitioner is its low cost, its ease and quickness of use, and the simplicity of the results — the final score is represented as a single number. So, the tool is suited be used by ergonomics teams to make recommendations to management. McAtamney and Corlett (1993) state that in their experience, "Managers were quick to recognize and remember the grand scores and their associated action levels." They also indicated that the tools lend themselves well to reevaluating changes in musculoskeletal loading after changes were made. They recommend the tool be used as a screening method or incorporated into a wider group of assessment tools.

34.4.3.2 Job Safety Analysis

The Job Safety Analysis (JSA), an analysis tool that has been commonly used in industry for years to identify and rectify safety hazards in the workplace, also can be used to identify MSD risk factors and potential intervention strategies (Peterson, 2003). The JSA process involves breaking a task into individual steps. For each step, potential hazards (including MSD risk factors) are identified and then work practices, procedures, or engineering controls are listed to ameliorate the hazards.

The process used when completing the JSA is critical to its success. It is imperative that workers doing the job are involved to ensure that the actual job steps are included and that recommendations are feasible. Worker involvement also increases acceptance of recommendations. Furthermore, solutions should focus on eliminating hazards and risk factors; administrative solutions such as worker training, recommended procedures, and job rotation should be considered as solutions only until engineering changes eliminate worker exposure to hazards can be implemented. Solutions should also be specific. According to Peterson (2003), "Solutions that merely state, 'Be more alert' or 'Use more caution' or something similar, are worthless."

Once the JSA is completed, it is important to make them readily available to workers performing the job — they do little good if filed in a drawer, which is all to often the case. Also, as the job demands change (i.e., after a recommended control is implemented), the JSA should be updated. It should be noted that although quantitative methods may be used to describe risk factors, JSAs do not provide a quantitative measure of risk factors. Therefore, JSAs alone cannot be used to predict injury, nor can they alone be used to prioritize tasks. However, JSAs do provide a systematic method to identify hazards, including MSD risk factors, and to develop potential solutions.

34.4.3.3 Checklists

Many checklists are available as an aid to identify MSD risk factors. Checklists, when used in ergonomics, are essentially lists of risk factors or workplace conditions that are believed to increase the risk of MSDs. They help the user to ensure that they have addressed areas of concern, and have not overlooked potential risk factors. Hence, they are particularly useful for the novice. In general, checklists indicate the presence of a risk, or if the level of risk is above a certain threshold level (e.g., if the weight lifted is greater than 20 lbs or if the neck angle is greater than $20°$). Consequently, checklists generally do not provide a quantitative level of risk and, therefore, cannot be used to predict risk of injury.

In their draft ergonomics standard for prevention of work-related MSDs developed in 1995, OSHA provided a checklist to be used as a screening tool (Schneider, 1995). In addition to indicating the presence of a risk factor, the tool also assigns a score between 0 and 3, depending on the duration of worker exposure and the severity of the identified risk factor. The tool is divided into "upper extremity," "back and legs," and "environmental" sections. Score totals are obtained for each section, and then combined for a total score. A slightly modified version of this MSD Risk Factor Screening Tool is shown in Table 34.1.

Although the tool has not been validated for its original intended use of providing a threshold value, the screening tool can provide a relative measure of worker exposure to MSDs. It should also be noted that interrater reliability is relatively poor, so comparisons made between raters may not be valid. The numerical score obtained from the screening tool can help prioritize tasks and can also be used to ensure that risk of injury has been reduced after a design change.

As an example, the tool was used as to provide an assessment of a window assembly task before and after ergonomic intervention. Figure 34.4 shows the window assembly task being performed before the ergonomic interventions were implemented. The upper extremity score was 19, the back and legs score was 6, and the environmental score was 0, for a total score of 25. Ergonomic interventions focusing on reducing upper extremity MSD risk factors were then implemented. The interventions consisted of: (1) providing automated screw feeders, (2) using in-line tools, and (3) providing an arm to counteract the tool torque. The redesigned task is shown in Figure 34.5. The follow-up screening tool analysis score was 11 for upper extremity, 6 for back and legs, and 0 for environmental. Thus, the screening tool indicated that we have reduced the upper extremity MSD risk factors as intended.

34.5   Conclusions

In order to get the most from data collected using ergonomic assessment tools, the analysis tool user must understand some fundamental issues associated with data collection. First, there may be some intersubject variation in the way tasks are performed, either due to variances in processes or variances in worker performance. So, it is important to observe the task for a long enough period to account for these variations. The length of time will depend on the task cycle time and any variation cycles times. Second, there will also likely be intrasubject variation; work practices can vary between

TABLE 34.1   MSD Risk Factor Screening Tool.

MSD Risk Factor Screening Tool

Task Title: _____   Company:

Name of Analyst:_____   Evaluation Date:

Brief Task Description:

_____

_____

_____

_____

Upper Extremity					
A	**B**	C	D	**E**	**F**
Risk Factor Category	**Risk Factors**	**2 to 4 Hours**	**4+ to 8 Hours**	**8+ Hours** *Add 0.5 per hour*	**Score**
Repetition (Finger, Wrist, Elbow, Shoulder, or Neck Motions)	**1. Identical or Similar Motions Performed Every Few Seconds** *Motions or motion patterns that are repeated every 15 seconds or less. (Keyboard use is scored below as a separate risk factor.)*	1	3		
	2. Intensive Keying *Scored separately from other repetitive tasks in the repetition category; includes steady pace, as in data entry.*	1	3		
	3. Intermittent Keying *Scored separately from other repetitive tasks. Keyboard or other input activity is regularly alternated with other activities for 50 to 75 percent of the work.*	0	1		
Hand Force (Repetitive or Static)	**1. Grip More Than 10-Pound Load** *Holding an object weighing more than 10 pounds or squeezing hard with hand in a power grip.*	1	3		
	2. Pinch More Than 2 Pounds *Pinch force of 2+ pounds as in the pinch used to open a small binder clip with the tips of fingers.*	2	3		
Awkward Postures	**1. Neck: Twist / Bend** *Twisting neck to either side more that 20°, bending neck forward more than 20° as in viewing a monitor, or bending neck backward more than 5°.*	1	2		
	2. Shoulder: Unsupported Arm or Elbow Above Mid-Torso Height *Arm is unsupported if there is not an arm rest when doing precision finger work, or when the elbow is above mid-torso height.*	2	3		
	3. Forearm: Rapid Rotation *Rotating the forearm or resisting rotation from a tool. An example of forearm rotation is using a manual screwdriver.*	1	2		
	4. Wrist: Bend / Deviate **Wrist bends that involve more than 20° of flexion (bending the wrist palm down) or more than 30° of extension (bending the wrist back). Bending can occur during manual assembly and data entry.**	2	3		
	5. Fingers *Forceful gripping to control or hold an object, such as click-and-drag operations with a computer mouse or deboning with a knife.*	0	1		

(Table Continued)

TABLE 34.1 (*Continued*)

Upper Extremity					
A	**B**	**C**	**D**	**E**	**F**
Risk Factor Category	**Risk Factors**	**2 to 4 Hours**	**4+ to 8 Hours**	**8+ Hours** *Add 0.5 per hour*	**Score**
	6. **Extended arm reaches**	1	2		
	7. **Reaching overhead (above shoulder level)**	1	2		
	8. **Reaching behind the torso**	1	2		
Contact Stress	1. **Hard/Sharp Objects Press Into Skin** *Includes contact of the palm, fingers, wrist, elbow, or armpit.*	1	2		
	2. **Using the Palm of the Hand as a Hammer**	2	3		
Vibration	1. **Localized Vibration** *Vibration from contact between the hand and a vibrating object, such as a power tool.*	1	2		
	TOTAL UPPER EXTREMITY SCORE:				

Back and Legs					
Awkward Postures (Repetitive or Static)	1. **Mild Forward or Lateral Bending of Torso More Than 20° But Less Than 45°**	1	2		
	2. **Severe Forward Bending of Torso More Than 45°**	2	3		
	3. **Backward Bending of Torso**	1	2		
	4. **Twisting Torso**	2	3		
	5. **Prolonged Sitting Without Adequate Back Support** *Back is not firmly supported by a back rest for an extended period*	1	2		
	6. **Standing Stationary or Inadequate Foot Support While Seated** *Stand in one place (an assembly line or check stand) without sit/stand option or walking, or feet are not firmly supported when sitting.*	0	1		
	7. **Kneeling / Squatting**	2	3		
	8. *Repetitive Ankle Extension / Flexion* *Using a foot pedal to start or stop a machine cycling (as in sewing machine operations).*	1	2		
Contact Stress	1. **Hard / Sharp Objects Press into Skin** *Includes contact against the leg.*	1	2		
	2. **Using the Knee as a Hammer or Kicker**	2	3		
Vibration	1. **Sitting/Standing on Vibrating Surface (Without Vibration Dampening)**	1	2		
Push/Pull	1. **Moderate Load** *Force needed to push / pull a shopping cart full of apples.*	1	2		
	2. **Heavy Load** *Force need to push / pull a two-drawer, full file cabinet across a carpeted room.*	2	3		
Manual Materials Handling — Load	1. **Weight** *Load being handled is more than 20 pounds.* **(Write actual weight of maximum load in box to right.)** **Actual Weight (lbs.)** _____	2	3		
	2. **Distance** *Horizontal distance from the mid-point between the ankles to center of the hand is greater than 10 inches.* **(Write actual maximum distance in box to right.)** **Actual Distance (in.)** _____	2	3		

(*Table Continued*)

TABLE 34.1 *(Continued)*

Back and Legs						
A Risk Factor Category	<u>B</u> Factors　　　　　　　　　Risk		<u>C</u> 2 to 4 Hours	<u>D</u> 4+ to 8 Hours	**E** 8+ Hours *Add 0.5 per hour*	**F** Score
Manual Materials Handling—Frequency	**1. Lifting Frequency** *Lifting frequency is between 1 and 5 times per minute. (**Write actual lifting frequency in the box to right.**)*	*Lifting Frequency* _____	1	1		
	2. Lifting Frequency *Lifting frequency is 5 or more times per minute.*		2	3		
TOTAL BACK AND LEGS SCORE:						

Environmental Worksheet						
A Risk Factor Category	**B** Risk Factors		<u>C</u> 2 to 4 Hours	<u>D</u> 4+ to 8 Hours	**E** 8+ Hours *Add 0.5 per hour*	**F** Score
Environment	**1. Lighting (Poor Illumination / Glare)** *Inability to see clearly (e.g. glare on a computer monitor).*		0	1		
	2. Cold Temperature *Air temperature less than 60°F for sedentary work, 40°F for light work, 20°F for moderate/heavy work; cold exhaust blowing on hands.*		0	1		
TOTAL ENVIRONMENTAL SCORE:						
Total Score: **(Upper Extremity + Back and Legs + Environmental)**						

workers. Therefore, the observer should sample enough workers to make a valid assessment. Oftentimes, it is feasible to collect data for each person performing the task. Other times a judgment on the sample size must be made based upon the amount of variability between workers and the number of workers performing the task. Third, there is likely to be interrater (or observer) variability, and this variability

FIGURE 34.4　Window assembly before ergonomic intervention.

FIGURE 34.5 Window assembly task after ergonomic intervention.

will increase the more subjective the measure being used. If the interrater variability is high, then comparison of analysis results performed by different raters is not advised. One way to reduce the amount of interrater variability is to gain consensus from a group or ergonomics team when recording subjective measures. Doing so will tend to make assessments more consistent and will facilitate communication within the team. A more complete description of the issues associated with making ergonomic assessments in the field can be found in Haines and McAtamney (1995).

Another useful tool that can be used to supplement each of the analysis tools mentioned above is videotaping. Videotaping provides a visual record of the task; this record is particularly useful when comparing jobs before and after implementing ergonomic interventions. Use of videotape also allows ergonomics teams to make assessments and discuss issues together, with minimal disruption of the work environment. Videotape also allows slow motion playback, zoom, and multiple viewing, enabling the observer to pick up on quick or subtle motions that may not be seen in real time. Some guidelines for recording work activities on videotape are presented in Grant (1996).

A myriad of analysis tools exist to help the ergonomic practitioner to identify MSD risk factors, to prioritize candidate tasks for resources, and to measure whether ergonomic intervention strategies have had the desired effect. Some tools that have been demonstrated to be practical for the ergonomics practitioner were presented in this chapter. None of the tools is perfect; each has its advantages and its drawbacks. In order to use the tools appropriately, the practitioner must understand the limitations of the assessment tools that they are using. As long as the strengths and weaknesses of the tools are kept in mind, they can provide the ergonomics team and the ergonomics practitioner with valuable data to prioritize projects, measure successes, and gain support from management to make the workplace safer for the worker.

References

ACGIH (2003). *2003 TLVs and BEIs: Threshold Limit Values for Chemical Substances and Physical Agents & Biological Exposure Indices*, American Conference of Governmental Industrial Hygienists, Cincinnati, Ohio.

Bernard, B.B. (1997). Musculoskeletal disorders and workplace risk factors: a critical review of epidemiologic evidence for work-related musculoskeletal disorders of the neck, upper extremity, and low back, DHHS (NIOSH) Publication #97-141, NIOSH, Cincinnati, Ohio.

Burdorf, A. (1992). Exposure assessment of risk factors for disorders of the back in occupational epidemiology, *Scand J Work Environ Health*, 18:1–9.

Corlett, E.N. and Bishop, R.P. (1976). A technique for assessing postural discomfort, *Ergonomics*, 19:175–182.

Dempsey, P.G. and Fathallah, F.A. (1999). Application issues and theoretical concerns regarding the 1991 NIOSH equation asymmetry multiplier, *Int Appl Ergon*, 23:181–191.

Grant, K.A. (1996). Job analysis, in *Occupational Ergonomics: Theory and Applications* Ed. by Battacharya A. and McGlothlin J.D., Marcel Dekker, Inc., New York.

Haines, H. and McAtamney, L. (1995). Undertaking an ergonomics study in industry, in *Evaluation of Human Work: A Practical Ergonomics Methodology*, 2nd edn. Ed. by Wilson, J.R. and Corlett, E.N., Taylor & Francis, London.

Hamrick, C.A. (2000). CTDs and ergonomics in Ohio, *Proceedings of the IEA 2000/HFES 2000 Congress*, 5-111–5-114.

Marras, W.S., Lavender, S.A., Leurgans, S., Rajulu, S.L., Allread, W.G., Fathallah, F.A., and Ferguson, S.A. (1993). The role of dynamic three dimensional trunk motion in occupationally-related low back disorders: the effects of workplace factors, trunk position and trunk motion characteristics on injury, *Spine*, 18(5):617–628.

Marras, W.S., Fine, L.J., Ferguson, S.A., and Waters, T.R. (1999). The effectiveness of commonly used lifting assessment methods to identify industrial jobs associated with elevated risk of low-back disorders, *Ergonomics*, 42(1):229–245.

Marras, W.S., Allread, W.G., Burr, D.L., and Fathallah, F.A. (2000). Validation of a low back disorder risk model: a prospective study of ergonomic interventions associated with manual materials handling jobs, *Ergonomics*, 43(11):1866–1886.

Marras, W.S., Davis, K.G., Ferguson, S.A., Lucas, B.R., and Gupta, P. (2001). Spine loading characteristics of patients with low back pain compared to asymptomatic individuals, *Spine*, 26(23):2566–2574.

McAtamney, L. and Corlett, E.N. (1993). RULA: a survey method for the investigation of work-related upper limb disorders, *Appl Ergon*, 24(2):91–99.

Moore, J.S. and Garg, A. (1995). The Strain Index: a proposed method to analyze jobs for risk of distal upper extremity disorders, *Am Ind Hyg Assoc J*, 56:443–458.

NIOSH (National Institute for Occupational Safety and Health) (1981). *Work Practices Guide for Manual Lifting*, DHHS (NIOSH) Publication No. 81–122, U.S. Department of Health and Human Services (NIOSH), Cincinnati, Ohio.

NIOSH (National Institute for Occupational Safety and Health) (1994). *Applications Manual for the Revised NIOSH Lifting Equation*, DHHS (NIOSH) Publication No. 94-110, U.S. Department of Health and Human Services (NIOSH), Cincinnati, Ohio.

NIOSH (National Institute for Occupational Safety and Health) (1997). *Elements of Ergonomics Programs: A Primer based on Workplace Evaluations of Musculoskeletal Disorders*, DHHS (NIOSH) Publication No. 97-117, U.S. Department of Health and Human Services (NIOSH), Cincinnati, Ohio.

Peterson, D. (2003). *Techniques of Safety Management: A Systems Approach*, 4th Edition, American Society of Safety Engineers, Des Plaines, Illinois.

Rucker, N. and Moore, J.S. (2002). Predictive validity of the strain index in manufacturing facilities, *Appl Occup Environ Hyg*, 17(1):63–73.

Saldaña, N. (1996). Active surveillance of work-related musculoskeletal disorders: an essential component in ergonomics programs, in *Occupational Ergonomics: Theory and Applications* Ed. by Battacharya A. and McGlothlin J.D., Marcel Dekker, Inc., New York.

Schneider, S. (ed.) (1995). OSHA's draft standard for prevention of work-related musculoskeletal disorders, *Appl Occup Environ Hyg*, 10(8): 665–674.

Sinclair, M.A. (1995). Subjective assessment, in *Evaluation of Human Work: A Practical Ergonomics Methodology*, 2nd edn. Ed. by Wilson, J.R. and Corlett, E.N., Taylor & Francis, London.

Snook, S.H. (1978). The design of manual handling tasks, *Ergonomics*, 21:963–985.

Snook, S.H. and Ciriello, V.M. (1991). The design of manual handling tasks: revised tables of maximum acceptable weights and forces, *Ergonomics*, 34(9): 1197–1213.

Tanaka, S. (1996). Record-based (passive) surveillance for cumulative trauma disorders, in *Occupational Ergonomics: Theory and Applications* Ed. by Battacharya A. and McGlothlin J.D., Marcel Dekker, Inc., New York.

United States Department of Labor (2003). Bureau of Labor Statistics (www.bls.com), Public Data Query, December 16, 2003.

United States Department of Labor (2004). OSHA Forms for Recording Work-Related Injuries and Illnesses, U.S. Department of Labor, Washington, D.C.

Waters, T.R., Putz-Anderson, V., Garg, A., and Fine, L.J. (1993). Revised NIOSH equation for the design and evaluation of manual lifting tasks, *Ergonomics*, 36(7):749–776.

Waters, T.R., Baron, S.L., and Kemmlert, K. (1998). Accuracy of measurements for the revised NIOSH lifting equation, *Appl Ergon*, 29(6):433–438.

Waters, T.R., Baron, S.L., Piacitelli, L.A., Anderson, V.P., Skov, T, Haring-Sweeney, M., Wall, D.K., and Fine, L.J. (1999). Evaluation of the revised NIOSH lifting equation: a cross-sectional epidemiologic study, *Spine*, 24(4):386–395.

35

Low Back Injury Risk Assessment Tools

Gary Mirka
Gwanseob Shin
North Carolina State University

35.1 Introduction

Over the last three decades, several low back injury risk assessment tools have been developed to provide ergonomics practitioners the ability to evaluate the relative risk posed by manual materials handling (MMH) tasks. The Work Practices Guide for Manual Lifting (NIOSH, 1981) and the Revised NIOSH Lifting Equation (Waters et al., 1993, 1994) are two well-established methods developed by the National Institute for Occupational Safety and Health (NIOSH). The Lumbar Motion Monitor (LMM) risk assessment model (Marras et al., 1993) and the Three-Dimensional Static Strength Prediction Program™ (3DSSPP) were developed by researchers at The Ohio State University and the University of Michigan, respectively. The goals of this paper are to compare and contrast these existing assessment models, develop the motivation for a hybrid modeling technique and identify gaps in our current low back injury risk assessment techniques for other high-risk activities.

35.2 NIOSH Lifting Equations

In 1981, the NIOSH published the Work Practices Guide for Manual Lifting (NIOSH, 1981) to provide industry with an assessment tool aimed at identifying jobs that carry an increased risk of low back injury. The general approach was to use basic data from the biomechanical, epidemiological, psychophysical, and physiological literature to create an assessment tool to quantify risk during MMH tasks. The mechanics of the tool require that the analyst gather important task-related data that describes the lifting task (i.e., the vertical location of the load, the distance of the load away from the spine, the required frequency of the lifting task and the vertical distance of travel of the load), and then use these data in a multiplicative model that derives an action limit (AL) value and a maximum permissible limit (MPL) value using the classic industrial hygiene approach. If the actual weight that the worker is asked to lift is below this AL level, the task is deemed acceptable. If the actual weight exceeds the MPL (calculated as three times the AL), significant risk exists and the task should be carefully evaluated and redesigned.

After over a decade of using this assessment tool, researchers at NIOSH recognized the limitations of this tool in assessing the variety of realistic working conditions faced by those performing manual materials handling tasks. Specifically, they identified the inability of the 1981 equation to assess the risk of asymmetric work postures and varied kinds of coupling between the lifter and the load being lifted. So in 1993 they published a Revised NIOSH Lifting Equation (Waters et al., 1993). The workplace variables considered in this revised equation include: vertical position of load, horizontal distance between the load and the spine, frequency of lifting, vertical travel distance of the load, asymmetric posture of torso, and coupling quality between the lifter and the object being lifted. As in the 1981 equation, these measures are then combined in a multiplicative model but instead of calculating an AL this model calculated a value called the recommended weight limit, or RWL, which describes a weight that can be lifted safely by a majority of the working population. The ratio of the actual weight being lifted in the job to this RWL is a value called the lifting index (LI). LI values greater than 1 are said to place some workers at increased risk, while values greater than 3 are said to be a potential problem for a majority of healthy industrial workers (Waters et al., 1994). There are a number of stated assumptions that should be considered when applying this assessment technique. Among these assumptions are that the workers perform only two-handed lifts, they work for no more than 8 h, they are not lifting or lowering objects faster than 75 cm/sec, and they are lifting in a relatively unrestricted work environment.

There have been several studies that have been conducted to evaluate the effectiveness of the NIOSH method in predicting the reporting of low back pain/discomfort. In a study of 97 MMH jobs, Wang et al. (1998) report a monotonically increasing relationship between the severity ratings of low back discomfort and the NIOSH LI. Their results showed that for jobs with a severity rating of 0 (on a 0–5 point scale) the mean LI was 0.8 while for jobs with a mean severity rating of 5 the mean LI was 4.1 with intermediate points following the trend. Another significant result from this study was that 42 of the 97 evaluated jobs had an RWL = 0, a result that the authors attribute to having tasks wherein lifting frequencies and/or horizontal distances exceeded those allowed by the NIOSH modeling methodology. In another study (Waters et al., 1999), 50 jobs in four different industrial facilities, were evaluated and the authors report that the unadjusted prevalence odds ratio for reported low back pain were 1.14, 1.54, and 2.54 for LIs of 0–1, 1–2, and 2–3, respectively. Interestingly, they report an unadjusted prevalence odds ratio of 1.63 for jobs with an LI of >3, and note potential selection and survivor effects may have influenced the results of their analysis. In addition to these studies that have considered the relationships between the NIOSH assessments and reporting of discomfort there have also been several studies that have considered "usability" aspects of the assessment tools. The reader is referred to the following articles for further information: Dempsey (2002), Dempsey and Fathallah (1999), and Waters et al. (1998).

While the NIOSH method is straightforward in application and has great utility in many industrial environments the static representation of the workplace does not take into account some of the human performance issues that have been implicated in the low back injury process. Specifically, in performing an MMH task, the three-dimensional postures, velocities and accelerations have been shown to play a role in the development of low back injuries (Marras et al., 1995). It should be noted that the NIOSH modeling approach did consider dynamics from physiological and psychophysical perspectives, but in many cases trunk motion plays a direct role in biomechanical loading and therefore is a facet of risk that is not directly addressed in the NIOSH approach.

35.3 LMM Risk Assessment Model

A risk assessment tool developed by Marras et al. (1993) recognized the importance of these lifting dynamics in the development of a low back injury. This assessment tool made use of a device called the Lumbar Motion Monitor (LMM) (Figure 35.1) that was developed to capture the instantaneous position, velocity, and acceleration of the lumbar spine in the three cardinal planes of human motion. Their approach to developing an assessment tool was to use this tri-axial goniometric device to capture the trunk kinematic profiles of workers performing their normal work tasks and then to relate these kinematic characteristics (along with a cadre of other task descriptors such as lifting frequency, moments about the spine created by the load, job satisfaction, the static workplace variables from the NIOSH Lifting Equation, etc.) to the historical incidence of low back injuries. They sampled 403 industrial jobs and then used multiple logistic regression techniques to form a relationship between historical injury data and the task parameters. Their results showed that five parameters were adequate to distinguish between the high- and low-risk jobs in this data set: lift rate, maximum sagittal angle, average twisting velocity, maximum lateral velocity, and maximum moment (model odds ratio of 10.7). The result of their work is a low back injury risk assessment model that takes as inputs these five task variables and the output is a single value that describes the probability of high-risk group membership (PHRGM) for that job.

The principal strength of this model is that it is based on the empirical relationship between outcome measures (injury and job turnover rates) with quantifiable job characteristics, including human performance-related variables. With this approach comes the ability to begin to consider the role that individual differences (i.e., lifting and MMH techniques) may play in the etiology of low back injury. While this model was able to overcome the static biomechanical modeling limitations of the NIOSH Lifting Guides, a limitation to the generalizabilty of this model is that it was developed using data collected from a sample of jobs where workers performed "repetitive jobs without job rotation." Since this was an empirical model, the specific job dataset that was used to develop the relationship between work characteristics and risk will have a great influence on the model output. Since nonrepetitive jobs were not included in the dataset, certain characteristics of these types of jobs may not be represented in this model's predictions. Further, because of the special emphasis placed on the variables describing trunk dynamics that resulted from this sample of jobs, some high-risk activities, such as lifting heavy loads in awkward, static postures will often escape identification.

FIGURE 35.1   The Lumbar Motion Monitor.

35.4 Three-Dimensional Static Strength Prediction Program Model

The Three-Dimensional Static Strength Prediction Program (3DSSPP) model developed by researchers at the University of Michigan is a biomechanical modeling system that can be used to compute the moments/forces acting about/on the joints comprising the kinematic chain. The inputs to this model are the major joint angles and the direction and magnitude of the force exerted by the hands. Once this three-dimensional biomechanical model is developed, the three-dimensional moments about the L5/S1 joint and spine compression can be calculated (Figure 35.2). Further, the three-dimensional moments can be compared with the data from the human strength capacity database so that an estimation of the percentage of a population capable of exerting these moments can be generated. Therefore, two of the measures relevant to the understanding of low back injury risk that can be derived from this modeling approach are the compression in the spine and the percent of the population that have the strength capacity to exert the required moments about the spine.

Several studies have illustrated the importance of documenting the relationship between a person's strength capacity and the physical demands placed on them during work (Chaffin and Park, 1973; Chaffin, 1974). Chaffin (1974) illustrated a sharp increase in low back pain incidence rates when the job demands required forces exceeding the workers' strength capacity. This author showed that jobs whose average lifting strength ratio (LSR — defined as the ratio of the heaviest weight lifted to the average strength of people asked to perform that lift) exceeded 1.0 had a job related low back incidence rate (low back incidences/1000 man-weeks) of ~2.3. This is as compared to ~0.75 and ~0.65 for jobs whose average LSR was between 0–0.5 and 0.5–1 respectively. Chaffin and Park (1973) performed a very

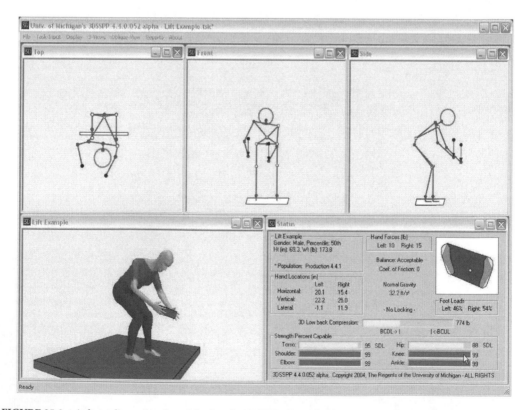

FIGURE 35.2   A three-dimensional model using the 3DSSPP. (Provided by the Center for Ergonomics, University of Michigan.)

similar analysis but used the strength capacity of a large/strong man in the denominator of the LSR, thus normalizing the data such that it would be very unlikely to encounter a job have lifting requirements with an LSR exceeding 1.0. Their results showed that for jobs with a LSR between 0.8 and 1.0 the job-related low back incidence rate approached 4 while the incidence rates associated with lower LSRs were less than 2.

The strength of the 3DSSPP approach is in its ability to assess risks associated with one time exertions, because it compares directly the required moments of the task with population strength data. Another strength of the model is its ability to estimate spine compression values that can be compared with established load limits to assess relative risk. The limitations of this approach are in its ability to quantify risk in jobs that are highly repetitive in nature but do not have torque or spine compression forces that approach human strength capabilities or spine compression load limits.

35.5 Discussion of Similarities and Differences

There are several similarities between these three assessment tools that should be identified. First, and not surprisingly, each considers the moment about the spine to be a central factor in quantifying risk. The three tools do, however, take different approaches to using this moment value. The 3DSSPP model uses the computed moment value in comparison to the moment generating capabilities of the population. The LMM model uses peak moment value as one of five parameters in the multiple logistic model, weighing it equally with the five other parameter. Finally, the NIOSH Lifting Equation considers the components of the moment (i.e., the moment arm and the load) at two different stages of the analysis. The moment arm is considered in the process of calculating the RWL (through the horizontal multiplier), while the load magnitude is not considered until the LI is computed. A second characteristic that these models share is an appreciation for the importance of the trunk posture during the lifting activity. In the LMM model, this comes in the form of the peak sagittal flexion parameter. In the 3DSSPP and NIOSH Lifting Equation model this is considered more indirectly. In the 3DSSPP model, the posture is reflected in the moment created by the mass of the torso in the static model, while in the NIOSH model these postural effects are considered in the vertical multiplier (quantifying the degree of sagittal flexion required) and asymmetry multiplier (describing the required trunk motion in the sagittal plane). One limitation that all three assessment tools share is their limited ability to address highly variable biomechanical requirements seen in some industries, the warehousing and construction industries being two that are notorious for being high-risk industries for back injuries. Gaining an appreciation for the range of stress levels that are experienced by the worker and the relative frequency of experiencing these levels could provide valuable insight into the cumulative and acute trauma risk posed. This particular limitation is addressed in the hybrid modeling approach described later in this chapter.

There are some important differences in the conceptual approaches that are noteworthy as well. Based on our current understanding of the etiology of occupation-related low back disorders it is clear that each of these assessment tools addresses an important facet of the low back injury risk paradigm, but that none of these models individually are able to identify all high risk activities. This notion is supported by Lavender et al. (1999) that showed poor correlation between the estimates of low back disorder risk that were produced by these three tools when assessing a variety of MMH tasks. These authors used each of these three assessment tools (along two variations of the United Auto Workers (UAW) – General Motors Ergonomic Risk Factor Checklist) to evaluate a mix of 93 randomly selected production jobs performed by 178 autoworkers. These authors showed that the intercorrelations between methods ranged from 0.21 to 0.80. For the three assessment tools considered in the paper the correlations were 0.54, 0.39, and 0.21 for LMM–NIOSH, LMM–3DSSPP, and NIOSH–3DSSPP comparisons, respectively. These authors conclude by noting that some of the differences in this assessment techniques may lie in their differential treatment of acute vs. cumulative loading characteristics.

Mirka et al. (2000) also highlighted some of the differences in these assessment tools but focused less on the acute vs. cumulative trauma risk perspective and more on the mechanics of the tools themselves.

In their study these three assessment tools were used in the analysis of construction workers in the home building industry to identify specific work tasks to be addressed through subsequent intervention research (Mirka et al., 2003). In this assessment a set of jobs was identified by each technique and the results showed little overlap in the specific tasks that were identified for intervention. In their discussion of these differences, the focus was less on the acute vs. cumulative loading distinction, but more on those task characteristics that the individual assessment tools keyed on in making their risk assessments. Specifically, it was noted that the NIOSH equation is very sensitive to the three-dimensional location of a hand-held load while spine compression is more sensitive to the deviation of the center of mass of the torso from its neutral position. The two extremes of this spectrum would be: (1) a person bending to the ground with little or no load in the hands (such as using a tape measure to mark off a wall location) and (2) a person holding a moderate weight in his or her hands at shoulder height with arms extended. In the first case, the LI would be very near zero due to the light hand-held load while there would be a significant amount of spine compression due to the body mass. In the second case, the LI could be very high due to the extreme position of the load relative to the lifter while the spine compression is relatively modest due to the neutral position of the torso. It is believed that the origin of these differences can be found in the differences between the pure biomechanical approach of the spine compression assessment and the contributions of the physiological and psychophysical aspects of lifting which played a role in the development of the NIOSH Lifting Equation. Finally, the LMM model approach is unique in its ability to consider the dynamic nature of the work demands in developing its assessment in overall risk. Since force is equal to mass times acceleration, it is logical to consider the additional forces encountered during dynamic lifting activities and this model is the only one that captures these important characteristics. This "human performance" aspect of this tool is somewhat unique among the three, however, the 3DSSPP does allow for the human performance aspect to be considered through the posture assumed by the worker.

35.6   A Hybrid Assessment Tool

The summary of the strengths and limitations of these three assessment tools in the pervious section points to the need for a tool that is able to make use the strengths of each. Further, as mentioned earlier, one limitation that each of these models share is the limited ability to characterize the risk posed by jobs with highly variable biomechanical demands. It is these concepts that are emphasized in the hybrid assessment tool known as CABS (Continuous Assessment of Back Stress) (Mirka et al., 2000).

The basic concept behind the CABS approach is that physically demanding work can be broken down into a series of individual subtasks, each of which can be adequately modeled using the three aforementioned assessment tools (Figure 35.3 and Figure 35.4). Using video analysis the cumulative amount of time spent performing each of these individual subtasks can be derived thus allowing a time-weighted histogram of the risk assessment metrics (NIOSH LI, PHRGM, spine compression, etc.) to be created (e.g., Figure 35.5). While the time-weighted histograms that are the output from this approach can be used to identify the highly visible right-hand tail high-risk activities (potential acute trauma risk activities), they can also help to identify the less prominent high-risk activities in the central portion of these distributions (potentially contributing through their percent of duty cycle characteristic to the cumulative trauma risk). In our construction industry application, there were several tasks that involved work at or near ground level that included little or no hand-held load. The contribution of these jobs to the histograms came in the form of the peaks in the central regions of the distributions of the spine compression. If we were to only prioritize for intervention based on a pure ordinal ranking of the biomechanical stress indices, these activities would not be set as priority items even though there is literature that highlights the risks in these postures (e.g., McGill, 1997; Adams and Dolan, 1995). Further, if these tasks were only infrequent and/or short-duration activities they would not have considered for intervention. It is only when one tabulates the significant amount of time that construction workers

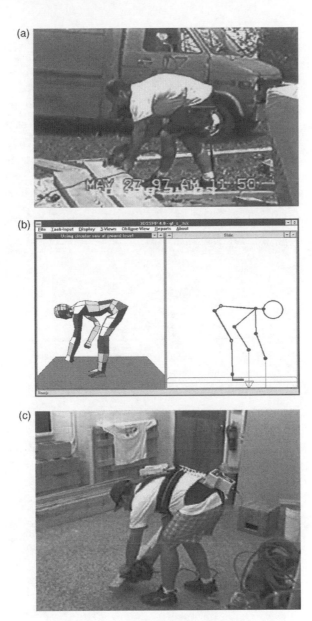

FIGURE 35.3 Three phases of modeling: (a) video capture, (b) stick figure representation for the 3DSSPP and NIOSH models, and (c) laboratory LMM simulations. (Sawing at ground level.)

spent in these postures that these activities move into the priority rankings. This illustrates that the time-weighted probabilistic representation of the biomechanical stresses gives an important insight into some of the activities performed by these workers that would have gone unnoticed with more traditional task analysis procedures.

35.7 Additional Low Back Risk Assessment Tool Needs

The main focus of these previous assessment tools has been on quantifying risk in traditional MMH tasks, that is, moving materials from one location to another. Another occupational activity that is

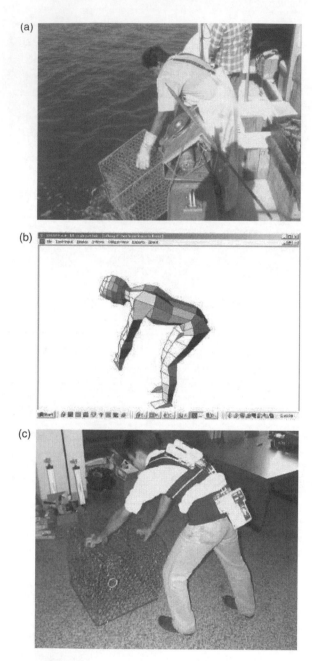

FIGURE 35.4 Three phases of modeling: (a) video capture, (b) stick figure representation for the 3DSSPP and NIOSH models, and (c) laboratory LMM simulations. (Lifting a crab pot into the deck of a boat.)

receiving more attention in the research literature is the prolonged stooped posture. There are a number of industries (agriculture and construction to name two) that require these prolonged stooped postures as part of the standard work activity and our current ability to quantify the risk associated with these postures is quite limited. In the stooped posture, the mass of the upper body is often supported through a passive mechanism that includes tension in the posterior spinal ligaments and fascia as well as compression in the spine itself. If the stooped posture is maintained, the viscoelastic properties of these passive tissues needs to be considered, which means that the time spent in the posture and the

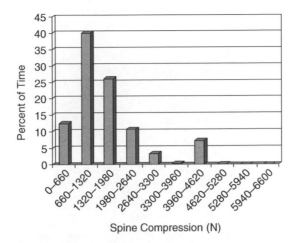

FIGURE 35.5 Distribution of probability of high-risk group membership values for framing carpenters.

required trunk posture are two important variables to consider when evaluating these work activities. Further, as stooping period continues, the passive tissues in the lumbar spine lose their stiffness and change their original dimensions, and micro-damage can occur in these tissues. These responses can result in the increase of the laxity and range of motion of the lumbar spine, and consequently, degrade the spinal stability.

A number of injury mechanisms have been proposed to understand the relationship between stooped postures and low back injury risk. A mechanism of lumbar spine instability has been suggested by *in vivo* experiments and biomechanical modeling studies. Cholewicki and McGill (1996) studied the lumbar spine stability during three-dimensional dynamic tasks and observed that there was a sufficient stability safety margin during tasks that demand a high muscular effort, whereas lighter tasks had a potential hazard of spine buckling and the risk increased if passive tissues lose their stiffness, which is the response of prolonged stooping. Another mechanism for low back disorder related to prolonged stooping has been recently introduced by Solomonow et al. (2003). These authors examined muscle activity patterns of multifidus and micro-damage in the L4/L5 supraspinous ligaments of *in vivo* feline lumbar spine during 20 min constant creep loading and 7 h recovery period. In the creep loading period, the multifidus showed exponential decrease over time and random spasms, suggesting possible decrease in the stability of the lumbar spine due to reduced muscle force and the development of inflammation in ligaments. The damage and reduced stiffness in the ligaments were not fully recovered even after 24 h. This indicates that the lumbar spine requires greater activation of multifidus muscles to maintain the stability and protect the damaged tissues even after a full day's rest. The concern might be that a similar task performed on the following day may cause continued creep deformation and severe tissue damage in the ligaments because of cumulative exposures to creep loading.

Risks associated with prolonged stooping include decreasing stability in lumbar spine structure, increasing muscle exertion level, and micro-damage in passive tissues, and these are time-dependent (stooping time and recovery time) and mainly driven by changes in physical characteristics of the posterior spinal ligaments. Assessment of these risks is quite essential to understand and prevent low back disorder from work-related prolonged stooping. No risk assessment tool of prolonged stooping has yet been developed but is the subject of on-going development research in our laboratory. The approach being pursued is to use finite element analysis (FEA) to develop a time-dependent biomechanical evaluation of the system (Figure 35.6). Mechanical changes of passive tissues in the lumbar spine under creep loading (prolonged stooping and recovery) can be simulated by modeling the lumbar spine using three-dimensional FEA technique. Nonlinear and viscoelastic material properties of ligaments and disc components have been investigated in *in vitro* experiments, and those data can be input into the FEA model.

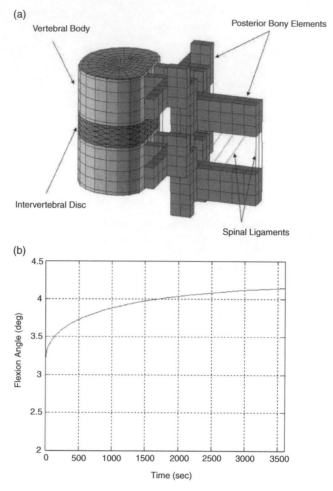

FIGURE 35.6 Finite element analysis: (a) three-dimensional FEA model of a L3/L4 motion segment; (b) predicted creep response of sagittal flexion angle during 1 h static moment loading.

The FEA model can simulate prolonged stooping and recovery, and the results (deformation and stress of passive tissues) can be used to estimate the changes in lumbar spinal stability and micro-damage as a function of time and posture. In the final assessment model the user would input the duty cycle characteristics of the alternating stooping and upright postures as well as the trunk flexion angles assumed during the stooping phase and the output would be a measure of the changes in a spinal stability index and predicted degrees of micro-damage occurring in the passive tissues.

References

Adams, M. and Dolan, P. (1995). Recent advances in lumbar spinal mechanics and their clinical significance, *Clinical Biomechanics*, 10(1):3–19.

Chaffin, D. (1974). Human strength capability and low-back pain, *Journal of Occupational Medicine*, 16:248–254.

Chaffin, D. and Park, K. (1973). A longitudinal study of low-back pain as associated with occupational weight lifting factors, *American Industrial Hygiene Association Journal*, 34:513–524.

Cholewicki, J. and McGill, S. (1996). Mechanical stability of the in vivo lumbar spine: implications for injury and chronic low back pain, *Clinical Biomechanics*, 11:1–15.

Dempsey, P. (2002). Usability of the Revised NIOSH Lifting Equation, *Ergonomics*, 45(12):817–828.

Dempsey, P. and Fathallah, F. (1999). Application issues and theoretical concerns regarding the 1991 NIOSH Equation Asymmetry Multiplier, *International Journal of Industrial Ergonomics*, 23(3):181–191.

Lavender, S., Oleske, D., Nicholson, L., Andersson, G., and Hahn, J. (1999). Comparison of five methods used to determine low back disorder risk in a manufacturing environment, *Spine*, 24:1441–1448.

Marras, W., Lavender, S., Leurgans, S., Rajulu, S., Alread, G., Fathallah, F., and Ferguson, S. (1993). The role of dynamic three-dimensional trunk motion in occupationally related low back disorders: the effects of workplace factors, trunk position and trunk motion characteristics on risk of injury, *Spine*, 18:617–628.

Marras, W., Lavender, S., Leurgans, S., Fathallah, F., Ferguson, S., Allread, G., and Rajulu, S. (1995). Biomechanical risk factors for occupationally related low back disorders, *Ergonomics*, 38:377–410.

McGill, S. (1997). The biomechanics of low back injury: implications on current practice in industry and the clinic, *Journal of Biomechanics*, 30(5):465–476.

Mirka, G., Kelaher, D., Nay, D., and Lawrence, B. (2000). Continuous assessment of back stress (CABS): a new approach to quantifying acute and cumulative low back stress, *Human Factors*, 42(2): 209–225.

Mirka, G., Monroe, M., Nay, D., Lipscomb, H., and Kelaher, D. (2003). Ergonomic interventions for the reduction of low back stress in framing carpenters in the home building industry, *International Journal of Industrial Ergonomics*, 31(6):397–409.

National Institute for Occupational Safety and Health (1981) *Work Practices Guide For Manual Lifting*, DHHS (NIOSH) Publication No. 81-122.

Solomonow, M., Baratta, R., Zhou, B., Burger, E., Zieske, A., and Gedalia, A. (2003). Muscular dysfunction elicited by creep of lumbar viscoelastic tissue, *Journal of Electromyography and Kinesiology*, 13(4):381–396.

Wang, M., Garg, A., Chang, Y., Shih, Y., Yeh, W., and Lee, C. (1998). The relationship between low back discomfort ratings and the NIOSH Lifting Index, *Human Factors*, 40:509–515.

Waters, T., Putz-Anderson, V., Garg, A., and Fine, L. (1993). Revised NIOSH equation for the design and evaluation of manual lifting tasks, *Ergonomics*, 36:749–776.

Waters, T., Putz-Anderson, V., and Garg, A. (1994). *Applications Manual for the Revised NIOSH Lifting Equation*, DHHS (NIOSH) Publication No. 94–110.

Waters, T., Baron, S., and Kemmlert, K. (1998). Accuracy of measurements for the Revised NIOSH Lifting Equation, *Applied Ergonomics*, 29(6):433–438.

Waters, T., Baron, S., Piacitelli, L., Andedrson, V., Skov, T., Haring-Sweeney, M., Wall, D., and Fine, L. (1999). Evaluation of the Revised NIOSH Lifting Equation: a cross-sectional epidemiologic study, *Spine*, 24:386–394.

36

Cognitive Task Analysis — A Review

Paul Salmon
Neville Stanton
Guy Walker
Damian Green
Brunel University

36.1 Introduction to Cognitive Task Analysis

Due to an increase in the use of technology and complicated procedures, operators of complex dynamic systems face increasing demands upon their cognitive skills and resources. During the design, development, and evaluation of these systems, designers require an analysis of the cognitive skills and demands associated with the operation of the system in question, in order to develop and design interventions, allocate tasks, develop training procedures, and evaluate operator competence and performance. While traditional task analysis outputs are sufficient when seeking a physical step-by-step description of the component task steps required during task performance, they are often inadequate when a description of the cognitive processes involved is required. When designing, supporting, and evaluating task performance in today's complex systems, a specific breakdown of the cognitive processes required during task performance is a necessity.

The past three decades has seen the emergence of cognitive task analysis (CTA) and, as a result, a number of techniques and approaches have been developed in order to aid the human factors (HF) practitioner in determining, describing, and analyzing the cognitive processes required during system operation. According to Schraagen et al. (2000), CTA represents an extension of traditional task analysis techniques and CTA techniques are used to describe the knowledge, thought processes, and goal structures underlying observable task performance. Militello and Hutton (2000) describe CTA techniques as those that focus upon describing and representing the cognitive elements that underlie goal generation, decision-making, and judgments. Redding (1989) indicated that the following components are essential to CTA: assessing individual abilities, assessing changes in knowledge base, identifying task components, identifying differences between novices and experts, identifying the conceptual and procedural knowledge of similar components, and specifying the conditions that best facilitate progression from one knowledge state to another. CTA techniques are used for a number of different purposes, and CTA outputs are used, among other things, to inform the design of procedures and processes, the design of new technology and systems, allocation of functions, the development of training procedures and interventions, and the evaluation of individual and team performance within complex systems.

Flanagan (1954) first probed the decisions and actions made by pilots in near accidents using the critical incident technique (CIT). However, the term "cognitive task analysis" did not appear until the early 1980s when it began to be used in research texts. According to Hollnagel (2003), the term was first used in 1981 to describe approaches to the understanding of the cognitive activities required of man–machine systems. Since then, the focus on the cognitive processes employed by system operators has increased, and CTA applications are now on the increase, particularly in complex, dynamic environments such as those seen in the nuclear power, defence, and emergency services domains. Various CTA techniques have been subject to widespread use over the past two decades, with applications in a number of domains, such as fire fighting (Militello and Hutton, 2000), aviation (O'Hare et al., 2000), emergency services (O'Hare et al., 2000), command and control (Chin et al., 1999), military operations (Klein, 2000), naval maintenance (Schaafstal and Schraagen, 2000), and even white-water rafting (O'Hare et al., 2000).

There are a plethora of CTA approaches available to the HF practitioner. The Cognitive Task Analysis Resource Website (www.ctaresource.com) lists over 100 CTA-related techniques designed to evaluate and describe the cognitive aspects of task performance. Roth et al. (2002) suggest that there are three different approaches to CTA. The first approach involves analyzing the domain in question in terms of goals and functions, in order to determine the cognitive demands imposed by the tasks performed. The second approach involves the use of empirical techniques, such as observation and interview techniques, in order to determine how the users perform the tasks under analysis, allowing a specification of the knowledge requirements and strategies involved. The third and more recent approach involves developing computer models that can be used to simulate the cognitive activities required during the task under analysis. Traditional CTA approaches use a combination of traditional knowledge elicitation methods such as observation, semistructured and structured interviews, and questionnaires in order to retrospectively elicit data regarding the mental processes used by system operators during task performance. For example, the cognitively orientated task analysis (COTA) framework is a collection of procedures, including verbal protocol analysis and interviews, that are used to describe the expertise involved during task performance (DuBois and Shalin, 2000). The critical decision method (CDM) (Klein et al., 1989) uses direct observation and semistructured interviews in order to analyze the cognitive processes underlying decision-making in complex environments. The applied cognitive task analysis (ACTA) approach (Militello and Hutton, 2000) uses sets of specific cognitive probes designed to elicit information regarding the cognitive processes employed during task performance. Finally, a more recent theme within CTA is the development of software packages that are designed to automate large portions of the typically lengthy CTA process and also to simulate the cognitive activities required during task performance. Software techniques such as the man–machine integration design and analysis system (MIDAS) attempt to simulate the cognitive processes required during task performance. There are a number of CTA software packages available, such as MIDAS, MicroSaint, the work domain analysis workbench (WDAW), the cognitive activity analysis toolset (CAATS), and the decompose, network, and assess (DNA) tool.

36.2 Cognitive Task Analysis of Teams

An increased use of teams in complex systems has resulted in the emergence of TCTA applications. A number of CTA techniques have been developed specifically to analyze the cognitive processes used during collaborative, team-based scenarios. According to Savoie (1998) (cited by Salas, 2004), the use of teams has risen dramatically with reports of "team presence" by workers rising from 5% in 1980 to 50% in the mid-1990s. Cooke (2004) suggests that an increase in task cognitive complexity caused by an increased use of technology has led to an increased requirement for teamwork. Salas (2004) defines a team as consisting of two or more people, dealing with multiple information resources, who work to accomplish some shared goal and suggests that while there are a number of advantages associated with the use of teams, there are also a number of disadvantages. Cooke (2004) suggests that teams are required to detect and interpret cues, remember, reason, plan, solve problems, acquire knowledge, and make decisions as an integrated and coordinated unit. Team performance comprises two components of behavior, taskwork and teamwork. Teamwork represents those instances where individuals interact or coordinate behavior in order to achieve tasks that are important to the team's goals, while taskwork describes those instances where individuals in a team setting are performing individual tasks separate from their team counterparts. Consequently, various analyses of team performance in industrial settings has led to the conclusion that team performance is extremely complex to understand and often flawed (Salas, 2004). Traditional CTA approaches such as CDM and GOMS tend to focus upon individual task performance, whereas team cognitive task analysis (TCTA) techniques focus on the cognitive components required during collaborative performance. The output of TCTA techniques is typically used to aid the design of team-based technology, the development of team training procedures, task allocation within teams, and also the organization of teams and selection of team members. Klein (2000) suggests that the main reason for conducting a TCTA is to improve the team in questions performance, and that this is achieved through using the CTA outputs to restructure the team, change the size of the team, to design better information technology, information management strategies, human computer interfaces, decision support systems, and communications, and also to develop novel team training methods.

According to Blickensderfer et al. (2000), TCTA differs from individual CTA in two ways. Firstly, a TCTA must identify, define, and describe the cognitive processes and knowledge associated with the teamwork processes (e.g., communication and coordination) involved in the task or scenario under analysis. Secondly, a TCTA technique must be able to determine the team knowledge required during the task or scenario under analysis. Klein (2000) identified a set of cognitive processes that are required during the performance of teamwork tasks. The team cognitive processes are outlined as follows (Klein, 2000):

- Control of attention — refers to the way a team engages in information management; for example, information seeking, information communication, and allocation of attention
- Shared situation awareness — refers to the degree to which the members have the same interpretation of ongoing events
- Shared mental models — refers to the extent that members have the same understanding of for the dynamics of key processes; for example, roles and functions of each team member, nature of the task, and use of equipment
- Application of strategies and heuristics to make decisions, solve problems, and plan
- Metacognition — refers to the process of self-monitoring in terms of difficulties encountered and also the limitations and vulnerabilities of the team. Effective teams are able to self-monitor their performance and shift strategies as appropriate

According to Klein (2000), a TCTA should capture and describe these five processes and then represent the findings to others so that they can be effectively used. Blickensderfer et al. (2000) conducted a review of existing CTA techniques that could potentially be applied to the analysis of team performance and highlighted a number of techniques that showed promise for use in the CTA of team performance. However, Blickensderfer et al. (2000) suggest that a thorough analysis of the cognitive components underlying team performance is lacking. Although a number of TCTA techniques do exist, a universally accepted approach to team CTA is yet to emerge.

36.3 CTA in C4i Environments

C4i (command control, communication, computers, and intelligence) systems consist of both human and technological agents and are designed to gather information and facilitate the accurate communication of this information between multiple agents dispersed across multiple locations (Salmon et al., 2004a). The tasks performed in C4i systems consist of both individual and team-based tasks, and involve multiple agents dispersed across multiple geographical locations. Effective communication and collaboration is key to achieving task goals. CTA approaches are required throughout the CADMID life-cycle of any C4i concept for a number of different purposes, such as interface design and specification, allocation of functions, evaluation of performance, and the design of processes. Due to the dynamic, dispersed and collaborative nature of C4i environments, any CTA approach used must be capable of dealing with both individuals and teams of multiple agents. Consider the example of an everyday command and control task. Command and control scenarios contain multiple agents each pursuing individual goals while simultaneously working to achieve the goals of their team. The commander at the command center issues orders and requires an accurate awareness of the global situation as it unfolds. The commanders on the ground not only receive command information and then disperse it throughout the team, they also relay information regarding the situation back to the command center. The agents "on the ground" have to act on the orders issued by the commanders, and also relay information back and forth between each other and also the commanders on the ground, while also conducting individual and/or team-based tasks. Each agent may possess both individual and shared goals and use very different cognitive processes during the performance of a particular task. The challenge facing any CTA application in C4i environments is to capture and describe not only the cognitive processes used by each agent within the team, but also the interactions between agents (coordination, communication, etc.) and the cognitive processes employed by the team as a whole.

36.4 The Problem with CTA

While its usefulness is assured, the concept of CTA is beset with a number of problems. Despite the increasing discussion of CTA in the open literature, it is apparent that there are a limited number of practical applications of the various CTA approaches available. Overall, the reported use of CTA techniques is limited, especially when compared to the use of other HF techniques, such as traditional task analysis applications (e.g., hierarchical task analysis). According to the literature, the cost in terms of time and money are the main reasons behind this limited use. A huge amount of data is generated during a typical CTA analysis. Before any data analysis can occur, observational data require coding and interview data transcribing, both of which are time-consuming procedures. When this is added to the time spent in collecting the data and then the time spent evaluating and analyzing the data, the amount of time invested is huge. According to Seamster et al. (2000), the use of CTA techniques has been limited due to the high financial and time costs and also a limited availability of adequately qualified researchers. Similarly, Shute et al. (2000) also report that CTA techniques have in the past been criticized due to the high expense (time, personnel, and money) involved.

It is also apparent that often there are shortages in available personnel who possess the appropriate skills required to adequately conduct cognitive task analyses. Seamster et al. (2000) suggest that the use of CTA techniques has been limited due to a limited availability of adequately qualified researchers (as well as the high financial and time costs). According to Shute et al. (2000), a vast number of appropriately skilled personnel are required if a CTA is to be conducted properly. A typical CTA would normally require the provision of a multidisciplinary team over a lengthy period of time. For example, a CTA of a military-based task would, at least, require the following personnel:

- Cognitive psychologists
- Human factors engineers
- System designers
- Research staff
- Various military subject matter experts

Of course, it is often difficult to assemble such multidisciplinary teams, let alone gather them together at one location for any period of time. The problems of cost and personnel shortages ensure that the process of merely getting to the stage where a CTA analysis can actually commence is a very difficult one. As a result, the problems associated with the cost, time invested, and personnel required may far outweigh the benefits associated with conducting the CTA in the first place. Consequently, organizations may be put off conducting CTA-type analyses by the cost-effectiveness issue alone, let alone the other problems associated with it.

Alongside the resource intensiveness of CTA techniques, there are also a number of associated methodological concerns. An analysis of the literature reveals a common problem associated with the format and presentation of the results of cognitive task analyses. It is apparent that once a CTA has been conducted and the results obtained, exactly what the results mean in relation to the problem goals is often difficult to understand or is often misinterpreted. It is also evident that it is often not clear what to do with the results of a CTA. Shute et al. (2000) highlight the imprecise and vague nature of CTA techniques. It seems that the great amount of resources that are invested in a CTA effort are often wasted as the output fails to be interpreted adequately by the system designers and their counterparts. Potter et al. (2000) describe a bottleneck that occurs during the transition from CTA to system design, and suggest that the information gained during the CTA must be effectively translated into design requirements and specifications.

Potter et al. (2000) conducted a review of CTA techniques in order to evaluate the current state of practice in CTA. They discovered that there was a wide diversity in the CTA techniques employed, the type of information generated, and also the manner in which the information is presented. According to Potter et al. (2000), this diversity has led to confusion as to what CTA actually refers to, what results are expected from CTA, and how the results will effect system development and evaluation. It was also concluded that typical CTA approaches are labor intensive and generate huge amounts of data, which, of course, leads to a lengthy transcription process.

Schraagen et al. (2000) conducted a review of existing CTA techniques and computer-based CTA tools. The review indicated that although there were a large number of CTA techniques available, they were generally limited. It was also concluded that there is limited guidance available in assisting practitioners in the selection of the most appropriate CTA techniques, in how to use the available CTA techniques, and also how to use the output of the CTA. As a result of the review of existing CTA techniques, Schraagen et al. (2000) identified the following issues surrounding CTA that require further investigation:

1. Validity and reliability of CTA techniques
2. CTA for novel tasks and systems
3. CTA for tasks that are difficult to verbalize (e.g., spatial tasks)
4. Requirements for analysts (e.g., level of training required)
5. Conditions under which CTA techniques should be employed
6. Relation between the purpose of the CTA and the results of the activity
7. Team CTA
8. Relationship between CTA techniques and theories of cognition
9. Individual differences and their implications for solutions
10. Effects of environmental stressors on the conduct of CTA
11. Defining and selecting expertise

12. Human operator/performance modeling systems
13. CTA for safety critical systems, human reliability analysis

36.5   Current Trends and the Future of CTA

As a result of the high resources invested when conducting CTA projects, the current trend appears to be the automation of the CTA process or, at least, part of the CTA process. The most common approach to this is the use of CTA software packages that remove some of the more resource intensive aspects of CTA. Computer models such as MIDAS attempt to simulate the cognitive processes required during task performance. At present there are numerous software packages that claim to be CTA orientated. Of course, while the provision of a software package that effectively automates part of the CTA process is useful, the extent to which this removes the burden of cost and time invested is questionable. Typically, software packages are expensive to purchase, and a lengthy training process is often required before analysts become proficient enough to actually use them. Furthermore, the extent to which a software package can remove the time-consuming process of data collection for CTA is questionable, and so observation and interviews may still be required, which are in themselves extremely time consuming. A universally accepted CTA software package is yet to emerge, and it remains to be seen whether one will. In the past, numerous attempts were made to automate the hierarchical task analysis (HTA) (Annett et al., 1971) procedure through the provision of software (Bass et al., 1995), and as yet none of these have succeeded in removing the traditional pen and paper approach.

As described previously in this report, TCTA is another area that is currently receiving considerable attention from the HF research community. The majority of work in complex and dynamic systems is now carried out using teams of multiple actors with varying skills. Complex, dynamic tasks are carried out by teams in a wide range of domains, such as aviation, ATC, the military, control room operation, and emergency services. According to Lesgold (2000), CTA of tasks that are performed by groups requires considerable additional work. The provision of valid and reliable CTA techniques for use in analyzing team processes poses a great challenge to the HF community, and such approaches are required for the analysis of tasks in the military, emergency services, and nuclear power domains.

Finally, one emerging theme currently receiving attention in the literature is the provision of a framework or toolkit of CTA techniques used by the HF practitioner in order to elicit specific information regarding the cognitive components of the task under analysis. Potter et al. (2000) suggest that there is more to CTA than the application of a single CTA technique, and that understanding a particular field relies upon multiple converging techniques. The toolkit or framework approach is not a new one, and has in the past been used in other psychological fields such as the prediction of human error (Kirwan, 1998a,b) and the measurement of mental workload.

36.6   CTA Methods Review

The CTA methods review was conducted as part the Human Factors Integration Defence Technology Center (HFI DTC) for "Human Factors design and evaluation methods review" (Salmon et al., 2004b). The CTA methods review was conducted over three stages. Firstly, an initial literature review of existing CTA methods and techniques was conducted. Secondly, a screening process was employed in order to remove any unsuitable techniques from the review. Thirdly, the CTA techniques selected for review were analyzed using a set of predetermined HF methods criteria. Each stage of the CTA methods review is described in more detail subsequently.

36.6.1   Stage 1: Initial Literature Review of Existing CTA Methods

A literature review was conducted in order to create a database of existing CTA techniques. The purpose of this literature review was to provide the authors with a comprehensive database of available CTA techniques and their associated source(s), author(s), and availability. For the purposes of this review, the following categories were used to determine the availability of the techniques identified in the literature review:

1. *Off the shelf* — includes CTA techniques that can be purchased at a financial cost. Once purchased the technique can be used freely by the owner(s)
2. *Proprietary* — includes CTA techniques that have been developed by HF consultancies and other organizations. For a financial cost, the creators will conduct a CTA analysis using the technique in question
3. *Free* — includes CTA techniques that are freely available in the public domain and can be used without the author's permission
4. *Software* — includes software-based CTA techniques and add-ons that can be purchased at a financial cost

The result of this initial literature review was a database of over 50 CTA techniques. The CTA technique database is presented in Table 36.1.

TABLE 36.1   CTA Technique Database and Descriptions

Technique	Source	Availability
Applied cognitive task analysis (ACTA)	Militello and Hutton (2000)	Free
Activity sampling	Various	Free
Applied cognitive work analysis (ACWA)	www.ctaresource.com	Free
APEX	www.andrew.cmu.edu/~bj07/apex/ index.html	Software tool
ATLAS	www.humaneng.co.uk	Software tool
Cognitive analysis tool (CAT)	www.ctaresource.com	Software tool
Critical decision method (CDM)	Klein et al. (1989), www.decisionmaking.com	Free and proprietary
Critical incident technique (CIT)	Flanagan (1954)	Free
The Cloze technique	www.ctaresource.com	Free
Cognitive walkthrough	Polson et al. (1992)	Free and proprietary
Cognitive work analysis (CWA)	Vicente (1999)	Free
Cognitive objects within a graphical environment (COGENT)	Zachary et al. (2000)	Software tool
Cognition as a network of tasks (COGNET)	www.ctaresource.com	Software tool
Concept mapping	www.ctaresource.com	Free
Content analysis	www.ctaresource.com	Free
Cognitively oriented task analysis (COTA)	DuBois and Shalin (2000)	Free and proprietary
Cognitive systems engineering (CSE)	www.csecenter.dk	Free and proprietary
Cognitive task load analysis (CTLA)	Neerincx (2003)	Free
Decompose, network and assess method (DNA)	Shute et al. (2000)	Software tool
Team decision requirements exercise (DRX)	Klinger and Hahn (2004)	Free and proprietary
Event tree analysis	Kirwan and Ainsworth (1992)	Free
Storyboarding	Various	Free
Functional analysis system technique (FAST)		Software tool
Goals, operators, methods and selection rules (GOMS)	Card et al. (1983)	Free and proprietary
Hierarchical task analysis (HTA)	Annett et al. (1971)	Free proprietary
HTA(T) — team hierarchical task analysis	Annett (2004)	Free and proprietary
IGEN	www.ctaresource.com	Software tool
Interviews	Various	Free and proprietary

(*Table continued*)

TABLE 36.1   *Continued*

Technique	Source	Availability
Interruption analysis	www.ctaresource.com	Free
Micro Saint	www.maad.com	Software tool
The man–machine integration design and analysis system (MIDAS)	http://caffeine.arc.nasa.gov/midas/ Research_Team.html	Software tool
Minimal scenario technique	www.ctaresource.com	Free
Observation	Various	Free and proprietary
Operator model architecture (OMAR)	www.omar.bbn.com	Software tool
Precursor, action, result and interpretation (PARI)	Hall et al. (1995)	Free
Questionnaires	Various	Free and off the shelf
Sub-goal template methodology (SGT)	Ormerod (2000)	Free and proprietary
Task analysis for knowledge descriptions (TAKD)	Diaper and Johnson (1989)	No longer used, free
Team cognitive task analysis (TCTA)	Klein (2000)	Free and proprietary
Team task analysis	Burke (2004)	Free
Repertory grid analysis	Baber (2004)	Free
Role play/scenarios	Various	Free
Shadowing	www.ctaresource.com	Free
Table top analysis	www.ctaresource.com	Free
Talkthrough analysis	Kirwan and Ainsworth (1992)	Free
Twenty questions	www.ctaresource.com	Free
Verbal protocol analysis (VPA)	Walker (2004)	Free
Walkthrough analysis	Kirwan and Ainsworth (1992)	Free
Work domain analysis	Vicente (1999)	Free

36.6.2   Stage 2: Initial Methods Screening

Before the CTA techniques were subjected to further analysis, a screening process was employed in order to discard any techniques that were deemed unsuitable for review with respect to their use during the design, development, and evaluation of C4i systems. Techniques were classed as unsuitable for review if they fell into the following categories:

- *Unavailable* — Those techniques subject to review should be freely available in the public domain. Any techniques that were unavailable were rejected
- *Software* — Software-based techniques are time consuming to acquire (process of ordering and delivery) and often require a lengthy training process. Any CTA software tools (e.g., ATLAS) were rejected
- *Inapplicable* — The applicability of each technique to the analysis of C4i scenarios was evaluated. Those techniques deemed unsuitable for the use in the analysis of behavior in armed forces were rejected
- *Misclassification* — It was felt that a number of the techniques encountered in CTA literature that were labeled as CTA techniques were in fact not specifically designed for use in CTA applications

As a result of the methods screening process, 18 of the CTA techniques identified were selected for further analysis as part of the methods review. The CTA techniques reviewed are presented in Table 36.2.

36.6.3   Stage 3: Methods Review

The 18 CTA techniques were then analyzed using the set of predetermined criteria outlined subsequently. The criteria were designed not only to establish which of the techniques were the most suitable for use in the evaluation of U.K. armed forces behavior, but also to aid the HF practitioner in the selection and use

TABLE 36.2 CTA Techniques Subject to Methods Review

Method	Author/Source
ACTA — Applied cognitive task analysis	Militello and Hutton (2000)
CDM — Critical decision method	Klein et al. (1989)
CIT — Critical incident technique	Flanagan (1954)
Cognitive walkthrough	Polson et al. (1992)
COTA — Cognitively oriented task analysis	DuBois and Shalin (2000)
CTLA — Cognitive task load analysis	Neerincx (2003)
CWA — Cognitive work analysis	Rasmussen et al. (1994), Vicente (1999)
DRX — Decision requirements exercise	Klinger and Hahn (2004)
GOMS — Goals, operators, methods and selection rules	Card et al. (1983)
HTA — Hierarchical task analysis	Annett et al. (1971)
HTA(T) — Team hierarchical task analysis	Annett (2005)
Interviews	Various
Observation	Various
SGT — Sub-goal template methodology	Ormerod (2000)
TCTA — Team cognitive task analysis	Klein (2000)
TTA — Team task analysis	Burke (2004)
VPA — Verbal protocol analysis	Walker (2004)
Walkthrough/talkthrough analysis	Kirwan and Ainsworth (1992)

of the appropriate methods. The output of the analysis is designed to act as a CTA techniques manual, aiding practitioners in the use of the CTA techniques reviewed:

1. *Name and acronym* — The name of the technique and its associated acronym
2. *Author(s), affiliations(s) and address(es)* — *The names, affiliations, and addresses of the authors are provided to assist with citation and requesting any further help in using the technique*
3. *Availability* — The availability of the technique is specified. Techniques are classed as free, off the shelf, proprietary, or as software add-ons or plug-ins
4. *Background and applications* — This section introduces the method, its origins and development, the domain of application of the method, and also application areas that it has been used in
5. *Domain of application* — Describes the domain that the technique was originally developed for and applied in
6. *Team/individual technique* — Denotes which aspects of performance the technique caters for
7. *Experts required* — This section attempts to clarify whether SMEs are required as either analysts or participants
8. *Procedure and advice* — This section describes the procedure for applying the method as well as general points of expert advice
9. *Flowchart* — A flowchart is provided, depicting the methods procedure
10. *Advantages* — Lists the advantages associated with using the method in the design of C4i systems
11. *Disadvantages* — Lists the disadvantages associated with using the method in the design of C4i systems
12. *Example* — An example, or examples, of the application of the method are provided to show the methods output
13. *Related methods* — Any closely related methods are listed, including contributory and similar methods
14. *Approximate training and application times* — Estimates of the training and application times are provided to give the reader an idea of the commitment required when using the technique
15. *Reliability and validity* — Any evidence on the reliability or validity of the method are cited
16. *Tools needed* — Describes any additional tools required when using the method
17. *Bibliography* — A bibliography lists recommended further reading on the method and the surrounding topic area

A summary of the CTA methods review is presented in Table 36.3. A brief description of the methods reviewed is presented below.

36.6.3.1 Applied Cognitive Task Analysis (Militello and Hutton, 2000)

Applied cognitive task analysis (ACTA) uses specific interview techniques in order to determine the cognitive skills and demands associated with a particular task or scenario. ACTA was developed as part of a Navy Personnel Research and Development Center funded project as a solution to the inaccessibility and difficulty associated with using existing cognitive task analysis type methods (Militello and Hutton, 2000). The ACTA procedure consists of the following components:

1. *Task diagram interview.* The task diagram interview is used to provide an overview of the task under analysis. The task diagram interview also allows the analyst to identify any cognitive aspects of the task that require further analysis.
2. *Knowledge audit.* The knowledge audit allows the analyst to determine the expertise required for each part of the task. During the knowledge audit interview, the analyst probes subject matter experts (SMEs) for specific examples of expertise that were employed during task performance.
3. *Simulation interview.* The simulation interview allows the analyst to probe specific cognitive aspects of the task under analysis. The analyst probes SMEs using specific probes that are designed to elicit information regarding the decisions made and the associated actions and information requirements.
4. *Cognitive demands table.* Once the task diagram, knowledge audit, and simulation interviews are complete, the cognitive demands table is used to group and sort the data.

The ACTA probes are presented in Table 36.4.

36.6.3.2 Cognitive Walkthrough (Polson et al., 1992)

The cognitive walkthrough technique is used to evaluate the usability of user interfaces. Based upon traditional design walkthrough techniques and a theory of exploratory learning (Polson and Lewis, 1990), the technique focuses upon the usability of an interface, in particular the ease of learning associated with the interface. A set of criteria is used to evaluate interfaces and their associated tasks. The criteria focus on the cognitive processes required to perform the tasks in question (Polson et al., 1992). The cognitive walkthrough procedure involves the analyst "walking" through each user action involved in a task step. The analyst then considers each criterion and the effect the interface has upon the user's goals and actions.

36.6.3.3 Critical Decision Method (Klein et al., 1989)

Probably the most commonly recognized and used of the CTA approaches, the critical decision method (CDM) is a semistructured interview technique that uses a set of cognitive probes in order to elicit information regarding expert decision-making. According to the authors, the technique can serve to provide knowledge engineering for expert system development, identify training requirements, generate training materials, and evaluate the task performance impact of expert systems (Klein et al., 1989). The CDM technique was developed in order to study naturalistic decision-making strategies of experienced personnel. CDM involves interviewing SMEs and probing them regarding the decision-making strategies employed during a particular incident, task, or scenario. CDM has been applied in a number of domains involving the operation of complex and dynamic systems, including fire fighting, military, and paramedics (Klein et al., 1989) and white water rafting (O'Hare et al., 2000). A set of CDM probes are presented in Table 36.5. A CDM analysis was conducted for a civil energy distribution task in order to analyse C4i activity (Salmon et al., 2004c). As a result of the CDM, the scenario was divided into four key phases: "first issue of instructions," "deal with switching requests," "perform isolation" and "report back to NOC." The CDM output for the "perform isolation" phase is presented in

TABLE 36.3 Summary of CTA Methods Review

Method	Domain	Team or Individual	Experts Required	Training Time	App. Time	Related Methods	Tools Needed	Validation Studies	Advantages	Disadvantages
ACTA	Generic	Individual	Yes	Med–high	High	Interviews Critical decision method	Pen and paper Audio recording equipment	Yes	Requires fewer resources than traditional cognitive task analysis techniques Provides the analyst with a set of probes	Great skill is required on behalf of the analyst for the technique to achieve its full potential Consistency/reliability of the technique is questionable Time consuming in its application
Cognitive walkthrough	Generic	Individual	Yes	High	High	HTA	Pen and paper Video and audio recording equipment	Yes	Has a sound theoretical underpinning (Normans action execution model) Offers a very useful output.	Requires further validity and reliability testing Time consuming in application Great skill is required on behalf of the analyst for the technique to achieve its full potential
Critical decision method	Generic	Individual	Yes	Med–High	High	Critical incident technique	Pen and paper Audio recording equipment	Yes	Can be used to elicit specific information regarding decision-making in complex environments Seems suited to C4i analysis Various cognitive probes are provided	Reliability is questionable There are numerous problems associated with recalling past events, such as memory degradation Great skill is required on behalf of the analyst for the technique to achieve its full potential
Team cognitive task analysis	Generic (military)	Team	Yes	High	High	Observation Interviews Critical decision method	Pen and paper Video and audio recording equipment	Yes	Can be used to elicit specific information regarding team decision-making in complex environments Seems suited to use in the analysis of C4i activity Output can be used to develop effective team decision-making strategies	Reliability is questionable Resource intensive High level of training and expertise is required in order to use the technique properly

(Table continued)

TABLE 36.3 *Continued*

Method	Domain	Team or Individual	Experts Required	Training Time	App. Time	Related Methods	Tools Needed	Validation Studies	Advantages	Disadvantages
Critical incident technique	Generic	Individual	Yes	Med–high	High	Critical decision method	Pen and paper Audio recording equipment	Yes	Can be used to elicit specific information regarding decision-making in complex environments; Seems suited to C4i analysis	Reliability is questionable; There are numerous problems associated with recalling past events, such as memory degradation; Great skill is required on behalf of the analyst for the technique to achieve its full potential
Team task analysis	Generic	Team	Yes	Med	Med	Coordination demand analysis Observation	Pen and paper	No	Output specifies the knowledge, skills, and abilities required during task performance; Useful for team training procedures; Specifies which of the tasks are team based and which are individual based	Time consuming in application; SME's are required throughout the procedure; Great skill is required on behalf of the analyst(s)
Decision requirements exercise	Generic (military)	Team	Yes	Med	Med–high	Critical decision method Observation	Pen and paper Video and audio recording equipment	No	Output is very useful, offering an analysis of team decision-making in a task or scenario; Based upon actual incidents, removing the need for simulation; Seems suited to use in the analysis of C4i activity	Data are based upon past events, which may be subject to memory degradation; Reliability is questionable; May be time consuming
CWA — Cognitive work analysis	Generic	Team and Individual	Yes	High	High	Abstraction hierarchy Decision ladders Information flow maps	Pen and paper Video audio recording equipment	No	Offers an extremely exhaustive analysis of the system in question; Has been used extensively in a number of domains such as command and control and the military; Can be used at any stage in the CADMID or design lifecycle	Extremely resource intensive, e.g., high time, cost, and resources intensive; May be laborious and unwieldy for complex systems

Technique	Type	Analyst					Methods	Tools	Advantages	Disadvantages
COTA — Cognitively oriented task analysis	Generic	Individual	Yes	High	High	No	Interviews Observation Plan goal graphs	Pen and paper Video audio recording equipment	Provides a description of the knowledge required during task performance COTA is only applied to tasks "in the field"	The COTA procedure is a time-consuming one, and the analysis may become unwieldy for large, complex tasks Limited evidence of the techniques usage and validation COTA only describes the knowledge required and does not consider the cognitive processes employed during task performance
Interviews	Generic	Individual	Yes	Med–high	High	Yes	Interviews Critical decision method	Pen and paper Audio recording equipment	Flexible technique that can be used to assess anything from usability to error Interviewer can direct the analysis Can be used to elicit data regarding cognitive components of a task	Data analysis is time consuming and laborious Reliability is difficult to assess Subject to various source of bias
Observation	Generic	Team and individual	Yes	Low	High	Yes	Acts as an input to various HF methods, e.g., HTA	Pen and paper Video and audio recording equipment	Can be used to elicit specific information regarding decision-making in complex environments Acts as the input to numerous HF techniques such as HTA Suited to the analysis of C4i activity	Data analysis procedure is very time consuming Coding data is also laborious Subject to bias
CTLA — Cognitive task load analysis	Generic	Individual	Yes	High	High	No	Operator sequence diagrams Observation Interviews Questionnairs	Pen and paper Video and audio recording equipment	Based upon sound theoretical underpinning Can be used to identify high workload tasks	Complex technique requiring a high level of training Extremely resource intensive, e.g., high time, cost, and resources intensive; also may be laborious and unwieldy for complex systems Limited evidence of the techniques usage

(Table continued)

TABLE 36.3 *Continued*

Method	Domain	Team or Individual	Experts Required	Training Time	App. Time	Related Methods	Tools Needed	Validation Studies	Advantages	Disadvantages
HTA — Hierarchical task analysis	Generic	Individual	No	Med	Med	HEI Task analysis	Pen and paper	Yes	HTA output feeds into numerous HF techniques Has been used extensively in a variety of domains Provides an accurate description of task activity	Provides mainly descriptive information Cannot cater for the cognitive components of task performance Can be time consuming to conduct for large, complex tasks
HTA(T)	Generic	Team	No	Med	Med	HEI Task analysis	Pen and paper	Yes	Team HTA based upon extensively used HTA technique Caters for team-based tasks	Limited use
GOMS — Goals, operators, methods and selection rules	HCI	Individual	No	Med–high	Med–high	NGOMSL CMN-GOMS KLM CPM-GOMS	Pen and paper	Yes No outside of HCI	Provides a hierarchical description of task activity	May be difficult to learn and apply for non-HCI practitioners Time consuming in its application Remains unvalidated outside of HCO domain
VPA — Verbal protocol analysis	Generic	Individual	Yes	Low	High	Walkthrough analysis	Audio recording equipment Observer software PC	Yes	Rich data source Verbalizations can give a genuine insight into cognitive processes Easy to conduct, providing the correct equipment is used	The data analysis process is very time consuming and laborious It is often difficult to verbalize cognitive behaviour Verbalizations intrude upon primary task performance
The sub-goal template method	Generic	Individual	No	Med	High	HTA	Pen and paper	No	The output is very useful. Information requirements for the task under analysis are specified	Techniques required further testing regarding reliability and validity Can be time consuming in its application
Walkthrough analysis	Generic	Individual	Yes	Low	Low	Talkthrough analysis	Pen and paper	No	Quick and easy to use involving little training and cost Allows the analyst(s) to understand the physical actions involved in the performance of a task Very flexible	SMEs required Access to the system under analysis is required Reliability is questionable

TABLE 36.4 ACTA Probes

	Basic probes
Past and future	Is there a time when you walked into the middle of a situation and knew exactly how things got there and where they were headed?
Big picture	Can you give me an example of what is important about the big picture for this task? What are the major elements you have to know and keep track of?
Noticing	Have you had experiences where part of a situation just "popped" out at you; where you noticed things going on that others did not catch? What is an example?
Job smarts	When you do this task, are there ways of working smart or accomplishing more with less — that you have found especially useful?
Opportunities/improvising	Can you think of an example when you have improvised in this task or noticed an opportunity to do something better?
Self-monitoring	Can you think of a time when you realized that you would need to change the way you were performing in order to get the job done?
	Optional probes
Anomalies	Can you describe an instance when you spotted a deviation from the norm, or knew something was amiss?
Equipment difficulties	Have there been times when the equipment pointed in one direction but your own judgment told you to do something else? Or when you had to rely on experience to avoid being led astray by the equipment?
	Simulation interview probes

As the (job you are investigating) in this scenario, what actions, if any, would you take at this point in time?
What do you think is going on here? What is your assessment of the situation at this point in time?
What pieces of information led you to this situation assessment and these actions?
What errors would an inexperienced person be likely to make in this situation?

Source: Militello, L. G., and Hutton, J. B. (2000). In J. Annett and N. A. Stanton (Eds), *Task Analysis*, pp. 90–113. London: Taylor & Francis. With permission.

Table 36.6. From the CDM analyses, it is possible to develop a propositional network that represents the ideal collection of knowledge objects for the scenario. A propositional network for the scenario is presented in Figure 36.1.

36.6.3.4   Critical Incident Technique (Flanagan, 1954)

The technique that most associate with the origins of CTA, the critical incident technique (CIT) was first used by Flanagan (1954) to analyze aircraft incidents that almost led to accidents and has since been used extensively and developed in the form of CDM (Klein et al., 1989). The CIT involves using an interview-based approach in order to facilitate operator recall of critical events or incidents, including the actions and decisions made by themselves and colleagues and why they made them. Examples of the CIT probes used by Flanagan (1954) are:

- Describe what led up to the situation
- Exactly what did the person do or not do that was especially effective or ineffective
- What was the outcome or result of this action?
- Why was this action effective or what more effective action might have been expected?

36.6.3.5   Cognitively Oriented Task Analysis (DuBois and Shalin, 2000)

The cognitively oriented task analysis (COTA) approach comprises a set of procedures that are used to describe the expertise that supports overall job performance (DuBois and Shalin, 2000). According to DuBois and Shalin (2000), the COTA approach comprises three phases: planning, describing job expertise, and developing CTA products. The planning phase has three main aims, which are to define the project goals, resources, and constraints, to develop a preliminary description of work context and tasks performed, and to specify a sampling plan for conducting the required knowledge elicitation

TABLE 36.5 CDM Probes

Goal Specification	What Were Your Specific Goals at the Various Decision Points?
Cue identification	What features were you looking for when you formulated your decision? How did you that you needed to make the decision? How did you know when to make the decision?
Expectancy	Were you expecting to make this sort of decision during the course of the event? Describe how this affected your decision-making process
Conceptual	Are there any situations in which your decision would have turned out differently? Describe the nature of these situations and the characteristics that would have changed the outcome of your decision?
Influence of uncertainty	At any stage, were you uncertain about either the reliability of the relevance of the information that you had available? At any stage, were you uncertain about the appropriateness of the decision?
Information integration	What was the most important piece of information that you used to formulate the decision?
Situation awareness	What information did you have available to you at the time of the decision?
Situation assessment	Did you use all of the information available to you when formulating the decision? Was there any additional information that you might have used to assist in the formulation of the decision?
Options	Were there any other alternatives available to you other than the decision you made?
Decision blocking — stress	Was their any stage during the decision-making process in which you found it difficult to process and integrate the information available? Describe precisely the nature of the situation?
Basis of choice	Do you think that you could develop a rule, based on your experience, which could assist another person to make the same decision successfully? Why/why not?
Analogy/generalization	Were you at any time reminded of previous experiences in which a similar decision was made? Were you at any time, reminded of previous experiences in which a different decision was made?

Source: O'Hare, D., Wiggins, M., Williams, A., and Wong, W. (2000). In J. Annett and N. A. Stanton (Eds), *Task Analysis*, pp. 170–190. London: Taylor & Francis. With permission.

activities. The COTA approach uses interviews with relevant SMEs in order to achieve these aims. In describing job expertise, the COTA approach uses videotaped protocol analysis of task performance in order to determine and describe the knowledge required during task performance. The final stage of a COTA analysis, developing CTA products, involves transforming the knowledge representation into appropriate inputs for the specified application.

36.6.3.6 Cognitive Task Load Analysis (Neerincx, 2003)

Cognitive task load analysis (CTLA) is used to assess or predict the cognitive load of a task or set of tasks imposed upon an operator. CTLA is typically used early in the design process is based upon a model of cognitive task load (Neerincx, 2003) that describes the effects of task characteristics upon operator mental workload. According to the model, cognitive (or mental) task load is comprised of percentage time occupied, level of information processing, and the number of task set switches exhibited during the task. According to Neerincx (2003), the operator should not be occupied by one task for more than 70 to 80% of the total time. The level of information processing is defined using the SRK framework (Rasmussen, 1986). Finally, task set switches are defined by changes of applicable task knowledge on the operating and environmental level exhibited by the operators under analysis (Neerincx, 2003). The three

TABLE 36.6   Phase 3: Perform Isolation

Goal specification	Ensure it is safe to perform local isolation
	Confirm circuits/equipment to be operated
Cue identification	Telecontrol displays/circuit loadings
	Equipment labels
	Equipment displays
	Other temporary notices
Expectancy	Equipment configured according to planned circuit switching
	Equipment will function correctly
Conceptual model	Layout/type/characteristics of circuit
	Circuit loadings/balance
	Function of equipment
Uncertainty	Will equipment physically work as expected (will something jam, etc.)
	Other work being carried out by other parties (e.g., EDF)
Information	Switching log
	Visual and verbal information from those undertaking the work
Situation awareness	Physical information from apparatus and telecontrol displays
Situation assessment	All information used
Options	Inform NOC that isolation cannot be performed/other aspects of switching
	instructions cannot be carried out
Stress	Some time pressure
	Possibly some difficulties in operating or physically handling the equipment
Choice	Yes — proceduralized within equipment types. Occasional nonroutine activities
	required to cope with unusual/unfamiliar equipment, or equipment not owned
	by NGT
Analogy	Yes — often. Except in cases with unfamiliar equipment

Source: Salmon, P. M., Stanton, N. A., and Walker, G. (2004c). Work Package 1.1.3: NGT Switching Scenario Report. Defence Technology Centre for Human Factors Integration Report.

variables (time occupied, level of information processing, and task set switches) are combined to determine the level of cognitive load imposed by the task. High ratings for the three variables equal a high cognitive load imposed on the operator by the task.

36.6.3.7   Cognitive Work Analysis (Vicente, 1999)

Cognitive work analysis (CWA) (Vicente, 1999) is a framework that is used to model and analyze decision-making in complex environments. A CWA involves five main stages. These are work domain analysis, activity analysis, strategies analysis, socio-organizational analysis, and worker competencies analysis. The work domain analysis involves describing the work system under analysis in terms of its functions and constraints using an abstraction decomposition space (ADS). The activity analysis component involves identifying the tasks that need to be performed in the work domain under analysis. Strategies analysis involves identifying the mental strategies that the agents involved may use during task performance in the domain under analysis. Social organization analysis involves identifying exactly how the work is distributed among the agents and artifacts within the system under analysis. Finally, worker competencies analysis involves the identification of the competencies that the agents involved are required to possess in order to perform the tasks in the work domain. The worker competencies are classified using Rasmussen's skill, rule, knowledge (SRK) framework. The CWA is an exhaustive procedure that utilizes a number of different methods, including an abstraction hierarchy, decision ladders, flowcharts, and the SRK classification. As part of a work domain analysis, abstraction and decomposition hierarchies are used to develop an "abstraction decomposition space." An ADS is comprised of an abstraction hierarchy and a decomposition hierarchy. The abstraction hierarchy consists of five levels of abstraction, ranging from the most abstract level of purposes to the most concrete level of form (Vicente, 1999). The decomposition hierarchy (the top row in the abstraction-decomposition space) comprises five levels of resolution, ranging from the coarsest level of total system to the finest level of component. According to Vicente (1999), each of the five levels represents a different level of granularity with respect to the system in

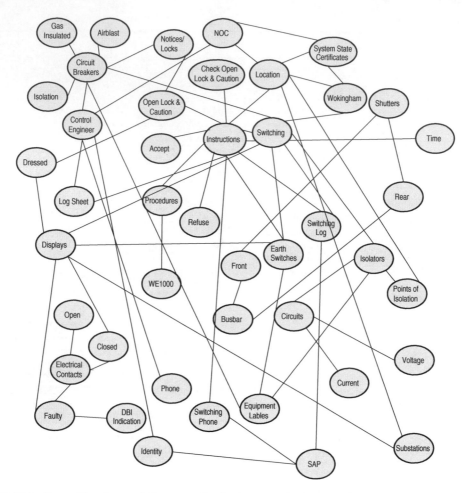

FIGURE 36.1   Propositional network for energy distribution scenario. (*Source:* Salmon, P. M., Stanton, N. A., and Walker, G. (2004c). Work Package 1.1.3: NGT Switching Scenario Report. Defence Technology Centre for Human Factors Integation Report.)

question (moving from left to right across the decomposition hierarchy is the equivalent of zooming into the system). An ADS can be used to develop so-called problem solving trajectories (Vicente, 1999) for tasks. Problem solving trajectories demonstrate how actors switch between different models of a particular work system in order to match their task demands. For example, Vicente (1999) describes how verbal protocol analysis (VPA) data of electronic technicians engaged in troubleshooting computer equipment were mapped onto an ADS in order to indicate the different ways (e.g. abstractions and decompositions) in which technicians were thinking about the work domain. An example of a task problem solving trajectory for a rail signaling task is presented in Figure 36.2 (Walker and Stanton, 2004). Each component task step was mapped onto an ADS to demonstrate how the platform staff and signaler were thinking about the system when performing each task.

Task list for rail signaler task:

1. Platform staff have train for detachment
2. Platform staff call signal box
3. Signaler states who they are
4. Platform staff state who they are and location
5. Platform staff justify why line was requested
6. Platform staff state which line is to be blocked

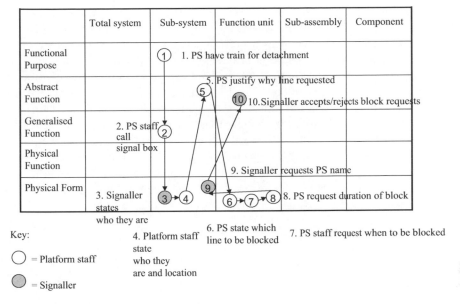

FIGURE 36.2 Decomposition space for rail signaling task.

7. Platform staff request when to be blocked
8. Platform staff request duration of block
9. Signaler requests platform staff name
10. Signaler accepts/rejects block requests

36.6.3.8 GOMS (Card et al., 1983)

The GOMS technique is part of a family of HCI techniques that is used to provide a description of human performance in terms of the user's goals, operators, methods, and selection rules. GOMS attempts to define the user's goals, decompose these goals into subgoals, and demonstrate how the goals are achieved through user interaction. GOMS can be used to provide a description of how a user performs a task, to predict performance times, and to predict human learning. The GOMS techniques are based upon the assumption that the user's interaction with a computer is similar to solving problems. Problems are broken down into subproblems, and these subproblems are broken down further. Four basic components of human interaction are used within the GOMS technique. These are as follows:

1. *Goals* — The goal represents exactly what the user wishes to achieve through the interaction. The goals are decomposed until an appropriate stopping point is achieved
2. *Operators* — The operators are the motor or cognitive actions that the user performs during the interaction. The goals are achieved through performing the operators
3. *Methods* — The methods describe the user's procedures for accomplishing the goals in terms of operators and subgoals. Often there are more than one set of methods available to the user
4. *Selection rules* — When there is more than one method for achieving a goal available to a user, selection rules highlight which of the available methods should be used

36.6.3.9 Hierarchical Task Analysis (Annett et al., 1971)

Hierarchical task analysis (HTA) is perhaps the most widely used of all HF techniques. Developed in order to analyze complex tasks, such as those found in the chemical processing and power generation industries (Annett, 2004), HTA involves describing tasks by breaking them down into a hierarchy of goals, operations, and plans. Tasks are broken down into hierarchical set of task goals and component subtasks and the plans used to dictate the performance of the tasks and subtasks are specified. One of the main reasons behind the enduring popularity of the technique is its flexibility and scope for further analysis of the subgoal hierarchy that it offers to the HF practitioner. The majority of HF analysis methods either require an initial HTA of the tasks under analysis as their input, or at least are made significantly easier through the provision of an HTA. HTA acts as an input into various HF analyses, such human error identification (HEI), allocation of function, workload assessment, interface design and evaluation, and many more. Consequently, HTA has been applied in a number of domains, including the process control and power generation industries (Annett, 2004), emergency services (Baber et al., 2004) military applications (Kirwan and Ainsworth, 1992; Ainsworth and Marshall, 2000), civil aviation (Marshall et al., 2003), driving (Walker, 2005) public technology (Stanton and Stevenage, 1998) and even retail (Shepherd, 2001). A HTA of the fire brigade training scenario "Hazardous chemical spillage at remote farmhouse" (Baber et al, 2004) is presented in Figure 36.3. Annett (2004) describes the application of HTA to a team-based naval warfare task, the purpose of which was to identify and measure team skills critical to successful antisubmarine warfare. An extract from the team HTA is presented in Figure 36.4. The analysis is also presented in tabular format in Table 36.7.

36.6.3.10   Verbal Protocol Analysis

Verbal protocol analysis (VPA) is also commonly used during CTA applications in order to provide an insight into the processes, cognitive and physical, that an individual employs during task performance. VPA involves creating a written transcript of operator behavior as they perform the task under analysis while verbally "thinking aloud." VPA has been used extensively as a means of gaining an insight into the cognitive aspects of complex behaviours. Walker (2004) reports the use of VPA in areas such as steel melting, Internet usability, and driving (Walker et al., 2001).

The subgoal template (SGT) method is a development of HTA (Annett et al., 1971) that is used to specify information requirements to system designers. The SGT technique was initially devised as a means of redescribing the output of HTA, in order to specify the relevant information requirements for the task or system under analysis (Ormerod, 2000). Although the technique was originally designed for use in the process control industries, Ormerod and Shepherd (2003) describe an adaptation that can be used in any domain. The technique itself involves redescribing an HTA for the tasks under analysis in terms of information handling operations (IHOs), SGT task elements, and the associated information requirements. The SGT task elements used are presented in Table 36.8 and Table 36.9.

Ormerod and Shepherd (2003) present a modified set of task elements, presented in Table 36.9.

36.6.3.11   Team Decision Requirements Exercise (Klinger and Hahn, 2004)

The team decision requirements exercise (DRX) (Klinger and Hahn, 2004) is an adaptation of the critical decision method (Klein et al., 1989) that is used to highlight critical decisions made by teams during task performance, and also to analyze the factors surrounding decisions; e.g., why the decision was made, how it was made, what factors affected the decision, etc. The DRX technique was originally used during the training of nuclear power control room crews, as a debriefing tool (Klinger and Hahn, 2004). Typically, a decision requirements table is constructed, and a number of critical decisions are analyzed within a group-interview-type scenario. A set of DRX probes are presented subsequentyly (Klinger and Hahn 2004). A DRX analysis of the fire training scenario "Hazardous chemical spillage at remote farmhouse" (Baber et al., 2004) was conducted in order to determine the factors surrounding the decisions made during the scenario. An extract of a DRX analysis of a fire service training scenario is presented in Table 36.10.

> Why was the decision difficult?
>> What is difficult about making this decision?
>> What can get in the way when you make this decision?
>> What might a less-experienced person have trouble with when making this decision?
> Common errors
>> What errors have you seen people make when addressing this decision?
>> What mistakes do less-experienced people tend to make in this situation?
>> What could have gone wrong (or did go wrong) when making this decision?

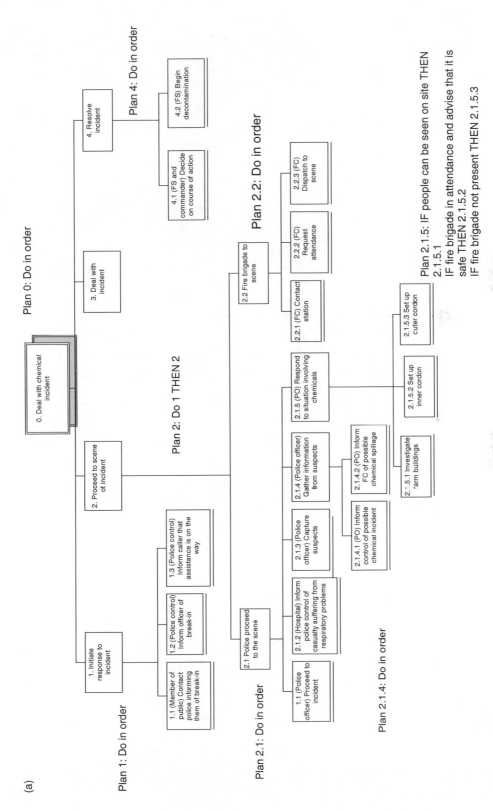

FIGURE 36.3 HTA for the fire brigade training scenario "Deal with chemical incident."

(b)

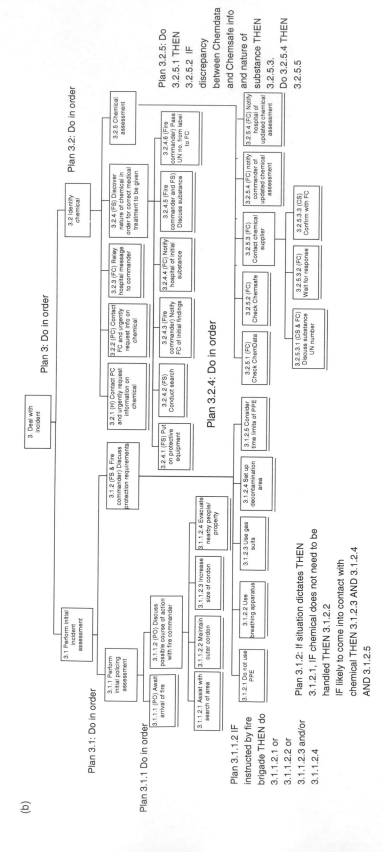

FIGURE 36.3 Continued.

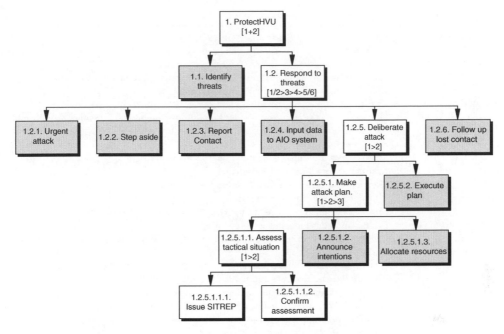

FIGURE 36.4 Extract from an analysis of an antisubmarine warfare team task. (*Source*: Annett, J. (2004). In N. A. Stanton, A. Hedge, K. Brookhuis, E. Salas, and H. Hendrick (Eds), *Handbook of Human Factors Methods*. Boca Raton, FL: CRC Press. With permission.)

Cues and factors

What cues did you consider when you made this decision?

What were you thinking about when you made the decision?

What information did you use to make the decision?

What made you realize that this decision had to be made?

Strategies

Is there a strategy you used when you made this decision?

What are the different strategies that can be used for this kind of decision?

How did you use various pieces of information when you made this decision?

Information sources

Where did you get the information that helped you make this decision?

Where did you look to get the information to help you here?

What about sources, such as other team members, individuals outside the team, technologies, and mechanical indicators, and even tools like maps or diagrams?

Suggested changes

How could you do this better next time?

What would need to be changed with the process or the roles of team members to make this decision easier next time?

What will you pay attention to next time to help you with this decision?

36.6.3.12 Team Cognitive Task Analysis (Klein, 2000)

Team cognitive task analysis (TCTA) is used to describe the cognitive skills that a team or group of individuals are required to undertake in order to perform a particular task or set of tasks. Klein (2000) describes an approach to TCTA that caters for the following team cognitive processes:

- Control of attention
- Shared situation awareness

TABLE 36.7　Tabular Form of Selected ASW Team Operations

1. Protect highly valued unit (HVU) [1 + 2]	*Goal*: ensure safe and timely arrival of HVU
	Teamwork: PWO in unit gaining initial contact with threat assumes tactical command and follows standard operating procedures in this role
	Plan: continues to monitor threats [1] whilst responding to identified threat
	Criterion measure: safe and timely arrival of HVU
1.2. Respond to threats [1/2 > 3 > 4 > 5/6]	*Goal*: respond to threat according to classification
	Teamwork: PWO selects response based on information provided by other team members
	Plan: If threat is immediate (e.g., torpedo) go to urgent attach [1.2.1] else execute 2,3,4 and 5 or 6
	Criterion measure: appropriate response with minimal delay
1.2.5. Deliberate attack [1 > 2]	*Goal*: get weapon in water within 6 min
	Teamwork: see further breakdown below
	Plan: make attack plan then execute
	Criterion measure: time elapsed since classification and/or previous attack
1.2.5.1. Make attack plan [1 > 2 > 3]	*Goals*: plan understood and accepted by team
	Teamwork: information regarding tactical situation and resources available from team members to PWO
	Plan: assess tactical situation; announce intentions; allocate resources
	Criterion measure: accurate information provided
1.2.5.1.1. Assess tactical situation [1 > 2]	*Goal*: arrive at correct assessment of tactical situation
	Teamwork: PWO must gather all relevant information by up-to-date status reports from own team and sensors and other friendly forces
	Plan: issue SITREP then confirm assessment
	Criterion measures: correct assessment; time to make assessment
1.2.5.1.1.1. Issue SITREP	*Goal*: to ensure whole team is aware of threat situation and to provide an opportunity for other team members to check any omissions or errors in tactical appreciation
	Teamwork: PWO issues situation report (SITREP) at appropriate time; all team members check against information they hold
	Criterion measure: all team members have accurate tactical information
1.2.5.1.1.2. Confirm tactical assessment	*Goal*: construct an accurate assessment of the threat and of resources available to meet it
	Teamwork: final responsibility lies with the PWO but information provided by and discussion with other team members essential to identify and resolve any inconsistencies
	Criterion measure: accurate assessment in light of information and resources available

Source: Annett, J. (2004). In N. A. Stanton, A. Hedge, K. Brookhuis, E. Salas, and H. Hendrick (Eds), *Handbook of Human Factors Methods*. Boca Raton, FL: CRC Press. With permission.

- Shared mental models
- Application of strategies/heuristics to make decisions, solve problems, and plan
- Metacognition

According to Klein (2000), a TCTA is provides a way of capturing each of these processes and representing the findings to others. The output of TCTA can be used to improve team performance through informing team training, team design, and team procedures. A study of marine corps command posts was conducted by Klein et al. (1996) as part of an exercise to improve the decision-making process in command posts. Three data collection phases were used during the exercise. Firstly, four regimental exercises were observed and any decision-making related incidents were recorded. As a result, over 200 critical decision-making incidents were recorded. Secondly, interviews with command post personnel were conducted in order to gather more specific information regarding the incidents recorded during the observation. Thirdly, a simulated decision-making scenario was used to test participant responses. Klein et al. (1996) presented 40 decision requirements, including details regarding the decision, reasons for difficulty in making the decision, errors and cues and strategies used for effective decision-making. The decision requirements were categorized into the following groups: building and maintaining situational awareness, managing information, and deciding on a plan. Furthermore, a list of 30 "barriers" to effective decision-making were also presented. A summary of the barriers identified is presented in Table 36.11.

TABLE 36.8 SGT Task Elements

Code	Label	Information Requirements
		Action elements
A1	Prepare equipment	Indication of alternative operating states, feedback that equipment is set to required state
A2	Activate	Feedback that the action has been effective
A3	Adjust	Possible operational states, feedback confirming actual state
A4	De-activate	Feedback that the action has been effective
		Communication elements
C1	Read	Indication of item
C2	Write	Location of record for storage and retrieval
C3	Wait for instruction	Projected wait time, contact point
C4	Receive instruction	Channel for confirmation
C5	Instruct or give data	Feedback for receipt
C6	Remember	Prompt for operator-supplied value
C7	Retrieve	Location of information for retrieval
		Monitoring elements
M1	Monitor to detect deviance	Listing of relevant items to monitor, normal parameters for comparison
M2	Monitor to anticipate change	Listing of relevant items to monitor, anticipated level
M3	Monitor rate of change	Listing of relevant items to monitor, template against which to compare observed parameters
M4	Inspect plant and equipment	Access to symptoms, templates for comparison with acceptable tolerances if necessary
		Decision-making elements
D1	Diagnose problems	Information to support trained strategy
D2	Plan adjustments	Planning information from typical scenarios
D3	Locate containment	Sample points enabling problem bracketing between a clean input and a contaminated output
D4	Judge adjustment	Target indicator, adjustment values
		Exchange elements
E1	Enter from discrete	Item position and delineation, advance descriptors, choice recovery
E2	Enter from continuous range	Choice indicator, range/category delineation, advance descriptors, end of range, range recovery
E3	Extract from discrete range	Information structure (e.g., criticality, weight, frequency structuring), feedback on current choice
E4	Extract from continuous range	Available range; information structure (e.g., criticality, weight, frequency structuring), feedback on current choices
		Navigation elements
N1	Locate a given information set	Organization structure cues (e.g., screen set/menu hierarchy, catalog, etc.), choice descriptor conventions, current location, location relative to start, selection indicator
N2	Move to a given location	Layout structure cues (e.g., screen position, menu selection, icon, etc.), current position, position relative to information coordinates, movement indicator
N3	Browse an information set	Information (e.g., screen/menu hierarchy, catalog, etc.), organization cues, information scope, choice points, current location, location relative to start, selection indicator

Source: Ormerod, T. C. (2000). In J. M. Schraagen, S. F. Chipman, and V. L. Shalin (Eds), *Cognitive Task Analysis*, pp. 181–200. Hillsdale; NJ: Lawrence Erlbaum Associates. With permission.

From the simulated decision-making exercise, it was found that the experienced personnel (colonels and lieutenant colonels) required only 5 to 10 min to understand a situation. However, majors took over 45 min to study and understand the same situation (Klein et al., 1996). In conclusion, Klein et al. (1996) reported that there were too many personnel in the command post, which made it more difficult to

TABLE 36.9 Modified SGT Task Elements

SGT	Task Elements	Context for Assigning SGT and Task Element	Information Requirements
Act		Perform as part of a procedure or subsequent to a decision made about changing the system	Action points and order; current, alternative, and target states; preconditions, outcomes, dependencies, halting, recovery indicators
	A1 Activate	Make subunit operational: switch from off to on	Temporal/stage progression, outcome activation level
	A2 Adjust	Regulate the rate of operation of a unit maintaining "on" state	Rate of state of change
	A3 Deactivate	Make subunit nonoperational: switch from on to off	Cessation descriptor
Exchange		To fulfill a recording requirement To obtain or deliver operating value	Indication of item to be exchanged, channel for confirmation
	E1 Enter	Record a value in a specified location	Information range (continuous, discrete)
	E2 Extract	Obtain a value of a specified parameter	Location of record for storage and retrieval; prompt for operator
Navigate		To move an informational state for exchange, action, or monitoring	System/state structure, current relative location
	N1 Locate	Find the location of a target value or control	Target information, end location relative to start
	N2 Move	Go to a given location and search it	Target location, directional descriptor
	N3 Explore	Browse through a set of locations and values	Current/next/previous item categories
Monitor		To be aware of system states that determine need for navigation, exchange, and action	Relevant items to monitor; record of when actions were taken; elapsed time from action to the present
	M1 Monitor to detect deviance	Routinely compare system state against target state to determine need for action	Normal parameters for comparison
	M2 Monitor to anticipate cue	Compare system state against target state to determine readiness for known action	Anticipated level
	Monitor transition	Routinely compare state of change during state transition	Template against which to compare observed parameters

Source: Ormerod, T. C. and Shepherd, A. (2003). In D. Diaper and N. A. Stanton (Eds), *The Handbook of Task Analysis for Human–Computer Interaction*, pp. 347–366. Hillsdale, NJ: Lawrence Erlbaum Associates. With permission.

complete the job in hand. Klein et al. (1996) suggested that reduced staffing at the command posts would contribute to speed and quality improvements in the decisions made.

36.6.3.13 Team Task Analysis

Team task analysis (TTA) is a technique that is used to provide a description of tasks distributed across a team and the requirements associated with the tasks in terms of operator knowledge, skills, and abilities. Typically, TTA is used to inform team task design, team training procedures, and team performance measurement. TTA aims to analyze team-based scenarios by gathering data regarding teamwork and taskwork. The TTA procedure is an exhaustive one, involving a requirements analysis, task analysis, a coordination demands analysis, the identification of teamwork tasks, and the specification of the knowledge, skills, abilities, and attitudes required to perform the tasks.

36.6.3.14 Interview and Observational Analysis

Finally, interviews and observations are examples of traditional knowledge elicitation techniques that are typically used during CTA applications. Observations of the tasks or scenarios under analysis are

TABLE 36.10 Extract of Decision Requirements Exercise for Hazardous Chemical Incident

Decision	What Did You Find Difficult When Making this Decision?	What Cues Did You Consider When Making this Decision?	Which Information Sources Did You Use When Making this Decision?	Were Any Errors Made while Making this Decision?	How Could You Make a Decision More Efficiently Next Time?
Level of protection required when conducting search activity	The level of protection required is dependent upon the nature of the chemical hazard within farmhouse. This was unknown at the time. There was also significant pressure from the hospital for positive ID of the substance	Urgency of diagnosis required by hospital. Symptoms exhibited by child in hospital. Time required to get into full protection suits	Correspondence with hospital personnel. Police officer. Fire control	Initial insistence upon full suit protection before identification of chemical type	Diagnose chemical type prior to arrival, through comms with farmhouse owner. Consider urgency of chemical diagnosis as critical
Determine type of chemical substance found and relay information to hospital	The chemical label identified substance as a liquid, but the substance was in powder form	Chemical drum labels. Chemical form, e.g., powder, liquid. Chemdata information. Chemsafe data	Chemical drum. Chemdata database. Fire control (chemsafe database)	Initial chemical diagnosis made prior to confirmation with chemdata and chemsafe databases	Use chemdata and chemsafe resources prior to diagnosis. Contact farmhouse owner enroute to farmhouse

TABLE 36.11　Summary of Decision-Making Barriers

Decision Requirements Category	Barriers
Building and maintaining SA	Information presented on separate map-boards
	Map-boards separated by location, furniture, and personnel
	System of overlays archaic and cumbersome
	Over-reliance upon memory while switching between maps
	Erroneous communication
Managing information	Sending irrelevant messages
	Inexperienced personnel used to route information
	Commanders critical information requirements (CCIR) concept misapplied
Deciding on a plan	Communication systems unreliable
	Too many personnel to coordinate information with

Source: Adapted from Klein, G. (2000). In J. M. Schraagen, S. F. Chipman, and V. L. Shalin (Eds), *Cognitive Task Analysis*, pp. 417–431. Hillsdale, NJ: Lawrence Erlbaum Associates. With permission.

normally conducted prior to interviews with the system personnel involved. Techniques such as CDM, ACTA, and CWA utilize interview and observational techniques during their application. Interview and observational techniques can also be used either alone or in conjunction with other techniques as part of a CTA effort.

36.7　Summary and Recommendations for CTA in C4i Systems

In summary, the methods review highlighted a number of pertinent issues regarding CTA in C4i systems. These issues are summarized as follows:

1. *Lack of published work regarding CTA applications in C4i environments* — There is a lack of published work regarding CTA applications in C4i environments.
2. *Lack of a standardized CTA approach* — An universally accepted CTA technique is yet to emerge in the general HF research community. There are a huge number of CTA techniques available, and while a small minority are the most popular (CDM, CWA, etc.), a standardized technique is yet to emerge.
3. *CTA techniques using traditional knowledge elicitation techniques (interviews, observations) are the most commonly used* — The methods review revealed that the most commonly used CTA techniques are those that utilize traditional knowledge elicitation techniques such as interviews and observations. Techniques such as CDM (Klein, 2000) and CWA (Rasmussen et al., 1994) are currently the most popular CTA approaches.
4. *High cost of CTA analyses* — One of the major problems associated with the application of CTA techniques is the high cost involved. CTA analyses are expensive, time consuming, and require a number of experienced personnel.
5. *Lack of guidance regarding selection and usage of CTA techniques* — It is apparent that there is a distinct lack of guidance for HF practitioners in the selection and application of the most appropriate CTA technique. Although CTA analyses are required for a number of purposes (performance evaluation, design evaluation, development of training procedures, allocation of function) there is limited guidance regarding the appropriateness of the various techniques available. Similarly, once the appropriate technique is selected, there is typically little guidance regarding the application of the technique.
6. *Lack of adequately qualified/experienced personnel* — According to Seamster et al. (2000), the use of CTA techniques has been limited due to a limited availability of adequately qualified researchers (as well as the high financial and time costs). Shute et al. (2000) also point out that a vast number of appropriately skilled personnel are required if a CTA is to be conducted properly.

7. *Poor interpretation of CTA outputs* — It is apparent that there are problems associated with the interpretation of the outputs offered from CTA analyses. Exactly what the outputs mean in relation to the project goals is often misinterpreted. As a result, the applicability and benefits of the approaches are often misunderstood. Potter et al. (2000) describe a bottleneck that occurs during the transition from CTA to system design, and suggest that the information gained during the CTA must be effectively translated into design requirements and specifications.

8. *Limited reliability and validity evidence associated with CTA techniques* — The problem of the validation of HF techniques remains a primary concern when considering HF method applications. As with other method types such as human error identification (HEI), situation awareness (SA) measurement, and mental workload (MWL) assessment, the literature revealed a distinct lack of CTA validation studies.

9. *Lack of TCTA techniques* — At the moment, there appears to be only a limited number of CTA techniques that can be applied in team settings. The provision of TCTA techniques is a critical requirement when considering the application of CTA in C4i environments. Exactly how to measure the cognitive processes involved in team performance and collaboration requires further investigation.

10. *Lengthy training requirements associated with CTA techniques* — Typical CTA approaches require a lengthy training procedure, and practitioners are required to be proficient in a number of different techniques (e.g., knowledge elicitation, task analysis, etc.) in order to use the techniques efficiently.

11. *Requirement of a multidisciplinary team in order to conduct CTA analyses effectively* — A typical CTA analysis effort requires a multidisciplinary team. According to Shute et al. (2000), a vast number of appropriately skilled personnel are required if a CTA is to be conducted properly. It is often difficult to assemble such multidisciplinary teams, let alone gather them together at one location for any period of time.

Despite the problems associated with the use of CTA techniques, their provision during the design lifecycle is essential. The original aim of the CTA methods review was to evaluate the suitability of existing CTA methods for use during the design, development, and evaluation of C4i systems. The methods review highlighted a subset of CTA techniques that appear to be suitable for use in C4i environments. These techniques are presented in Table 36.12.

Furthermore, the methods review highlighted that the different CTA approaches are more suitable at different stages of the CADMID design lifecycle. Thus, it is tentatively concluded that the selection of the most appropriate CTA approach is dependent upon the stage in the CADMID (Concept, Assessment, Demonstration, Manufacture, In-service, and Disposal) lifecycle that the CTA analysis is required. The CADMID cycle (SEA Technologies, 2002) refers to the series of stages that are designed to take the project in question through its whole life, from the initial concept stage to its disposal. There is a requirement for human factors integration (HFI) input at each stage in the CADMID cycle. A brief description of each of the CADMID stages is given subsequentlyly (adapted from Anon, 2002):

- *Concept* — The concept stage represents the beginning of any project, and requires an analysis of the human issues related to the proposed project, and also an assessment of the associated risks and requirements. The concept process is normally referred to as early human factors analysis (EHFA).
- *Assessment* — During the assessment stage, more detailed work is conducted in order to quantify the HFI issues and risks. Information regarding the user tasks, working conditions, and expected performance is normally required during this stage. Issues such as manpower reduction, workload, and performance shortfalls are also assessed during this stage. The main objective during the assessment stage is to focus on the human-related issues arising from the project in question.
- *Demonstration* — During the demonstration stage, specifications are refined to ensure robust HFI content, and clear human performance targets are specified.

TABLE 36.12 CTA Techniques Recommended for Use in C4i Applications

Method	Author/Source
ACTA — Applied cognitive task analysis	Militello and Hutton (2000)
CDM — Critical decision method	Klein et al. (1989)
CWA — Cognitive work analysis	Rasmussen et al. (1994), Vicente (1999)
DRX — Decision requirements exercise	Klinger and Hahn (2004)
HTA — Hierarchical task analysis	Annett et al. (1971)
HTA(T) — Team hierarchical task analysis	Annett (2005)
SGT — Subgoal template methodology	Ormerod (2000)
TCTA — Team cognitive task analysis	Klein (2000)
TTA — Team task analysis	Burke (2004)
VPA — Verbal protocol analysis	Walker (2004)
Interviews	Various
Observation	Various

- *Manufacture* — During the manufacture stage, the MOD ensures integration with training development, tactics development, support strategy, etc. End-user trials are also conducted in order to build confidence in equipment operability, which leads to acceptance and subsequent hand over.
- *In-Service* — Declaration of in-service date (ISD) follows the demonstration of effective integration of the equipment with the human component (personnel, procedures, support, and training games) under operational conditions. While in-service, HFI evaluation helps to identify any human-related performance shortfalls or failures of human–equipment integration.
- *Disposal* — The aim of the disposal stage is to dispose of the equipment safely, efficiently, and effectively.

Table 36.13 presents the CADMID lifecycle phases, their associated HFI aims, objectives and activities, and recommended CTA techniques for each stage.

The concept stage involves the development of a description of the users of the proposed system and also a high-level task description. The CWA approach (Rasmussen et al., 1994) is useful at this stage as it involves conducting an analysis of the tasks involved in the work domain, an analysis of the allocation of work, and also worker competency and mental strategies involved. HTA and HTA(T) offers a description of the tasks involved in the system under analysis. The assessment stage involves identifying the human-related system requirements, analyzing human performance, reviewing the human issues involved, and conducting user interface modeling. Again, the CWA approach is useful at this stage as it offers an exhaustive breakdown of the work involved, the competencies and strategies required, and also the allocation of work within the system. The CWA approach has also been commonly used during the design of user interfaces in the past. HTA and HTA(T) are also required here in order to determine the key human-related system requirements. The demonstration stage is perhaps the CADMID stage where there is the most scope for CTA applications. The demonstration phase involves the analysis and evaluation of design concepts, the definition of human performance and operability targets, the coordination of the human aspects of system design, training analysis and support analysis, and dealing with HFI issues as they arise during the design process. Thus, there is potential for the applications of various CTA approaches during the demonstration stage. HTA, HTA(T), CWA, CDM, TCTA, TTA, and the DRX could all potentially be applied during the demonstration stage, particularly for the analysis and evaluation of the design concepts. Techniques such as the CDM and TCTA are particularly useful when considering the evaluation and definition of human performance and also the specification of training and support procedures. The manufacture stage also has scope for CTA applications, as it involves further HFI evaluation and analysis. As such, numerous CTA techniques could potentially be applied, including CDM, TCTA, TTA, DRX, interviews, and observation. The in-service stage involves the demonstration of effective integration of the equipment with the human component (personnel, procedures, support, and training games) under operational conditions. To support further performance

TABLE 36.13 Recommended CTA Techniques for Each CADMID Stage

CADMID Phase	HFI Objectives	HFI Activities	Recommended CTA Techniques
Concept	Contribute to the development of user requirements document Nominate HFI focus Identify HFI issues Identify human implications of technology and procurement options	Conduct early human factors Analysis Develop description of users Develop high-level task description Specify need for detailed HF analysis Conduct outline manning analysis User interface prototyping Specify outline operability targets	HTA HTA(T) Cognitive work analysis
Assessment	Identify key human-related system requirements Identify human impact on cost and performance for different options Explore and quantify human-related trade-offs as part of option assessment process Undertake analysis and evaluation to quantify and mitigate human-related risks to overall performance and costs	Conduct detailed HF analysis Quantify human performance and cost of options Review and update list of human issues Conduct user interface modeling Develop HFI aspects of solutions Specify detailed operability targets Plan HFI work for demonstration	HTA HTA(T) Cognitive work analysis
Demonstration	Evaluate HFI aspects of proposed solutions. Predict human-related performance where necessary Demonstrate ability to deliver operability. Coordinate human aspects of system design, training analysis, and support analysis. Conduct progressive operability trials Contribute HFI criteria for solution and contractor assessment Undertake or commission HFI analysis or evaluation	Further analysis, evaluation, trials, and design support working with contractor prototypes Ensure HFI content of specifications is robust Define human performance and operability targets Coordinate human aspects of system design, training analysis, and support analysis Resolve HFI issues as they arise throughout design Refine HFI acceptance strategy Undertake progressive operability trials	HTA HTA(T) Cognitive work analysis Critical decision method Team cognitive task analysis Team task analysis Decision requirements Exercise
Manufacture	Coordinate development of human-related components of capability, i.e., manpower, training, support Support contractor with input to continued HFI evaluation Monitor and assess operability trials Transfer HFI information for use in the support phase Support HFI evaluation of field trials	Continued input to HFI evaluation Resolve HFI issues as they arise Monitor and assess operability trials Support evaluation of operational work-up trials Transfer HFI information for use in the support phase	Critical decision method Team cognitive task analysis Team task analysis Decision requirements exercise Interviews Observation
In-service	Initiate in-service HFI audit. Identify issues requiring action Monitor exercise and operational reports for evidence of HFI issues Monitor efficiency and effectiveness of maintenance, support, operation, etc. Rerun procedures used during main procurement, at a reduced scale, and drawing on stored HFI data	Develop in-service HF issues log Maintain and initiate mitigating actions to resolve HF issues Ensure health and safety requirements are defined, understood, and met	Critical decision method Team cognitive task analysis Team task analysis Decision requirements exercise Interviews Observation
Disposal	Assess HF issues of the proposed disposal process. Take action as required	Assess HF issues of the proposed disposal process. Take action as required	CWA HTA, HTA(T)

evaluation, techniques such as CDM, TCTA, TTA, and DRX are required. Finally, the disposal stage involves the assessment of the HF issues involved during the disposal of the system in question, in order to ensure its efficient and safe disposal. HTA and HTA(T) could potentially be used to offer a step-by-step description of the proposed disposal process, while the CWA technique could be used to determine the work, organization, competencies, and strategies involved during the proposed disposal process.

It is apparent that there is scope for much further work in the area of CTA. Further investigation into the development of guidelines for the selection and usage of the various CTA techniques, the validation of CTA techniques, the interpretation of CTA outputs, and the development of software assistance is required. It is clear that for the future of CTA to be assured, a great deal of work is required. Without this, the usefulness of CTA techniques may well be questioned.

Acknowledgment

The work reported in this paper was carried out under the U.K. Defence Technology Centre (DTC) for Human Factors Integration (HFI), funded by the U.K. Ministry of Defence and the Defence Science and Technology Laboratory (DSTL).

References

Ainsworth, L. and Marshall, E. (2000). Issues of quality and practicability in task analysis: preliminary results from two surveys. In J. Annett and N. A. Stanton (Eds), *Task Analysis* pp. 79–90. Boca Raton, FL: CRC Press.

Annett, J. (2005). Hierarchical task analysis (HTA). In N. A. Stanton, A. Hedge, K. Brookhuis, E. Salas, and H. Hendrick (Eds), *Handbook of Human Factors Methods*. Boca Raton, FL: CRC Press.

Annett, J., Duncan, K. D., Stammers, R. B., and Gray, M. (1971). *Task Analysis*. London: HMSO.

Baber, C. (2005). Repertory grid for product evaluation. In N. A. Stanton, A. Hedge, K. Brookhuis, E. Salas, and H. Hendrick (Eds), *Handbook of Human Factors Methods*. Boca Raton, FL: CRC Press.

Baber, C., Walker, G., Salmon, P., and Stanton, N. A. (2004). Observation Study Conducted at the Fire Service Training College. Human Factors Integration Defence Technology Centre Report.

Bass, A., Aspinal, J., Walter, G., and Stanton, N. A. (1995). A software toolkit for hierarchical task analysis. *Applied Ergonomics, 26* (2) 147–151.

Blickensderfer, E., Cannon-Bowers, J. A., Salas, E., and Baker, D. P. (2000). Analysing knowledge requirements in team tasks. In J. M. Schraagen, S. F. Chipman, V. L. Shalin, (Eds), *Cognitive Task Analysis*, pp. 431–447. Hillsdale, NJ: Lawrence Erlbaum Associates.

Burke, C. S. (2005). Team task analysis. In N. A. Stanton, A. Hedge, K. Brookhuis, E. Salas, and H. Hendrick (Eds), *Handbook of Human Factors Methods*. Boca Raton, FL: CRC Press.

Card, S. K., Moran, T. P., and Newell, A. (1983). *The Psychology of Human Computer Interaction*. Hillsdale, NJ: Lawrence Erlbaum Associates.

Chin, M., Sanderson, P., and Watson, M. (1999). Cognitive work analysis of the command and control work domain. *Proceedings of the 1999 Command and Control Research and Technology Symposium* (CCRTS), June 29–July 1, Newport, RI, Volume 1, pp. 233–248.

Cooke, N. J. (2004). Measuring team knowledge. In N. A. Stanton, A. Hedge, K. Brookhuis, E. Salas, and H. Hendrick (Eds), *Handbook of Human Factors Methods*. Boca Raton, FL: CRC Press.

Diaper, D. and Johnson, P. (1989). Task analysis for knowledge descriptions: Theory and application in training. In J. Long and A. Whitefield (Eds), *Cognitive Ergonomics in Human Computer Interaction*, pp. 191–224. Cambridge, UK: Cambridge University Press.

DuBois, D. and Shalin, V. L. (2000). Describing job expertise using cognitively-oriented task analysis. In J. M. Schraagen, S. F. Chipman, and V. L. Shalin (Eds), *Cognitive Task Analysis*, pp. 41–56. Hillsdale, NJ: Lawrence Erlbaum Associates.

Flanagan, J. C. (1954). The critical incident technique. *Psychological Bulletin, 51,* 327–358.

Hall, E., Gott, S., and Pokomy, R. (1995). *A Procedural Guide to Cognitive Task Analysis: The PARI Methodology.* Technical Report AL/HR-TR-1995-0108. Brooks AFB, TX: Armstrong Laboratory.

Hollnagel, E. (2003). *Handbook of Cognitive Task Design.* Hillsdale, NJ: Lawrence Erlbaum Associates.

Kirwan, B. (1998a). Human error identification techniques for risk assessment of high-risk systems — Part 1: review and evaluation of techniques. *Applied Ergonomics, 29,* 157–177.

Kirwan, B. (1998b). Human error identification techniques for risk assessment of high-risk systems — Part 2: towards a framework approach. *Applied Ergonomics, 5,* 299–319.

Kirwan, B. and Ainsworth, L. K. (1992). *A Guide to Task Analysis.* London: Taylor & Francis.

Klein, G. (2000). Cognitive task analysis of teams. In J. M. Schraagen, S. F. Chipman, and V. L. Shalin (Eds), *Cognitive Task Analysis,* pp. 417–431. Hillsdale, NJ: Lawrence Erlbaum Associates.

Klein, G. A., Calderwood, R., and MacGregor, D. (1989). Critical decision method for eliciting knowledge. *IEEE Transactions on Systems, Man and Cybernetics, 19* (3), 462–472.

Klein, G., Schmitt, J., McCloskey, M., Heaton, J., Klinger, D., and Wolf, S. (1996). *A Decision-Centred Study of the Regimental Command Post.* Fairborn, OH: Klein Associates.

Klinger, D. W. and Hahn, B. B. (2005). Team decision requirement exercise: making team decision requirements explicit. In N. A. Stanton, A. Hedge, K. Brookhuis, E. Salas, and H. Hendrick. (Eds), *Handbook of Human Factors Methods.* Boca Raton, FL: CRC Press.

Lesgold, A. (2000). On the future of cognitive task analysis. In J. M., Schraagen, S. F. Chipman, and V. L. Shalin (Eds.), *Cognitive Task Analysis,* pp. 135–146. Hillsdale, NJ: Lawrence Erlbaum Associates.

Marshall, A., Stanton, N., Young, M., Salmon, P., Harris, D., Demagalski, J., Waldmann, T., and Dekker, S. (2003). Development of the human error template — a new methodology for assessing design induced errors on aircraft flight decks. DTI ErrorPred Report.

Militello, L. G. and Hutton, J. B. (2000). Applied cognitive task analysis (ACTA): a practitioner's toolkit for understanding cognitive task demands. In J. Annett & N. A Stanton (Eds), *Task Analysis,* pp. 90–113. London: Taylor & Francis.

Neerincx, M. A. (2003). Cognitive task load analysis: allocating tasks and designing support. In E. Hollnagel (Ed), *Handbook of Cognitive Task Design,* pp. 281–305. Hillsdale, NJ: Lawrence Erlbaum Associates.

O'Hare, D., Wiggins, M., Williams, A., and Wong, W. (2000). Cognitive task analyses for decision centred design and training. In J. Annett and N. Stanton (Eds), *Task Analysis,* pp. 170–190. London: Taylor & Francis.

Ormerod, T. C. (2000). Using task analysis as a primary design method: the SGT approach. In J. M. Schraagen, S. F. Chipman, and V. L. Shalin (Eds), *Cognitive Task Analysis,* pp. 181–200. Hillsdale, NJ: Lawrence Erlbaum Associates.

Ormerod, T. C. and Shepherd, A. (2004). Using task analysis for information requirements specification: the sub-goal template (SGT) method. In D. Diaper and N. Stanton (Eds), *The Handbook of Task Analysis for Human–Computer Interaction,* pp. 347–366. Hillsdale, NJ: Lawrence Erlbaum Associates.

Polson, P. G. and Lewis, C. (1990). Theory-based design for easily learned interfaces. *Human Computer Interaction, 5* (2 & 3), 191–220.

Polson, P. G., Lewis, C., Rieman, J., and Wharton, C. (1992). Cognitive walkthroughs: a method for theory based evaluation of user interfaces. *International Journal of Man–Machine Studies, 36,* 741–773.

Potter, S. S., Roth, E. M., Woods, D. D., and Elm, W. (2000). In J. M. Schraagen, S. F. Chipman, and V. L. Shalin (Eds), *Cognitive Task Analysis,* pp. 317–340. Hillsdale, NJ: Lawrence Erlbaum Associates.

Rasmussen, J., Pejtersen, A., and Goodstein, L. P. (1994). *Cognitive Systems Engineering.* New York: John Wiley & Sons.

Redding, R. E. (1989). Perspectives on cognitive task-analysis: the state of the state of the art. In *Proceedings of the Human Factors Society, Vol. 33,* pp. 1353–1357. Santa Monica, CA: Human Factors Society.

Roth, E. M., Patterson, E. S., and Mumaw, R. J. (2002). Cognitive engineering: issues in user-centred system design. In J. J. Marciniak (Ed), *Enclyopedia of Software Engineering*, 2nd edn, pp. 163–179. New York: John Wiley & Sons.

Salas, E. (2005). Team methods. In N. A. Stanton, A. Hedge, K. Brookhuis, E. Salas, and H. Hendrick. (Eds), *Handbook of Human Factors Methods*. Boca Raton, FL: CRC Press.

Salmon, P. M., Stanton, N. A., Walker, G., and Green, D. (2004a). Situation awareness measurement: A review of applicability for C4i environments. *The Journal of Applied Ergonomics* (In Press).

Salmon, P. M., Stanton, N. A., Walker, G., and Green, D. (2004b). Human Factors Design and Evaluation Methods Review. Defence Technology Centre for Human Factors Integration, Report No. HFIDTC/WP1.3.2/1.

Salmon, P. M., Stanton, N. A., and Walker, G. (2004c). Work Package 1.1.3: NGT Switching Scenario Report. Defence Technology Centre for Human Factors Integration Report.

Savoie, E. J. (1998). Tapping the power of teams. In R. S. Tindale, L. Heath, et al. (Eds), *Theory and Research on Small Groups. Social Psychological Applications to Social Issues*, vol. 4, pp. 229–244. New York: Plenum Press.

Schaafstal, A. and Schraagen, J. M. (2000). Training of troubleshooting: a structured, task analytical approach. In J. M. Schraagen, S. F. Chipman, and V. L. Shalin (Eds), *Cognitive Task Analysis*, pp. 57–71. Hillsdale, NJ: Lawrence Erlbaum Associates.

Schraagen, J. M., Chipman, S. F., and Shalin, V. L. (2000). *Cognitive Task Analysis*. Hillsdale, NJ: Lawrence Erlbaum Associates.

Schraagen, J. M., Ruisseau, J. I., Graff, N., Annett, J., Strub, M. H., Sheppard, C., Chipman, S. E., Shalin, V. L., and Shute, V. L. (2000). Cognitive Task Analysis. RTO Technical Report 24.

Seamster, T. L., Redding, R. E., and Kaempf, G. L. (2000). A skill-based cognitive task analysis framework. In J. M. Schraagen, S. F. Chipman, and V. L. Shalin (Eds), *Cognitive Task Analysis*, pp. 135–146. Hillsdale, NJ: Lawrence Erlbaum Associates.

SEA Technologies (2002). *STGP 10 HFI Management Guide*. Bristol: SEA.

Shepherd, A. (2001). *Hierarchical Task Analysis*. London: Taylor & Francis.

Shute, V. J., Torreano, L. A., and Willis, R. E. (2000). DNA: providing the blueprint for instruction. In J. M. Schraagen, S. F. Chipman, and V. L. Shalin (Eds), *Cognitive Task Analysis*, pp. 71–86. Hillsdale, NJ: Lawrence Erlbaum Associates.

Stanton, N. A. and Stevenage, S. V. (1998). Learning to predict human error: issues of acceptability, reliability and validity. *Ergonomics*, 41 (11), 1737–1756.

Stanton, N. A., Hedge, A., Brookhuis, K., Salas, E., and Hendrick, H. (2005). *Handbook of Human Factors Methods*. Boca Raton, FL: CRC Press.

Vicente, K. J. (1999). *Cognitive Work Analysis: Towards Safe, Productive, and Healthy Computer-Based Work*. Mahwah, NJ: Lawrence Erlbaum Associates.

Walker, G. H. (2005). Verbal protocol analysis. In N. A. Stanton, A. Hedge, K. Brookhuis, E. Salas, and H. Hendrick (Eds), *Handbook of Human Factors Methods*. Boca Raton, FL: CRC Press.

Walker, G. H. and Stanton, N. A. (2004). Work Package 1.1.3: EAST Analysis of a Railway Signalling Task. Defence Technology Centre for Human Factors Integration Report.

Walker, G. H., Stanton, N. A., and Young, M. S. (2001). An on-road investigation of vehicle feedback and its role in driver cognition: Implications for cognitive ergonomics. *International Journal of Cognitive Ergonomics*, 5 (4), 421–444.

Zachary, W. W., Ryder, J. M., and Hicinbothom, J. H. (2000). Building Cognitive Task Analyses and Models of a Decision-Making Team in a Complex Real-Time Environment.

37

Subjective Scales of Effort and Workload Assessment

B. Sherehiy
W. Karwowski
University of Louisville

37.1   Subjective Workload Assessment

Workload assessment techniques are traditionally divided into four major categories: subjective, physiological, performance based, and analytical methods (Meshkati et al., 1992; Wierwille and Eggemeier, 1993). The performance based measurement assumes that any increase in task difficulty will lead to an increase in demands, which, in turn, will result in degrading performance (Young and Stanton, 2004). Physiological indexes assume that the mental workload can be measured by means of the level of physiological activation (Rubio et al., 2004). Subjective procedures are based on operator judgments of the workload associated with performance of a task or system function. Subjective procedures assume that an energy expenditure associated with task performance is linked to the perceived effort, and can be appropriately assessed by individuals. Thus, the subjective measures reflect direct opinion of an operator about the effort required to accomplish task in the context of the task environment, and his or her skill and experience (Tsang and Wilson, 1997).

Subjective methods are very important tools in system evaluation and workload assessment, and are extensively used to assess workload in a wide rage of applications (Tsang and Wilson, 1997; Annet, 2002; Rubio et al., 2004). Among subjective (or self-reported) methods one can distinguish such techniques as interviews, questionnaires, ranking and ratings scales, and critical incidence techniques (Sinclair, 1995). In the field of subjective workload assessment, the rating scales are used mostly to collect the data. Typically, the individual is asked to rate the level of effort associated with performed task on a scale. Three main types of rating scales can be distinguished, that is, simple rating scale, Thurstone's

Paired Comparison Technique, and the Likert Scale (Sinclair, 1995). The simple rating method consists of a set of questions related to workload or task attributes, and a scale on which those questions are rated. The rating scale is represented by the 100 mm line, usually with subdivisions and labels at the end of the scale. The intermediate labels or numbers are often assigned to some or all subdivisions. Thurstone's Paired Comparison Technique asks the subject to compare two entities (e.g., two tasks) or every combination of all measured entities, and decide which one is larger or smaller. Each pair of compared entities can be also evaluated on the rating scale with assigned numbers. The Likert Scale has an odd number of discrete options and consisting of a range from 1 to 5 (or 7), with labels from "strongly disagree" to "strongly agree," respectively.

There are several common approaches applied in subjective workload measurements (Tsang and Vidulich, 1994): (1) unidimensional versus multidimensional rating scale; (2) immediate versus retrospective assessment procedure; and (3) absolute estimation versus relative judgments approach. Unidimensional scale assesses only one dimension of workload or focuses on the overall workload level, while multidimensional scale evaluates several aspects or components of workload. The ratings can be obtained immediately after performance of each condition or retrospectively after experiencing all task conditions. The simple or direct rating approach is called the absolute magnitude estimation, in contrast to the relative judgment approach where as in Thurstone's Paired Comparison Technique, the subject is asked to assess the task condition in reference to the single standard or multiple task conditions.

Each subjective method should demonstrate several properties in order to be accepted as good measurement tool. These properties are: validity, reliability, sensitivity, diagnosticity, intrusiveness, transferability, and ease of implementation and subject acceptability (Eggemeier et al., 1991; Wierwille and Eggemeier, 1993). These criteria should be applied when the suitability of workload assessment method is considered for evaluation of a particular type of workload, or for a particular type of environment, task, or system.

Reliability is the degree of precision to which a method or instrument is able to measure what it measures. Reliability can be assessed as homogeneity, consistency or stability of measurement, or in the case of two or more raters as the interrater reliability (ISO/FDIS 10075-3, 2002).

Validity is the degree to which a method or instrument is able to measure what it is intended to measure (ISO/FDIS 10075-3, 2002). Thus, in case of mental workload, the measures should reflect differences in cognitive demands, but not changes in other variables such as physical workload, which are not necessarily associated with mental workload.

Sensitivity refers to how well a technique can distinguish the differences in levels of load required to accomplish a task (Wierwille and Eggemeier, 1993). With regard to workload, this criterion is primary among other criteria, since it is important to access differences in the workload imposed by a task or system.

Diagnosticity is the ability to distinguish the type of workload, or ability to attribute it to a particular aspect of a performed task (Tsang and Wilson, 1997). Thus, the diagnosticity can help to determine which elements or aspects of the task caused workload.

Intrusiveness is related to the fact that application of measurement technique can interfere with the task performance, which is evaluated, and can cause performance changes that are not related to the task itself (Wierwille and Eggemeier, 1993). In order to avoid contamination of workload measures, it is desirable to minimize intrusiveness of the measurement method.

Transferability refers to the possibility of applying a given technique in different environments or tasks. Implementation requirements related to such issues are: ease of data collection, robustness of the measurement instruments, and overall data quality control (Wierwille and Eggemeier, 1993). Finally, subject acceptability refers to the subject's perception of the measurement procedure.

In reference to criteria presented earlier, the subjective workload methods are described as sensitive to different levels of the workload. However, since many research studies showed rather low diagnosticity of the measures, it was concluded that the subjective rating scales can be used only as global measure of workload (Eggemeier and Wilson, 1991). There is some evidence that the multidimensional workload rating scales present better diagnostic properties (Tsang, 2001). Subjective measures appear to be reliable and have concurrent validity with performance measures. In most applications, the rating scales are completed after the task performance and, therefore, are considered as nonintrusive methods. Furthermore, subjective methods have high subject acceptance and are easy to implement.

37.2 Psychophysical Scaling

Psychophysics is the study of the relationship between the physical qualities of a stimulus and the perception of those qualities (Stevens, 1974). The psychophysical approach assumes that humans are able to perceive the strain generated in the body by a given work task and to make absolute and relative judgments about perceived effort (Kroemer, et al., 1994). Subjective measurement techniques rely on an individual's ability to relate their sensations to some quantitative measure (Noble and Robertson, 1996). Psychophysical scaling is a technique which allows a subject to assign a numerical value to the subjective magnitude of a stimulus.

Weber in 1838 and Fechner in 1860 established formal relationships between physical stimulus and its perceptual sensation (Kroemer et al., 1994). The Weber–Fechner Law states that a relative increase in external energy corresponds to a fixed increment in subjective intensity (Stevens, 1974). Thus this law showed that the relationship between the stimulus and the sensation can be defined by a constant, and equal physical ratios produce equal psychological differences. According to Fechner, a sensation grows as a logarithm of the stimulus:

$$S = k \log I$$

where S is the strength of sensation, k is a constant, and I is the intensity of the stimulus.

Stevens (1974) argued that the approach of Fechner is "indirect" because it defines ability to discriminate among stimuli as the basic unit of sensation. Stevens developed a ratio scaling technique, for example, magnitude estimation (ME), where the subjects were asked to "directly" assign numbers to stimuli to represent the magnitudes of their sensations. Stevens (1974) modified the Fechner Law and discovered more general psychophysical law which holds in all sense modalities. Stevens's power law states that the strength of sensation (S) and the intensity of its physical stimulus (I) is related by the power function:

$$S = k * I^n$$

where S is the strength of sensation, k is a constant, I is the intensity of the stimulus, and n is the slope of the line that represents the power function when plotted on log–log coordinates. According to the aforementioned power law, equal stimulus ratios produce equal sensation ratios.

37.3 Scales of Perceived Physical Effort

37.3.1 Borg's Scales for Perceived Exertion

Borg (1982) developed formal techniques to rate the perceived exertion associated with different kinds of efforts. He found that the perception of both muscular effort and force obey the psychophysical function, where sensation magnitude (S) grows as a power function of the stimulus. Contrary to the Sevens approach, according to which rating scale should be a "ratio scale," Borg developed a "category scale" for rating of perceived exertion (RPE). According to Borg (2001), the category scale takes into account interindividual subjective differences, which was not possible with conventional ratio scaling method. Category scales are equally partitioned continuous rating scales that designate categories with adjectives or a finite set of numbers (Noble and Robertson, 1996). Scales with ratio properties require individuals to estimate subjective intensities on a scale with an absolute zero and equivalent scale steps. This is in contrast to the interval scale, which has equal steps but no true zero (Borg, 2001).

TABLE 37.1 Borg's 15-Point Scale for Ratings of Perceived Exertion

6	
7	Very, very light
8	
9	Very light
10	
11	Fairly light
12	
13	Somewhat hard
14	
15	Hard
16	
17	Very hard
18	
19	Very, very hard
20	

Source: Adapted from Borg, G.A. (1982). *Med Sci Sports Exerc*, 14, 377–381. With permission.

The RPE scale was developed according to the principles of quantitative semantics, which deals with the meaning of words and the quantitative relationship between verbal expressions (Noble and Robertson, 1996; Borg and Borg, 2002). Borg chose verbal expressions for the scale that were not only well defined but also had a comparatively equal meaning for subjects. The original RPE scale was developed to be linearly related to heart rate, and based on subjective force estimates provided by subjects while performing short-time work on a bicycle ergometer with a stepwise increase in workload. In order to relate the scale linearly with heart rate, the scale number were arranged to conform to the absolute heart rate quantities. Since heart rate ranges from 60 to 200 in a healthy group of adults, the scale numbers were set from 6 to 20 (Table 37.1). Close relationship with the pulse rate and application of quantitative semantics support the notion that the RPE scale is a category scale with interval properties, that is, the scale not only rank sensations but, also satisfies the equal interval criterion (Noble and Robertson, 1996).

Since the original 15-point RPE scale does not reflect the more exponential growth of perceptual responses as intensity increases, Borg (1982) developed a Category–Ratio (CR10) scale for perceived exertion (Table 37.2). This scale follows the growth of perceptual responses, and the exponential growth of nonlinear physiological responses such as lactate accumulation and pulmonary ventilation. This scale also combined advantages of both scales, that is, category-rating and ratio scale (Borg, 2002). CR scale can be useful for identifying local sensations of effort during high-intensity exercise. However, a valid determination of direct intensity levels with this scale is not possible. Thus, Borg (1982) did not recommend use of this scale in place of the 15-point RPE scale for general exercise testing. The CR10 scale has 12 categories, with values ranging from 0 to 10, with "0" for "nothing at all" and 0.5 for "extremely weak (just noticeable)". In order to prevent a ceiling affect, the term "maximal" was placed outside of the scale, to which any number can be assigned by subject (Noble and Robertson, 1996; Borg and Borg, 2002).

The original Borg RPE scale has been shown to be an accurate and reliable instrument to measure perceived exertion in a number of investigations. The use of the Borg scale was validated on several tasks such as bicycle ergometer, walking, stool stepping, and walking treadmill. Various physiological measures of physical exertion: heart rate, blood and muscle lactate concentration, oxygen uptake, ventilation and respiration rate have been used as criterion in the validation research (Noble and Robertson, 1996; Russell, 1997; Chen et al., 2002). The validation against heart rate showed correlation coefficients in the range of $r = 0.80$–0.90 (Noble and Robertson, 1996; Hampson et al., 2001; Chen, 2002). The reliability coefficients established by test–retest method were 0.78 and 0.90 for bicycle ergometer tasks (Skinner et al., 1973; Noble and Robertson, 1996), 0.76 for the oscillating test, 0.76 for stool stepping exercise, and 0.76 for treadmill walking test (Noble and Robertson, 1996). In addition, validation

TABLE 37.2 Borg's CR10 Scale with "P" Representing Perception

0	Nothing at all	"No P"
0.3		
0.5	Extremely weak	Just noticeable
1	Very weak	
1.5		
2	Weak	Light
2.5		
3	Moderate	
4		
5	Strong	Heavy
6		
7	Very strong	
8		
9		
10	Extremely strong	"Max P"
●	Absolute maximum	Highest possible

Source: Adapted from Borg, G.A. (1982). *Med Sci Sports Exerc*, 14, 377–381. With permission.

studies showed that different levels of physical activity, gender, age and conditions such as asthma, blindness, and cardiac disease do not affect the validity of this scale (Hampson et al., 2001). The CR10 scale has also been correlated with physiological measures, such as heart rate, blood lactate, and muscle lactate, supporting the validity of the scale (Borg, 1985; Borg et al., 1987; Noble and Robertson, 1996).

Most of the differences in RPE responses are attributed to physiological factors (oxygen uptake and heart rate, ventilation and respiration rate, electromyographic (EMG) measures, perspiration rate, blood lactate) (Borg, 2001). However, quite large variance in RPE responses may be explained by psychological factors such as personality and rating behavior, emotional and motivational factors. Morgan (1973) demonstrated that subjects who are neurotic, anxious, and depressed have difficulties with accurate rating of perceived exertion. It was also revealed that preferred exercise intensity was higher for extraverted subjects than for introverted subjects, and supported the hypothesis that extraverts suppress painful stimuli (Morgan, 1973).

The overwhelming focus on these scales application has been the use of RPE for regulation of exercise intensity and exercise prescription (Russell, 1997). The main application of scales of perceived exertion in human factors research is in the assessment of physical workload (Borg, 2001). The RPE scale is often used to evaluate the exertion and difficulty of a work task in order to compare different performance techniques or to adjust the task to specific population group (age, gender, or handicapped). Perceived exertion has been used in different context to evaluate the level of physical effort required to complete different tasks (Shepard, 1994). The Borg CR10 scale was also used to rate physical exertions of truck drivers during heavy operations (Johansson and Borg, 1993). In a package delivery industry, the scale of perceived exertion was used to estimate the amounts of load that correspond to various levels of load heaviness (Genaidy et al., 1998). Psychophysical assessments were also applied to investigate the perceived physical effects of resident-transferring methods on nursing assistants, and to determine which method minimizes psychophysical stress (Zhuang et al., 2000). Ratings of perceived exertions also have been used to determine appropriate work design parameters, for example, efficient working postures, task frequency, work capacity, and tool assessment (Pandolf et al., 1978; Legg et al., 1997; Olendorf and Drury, 2001).

37.3.2 Other Scales of Perceived Effort

37.3.2.1 Fleishman Index of Perceived Physical Effort

Hogan and Fleishman (1979) analyzed perception of physical effort for a wide range of occupational tasks. Based on the Borg RPE scale, they developed a category scale for assessment of perceived effort during job task performance. The proposed index (Table 37.3) of perceived physical effort is a

TABLE 37.3　Fleishman Index of Perceived Physical Effort

1	Very, very light
2	
3	
4	Somewhat hard
5	
6	
7	Very, very hard

seven-point category scale with values from 1 to 7, each number anchored with verbal descriptions from 1 = very, very light, to 7 = very, very hard, and 4 = somewhat hard. In the first study (Hogan and Fleishman, 1979) the proposed index was applied to assessment of 37 occupational and 62 recreational tasks whose metabolic costs had been previously determined. In the second study (Hogan et al., 1980) energy expenditure for 24 manual handling tasks were measured and perceived effort were assessed.

Both studies supported validity and reliability of the physical effort scale. The results showed high interrater agreement (0.83) among subjects concerning the perceived effort needed to perform the assessed tasks. The ratings of perceived effort were highly related (0.83–0.88) to estimated metabolic costs. Subjects were also able to make distinction between tasks at all ranges of physical effort. Similarities in ratings of perceived exertion suggest that group ratings of perceived effort can be used to accurately reflect metabolic costs of task performance (Fleishman et al., 1984), as opposed to individual ratings that only represent the effort perceived by one person and do not always accurately represent perception of population.

37.3.2.2　Pittsburgh Perceived Exertion Scale

The Pittsburgh perceived exertion scale (Table 37.4) is a nine-point category scale with terminal expressions of 2 and 8 on the scale (Noble and Robertson, 1996). These numbers were used intentionally in order to avoid the "end effect" of category scales, when subjects tend to use the middle of the scale. When the absolute ends of the scale are not verbally anchored, the use of the terminal expressions increases. Ratings on this scale have been shown to correlate well with the heart rate and power output (Noble and Robertson, 1996).

37.3.3　Subjective Scales for Postural Assessment

37.3.3.1　Body Part Discomfort Scale

The Body Part Discomfort (BPD) Scale (Corlett and Bishop, 1976) has been adopted technique for assessment of postural discomfort. Corlett and Bishop (1976) analyzed the relationship between the perceived discomfort level and holding time, that is, the amount of time a subject could hold different postures before voluntary termination of the posture. This research demonstrated that if a force was exerted

TABLE 37.4　Pittsburgh Perceived Exertion Scale

1	
2	Not at all stressful
3	
4	
5	
6	
7	
8	Very, very stressful
9	

for as long as possible, and if estimates of discomfort level were made at intervals during the holding time, the growth of discomfort was linearly related to holding time regardless of the level of force being exerted (Corrlet, 1995). Therefore, it was concluded that perception of discomfort level can be used as a linear scale. In this technique the severity of discomfort is evaluated on a five- or seven-point ordinal scale anchored at 0 and 5(7) categories by "no discomfort" and "extreme discomfort."

In order to identify the body areas where discomfort is experienced the body map divided into segments is used. The original body chart used in Corlett and Bishop (1976) has been modified by many researchers to fit their particular purposes (Olendorf and Drury 2001; Drury et al., 1989). Coury and Drury (1982) and Drury et al. (1989) developed three summary measures for the Body Part Discomfort Scale: BPD frequency (BPDF), BPD severity (BPDS), and BDP frequency severity (BPDFS). BPDF is calculated as the number of body parts rated greater than zero, whereas BPDS is the average of all nonzero ratings. BPDFS is the product of the BPDF and BPDS.

The Body Part Discomfort Scale (Corrlett and Bishop, 1976) has been validated in a many studies. Significant relationship was found between the applied biomechanical torque and discomfort (Boussenna et al., 1982). The Boussenna et al. (1982) study also supported relationship between holding time and perceived discomfort level, that is postures that caused greater discomfort than others led to shorter holding times. Jung and Choe (1996) demonstrated that a psychophysical scale of discomfort correctly reflects physiological muscle activity obtained with EMG recordings. The cross-validation of this technique with RPE scale and OWAS showed high intercorrelation of results (Drury et al., 1989; Liao and Drury, 2000; Olendorf and Drury, 2001). The reliability of body discomfort scale was demonstrated with high intersubject agreement correlation in the study on the postural loading effect at joints (Boussenna et al., 1982). However, the research results on sensitivity of the Body Part Discomfort Scale are equivocal. Bonney et al. (1990) revealed that perceived discomfort discriminate 20°changes in back postures from one another. In the study by Kumar et al. (1999), the BDPR was unable to differentiate between task variables, in comparison to RPE and visual scale (VS) that were sensitive to changes in the lifting tasks, which were of a short duration, continuous, and not biomechanically demanding.

37.3.3.2 Nordic Questionnaire for the Musculoskeletal Symptoms

The Nordic questionnaire (NQ) consists of structured, forced binary or multiple-choice questions used to indicate general and specific information regarding pain in various regions of the body (Kuorinka et al., 1987). It can be used in self-assessment and in interviews. The questionnaire is enhanced with generic body diagrams used for identifying painful regions of the body. There are two types of the questionnaire: (1) a general questionnaire, and (2) specific ones focused on the low back and neck/shoulders. The main purpose of the questionnaire is the screening of musculoskeletal disorders in the context of ergonomics. The localization of symptoms may reveal their causes in terms of loading analyzed in reference to the daily activities, job tasks, and work environment (Dickinson et al., 1992).

37.4 Subjective Scales of Mental Workload Measurement

37.4.1 NASA Task Load Index

The NASA Task Load Index (TLX) (Hart and Staveland, 1988) is a multidimensional subjective workload rating technique. In TLX workload is defined as the "cost incurred by human operators to achieve a specific level of performance." (Rehman, 1995). The NASA TLX uses six dimensions to assess subjective workload: (1) mental demand, (2) physical demand, (3) temporal demand, (4) performance, (5) effort, and (6) frustration. Rating scales description is presented in Table 37.5. Each of the dimensions is rated on the 20-step bipolar scale (Figure 37.1).

The overall workload score is based on the weighted average of ratings on the six workload dimensions. The evaluation procedure with TLX consists of two stages. First, during the scale development procedure, each subject evaluates all possible paired comparisons of the six dimensions, in order to specify the member of each pair that has larger contribution to the workload of assessed task. Values obtained

TABLE 37.5   Rating Scales Description of NASA Task Load Index

Title	Endpoints	Descriptions
Mental demand	Low/high	How much mental and perceptual activity was required (e.g., thinking, deciding, calculating, remembering, looking, searching, etc.)? Was the task easy or demanding, simple or complex, exacting or forgiving?
Physical demand	Low/high	How much physical activity was required (e.g., pushing, pulling, turning, controlling, activating, etc.)? Was the task easy or demanding, slow or brisk, slack or strenuous, restful or laborious?
Temporal demand	Low/high	How much time pressure did you feel due to the rate or pace at which the tasks or task elements occurred? Was the pace slow and leisurely or rapid and frantic?
Effort	Low/high	How hard did you have to work (mentally and physically) to accomplish your level of performance?
Performance	Low/high	How successful do you think you were in accomplishing the goals of the task set by the experimenter (or yourself)? How satisfied were you with your performance in accomplishing these goals?
Frustration level	Low/high	How insecure, discouraged, irritated, stressed, and annoyed versus secure, gratified, content, relaxed, and complacent did you feel during the task?

from paired comparisons, is used for calculation of the weights for each dimension. These weights represent relative importance associated with each dimension for the workload of rated task. During the event scoring procedure, subjects rate performed task on each of the six workload scales. Workload evaluation with TLX is conducted immediately after completion of task.

Several studies reported that NASA TLX was a valid and reliable measure of workload (Hart and Staveland, 1988; Hill et al., 1992). Hart and Staveland (1988) stated that the TLX provides a sensitive indicator of the overall workload as it differentiates among tasks of various cognitive and physical demands. They also concluded that the weights determined for each TLX dimension reflect diagnostic information about the sources of loading within a task. Instrument sensitivity has been tested in a wide variety of multitask environments, including flight simulators (Battiste and Bortolussi, 1988; Tsang and Johnson, 1989), actual flight (Shively et al., 1987), air combat (Hill et al., 1989; Bittner et al., 1989). These studies showed that TLX ratings significantly discriminated the flight segments (Shively et al., 1987; Battiste and Bortolussi, 1988) and between low- and high-workload scenarios (Battiste and Bortolussi, 1988; Tsang and Johnson, 1989).

37.4.2   Subjective Workload Assessment Technique

The Subjective Workload Assessment Technique (SWAT) is based on the additive multidimensional model of mental workload (Reid and Nygren, 1988). SWAT measures workload on three dimensions: (1) time load, (2) mental load, and (3) psychological stress load. The description of workload dimensions is provided in Table 37.6. These workload dimensions are rated on a three-point scale (low, medium, and high).

Application of SWAT consists of two phases. During the scale development phase, each subject rank orders all 27 possible combinations at three levels of each of the three dimensions. Next, the conjoint

FIGURE 37.1   Rating scale used in NASA task load index.

TABLE 37.6 The Description of Workload Dimensions of Subjective Workload Assessment Technique (SWAT)

I. *Time load*
1. Often have spare time. Interruptions or overlap among activities occur infrequently or not at all
2. Occasionally have spare time. Interruptions or overlap among activities occur frequently
3. Almost never have spare time. Interruptions or overlap among activities are very frequent, or occur all the time

II. *Mental effort load*
1. Very little conscious mental effort or concentration required. Activity is almost automatic, requiring little or no attention
2. Moderate conscious mental effort or concentration required. Complexity of activity is moderately high due to uncertainty, unpredictability, or unfamiliarity. Considerable attention required
3. Extensive mental effort and concentration are necessary. Very complex activity requiring total attention

III. *Psychological stress load*
1. Little confusion, risk, frustration, or anxiety exists and can be easily accommodated
2. Moderate stress due to confusion, frustration, or anxiety noticeably adds to workload. Significant compensation is required to maintain adequate performance
3. High to very intense stress due to confusion, frustration, or anxiety
4. High to extreme determination and self-control required

measurement methodology is applied to develop the best-fitting scale for perceived workload. SWAT has the capability to account for individual differences by grouping the subjects according to the dimensions they emphasize most in their ratings (Meshkati et al., 1992). A separate workload scale can be derived for each subgroup. Although the individuals are not asked to evaluate the importance of each of the three SWAT dimensions, the estimates of the relative importance are obtained as a function of rescaling of rank-ordered data by conjoint scaling procedure (Nygren, 1991). The obtained scale is used in the event scoring phase to assess the workload associated with performed task.

Mental workload assessment with SWAT was extensively tested in diverse environments, including military flight scenarios and commercial air travel (Nataupsky and Abbott, 1987; Battiste and Bortolussi, 1988), nuclear plant simulations (Beare and Dorris, 1984), military tank simulators (Whitaker, Peters and Garinther, 1989), different systems of air defense (Bittner et al., 1989), and remote control vehicles (Byers et al., 1988). SWAT demonstrated sensitivity to variations in mental workload during a variety of tasks, including visual display monitoring, memory tasks, and manual control (Rubio, et al., 2004). It was also found that the three SWAT rating scales are differently sensitive to the tasks demands. Therefore, it was suggested that the individual scales have differential diagnosticity in assessing workload, and individual scale information should be retained and separately examined as workload components (Moroney et al., 1995).

37.4.3 Modified Cooper–Harper Scale

The Cooper–Harper scale is one of the first standardized scales for measuring workload, was originally developed to evaluate handling qualities of the aircraft (Cooper and Harper, 1969). It was concluded that this technique is well suited to evaluate other manual control tasks as well (Moray, 1982; Skipper et al., 1986). In order to make the instrument applicable to a wider variety of tasks, the modified scale was developed (Modified Cooper–Harper: MCH scale) (Wierwille and Casali 1983). The modification was done in order to assess workload associated with cognitive functions, such as perception, monitoring, evaluation, communications, and problem solving. The flow diagram of the original technique was retained, but the verbal descriptors and the rating scale range were changed. The MCH scale consists of a decision tree and a unidimensional 10-point rating scale that ranges from easy (1) to impossible (10). Figure 37.2 presents the MCH decision tree and rating scale.

The MCH scale is especially appropriate for evaluation of tasks with perceptual, mediational, and communications activities (Casali and Wierwille, 1983, 1984). It was argued that application of the decision tree flowchart in the subjective rating scale may reduce the variability due to its tighter structure, whereas conventional scales such as, bipolar leave too many of the scale levels open to operator judgment and selection variability (Skipper et al., 1986). However, the decision tree scales can provide only ordinal

Difficulty Level	Operator Demand Level	Rating
Very easy, Highly desirable	Operator mental effort is minimal and desired performance is easily attainable	1
Easy, desirable	Operator mental effort is low and desired performance is attainable	2
Fair, mild difficult	Acceptable operator mental effort is required to attain adequate system performance	3
Minor but annoying difficulty	Moderately high operator mental effort is required to attain adequate system performance	4
Moderately objectionable difficulty	High operator mental effort is required to attain adequate system performance	5
Very objectionable but tolerable	Maximum operator mental effort is required to attain adequate system performance	6
Major difficulty	Maximum operator mental effort is required to bring errors to moderate level	7
Major difficulty	Maximum operator mental effort is required to avoid large or numerous errors	8
Major difficulty	Intense operator mental effort is required to accomplish task, but frequent or numerous errors persist	9
Impossible	Instructed task could not be accomplished reliably	10

FIGURE 37.2   Modified Cooper–Harper rating scale.

estimates of mental workload. Validation research showed that the MCH scale reflects differences in both performance and workload, and is sensitive to variations in controls, displays, and aircraft stability (Rehman, 1995). The MCH was successfully applied to workload evaluation in many flight simulation experiments (Casali and Wierwille, 1983, 1984; Wierwille et al., 1985; Skipper et al., 1986). The MCH scale was able to discriminate between the low, moderate, and high communication loads and mental loads in such tasks as different hazard detection conditions were also observed (Casali and Wierwille, 1984) and navigation dilemmas (Wierwille et al., 1985). Significant increase of MSH ratings with increased danger conditions (Casali and Wierwille, 1984). The applications of the MCH scale in such environments as remotely piloted vehicle system (Byers et al., 1988) and generic air defense system (Bittner et al., 1989) confirmed sensitivity of the scale. It was concluded that the MCH scale provides consistent and sensitive ratings of workload across a range of tasks (Wierwille et al., 1985; Skipper et al., 1996). However, some studies showed that the MCH is less sensitive than the NASA TLX or the overall workload scale (Hill et al., 1992).

37.4.4   Workload Profile

The Workload Profile (WP) is a relatively new multidimensional workload assessment technique (Tsang and Velazquez, 1996). The workload dimensions assessed by this instrument are based on the resource dimensions described in multiple resource model of Wickens (Rubio et al., 2004). The workload dimensions represent demands that can be imposed by a task, including perceptual/central processing, response selection and execution, spatial processing, verbal processing, visual processing, auditory processing, manual output, and speech output. This instrument aims to combine the advantages of secondary task performance based procedures (high diagnosticity) and subjective techniques (high subject acceptability and low implementation requirements and intrusiveness) (Rubio et al., 2004). During workload assessment with WP, the subjects are asked to provide the proportion of attentional resources used after experiencing all of the tasks to be evaluated. Subjects are provided with the definition of each dimension at the time of their rating. Each dimension is rated with the number between 0 and 1 to reflect the proportion of attentional resources used in each task. A rating 0 means that tasks placed no demands on the dimension rated, and 1 means the maximum attentional demands (Tsang and Velazquez, 1996).

The WP procedure performance was investigated by two studies (Tsang and Velazquez, 1996; Rubio et al., 2004). Tsang and Velazquez (1996) established instrument reliability with test–retest method and concurrent validity in reference to task performance. Both studies showed that WP ratings are sensitive to the task demand manipulations. However, the properties of WP demand more detailed and extensive research (Tsang and Velazquez, 1996; Rubio et al., 2004).

37.4.5 Other Methods of Mental Workload Assessment

37.4.5.1 Overall Workload Scale

The Overall Workload (OW) scale is a bipolar scale requiring subjects to provide a single workload rating. The OW scale is easy to use, but is less valid and reliable than NASA TLX or AHP ratings (Vidulich and Tsang, 1986). Hill et al. (1992) reported that OW was consistently more sensitive to workload and had greater operator acceptance than the MCH rating scale or the SWAT.

37.4.5.2 Bedford Scale

The Bedford scale is a unidimensional rating scale designed to identify operator's spare mental capacity while completing a task. The Bedford scale is a modification of the Cooper–Harper scale (Rehman, 1995). The single dimension is assessed using a hierarchical decision tree that guides the operator through a 10-point rating scale, each point of which is accompanied by a descriptor of the associated level of workload. It is simple, quick, and easy to apply *in situ* to assess task load in high workload environments, but it does not have a diagnostic capability. The Bedford scale provides a good measure of spare capacity (Tsang and Johnson, 1987), and demonstrates high sensitivity to the manipulations of task demands (Tsang and Velazquez, 1996). However, other research showed that the scale is not sensitive to differences in either control configurations or combat conditions (Rehman, 1995).

37.4.5.3 Defence Research Agency Workload Scale

The Defence Research Agency Workload Scale (DRAWS) (Table 37.7) is a multidimensional tool designed to perform subjective assessment of the operators' workload. This instrument evaluates workload along several dimensions, including input demand, central demand, output demand, and time pressure. Input demand is a load associated with the acquisition of information from external sources, central demand is the load associated with interpreting information and deciding on action.

TABLE 37.7 Defence Research Agency Workload Scales (DRAWS)

(a) How much demand was imposed by the acquisition of information from external sources (e.g., from a visual display or auditory signals)?

(b) How much demand was imposed by the mental operations (e.g., memorization, calculation, decision making) required by the task?

(c) How much demand was imposed by the responses (e.g., keypad entries, control adjustments, vocal utterances) required by the task?

(d) How much demand was imposed by time pressure?

Output demand is the load associated with the responses required by the task. Time pressure is the load associated with the speed at which tasks must be performed. The DRAWS tool requires people to provide ratings for each description of demand on a scale of 0–100, where 0 means no demand and 100 the maximum demand (HIFAdata, 2002).

37.4.5.4 Rating Scale Mental Effort

The Rating Scale Mental Effort (RSME) is a unidimensional rating scale. This scale ranges from 0 to 150 and has nine descriptive indicators along its axis (e.g., "not effortful" and "awfully effortful"). Validation of this technique ensured that the meanings of the verbal labels are the same for different people (Verwey and Veltman, 1996).

37.4.5.5 Analytical Hierarchy Process

The Analytical Hierarchy Process (AHP) (Saaty, 1980) is a relative, retrospective, and redundant technique of mental workload assessment. The AHP procedure based on the Gopher's psychophysical scaling approach and uses relative judgments for workload assessment (Vidulich et al., 1991). However, there is no single reference task, each task is compared with all other tasks. It is also a fully retrospective technique, where all comparisons are made after the rater completed all tasks. A 17-point rating scale is used to evaluate all possible pairs of tasks comparisons. The scores are inputs, the in-judgment matrix, in which each row or column represents the workload dominance of one task relative to all of the other. Research showed that AHP ratings were sensitive to changes in difficulty among 10 procedural elements performed during in-flight aircraft testing (Vidulich et al., 1991).

37.4.5.6 Subjective Workload Dominance Technique

The Subjective Workload Dominance (SWORD) technique uses a series of relative judgments comparing the workload of different task conditions (Vidulich et al., 1991). The SWORD technique is an implementation of the AHP approach designed specifically for subjective workload assessment. There are three required steps: (1) a rating scale listing all possible pairwise comparisons of the tasks performed must be completed, (2) a judgment matrix comparing each task to every other task must be filled in with each subject's evaluation, and (3) ratings must be calculated using a geometric means approach. Vidulich and Tsang (1986) showed that *SWORD* is a sensitive and reliable workload measure.

37.4.5.7 Comparison between Different Subjective Rating Scales for Mental Workload

The most common approach applied in subjective workload assessment has been the combination of absolute magnitude estimation with immediate presentation (Tsang and Vidulich, 1994). Two most popular techniques — the NASA TLX and SWAT — use the absolute estimation approach, that is, both techniques require independent assessments of each condition on abstract scale dimension. Also, both techniques (the NASA TLX and SWAT) are designed to be used immediately after the performance of rated task. Immediate procedure has been promoted as protection against loss of information from short-term memory (Reid and Nygren, 1988). The raw ratings from these instruments are intended to be based on fresh memory and free from contamination from the raters hypothesis concerning tasks ordering. Relative judgment approaches (SWORD, AHP) to subjective mental workload assessments are rather uncommon. Some research (Vidulich and Tsang, 1986; Tsang and Vidulich, 1994) compared the relative-retrospective method (AHP) to the absolute-immediate evaluation method (NASA TLX). It was found that the retrospective rating has higher test–retest reliability and sensitivity than the immediate ratings (Vidulich et al., 1991). The investigation of the relative-immediate and relative–respective approach in SWORD technique (Tsang and Vidulich, 1994) showed that relative-immediate approach had the lowest reliability, concurrent validity, and sensitivity.

Vidulich and Tsang (1986) compared the SWAT and the NASA-Bipolar methods ratings with respect to performance ratings. These methods were applied to workload assessment in the tracking and spatial transformation tasks with different levels of difficulty, different input/output configurations, and various degree of resources competition. The evaluation of construct validity of the scales was based on their ability to differentiate the levels of task difficulty. The concurrent validity was established by the correlation between the subjective workload and performance ratings. Both techniques showed similar sensitivity to different tasks manipulations and task difficulty. However, both techniques were not able to detect resource competition effects in dual tasks performance, demands of response execution processing, and the dynamics of the difficulty changes.

Hill et al. (1992) compared four subjective workload ratings scales: NASA TLX, OW, SWAT, and MCH scale. These techniques were compared in reference to the following criteria: (1) sensitivity (measured by factor validity), (2) operator acceptance, (3) resource requirements, and (4) special procedures. The results showed that NASA TLX had the highest factors validity (the greatest correlation with the operator workload factor), while the OW had the second highest average factor validity. The TLX was liked best by the operators, and OW was the easiest to complete. Verwey and Veltman (1996) compared sensitivity and diagnosticity of several workload assessment methods such as SWAT, RMSE, workload secondary task, ratings, heart rate, and eyebinks. The workload assessment techniques were compared for the short periods of elevated visual and mental workload during driving. The results revealed that the secondary performance technique, SWAT, and RSME were sensitive to the visual workload peaks. Secondary performance technique and RSME were also sensitive to the mental workload peaks, while SWAT was less sensitive to the mental workload peaks.

Rubio et al. (2004) evaluated psychometric properties (sensitivity, diagnosticity, and validity) of three instruments: the NASA TLX, SWAT, and WP. Two laboratory tasks were evaluated: Sternberg's memory searching tasks and tracking tasks. The sensitivity and diagnosticity of WP ratings were higher than NASA TLX and SWAT ratings. The assessment of concurrent validity by correlation of ratings with performance showed that the NASA TLX ratings had higher correlation than SWAT and WP.

Nygren (1991) made a theoretical analysis of the psychometric properties of NASA TLX and SWAT. He concluded that psychometric properties of the unidimesional scale (such as Cooper–Harper or OW) make them less sensitive to differences in workload than either TLX or SWAT, and relatively more variable than TLX and SWAT scores. According to Nygren (1991), the advantage of SWAT over TLX is that it is a psychological model of subjective judgment, which may be oversimplified, but may also reflect cognitive mechanisms and biases that actually affect the process of mental workload judgments. The advantage of TLX over SWAT is based on the general linear model, where six dimensions derived from extensive multivariate analyses based on numerous studies and different workload domains. SWAT is more relevant for empirical testing of the appropriateness of particular additive model of workload assessment. TLX has greater potential for solving workload problems in many applied settings, by accurately predicting operator workload levels across a variety of tasks (Nygren, 1991). The results of numerous validation studies indicate that MCH, Bedford scale, SWAT, and TLX procedures represent globally sensitive measures of operator workload (Wierwille and Eggemeier, 1993). Since, both SWAT and TLX are multidimensional scale and they can therefore provide some diagnostic information concerning causes of workload represented by the subscales (Moroney et al., 1995).

37.5   Summary

Several authors concluded that subjective ratings scales are among the simplest and most efficient of workload estimation instruments that can be used for ergonomics applications (Skipper et al. 1986; Nygren, 1991; Wierwille and Eggemeier, 1993; Tsang, 2001). These scales are the most sensitive, most transferable, and least intrusive techniques for workload estimation, are easy to administer, and require little effort or no equipment. Several scales demonstrated global sensitivity and thus can provide appropriate workload indicators in test and evaluation situations. Some authors stated that subjective rating scales constitute the most relevant method of mental workload assessment, since" subjective scaling is the most direct measure of such subjective experience" (Sheridan, 1980). However, the subjective rating scales have also some serious disadvantages. The source of the resource demands is hard to introspectively diagnose within a dimensional framework. Subjects may not be able to distinguish mental demands from other type of demands such as physical or manipulative (O'Donnell and

Eggemeier, 1986). Additionally, the levels of fatigue and emotional states can have significant effect on the workload ratings. Furthermore, operators who provide ratings may have been adapted to particular system and learned to compensate for its deficiencies, thus the rating will not reflect properly the workload imposed by a task or a system (Skipper et al., 1986). Finally, all subjective methods are prone to biases due to central tendency, halo, and leniency effects. Careful instruction and properly performed measurement procedures can minimize the possibility of bias occurrence and ensure accurate ratings.

References

Annett, J. 2002. Subjective rating scales: science or art. *Ergonomics*, 45, 966–987.

Battiste, V. and Bortolussi, M. 1988. Transport pilot workload: a comparison of two objective techniques. In *Proceedings of the Human Factors Society 32nd Annual Meeting*, 150–154.

Beare, A. and Dorris, R. 1984. The effects of supervisor experience and the presence of a shift technical advisor on the performance of two-man crews in a nuclear power plant simulator. In *Proceedings of the Human Factors Society, 28th Annual Meeting*, 242–246. Santa Monica, CA: Human Factors Society.

Bittner, A.V., Byers, J.C., Hill, S.G., Zaklad, A.L., and Christ, R.E. 1989. Generic workload ratings of a mobile air defense system (LOS-F-H). In *Proceedings of the 33rd Annual Meeting of the Human Factors Society*, 1476–1480. Santa Monica, CA: Human Factors Society.

Bonney, R., Weisman, G., Haugh, L.D., and Finkelstein, J. 1990. Assessment of postural discomfort. In *Proceedings of the Human Factors Society 34th Annual Meeting*, 684–687. Orlando, FL, 2–5 October.

Borg, G.A. 1982. Psychophysical bases of perceived exertion. *Med Sci Sports Exerc*, 14, 377–381.

Borg, G. 1985. *An Introduction to Borg's RPE-scale*. Ithaca, NY: Mouvement.

Borg, G. 2001. Rating scales for perceived physical effort and exertion. In *International Encyclopedia of Ergonomics and Human Factors*, Karwowski, W. (ed.), 538–542. New York: Taylor & Francis.

Borg, G., Hassmen, P., and Lagerstrom, M. 1987. Perceived exertion related to heart rate and blood lactate during arm and leg exercise. *Eur J App Physiol*, 65, 679–685.

Borg, E. and Borg, G. 2002. A comparison of AME and CR100 for scaling perceived exertion. *Acta Psychol*, 109, 157–175.

Boussenna, M., Corlett, E.N., and Pheasant, S.T. 1982. The relation between discomfort and postural loading at the joints. *Ergonomics*, 25, 315–322.

Byers, J.C., Bittner, A.C., Hill, S.G., Zaklad, A.L., and Christ, R.E. 1988. Workload assessment of a remotely piloted vehicle (RPV) system. In *Proceedings of the Human Factors Society 32nd Annual Meeting*, 1145–1149. Santa Monica, CA: Human Factors Society.

Casali, J.G. and Wierwille, W.W. 1983. A comparison of rating scale, secondary task, physiological, and primary task workload estimation techniques in a simulated flight emphasizing communications load. *Human Factors*, 25(6), 623–641.

Casali, J.G. and Wierwille, W.W. 1984. On the measurement of pilot perceptual workload: a comparison of assessment techniques addressing sensitivity and intrusion issues. *Ergonomics*, 27, 1033–1050.

Chen, M.J., Fan, X., and Moe, S.T. 2002. Criterion-related validity of the Borg ratings of perceived exertion scale in healthy individuals: a meta-analysis. *J Sports Sci*, 20, 873–899.

Cooper, G.E. and Harper, R.P. 1969. The use of pilot ratings in the evaluation of aircraft handling qualities. NASA Ames Technical Report NASA TN-D-5153. Moffett Field, CA: NASA Ames Research Center.

Corlett, E.N. and Bishop, R.P. 1976. A technique for measuring postural discomfort. *Ergonomics*, 9, 175–182.

Corlett, E.N. 1995. The evaluation of posture and its effects. In Wilson J.R. and Corlett E.N. (eds), 662–713, London: Taylor and Francis.

Coury, B.G. and Drury, C.G. 1982. Optimum handle positions in a box-holding task. *Ergonomics*, 25, 645–662.

Dickinson, C.E., Campion, K., Foster, A.F., Newman, S.J., O'Rourke, A.M.T., and Thomas, P.G. 1992. Questionnaire development: an examination of the Nordic Musculoskeletal Questionnaire. *Appl Ergon*, 23, 197–201.

Drury, C.G., Deeb, J.M., Hartman, B., Woolley, S., Drury, C.E., and Gallagher, S. 1989. Symmetric and asymmetric manual materials handling. Part 1: Physiology and psychophysics. *Ergonomics*, 32, 467–489.

Eggemeier, F.T., Wilson, G.F., Kramer, A.F., and Damos, D.L. 1991. General considerations concerning workload assessment in multi-task environments. In *Multiple Task Performance* Damos, D.L. (ed.), 207–216. London: Taylor & Francis.

Eggemeier, F.T. and Wilson, G.F. 1991. Performance-based and subjective assessment of workload in multi-task environments. In *Multiple Task Performance*, Damos, D.L. (ed.), 217–278. Washington, DC: Taylor & Francis.

Fleishman, E.A., Gebhardt, D.L., and Hogan, J.C. 1984. The measurement of effort. *Ergonomics*, 27(9), 947–954.

Genaidy, A.M., Karwowski, W., Christensen, D.M., Vogiatzis, C., Deraiseh, N., and Prins, A. 1998. What is 'heavy'? *Ergonomics*, 41(4), 420–432.

Hampson, D.B., Gibson, A.S.C., Lambert, M.I., and Noakes, T.D. 2001. The Influence of sensory cues on the perception of exertion during exercise and central regulation of exercise performance. *Sports Med*, 31, 935–952.

Hart, S.G. and Staveland, L.E. 1988. Development of NASA-TLX (Task Load Index): results of empirical and theoretical research. In *Human Mental Workload*, Hancock P.A. and Meshkati, N. (eds.), 139–183. Amsterdam: Elsevier Science Publishers.

HIFAdata. 2002. Human Factor Integration in Future database website. http://www.eurocontrol.int/eatmp/hifa/hifa/HIFAdata.html.

Hill, S.G., Byers, J.C., Zaklad, A.L., and Christ, R.E. 1989. Subjective workload assessment during 48 continuous hours of LOS-F-H operations. In *Proceedings of the Human Factors Society 33rd Annual Meeting*, 1129–1133. Santa Monica, CA: Human Factors Society.

Hill, S.G., Iavecchia, H.P., Byers, J.C., Bittner, A.C., Zaklad, A.L., and Christ, R.E. 1992. Comparison of four subjective workload rating scales. *Human Factors*, 34, 429–439.

Hogan, J.C. and Fleishman, E.A. 1979. An index of the physical effort required in human task performance. *J Appl Psychol* 64, 197–204.

Hogan, J.C., Ogden, G., and Fleishman, E.A. 1980. Reliability and validity of methods for evaluating perceived physical effort. *J Appl Psychol* 65, 672–679.

ISO/FDIS 10075-3, 2002, Ergonomic principles related to mental workload — Part 3: Principles and requirements concerning methods for measuring and assessing mental workload. International Standardization Organization.

Johansson, S.-E. and Borg, G. 1993. Perception of heavy work operations by tank truck drivers. *Appl Ergon*, 24, 421–426.

Jung, E.S. and Choe, J. 1996. Human reach posture prediction based on psychophysical discomfort. *In J Ind Ergon*, 18, 173–179.

Kroemer, K.H.E., Kroemer, H.B., and Kroemer-Elbert, K.E. 1994. *Ergonomics: How to Design for Ease and Efficiency*. Englewood Cliffs, NJ: Prentice Hall.

Kumar, S., Narayan, Y., and Bjornsdottir, S. 1999. Comparison of the sensitivity of three psychophysical techniques to three manual materials handling task variables. Ergonomics, 42, 61–73.

Kuorinka, I., Jonsson, B., Kilbom, A., Vinterberg, H., Biering-Sorensen, F., Andersson, G., and Jorgensen, K. 1987. Standardized Nordic Questionnaires for the analysis of musculoskeletal symptoms. *Appl Ergon*, 18, 233–237.

Legg, S.J., Perko, L., and Campbell, P. 1997. Subjective perceptual methods for comparing backpacks. Ergonomics, 40, 809–817.

Liao, M.-H. and Drury, C.G. 2000. Posture, discomfort and performance on a VDT task. *Ergonomics*, 43, 345–359.

Meshkati, N., Hancock, P., and Rahimi, M. 1992. Techniques in mental workload assessment. In *Evaluation of Human Work. A Practical Ergonomics Methodology*, Wilson, J. and Corlett, E. (eds.), 605–627. London: Taylor and Francis.

Moray, N. 1982. Subjective mental workload. *Human Factors*, 24, 25–40.

Morgan, W.P. 1994. Psychological components of effort sense. *Med Sci Sports Exerc*, 26, 1071–1077.

Moroney, W.F., Biers, D.W., and Eggemeier, F.T. 1995. Some measurement and methodological considerations in the application of subjective workload measurement techniques. *Int J Aviat Psychol*, 5, 87–106.

Nataupsky, M. and Abbott, T.S. 1987. Comparison of workload measures on computer-generated primary flight displays. In *Proceedings of the Human Factors Society 31st Annual Meeting*, 548–552. Santa Monica, CA: Human Factors Society.

Noble, B.J. and Robertson, R.J., 1996. *Perceived Exertion*. Champaign, IL: Human Kinetics Nygren, T.E. 1991. Psychometric properties of subjective workload measurement techniques: implications for their use in the assessment of perceived mental workload. *Human Factors*, 33, 17–33.

O'Donnell, R.D. and Eggemeier, F.T. 1986. Workload assessment methodology. In Boff, K.R., Kaufman, L. and Thomas, J.P. (eds.). *Handbook of perception and human performance*, Volume II: Cognitive Processes and Performance, 42/1–42/49. New York: John Wiley & Sons.

Olendorf, M.R. and Drury, C.G. 2001. Postural discomfort and perceived exertion in standardized box-holding postures. *Ergonomics*, 44, 1341–1367.

Pandolf, K.B. 1983. Advances in the study and application of perceived exertion. *Exerc Sport Sci Rev*, 11, 118–158.

Reid, G.B. and Nygren, T.E. 1988. The subjective workload assessment technique: a scaling procedure for measuring mental workload. In Hancock, P.A. and Meshkati, N. (eds.), *Human Mental Workload*, 185–218. Amsterdam: Elsevier Science Publishers.

Rehman, A.J. 1995. *Handbook of Human Performance Measures and Crew Requirements for Flightdeck Research*. Technical report, DOT/FAA/CT-TN95/49.

Rubio, S., Diaz, E., Martin, J., and Puente, J.M. 2004. Evaluation of subjective mental workload: a comparison of SWAT, NASA-TLX, and Workload Profile methods. *Appl Psychol Int Rev*, 53, 61–86.

Russell, W.D. 1997. On the current status of rated perceived exertion. *Percept Motor Skills*, 84, 799–808.

Saaty, T.L. 1980. *Multicriteria Decision Making: The Analytic Hierarchy Process*. New York: McGraw Hill Company.

Shephard, R.J. 1994. Perception of effort in the assessment of work capacity and the regulation of the intensity of effort. *Int J Ind Ergon*, 13, 67–80.

Sheridan, T.B. 1980. Mental workload, what is it? Why bother with it? *Human Factor Soc Bull*, 23, 1–2.

Shively, R., Battiste, V., Matsumoto, J., Pepiton, D., Bortolussi, M., and Hart, S.G. (1987). In-flight evaluation of pilot workload measures for rotorcraft research. In *Proceedings of the Fourth Symposium on Aviation Psychology*, 637–643. Columbus, OH: Department of Aviation, The Ohio State University.

Sinclair, M.A. 1995. Subjective assessment. In Wilson, J.R. and Corlett, E.N. (eds.), *Evaluation of Human Work*, 69–100. London: Taylor & Francis.

Skinner, J.S., Hustler, R., Bergsteinova, V., and Buskirk, E.R. 1973. The validity and reliability of a rating scale of perceived exertion. *Med Sci Sports*, 5, 94–96.

Skipper, J.H., Rieger, C.A., and Wierwille, W.W. 1986. Evaluation of decision tree rating scales for mental workload estimation. *Ergonomics*, 29, 585–599.

Stevens, S.S. (1974). perceptual magnitude and its measurement. In *Handbook of Perception: Psychophysical Judgment and Measurement*, Carterette C. and Friedman, M.P. (eds.), 361–389. New York: Academic Press.

Tsang, P.S. and Johnson, W.W. 1989. Cognitive demands in automation. *Aviat Space Environ Med*, 60, 130–135.

Tsang, P.S. and Velazquez, V.L. 1996. Diagnosticity and multidimensional subjective workload ratings. *Ergonomics*, 39, 358–381.

Tsang, P.S. and Vidulich, M.A. 1994. The roles of immediacy and redundancy in relative subjective workload assessment. *Human Factors*, 36, 503–513.

Tsang, P.S. and Wilson, G.F. 1997. Mental workload. In *Handbook of Human Factors and Ergonomics*, Salvendy, G. (ed.). New York: John Wiley & Sons.

Tsang, P.S., 2001. Mental workload. In *International Encyclopedia of Ergonomics and Human Factors*, Karwowski, W. (ed.). New York, Taylor & Francis.

Verwey, W.B. and Veltman, H.A. 1996. Detecting short periods of elevated workload: a comparison of nine workload assessment techniques. *J Exp Psychol Appl*, 2, 270–285.

Vidulich, M.A. and Tsang, P.S. 1986. Techniques of subjective workload assessment: a comparison of SWAT and the NASA-Bipolar methods. *Ergonomics*, 29(11), 1385–1398.

Vidulich, M.A., Ward, G.F., and Schueren, J. 1991. Using subjective Workload Dominance (SWORD) technique for projective workload assessment. *Human Factors*, 33, 677–692.

Whitaker, L., Peters, L., and Garinther, G. 1989. Tank crew performance: effects of speech intelligibility in target acquisition and subjective workload assessment. In *Proceedings of the Human Factors Society 33rd Annual Meeting*, 1411–1413. Santa Monica, CA: Human Factors Society.

Wierwille, W. W. and Eggemeier, F.T. 1993. Recommendations for mental workload measurement in a test and evaluation environment. *Human Factors*, 35, 263–281.

Wierwille, W. W., Rahimi, M., and Casali, J.G. 1985. Evaluation of 16 measures of mental workload using a simulated flight task emphasizing mediational workload. *Human Factors*, 21, 499–502.

Wierwille, W.W. and Casali, J.G. 1983. A validated rating scale for global mental workload measurement applications. In *Proceedings of the Human Factors Society 27th Annual Meeting*, 129–133. Santa Monica, CA: Human Factors Society.

Zhuang, Z., Stobbe, T.J., Collins, J.W., Hsiao, H., and Hobbs, G.R. 2000. Psychophysical assessment of assistive devices for transferring patients/residents. *Appl Ergon*, 31, 35–44.

38

Rest Allowances*

Stephan Konz
Kansas State University

38.1 Introduction

The chapter is divided into two major parts. The first part (Section 38.2 to Section 38.6) gives numerical allowances for time standards for various tasks from the International Labor Organization (ILO). Although the ILO values are logical, the values are subject to many questions.

The second part (Section 38.7 and Section 38.8) gives general guidelines for fatigue.

38.2 Time Standard Concept

Beginning in the 1920s, the concept of a time standard for repetitive production work was initiated:

$$\text{Normal time} = (\text{Observed time})(\text{Rating})$$

That is, the time study technician recorded the time the worker took to do the job (observed time) and multiplied the observed time by a pace rating to determine the normal time of an experienced operator to do the job. Standard data tables from systems such as MTM are normal time.

---

*This chapter is a concise version of material from Konz and Johnson (2004).

However, it was recognized that, in addition to normal time, additional time was needed for "allowances" — typically divided into machine allowances, personal allowances, delay allowances, and fatigue allowances. Some firms express the allowances as a percentage of shift time and some as a percentage of work time.

For shift time:

$$\text{Standard time} = \frac{\text{Normal time}}{1 - \text{Allowances}}$$

For work time:

$$\text{Standard time} = \text{Normal time}(1 + \text{Allowances})$$

38.3   Personal Allowances

Personal allowances are given for such things as going to the toilet, getting a drink of water, smoking, and so on. They do not vary with the task — they are the same for all tasks in the firm. There is no scientific or engineering basis for the percentage to be given. A value of about 5% (24 min in a 480-min day) seems typical.

Most firms have standardized break periods (coffee breaks); for example, 15 min during the first part of the shift and the same during the second part. It is not clear whether firms consider this part of the personal allowance or in addition to it.

The mid-shift break (lunch) is another break time; it typically is not considered part of allowances.

Some firms also give additional allowances to workers for cleanup (either of the person or the machine), putting on or taking off protective clothing, or for travel. In mines, the travel allowance is called portal-to-portal pay; pay begins when the worker crosses the mine portal even though the worker will not arrive at the work site until some time later.

Note that some allowances (such as cleanup time at the end of the shift) are given per shift and not given as a percentage of time/unit.

38.4   Delay Allowances

Delay allowances should vary with the task, not the operator. They compensate for machine breakdowns, interrupted material flow, machine maintenance and cleaning, etc. For "long" delays (say >30 min), the operator clocks out and works on something else during the clocked out time. Delays usually permit the operator to take some personal time and reduce fatigue; that is, they also serve as personal allowances and fatigue allowances.

38.5   Machine Allowances

With increasing capabilities of servomechanisms and computers, many machines operate semiautomatically (operator's primary duty is to load/unload the machine) or automatically (machine loads, processes, and unloads). During the machine time of the work cycle, the operator may be able to drink coffee (personal allowance), talk to the supervisor (delay allowance), or recover from fatigue. Thus, as a general rule, allowances are to be given for fatigue only for the portion of the work cycle outside the machine time.

38.6   ILO Fatigue Allowances

The following discusses the fatigue allowances developed for the International Labor Organization (ILO, 1992). They were supplied by a British consulting firm.

The use of the ILO values is complex. Remembering that fatigue allowances are given only for work time (not machine time), the applicable fatigue allowance points are to be totaled. Then, using Table 38.1, the points are converted to percentage of time. A large fatigue allowance is an indication that there is a large potential for improved ergonomics.

Fatigue allowances will be divided into physical, mental, and environmental.

38.6.1 Physical Fatigue Allowances

There are four subcategories of physical fatigue allowances: physical fatigue, body posture, short cycle, and restrictive clothing:

1. *Physical fatigue.* Table 38.2 has three fatigue categories: push (including foot pedal as well as carrying a load on the back), carry (hand/arm carry), and lifting/lowering. All three categories are straight lines when force is plotted versus percentage of allowances.

   Carrying items on the back is rare in present industrial practice. For pedal force, the recommended maximum pedal force is 90 N (9 kg) (Konz and Johnson, 2004). However, a 9-kg pedal force seems excessive in present industrial practice — especially for a repetitive operation. Carrying items in the hand/arms probably would be intermittent rather than continuous so the stress would be on local muscle systems rather than a cardiovascular load. Mital et al. (1993) give recommended weights for intermittent carrying for one- and two-handed carrying. They do not give any recommendations for fatigue of carrying but do recommend reducing the weight carried by 30% if the carrying is done frequently. The ILO lifting allowances probably are best compared versus the National Institute of Occupational Safety and Health (NIOSH) lifting guidelines (see Chapter 46). The ILO uses only the single variable of weight lifted while NIOSH considers variables of lift origin, lift destination, frequency of move, angle, and container design to determine the permissible weight to be lifted. However, NIOSH does not give fatigue allowances for lifting.

2. *Body posture.* Table 38.3 gives posture allowances. Sitting easily is 0 points with 10 points the maximum in present industrial practice. Activities given an allowance include standing, carrying

TABLE 38.1 Conversion from Points Allowance to Percentage Allowance for ILO

	Points									
	0	1	2	3	4	5	6	7	8	9
0	10	10	10	10	10	10	10	11	11	11
10	11	11	11	11	11	12	12	12	12	12
20	13	13	13	13	14	14	14	14	15	15
30	15	16	16	16	17	17	17	18	18	18
40	19	19	20	20	21	21	22	22	23	23
50	24	24	25	26	26	27	27	28	28	29
60	30	30	31	32	32	33	34	34	35	36
70	37	37	38	39	40	40	41	42	43	44
80	45	46	47	48	48	49	50	51	52	53
90	54	55	56	57	58	59	60	61	62	63
100	64	65	66	68	69	70	71	72	73	74
110	75	77	78	79	80	82	83	84	85	87
120	88	89	91	92	93	95	96	97	99	100
130	101	103	105	106	107	109	110	112	113	115
140	116	118	119	121	122	123	125	126	128	130

Note: The second column (0) gives the 10 s, and the remaining columns give the units. Thus, 30 points (0 column) = 15%; 31 points (1 column) = 16%; 34 points = 17%.

The percentage allowance is for manual work time (not machine time) and includes 5% personal time for coffee breaks. Note that 0 points gives 10% allowance. In addition, at "low" points (say 0 to 20), it takes about six points to get an additional 1% allowance while at "high" points (say 100), one point gives about an additional 1% allowance.

Source: International Labor Office. 1992. *Introduction to Work Study,* 4th ed. Geneva, Switzerland: ILO. With permission.

TABLE 38.2 Pushing/Back Carrying, Arm Carrying, and Lifting Force Allowances

Weight or Force, kg	Push/Back Carry Points	Arm Carry Points	Lifting Points
1	0	0	0
2	5	5	10
3	8	9	15
4	10	13	18
5	12	15	21
6	14	17	23
7	15	20	26
8	17	21	29
9	19	24	32
10	20	26	34
11	21	29	37
12	23	31	40
13	25	33	44
14	26	34	46
15	27	36	50
16	28	39	50
17	30	40	53
18	32	42	56
19	33	44	58
20	34	46	60

Note: Push includes foot pedal push and carry on the back. Arm carry includes hand carry and swinging arm movements. Weight is averaged over time. A 15-kg load lifted for 33% of a cycle is 5 kg.

Source: International Labor Office. 1992. *Introduction to Work Study*, 4th ed. Geneva, Switzerland: ILO. With permission.

a load, and various major body movements such as climbing a ladder, lifting, and shoveling. For an ergonomics analysis of the effects of posture, see the Rapid Upper Limb Assessment (RULA) technique in Chapter 42.

3. *Short cycle.* Table 38.4 gives short-cycle allowances to permit muscles some time to recover from highly repetitive movements. Since it is for cycle times of 3 to 10 sec, this really involves only hand–arm movements and probably, even more narrowly, hand–finger movements. In a factory, this would be "highly repetitive bench assembly."

4. *Restrictive clothing.* Table 38.5 gives allowances due to clothing restrictions.

TABLE 38.3 Short-Cycle Allowances

Points	Cycle Time, sec	Cycle Time, min	Cycles/minute
1	10.0	0.167	6.0
2	9.0	0.150	6.7
3	8.0	0.133	7.5
4	7.2	0.120	8.3
5	6.0	0.100	10.0
6	5.0	0.083	12.0
7	4.2	0.070	14.3
8	3.6	0.060	16.7
9	3.0	0.050	20.0
10	<3.0	<0.050	>20.0

Source: International Labor Office. 1992. *Introduction to Work Study*, 4th ed. Geneva, Switzerland: ILO. With permission.

TABLE 38.4 Posture Allowance

Points	Activity
0	Sitting easily
2	Sitting awkwardly or mixed sitting and standing
4	Standing or walking freely
5	Ascending or descending stairs, unladen
6	Standing with a load; walking with a load
8	Climbing up and down ladders; some bending, lifting, stretching, or throwing
10	Awkward lifting; shoveling ballast to container
12	Constant bending, lifting, stretching, or throwing
16	Coal mining with pickaxes; lying in a low seam

Source: International Labor Office. 1992. *Introduction to Work Study*, 4th ed. Geneva, Switzerland: ILO. With permission.

38.6.2 Mental Fatigue Allowances

The ILO (1992) has two mental fatigue allowances: concentration/anxiety and monotony.

1. *Concentration/anxiety.* Table 38.6 gives allowances for concentration/anxiety
2. *Monotony.* Table 38.7 gives allowances for monotony

38.6.3 Environmental Fatigue Allowances

The ILO (1992) has four environmental fatigue allowances: climate; dust, dirt, and fumes; noise and vibration; and eye strain (Konz and Johnson, 2004):

1. *Climate.* Table 38.8 gives allowances for climate, subdivided into temperature/humidity, wet, and ventilation. For more on temperature/humidity, see Konz and Johnson (2004)
2. *Dirt, dust, and fumes.* Table 38.9 gives allowances for dirt, dust, and fumes
3. *Noise and vibration.* Table 38.10 gives allowances for noise and vibration. For more on noise and vibration, see Chapter 31 and Chapter 33
4. *Eye strain.* Table 38.11 gives allowance for eye strain. For more on eyes, vision, and illumination, see Konz and Johnson (2004)

TABLE 38.5 Restrictive Clothing Allowances

Points	Clothing
1	Thin rubber (surgeon's) gloves
2	Household rubber gloves; rubber boots
3	Grinder's goggles
5	Industrial rubber or leather gloves
8	Face mask (e.g., for spray painting)
15	Asbestos suit or tarpaulin coat
20	Restrictive protective clothing and respirator

Notes: ILO (1992) considers clothing effects on dexterity as well as clothing weight in relation to effort and movement. Also consider whether the clothing affects vision or breathing.

Source: International Labor Office. 1992. *Introduction to Work Study*, 4th ed. Geneva, Switzerland: ILO. With permission.

TABLE 38.6   Concentration/Anxiety Allowances

Points	Degree
0	Routine simple assembly; shoveling ballast
1	Routine packing, washing vehicles, wheeling trolley down clear gangway
2	Feed press tool (hand clear of press); topping up battery
3	Painting walls
4	Assembling small and simple batches (performed without much thinking); sewing machine work (automatically guided)
5	Assembling warehouse orders by trolley; simple inspection
6	Load/unload press tool; hand feed into machine; spray painting metalwork
7	Adding up figures; inspecting detailed components
8	Buffing and polishing
10	Guiding work by hand on sewing machine; packing assorted chocolates (memorizing patterns and selecting accordingly); assembly work too complex to become automatic; welding parts held in a jig
15	Driving a bus in heavy traffic or fog; marking out in detail with high accuracy

Note: ILO (1992) considers what would happen if the operator were to relax attention, responsibility, need for exact timing, and accuracy or precision required.

Source: International Labor Office. 1992. *Introduction to Work Study*, 4th ed. Geneva, Switzerland: ILO. With permission.

38.6.4   Overview of ILO Fatigue Allowances

The ILO fatigue allowances were developed by a British consulting firm, which did not publish any detailed justification or explanation of how the values were determined. They give a rational structure of allowances but the allowances certainly are subject to question.

An important point is that they vary with only the job — not the person. They do not depend on whether the worker is male or female, age 20 or 60 yr, or weighs 50 or 100 kg.

Another point is the appearance of inadequate ranges for many of the factors. For example, for posture, sitting easily is 0 points while climbing up and down ladders is 8 points. For monotony, two people on jobbing work is 0 points while adding similar columns of figures is 8 points. As mentioned in Table 38.1, a point is not a percentage allowance. At low values of points (say up to a total of 14 points), only a couple of additional percent times are given.

The length of the work time per day, the days per week, and the hours per year are not specified. Presumably, it is 8 h/day, 5 days/week, and approximately 1900 h/yr. What the allowances would be for longer or shorter times is not given.

The allowances consider recovery from fatigue to be complete by the start of the shift on the next day. They do not consider the extra recovery time of weekends, holidays, or vacations. In addition, the

TABLE 38.7   Monotony Allowances

Points	Degree
0	Two people on jobbing work
3	Cleaning own shoes for 0.5 h on one's own
5	Operator on repetitive work; operator working alone on nonrepetitive work
6	Routine inspection
8	Adding similar columns of figures
11	One operator working alone on highly repetitive work

Note: ILO (1972) considers the degree of mental stimulation and if there is companionship, competitive spirit, music, and so on.

Source: International Labor Office. 1992. *Introduction to Work Study*, 4th ed. Geneva, Switzerland: ILO. With permission.

TABLE 38.8 Climate Allowances

Humidity	Points for "temperature/humidity"		
	Temperature		
	Up to 24°C	24–32°C	Over 32°C
Up to 75	0	6–9	12–16
76–85	1–3	8–12	15–26
Over 85	4–6	12–17	20–36

Points for "wet"

Points	Wet
0	Normal factory operations
1	Outdoor workers (e.g., postal delivery)
2	Working continuously in the damp
4	Rubbing down walls with wet pumice block
5	Continuous handling of wet articles
10	Laundry washhouse, wet work, steamy, floor running with water, hands wet

Points for "ventilation"

Points	Ventilation
0	Offices; factories with "office-type" conditions
1	Workshop with reasonable ventilation but some drafts
3	Drafty workshops
14	Working in sewer

Note: ILO (1992) considers temperature/humidity, wet, and ventilation. For temperature/humidity, use the average environmental temperature. For wet, consider the cumulative effect over a long period. For ventilation, consider quality/freshness of air and its circulation by air conditioning or natural movement.

Source: International Labor Office. 1992. *Introduction to Work Study*, 4th ed. Geneva, Switzerland: ILO. With permission.

allowances were developed during a time in which Europeans worked longer hours per year than at present; presently they work 1500 to 1600 h/yr.

Finally, the ILO allowances consider only the duration of the rest and not what happens during the rest. Each minute of rest is considered equally valuable.

The following will give some general guidelines to the temporal aspects of fatigue — especially from Chapter 21, Temporal Ergonomics, of Konz and Johnson (2004).

38.7 Temporal Aspects of Fatigue

Ashberg and Gamberale (1998) divide fatigue into five factors:

1. Physical exertion (e.g., bicycle ergometer work; described as warm, sweaty, out of breath, breathing heavily, palpitations)
2. Physical discomfort (e.g., static load on small muscle groups; as tense muscles, aching, numbness, hurting, stiff joints)
3. Lack of energy (mental + physical; adjectives such as exhausted, spent, overworked, worn out, drained)
4. Lack of motivation (mental; described as listless, passive, indifferent, uninterested; lack of initiative)
5. Sleepiness (mental; described as sleepy, yawning, drowsy, falling asleep, lazy)

Jobs will have different combinations of fatigue; often varying during the shift. A VDT operator may have lack of energy and two different kinds of physical discomfort (static loading of the back from posture and repetitive strain on the fingers).

TABLE 38.9   Dirt, Dust, and Fumes Allowances

Points	Dust
0	Office, normal light assembly, press shop
1	Grinding or buffing with good extraction
2	Sawing wood
4	Emptying ashes
6	Finishing weld
10	Running coke from hoppers into skips or trucks
11	Unloading cement
12	Demolishing building
	Dirt
0	Office work, normal assembly operations
1	Office duplicating
2	Garbage collector
4	Stripping internal combustion engine
5	Working under old motor vehicle
7	Unloading bags of cement
10	Coal mining; chimney sweeping with brushes
	Fumes
0	Lathe turning with coolants
1	Emulsion paint, gas cutting, soldering with resin
5	Motor vehicle exhaust in small commercial garage
6	Cellulose painting
10	Molder procuring metal and filling mold

Notes: For dust, consider both volume and nature of the dust. The dirt allowance covers "washing time" where this is paid for (e.g., 3 min for washing). Do not allow both time and points. For fumes, consider the nature and concentration, odor, whether toxic or injurious to the health, irritating to the eyes, nose, and throat.

Source: International Labor Office. 1992. *Introduction to Work Study*, 4th ed. Geneva, Switzerland: ILO. With permission.

Finkleman (1994) analyzed 3700 people who had reported fatigue in their work. Physically demanding jobs had less fatigue than jobs with low physical demand! Significant predictors of fatigue included job pay, job control, and supervisor quality — emphasizing the importance of lack of motivation.

Fatigue generally can be overcome by rest (recovery). Resting time can be "off work" (evenings, weekends, holidays, vacations) and "at work." "At work" is further divided into "formal breaks" (lunch, coffee) and "informal breaks" (work interruptions, training), microbreaks (short breaks of a minute or less), and "working rest" (doing a different task, using a different part of the body, such as answering the phone instead of keying data).

The following are given as axioms:

- Most jobs have peaks and valleys of demand within the shift — that is, the "load" is not constant
- Fatigue increases exponentially with time
- Rest is more beneficial if it occurs before the muscle (cardiovascular system, brain) has "too much" fatigue
- The value of a rest declines exponentially with time
- Different parts of the body have different recovery rates

The "body" will be divided into three parts: (1) cardiovascular system, (2) musculoskeletal system, and (3) the brain.

TABLE 38.10   Noise and Vibration Allowances

Points	Noise
	Noise
0	Working in a quiet office, no distracting noise; light assembly work
1	Work in a city office with continual traffic noise outside
2	Light machine shop; office or assembly shop where noise is a distraction
4	Woodworking machine shop
5	Operating steam hammer in forge
9	Riveting in a shipyard
10	Road drilling
	Vibration
1	Shoveling light materials
2	Power sewing machine; power press or guillotine if operator is holding the material; cross-cut sawing
4	Shoveling ballast; portable power drill operated by one hand
6	Pickaxing
8	Power drill (two hands)
15	Road drill on concrete

Note: For noise, ILO (1992) considers whether the noise affects communication, is a steady hum or a background noise, is regular or occurs unexpectedly, and is irritating or soothing. For vibration, consider the impact of vibration on the body, limbs, or hands and the addition to mental effort as a result, or to a series of jars and shocks.

Source: International Labor Office. 1992. Introduction to Work Study, 4th ed. Geneva, Switzerland: ILO. With permission.

38.7.1   Cardiovascular System

The cardiovascular system is fatigued during "heavy" work — typically manual handling. A key question is the duration of the work. Mital (1984a, b) reported that male material handlers could sustain, without overexertion, for 8-h workdays, 29% of their maximal oxygen uptake; for 12 h shifts, the value declined to 23%. Moreover, Mital et al. (1994) found workers in an aircargo firm's package-handing area, working 2-h shifts, worked at 40 to 54% of their treadmill aerobic capacity.

38.7.2   Musculoskeletal System

This section is divided into static work, dynamic work, and VDT work.

TABLE 38.11   Eye Strain Allowances

Points	Degree
0	Normal factory work
2	Inspection of easily visible faults; sorting distinctively colored articles by color; factory work in poor lighting
4	Intermittent inspection for detailed faults; grading apples
8	Reading a newspaper in a bus
10	Continuous visual inspection (cloth from a loom)
14	Engraving using an eyeglass

Notes: ILO (1992) considers the lighting conditions, glare, flicker, illumination, color, and closeness of work and for how long the strain is endured.

Source: International Labor Office. 1992. *Introduction to Work Study*, 4th ed. Geneva, Switzerland: ILO. With permission.

38.7.2.1 Static Work

Pure static work is not common in most tasks. Usually there are small movements or the body parts are partially supported. There is a great difference in the fatigue resistance of different muscles. For example, using the same protocol, endurance time of the soleus was seven times greater than that of the quadriceps (Bigland-Ritchie et al., 1986).

38.7.2.2 Dynamic Work

In dynamic work, the muscles automatically create micropauses — in contrast to the constant load of static work. For example, when reaching out with the arm, the set of muscles for reaching out work while the set of muscles for reaching in have a rest.

In general, short breaks often are better than long breaks occasionally. For example, Bhatia and Murrell (1969) studied industrial workers who had either six breaks of 10 min or four breaks of 15 min; six breaks of 10 min was preferred.

38.7.2.3 VDT Work

VDT work combines static load on the shoulders and back with dynamic work on the fingers and mental stress. Kadefors and Laubli (2002) emphasize that breaks from computer work must be frequent and allow for mental relaxation as mental load activates the same muscle motor units as does keying.

In general, the literature suggests continuous computer work should have a break after approximately an hour. The break length should be approximately 10 to 15 min after 50 to 60 min of work. Microbreaks (say 3 min long) seem worthwhile. Computer programs are now available which count the time or keystrokes since the last break and then recommend when a break should occur. (The physiological concept underlying the breaks is that the tense muscles squeeze the blood out of the capillaries, reducing oxygen input and removal of waste products. The breaks [perhaps combined with exercises] allow reestablishment of the blood supply.) This brings up the concepts of passive, active, and working rest. Passive rest would be sitting at the workstation, perhaps conversing with coworkers. Active rest could be walking to another area, perhaps just to have a beverage or perhaps to do some exercises. Working rest is other work activity such as going to the printer or answering the phone. In general, active or working rest gives better recovery. Working rest has an advantage to the organization that work is done during the rest time.

38.7.3 Brain

Two aspects involving the brain are optimum stimulation and concentration and attention.

38.7.3.1 Optimum Stimulation

Many people have pointed out that fatigue is not just physiological but has a strong psychological (lack of motivation) aspect (Finkleman, 1994). Finkleman concluded that an important predictor of fatigue was processing either too much information (overload, and thus fatigue) or too little information (underload, and thus boredom).

38.7.3.2 Concentration and Attention

There are some sedentary jobs with mental activity; three examples are simultaneous translation, gambling, and education. Two examples of mental underload are monitoring/vigilance and controlling a vehicle at night.

Simultaneous translation usually is done by pairs of translators — one translates while the other rests. In Quebec (French/English), they switch every 30 min; in Japan (Japanese/English), they switch every 20 min. Sign language translators for the hearing-impaired switch every 20 min. In casinos, blackjack dealers (standing work with intense concentration and finger activity) work for 60 min and then have a 20 min nonworking rest. The typical schedule at universities in the United States is 50 min of lecture with a 10 min break before the next class. However, the typical student schedule is 15 class hours per week, so there is considerable recovery time between classes. Professors tend to have 6 to 12 teaching "hours" (each of 50 min) each week; typically they teach for 32 weeks/yr. U.S. high school students typically have six to seven periods/day; each period is 60 min, which includes a 5 min break. Often one period is physical education; lunch is an additional break. Typically they go to school <190 days/yr.

Example underload tasks (monitoring/vigilance) are process control monitoring, hospital monitoring, and industrial inspection. The Accreditation Council for Graduate Medical Education, in an effort to reduce medical errors, set new standards for doctors-in-training. They are now limited to 80 h workweeks and they must get at least 10 h of rest between shifts. They are not be to be on duty for more than 24 h continuously. Craig (1985) summarized some vigilance studies. Changes that reduced boredom were beneficial; one example was loading/unloading (10 min) as well as inspecting coins (14 min). Another example was 30 min of inspection followed by 60 min of other tasks.

Driving/piloting at night is an underload problem. Buck and Lamonde (1993) reported locomotive drivers had sleep problems — especially close to 0300 and 1500 h. Fatigue may be more of a problem when the driver is "externally scheduled" — for example, when tired, a bus driver or a pilot cannot stop and take a break, whereas a car or truck driver, being self-paced, can stop. Japanese taxi drivers typically work a 16 h day and then have the following 1 or 2 days off.

38.7.4 Sleep/Biological Clock

Three divisions are sleep, biological clock (circadian rhythm), and countermeasures.

38.7.4.1 Sleep

Sleep restores the functions of the brain. Only sleep allows some form of cerebral shutdown. Females and people over 50 yr are more prone to night-sleep problems. Sleep deprivation of 24 to 48 h primarily affects motivation to perform rather than ability to perform; thus uninteresting, undemanding, simple tasks are the most affected.

38.7.4.2 Circadian Rhythm

Various physiological functions vary in a circadian (24 h) rhythm. Both internal temperature and potassium peak during the day and bottom at night. Cortisol (the "wake-up" hormone) peaks around 9 a.m.; melatonin (the "go-to-sleep" hormone) peaks around 2 p.m. Alertness peaks in the late afternoon and bottoms during 2300 to 0600 h.

38.7.4.3 Countermeasures

Three sleep-deprivation countermeasures are rest, drugs/food, and environmental stimulation.

Naps are a potential supplement to sleep at night. However, people tend to think of naps as inadequate opportunity to sleep rather than an opportunity for partial recuperation. A nap problem is sleep inertia — poor performance for about 15 to 30 min after being awakened.

The most common drug to decrease sleepiness/increase alertness is caffeine. Amphetamines also work well. Caffeine has a metabolic half-life of 3 to 7 h. Pregnancy increases half-life (to as much as 18 h) while smoking decreases caffeine's half-life. A pharmacologically active dose, depending on the individual, is about 3 mg/kg of body weight. Truck drivers often recommend eating — possibly while still driving. Eat a moderate intake of carbohydrates rather than fats or proteins. Another possibility is having a drink with caffeine.

Environmental stimulation can be exposure to bright light such as daylight. (Normal indoor lighting is too dim to stimulate.) In a vehicle, modify the ventilation or decrease isolation by listening to a radio/CD or talking on a cell phone. Physical activity (say walking 100 m) can be useful.

38.8 Fatigue Guidelines

The seven guidelines are divided into fatigue prevention (guidelines 1 to 3) and fatigue reduction (guidelines 4 to 7).

38.8.1 Guideline 1: Have a Work Scheduling Policy

The problem is insufficient rest. Two aspects are: (1) too many work hours and (2) work hours at the wrong time.

38.8.1.1 Too Many Hours

Count all hours in "duty time." For example, jobs such as train crews and flight crews often have waiting and preparation time required before and after the "primary" job. There may be, especially for supervisors, "shift turnover" time (people from the previous shift stay to communicate with the new shift).

People may work overtime when other individuals have to be replaced (illness, absenteeism). The resulting shift can be 12 or even 16 h long. There probably should be some organizational restriction on *prolonged* overtime (say over 12 h/day and over 55 h/week). Lack of sleep can increase if the individual has a long commute time or moonlights.

38.8.1.2 Work Hours at the Wrong Time

Lack of sleep can result from sleeping at the "wrong time" or having irregular hours of work. Table 38.12 gives tips for day sleeping.

38.8.2 Guideline 2: Optimize Stimulation During Work

There may be too much stimulation (overload) or too little (boredom). Stimulation comes from both the task and the environment.

- *Too much stimulation.* The usual solution is to reduce environmental stimulation. For example, for office tasks, increase visual and auditory privacy
- *Too little stimulation.* Tasks are more stimulating if there is physical activity. Add variety within the task or schedule a variety of tasks

Add environmental stimulation by (1) encouraging conversation with others (this may require two-way radios for those who are physically isolated), (2) varying the auditory environment (talk radio, stimulating music), (3) varying the visual environment (e.g., windows with a view), or (4) varying the climate (change temperature, air velocity). Also consider chemicals such as caffeine.

TABLE 38.12   Tips for Day Sleeping

Develop a good sleeping environment (dark, quiet, cool, with a bed). Have it *dark* (e.g., opaque curtains). Have it *quiet* since it is difficult to go back to sleep when daytime sleep is interrupted. Minimize changes in noise volume. Consider earplugs, unplugging bedroom phones, turning down the phone volume in other rooms, reducing TV volume in other rooms, using a fan to make noise. Train your children. Have the sleeping area *cool*. The *bed* normally is OK but may be poor if the sleeper is not sleeping at home (e.g., is part of an "augmented crew" for trucks, aircraft). Then provide a good mattress and enough space

Plan your sleeping time. Tell others your schedule (minimize interruptions). Consider sleeping in two periods (5 to 6 h during the day and 1 to 2 h in the late evening before returning to work). Less daytime sleep and more late evening sleep not only make it easier to sleep but also may give a better fit with family/social activities. Night workers should go to sleep as soon as they get home because the sooner they go to bed, the less adjustment their biological clock must make.

Have a light (not zero or heavy) meal before sleep. Avoid liquid consumption, as it increases the need to urinate (which wakes you up). Avoid caffeine. A warm drink before your bedtime (perhaps with family members starting their day) may help meet your social needs. Avoid foods that upset your stomach — and thus wake you up

If under emotional stress, relax before going to bed. One possibility is light exercise

Source: Konz, S. and Johnson, S. 2004. *Work Design: Occupational Ergonomics*, 6th ed. Scottsdale, AZ: Holcomb-Hathaway. With permission.

38.8.3   Guideline 3: Minimize the Fatigue Dose

The "dose" of fatigue may become too great to overcome easily. Two aspects are intensity and work/rest schedule.

38.8.3.1   Intensity

Good ergonomics practice reduces high stress on the individual. For example, use machines and devices to reduce hold-and-carry activities. Static work is especially stressful, as specific muscles are activated continuously and the alternation of muscles that occurs in dynamic work is not present.

38.8.3.2   Work/Rest Schedule

The effect of fatigue increases exponentially (not linearly) with time. Thus, it is important to get rest before the fatigue gets too high. The normal approach is to schedule a break. Another approach is to use part-time workers (as with sorting express packages).

38.8.4   Guideline 4: Use Work Breaks

The problem with a conventional break is that there is no productivity during the break. A solution is to use a different part of the body to work while resting the fatigued part.

On an automatic or semiautomatic machine, the operator may be able to rest during the automatic part of the cycle (machine time). Job rotation has the worker periodically shift tasks. Fatigue recovery is best if the alternative work uses a distinctively different part of the body — for example, loading/unloading a truck versus driving a truck; word processing versus answering a telephone.

Not quite as good, but still beneficial, is alternating similar work, as there would be differences in body posture, force requirements, mental activity, and so on. For example, inspectors could periodically shift work with other inspectors. Assembly teams could shift jobs every 30 min. Checkout clerks could use a left-hand station and then a right-hand station. In a warehouse, workers could pick cases from pallets to a conveyor for 4 h and unload the conveyor to trucks for 4 h. Job rotation, in addition to reducing fatigue, reduces the feeling of inequity among workers, as everyone shares the good and bad jobs; it also reduces boredom and fatigue. Job rotation requires cross-trained people (able to do more than one thing). But this gives scheduling flexibility to management.

38.8.5   Guideline 5: Use Frequent Short Breaks

The problem is how to divide break time. The key to the solution is that fatigue recovery is exponential. The time at which the y-axis response of an exponential curve reaches 50% is called its half-life (x-axis value). For a curve with a half-life of 9 min, if recovery is "complete" in 60 min, it takes only 4 min to drop from 100% fatigue to 75% fatigue but it takes 42 min to drop from 25% fatigue to 0% fatigue. Thus, give breaks in small segments. Ideally, the break should be in the middle of the work period rather than at the beginning or end. Operator-controlled breaks are better than fixed-period breaks as they allow for individual differences among workers. Some production is lost for each break. Reduce this loss by not turning the machine off and on, by taking the break near the machine, and the like.

38.8.6   Guideline 6: Maximize the Recovery Rate

The problem is to recover as quickly as possible — in technical terms, reduce the fatigue half-life. For environmental stressors, reduce contact with the stressor. For heat stress, use a cool recovery area. Use a quiet area to recover from noise, no glare to recover from glare, no vibration to recover from vibration. For muscle stressors, it helps to have a good circulatory system. That is, a person in good physical shape will recover faster than a person in poor shape. Active rest seems to be better than passive rest. The active rest may be just walking to the coffee area (blood circulation in the legs improves dramatically within about ten steps). For working rest, to encourage walking and using alternate muscle groups, consider cell layout and having the operator do the material handling for the workstation (obtaining supplies or disposal of finished components).

38.8.7 Guideline 7: Increase the Recovery/Work Ratio

When the problem is insufficient time to recover, the solution is to increase the recovery time or decrease the work time. For example, if a specific muscle is used for 8 h/day, there are 16 h to recover; hence, 2 h of recovery/1 h of work. But, if the two arms are alternated so one arm is used 4 h/day, then there is 20 h to recover; 5 h of recovery/1 h of work. Overtime, moonlighting, or 12-h shifts can cause problems. Working 12 h/day gives 12 h of recovery, so there is 1 h of recovery/1 h of work.

Consider all break time, both paid and unpaid. In particular, consider machine time and job rotation as well as coffee and lunch breaks. Holidays, weekends, and vacations are valuable in reducing long-term fatigue (long half-life) where there still is a fatigue effect at the start of the day.

References

Ahsberg, E. and Gamberale, F. 1998. Perceived fatigue during physical work: an experimental evaluation of a fatigue inventory. *International Journal of Industrial Ergonomics*, 21: 177–131.

Bhatia, N. and Murrell, K. 1969. An industrial experiment in organized rest pauses. *Human Factors*, 11 (2): 167–174.

Bigland-Ritchie, B., Furbach, F., and Woods, J. 1986. Fatigue of intermittent submaximal voluntary contractions: central and peripheral factors. *Journal of Applied Physiology*, 61 (2): 421–429.

Buck, L. and Lamonde, F. 1993. Critical incidents and fatigue among locomotive engineers. *Safety Science*, 16: 1–18.

Craig, A. 1985. Field studies of human inspection: the application of vigilance research. In *Hours of Work*, Folkard, S. and Monk, T. Eds., John Wiley & Sons: New York.

Finkleman, J. 1994. A large database study of the factors associated with work-induced fatigue. *Human Factors*, 36 (2): 232–243.

International Labor Office. 1992. *Introduction to Work Study*, 4th ed. Geneva, Switzerland: ILO.

Kadefors, R. and Laubli, T. 2002. Muscular disorders in computer users: an introduction. *International Journal of Industrial Ergonomics*, 30: 203–210.

Konz, S. and Johnson, S. 2004. *Work Design: Occupational Ergonomics*, 6th ed. Scottsdale, AZ: Holcomb-Hathaway.

Mital, A. 1984a. Comprehensive maximum acceptable weights of lift database for regular 8-hour work shifts. *Ergonomics*, 27: 1127–1138.

Mital, A. 1984b. Maximum weights of lift acceptable to male and female industrial workers for extended work shifts. *Ergonomics*, 27: 1115–1126.

Mital, A., Hamid, F., and Brown, M. 1994. Physical fatigue in high and very high frequency manual materials handling: perceived exertion and physiological factors. *Human Factors*, 36 (2): 219–231.

Mital, A., Nicholson, A., and Ayoub, M.M. 1993. *A Guide to Manual Material Handling*. London: Taylor & Francis.

Further Reading

Konz, S. and Johnson, S. 2004. *Work Design: Occupational Ergonomics*, 6th ed. Scottsdale, AZ: Holcomb-Hathaway. This popular textbook has an extensive coverage of fatigue and allowances as well as design guidelines for occupational ergonomics.

39

Wrist Posture in Office Work

Mircea Fagarasanu
Shrawan Kumar
University of Alberta

39.1 Introduction — Ergonomic Relevance

In the recent past, cumulative trauma disorders (CTDs) have been the fastest growing occupational health problem. Compared to 1981 when only 24% of all occupational musculoskeletal disorders were CTD, in 1992 almost 66% of all work related illnesses reported in the United States were attributed to this category (Bureau of Labor Statistics, 1994). Since changes have occurred in many jobs during recent years (characterized by less force demands and increased mental load, higher social stress leading to a sustained increase in muscle load) (Viikari-Juntura and Riihimaki, 1999), this trend is expected to continue. This trend is even more visible in office work. Hence, ergonomic intervention becomes very important.

In the last century, the keyboards have constituted a constant presence in the office setting. While at the beginning designers addressed mostly the mechanical aspect, the next period (20–25 yr) emphasized increasing performance. Lately typist fatigue, muscular strain, perceived pain, and ergonomics interventions have been the focus. Currently, the computer keyboard is the primary input device for data entry tasks. The office work-related musculoskeletal problems are due to the fact that although the keyboard is often a nonadjustable device, all the computer users regardless of age, gender, performance, and anthropometric characteristic use it.

Carpal tunnel syndrome (CTS) is the "chief occupational hazard of the 90s" — disabling workers in epidemic proportions (U.S. Department of Labor, 1999). Currently, over 8 million Americans suffer from CTS (U.S. Department of Labor, 1999). In United States alone, approximately 260,000 carpal tunnel release operations are performed each year, with almost half of the cases considered to be work related (NCHS, 2000). Among all work-related upper extremity disorders (WRUEDs), CTS has the biggest impact in the office workers' health. Also, the medical and nonmedical costs for CTS are the highest among all upper extremity musculoskeletal disorders. Since 66% of the entire population spends 33% of their time at work (WHO, 1995 cited by Kumar et al., 1997), and the incidence of CTS is on the rise (Hedge and Powers, 1995) an association may be argued. Almost 25% of all cases of work-related CTS reported in 1994 were attributed to repetitive typing or key entry of data (Szabo, 1998). The loss in productivity occurs before (less typing speed), during, and after (days of hospitalization) the treatment of CTS (Moore, 1992).

39.2   Anatomic vs. Physiologic Wrist Neutral Zone

The neutral zone is defined as "the part of the range of physiological motion, measured from the neutral position, within which the motion is produced with a minimal internal resistance" (Kumar and Panjabi, 1995).

Previous studies define the wrist neutral position for radial/ulnar deviation plane of motion as the axis where the line that is in continuation of the middle finger (finger III) and the forearm's long axis. In fact, the wrist has already an ulnar deviation of 4°–6° in the anatomical neutral position. This point of view is supported by the findings that the intracarpal pressure is lowest when the hand is in slight pronation, 3°–5° ulnar deviation, 2°–3.5° flexion, and 45° metacarpophalangeal (finger) flexion (Hedge and Powers, 1995). Marklin et al. (1999) also assessed that ulnar deviation of 10° does not increase the carpal tunnel pressure (CTP). Also, Fagarasanu et al. (2004a) assessed the 5°–7° ulnar deviation and 7°–9° extension as being the wrist's neutral posture.

39.3   Wrist Neutral Zone — Experimental Evidence

The physiologic wrist neutral posture was assessed in one of our previous studies (Fagarasanu et al., 2004a), in which blindfolded subjects were asked to position their wrist in the neutral posture starting from a randomly chosen wrist deviated postures (45° flexion and extension, 30° ulnar deviation, and end of range of motion for radial deviation). Measurement of electromyographic (EMG) forearm muscle activity in both deviated and neutral wrist positions was also done.

During the experiment, wrist motion was measured using a calibrated custom-made electrogoniometer. It consisted of two mobile plastic arms articulated with a central high-precision potentiometer (Figure 39.1). For the EMG forearm muscle activity, DelSys Bagnoli[TM] (Boston, USA) EMG system (active surface electrodes, electrode cables, preamplifiers, and amplifiers) was used. Subjects were seated upright into a straight-back chair with feet flat on the floor and looking straight ahead. The forearm was rested on the table, being fully pronated (the forearm volar side was parallel to the table) when wrist deviation in the ulnar–radial deviation plane was measured and semipronated (the forearm lateral side was parallel to the table) for the flexion–extension plane (Figure 39.1).

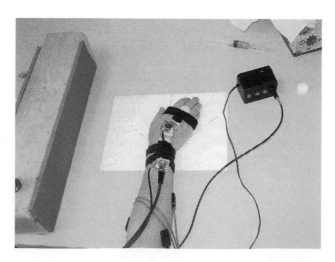

FIGURE 39.1   Experimental set-up.

The electrogoniometer was adjusted across the wrist with the goniometer's arms aligned to the long axes of the hand and the lower arm. For radial and ulnar deviation assessment, the electrogoniometer's fulcrum was centered over the middle of the dorsal aspect of the wrist over the capitate. The proximal arm was aligned with the dorsal midline of the forearm, using the lateral epicondyle of the humerus for reference and the distal arm was aligned with the dorsal midline of the third metacarpal bone. For flexion and extension measurements, the fulcrum of the electrogoniometer was centered over the radial aspect of the wrist (trapezium level) with the proximal arm aligned with the medial side of the radius and the distal arm aligned with the midline of the second metacarpal bone. The device was adhered using Velcro closures. For EMG recording, after forearm preparation, the EMG electrodes were applied 5–7 cm distal to the line connecting the medial epicondyle and biceps tendon for flexor carpi radialis (FCR), above the shaft of ulna in the middle of forearm for extensor carpi ulnaris (ECU), at 2–3 cm volar to ulna at the junction of the upper and middle thirds of the forearm for flexor carpi ulnaris (FCU), and at 3 cm medio-distal to lateral epicondyle for extensor carpi radialis (ECR).

Subjects were blindfolded. After completing the maximal isometric contractions (MICs), volunteers were asked to bring the passively deviated wrist (45° flexion, 45° extension, 30° ulnar deviation, and at the end of range of motion for radial deviation) in the subjective neutral position. The sequence was randomized in order to avoid the carry-over effect. Each condition was repeated once (two trials). Between conditions, a 2-min resting period was given. The forearm muscles' EMG activity was measured in both deviated and neutral wrist positions.

39.3.1 Self-Selected Wrist Position

The results indicated that all subjects consistently positioned their wrist in 5°–7° ulnar deviation and 7° to 9° extension. Males tend to adopt more deviated postures (8°–9° extension and 7° ulnar deviation compared to 7°–8° extension and 6° ulnar deviation for females) while keeping the wrist in the neutral posture, but the differences between genders were not statistically significant. Also, no significant differences in terms of wrist position in the neutral posture were found between the left and right sides for both genders. The recorded postures had a significant effect on all four forearm muscles, causing a 66–75% decrease in muscle activity. Additional training and device and workstation targeted design modifications would be able to reduce the musculoskeletal disorders risk factors. This information may be valuable for physical therapists and surgeons. Also, office and industrial workstations redesign and ergonomic interventions (e.g., job/device design/redesign, final wrist joint position following reconstructive interventions) may benefit from the outcome.

O'Driscoll et al. (1992) reported the same self-selected wrist posture with moderate extension and ulnar deviation. Also, our results are supported by Hedge and Powers (1995) who demonstrated that the lowest CTP is recorded when the hand is 5° ulnar deviated. It is suggested that performing tasks with a wrist position within neutral zone would reduce carpal tunnel pressure, helping those with a diagnosis of CTS or exposed to increased risk.

Working with extreme wrist deviation poses significant risk for CTS development. Extreme wrist extension causes the finger flexors' tendons to slide in the area between volar carpal ligament and the carpal bones increasing tissue crowding. Also, wrist flexion causes the tunnel elements to be close together on the volar side of the wrist and spread apart on the dorsal side. Furthermore, the flexor retinaculum presses the flexor tendons and bursae against the head of the radius. Although the carpal tunnel cross section decreases in ulnar and radial deviations, it is not so acute owing to constrained range of motion to cause significant problem.

39.3.2 EMG Muscle Activity

When compared to values obtained while keeping the wrist deviated, in the neutral zone muscle activity was significantly lower ($P < 0.05$). For both males and females, all forearm muscles demonstrated a similar pattern with ECR being the most active (9.2–11.1% of MVC), followed by ECU, FCR, and FCU with normalized average EMG values varying between 7.7–9.3%, 6.9–8.4%, and 4.8–8.5% of MVC, respectively (Figure 39.2). Hence, a drop of up to 75% in the muscle activity in the neutral zone when compared to normalized average EMG values in wrist-deviated postures (16.5–38.3% of MVC) was found. Although not significant, females presented higher %MVC EMG values for all muscles.

While keeping the wrist in ulnar deviation, the maximum activity was observed for ECU (26.9–35.7% of MVC) and FCU (16.5–29.1% of MVC). ECR was the most active muscle in both wrist radial deviation (25.5–36.8% of MVC) and extension (29.4–38.3% of MVC). FCR (19.9–26.8% of MVC) and FCU (18.3–23.6% of MVC) were the most active muscles while the wrist was maintained in flexion. FCR in wrist radial deviation (19.8–24.2% of MVC) and ECU in wrist extension (17.3–34.1% of MVC) were the second most active muscles after FCU.

The proven coactivation of wrist muscles was also noted by Hoffman and Strick (1999). This coactivation included both synergists and antagonist muscles. Since wrist extensors have smaller moment arms compared to flexors, larger forces will be required by extensors to maintain the wrist posture (Keir et al., 1996) posing this group of muscles to elevated risk of injury while performing with in wrist flexion posture. Passive muscle forces in antagonist muscles may further increase the risk. The deviated joints cause muscle overstretch, thus pose a greater risk for musculoskeletal injury.

EMG activity levels between 8% and 17% of MVC were recorded for muscles acting as secondary effectors (FCU in extension and radial deviation, FCR in extension and ulnar deviation, ECR in flexion and ulnar deviation, and ECU in flexion and radial deviation). These levels demonstrate their concomitant dual role in wrist stabilization and force exertion. Prolonged muscle loading promoted fatigue. As a result, due to lack of rest, the risk of musculoskeletal injury is increased (Kumar, 2001). Also, Drury et al. (1985) noted an important increase in EMG at extreme wrist deviations, whereas the muscle activity for wrist angles between 5° radial deviation and 10° ulnar deviation was low and almost constant.

ECR was the most active muscle in both radial deviation (25.5–36.8%) and extension (29.4–38.3%) making it at risk in activities that require this wrist deviation concomitantly (e.g., use of computer mouse). The higher prevalence of epicondylitis on the extensor side can be explained by the ECR's role as wrist stabilizer and primary effector in wrist extension and radial deviation. These values were obtained in passively deviated wrist postures and any active contractions would require significantly greater muscle activity, increasing the risk for musculoskeletal injuries even more.

Simultaneous recordings of forearm muscle activity and CTP are needed in order to see if the selected wrist posture corresponds to the lowest values for both EMG and CTP. Although during rest, forearm muscle activity and CTP are low, some office tasks require awkward upper extremity postures, significantly changing the required muscle activity.

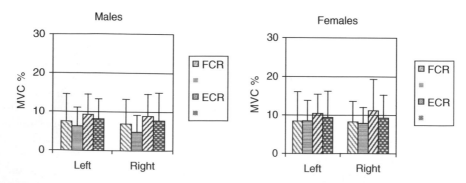

FIGURE 39.2 The forearm muscles normalized average EMG (% isometric MVC) for both genders in the self-selected neutral position (FCU, flexor carpi ulnaris; FCR, flexor carpi radialis; ECR, extensor carpi radialis; ECU, extensor carpi ulnaris).

39.4 Keyboarding and Upper Extremity

39.4.1 Risk Factors for CTS

The neck region including shoulder and upper extremity are at a greater relative risk for developing musculoskeletal problems in a video display terminal (VDT) task (Sauter et al., 1991). The highest risks are for hand, wrist, and arm (Rempel et al., 1999). Sauter et al. (1991) carried out a study with 932 VDT users and assessed discomfort levels of 13% and 12% for wrist and right hand, respectively. They also indicated the keyboard height as being most important variable for arm discomfort. Owing to the close relationship between arm abduction and arm pronation and wrist ulnar deviation, wrist position is indirectly affected by the arm posture (Harvey and Peper, 1997; Marklin and Simoneau, 2001). Keyboarding is associated with a large palette of risk factors that are present in such simultaneous and cumulative levels only in office work. Excessive wrist deviation in the extension/flexion movement plane is closely correlated with the type of keyboard used (slope angle). In addition, ulnar deviation occurs directly due to the need to reach the far left or right keys (Werner et al., 1997; Marklin et al., 1999) and indirectly as a compensation of the arm abduction.

The conventional QWERTY keyboard (named after the first six letters of the left portion of the top alphabet row) is the best-known and popular keyboard among VDT users. It has a slope (keyboard inclination in sagital plane) ranging from 0° to 15°, slant angle (the angle between the two groups of keys measured in horizontal plane) of 0°, and tilt angle (lateral inclination of the keys) of 0°. The fact that the QWERTY layout poses a great risk of injury on the users, may be due to its initial design purpose (it has been designed for mechanical typing machines, where mechanical linkage jam was an issue). Although there have been many attempts to adapt the conventional keyboard design, none has replaced it. The most comprehensive description of the needed modifications was made by Dvorak (1943) on the basis of his analysis that the following defects exist in the QWERTY design: overloading of the weaker left hand in a right-handed person, too little typing on the home row, overworking certain fingers and not assigning enough work to others, and fingers are required to execute an excessive amount of jumping back and forth from row to row.

The aforementioned suggestions have been proven by many studies ever since. When using traditional QWERTY key layout, both forearms are pronated and both wrists are in ulnar deviation and extension (Hedge and Powers, 1995; Smith et al., 1998; Marklin et al., 1999; Simoneau et al., 1999; Liao and Drury, 2000; Visser et al., 2000; Marklin and Simoneau, 2001). The forearm pronation mean is between 69° and 79° with right pronation significantly greater than left pronation $65.6° \pm 8.3°$ and $62.2° \pm 10.6°$ ($F = 12.28$, $p < 0.01$), respectively. Also, there was a significantly greater left hand ulnar deviation when compared to the right side ($15.0° \pm 7.7°$ compared with $10.1° \pm 7.2°$, $F = 41.57$, $p < 0.01$) and left extension exceeded the right one ($21.2° \pm 8.8°$ compared with $17.0° \pm 7.4°$, $F = 23.24$, $p < 0.01$) (Simoneau et al., 1999). Variation in arm/forearm/wrist muscle load and typing performance's fluctuation due to various hand sizes have not been assessed. Distribution and frequency of use of alphabetic, numeric, or special keys, like *CapsLock, Tab,* and *Shift* for left hand (Marklin and Simoneau, 2001) are accountable for the differences in hand posture between sides. Another reason for the difference is that 58% of letters typed in English text are typed with the left hand.

Users with big hands are forced to adopt more pronounced finger flexion and wrist extension increasing tendon travel (Treaster and Marras, 2000). For 1 h of continuous typing, the tendon travel ranges from 30 to 59 m (Nelson et al., 2000). Repetitive sliding of tendons within their sheaths will increase the friction that is a major trigger for the disorders of the tendons, their sheaths or adjacent nerves (Moore, 1992). Although the prevalence of CTS is higher in females (Armstrong and Chaffin, 1979), males have a greater tendon travel (Treaster and Marras, 2000). Additional research is needed in order to clarify the causal relationship between gender attributes and CTS pathogenesis. Also, ergonomics interventions should consider the differences between postures of right and left hands as well as the particularities of special group of users.

Training work sessions in a particular wrist position should be made after the wrist neutral position has been determined. Owing to the fact that static load is an important factor for musculoskeletal disorders development, even after the safe margins for wrist deviation are known (neutral zone), wrist positions should be alternated within its limits. Serina et al. (1999), studying the typists' posture when using a standard flat QWERTY keyboard, noted that typing on a keyboard in an adjusted workstation, forces the users to spend about 75% of the working time with the wrist in greater than 15° extension and 28% and 9% of their time with a wrist extension greater than 30° for the left and right hand, respectively. Ergonomic assessment of hazardous postures should precede the design and introduction of alternate keyboards. Otherwise, elevated CTS prevalence and complaints will follow.

39.4.1.1 Wrist Extreme Postures

CTP, an important factor for CTS' pathogenesis (Phalen and Kendrick, 1957; Szabo, 1989a, b; Seradge et al., 1995; Keir et al.,1998, 1999), is lowest when wrist is in neutral position, hand is relaxed with fingers flexed at 30° and forearm in a semipronated position (Werner et al., 1997). The aforementioned postures are rarely reached during a typing task. The typing posture is usually that in which the arm is abducted and pronated, wrist extended, ulnar deviated and fingers extended in order to fit the keyboard. All these working positions determine an elevated CTP (Werner et al., 1997).

Wrist extension has a greater effect than ulnar deviation on carpal tunnel pressure (Marklin et al., 1999). During typing the hand and wrist adopt awkward postures that increased CTP exceeding the upper safe limit. CTP is increased by the following factors: movement of lumbrical muscles into the carpal tunnel, changes in cross-sectional area (affected by wrist position), and folding of skin at the distal palm. Werner and Armstrong (1997) noted that wrist extension stretches flexor tendons and median nerve, exerting pressure on their dorsal face. During wrist flexion, the flexor digitorum tendons are pushed against the palmar side of the carpal tunnel, causing pressure on both the tendons and the flexor retinaculum. Owing to its close relationship with flexor retinaculum and the flexor tendons, the pressure exerted on the median nerve will rise (De Krom et al., 1990; Szabo, 1998). Overload of the flexor muscles due to lack of rest, leads to an imbalance between flexor and extensor muscles causing elevated pressure on the palmar surface of the carpal tunnel (Ostrem, 1995). This increased pressure exaggerates the already existing elevated CTP, exposing the tissues to greater risk. When compared to wrist flexion, wrist extension is encountered more often during data entry tasks. It causes the tendons to be displaced against the dorsal side of the carpal tunnel and the head of the radius, leading to high pressure on the tendons. When the wrist adopts extreme postures, the resultant high pressure leads to in endoneurial edema and microscopic pathological changes (Cullum and Molloy, 1994).

The flexion of the fingers will lead to an increase in the CTP (Keir et al., 1998). Finger flexion is very important for CTP. The fingers are constantly fitted to the keyboard during typing. Straight postures and elevates CTP compared with the relaxed finger posture (Keir et al., 1998) are present. Besides the presence of elevated CTP in patients with CTS, Szabo (1989b) noted that postexercise, the increased CTP inertia in CTS patients leads to elevated risk of nerve injury. Werner et al. (1983) and Braun (1988) noted the same comportment for CTP, indicating elevated CTP and increased sensory impairment in patients with CTP postactive motion of the wrist.

The CTP is not uniformly distributed in the carpal tunnel. It is higher in the distal portion of the carpal tunnel and that is why the sensory conduction velocity action potential amplitude is affected more in this portion (Keir et al., 1998).

39.4.1.2 Repetition

Extreme postures and high-repetitive actions (38–40 per minute per finger) are frequently required during computer tasks. This value exceeds the highest acceptable frequency in a repetitive motion (frequency of 30 per minute) (Bergamasco et al., 1998). Cumulative load is a risk factor for causation of musculoskeletal injuries (Kumar, 1990, 2001). The adjacent tendons are sliding one against the other with the friction force being proportional to the tension in the tendon and inversely proportional to the radius curvature (Hadler, 1987). Serina et al. (1999) noted that velocities during typing in flexion/extension plane are similar to velocities measured in workers involved in industrial activities with great risk for CTS. Nerve compression due to thickening of the flexor tendon sheaths has been proven by Yamaguchi et al. (1965) who found greater fibrosis and edema in the tendon sheaths in CTS patients compared with controls. Highest velocity and accelerations occurred in flexion/extension and radial/ulnar deviation movements (Serina et al., 1999).

39.4.2 The Effect of Alternative Design on CTS Risk Factors

Previous research and design efforts to reduce the risk factors level have focused on re-shaping the standard keyboard, or making it more adjustable, while keeping its basic design and QWERTY layout. This would ensure a smooth transition to the new keyboard design. Improved hand and arm postures, without the need for learning a whole new typing skill, are achieved. Split keyboards are the most commonly seen by most and are typically the least expensive of the alternative keyboards. They have a set horizontal split angle and possibly a slight center raise or "tenting" of the left and right hand key segments. They have been used in many studies (Hedge and Powers, 1995; Harvey and Peper, 1997; Smith et al., 1998; Marklin et al., 1999; Lincoln et al., 2000; Marklin and Simoneau, 2001) along with QWERTY or other alternative keyboards.

Smith et al. (1998), in a comparative study between split and conventional keyboards, noted that a split keyboard allows the hand, wrist and arms to be maintained in more neutral positions. Both right and left ulnar deviation and pronation were reduced. Keeping the wrist within the neutral zone for a longer period of time, promotes decreased force applied on carpal bones, ligaments, and tendon sheaths (Armstrong and Chaffin, 1979; Armstrong et al., 1984; Marklin et al., 1999). Mitigated CTP, the major trigger for CTS follows. Marklin and Simoneau (2001), while assessing split computer keyboards, showed that wrist ulnar deviation ranged from $7.0°$ to $8.5°$ for the left wrist and from $2.7°$ to $5.0°$ for the right wrist for alternative keyboards as compared to $15°-30°$ for both hands for conventional keyboards. Using the split keyboard under correct settings would ensure a reduction in ulnar deviation equal to half of the split angle. Another advantage of split keyboards have been cited by Treaster and Marras (2000) who determined that alternative keyboard design can affect tendon travel by as much as 11%, reducing the thickening process of the tendon sheaths.

When using a split keyboard the problem of wrist-extended posture is still present (Hedge and Powers, 1995). Also, placing the mouse in a more lateral position will require elevated arm abduction (Harvey and Peper, 1997). Treaster and Marras (2000) noted that volunteers failed to set the split keyboard in order to reduce the tendon travel. Educational and ergonomic programs are needed to increase the awareness among VDT users regarding the safe postures that are required while typing.

Negative slope keyboard support (NSKS) reduces the wrist extension problem mitigating it from $13°$ extension to $1.2°$ flexion (Hedge and Powers, 1995). Subjects responded very favorably to the NSKS system. A downside is that the ulnar deviation remains the same or it is even greater because of the active process of fitting the finger to reach the same point. Despite the fact that the keyboards of many computers are flat, almost none of the conventional computer keyboards used on a flat work surface actually has a $0°$ slope, and therefore, a much more ergonomic keyboard would be an NSKS with a split angle.

Participants have had the ability to rapidly adapt to the changes in keyboard design (Hedge and Powers, 1995; Smith et al., 1998; Marklin et al., 1999; Zecevic et al., 2000; Marklin and Simoneau, 2001). The average speed for alternative keyboards was reduced with 10% when compared to the speed for conventional keyboards (Marklin et al., 1999; Marklin and Simoneau, 2001). The resultant typing performance is even more remarkable if we take into account the training time, which was very short (Smith et al., 1998). Although the aforementioned studies have not had too many subjects, and not all the alternative designs have been included, there is sufficient evidence to support the superiority of alternative keyboards over conventional ones.

The effect of alternative keyboard design on typing variables have been assessed by Fagarasanu et al. (2004b) who conducted an experiment in which wrist motion, forearm muscle activity, applied

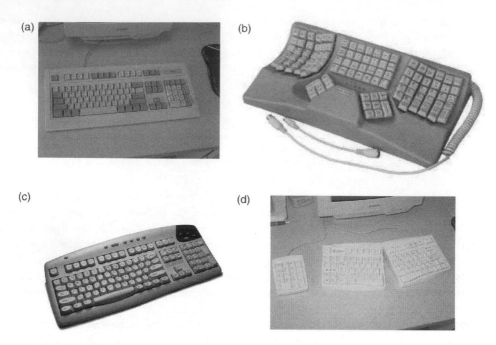

FIGURE 39.3 The keyboards that have been used in the experiment: (a) conventional; (b) maltron; (c) prosper street technologies; and (d) goldtouch.

force, and performance were measured during a standardized typing task. Four different keyboards (one conventional and three alternative) were used in this study (Figure 39.3). The conventional keyboard was a Fujitsu 105-keys traditional QWERTY layout with 5° positive slope. The alternative keyboards were: Maltron E-Type (fixed split design, tilted keys and pads, straight vertical key columns, central number pad, and slightly modified layout such as thumb keys for *Enter, Space*, and *Backspace*), Goldtouch Adjustable Ergonomic Keyboard (adjustable split angle and lateral slope with lacking ball and socket latch mechanism), and Prosper Street Technologies (PST) LLC Wave Keyboard (QWERTY slim design with row vertical curves for longer fingers). Although the Goldtouch® keyboard lateral inclination could have influenced both pronation and ulnar deviation, in order to be able to assess the impact of split angle design on typing posture, the authors chose a fixed split angle of 25° and a 0° lateral slope. Also, wrist motion and number of wrist repetition greater than 10° (changes in wrist movement) were measured bilaterally using two SG 65 Biometrics Ltd electrogoniometers. The EMG forearm muscle activity was measured using DelSys Bagnoly™ EMG system. In order to record the overall applied typing force, an AMTI force plate was placed under the keyboards.

The results of the study indicate that for all four keyboards, the wrist was ulnarly deviated and extended for typing. Table 39.1 presents the wrist deviation angles for extension/flexion and ulnar/radial deviation planes while typing on each of the four tested keyboards. When compared to conventional keyboard, Maltron and Goldtouch keyboards significantly reduced the wrist ulnar deviation for

TABLE 39.1 Means and Standard Deviations for Average Wrist Angles on Four Different Keyboard Designs

Keyboard	Extension (+)/Flexion (−)		Ulnar (+)/Radial (−) Deviation	
	Left	Right	Left	Right
Conventional	21.80 (4.89)	21.73 (6.12)	15.67 (3.63)	16.91 (4.33)
Goldtouch	25.76 (4.67)	23.56 (5.87)	0.55 (6.44)	3.62 (5.61)
Maltron	15.01 (5.55)	13.29 (5.45)	4.69 (4.11)	7.00 (4.80)
PST	21.00 (5.46)	21.78 (4.45)	14.04 (4.81)	15.58 (3.32)

both left and right sides ($p < 0.001$). The PST keyboard required $9°$–$13°$ more ulnar deviation than Maltron and Goldtouch keyboards for both LUR and RUR planes ($p < 0.001$). The Goldtouch keyboard design forced 80% of subjects to type with the left or right wrist in greater than $20°$ extension as compared to 70% for the conventional and PST and 30% for Maltron. Within keyboards, no significant differences were found between sides for all four planes. In order to reduce the risk of musculoskeletal problems at this level, wrist extension should be lowered at $5°$–$10°$. A split design QWERTY layout with $25°$–$30°$ split angle, $0°$ lateral slope, and horizontal or negative slope is needed. While lateral slope tends to decrease forearm pronation, it also reduces typing productivity and user's acceptance. The reduction in both ulnar deviation and shoulder external rotation due to split angle design promote a safe forearm pronation while typing.

In terms of wrist excursions greater than $10°$, Maltron caused decreased wrist repetition for REF, LUR, and RUR planes. Table 39.2 presents the repetitive values for all four keyboards tested. No significant differences were found between the conventional, Goldtouch, and PST keyboards for all four movement planes. For the Maltron keyboard, the key-column vertical curvature and the thumb keys for *Enter, Backspace, Delete*, and other frequently used keys reduced the hand movement in the extension–flexion plane. For the ulnar–radial deviation plane, the wrist repetitive movements over $10°$ were reduced by the presence of the central numeric pad, which could be used by either hand, as preferred. Also, straight vertical key-columns reduced wrist excursions. Some of these design features should be further evaluated and, if valid, adopted by other keyboards.

Compared with the conventional keyboard, only Maltron had a statistically significant difference in applied force ($p < 0.001$). The mean typing force for participants using the conventional keyboard was 1.91 N (SD = 1.05), as compared to Maltron ($M = 5.84$; SD = 4.16). The mean applied typing forces for Goldtouch and PST were 0.97 N (SD = 0.52) and 1.28 N (SD = 0.85), in that order, which were 4.87 and 4.56 N, respectively, lower than the Maltron's average force ($p < 0.001$). The overall applied typing force recorded while using the conventional and Maltron keyboards exceeded the ANSI/HFS recommendations (0.5–1.5 N). These results are similar to those of Rempel et al. (1994), who using a piezoelectric load cell determined that the subject's mean peak force ranged between 1.6 and 5.3 N. The differences in overall typing force can be explained by important variations in keyswitch characteristics (key travel distance, over travel distance, stiffness, and keyswitch make force).

In terms of EMG muscle activity while typing, no significant differences were found between the six recorded muscles (ECU, FCR, FCU bilaterally) for all four keyboards.

Table 39.3 presents the average values for typing speed and accuracy for each keyboard. The Maltron keyboard was associated with significantly lower performance compared to other three keyboards for both typing speed and error rate ($p < 0.001$). While the conventional and PST keyboard were statistically similar in terms of accuracy, Goldtouch keyboard showed significantly higher error rate than the conventional, with 89% level of confidence. For the Maltron keyboard, the productivity was significantly reduced (58% decrease in typing speed and 149% increase in error rate, when compared to the conventional keyboard). On the Goldtouch keyboard, subjects reached 86% and on the PST keyboard 90%, of their typing speed on the traditional keyboard. Also, the error rate for these keyboards was statistically identical when compared to the conventional design. The decrease in productivity of 58% for the

TABLE 39.2 Mean Values and Standard Deviations for Wrist Repetition $>10°$ Per Minute for All Planes and Keyboards

Keyboard	Movement Plane			
	LEF	REF	LUR	RUR
Conventional	51 (13)	44 (15)	9 (3)	12 (5)
Goldtouch	43 (10)	39 (14)	12 (6)	10 (5)
Maltron	26 (8)	29 (12)	7 (3)	10 (5)
PST	46 (14)	41 (12)	9 (6)	12 (4)

TABLE 39.3 Typing Speed Words Per Minute (WPM) and Accuracy (*Backspace* Strokes per 100 Typed Words) for Different Keyboards

Keyboard	Typing speed	Typing accuracy
Conventional	69.67 (19.61)	7.61 (4.09)
Goldtouch	58.92 (21.40)	11.38 (5.37)
Maltron	29.26 (8.86)	19.29 (5.88)
PST	62.37 (17.28)	8.39 (3.72)

Maltron keyboard represents an important impediment for its acceptance. Table 39.4 summarizes the effect of alternative keyboard designs on the tested typing variables.

The current data demonstrate that after a relative short practice session typists were able to adjust their posture, performing as well with some of the tested alternative keyboards as with the conventional keyboard. This study indicated that keyboard design had an important effect on typing in terms of musculoskeletal diseases risk factors.

39.4.3   Training Effect on Typing on Ergonomic Keyboards

Immediate interests such as performance preservation and high training costs have delayed the introduction of ergonomic keyboard designs. In the literature there is a lack of data regarding the improvement in typing parameters induced by subjects' adaptation to the new designs.

In an attempt to reduce the impact of extensive office work on CTD incidence, a wide variety of alternative keyboard designs have been developed. Although previous studies noted an alleviation of tendon travel, wrist deviation, and forearm pronation, when typing on alternative designs, urgent interests including replacement costs and early decreased performance deferred the introduction of alternative keyboards. Also, dramatic design modifications caused unbending resistance from companies and data entry personnel. Zacevic et al. (2000) noted that after 10 h of training, the decline in productivity was 10% for the FIXED keyboard (split angle of 12° and a lateral slope of 10°) and 20% for the OPEN keyboard (split angle of 15° and lateral inclination of 42°) when compared to the standard keyboard. The majority of studies included in their experimental design very short training sessions, leading to biased outcomes. Swanson et al. (1999), in a study that assessed the impact of different keyboard designs on performance and comfort, described typing performance for alternative keyboards as a curve with an initial decline that is recuperated through the session.

In a study that determined the training effect on wrist posture and repetition, overall applied force, typing performance in terms of number of typed words per minute and number of mistakes per 100 typed words, and EMG forearm muscles activity for two different ergonomic keyboards, Fagarasanu et al. (2004c) noted important variations in studied typing variables. Twenty subjects underwent 8 h of training on each of the alternative keyboards.

The typing speed was significantly improved by training. Typing speed improved by 48% for both Goldtouch and Maltron keyboards. For the Goldtouch keyboard, the accuracy rate for the trained

TABLE 39.4 Changes in Typing Parameters for the Tested Alternative Keyboards when Compared with the Conventional Design

Keyboard	Wrist posture		Wrist Repetition	Applied Force	Muscle Activity	Typing Performance	
	Ulnar deviation	Extension				Pm	Error Rate
PST	↔	↔	↔	↓	↔	↔	↔
Goldtouch	↓	↔	↔	↓	↔	↔	↔
Maltron	↓	↓	↓	↑	↔	↓	↑

Note: ↔ = no statistically significant difference; ↓ = statistically significant decrease; ↑ = statistically significant increase.

group was 11.3 errors per 100 typed words (SD = 5.37), as compared to 15.3 (SD = 3.14) for the untrained subjects ($p < 0.039$). Training on the Maltron keyboard significantly reduced the error rate from 26.5 (SD = 7.40) to 19.2 (SD = 5.88) ($p < 0.007$). The fact that trained participants were able to type at 89% of their baseline typing speed when using the Goldtouch keyboard constitutes strong evidence that with additional experience alternative keyboards could easily replace the widespread traditional design without any loss in productivity, or perhaps even a gain. Our results differ from those reported by Treaster and Marras (2000), who noted a decrease of only 14% when typing on the Kinesis[TM] keyboard (similar design to Maltron). With regard to typing performance, Smith et al. (2000) noted values similar to those recorded for the traditional keyboard after only 2 h of training on a split angle keyboard. Previous research indicated that the initial decline in typing productivity has been recuperated after 2 days of training.

For the Maltron keyboard, the more ergonomic working postures found in untrained participants (less wrist ulnar deviation and extension) when compared with the traditional design, were maintained after the training session. Working on the Goldtouch keyboard promoted mitigated ulnar deviation in both untrained and trained groups. Although training decreased wrist repetition (number of wrist excursions greater than $10°$) with two to six repetitions per minute, the difference was not significant ($p > 0.05$).

The 10% decrease in wrist repetition represents an important decrease in risk factors associated with prolonged typing (e.g., tendon travel, tendon sheaths friction). After training, for one day of work the decline in wrist repetition would be of 2400 movements per movement plane (5 repetitions/min × 60 min per hour × 8 h of work). For one hand (both ulnar–radial deviation and flexion–extension planes) a total of approximately 4800 unnecessary wrist movements would be avoided through training. For the Maltron keyboard, training induced values below 30 per minute, which is the recommended highest acceptable frequency in a repetitive motion.

In terms of applied force, training significantly reduced the overall applied force for both Goldtouch ($p < 0.022$) and Maltron ($p < 0.031$) keyboards. The mean typing force was reduced by 58% for Goldtouch and by 42% for the Maltron keyboard. For the Goldtouch keyboard the training session was enough in order to reduce the typing force below the values for the conventional design (from 2.27 to 0.97 N, compared with 2.17 N recorded for the conventional one).

The decrease in overall applied force following training could be explained by hesitancy alleviation. Working under time pressure, especially with keyboard designs totally different than the one subjects are used to, spending more time in order to find the right keys leads to higher key stroke force (increased finger velocities) when the key is found. Training makes devices more familiar, eliminating unnecessary actions.

A synthesis of the effects of training on the studied variables is presented in Table 39.5.

39.5 Conclusions and Summary

Office work is associated with an elevated level of musculoskeletal risk factors. The upper extremity is forced to be positioned in deviated postures with concomitant repetitive motions. Wrist neutral zone should not be viewed as an arbitrary region of this particular range of motion. Low muscle activity and CTP are present in activities carried out within these boundaries. The neutral zone varied between $7°$ and $9°$ extension and between $5°$ and $7°$ ulnar deviation. Significantly lower EMG muscle activity was recorded while the wrist was positioned within neutral zone as compared to deviated

TABLE 39.5 The Effect of Training on Studied Variables

					Typing Performance	
Keyboard	Wrist Posture	Wrist Repetition	Applied Force	EMG Activity	WPM	Accuracy
Goldtouch	↔ $p > 0.05$	↓ $p > 0.05$	↓ *$p = 0.022$	↔ $p > 0.05$	↑ *$p = 0.027$	↓ *$p = 0.039$
Maltron	↔ $p > 0.05$	↓ $p > 0.05$	↓ *$p = 0.031$	↔ $p > 0.05$	↑ *$p = 0.008$	↓ *$p = 0.007$

postures. Encouraging workers to perform with wrist positions within neutral zone as it could reduce job-associated musculoskeletal disorders risks.

Typing on alternative keyboards improves upper extremity posture, thus eliminating, or at least reducing the risk factors associated with awkward posture while typing. Further, wrist repetition and overall applied typing force are reduced without a significant effect on typing performance. When designing alternative keyboards, one should keep in mind the trade-off between drastic design modifications and typing performance. In any new design, a balance between new features and their effect on work performance should be in the designer's mind. Not only are ergonomic keyboards able to meet the immediate requirements such as performance, typing speed, and short training time, but they also promote safer hand postures. Additional research is mandatory in order to see if prolonged office work on alternative keyboards supports these findings. The current results show that with additional experience, alternative keyboards represent a valid alternative for the conventional keyboard design.

References

Armstrong TJ, Castelli WA, Evans FG, Diaz-Perez R. 1984. Some histological changes in carpal tunnel contents and their biomechanical implications. *Journal of Occupational Medicine* 26: 197–201.

Armstrong TJ, Chaffin DB. 1979. Carpal tunnel syndrome and selected personal attributes. *Journal of Occupational Medicine* 21: 481–486.

Bergamasco R, Girola C, Colombini D. 1998. Guidelines for designing jobs featuring repetitive tasks. *Ergonomics* 41: 1364–1383.

Braun RM. 1988. The dynamic diagnosis of carpal tunnel syndrome. *4th Congress of Peripheral Neuropathy Association of America*, Halifax, Nova Scotia, Canada.

Bureau of Labor Statistics. 1994. CTD News. http://ctdnews.com/bls.html.

De Krom MCTFM, Kester ADM, Knipschild PG, Spaans F. 1990. Risk factors for carpal tunnel syndrome. *American Journal of Epidemiology* 132: 1102–1110.

Drury CG, Begbie K, Ulate C, Deeb JM. 1985. Experiments on wrist deviation in manual materials handling. *Ergonomics* 28: 577–589.

Dvorak A. 1943. There is a better typewriter keyboard. *National Business Education Quarterly* 12: 51–58, 66.

Fagarasanu M, Kumar S, Narayan Y. 2004a. Measurement of angular wrist neutral zone and forearm muscle activity. *Journal of Clinical Biomechanics* (in press).

Fagarasanu M, Kumar S, Narayan Y. 2004b. An ergonomic comparison of four computer keyboards. *Applied Ergonomics* (in press).

Fagarasanu M, Kumar S, Narayan Y. 2004c. The training effect of typing on two ergonomic keyboards. *International Journal of Human–Computer Studies* (pending).

Hadler NM. 1987. *Clinical Concepts in Regional Musculoskeletal Illness*. Grune & Stratton, Inc., Harcourt Brace Jovanovich, Publishers.

Harvey R, Peper E. 1997. Surface electromyography and mouse use position. *Ergonomics* 40: 781–789.

Hedge A, Powers JR. 1995. Wrist postures while keyboarding: effects of a negative slope keyboard system and full motion forearm supports. *Ergonomics* 38: 508–517.

Hoffman DS, Strick PL. 1999. Step-tracking movements of the wrist. IV. Muscle activity associated with movements in different directions. *Journal of Neurophysiology.* 81: 319–333.

Keir PJ, Bach JM, Rempel DM. 1996. Effects of finger posture on carpal tunnel pressure during wrist motion. *Journal of Hand Surgery* 23A(6): 1004–1009.

Keir PJ, Bach JM, Rempel DM. 1998. Effects of finger posture on carpal tunnel pressure during wrist motion. *The Journal of Hand Surgery* 23A: 1004–1009.

Keir P, Bach J, Rempel D. 1999. Effects of computer mouse design and task on carpal tunnel pressure. *Ergonomics* 42: 1350–1360.

Kumar S. 1990. Analysis of selected high risk operations in a garmet industry. In: *Advances in Industrial Ergonomics and Safety II*, Das B. Ed. 227–236. Taylor & Francis, London.

Kumar S. 2001. Theories of musculoskeletal injury causation. *Ergonomics* 44: 17–47.

Kumar S. Panjabi M. 1995. In vivo axial rotations and neutral zones of he thoracolumbar spine. *Journal of Spinal Disorders* 8(4): 253–263.

Kumar S, Narayan Y, Chouinard K. 1997. Effort reproduction accuracy in pinching, gripping, and lifting among industrial males. *Ergonomics* 20: 109–119.

Liao MH, Drury CG. 2000. Posture, discomfort and performance in a VDT task. *Ergonomics* 43: 345–359.

Lincoln AE, Vernick JS, Ogaitis S, Smith GS, Mitchell CS, Agnew J. 2000. Interventions for the primary prevention of work-related carpal tunnel syndrome. *American Journal of Preventive Medicine* 18: 37–50.

Marklin RW, Simoneau GG. 2001. Effect of setup configurations of split computer keyboard on wrist angle. *Physical Therapy* 81: 1038–1048.

Marklin RW, Simoneau GG, Monroe JF. 1999. Wrist and forearm posture from typing on split and vertically inclined computer keyboards. *Human Factors* 41: 559–569.

Moore SJ. 1992. Carpal tunnel syndrome. *Occupational Medicine. State of the Art Reviews* 7: 741–763.

Nelson JE, Treaster DE, Marras WS. 2000. Finger motion, wrist motion and tendon travel as a function of keyboard angles. *Clinical Biomechanics* 15: 489–498.

O'Driscoll SW, Horii E, Ness R, Richards RR, An K-N. 1992. The relationship between wrist position, grasp size, and grip strength. *Journal of Hand Surgery* 17A: 169–177.

Phalen GS, Kendrick JI. 1957. Compression neuropathy of the median nerve in the carpal tunnel. *Journal of the American medical Association* 164: 524–530.

Rempel DR, Tittiranonda P, Burasteno S, Hudes M, So Y. 1999. Effect of keyboard keyswitch design on hand pain. *Journal of Occupational and Environmental Medicine* 41: 111–119.

Sauter SL, Schleifer M, Knutson S. 1991. Work posture, workstation design, and musculoskeletal discomfort in a VDT data entry task. *Human Factors* 33: 151–167.

Seradge H, Jia YC, Owens W. 1995. In vivo measurement of carpal tunnel pressure in the functioning hand. *Journal of Hand Surgery* 20A: 855–859.

Serina E, Tal R, Rempel D. 1999. Wrist and forearm postures and motions during typing. *Ergonomics* 42: 938–951.

Simoneau GG, Marklin RW, Monroe JF. 1999. Wrist and forearm postures of users of conventional computer keyboards. *Human Factors* 41: 413–424.

Smith MJ, Karsh BT, Conway FT, Cohen WJ, James CA, Morgan JJ, Sanders K, Zehel DJ. 1998. Effects of a split keyboard design and wrist rest on performance, posture, and comfort. *Human Factors* 40: 324–336.

Szabo RM. 1998. Carpal tunnel syndrome as a repetitive motion disorder. *Clinical Orthopaedics and Related Research* 351: 78–89.

Szabo RM. 1989a. *Nerve Compression Syndromes, Diagnostic and Treatment.* SLACK Incorporated, Thorofare, NJ.

Szabo RM. 1989b. Stress carpal tunnel pressures in patients with carpal tunnel syndrome and normal patients. *Journal of Hand Surgery* 14A: 624–627.

Treaster DE, Marras WS. 2000. An assessment of alternate keyboards using finger motion, wrist motion and tendon travel. *Clinical Biomechanics* 15: 499–503.

U.S. Department of Labor. 1999. Notes: CTD News. http://ctdnews.com/bls.html.

Viikari-Juntura E, Riihimaki H. 1999. New avenues in research on musculoskeletal disorders. *Scandinavian Journal of Work, Environment and Health* 25: 564–568.

Visser B, de Korte E, van der Kraan I, Kuijer P. 2000. The effect of arm and wrist supports on the load of the upper extremity during VDU work. *Clinical Biomechanics* 15: S34–S38.

Wahlstrom J, Svensson J, Hagberg M, Johnson P. 2000. Differences between work methods and gender in computer mouse use. *Scandinavian Journal of Work Environment and Health* 26: 390–397.

Werner CO, Elmquist D, Ohlin P. 1983. Pressure and nerve lesion in the carpal tunnel. *Acta Orthopaedica Scandinavica* 54: 312–316.

Werner R, Armstrong TJ, Aylard MK. 1997. Intracarpal canal presure: the role of finger, hand, wrist and forearm position. *Clinical Biomechanics* 12: 44–51.

Werner RA, Armstrong TJ. 1997. Carpal tunnel syndrome — ergonomic risk factors and intracarpal canal pressure. *Physical Medicine and Rehabilitation Clinics of North America* 8: 555–569.

Yamaguchi DM, Lipscomb PR, Soule EH. 1965. Carpal tunnel syndrome. *Minnesota Medicine* January: 22–23.

Zecevic A, Miller DI, Harburn K. 2000. An evaluation of the ergonomics of three computer keyboards. *Ergonomics* 43: 55–72.

Further Reading

Armstrong TJ, Castelli WA, Evans FG, Diaz-Perez R. 1984. Some histological changes in carpal tunnel contents and their biomechanical implications. *Journal of Occupational Medicine* 26: 197–201.

Chen C, Burastero S, Tittiranonda P, Hollerbach K, Shih M, Denhov R, 1994. Quantitative evaluation of four computer keyboards: wrist posture and typing performance. *Proceedings of the Human Factors and Ergonomics Society 38th Annual Meeting.* Santa Monica, CA, pp. 1094–1098.

Fernandez JE, Dahalan JB, Halpern CA, Viswanath V, 1991. The effect of wrist posture on pinch strength. *Proceedings of the Human Factors 35th Annual Meeting*, Human Factors Society, San Francisco, CA, pp. 748–752.

Hertting-Thomasius R, Steidel F, Prokop M, Lettow H. 1992. On the introduction of ergonomically designed keyboards. *Abstracts Book: Work with Display Units WWDU '92.* Luczak, H Cakir, A Cakir G Eds. Tachnische Universitat Berlin, Berlin p. P-2.

Kapandji IA, 1982. *The Physiology of the Joints, Annotated Diagrams of the Mechanics of the Human Joints.* Churchill Livingstone, Edinburgh.

Pryce J. 1980. The wrist position between neutral and ulnar deviation that facilitates the maximum power grip strength. *Journal of Biomechanics* 13: 505–511.

Tittiranonda P, Rempel D, Armstrong T, Burastero S, 1999. Workplace use of an adjustable keyboard: Adjustment preferences and effect on wrist posture. *American Industrial Hygiene Association Journal,* 60(3): 340–349.

Yoshitake R, Ise N, Yamada S, Tsuchiya K. 1997. An analysis of users' preference on keyboards through ergonomic comparison among four keyboards. *Applied Human Science* 16(5): 205–211.

Weiss ND, Gordon L, Bloon T, So Y, Rempel DM, 1995. Position of the wrist associated with the lowest carpal tunnel pressure: implications for splint design. *Journal of Bone and Joint Surgery* 77A(11): 1695–1699.

PLIBEL — A Method Assigned for Identification of Ergonomics Hazards

Kristina Kemmlert

National Board of Occupational
Safety and Health

40.1 Introduction

The Swedish Work Environment Act stipulates that the employer shall investigate occupational injuries, draw up action plans, and organize and evaluate job modifications. Hence, it is also of interest for the Labor Inspectorate to study conditions and improvements at workplaces.

A method for the identification of musculoskeletal stress factors that may have injurious effects (PLIBEL) was designed to meet such needs[3–5,7] (Figure 40.1). PLIBEL has been used in several studies, in practical on-site ergonomics work, and also in education. It has been presented in various parts of the world and translated into several languages.[3,6,8,9,11]

PLIBEL is a simple checklist screening tool intended to highlight musculoskeletal risks in connection with workplace investigations. Time aspects, environmental factors, and organizational factors also have to be considered as modifying factors.

The checklist was designed so that items ordinarily checked in a workplace assessment of ergonomics hazards would be listed and linked to five body regions (Figure 40.1). Only specific work characteristics, defined and documented as ergonomics hazards in scientific papers or textbooks, are listed (Figure 40.2 and Figure 40.3). Whenever a question is irrelevant to a certain body region, as documentation has not been found in the literature, it is represented by a gray field in the checklist.

The list was made in 1986 and new references have since then been read continuously and the list updated. Mostly, these only add knowledge to the primary list, which accordingly has not been changed. Only one, concerning hips, knees, feet, and the lower spinal region, has the kind of new information searched for and has therefore been added to the documented background (Figure 40.2).

Method for the identification of musculo-skeletal stress factors which may have injurious effects-PLIBEL

Kemmlert, K. Kilbom, Å. (1986) National Board of Occupational Safety and Health, Research Department, Work Physiology Unit, 171 84 Solna, Sweden.

Body regions (columns): neck/shoulders, upper part of back — elbows, forearms hands — feet — knees and hips — low back

1. Is the walking surface uneven, sloping, slippery or non resilient?
2. Is the space too limited for work movements or work materials?
3. Are tools and equipment unsuitably designed for the worker or the task?
4. Is the working height incorrectly adjusted?
5. Is the working chair poorly designed or incorrectly adjusted?
6. (If the work is performed whilst standing): Is there no possibility to sit and rest?
7. Is fatiguing foot-pedal work performed?
8. Is fatiguing leg work performed e. g.:
   a) repeated stepping up on stool, step etc.?
   b) repeated jumps, prolonged squatting or kneeling?
   c) one leg being used more often in supporting the body?
9. Is repeated or sustained work performed when the back is:
   a) mildly flexed forward?
   b) severely flexed forward?
   c) bent sideways or mildly twisted?
   d) severely twisted?
10. Is repeated or sustained work performed when the neck is:
    a) flexed forward?
    b) bent sideways or mildly twisted?
    c) severely twisted?
    d) extended backwards?
11. Are loads lifted manually? Notice factors of importance as:
    a) periods of repetitive lifting
    b) weight of load
    c) awkward grasping of load
    d) awkward location of load at onset or end of lifting
    e) handling beyond forearm length
    f) handling below knee height
    g) handling above shoulder height
12. Is repeated, sustained or uncomfortable carrying, pushing or pulling of loads performed?
13. Is sustained work performed when one arm reaches forward or to the side without support?
14. Is there repetition of:
    a) similar work movements?
    b) similar work movements beyond comfortable reaching distance?
15. Is repeated or sustained manual work performed? Notice factors of importance as:
    a) weight of working materials or tools
    b) awkward grasping of working materials or tools
16. Are there high demands on visual capacity?
17. Is repeated work, with forearm and hand, performed with:
    a) twisting movements?
    b) forceful movements?
    c) uncomfortable hand positions?
    d) switches or keyboards?

Method of application.

* Find the injured body region
* Follow white fields to the right
* Do the work tasks contain any of the factors described?
* If so, tick where appropriate

Also take these factors into consideration:

a) the possibility to take breaks and pauses
b) the possibility to choose order and type of work tasks or pace of work
c) if the job is performed under time demands or psychological stress
d) if the work can have unusual or unexpected situations
e) presence of cold, heat, draught, noise or troublesome visual conditions
f) presence of jerks, shakes or vibrations

FIGURE 40.1   The PLIBEL form.

Item	Neck/shoulders, upper part of back	Elbows, fore-arms, hands	Hips, knees, feet	Low back
1			40 49 63	49 63
2	11 12	11 12	2	2 15 30
3	1 3 7 9 11 12 21	7 9 18 60	30	3 9 15
4	1 3 9 11 12 21			3 9 15 30
5	3 9 11 12 21			3 9 30
6			15 28 38 49	28
7			38	
8a			38	44
8b			8 19 38	28
8c			8 38	39
9a	36			15 26 39 56
9b	36			15 26 39 48 56
9c	36			15 26 39 48 56
9d	36			15 26 39 48 56
10a	10 11 12 14 36 57 62			
10b	12 36 55 62			
10c	12 36 55 62			
10d	29 62			
11a	23 43 47 55 58 61			48 56
11b	21 23 34 43 57			48 56
11c	7			53
11d	44			15 26 48 56
11e	21 57			15 38 48
11f	53			15 26 56
11g	21 34 41 47 57			48 56
12	17 54	23		48 54 56
13	1 4 11 14 29 47 62			
14a	4 11 14 16 21 34 47	16 23 31 52 58 61		
14b	1 4 14 21 29 62	1 4 14 21 55 62		
15a	1 21 34 41 57 62	1 18 31 52 55 58		
15b	1 7	18 23 31 52 58 60		
16	1 11 22 33			
17a		23 31 52 58 60 61		
17b		16 23 31 52 60 61		
17c		16 23 31 52 60 61		
17d		11 55		

Original papers

1 Aarås 1987
2 Anderson 1984
3 Bhatnager et al. 1985
4 Bjelle et al. 1981
5 Bovenzi 1991
6 Chen 1991
7 Drury 1985
8 Felson 1988
9 Grandjean et al.1983
10 Harms-Ringdahl 1986
11 Hünting et al. 1980
12 Hünting et al. 1981
13 Johansson and Aronsson 1980
14 Jonsson et al. 1988
15 Keyserling et al. 1988
16 Kilbom 1994 a
17 Kilbom et al.1984
18 Kilbom et al.1991
19 Kivimäki et al. 1992
20 Kjellberg et al. 1992

21 Kvarnström 1983
22 Laville 1968
23 Luopajärvi et al. 1979
24 Magnusson 1991
25 Mathiassen 1993
26 Punnett et al. 1991
27 Riihimäki 1990
28 Ryan 1989
29 Sakakibara et al. 1987
30 Shute and Starr 1984
31 Silverstein et al. 1986
32 Sköldström 1987
33 Starr et al. 1985
34 Stenlund et al. 1993
35 Sundelin 1992
36 Tola et al. 1988
37 Van den Bossche and Lahaye 1984
38 Vingård et al. 1991
39 Vink et al. 1992
40 Winkel and Ekblom 1982
41 Örtengren et al. 1991

Review papers and textbooks

42 Bongers et al. 1993
43 Chaffin 1973
44 Chaffin and Andersson 1984
45 Enander 1984
46 Gemne et al. 1993
47 Hagberg and Wegman 1987
48 Jørgensen and Biering-Sørensen 198
49 Jørgensen et al. 1993
50 Karasek and Theorell 1990
51 Lloyd 1986
52 Mital and Kilbom 1992
53 Mital et al. 1993
54 Pedersen et al. 1992
55 Rempel et al. 1992
56 Riihimäki 1991
57 Sommerich et al. 1993
58 Stock 1991
59 Sundström-Frisk 1990
60 Tichaner 1978
61 Wallace and Buckle 1987
62 Winkel and Westgaard 1992
63 Hansen, Winkel and Jörgensen 1998

FIGURE 40.2 Documented background for PLIBEL. References, as numbered in the footnote, are given for each risk factor in relation to body regions as in the PLIBEL form. Note, however, that in this presentation the distribution is by four body regions. Hips, knees, and feet are combined in the table.

Also take these factors into consideration

The possibility to take breaks and pauses	21	25	35	50	57	59	61		
The possibility to choose order and type or work tasks or pace of work	13	21	35	50	57				
If the job is performed under time demands or psychological stress	13	21	35	42	50	59	61		
If the work can have unusual or unexpected situations	34	38	50	56					
Presence of:									
Cold	6	45	51						
Heat	32	53							
Draught	35	36							
Noise	20								
Troublesome visual conditions	1	11	22	33	61				
Jerks, shakes, or vibration	5	24	34	37	46	48	52	54	56

FIGURE 40.3   Documented background for modifying factors (for references, see footnote to Figure 40.2).

40.2   Procedure

A workplace assessment using PLIBEL starts with an introductory interview with the employee and preliminary observation. Representative parts of the job, the tasks that are conducted for most of the working hours, and tasks that the worker or the observer look upon as particularly stressful to the musculoskeletal system are chosen for the assessments. Thus, several PLIBEL forms may have to be filled in for each employee.

The assessments should be related to the capacity of the individual observed. Unusual or personal ways of doing a task are also recorded.

When an ergonomics hazard is observed, the numbered area on the form is checked or a short note is made. In the concluding report, where the crude dichotomous answers are arranged in order of importance, quotations from the list of ergonomics hazards may be used. Modifying factors, duration, and quantities of environmental or organizational factors are then taken into consideration (Figure 40.1).

Usually PLIBEL is used to identify musculoskeletal injury risk factors for a specific body region, and only questions relevant to that body region need be answered. A more general application may also be feasible. Here, the whole list is used, and the result can be referred to one or more body regions.

To use PLIBEL, first locate the injured body region, then follow the white fields to the right and check any observed risk factors for the work task. The continued assessment is more difficult, as it requires consideration of questions a–f. These can either upgrade or minimize the problem. Additional evident risks, not mentioned in the checklist, are noted and addressed. For example, there are no duration criteria for a PLIBEL record, and so cumbersome but short-lasting or rare events can also be recorded. In fact, the purpose of the interview with the worker, which precedes the observation, is to make such aspects of the task manifest.

A participatory approach of this kind has also been suggested by other authors, for example, Drury[2], who recommend that observes talk to operators to get a feel for what is important. If only "normal" subjects and work periods are chosen for assessments, many of the unusual conditions, which may constitute main hazards, may be missed.

A handbook (unpublished material) has been compiled to provide the scientific background for each item and help identify the cut-off point for "yes" or "no" answers. This facilitates the assessments, which are to be performed by knowledgable and experienced observers. To make the checklist easy to handle and applicable in many different situations, the questions are basic.

The analysis of possible ergonomics hazards is done at the workplace and only relevant risk information from the assessment is considered. The issues identified as risks are arranged in order of importance. The concluding report gives an interpretation of the ergonomics working conditions, starting with the most tiresome movements and postures.

The PLIBEL method is a general assessment method and is not intended for any specific occupation or task. It observes a part, or the whole, of the body and summarizes the actual identification of ergonomics hazards in a few sentences.

It is simple and designed for primary checking. For labor inspectors and others looking at many tasks every day, it is certainly enough to be equipped and well acquainted with the checklist.

PLIBEL is an initial investigative method for the workplace observer to identify ergonomics hazards, and it can be supplemented by other measurements, for instance, weight and time, or quotations/observations from other studies.

Although it is tempting to add up items, to obtain a simple and quantitative measure of ergonomics conditions after a workplace assessment, PLIBEL should not be modified or used in this way. Different ergonomics hazards do not have an equal influence, and certain problems can appear with more than one hazardous factor in the checklist.

Many other methods are intended for a specific occupation or body region and can record more detailed answers. If necessary, these more specific data can easily be used to supplement the PLIBEL questions.

40.3 Example

PLIBEL analysis of the task shown in Figure 40.4 reveals that it entails a risk of musculoskeletal stress to the lower region of the back, due to the nonresilient walking surface, the unsuitably designed tools and equipment, and the lack of any possibility to sit and rest. Repetitive and sustained work is performed with the back flexed slightly forward, bent sideways, and slightly twisted. Loads are repeatedly lifted manually and often above shoulder height. Note that the text order has been expressed by giving the most exposed body region and the environmental and instrumental conditions first. The following phrase gives "the answers" from the body, followed by a description of the tiresome, and perhaps individual, way of performing the task.

40.4 Reliability and Validity

A reliability and validity study of the method has been performed according to Carmines and Seller.[1] It was tested for construct validity, criterion validity, reliability, and applicability.[3] The agreement

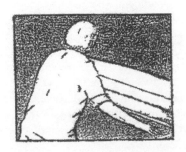

FIGURE 40.4 Example of a task posing ergonomics hazards that was analyzed using PLIBEL. (From Prevention of Occupational Musculosketetal Injuries. *Scand Journ. Of Rehab. Medicine.* 35, 1996. With premission.)

between matching items was considerable and the interobserver reliability yielded kappa values expressing a fair to moderate agreement on the following questions:

1. Is the content of PLIBEL and the set of items consistent with theoretically derived expectations?
2. Can occurrence of the criterion (ergonomics hazard) be validated by comparison with another method?
3. Are the results from different users of the PLIBEL method consistent when observing the same working situation?
4. How has the method been used? What are the experiences?

PLIBEL was written in Swedish, but has also been translated into several languages: Dutch,[9] English,[3] French[6] Greek,[8] and Spanish.[11]

To provide a reference instrument for PLIBEL, an inventory of available scientific literature on occupational risk factors for musculoskeletal disorders was performed. Original papers, review papers, and textbooks were studied.

After a thorough review of the literature, the German ergonomics job analysis procedure AET (Arbeitswissenshaftliche Erhebungsverfahren zur Tätigkeitsanalyse) was chosen as the referent instrument for field testing.[10]

Like PLIBEL, AET is applicable to all sorts of occupational tasks, and covers workplaces, tools and objects, degree of repetitiveness, work organization, cognitive demands, and also environmental factors such as visual conditions, noise, and vibration.

However, while AET analyzes all components in a man-at-work system, PLIBEL focuses on one extreme phenomenon, that is, the occurrence of an ergonomics hazard.

Two researchers, each of whom had been practising AET and PLIBEL, respectively, clearing very many workplace assessments, identified 18 matching items in the two methods. For PLIBEL only dichotomous answers are used, whereas multilevel codes, in steps zero to five, are applied in AET. For each of the items, the corresponding level between the two methods was identified. The two observation methods were then used simultaneously for observations on a total of 25 workers, men and women in different tasks.

When comparing the results of PLIBEL and AET, the agreement between matching items was considerable. However, the modifications of AET scores for a dichotomous coding could not completely eliminate the differences between the methods. In concordance with its purposes, PLIBEL was more sensitive to ergonomics hazards.

40.5 Conclusion

PLIBEL was designed to meet the need of a standardized and practical method for the identification of ergonomics hazards and for a preliminary assessment of risk factors. An ergonomics screening tool, for the assessment of ergonomics conditions at workplaces, has been suggested as a feasible instrument by other researchers.

Moreover, it is valuable to have a systematic way of assessment when doing follow-ups and when analyzing how intervention after occupational musculoskeletal injuries could be made more effective.

PLIBEL follows standards and regulations of the day, and though it is a self-explanatory, subjective assessment method, registering only at a dichotomous level, it requires a solid understanding of ergonomics. For using the method skilfully, a certain degree of practice is firmly recommended.

To see that a situation is awkward is not difficult, nor is it difficult to find such a situation with the aid of the checklist. PLIBEL is quick to use and easy to understand, and users will become familiar with this within hours. However, although PLIBEL is a self-explanatory subjective assessment method, making dichotomous judgments about risks, it requires a solid ergonomics understanding, and using the method skilfully requires practice.

Paper, pencil, a folding rule, and a camera are sufficient for ordinary workplace observations and for initial identification of ergonomics hazards.

Observational findings have provided a base for recommended improvements, for discussion of ergonomics problems, and for worksite education. Moreover, PLIBEL has been used for ergonomics education both in industry and in the Swedish school system.

References

1. Carmines, E.G. and Seller, R.A. (1979). *Reliability and Validity Assessment*, Sage Publications, London.
2. Drury, C.G. (1990). Methods for direct observation of performance, in: Wilson, J.R. and Corlett, E.N. (eds), *Evaluation of Human Work*, Taylor & Francis, London, pp. 35–57.
3. Kemmlert, K. (1995). A method assigned for the identification of ergonomic hazards — PLIBEL, *Applied Ergonomics*, 26(3), 199–211.
4. Kemmlert, K. (1996a). Prevention of occupational musculo-skeletal injuries. Labour Inspectorate investigation, *Scandinavian Journal of Rehabilitaiton Medicine*, 35 (Suppl.), 1–34.
5. Kemmlert, K. (1996b). New analytic methods for the prevention of work-related musculoskeletal injuries, in: *Fifteen Years of Occupational Accident Research in Sweden*, Rådet för arbetslivsforskning, Stockholm, pp. 176–186.
6. Kemmlert, K. and Kilbom, Å. (1996c). La check-list, in: D'Hertefelt, H, Bentein, K, and Willcox, M, *Le corps au travail*, Bruxelles, INRCT, pp. 224–226.
7. Kemmlert, K. (1997), *On the Identification and Prevention of Ergonomic Risk Factors*, Solna, Arbetslivsinstitutet. Doctoral thesis, Luleå University of Technology.
8. Lomi, C. (2002). *ΜΕΘΟΔΟΣΓΙΑ ΤΟΝ ΠΡΟΣΔΙΟΡΙΣΜΟ ΤΩΝ ΠΑΡΑΓΟΝΤΩΝ ΚΑΤΑ- ΠΟΝΗΣΗΣ ΤΟΥ ΜΥΟΣΚΕΛΕΤΙΚΟΥ ΣΥΣΤΗΜΑΤΟΣ ΠΟΥ ΜΗΟΡΕΙ ΝΑ ΕΙΝΑΙ ΒΛΑΠΤΙΚΗ — PLIBEL*. [Method for the identification of musculo-skeletal stress factors which may have injurious effects—PLIBEL] *ΥΓΙΕΙΝΗ ΚΑΙ ΑΣΦΑΛΕΙΑ ΤΗΣ ΕΡΓΑΣΙΑΣ*, 9, 5–12 [Occupational hygiene and safety, Vol. 9, pp. 5–12].
9. Ollongren, G., Debout, J., and Desiron, H. (1990). *Van klacht naar problemformulering* [From complaints to the wording of a problem], Postuniversitaire opleiding bedrijfsergonomie en arbeidshygiëne, Universiteit Antwerpen [University training in occupational health and hygiene, University of Antwerp].
10. Rohmert, W. and Landau, K. (1983). *A New Technique for Job Analysis*, Taylor & Francis, London.
11. Serratos-Perés, N. and Kemmlert, K. (1998). Assessing ergonomic conditions in industrial operations, *Asian-Pacific Newsletter on Occupational Health and Safety*, 5(3), 67–69.

41

The ACGIH TLV® for Hand Activity Level

Thomas J. Armstrong

University of Michigan

41.1 Introduction

Musculoskeletal disorders (MSDs) continue to be a major cause of disability and lost work in many industries involving hand-intensive activities (BLS, 2003). MSDs are "multifactorial," which means that they may be caused or aggravated by multiple factors. Factors that pertain to the individual, such as weight, age, and leisure activities, are referred to as "personal" factors. Factors that pertain to the job, such as work posture, force required to handle materials and use tools, and recovery time, are referred to as "work-related" factors. There is a growing body of literature that demonstrates exposure to work factors results in an increased risk of MSDs (NRC, 1999; NRC and IOM, 2001). MSDs that result from exposure to work factors are referred to as "work-related" MSDs or "WMSDs." This chapter describes the American Council of Industrial Hygienists Threshold Limit Value (ACGIH TLV®) for monotask hand work and reviews the basic concepts underlying general development of TLVs. It discusses specifically some of the studies on which the TLV for monotask handwork was based and some studies of the TLV that have been reported since it was proposed. The chapter concludes with the concepts and methods that are used to apply this TLV.

41.2 ACGIH TLVs

ACGIH TLVs are not standards, they are guidelines. ACGIH is a scientific organization that has established committees to review existing published, peer-reviewed literature. The committees recommend and the ACGIH Board of Directors approves the publication of guidelines known as Threshold Limit Values for use by industrial hygienists in making decisions regarding safe levels of exposure to various physical agents encountered in the workplace. Users of their guidelines are cautioned that there may be other factors in addition to those specified by the TLVs that should be considered in evaluating specific workplaces. For this reason, professional judgment is necessary for applying and interpreting TLVs.

The TLV for hand activity is intended to protect most workers from MSDs for the hand, wrist, and forearm. Workers with predisposing health conditions such as arthritis, endocrinological disorders, obesity, pregnancy, old age, or previous injuries may be affected by exposures below the TLV. It is likely that exposures below the TLV may produce some discomfort, but it should not persist from day to day or interfere with activities of work or daily living. The ACGIH TLV is designed to prevent only work-related cases. Employers should educate workers about non-work-related causes of musculoskeletal disorders and how they can minimize their risk.

41.3   The Basis for a TLV on Hand Activity Level

ACGIH "Documentation" is the official source of information on the basis for ACGIH TLVs (ACGIH, 2005a). Here we will briefly describe some of the very early studies that called attention to the hand–wrist–forearm MSD as an occupational health problem, some of the key references that relate to specific aspects of the TLV, and some studies that have been reported since the TLV was proposed.

Obolenskaja and Goljanitzki (1927) suggested work rates of 7600 to 12,000 exertions per shift were a factor in 189 cases of tenosynovitis among 700 tea packers. Hammer (1934) suggested human tendons do not tolerate more than 1500 to 2000 exertions per hour. Although these were not controlled epidemiological studies, they demonstrated a long standing concern about MSDs in manual work and introduced the concept of a dose–response relationship between repetitive hand work and hand–wrist–forearm MSDs. Since these studies, there have been many controlled studies that support this relationship.

Work by Leclerc et al. (1997), Fransson-Hall et al. (1995), and Faucett and Rempel et al. (1994) indicated that there is elevated risk of hand–wrist–forearm MSDs with exposures to repetitive hand intensive work for four or more hours per shift. Therefore, the TLV applies to exposures of four or more hours.

Latko et al. (1999), Luopajarvi et al. (1979), Silverstein et al. (1987), Armstrong et al. (1987), Fransson-Hall et al. (1995), Roquelaure et al. (1997), Leclerc et al. (1998), and Marras and Schonmarklin (1993) supported a exposure–response relationship between repetitive work and hand–wrist–forearm MSDs. Chaing et al. (1993) showed a relationship between force and carpal tunnel syndrome. Silverstein et al. (1987), Armstrong et al. (1987), and Roquelaure et al. (1997) demonstrated an interaction between force and repetition. Latko et al. (1999) reported that the prevalence of nonspecific discomfort, symptoms of carpal tunnel syndrome and tendonitis, and nerve conduction findings were all strongly related to repetitive hand work. Latko et al. defined repetition on a scale from zero to ten where zero was "hands idle most of the time; no regular exertions" and ten was "rapid steady motion or continuous exertion; difficulty keeping up." Additional anchor points were described at two-point intervals based on the frequency, speed of motion, and recovery time (see Figure 41.1a). The Latko et al. (1997) scale was renamed "hand activity level" or HAL to emphasize that this TLV applies to the hand and to emphasize activity in the broadest sense (ACGIH, 2005a,b).

Force is expressed as the "peak finger force," which is defined as the 90th percentile of the force the worker exerts with his or her fingers over the work shift. The term finger is included with force because the TLV applies to the hand, wrist, and forearm. The fingers are the primary link between work objects and tendons and muscles in the hand, wrist, and forearm (Armstrong and Chaffin, 1979). Force is normalized on a zero to ten scale as described by Latko et al. (1999) for direct comparison with the HAL.

Peak force is used as opposed to an average force. Average forces above 15% of maximum strength or 1.5 on a zero to ten scale generally cannot be sustained for prolonged exertions (Rohmer, 1973; Bystrom and Fransson-Hall, 1994). Repetitive work is typically composed of patterns of varying force levels (Armstrong and Chaffin, 1979; Armstrong et al., 1982). Periods of high force are typically offset by periods of low force. Average values mask the magnitude of the peak forces; however, peak forces may vary significantly from one cycle to the next and are subject to random fluctuations. Silverstein et al. (1987) utilized an adjusted force that was expressed as: average force × (1+ coefficient of variation squared). For a coefficient variation of 0.9 reported by Silverstein (1985), the adjusted force was very

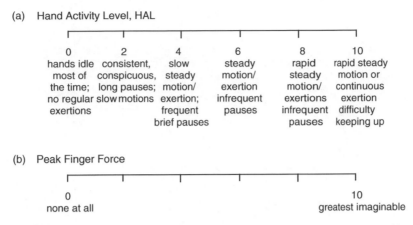

FIGURE 41.1 (a) Visual analog scale for estimating HAL from observations (Latko et al., 1997; ACGIH, 2005). (b) Peak finger force can be estimated from observations of a job and using the visual analog scale given in (a). (From Latko, et al., 1999; ACGIH, 2005.)

close to a 97.5 percentile. For purposes of the TLV, peak finger force is defined as a 90th percentile value. Peak forces also are consistent with the measurements used by Roquelaure et al. (1997). Assessment of 90th percentile peak finger forces is discussed next.

The maximum acceptable hand force was established as zero for an HAL of ten. By definition an HAL of ten is the "fastest possible: continuous exertions, rapid motion difficulty keeping up." At this point, external forces are assumed to be zero. However, there are likely to be very high inertial forces on tendons and muscles due to the rapid motions (Marras and Schonmarklin, 1993). The cyclical motions of the wrist translate into constant accelerations and decelerations. Wrist motions can be described as a series of sine functions (Radwin and Lin, 1993). The angular velocity of the joints is proportional to the amplitude and the frequency of motion, and the angular acceleration is proportional to the amplitude and the frequency squared. The force required to overcome the inertia of the fingers and the wrist is equal to the product of angular acceleration and the moment of inertia of the hand segments. Thus, "rapid steady motion or continuous exertion difficulty keeping up" is associated with high inertial forces that would be transmitted to the finger and wrist tendons and the forearm muscles.

The TLV assumes a simple linear relationship in which peak finger force decreases from an value of seven for a HAL value of one to a peak finger force of zero for an HAL value of zero and is shown in Figure 41.2. The relationship between peak finger force and HAL can be described mathematically as

$$\text{Peak finger force} \leq 70/9 - 7 * \text{HAL}/9$$

A nonlinear relationship is possible between peak finger force and HAL, but at the time the TLV was developed there were insufficient epidemiological data to rationalize a more complex function.

There was concern that a significant number of workers might still be at risk at exposure levels below the TLV, so an action limit was recommended. In contrast to the TLV, which should not be exceeded, the action limit may be exceeded, but it triggers a proactive program that includes: training, job and health surveillance, and medical management. The action limit goes from an HAL and peak finger force of one and five to ten and zero, respectively. The action limit can be described mathematically as

$$\text{Peak finger force} \leq 50/9 - 5 * \text{HAL}/9$$

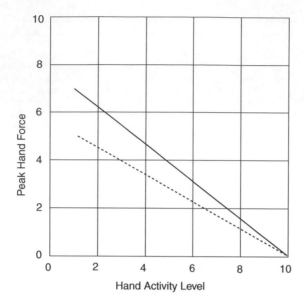

FIGURE 41.2 The ACGIH TLV for monotask hand work. The solid line depicts combinations of hand activity level and peak finger force that should not be exceeded to prevent excessive risk of hand and wrist MSDs. The dashed line is an action limit at which some people may be elevated risk. The action limit should trigger additional job analysis and worker training. (From ACGIH (2005a). *Documentation of the TLVs and BEI with Other Worldwide Occupational Exposure Values 2005.* Cincinnati, OH: ACGIH Worldwide; ACGIH (2005b). *Hand Activity Level. 2005 Threshold Limit Values for Chemical Substances and Physical Agents and Biological Exposure Limits.* Cincinnati, OH: ACGIH, pp. 112–117. With permission.)

There is strong biomechanical support for the contribution of non-neutral wrist postures to hand–wrist–forearm MSDs (Armstrong and Chaffin, 1979; Moore et al., 1991; Armstrong et al., 1993; Moore, 2002; NRC, 1999; NRC and IOM 2001; Clark et al., 2004). In contrast, Marras and Schoenmarklin (1993) showed that angular wrist velocity and acceleration may be more important than posture. Angular velocity and acceleration are captured with the speed consideration in HAL (see Figure 41.1a). Presently, posture is deferred to professional judgment.

There is a growing body of literature concerned with organizational issues (NRC, 1999; NRC and IOM, 2001). The mechanism by which organizational factors contribute to MSDs is not clear or at least not clear enough to specify a TLV for these factors at this time.

Several studies of the TLV have been reported since it was proposed. Franzblau et al. (2005) examined the prevalence of symptoms and specific disorders among 908 workers from seven different job sites in relation to the TLV. Worker exposures were categorized as below the action limit, between TLV and the action limit, or below the action limit. Elbow/forearm tendonitis was found to be significantly related to TLV category as were all measures of carpal tunnel syndrome. Still, there was a substantial prevalence of symptoms and specific disorders below the TLV action limit. These results suggest that adherence to the TLV and action limit will reduce, but not eliminate symptoms and/or upper extremity MSDs and that a control program may still be necessary.

Gell et al. (2005) conducted a prospective study of 432 industrial and clerical workers over a period of 5.4 yr. Incident cases were defined as diagnosed with CTS in workers who had no history of CTS at the beginning of the study. There was elevated incidence of new CTS cases among workers whose jobs exceed the TLV (relative risk 1.6); however, the relationship was not statistically significant at $p < 0.05$.

Werner et al. (2005) studied upper extremity pain in a cohort of 501 active workers from four industrial and three clerical work sites for an average of 5.4 yr. Cases were defined as workers who were asymptomatic or had a discomfort score of two or less at baseline testing, but reported a discomfort score of four or above on a ten point visual analog scale at follow-up. Controls were defined as those who reported pain of two or less at follow-up. The peak finger force was close to or exceeded the TLV for 65% of cases versus only 39% of control ($p = 0.01$). The peak finger force was higher among cases (3.4 versus 2.9, $p = 0.04$). Significant relationships were also reported between upper extremity discomfort and both peak and average finger forces and both peak and average wrist postures.

These studies provide additional support for the TLV, but suggest that the action limit for a proactive control program may be set too high. Users should consider lowering it to a peak finger force of 3 for an HAL value of 1. The Werner et al. (2005) study also supports the concern about non-neutral wrist postures.

41.4 Applying the ACGIH TLV

Key points of the TLV on HAL.

- Considers repetition and force and applies to monotask hand work performed for four or more hours per day
- The term "hand activity level" or "HAL" is used to refer to repeated and sustained exertions of the hand
- Peak finger force is the 90th percentile force value exerted with the fingers
- Factors besides repetition and force, such as posture, contact stresses, vibration and psychosocial stresses, are deferred to professional judgment of the analyst
- The TLV specifies an exposure level that presents a significant risk of work-related MSDs and it should not be exceeded
- The action limit specifies exposure limits that may present a risk to some people — particularly if combined with other work or personal factors. The action limit should trigger a proactive program
- The TLV should be applied to the right and left hands separately

41.4.1 Monotask Hand Work

The ACGIH TLV is intended for monotask hand work for four or more hours per shift (ACGIH, 2005a,b). Many will think of a task as a way of organizing groups of work elements according to their purpose. In this case, monotask work implies that there is a predictable pattern of work elements throughout the work shift.

41.4.2 Hand Activity Level

HAL was adopted from Latko et al. (1997) and considers exertion frequency, recovery time, and the speed of motion. The term hand was selected to emphasize that the TLV is concerned primarily with the actions of the hand and wrist as opposed to other body parts. The term "activity" was selected over "repetition" to emphasize both dynamic and static exertions that may contribute to MSDs in the hand, wrist, and forearm. The term "level" was selected because the HAL is expressed on a relative zero to ten point scale shown in Figure 41.1(a).

Latko et al. (1997) showed that the HAL was closely related to the frequency of exertion and the recovery time in the work cycle. These data were used by the ACGIH Physical Agents committee to develop a table of HALs versus frequency and duty cycle. Table 41.1 was adapted from ACGIH (2005a). Duty cycle has been replaced with recovery time, the complement of duty cycle. Recovery time will be more familiar to those concerned with setting up production standards. Work standards use allowances that are generally expressed as a percentage of work time. A second modification of Table 41.1 includes upper and lower limits for each exertion frequency range. There are some gaps in the ACGIH table because of the range of the available data. As many readers will inevitably encounter values outside of the range of the observed data, values are suggested for extrapolation. Extrapolated values are indicated with an asterisk and are the opinion of the author and do not reflect the views of ACGIH.

TABLE 41.1 HAL can be Estimated from Exertion Frequency or Period and Work Cycle Recovery Time

Freq	Period	Recovery				
		0–20%	20–40%	40–60%	60–80%	80–100%
0.12/s (0.09–0.18)	8.0s (5.66–11.31)	6*	4*	3*	1	1
0.25/s (0.18–0.35)	4.0s (2.83–5.66)	6*	4*	3	2	2
0.5/s (0.35–0.71)	2.0s (1.41–2.83)	6	5	5	4	3
1.0/s (0.71–1.41)	1.0s (0.71–1.41)	7	6	5	5	4
2.0/s (1.41–2.83)	0.5s (0.35–0.71)	8	7	6	5	4*

*Values Extrapolated by author – not From ACGIH.
Source: Adapted from ACGIH, 2005.

The concept of an exertion is important in ratings based on observations and on calculations. An exertion is a single movement or exertion of force. Figure 41.3 shows time-based plots of finger forces and postures for two cycles of a turkey thigh boning job reported by Armstrong et al. (1982). In this case, an exertion entails positioning the knife and performing the cut. The force plots are estimated from surface EMG measurements of the forearm. It can be seen from the force plot for the right hand (Figure 41.3a), that over the course of each 7.6 sec cycle, there are approximately four peaks that correspond with each cut, plus a fifth exertion to control the knife between thighs. It can be seen from the plot of the wrist posture, in Figure 41.3c and Figure 41.3d, that these cuts also involve flexion and ulnar deviation of the wrist. The average exertion frequency for the right hand is 0.66 exertions per second, computed by dividing five exertions by 7.6 sec. The left hand is used to hold the thigh and pull the

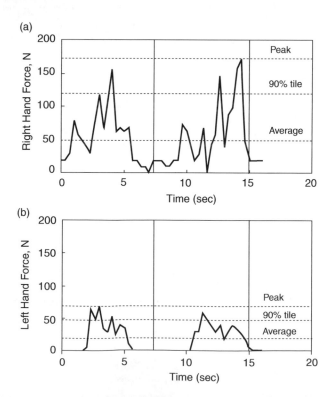

FIGURE 41.3 Time-based plots of hand force, a and b, wrist flexion/extension, c and d, and wrist ulnar/radial deviation, e and f, for the right (a, c, e) and left (b, d f) hands for two cycles of turkey thing boning. (Adapted form Armstrong T., J. Foulke, B. Joseph and S. Goldstein (1982). *Am Ind Hyg Assoc J* **43**(2): 103–116. With permission.)

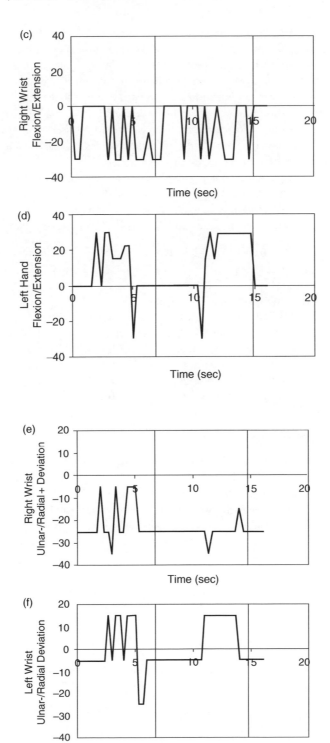

FIGURE 41.3 *Continued.*

meat away from the bone. Although the force peaks are not as conspicuous, four exertions per thigh are required for each 7.6 sec cycle. The left hand can rest between successive thighs. The average exertion frequency for the left hand is 0.5 exertions per second.

Other examples of an exertion include swinging a hammer, twisting a screwdriver, driving a threaded fastener with a power screwdriver, getting and placing a part, folding the flap of a box, ejecting the contents of a pipette, activating a control on a machine, pressing a key on a keyboard, etc. All of these examples entail a single movement or exertion of force. It is not always necessary to make a plot as shown in Figure 41.3. Exertions can be counted from detailed job descriptions used by engineers to establish production standards or from video recordings.

In addition to exertion frequency, the force plots also provide information about recovery time so that Table 41.1 can be used to estimate the HAL. There is essentially no recovery time for the right hand because of the exertion to hold the knife in between successive cuts and thighs. The recovery time can be calculated by adding up the times that the hand is at rest or estimated from job observations:

$$\text{Rest: } (2.0 - 0) + (10.6 - 5.6) = 7 \, \text{sec}$$
$$\text{Work: } 2.0 - 5.6; \ 10.6 - 15 = (5.6 - 2.0) + (15 - 10.6) = 8 \, \text{sec}$$

In this case the recovery time was found to be 47%.

Using Table 41.1, the HAL can be estimated for the right hand with 0.66 exertions per sec and 0% recovery time as 6, and for the left hand with 0.5 exertions per sec and 47% recovery time as 5.

HALs also can be estimated using the visual analog scale shown in Figure 41.1a (Latko, 1997). The HAL of six for the right hand corresponds to "slow steady motion per exertions frequent brief pauses." The HAL of five for the left hand is in between six "steady motion per exertion infrequent pauses" and four "slow steady motion per exertions frequent brief pauses."

41.4.3 Peak Finger Force

The finger force varies with time and work elements as shown in Figure 41.3a and Figure 41.3b for the turkey boning job. Peak finger force is expressed as a 90th percentile on a relative zero to ten scale (see Figure 41.1b). The upper percentiles are more sensitive to jobs with high force variations than are average values, but are not as sensitive to random values as are absolute peaks. The relative zero to ten scale adjusts for posture variations and can be assessed using the Borg (1990) scale or observer ratings (as shown in Figure 41.1b. In cases where force is measured in conventional force units, it is necessary to divide by the posture-specific strength. Peak finger force on a zero to ten scale can be calculated based on a 90th percentile finger force and posture-specific hand strength. Figure 41.4 shows frequency histograms for the force plots shown in Figure 41.3a and Figure 41.3b. Using the cumulative frequency curve the 90th finger forces are estimated for the right and left hands as 118 and 52 N, respectively. The can be expressed on a zero to ten scale by dividing them by the posture-specific strength and multiplying the result by ten:

$$\text{Peak finger force } (0-10) = 90\text{th percentile finger force (N)/strength (N)} \times 10$$

The knife and thigh in the thigh boning example are held in a power grip posture. Based on a survey of 40–45-yr-old suburban and rural females by Mathiowetz et al. (1985), average right and left hand grip strengths are estimated as 314 ± 60 and 278 ± 61 N respectively:

$$\text{Right hand: peak finger force} = 118 \, \text{N}/314 * 10 = 3.8$$
$$\text{Left hand: peak finger force} = 52 \, \text{N}/278 * 10 = 1.9$$

Strength can be adjusted from males or females, young or old workers, or higher or lower percentiles, depending on the population of interest.

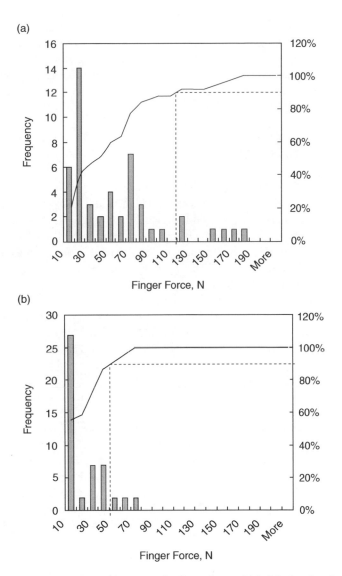

FIGURE 41.4 Frequency and cumulative histograms for finger force of (a) right hand and of (b) left hand from Figure 41.3(a) and Figure 41.3(b). The dashed lines show the 90th percentile forces.

Peak finger force can be estimated by the observer (Latko et al., 1995). Force estimates can also be obtained from the workers themselves. Marshall et al. (2004) showed that worker force ratings can be improved significantly if they are first asked to make a maximal exertion on a force gauge. The maximum force values can be compared with the relative worker ratings to estimate forces in pounds or Newtons. Bao and Silverstein (2005) described the use of force matching in which workers indicate the task force requirements by squeezing a force dynamometer in a similar position. They reported good agreement between actual and observed forces, but that clear instructions are important.

In some cases, forces can be estimated from the weight of a work object using biomechanical analysis. For example, the finger force required to lift a tool box in a hook grip posture will be equal to the weight of the tool box. The finger force required to hold a work object in a pinch posture will be equal to the weight of the object divided by two times the coefficient of friction. The finger force required to one part against another may be difficult to measure. Biomechanical analysis of these exertions are complex and beyond the scope of this discussion.

Surface electromyography (EMG) can be used to estimate hand forces in some settings (Armstrong et al., 1982; Jonsson, 1988; Matiassen and Winkel, 1991). Finally, some tasks may be suitable to the use of electronic force gauges for measurement of forces. Armstrong et al. (1994) describe the use of force gauges under a keyboard for measuring reaction forces as subjects type on a keyboard. The method selected for assessing force and repetition will depend on the desired level of quality, type of job, and available resources.

41.5   Applying the TLV

As described earlier, the HAL and peak finger force can be determined from observations using the scales provided in Figure 41.1a and Figure 41.1b. This method is particular well suited for plant walk through inspections, but ratings also may be performed from video recordings. The HAL and peak force values can be determined as described earlier using observer ratings or methods analysis for HAL. Observer ratings also can be used for peak finger force and other ergonomics stresses such average force, peak and average postures, and contact stress. Observer ratings are particularly useful for identifying jobs for follow-up analyses or during walk through surveys.

Latko (1997) describes a procedure in which a team of two or more people observed video recordings of jobs, made independent ratings, and then discussed them until they achieved a consensus of ± 0.5 for each score. This process was very time consuming and not well suited for in-plant surveys. In response, Latko et al. (1997) recommended that jobs first be rated by two or more observers. Those ratings can then be compared and discussed until agreement is achieved within one point on a ten-point scale. Furthermore, Ebersole and Armstrong (2002) described a procedure in which a pair of analysts observed and rated jobs independently on the plant floor or from videos. They then discussed their results and re-rated the jobs, but consensus was not forced.

This last procedure was used by six observers working in pairs to rate 410 auto-assembly jobs (Ebersole and Armstong, 2002). Observers first rated the jobs independently. They then compared and discussed their findings before re-rating the jobs. They were not required to achieve consensus. The raters were periodically reassigned to a new partner. One hundred percent of the HAL ratings were all within 2 points of each other on the 10-point scale. In fact, 91% of the ratings were within 1 point, which is the definition of consensus by this study. The agreement for peak finger force ratings were not as good as for HAL. Force ratings were within two points in 91.7% of the cases; 71.7% of the initial force values were within one point of each other. These agreements are acceptable for plant surveys. For some jobs, a time-based analysis of methods, forces, and postures as described by Armstrong et al. (2003) could provide useful details and increased accuracy.

41.6   Determining Compliance with the TLV

Compliance with the TLV is determined by locating the observed HAL and peak finger force in Figure 41.2 or by comparing the observed and acceptable peak finger forces for the observed HALs as described in the example described earlier. The turkey boning job in which workers used a knife to debone 3780 turkey thighs per shift was used to illustrate the concept of an exertion (Armstrong et al., 1982). The HALs for the right and left hands was estimated as six and five, respectively, from the force plots shown in Figure 41.3. The peak finger forces were estimated as four and two for the right and left hands, respectively, from the cumulative force frequency histograms shown in Figure 41.4 and average grip strengths for females in their early 40s by Mathiowetz et al. (1985). These values are located on a plot of peak finger force versus HAL with vertical and horizontal lines as shown in Figure 41.5. The lines are deliberately one HAL unit and one force unit wide to remind the user that these are ten point scales. Users should be cautious when the intersection of the lines is close to the TLV or action limit. In this case, it can be seen that the intersection of these lines for

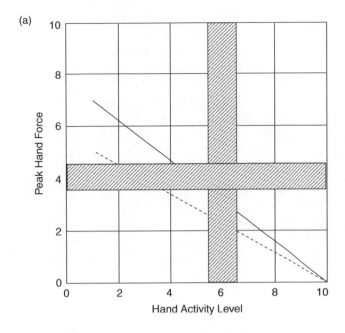

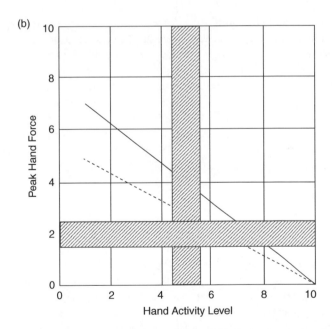

FIGURE 41.5 Applications of the TLV for the right hand and left hand for boning 3780 turkey thighs per shift. The right hand exceeds the TLV; the left hand is below the action limit.

the right hand is clearly above the TLV. In this case, the intersection of the left hand lines is below the action limit.

The TLV also can be evaluated computationally as shown previously:

$$\text{Peak finger force} \leq 70/9 - 7 \times \text{HAL}/9$$

For the right hand of the thigh boner, the acceptable finger force is

$$\text{Peak finger force} \leq 7.77 - 0.77 * 6$$
$$\text{Peak finger force} \leq 3.1$$

Since the observed force was four, the TLV is clearly exceeded by the right hand. In addition to the force and posture, it can be seen from Figure 41.3(a) and Figure 41.3(e) that there is also repeated wrist flexion and deviation. Engineering controls need to be considered to reduce the finger forces and the postural stresses on the wrist. Administrative controls also should be implemented as long as the peak finger force and HAL values are near the action limit.

Similarly, the TLV is not exceeded by the left hand, but was slightly above the action limit.

TLV
   Peak finger force $\leq 7.77 - 0.77*5$
   Peak finger force ≤ 3.8
Action limit:
   Peak finger force $\leq 5.55 - 0.55*5$
   Peak finger force ≤ 2.8

This example was based on the average strength of a female in her early 40s. Different results would be obtained for workers with different strengths. It is important that strength values representative of all of the workers be considered — male–female, young–old, and lower percentiles in each category.

Administrative controls should be continued as long as exposures are close to the action limit. Administrative controls should include educating workers about the symptoms of hand–wrist–forearm MSDs and where to go to get help if symptoms occur. Workers also should be instructed in best work practices, and how to adjust their equipment. Supervisors should be instructed about the symptoms of MSDs, best work practices, and work station setup so that they can assist workers. Controls should include surveillance of worker symptoms, injuries, and illnesses to identify both reported and unreported problems so that appropriate interventions can be implemented. Surveillance should include further analysis of the job to identify causes for high forces and posture stresses so that they can be addressed. Finally, there should be some kind of administrative structure to manage these control efforts, for example, an ergonomics team or safety committee.

41.7   Summary

The ACGIH TLV is a tool for assessing the risk of hand–wrist–forearm MSDs. The TLV applies to mono-task hand work performed for four or more hours per day. The TLV considers HAL and peak finger force, but recommends that other work factors, such as wrist posture, contact stresses, and hand–arm vibration, also be considered — particularly when exposures are close to the TLV. Also, workers should be advised to discuss personal factors, such as weight, chronic diseases, pregnancy, and past injuries, with their health-care provider. Peak finger force and HAL can be determined from observations or calculated from time-based measurements. Time-based measurements may be obtained from sequential analysis of video recordings or electronically from electromyography, force gauges, and goniometry. Exceeding the TLV should initiate an engineering control change process. Those jobs exceeding the action limit should have administrative or engineering controls applied.

References

ACGIH (2005a). *Documentation of the TLVs and BEI with Other Worldwide Occupational Exposure Values 2005*. Cincinnati, OH: ACGIH Worldwide.

ACGIH (2005b). *Hand Activity Level. 2005 Threshold Limit Values for Chemical Substances and Physical Agents and Biological Exposure Limits.* Cincinnati, OH: ACGIH, pp. 112−117.

Armstrong, T. and D. Chaffin (1979). Some biomechanical aspects of the carpal tunnel. *J Biomech*, **12**(7): 567−570.

Armstrong, T., J. Foulke, B. Joseph and S. Goldstein (1982). Investigation of cumulative trauma disorders in a poultry processing plant. *Am Ind Hyg Assoc J* **43**(2): 103−116.

Armstrong, T., L. Fine, S. Goldstein, Y. Lifshitz and B. Silverstein (1987). Ergonomics considerations in hand and wrist tendinitis. *J Hand Surg [Am]* **12**(5 Pt 2): 830−837.

Armstrong, T., P. Buckle, L. Fine, M. Hagberg, B. Jonsson, A. Kilbom, I. Kuorinka, B. Silverstein, G. Sjogaard and E. Viikari-Juntura (1993). A conceptual model for work-related neck and upper-limb musculoskeletal disorders. *Scand J Work Environ Health* **19**(2): 73−84.

Armstrong, T., J. Foulke, B. Martin, J. Gerson and D. Rempel (1994). Investigation of applied forces in alphanumeric keyboard work. *Am Ind Hyg Assoc J* **55**(1): 30−35.

Armstrong, T., W. Keyserling, D. Grieshaber, M. Ebersole and E. Lo (2003). *Time based job analysis for control of work related musculoskeletal disorders.* 15th Triennial Congress of the International Ergonomics Association, Seoul, South Korea.

Bao, S. and B. Silverstein (2005). Estimation of hand force in ergonomic job evaluations. *Ergonomics* **48**(3): 288−301.

BLS (2003). *Lost-Worktime Injuries and Illnesses: Characteristics and Resulting Time Away from Work, 2001.* Washington, DC: Bureau of Labor Statistics, United States Department of Labor.

Borg, G. (1990). Psychophysical scaling with applications in physical work and the perception of exertion. *Scand J Work Environ Health* **16**(Suppl 1): 55−58.

Bystrom, S. and C. Fransson-Hall (1994). Acceptability of intermittent handgrip contractions based on physiological response. *Hum Factors* **36**(1): 158−171.

Chiang, H.-C., Y.-C. Ko, S.-S. Chen, H.-S. Yu and T. Wu (1993). Prevalence of shoulder and upper-limb disorders among workers in the fish-processing industry. *Scand J Work Environ Health* **19**(2): 126−131.

Clark, B., T. Al-Shatti, A. Barr, M. Amin and M. Barbe (2004). Performance of a high-repetition, high-force task induces carpal tunnel syndrome in rats. *J Orthop Sports Phys Ther* **34**(5): 244−253.

Ebersole, M. and T. Armstrong (2002). *Inter-Rater Reliability for Hand Activity Level (HAL) and Force Metrics.* Baltimore, MD: Human Factor and Ergonomics Society.

Faucett, J. and D. Rempel (1994). VDT-related musculoskeletal symptoms: Interactions between work posture and psychosocial work factors. *Am J Ind Med* **26**(5): 597−612.

Fransson-Hall, C., S. Bystrom and A. Kilbom (1995). Self-reported physical exposure and musculoskeletal symptoms of the forearm-hand among automobile assembly-line workers. *J Occup Environ Med* **37**(9): 1136−1144.

Franzblau, A., T. Armstrong, R. Werner and S. Ulin (2005). A cross-sectional assessment of the ACGIH TLV for hand activity level. *J Occup Rehabil* **15**(1): 57−67.

Gell, N., R. Werner, A. Franzblau, S. Ulin and T. Armstrong (2005). A longitudinal study of industrial and clerical workers: incidence of carpal tunnel syndrome and assessment of risk factors. *J Occup Rehabil* **15**(1): 47−55.

Hammer, A. (1934). Tenosynovitis. *Med Rec* **140**: 353−355.

Jonsson, B. (1988). The static load component in muscle work. *Eur J Appl Physiol Occup Physiol* **57**(3): 305−310.

Latko, W. (1997). Development and evaluation of an observational method for quantifying exposure to hand activity and other physical stressors in manual work. Doctoral Dissertation: *Industrial and Operations Engineering.* University of Michigan, Ann Arbor, MI, 188.

Latko, W., T. Armstrong, A. Franzblau and S. Ulin (1995). *Comparison of three methods for assessing repetition in manual work.* Montreal: PREMUS.

Latko, W., T. Armstrong, J. Foulke, G. Herrin, R. Rabourn and S. Ulin (1997). Development and evaluation of an observational method for assessing repetition in hand tasks. *Am Ind Hyg Assoc J* **58**(4): 278–285.

Latko, W., T. Armstrong, A. Franzblau, S. Ulin, R. Werner and J. Albers (1999). Cross-sectional study of the relationship between repetitive work and the prevalence of upper limb musculoskeletal disorders. *Am J Ind Med* **36**(2): 248–259.

Leclerc, A., P. Franchi, M. Cristofari, B. Delemotte, P. Mereau, C. Teyssier-Cotte and A. Touranchet (1998). Carpal tunnel syndrome and work organisation in repetitive work: a cross sectional study in France. Study Group on Repetitive Work. *Occup Environ Med* **55**(3): 180–187.

Luopajarvi, T., I. Kuorinka, M. Virolainen and M. Holmberg (1979). Prevalence of tenosynovitis and other injuries of the upper extremities in repetitive work. *Scand J Work Environ Health* **5** (Suppl 3): 48–55.

Marshall, M., T. Armstrong and M. Ebersole (2004). Verbal estimation of peak exertion intensity. *Human Factors and Ergonomics* **46**(4): 697–710.

Marras, W. and R. Schoenmarklin (1993). Wrist motions in industry. *Ergonomics* **36**(4): 341–351.

Mathiassen, S. and J. Winkel (1991). Quantifying variation in physical load using exposure-vs-time data. *Ergonomics* **34**(12): 1455–1468.

Mathiowetz, V., N. Kashman, G. Volland, K. Weber, M. Dowe and S. Rogers (1985). Grip and pinch strength: normative data for adults. *Arch Phys Med Rehabil* **66**(2): 69–74.

Moore, J. (2002). Biomechanical models for the pathogenesis of specific distal upper extremity disorders. *Am J Ind Med* **41**(5): 353–369.

Moore, A., R. Wells and D. Ranney (1991). Quantifying exposure in occupational manual task with cumulative trauma disorder potential. *Ergonomics* **34**(12): 1433–1453.

NRC (1999). *Work-Related Musculoskeletal Disorders: A Review Of The Evidence.* Washington: DC: 4 National Academy Press.

NRC and IOM (2001). *Musculoskeletal Disorders and the Workplace: Low Back and Upper Extremities.* Washington, DC: National Academy Press.

Obolenskaja, A. and I. Goljanitzki (1927). Die seröse tendovaginitis in der Klinic und im Experiment. *Dtsch Z Chir Leipz* **201**: 388–399.

Radwin, R. and M. Lin (1993). An analytical method for characterizing repetitive motion and postural stress using spectral analysis. *Ergonomics* **36**(4): 379–389.

Rohmert, W. (1973). Problems in determining rest allowances. *Appl Ergon* **4**(2): 91–95.

Roquelaure, Y., S. Mechali, C. Dano, S. Fanello, F. Benetti, D. Bureau, J. Mariel, Y.H. Martin, F. Derriennic and D. Penneau-Fontbonne (1997). Occupational and personal risk factors for carpal tunnel syndrome in industrial workers. *Scand J Work Environ Health* **23**(5): 364–369.

Silverstein, B. (1985). The prevalence of upper extremity cumulative trauma disorders in industry. Doctoral Disseration: *Department of Epidemiology,* The University of Michigan, Ann Arbor, MI, p. 198.

Silverstein, B., L. Fine and T. Armstrong (1987). Occupational factors and carpal tunnel syndrome. *Am J Ind Med* **11**(3): 343–358.

Werner, R.A., A. Franzblau, N. Gell, S.S. Ulin and T.J. Armstrong (2005). Predictors of upper extremity discomfort: a longitudinal study of industrial and clerical workers. *J Occup Rehabil* **15**(1): 27–35.

42

REBA and RULA: Whole Body and Upper Limb Rapid Assessment Tools

Sue Hignett
Loughborough University

Lynn McAtamney
National Occupational Health and Safety Commission

42.1 Introduction

REBA (Rapid Entire Body Assessment) and RULA (Rapid Upper Limb Assessment) provide a quick analysis of the demands on a person's musculoskeletal system when performing a specific task. Both tools are required to be used as part of a full ergonomic workplace assessment and have proved popular in providing a simple, visual indication as to the level of risk and need for action associated with the task. REBA was developed to provide a quick and easy observational postural analysis tool for whole body activities (static and dynamic) giving a musculoskeletal risk action level (Hignett and McAtamney, 2000). RULA was developed earlier (McAtamney and Corlett, 1993) to provide a rapid objective measure of musculoskeletal risk caused by mainly sedentary tasks where upper body demands were high. Both tools use body part diagrams to assist with the coding of joint angles and body postures, with additional coding for load/force, coupling and muscle activity. They both produce risk level scores on a given scale to indicate whether the risk is negligible through to very high.

There are several postural analysis tools available for ergonomic and occupational health practitioners, many of which are included in this publication. In order to differentiate between the tools, consider the following questions:

1. Task
   a. Which area of the body is being assessed, for example, whole body or upper limb?

  b. Does the activity include static and dynamic postures?

 2. Sensitivity and generality

  a. How detailed will the assessment be?

  b. Will the same postural analysis tool be used for a range of tasks in several industrial settings?

When choosing between REBA and RULA consider the task demands and type of assessment required as suggested in Table 42.1. RULA is generally used if the person is sitting, standing still or in an otherwise sedentary position and mainly using the upper body and arms to work, for all other tasks REBA should be used. In all applications of REBA and RULA, users should receive training or be confidently skilled in the tool before using it although no previous ergonomic skills are required.

REBA was initially designed to provide a pen-and-paper postural analysis tool that could either be used in the field by direct observation or with still/video photographs. It has been further developed and there are now simple computer programmes available which support the coding and analysis (Janik et al., 2002). As a pen-and-paper tool, it was designed to have wider application than more complex postural analysis tools (e.g., NIOSH, Waters et al., 1993) and so was developed using examples from electricity, health care, and manufacturing industries. However, it also has more sensitivity and anatomical (body part) detail than other postural analysis tools (e.g., OWAS, Karhu et al., 1977).

RULA was developed to provide postural analysis where work placed physical demands on the trunk, neck, and upper limbs in particular and therefore in tasks where work-related upper limb disorders, cumulative trauma disorders, and similar problems are a concern. RULA assesses the posture, force and movement associated with sedentary tasks such tasks include screen based or computer tasks, manufacturing or retail tasks where the worker is seated or standing without moving about. The main applications of RULA are to measure the musculoskeletal risk, usually as part of a broader ergonomic investigation and then:

 1. Compare the effects of a current and modified workstation designs

 2. Evaluate outcomes such as productivity or suitability of equipment

 3. Educate workers about musculoskeletal risk created by different working postures

RULA has been used extensively in studies including:

- Manufacturing where Gutierrez (1998) evaluated assembly improvements in postures of electronics factory workers using redesigned workstations
- Computer equipment assessments (Hedge et al., 1995)
- Office-based tasks. Leuder (1996) modified RULA (http://www.humanics-es.com/rulacite.htm) to assess broader risks associated with office-based tasks including glare on the computer screen (see http://www.humanics-es.com/files/rula.pdf). Whilst the modified tool has not been validated, it provides useful information on workstation risks
- Cost benefits of improving workstation design. Axelsson (1997) found a correlation between high RULA scores and a higher proportion of products that were discarded as defective at that workstation. As part of their macro ergonomic management program the subsequent improvements to the identified high-risk workstations produced a 39% drop in quality deficiencies representing a cost saving of $25,000 per year.

A web-based RULA assessment tool is available free of charge at http://www.ergonomics.co.uk.

TABLE 42.1 Choosing between REBA and RULA

Task Demands	Overview Assessment	Detailed Assessment
Whole of body activity (static or dynamic)	Use REBA	Specific assessment tools (REBA useful as overview)
Mainly upper body activity	Use RULA	Specific assessment tools (RULA useful as overview)

42.2 REBA: The Postures

The baseline posture is the functional anatomically neutral posture as defined by the American Academy of Orthopedic Surgeons (1965). Increasing scores are allocated as the posture moves away from the neutral position based on biomechanical studies from the literature. The body parts are grouped into:

Group A: trunk, neck, and legs (Figure 42.1 and Table 42.2)
Group B: upper arms, lower arms, and wrists (Figure 42.2 and Table 42.3)

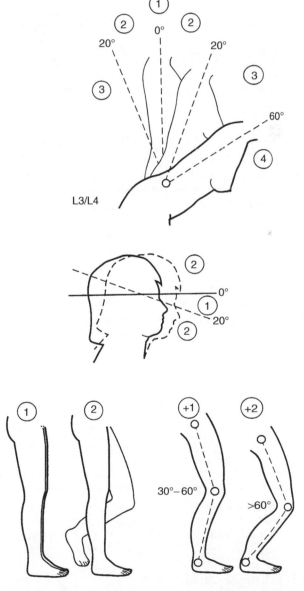

FIGURE 42.1 Group A postures. (From Hignett, S. 1998. In: Pitt-Brooke, J., Raid, H., Lockwood, J., and Kerr K., (Eds.), *Rehabilitation of Movement. Theoretical Basis of Clinical Practice.* London: W.B. Saunders Company Ltd. Chapter 13, pp. 480–486.)

TABLE 42.2 Group A Definitions

Movement	Score	Change score
Trunk Postures		
Upright	1	+1 if twisting of side flexed
0–20° flexion 0–20° extension	2	
20–60° flexion >20° extension	3	
>60 flexion	4	
Neck Postures		
0–20° flexion	1	+1 if twisting or side flexed
>20° flexion or extension	2	
Leg Postures		
Position		
Bilateral weight-bearing, walking, or sitting	1	+1 if knee(s) between 30 and 60° flexion
Unilateral weight-bearing, feather weight-bearing, or an unstable posture	2	+2 if knee(s) >60° flexion (N.B. not for sitting)

The postures are scored by observing the task by video, photograph or in real time and allocating scores for the body parts (Group A and Group B) load/force, coupling, and activity as shown in Figure 42.3. These data are recorded on the Score Sheet (Figure 42.4). The choice of right or left arm is usually driven by availability (what can be observed); however, it is also possible to score both sides and then choose the highest score to take forward in the score sheet (see examples in Figure 42.5 and Figure 42.6). The scores are then transformed via Table A (Table 42.4) and Table B (Table 42.5) into SCORE A and SCORE B. At this stage, the additional scores are added for load/force (Table 42.4), with an additional score for shock or rapid build-up of force (e.g., catching a load), and coupling (Table 42.5). The coupling score uses four levels (good, fail, poor, and unacceptable) to give an indication of the interface between the person and the load and allows for both manual and other body region interfaces.

SCORE A and SCORE B are then entered onto Table C (Table 42.6) to produce SCORE C. At this stage, the Activity Score (Table 42.7) is added to give additional scores for:

- One or more body parts are static
- Repeated small range actions
- Large range changes in postures or unstable base

This gives a final REBA SCORE, which is then interpreted into an Action Level using Table 42.7. The five action levels give an indication of the urgency of avoiding or reducing the risk of the assessed posture.

42.3 Examples

Two examples are given in Figure 42.5 and Figure 42.6. The scoring rationale is outlined in the following sections. Both of these examples were part of an initial ergonomic workplace assessment and have been subsequently changed. The medical notes area was relocated in a specially designed building, with appropriate height racking and safety stepladders. The pediatric cot was redesigned with a U.K. manufacturer and has been replaced with an electric cot with redesigned cot sides.

42.3.1 Example 1: Filing Medical Notes

Group A

Trunk = 2 (0°–20° extension; twisting/side flexion more than 20°)
Neck = 2 (greater than 20° extension; twisting/side flexion less than 20°)

Legs = 1 (bilateral weight-bearing, with knees extended (no flexion))
Load/Force = 0 (less than 5 kg, with no shock or rapid build-up of force)

Group B

Upper Arms = 4 + 1 + 1 (greater than 90° flexion with medial rotation and a raised shoulder)
Lower arms = 2 (less than 60° flexion)
Wrist = 2 (greater than 15° extension, with no deviation/twist)
Coupling = 2 (poor, not acceptable although possible)
SCORE A = 3
SCORE B = 11 (9 + 2)

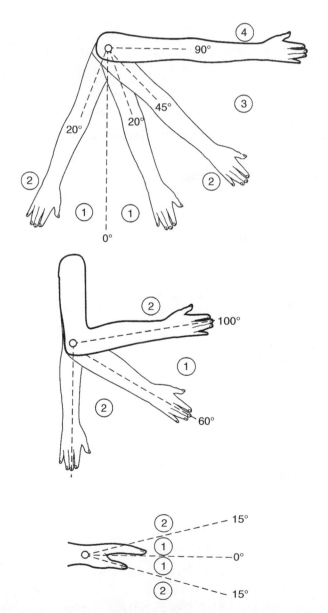

FIGURE 42.2 Group B postures. (From Hignett, S. 1998. In: Pitt-Brooke, J., Raid, H., Lockwood, J., and Kerr K., (Eds.), *Rehabilitation of Movement. Theoretical Basis of Clinical Practice.* London: W.B. Saunders Company Ltd. Chapter 13, pp. 480–486.)

TABLE 42.3   Group B Definitions

Position	Score	Change score
Upper Arms		
20°extension to 20° flexion	1	+1 if arm is: abducted and/or rotated
>20° extension 20–45° flexion	2	+1 if shoulder is raised
45–90° flexion	3	−1 if leaning, supporting weight of arm or if posture
>90° flexion	4	is gravity-assisted
Lower Arms		
60°–100° flexion	1	
<60° flexion	2	
> 100° flexion		
Wrist		
0°–15° flexion/extension	1	+1 if wrist is deviated or twisted
>15° flexion/extension	2	

SCORE C = 8

Activity Score = +1 (unstable base)

REBA SCORE = 9 (Action level 3. High risk level, action is necessary soon)

42.3.2   Example 2: Raising a Cot Side

Group A

Trunk = 4+1 (greater than 60° flexion with twisting/side flexion greater than 20°)

Neck = 2+1 (greater than 20° extension with twisting greater than 20°)

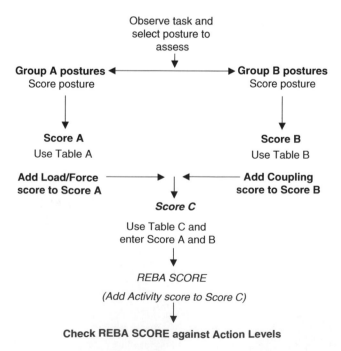

FIGURE 42.3   Using REBA.

Legs = 1+1 (bilateral weight bearing with knees flexed between 30 and 60°)
Load/Force = 1 (less than 5 kg force exerted but with a rapid build-up of force (jerk))

Group B

Upper arms = 3 + 1−1 (between 45 and 90° flexion with medial rotation and a gravity-assisted posture
        (−1))
Lower arms = 2 (less than 60° flexion)
Wrist = 1 (between 0 and 15° extension, with no deviation/twist)
Coupling = 1 (fair; hand hold acceptable, but not ideal)
SCORE A = 9 (8 + 1)
SCORE B = 5 (4 + 1)
SCORE C = 10

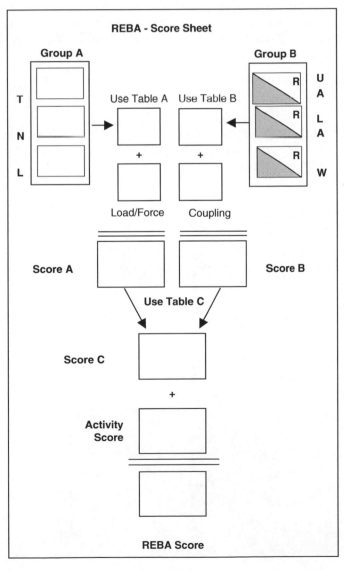

FIGURE 42.4 Score Sheet.

Activity Score = +1 (rapid large change in posture)
REBA SCORE = 11. Action level 4. Very high risk level, action is necessary NOW

42.4 Reliability and Validity

42.4.1 Reliability

In the development stages of REBA, the reliability was tested by using over 600 examples of postures from the electricity, health care, and manufacturing industries. The examples were coded by 14 professionals (occupational therapists, physiotherapists, nurses, and ergonomists). The reliability of the upper arm posture score (56%) was excluded due to the addition of the gravity-assisted code during the reliability testing. The results were analyzed by body part (Figure 42.7) with an agreement rate of between 62 and 85%.

This was felt to be a satisfactory interrater reliability for this stage of development with such a large group. In comparison, Suurnäkki et al. (1988) reported achieving a reliability range of 74–99% with OWAS with only six experts (Corlett, 1998). Hignett (1998) reported a 96% agreement with only two participants in an interrater reliability study.

42.4.2 Validity

REBA has very good face validity and is widely used internationally. External validity has been achieved through use in a range of industries and it considered to present generalizable results within the context of a full ergonomics workplace assessment.

Filling Medical Notes

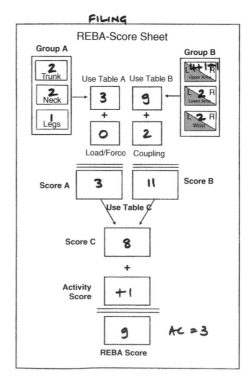

FIGURE 42.5 Filing medical notes.

Raising a Paediatric Cot Side

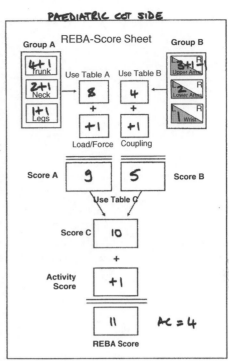

FIGURE 42.6 Raising a pediatric cot side.

42.5 Using RULA

It is recommended that RULA is used as part of an ergonomic investigation and that users of the tool have undertaken some training to become familiar with the tool first.

Using RULA takes three steps.

Step 1: The posture or postures for assessment are selected
Step 2: The postures are scored using the scoring sheet, body part diagrams
Step 3: The scores are put through tables and converted to a score of one to seven which are clustered into four action levels

Table A and Load/Force

Table A

Trunk	Neck 1				Neck 2				Neck 3			
	Legs											
	1	2	3	4	1	2	3	4	1	2	3	4
1	1	2	3	4	1	2	3	4	3	3	5	6
2	2	3	4	5	3	4	5	6	4	5	6	7
3	2	4	5	6	4	5	6	7	5	6	7	8
4	3	5	6	7	5	6	7	8	6	7	8	9
5	4	6	7	8	6	7	8	9	7	8	9	9

Load/Force			
0	1	2	+1
<5 kg	5–10 kg	>10 kg	Shock or rapid build up of force

TABLE 42.5 Table B and Coupling

Table B

Upper Arm	Lower Arm 1			Lower Arm 2		
	Wrist					
	1	2	3	1	2	3
1	1	2	2	1	2	3
2	1	2	3	2	3	4
3	3	4	5	4	5	5
4	4	5	5	5	6	7
5	6	7	8	7	8	8
6	7	8	8	8	9	9

Coupling

0	1	2	3
Good	Fair	Poor	Unacceptable
Well-fitting handle and a mid-range power grip	Hand hold acceptable but not ideal or coupling is acceptable via another part of the body	Hand hold not acceptable although possible	Awkward, unsafe grip, no handles Coupling is unacceptable using other parts of the body

42.5.1   Step 1 — Observing and Selecting the Postures to Assess

A RULA assessment is taken as a snapshot in the work cycle. The user is required to observe and make a judgement as to the posture for assessment. This may be the most frequent, most sustained or worst posture depending on the purpose of the assessment. It can be useful to estimate the proportion of time spent in the various postures being evaluated (McAtamney and Corlett, 1993).

42.5.2   Step 2 — Scoring and Recording the Posture

Assess left, right, or both sides using the scoring sheet. Score the posture of each body part using the free software found at http://www.ergonomics.co.uk/Rula/Ergo/index.html or the paper version

TABLE 42.6 Table C and Activity Score

Table C

Score A	Score B											
	1	2	3	4	5	6	7	8	9	10	11	12
1	1	1	1	2	3	3	4	5	6	7	7	7
2	1	2	2	3	4	4	5	6	6	7	7	8
3	2	3	3	3	4	5	6	7	7	8	8	8
4	3	4	4	4	5	6	7	8	8	9	9	9
5	4	4	4	5	6	7	8	8	9	9	9	9
6	6	6	6	7	8	8	9	9	10	10	10	10
7	7	7	7	8	9	9	9	10	10	11	11	11
8	8	8	8	9	10	10	10	10	10	11	11	11
9	9	9	9	10	10	10	11	11	11	12	12	12
10	10	10	10	11	11	11	11	12	12	12	12	12
11	11	11	11	11	12	12	12	12	12	12	12	12
12	12	12	12	12	12	12	12	12	12	12	12	12

Activity Score

+1	1 or more body parts are static, e.g., held for longer 1 min
+1	Repeated small range actions. e.g., repeated more than four times per minute (not including walking)
+1	Action causes rapid large changes in posture or an unstable base

TABLE 42.7   Action Levels

Action Level	REBA Score	Risk Level	Action (Including Further Assessment)
0	1	Negligible	None necessary
1	2–3	Low	May be necessary
2	4–7	Medium	Necessary
3	8–10	High	Necessary soon
4	11–15	Very high	Necessary NOW

(McAtamney and Corlett, 1993) that is on the web at http://ergo.human.cornell.edu/ahRULA.html. Use the diagrams to score the posture for each body part, along with the forces/loads and the muscle use required for that particular posture.

42.5.3   Step 3 — Action Level

Follow the score sheet to calculate the posture scores for Groups A and B if using the paper version (the software version does this for you). Use the calculation button on the software or use Table C to calculate the Grand Score. The grand score can be compared to the Action Level List.

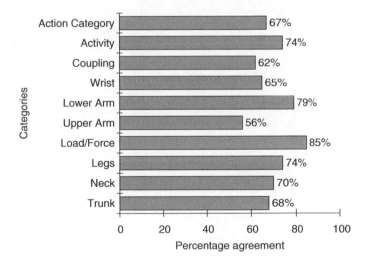

FIGURE 42.7   Percentage agreement for REBA codes.

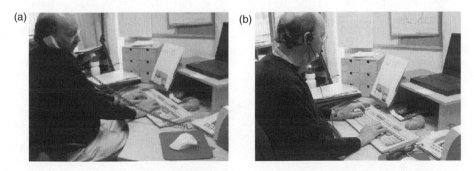

FIGURE 42.8   RULA example. (a) Telephone held on shoulder. RULA Score of 5, (b) Headset used. RULA Score of 1.

42.6 Example

A habitual posture in organizations where headsets are not used is holding a phone on the shoulder. The nerves and blood vessels in the side of the neck can be placed under high loading and there is a direct impact on shoulder/neck discomfort after prolonged exposure. The RULA score is five when holding a phone as shown in Figure 42.8. This changes to a RULA score one simply with the provision of a headset.

References

American Academy of Orthopedic Surgeons. 1965. *Joint Motion: Method of Measuring and Recording.* Edinburgh: Churchill Livingstone.

Axelsson, J.R.C. 1997. RULA in action: enhancing participation and continuous improvements. In: Seppala, P., Luopajarvi T., Nygard, C.H., and Mattila, M. (Eds.) *From Experience to Innovation — IEA '97. Proceedings of the 13th Triennial Congress of the International Ergonomics Association,* Tampere, Finland, June 29–July 4, 1997. Helsinki: Finnish Institute of Occupational Health, Volume 4, 251–253.

Corlett, E.N. 1998. Personal communication.

Gutierrez, A.M.J.A. 1998. A workstation design for a Philippine semiconductor. In: Bishu, R., Karwowski, W., and Goonetilleke, R.S. (Eds.), *ERGON-AXIA '98. Proceedings of the First World Congress on Ergonomics for Global Quality and Productivity,* Clear Water Bay, Hong Kong, July 8–11, 1998, Hong Kong: Hong Kong University of Science and Technology, pp. 133–136

Hignett, S. 1998. Ergonomics. In: Pitt-Brooke, J., Reid, H., Lockwood J., and Kerr, K. (Eds.) *Rehabilitation of Movement. Theoretical Basis of Clinical Practice.* London: W.B Saunders Company Ltd, Chapter 13, pp. 480–486.

Hignett, S. and McAtamney, L. 2000. Rapid Entire Body Assessment (REBA). *Applied Ergonomics,* 31, 201–205

Janik, H., Münzbergen, E., and Schultz, K. 2002. REBA-verfahren (Rapid Entire Body Assessment) auf einem Pocket Computer. *Proceedings of 42. Jahrestagung der Deutschen Gesellschaf für Arbeitsmedizin un Umweltmedizi*n e. V. (DGAUM). April 10–13, 2002, Munich, V25.

Karhu, O., Kansi, P., and Kuorinka, I. 1977. Correcting working postures in industry: a practical method for analysis. *Applied Ergonomics,* 8(4), 199–201.

Leuder, R. 1996. A proposed RULA for computer users. *Proceedings of the Ergonomics Summer Workshop,* U.C. Berkeley, August 8–9.

McAtamney, L. and Corlett, E.N. 1993. RULA: a survey method for the investigation of work-related upper limb disorders. *Applied Ergonomics,* 24, 91–99.

Suurnäkki, T., Louhevaara, V., Karhu, O., Kuorinka, I., Kansi, P., and Peuraniemi, A. 1988. Standardised observation method for the assessment of working postures: the OWAS method. In: Adams, A.S., Hall, R.R., McPhee, B.J., and Oxenburgh, M.S. (Eds.), *Designing a Better World. Ergonomics International '88. Proceedings of the 10th Congress of the International Ergonomics Association,* August 1–5, 1988, Sydney, Australia.

Waters, T.R., Putz-Anderson, V., Garg, A., and Fine, L.J. 1993. Revised NIOSH equation for the design and evaluation of manual lifting tasks. *Ergonomics,* 36(7), 749–776.

43

An Assessment Technique for Postural Loading on the Upper Body (LUBA)

Dohyung Kee
Keimyung University

Waldemar Karwowski
University of Louisville

43.1 Posture Classification Scheme

The relationship between awkward working postures and the risk of musculoskeletal disorders (MSDs) has been widely studied in the past. For example, van Wely (1969) discussed the relation between inadequate work postures and probable sites of pain, and Armstrong et al. (1993) reported a comprehensive review of epidemiological studies examining the relationship between work postures and (MSDs). Heinsalmi (1986) and Burdof et al. (1991) pointed out that a significant relationship was found between poor working postures and musculoskeletal-related lost workdays or low-back disorders. Bhatnager et al. (1985) indicated that a working posture affects postural discomfort and inspection performance for printed board reproductions.

Awkward, extreme, and repetitive postures can increase the risk of musculoskeletal disorders. Therefore, cost-effective quantification of the magnitude for physical exposure to poor working postures is important and needed, if the potential for injury as a result of postures is to be reduced (Andrew et al., 1998). Since development of the posturegram, a technique for numerically defining a posture proposed by Priel (1974), various postural classification methods have been developed to identify and quantify postural stress during work. These schemes can be classified into two basic categories depending upon the methods used for quantifying postural stresses: instrument-based and observational techniques. The latter is more widespread in industry, because it does not interfere with the worker during observations, and does not require use of expensive equipment for estimating the angular deviation of a body from the neutral position (Genaidy et al., 1994). The instrument-based method using bioinstrumentation such as electromyography has been rarely used as a major tool for quantifying postural stresses in industry, because it is expensive, obtrusive, and limited due to the nature of the production process in industrial sites. Most postural classification schemes developed are the observation methods. These include the Posture Targeting (Corlett et al., 1979), OWAS (Karhu et al., 1977), PATH (Buchholz et al., 1996), and RULA (McAtamney and Corlett, 1993).

Depending upon the grouping methods for joint motions involved in a classified posture, Genaidy et al. (1994) categorized the postural classification approaches used in observational techniques into: macropostural, micropostural, and postural–work activity classifications. The macropostural classification groups more than one non-neutral posture around a joint into one category, while the macropostural classification is more detailed than the previous method. The postural–work activity classification combines postures and work activities (Genaidy et al., 1994).

Although the existing methods have proved useful for quantification of postural stresses in field studies, and contributed to preventing work-related MSDs, they have many disadvantages. First, many of the observational classification schemes are not based on experimental data. Second, the existing methods have been developed for specific application purposes, and, consequently, are not generic in many respects. Third, many methods deal with only a few representative joint motions, as they focus on specific joint motions frequently linked to MSDs. Another problem is that only a few schemes (including OWAS and RULA) utilize specific evaluation criteria for the classified postures, which provide information on any corrective actions to be undertaken for reducing postural burden at work. In addition, the evaluation criteria provided by RULA and OWAS were not based on experimental results, but rather relied on the rankings provided by ergonomists and occupational physiotherapists using biomechanical and muscle function criteria (McAtamney and Corlett, 1993), or the subjective rankings provided by experienced steel workers (Karhu et al., 1977), respectively.

43.2 Objectives

To complement the restrictions of the existing methods, this chapter presents an observational and macropostural technique for postural loading on the upper body assessment (LUBA). The method is based on experimental data for perceived discomfort, expressed as numerical ratio scores for a set of joint motions, including the hand, arm, neck, and back. This technique is applied to the seated posture or standing posture with the lower limb well supported in an evenly balanced posture.

Perceived discomforts were gathered for varying postures of five joints in the upper body (Table 43.1), which included almost every possible joint motion occurring in the sitting and standing postures. Perceived discomforts were measured at five levels of range of joint motion (ROM) in each motion: 0 (neutral), 25, 50, 75, and 100% of ROM, respectively. The magnitude estimation was adopted for measuring discomfort scores. It has the advantage of providing data with the characteristics of the interval or ratio scale that can be applied to quantitative statistical techniques.

43.3 LUBA

43.3.1 Relative Discomfort Scores by Joint Motions

Each joint motion class was assigned a numerical relative discomfort score on the basis of discomfort value for the neutral position of elbow flexion, which was the least stressful of all the joint motions investigated. A relative discomfort score of 1.0 was assigned to the neutral position of elbow flexion, with a higher number indicating that a given joint motion class is more stressful. The relative discomfort score is the ratio scale that could be applied to four arithmetic rules such as addition, subtraction, multiplication, and division. In other words, the back flexion of $>60°$ with the relative discomfort score of 10 has ten times the discomfort magnitude of the elbow flexion of 0 to 45°, the relative discomfort score of

TABLE 43.1   Joint Motions Measured in this Study

	Joint Motion	
Joint	Sitting Posture	Standing Posture
Wrist	Flexion	Flexion
	Extension	Extension
	Radial deviation	Radial deviation
	Ulnar deviation	Ulnar deviation
Elbow	Flexion	Flexion
	Supination	Supination
	Pronation	Pronation
Shoulder	Flexion	Flexion
	Extension	Extension
	Adduction	Adduction
	Abduction	Abduction
	Medial rotation	Medial rotation
	Lateral rotation	Lateral rotation
Neck	Flexion	Flexion
	Extension	Extension
	Rotation	Rotation
	Lateral bending	Lateral bending
Lower back	Flexion	Flexion
	Rotation	Extension
	nm	Rotation
	Lateral bending	Lateral bending

nm: not measured.

which is 1.0. The developed relative discomfort score scheme is presented in Table 43.2 through Table 43.6, which are classified by the joints involved in motions.

As shown in Table 43.2 through Table 43.6, the relative discomfort scores are almost identical for both sitting and standing postures, and increase drastically when the joints approach the limit of their ROM. The results also showed that back movements were perceived as more stressful than any other joint motion. Specifically, the back extension of $>30°$ (with a relative discomfort score of 15) in the standing posture was the most stressful of all the joint motions examined.

TABLE 43.2   Postural Classification Scheme for the Wrist

	Posture and Discomfort Score			
	Sitting Posture		Standing Posture	
Joint Motions	Class	Relative Discomfort Score	Class	Relative Discomfort Score
Flexion	0–20°	1	0–20°	1
	20–60°	2	20–60°	2
	>60°	5	>60°	5
Extension	0–20°	1	0–20°	1
	20–45°	2	20–45°	2
	>45°	7	>45°	7
Radial deviation	0–10°	1	0–10°	1
	10–30°	3	10–30°	3
	>30°	7	>30°	7
Ulnar deviation	0–10°	1	0–10°	1
	10–20°	3	10–20°	3
	>20°	6	>20°	6

TABLE 43.3   Postural Classification Scheme for the Elbow

| | Posture and Discomfort Score | | | |
| | Sitting Posture | | Standing Posture | |
Joint Motions	Class	Relative Discomfort Score	Class	Relative Discomfort Score
Flexion	0–45°	1	0–45°	1
	45–120°	2	45–120°	3
	>120°	5	>120°	5
Pronation	0–70°	2	0–70°	2
	>70°	7	>70°	7
Supination	0–90°	2	0–90°	2
	>90°	7	>90°	7

43.3.2   Posture Evaluation Procedures

Procedures used for application of the postural classification scheme consisted of five steps: first, the operator was videotaped in order to record working postures during several work cycles. This was done for selecting of the tasks and postures to be assessed in the next step. Typically, a camera should be positioned at an angle to the operator so that three-dimensional working postures can be identified during playback. Several working cycles should be recorded because postures can vary from cycle to cycle depending upon the nature and demands of the job (Keyserling, 1986), and many of the unusual conditions that may constitute main hazards should be included (Kemmlert, 1995). Second, target postures of the recorded job were chosen for assessment, based on the posture holding time or possible postural stresses. The postures to be assessed may be those that are held for the greatest amount of the work cycle,

TABLE 43.4   Postural Classification Scheme for the Shoulder

| | Posture and Discomfort Score | | | |
| | Sitting Posture | | Standing Posture | |
Joint Motions	Class	Relative Discomfort Score	Class	Relative Discomfort Score
Flexion	0–45°	1	0–45°	1
	45–90°	3	45–90°	3
	90–150°	6	90–150°	6
	>150°	11	>150°	11
Extension	0–20°	1	0–20°	1
	20–45°	4	20–45°	3
	45–60°	9	45–60°	6
	>60°	13	>60°	10
Adduction	0–10°	1	0–10°	1
	10–30°	2	10–30°	2
	>30°	8	>30°	8
Abduction	0–30°	1	0–30°	1
	30–90°	3	30–90°	3
	>90°	10	>90°	7
Medial rotation	0–30°	1	0–30°	1
	30–90°	2	30–90°	2
	>90°	7	>90°	5
Lateral rotation	0–10°	1	0–10°	1
	10–30°	3	10–30°	2
	>30°	7	>30°	5

TABLE 43.5   Postural Classification Scheme for the Neck

| | Posture and Discomfort Score | | | |
| | Sitting Posture | | Standing Posture | |
Joint Motions	Class	Relative Discomfort Score	Class	Relative Discomfort Score
Flexion	0–20°	1	0–20°	1
	20–45°	3	20–45°	3
	>45°	5	>45°	5
Extension	0–30°	1	0–30°	1
	30–60°	6	30–60°	4
	>60°	12	>60°	9
Lateral bending	0–30°	1	0–30°	1
	30–45°	3	30–45°	2
	>45°	10	>45°	7
Ulnar deviation	0–30°	1	0–30°	1
	30–60°	2	30–60°	2
	>60°	8	>60°	8

or those that the worker itself (or the observer) considers as stressful to the muculoskeletal system. Third, each joint motion observed in the selected postures is assigned a relative discomfort score according to the above classification scheme. It is recommended that only the right or left of the upper body, which seems to be more stressful of two body sides, be assessed at a time.

Figure 43.1 makes it easy and fast for the posture analysts to classify working postures, where they just tick the corresponding item for each joint motion. After completion of posture classification, the postural load for the selected posture can be obtained by summing up the respective discomfort score values ticked in Figure 43.1. Fourth, the following equation is used to calculate the postural load index for joint motions deviated from their neutral positions in the chosen postures, that is, for joint motions

TABLE 43.6   Postural Classification Scheme for the Back

| | Posture and Discomfort Score | | | |
| | Sitting Posture | | Standing Posture | |
Joint Motions	Class	Relative Discomfort Score	Class	Relative Discomfort Score
Flexion	0–20°	1	0–30°	1
	20–60°	3	30–60°	3
	>60°	10	60–90°	6
			>90°	12
Extension	nm	nm	0–10°	1
			10–20°	4
			20–30°	8
			>30°	15
Lateral bending	0–10°	1	0–10°	1
	10–20°	3	10–20°	4
	20–30°	9	20–30°	9
	>30°	13	>30°	13
Rotation	0–20°	1	0–20°	1
	20–30°	2	20–60°	3
	30–45°	7	>60°	10
	>45°	11		

nm: not measured.

Department:			Task:		Operator:		
Analyst name:					Date:		
Joint	Motion	Class	Score	Motion	Class	Score	
Wrist	Flexion	0–20°	1 ___	Extension	0–20°	1 ___	
		20–60°	2 ___		20–45°	2 ___	
		>60°	5 ___		>45°	7 ___	
	Radial	0–10°	1 ___	Ulnar	0–10°	1 ___	
	deviation	10–30°	3 ___	deviation	10–20°	3 ___	
		>30°	6 ___		>20°	7 ___	
Elbow	Flexion	0–45°	1 ___	Supination	0–90°	2 ___	
		45–120°	2 _v_		>90°	7 ___	
		>120°	5 ___				
	Pronation	0–70°	2 ___				
	deviation	>70°	7 ___				
Shoulder	Flexion	0–45°	1 ___	Extension	0–20°	1 ___	
		45–90°	3 _v_		20–45°	4 ___	
		90–150°	6 ___		45–60°	9 ___	
		>150°	11 ___		>60°	13 ___	
	Adduction	0–10°	1 ___	Abduction	0–30°	1 ___	
		10–30°	2 _v_		30–90°	3 ___	
		>30°	8 ___		>90°	10 ___	
	Medial	0–30°	1 ___	Lateral	0–10°	1 ___	
	rotation	30–90°	2 ___	rotation	10–30°	3 ___	
		>90°	7 ___		>30°	7 ___	
Neck	Flexion	0–20°	1 ___	Extension	0–30°	1 ___	
		20–45°	3 ___		30–60°	6 ___	
		>45°	5 ___		>60°	12 ___	
	Lateral	0–30°	1 ___	Rotation	0–30°	1 ___	
	bending	30–45°	3 ___		30–60°	2 ___	
		>45°	10 ___		>60°	8 ___	
Back	Flexion	0–20°	1 ___	Extension	Not included		
		20–60°	3 ___				
		>60°	10 ___				
	Lateral	0–10°	1 ___	Rotation	0–20°	1 ___	
	bending	10–20°	3 _v_		20–30°	2 _v_	
		20–30°	9 ___		30–45°	7 ___	
		>30°	13 ___		>45	11 ___	
Postural load = 12							

FIGURE 43.1   Checklist for evaluating postures.

having relative discomfort scores of two or more. The postural load index is calculated for the left or right arm/hand, the neck, and back motions. Only the right or left arm/hand is assessed at a time when calculating postural load index. Finally, based on the postural load index, the posture is evaluated using the criterion of four action categories in the following in terms of whether the posture is acceptable or any correction actions are needed:

$$\text{Postural load index} = \sum_{j=1}^{n} \sum_{i=1}^{mj} Sij$$

where i is the ith joint motion, j is the jth joint, n is the number of joints involved, m_j is the number of joint motions studied in the jth joint, S_{ij} is the relative discomfort score of the ith joint motion in the jth joint (here, $S_{ij} = 0$ if a corresponding relative discomfort score is 1.0).

The four action categories are as follows:

Category I: Postures with the postural load index of 5 or less. This category of postures is acceptable, except in special situations such as repeating and sustaining them for long periods of time, etc. No corrective actions are needed

Category II: Postures with the postural load index from 5 to 10. This category of postures requires further investigation and corrective changes during the next regular check, but immediate intervention is not needed

Category III: Postures with the postural load index from 10 to 15. This category of postures requires corrective action through redesigning workplaces or working methods soon

Category IV: Postures with the postural load index of 15 or more. This category of postures requires immediate consideration and corrective action

43.4 Application Example

A working posture was selected in an electromechanics manufacturing company of Korea (Figure 43.2), and assessed using LUBA. Of two body sides, the right side was evaluated, because it was thought to be more stressful. The postural load score was 12, the details of which are shown in Figure 43.1. This refers to a LUBA action category of 3, which means that the posture requires corrective action soon. The high score of 12 was mainly attributed to back lateral bend and rotation. For eliminating the motions or reducing their degree, the roller at the left side of the worker should be positioned as closer to the worker as possible. The improvement results in the postural score of 7 or less, which indicates that immediate ergonomics intervention is not needed.

FIGURE 43.2 Example of working posture.

43.5 Conclusions

A technique for postural loading on the upper body assessment was presented based on the new experimental data for perceived discomfort values for a set of joint motions, including the hand, arm, neck, and back. Each postural class was assigned a relative discomfort score with the characteristics of the ratio scale relative to the perceived discomfort for the neutral position of elbow flexion. The ratio discomfort score makes it easy to quantitatively evaluate postural stresses for varying postures and to compare them across different postures. It is expected that the postural classification scheme based on consideration of perceived discomfort can be used as a valuable tool for assessing postural stresses and preventing posture-related MSDs.

References

Andrews, D.M., Norman, R.W., Wells, R.P., Neumann, P., 1998. Comparison of self-report and observer methods for repetitive posture and load assessment. *Occupational Ergonomics* 1(3), 211–222.

Armstrong, T.J., Buckle, P., Fine, L.J., Hagberg, M., Jonsson, B., Kilbom, A., Kuorinka, I.A.A., Silverstein, B.A., Sjogaard, G., Viikari-Juntura, E.R.A., 1993. A conceptual model for work-related neck and upper-limb musculoskeletal disorders. *Scandinavian Journal of Work, Environment, and Health* 19, 73–74.

Bhatnager, V., Drury, C.G., Schiro, S.G., 1985. Posture, postural discomfort, and performance. *Human Factors* 27(2), 189–199.

Buchholz, B., Paquet, V., Punnett, L., Lee, D., Moir, S., 1996. PATH: a work sampling-based approach to ergonomics job analysis for construction and other non-repetitive work. *Applied Ergonomics* 27(3), 177–187.

Burdorf, F.J., Govaert, G., Elders, L., 1991. Postural load and back pain of workers in the manufacturing of prefabricated concrete elements. *Ergonomics* 34(7), 909–918.

Corlett, E.N., Madeley, S.J., Manencia, I., 1979. Posture Targetting: a technique for recording working postures. *Ergonomics* 22(3), 357–366.

Genaidy, A.M., Al-Shedi, A.A., Karwowski, W., 1994. Postural stress analysis in industry. *Applied Ergonomics* 25, 77–87.

Heinsalmi, P., 1986. Method to measure working posture loads at working site (OWAS). In: Corlett, E.N., Wilson, J., Manencia, I. Eds., *The Ergonomics of Working Postures*. Talyor & Francis, London, pp. 100–104.

Karhu, O., Kansi, P., Kuorinka, I., 1977. Correcting working postures in industry: a practical method for analysis. *Applied Ergonomics* 8(4), 199–201.

Kemmlert, K., 1995. A method assigned for the identification of ergonomics hazards — PLIBEL. *Applied Ergonomics* 36(3), 199–211.

Keyserling, W.M., 1986. Postural analysis of trunk and shoulders in simulated real time. *Ergonomics* 29(4), 569–583.

McAtamney, L., Corlett, E.N., 1993. RULA: a survey method for the investigation of work-related upper limb disorders. *Applied Ergonomics* 24(2), 91–99.

Priel, V.Z., 1974. A numerical definition of posture. *Human Factors* 16, 576–584.

van Wely, P., 1969. Design and disease. *Applied Ergonomics* 1, 262–269.

44

The Washington State SHARP Approach to Exposure Assessment

Stephen Bao

Barbara Silverstein

Ninica Howard

Peregrin Spielholz
Washington State Department of Labor & Industries

44.1 Introduction

Understanding the relationships between workplace exposure parameters and the health outcomes of the musculoskeletal system is the basis for preventing and reducing work-related musculoskeletal disorders. Quantification of exposure parameters is critical in epidemiological studies as well as ergonomics applications. Methods used for the exposure assessment vary depending on the purpose of the applications and feasibilities of using these methods. This chapter discusses the various exposure parameters at workplaces related to work-related upper extremity disorders, measurement strategies, and some exposure assessment methods used in epidemiological studies. The exposure assessment approach used by SHARP in a large prospective study of upper extremity musculoskeletal disorders is presented and discussed.

44.2 Exposure Parameters, Measurement Strategy, and Measurement Methods

Workplace exposure parameters associated with the development of upper extremity musculoskeletal disorders include work organization variables and various physical exposures of the jobs. The National Institute for Occupational Safety and Health (NIOSH) published a critical review of the evidence for work-related musculoskeletal disorders of the neck, upper extremity, and low back and summarized major findings on the various exposure parameters (NIOSH, 1997). In 2001, the National Research Council published a comprehensive review of the evidence on work-related musculoskeletal disorders in which they concluded that repetition, force, and vibration, as well as high job demands and job stress, were particularly important risk factors for upper extremity disorders (Panel on Musculoskeletal Disorders and the Workplace, 2001). They also found that modification of these factors could substantially reduce the risk for these disorders.

The way work is organized and performed often determines subsequent physical/mechanical and psychological job demands on individual workers. Work organization also encompasses the organizational practices and production methods that affect job design. These include the temporal aspects of work (e.g., work–rest schedules, work shifts, hours of work, work pacing), job content (e.g., repetitiveness of tasks, use of skills, vigilance, participation in decision-making), compensation arrangements (salary, hourly, quota, piece rate), work status (fulltime, part-time, seasonal, temporary), social interactions (isolated, various levels of team work), task (single, rotating, multiple), and opportunities for development (Kasl, 1992; Sauter and Swanson, 1996). Consideration of work organization provides information at the group level and enables multilevel analysis. Work organizational observational exposure assessment methods that are potentially relevant for assessing relationships with musculoskeletal disorders include those of Rohmert and Landau (1983), Ergonomic Workplace Analysis (Ahonen et al., 1996), Meaning of Work (MOW International Research Team, 1987), and the Occupational Stress Index (Belkic et al., 1995).

Typically, physical exposures identified in workplaces include forceful exertions (Stetson et al., 1993; Fransson-Hall et al., 1996; Roquelaure et al., 1997), such as gripping a high force demanding hand tool, lifting a heavy object, pushing a fully loaded cart. Non-neutral postures of hands and upper extremities (Frost and Andersen, 1999; Punnett et al., 2000; Viikari-Juntura et al., 2000), such as bending the wrist when using a hand tool and raising the hand above the head when performing a task, increase force requirements. Highly repetitive motions of the hand, wrist, and upper arms (Veiersted and Westgaard, 1993; Blanc et al., 1996; Nordstrom et al., 1997; Punnett, 1998) are found in hand-intensive jobs such as assembly and data entry. Some other physical demand parameters at workplaces include hand–arm vibration, wearing gloves, and some environmental conditions such as extremely cold or hot temperatures. Work organization parameters such as work methods, social content, and task pacing may also influence the development of work-related upper extremity disorders.

Different measurement strategies may be used to meet the various needs of the exposure assessments. For example, most cross-sectional epidemiological studies measure exposure parameters at a certain point in time, while most prospective epidemiological studies require the quantification of the exposures for the days, weeks, and years on the job. Thus, data collected for prospective studies should make it possible to calculate cumulative exposures. For example, a worker performs two different tasks in a workday. Exposure from both tasks should be measured and the compound exposure for the whole day should be calculated depending on the task distribution (time spent on the two tasks). If the worker's exposure is changed during the course of the study (e.g., job changes), a new exposure measurement should be performed, and the accumulated exposure is then calculated. If the purpose of the measurement is to assess exposure differences among two or several conditions, measurement can be done for each of the conditions.

When considering the measurement strategy to be used, one should also consider the three main dimensions of physical exposure: amplitude, frequency, and duration, rather only one single dimension. This is because the physiological significance is dependent on the combination of these exposure dimensions. Therefore, exposure quantification should include the measurement of exposure amplitude (e.g., level of hand force, degree of a joint angle), exposure frequency (e.g., number of exertions per minute), and exposure duration (e.g., length of time in hours). Other aspects of physical exposure may also need to be quantified, such as duty cycle and speed.

After deciding on the measurement strategy, one should consider the selection of exposure quantification methods to be used. There are numerous exposure quantification methods available.

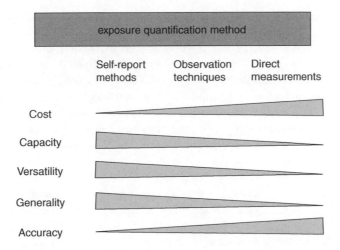

FIGURE 44.1   General characteristics of the three types of exposure quantification methods. (After Winkel, J. and Mathiassen, S.E. (1994) *Ergonomics* 37: 979–988. With permission.)

However, different methods may have different specificity, validity, and reproducibility. Depending on the types of methods, they may also require users with varied expertise. The time and cost required for data collection and analysis could also vary significantly.

Although risk estimation from job titles has been used as the normal exposure quantification in many epidemiological studies, they may only give vague or invalid exposure assessment in musculoskeletal epidemiological studies. This is because workers under the same job title could perform very different activities due to differing technologies and machines used, individual work techniques, and work organizational differences between different companies. Commonly used exposure quantification methods in work-related musculoskeletal studies can usually be divided into the following three categories: (1) self-report questionnaires; (2) observational methods; and (3) direct measurement techniques. Each of these has its own strengths and weaknesses and may be used in different applications depending on the study purposes. Figure 44.1 illustrates some general characteristics of the different methods, and can serve as a guide for selection. In general, direct measurement with instrumentations gives the most specific and accurate exposure estimation, but involves significant costs. This method would be impractical for individual exposure assessment in very large populations of large-scale epidemiological studies because of the significant resources and expertise that would be required. Self-reporting using questionnaires or interview methods can access large populations with reasonable cost, but the data, in general, have low validity with respect to exposure level and variation. Observational methods are usually considered to be in between the direct and self-reporting methods in terms of the different characteristics. The following sections discuss some of the commonly used methods of the three groups.

44.2.1   Self-Report Questionnaires

Self-report questionnaires are appealing due to their relative ease of administration and low cost in comparison to direct measurement methods. The questionnaires can be self-administered or through interviews. They could be used among large population samples within a short period of time and with relatively low cost. However, the obvious drawback is the resulting information may be potentially less reliable and more easily influenced by environmental and personal factors. This type of method also has low specificity so that they may not be able to detect differences between groups, which could otherwise be detected by different exposure measurement methods. Self-report questionnaires have been used with other methods in industry by several researchers with varying levels of success (Baty et al., 1986; Burdorf and Laan, 1991; Holmstrom et al., 1992; Wiktorin et al., 1993; Punnett, 1998).

Self-report questionnaires tend to be both reproducible and relatively close to observational and direct measurement for gross levels of activity (Baty et al., 1986; Wiktorin et al., 1993, 1996). Wiktorin et al. (1996) reported acceptable reproducibility in the 0.5 to 0.8 range for questions relating to overall physical activity, whole body working postures, and specific leisure activities. Some success has been reported in subjective estimates of impulses or forces on the hands (Freivalds and Eklund, 1993). However, both reproducibility and reliability dropped significantly compared to other measurement methods when specific questions relating to bent postures and levels of loads were asked (Holmstrom et al., 1992; Wiktorin et al., 1993). Viikari-Juntura et al. (1996) reported moderate correlation (0.42 to 0.55) between self-reports and observations of physical workload factors such as frequency of manual handling, duration of trunk flexion, neck rotation, hands above shoulders, and squatting or kneeling. The correlations, in general, were higher for those without low back pain than for those with low back pain. Pope et al. (1998) compared results from a self-report questionnaire on physical demands (postures, manual handling, repetitive upper limb movements) to direct simultaneous observations in six different occupational settings. Agreement was good for most of the manual material-handling activities. However, minutes of repetitive arm and wrist movements appeared to be the least accurate (overestimates). These authors concluded that dichotomous recall is satisfactory (i.e., ever, never) and that exposure magnitude recall can also be satisfactory for some risk factors.

Toomingas et al. (1997) tested the hypothesis that those who rate health outcomes high on self-reports would also rate exposures high on self-reports, thereby biasing risk estimates. Conducting separate analyses by age, gender, and socioeconomic status, correlations were close to zero for fixed and nonfixed stimuli, including symptoms and physical exposures, indicating no systematic differences by rating behavior. Punnett (1998) reported consistent findings of good comparability in estimates, when comparing self-reported physical exposures with observations by researchers blinded to health status. Kerr et al. (2001) reported good agreement between back injury cases and job-matched controls on self-reported physical demands of the job (ICC = 0.6), suggesting a lack of symptoms-related bias in estimates. Bernard et al. (1994) compared observational analysis to self-reports of exposure among symptomatic newspaper workers and referents. Both groups reported a longer duration of typing time (approximately 50% more) than the observational analysis. Similar results by Spielholz et al. (1999) showed consistent overestimation of upper extremity risk factors by most individuals. These studies indicate that self-reports may provide valuable information regarding task duration/frequency and whole body postures but are generally neither accurate nor reliable for measurement of hand/arm exposure to risk factors in terms of duration and frequency.

44.2.2 Observational Methods

Observational measurement methods are frequently used in field studies as a compromise between questionnaire and direct measurement methods. Observational methods present the best compromise for individual exposure assessment in large-scale epidemiological studies. Observational measurement systems are usually categorized into two types: event based and time based.

Event-based methods such as the NIOSH lifting equation (NIOSH, 1994), the rapid upper limb assessment method or RULA (McAtamney and Corlett, 1993), and the rapid entire body assessment method or REBA (Hignett and McAtamney, 2000) are applied to the complete event of a task or subtask and give a score or index to represent the risk level. This type of method, though widely used by ergonomics practitioners for its simplicity, does not provide detailed information on changes in exposure during task performance. Often these types of methods are used in field-based risk assessment. In addition, no one measurement method has been widely accepted as a standard, although several are currently being developed to fill this role.

One event-based method used as a standard is the ACGIH hand activity level (HAL) threshold limit value (ACGIH, 2001). This method adopts the previous work of Latko et al. (1997) to set levels of physical exposure for the hand and the wrist. The HAL is applicable to single-task jobs, although some approaches have been attempted to extend its use in multiple-task situations. The RULA method (McAtamney and Corlett, 1993) is used to assess the postures of the neck, trunk, and upper limbs, muscular effort, and the external loads on the body. This postural exposure assessment system has been used in several different formats and adopted for use in many different types of industries (Lueder, 1996; Hignett and McAtamney, 2000).

Time-based methods such as OWAS (Karhu et al., 1981), VIRA (Persson and Kilbom, 1983; and Kilbom et al., 1986), ARBAN (Holzmann, 1982), and PEO (Fransson et al., 1991) require the analyst to observe the job performance continuously or at specific time samples during the task performance. The analyst records the exposure changes based on predefined categories, such as, hand with weight versus hand without weight, and neck flexion between $0°$ and $20°$ versus greater than $20°$. Observations can be performed on-site with a computer or off-site where video-tapes are analyzed. Advantages of the time-based methods are that they more closely represent the true exposure during the task performance. The disadvantage is that it is time consuming and may also limit the number of exposure parameters that an analyst can observe if the method is used on-site.

Video-based off-site techniques often use categorical scoring of body positions, movement frequency, type of grip, and force based on either sampled or real-time recording (Karhu et al., 1977; Corlett et al., 1979; Holzman-Voigt, 1979; Kemmlert and Kilbom, 1986; Keyserling, 1986; Armstrong et al., 1982). The method employed by Armstrong et al. (1982), for example, sampled postures several times a second and classified wrist postures into five categories: (1) neutral, (2) flexion, (3) extreme flexion, (4) extension, and (5) extreme extension. In general, video-based analysis may be the most appropriate observational method for risk factor quantification and definition of work activities for large-scale epidemiological studies because it allows the analysts sufficient time to estimate the postures of the various body parts and provides the possibility to reanalyze the data for quality control purpose.

With the availability of newer computer technologies, time studies of task performance and postural analysis can now be carried out on computers. A recently developed multimedia video task analysis (MVTA) system (Yen and Radwin, 1995) is able to set accurate time codes on videotapes and perform time analysis on various time-based events (e.g., tasks, postures, and hand exertions). With its flexible design, users can set their own parameters to be studied (e.g., tasks, wrist flexion and extension postures, hand exertions) and define their own categories of the different parameters (e.g., for the parameter of task with two levels: computer keyboarding and writing notes; for the parameter of wrist flexion/extension with four categories: flexion 0 to $30°$, flexion $>30°$, extension 0 to $30°$, and extension $>30°$). A drawback of this type of analysis is that the analyst has no control on the angles of observation, and has to depend on the quality of the videotapes. Therefore, to obtain reliable and adequate exposure information, it is important to take good-quality video. Another disadvantage with the computer-based observation systems is that one cannot obtain direct measurements such as object weight and forces required to operate a tool while the analyst is sitting in his or her laboratory. In contrast, when the observation is done on-site, the analyst can most often communicate or interact with the operator to obtain the information. Therefore, if the analysis is performed off-site, it is important to obtain the required information on-site and be prepared for use in the off-site analysis.

Falling within the scope of observational field methods are methods based on workloads. These methods define a system of quantifying an overall load score (Helliwell et al., 1992) or classify workers into classes based on work levels (Nathan et al., 1993). The Strain Index developed by Moore and Garg (1995) identified six risk factors, each given a categorical 1 to 5 score, that give an overall severity index (SI) score when multiplied together. This tool has been used in meatpacking and has shown data that support its validity in predicting morbidity (Moore and Garg, 1995). Although the Strain Index method was originally designed for single-task jobs, the authors have made attempts to extend this method to multiple-task jobs.

Force quantification often presents a problem in observational methods. Hand force cannot be seen. Consequently, it must be estimated, which can be achieved using several methods. A simple dichotomous classification of either high or low force has commonly been used, typically using manipulation of a

4-kg object or its equivalent force for power grip or 1-kg object or its equivalent force for pinch grip as the determinants of class (Silverstein et al., 1987; Stetson et al., 1991). Several researchers have used a modified Borg scale, which classifies an expert estimate on a 10-point categorical scale ranging from zero to maximal exertion (Borg et al., 1985; Lloyd et al., 1991). Another approach is to estimate the tendon force based on hand geometry, assumed friction, and object weight (Helliwell et al., 1992). Despite the provision of an actual force value, this method relies on the estimation of every factor and may not be any more reliable than scaling techniques. A continuous method used by Latko (1997) employs expert consensus rating of average and peak force on 10-cm visual–analog scales. Reproducibility estimates of this method have been between 0.6 and 0.8, showing promise as a continuous scaling method.

Psychophysical studies use a subject's perception of sensation to measure a factor of interest. This has been applied in the field of exposure observation. Snook et al. (1995) developed guidelines for hand/wrist flexion and extension based on psychophysical studies. Analysts observe the hand/wrist postures during task performance and give subjective ratings on the postures. Previously, Snook (1978) also used perception of object weight to develop acceptable guidelines for lifting based on location, lift frequency, and weight. A more recent method developed by Latko et al. (1997) employed expert group rating of physical components of work on visual–analog scales. This study evaluated the use of rating several risk factor exposure metrics on continuous visual–analog line scales. The technique shows great promise in terms of reproducibility and reliability of quantifying hand activities (Latko et al., 1997). These scales have been incorporated into primary measurement methods of the American Conference of Government Industrial Hygienists (ACGIH, 2001) TLV on hand activity.

Although there have been major advances in observational methods and they have been widely used in musculoskeletal epidemiological studies, some drawbacks exist. One of the major drawbacks is that the observational methods are based on the subjective judgment of the individual analyst. Some variations within and between analysts are unavoidable. Measures to reduce such variation should be taken. Another common problem associated with observational methods in epidemiological studies is that different researchers have used different predefined exposure categories. This makes it difficult to compare results from different studies.

44.2.3   Direct Measurement Methods

Direct measurement aims to provide the standard by which the validity and reliability of all other methods are measured. However, much work remains in developing accurate systems that can be used in the field to measure posture, motion, and force. The two most commonly used methods to measure posture and motion, electrogoniometry and video-based motion tracking systems, have only recently been used in field studies for the upper extremities (Hägg et al., 1997). This is in large part due to issues of feasibility and measurement error.

Video motion tracking systems, which typically operate by using computer-aided edge detection to follow markers placed on a worker, are not widely used in the field due to feasibility issues. In order to perform unobstructed tracking in three dimensions, the worker must be in view of three cameras. This, however, does not eliminate the analysis of many obstructed-view estimations and makes recording of dynamic work or work inside enclosures impossible, as work is often performed outside the field of view of the cameras.

Electrogoniometers, physically placed on the hand/wrist and forearm, do not have the obstructed-view problems associated with video tracking, and systems have been developed which may easily be used in the field (Moore et al., 1991). Continuous angle recordings may be analyzed to determine the length of time in specified body postures, repetitiveness of motion, and angular velocity and acceleration (Marras and Schoenmarklin, 1993; Radwin et al., 1993). However, measurement error largely due to cross-talk between recording channels has been a pervasive problem in past studies (Moore et al., 1991; Buchholz and Wellman, 1997). Cross-talk can be thought of as the bias created by the distortion of the resistive strain gauges in the electrogoniometers by movement on one or both of the axes.

Several researchers have evaluated electrogoniometers and the introduction of significant cross-talk in flexion/extension measurements and deviation measurements from extreme forearm rotation (Armstrong et al., 1993; Smutz et al., 1994; Buchholz and Wellman, 1997; Roberts, 1997).

Researchers have developed procedures that may reduce errors caused by cross talk (e.g., Smutz et al., 1994; Buchholz and Wellman, 1997; Roberts, 1997). These results show promise for the use of an electrogoniometer and electrotorsiometer in tandem to measure motions and allow for error correction. Researchers appear to agree that an electrotorsiometer and electrogoniometer can be used in a telemetric system to perform angle measurements with errors less than 5° (Armstrong et al., 1993; Smutz et al., 1994; Buchholz and Wellman, 1997).

Force exerted by muscle groups is commonly measured by force transducers placed in the line of action or by the use of surface electromyography (EMG) (Armstrong and Chaffin, 1979). Force transducers may provide accurate information if specific conditions exist where their placement does not affect the work.

EMG has become relatively easy to perform in the field with the use of disposable surface electrodes and portable measurement devices (Winkel and Gard, 1988; Hägg et al., 1997). Typically, EMG data may be measured and analyzed for either physical signs of fatigue or for comparison of static force levels (NIOSH, 1992). Electrodes transmit motor unit action potentials from the underlying muscles. These signals, when root-mean-square (RMS) transformed or integrated, have shown a linear or exponential relationship ($r^2 > 0.90$) to developed static force (NIOSH, 1992).

Direct estimation, an alternative to EMG for calculation of hand force, can be classified as a direct measurement method. Field practitioners and consultants commonly use this method to obtain job force requirements. Direct estimation can be done simply by measuring the force requirements of a tool or piece of equipment with a force gauge. When this is not possible, estimation of force can also be accomplished through the reproduction of the exertion on a force gauge in the same orientation and type of grip as performed by a worker. Kingdon and Wells (2000) conducted a laboratory study on the accuracy of matching a manual gripping force using a hand dynamometer. Initial findings have shown that force matching may be relatively accurate and consistent at lower force levels. A study published by the Safety & Health Assessment & Research for Prevention (SHARP) program on 113 government workers showed that with the use of a hand dynamometer, the force matching method can be quite accurate and consistent in estimations of power grip force and pinch grip force (Bao and Silverstein, 2005). These results support the use of this method as an alternative to more time-consuming and expensive instrumentation techniques for quantifying hand force levels in large epidemiological studies.

In conclusion, no method is perfect and different methods may be used in different situations for different purposes. Direct measurement using current techniques represent the most accurate and reliable exposure assessment method. Work by Spielholz et al. (1999) comparing self-reports, video observation, and direct measurement showed that video observation may have approximately 30% more error than direct measurement in some risk factor measurements. Direct measurement would be the preferred method given unlimited resources; however, modern video observation techniques have the advantage of providing larger numbers of evaluated participants due to less time-consuming data collection and analysis.

Due to the population size requirement of most epidemiological studies, direct measurement of all participants would require resources well beyond what is available from a granting agency. Video-based observational assessment in combination with direct measurement and estimation of forces is the only method that would allow measurement of all participants with an acceptable level of accuracy and reliability. Additionally, discrimination calculations by Spielholz et al. (1999) show that the estimated tenfold increase in number of measured participants (100 to 1000) possible with video-based techniques over direct measurement will give a more accurate exposure assessment at the group level despite the increased measurement error. For these reasons, video-based observation and direct force estimation techniques used previously by the SHARP (1999) program were chosen as the primary exposure assessment method in a large prospective study of upper extremity musculoskeletal disorders conducted by SHARP (referred to as the SHARP Study in the subsequent text).

44.3 SHARP Study Exposure Assessment Methods

The "SHARP Study" involves health assessments (structured interview, physical examination, nerve conduction studies, and psychosocial questionnaires) and exposure assessments (collected by different teams blinded to either health or exposure status) of workers at 13 different worksites, collected at baseline and 4-month intervals over 3 yr. Work organizational factors are collected at the departmental level by the exposure assessment team. If there is a significant job change, exposure assessment is repeated. In the following sections, the exposure assessment methods used in the SHARP Study will be presented and discussed. The discussion starts with job sampling followed by on-site data collection, and then discussions of the various exposure assessment methods (e.g., job analysis, posture analysis, repetitive exertion analysis, repetitive movement analysis, and work organization measurement).

44.3.1 Job Sampling

It is impractical to follow a worker for a whole workday to document his or her exposure at the job for large-scale epidemiological studies with a large population. If one has to follow each of 1000 subjects for 8 h to document exposure, the total measurement time for just the baseline would be 8000 h. This is not feasible for most epidemiological studies. However, it is also important to obtain exposure measurements for individual workers. Many previous studies used a group exposure measurement approach. This is done by measuring exposure among a small number of subjects in a specific job or job category and assuming the same exposure for the whole group. The large within-group variance in some exposure parameters could unravel specific risk factors for work-related musculoskeletal disorders (Fallentin et al., 2001). Individual exposure assessment is the alternative. For this type of assessment some sort of job sampling method is necessary to obtain exposure measurement over a short period of time during a representative workday. Job sampling has been a common procedure in industrial engineering time studies. The requirement for job sampling is that exposure results taken from some short periods of time can be used to represent the whole day exposure with reasonable accuracy.

Job samples should be taken from a typical workday where the employee performs his or her usual type of work at a normal pace without any restrictions by process limitations. This is usually confirmed at the beginning of the measurement period by asking supervisors and workers.

In the field of industrial engineering, the length of observation or the number of cycles that should be studied in order to arrive at an equitable standard is a subject that has caused considerable discussion among time study analysts as well as union representatives (Niebel, 1988). Since the activity of the job, as well as its cycle time, directly influences the number of cycles that can be studied from an economical standpoint, one cannot be completely governed by the sound statistical practice that demands a certain size sample based on the dispersion of the individual element readings. There are no generally agreed upon criteria for recording length or number of cycles in epidemiological studies. Most often, common practice, personal judgment, and available resources play a more important role than strict statistical procedures in determining the number and length of job samples. In the SHARP Study, the recording time was determined after reviews of other similar studies and feasibility considerations based on available resources.

As in many industries, workers can be assigned to different tasks during a workday. For example, an electronic assembly worker can be assigned to an assembly task for 6 h and a packaging task for 2 h in an 8-h workday. Therefore, job samples are usually taken at the task level, instead of at the job level. When a task is cyclic (i.e., repetition of the same activities), the exposure variation within a single task is usually smaller than between different tasks. Some tasks may have no repetitive pattern, that is, activities may not be repeated regularly and can happen at different times of the day without any specific patterns. This type of task can be labeled as a noncyclic task. Exposure variation within this type of task is usually large. Exposure measurement obtained from a continuous job sample of this type of task may not represent the real exposure of that task. Therefore, a different approach may need to be considered. A common sampling practice for noncyclic tasks is to take a number of random samples during the task performance period.

The job sampling approach used in the SHARP Study is based on whether the job is a single-task job or a multiple-task job, and the tasks are cyclic or noncyclic (Figure 44.2). The first step is to determine whether the job is a single-task job or a multiple-task job and, secondly, to determine if each task is cyclic or noncyclic. For a cyclic single-task job, a continuous 15-min job sample is taken for the exposure measurement purpose. For a noncyclic single-task job, three 5-min job samples are taken randomly during the workday. This attempts to obtain good equitable exposure measures for this type of job and capture the fluctuations of the exposure during a workday. The total job sample length for a single task job is 15 min for both cyclic and noncyclic task jobs.

For multiple-task jobs (more than one task is performed during a workday), a different job sampling method is used for both cyclic and noncyclic tasks. A 10-min job sample is taken for each of the cyclic tasks and two 5-min job samples are randomly taken from each of the noncyclic tasks in a multiple-task

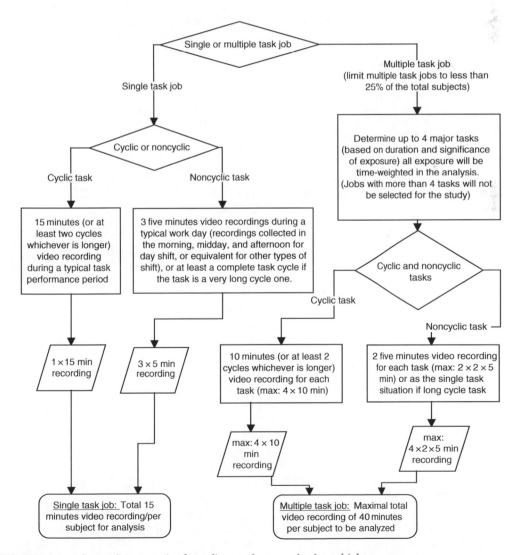

FIGURE 44.2 Job sampling strategies depending on the type of tasks and jobs.

job. Depending on the number of tasks, the total length of the job samples is longer in a multiple-task job than that in a single-task job.

For practical purposes, it may be necessary to limit the number of multiple-task jobs in a large epidemiological study as the amount of time needed for data collection, processing, and analysis is much longer in multiple-task jobs than that in single-task jobs. In many industries where manual activities are dominant, a common practice is to rotate workers between different tasks. Rotations commonly occur at break times. In the SHARP Study, jobs with more than four tasks were excluded.

44.3.2 On-Site Data Collection

After the job sampling strategy is determined, jobs are documented and data are collected on site. Depending on the types of analysis, data collection can be done through interviewing workers, supervisors, and other plant personnel, examining production data logs, performing on-site observations, video filming jobs for further analysis, and collecting direct measurements of relevant parameters. Often a combination of these methods is used. As an example, an on-site data collection form used in the SHARP Study is shown in Figure 44.3.

Before taking any exposure measurements, work hours and days, job change information, and task distribution information (task rotation schedule and number of hours at each task) are obtained from

---

Field Physical Exposure Data Collection Form

Location: _____ ; Physical location: _____

Subject ID: _____ Analyst(s): _____ Time: _____

Period ID: _____ Initial job category: _____

Age: _____ ; Gender: ☐ Male ☐ Female; Height: _____ in.; Weight: _____ lb

Hand (write): ☐ Right ☐ Left ☐ Both;   Hand (work): ☐ Right ☐ Left ☐ Both;

Current job: _____ Shift hours: _____ Work days: _____

Task ID	Task Activity	Duration (h/d)	Expected time(sec)	Cyclic	Note
				Y / N	
				Y / N	
				Y / N	
				Y / N	
				Y / N	

Break times: _____

Power tool information (PT)

Tool ID	Power Tool/Model	Weight	Grip span(s)/Handle Diameter(s)	Note
PT				
PT				
PT				
PT				
PT				

On site direct measurement

Task: _____ ; Task ID: _____

Major object handled

Object	Measure (lb)	Estimated[1]	Sub-task description

Significant force measurement

Force type	Measure (lb)	Est[1]	Not measurable?	Sub-task description
☐ push ☐ pull			☐	
☐ push ☐ pull			☐	
☐ push ☐ pull			☐	
☐ push ☐ pull			☐	

Significant hand force measurement

Measmnt 1 (lb)	Measmnt 2 (lb)	Measmnt 3 (lb)	RPE	Est[1]	Grip type	Not measurable?	Sub-task description
						☐	
						☐	
						☐	
						☐	

Other significant force measurement

Force type (please describe)	Measurement (lb)	Est[1]	Not measurable?	Sub-task description

Other estimated and observed measurements

	Task average		Note
	Left	Right	
HAL[2]			
Duration of exertion (% of cycle)[3]			
Efforts/min[4]			
Hand/wrist posture[5]			
Speed of work[6]			
Duration per day[7]			

	Check (✓)if yes		Estimated frequency[8]	Sub-task description
	Left	Right		
Power tool use	☐ id: ___	☐ id: ___		
Contact stress	☐	☐		
Jerking	☐	☐		
Impact action	☐	☐		

1. Estimated (Est) intensity of force: 1 (low) to 10 (high)
2. HAL

0–handle idle most of the time, no regular exertions	2–consistent, conspicuous, long pauses or very slow motions	4–slow, steady motion/exertions, frequent brief pauses	6–steady motion/exertion; infrequent pause	8–rapid, steady motion/exertions; no regular pauses	10–rapid, steady motions/difficulty keeping up or continuous exertion

3. Duration of exertion (% of cycle)

1 - <10%	2 – 10 to 29%	3 – 30 to 49%	4 – 50 to 79%	5 - >80

4. Efforts/min

1 - <4	2 – 4 to 8	3 – 9 to 14	4 – 15 to 19	5 - ≥20

5. Hand/wrist posture

1 – very good	2 – good	3 – fair	4 – bad	5 – very bad

6. Speed of work

1 – very slow	2 – slow	3 – fair	4 – fast	5 – very fast

7. Duration per day

1 - <1	2 – 1 to 2	3 – 2 to 4	4 – 4 to 8	5 - ≥8

8. Estimated frequency: 1-very infrequent, 2-infrequent, 3-average, 4-frequent, 5-very frequent

FIGURE 44.3   Sample of SHARP study's on-site data collection form.

interviewing workers and supervisors. Jobs are then video filmed according to the job sampling strategy discussed previously. Two synchronized cameras are used in order to capture both sides of the body while the worker is performing tasks. The camera crews should be well coordinated so that when the worker moves the cameras should be moved accordingly in order for at least one camera to capture both sides of the body. This will help the off-site data processing in the laboratory.

During the observation period, forces applied in the task are noted and later measured. As it is not feasible to measure all forces that the worker applies in the task, a subjective determination of "significant force" is made. Operationally, when one of the ergonomists considers that the force is obvious and may be of importance to the exposure, the force data will be collected. This is similar to most ergonomics consultations where an ergonomist takes measurements he or she thinks necessary. Conceptually, a "significant force" is defined as a lifting force of ≥ 0.9 kg, a pinch grip force of ≥ 0.9 kg, a power grip force of ≥ 4.4 kg, and a push/pull force of ≥ 4.4 kg. The force value is not known until measured. Therefore, in practice, forces that are lower than the defined levels are sometimes measured. A lifting force is measured by the object weight. This is typically measured by using a force gauge or a weight scale. Object weights can also be obtained from the company. A push/pull force is also measured using a force gauge. For practical purposes, no distinction is made between push and pull forces, though they may have different physiological impacts. Additionally, both lifting weight and push/pull force are also estimated by an ergonomist using a 1 to 10 rating scale. A pinch or power grip hand force is measured using the force matching method (Bao and Silverstein, 2005). This is done by asking a subject to recreate the amount of force he or she uses in the task on a force dynanometer using similar hand/wrist postures. This process is repeated three times, and the median of the three is used in the analysis. Borg ratings by subjects and researchers are also collected for force applications (CR-10, Borg, 1982). Different measurement methods are used for the same exposure parameters in order to study the differences and similarities between the different methods.

Other observed parameters, such as the HAL, and parameters for computing the Strain Index (i.e., duration of exertion, efforts/min, hand/wrist posture, speed of work, and duration per day) are also collected during the on-site data collection period. This allows several event-based exposure estimations to be made.

If the worker uses vibrating tools, the tool information is collected. This is used to crudely estimate vibration exposure to the worker.

It is important to ensure that the data collection process of the on-site analysis does not interfere with the normal performance of the task. Some workers may have the tendency to modify their performance in front of video cameras. This must be discouraged. At the end of the data collection, it is also important to check the completeness of the data and be sure all data are collected properly.

44.3.3 Job Analysis (Significant Force Analysis)

Job analysis is done in the laboratory and is based on the video recording and data collected at the worksite. The purpose of this analysis is to obtain the frequency and duration of significant forces. This can be done by performing time studies on the recorded tasks. A software program called MVTA (multi-video task analysis), developed by the University of Wisconsin (Yen and Radwin, 1995), is used in the SHARP Study. A typical data processing screen is shown in Figure 44.4. The record shows a time line where a certain event (activities of various significant forces) occurred. Significant forces are listed in the right panel. The video window shows the recorded task performance. The analyst can use any video clip from the two synchronized cameras to obtain the best view for the analysis. Time-line marks are inserted at the time when significant forces occur. The analyst may often play the video at normal speed first in order to understand the task activity contents and then play the video in slow motion mode to set the event marks.

After the data processing, a time-study report can be generated. For instance in Table 44.1 it is shown that at recorded cycle #3, the worker lifted an object of 56 lb for a duration of 142 frames (or 4.7 sec) and spent 286 frame time (or 9.5 sec) performing other activities where no significant force was applied.

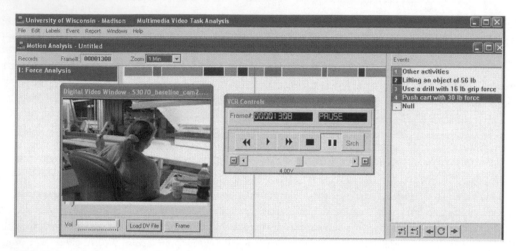

FIGURE 44.4   A typical time study on significant forces.

With the time-study report, a summary statistics report can be produced where average duration, percentage of time, and frequency of the various significant forces can be calculated as follows:

$$\text{Average duration (sec)} = \frac{\sum \text{Duration of individual exertions}}{\text{Total number of exertion cycles}}$$

$$\% \text{ Time} = \frac{\sum \text{Duration of individual exertions}}{\text{Total recording time}} \times 100\%$$

$$\text{Frequency of exertion (times/min)} = \frac{\text{Total number of exertions}}{\text{Total recording time}}$$

When the worker is involved in the use of vibrating tools during task performance, the vibration exposure needs to be quantified. Theoretically, vibration levels should be measured during task performance. However, vibration measurement requires sophisticated instrumentation and enormous resources.

TABLE 44.1   A Typical Time-Study Report of Significant Forces

Time-Study Report	
Event #	Event Elements
1	Other activities
2	Lifting an object of 56 lb
3	Use a drill with 16 lb grip force
4	Push cart with 30 lb force

Time Units in Frames

Cycle	Other Activities	Lifting an object of 56 lb	Use a Drill with 16 lb Grip Force	Push Cart with 30 lb Force
1	270			42
2	198		61	
3	286	142		
...	...	...	...	...
66	92	75		
67	110			124
68	272		38	
69	212	49		

In the SHARP Study, it is not possible to perform accurate vibration measurements while also collecting a large amount of other physical exposure parameters for each of the subjects. The alternative is to obtain the declared vibration values of the tools that the workers use, and then perform a time study on the video recordings to measure the actual time that the vibration tools are activated. This estimation may not reflect the real vibration exposure of the workers, as other factors such as tool balancing, work surface conditions, and individual work techniques can influence the true vibration level, but it can give an estimation of the vibration exposure.

44.3.4   Posture Analysis, Event Based vs. Time Based

Posture analysis is based on observations of recorded tasks as well. There are two types of posture analyses, "event based" and "time based." For "event-based" posture analysis the overall postures (the most common posture and the worst posture) for the different body parts when performing a specific task are determined. In "time-based" posture analysis, postures are measured at a particular time for a specific task. There are two types of analyses for the time-based posture analysis: (1) continuous observation and (2) time-sampled observation.

In the continuous observation, the analyst observes the postures of the different body parts continuously, and marks down the changes whenever the body part moves from one predefined angular category to another. This analysis allows the determination of the distribution of the angles of the different body parts and the movement frequency between the different predefined angular categories. However, this type of analysis is very time consuming, particularly when there are several predefined angular categories for each body part. Also, it is very difficult to observe several body parts simultaneously. Therefore, the analyst must play the video several times in order to complete the analyses for the different body parts. In the time-sampled observation of the time-based posture analysis, the analyst observes the postures at a number of preselected times during the task performance, and a distribution of the postures is calculated. Although this method significantly reduces the data processing time, it is not possible to obtain information about repetitive movements of the different body parts in combination with the postures.

In the SHARP Study, both event-based and time-based posture analyses are used. One of the reasons for using both methods is to compare results obtained by the two methods. However, due to the large amount of data processing and analysis in this project, only the time-sampled observation method is used for the time-based posture analysis, rather than the continuous observation approach.

In the event-based posture analysis, predefined postures are used (Table 44.2). Posture distribution results can then be calculated and used for epidemiological modeling. An illustrative distribution result is shown in Figure 44.5, where job A seems to have more wrist extensions and flexion compared to job B and workers at job B maintain more neutral wrist postures compared to workers at job A. Depending on the need of the epidemiological analysis, some of the predefined angular categories may be consolidated.

Using the event-based posture results, one can also calculate certain indices for the different body parts, such as the RULA scores. To obtain the final RULA score, apart from the posture results, additional information such as forces and muscle use should also be obtained. Details on the computation of RULA scores can be found in relevant articles (McAtamney and Corlett, 1993).

In the time-based posture analysis, postures of the same body regions used in the event-based posture analysis are measured. However, instead of giving overall estimated posture values for an entire task (event), postures are estimated at certain points of time during a task performance. In the SHARP Study, postures are estimated for numerous randomly selected frames during a 15-min task recording (75 frames for a single-task job, 80 frames for a two-task job, 90 frames for a three-task job, and 100 frames for a four-task job). To lower individual analyst variation, the frames are assigned to two analysts for processing.

One of the potential problems with predefined angular categories is that the analyst may be biased by the nature of the job. For instance, when a posture is on the threshold of two predefined angular categories (e.g., posture is approximately $30°$ while the categories are 0 to $30°$ and 30 to $60°$), the

TABLE 44.2   Predefined Angular Categories for the Different Body Parts

Trunk	Trunk flexion–extension	Trunk lateral flexion	Trunk twisting
	<0° (extension)	0 to 10°	0 to 10°
	0 to 20° (flexion)	10 to 30°	10 to 45°
	20 to 60°	>30°	>45°
	>60°		
Neck	Neck flexion–extension	Neck lateral flexion	Neck twisting
	<0° (extension)	0 to 10°	0 to 10°
	0 to 20° (flexion)	10 to 30°	10 to 45°
	>20°	>30°	>45°
Upper arms	Upper arm flexion–extension	Upper arm abduction–adduction	Upper arm rotation
	<0° (extension)	<0° (adduction)	<0° (outward)
	0 to 20° (flexion)	0 to 30° (abduction)	0 to 15° (inward)
	20 to 45°	30 to 60°	15 to 45°
	45 to 90°	60 to 90°	>45°
	>90°	>90°	
Shoulders and elbows	Shoulder raise	Arm supported	Elbow flexion
	Yes or no	Yes or no	<0° (extension)
			0 to 20° (flexion)
			20 to 60°
			60 to 100°
			>100°
Forearms and wrists	Forearm rotation	Wrist flexion–extension	Wrist ulnar–radial deviation
	−180 to −90° (supernation)	<−45° (extension)	<−15° (radial)
	−90 to 0° (supernation)	−45 to −15°	−15 to −5°
	0 to 90° (pronation)	−15 to 0°	−5 to 0°
	90 to 180° (pronation)	0 to 15° (flexion)	0 to 10° (ulnar)
		15 to 45°	10 to 20°
		>45°	>20°

analyst may assign the worse posture category to the subject when he or she thinks the job is hazardous, or vice versa. To overcome this problem, in the SHARP Study, a continuous angular scale was used during data processing (posture estimate) and then the data were categorized later during the analysis. In order to make the continuous scale for the posture estimate, a special data processing program has been created. One of the data processing screens is shown in Figure 44.6. In this screen the worker is shown from two camera angles (pictures are just for illustrative purpose in the figure and does not represent the actual analysis) at a preselected video frame. The analyst observes the posture and estimates the approximate locations of the body parts by clicking on the posture diagrams. The continuous angle data are automatically entered into a database.

From the raw posture data, posture distributions can be computed based on pre-defined angle categories (e.g., Job A: 15% of time wrist posture is >45° extended, 67% of time wrist posture is

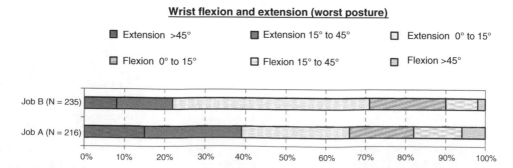

FIGURE 44.5   Distribution results of event-based posture analysis (*n* — number of subjects).

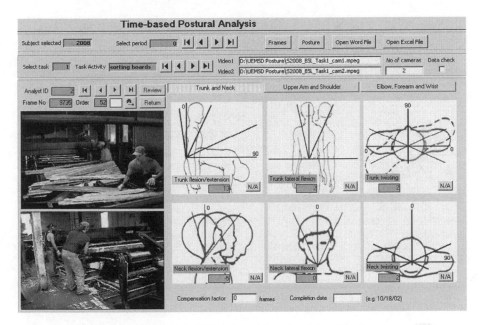

FIGURE 44.6 Time-based posture analysis using continuous angular scales (data do not represent the actual angle measurements).

between 45° extension and 15° flexion, and 18% of time wrist posture is >15° flexion; Job B: 8% of time wrist posture is >45° extended, 82% of time wrist posture is between 45° extension and 15° flexion, and 10% of time wrist posture is >15° flexion). Depending on the epidemiological analysis needs, alternatively, the descriptive statistics for individual subjects such as mean, median, 95% percentile of the postures can also be calculated.

44.3.5 Repetitive Exertion Analysis

In the SHARP Study, hand exertion is defined as any physical effort of the hand when performing a task activity. This is different from the significant force applications, which may not include all hand exertions. This analysis is another way to capture exposure on hand activities. The purpose of this analysis is to calculate the frequency of exertion (or cycle time of exertion), duration of exertion, and exertion duty cycles (% exertion). In the SHARP Study, this analysis is done using the MVTA time-study program (Yen and Radwin, 1995). In this analysis, completed in the laboratory, each hand is analyzed separately. Videotapes may be played at normal speed in order for the analysts to understand the contents of the task activities and played in slow motion mode in order for the analyst to capture the exertions. To reduce processing time, particularly when the tasks are of a very repetitive nature, the analysis is done on five to six randomly selected 1-min intervals of the recorded task instead of the whole video recording.

A typical repetitive exertion MVTA analysis window is shown in Figure 44.7. The left and right hands are analyzed separately. From this analysis, a raw data report, containing durations of individual exertion and nonexertion events, can be generated. The raw data report can be used to calculate the following:

$$\text{Average duration of exertion (sec)} = \frac{\sum \text{Duration of individual exertion}}{\text{Number of exertion cycles}}$$

$$\text{Average cycle time (sec)} = \frac{\text{Total recording time}}{\text{Number of exertion cycles}}$$

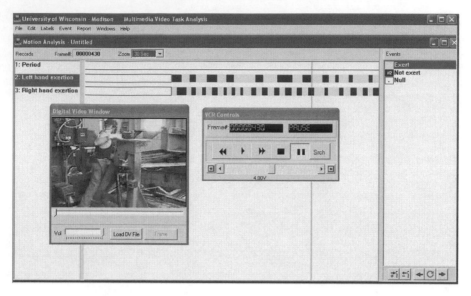

FIGURE 44.7 A typical repetitive exertion analysis screen in MVTA.

or

$$\text{Average frequency of exertion (times/ min)} = \frac{\text{Number of exertion cycles}}{\text{Total recording time}}$$

Average duty cycle (%)

$$= \frac{\sum \left[\text{Individual exertion/(Individual exertion + individual nonexertion)} \right]}{\text{Number of exertion cycles}} \times 100\%$$

44.3.6 Repetitive Movement Analysis

Repetitive movement analysis quantifies the frequency of movements of a specific joint. In the SHARP Study, shoulder movements are quantified through a time study of the recorded tasks using the MVTA program (Yen and Radwin, 1995). Two events are defined to distinguish the directions of the shoulder (the upper arm) movements. A complete movement cycle is defined as movement of the upper arm from one direction to another. Again the left and right shoulders are analyzed separately. Average frequency of the shoulder movement can then be computed based on the raw data report generated by the MVTA program:

$$\text{Average frequency (times/ min)} = \frac{\text{Number of movement cycles}}{\text{Total recording time}}$$

44.3.7 Work Organization Measurement

The SHARP Study is associated with the NIOSH Research Consortium on Work-related Musculoskeletal Disorders, along with other U.S. research groups. One of the first collaborative efforts of the consortium was to develop an observational assessment tool for work organization characteristics adapting elements from AET (Rohmert and Landau, 1983), EWA (Ahonen et al., 1996), and VERA (Volpert et al., 1989). SHARP has been using this checklist (Figure 44.8) at the departmental level during baseline exposure assessment. Prior to specific physical job demand parameter observations, supervisors and workers are queried about the number and type of tasks and their usual duration, rotation patterns, and upset conditions. The ergonomist observes the department that the worker is in for environmental factors, demographic segregation, work method, social content, pacing, positioning, preparation for action, flexibility, attentiveness demands (adapted from EWA; Ahonen et al., 1996), responsibility for the safety of

Organization of Work Observational Rating Sheet			
Worksite_____ Department_____ Job			
Date ___/___/____ Analyst_____ _____			
Gender mix overall job:	❏Male[0]	❏Female[1] ❏Mixed[2]	
Gender segregation by sub-task?	❏ No[0] ❏ Yes[1] ➔ *Specify*		
Other demographic segregation?	❏ No[0] ❏ Yes[1] ➔ *Specify*		
Ambient Environment			
Temperature:		_____	
Humidity:		_____	
Illumination:		❏ Adequate ❏ Too Bright ❏ Too low for task	
Noise:		❏ Normal talk ❏ Yell ❏ Hearing protection	
Housekeeping:	❏ Very Good	❏ Good	❏ Bad	❏ Very Bad
	well controlled storage and cleanliness	neat and orderly; some clutter but generally contained	housekeeping is fair; some clutter on pedestrian traffic routes	dirty and poor storage of materials

Task-Level Work Organization	
Labor Content:	a) ❏ Direct production / service delivery[1] ❏ Indirect production / service delivery[2] b) ❏ Non-skilled[1] ❏ Semi-skilled[2] ❏ Skilled manual[3] ❏ Skilled trade[4] ❏ Professional[5]
Job Type:	❏ Temporary Workers[1] ❏ Hourly, Full-time Workers[2] ❏ Salaried, Full-time Workers[3]
Work Method:	❏ Assembly Line[1] ❏ Work Cells[2] ❏ Desk Work[3]
Posture Type:	❏ Primarily dynamic posture[1] ❏ Primarily static postures[2] ❏ Combination of dynamic and static postures[3]
Social Content:	❏ Individual[1] ❏ Work team, minimal coordination[2] ❏ Work team, mod. coordination[3] ❏ Work team, high coordination *(planning, decision making)*[4]
Positioning:	❏ Single-task, single activity[1] *(e.g. 1 station, only insert screws)* ❏ Single-task, multiple activities[3] *(e.g. 1 station, operates multiple machines)* ❏ Multiple tasks[2] *(e.g., rotate through stations, perform different tasks throughout day)*
Pacing:	❏ Self[1] ❏ Social/Peer[2] ❏ Machine[3] ❏ Line[4] ❏ Piece rate[5] ❏ Quota[6]
Pacing Control:	❏ None[1] ❏ Manual over-ride[2] ❏ Event triggered[3] ❏ Work ahead *(e.g., up-line)*[4] ❏ Inventory "buffer"[5] ❏ Material staging *(e.g., sub-assembly)*[6] ❏ Regular informal breaks *(e.g., relief staff coverage)*[7]
Prep for Action:	❏ None[1] ❏ Maintain "working posture" between work "events"[2] ❏ Part handling[3] ❏ Maintain grasp of tool/other between work "events"[4]

FIGURE 44.8 A sample of work organization checklist used in the SHARP study.

Job Rotation:	☐ No[0] ☐ Yes[1] ➔ ☐ Hourly[1] ☐ Other[2] *(specify)* _____
	➔ Number of jobs in rotation schedule: _____ *(# jobs)*
	➔ Use of different muscle groups? ☐ No[0] ☐ Some[1] ☐ Mostly[2]
Recent Dept Changes:	☐ None[1] ☐ Work methods, all new duties[2] ☐ Job enlargement[3] *(i.e. vertical job additions)* ☐ Job Rotation[4] *(i.e. horizontal job additions)*

Work Group		Comments
Flexible work hours	☐Yes ☐Somewhat ☐No	
Flexible work arrangements	☐Yes ☐Somewhat ☐No	
Formal break schedule	☐Yes ☐Somewhat ☐No	
Informal break possibilities	☐Yes ☐Somewhat ☐No	
Extended work hours	☐8 hr ☐10 hr ☐12 hr ☐Other	
Shift work	☐Yes ☐Somewhat ☐No	

Attentiveness Demands (adapted from EWA)			
☐[1] Superficial	☐[2] Average	☐[3] Somewhat High	☐[4] Very High
handling materials that are not delicate, stamping papers	positioning parts, passive monitoring	proof reading, inspection, active monitoring with some consequence	use of adjusting and measuring instruments; drawing maps; active monitoring with critical consequence

Responsibility for Safety of Others					
☐[1] Does not apply	☐[2] Very Limited	☐[3] Limited	☐[4] Average	☐[5] Significant	☐[6] Very Significant
	worker bears little responsibility for safety of others	worker responsible for safety of others within narrow limits (e.g., proper operation of punch press, lathe)	Worker must maintain vigilance to ensure others are not injured by his/her actions (e.g., operating a crane, auto)	Continual care to ensure safety of others (e.g., handling explosives)	Safety of other depends mainly on the corrective actions of worker (e.g., pilot, surgeon)

Job Content *			
Rating	AET (Landau & Rohmert)	Ergonomic Workplace Analysis	Action Regulation
☐[1] Very minor structural restraints	Worker determines the organization of work, e.g., freelance artist	Work method is in no way restricted to the requirements of a machine, process or the production method or pace.	Establish new working processes - the goal of the actions is not determined in advance.
☐[2] Little structural restraint	Personal freedom of action to organize work to meet general requirements of the job, e.g., sales representative or a scientist		Coordinating several working processes - planning of several sub-goals is needed and their coordination needs to be considered.
☐[3] Average structural restraints	e.g., activities of teachers, administrative workers	Work method is occasionally restricted; demands concentration on the task for a certain time.	Sub-goal planning - a rough planning of sequence is needed and each activity needs its own planning with the sequent activity depending upon the result of the first.
☐[4] Strong structural restraints	Performs a largely predetermined sequence of tasks		Action planning - a sequence of work steps needs to be planned and different circumstances need to be considered.
☐[5] Very strong structural restraints	The sequence of tasks is precisely determined, e.g., parts assembly.	Work method is completely restricted by a machine, process or work group.	Sensory motoric - no conscious planning is needed and occasionally different tools might be used.

FIGURE 44.8 *Continued.*

others, and job content. During subsequent visits, changes to the work organization are documented using the checklist.

The annual worker questionnaire interview inquires about duration on the current job, shift, hours and days of work, overtime schedules, the number of days off in the last week and month, changes in the previous 4 months in tools/equipment, parts made, workstation/area, tasks, rotation pattern, pace. Changes in these parameters are queried every 4 months. Annual self-administered questionnaires include questions about job demands, decision latitude, social support, job satisfaction, job security, and motivation.

It should be noted that there is high correlation between this observational assessment of work organization and worker assessment of psychosocial demands. There is also high correlation between job content identified on this checklist and HAL.

44.4 Summary

The SHARP Study uses a variety of exposure assessment approaches at the individual level and the more global level, incorporating worker estimates (force matching, job content, psychosocial demands, social support, work scheduling), direct measurement (significant push/pull, lifting, pinch and grip forces), observational methods of physical demands (upper extremity postures and motions in terms of amplitude, frequency, and duration), and departmental-level measures of work organization. These data collection methods allow us to input variables into existing event-based and time-based exposure assessment methods (RULA, REBA, HAL, Strain Index) as well as to add precision to estimates of risk based on more quantitative time-based methods. It is hoped that these detailed exposure assessment methods used by ourselves and others in prospective studies will ultimately lead to more efficient and easy-to-use exposure assessment methods for practitioners.

References

ACGIH (2001) Hand activity level. In *TLVs and BEIs — Threshold Limit Values for Chemical Substances and Physical Agents*. Cincinnati, OH: ACGIH.

Ahonen, A., Launis, M., and Kuorinka, T. (1996) *Ergonomic Workplace Analysis*. Helsinki, Finland: Finnish Institute of Occupational Health.

Armstrong, T. J. and Chaffin, D. B. (1979) Carpal tunnel syndrome and selected personal attributes, *J Occup Med* 21 (7): 481–486.

Armstrong, T. J., Dunnigan, J., Ulin, S., and Foulke, J. (1993) Evaluation of a biaxial flexible wire electrogoniometer for measurement of wrist posture. Paper presented at the 24th Congress of the International Commission on Occupational Health.

Armstrong, T. J., Foulke, J. A., Joseph, B. S., and Goldstein, S. A. (1982) Investigation of cumulative trauma disorders in a poultry processing plant, *Am Ind Hyg Assoc J* 43 (2): 103–116.

Bao, S. and Silverstein, B. (2005) Estimation of hand force in ergonomic job evaluations, *Ergonomics* 48: 288–301.

Baty, D., Buckle, P. N., and Stubbs, D. A. (1986) Posture recording by direct observation, questionnaire assessment and instrumentation: a comparison based on a recent field study. In *The Ergonomics of Working Postures*. Edited by N. Corlett and J. Wilson. London: Taylor & Francis.

Belkic, K., Savic, C., Theorell, T., and Cizinsky, S. (1995) *Work Stressors and Cardiovasuclar Risk: Assessment for Clinical Practice. Part I*. Stockholm (Sweden): Stress Research Reports, National Institute for Psychosocial Factors and Health. Section for Stress Research, Karolinska Institute, WHO Psychosocial Center.

Bernard, B. P., Sauter, S. L., Fine, L. J., Petersen, M., and Hales, T. R. (1994) Job task and psychosocial risk factors for work-related musculoskeletal disorders among newspaper employees, *Scand J Work Environ Health* 20: 417–426.

Blanc, P., Faucett, J., Kennedy, J. J., Cisternas, M., and Yelin, E. (1996) Self-reported carpal tunnel syndrome: predictors of work disability form the National Health Interview Survey Occupational Health Supplement, *Am J Ind Med* 30 (3): 362–368.

Borg, G., Ljundggren, G., and Ceci, R. (1985) The increase of perceived exertion, aches and pain in the legs, heart rate and blood lactate during exercise on a bicycle ergometer, *Eur J Appl Physiol* 54: 343–349.

Borg, G. A. V. (1982) Psychophysical bases of perceived exertion, *Med Sci Sports Exer* 14 (5): 377–381.

Buchholz, B. and Wellman, H. (1997) Practical operation of a biaxial goniometer at the wrist joint, *Hum Factors* 39: 119–129.

Burdorf, A. and Laan, J. (1991) Comparison of methods for the assessment of postural load on the back, *Scand J Work Environ Health* 17: 425–429.

Corlett, E. N., Madeley, S. J., and Manenica, I. (1979) Posture targetting: a technique for recording working postures, *Ergonomics* 22 (3): 357–366.

Fallentin, N., Juul-Kristensen, B., Mikkelsen, S., Andersen, J. H., Bonde, J. P., Frost, P., and Endahl, L. (2001) Physical exposure assessment in monotonous repetitive work — the PRIM study, *Scand J Work Environ Health* 27 (1): 21–29.

Fransson, C., Gloria, R., Kilbom, Å., Karlqvist, L., Nygård, C-H., Wiktorin, C., Winkel, J., and the Stockholm MUSIC I Study Group. (1991) Presentation and evaluation of a portable ergonomic observation method (PEO). In *11th Congress International Ergonomics Association: Designing for Everyone.* Edited by Y. Quéinnec and Daniellou. London: Taylor & Francis.

Fransson-Hall, C., Bystrom, S., and Kilbom, A. (1996) Characteristics of forearm-hand exposure in relation to symptoms among automobile assembly line workers, *Am J Ind Med* 29 (1): 15–22.

Freivalds, A. and Eklund, J. (1993) Reaction torques and operator stress while using powered nutrunners, *Appl Ergon* 24 (3): 158–164.

Frost, P. and Andersen, J. H. (1999) Shoulder impingement syndrome in relation to shoulder intensive work, *Occup Environ Med* 56 (7): 494–498.

Helliwell, P. S., Mumford, D. B., Smeathers, J. E., and Wright, V. (1992) Work related upper limb disorder: the relationship between pain, cumulative load, disability, and psychological factors, *Ann Rheum Dis* 51: 1325–1329.

Hignett, S. and McAtamney, L. (2000) Rapid entire body assessment (REBA), *Appl Ergon* 31: 201–205.

Holmstrom, E. B., Lindell, J., and Mortiz, U. (1992) Low back and neck/shoulder pain in construction workers: occupational workload and psychosocial risk factors Part 1: relationship to low back pain, *Spine* 17 (6): 663–671.

Holzman-Voigt, P. (1979) ARBAN: a method for ergonomic analysis of work sites, *Arh Hig Rada Toksikol* 30: 82–86.

Holzmann, P. (1982) ARBAN — a new method for analysis of ergonomic effort, *Appl Ergon* 13 (2): 82–86.

Hägg, G. M., Oster, J., and Bystrom, S. (1997) Forearm muscular load and wrist angle among automobile assembly line workers in relation to symptoms, *Appl Ergon* 28 (1): 41–47.

Karhu, O., Härkönen, R., Sorvali, P., and Vespsäläijnen, P. (1981) Observing working postures in industry: examples of OWAS application, *Appl Ergon* 12: 13–17.

Karhu, O., Kansi, P., and Kuorinka, I. (1977) Correcting working postures in industry: a practical method for analysis, *Appl Ergon* 8 (4): 199–201.

Kasl, S. V. (1992) Surveillance of psychological disorders in the workplace panel. In *Work and Well-being: An Agenda for the 1990s.* Edited by G. P. Keita, and S. L. Sauter. Washington, DC: American Psychological Association, pp. 73–95.

Kemmlert, K. and Kilbom, Å. (1986) *Method for the Identification of Musculo-skeletal Stress Factors which may Have Injurious Effects — PLIBEL.* Solna, Sweden: National Board of Occupational Safety and Health, Research Department, Work Physiology Unit.

Kerr, M. S., Frank, J. W., Shannon, H. S., Norman, R. W., Wells, R. P., Neumann, W. P., and Bombardier, C. (2001) Biomechanical and psychosocial risk factors for low back pain at work, *Am J Public Health* 91 (7): 1069–1075.

Keyserling, W. M. (1986) Postural analysis of the trunk and shoulders in simulated real time, *Ergonomics* 29: 569–583.

Kilbom, Å., Persson, J., and Jonsson, B. (1986) Risk factors for work-related disorders of the neck and shoulder with special emphasis on working postures and movements. Edited by N. Corlett, J. Wilson, and I. Manenica. London: Taylor & Francis.

Kingdon, K. and Wells, R. (2000) Accuracy of force matching using a hand dynamometer, Unpublished work from University of Waterloo, Ontario, Canada.

Latko, W. (1997) Development and evaluation of an observational method for quantifying exposure to hand activity and other physical stressors in manual work, Ph.D. Dissertation, The University of Michigan.

Latko, W. A., Armstrong, R. J., Foulke, J. A., Herrin, G. D., Ranbourn, R. A., and Ulin, S. S. (1997) Development and evaluation of an observation method for assessing repetition in hand tasks, *Am Ind Hyg Assoc J* 58 (4): 278–285.

Lloyd, A., Gandevia, S., and Hales, J. (1991) Muscle performance, voluntary activation, twitch properties and perceived effort in normal subjects and patients with the chronic fatigue syndrome, *Brain* 114: 85–98.

Lueder, R. (1996) A proposed RULA for computer users. Proceedings of the Ergonomics Summer Workshop, August 8–9, UC Berkeley Center for Occupational and Environmental Health.

Marras, W. S. and Schoenmarklin, R. W. (1993) Wrist motions in industry, *Ergonomics* 36: 341–351.

McAtamney, L. and Corlett, E. N. (1993) RULA: a survey method for the investigation of work-related upper limb disorders, *Appl Ergon* 24: 91–99.

Moore, J. S. and Garg, A. (1995) The strain index: a proposed method to analyze jobs for risk of distal upper extremity disorders. *Am Ind Hyg Assoc J* 56: 443–458.

Moore, A., Wells, R., and Ranney, D. (1991) Quantifying exposure in occupational manual task with cumulative trauma disorder potential, *Ergonomics* 34 (12): 1433–1453.

MOW International Research Team (1987) *The Meaning of Working.* London: Academic Press.

Nathan, P. A., Keniston, R. C., Meadows, K. D., and Lockwood, R. S. (1993) Validation of occupational hand use categories, *J Occup Med* 35: 1034–1042.

National Institute for Occupational Safety and Health (NIOSH) (1992) *Selected Topics in Surface Electromyography for Use in the Occupational Setting: Expert Perspectives.* DHHS (NIOSH) Publication No. 91–100.

National Institute for Occupational Safety and Health (NIOSH) (1997) *Musculoskeletal Disorders and Workplace Factors: A Critical Review of Epidemiologic Evidence for Work-Related Musculoskeletal Disorders of the Neck, Upper Extremity, and Low Back.* DHHS (NIOSH).

National Institute for Occupational Safety and Health (NIOSH) (1994) *Revised NIOSH Lifting Equation.* Cincinnati, OH: U.S. Department of Health and Human Services, Public Health Service, Centers for Disease Control, National Institute for Occupational Safety and Health.

Niebel, B. W. (1988) *Motion and Time Study*, 8th edn. Homewood, IL: Irwin.

Nordstrom, D. L., Vierkant, R. A., DeStefano, F., and Layde, P. M. (1997) Risk factors for carpal tunnel syndrome in a general population, *Occup Environ Med* 54: 734–740.

Panel on Musculoskeletal Disorders and the Workplace; Commission on Behavioral and Social Sciences and Education; National Research Council, and Institute of Medicine (2001) *Musculoskeletal Disorders and the Workplace: Low Back and Upper Extremities.* Washington, DC: National Academy Press.

Persson, J. and Kilbom, Å. (1983) VIRA — en enkel videofilmteknik för registrering och analys av arbetsställningar och rörelser, Vol. 10, 23 pp.

Pope, D. P., Silman, A. J., Cherry, N. M., Pritchard, C., and Macfarlane, G. J. (1998) Validity of self-completed questionnaire measuring the physical demands of work, *Scand J Work Environ Health* 24 (5): 376–385.

Punnett, L. (1998) Ergonomic stressors and upper extremity disorders in vehicle manufacturing: cross sectional exposure–response trends, *Occup Environ Med* 55: 414–420.

Punnett, L., Fine, L. J., Keyserling, W. M., Herrin, G. D., and Chaffin, D. B. (2000) Shoulder disorders and postural stress in automobile assembly work, *Scand J Work Environ Heal* 26 (4): 283–291.

Radwin, R. G., Lin, M. L., and Yen, T. Y. (1993) Exposure assessment of biomechanical stress in repetitive manual work using spectral analysis. In Proceedings of the Human Factors and Ergonomics Society 37th Annual Meeting, pp. 669–693.

Roberts, R. (1997) Calibration procedure for the Penny and Giles Z110 torsiometer. University of Massachusetts Lowell, unpublished master's project.

Rohmert, W. and Landau, K. (1983) *A New Technique for Job Analysis.* London: Taylor & Francis.

Roquelaure, Y., Mechali, S., Dano, C., Fanello, S., Benetti, F., Bureau, D., Mariel, J., Martin, Y.-H., Derriennic, F., and Penneau-Fontbonne, D. (1997) Occupational and personal risk factors for carpal tunnel syndrome in industrial workers, *Scand J Work Environ Health* 23: 364–369.

Sauter, S. L. and Swanson, N. G. (1996) An ecological model of musculoskeletal disorders in office work. In *Beyond Biomechanics: Psychosocial Aspects of Musculoskeletal Disorders in Office Work*. Edited by S. D. Moon and S. L. Sauter. London: Taylor & Francis.

SHARP (1999) Ergonomics evaluation report on a trimmer operator's job. Technical Report 22-4-1999. Washington State Department of Labor and Industries, Olympia, WA.

Silverstein, B., Fine, L. J., and Armstrong, T. J. (1987) Occupational factors and carpal tunnel syndrome, *Am J Ind Med* 11: 343–358.

Smutz, P., Serina, E., and Rempel, D. (1994) A system for evaluating the effect of keyboard design on force, posture, comfort, and productivity, *Ergonomics* 37 (10): 1649–1660.

Snook, S. H. (1978) The design of manual handling tasks, *Ergonomics* 21 (12): 963–985.

Snook, S. H., Vaillancourt, D. R., Ciriello, V. M., and Webster, B. S. (1995) Psychophysical studies of repetitive wrist flexion and extension, *Ergonomics* 38 (7): 1488–1507.

Spielholz, P., Silverstein, B. A., and Stuart, M. (1999) Reproducibility of a self-report questionnaire for upper extremity musculoskeletal disorder risk factors, *Appl Ergon* 30: 429–433.

Stetson, D. S., Keyserling, W. M., Silverstein, B. A., and Leonard, J. A. (1991) Observational analysis of the hand and wrist: a pilot study, *Appl Occup Environ Hyg* 6 (11): 927–937.

Stetson, D. S., Silverstein, B. A., Keyserling, W. M., Wolfe, R. A., and Albers, J. W. (1993) Median sensory distal amplitude and latency: comparisons between nonexposed managerial/professional employees and industrial workers, *Am J Ind Med* 24: 175–189.

Toomingas, A., Theorell, T., Michelsen, H., Nordemar, R., and Stockholm MUSIC I Study Group. (1997) Associations between self-rated psychosocial work conditions and musculoskeletal symptoms and signs, *Scand J Work Environ Health* 23: 130–139.

Veiersted, K. B. and Westgaard, R. H. (1993) Development of trapezius myalgia among female workers performing light manual work, *Scand J Work Environ Health* 19: 277–283.

Viikari-Juntura, E., Martikainen, R., Luukkonen, R., Mutanen, P., Takala, E., and Riihimaki, H. (2000) A longitudinal study of work-related and individual risk factors of radiating neck pain. People and Work Research Report, Helsinki, Finland.

Viikari-Juntura, E., Rauas, S., Marikainen, R., Kuosma, E., Riihimaki, H., Takala, E., and Saarenmaa, K. (1996) Validity of self-reported physical work load in epidemiologic studies on musculoskeletal disorders, *Scand J Work Environ Health* 22: 251–259.

Volpert, W., Kötter, W., Gohde, H.-E., and Weber, W. G. (1989) Psychological evaluation and design of work tasks: two examples, *Ergonomics* 32 (7): 881–890.

Wiktorin, C., Karlqvist, L., Winkel, J., and Stockholm MUSIC I Study Group, (1993) Validity of self-reported exposures to work postures and manual materials handling, *Scand J Work Environ Health* 19: 208–214.

Wiktorin, C., Selin, K., Ekenvall, L., Kilbom, A., and Afredsson, L. (1996) Evaluation of perceived and self-reported manual forces exerted in occupational materials handling, *Appl Ergon* 27 (4): 231–239.

Winkel, J. and Gard, G. (1988) An EMG-study of work methods and equipment in crane coupling as a basis for job redesign, *Appl Ergon* 19 (3): 178–184.

Winkel, J. and Mathiassen, S. E. (1994) Assessment of physical work load in epidemiologic studies: concepts, issues and operational considerations, *Ergonomics* 37: 979–988.

Yen, T. Y. and Radwin, R. G. (1995) A video-based system for acquiring biomechanical data synchronized wit arbitrary events and activities, *IEEE Transactions on Biomedical Engineering* 42 (9): 944–948.

Upper Extremity Analysis of the Wrist

Andris Freivalds
Pennsylvania State University

45.1 Anatomy of the Hand and Wrist

45.1.1 Bones of the Hand and Wrist

The human hand has 27 bones divided into three groups: 8 carpal bones in the wrist, 5 metacarpal bones, and 14 phalanges of the fingers. The carpal bones are arranged in two rows and have names reflecting their shapes (Figure 45.1). The bones of the distal row, from the lateral side to the medial side, include the trapezium (four sided with two parallel sides), the trapezoid (four sided), the capitate (the central bone), and the hamate (hook shaped). These four bones fit together, tightly bound by inter-osseous ligaments to form a relatively immobile unit that articulates with the metacarpals to form the carpometacarpal (CMC) joint. The bones of the proximal row include the scaphoid (boat shaped), the lunate (half-moon shaped), the triquetrum (triangle shaped), and the pisiform (pea shaped). The proximal surfaces of the scaphoid, lunate, and triquetrum form a biconvex elliptical surface which articulates with the biconcave surface of the distal extremity of the radius. The articulation between the proximal and distal rows is term the midcarpal joint while articulations between adjacent bones are called intercarpal joints.

The five metacarpal bones are cylindrical in shape and articulate proximally with the distal carpal bones and distally with the proximal phalanges of the digits. The base of the metacarpal bones of the index and middle fingers are linked together tightly and articulate little with the trapezoid and the capitate bones in the CMC joint. On the other hand, the CMC joint for metacarpals of the ring and little fingers with the hamate allow up to $10°-15°$ and $20°-30°$ of flexion/extension, respectively. The arched shafts of the metacarpal bones form the palm and the distal ends are spherical in shape, allowing

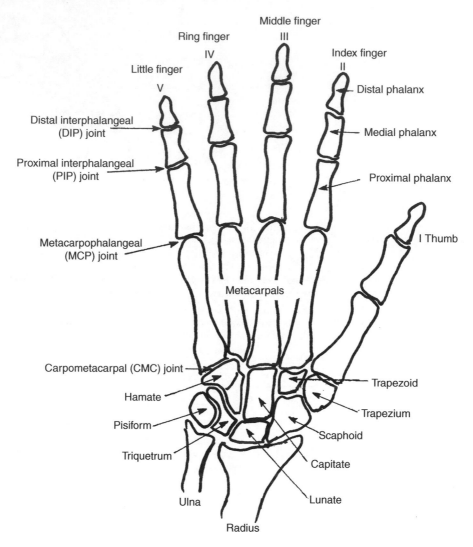

FIGURE 45.1   Bones and joints of the right hand (palmar view).

articulation with the base of the corresponding phalanges. Interosseous muscles and extensor tendons run along the concave side and the large and smooth dorsal surface of the shaft, respectively, while the distal ends have a grooved volar surface for the flexor tendons (Nordin and Frankel, 2001).

There are three phalanges for each digit and two for the thumb for a total of 14 bones. They are labeled proximal, middle, and distal phalanges (with the middle one missing in the thumb), according to their positions and become progressively smaller. The heads of the proximal and middle phalanges are bicondylar, facilitating flexion, extension, and circumduction. The shafts are semicircular in cross section (the palmar surface being almost flat), as opposed to the cylindrical metacarpals. The axes of the distal phalanges of the index, ring, and little fingers are respectively deviated ulnarly, radially, and radially from the axes of the middle phalanges (Gigis and Kuczynski, 1982).

45.1.2   Joints of the Hand and Wrist

There are four joints in each finger, in sequence from the proximal to distal: carpometacarpal (CMC), metacarpophalangeal (MCP), proximal interphalangeal (PIP), and distal interphalangeal (DIP) joints

(Figure 45.1). The CMC joints are formed by the bases of the four metacarpals and the distal carpal bones and are stabilized by interosseous ligaments to form a relatively immobile joint. However, a major function of the CMC joint is to form the hollow of the palm and allow the hand and digits to conform to the shape of the object being handled (Norkin and Levangie, 1992).

The MCP joints are composed of the convex metacarpal head and the concave base of the proximal phalanx and stabilized by a joint capsule and ligaments. Flexion of 90° and extension of 20°–30° from neutral take place in the sagittal plane. The range of flexion differs among fingers (and individuals) with the index finger having the smallest flexion angle of 70° and the little finger showing the largest angle of 95° (Batmanabane and Malathi, 1985). Radial and ulnar deviation of approximately 40°–60° occurs in the frontal plane, with the index finger showing up to 60° abduction and adduction, middle and ring fingers up to 45°, and the little finger about 50° of mostly abduction (Steindler, 1955). The range of motion at the MCP joint decreases as the flexion angle increases because of the bicondylar metacarpal structure (Youm et al., 1978; Schultz et al. 1987). There is also some axial rotation of the fingers from a pronated to a supinated position as the fingers are extended. In the reverse motion, the fingers crowd together as they go into flexion (Steindler, 1955).

The IP joints, being hinge joints, exhibit only flexion and extension. Each finger has two IP joints, the PIP and the DIP joints, except the thumb, which has only one. Volar and collateral ligaments, connected with expansion sheets of the extensor tendons, prevent any side to side motion. The largest flexion range of 100°–110° is found in the PIP joints; while a smaller flexion range of 60°–70° is found in the DIP joints. Hyperextension or extension beyond the neutral position, due to ligament laxity, can also be found in both DIP and PIP joints (Steindler, 1955).

45.1.3 Muscles of the Forearm, Hand, and Wrist

The muscles producing movement of the fingers are divided into two groups: extrinsic and intrinsic, based on the origin of the muscles. The extrinsic muscles originate primarily in the forearm, while the intrinsic muscles originate primarily in the hand. Therefore, the extrinsic muscles are large and provide strength, while the intrinsic muscles are small and provide precise coordination for the fingers. Each finger is innervated by both sets of muscles, requiring good coordination for hand movement.

The extrinsic muscles are divided into flexors found primarily on the anterior forearm (Figure 45.2) and extensors found primarily on the posterior forearm (Figure 45.3). Most of the flexors originate from the medial epicondyle of the humerus while most of extensors originate from the lateral epicondyle of the humerus. Both sets of muscles insert on the carpal bones, metacarpals, or phalanges. Each group can be further divided into superficial and deep groups of muscles as categorized in greater detail in Table 45.1.

The intrinsic muscles are divided into three groups: the thenar, the hypothenar, and the midpalmar muscle groups. The thenar group acts on the thumb and comprises the thenar eminence at the base of the thumb. The hypothenar group acts on the little finger and comprises the hypothenar eminence at the base of the medial palm. The midpalmar muscles act on all of the phalanges except the thumb. Primarily located on the palmar side, these intrinsic muscles allow for the independent flexion/extension and abduction/adduction of each of the phalanges giving rise to precise finger movements. These muscles are shown in Figure 45.4 and categorized in Table 45.2.

45.1.4 The Flexor Digitorum Profundus and Flexor Digitorum Superficialis

The flexor digitorum profundus (FDP) and flexor and digitorum superficialis (FDS) are the main finger flexor muscles and are involved in most repetitive work. Using EMG, Long et al. (1970) identified the FDP as the muscle performing most of the unloaded finger flexion, while the FDS came into play when additional strength was needed, with the FDP comprising about 12% of the total muscle capability below the elbow. There is significant variation in force contributions of the FDS tendons for each finger

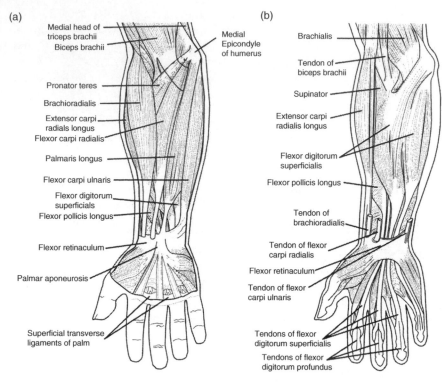

FIGURE 45.2   Anterior muscles of the right hand: (a) superficial layer, (b) middle layer. (Adapted from Spence, A.P. (1990). *Basic Human Anatomy*, 3rd ed. Redwood City, CA: Benjamin/Cummings. With permission.)

(0.9–3.4%), while the FDP tendons provide a relatively constant force contribution to each finger (2.7–3.4%). This results in a relatively large range of force ratios, from 1.5 to 3. Average FDP resting tendon fiber length is slightly shorter than for the average FDS tendon. Table 45.3 provides a summary of FDP and FDS tendon characteristics.

The FDP originates from the proximal anterior and medial surface of the ulna and inserts into the base of the distal phalanx (Figure 45.5). In the midforearm, the muscle divides into two bellies: the radial and the ulnar. The radial part inserts into the index finger, while the ulnar part inserts into middle, ring, and little fingers. Consequently, the latter three fingers tend to move together, while the index finger can function independently of the others. The FDP tendon passes along the finger through a series of pulleys, which maintain a reasonably constant moment arm for flexing or extending the finger. Before inserting into the distal phalanx, the FDP passes through a split in the FDS tendon (Fahrer, 1971; Steinberg, 1992; Brand and Hollister, 1993).

45.1.5 Flexor Tendon Sheath Pulley Systems

The tendon sheath is a double-walled tube, surrounding the tendons and containing synovial fluid. The synovial sheath provides both a low-friction gliding as well as a nutritional environment for the flexor tendon. The flexor tendon sheath, sometimes termed the fibro-osseous tunnel, begins at the neck of the metacarpal phalanx, ends at the distal interphalangeal joint, and is held against the phalanges by pulleys. These pulleys primarily act to prevent tendon bowstringing across the joints during flexion but also maintain a relatively constant moment arm.

The pulleys can be divided into three types based on their locations: a palmar aponeurosis pulley, five annular (ring shaped) pulleys (A1, A2, A3, A4, and A5) and three cruciate (cross-like) pulleys (C1, C2, and C3). The A2 and A4 pulleys are located on the proximal and middle phalanges, while the A1, A3, and

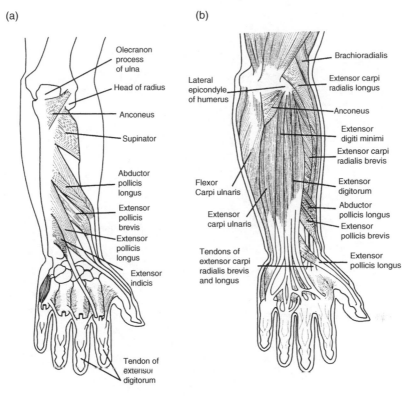

FIGURE 45.3 Posterior muscles of the right hand: (a) deep layer, (b) superficial layer. (Adapted from Spence, A.P. (1990). *Basic Human Anatomy,* 3rd ed. Redwood City, CA: Benjamin/Cummings. With permission.)

A5 pulleys are located at the palmar surface of the MCP, PIP, and DIP joints (Figure 45.6). The A2 and A4 pulleys are most important for normal function and a stable joint, with the A3 and other pulleys coming into play when the A2 and A4 have been damaged (Manske and Lesker, 1977; Idler, 1985; Lin et al., 1990). Such damage to the pulleys can occur in extreme activities, in which much of the body

TABLE 45.1 Extrinsic Muscles of the Hand and Wrist

Group	Layer	Name	Nerve	Function
Anterior	Superficial	Flexor carpi radialis	Median	Flexes and adducts hand
		Palmaris longus	Median	Flexes hand
		Flexor carpi ulnaris	Ulnar	Flexes and adducts hand
	Middle	Flexor digitorum superficialis	Median	Flexes phalanges and hand
	Deep	Flexor digitorum profundus	Median	Flexes phalanges and hand
Posterior	Superficial	Extensor carpi radialis longus	Radial	Extends and abducts hand
		Extensor carpi radialis brevis	Radial	Extends hand
		Extensor digitorum	Radial	Extends little finger
		Extensor digiti minimi	Radial	Extends little finger
		Extensor carpi ulnaris	Radial	Extends and adducts hand
	Deep	Abductor pollicis longus	Radial	Abducts thumb and hand
		Extensor pollicis brevis	Radial	Extends thumb
		Extensor pollicis longus	Radial	Extends thumb
		Extensor indicis	Radial	Extends index finger

Source: Adapted from Spence, A.P. (1990). *Basic Human Anatomy,* 3rd ed. Redwood City, CA: Benjamin/Cummings and Tubiana, R. (ed.) (1981). *The Hand,* Vol. 1. Philadelphia, PA: Saunders. With permission.

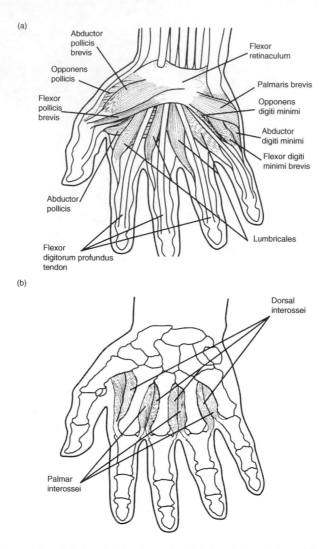

FIGURE 45.4   Intrinsic muscles of the hand: (a) palmar view, (b) dorsal view. (Adapted from Spence, A.P. (1990). *Basic Human Anatomy*, 3rd ed. Redwood City, CA: Benjamin/Cummings. With permission.)

TABLE 45.2   Intrinsic Muscles of the Hand and Wrist

Group	Name	Nerve	Function
Thenar muscles	Abductor pollicis brevis	Median	Abducts thumb
	Opponens pollicis	Median	Pulls thumb toward little finger
	Flexor pollicis brevis	Median	Flexes thumb
	Adductor pollicis	Ulnar	Adducts thumb
Hypothenar muscles	Palmaris brevis	Ulnar	Folds skin on ulnar side of palm
	Abductor digiti minimi	Ulnar	Abducts little finger
	Flexor digiti minimi	Ulnar	Flexes little finger
	Opponens digiti minimi	Ulnar	Pulls little finger toward thumb
Midpalmar muscles	Lumbricales	Median, ulnar	Flex proximal phalanges
	Dorsal interossei	Ulnar	Abduct fingers
	Palmar inerossei	Ulnar	Adduct fingers

Source: Adapted from Spence, A.P. (1990). *Basic Human Anatomy*, 3rd ed. Redwood City, CA: Benjamin/Cummings and Tubiana, R. (ed.) (1981). *The Hand*, Vol. 1. Philadelphia, PA: Saunders. With permission.

TABLE 45.3 General Characteristics of the FDP and FDS Tendons

| Finger | Resting Fiber Length (cm)[a] | | Moment Arm (cm)[b] | | | | | | Relative Force of Finger Flexors (%)[c] | | |
| | FDP | FDS | DIP joint | | PIP joint | | MCP joint | | FDP | FDS | FDP/FDS |
			FDP	FDS	FDP	FDS	FDP	FDS			
Index	6.6	7.2	0.65	—	0.98	0.83	1.01	1.21	2.7	2.0	1.35
Middle	6.6	7.0	0.70	—	1.07	0.87	1.16	1.40	3.4	3.4	1.00
Ring	6.8	7.3	0.68	—	1.04	0.85	1.04	1.30	3.0	2.0	1.50
Little	6.2	7.0	0.60	—	0.85	0.74	0.89	0.98	2.8	0.90	3.11

[a]Brand et al. (1981).
[b]Ketchum et al. (1978).
[c]Brand and Hollister (1993).

weight is supported by the fingers, such as rock climbing. Although, the A3 pulley is relatively weaker and closer to the PIP joint, it is more flexible and stretches, transferring the load to the A2 and A4 pulleys, which then fail first (Marco et al., 1998).

45.1.6 Wrist Mechanics

The seven main muscles involved in wrist and hand motion are flexor carpi radialis (FCR), flexor carpi ulnaris (FCU), flexor digitorum profundus (FDP), flexor digitorum superficialis (FDS), extensor carpi radialis brevis (ECRB), extensor carpi radialis longus (ECRL), and extensor carpi radialis ulnaris (ECU) (Garcia-Elias et al., 1991). FCR, FCU, ECRB, ECRL, and ECU's primary function is to move the wrist, while FDP and FDS are secondary wrist movers. FDP and FDS's primary function is to flex and extend the fingers and secondarily to rotate the wrist. The FDP and FDS pass through carpal tunnel. The primary muscles and tendons involved with wrist movements of flexion, extension, radial and ulnar deviation planes are listed below (An et al., 1981):

Flexion: FCR and FCU
Extension: ECRB, ECRL, and ECU
Radial deviation: FCR, ECRB, and ECRL
Ulnar deviation: FCU, and ECU

The parameters that have been commonly used to describe the muscles are muscle fiber length (FL) and physiological cross-sectional area (PCSA). Muscle length is related to mechanical potential for tendon excursion, and the maximum tension of the muscle to its PCSA (An et al., 1991). In moving the wrist, each tendon across the wrist joint slides a certain distance to execute the movement, and the tendon excursion and moment arm at various wrist joint angles can be measured and derived through experiments (Armstrong and Chaffin, 1978; An et al., 1991). Muscle parameters, tendon excursions, and moment arms at wrist joint (An et al., 1981; Lieber et al., 1990; An et al., 1991) are summarized in Table 45.4. The magnitudes of tendon excursion were measured over a 100° range of motion in the

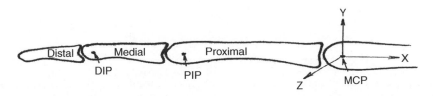

FIGURE 45.5 FDP and FDS tendons of a typical digit.

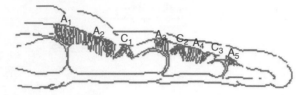

FIGURE 45.6   Pulley structure of the finger. (Adapted from Nordin, M. and Frankel, V.H. (2001). *Basic Mechanics of the Musculoskeletal System*, 3rd ed. Philadelphia, PA: Lippincott Williams & Wilkins. With permission.)

flexion–extension plane and a 50° range of motion in the radial–ulnar deviation plane in forearm neutral position.

Table 45.4 reveals that the FCR, FCU, and ECRB provide larger tendon excursion during flexion and extension movement than ECRL and ECU, while ECRL and ECU have greater tendon excursion during radial and ulnar deviation movement. The results also demonstrate that the FCR and FCU are prime muscles for flexion, ECRB for extension, ECRL for radial deviation, and ECU for ulnar deviation. In spite of the three-dimensional orientation of the wrist tendons to the rotation axes and the complexity of carpal bone motion, Table 45.4 indicates that the moment arms of wrist motion are maintained fairly consistently and correspond well with the anatomical location of the tendons. According to An et al. (1991), these findings are related to the anatomical considerations; the extensor retinaculum ensures a consistent relationship of the wrist extensors (ECRB, ECBL, and ECU) to the rotation axes, while the FCR is firmly fixed in the fibro-osseous groove, and the FCU infixed on the pisiform.

45.1.7   Select Anthropometry Data

For any sort of biomechanical modeling it is necessary to have a variety of key anthropometric properties, such as segment link lengths, segment weights, the location of the center of gravity, the location of the center of joint rotation, the range of motion for each joint, and muscle insertion points. Not all of this information has been measured or documented at the level of individual phalanges, but the following tables may provide some useful data: Table 45.5 — phalange lengths, Table 45.6 — interphalangeal joint dimensions, Table 45.7 — joint center locations, and Table 45.8 — tendon insertion points. The latter two tables were adapted from the data of An et al. (1979) based on three separate coordinate systems located at the center of joint rotation for each phalanx as shown in Figure 45.7. Note the x-axis is in the axial direction of the phalanx, with the positive direction pointing proximally. The y-axis is perpendicular with the positive direction pointing dorsally. The positive z-axis points radially for the right hand. However, any deviations in the z-direction were generally minimal and were omitted for simplicity. The tendon insertion points of Table 45.8 correspond approximately to the location of the five annular pulleys. This still allows for adequate modeling of finger flexion in the $x–y$ plane. Other, more functional,

TABLE 45.4   Physiological and Mechanical Properties of Wrist Joint Muscles and Tendons

Muscle and Tendon	Physiological Size		Tendon Excursion (mm)		Moment Arm (mm)[a]	
	Length (cm)	PCSA (cm^2)	F/E plane	R/U plane	F/E plane	R/U plane
FCR	10.9–12.4	2.0	25 ± 4	7 ± 1	+15 ± 3	+8 ± 2
FCU	15.2–15.4	3.2–3.4	28 ± 4	12 ± 3	+16 ± 3	−14 ± 3
ECRB	13.8–15.8	2.7–2.9	20 ± 3	11 ± 1	−12 ± 2	+13 ± 2
ECRL	11.8–18.3	1.5–2.4	12 ± 3	17 ± 1	−7 ± 2	+19 ± 2
ECU	13.6–14.9	2.6–3.4	10 ± 2	16 ± 2	−6 ± 1	−17 ± 3

[a] + Denotes flexion and radial deviation, and − denotes extension and ulnar deviation.

Sources: Adapted from, An, K.N., Hui, F.C., Morrey, B.F., Linscheid, R.L., and Chao, E.Y.S. (1981). *Journal of Biomechanics*, 14:659–669; Lieber, R.L., Fazeli, B.M. and Botte, M.J. (1990). *Journal of Hand Surgery*, 15:244–250; An, K.N., Horii, E., and Ryu, J. (1991). *Biomechanics of the Wrist Joint*, New York: Springer Verlag, pp. 157–169. With permission.

TABLE 45.5 Phalange Lengths as Percentage of Hand Length for Males and Females

Phalanx	Proximal	Medial	Distal
Thumb	17.1	—	12.1
Index	21.8	14.1	8.6
Middle	24.5	15.8	9.8
Ring	22.2	15.3	9.7
Little	17.7	10.8	8.6

Sources: From Davidoff, N.A. MS Thesis, Pennsylvania State University and Davidoff, N.A. and Freivalds, A. (1993). *International Journal of Industrial Ergonomics*, 12: 255–264. With permission.

data on the hand can be found in Garrett (1971) and detailed data on muscle moment arms and tendon excursions for the index finger can be found in An et al. (1983).

45.2 Models of the Hand and Wrist

45.2.1 Static Tendon Pulley Models

Landsmeer (1960, 1962) developed three biomechanical models for finger flexor tendon displacements, in which the tendon–joint displacement relationships are determined by the spatial relationships between the tendons and joints. In Model I (Figure 45.8a), he assumed that the tendon is held securely against the curved articular surface of the proximal bone of the joint, and the proximal articular surface can be described as a trochlea. Such a model is particularly useful in describing extensor muscles. The tendon displacement relationship is described by:

$$x = R\theta \tag{45.1}$$

where x is the tendon displacement, R is the distance from the joint center to the tendon, and θ is the joint rotation angle.

However, if the tendon is not held securely, it may be displaced from the joint when the joint is flexed and will settle in a position along the bisection of the joint angle (see Figure 45.8b). Model II is useful for describing tendon displacement in intrinsic muscles as:

$$x = 2R \sin(\theta/2) \tag{45.2}$$

TABLE 45.6 Interphalangeal Joint Dimensions — Mean and Standard Deviations (in parentheses) in mm

Joint	Breadth		Thickness	
	Male	Female	Male	Female
IP (I)	22.9 (3.8)	19.1 (1.3)	20.1 (1.5)	16.8 (1.0)
PIP (II)	21.3 (1.3)	18.3 (1.0)	19.6 (1.3)	16.3 (1.0)
DIP (II)	18.3 (1.3)	15.5 (1.0)	15.5 (1.3)	13.0 (1.0)
PIP (III)	21.8 (1.3)	18.3 (1.0)	20.1 (1.5)	16.8 (1.0)
DIP (III)	18.3 (1.3)	15.2 (1.0)	16.0 (1.3)	13.2 (1.0)
PIP (IV)	20.1 (1.3)	18.3 (1.0)	18.8 (1.3)	15.8 (1.0)
DIP (IV)	17.3 (1.0)	14.5 (0.8)	15.2 (1.3)	12.5 (0.8)
PIP (V)	17.8 (1.5)	14.5 (0.8)	16.8 (1.3)	14.0 (1.0)
DIP (V)	15.8 (1.3)	13.2 (0.8)	13.7 (1.3)	11.4 (0.8)

Note: I = thumb, II = index finger, III = middle, IV = ring, V = little.
Source: From Garrett, J.W. (1970a,b). AMRL-TR-69-26 and AMRL-TR-69-42. Wright Patterson AFB, OH: Aerospace Medical Research Laboratory. With permission.

TABLE 45.7   Location of Finger Joint Centers from Distal End of Phalanx. DIP Distances as percentage of Medial Phalanx Length, PIP and MCP Distances as Percentage of Proximal Phalanx Length

Finger	DIP	PIP	MCP
Index	18	13	20
Middle	15	12	20
Ring	13	12	19
Little	17	14	24

Source: Adapted from An, K.N., Chao, E.Y., Cooney, W.P., and Limscheid, R.L. (1979). *Journal of Biomechanics*, 12:775–788. With permission.

Landsmeer's (1960) Model III depicts a tendon running through a tendon sheath held securely against the bone, which allows the tendon to curve smoothly around the joint (Figure 45.8c). The tendon displacement is described by:

$$x = 2\left[y + \frac{1}{2}\theta\left(d - y/\tan\frac{1}{2}\theta\right)\right] \tag{45.3}$$

where y is the tendon length to joint axis measured along the long axis of the bone and d is the distance of tendon to the long axis of the bone.

For small angles of flexion ($\theta < 20°$), tan θ is almost equal to θ, and Equation (45.3) simplifies to:

$$x = d\,\theta \tag{45.4}$$

Armstrong and Chaffin (1979) proposed a static model for the wrist based on Landsmeer's (1962) tendon Model I and LeVeau's (1977) pulley-friction concepts (Figure 45.9). Armstrong and Chaffin (1978) found that, when the wrist is flexed, the flexor tendons are supported by flexor retinaculum on the volar side of the carpal tunnel. When the wrist is extended, the flexor tendons are supported by the carpal bones. Thus, deviation of the wrist from neutral position causes the tendons to be displaced against and past the adjacent walls of the carpal tunnel. They assumed that a tendon sliding over a curved surface is analogous to a belt incurring friction forces while wrapped around a pulley. The radial reaction force on the ligament or the carpal bones, F_R, can be characterized as follows:

$$F_R = 2F_T\,e^{\mu\Theta}\,\sin(\theta/2) \tag{45.5}$$

TABLE 45.8   Representative Tendon Insertion Distances for Each Finger with Respect to that Joint Center Coordinate System (Figure 45.7). DIP distances as percentage of Medial Phalanx Length, PIP and MCP Distances as percentage of Proximal Phalanx Length

		Index Finger				Middle Finger				Ring Finger				Little Finger			
		Distal		Proximal		Distal		Proximal		Distal		Proximal		Distal		Proximal	
Joint	Muscle	x	y	x	y	x	y	x	y	x	y	x	y	x	y	x	y
DIP	FDP	−18	−13	25	−22	−19	−12	26	−22	−19	−13	26	−22	−18	−15	2	−20
PIP	FDP	−22	−14	18	−17	−27	−15	22	−14	−28	−16	23	−16	−24	−15	20	−18
	FDS	−22	−11	18	−13	−27	−11	22	−12	−28	−13	23	−13	−24	−13	20	−15
MCP	FDP	−24	−17	13	−27	−30	−18	16	−26	−31	−16	17	−26	−27	−20	15	−28
	FDS	−24	−22	13	−31	−30	−18	16	−31	−31	−19	17	−29	−27	−24	15	−32

Source: Adapted from An, K.W., Chao, E.Y., Cooney, W.P., and Limscheid, R.L. (1979). *Journal of Biomechanics*, 12: 775–788. With permission.

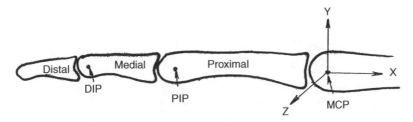

FIGURE 45.7 Finger coordinate system. (Adapted from An, K.W., Chao, E.Y., Cooney, W.P., and Linscheid, R.L. (1979). *Journal of Biomechanics*, 12:775–788. With permission.)

where F_R is the radial reaction force, F_T is the tendon force or belt tension, μ is the coefficient of friction between tendon and supporting tissues, and θ is the wrist deviation angle (in radians).

The resulting normal forces on the tendon exerted by the pulley surface can be expressed per unit arc length as:

$$F_N = \frac{2F_T\, e^{\mu\Theta}\, \sin(\theta/2)}{R\theta} \tag{45.6}$$

where F_N is the normal forces exerted on tendon and R is the radius of curvature around supporting tissues.

For small coefficient of frictions, comparable to what is found in joints ($\mu < 0.04$) and for small angles of θ, Equation (45.6) reduces to the simple expression:

$$F_N = F_T/R \tag{45.7}$$

Thus, F_N is a function of only the tendon force and the radius of curvature. As the tendon force increases or the radius of curvature decreases (e.g., small wrists), the normal supporting force exerted on tendon increases. F_R, on the other hand, is independent of radius of curvature but is dependent on the wrist deviation angle.

This tendon-pulley model provides a relatively simple mechanism for calculating the normal supporting force exerted on tendons that are a major factor in work-related musculoskeletal disorders (WRMSDs). However, this model does not include the dynamic components of wrist movements such as angular velocity and acceleration, which might be risk factors in WRMSDs.

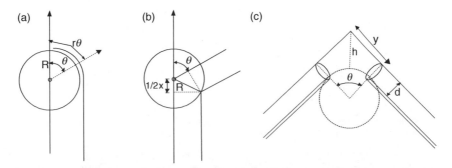

FIGURE 45.8 Landsmeer's tendon models. (a) Model I, (b) Model II, (c) Model III, (Adapted from Landsmeer, J.M.F. (1962). *Annals of Rheumatoid Diseases*, 21:164–170. With permission.)

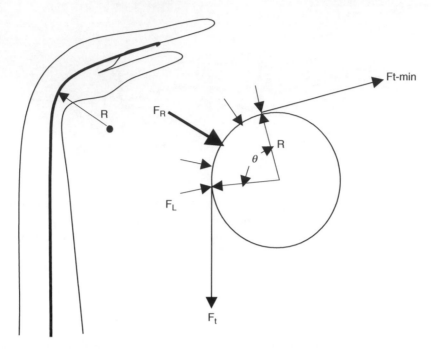

FIGURE 45.9 Wrist tendon pulley model. (From Chaffin, D.B., Andersson, G.B.J., and Martin, B.J. (1999). *Occupational Biomechanics*, 3rd ed., New York: John Wiley & Sons. With permission.)

45.2.2 Dynamic Tendon-Pulley Models

Schoenmarklin and Marras's (1990) dynamic biomechanical model extended Armstrong and Chaffin's (1979) static model to include dynamic components of angular acceleration (Figure 45.10). This model is two-dimensional in that only the forces in flexion and extension plane are analyzed. It also investigates the effects of maximum angular acceleration on the resultant reaction force that the wrist ligaments and carpal bones exert on tendons and their sheaths.

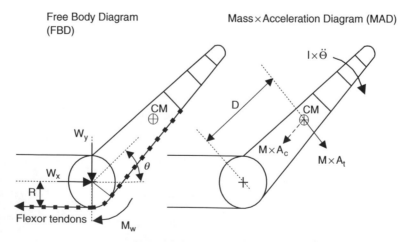

FIGURE 45.10 Dynamic wrist tendon pulley model. (Adapted from Schoenmarklin, R.W. and Marras, W.P. (1990). *Proceedings of the Human Factors Society 34th Annual Meeting*. Santa Monica, CA: Human Factors Society, 809–809. With permission.)

Key forces and movements in the model include the reaction force at the center of the wrist (W_x and W_y), the couple or moment (M_w) required to flex and extend the wrist, and the inertial force ($M \times A_c$ and $M \times A_t$) and inertial moment $I \times \ddot{\Theta}$ acting around the hand's center of mass. For equilibrium, the magnitude of moment around the wrist in the free body diagram must equal the magnitude of moment acting around the hand's center of mass in the moment acceleration diagram:

$$F_t \times R = (M \times A_t + M \times A_c) \times D + I \times \ddot{\Theta} \tag{45.8}$$

where M is the mass, A_t is the tangential acceleration, A_c is the centripetal acceleration, F_T is the tendon force, I is the moment of inertia of the hand in flexion and extension, and $\ddot{\Theta}$ is the angular acceleration.

Thus, the hand is assumed to accelerate from a stationary posture, so, the angular velocity is theoretically zero, resulted in zero centripetal force ($A_c = V^2/R = 0$). Then,

$$F_t \times R = (M \times A_t) \times D + I \times \ddot{\Theta} \tag{45.9}$$

$$F_t \times R = (M \times D \times \ddot{\Theta}) \times D + I \times \ddot{\Theta} \tag{45.10}$$

$$F_t = \frac{(M \times D^2 + I) \times \ddot{\Theta}}{R} \tag{45.11}$$

$$F_R = 2 \times \left[\frac{(M \times D^2 + I) \times \ddot{\Theta}}{R} \right] \times \sin\left(\frac{\theta}{2}\right) \tag{45.12}$$

where R is the radius of curvature of the tendon, D is the distance between the center of mass of hand and wrist, M is the weight of hand, and θ is the wrist deviation angle.

Equation (45.9)–Equation (49.12) indicate that the resultant reaction force, F_R, is a function of angular acceleration, radius of curvature, and wrist deviation. Thus, exertion of wrist and hand with greatly angular acceleration and deviated wrist angle would result in greater total resultant reaction forces on the tendons and supporting tissues than exertions with small angular acceleration and neutral wrist position. According to Armstrong and Chaffin (1979), increases in resultant reaction force would increase the supporting force that the carpal bones and ligaments exert on the flexor tendons, therefore increasing the chance of inflammation and risk of carpal tunnel syndrome (CTS). Therefore, these results might provide theoretical support to why angular acceleration variable can be considered a risk factor of WRMSDs.

The advantage of Schoenmarklin and Marras (1990) model is that it does include the dynamic variable of angular acceleration into assessment of resultant reaction force on the tendons. But the model is two-dimensional, and it does not consider the coactivation of antagonistic muscles in wrist joint motions. This points to the need for further model development to account for additional physiological factors.

45.2.3 Complex Tendon Models

Any model that incorporates more than the one muscle–tendon unit of the aforementioned models, all of sudden, becomes much more complicated, due to the fact that the number of unknown muscle forces exceeds the number of equilibrium or constraint equations. This is known as the statically indeterminate problem. The two main approaches utilized in solving this problem are reduction methods and optimization methods.

The main objective of the reduction method is to reduce the number of excessive variables until the number of unknown forces is equal to the number of required equilibrium equations eliminating static indeterminacy.

Smith et al. (1964) initiated mathematical analyses of the finger tendon forces to find the effects of the flexor tendons acting on metacarpal phalangeal (MCP) joint deformed by rheumatoid arthritis. They used a two-dimensional model to analyze the MCP joint and muscle forces of the index finger during tip pinch. To reduce the number of unknown muscle forces, the following assumptions were applied: (1) the sum of the interosseous (I) forces is treated as a single force $2I$; (2) half of the interosseous forces of I act at the PIP joint and the other half act at the DIP joint; (3) the lumbrical (L) is much smaller than the interosseous (I), as much as $1/3I$. They solved the three moment equations using these assumptions and anthropometrical data of the index finger obtained from a cadaver hand in a tip pinch position. They reported the tendon forces normalized to the external force F, as $3.8F$, $2.5F$, $0.9F$, and $0.3F$ for the FDP, FDS, I, and L, respectively. They also found a value of $7.5F$ for the MCP joint force. The results indicate that the flexor tendons are dominant and the forces are many times larger than the intrinsic muscle forces during tip pinch.

Chao et al. (1976) presented a comprehensive analysis of the three-dimensional tendon and joint forces of the fingers in pinch and power grip functions. Kirschner wires were drilled through the phalanges to fix the finger configuration in the desired position and different surgical wires were inserted into the tendon and muscles of hand specimens of the cadaver to identify different tendons on x-ray film. The exact orientations of finger digits and the locations of the tendons were defined by biplanar x-ray analysis. Through a free-body analysis, 19 independent equations were obtained for 23 unknown joint and tendon forces. Using the permutation–combination principle of setting any four of the nine tendon forces equal to zero solved the indeterminate problem. The selection of these tendons was based primarily on electromyographic (EMG) responses and physiological assessment. They found that high constraint forces and moments at the DIP and PIP joints were found during pinches, whereas large magnitudes of constraint forces at the MCP joint were found during power grips. The total of the intrinsic muscles (RI, UI, and LI) produced a greater force than the total of the flexor tendons (FDP and FDS) in both pinch and power grip actions.

An alternate method using a typical optimization technique was suggested by Seireg and Arvikar (1973) and Penrod et al. (1974). In this approach, force equilibrium equations and anatomical constraint relationships were used for the equality constraints and the physiological limits on the tendon, muscle, and joint forces were applied as the inequality constraints. In addition, the most important factor in this method is optimal criteria that correspond to the objective function of the formulation. The possible solutions can vary based on the optimal criteria selected.

Chao and An (1978a) studied the middle finger during tip pinch and power grip actions, with an aid of three-dimensional analysis. They analyzed the same problem using the optimization and linear programming (LP) technique of Chao et al. (1976) instead of the previously described EMG and permutation-combination method. The predicted middle finger muscle and joint forces were very similar to those of the previous study (Penrod et al., 1974), except for the intrinsic muscle forces whose predicted values were considerably lower. They found that the highest joint contact forces for all three joints occurred for pinch grip rather than power grip. They also found that the main flexors (FDP and FDS) were most active in both pinch and power grip functions, whereas the intrinsic muscles were less active in power grip than in pinch.

An et al. (1985) also applied LP optimization techniques to solve the indeterminate problem of a three-dimensional analytic hand model. The ranges of muscle forces of the index finger under isometric hand functions, such as tip pinch, lateral key pinch, power grip and other functional activities were analyzed. FDP and FDS carried high tendon forces compared with other muscles in most hand functions, although the predicted FDS force was zero in a pinch grip. The long extensors (LE) and two intrinsic muscles contributed large forces in the key pinch. The large force of these intrinsic muscles in pinch action can be explained by the role of these muscles maintaining balance and stabilization of the MCP joint. The joint constraint forces for each finger were also studied. The Chao et al. (1976) study showed a trend for joint constraint forces in which the DIP joint had the lowest force and the force progressively increased for the PIP joint and was largest at the MCP joint. An et al. (1985) showed the same trend in lateral pinch functions.

Chao and An (1978b) used a graphical presentation with a combined permutation and optimization technique to solve the statically indeterminate tendon force problem. They analyzed the maximum tip pinch force of the index finger as a function of external force directions ($0°$, $30°$, and $45°$) and the

DIP joint flexion angles ($10° - 50°$). The results showed that the pinch strength relied on the direction of applied external force as well as on the finger joint configuration. The tendon forces of the index finger were also studied with the same finger posture as that in Chao et al. (1976) study, but only one angle ($45°$) of the external force was assumed. Also, the predicted extrinsic extensor tendon force was considerably larger than in their previous studies.

Weightman and Amis (1982) presented a good critical review for the previously published studies and applied their two-dimensional finger model to the analysis of resultant joint forces and muscle tensions in various pinch actions. To create a statically determinate problem, all joints were assumed to be pin joints with a fixed center of rotation during flexion. The relationships of the intrinsic muscle forces were assumed identical to those of Chao and An (1978a), except that the long extensor muscles forces dropped to zero. They also used the physiological cross-sectional area (PCSA) of the muscles to define the force distributions in the intrinsic muscles. Their results compared to other previously published studies with a good correlation of both muscle and joint force predictions. Based on these comparisons, they verified that a two-dimensional finger model could be valid for analyzing two-dimensional finger actions, even though realistically any finger motion is still three dimensional.

45.2.4   A Two-Dimensional Hand and Wrist Model

From a biomechanical perspective, the extrinsic finger flexors, FDP and FDS comprise the main sources of power for finger flexion in grasping type motions, especially the power grip. Also, since most of the tendon pulley attachments are in line with the long axis (i.e., x-axis) of the phalanges and there are small lateral force components (i.e., along the z-axis), only two axes, x and y, need to be defined. Therefore, a simple two-dimensional model utilizing those two tendons should be sufficient for most applications.

To further define the model, several other assumptions need to be made. These are as follows:

1. The effects of intrinsic muscles and extensors on the finger flexion can be neglected since these muscles will typically be in a relaxed state during the normal range of motion for a power grip (Armstrong, 1976; Cailliet, 1994)
2. All of the interphalangeal and metacarpal joints (DIP, PIP, and MP joints) are assumed to be pure hinge joints, allowing only flexion and extension
3. Anatomic analysis shows that the FDS is inserted by two slips to either side of the proximal end of the middle phalanx (Steinberg, 1992; Cailliet, 1994). It is assumed that each FDS tendon is inserted to the palmar side of the proximal end of the middle phalange, parallel to the long axis. In a two-dimensional biomechanical model, the effect of having two splits inserted along the sides of the bone is the same as having one tendon inserted on the palmar side of the proximal end of the middle phalange
4. Tendons and tendon sheaths are modeled as a frictionless cable and pulley system. Therefore, a single tendon passing through several joints maintains the same tensile force (Chao et al., 1976)
5. The externally applied forces are assumed to be a single unit-force exerted at the midpoint pulp of a distal phalange for pinch or by three unit-forces applied normally at the midpoint of each phalange and metacarpal bone for grasp as shown in Figure 45.10. The direction of the force is assumed to be perpendicular to the long axis of the bone
6. The weight of the bones together with other soft tissues on the hand are assumed to be negligible
7. Due to indeterminacy, the tendon force ratio of FDP to FDS at the each phalange: is assumed to be 3:1, that is, $\alpha = 0.333$ (Marco et al., 1998)
8. FDP and FDS tendon moment arms (in mm) are estimated for DIP, PIP, and MCP joints of different thickness equations from the equations of Armstrong (1976):

$$PR_{ik} = 6.19 - 1.66X_1 - 4.03X_2 + 0.225X_3 \qquad (45.13)$$

$$SR_{ik} = 6.42 + 0.10X_1 - 4.03X_2 + 0.225X_3 \qquad (45.14)$$

where: PR_{ik} is the FDP moment arm for the ith finger and kth joint, SR_{ik} is the FDS moment arm for the ith finger and kth joint, $X_1 = 1$ for PIP and 0 for all others, $X_2 = 1$ for DIP and 0 for all others, X_3 is the joint thickness (mm) from Table 45.5.

Consequently, the pertinent equations are as follows:

$$PR_{DIP} = 2.16 + 0.225X_3 \tag{45.15}$$

$$PR_{PIP} = 4.53 + 0.225X_3 \tag{45.16}$$

$$SR_{PIP} = 6.52 + 0.225X_3 \tag{45.17}$$

$$PR_{MCP} = 6.19 + 0.225X_3 \tag{45.18}$$

$$SR_{MCP} = 6.42 + 0.225X_3 \tag{45.19}$$

Four Cartesian coordinate systems are established to define the locations and orientations of the tendons and to describe the joint configuration (Figure 45.11). There are two coordinate systems for both the middle and proximal phalanges and only one system for the distal and metacarpal phalanges. The y-axis is defined along the long axis of the each phalanx, from the proximal end to the distal end. The x-axis is defined as perpendicular to the long axis of each phalanx and in the palmar–dorsal plane, from the palmar side to the dorsal side of the finger bone. Both x- and y-axes have their origins at the center of the proximal end of phalanx. Note that these definitions are different from that used by An et al. (1979) in Figure 45.7, Table 45.7 and Table 45.8.

In terms of notation, subscript i refers to fingers, with $1-4$ for the index, middle, ring, and little finger, respectively, subscript j refers to joints, with $1-4$ for the DIP, PIP, MP, and wrist joints, respectively, while subscript k refers to phalanges, with $1-4$ for the distal, middle, proximal phalanges, and the metacarpal bone, respectively.

In terms of model input values, the external force on each phalange of each finger is indicated by $F(i,k)$. The finger joint flexion angles, measured with reference to straight fingers as the hand is lying flat, are indicated by (i,θ_j). The length of each phalanx for each finger is indicated by $L(i,k)$.

For output variables, the FDP tendon force for each phalanx of each finger is indicated by $TP(i,k)$. The FDS tendon force for each phalanx of each finger is indicated by $TS(i,k)$. Finally, joint constraint forces along the X_k- and Y_k- axes are indicated by $Rx_k(i,j)$, and $Ry_k(i,j)$, respectively.

To solve for the aforementioned unknown model output variables, a static equilibrium analyses of each phalanx must be performed. Specifically, the summation of forces acting on each phalanx in the X- and

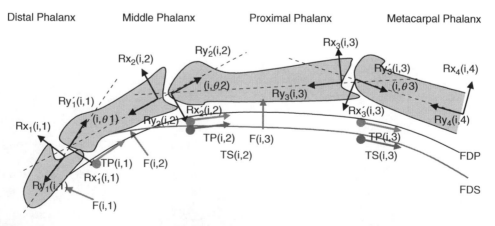

FIGURE 45.11 Simple two-dimensional hand model. For symbols and notation, see text. (Reproduced from Kong, Y.K. Ph.D. Dissertation, Pennsylvania State University. With permission.)

Y-axes must be zero. Similarly, the summation of all moments acting on each phalanx must also be equal to zero. The resulting equations for the distal phalanx are:

$$TP(i,1) = 0.5 \, L(i,1) \, F(i,1)/PR(i,1) \tag{45.20}$$
$$Ry_1(i,1) = TP(i,1) \cos (i,\theta_1) \tag{45.21}$$
$$Rx_1(i,1) = TP(i,1) \sin (i,\theta_1) - F(i,1) \tag{45.22}$$

For the middle phalanx, they are:

$$TP(i,2) = \frac{[0.5 \, F(i,2) - Rx_1(i,1) \cos (i,\theta_1) + Ry_1(i,1) \sin (i,\theta_1)]}{\alpha SR(i,2) + PR(i,2)} L(i,2) \tag{45.23}$$
$$TS(i,2) = \alpha TP(i,2) \tag{45.24}$$
$$Ry_2(i,2) = Rx_1(i,1) \sin (i,\theta_1) + Ry_1(i,1) \cos (i,\theta_1) + (\alpha + 1) \, TP(i,2) \cos (i,\theta_2) \tag{45.25}$$
$$Rx_2(i,2) = Rx_1(i,1) \cos (i,\theta_1) - Ry_1(i,1) \sin (i,\theta_1) + (\alpha + 1) \, TP(i,2) \sin (i,\theta_2) - F(i,2) \tag{45.26}$$

For the proximal phalanx, they are:

$$TP(i,3) = \frac{[0.5 \, F(i,3) - Rx_2(i,2) \cos (i,\theta_2) + Ry_2(i,2) \sin (i,\theta_2)]}{\alpha SR(i,3) + PR(i,3)} L(i,3) \tag{45.27}$$
$$TS(i,3) = \alpha TP(i,3) \tag{45.28}$$
$$Ry_3(i,3) = Rx_2(i,2) \sin (i,\theta_2) + Ry_2(i,2) \cos (i,\theta_2) + (\alpha + 1) \, TP(i,3) \cos (i,\theta_3) \tag{45.29}$$
$$Rx_3(i,3) = Rx_2(i,2) \cos (i,\theta_2) - Ry_2(i,2) \sin (i,\theta_2) + (\alpha + 1) \, TP(i,3) \sin (i,\theta_3) - F(i,3) \tag{45.30}$$

One interesting application of such a biomechanical hand model is to identify the optimum handle size for gripping so as to minimize tendon forces. However, in a typical power grip, there are two alternate ways in which the geometry of a cylindrical handle surface and phalange contacts can be defined. In Grip I (Figure 45.12), the point of contact between the distal phalange (L_1) and a handle is assumed to be at the middle point of the distal phalange, that is, the distal phalange is divided into two equal lengths ($L_1/2$). The bisector of the DIP angle establishes a right triangle with the distal phalanx as the base and the altitude passing through the contact point to the center of handle. Through trigonometry, the DIP joint angle then becomes:

$$\theta_1' = 2 \tan^{-1}[2(R + D_1)/L_1] \tag{45.31}$$

where: θ_1' is the DIP joint angle, R is the radius of the cylindrical handle, and D_1 is the thickness of the distal phalanx.

The second contact point (between the middle phalange (L_2) and the handle) divides the middle phalange into two unequal lengths, one being $L_1/2$ (due to the DIP bisector) and the other being $L_2 - L_1/2$. The bisector of the PIP angle establishes a right triangle with the medial phalanx as the base and the altitude again passing through the contact point to the center of handle. Through trigonometry, the PIP joint angle then becomes:

$$\theta_2' = 2 \tan^{-1} \frac{2(R + D_2)}{2L_2 - L_1} \tag{45.32}$$

where θ_2' is the PIP joint angle and D_2 is the thickness of the medial phalanx.

The third contact point with the handle also divides the proximal phalange (L_3) into two parts, one being the same length as the proximal part of middle phalange ($L_2 - L_1/2$) and the other being

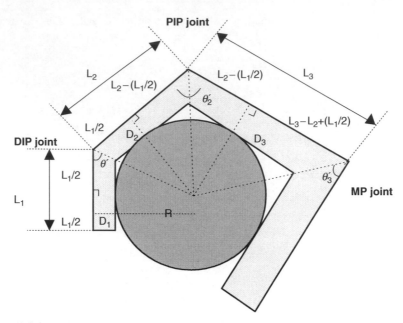

FIGURE 45.12   Schematic diagram of Grip I. (Reproduced from Kong, Y. K. (2001)   Ph.D Dissertation. Pennsylvania State University. With permission.).

$(L_3 - L_2 + L_1/2)$. The bisector of the MP angle creates another right triangle:

$$\theta_3' = 2 \tan^{-1} \frac{2(R + D3)}{2L_3 - 2L_2 + L_1} \tag{45.33}$$

where: θ_3' is the MP joint angle.

In the case of Grip II (see Figure 45.13), the second contact point is assumed to divide the medial phalanx into two equal lengths $(L_2/2)$. The perpendicular bisector of the medial phalanx forms two identical right triangles, yielding equal DIP and PIP joint angles (θ_1' and θ_2'):

$$\theta_1' = \theta_2' = 2 \tan^{-1}[2(R + D_2)/L_2] \tag{45.34}$$

The third contact point divides the proximal phalange into two unequal lengths. One is $L_2/2$ while the other is $L_3 - L_2/2$. The MP joint angle (θ_3') can then be estimated as:

$$\theta_3' = 2 \tan^{-1} \frac{2(R + D_3)}{2L_3 - L_2} \tag{45.35}$$

Based on this biomechanical hand model, tendon forces for each finger and in total were calculated for both types of grip and for 11 cylindrical handles with diameters ranging from 10 to 60 mm. For Grip I, tendon force were minimized at 30–35, 38–43, 40–45, and 25–30 mm for index, middle, ring, and little fingers, respectively. The total of tendon force for all fingers was minimized for an approximately 40 mm cylindrical handle (see Figure 45.14). As the size of the handle deviated above or below 40 mm, the total tendon forces increased.

For Grip II, tendon force were minimized at 23–28, 28–33, 28–33, and 20–25 mm for index, middle, ring, and little fingers, respectively. The total of tendon force for all fingers was minimized for an approximately 28 mm cylindrical handle (see Figure 45.15). The combined results for each type of grip are

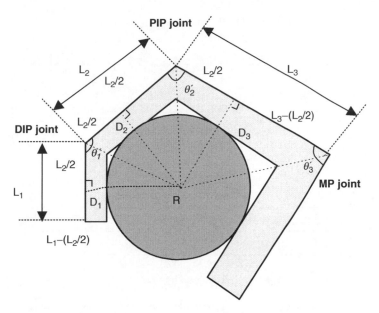

FIGURE 45.13 Schematic diagram of Grip II. (Reproduced from Kong, Y.K. (2001). Ph.D. Dissertation. Pennsylvania State University. With permission.)

summarized in Table 45.9. As noted previously, as the size of the handle deviates from the optimum size, tendon forces increase. This is an important principle that should be utilized in the design of hand tools. Also, in either type of grip, the optimal handle sizes depends greatly on which finger is considered. Therefore, a purely traditional cylindrical handle cannot provide optimality for all fingers simultaneously and alternate handle shapes, such as the double frustum, need to be considered. Further details can be found in Kong (2001) and Kong, et al. (2004), with applications to meat hook handles in Kong and Freivalds (2003).

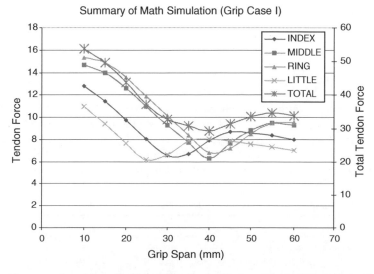

FIGURE 45.14 Finger tendon forces for Grip I. (Reproduced from Kong, Y.K. (2001). Ph.D. Dissertation. Pennsylvania State University. With permission.)

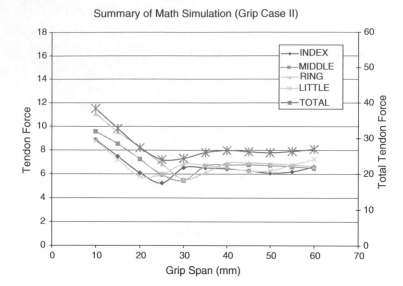

FIGURE 45.15   Finger tendon forces for Grip II (Reproduced from Kong, Y.K. (2001). From Ph.D. Dissertation. The Pennsylvania State University. With permission.)

45.3 Direct Measurement Studies

Directly measured tendon forces under isometric finger function were first reported by Bright and Urbaniak (1976). They developed a strain gauge to measure the tendon forces in both tip pinch and power grip actions during operative procedures. Flexor tendon forces were found to be in the range of 40–200 and 12.5–150 N for the FDP and FDS, respectively, in power grip action, while 25–125 N for the FDP and 10–75 N for the FDS in pinch action. Since they directly measured the tendon forces only, they did not report the actual applied pinch and power grip force and the ratio of tendon force to the externally applied force.

Schuind et al. (1992) directly measured the flexor tendons (flexor pollicis longus, or FPL, FDP, and FDS) during various finger functions. They developed an S-shaped tendon force transducer and measured the flexor tendon forces in pinch and power grip functions. Also, a pinch dynamometer was used to record the applied loads in pinch action. The tendon forces showed proportionality to the externally applied forces. To compare their results with the previously published mathematical finger models, they normalized their tendon forces, as a ratio of the tendon force to the applied forces. In tip pinch, the ratios were 3.6*F*, 7.92*F*, and 1.73*F*, for the externally applied force *F*, for the FPL, FDP, and FDS, respectively. In lateral pinches, the ratios were 3.05*F*, 2.9*F*, and 0.71*F* for the FPL, FDP, and FDS, respectively. Although the FDP and FPL showed high forces during tip and lateral pinch, the maximal values recorded are probably on the lower side of the potential forces and it could be explained by the significantly weaker pinch and power grip forces during carpal tunnel surgery due to the denervation or partial anesthesia of the sensory area of the median nerve. However, the magnitude of tendon forces was similar to values reported by Bright and Urbaniak (1976), although direct comparison is not possible since the applied force was not recorded in their study.

TABLE 45.9   Summary of Optimal Cylindrical Handle Sizes

Finger	Grip I	Grip II	Combined Grip
Index	30	25	25–30
Middle	40	30	30–40
Ring	40	30	30–40
Little	25	20	20–25

In another *in vivo* tendon force measurement study, Dennerlein et al. (1998) measured only the FDS tendon forces of the middle finger at three finger postures, which ranged from extended to flexed pinch postures, using a gas-sterilized tendon force transducer (Dennerlein et al., 1997) and a single axis load cell. The investigation was centered upon the average ratio of the FDS tendon tension to the externally applied force. The average ratio ranged from 1.7*F* to 5.8*F*, with a mean of 3.3*F*, in the study. Tip pinches with the DIP joint flexed were also studied with the tendon-to-tip force ratio being 2.4*F*. These ratios were compared with the results of their own three-finger models as well as other contemporarily published isometric tendon force models. These ratios were larger than those of other studies. The average values were also slightly higher than that (1.73*F*) of Schuind et al.'s (1992) *in vivo* tendon force measurement study. It was found that the tendon force ratios and muscle strength varied substantially from individual to individual, although the ratio of force from tendon to tendon was relatively constant within the same limb for all studies (Ketchum et al., 1978; Brand et al., 1981; Dennerlein et al., 1998).

Although the intrinsic muscles are more active in pinch action than in power grip action, the relative magnitudes of the main flexor tendon forces (such as FDP and FDS) are usually high in both actions. These *in vivo* tendon forces of the flexors are presented in Table 45.10 based on the previous studies. In general, these averages and ranges of tendon forces are very similar with a few exceptions. Schuind et al. (1992) showed lower FDS tendon forces in power grip action than those of other types of grips. The tendon force ranges of Brand et al. (1981) show similar magnitudes with Bright and Urbaniak (1976) power grips. Tip pinch action studies by Ketchum et al. (1978) and Bright and Urbaniak (1976) also show the similar ranges of tendon forces. Schuind et al. (1992) represented the significant differences between FDP and FDS tendon forces in both pinch and power grip, whereas the others showed that the force of FDP tendon was only slightly larger than that of the FDS tendon. These discrepancies can be explained by the different finger postures utilized in each study, since each finger could have various functional muscle capacities depending upon its joint configuration (Chao and An, 1978b).

45.4   Data-Driven WRMSD Risk Index

Detailed data on hand and wrist postures can be used as input for assessing risk for upper limb WRMSDs. On such risk index was developed by Seth et al. (1999) using wrist posture data fed directly from the Yun et al. (1997) integrated touch glove system and WRMSD data from industry to yield a predicted incidence rate. The risk index is presented here in a step-by-step approach and is available at http://www.ie.psu.edu/courses/ie552/CTDriskindex.htm.

Wrist deviation and arm rotation cause a significant grip strength decrement as compared to a neutral wrist posture. These effects, from the data of Terrell and Purswell (1976), Imrahn (1991), and Hallbeck et al. (1992), were expressed as equations for maximum voluntary contraction (MVC) for power grip and various pinches produced in five wrist positions (neutral, flexion, extension, radial and ulnar deviation) and in three arm rotations (pronation, midposition, and supination) (see Table 45.11). Inputting the appropriate parameters yields the variable "Force Capacity Wrist." Similarly, force decrements occur for both power and pinch grip depending on the grip span utilized. This effect again can be quantified

TABLE 45.10   *In Vivo* FDP and FDS Tendon Forces (kg)

Study	Finger Configuration	FDP	FDS
Brand et al. (1981)	—	14.9 (13.5−17.0)	10.4 (4.5−17)
Bright and Urbaniak (1976)	Tip pinch	2.5−12.5	1.0−7.5
	Power grip	4.0−20.0	1.25−15.0
Dennerlein et al. (1998)	Tip pinch	—	3.3 (1.7−5.8)
Ketchum et al. (1978)	MCP joint flexion	5.7 (5.27−6.18)	6.12 (3.73−7.63)
Schuind et al. (1992)	Tip pinch	8.3 (2.0−12.0)	1.9 (0.3−3.5)
	Power grip	4.0 (1.9−6.4)	0.6 (0.0−0.9)

TABLE 45.11   Force Capacity Wrist Equations

Grip Type	Wrist Position	%MVC as a Function of Wrist Angle ($A = °$)
Power	Flexion/extension/pronation	$-0.0113A^2 + 0.1826A + 88$
Power	Flexion/extension/mid-position	$-0.0114A^2 + 0.1308A + 99$
Power	Flexion/extension/supination	$-0.0112A^2 + 0.0979A + 100$
Power	Radial/ulnar deviation/pronation	$-0.0337A^2 + 0.025A + 99$
Power	Radial/ulnar deviation/mid-position	$-0.0538A^2 + 0.275A + 88$
Power	Radial/ulnar deviation/supination	$-0.0338A^2 + 0.075A + 100$
Lateral pinch	Flexion/extension/supination	$-0.0124A^2 + 0.1905A + 100$
Lateral pinch	Radial/ulnar deviation/pronation	$-0.0409A^2 + 0.1525A + 100$
Chuck pinch	Flexion/extension/supination	$-0.0158A^2 + 0.2191A + 98$
Chuck pinch	Radial/ulnar deviation/Pronation	$-0.0626A^2 + 0.1275A + 98$
Two-Point Pinch	Flexion/extension/supination	$-0.0114A^2 + 0.1423A + 72.4$
Two-Point Pinch	Radial/ulnar deviation/pronation	$-0.00535A^2 + 72.4$

from the data of Greenberg and Chaffin (1976) and Petrofsky et al. (1980) to yield average grip span strength decrements for spans ranging from 0 to 11 cm as shown in Table 45.12. Inputting the appropriate parameters yields the variable "Grip Span Force."

Repeated exertions will result in muscle fatigue and reduced capacity for further exertions. This effect can be quantified from the data of Schutz (1972) as the maximum "%MVC Allowed" for a particular wrist motion based on the exertion time for the motion and the time between exertions or motions. If the "%MVC Required" performing the motion is greater than that allowed as calculated below (%MVC Allowed), then a penalty is assessed to that motion.

$$\%\text{MVC Allowed} = [0.9503\,\text{TW}^{-0.394} \times \text{TR}^{0.2246} \times \text{MW}^{0.258} \times 0.475] \times 100 \qquad (45.36)$$

where: TW is the work time of grip or motion (min), TR is the rest time between grips or motions (min), and MW is the time worked during one shift (min).

For each motion a "Force Capacity" is calculated as the product of "Force Capacity Wrist" and "Grip Span Force":

$$\text{Force Capacity} = \text{Force Capacity Wrist} \times \text{Grip Span Force}/100 \qquad (45.37)$$

%MVC Required$_{\text{Adj}}$ is calculated based on %MVC Required performing the motion and the previously calculated Force Capacity. %MVC Required is typically found by dividing the subjective job force requirement to the operator's MVC.

$$\%\text{MVC Required}_{\text{Adj}} = \%\text{MVC Required}/\text{Force Capacity} \times 100 \qquad (45.38)$$

The %MVC Required$_{\text{Adj}}$ is then compared to the %MVC Allowed for Equation (45.36). If the %MVC Allowed is greater than the %MVC Required$_{\text{Adj}}$, then no penalty is assessed to that hand motion and that motion will have a "Force Frequency Score" (FFS) of 1. If the %MVC Allowed is less than the %MVC

TABLE 45.12   Grip Span Force Equations

Grip Type	% MVC as a Function of Grip Span Width ($G = $ cm)
Power	$-0.3624G^3 + 4.6865G^2 - 3.6186G + 14.4$
Lateral pinch	$-0.8308G^2 + 3.0288G + 97.59$
Chuck pinch	$-0.2363G^2 - 0.7093G + 100.22$
Two-point pulp pinch	$-0.3931G^2 + 1.6383G + 99.22$

Required$_{\text{Adj}}$, then the FFS$_i$ for that individual motion is calculated as follows:

$$\text{FFS}_i = \%\text{MVC Required}_{\text{Adj}}/\%\text{MVC Allowed} \times n \tag{45.39}$$

where n is the number of hand motions per job cycle.

The FFS$_i$ values for each individual hand motion are then summed to obtain an overall force frequency score:

$$\text{FFS} = \sum \text{FFS}_i \times N/10,000 \tag{45.40}$$

where N is the number of job cycles per shift.

The scaling factor of 10,000 is the National Institute of Occupational Safety and Health (NIOSH) (Hales et al., 1989) maximum recommended number of damaging wrist motions that can be performed in an 8-h shift. The model then assesses which hand has a higher overall FFS and uses that value for further calculations.

Gross torso posture of a worker while performing a job is important because awkward postures can lead to fatigue. The resulting fatigue, defined as "%Endurance Capacity", can be quantified in a manner similar to Equation (45.36)

$$\%\text{Endurance Capacity} = \{1 - [1.1 \times F_{\text{MVC}}^{1.863} \times \text{TW}^{0.734} \times \text{TR}^{-0.413} \times \text{MW}^{0.481}]\} \times 100 \tag{45.41}$$

where TW is the work time posture held (min), TR is the rest time between postures (min), MW is the time worked during one shift (min), and F_{MVC} is the force/load held normalized by 51 lbs, from the NIOSH Lifting Equation of Waters et al. (1993).

Five postures are considered: neck and back flexion, elbow and shoulder flexion, and shoulder abduction. For simplicity sake (and difficulty interpreting angles from job videotapes), *Points* are assigned to various angles as follows: for back flexion $0° - 10° = 0$, $10° - 20° = 1$, $20° - 45° = 2$, and $>45° = 3$ points; for neck flexion $0° - 30° = 0$, $30° - 45° = 1$, $45° - 60° = 2$, and $>60° = 3$ points; for elbow flexion/extension (with a 90° bent elbow being considered the neutral posture) 10° flexion to 30°extension $= 0$, everything else is 1 point; for shoulder flexion $0° - 20° = 0$, $20° - 45° = 1$, $45° - 90° = 2$ and $>90° = 3$ points; and for shoulder abduction $0° - 30° = 0$, $30° - 60° = 1$, $60° - 90° = 2$, $>90° = 3$ points.

The "Posture Score" for each body part is obtained by multiplying the point value by the %Endurance Capacity (Equation 45.41) and dividing the product into 50, which is considered as the limit for acceptable fatigue or endurance:

$$\text{Posture Score} = 50/(\%\text{Endurance Capacity} \times \text{Points}) \tag{45.42}$$

The "Overall Posture Score" (OPS) is the maximum of individual Posture Scores for each joint.

The final risk score is a weighted average (coefficients obtained from regression of a subset of jobs) of the FFS (Equation 45.40) and the OPS in the form of a predicted incidence rate (IR) normalized to 200,000 exposure hours:

$$\text{IR} = 3.41\,\text{FFS} + 1.87\,\text{OPS} \tag{45.43}$$

Regression of predicted incidence rates against the actual incidence rates experienced on 24 industrial jobs in the garment and the printing industry (involving a total of 288 workers) yielded a significant ($p < 0.001$) linear regression with $r^2 = 0.52$. As a comparison, Moore and Garg's (1995) Strain Index yielded a nonsignificant ($p = 0.2$) regression with $r^2 = 0.17$. The only limitations found were for very short cycle jobs (typically under 4-sec cycle times), in which any error in miscounting motions could be amplified into a large error for the final predicted incidence rate. Note that the roughly 48%

of unaccounted variance was thought to be due individual differences such as gender, age, physical fitness, etc. and the psychosocial risk factors.

Novice ergonomists (university graduate students) required at least several trials in becoming proficient with the risk assessment model. However, by the fifth trial, average time required for job analysis had decreased to 12 min and test–retest reliability was up to $r^2 = 0.99$. A simplified version of this risk index has been developed into a paper-and-pencil checklist for use in industry (see Figure 45.16), with

CTD Risk Index

Job Title:	VCR Counter No.:	Date:
Job Description:	Department:	Analyst:

Cycle Time (in minutes; obtain from videotape)		①
# Cycle/Day = $\dfrac{(480-Lunch-Breaks)}{CycleTime}$ =	②a	③ Larger of ②a or ②b:
# Parts / Day (if known)	②b	
# Handmotions / Cycle		④
# Handmotions / Day (③ x ④)		⑤

Frequency Factor (Divide ⑤ by 10,000) =

(Circle appropriate condition)	Points			
	0	1	2	3
Working Posture	Sit	Stand		
Hand Posture 1: Pulp Pinch	No	Yes		
Hand Posture 2: Lateral Pinch	No	Yes		
Hand Posture 3: Palm Pinch	No	Yes		
Hand Posture 4: Finger Press	No	Yes		
Hand Posture 5: Power Grip	Yes	No		
Type of Reach	Horizontal	Up/Down		
Hand Deviation 1: Flexion	No	Yes		
Hand Deviation 2: Extension	No	Yes		
Hand Deviation 3: Radial Dev.	No	Yes		
Hand Deviation 4: Ulnar Dev.	No	Yes		
Forearm Rotation	Neutral	In/Out		
Elbow Angle	=90°	≠90°		
Shoulder Abduction	0	<45°	<90°	>90°
Shoulder Flexion	0	<90°	<180°	>180°
Back/Neck Angle	0	<45°	<90°	>90°
Balance	Yes	No		

Total the Points for the Circled Conditions ⑥

Posture Factor (Divide ⑥ by 10) =

Grip or Pinch Force Used on Task	⑦	lbs.	⑨Divide⑦ by⑧:
Max Grip or Pinch Force	⑧	lbs.	

Force Factor (Divide ⑨ by .15) =

(Circle appropriate condition)	Points			
	0	1	2	3
Sharp Edge	No	Yes		
Glove	No	Yes		
Vibration	No	Yes		
Type of Action	Dynamic	Intermittent	Static	
Temperature	Warm	Cold		

Total the Points for the Circled Conditions ⑩

Miscellaneous Factor (Divide ⑩ by 3) =

CTD Risk Index = .3 x (Frequency + Posture + Force Factors) + .1 x (Miscellaneous Factor)

CTD Risk Index = .3 x (+   +   **) + .1 x (**   **)**   =

FIGURE 45.16 A simple carpal tunnel disorder risk index (Reproduced from Niebel, B. and Freivalds, A. (2003). *Methods, Standards, and Work Design.* New York: McGraw-Hill. With permission.)

values greater than 1.0 indicating risk for injury (Niebel and Freivalds, 2003). However, it has not been validated or checked for reliability.

45.5 Conclusions

A variety of models and analysis tools have been presented for the hand and wrist. For all these, the main risk factors for WRMSDs are high external forces and awkward wrist postures that produce high tendon forces. In addition, high velocities and accelerations, resulting from frequent rapid movements, can further increase the risk for injury.

References

An, K.W., Chao, E.Y., Cooney, W.P., and Linscheid, R.L. (1979). Normative model of human hand for biomechanical analysis. *Journal of Biomechanics*, 12:775–788.

An, K.N., Hui, F.C., Morrey, B.F., Linscheid, R.L., and Chao, E.Y.S. (1981). Muscle across the elbow joint: a biomechanical analysis. *Journal of Biomechanics*, 14:659–669.

An, K.N., Ueba, Y., Chao, E.Y., Cooney, W.P., and Linscheid, R.L. (1983). Tendon excursion and moment arm of index finger muscles. *Journal of Biomechanics*, 16:419–425.

An K.N., Chao E.Y., Cooney, W.P., and Linscheid R.L. (1985). Forces in the normal and abnormal hand. *Journal of Orthopedic Research*, 3:202–211.

An, K.N., Horii, E., and Ryu, J. (1991). Muscle function. In An, K.N., Berger, R.A. and Cooney, W.P. (Eds.), *Biomechanics of the Wrist Joint*, New York: Springer Verlag, pp. 157–169.

Armstrong, T.J. (1976). Circulatory and Local Muscle Responses to Static Manual Work. Ph.D. Dissertation, Ann Arbor, MI: University of Michigan.

Armstrong, T.J. and Chaffin, D.B. (1978). An investigation of the relationship between displacements of the finger and wrist joints and the extrinsic finger flexor tendons. *Journal of Biomechanics*, 11:119–128.

Armstrong, T.J. and Chaffin, D.B. (1979). Some biomechanical aspects of the carpal tunnel. *Journal of Biomechanics*, 12:567–570.

Batmanabane, M. and Malathi, S. (1985). Movements at the carpometacarpal and metacarpophalangeal joints of the hand and their effect on the dimensions of the articular ends of the metacarpal bones. *Anatomical Record*, 213:102–110.

Brand, P.W. and Hollister, A. (1993). *Clinical Mechanics of the Hand*, 2nd ed. St. Louis, MO: Mosby Year Book, Inc.

Brand, P.W., Beach, R.B., and Thompson, D.E. (1981). Relative tension and potential excursion of muscles in the forearm and hand. *Journal of Hand Surgery*, 6:209–218.

Bright, D.S. and Urbaniak, J.S. (1976). Direct measurements of flexor tendon tension during active and passive digit motion and its application to flexor tendon surgery. *Transactions of the 22nd Annual Orthopedic Research Society*, 240.

Cailliet, R. (1994). *Hand Pain and Impairment*. 3rd ed. Philadelphia, PA: Davis.

Chaffin, D.B., Andersson, G.B.J., and Martin, B.J. (1999). *Occupational Biomechanics*, 3rd ed., New York: John Wiley & Sons.

Chao, E.Y.S. and An, K.N. (1978a). Determination of internal forces in human hand. *Journal of Engineering Mechanics Division ASCE*, 104:255–272.

Chao, E.Y.S. and An, K.N. (1978b). Graphical interpretation of the solution to the redundant problem in biomechanics. *Journal of Biomechanics*, 11:159–167.

Chao, E.Y.S., Opgrande, J.D., and Axmear, F.E. (1976). Three-dimensional force analysis of finger joints in selected isometric hand functions. *Journal of Biomechanics*, 9:387–396.

Davidoff, N.A. (1990). The Development of a Graphic Model of a Human Hand, MS Thesis, University Park, PA: Pennsylvania State University.

Davidoff, N.A. and Freivalds, A. (1993). A graphic model of the human hand using CATIA. *International Journal of Industrial Ergonomics*, 12:255–264.

Dennerlein, J.T., Miller, J., Mote, C.D. Jr., and Rempel, D.M. (1997). A low profile human tendon force transducer: the influence of tendon thickness on calibration. *Journal of Biomechanics*, 30:395–397.

Dennerlein, J.T., Diao, E., Mote, C.D. Jr., and Rempel, D.M. (1998). Tensions of the flexor digitorum superficialis are higher than a current model predicts. *Journal of Biomechanics*, 31:295–301.

Fahrer, M. (1971). Considerations on functional anatomy of the flexor digitorum profundus. *Annales de Chirurgie*, 25:945.

Garcia-Elias, M., Horii, E., and Berger, R.A. (1991). Individual carpal bone motion. In: An, K-N., Berger, R.A. and Cooney, W.P. (Eds.). *Biomechanics of the Wrist Joint.* New York: Springer-Verlag, pp. 61–75.

Garrett, J.W. (1970a). *Anthropometry of the Air Force Female Hand*, AMRL-TR-69–26, Wright Patterson AFB, OH: Aerospace Medical Research Laboratory.

Garrett, J.W. (1970b). *Anthropometry of the Hands of Male Air Force Flight Personnel*, AMRL-TR-69–42, Wright Patterson AFB, OH: Aerospace Medical Research Laboratory.

Garrett, J.W. (1971). The adult human hand: Some anthropometric and biomechanical considerations. *Human Factors*, 13:117–131.

Gigis, P.I. and Kuczynski, K. (1982). The distal interphalangeal joints of the human fingers. *Journal of Hand Surgery*, 7:176–182.

Greenberg, L. and Chaffin, D.B. (1976). *Workers and Their Tools: A Guide to the Ergonomics of Hand Tools and Small Presses.* Midland, MI: Pendell Pub.

Hales, T., Habes, D., Fine, L., Hornung, R., and Boiano, J. (1989). *Health Hazard Evaluation Report, Bennett Industries, Peotone, IL*, HETA 89-146-2049, Cincinnati, OH: National Institute of Occupational Safety and Health.

Hallbeck, M.S., Kamal, M.S., and Harmon, P.E. (1992). The effects of forearm posture, wrist posture and hand on three peak pinch force types. *Proceedings of the Human Factors and Ergonomics Society 36th Annual Conference*, Santa Monica, CA: Human Factors and Ergonomics Society, pp. 801–805.

Idler, R.S. (1985). Anatomy and biomechanics of the digital flexor tendons. *Hand Clinic*, 1:3–12.

Imrahn, S.N. (1991). The influence of wrist posture on different types of pinch strength. *Applied Ergonomics*, 22:379–384.

Ketchum, L.D., Thompson, D., Pocock, G., La, C., and Wallingford, D. (1978). A clinical study of forces generated by the intrinsic muscles of the index finger and the extrinsic flexor and extensor muscles of the hand. *The Journal of Hand Surgery*, 3:571–578.

Kong, Y.K. (2001). Optimum Design of Handle Shape Through Biomechanical Modeling of Hand Tendon Forces. Ph.D. Dissertation, University Park, PA: Pennsylvania State University.

Kong, Y.K. and Freivalds, A. (2003). Evaluation of meat hook handle shapes. *International Journal of Industrial Ergonomics*, 32:13–23.

Kong, Y., Freivalds, A., and Kim, S.E. (2004). Evaluation of handles in a maximum gripping task. *Ergonomics*, 47.

Landsmeer, J.M.F. (1960). Studies in the anatomy of articulation. *Acta Morphologica Nederlands*, 3–4:287–303.

Landsmeer, J.M.F. (1962). Power grip and precision handling. *Annals of Rheumatoid Diseases*, 21:164–170.

LeVeau, B. (1977). *William and Lissner Biomechanics of Human Motion.* Philadelphia, PA: W.B. Saunders.

Lieber, R.L., Fazeli, B.M., and Botte, M.J. (1990). Architecture of selected wrist flexor and extensor muscles. *Journal of Hand Surgery*, 15:244–250.

Lin, G.T., Cooney, W.P., Amadio, P.C., and An, K.N. (1990). Mechanical properties of human pulleys. *Journal of Hand Surgery*, 15B:429–434.

Long, C., Conrad, P.W., Hall, E.A., and Furler, S.L. (1970). Intrinsic–extrinsic muscle control of the hand in power grip and precision handling: an electromyographic study. *Journal of Bone and Joint Surgery*, 52A:853–867.

Manske, P.R. and Lesker, P.A. (1977). Strength of human pulleys. *Hand*, 9:147–152.

Marco, R.A.W., Sharkey, N.A., and Smith, T.S. (1998). Pathomechanics of closed rupture of the flexor tendon pulleys in rock climbers. *The Journal of Bone and Joint Surgery*, 80A:1012–1019.

Moore, J. S. and Garg, A. (1995). The strain index: a proposed method to analyze jobs for risk of distal upper extremity disorders. *American Industrial Hygiene Association Journal*, 56:443–458.

Niebel, B. and Freivalds, A. (2003). *Methods, Standards, and Work Design.* New York: McGraw-Hill.

Nordin, M. and Frankel, V.H. (2001). *Basic Biomechanics of the Musculoskeletal System*, 3rd ed. Philadelphia, PA: Lippincott Williams & Wilkins.

Norkin, C.C. and Levangie, P.K. (1992). *Joint Structure and Function: A Comprehensive Analysis*, Philadelphia, PA: Davis.

Penrod, D.D., Davy, D.T., and Singh, D.P. (1974). An optimization approach to tendon force analysis. *Journal of Biomechanics*, 7:123–129.

Petrofsky, J.S., Williams, C., Kamen, G., and Lind, A.R. (1980). The effect of handgrip span on isometric exercise performance. *Ergonomics*, 23:1129–1135.

Schoenmarklin, R.W. and Marras, W.P. (1990). A dynamic biomechanical model of the wrist joint. *Proceedings of the Human Factors Society 34th Annual Meeting*, Santa Monica, CA: Human Factors Society, pp. 805–809.

Schuind, F., Garcia-Elias, M., Cooney W.P. III, and An, K.N. (1992). Flexor tendon forces: *In vivo* measurements. *The Journal of Hand Surgery*, 17A:291–298.

Schultz, R.B., Stroave, A., and Krishnamurthy, S. (1987). Metacarpophalangeal joint motion and the role of the collateral ligaments. *International Orthopaedics*, 11:149–155.

Schutz, R.K. (1972). Cyclic Work–Rest Exercise's Effect on Continuous Hold Endurance Capability, Ph.D. Dissertation, Ann Arbor, MI: University of Michigan.

Seireg, A. and Arvikar, R.B. (1973). A mathematical model for evaluation of forces in lower extremities of musculoskeletal system. *Journal of Biomechanics*, 6:313–326.

Seth,V., Weston, R.L., and Freivalds, A. (1999). Development of a cumulative trauma disorder risk assessment model for the upper extremities. *International Journal of Industrial Ergonomics*, 23:281–291.

Smith, E.M., Juvenile, R.B., Bender, L.F., and Pearson, J.R. (1964). Role of the finger flexors in rheumatoid deformities of the metacarpophalangeal joints. *Arthritis and Rheumatism*, 7:467–480.

Spence, A.P. (1990). *Basic Human Anatomy*, 3rd ed. Redwood City, CA: Benjamin/Cummings Pub. Co.

Steinberg, D.R. (1992). Acute flexor tendon injuries. *Orthopaedics Clinics of North America*, 23:125–140.

Steindler, A. (1955). *Kinesiology of the Human Body under Normal and Pathological Conditions*. Springfield, IL: Thomas.

Terrell, R. and Purswell, J.L. (1976). The influence of forearm and wrist orientation on static grip strength as a design criterion for hand tools. *Proceedings of the Human Factors and Ergonomics Society 36th Annual Conference*, Santa Monica, CA: Human Factors and Ergonomics Society, pp. 28–32.

Tubiana, R. (Ed.) (1981). *The Hand*, Vol. 1, Philadelphia, PA: Saunders.

Waters, T.R., Putz-Anderson, V., Garg, A., and Fine, L. (1993). Revised NIOSH equation for the design and evaluation of manual lifting tasks. *Ergonomics*, 36:749–776.

Weightman, B. and Amis, A.A. (1982). Finger joint force predictions related to design of joint replacements. *Journal of Biomedical Engineering*, 4:197–205.

Youm, Y., McMurty, R.Y., Flatt, A.E., and Gillespie, T.E. (1978). Kinematics of the wrist. *Journal of Bone and Joint Surgery*, 60A:424–431.

Yun, M.H., Cannon, D., Freivalds, A., and Thomas, G. (1997). An instrumented glove for grasp specification in virtual-reality based point-and-direct telerobotics. *IEEE Transactions on Systems, Man, and Cybernetics, Part B: Cybernetics*, 27:835–846.

46

Revised NIOSH Lifting Equation

Thomas R. Waters

National Institute for
Occupational Safety and
Health

46.1 Introduction

This chapter provides information about a revised equation developed by the National Institute for Occupational Safety and Health (NIOSH) for assessing the physical demands of certain two-handed manual lifting tasks, which was described by Waters et al. (1993). The chapter contains sections describing what factors need to be measured, how they should be measured, what procedures should be used, and how the results can be used to ergonomically design new jobs or make decisions about redesigning existing jobs that may be hazardous. The chapter defines all pertinent terms and presents the mathematical formulae and procedures needed to properly apply the NIOSH lifting equation. Several example problems are also provided to demonstrate how the equations should be used. An expanded, more detailed version of this chapter is contained in a NIOSH report entitled *Applications Manual for the Revised NIOSH Lifting Equation* (Waters et al., 1994).

Historically, the NIOSH has recognized the problem of work-related back injuries, and published the *Work Practices Guide for Manual Lifting* (WPG) in 1981 (NIOSH, 1981). The NIOSH WPG contained a summary of the lifting-related literature before 1981; analytical procedures and a lifting equation for calculating a recommended weight for specified two-handed, symmetrical lifting tasks; and an approach for controlling the hazards of low back injury from manual lifting. The approach to hazard control was coupled to the action limit (AL), a resultant term that denoted the recommended weight derived from the lifting equation.

In 1985, the NIOSH convened an ad hoc committee of experts who reviewed the current literature on lifting, including the NIOSH WPG.[1] The literature review was summarized in a document containing updated information on the physiological, biomechanical, psychophysical, and epidemiological aspects of manual lifting. Based on the results of the literature review, the ad hoc committee recommended criteria for defining the lifting capacity of healthy workers. The committee used the criteria to formulate the revised lifting equation.[2] Subsequently, NIOSH staff developed the documentation for the equation and played a prominent role in recommending methods for interpreting the results of the lifting equation.

The rationale and criterion for the development of the revised NIOSH lifting equation are provided in a journal article entitled: "Revised NIOSH equation for the design and evaluation of manual lifting tasks" (Waters et al., 1993). We suggest that those users who wish to achieve a better understanding of the data and decisions that were made in formulating the revised equation consult the article by Waters et al., 1993. The 1993 article provides an explanation of the selection of the biomechanical, physiological, and psychophysical criterion, as well as a description of the derivation of the individual components of the revised lifting equation. For those individuals, however, who are primarily concerned with the use and application of the revised lifting equation, provides a more complete description of the method and limitations for using the revised equation. The applications manual for the revised NIOSH Lifting Equation (Waters et al., 1994).

Although there are limited data examining the validity or effectiveness of the revised lifting equation to identify lifting jobs with increased risk of low back disorders, the recommended weight limits derived from the revised equation are consistent with, or lower than, those generally reported in the literature as being safe for workers (Waters et al., 1993, Table 2, Table 4, and Table 5). Moreover, the proper application of the revised equation is more likely to protect healthy workers for a wider variety of lifting tasks than methods that rely on only a single task factor or single criterion. A later section of this chapter provides a summary of studies examining the effectiveness of the NIOSH equation to identify manual lifting jobs with increased risk of lifting-related low back pain (LBP).

Finally, it should be stressed that the NIOSH lifting equation is only one tool in a comprehensive effort to prevent work-related LBP and disability. Some examples of other approaches are described elsewhere (ASPH/NIOSH, 1986). Moreover, lifting is only one of the causes of work-related LBP and disability. Other causes that have been hypothesized or established as risk factors include whole body vibration, static postures, prolonged sitting, and direct trauma to the back. Psychosocial factors, appropriate medical treatment, and job demands also may be particularly important in influencing the transition of acute low back pain to chronic disabling pain (see chapter entitled "Manual Materials Handling").

46.2   Definition of Terms

This section provides the basic technical information needed to properly use the revised lifting equation to evaluate a variety of two-handed manual lifting tasks. Definitions and data requirements for the revised lifting equation are also provided.

---

[1]The ad hoc 1991 NIOSH Lifting Committee members included: M.M. Ayoub, Donald B. Chaffin, Colin G. Drury, Arun Garg, and Suzanne Rodgers. NIOSH representatives included Vern Putz-Anderson and Thomas R. Waters.

[2]For this document, the revised 1991 NIOSH lifting equation will be identified simply as "the revised lifting equation." The abbreviation WPG will continue to be used as the reference to the earlier NIOSH lifting equation, which was documented in a publication entitled *Work Practices Guide for Manual Lifting*.

46.2.1   Recommended Weight Limit

The recommended weight limit (RWL) is the principal product of the revised NIOSH lifting equation. The RWL is defined for a specific set of task conditions as the weight of the load that nearly all healthy workers could perform over a substantial period of time (e.g., up to 8 h) without an increased risk of developing lifting-related LBP. By "healthy workers," we mean workers who are free of adverse health conditions that would increase their risk of musculoskeletal injury.

The concept behind the revised NIOSH lifting equation is to start with a recommended weight that is considered safe for an "ideal" lift (i.e., load constant equal to 51 lb) and then reduce the weight as the task becomes more stressful (i.e., as the task-related factors become less favorable). The precise formulation of the revised lifting equation for calculating the RWL is based on a multiplicative model that provides a weighting (multiplier) for each of six task variables, which include the: (1) *horizontal* distance of the load from the worker (*H*); (2) *vertical* height of the lift (*V*); (3) vertical *displacement* during the lift (*D*); (4) angle of *asymmetry* (*A*); (5) *frequency* (*F*) and duration of lifting; and (6) quality of the hand-to-object *coupling* (*C*). The weightings are expressed as coefficients that serve to decrease the load constant, which represents the maximum recommended load weight to be lifted under ideal conditions. For example, as the horizontal distance between the load and the worker increases from 10 in., the recommended weight limit for that task would be reduced from the ideal starting weight.

The RWL is defined as follows:

$$RWL = LC \times HM \times VM \times DM \times AM \times FM \times CM$$

where

	Metric	U.S. Customary
LC, Load constant	23 kg	51 lb
HM, Horizontal multiplier	$(25/H)$	$(10/H)$
VM, Vertical multiplier	$1 - (0.003 \mid V - 75\mid)$	$1 - (0.0075 \mid V - 30\mid)$
DM, Distance multiplier	$0.82 + (4.5/D)$	$0.82 + (1.8/D)$
AM, Asymmetric multiplier	$1 - (0.0032\,A)$	$1 - (0.0032\,A)$
FM, Frequency multiplier	From Table 46.5	From Table 46.5
CM, Coupling multiplier	From Table 46.7	From Table 46.7

The term "task variables" refers to the measurable task-related measurements that are used as input data for the formula (i.e., *H*, *V*, *D*, *A*, *F*, and *C*), whereas the term "multipliers" refers to the reduction coefficients in the equation (i.e., HM, VM, DM, AM, FM, and CM).

46.2.2   Measurement Requirements

The following list briefly describes the measurements required to use the revised NIOSH lifting equation. Details for each of the variables is presented later in this chapter (Section 46.4):

   H — Horizontal location of hands from midpoint between the inner ankle bones. Measure at the origin and the destination of the lift (cm or in.)

   V — Vertical location of the hands from the floor. Measure at the origin and destination of the lift (cm or in.)

   D — Vertical travel distance between the origin and the destination of the lift (cm or in.)

   A — Angle of asymmetry — angular displacement of the load from the worker's sagittal plane. Measure at the origin and destination of the lift (°)

F—Average frequency rate of lifting measured in lifts/min. Duration is defined to be ≤1 h, ≤2 h, or ≤8 h assuming appropriate recovery allowances (Table 46.5)

C—Quality of hand-to-object coupling (quality of interface between the worker and the load being lifted). The quality of the coupling is categorized as good, fair, or poor, depending upon the type and location of the coupling, the physical characteristics of load, and the vertical height of the lift

46.2.3 Lifting Index

The Lifting Index (LI) is a term that provides a relative estimate of the level of physical stress associated with a particular manual lifting task. The estimate of the level of physical stress is defined by the relationship of the weight of the load lifted and the recommended weight limit. The LI is defined by the following equation:

$$LI = \frac{\text{Load weight}}{\text{Recommended weight limit}} = \frac{L}{\text{RWL}}$$

where load weight (L) is equal to the weight of the object lifted (lb or kg).

46.2.4 Miscellaneous Terms

Lifting task — Defined as the act of manually grasping an object of definable size and mass with two hands, and vertically moving the object without mechanical assistance.

Load weight (L) — Weight of the object to be lifted, in pounds or kilograms, including the container.

Horizontal location (H) — Distance of the hands away from the midpoint between the ankles, in inches or centimeters (measure at the origin and destination of lift). See Figure 46.1.

Vertical location (V) — Distance of the hands above the floor, in inches or centimeters (measure at the origin and destination of lift). See Figure 46.1.

Vertical travel distance (D) — Absolute value of the difference between the vertical heights at the destination and origin of the lift, in inches or centimeters.

Angle of asymmetry (A) — The angular measure of how far the *object* is displaced from the front (midsagittal plane) of the worker's body at the beginning or ending of the lift, in degrees (measure at the origin and destination of lift) see Figure 46.2. The asymmetry angle is defined by the location of the load relative to the worker's midsagittal plane, as defined by the neutral body posture, rather than the position of the feet or the extent of body twist.

Neutral body position — Describes the position of the body when the hands are directly in front of the body and there is minimal twisting at the legs, torso, or shoulders.

Frequency of lifting (F) — Average number of lifts per minute over a 15-min period.

Duration of lifting — Three-tiered classification of lifting duration specified by the distribution of worktime and recoverytime (work pattern). Duration is classified as either short (1 h), moderate (1 to 2 h), or long (2 to 8 h), depending on the work pattern.

Coupling classification — Classification of the quality of the hand-to-object coupling (e.g., handle, cut-out, or grip). Coupling quality is classified as good, fair, or poor.

Significant control — Significant control is defined as a condition requiring "precision placement" of the load at the destination of the lift. This is usually the case when (1) the worker has to regrasp the load near the destination of the lift, (2) the worker has to momentarily hold the object at the destination, or (3) the worker has to carefully position or guide the load at the destination.

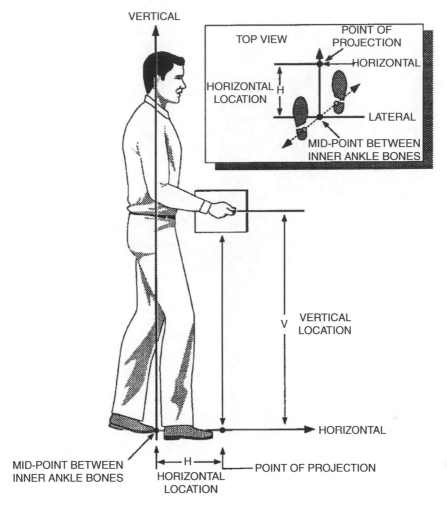

FIGURE 46.1 Graphic representation of hand location.

46.3 Limitations of Equation

The lifting equation is a tool for assessing the physical stress of two-handed manual lifting tasks. As with any tool, its application is limited to those conditions for which it was designed. Specifically, the lifting equation was designed to meet specific lifting-related criteria that encompass biomechanical, physiological, and psychophysical assumptions and data used to develop the equation. To the extent that a given lifting task accurately reflects these underlying conditions and criteria, this lifting equation may be appropriately applied.

The following list identifies a set of work conditions in which the application of the lifting equation could either underestimate or overestimate the extent of physical stress associated with a particular work-related activity. Each of the following task limitations also highlights research topics in need of further research to extend the application of the lifting equation to a greater range of real world lifting tasks.

The revised NIOSH lifting equation does not apply if any of the following occur:

- Lifting/lowering with one hand
- Lifting/lowering for over 8 h
- Lifting/lowering while seated or kneeling

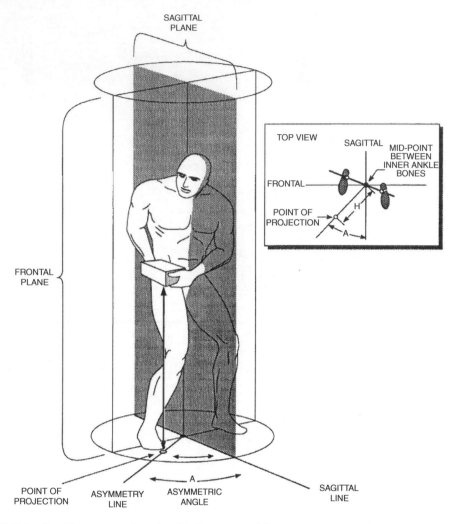

FIGURE 46.2 Graphic representation of angle of asymmetry (*A*).

- Lifting/lowering in a restricted work space
- Lifting/lowering unstable objects
- Lifting/lowering while carrying, pushing, or pulling
- Lifting/lowering with wheelbarrows or shovels
- Lifting/lowering with "high-speed" motion (faster than about 30 in./sec)
- Lifting/lowering with unreasonable foot/floor coupling (<0.4 coefficient of friction between the sole and the floor)
- Lifting/lowering in an unfavorable environment (temperature significantly outside 66 to 79°F [19 to 26°C] range; relative humidity outside 35 to 50% range)

46.4   Obtaining and Using the Data

46.4.1   Horizontal Component

46.4.1.1   Definition and Measurement

Horizontal location (*H*) is measured from the midpoint of the line joining the inner ankle bones to a point projected on the floor directly below the midpoint of the hand grasps (i.e., load center), as defined by the large middle knuckle of the hand (Figure 46.1). Typically, the worker's feet are not aligned with the midsagittal plane, as shown in Figure 46.1, but may be rotated inward or outward. If this is the case, then the midsagittal plane is defined by the worker's neutral body posture as defined above. If significant control is required at the destination (i.e., precision placement), then H should be measured at both the origin and destination of the lift. Also, if the worker leans over on one foot during lifting, concentrating nearly all of their support on one foot, while using the other leg and foot as a counterbalance so that they can reach out further to pick up the load, the H variable is measured from a point directly below the weight bearing foot, rather than the midpoint between the ankles. In cases where it is not clear that the weight is concentrated primarily on one foot, the point between the ankles should still be used as the reference point for measurement of the horizontal location (H). It also important to note that it has also come to our attention that users sometimes overestimate the magnitude of the horizontal location (H) and the asymmetric angle (A) for some types of lifts because they mistakenly measure the task variables at the incorrect location for the origin of the lift. This may occur when the lifters stand with the side of their body next to a table or shelf and reach over to slide the object horizontally toward the front of the body as they begin the lift. When the lift is performed this way, the load actually moves horizontally toward the front of the body before it actually begins to move vertically. When this type of lift is analyzed, the task variables should be measured at the actual location where the object first begins to move upward (liftoff point), rather than at the point where the object first begins to move horizontally. This change will generally result in smaller H values than would have been determined if the measurements had been taken at the point where the object first began to move horizontally rather than vertically.

Horizontal distance (H) should be measured. In those situations where the H value cannot be measured, then H may be approximated from the following equations:

Metric (all distances in cm)	U.S. Customary (all distances in in.)
$H = 20 + W/2$ for $V \geq 25$ cm	$H = 8 + W/2$ for $V \geq 10$ in.
$H = 25 + W/2$ for $V < 25$ cm	$H = 10 + W/2$ for $V < 10$ in.

where W is the width of the container in the sagittal plane and V is the vertical location of the hands from the floor.

46.4.1.2 Horizontal Restrictions

If the horizontal distance is less than 10 in. (25 cm), then H is set to 10 in. (25 cm). Although objects can be carried or held closer than 10 in. from the ankles, most objects that are closer than this cannot be lifted without encountering interference from the abdomen or hyperextending the shoulders. While 25 in. (63 cm) was chosen as the maximum value for H, it is probably too large for shorter workers, particularly when lifting asymmetrically. Furthermore, objects at a distance of more than 25 in. from the ankles normally cannot be lifted vertically without some loss of balance.

46.4.1.3 Horizontal Multiplier

The horizontal multiplier (HM) is $10/H$ for H measured in inches, and HM is $25/H$ for H measured in centimeters. If H is less than or equal to 10 in. (25 cm), the multiplier is 1.0. HM decreases with an increase in H value. The multiplier for H is reduced to 0.4 when H is 25 in. (63 cm). If H is greater than 25 in., then HM = 0. The HM value can be computed directly or determined from Table 46.1.

TABLE 46.1   Horizontal Multiplier

H (in.)	HM	H (cm)	HM
≤10	1.00	≤25	1.00
11	0.91	28	0.89
12	0.83	30	0.83
13	0.77	32	0.78
14	0.71	34	0.74
15	0.67	36	0.69
16	0.63	38	0.66
17	0.59	40	0.63
18	0.56	42	0.60
19	0.53	44	0.57
20	0.50	46	0.54
21	0.48	48	0.52
22	0.46	50	0.50
23	0.44	52	0.48
24	0.42	54	0.46
25	0.40	56	0.45
>25	0.00	58	0.43
		60	0.42
		63	0.40
		>63	0.00

46.4.2   Vertical Component

46.4.2.1   Definition and Measurement

Vertical location (V) is defined as the vertical height of the hands above the floor. V is measured vertically from the floor to the midpoint between the hand grasps, as defined by the large middle knuckle. The coordinate system is illustrated in Figure 46.1.

46.4.2.2   Vertical Restrictions

The vertical location (V) is limited by the floor surface and the upper limit of vertical reach for lifting (i.e., 70 in. or 175 cm). The vertical location should be measured at the origin and the destination of the lift to determine the travel distance (D).

46.4.2.3   Vertical Multiplier

To determine the vertical multiplier (VM), the absolute value or deviation of V from an optimum height of 30 in. (75 cm) is calculated. A height of 30 in. above the floor level is considered "knuckle height" for a worker of average height (66 in. or 165 cm). The VM is $(1- (0.0075 \mid V - 30\mid))$ for V measured in inches, and VM is $(1 - (0.003 \mid V - 75\mid))$ for V measured in centimeters.

When V is at 30 in. (75 cm), the VM is 1.0. The value of VM decreases linearly with an increase or decrease in height from this position. At floor level, VM is 0.78, and at 70 in. (175 cm) height VM is 0.7. If V is greater than 70 in., then VM $= 0$. The VM value can be computed directly or determined from Table 46.2.

46.4.3   Distance Component

46.4.3.1   Definition and Measurement

The distance variable (D) is defined as the vertical travel distance of the hands between the origin and destination of the lift. For lifting, D can be computed by subtracting the vertical location (V) at the origin of the lift from the corresponding V at the destination of the lift (i.e., D is equal to V at the destination minus V at the origin). For a lowering task, D is equal to V at the origin minus V at the destination.

TABLE 46.2   Vertical Multiplier

V (in.)	VM	V (cm)	VM
0	0.78	0	0.78
5	0.81	10	0.81
10	0.85	20	0.84
15	0.89	30	0.87
20	0.93	40	0.90
25	0.96	50	0.93
30	1.00	60	0.96
35	0.96	70	0.99
40	0.93	80	0.99
45	0.89	90	0.96
50	0.85	100	0.93
55	0.81	110	0.90
60	0.78	120	0.87
65	0.74	130	0.84
70	0.70	140	0.81
>70	0.00	150	0.78
		160	0.75
		170	0.72
		175	0.70
		>175	0.00

46.4.3.2 Distance Restrictions

The distance variable (D) is assumed to be at least 10 in. (25 cm), and no greater than 70 in. (175 cm). If the vertical travel distance is less than 10 in. (25 cm), then D should be set to the minimum distance of 10 in. (25 cm).

46.4.3.3 Distance Multiplier

The distance multiplier (DM) is $(0.82 + (1.8/D))$ for D measured in inches, and DM is $(0.82 + (4.5/D))$ for D measured in centimeters. For D less than 10 in. (25 cm) D is assumed to be 10 in. (25 cm), and DM is 1.0. The DM, therefore, decreases gradually with an increase in travel distance. The DM is 1.0 when D is set at 10 in., (25 cm); DM is 0.85 when $D = 70$ in. (175 cm). Thus, DM ranges from 1.0 to 0.85 as the D varies from 0 in. (0 cm) to 70 in. (175 cm). The DM value can be computed directly or determined from Table 46.3.

TABLE 46.3   Distance Multiplier

D (in.)	DM	D (cm)	DM
≤10	1.00	≤25	1.00
15	0.94	40	0.93
20	0.91	55	0.90
25	0.89	70	0.88
30	0.88	85	0.87
35	0.87	100	0.87
40	0.87	115	0.86
45	0.86	130	0.86
50	0.86	145	0.85
55	0.85	160	0.85
60	0.85	175	0.85
70	0.85	>175	0.00
>70	0.00		

46.4.4   Asymmetry Component

46.4.4.1   Definition and Measurement

Asymmetry refers to a lift that begins or ends outside the midsagittal plane (see Figure 46.2). In general, asymmetric lifting should be avoided. If asymmetric lifting cannot be avoided, however, the recommended weight limits are significantly less than those limits used for symmetrical lifting.[3]

An asymmetric lift may be required under the following task or workplace conditions:

1. The origin and destination of the lift are oriented at an angle to each another
2. The lifting motion is across the body, such as occurs in swinging bags or boxes from one location to another
3. The lifting is done to maintain body balance in obstructed workplaces, on rough terrain, or on littered floors
4. Productivity standards require reduced time per lift

The asymmetric angle (A), which is depicted graphically in Figure 46.2, is operationally defined as the angle between the asymmetry line and the midsagittal line. The *asymmetry line* is defined as the line that joins the midpoint between the inner ankle bones and the point projected on the floor directly below the midpoint of the hand grasps, as defined by the large middle knuckle. The *sagittal line* is defined as the line passing through the midpoint between the inner ankle bones and lying in the midsagittal plane, as defined by the neutral body position (i.e., hands directly in front of the body, with no twisting at the legs, torso, or shoulders). *Note*: The asymmetry angle is not defined by foot position or the angle of torso twist, but by the location of the load relative to the worker's midsagittal plane.

In many cases of asymmetric lifting, the worker will pivot or use a step turn to complete the lift. Since this may vary significantly between workers and between lifts, we have assumed that no pivoting or stepping occurs. Although this assumption may overestimate the reduction in acceptable load weight, it will provide the greatest protection for the worker.

The asymmetry angle (A) must always be measured at the origin of the lift. If significant control is required at the destination, however, then angle A should be measured at both the origin and the destination of the lift. Remember that A should be measured at the liftoff point, when the load actually begins to move upward, rather than at the point when the object begins to move horizontally. This is often easiest to see when the job is videotaped and the videotape is played back at a slow speed or frame by frame.

46.4.4.2   Asymmetry Restrictions

The angle A is limited to the range from 0 to 135°. If $A > 135°$, then the asymmetric multiplier (AM) is set equal to zero, which results in a RWL of zero, or no load.

46.4.4.3   Asymmetric Multiplier

The AM is $1 - (0.0032A)$. The AM has a maximum value of 1.0 when the load is lifted directly in front of the body. The AM decreases linearly as the angle of asymmetry (A) increases. The range is from a value of 0.57 at 135° of asymmetry to a value of 1.0 at 0° of asymmetry (i.e., symmetric lift). If A is greater than 135°, then AM = 0, and the load is zero. The AM value can be computed directly or determined from Table 46.4.

---

[3]It may not always be clear if asymmetry is an intrinsic element of the task or just a personal characteristic of the worker's lifting style. Regardless of the reason for the asymmetry, any observed asymmetric lifting should be considered an intrinsic element of the job design and should be considered in the assessment and subsequent redesign. Moreover, the design of the task should not rely on worker compliance, but rather the design should discourage or eliminate the need for asymmetric lifting.

TABLE 46.4 Asymmetric Multiplier

A (°)	AM
0	1.00
15	0.95
30	0.90
45	0.86
60	0.81
75	0.76
90	0.71
105	0.66
120	0.62
135	0.57
>135	0.00

46.4.5 Frequency Component

46.4.5.1 Definition and Measurement

The frequency multiplier is defined by (a) the number of lifts per minute (frequency), (b) the amount of time engaged in the lifting activity (duration), and (c) the vertical height of the lift from the floor. Lifting frequency (F) refers to the average number of lifts made per minute, as measured over a 15-min period. Because of the potential variation in work patterns, analysts may have difficulty obtaining an accurate or representative 15-min work sample for computing the lifting frequency (F). If significant variation exists in the frequency of lifting over the course of the day, analysts should employ standard work sampling techniques to obtain a representative work sample for determining the number of lifts per minute. For those jobs where the frequency varies from session to session, each session should be analyzed separately, but the overall work pattern must still be considered. For more information, most standard industrial engineering or ergonomics texts provide guidance for establishing a representative job sampling strategy (e.g., Eastman Kodak Company, 2004).

46.4.5.2 Lifting Duration

Lifting duration is classified into three categories based on the pattern of continuous work-time and recovery-time (i.e., light work) periods. A continuous work-time (WT) period is defined as a period of uninterrupted work. Recovery time (RT) is defined as the duration of light work activity following a period of continuous lifting. Examples of light work include activities such as sitting at a desk or table, monitoring operations, light assembly work, etc. The three categories are short duration, moderate duration and long duration:

1. *Short-duration* defines lifting tasks that have a work duration of 1 h or less, followed by a recovery time equal to 1.0 times the work time (i.e., at least a 1.0 recovery time to work time ratio [RT/WT]). (*Note*: the RT/WT ratio has been changed from 1.2 to 1.0 since the equation was originally published.)

   For example, to be classified as short duration, a 45-min lifting job must be followed by at least a 45-min recovery period prior to initiating a subsequent lifting session. If the required recovery time is not met for a job of 1 h or less, and a subsequent lifting session is required, then the total lifting time must be combined to correctly determine the duration category. Moreover, if the recovery period does not meet the time requirement, it is disregarded for purposes of determining the appropriate duration category.

   As another example, assume a worker lifts continuously for 30 min, then performs a light work task for 10 min, and then lifts for an additional 45-min period. In this case, the recovery time between lifting sessions (10 min) is less than 1.0 times the initial 30-min work time (36 min). Thus, the two work times (30 and 45 min) must be added together to determine the duration.

Since the total work time (75 min) exceeds 1 h, the job is classified as moderate duration. On the other hand, if the recovery period between lifting sessions was increased to 30 min, then the short-duration category would apply, which would result in a larger FM value.

A special procedure has been developed for determining the appropriate lifting frequency (F) for certain repetitive lifting tasks in which workers do not lift continuously during the 15-min sampling period. This occurs when the work pattern is such that the worker lifts repetitively for a short time and then performs light work for a short time before starting another cycle. For work patterns such as this, the lifting frequency (F) may be determined as follows, as long as the actual lifting frequency does not exceed 15 lifts per minute:

(i) Compute the total number of lifts performed for the 15-min period (i.e., lift rate times work time)

(ii) Divide the total number of lifts by 15

(iii) Use the resulting value as the frequency (F) to determine the FM from Table 46.5

For example, if the work pattern for a job consists of a series of cyclic sessions requiring 8 min of lifting followed by 7 min of light work, and the lifting rate during the work sessions is ten lifts per minute, then the frequency rate (F) that is used to determine the FM for this job is equal to $(10 \times 8)/15$ or 5.33 lifts/min. If the worker lifted continuously for more than 15 min, however, then the actual lifting frequency (10 lifts/min) would be used.

When using this special procedure, the duration category is based on the magnitude of the recovery periods *between* work sessions, not *within* work sessions. In other words, if the work pattern is intermittent and the special procedure applies, then the intermittent recovery periods that occur during the 15-min sampling period are not considered as recovery periods for purposes of determining the duration category. For example, if the work pattern for a manual lifting job was composed of repetitive cycles consisting of 1 min of continuous lifting at a rate of 10 lifts/min, followed by 2 min of recovery, the correct procedure would be to adjust the frequency according to the special procedure (i.e., $F = [10 \text{ lifts/min} \times 5 \text{ min}]/15 \text{ min} = 50/15 = 3.4$ lifts/min).

TABLE 46.5 Frequency Multiplier Table

Frequency[a] Lifts/min (F)	≤1 h		>1 but ≤2 h		>2 but ≤8 h	
	$V^b < 30$	$V \geq 30$	$V < 30$	$V \geq 30$	$V < 30$	$V \geq 30$
≤0.2	1.00	1.00	0.95	0.95	0.85	0.85
0.5	0.97	0.97	0.92	0.92	0.81	0.81
1	0.94	0.94	0.88	0.88	0.75	0.75
2	0.91	0.91	0.84	0.84	0.65	0.65
3	0.88	0.88	0.79	0.79	0.55	0.55
4	0.84	0.84	0.72	0.72	0.45	0.45
5	0.80	0.80	0.60	0.60	0.35	0.35
6	0.75	0.75	0.50	0.50	0.27	0.27
7	0.70	0.70	0.42	0.42	0.22	0.22
8	0.60	0.60	0.35	0.35	0.18	0.18
9	0.52	0.52	0.30	0.30	0.00	0.15
10	0.45	0.45	0.26	0.26	0.00	0.13
11	0.41	0.41	0.00	0.23	0.00	0.00
12	0.37	0.37	0.00	0.21	0.00	0.00
13	0.00	0.34	0.00	0.00	0.00	0.00
14	0.00	0.31	0.00	0.00	0.00	0.00
15	0.00	0.28	0.00	0.00	0.00	0.00
>15	0.00	0.00	0.00	0.00	0.00	0.00

[a]For lifting less frequently than once per 5 min, set $F = 0.2$ lifts/min.
[b]Values of V are in inches.

The 2-min recovery periods would not count toward the RT/WT ratio, however, and additional recovery periods would have to be provided as described earlier.

2. Moderate duration defines lifting tasks that have a duration of more than 1 h, but not more than 2 h, followed by a recovery period of at least 0.3 times the work time (i.e., at least a 0.3 recovery time to work time ratio [RT/WT]).

    For example, if a worker continuously lifts for 2 h, then a recovery period of at least 36 min would be required before initiating a subsequent lifting session. If the recovery time requirement is not met, and a subsequent lifting session is required, then the total work time must be added together. If the total work time exceeds 2 h, then the job must be classified as a long-duration lifting task.

3. Long duration defines lifting tasks that have a duration of between 2 and 8 h, with standard industrial rest allowances (e.g., morning, lunch, and afternoon rest breaks).

    Note: no weight limits are provided for more than 8 h of work.

The difference in the required RT/WT ratio for the short-duration (less than 1 h) category, which is 1.0, and the moderate duration category (1 to 2 h), which is 0.3, is due to the difference in the magnitudes of the frequency multiplier values associated with each of the duration categories. Since the moderate category results in larger reductions in the RWL than the short category, there is less need for a recovery period between sessions than for the short-duration category. In other words, the short-duration category would result in higher weight limits than the moderate duration category, so larger recovery periods would be needed.

46.4.5.3 Frequency Restrictions

Lifting frequency (F) for repetitive lifting may range from 0.2 lifts/min to a maximum frequency that is dependent on the vertical location of the object (V) and the duration of lifting (Table 46.5). Lifting above the maximum frequency results in an RWL of 0.0. (Except for the special case of discontinuous lifting discussed earlier, where the maximum frequency is 15 lifts/min.)

46.4.5.4 Frequency Multiplier

The FM value depends upon the average number of lifts per minute (F), the vertical location (V) of the hands at the origin, and the duration of continuous lifting. For lifting tasks with a frequency less than 0.2 lifts/min, set the frequency equal to 0.2 lifts/min. Otherwise, the FM is determined from Table 46.5.

46.4.6 Coupling Component

46.4.6.1 Definition and Measurement

The nature of the hand-to-object coupling or gripping method can affect not only the maximum force a worker can or must exert on the object, but also the vertical location of the hands during the lift. A "good" coupling will reduce the maximum grasp forces required and increase the acceptable weight for lifting, while a "poor" coupling will generally require higher maximum grasp forces and decrease the acceptable weight for lifting.

The effectiveness of the coupling is not static, but may vary with the distance of the object from the ground, so that a good coupling could become a poor coupling during a single lift. The entire range of the lift should be considered when classifying hand-to-object couplings, with classification based on overall effectiveness. The analyst must classify the coupling as good, fair, or poor. The three categories are defined in Table 46.6. If there is any doubt about classifying a particular coupling design, the more stressful classification should be selected.

The decision tree shown in Figure 46.3 may be helpful in classifying the hand-to-object coupling.

46.4.6.2 Coupling Multiplier

Based on the coupling classification and vertical location of the lift, the coupling multiplier (CM) is determined from Table 46.7.

TABLE 46.6 Hand-to-Container Coupling Classification

Good	Fair	Poor
For containers of optimal design, such as some boxes, crates, etc., a "good" hand-to-object coupling would be defined as handles or hand-hold cut-outs of optimal design[a–c]	For containers of optimal design, a "fair" hand-to-object coupling would be defined as handles or hand-hold cut-outs of less than optimal design[a–d]	Containers of less than optimal design or loose parts or irregular objects that are bulky, hard to handle, or have sharp edges[e]
For loose parts or irregular objects, which are not usually containerized, such as castings, stock, and supply materials, a "good" hand-to-object coupling would be defined as a comfortable grip in which the hand can be easily wrapped around the object[f]	For containers of optimal design with no handles or hand-hold cut-outs or for loose parts or irregular objects, a "fair" hand-to-object coupling is defined as a grip in which the hand can be flexed about 90°[d]	Lifting nonrigid bags (i.e., bags that sag in the middle)

[a]An optimal handle design has 0.75 to 1.5 in. (1.9 to 3.8 cm) diameter, ≥ 4.5 in. (11.5 cm) length, 2 in. (5 cm) clearance, cylindrical shape, and a smooth, nonslip surface.

[b]An optimal hand-hold cut-out has the following approximate characteristics: ≥ 1.5 in. (3.8 cm) height, 4.5 in. (11.5 cm) length, semioval shape, ≥ 2 in. (5 cm) clearance, smooth nonslip surface, and ≥ 0.25 in. (0.60 cm) container thickness (e.g., double thickness cardboard).

[c]An optimal container design has ≤ 16 in. (40 cm) frontal length, ≤ 12 in. (30 cm) height, and a smooth nonslip surface.

[d]A worker should be capable of clamping the fingers at nearly 90° under the container, such as required when lifting a cardboard box from the floor.

[e]A container is considered less than optimal if it has a frontal length > 16 in. (40 cm), height > 12 in. (30 cm), rough or slippery surfaces, sharp edges, asymmetric center of mass, unstable contents, or requires the use of gloves.

[f]A worker should be able to comfortably wrap the hand around the object without causing excessive wrist deviations or awkward postures, and the grip should not require excessive force.

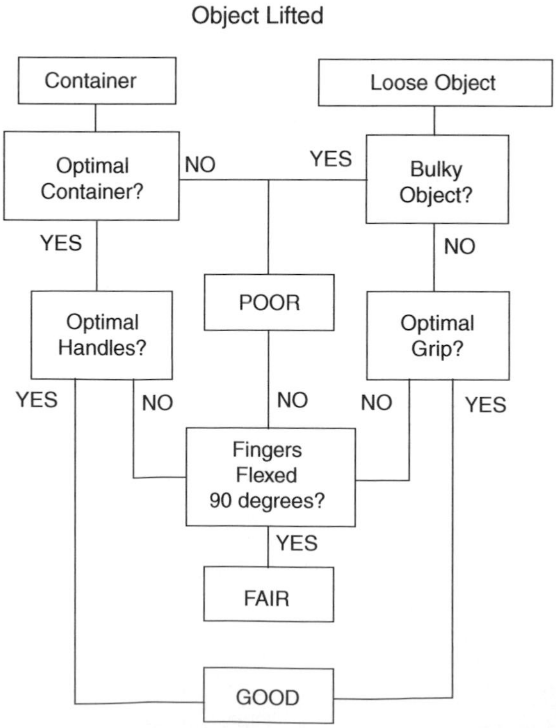

FIGURE 46.3 Decision tree for coupling quality.

TABLE 46.7 Coupling Multiplier

	Coupling Multiplier	
Coupling Type	$V < 30$ in. (75 cm)	$V \geq 30$ in. (75 cm)
Good	1.00	1.00
Fair	0.95	1.00
Poor	0.90	0.90

46.5   Procedures

Prior to data collection, the analyst must decide (1) if the job should be analyzed as a single-task or multi-task manual lifting job and (2) if significant control is required at the destination of the lift. This is necessary because the procedures differ, depending on the type of analysis required.

A manual lifting job may be analyzed as a single-task job if the task variables do not differ from task to task, or if only one task is of interest (e.g., single most stressful task). This may be the case if one of the tasks clearly has a dominant effect on strength demands, localized muscle fatigue, or whole body fatigue. On the other hand, if the task variables differ significantly between tasks, it may be more appropriate to analyze a job as a multi-task manual lifting job. A multi-task analysis is more difficult to perform than a single-task analysis because additional data and computations are required. The multi-task approach, however, will provide more detailed information about specific strength and physiological demands.

For many lifting jobs, it may be acceptable to use either the single- or multi-task approach. The single-task analysis should be used when possible, but when a job consists of more than one task and detailed information is needed to specify engineering modifications, then the multi-task approach provides a reasonable method of assessing the overall physical demands. The multi-task procedure is more complicated than the single-task procedure, and requires a greater understanding of assessment terminology and mathematical concepts. Therefore, the decision to use the single- or multi-task approach should be based on: (1) the need for detailed information about all facets of the multi-tasked lifting job; (2) the need for accuracy and completeness of data regarding assessment of the physiological demands of the task; and (3) the analyst's level of understanding of the assessment procedures.

The decision about control at the destination is important because the physical demands on the worker may be greater at the destination of the lift than at the origin, especially when significant control is required. When significant control is required at the destination, for example, the physical stress is increased because the load will have to be accelerated upward to slow down the descent of the load. This acceleration may be as great as the acceleration at the origin of the lift and may create high loads on the spine. Therefore, if significant control is required, then the RWL and LI should be determined at both locations and the lower of the two values will specify the overall level of physical demand.

To perform a lifting analysis using the revised lifting equation, two steps are undertaken: (1) data is collected at the worksite as described in Step 1 (described next); and (2) the RWL and LI values are computed using the single- or multi-task analysis procedures described in Step 2 (described later).

46.5.1   Step 1: Collect Data

The relevant task variables must be carefully measured and clearly recorded in a concise format. As mentioned previously, these variables include the horizontal location of the hands (H), vertical location of the hands (V), vertical displacement (D), asymmetric angle (A), lifting frequency (F), and coupling quality (C). A job analysis worksheet, as shown in Figure 46.4 for single-task jobs or Figure 46.5 for multi-task jobs, provides a simple form for recording the task variables and the data needed to calculate the RWL and LI values. A thorough job analysis is required to identify and catalog each independent lifting task that comprises the worker's complete job. For multi-task jobs, data must be collected for each individual task.

JOB ANALYSIS WORKSHEET

DEPARTMENT _____   JOB DESCRIPTION
JOB TITLE _____
ANALYST'S NAME _____
DATE _____

STEP 1. Measure and record task variables

Object Weight (lbs)		Hand Location (in)				Vertical Distance (in)	Asymmetric Angle (degrees)		Frequency Rate lifts/min	Duration (HRS)	Object Coupling
		Origin		Dest			Origin	Destination			
L (AVG.)	L (Max.)	H	V	H	V	D	A	A	F		C

STEP 2. Determine the multipliers and compute the RWLs

RWL = LC × HM × VM × DM × AM × FM × CM

ORIGIN   RWL = [51] × [ ] × [ ] × [ ] × [ ] × [ ] × [ ] =   Lbs

DESTINATION   RWL = [51] × [ ] × [ ] × [ ] × [ ] × [ ] × [ ] =   Lbs

STEP 3. Compute the LIFTING INDEX

ORIGIN   LIFTING INDEX = $\dfrac{\text{OBJECT WEIGHT (L)}}{\text{RWL}}$ = _____ = [ ]

DESTINATION   LIFTING INDEX = $\dfrac{\text{OBJECT WEIGHT (L)}}{\text{RWL}}$ = _____ = [ ]

FIGURE 46.4   Single-task analysis sheet.

MULTI-TASK JOB ANALYSIS WORKSHEET

DEPARTMENT _____   JOB DESCRIPTION
JOB TITLE _____
ANALYST'S NAME _____
DATE _____

STEP 1. Measure and Record Task Variable Data

Task No.	Object Weight (lbs)		Hand Location (in)				Vertical Distance (in)	Asymmetry Angle (degs)		Frequency Rate lifts/min	Duration Hrs	Coupling
	L (Avg.)	L (Max.)	Origin		Dest.			Origin	Dest.			
			H	V	H	V	D	A	A	F		C
_____												
_____												
_____												
_____												

STEP 2. Compute multipliers and FIRWL, STRWL, FILI, and STLI for Each Task

Task No.	LC × HM × VM × DM × AM × CM	FIRWL × FM	STRWL	FILI = L/FIRWL	STLI = L/STRWL	New Task No.	F
51							
51							
51							
51							
51							

STEP 3. Compute the Composite Lifting Index for the Job   (After renumbering tasks)

CLI = STLI$_1$ + $\triangle$ FILI$_2$ + $\triangle$ FILI$_3$ + $\triangle$ FILI$_4$ + $\triangle$ FILI$_5$

FILI$_2$(1/FM$_{1,2}$ − 1/FM$_1$)	FILI$_3$(1/FM$_{1,2,3}$ − 1/FM$_{1,2}$)	FILI$_4$(1/FM$_{1,2,3,4}$ − 1/FM$_{1,2,3}$)	FILI$_5$(1/FM$_{1,2,3,4,5}$ − 1/FM$_{1,2,3,4}$)

CLI = [ ]

FIGURE 46.5   Multi-task analysis sheet.

46.5.2 Step 2: Single- and Multi-Task Procedures

46.5.2.1 Single-Task Procedure

46.5.2.1.1 Compute the Recommended Weight Limit and Lifting Index

Calculate the RWL at the origin for each lift. For lifting tasks that require significant control at the destination, calculate the RWL at both the origin and the destination of the lift. The latter procedure is required if (1) the worker has to regrasp the load near the destination of the lift, (2) the worker has to momentarily hold the object at the destination, or (3) the worker has to position or guide the load at the destination. The purpose of calculating the RWL at both the origin and destination of the lift is to identify the most stressful location of the lift. Therefore, the lower of the RWL values at the origin or destination should be used to compute the LI for the task, since this value would represent the limiting set of conditions.

The assessment is completed on the single-task worksheet by determining the LI for the task of interest. This is accomplished by comparing the actual weight of the load (L) lifted with the RWL value obtained from the lifting equation.

46.5.2.2 Multi-Task Procedure

1. Compute the frequency-independent recommended weight limit (FIRWL) and single-task recommended weight limit (STRWL) for each task
2. Compute the frequency-independent lifting index (FILI) and single-task lifting index (STLI) for each task
3. Compute the composite lifting index (CLI) for the overall job

46.5.2.2.1 Compute the Frequency-Independent Recommended Weight Limits

Compute the FIRWL value for each task by using the respective task variables and setting the FM to a value of 1.0. The FIRWL for each task reflects the compressive force and muscle strength demands for a single repetition of that task. If significant control is required at the destination for any individual task, the FIRWL must be computed at both the origin and the destination of the lift, as described earlier for a single-task analysis.

46.5.2.2.2 Compute the Single-Task Recommended Weight Limit

Compute the STRWL for each task by multiplying its FIRWL by its appropriate FM. The STRWL for a task reflects the overall demands of that task, assuming it was the only task being performed. Note that this value does not reflect the overall demands of the task when the other tasks are considered. Nevertheless, this value is helpful in determining the extent of excessive physical stress for an individual task.

46.5.2.2.3 Compute the Frequency-Independent Lifting Index

The FILI is computed for each task by dividing the *maximum* load weight (L) for that task by the respective FIRWL. The maximum weight is used to compute the FILI because the maximum weight determines the maximum biomechanical loads to which the body will be exposed, regardless of the frequency of occurrence. Thus, the FILI can identify individual tasks with potential strength problems for infrequent lifts. If any of the FILI values exceed a value of 1.0, then job design changes may be needed to decrease the strength demands.

46.5.2.2.4 Compute the Single-Task Lifting Index

The STLI is computed for each task by dividing the *average* load weight (L) for that task by the respective STRWL. The average weight is used to compute the STLI because the average weight provides a better representation of the metabolic demands, which are distributed across the tasks, rather than dependent on individual tasks. The STLI can be used to identify individual tasks with excessive physical demands (i.e., tasks that would result in fatigue). The STLI values do not indicate the relative stress of the individual tasks in the context of the whole job, but the STLI value can be used to prioritize the individual tasks according to the magnitude of their physical stress. Thus, if any of the STLI values exceed a value of 1.0, then ergonomics changes may be needed to decrease the overall physical demands of the task. Note that it may be possible to have a job in which all of the individual tasks have an STLI less than 1.0 and still be physically demanding due to the combined demands of the tasks. In cases where the FILI exceeds the STLI for any task, the maximum weights may represent a significant problem and careful evaluation is necessary.

46.5.2.2.5 Compute the Composite Lifting Index

The assessment is completed on the multi-task worksheet by determining the CLI for the overall job. The CLI is computed as follows:

1. The tasks are renumbered in order of decreasing physical stress, beginning with the task with the greatest STLI down to the task with the smallest STLI. The tasks are renumbered in this way so that the more difficult tasks are considered first.
2. The CLI for the job is then computed according to the following formula:

$$CLI = STLI_1 + \sum \Delta LI$$

where

$$\sum \Delta LI = \left(FILI_2 \times \left(\frac{1}{FM_{1,2}} - \frac{1}{FM_1} \right) \right)$$

$$+ \left(FILI_3 \times \left(\frac{1}{FM_{1,2,3}} - \frac{1}{FM_{1,2}} \right) \right)$$

$$+ \left(FILI_4 \times \left(\frac{1}{FM_{1,2,3,4}} - \frac{1}{FM_{1,2,3}} \right) \right)$$

$$\vdots$$

$$+ \left(FILI_n \times \left(\frac{1}{FM_{1,2,3,4,\cdots,n}} - \frac{1}{FM_{1,2,3,\cdots,(n-1)}} \right) \right)$$

Note that (1) the numbers in 46.5 subscript refer to the new task numbers and (2) the FM values are determined from Table 46.5, based on the sum of the frequencies for tasks listed in subscript.

The following example is provided to demonstrate this step of the multi-task procedure. Assume that an analysis of a typical three-task job provided the results shown in Table 46.8.

To compute the CLI for this job, the tasks are renumbered in order of decreasing physical stress, beginning with the task with the greatest STLI down to the task with the smallest STLI. In this case, as shown in Table 46.8, the task numbers do not change. Next, the CLI is computed according to the formula given earlier. The task with the greatest CLI is Task 1 (STLI = 1.6). The sum of the frequencies for Tasks 1 and 2 is 1+2 or 3, and the sum of the frequencies for Tasks 1, 2, and 3 is $1 + 2 + 4$ or 7. Then, from Table 46.5, FM_1 is 0.94, $FM_{1,2}$ is 0.88, and $FM_{1,2,3}$ is 0.70. Finally, the CLI $= 1.6 + 1.0(1/0.88 - 1/0.94) + 0.67(1/0.70 - 1/0.88) = 1.6 + 0.07 + 0.20 = 1.9$. Note that the FM values were based on the sum of the frequencies of the subscripts, the vertical height, and the duration of lifting.

TABLE 46.8 Computations from Multi-Task Example

Task #	Load Weight (L)	Task Frequency (F)	FIRWL	FM	STRWL	FILI	STLI	New Task #
1	30	1	20	0.94	18.8	1.5	1.6	1
2	20	2	20	0.91	18.2	1.0	1.1	2
3	10	4	15	0.84	12.6	0.67	0.8	3

46.6 Applying the Equations

46.6.1 Using the RWL and LI to Guide Ergonomics Design

The RWL and LI can be used to guide ergonomics design in several ways:

1. The individual multipliers can be used to identify specific job-related problems. The relative magnitude of each multiplier indicates the relative contribution of each task factor (e.g., horizontal, vertical, frequency, etc.).
2. The RWL can be used to guide the redesign of existing manual lifting jobs or to design new manual lifting jobs. For example, if the task variables are fixed, then the maximum weight of the load could be selected so as not to exceed the RWL; if the weight is fixed, then the task variables could be optimized so as not to exceed the RWL.
3. The LI can be used to estimate the relative magnitude of physical stress for a task or job. The greater the LI, the smaller the fraction of workers capable of safely sustaining the level of activity. Thus, two or more job designs could be compared.
4. The LI can be used to prioritize ergonomics redesign. For example, a series of suspected hazardous jobs could be rank ordered according to the LI and a control strategy could be developed according to the rank ordering (i.e., jobs with lifting indices above 1.0 or higher would benefit the most from redesign).

46.6.2 Rationale and Limitations for LI

The NIOSH RWL equation and LI are based on the concept that the risk of lifting-related LBP increases as the demands of the lifting task increase. In other words, as the magnitude of the LI increases, (1) the level of the risk for a given worker would be increased and (2) a greater percentage of the workforce is likely to be at risk for developing lifting-related LBP. The shape of the risk function, however, is not known. Without additional data showing the relationship between LBP and the LI, it is impossible to predict the magnitude of the risk for a given individual or the exact percentage of the work population who would be at an elevated risk for LBP.

To gain a better understanding of the rationale for the development of RWL and LI, consult the paper Waters et al. (1993). This article provides a discussion of the criteria underlying the lifting equation and of the individual multipliers. This article also identifies both the assumptions and uncertainties in the scientific studies that associate manual lifting and low back injuries.

46.6.3 Job-Related Intervention Strategy

The LI may be used to identify potentially hazardous lifting jobs or to compare the relative severity of two jobs for the purpose of evaluating and redesigning them. From the NIOSH perspective, it is likely that lifting tasks with an LI > 1.0 pose an increased risk for lifting-related LBP for some fraction of the workforce (Waters et al., 1993). Hence, to the extent possible, lifting jobs should be designed to achieve an LI of 1.0 or less.

Some experts believe, however, that worker selection criteria may be used to identify workers who can perform potentially stressful lifting tasks (i.e., lifting tasks that would exceed an LI of 1.0) without significantly increasing their risk of work-related injury above the baseline level. Those who endorse the use of selection criteria believe that the criteria must be based on research studies, empirical observations, or theoretical considerations that include job-related strength testing or aerobic capacity testing. Even these experts agree, however, that many workers will be at a significant risk of a work-related injury when performing highly stressful lifting tasks (i.e., lifting tasks that would exceed an LI of 3.0). Also, "informal" or "natural" selection of workers may occur in many jobs that require repetitive lifting tasks. According to some experts, this may result in a unique workforce that may be able to work above a lifting index of 1.0, at least in theory, without substantially increasing their risk of low back injuries above the baseline rate of injury.

46.7 Example Problems

Two example problems are provided to demonstrate the proper application of the lifting equation and procedures. The procedures provide a method for determining the level of physical stress associated with a specific set of lifting conditions, and assist in identifying the contribution of each job-related factor. The examples also provide guidance in developing an ergonomics redesign strategy. Specifically, for each example, a job description, job analysis, hazard assessment, redesign suggestion, illustration, and completed worksheet are provided.

A series of general design/redesign suggestions for each job-related risk factor are provided in Table 46.9. These suggestions can be used to develop a practical ergonomics design/redesign strategy.

46.7.1 Loading Supply Rolls, Example 1

46.7.1.1 Job Description

With both hands directly in front of the body, a worker lifts the core of a 35-lb roll of paper from a cart, and then shifts the roll in the hands and holds it by the sides to position it on a machine, as shown in Figure 46.6. Significant control of the roll is required at the destination of the lift. Also, the worker must crouch at the destination of the lift to support the roll in front of the body, but does not have to twist.

46.7.1.2 Job Analysis

The task variable data are measured and recorded on the job analysis worksheet (Figure 46.7). The vertical location of the hands is 27 in. at the origin and 10 in. at the destination. The horizontal location of the hands is 15 in. at the origin and 20 in. at the destination. The asymmetric angle is $0°$ at both the origin and the destination, and the frequency is 4 lifts/shift (i.e., less than 0.2 lifts/min for less than 1 h — see Table 46.5).

Using Table 46.6, the coupling is classified as poor because the worker must reposition the hands at the destination of the lift and they cannot flex the fingers to the desired $90°$ angle (e.g., hook grip). No asymmetric lifting is involved (i.e., $A = 0$), and significant control of the object is required at the destination of the lift. Thus, the RWL should be computed at both the origin and the destination of the lift. The multipliers are computed from the lifting equation or determined from the multiplier tables (Table 46.1 through Table 46.5, and Table 46.7). As shown in Figure 46.7, the RWL for this activity is 28.0 lb at the origin and 18.1 lb at the destination.

TABLE 46.9 General Design/Redesign Suggestions

If HM is less than 1.0	Bring the load closer to the worker by removing any horizontal barriers or reducing the size of the object. Lifts near the floor should be avoided; if unavoidable, the object should fit easily between the legs
If VM is less than 1.0	Raise/lower the origin/destination of the lift. Avoid lifting near the floor or above the shoulders
If DM is less than 1.0	Reduce the vertical distance between the origin and the destination of the lift
If AM is less than 1.0	Move the origin and destination of the lift closer together to reduce the angle of twist, or move the origin and destination further apart to force the worker to turn the feet and step, rather than twist the body
If FM is less than 1.0	Reduce the lifting frequency rate, reduce the lifting duration, or provide longer recovery periods (i.e., light work period)
If CM is less than 1.0	Improve the hand-to-object coupling by providing optimal containers with handles or handhold cutouts, or improve the handholds for irregular objects
If the RWL at the destination is less than at the origin	Eliminate the need for significant control of the object at the destination by redesigning the job or modifying the container/object characteristics

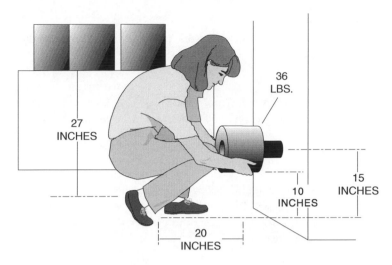

FIGURE 46.6   Loading supply rolls, example 1.

46.7.1.3   Hazard Assessment

The weight to be lifted (35 lb) is greater than the RWL at both the origin and destination of the lift (28.0 lb and 18.1 lb, respectively). The LI at the origin is 35/28.0 lb or 1.3, and the LI at the destination is 35/18.1 lb or 1.9. These values indicate that this job is only slightly stressful at the origin, but moderately stressful at the destination of the lift.

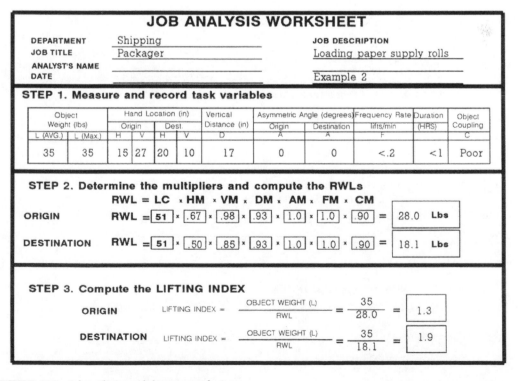

FIGURE 46.7   Job analysis worksheet, example 1.

46.7.1.4 Redesign Suggestions

The first choice for reducing the risk of injury for workers performing this task would be to adapt the cart so that the paper rolls could be easily pushed into position on the machine, without manually lifting them.

If the cart cannot be modified, then the results of the equation may be used to suggest task modifications. The worksheet displayed in Figure 46.7 indicates that the multipliers with the smallest magnitude (i.e., those providing the greatest penalties) are 0.50 for the HM at the destination, 0.67 for the HM at the origin, 0.85 for the VM at the destination, and 0.90 for the CM value. Using Table 46.9, the following job modifications are suggested:

1. Bring the load closer to the worker by making the roll smaller so that the roll can be lifted from between the worker's legs. This will decrease the *H* value, which in turn will increase the HM value.
2. Raise the height of the destination to increase the VM.
3. Improve the coupling to increase the CM.

If the size of the roll cannot be reduced, then the vertical height (*V*) of the destination should be increased. Figure 46.8 shows that if *V* was increased to about 30 in., then VM would be increased from 0.85 to 1.0; the *H* value would be decreased from 20 in. to 15 in., which would increase HM from 0.50 to 0.67; the DM would be increased from 0.93 to 1.0. As shown in Figure 46.8, the final RWL would be increased from 18.1 to 30.8 lb, and the LI at the destination would decrease from 1.9 to 1.1.

In some cases, redesign may not be feasible. In these cases, use of a mechanical lift may be more suitable. As an interim control strategy, two or more workers may be assigned to lift the supply roll.

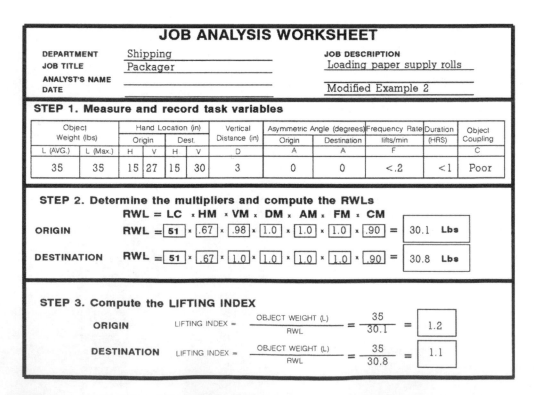

FIGURE 46.8 Modified job analysis worksheet, example 1.

46.7.1.5   Comments

The horizontal distance (*H*) is a significant factor that may be difficult to reduce because the size of the paper rolls may be fixed. Moreover, redesign of the machine may not be practical. Therefore, elimination of the manual lifting component of the job may be more appropriate than job redesign.

46.7.2   Dish-Washing Machine Unloading, Example 2

46.7.2.1   Job Description

A worker manually lifts trays of clean dishes from a conveyor at the end of a dish-washing machine and loads them on a cart as shown in Figure 46.9. The trays are filled with assorted dishes (e.g., glasses, plates, bowls) and silverware. The job takes between 45 min and 1 h to complete, and the lifting frequency rate averages 5 lifts/min. Workers usually twist to one side of their body to lift the trays (i.e., asymmetric lift) and then rotate to the other side of their body to lower the trays to the cart in one smooth continuous motion. The maximum amount of asymmetric twist varies between workers and within workers; however, there is usually equal twist to either side. During the lift the worker may take a step toward the cart. The trays have well designed handhold cutouts and are made of lightweight materials.

46.7.2.2   Job Analysis

The task variable data are measured and recorded on the job analysis worksheet (Figure 46.10). At the origin of the lift, the horizontal distance (*H*) is 20 in., the vertical distance (*V*) is 44 in., and the angle of asymmetry (*A*) is 30°. At the destination of the lift, *H* is 20 in., *V* is 7 in., and *A* is 30°. The trays normally weigh from 5 to 20 lb, but for this example, assume that all of the trays weigh 20 lb.

Using Table 46.6, the coupling is classified as "good." Significant control is required at the destination of the lift. Using Table 46.5, the FM is determined to be 0.80. As shown in Figure 46.10, the RWL is 14.4 lb at the origin and 13.3 lb at the destination.

46.7.2.3   Hazard Assessment

The weight to be lifted (20 lb) is greater than the RWL at both the origin and destination of the lift (14.4 and 13.3 lb, respectively). The LI at the origin is 20/14.4 or 1.4 and the LI at the destination is 1.5. These results indicate that this lifting task would be stressful for some workers.

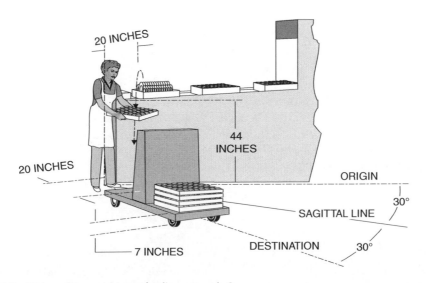

FIGURE 46.9   Dish-washing machine unloading, example 2.

JOB ANALYSIS WORKSHEET

DEPARTMENT Food Service

JOB TITLE Cafeteria Worker

ANALYST'S NAME

DATE

JOB DESCRIPTION
Unloading a dish-washing
machine
Example 5

STEP 1. Measure and record task variables

Object Weight (lbs)		Hand Location (in)				Vertical Distance (in)	Asymmetric Angle (degrees)		Frequency Rate lifts/min	Duration (HRS)	Object Coupling
		Origin		Dest.			Origin	Destination			
L (AVG.)	L (Max.)	H	V	H	V	D	A	A	F		C
20	20	20	44	20	7	37	30	30	5	< 1	Good

STEP 2. Determine the multipliers and compute the RWLs

$$RWL = LC \times HM \times VM \times DM \times AM \times FM \times CM$$

ORIGIN $RWL = \boxed{51} \times \boxed{.50} \times \boxed{.90} \times \boxed{.87} \times \boxed{.90} \times \boxed{.80} \times \boxed{1.0} = \boxed{14.4 \text{ Lbs}}$

DESTINATION $RWL = \boxed{51} \times \boxed{.50} \times \boxed{.83} \times \boxed{.87} \times \boxed{.90} \times \boxed{.80} \times \boxed{1.0} = \boxed{13.3 \text{ Lbs}}$

STEP 3. Compute the LIFTING INDEX

ORIGIN LIFTING INDEX = $\dfrac{\text{OBJECT WEIGHT (L)}}{\text{RWL}} = \dfrac{20}{14.4} = \boxed{1.4}$

DESTINATION LIFTING INDEX = $\dfrac{\text{OBJECT WEIGHT (L)}}{\text{RWL}} = \dfrac{20}{13.3} = \boxed{1.5}$

FIGURE 46.10   Job analysis worksheet, example 2.

46.7.2.4   Redesign Suggestions

The worksheet shows that the smallest multipliers (i.e., the greatest penalties) are 0.50 for the HM, 0.80 for the FM, 0.83 for the VM, and 0.90 for the AM. Using Table 46.9, the following job modifications are suggested:

1. Bring the load closer to the worker to increase HM
2. Reduce the lifting frequency rate to increase FM
3. Raise the destination of the lift to increase VM
4. Reduce the angle of twist to increase AM by either moving the origin and destination closer together or moving them further apart

Since the horizontal distance (H) is dependent on the width of the tray in the sagittal plane, this variable can only be reduced by using smaller trays. Both the DM and VM, however, can be increased by lowering the height of the origin and increasing the height of the destination. For example, if the height at both the origin and destination is 30 in., then VM and DM are 1.0, as shown in the modified worksheet (Figure 46.11). Moreover, if the cart is moved so that the twist is eliminated, the AM can be increased from 0.90 to 1.00. As shown in Figure 46.11, with these redesign suggestions the RWL can be increased from 13.3 to 20.4 lb, and the LI values are reduced to 1.0.

46.7.2.5   Comments

This analysis was based on a 1-h work session. If a subsequent work session begins before the appropriate recovery period has elapsed (i.e., 1.0 h), then the 8-h category would be used to compute the FM value.

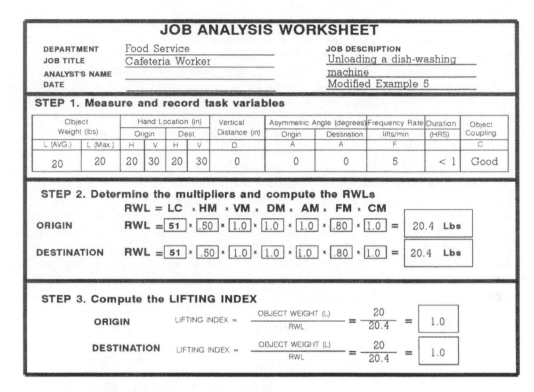

FIGURE 46.11 Modified job analysis worksheet, example 2.

46.8 Validation of the Revised NIOSH Lifting Equation

Several studies have been conducted examining the effectiveness of the revised NIOSH lifting equation to identify jobs with increased risk of lifting-related low back disorders. Waters et al. (1999) conducted a cross-sectional study to investigate whether there was a significant relationship between risk of LBP and exposure to jobs with various LI values. In the study, the 1-yr prevalence of LBP was determined for workers employed in jobs with varying LI values. Fifty jobs at four industrial sites were evaluated with the NIOSH lifting equation. A symptom and occupational history questionnaire was administered to 204 people employed in those lifting jobs and 80 people employed in nonlifting jobs at the four sites. Jobs were categorized by exposure according to the following LI values: LI = 0 (unexposed), $0.0 < LI \leq 1.0$, $1.0 < LI \leq 2.0$, $2.0 < LI \leq 3.0$, and LI > 3.0. Regression analysis was used to determine whether there was a correlation between the LI and reported LBP. The authors found that as the LI increased from 0.0 to 3.0, the odds of LBP increased, with a peak and statistically significant odds ratio (OR = 2.45) occurring in the category of jobs with an LI value between 2.0 and 3.0. For the group of jobs with an LI value greater than 3.0, the OR value was lower (OR = 1.63) than for the 2.0 to 3.0 group and was not statistically different from the nonlifting group, but the authors explained that this finding was most likely due to a combination of worker selection, a survivor effect, and high turnover in the higher risk jobs. In a study of worker turnover rates on physically demanding jobs, Lavender and Marras (1994) showed that high turnover rate was a good indicator of high risk for LBP. In that study, the authors attributed lower than expected injury rates in jobs with high turnover to the "healthy worker effect." Based on the overall findings, Waters et al. (1999) concluded that, "although LBP is a common disorder, the lifting index appears to be a useful indicator for determining the risk of low back pain caused by manual lifting."

In a study designed to investigate lifting-related musculoskeletal disorders in the metal processing industry in China, Xiao et al. (2004) used the revised NIOSH equation to analyze the risk factors for

LBP and to study the validity and feasibility of using the NIOSH lifting equation in China. The NIOSH equation was used to evaluate lifting risk for 69 workers mainly involved in manual materials handling (MMH) (job A) and 51 machinery workers who worked in less demanding MMH tasks (job B). The prevalence of LBP lasting for more than a week due to lifting were 26.09 and 5.88% for jobs A and B, respectively. The NIOSH LI was estimated to be 2.4 for job A, and $0 < LI < 1$ for job B. The authors concluded that the NIOSH equation is an important tool in assessing characteristics and risk factors of LBP for MMH tasks.

In another study, Marras et al. (1999) examined the relationship between low-back disorders and various risk factors for LBP due to manual lifting. One of the objectives of the study was to evaluate the validity and effectiveness of the revised NIOSH lifting equation to correctly identify jobs with varying levels of risk of low back disorder, where job risk was defined according to historical records of low back disorder injuries. High-risk jobs were defined as jobs in which more than 12 low back disorders were recorded per 100 exposed workers (mean of 22 injuries/100 exposed workers), medium-risk jobs were defined as jobs in which between 1 and 12 low back disorders were recorded per 100 exposed workers, and low-risk jobs were defined as jobs in which no low back disorder was recorded per 100 exposed workers. The results indicated that the when the average horizontal distance was used, the NIOSH equation was predictive of risk of low back disorder, resulting in an odds ratios of 3.1 (95% CI, 2.6 to 3.8) for high-risk jobs compared to low-risk jobs. When the maximum horizontal distance was used, the odds ratio was increased to 4.3 for high-risk jobs compared to low-risk jobs.

In another study, conducted by Lee et al. (1996), researchers examined whether the NIOSH lifting equation was applicable for an Asian population. The application of the NIOSH equation for establishing weight limits for Korean workers was examined using the psychophysical method. The study population consisted of 53 male college students and 16 male field workers. The subjects were required to perform six different lifting tasks in the sagittal plane, at various lifting frequencies and heights, for 8 h. The RWLs for each lift were calculated using the NIOSH equation. The psychophysical method, in which subjects were allowed to adjust the weight of lift during a 20-min period, was also used to estimate the maximum acceptable weight of lift (MAWL) for each lifting task. While students generally had larger body sizes than the worker population, workers were generally stronger than the students. Within each group, neither the frequency nor the vertical height of lift was significantly related to the differences between the NIOSH Lifting Equation (NLE) and psychophysically-based weight limits. The MAWLs of the workers were significantly higher than those of the students. While this difference increased with increasing lift frequency, it was not sensitive to lift height. When the data were adjusted to represent the entire Korean young male population, no significant differences were observed between the NIOSH recommended weights of load and the adjusted MAWLs. Although the load constant of the NIOSH equation was 23 kg, that of the students was 20.24 kg and that of the workers was 25.05 kg. The authors conclude that the NIOSH weight limit equation is well suited for young, healthy Korean males.

Recently, Hidalgo et al. (1995, 1997) conducted a study designed to evaluate the validity of the psychophysical, biomechanical, and physiological criteria used in establishing the NIOSH lifting equation (Waters et al., 1991). The criteria used to develop the equation were cross-validated against the data published by different researchers in the scientific literature. Assessment of the 1991 NIOSH lifting equation indicated that there are differences between the NIOSH equation values and the psychophysical limits for some types of lifts and that the RWL likely would protect about 85% of the female population and 95% of the male population. The authors, however, noted that the 3.4 kN limit for compression on the lumbosacral joint may be too high to protect all workers and that the energy expenditure limits used in development of the RWL index can be sustained by 57 to 99% of worker population when compared to the physiological limits based on previous fatigue studies. The authors concluded that the results of the cross-validation for psychophysical criterion confirmed the validity of assumptions made in the 1991 NIOSH revised lifting equation, but that the results of cross-validation for the biomechanical and physiological criteria were not in total agreement with the 1991 NIOSH model. They did not, however, actually evaluate whether the equation would protect workers or not in the study.

Sesek et al. (2003) conducted a study designed to investigate the ability of the revised NIOSH lifting equation to measure the risk of low back injury using employee health outcomes to identify high-risk manual lifting jobs. In addition to the revised NIOSH lifting equation, a slightly modified version of the equation was evaluated, in which some factors were removed from the equation for simplification. The authors found that, without the modifications, the revised NIOSH lifting equation was able to predict back injuries with odds ratios of 2.1 (95% CI, 1.0 to 4.43) and 4.0 (95% CI, 1.5 to 10.3) for lifting indices of 1.0 and 3.0, respectively. They reported that simplifying the lifting equation by removing several variables did not significantly reduce the predictive performance of the equation. When the authors modified the equation, they found that the modified NIOSH lifting equation was able to predict back injuries with odds ratios of 2.2 (95% CI, 1.0 to 4.6) and 5.3 (95% CI, 1.5 to 19.1) for lifting indices of 1.0 and 3.0, respectively. The authors concluded that these modifications to the NIOSH lifting equation show promise for increasing both the usability and utility of the lifting equation.

An epidemiological study, conducted by Wang et al. (1998), evaluated the relation between low back discomfort ratings and use of the revised NIOSH lifting equation to assess the risk of MMH tasks. In the study, the authors surveyed 97 MMH workers on site in 15 factories and designed a questionnaire to systematically collect job-related information. Approximately 90% of the workers had suffered various degrees of lower back discomfort, and 80% had sought medical treatment. The survey showed that 42 of the 97 jobs analyzed had a RWL of 0, which was attributed to either a horizontal distance or a lifting frequency that exceeded the bounds of the NIOSH LI. Based on the results of the study, the authors suggested that the limits for horizontal distance and maximum allowable frequency may be too stringent to accommodate many existing MMH jobs. The authors also reported that for the remaining 55 jobs the significant positive correlation obtained between the LI and the severity of low back discomfort suggests that the LI is reliable in assessing the potential risk of low back injury in MMH. The authors concluded that their results provide useful information on the application of the NIOSH lifting guide to the assessment of LBP.

In a study examining the effectiveness of a training course to provide instruction on the proper use of the NIOSH lifting equation, Waters et al. (1998), trained a group of nonergonomists to use the revised NIOSH lifting equation and then tested them 8 weeks post-training to evaluate their knowledge in making the measurements needed to use the equation. Twenty-seven individuals from NIOSH participated in a 1-day training session on the use of the NIOSH lifting equation. The participants were subsequently tested on a simulated lifting task 8 weeks later to determine their accuracy in measuring the variables. Analysis of the results indicated that (1) interobserver variability was small, especially for the most important factor (i.e., horizontal distance); (2) individuals can be trained to make measurements with sufficient accuracy to provide consistent recommended weight limit and lifting index values; and (3) measurement of the coupling and asymmetric variables were the least accurate and additional training should be provided to clarify these factors.

References

ASPH/NIOSH. 1986. *Proposed National Strategies for the Prevention of Leading Work-Related Diseases and Injuries: Part 1* (Published by the Association of Schools of Public Health under a cooperative agreement with the National Institute for Occupational Safety and Health).

Eastman Kodak, 2004. *Ergonomic Design for People at Work*, 2nd Ed. (John Wiley & Sons, NJ).

Hidalgo, J., Genaidy, A., Karwowski, W., Christensen, D., Huston, R., and Stambough, J. 1995. A cross-validation of the NIOSH limits for manual lifting. *Ergonomics* 38(12): 2455–2464.

Hidalgo, J., Genaidy, A., Karwowski, W., Christensen, D., Huston, R., and Stambough J. 1997. A comprehensive lifting model: beyond the NIOSH lifting equation. *Ergonomics*, 40(9): 926–927.

Lavender, S.A. and Marras, W.S. 1994. The use of turnover rate as a passive surveillance indicator for potential low back disorders. *Ergonomics* 37(6): 971–978.

Lee, K.S., Park, H.S., and Chun, Y.H. 1996. The validity of the revised NIOSH weight limit in a Korean young male population: a psychophysical approach. *International Journal of Industrial Ergonomics* 18(2/3): 181–186.

Marras, W., Fine, L., Ferguson, S., and Waters, T. 1999. The effectiveness of commonly used lifting assessment methods to identify industrial jobs associated with elevated risk of low-back disorders. *Ergonomics*, 42(1): 229–245.

NIOSH. 1981. *Work Practices Guide for Manual Lifting.* NIOSH Technical Report No. 81-122 (U.S. Department of Health and Human Services, National Institute for Occupational Safety and Health, Cincinnati, OH).

NIOSH. 1991. *Scientific Support Documentation for the Revised 1991 NIOSH Lifting Equation: Technical Contract Reports, May 8, 1991* (U.S. Department of Health and Human Services, National Institute for Occupational Safety and Health, Cincinnati, OH). Available from the National Technical Information Service (NTIS No. PB-91-226-274).

Sesek, R., Gilkey, D., Drinkaus, P., Bloswick, D.S., and Herron, R. 2003. Evaluation and quantification of manual materials handling risk factors. *International Journal of Occupational Safety and Ergonomics.* 9(3): 271–287.

Wang, M.J., Garg, A., Chang, Y.C., Shih, Y.C., Yeh, W.Y., and Lee, C.L. 1998. The relationship between low back discomfort ratings and the NIOSH lifting index. *Human Factors and Ergonomics* 40(3): 509–515.

Waters, T.R., Barron, S.L., Piacitelli, L.A., Anderson, V.P., Skov, T., Haring-Sweeney, M., Wall, D.K., and Fine, L.J. 1999. Evaluation of the revised NIOSH Lifting Equation. Spine 24(4): 386–394.

Waters, T.R., Putz-Anderson, V., Garg, A., and Fine, L.J. 1993. Revised NIOSH equation for the design and evaluation of manual lifting tasks. *Ergonomics* 36(7): 749–776.

Waters, T.R., Putz-Anderson, V., and Garg, A. 1994. *Applications Manual for the Revised NIOSH Lifting Equation.* National Institute for Occupational Safety and Health, Technical Report. DHHS(NIOSH) Pub. No. 94-110. Available from the National Technical Information Service (NTIS). NTIS document number PB94-176930 (1-800-553-NTIS).

Waters, T.R., Baron, S.L., and Kemmlert, K. 1998. Accuracy of measurements for the revised NIOSH lifting equation. National Institute for Occupational Safety and Health. *Applied Ergonomics.* 29(6): 433–438.

Xiao, G.B., Lei, L., Dempsey, P., Ma, Z.H., and Liang, Y.X. 2004. *Zhonghua Lao Dong Wei Sheng Zhi Ye Bing Za Zhi.* Study on lifting-related musculoskeletal disorders among workers in metal processing. 22(2): 81–85.

47

Psychophysical Approach to Task Analysis

Patrick G. Dempsey
*Liberty Mutual Research
Institute for Safety*

47.1 Introduction

One approach to the prevention of work-related musculoskeletal disorders (WRMSDs) is the psychophysical approach. This approach seeks to provide limits and guidelines for manual work that represent "maximum acceptable" workloads that minimize the injury potential of the work. Most often, these data reflect maximum acceptable weights or forces acceptable to different population percentages given other constant task parameters such as frequency. Limited data have been collected that examine maximum acceptable frequencies when the force or load parameter is held constant. It is often quite difficult to assess whether forces exerted or loads handled in the workplace exceed what is thought to pose excess risk, and the psychophysical approach is one approach to this problem.

Like all assessment methods, there are advantages and disadvantages to consider when selecting an analysis tool. These will be discussed, as will sources of data for workplace application. No single approach is applicable to all workplace analyses, and other tools such as biomechanical and physiologic models for analyzing materials handling tasks, can be used as alternate or supplemental analyses.

47.1.1 Chapter Goal and Outline

The goal of this chapter is to provide the reader with information concerning basic theory behind the psychophysical approach, the availability of data for designing manual tasks, the methods of applying the data, and the limitations of the data. The primary focus will be on the application of psychophysical techniques rather than on empirical methodologies or a literature review of the theoretical underpinnings of the psychophysical approach. Where necessary, theoretical and empirical results will be used to justify specific application techniques or to explain caveats of psychophysical data. Manual materials handling (MMH) and upper extremity intensive (UEI) tasks have been the focus of psychophysical research, and each will be considered separately due to the disparity in application methodologies.

Readers interested in the empirical and theoretical aspects of the psychophysical approach to MMH task design are referred to Snook[1-3] and Ayoub and Mital[4] for further reading. For specific information on the comparisons of the psychophysical approach to the biomechanical and physiological approaches to MMH task design, the reader should consult Ayoub[5], Mital et al.[6] and Nicholson[7]. Less thorough information on the empirical and theoretical aspects of the psychophysical approach to UEI task design is available. However, there are useful discussions in Kim et al.[8] and Fernandez et al.[9]

47.1.2   Introduction to Psychophysics

Psychophysics is a branch of psychology dealing with the relationships between stimuli and sensations. These relationships can be best described by the psychophysical power law. The psychological magnitude (sensation) ψ grows as a power function of the physical magnitude ϕ (stimulus) in the following manne:[10]

$$\psi = k\phi^n$$

The value of the constant k depends on the units of measure, while the exponent n has a value that varies for different sensations. The value of n may be lower than 1 for stimuli such as smell and brightness, or as high as 3.5 for electric shock.[10] Ljungberg et al.[11] found a value of $n = 1.86$ for a simulated brewery lifting task whereas Gamberale et al.[12] found a value of 2.43 for a similar task. Gamberale et al. attributed the larger value in the latter study to more demanding lifting cycles.

In MMH experiments, the subject adjusts the magnitude of the stimulus (weight, force, or frequency) to correspond to a sensation which is "dictated" by the instructions given by the experimenter, that is, "without straining yourself, or becoming unusually tired, weakened, overheated, or out of breath".[13] Adaptations of these instructions have been used to study UIE tasks. For UEI tasks, the instructions are directed at having subjects select workloads that do not result in "unusual discomfort in the hands, wrists or forearms".[14] Subjects are also monitored during experimentation for signs of soreness, stiffness, and numbness.

47.1.3   The Current State of Psychophysical Data

Psychophysical data are currently available for designing MMH tasks as well as UEI tasks. For MMH tasks, there are fairly extensive data available for maximum acceptable weights and forces for lifting, lowering, pushing, pulling, holding, and carrying tasks. Research of the psychophysical approach to MMH task design has spanned approximately three decades, and current work continues to expand the range of task conditions for which psychophysical data are available.

For UEI tasks, data are available for maximum acceptable frequencies and forces for a variety of tasks. It should be noted that there are considerably more data available for designing MMH tasks than UEI tasks. The increased attention to WRMSDs in the past decade or so led researchers to adapt psychophysical techniques used in MMH research to the study of UEI tasks. In part, this was done because of the lack of quantitative guidelines for forces, durations, and postures associated with UEI work. Although there is not extensive data, there are data that can be used to design manual work involving the upper extremities.

47.2 The Psychophysical Approach to Designing Manual Materials Handling Tasks

One of the first approaches to the control of MMH injuries through specifying task limits was the psychophysical approach. Applications in the military which relied on subjective estimates of load handling limits[15,16] were followed by the psychophysical approach being utilized to set industrial materials handling limits.[17] The methodologies and database developed by Snook and his colleagues[1,17–20] at the Liberty Mutual Research Center have been used by researchers and practitioners for the past several decades. This has resulted in the availability of a wide range of data available for designing MMH tasks.[6,20–23]

47.2.1 Setting Weight and Force Limits

The application of the psychophysical approach to MMH task design is performed by using databases in the literature which provide limits specific to task conditions such as frequency, pushing or pulling distance, and load dimensions. Figure 47.1 presents a general model of the procedures associated with applying psychophysical data.

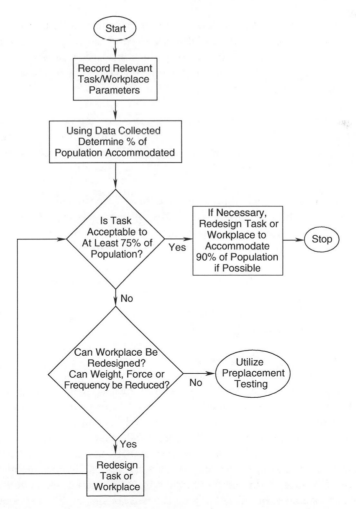

FIGURE 47.1 General model of the psychophysical approach to analyzing manual tasks.

To use a database, the user records the relevant task parameters, then finds the value in the database applicable to the specific task. Since it is impossible to collect data for all combinations of tasks, one can either use interpolation to find the appropriate value, or one can use the closest value. In the latter case, the user should use the lower of the values in between which the task parameters fall. Next, the analyst determines if a task is acceptable. At a minimum, an MMH task should be designed to accommodate 75% of the population;[1] however, one should strive to accommodate at least 90% of the population whenever possible. If females perform the task, then the design should be based upon accommodating the female workers.

When a task does not accommodate at least 75% of the population, or if the task does accommodate 75% of the population and minor changes can be made to increase the acceptability of the task at little or no cost, then the task and workplace should be redesigned to accommodate at least 90% of the population. Options to increase the acceptability of a task will be described in more detail later. If a task cannot be redesigned to accommodate at least 75% of the population, then preplacement testing should be considered.

47.2.1.1 Psychophysical Data

Most psychophysical databases present maximum acceptable weights or forces. There is limited information available on maximum acceptable frequencies.[24–26] The data described in detail here are all related to force and weight data.

The largest and most comprehensive single database for designing MMH tasks is that of Snook and Ciriello.[20] The data were collected from industrial subjects. The database includes maximum acceptable weights for lifting and lowering tasks, maximum acceptable initial and sustained forces for pushing and pulling tasks, and maximum acceptable weights for carrying tasks. The database provides values for males and females, as well as values that accommodate 10, 25, 50, 75, and 90% of the respective populations.

For lifting and lowering tasks, Snook and Ciriello's[20] database accommodates frequencies between one lift/lower every 8 h and 12 lifts/lowers per minute. Box dimensions in the sagittal plane between 34 and 75 cm are accommodated, as are lifts and lowers with vertical distances between 25 and 76 cm. Finally, data for floor to knuckle height, knuckle to shoulder height, and shoulder height to arm reach ranges of lift/lower are available.

Snook and Ciriello[20] presented maximum acceptable forces for pushing and pulling tasks, including forces required to start an object in motion (initial force) and forces required to keep an object in motion (sustained forces). Vertical handle heights between 57 and 135 cm are accommodated by the data. Frequencies for pushing/pulling are the same as those for lifting. The push/pull distances range between 2.1 and 61.0 m.

Finally, Snook and Ciriello's[20] database also contains maximum acceptable weight of carry values for the same frequencies as the other MMH tasks mentioned above. Carry distances of 2.1, 4.3, and 8.5 m are available with vertical heights of 79 and 111 cm for males and 72 and 105 cm for females. Snook and Ciriello's[20] data are presented in Table 47.1–Table 47.9.

47.2.1.1.1 Modified Psychophysical Databases

Mital et al.[6] modified Snook and Ciriello's[20] database to conform with biomechanical, physiological, and epidemiological criteria. Mital et al. also present data from other sources for additional types of MMH tasks. Snook and Ciriello's[20] database was modified to satisfy the following criteria: spinal compression limits of 2689 and 3920 N for females and males, respectively, in consideration of the results of Liles et al.;[27] an intra-abdominal pressure limit of 90 mmHg; and an energy expenditure limit of 21–23% of treadmill aerobic capacity or 28–29% of bicycle aerobic capacity. Mital et al.[6] also present maximum acceptable frequencies of lift for one-handed lifting, acceptable holding time data, and maximum acceptable forces and weights of lift for MMH tasks performed in nonstandard postures such as lying down and kneeling. In order to extend the range of applicability of Snook and Ciriello's[20] database, Mital et al.[6] presented multipliers to adjust lifting data for the following factors: work duration,

TABLE 47.1 Maximum Acceptable Weight of Lift for Males (kg)

Width[a]	Distance[b]	Percent[c]	Floor level to knuckle height One lift every								Knuckle height to shoulder height One lift every								Shoulder height to arm reach One lift every							
			sec			min				h	sec			min				h	sec			min				h
			5	9	14	1	2	5	30	8	5	9	14	1	2	5	30	8	5	9	14	1	2	5	30	8
76	76	90	6	7	9	11	13	14	14	17	8	10	12	13	14	14	16	17	6	8	9	10	10	11	12	13
		75	9	11	13	16	19	20	21	24	10	14	16	18	18	19	21	23	8	10	12	14	14	14	16	17
		50	12	15	17	22	25	27	28	32	13	17	20	22	23	24	26	29	10	13	15	17	17	18	20	22
		25	15	18	21	28	31	34	35	41	16	21	24	27	27	28	32	35	11	16	18	21	21	22	24	27
		10	18	22	25	33	37	40	41	48	19	24	28	31	32	33	37	40	14	18	21	24	24	25	28	31
75	51	90	6	8	9	12	13	15	15	17	8	11	13	15	15	16	18	19	6	8	9	12	12	12	14	15
		75	9	11	13	17	19	21	22	25	11	15	17	20	20	21	23	25	8	11	12	15	15	16	18	20
		50	13	15	18	23	26	28	29	34	14	19	21	25	25	26	29	32	10	14	16	19	20	20	23	25
		25	16	19	22	29	33	35	36	42	17	23	26	30	31	32	36	39	13	17	19	23	24	25	27	30
		10	19	22	26	34	38	42	43	50	20	26	30	35	36	37	41	45	15	19	22	27	27	29	32	35
	25	90	8	9	11	13	15	16	17	20	10	13	15	18	18	19	21	23	7	10	11	14	14	14	16	18
		75	11	13	15	19	22	24	24	28	13	17	20	23	24	25	27	30	10	13	15	18	18	19	21	23
		50	15	18	21	26	29	32	33	38	17	22	25	30	30	31	35	38	12	16	19	23	23	24	27	29
		25	18	22	26	33	37	40	41	48	20	27	30	36	36	38	42	46	15	20	22	28	28	29	32	35
		10	22	26	31	38	44	47	49	57	23	31	35	42	42	44	49	53	17	23	26	32	32	34	38	41
49	76	90	7	8	10	13	15	16	17	20	8	10	12	13	14	14	16	17	7	9	10	12	12	13	14	16
		75	10	12	14	19	22	24	24	28	10	14	16	18	18	19	21	23	9	11	13	16	16	17	19	21
		50	14	16	19	26	29	32	33	38	13	17	20	22	23	24	26	29	11	15	17	20	21	21	24	26
		25	17	20	24	33	37	40	41	48	16	21	24	27	27	28	32	35	13	18	20	25	25	26	29	31
		10	20	24	28	38	43	47	48	57	19	24	28	31	32	33	37	40	15	21	23	28	28	30	33	36
	51	90	7	9	10	14	16	17	18	20	8	11	13	15	15	16	18	19	7	10	11	14	14	14	16	18
		75	10	13	15	20	23	25	25	30	11	15	17	20	20	21	23	25	10	13	15	18	18	19	21	23
		50	14	17	20	27	30	33	34	40	14	19	21	25	25	26	29	32	12	16	19	23	23	24	27	29
		25	18	21	25	34	38	42	43	50	17	23	26	30	31	32	36	39	15	20	22	28	28	29	32	35
		10	21	25	29	40	45	49	50	59	20	26	30	35	36	37	41	45	17	23	26	32	32	34	38	41
	25	90	8	10	12	16	18	19	20	23	10	13	15	18	18	19	21	23	9	11	12	16	16	17	19	21
		75	12	15	17	23	26	28	29	33	13	17	20	23	24	25	27	30	11	14	16	21	21	22	25	27
		50	16	20	23	30	34	37	38	45	17	22	25	30	30	31	35	38	14	18	21	27	28	28	32	35
		25	21	25	29	38	43	47	48	56	20	27	30	36	36	38	42	46	16	22	25	33	33	34	38	42
		10	24	29	34	45	51	56	57	67	23	31	35	42	42	44	49	53	19	25	29	38	38	40	44	48

(Table continued)

TABLE 47.1 *Continued*

Width[a]	Distance[b]	Percent[c]	Floor level to knuckle height One lift every									Knuckle height to shoulder height One lift every									Shoulder height to arm reach One lift every								
			sec			min				h	sec			min				h	sec			min				h
			5	9	14	1	2	5	30	8	5	9	14	1	2	5	30	8	5	9	14	1	2	5	30	8
34	76	90	8	10	11	15	17	19	19	23	8	11	13	15	15	16	18	19	8	10	12	14	14	15	16	18
		75	12	14	17	22	25	28	28	33	11	15	17	20	20	21	23	25	10	14	16	18	19	19	22	24
		50	16	19	22	30	34	37	38	44	14	19	21	25	25	26	29	32	13	17	20	23	24	25	27	30
		25	20	24	28	37	42	47	47	55	17	23	26	30	31	32	36	39	16	21	24	28	29	30	33	36
		10	24	29	33	44	50	54	56	65	20	26	30	35	36	37	41	45	18	24	28	33	33	34	38	42
	51	90	9	10	12	16	18	20	20	24	9	12	14	17	17	18	20	22	8	11	13	16	16	17	18	20
		75	12	15	18	23	26	28	29	34	12	16	18	22	23	23	26	29	11	14	17	21	21	22	24	26
		50	17	20	24	31	35	38	39	46	15	20	23	28	29	30	33	36	14	18	21	26	27	28	31	34
		25	21	25	30	39	44	48	49	57	18	24	27	34	35	36	40	44	17	22	25	32	32	33	37	41
		10	25	30	35	46	52	57	58	68	21	28	32	40	40	42	46	51	19	26	29	37	37	39	43	47
	25	90	10	12	14	18	20	22	23	27	11	14	16	20	20	21	23	26	10	13	15	19	19	19	22	24
		75	15	18	21	26	30	32	33	38	14	18	21	26	27	28	31	34	13	17	20	24	25	26	29	31
		50	20	24	28	35	40	43	44	52	18	23	27	33	34	35	39	43	16	22	25	31	31	33	36	40
		25	26	30	35	44	50	54	55	65	21	28	32	40	41	42	47	52	20	26	30	37	38	39	44	46
		10	29	35	41	52	59	64	66	76	25	33	37	47	47	49	55	60	23	30	35	43	44	45	51	55

[a]Box width (the dimension away from the body) (cm).

[b]Vertical distance of lift (cm).

[c]Percentage of industrial population.

Note: Italicized values exceed 8 h physiological criteria.

Source: Reprinted from Snook, S.H. and Ciriello, V.M. *Ergonomics*, 34, 1197–1213, 1991. With permission.

TABLE 47.2 Maximum Acceptable Weight of Lift for Females (kg)

Width[a]	Distance[b]	Percent[c]	Floor level to knuckle height — One lift every								Knuckle height to shoulder height — One lift every								Shoulder height to arm reach — One lift every							
			sec 5	sec 9	sec 14	min 1	min 2	min 5	min 30	h 8	sec 5	sec 9	sec 14	min 1	min 2	min 5	min 30	h 8	sec 5	sec 9	sec 14	min 1	min 2	min 5	min 30	h 8
76	76	90	5	6	7	7	8	8	9	12	5	6	7	9	9	9	10	12	4	5	5	6	7	7	7	8
		75	7	8	9	9	10	10	11	14	6	7	8	10	11	11	12	14	5	6	6	7	8	8	8	10
		50	8	10	10	11	12	12	13	17	7	8	9	11	12	12	13	16	6	7	7	8	9	9	10	11
		25	9	11	12	13	14	14	15	21	8	9	10	13	14	14	15	18	7	7	8	9	10	10	11	13
		10	11	13	14	14	15	16	17	23	9	10	11	14	15	15	17	20	7	8	9	10	11	11	12	14
75	51	90	6	7	8	8	9	9	10	14	6	7	8	9	10	10	11	13	5	6	7	7	7	7	8	9
		75	7	9	9	10	11	11	13	17	7	8	9	11	12	12	13	15	6	7	8	8	9	9	9	11
		50	9	10	11	12	13	14	15	21	9	9	11	13	14	14	15	17	7	8	8	9	10	10	11	13
		25	10	12	13	15	16	16	18	24	10	11	12	16	16	16	17	20	8	9	10	10	11	11	12	14
		10	11	14	15	17	18	18	20	27	11	12	14	18	17	17	19	22	9	10	11	12	13	13	14	16
	25	90	6	8	8	9	9	9	11	14	6	7	8	10	11	11	12	14	5	6	7	8	8	8	9	10
		75	8	10	11	11	12	12	13	18	7	8	9	12	13	13	14	17	6	7	8	9	9	9	10	12
		50	10	12	13	13	14	14	16	21	9	10	11	14	15	15	16	19	7	8	9	10	11	11	12	14
		25	11	14	15	15	16	17	19	25	10	11	12	16	17	17	19	22	8	9	10	12	12	12	14	16
		10	13	16	17	17	18	19	21	29	11	12	14	18	19	19	21	24	9	10	11	13	14	14	15	17
49	76	90	5	6	7	8	8	8	9	13	5	6	7	9	9	9	10	12	4	5	5	7	7	7	8	9
		75	7	8	9	10	10	10	12	16	6	7	8	10	11	11	12	14	5	6	6	8	8	8	9	11
		50	8	10	10	12	12	13	14	19	7	8	9	11	12	12	13	16	6	7	7	9	10	10	11	12
		25	9	11	12	14	15	15	17	22	8	9	10	13	14	14	15	18	7	7	8	10	11	11	12	14
		10	11	13	14	15	17	17	19	25	9	10	11	14	15	15	17	20	7	8	9	11	12	12	13	15
	51	90	6	7	8	9	10	10	11	15	6	7	8	9	10	10	11	13	5	6	7	7	8	8	9	10
		75	7	9	9	11	12	12	14	18	7	8	9	11	12	12	13	15	6	7	8	9	9	9	10	12
		50	9	10	11	13	15	15	16	22	9	9	11	13	14	14	15	17	7	8	8	10	11	11	12	14
		25	10	12	13	16	17	17	19	26	10	11	12	14	16	16	17	20	8	9	10	11	12	12	13	15
		10	11	14	15	18	19	19	22	30	11	12	14	16	17	17	19	22	9	10	11	13	14	14	15	17
	25	90	6	8	8	9	10	10	11	15	6	7	8	10	11	11	12	14	5	6	7	8	9	9	10	11
		75	8	10	11	12	12	13	14	19	7	8	9	12	13	13	14	17	6	7	8	9	10	10	11	13
		50	10	12	13	14	15	15	17	23	9	10	11	14	15	15	16	19	7	8	9	11	12	12	13	15
		25	11	14	15	16	18	18	20	27	10	11	12	16	17	17	19	22	8	9	10	12	13	13	15	17
		10	13	16	17	19	20	21	23	31	11	12	14	18	19	19	21	24	9	10	11	14	15	15	16	19

(*Table continued*)

TABLE 47.2 *Continued*

Width[a]	Distance[b]	Percent[c]	Floor level to knuckle height One lift every								Knuckle height to shoulder height One lift every								Shoulder height to arm reach One lift every							
			sec			min				h	sec			min				h	sec			min				h
			5	9	14	1	2	5	30	8	5	9	14	1	2	5	30	8	5	9	14	1	2	5	30	8
76		90	7	8	9	9	10	10	11	15	6	7	8	9	10	10	11	13	5	6	7	8	9	9	10	11
		75	8	10	11	12	13	13	14	19	7	8	9	11	12	12	13	15	6	7	8	9	10	10	11	13
		50	10	12	13	14	15	16	17	23	9	9	11	13	14	14	15	17	7	8	9	11	12	12	13	15
		25	12	14	15	17	18	18	20	27	10	11	12	14	16	16	17	20	8	9	10	12	13	13	15	17
		10	13	16	18	19	20	21	23	31	11	12	13	16	17	17	19	22	9	10	11	14	15	15	16	19
51		90	7	9	9	11	12	12	13	18	8	8	9	10	11	11	12	14	7	7	8	9	10	10	11	12
		75	9	11	12	14	15	15	16	22	9	10	11	12	13	13	14	17	8	8	9	11	11	11	12	14
		50	11	13	14	16	18	18	20	27	10	11	13	14	15	15	17	19	9	10	11	12	13	13	14	17
		25	13	15	17	19	21	21	24	32	12	13	14	16	17	17	19	22	10	11	12	14	15	15	16	19
		10	14	18	19	22	24	24	27	36	13	14	16	18	19	19	21	24	11	12	14	15	16	16	18	21
34		90	8	10	11	11	12	12	14	19	8	8	9	12	12	12	14	16	7	7	8	10	11	11	12	14
		75	10	12	13	14	15	15	17	23	9	10	11	13	14	14	16	18	8	8	9	12	12	12	14	16
		50	12	15	16	17	18	19	21	28	10	11	13	16	17	17	18	21	9	10	11	13	14	14	16	18
		25	14	17	19	20	22	22	24	33	12	13	14	18	19	19	21	24	10	11	12	15	16	16	18	21
		10	16	20	21	23	25	25	28	38	13	14	16	19	21	21	23	27	11	12	14	17	18	18	20	23

[a]Box width (the dimension away from the body) (cm).
[b]Vertical distance of lift (cm).
[c]Percentage of industrial population.
Note: Italicized values exceed 8 h physiological criteria.
Source: Reprinted from Snook, S.H. and Ciriello, V.M. *Ergonomics*, 34, 1197–1213, 1991. With permission.

TABLE 47.3 Maximum Acceptable Weight of Lowering for Males (kg)

Width[a]	Distance[b]	Percent[c]	Knuckle height to floor level One lowering every								Shoulder height to knuckle height One lowering every								Arm reach to shoulder height One lowering every							
			sec			min				h	sec			min				h	sec			min				h
			5	9	14	1	2	5	30	8	5	9	14	1	2	5	30	8	5	9	14	1	2	5	30	8
75	76	90	7	9	10	12	14	15	16	20	10	11	14	14	15	15	16	19	6	7	9	9	10	10	11	13
		75	10	13	14	18	20	22	22	29	13	16	18	18	21	21	21	26	9	10	12	12	14	14	14	18
		50	14	17	19	23	27	29	30	38	18	20	24	24	27	27	28	34	11	13	15	16	18	18	19	23
		25	17	21	24	29	33	36	37	47	21	25	29	29	34	34	34	42	14	16	19	20	23	23	23	28
		10	20	25	28	34	39	42	44	56	25	29	34	34	39	39	39	49	16	19	22	23	26	26	27	33
	51	90	8	10	11	13	15	16	17	21	11	12	14	15	17	17	18	22	7	8	9	10	12	12	12	15
		75	11	14	15	18	21	23	23	30	14	17	20	21	24	24	24	30	9	11	13	14	16	16	16	20
		50	14	18	20	24	28	30	31	40	19	21	25	27	31	31	31	38	12	14	16	18	21	21	21	26
		25	18	22	25	30	34	37	39	49	23	26	31	33	38	38	38	47	15	17	20	22	25	25	26	32
		10	21	26	29	36	41	44	46	58	27	31	36	38	44	44	44	55	17	20	24	26	30	30	30	37
	25	90	9	11	12	15	17	18	19	24	12	14	17	18	21	21	21	26	8	9	11	12	14	14	14	17
		75	13	16	17	21	24	25	26	34	17	20	23	24	28	28	28	35	11	13	15	16	19	19	19	24
		50	17	21	23	27	31	34	35	45	22	25	30	32	36	36	37	45	14	16	19	21	24	24	25	31
		25	21	26	29	34	39	42	44	56	27	31	37	39	44	44	45	56	17	20	24	26	30	30	30	38
		10	24	31	34	40	46	49	51	66	31	36	43	45	52	52	52	65	20	23	28	30	35	35	35	44
49	76	90	8	10	11	15	17	18	19	24	10	11	14	14	15	15	16	19	7	8	10	11	12	12	12	15
		75	12	15	16	21	24	26	26	34	13	16	18	18	21	21	21	26	10	11	14	15	16	17	17	21
		50	15	19	21	27	31	34	35	45	18	20	24	24	27	27	28	34	13	15	17	19	22	22	22	27
		25	19	24	26	34	39	42	44	56	21	25	29	29	34	34	34	42	16	18	21	23	27	27	27	33
		10	25	28	31	40	46	49	51	65	25	29	34	34	39	39	39	49	18	21	25	27	31	31	31	39
	51	90	9	11	12	15	17	19	19	25	11	12	14	15	17	17	18	22	8	9	10	12	14	14	14	17
		75	12	15	17	22	25	26	28	35	14	17	20	21	24	24	24	30	10	12	14	16	19	19	19	24
		50	16	20	22	29	33	35	37	47	19	21	25	27	31	31	31	38	14	16	18	21	24	24	25	31
		25	20	25	27	36	41	44	46	58	23	26	31	33	38	38	38	47	17	19	23	26	30	30	30	37
		10	23	29	32	42	48	51	54	68	27	31	36	38	44	44	44	55	19	22	26	30	35	35	35	44
	25	90	10	13	14	17	20	21	22	28	12	14	17	18	21	21	21	26	9	10	12	14	16	16	16	20
		75	14	18	19	24	28	30	31	40	17	20	23	24	28	28	28	35	12	14	17	19	22	22	22	28
		50	19	24	26	32	37	40	41	54	22	25	30	32	36	36	36	45	16	18	22	25	29	29	29	36
		25	23	29	32	40	46	49	51	65	27	31	37	39	44	44	44	56	20	23	27	31	35	35	36	44

(Table continued)

TABLE 47.3 *Continued*

			Knuckle height to floor level One lowering every								Shoulder height to knuckle height One lowering every								Arm reach to shoulder height One lowering every							
			sec			min				h	sec			min				h	sec			min				h
Width[a]	Distance[b]	Percent[c]	5	9	14	1	2	5	30	8	5	9	14	1	2	5	30	8	5	9	14	1	2	5	30	8
		10	27	34	38	47	54	58	60	77	31	36	43	45	52	52	52	65	23	26	31	36	41	41	42	52
		90	10	12	13	17	19	21	21	27	11	12	14	15	17	17	18	22	9	10	12	12	14	14	14	18
		75	14	17	19	24	27	29	30	39	14	17	20	21	24	24	24	30	12	13	16	17	19	19	19	24
	76	50	18	23	25	32	36	39	40	51	19	21	25	27	31	31	31	38	15	17	21	22	25	25	25	31
		25	23	29	31	39	45	48	50	64	23	26	31	33	38	38	38	47	19	21	25	27	31	31	31	38
		10	27	34	37	46	53	57	59	75	27	31	36	38	44	44	44	55	22	25	30	31	36	36	36	45
		90	10	13	14	17	20	22	22	29	11	13	15	17	20	20	20	24	9	10	12	14	16	16	16	20
		75	14	18	20	25	28	30	32	40	15	18	21	23	27	27	27	33	12	14	17	19	22	22	22	27
34	51	50	19	24	26	33	37	40	42	53	20	23	27	30	35	35	35	43	16	19	22	24	28	28	28	35
		25	24	30	33	41	47	50	52	67	24	28	33	37	42	42	43	53	20	23	27	30	34	34	35	43
		10	28	35	38	48	55	59	62	78	28	33	39	43	49	49	50	62	23	27	31	35	40	40	40	50
		90	12	15	16	20	23	24	25	32	13	15	18	20	23	23	23	29	11	12	15	16	19	19	19	23
		75	17	21	23	28	32	34	36	46	18	21	25	27	31	31	32	39	15	17	20	22	26	26	26	32
	25	50	23	28	31	37	42	46	47	60	23	27	32	35	41	41	41	51	19	22	26	29	33	33	33	41
		25	28	35	38	46	53	57	59	75	29	33	39	42	50	50	50	63	23	27	32	35	41	41	41	51
		10	33	41	45	54	62	67	70	89	33	39	46	51	58	58	59	73	27	31	37	41	47	47	48	59

[a]Box width (the dimension away from the body) (cm).

[b]Vertical distance of lower (cm).

[c]Percentage of industrial population.

Note: Italicized values exceed 8 h physiological criteria.

Source: Reprinted from Snook. S. H. and Ciriello, V.M. *Ergonomics*, 34, 1197–1213, 1991. with permission.

TABLE 47.4 Maximum Acceptable Weight of Lowering for Females (kg)

Width[a]	Distance[b]	Percent[c]	Knuckle height to floor level (One lowering every)								Shoulder height to knuckle height (One lowering every)								Arm reach to shoulder height (One lowering every)							
			sec			min				h	sec			min				h	sec			min				h
			5	9	14	1	2	5	30	8	5	9	14	1	2	5	30	8	5	9	14	1	2	5	30	8
75	76	90	5	6	7	7	8	8	9	12	6	6	7	8	9	10	10	13	5	5	5	6	7	7	7	9
		75	6	8	8	9	10	10	11	14	7	8	8	11	11	12	12	15	5	6	6	7	8	9	9	11
		50	7	9	10	11	12	12	13	17	8	9	10	12	13	14	14	18	7	8	8	8	10	10	10	13
		25	9	11	12	12	14	14	15	20	9	11	11	13	15	17	17	21	8	9	9	10	11	12	12	15
		10	10	13	13	14	15	16	17	23	11	12	13	15	17	19	19	24	9	10	10	11	12	14	14	17
	51	90	6	7	7	8	9	10	10	14	7	8	8	9	10	11	11	14	5	6	6	6	7	8	8	10
		75	7	8	9	10	11	12	13	17	8	9	9	11	12	13	13	17	7	7	8	8	9	10	10	12
		50	8	10	11	12	14	14	15	20	10	11	11	13	15	16	16	20	8	9	9	9	11	12	12	15
		25	10	12	13	14	16	17	18	24	11	13	13	15	17	19	19	23	9	10	11	11	12	13	13	17
		10	11	13	14	16	18	19	20	27	13	15	15	17	19	21	21	26	10	12	12	12	14	15	15	19
	25	90	6	8	8	9	10	10	11	14	7	8	8	10	11	12	12	15	5	6	6	7	8	9	9	11
		75	8	10	10	11	12	12	13	17	8	9	9	12	13	15	15	19	7	7	8	9	10	11	11	13
		50	9	11	12	13	14	15	16	21	10	11	11	14	16	18	18	22	8	9	9	10	12	13	13	16
		25	11	13	14	15	17	17	19	25	11	13	13	16	19	20	20	26	9	10	11	12	13	15	15	19
		10	12	15	16	17	19	20	21	28	13	15	15	19	21	23	23	29	10	12	12	13	15	17	17	21
49	76	90	5	6	7	8	8	9	10	13	6	6	7	8	9	10	10	13	5	5	5	6	7	8	8	10
		75	6	8	8	9	10	11	12	16	7	8	8	10	11	12	12	15	5	6	6	8	9	9	9	12
		50	8	9	10	11	13	13	14	19	8	9	10	12	13	14	14	18	7	8	8	9	10	11	11	14
		25	9	11	12	13	15	16	17	22	9	11	11	13	15	17	17	21	8	9	9	11	12	13	13	16
		10	10	13	13	15	17	18	19	25	11	12	13	15	17	19	19	24	9	10	10	12	13	15	15	19
	51	90	6	7	7	9	10	10	11	15	7	8	8	9	10	11	11	14	5	6	6	7	8	9	9	11
		75	7	8	9	11	12	13	13	18	8	9	9	11	12	13	13	17	7	7	8	8	10	10	10	13
		50	8	10	10	13	15	15	16	22	10	11	11	13	15	16	16	20	8	9	9	10	11	13	13	16
		25	10	12	13	15	17	18	19	26	11	13	13	15	17	19	19	23	9	10	11	12	13	15	15	18
		10	11	13	14	17	19	20	22	29	13	15	15	17	19	21	21	26	10	12	12	13	15	16	16	21
	25	90	6	8	8	9	10	11	12	15	7	8	8	10	11	12	12	15	5	6	6	8	9	9	9	12
		75	8	10	10	11	13	13	14	19	8	9	9	12	13	15	15	19	7	7	8	9	10	12	12	14
		50	9	11	12	14	15	16	17	23	10	11	11	14	16	18	18	22	8	9	9	11	13	14	14	17
		25	11	13	14	16	18	19	20	27	11	13	13	16	19	20	20	26	9	10	11	13	15	16	16	20
		10	12	15	16	18	20	21	23	30	13	15	15	19	21	23	23	29	10	12	12	15	16	18	18	23

(*Table continued*)

TABLE 47.4 Continued

Width[a]	Distance[b]	Percent[c]	Knuckle height to floor level — One lowering every								Shoulder height to knuckle height — One lowering every								Arm reach to shoulder height — One lowering every							
			sec			min				h	sec			min				h	sec			min				h
			5	9	14	1	2	5	30	8	5	9	14	1	2	5	30	8	5	9	14	1	2	5	30	8
	76	90	6	8	9	9	10	11	12	15	7	8	8	9	10	11	11	14	6	6	7	8	9	9	9	12
		75	8	10	11	11	13	13	14	19	8	9	9	11	12	13	13	17	7	8	8	9	10	11	11	14
		50	10	12	13	14	15	16	17	23	10	11	11	13	15	16	16	20	8	9	10	11	13	14	14	17
		25	11	14	15	16	18	19	20	27	11	13	13	15	17	19	19	23	9	11	11	13	15	16	16	20
		10	13	16	17	18	20	21	23	30	12	14	15	17	19	21	21	26	11	12	13	14	16	18	18	23
34	51	90	7	9	9	11	12	13	14	18	8	9	9	10	11	12	12	15	7	8	8	8	10	11	11	13
		75	9	11	11	13	15	16	17	22	9	11	11	12	14	15	15	19	8	9	10	10	12	13	13	16
		50	10	13	14	16	18	19	20	27	11	13	13	14	16	18	18	22	10	11	11	12	14	15	15	19
		25	12	15	16	19	21	22	24	31	13	15	15	17	19	21	21	26	11	13	13	14	16	18	18	22
		10	14	17	18	21	24	25	27	35	16	17	17	19	21	23	23	29	13	15	15	16	18	20	20	25
	25	90	8	10	10	11	13	13	14	19	8	9	9	11	12	13	13	17	7	8	8	9	11	12	12	15
		75	10	12	13	14	15	16	17	23	9	11	11	13	15	16	16	21	8	9	10	11	13	14	14	18
		50	12	14	15	17	19	20	21	28	11	13	13	16	18	20	20	25	10	11	11	14	15	17	17	21
		25	14	17	18	20	22	23	24	33	13	15	15	18	21	23	23	29	11	13	13	16	18	19	19	24
		10	15	19	20	22	25	26	28	37	15	17	17	21	23	26	26	32	13	15	15	18	20	22	22	28

[a]Box width (the dimension away from the body) (cm).
[b]Vertical distance of lower (cm).
[c]Percentage of industrial population.
Note: Italicized values exceed 8 h physiological criteria.
Source: Reprinted from Snook, S. H. and Ciriello, V. M. Ergonomics, 34, 1197–1213, 1991. With permission.

TABLE 47.5 Maximum Acceptable Forces of Push for Males (kg)

Height[a]	Percent[b]	2.1 m push One push every							7.6 m push One push every							15.2 m push One push every							30.5 m push One push every					45.7 m push One push every					61.0 m push One push every			
		sec		min				h	sec		min				h	sec		min				h	min				h	min				h	min			h
		6	12	1	2	5	30	8	15	22	1	2	5	30	8	25	35	1	2	5	30	8	1	2	5	30	8	1	2	5	30	8	2	5	30	8
Initial forces[c]																																				
144	90	20	22	25	25	26	26	31	14	16	21	21	22	22	26	16	18	19	19	20	21	25	15	16	19	19	24	13	14	16	16	20	12	14	14	18
	75	26	29	32	32	34	34	41	18	20	27	27	28	28	34	21	23	25	25	26	27	32	19	21	25	25	31	16	18	21	21	26	16	18	18	23
	50	32	36	40	40	42	42	51	23	25	33	33	35	35	42	26	29	31	31	33	33	40	24	27	31	31	38	20	23	26	26	33	20	22	22	28
	25	38	43	47	47	50	51	61	27	3	40	40	42	42	51	31	35	37	37	40	40	48	28	32	37	37	46	24	27	32	32	39	23	27	27	34
	10	44	49	55	55	58	58	70	31	35	46	46	48	49	58	36	40	43	43	45	46	55	32	37	42	48	53	28	31	36	36	45	27	31	31	39
95	90	21	24	26	26	28	28	34	16	18	23	23	25	25	30	18	21	22	22	23	24	28	17	19	22	22	28	14	16	19	19	23	14	16	16	20
	75	28	31	34	34	36	36	44	21	23	30	30	32	32	39	24	27	28	28	30	30	36	21	24	28	28	36	18	21	24	24	30	18	21	20	26
	50	34	38	43	43	45	45	54	26	29	38	38	40	40	48	29	33	35	35	37	38	45	27	30	35	35	45	23	26	30	30	37	22	26	26	32
	25	41	46	51	51	54	55	65	31	35	45	45	48	48	58	35	40	42	42	45	45	54	32	36	42	42	54	27	31	36	36	45	27	31	31	38
	10	47	53	59	59	62	63	75	35	40	52	52	55	56	66	40	46	49	49	52	52	62	37	41	48	48	62	32	36	41	41	52	31	35	35	44
64	90	19	22	24	24	25	26	31	13	14	20	20	21	21	26	15	17	19	19	20	20	26	14	16	19	19	24	12	14	16	16	20	12	14	14	17
	75	25	28	31	31	33	33	40	16	19	26	26	27	28	33	19	21	24	24	26	26	33	18	21	24	24	30	16	18	21	21	26	15	18	18	22
	50	31	35	39	39	41	41	50	20	23	32	32	34	35	41	23	27	30	30	32	33	41	23	26	30	30	37	20	22	26	26	32	19	22	22	28
	25	38	42	46	46	49	50	59	25	28	39	39	41	41	50	28	32	36	36	39	39	50	28	31	36	36	45	24	27	31	31	39	23	26	26	33
	10	43	48	53	53	57	57	68	28	32	45	45	47	48	57	32	37	42	42	44	45	57	32	36	41	41	52	27	31	36	36	44	26	30	30	38
Sustained forces[d]																																				
144	90	10	13	15	16	18	18	22	8	9	13	13	15	16	18	8	9	11	11	12	13	16	8	10	12	13	16	7	8	10	11	13	7	8	9	11
	75	13	17	21	22	24	25	30	10	13	17	18	20	21	25	11	13	15	16	18	18	22	11	13	16	18	21	10	11	13	15	18	9	11	13	15
	50	17	22	27	28	31	32	38	13	16	22	23	26	27	32	14	17	20	20	23	23	28	15	17	20	23	28	12	14	17	19	23	12	14	16	19
	25	21	27	33	34	38	40	47	16	20	28	29	32	33	39	17	20	24	25	28	29	34	18	21	25	29	34	15	18	21	24	28	15	17	20	24
	10	25	31	38	40	45	46	54	19	23	32	33	38	39	46	20	24	28	29	33	34	40	21	25	29	33	39	18	21	24	28	33	19	20	23	28
	90	*10*	*13*	*16*	*17*	*19*	*19*	*23*	*8*	*10*	*13*	*13*	*15*	*15*	*18*	*8*	*10*	*11*	*12*	*13*	*13*	*16*	*8*	*10*	*12*	*13*	*16*	*7*	*8*	*9*	*11*	*13*	*7*	*8*	*9*	*11*
	75	*14*	*18*	*22*	*22*	*25*	*26*	*31*	*11*	*13*	*17*	*18*	*20*	*21*	*25*	*11*	*13*	*15*	*16*	*18*	*18*	*21*	*11*	*13*	*16*	*18*	*21*	*9*	*11*	*13*	*15*	*18*	*9*	*11*	*12*	*15*

(Table continued)

TABLE 47.5 *Continued*

Height[a]	Percent[b]	2.1 m push One push every							7.6 m push One push every							15.2 m push One push every							30.5 m push One push every					45.7 m push One push every					61.0 m push One push every			
		sec		min				h	sec		min				h	sec		min				h	min				h	min				h	min			h
		6	12	1	2	5	30	8	15	22	1	2	5	30	8	25	35	1	2	5	30	8	1	2	5	30	8	1	2	5	30	8	2	5	30	8
95	50	18	23	28	29	33	34	40	14	17	22	23	26	27	32	14	17	19	20	23	23	28	15	17	20	23	28	12	14	17	19	23	12	14	16	19
	25	22	28	34	35	40	41	49	17	21	27	29	32	33	39	18	21	24	25	28	29	34	18	21	25	28	34	15	18	21	24	28	15	17	20	23
	10	26	33	40	41	46	48	57	20	24	32	33	37	38	45	20	25	28	29	32	33	40	21	25	29	33	40	17	20	24	27	32	17	20	23	27
	90	10	13	16	16	18	19	23	8	10	12	13	14	15	18	8	10	11	11	12	13	15	8	9	11	13	15	7	8	9	11	13	7	8	9	10
	75	14	18	21	22	25	26	31	11	13	17	17	19	20	24	11	13	14	15	17	17	21	11	13	15	17	20	9	11	12	14	17	9	10	12	14
64	50	18	23	28	29	32	33	39	14	17	21	22	25	26	31	14	17	19	19	22	22	27	14	16	19	22	26	12	14	16	18	22	12	14	15	18
	25	22	28	34	35	39	41	48	17	21	26	27	31	32	37	18	21	23	24	27	28	33	17	20	24	27	32	14	17	20	23	27	14	17	19	22
	10	26	32	39	41	46	48	56	20	25	30	32	36	37	44	21	25	27	28	31	32	38	20	24	28	32	37	17	20	23	26	31	16	19	22	26

[a]Vertical distance from floor to hands (cm).
[b]Percentage of industrial population.
[c]The force required to get an object in motion.
[d]The force required to keep an object in motion.
Note: Italicized values exceed 8 h physiological criteria.
Source: Reprinted from Snook, S. H. and Ciriello, V.M. *Ergonomics*, 34, 1197–1213, 1991. With permission.

TABLE 47.6 Maximum Acceptable Forces of Push for Females (kg)

Height[a]	Percent[b]	2.1 m push (One push every)							7.6 m push (One push every)							15.2 m push (One push every)							30.5 m push (One push every)					45.7 m push (One push every)					61.0 m push (One push every)			
		sec		min				h	sec		min				h	sec		min				h	min				h	min				h	min			h
		6	12	1	2	5	30	8	15	22	1	2	5	30	8	25	35	1	2	5	30	8	1	2	5	30	8	1	2	5	30	8	2	5	30	8
Initial forces[c]																																				
135	90	14	15	17	18	20	21	22	15	16	16	16	18	19	20	12	14	14	14	15	16	17	12	13	14	15	17	12	13	14	15	17	12	12	14	15
	75	17	18	21	22	24	25	27	18	19	19	20	22	23	24	15	17	17	17	19	20	21	15	16	17	19	21	15	16	17	19	21	14	15	17	19
	50	20	22	25	26	29	30	32	21	23	23	24	26	27	29	18	20	20	20	22	23	25	18	19	21	22	25	18	19	21	22	25	17	18	20	22
	25	24	25	29	30	33	35	37	25	26	27	28	31	32	34	20	23	23	24	26	27	29	20	22	24	26	29	20	22	24	26	29	20	21	23	26
	10	26	28	33	34	38	39	41	28	30	30	31	34	36	38	23	26	26	26	29	31	32	23	25	27	29	33	23	25	27	29	33	22	24	26	29
89	90	14	15	17	18	20	21	22	14	15	16	17	19	19	21	11	13	14	14	16	16	17	12	14	14	15	18	12	14	15	16	18	12	12	14	16
	75	17	18	21	22	24	25	27	17	18	20	20	22	23	25	14	16	17	17	19	20	21	15	16	16	18	21	15	16	18	19	21	15	16	17	19
	50	20	22	25	26	29	30	32	20	21	23	24	27	28	30	16	19	20	21	23	24	25	18	20	20	21	26	18	20	21	23	26	18	19	20	23
	25	24	25	29	30	33	35	37	23	25	27	28	31	33	34	19	22	23	24	27	28	29	21	23	23	24	30	21	23	24	26	30	20	22	24	27
	10	26	28	33	34	38	39	41	26	28	31	32	35	37	39	22	24	26	27	30	31	33	24	26	26	28	33	24	26	28	30	33	23	25	26	30
57	90	11	12	14	14	16	17	18	11	12	14	14	16	16	17	9	11	12	12	13	14	15	11	12	12	13	15	11	12	12	13	15	10	11	12	13
	75	14	15	17	17	19	20	21	14	15	17	17	19	19	21	11	13	14	15	16	17	18	13	14	14	15	18	13	14	15	16	18	12	103	14	16
	50	16	17	20	21	23	24	25	16	18	20	21	23	24	25	14	15	17	18	19	20	21	15	17	17	18	22	15	17	18	19	22	16	16	17	19
	25	19	20	23	24	27	28	30	19	21	23	24	27	28	29	16	18	20	20	23	24	25	18	19	19	21	25	18	19	21	22	25	17	19	20	23
	10	21	23	26	27	30	31	33	22	23	26	27	30	31	33	18	20	22	23	25	26	28	20	22	22	23	28	20	22	23	25	28	19	21	23	25
Sustained forces[d]																																				
135	90	6	8	10	10	11	12	14	6	7	7	7	8	9	11	5	6	6	6	7	7	9	5	6	6	7	8	5	5	5	6	8	4	4	5	6
	75	9	12	14	14	16	17	21	9	10	11	11	12	13	16	7	8	9	9	10	11	13	7	8	8	9	12	7	8	8	8	11	6	6	6	9
	50	12	16	19	20	21	23	28	12	14	15	16	18	17	21	10	11	12	12	14	14	18	10	11	11	12	16	9	10	11	11	15	8	8	9	12
	25	16	20	24	25	27	29	36	15	17	18	18	20	22	27	12	14	15	16	17	18	22	13	14	14	15	21	11	13	13	13	19	10	10	11	15
	10	18	23	28	29	32	34	42	18	20	21	22	24	26	32	14	17	18	18	20	22	27	15	17	17	18	25	14	15	16	17	22	12	12	13	17
	90	6	7	9	9	10	11	13	6	7	8	8	9	9	11	5	6	6	6	7	8	10	5	6	6	7	9	5	6	6	6	8	4	4	5	6
	75	8	11	13	13	15	16	19	9	10	11	11	13	13	17	7	8	9	9	11	11	14	8	9	9	10	13	7	8	8	9	12	6	6	7	9

(Table continued)

TABLE 47.6 *Continued*

Height[a]	Percent[b]	2.1 m push — One push every sec 6	2.1m sec 12	2.1m min 1	2.1m min 2	2.1m min 5	2.1m min 30	2.1m h 8	7.6 m push — One push every sec 15	7.6m sec 22	7.6m min 1	7.6m min 2	7.6m min 5	7.6m min 30	7.6m h 8	15.2 m push — One push every sec 25	15.2m sec 35	15.2m min 1	15.2m min 2	15.2m min 5	15.2m min 30	15.2m h 8	30.5 m push — One push every min 1	30.5m min 2	30.5m min 5	30.5m min 30	30.5m h 8	45.7 m push — One push every min 1	45.7m min 2	45.7m min 5	45.7m min 30	45.7m h 8	61.0 m push — One push every min 2	61.0m min 5	61.0m min 30	61.0m h 8
89	50	11	15	18	18	20	21	26	12	13	15	15	17	18	22	9	11	13	13	14	15	19	10	12	12	13	17	10	11	11	12	16	8	9	9	12
	25	14	18	22	23	25	27	33	15	17	19	19	21	23	28	12	14	16	16	18	19	24	13	15	15	16	22	12	14	14	15	20	11	11	12	15
	10	17	22	26	27	30	32	39	17	20	22	23	25	27	33	14	17	19	19	21	23	28	16	18	18	19	26	14	16	17	18	24	13	13	14	18
	90	5	6	8	8	9	9	12	6	7	7	8	8	9	11	5	6	6	6	7	7	9	5	6	6	6	8	5	5	5	6	7	4	4	4	6
	75	7	9	11	12	13	14	17	8	10	10	11	12	12	15	7	8	9	9	10	10	13	7	8	8	9	12	7	7	8	8	11	6	6	6	8
57	50	10	13	15	16	17	18	23	11	13	14	14	16	17	21	9	11	12	12	13	14	17	10	11	11	12	16	9	10	10	11	15	8	8	8	11
	25	12	16	19	20	22	23	29	14	17	18	18	20	21	26	12	14	15	15	17	18	22	12	14	14	15	20	11	13	13	14	18	10	10	11	14
	10	15	19	23	23	26	28	34	17	20	21	21	23	25	31	14	16	17	18	20	21	26	15	16	17	18	24	13	15	16	16	22	12	12	13	17

[a]Vertical distance from floor to hands (cm).

[b]Percentage of industrial population.

[c]The force required to get an object in motion.

[d]The force required to keep an object in motion.

Note: Italicized values exceed 8 h physiological criteria.

Source: Reprinted from Snook, S. M. and Ciriello, V.M. *Ergonomics*, 34, 1197–1213, 1991. With permission.

TABLE 47.7 Maximum Acceptable Forces of Pull for Males (kg)

Column groups are "One pull every" for each push distance. Sub-interval columns: 2.1 m push = 6 sec, 12 sec, 1 min, 2 min, 5 min, 30 min, 8 h; 7.6 m push = 15 sec, 22 sec, 1 min, 2 min, 5 min, 30 min, 8 h; 15.2 m push = 25 sec, 35 sec, 1 min, 2 min, 5 min, 30 min, 8 h; 30.5 m push = 1 min, 2 min, 5 min, 30 min, 8 h; 45.7 m push = 1 min, 2 min, 5 min, 30 min, 8 h; 61.0 m push = 2 min, 5 min, 30 min, 8 h.

Height[a]	Percent[b]	2.1m 6s	2.1m 12s	2.1m 1m	2.1m 2m	2.1m 5m	2.1m 30m	2.1m 8h	7.6m 15s	7.6m 22s	7.6m 1m	7.6m 2m	7.6m 5m	7.6m 30m	7.6m 8h	15.2m 25s	15.2m 35s	15.2m 1m	15.2m 2m	15.2m 5m	15.2m 30m	15.2m 8h	30.5m 1m	30.5m 2m	30.5m 5m	30.5m 30m	30.5m 8h	45.7m 1m	45.7m 2m	45.7m 5m	45.7m 30m	45.7m 8h	61.0m 2m	61.0m 5m	61.0m 30m	61.0m 8h
Initial forces[c]																																				
144	90	14	16	18	18	19	19	23	11	13	16	16	17	18	21	13	15	15	15	16	17	20	12	13	15	15	19	10	11	11	13	16	10	11	11	14
	75	17	19	22	22	23	24	28	14	15	20	20	21	21	26	16	18	19	19	20	20	24	14	16	19	19	23	12	14	14	16	20	12	14	14	17
	50	20	23	26	26	28	28	33	16	18	24	24	25	26	31	19	21	22	22	24	24	29	17	19	22	22	27	15	16	16	19	24	14	16	16	20
	25	24	27	31	31	32	33	39	19	21	28	28	29	30	36	22	25	26	26	28	28	33	20	22	26	26	32	17	19	19	22	28	16	19	19	24
	10	26	30	34	34	36	37	44	21	24	31	31	33	33	40	24	28	29	29	31	31	38	22	25	29	29	37	20	22	22	25	31	18	21	21	27
95	90	19	22	25	25	27	27	32	15	18	23	23	24	24	29	18	20	21	21	23	23	28	16	18	21	21	26	14	16	16	18	23	13	16	16	19
	75	23	27	31	31	32	33	39	19	21	28	28	29	30	36	22	25	26	26	28	28	33	20	22	26	26	32	17	19	19	22	28	16	19	19	24
	50	28	32	36	36	39	39	47	23	26	33	33	35	35	42	26	29	31	31	33	33	40	24	27	31	31	38	20	23	23	27	33	20	23	23	28
	25	33	37	42	42	45	45	54	26	30	39	39	41	41	49	30	34	36	36	38	39	46	27	31	36	36	45	24	27	27	31	38	23	26	26	33
	10	37	42	48	48	51	51	61	30	33	43	43	46	47	56	33	38	41	41	43	44	52	31	35	40	40	50	27	30	30	35	43	26	30	30	37
64	90	22	25	28	28	30	30	36	18	20	26	26	27	28	33	20	23	24	24	26	26	31	18	21	24	24	30	16	18	18	21	26	15	18	18	22
	75	27	30	34	34	37	37	44	21	24	31	31	33	34	40	24	28	29	29	31	32	38	22	25	29	29	36	19	22	22	25	31	19	21	21	27
	50	32	36	41	41	44	44	53	25	29	37	37	40	40	48	29	33	35	35	37	38	45	27	30	35	35	43	23	26	26	30	37	22	26	26	32
	25	37	42	48	48	51	51	61	30	34	44	44	46	47	56	34	39	41	41	43	44	52	31	35	41	41	50	27	30	30	35	43	26	30	30	37
	10	42	48	54	54	57	58	69	33	38	49	49	52	53	63	38	43	46	46	49	49	59	35	39	46	46	57	30	34	34	39	49	29	34	34	42
Sustained forces[d]																																			
144	90	8	10	12	13	15	15	18	6	8	10	11	12	12	15	7	8	9	9	10	11	13	7	8	9	11	13	6	7	9	9	10	6	6	7	9
	75	10	13	16	17	19	20	23	8	10	13	14	16	16	19	9	10	12	12	14	14	17	9	10	12	14	16	7	9	11	11	14	7	8	10	11
	50	13	16	20	21	23	24	28	10	13	16	17	19	20	23	11	13	14	15	17	17	20	11	13	15	17	20	9	11	13	14	17	9	10	12	14
	25	15	20	24	25	28	29	34	12	15	20	20	23	24	28	13	15	17	18	20	21	24	13	15	18	20	24	11	13	15	17	20	11	12	14	17
	10	17	22	27	28	32	33	39	14	17	22	23	26	27	32	14	17	19	20	23	24	28	15	17	20	23	27	12	14	17	19	23	12	14	16	19
	90	10	13	16	17	19	20	24	8	10	13	14	16	16	19	9	10	12	12	14	14	17	9	10	12	14	17	7	9	12	12	14	7	9	10	12
	75	13	17	21	22	25	26	30	11	13	17	18	20	21	25	11	14	15	15	18	18	22	12	13	16	18	21	10	11	13	15	18	9	11	13	15

(Table continued)

TABLE 47.7 *Continued*

Height[a]	Percent[b]	2.1 m push (One pull every)							7.6 m push (One pull every)							15.2 m push (One pull every)							30.5 m push (One pull every)					45.7 m push (One pull every)					61.0 m push (One pull every)			
		6 s	12 s	1 min	2 min	5 min	30 min	8 h	15 s	22 s	1 min	2 min	5 min	30 min	8 h	25 s	35 s	1 min	2 min	5 min	30 min	8 h	1 min	2 min	5 min	30 min	8 h	1 min	2 min	5 min	30 min	8 h	2 min	5 min	30 min	8 h
95	50	16	21	26	27	31	32	37	13	17	21	22	25	26	31	14	17	19	19	22	23	27	14	17	19	22	26	12	14	16	19	22	12	14	16	18
	25	19	26	31	33	37	38	45	16	20	26	27	30	31	37	17	20	22	23	26	27	32	17	20	23	27	32	14	17	19	22	26	14	16	19	22
	10	22	29	36	37	42	43	51	18	23	29	31	34	36	42	19	23	26	27	30	31	37	19	23	27	31	36	16	19	22	25	30	16	19	21	25
64	90	11	14	17	18	20	21	25	9	11	14	15	17	17	20	9	11	12	13	15	15	18	9	11	13	15	18	8	9	11	12	15	8	9	10	12
	75	14	19	23	23	26	27	32	11	14	19	19	22	22	26	12	14	16	17	19	19	23	12	14	17	19	23	10	12	14	16	19	10	12	13	16
	50	17	23	28	29	32	34	40	14	18	23	24	27	28	33	15	18	20	21	23	24	28	15	18	21	24	27	13	15	17	20	23	12	14	16	20
	25	20	27	33	35	39	40	48	17	21	27	28	32	33	39	18	21	24	25	28	29	34	18	21	25	28	33	15	18	21	24	28	15	17	20	23
	10	23	31	38	40	45	46	54	19	24	31	32	37	38	45	20	24	27	28	32	33	39	21	24	28	32	38	17	20	24	27	32	17	20	23	27

[a]Vertical distance from floor to hands (cm).
[b]Percentage of industrial population.
[c]The force required to get an object in motion.
[d]The force required to keep an object in motion.
Note: Italicized values exceed 8 h physiological criteria.
Source: Reprinted from Snook, S. H. and Ciriello, V. M. *Ergonomics,* 34, 1197–1213, 1991. With permission.

TABLE 47.8 Maximum Acceptable Forces of Pull for Females (kg)

Height[a]	Percent[b]	2.1 m push — One push every							7.6 m push — One push every							15.2 m push — One push every							30.5 m push — One push every					45.7 m push — One push every					61.0 m push — One push every			
		6 sec	12 sec	1 min	2 min	5 min	30 min	8 h	15 sec	22 sec	1 min	2 min	5 min	30 min	8 h	25 sec	35 sec	1 min	2 min	5 min	30 min	8 h	1 min	2 min	5 min	30 min	8 h	1 min	2 min	5 min	30 min	8 h	2 min	5 min	30 min	8 h
Initial forces[c]																																				
135	90	13	16	17	18	20	21	22	13	14	16	16	18	19	20	10	12	13	14	15	16	17	12	13	14	15	17	12	13	14	15	17	12	13	14	15
	75	16	19	20	21	24	25	26	16	17	19	19	21	22	24	12	14	16	16	18	19	20	14	16	17	18	20	14	16	17	18	20	14	15	16	18
	50	19	22	24	25	28	29	31	19	20	22	23	25	26	28	14	16	19	19	21	22	24	17	18	20	21	24	17	18	20	21	24	16	18	19	21
	25	21	25	28	29	32	33	35	21	23	25	26	29	30	32	16	19	21	22	25	26	27	19	21	23	24	27	19	21	23	24	27	19	20	22	25
	10	24	28	31	32	36	37	39	24	26	28	29	32	34	36	18	21	24	25	27	29	30	22	24	25	27	31	22	24	25	27	31	21	23	24	27
89	90	14	16	18	19	21	22	23	14	15	16	17	19	19	21	10	12	14	14	16	17	18	13	14	15	16	18	13	14	15	16	18	12	13	14	16
	75	16	19	21	22	25	26	27	16	18	19	20	22	23	25	12	15	17	17	19	20	21	15	16	18	19	21	15	16	18	19	21	15	16	17	19
	50	19	23	25	26	29	30	32	19	21	23	24	26	27	29	14	17	19	20	22	23	25	18	19	21	22	25	18	19	21	22	25	17	18	20	22
	25	22	26	29	30	33	35	37	22	24	26	27	30	31	33	16	20	22	23	26	27	28	20	22	24	25	29	20	22	24	25	29	20	21	23	26
	10	25	29	32	33	37	39	41	25	27	30	30	33	35	37	18	22	25	26	29	30	32	23	25	26	28	32	23	25	26	28	32	22	24	25	29
57	90	15	17	19	20	22	23	24	15	16	17	18	20	20	22	11	13	15	15	17	18	19	13	14	15	17	19	13	14	15	17	19	13	14	15	17
	75	17	20	22	23	26	27	28	17	19	20	21	23	23	26	13	15	17	18	20	20	22	16	17	18	20	22	16	17	18	20	22	15	16	18	20
	50	20	24	26	27	30	32	33	20	22	24	25	28	29	30	15	18	20	21	23	23	26	18	20	22	23	26	18	20	22	25	26	17	18	19	21
	25	23	27	30	31	35	36	38	23	25	27	29	32	33	35	17	21	23	24	27	28	30	21	23	25	27	30	21	23	25	27	30	21	22	24	27
	10	26	31	34	35	39	40	43	26	28	31	32	35	37	39	19	23	26	27	30	31	33	24	26	28	30	34	24	26	28	30	34	23	25	27	30
Sustained forces[d]																																				
135	90	6	9	10	10	11	12	15	7	8	9	9	10	11	13	6	7	7	8	8	9	11	6	7	8	8	10	6	6	7	7	10	5	5	5	7
	75	8	12	13	14	15	16	20	9	11	12	12	13	14	18	7	9	9	10	11	12	15	8	9	10	10	14	8	9	9	9	14	7	7	7	10
	50	10	16	17	18	19	21	25	12	13	15	16	18	18	22	9	11	13	13	14	15	19	11	12	12	13	17	10	11	11	12	16	8	9	9	12
	25	13	19	21	23	25	25	31	14	16	18	19	21	22	27	11	14	15	16	17	19	23	13	15	15	16	21	12	13	14	14	19	10	11	11	15
	10	15	22	24	25	27	29	36	16	19	21	22	24	26	32	13	16	18	18	20	22	27	15	17	17	18	25	14	15	16	17	23	12	12	13	17
89	90	6	9	10	10	11	12	14	7	8	9	9	10	10	13	5	6	7	7	8	9	11	6	7	7	7	10	5	6	6	7	9	5	5	5	7
	75	8	12	13	13	15	16	19	9	10	11	11	12	13	17	7	8	10	10	11	12	14	8	9	9	10	13	7	8	9	9	12	6	7	7	9
	50	10	15	16	17	19	20	25	11	13	15	15	16	18	22	9	11	12	13	14	15	18	10	12	12	13	17	9	11	11	12	15	8	8	9	12

(Table continued)

TABLE 47.8 *Continued*

Height[a]	Percent[b]	2.1 m push One push every							7.6 m push One push every							15.2 m push One push every							30.5 m push One push every					45.7 m push One push every					61.0 m push One push every			
		sec		min				h	sec		min				h	sec		min				h	min				h	min				h	min			h
		6	12	1	2	5	30	8	15	22	1	2	5	30	8	25	35	1	2	5	30	8	1	2	5	30	8	1	2	5	30	8	2	5	30	8
	25	12	18	20	21	23	24	30	14	16	18	18	20	22	27	11	13	15	15	17	18	22	12	14	15	15	21	11	13	13	14	19	10	10	11	15
	10	14	21	23	24	26	28	35	16	18	21	21	23	25	31	13	15	17	18	20	21	26	15	16	17	18	24	13	15	16	16	22	12	12	13	17
	90	5	8	9	9	10	11	13	6	7	8	8	9	10	12	5	6	7	7	7	8	10	6	6	6	7	9	5	6	6	6	8	4	5	5	6
	75	7	11	12	12	13	14	18	8	9	11	11	12	13	16	7	8	9	9	10	11	13	7	8	8	9	12	7	8	8	8	11	6	6	6	9
57	50	9	14	15	16	17	18	23	10	12	13	14	15	16	20	8	10	11	12	13	14	17	9	11	11	12	16	9	10	10	11	14	8	8	8	11
	25	11	17	18	19	21	22	27	13	15	16	17	19	20	24	10	12	14	14	16	17	21	11	13	13	14	19	11	12	12	13	17	9	10	10	13
	10	13	20	21	22	24	26	32	15	17	19	20	22	23	28	12	14	16	16	18	20	24	13	15	16	16	22	12	14	14	15	20	11	11	12	16

[a]Vertical distance from floor to hands (cm).
[b]Percentage of industrial population.
[c]The force required to get an object in motion.
[d]The force required to keep an object in motion.
Note: Italicized values exceed 8 h physiological criteria.
Source: Reprinted from Snook, S. H. and Ciriello, V. M. *Ergonomics,* 34, 1197–1213, 1991. With permission.

TABLE 47.9 Maximum Acceptable Weight of Carry (kg)

Height[a]	Percent[b]	2.1 m carry — One carry every							4.3 m carry — One carry every							8.5 m carry — One carry every						
		sec 6	sec 12	min 1	min 2	min 5	min 30	h 8	sec 10	sec 16	min 1	min 2	min 5	min 30	h 8	sec 18	sec 24	min 1	min 2	min 5	min 30	h 8
Males																						
111	90	10	14	17	17	19	21	25	9	11	15	15	17	19	22	10	11	13	13	15	17	20
	75	14	19	23	23	26	29	34	13	16	21	21	23	26	30	13	15	18	18	20	23	27
	50	19	25	30	30	33	38	44	17	20	27	27	30	34	39	17	19	23	24	26	29	35
	25	23	30	37	37	41	46	54	20	25	33	33	37	41	48	21	24	29	29	32	36	43
	10	27	35	43	43	48	54	63	24	29	38	39	43	48	57	24	28	34	34	38	42	50
79	90	13	17	21	21	23	26	31	11	14	18	19	21	23	27	13	18	17	18	20	22	26
	75	18	23	28	29	32	36	42	16	19	25	25	28	32	37	17	20	24	24	27	30	35
	50	23	30	37	37	41	46	54	20	25	32	33	36	41	48	22	26	31	31	35	39	46
	25	28	37	45	46	51	57	67	25	30	40	40	45	50	59	27	32	38	38	42	48	56
	10	33	43	53	53	59	66	78	29	35	47	47	52	59	69	32	38	44	45	50	56	65
Females																						
105	90	11	12	13	13	13	13	18	9	10	13	13	13	13	18	10	11	12	12	12	12	16
	75	13	14	15	15	16	16	21	11	12	15	15	16	16	21	12	13	14	14	14	14	19
	50	15	16	18	18	18	18	25	12	13	18	18	18	18	24	14	15	16	16	16	16	22
	25	17	18	20	20	21	21	28	14	15	20	20	21	21	28	15	17	18	18	19	19	25
	10	19	20	22	22	23	23	31	16	17	22	22	23	23	31	17	19	20	20	21	21	28
72	90	13	14	16	16	16	16	22	10	11	14	14	14	14	20	12	12	14	14	14	14	19
	75	15	17	18	18	19	19	25	11	13	16	16	17	17	23	14	15	16	16	17	17	23
	50	17	19	21	21	22	22	29	13	15	19	19	20	20	26	16	17	19	19	20	20	26
	25	20	22	24	24	25	25	33	15	17	22	22	22	22	30	18	19	21	22	22	22	30
	10	22	24	27	27	28	28	37	17	19	24	24	25	25	33	20	21	24	24	25	25	33

[a] Vertical distance from floor to hands (cm).
[b] Percentage of industrial population.
Note: Italicized values exceed 8 h physiological criteria (see text).
Source: Reprinted from Snook, S. H. and Ciriello, V. M. *Ergonomics*, 34, 1197–1213, 1991. With permission.

limited headroom, asymmetrical lifting, load asymmetry, couplings, load placement clearance, and heat stress.

Mital[28] found that psychophysical data collected in short periods (i.e., 20–25 min) assuming a longer work period (8–12 h) should be reduced. Subsequently, Mital[22] presented psychophysical data for males and females performing lifting tasks for eight-hour work shifts based on the adjustments determined in the earlier study. The data were collected from 37 males and 37 females experienced in manual lifting. Mital[22] also presented a modified database representing the combined data from his study, Snook's[1] data, and data collected by Ayoub et al.[21] Although the modified database only accommodates lifting tasks, the combined sample size is considerable. Similarly, Mital[23] presented a psychophysical database for lifting tasks for males and females working 12-h shifts. The database represents values valid for 12 h based on adjustments of 8-h data.

47.2.1.1.2 Data for Nonstandard MMH Tasks

One advantage of the psychophysical approach is that it allows for the realistic simulation of many types of materials handling tasks. Several such examples will be provided to illustrate how psychophysics has been used to develop guidelines for specific applications.

Smith et al.[29] presented a psychophysical database for evaluating MMH tasks performed in unusual postures. Maximum acceptable weight data for 99 different tasks were presented, including data for one- and two-handed lifting and lowering tasks performed in postures such as lifting on one knee, lifting on two knees, lifting while lying down, etc. These data are particularly appropriate for occasional maintenance tasks which impose postural constraints on the operator.

Like maintenance, mining is comprised of many activities for which standard psychophysical data are not applicable to. Mining tasks are often performed under postural constraints, such as limited headroom. Gallagher[30] collected psychophysical data to address tasks performed under restricted headroom conditions and provided guidelines for tasks performed in low-seam coal mines requiring lifting while kneeling. Mining also requires handling of a variety of materials. Gallagher and Hamrick[31] provided psychophysical guidelines for the handling of rock dust bags, ventilation stopping blocks, and crib blocks.

47.2.1.1.3 Data for Return to Work from Low-Back Disability

Snook[3] suggested that a future direction for psychophysical research is including symptomatic subjects. Previous research and databases typically reflect values generated by healthy subjects, in some cases with no history of musculoskeletal disorders. Since low-back pain will never be completely prevented, data that indicate the loads and forces that can be handled by persons with musculoskeletal disorders can be used to prevent the associated disability. This information could be used to design workplaces that allow workers to work at sufficient levels of productivity in spite of low-back pain or similar disorders.

47.2.1.2 Assessment of Multiple Component MMH Tasks

More often than not, workers that perform MMH tasks perform a number of tasks, often in sequence. For example, a common combination task in industry is where a worker lifts material, carries it for some distance, then lowers the material. In such situations, all of the tasks should be evaluated.

Snook and Ciriello[20] recommend using the weight or force limit for the task with lowest percentage of the population accommodated as the design criterion for multiple component tasks. This recommendation was based upon the findings of Ciriello et al.[18,19] Thus, for a combination where the worker lifts materials, carries materials, and then lowers material, the limiting component task would be the lift. Snook and Ciriello[20] do caution that this method of analysis may result in violation of recommended energy expenditure criteria for some multiple component tasks.

Straker et al.[32] disagree with the multiple component methodology described in the preceding paragraph, stating that "combination tasks should probably be assessed as whole entities and not separated into components for analysis." However, Straker et al.[32] recommend no alternative methodology to assess multiple component tasks as "whole entities." In general, the design and evaluation of multiple component MMH jobs is one of the more underdeveloped areas of MMH research and practice.

The only other method for multitask assessment that incorporates MMH tasks in addition to lifting that the author was able to find was the method presented by Mital.[33] The data used with this methodology is from Mital et al.,[6] which is modified psychophysical data as described earlier. The method is similar to that developed by Jiang and Mital,[34] except that capacity is predicted using more contemporary data.

In general, this method requires that each MMH task is analyzed, and data regarding work duration, etc. are also needed. The analyst then determines the percentage of the population that the design should accommodate, which should be 75% or 90%. The next step involves calculating the recommended work rate (kg m/min) for the percentage of the population being analyzed using the Mital et al.[6] data. The actual work rate is divided by the recommended work rate, which yields the risk potential. Any risk potential values greater than 1 signal the need for task redesign. This method focuses on the individual components that are unacceptable, as with the Snook and Ciriello[20] method.

47.2.1.3 Example

An example of Snook and Ciriello's[20] multicomponent task assessment will be used to illustrate how psychophysical data are used to analyze MMH tasks. The analysis was performed with the CompuTask[TM] computer program. The set of tasks is fairly simple and includes a worker bending over and lifting a box, carrying the box 5 ft, and lowering the box to the floor. The set of tasks is performed three times per minute for 8 h. The relevant data that need to be collected as well as the analysis are shown in Figure 47.2. Aside from psychophysical results, physiological analyses and National Institute of Occupational Safety and Health (NIOSH) lifting equation computations (STRWL = single task recommended weight limit, FIRWL = frequency independent recommended weight limit, STLI = single task lifting index, FILI = frequency independent lifting index) are provided by the software.

The task with the lowest percentage of the population accommodated is the lifting task. This task accommodates 75% of the male population and <10% of the female population. Thus, the set of tasks is marginally acceptable for males and unacceptable for females. Also, the overall physiological evaluation shows that the task is not acceptable for 8 h. As was discussed earlier, the method of analysis being used may result in violation of energy expenditure criterion. Redesign efforts would be focused on eliminating the tasks through materials handling devices or a conveyor, or eliminating the need to lift and lower the boxes by increasing the vertical origin and destination of the lifting and lowering tasks, respectively.

47.2.2 Task and Workplace Design

An often overlooked use of psychophysical data is workplace design. In situations where unacceptable tasks cannot be eliminated, or loads, frequencies or forces cannot be reduced to "acceptable" levels, psychophysical results can be used to suggest workplace design changes to increase the percentage of the population that a task or job will accommodate. For example, a common problem is that the dimensions and weight of the load cannot be reduced. In situations such as this, the task and workplace can be redesigned to decrease the physical demands of the task.

The following list provides several examples of task and workplace redesign principles to reduce physical demands when altering the material being handled is infeasible or mechanical aids cannot alleviate the need to handle materials manually. Application of these principles will increase the percentage of the population a task will accommodate.

1. For lifting or lowering tasks, bring the load closer to the body
2. When lifting low loads, bring the vertical origin of the load as close to knuckle height as possible. Alternately, when lowering loads, the destination should be as close to knuckle height as possible. This principle will reduce bending. In general, try to avoid lifting or lowering to or from high and low locations

LIBERTY MUTUAL®

Shipping Department

Evaluation Results

COMPONENT	FREQUENCY ONE EVERY	FORCE (lbs)	HAND HEIGHT AT START (in.)	HAND HEIGHT AT END (in.)	DISTANCE MOVED	HAND DISTANCE FROM BODY (in.)	BODY MOTION TWIST	REACH	BEND	POPULATION PERCENTAGES MALE	FEMALE	SUGGESTED MAXIMUM DURATION (h) MALE/FEMALE
Lift	20 sec	34	5	33	28 in.	10	Yes	Moderate	Considerable	75	<10	8/EL
Carry	20 sec	34	33	33	5 ft.	10	No	Moderate	None	>90	77	8/8
Lower	20 sec	34	33	5	28 in.	10	No	Moderate	Considerable	83	<10	8/8
EVALUATION FOR ENTIRE TASK							Yes	Moderate	Considerable	75	<10	EL/EL

EL → Exceeds energy expenditure limit for the task duration specified

"POPULATION PERCENTAGES" are the percentages of the male and female population that can be expected to perform the task without excessive stress or excessive fatigue.

"SUGGESTED MAXIMUM DURATION" is the recommended continuous time the job can be performed during an 8-h workday before exceeding the Energy Expenditure (kcal/min) guidelines for males and females.

	STRWL	FIRWL	STLI	FILI
Component #1 - Lift	11.6	21.1	2.9	1.6
Component #3 - Lower	15.2	27.6	2.2	1.2

FIGURE 47.2 Example of psychophysical analysis of a multiple-component MMH job using CompuTask^{TM}.

3. Decrease the vertical distance that loads must be lifted/lowered and the distance which loads must be pushed, pulled or carried
4. Decrease the frequency of the task or increase the number or workers performing the task
5. For pushing and pulling tasks, provide equipment that provides the least resistance so that initial forces required to overcome inertia are as low as possible. Maintenance of mechanical assists is very important with regards to this principle
6. For all MMH tasks, provide good hand-to-object coupling when possible, that is, tote boxes with handles, carts with handle bars, etc
7. Decrease the duration over which the task is performed
8. Change pulling tasks to pushing tasks

47.3 The Psychophysical Approach to Designing Upper Extremity Tasks

The primary risk factors for WRMSDs of the upper extremity are fairly well known[35]. Task-related risk factors include posture, force, and repetition. Vibration and cold are task-related risk factors for some disorders such as carpal tunnel syndrome. Duration of the task and rest periods are also important since these factors affect the acceptability of task. Altering work–rest relationships can alter the acceptability of a particular combination of posture, force, and repetition.

There are few quantitative guidelines for limits of posture, force, and repetition. Although general guidelines suggest maintaining a neutral wrist posture and reducing the force requirements and frequency of a task, these guidelines do not indicate acceptable levels of the variables. Once ergonomic analyses and task redesign are done, a decision as to the acceptability of a task is difficult.

The application of the psychophysical approach to the design of UEI tasks was a response to the need for establishing quantitative guidelines with which to assess tasks. Currently, quantitative dose–response relationships developed with epidemiological techniques that provide relationships between individual risk factors and their interactions and the risk of upper extremity WRMSDs do not exist. In the absence of such relationships, psychophysical data will continue to be one option of setting task limits. The remainder of this section will provide an overview of the current state of psychophysical data as well as discussion of how these data are applied in the workplace.

One advantage of the psychophysical approach is that data can be developed which incorporate force, posture, and repetition into the development of data for different durations. This is important in that this approach allows for trade-offs between variables, that is, for some tasks, it is not always possible to modify all factors. The psychophysical approach has been extended to study specific tasks such as acceptable impact severity levels for an automotive trim installation.[36] Subjects performed five hand impacts per minute on a device that simulated the process of seating push pins during door trim panel installation, and subjects altered the impact to a level they felt could be performed without causing injury, numbness, or pain.

47.3.1 Setting Acceptable Force and Frequency Limits

Fernandez and his colleagues have collected maximum acceptable frequency data for several types of tasks include drilling,[37–40] riveting,[41,42] and tasks requiring pinch and power grasps.[43,44] In these studies, factors such as wrist posture and duration were incorporated into the experimental protocol to provide frequency limits for a variety of UEI tasks. In a similar study, Abu-Ali et al.[45] had subjects control the length of rest periods for task combinations of varying wrist postures, exertion periods, and power grip forces.

In order to use these data, one would record relevant task parameters, find the data relevant for a particular task in the database, and determine if the task is acceptable to the majority of the population, just as with MMH data. If the task is not acceptable, then the frequency would need to be reduced, the

TABLE 47.10   Maximum Acceptable Forces for Female Wrist Flexion (Power Grip) (N)

Percent of population	Repetition rate				
	2/min	5/min	10/min	15/min	20/min
90	14.9	14.9	13.5	12.0	10.2
75	23.2	23.2	20.9	18.6	15.8
50	32.3	32.3	29.0	26.0	22.1
25	41.5	41.5	37.2	33.5	28.4
10	49.8	49.8	44.6	40.1	34.0

Source: Reprinted from Snook, S. H. and Ciriello, V. M. *Journal of Occupational Medicine*, 16, 527–534, 1974. With permission.

duration would need to be reduced, or factors such as wrist deviation would need to be modified to increase the acceptability of a task.

While Fernandez and his colleagues chose frequency as the variable that subjects manipulate, Snook et al.[14] chose force as the manipulated variable. Also, Snook et al.[14] used a 7-h adjustment period which was much longer than the shorter (20–25 min) period used in the studies cited earlier. Snook et al.[14] studied tasks requiring wrist flexion with a power grasp, wrist flexion with a pinch grip, and extension with a power grasp. Frequencies between 2 and 20 repetitive motions per minute were studied. Subjects adjusted wrist torque during the experiment by manipulating the resistance offered by a magnetic particle brake. Aside from reporting the torques, forces were also reported which were computed by dividing the torques by the moment arms. The forces are reported in Table 47.10 through Table 47.12.

Ciriello et al.[46] reported the results of a psychophysical study of six hand movements. In addition to the motions studied by Snook et al.,[14] wrist extension with a pinch grip, ulnar deviation with a power grip, and a handgrip task (power grip) were studied. Ciriello et al.[47] investigated ulnar deviation with a power grip and the handgrip task, but included clockwise screwdriver motions using 31 and 40 mm handles and a 39 mm yoke handle. Counterclockwise screwdriver motions with a 31 mm handle completed the experimental conditions. Subjects selected 14–24% of maximum isometric torque for the different motions, at different frequencies.

The data reported by Snook et al.[14] and Ciriello et al.[46,47] would be applied in a manner similar to that described earlier in this section. However, these data are more generic than some of the data collected by Fernandez and his colleagues. The Ciriello et al.[47] data contain a mix of generic and task-specific data. For example, the data collected by Snook et al.[14] do not apply only to specific tasks such as drilling, as do some of the data from other studies mentioned.[37–40]

Krawczyk et al.[48] presented preferred weights for manual transfer tasks for transfer distances of 0.5 and 1.0 m and frequencies between 10 and 30 transfers per minute for an 8 h work duration. Thus, depending on the situation, one could adjust frequency or transfer distance for a particular weight of object being transferred.

TABLE 47.11   Maximum Acceptable Forces for Female Wrist Flexion (Pinch Grip) (N)

Percent of population	Repetition rate				
	2/min	5/min	10/min	15/min	20/min
90	9.2	8.5	7.4	7.4	6.0
75	14.2	13.2	11.5	11.5	9.3
75	19.8	18.4	16.0	16.0	12.9
75	25.4	23.6	20.6	20.6	16.6
10	30.5	28.2	24.6	24.6	19.8

Source: Reprinted from Snook, S. H. and Ciriello, V. M. *Journal of Occupational Medicine*, 16, 527–534, 1974. With permission.

TABLE 47.12   Maximum Acceptable Forces for Female Wrist Extension (Power Grip) (N)

	Repetition rate				
Percent of population	2/min	5/min	10/min	15/min	20/min
90	8.8	8.8	7.8	6.9	5.4
75	13.6	13.6	12.1	10.9	8.5
75	18.9	18.9	16.8	15.1	11.9
75	24.2	24.2	21.5	19.3	15.2
10	29.0	29.0	25.8	23.2	18.3

Source: Reprinted from Snook, S. H. and Ciriello, V. M. *Journal of Occupational Medicine*, 16, 527–534, 1974. With permission.

47.3.2   Tool and Workplace Design

Up to this point, the psychophysical results discussed have all involved experimentation where subjects control a variable. Another approach to the study of manual work using psychophysics is to elicit perceived exertion or discomfort ratings for subjects performing specific tasks. The results can then be used to select the task conditions with the lowest perceived exertion or discomfort.

Ulin et al.[49–51] utilized perceived exertions to study tool masses, tool shapes, and horizontal and vertical work locations. A very general summary of the results indicates that 114 cm was the preferred vertical height for driving screws with a variety of tools. At that height, a pistol-shaped tool was the most preferred. The ratings of perceived exertion with respect to the horizontal distance of the workpiece from the body indicated that the workpiece should not be greater than 38 cm from the front of the worker. The optimal workpiece location was a vertical height of 114 cm and a horizontal distance of 13 cm. Likewise, increasing the tool mass increased the ratings, as would be expected.

In general, the following recommendations are examples of means to increase the psychophysical acceptability of a UEI task:

1. Maintain a neutral wrist posture
2. Decrease the force requirements of a task
3. Decrease the duration over which a task is performed. Worker rotation can be very helpful
4. Decrease the frequency at which the task is performed. This will increase recovery time between exertions and help to prevent fatigue and possible WRMSDs
5. Reduce the horizontal distance between the workpiece and the worker
6. For work with hand tools performed while the operator is standing, position the workpiece at a vertical height of approximately 114 cm. Ideally, the location of the workpiece should be adjustable to accommodate different operators

47.4   Advantages and Disadvantages of the Psychophysical Approach

Like any approach to setting limits for manual work, there are advantages and disadvantages of the psychophysical approach. This approach is only one of several approaches available for designing MMH and UEI tasks, and the advantages and disadvantages of each approach should be examined to determine the best fit for a particular situation. The advantages and disadvantages of the psychophysical approach to the design of MMH and UEI tasks are given in the following text. When necessary, a distinction is made if the advantage or disadvantage is specific to MMH or UEI tasks.

The advantages of the psychophysical approach include:

- Psychophysics allows for the realistic simulation of industrial work[2]
- Currently, there is a considerable amount of psychophysical data for MMH tasks available that were collected from industrial workers. Many physiological models were developed from limited

samples of university students. Likewise, the representativeness of cadaver data used to set certain biomechanical criteria such as lumbosacral compression limits is questionable

- Psychophysical results are consistent with the industrial engineering concept of a "fair day's work for a fair day's pay"[2]
- Psychophysics can be used to study intermittent MMH tasks which are common in industry.[2] Such tasks are not amenable to physiological analyses
- Psychophysical results are very reproducible[2]
- For MMH tasks, psychophysical judgments take into account the whole job, and integrate biomechanical and physiological factors[52,53]
- Psychophysical results for MMH tasks appear to be related to low-back pain[1,2,27]
- For MMH tasks, psychophysical data apply to a wider array of tasks than either the biomechanical or physiological approach[54]
- For MMH tasks that must necessarily be performed under postural restrictions (i.e., maintenance work and mining), psychophysics is one technique that can be used to develop handling limits specific to the tasks being examined[54]
- The psychophysical approach is less costly and less time consuming to apply in industry than many of the biomechanical and physiological techniques[54]
- Currently, psychophysical data represent one of the only quantitative guides for the design of force limits for UEI work. In the absence of objective biomechanical or physiological criteria, psychophysics may be used to elicit acceptable task parameters for UEI work[8]

The disadvantages and limitations of the psychophysical approach include:

- Psychophysics is a subjective method[2]
- The assumption that the subjective workloads selected by subjects are below the threshold for injury has not been validated.[55] There is not extensive epidemiological support for psychophysical data for MMH tasks and no epidemiological support for using psychophysical data for the design of UEI data. However, the same is true of most of the other criteria currently in use for designing manual work
- Psychophysical results for high-frequency MMH tasks exceed energy expenditure criteria[2]
- Some psychophysical values for MMH tasks may violate the biomechanical spinal compression criterion of 3400 N.[56] However, this assumes that the spinal compression criterion of 3400 N is correct, for which there is not much support
- Psychophysics does not appear to be sensitive to bending and twisting while performing MMH tasks, both of which have been related to compensable low-back pain cases[2]
- The range of data for designing UEI tasks is somewhat limited at this time

47.5   Conclusions

Psychophysical data are one option available to the ergonomist for designing manual tasks and assessing whether or not a task or set of tasks needs to be redesigned. These data have been applied in the workplace for many years with considerable success. As with any assessment tool, there are advantages and limitations associated with psychophysics as discussed in the previous section. When used properly, psychophysical data provide the analyst with a tool applicable to a diverse set of tasks involving manual work.

References

1. Snook, S. H., The design of manual handling tasks, *Ergonomics*, 21, 963–985, 1978.
2. Snook, S. H., Psychophysical considerations in permissible loads, *Ergonomics*, 28, 327–330, 1985.
3. Snook, S. H., Future directions of psychophysical studies, *Scandinavian Journal of Work, Environment & Health*, 25, 13–18, 1999.

4. Ayoub, M. M. and Mital, A., *Manual Materials Handling*, Taylor & Francis, London, 1989.
5. Ayoub, M. M., Problems and solutions in manual materials handling: the state of the art, *Ergonomics*, 35, 713–728, 1992.
6. Mital, A., Nicholson, A. S, and Ayoub, M.M., *A Guide to Manual Materials Handling*, Taylor & Francis, London, 1993.
7. Nicholson, L. M., A comparative study of methods for establishing load handling capabilities, *Ergonomics*, 32, 1125–1144, 1989.
8. Kim, C. H., Marley, R. J., Fernandez, J. E., and Klein, M. G., Acceptable work limits for the upper extremities with the psychophysical approach, in *Proceedings of the 3rd Pan Pacific Conference on Occupational Ergonomics*, 1994, 312–316.
9. Fernandez, J. E., Fredericks, T. K., and Marley, R. J., The psychophysical approach in upper extremities work, in *Contemporary Ergonomics 1995*, Robertson, S. A., Ed., Taylor & Francis, London, 1995, 456–461.
10. Stevens, S. S., The psychophysics of sensory function, *American Scientist*, 48, 226–253, 1960.
11. Ljungberg, A. S., Gamberale, F., and Kilbom, Å., Horizontal lifting — physiological and psychological responses, *Ergonomics*, 25, 741–757, 1982.
12. Gamberale, F., Ljungberg, A. S., Annwall, G., and Kilbom, Å., An experimental evaluation of psychophysical criteria for repetitive lifting work, *Applied Ergonomics*, 18, 311–321, 1987.
13. Snook, S. H. and Ciriello, V. M., Maximum weights and work loads acceptable to female workers, *Journal of Occupational Medicine*, 16, 527–534, 1974.
14. Snook, S. H., Vaillancourt, D. R., Ciriello, V. M., and Webster, B. S., Psychophysical studies of repetitive wrist flexion and extension, *Ergonomics*, 38, 1488–1507, 1995.
15. Emanuel, I., Chaffee, J. W., and Wing, J, A study of human weight lifting capabilities for loading ammunition into the F-86H aircraft, WADC Technical Report 56–367, Wright Air Development Center, Wright-Patterson Air Force Base, Ohio, 1956.
16. Switzer, S. A., Weight-lifting capabilities of a selected sample of human subjects, Technical Document Report No. MRL-TDR-62-57, Aerospace Medical Research Laboratories, Wright-Patterson Air Force Base, Ohio, 1962.
17. Snook, S. H. and Irvine, C. H., The evaluation of physical tasks in industry, *American Industrial Hygiene Association Journal*, 27, 228–233, 1966.
18. Ciriello, V. M., Snook, S. H., and Hughes, G. J. Further studies of psychophysically determined maximum acceptable weights and forces, *Human Factors*, 35, 175–186, 1993.
19. Ciriello, V. M., Snook, S. H., Blick, A. C., and Wilkinson, P. L., The effects of task duration on psychophysically-determined maximum acceptable weights and forces, *Ergonomics*, 33, 187–200, 1990.
20. Snook, S. H. and Ciriello, V. M., The design of manual handling tasks: revised tables of maximum acceptable weights and forces, *Ergonomics*, 34, 1197–1213, 1991.
21. Ayoub, M. M., Bethea, N., Deivanayagam, S., Asfour, S., Bakken, G., Liles, D., Selan, J., and Sherif, M., Determination and modeling of lifting capacity, Final Report, HEW (NIOSH) Grant No. 5R01OH00545-02, 1978.
22. Mital, A., Comprehensive maximum acceptable weight of lift database for regular 8-hour work shifts, *Ergonomics*, 27, 1127–1138, 1984.
23. Mital, A., Maximum weights of lift acceptable to male and female industrial workers for extended work shifts, *Ergonomics*, 27, 1115–1126, 1984.
24. Fox, R. R., A psychophysical study of high-frequency lifting, Unpublished doctoral dissertation, Texas Tech University, Lubbock, Texas, 1993.
25. Nicholson, L. M. and Legg, S. J., A psychophysical study of the effects of load and frequency upon selection of workload in repetitive lifting, *Ergonomics*, 29, 903–911, 1986.
26. Snook, S. H. and Irvine, C. H., Maximum frequency of lift acceptable to male industrial workers, *American Industrial Hygiene Association Journal*, 29, 531–536, 1968.
27. Liles, D. H., Deivanayagam, S., Ayoub, M. M., and Mahajan, P., A job severity index for the evaluation and control of lifting injury, *Human Factors*, 26, 683–693 1984.

28. Mital, A., The psychophysical approach in manual lifting — a verification study, *Human Factors*, 25, 485–491, 1983.

29. Smith, J. L., Ayoub, M. M., and McDaniel, J. W., Manual materials handling capabilities in non-standard postures, *Ergonomics*, 35, 807–831, 1992.

30. Gallagher, S., Acceptable weights and physiological costs of performing combined manual handling tasks in restricted postures, *Ergonomics*, 34, 939–952, 1991.

31. Gallagher, S. and Hamrick, C.A., Acceptable workloads for three common mining materials, *Ergonomics*, 35, 1013–1031, 1992.

32. Straker, L. M., Stevenson, M. G., and Twomey, L. T., A comparison of risk assessment of single and combination manual handling tasks: 1. Maximum acceptable weight measures, *Ergonomics*, 39, 128–140, 1996.

33. Mital, A., Using "A Guide to Manual Materials Handling" for designing/evaluating multiple activity manual materials handling tasks, in *Proceedings of the IEA World Conference on Ergonomic Design, Interfaces, Products, and Information*, 1995, 550.

34. Jiang, B.C. and Mital, A., A procedure for designing/evaluating manual materials handling tasks, *International Journal of Production Research*, 24, 913–925, 1986.

35. Putz-Anderson, V. (Ed.), *Cumulative Trauma Disorders: A Manual for Musculoskeletal Disorders of the Upper Limb*, Taylor and Francis, London, 1988.

36. Potvin, J. R., Chiang, J., McKean, C., and Stephens, A., A psychophysical study to determine acceptable limits for repetitive hand impact severity during automotive trim installation, *International Journal of Industrial Ergonomics*, 26, 625–637, 2000.

37. Davis, P. J. and Fernandez, J. E., Maximum acceptable frequencies for females performing a drilling task in different wrist postures, *Journal of Human Ergology*, 23, 81–92, 1994.

38. Fernandez, J. E., Dahalan, J. B., and Klein, M. G., Using the psychophysical approach in hand-wrist work, in *Proceedings of the M.M. Ayoub Occupational Ergonomics Symposium*, Institute for Ergonomics Research, Lubbock, Texas, 1993, 63.

39. Kim, C. H. and Fernandez, J. E., Psychophysical frequency for a drilling task, *International Journal of Industrial Ergonomics*, 12, 209–218, 1993.

40. Vaidyanathan, V. and Fernandez, J. E., MAF for males performing drilling tasks, in *Proceedings of the Human Factors Society 36th Annual Meeting*, Human Factors Society, Santa Monica, California, 1992, 692–696.

41. Fredericks, T. K., The effect of vibration on maximum acceptable frequency for a riveting task, Unpublished doctoral dissertation, The Wichita State University, Wichita, Kansas, 1995.

42. Fredericks, T. K. and Fernandez, J. E., The effect of vibration on maximum acceptable frequency for a riveting task, in *Proceedings of the Konz/Purswell Occupational Ergonomics Symposium*, Institute for Ergonomics Research, Lubbock, Texas, 1995, 27.

43. Dahalan, J. B. and Fernandez, J. E., Psychophysical frequency for a gripping task, *International Journal of Industrial Ergonomics*, 12, 219–230, 1993.

44. Klein, M. G. and Fernandez, J. E., The effects of posture, duration, and force on pinching frequency, *International Journal of Industrial Ergonomics*, 20, 267–275, 1997.

45. Abu-Ali, M., Purswell, J. L., and Schlegel, R. E., Psychophysically determined work-cycle parameters for repetitive hand gripping, *International Journal of Industrial Ergonomics*, 17, 35–42, 1996.

46. Ciriello, V. M., Snook, S. H., Webster, B. S., and Dempsey, P. G., A psychophysical study of six hand movements. *Ergonomics*, 44, 922–936, 2001.

47. Ciriello, V. M., Webster, B. S., and Dempsey, P. G., Maximum acceptable torques of highly repetitive screw driving, ulnar deviation, and handgrip tasks for seven hour workdays. *American Industrial Hygiene Association Journal*, 63, 594–604, 2002.

48. Krawczyk, S., Armstrong, T. J., and Snook, S. H., Preferred weights for hand transfer tasks for an eight hour day, in *Proceedings of International Scientific Conference on Prevention of Work-related Musculoskeletal Disorders*, Hagberg, M., and Kilbom, Å., Eds., 1992, 157.

49. Ulin, S. S., Armstrong, T. J., Snook, S. H., and Franzblau, A., Effect of tool shape and work location on perceived exertion for work on horizontal surfaces, *American Industrial Hygiene Association Journal*, 54, 383–391, 1993.

50. Ulin, S. S., Armstrong, T. J., Snook, S. H., and Keyserling, W. M., Examination of the effect of tool mass and work postures on perceived exertion for a screw driving task, *International Journal of Industrial Ergonomics*, 12, 105–115, 1993.

51. Ulin, S. S., Snook, S. H., Armstrong, T. J., and Herrin, G. D., Preferred tool shapes for various horizontal and vertical work locations, *Applied Occupational and Environmental Hygiene*, 7, 327–337, 1992.

52. Haslegrave, C. M. and Corlett, E. N., Evaluating work conditions for risk of injury–techniques for field surveys, in *Evaluation of Human Work*, 2nd. ed., Wilson, J. R., and Corlett, E. N., Eds., Taylor & Francis, London, 1995, 892–920.

53. Karwowski, W. and Ayoub, M. M., Fuzzy modelling of stresses in manual lifting tasks, *Ergonomics*, 27, 641–649, 1984.

54. Ayoub, M. M. and Dempsey, P. G., The psychophysical approach to manual materials handling task design, *Ergonomics*, 42, 17–31, 1999.

55. Gamberale, F. and Kilbom, Å., An experimental evaluation of psychophysically determined maximum acceptable workload for repetitive lifting work, in *Ergonomics International 88*, Adams, A. S., Hall, R. R., McPhee, B. J., and Oxenburgh, M. S., Eds., Ergonomics Society of Australia, Sydney, 1988, 233–235.

56. Chaffin, D. B. and Page, G. B., Postural effects on biomechanical and psychophysical weight-lifting limits, *Ergonomics*, 37, 663–676, 1994.

48

Static Biomechanical Modeling in Manual Lifting

Don B. Chaffin
Charles B. Woolley
University of Michigan

48.1 Introduction

Though most manual tasks in industry involve significant body motions, it continues to be very helpful to evaluate specific exertions within a manual task by performing a static biomechanical analysis. Such analyses are normally performed by combining the postural information (body angles) obtained from a stopped frame video image (or photograph) of a worker, and measured forces exerted at the hands. The latter is often obtained with a simple handheld force gauge.

What follows is a description of a computerized static biomechanical model that has been developed and used over the last 30 yr to predict:

1. The percentage of men and women who would be capable of exerting specified hand forces in various work postures
2. The forces acting on various spinal motion segments

Since these two different output predictions have specific criterion values referenced in the NIOSH Work Practices Guide (NIOSH, 1981), they are often used by professional ergonomists to determine the relative risk of injury associated with the performance of a manual exertion of interest (Chaffin, 1988a). It also should be noted that the prediction of the percent of the population capable of performing a specific exertion required on a job is often crucial to the determination of a job specific strength test score for pre-employment and return to work purposes (Chaffin, 1996). Finally, because the biomechanical population strengths and low back stresses are predicted by a computerized model which runs on common personal computer platforms, this has meant that job and product designers and engineers have been able to simulate various expected high exertion tasks during the early part of the design process, and thus avoid costly prototype evaluations and retrofits when the products and processes become operational (Chaffin, 2001). It is this latter application of the biomechanical static strength prediction programs that provides perhaps the greatest potential benefit over other common job evaluation methods. Many other methods require a person to be observed and measured, sometimes with expensive instrumentation. This precludes the use of these empirical methods for use in prescriptive job design; where in the job exists only on paper or in a computer rendered drawing of the workspace. By interfacing a computerized biomechanical model of a person (as described in the following) into a computerized rendering of the workspace, the designer can quickly perform a large number of simulated exertions to determine the human consequences of altering a proposed job design, much like that being done to accommodate various sized individuals using computerized anthropometric manikins.

What follows is a brief description of the development of a static biomechanical strength modeling technology, including a couple illustrations of how it has been used to evaluate various manual lifting situations.

48.2   Development of Static Strength Prediction Programs

The general logic used to predict population static strengths in various jobs is depicted in Figure 48.1. In this model, specific muscle group strength data and spinal vertebrae failure data are used as the limiting values for the reactive moments at various body joints created when a person of a designated stature and body weight attempts an exertion (i.e., lifts, pushes, or pulls in a specific direction with one or both hands while maintaining a known posture).

This logic has been well described for a sagittal, coplanar static strength analysis in Chaffin et al. (1999). When wishing to perform an analysis in three dimensions, the body is represented as a set of links with known mass, as depicted in Figure 48.2. The load moments M_j are computed by the cross products of the unit distance vectors to each joint and the respective body segment weight and hand force vectors.

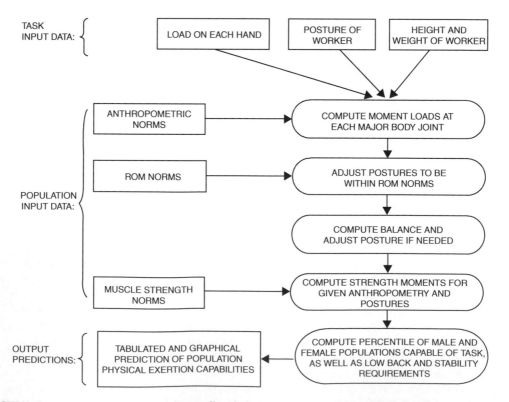

FIGURE 48.1   Biomechanical logic used to predict whole-body static exertion capabilities for given postures, hand force directions, and anthropometric groups.

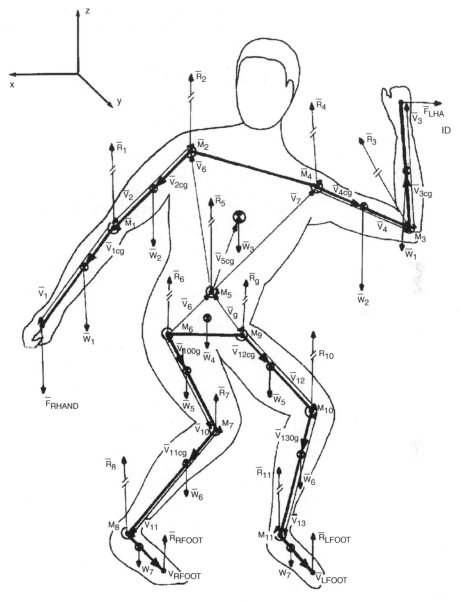

FIGURE 48.2 Three-dimensional distance, force, and moment vectors used in 12 link biomechanical model for strength prediction. (From Chaffin, D.B. and Erig, M. 1991. *IIE Trans.* 23(3):216–227. With permission.)

The static moment equilibrium equations for the elbow and shoulder in this linkage can be defined as:

$$\bar{F}_{R\,HAND} = F_{RX}i + F_{RY}j + F_{RZ}k$$

where $\bar{F}_{R\,HAND}$ is the right hand force with X, Y, Z unit vector (i, j, k) components for each, and F_{RX}, F_{RY}, and F_{RZ} are, respectively, the X, Y, and Z components of the right hand force.

$$\bar{V}_1 = (V_{1X}i + V_{1Y}j + V_{1Z}k) \cdot (\text{Lelink})$$

where $\bar{V}_1$ is the forearm length with link unit vectors, V_{1X}, V_{1Y}, V_{1Z} are, respectively, the X, Y, Z components of the forearm unit vector, and LeLink is the magnitude of forearm length from anthropometric data.

$$\overline{V_{1cg}} = (V_{1X}i + V_{1Y}j + V_{1Z}k) \cdot (\text{cglink})$$

where V_{1cg} is the cg distance from elbow to forearm center of gravity vector expressed in unit vector form, multiplied by the cglink, which is the magnitude of proximal distance to cg of forearm from anthropometric data.

$$\overline{M}_1 = (M_{1X}i + M_{1Y}j + M_{1Z}k)$$

where $\overline{M}_1$ is the elbow resultant moment with X, Y, Z unit vector components, and M_{1X}, M_{1Y}, M_{1Z}, are, respectively, the elbow moment about X, Y, Z axes and

$$\overline{M}_1 = (\overline{V}_1 * \overline{F}_{R\,HAND} + \overline{V}_{1cg} * \overline{W}_1)$$

where $\overline{W}_1 = 0i + 0j - W_{1Z}k$, (which is the forearm weight Vector), and

$$\overline{R}_1 = (R_{1X}i + R_{1Y}j + R_{1Z}k)$$

where $\overline{R}_1$ is the elbow joint reaction force vector with X, Y, Z unit vector components, and

$$\overline{M}_2 = \overline{M}_1 + \overline{V}_{2cg} * \overline{W}_2 + \overline{V}_2 * (-\overline{R}_1)$$

where $\overline{M}_2$ is the right shoulder resultant moment with $\overline{V}_{2cg}$ the upperarm center of gravity vector, $\overline{V}_2$ the upperarm link vector, and $\overline{W}_2$ the upperarm weight vector.

$$\overline{M}_2 = M_{2X}i + M_{2Y}j + M_{2Z}k$$

where $\overline{M}_{2X}$ is the right shoulder movement with X, Y, Z, unit vector components, $\overline{M}_{2X}$, $\overline{M}_{2Y}$, and $\overline{M}_{2Z}$, are respectively, the shoulder moment about the X, Y, and Z axes.

A recursive computational procedure is used to continue the analysis to compute external load moments and forces at the elbow and shoulder of the arm or arms doing the exertion, the lumbosacral joint, hip joints, and knee and ankle joints.

The size and mass of the person (linkage size) is most often specified as a select stratum of the population (i.e., a percentile of specific anthropometric dimensions is selected from population surveys). Thus, a small, medium, or large man or woman can be specified, or specific link anthropometry can be used if available. Link length-to-stature ratios from Drillis and Contini (1966) and link mass-to-bodyweight ratios from Dempster (1955) and Clauser et al. (1969) are used to simplify this procedure, if specific anthropometry is not available on a subject. Most often an average male or female anthropometry is chosen for assessing the strength requirements of a given task in industry.

The strength moment values used as population limit values in the program were measured by Stobbe (1982) for 25 men and 22 women employed in manual jobs in three different industries. These values have been combined with the earlier values from Chaffin and Baker (1970) and Schanne (1972) to form the statistical data for the population joint moment limits.

Once the size of the person has been specified or selected from a known anthropometric data source, the posture is entered with reference to either photographs or videos (or by manipulating a computer generated hominoid) and then the hand forces of interest are entered. The program then computes the load moments at each joint of the linkage, and compares each to the corresponding strength moment capability obtained from the previously measured populations. This provides a prediction of the percent of the population that is capable of producing the necessary strength moments at each joint.

The logic for computing the lumbar motion segment compression force is shown in Figure 48.3. Once the lumbar moment is computed (as described in the preceding), torso muscle contraction, which stabilizes the column, is estimated. Two models of torso muscle contraction will be described. In the sagittal plane low-back model by, Chaffin (1975), a single equivalent torso muscle contraction force is implemented. When the necessary reactive torso muscle force is added to body segment weights and hand forces (with a minor adjustment for abdominal pressure effects) a prediction of the compression force on the L5/S1 disc results, as shown in Figure 48.4.

When an asymmetric exertion (e.g., one-handed force, or twisted or laterally bent torso) is being analyzed, many different torso muscle actions and passive supporting tissue reactions need to be considered (Chaffin 1988b). The first step in such a procedure requires that the position, orientation, cross-sectional size, and length of the various connective tissues be modeled at the lumbar spinal level. A geometric torso model proposed by Nussbaum and Chaffin (1996) for this purpose is shown in Figure 48.5. This model includes estimates of specific tissue geometry acquired from various computed tomography (CT) scans (Tracy et al., 1989; Chaffin et al., 1990; Moga, et al., 1993), along with passive tissue reaction forces estimated by McCully and Faulkner (1983), Nachemson et al. (1979), Miller et al. (1986), and others.

The most important predictors of spinal column stress, however, are the muscle reaction forces required to stabilize the spine to external load moments. In three-dimensional (3D) torso models various approaches have been used to predict these required reactive muscle forces. Perhaps the most commonly cited torso biomechanical model for 3D static analysis is that developed by Schultz and

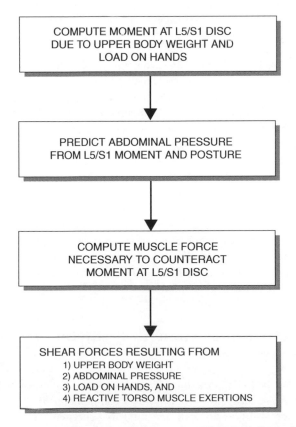

FIGURE 48.3   Logic for computing L5/S1 compression forces in 2D/3D Static Strength Prediction Program.

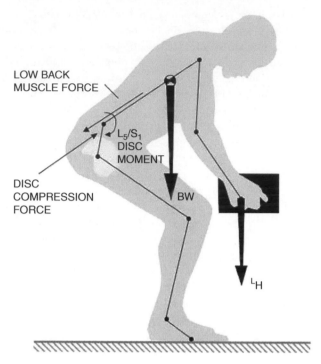

FIGURE 48.4 Simple low-back model of lifting for static coplanar lifting analyses. The load on the hands L_H and torso and arm weights BW act to create moments at the L5/S1 disc of the spine. The moments are resisted primarily by the back muscles. The high muscle forces required in such a task cause high disc compression forces.

Andersson (1981). It is depicted in Figure 48.6. Bean et al. (1988) developed a revised version of this model that provides a more efficient computational method for solving the linear programs used to simultaneously minimize the torso muscle contraction intensities and motion segment compression forces. The present 3D low-back model included in the computerized version described in the following text predicts the minimum muscle force contractile intensities required to meet the moment equilibrium requirements about the three orthogonal axes of rotation of the motion segment. Given a set of optimal forces so computed, the model further seeks to minimize the disc compression force. Because such an approach attempts to minimize *both* muscle intensity requirements and disc compression forces simultaneously, it is referred to as a "double linear optimization" approach.

Hughes and Chaffin (1995) proposed that a nonlinear objective function be used as the basis for selecting the various muscle reaction forces during a given exertion. They referred to this as the sum of the cubed muscle intensity objective. Nussbaum et al. (1996) also have proposed a neural network model to predict torso muscle actions. Raschke and Chaffin (1996) have proposed that the external moment is normally distributed about the torso, and activates several muscles simultaneously depending on the direction and magnitude of the external moment.

48.3 Computerization of Strength Prediction and Back Force Prediction Models

It should be clear from the preceding descriptions that the biomechanical models used for population strength and spinal motion segment force prediction are computationally intense, especially in the 3D form. For this reason a number of faculty, staff, and students associated with the Center for Ergonomics at the University of Michigan have worked to provide user-friendly, computer programs of the models.

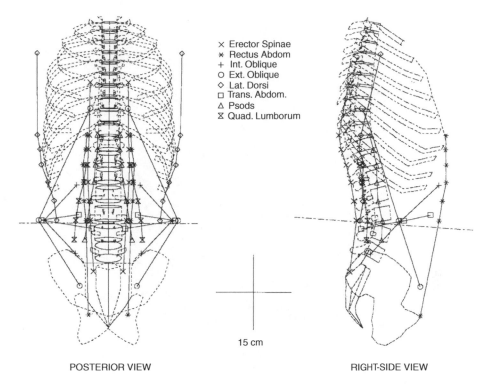

× Erector Spinae
∗ Rectus Abdom
+ Int. Oblique
○ Ext. Oblique
◇ Lat. Dorsi
□ Trans. Abdom.
△ Psods
⊠ Quad. Lumborum

15 cm

POSTERIOR VIEW   RIGHT-SIDE VIEW

FIGURE 48.5   Muscle geometry illustrated for a 50th percentile male. Muscles are treated as point-wise connections from origin to insertion (see text). An imaginary cutting plane, which bisects the L3/L4 motion segment, is also shown. (From Nussbaum, M.A. and Chaffin, D.B. 1996. *Clin. Biomech.* 11(1):25–34. With permission.)

These have been referred to as the University of Michigan's Two-Dimensional and Three-Dimensional Static Strength Prediction Programs™ (i.e., 2DSSPP™ and 3DSSPP™). Currently only one version is available, 3DSSPP, which includes both the sagittal plane and 3D low-back models as well as the strength model. The University of Michigan's Office of Technology Transfer has granted over 2500 individual licenses for use of these programs since 1984.

The 3DSSPP program requires more input data than the previous 2D version. 3D exertions often involve two hand forces, which can act in any direction. Also, a model of the human body in 3D has 12 body links (some with three postural angles). The main window of the 3DSSPP program is depicted in Figure 48.7. The input values (posture, hand forces, and anthropometry) are entered in dialogs available from the pull-down menus. Postures can be entered manually by specifying body link angles, by specifying hand locations and using the posture prediction feature, or graphically by selecting and dragging joints using a mouse entry device. The posture prediction feature uses an inverse kinematic model with preferred postural prediction capability and is included to allow the user to easily manipulate the figure. Orthogonal stick figures depicting the body posture, the hand location, and the hand force directions are provided across the top of the window. An oblique-view enfleshed hominoid, which along with the stick figures assist with the entry and adjustment of the posture, is provided in the lower-left area of the window and can be manipulated to appear in the same orientation as a photo, video, or digital image. The use of the 3D hominoid was found by Beck and Chaffin (1992) to allow postures to be accurately entered and represented in a computer.

Summary analysis results, including the predicted percent of the population having sufficient strength to perform the designated exertion, back compression, and balance status, are shown in tabular form in the lower right quadrant of the main window. From inspection of the percent capable predictions shown in Figure 48.7 for the analysis of a 50th percentile male lifting a 44-lb stock reel, it is obvious that hip

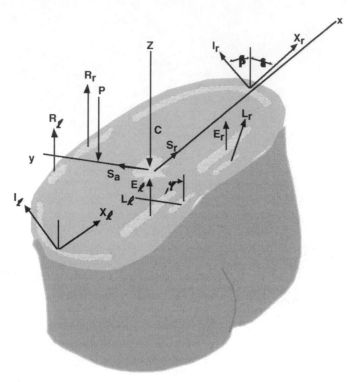

FIGURE 48.6   Schematic of 10-muscle model. (developed by Schultz and Andersson 1981. Analysis of loads on the lumbar spine. *Spine* 6(1): 76–82. With permission.)

strength is the most limiting muscle group strength (only 86% of men have sufficient hip strength to lift the reel). This value is below that recommended by NIOSH, which proposes that jobs should accommodate 99% of men's and 75% of women's strength (or 90% of a mixed gender population). The estimated 3D low-back compression force of 969 lbs also is above the 760 lbs recommended by NIOSH. Two balance analyses are performed: possible static equilibrium of moments at the ankles and an estimated center of pressure within the basis of support on the floor. The balance status indicates that the task as depicted provides acceptable balance.

Analysis output windows for these and other calculated results are available, and two are shown in Figure 48.8. These depict percent capable strength predictions (for 34 muscle functions) and lumbar compression forces at the L4L5 disc; the latter includes 3D predictions of a large number of individual muscle and spinal forces.

48.4   Validation of Strength and Back Force Prediction Models

The validation of the static strength predictions has been accomplished in three different studies. All three validations required using the models to simulate whole-body exertions and compare the percent capable predictions with the mean, 10, and 90 percentile strengths of a group of volunteers who performed the same tasks.

In the first validation Garg and Chaffin (1975) had 71 male Air Force personnel perform 38 different maximum arm exertions (i.e., lifts, pushes, pulls, etc.) in a variety of arm/torso postures while seated. They found the predicted strengths were highly correlated with the group strengths when performing the 38 upper-body tasks ($r = 0.93$–0.97). Chaffin et al. (1987) simulated 15 different whole-body exertions in the sagittal plane, which were also performed by both men and women from a variety of industries. In some of these tests over 1000 people performed the exertions, though on average about 200 people

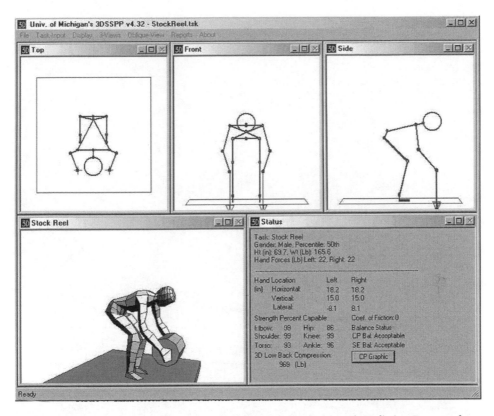

FIGURE 48.7 Main screen from the University of Michigan 3D Static Strength Prediction Program for personal computers shown lifting a 44-lb stock reel. (With permission, Regents of the University of Michigan.)

performed each one. Comparison of the strength prediction program with the group strength data revealed a very high correlation ($r = 0.92$) for sagittal plane symmetric exertions. This same study also included asymmetric simulations with the strength prediction program of 72 different one-arm exertions performed by five male Army personnel. The correlations ranged from $r = 0.71$ to $r = 0.83$. Unfortunately, in this latter comparison, exact postural and bracing conditions were not available to use in the simulations. This may have contributed to the lower correlations.

The last validation involved simulations of 56 one- and two-handed, whole-body exertions in 14 different symmetric, bent, and twisted-torso postures (Chaffin and Erig, 1991). The simulation results were compared with the group strengths of 29 young males. Photographs from several views were available to assist in replicating the postures used by these subjects. The results indicated that if care is taken to assure that the posture used in the model simulation is the same as that chosen by people performing the exertions, the prediction error standard deviation will be less than 6%.

In conclusion, it appears that the strength prediction models and population norms used in the present models are accurate in predicting the percent of the population capable of performing a large variety of different types of maximal static exertions. One caution should be noted, however. At present the strength norms used as limits in the models are based on male and female populations who are relatively young (i.e., 18–49 yr). To improve the models further, strength values are currently being gathered on older populations by these investigators. In this regard, one comparison involving 98 men and women with a mean age of 73 yr, showed a major decrease in strength performance in certain muscle functions. When these decreases were included in the 3DSSPP population data base, it was found that some exertions that could easily be performed by younger people were predicted to be impossible to perform by most older people (Chaffin et al., 1994).

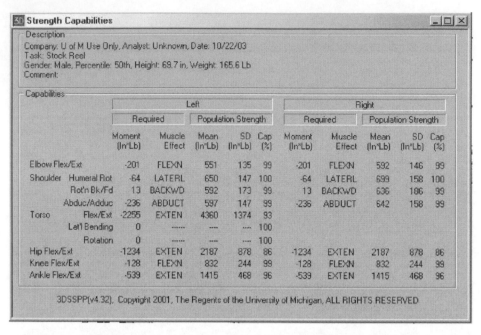

FIGURE 48.8 Analysis result widows from 3DSSPP™ for strength prediction capability and for 3D low-back compression. (With permission, Regents of the University of Michigan.)

Validation of the low back biomechanical model has been largely dependent on EMG estimates of muscle reactions in subjects performing controlled torso exertions. Hughes et al. (1994) discusses this procedure, and the results of comparisons with four different optimization procedures used to predict torso muscle responses to different external torso moment loads. Generally speaking, relatively high correlation ($r > 0.8$) are achieved when loading the torso approximately in the sagittal plane. With greater asymmetric or sudden loading, more complex muscle patterns result, sometimes with a 10–30%

antagonistic type of muscle response. These complex responses are often not well predicted ($r < 0.6$) by existing models. Thus, it is expected that existing models may underpredict the muscle-induced compression and shear forces on the spinal motion segments by as much as 30% during sudden (i.e., jerking) motions or lateral, asymmetric exertions. The newer neural network and geometric moment distribution models are yet to be thoroughly tested under complex loading conditions. They may be less sensitive to this cocontraction phenomenon than existing optimization models.

48.5 Final Comments

The development of biomechanically based, static strength and low back force prediction models provide powerful tools for performing ergonomic assessments of high physical effort jobs. With the advent of faster personal computers these models are gaining popularity not only as a means to evaluate existing manual tasks, but also to aid in the design of new workplaces and equipment, as well as in the specification of personnel selection and training programs.

The inclusion of inverse kinematics, behaviorally based posture prediction methods, improved human form graphics, and direct human image video input data, have made the use of these models relatively easy for the ergonomics practitioner. By simply clicking and pointing to parts of the body, one can adjust postures and have a comprehensive biomechanical assessment of a specific exertion (such as depicted in Figure 48.9).

With the widely recognized need to "design workplaces right the first time," computerized biomechanical strength prediction models will become even more useful. It is hoped that this presentation assists those interested in knowing more about both the scientific basis for this rapidly expanding technology, and the potential benefits and limitations inherent in it.

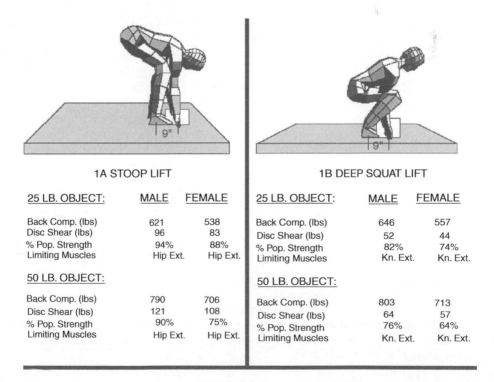

1A STOOP LIFT				1B DEEP SQUAT LIFT		
25 LB. OBJECT:	MALE	FEMALE		25 LB. OBJECT:	MALE	FEMALE
Back Comp. (lbs)	621	538		Back Comp. (lbs)	646	557
Disc Shear (lbs)	96	83		Disc Shear (lbs)	52	44
% Pop. Strength	94%	88%		% Pop. Strength	82%	74%
Limiting Muscles	Hip Ext.	Hip Ext.		Limiting Muscles	Kn. Ext.	Kn. Ext.
50 LB. OBJECT:				50 LB. OBJECT:		
Back Comp. (lbs)	790	706		Back Comp. (lbs)	803	713
Disc Shear (lbs)	121	108		Disc Shear (lbs)	64	57
% Pop. Strength	90%	75%		% Pop. Strength	76%	64%
Limiting Muscles	Hip Ext.	Hip Ext.		Limiting Muscles	Kn. Ext.	Kn. Ext.

FIGURE 48.9 Comparison of two different postures used to lift 25- and 50-lb objects from the floor close to the feet using the Michigan 3DSSPP™.

Acknowledgments

We wish to acknowledge NIH Grant AR-39599 for partial support of some of the work described in this presentation.

References

Bean, J.C., Chaffin, D.B., and Schultz A.B. 1988. Biomechanical model calculation muscle contraction forces: a double linear programming method. *J. Biomech.* 21(1): 59–66.

Beck, D.J. and Chaffin, D.B. 1992. An evaluation of inverse kinematics models for posture prediction. *Proceedings of CAES 1992*, Tampere, Finland. Center for Ergonomics, University of Michigan, Ann Arbor, MI.

Chaffin, D.B., 1975. On the validity of biomechanical models of the low back for weight lifting analysis. *ASME Proceedings.* 75-WA-Bio-1, New York: Amer. Soc. of Mech. Eng.

Chaffin, D.B., 1988a. A biomechanical strength model for use in industry. *Appl. Ind. Hyg.* 3(3): 79–86.

Chaffin, D.B., 1988b. Biomechanical modeling of the low back during load lifting. *Ergonomics* 31(5): 685–697.

Chaffin, D.B., 1996. Ergonomic basis for job-related strength testing. In *Disability Evaluations*, eds. S.L. Demeter, G.B.J. Andersson, and G.M. Smith, pp. 159–167. St. Louis, MO: American Medical Association, Mosby.

Chaffin, D.B. (ed.) 2001. *Digital Human Modeling for Vehicle and Workplace Design.* Warrendale, PA: Society of Automotive Engineers.

Chaffin, D.B. and Baker, W.H. 1970. A biomechanical model for analysis of symmetric sagittal plane lifting. *AIIE Trans.* 2(1): 16–27.

Chaffin, D.B. and Erig, M. 1991. Three-dimensional biomechanical static strength prediction model sensitivity to postural and anthropometric inaccuracies. *IIE Trans.* 23(3): 216–227.

Chaffin, D.B., Freivalds, A., and Evans, S.M. 1987. On the validity of an isometric biomechanical model of worker strengths. *IIE Trans.* 19(3): 280–288.

Chaffin, D.B., Andersson, G.B.J., and Martin, B.J. 1999. *Occupational Biomechanics*, 3rd ed. New York: John Wiley & Sons, Inc.

Chaffin, D.B., Redfern, M.S., Erig, M., and Goldstein, S.A. 1990. Lumbar muscle size and location measurements from CT scans of 96 older women. *Clin. Biomech.* 5(1): 9–16.

Chaffin, D.B., Woolley, C.B., Buhr, T., and Verbrugge, L. 1994. Age effects in biomechanical modeling of static lifting strengths. *Proceedings of Human Factors Society*, Nashville, TN.

Clauser, C.E., McConville, J.T., and Young, J.W. 1969. Weight, volume and center of mass of segments of the human body. AMRL-TR-69-70, Aerospace Med. Res. Lab., OH.

Dempster, W.T. 1955. Space requirements of the seated operator. WADC-TR-55-159, Aerospace Med. Res. Lab. Wright-Patterson AFB, OH.

Drillis, R. and Contini, R. 1966. *Body Segment Parameters.* BP-174-945, Tech. Rep. No. 1166.03, S. of Eng. and Sci., NYU, NY.

Garg, A.D. and Chaffin, D.B. 1975. A biomechanic computerized simulation of human strength. *AIIE Trans.* 14: 272–280.

Hughes, R.E. and Chaffin, D.B. 1995. The effect of strict muscle stress limits on abdominal muscle force predictions for combined torsion and extension loadings. *J. Biomech.* 28(5): 527–533.

Hughes, R.E., Chaffin, D.B., Lavender, S.A., and Andersson, G.B.J. 1994. Evaluation of muscle force prediction models of the lumbar trunk using surface electromyography. *J. Orthop. Res.* 12: 689–698.

McCully, K.K. and Faulkner, J.A. 1983. Length–tension relationship of mamallian diaphragm muscles. *J. Appl. Physiol.* 54: 1681–1686.

Moga, P.J., Erig, M., Chaffin, D.B., and Nussbaum, M.A. 1993. Torso muscle moment arms at intervertebral levels T10 through L5 from CT scans on eleven male and eight female subjects. *Spine* 18(15): 2305–2309.

Miller, J.A.A., Schultz, A.B., Warwick, D.N., and Spencer, D.L. 1986. Mechanical properties of lumbar spine motion segments under large loads. *J. Biomech.* 19: 79–84.

Nachemson, A.L., Schultz, A.B., and Berkson, M.H. 1979. Mechanical properties of human lumbar spine motion segments: influences of age, sex, disc level, and degeneration. *Spine* 4: 1–8.

National Institute for Occupational Safety and Health 1981. *Work Practices Guide for Manual Lifting*, Technical Report No. 81-122, U.S. Dept. of Health and Human Services (NIOSH), Cincinnati, OH.

Nussbaum, M.A. and Chaffin, D.B. 1996. Development and evaluation of a scalable and deformable geometric model of the human torso. *Clin. Biomech.* 11(1): 25–34.

Nussbaum, M.A., Chaffin, D.B., and Martin, B.J. 1996. A back-propagation neural network model of lumbar muscle recruitment during moderate static exertions. *J. Biomech.* 28(9): 1015–1024.

Raschke, U. and Chaffin, D.B. 1996. Trunk and hip muscle recruitment in response to external anterior lumbosacral shear and moment loads. *Clin. Biomech.* 11(3): 145–152.

Schanne, F.T. 1972. Three dimensional hand force capability model for a seated person. An unpublished Ph.D. dissertation, University of Michigan, Ann Arbor, MI.

Schultz, A.B. and Andersson, B.J.G. 1981. Analysis of loads on the lumbar spine. *Spine* 6(1): 76–82.

Stobbe, T.J. 1982. The development of a practical strength testing program in industry. An unpublished Ph.D. dissertation, University of Michigan, Ann Arbor, MI.

Tracy, M.F., Gibson, M.J., Szypryt, E.P., et al. 1989. The geometry of the muscles of the lumbar spine determined by magnetic resonance imaging. *Spine* 14: 186–193.

49

Industrial Lumbar Motion Monitor

William S. Marras
W. Gary Allread
The Ohio State University

49.1 Introduction

49.1.1 Occupational Back Injuries

Low back disorders (LBDs) are among the most commonly reported injuries. Reisbord and Greenland (1985) found that 18% of adults over age 25 annually report having frequent back pain. In occupational settings, Andersson (1997) reported that LBDs affect an estimated 80% of the population during their working career, and The National Center for Health Statistics (1977) has documented that LBDs are the prime reason for activity limitation in those 45 yr of age or younger. Guo et al. (1999) estimated that back pain accounts for 149 million lost workdays in the U.S. annually; 68% of these are associated specifically with work-related back pain. Cats-Baril (1996) has shown that LBDs cost society up to 100 billion dollars annually. Despite the prevalence and cost of these injuries, there are relatively few accurate methods available to predict the risk of occupationally related LBDs.

49.1.2   Tools for Analyzing Low Back Injury Risk

A wide range of tools is available to evaluate LBD risk in industrial jobs. These tools vary considerably in their applicability, complexity, and ease of use. When choosing the appropriate analysis technique it is important to match the capabilities of the tool to the characteristics of the job being evaluated.

Lifting equations, such as the NIOSH *Work Practices Guide*, and tables of acceptable load weight limits are easy to use. They require only a few measurements of the workplace and some observations of a lifting task to compute recommended lifting limits. However, they assume that all movements are slow and smooth. Two- and three-dimensional static biomechanical models also are readily available and can often be relatively easy to apply. However, these also do not take into account trunk motions and, thus, may not accurately reflect the job's level of risk.

Recent research has suggested that motion plays a role in LBD risk. Numerous epidemiological studies have specifically indicated that the risk of LBD increases when dynamic lifting occurs. Data from a retrospective study of over 4500 injuries (Bigos et al., 1986) found that there were greater reports of LBD with dynamic tasks as compared to awkward static tasks. Magora (1973) concluded that lateral bending and twisting were only significant risk factors when they occurred simultaneously with sudden (quick) movements. Punnet et al. (1991) studied non-neutral postures in automobile assembly plants and reported that postural stress to the back was more a function of dynamic than static tasks. All of these studies indicated that the risk of LBD increases with dynamic activity, especially when the body moves asymmetrically. Therefore, if a job being evaluated has a dynamic component it is important to choose an analysis method that accounts for the additional risk that motion imparts.

Video-based motion analysis systems offer one way to study the dynamic components of a job, but they have several drawbacks when used in an industrial setting. First, video assessments must take place in a calibrated space of usually no more than two to three cubic meters. Cameras must be carefully placed to obtain data for all three planes of motion, and time-consuming analysis is necessary to obtain usable data. In industrial environments, these motion analysis systems often are not practical, as tasks typically involve movement outside of the calibrated space and work areas often make setting up of cameras difficult. These limitations often make video-based systems impractical for routinely evaluating a large number of industrial jobs.

49.2   Development of the Industrial Lumbar Motion Monitor

49.2.1   Physical Description

The industrial lumbar motion monitor (iLMM) (Figure 49.1) was developed in the Biodynamics Laboratory at The Ohio State University's Institute for Ergonomics. Its development was in response to the need for a practical method of assessing the dynamic component of occupationally related LBD risk in industrial settings. The patented iLMM is a triaxial electrogoniometer that acts as a lightweight exoskeleton of the lumbar spine. It is positioned on the back of an individual directly in line with the spine and is attached using a waist belt at the pelvis and a harness worn over the shoulders. Four potentiometers at the base of the iLMM measure the instantaneous position of the spine (as a unit) in three-dimensional space relative to the pelvis. Position data from the potentiometers are recorded at 60 Hz, converted to a digital signal, and then recorded on a microcomputer. The data then are processed to calculate the position, velocity, and acceleration of the spine in each of the three planes of motion as a function of time.

49.2.2 Calibration and Measurement Accuracy

During data collection and analysis, each iLMM size is used with a matching calibration file. Before its first use, an iLMM is calibrated using a specially designed reference frame (Figure 49.2). During this calibration process, the iLMM is positioned on the frame, at 225 different positions in three-dimensional space, and the voltage outputs from each potentiometer are recorded. This calibration process eliminates any individual variability across iLMMs.

The iLMM was checked to ensure its accuracy and sensitivity using a video-based motion analysis system (Marras et al., 1992). During this measurement process, the predicted velocities and accelerations of the iLMM were compared relatively (to position accuracy) against the predicted values of the motion analysis system. The results of this process (Table 49.1) show high correlation coefficient values and significance ($r > 0.95$, $p < 0.0001$) for all three planes of motion. An independent group also has determined that the reproducibility of the iLMM is suitably high for range of motion and velocity, for the device to be used for evaluation in a clinical and research setting (Gill and Callaghan, 1996).

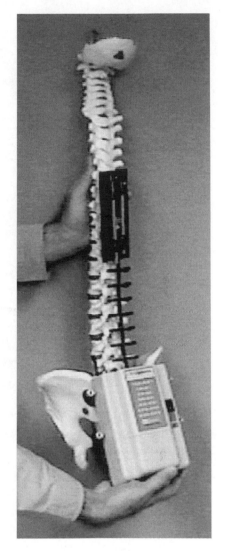

FIGURE 49.1 The industrial lumbar motion monitor (iLMM), compared with an anatomical model of the spine.

49.3 Development of the LBD Risk Model

An *in vivo* study was undertaken to determine quantitatively whether dynamic trunk motions in combination with workplace and environmental factors may better describe LBD risk in repetitive manual materials handling (MMH) tasks.

49.3.1 Approach

The initial study (Marras et al., 1993) involved an industrial surveillance of the trunk motions and workplace factors associated with repetitive MMH jobs having either high or low LBD rates. The approach used in this project was to: (1) identify industries having repetitive MMH production work; (2) examine the medical and health and safety records for these companies, to identify those repetitive MMH jobs that had high or a low LBD rates; (3) quantitatively monitor the trunk motions and

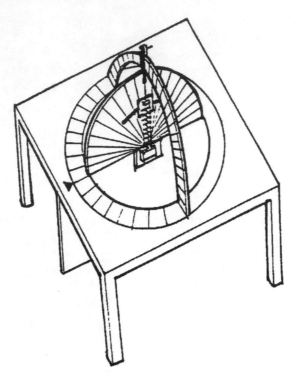

FIGURE 49.2   The iLMM's three-dimensional calibration and reference frame.

workplace factors associated with each of these jobs; and (4) evaluate the data, to determine which combination of gathered factors best distinguished between the high- and low-rate groups.

49.3.2   Study Design

This was a cross-sectional study of 403 industrial jobs from 48 manufacturing companies throughout the midwestern United States. Only repetitive jobs without job rotation were examined in this study. This was necessary to prevent the confounding effects created by alternating jobs. Jobs examined in this study were divided into two groups — having either high or low LBD rates — based on an examination of the company's injury and medical records. Whenever possible, company medical reports were used to categorize these rates, but in some cases only injury logs (i.e., OSHA 200 logs) were available. All medical reports, injury records, and logs were scrutinized to ensure they were as accurate as possible. The outcome measure derived from these records (LBD rate) consisted of the normalized incidence of

TABLE 49.1   Correlations of the Velocities and Accelerations of the Motion Analysis System in the Three Planes of Motion with the Velocities and Accelerations of the iLMM

Plane	Correlation*	
	Velocity	Acceleration
Lateral	0.95	0.95
Sagittal	0.99	0.96
Twisting	0.99	0.99

$^{*}p < 0.0001$.

reported occupationally related LBD. Occurrences of reported LBDs were considered regardless of the amount of restricted or lost time associated with the incident.

The dependent variable in this study consisted of two levels of job-related LBD rates. Low-rate group jobs were defined at those with at least 3 yr of records showing zero low-back injuries and zero turnover. Turnover was defined as the average number of employees leaving a job per year. High-rate group jobs were those associated with at least 12 injuries per 200,000 h of work exposure. (The average rate for this group was 26.0 injuries per 200,000 h.) The incidence rate for the high LBD group category corresponds to the 75th percentile value of risk for the 403 jobs examined. Of these jobs, 124 were categorized in the low injury rate group, and 111 were in the high-rate group. The remaining jobs (totaling 168) were in the medium injury rate group and were not used in this particular analysis.

The independent variables in this study consisted of workplace, individual, and trunk motion characteristics that were indicative of each job. The workplace and individual characteristics consisted of variables typically considered in current workplace guidelines for MMH (NIOSH, 1981; Putz-Anderson and Waters, 1991). Specifically, these variables were: (1) the maximum external horizontal distance of the load from the spine; (2) the weight of the object lifted; (3) the vertical height of the load at both the origin and destination of the lift; (4) the frequency of lifting (e.g., lift rate); (5) the asymmetric angle of the lift (as defined by NIOSH, 1981); (6) 12 measures of employee anthropometry; (7) employee injury history; (8) employee satisfaction; and (9) trunk motion. Trunk motions were obtained using the iLMM. These variables consisted of the trunk angular position, velocity, and acceleration characteristics (i.e., means, ranges, maximums, minimums, etc.) in each of the cardinal planes of movement. Selected trunk motion factors, along with selected workplace factors, were used to develop a quantitative model of occupational LBD risk.

49.3.3 Data Collection

Data about employee health, employment history, and anthropometry were collected. Next, the employee was fitted with the iLMM. A baseline reading from the monitor was then taken as the individual stood erect and rigid. The employee was asked to return to the job, wearing the iLMM for at least ten job cycles. Thus, the length of time the employee wore the monitor depended upon the cycle time of the job. Monitoring of back motions was initiated as the employee began the MMH task and concluded when the task was completed. Extraneous activities not involving MMH were not monitored. Signals from the iLMM were sampled at 60 Hz via an analog-to-digital converter and stored on a portable microcomputer. The data were further processed in the laboratory, to determine position, velocity, and acceleration of the trunk for each of the job cycles, as a function of time spent in the sagittal, lateral, and axial twisting planes of motion.

49.3.4 Analysis

The data were examined initially to determine whether the trunk motions gathered were repeatable. This analysis indicated that task-to-task variation was much larger than the variability due either to multiple cycles performed within a task or to different employees performing the same task (Allread et al., 2000). Hence, trunk motions were dictated largely by the design of the task, not by the individual, and repetitive task cycles resulted in motions that were fairly similar.

The various personal, environmental, and workplace factors from the database were analyzed using logistic regression techniques, to determine if any single factor could distinguish jobs associated with high LBD rates from those with low rates. The most powerful single variable was maximum external moment, which yielded an odds ratio of 5.17. Overall, however, the odds ratios were low, indicating that few of the individual variables discriminated well between the two injury rate groups. Of the trunk motion factors, the velocity variables generally produced greater odds ratios than maximum or minimum position, range of motion, or acceleration. Table 49.2 shows the descriptive statistics of the workplace and trunk motion factors for the high and low LBD rate groups.

Next, multiple logistic regression was used to predict the probability of high-risk group membership as a function of the values for several workplace and trunk motion factors. A five-variable model incorporating these factors was developed and further refined after examining a series of stepwise logistic regression models (containing different variables, such as velocity, acceleration) fitted to several intermediate data sets. A combination of five variables (external moment, lift rate, sagittal flexion, twisting velocity, and lateral velocity) was found to have the greatest odds of predicting high-risk

TABLE 49.2   Descriptive Statistics of the Workplace and Trunk Motion Factors in Each of the Risk Groups

Factors	High Injury Rate Jobs ($n = 111$)				Low Injury Rate Jobs ($n = 124$)			
	Mean	SD	Min	Max	Mean	SD	Min	Max
Workplace factors								
Lift rate (lifts/h)	175.89	8.65	15.30	900.00	118.83	169.09	5.40	1500.00
Vertical location at origin (m)	1.00	0.21	0.38	1.80	1.05	0.27	0.18	2.18
Vertical load location at destination (m)	1.04	0.22	0.55	1.79	1.15	0.26	0.25	1.88
Vertical distance traveled by load (m)	0.23	0.17	0.00	0.76	0.25	0.22	0.00	1.04
Average weight handled (N)	84.74	79.39	0.45	423.61	29.30	48.87	0.45	280.92
Maximum weight handled (N)	104.36	88.81	0.45	423.61	37.15	60.83	0.45	325.51
Average horizontal distance between load and L_5S_1 joint (N)	0.66	0.12	0.30	0.99	0.61	0.14	0.33	1.12
Maximum horizontal distance between load and L_5S_1 joint (N)	0.76	0.17	0.38	1.24	0.67	0.19	0.33	1.17
Average moment (N m)	55.26	51.41	0.16	258.23	17.70	29.18	0.17	150.72
Maximum moment (N m)	73.65	60.65	0.19	275.90	23.64	38.62	0.17	198.21
Job satisfaction	5.96	2.26	1.00	10.00	7.28	1.95	1.00	10.00
Trunk motion factors								
Sagittal plane								
Maximum extension position (°)	−8.30	9.10	−30.82	18.96	−10.19	10.58	−30.00	33.12
Maximum flexion position (°)	17.85	16.63	−13.96	45.00	10.37	16.02	−25.23	45.00
Range of motion (°)	31.50	15.67	7.50	75.00	23.82	14.22	399.00	67.74
Average velocity (°/sec)	11.74	8.14	3.27	48.88	6.55	4.28	1.40	35.73
Maximum velocity (°/sec)	55.00	38.23	14.20	207.55	38.69	26.52	9.02	193.29
Maximum acceleration (°/sec²)	316.73	224.57	80.61	1341.92	226.04	173.88	59.10	1120.10
Maximum deceleration (°/sec²)	−92.45	63.55	−514.08	−18.45	−83.32	47.71	−227.12	−4.57
Lateral plane								
Maximum left bend (°)	−1.47	6.02	−16.80	24.49	−2.54	5.46	−23.80	13.96
Maximum right bend (°)	15.60	7.61	3.65	43.11	13.24	6.32	0.34	34.14
Range of motion (°)	24.44	9.77	7.10	47.54	21.59	10.34	5.42	62.41
Average velocity (°/sec)	10.28	4.54	3.12	33.11	7.15	3.16	2.13	18.86
Maximum velocity (°/sec)	46.36	19.12	13.51	119.94	35.45	12.88	11.97	76.25
Maximum acceleration (°/sec²)	301.41	166.69	82.64	1030.29	229.29	90.90	66.72	495.88
Maximum deceleration (°/sec²)	−103.65	60.31	−376.75	0.00	−106.20	58.27	−294.83	0.00
Twisting plane								
Maximum CCW twist (°)	1.21	9.08	−27.56	29.54	−1.92	5.36	−30.00	11.44
Maximum CW twist (°)	13.95	8.69	−13.45	30.00	10.83	6.08	−11.20	30.00
Range of motion (°)	20.71	10.61	3.28	53.30	17.08	8.13	1.74	38.59
Average velocity (°/sec)	8.71	6.61	1.02	34.77	5.44	3.19	0.66	17.44
Maximum velocity (°/sec)	46.36	25.61	8.06	136.72	38.04	17.51	5.93	91.97
Maximum acceleration (°/sec²)	304.55	175.31	54.48	853.93	269.49	146.65	44.17	940.77
Maximum deceleration (°/sec²)	−88.52	70.30	−428.94	−5.84	−100.32	77.40	−325.93	−2.74

group membership. This combination of workplace and trunk motion factors formed the basis of the LBD risk model. The model was selected for its statistical importance of the predictors and for biomechanical plausibility. The model variables remained consistent when tested with the various intermediate data sets. The empirical stability of the model was checked by predicting the classification of 100 jobs, based on the preliminary model. This model resulted in an odds ratio of 10.6.

By averaging individual probability values for moment, lift rate, sagittal flexion, twisting velocity, and lateral velocity, the LBD risk model is able to predict the probability of high-risk group membership (LBD risk) for any repetitive job. A chart depicting this information is shown in Figure 49.3. It is important to understand that the predictive power of this model is a result of the *interaction* of these five variables. Individually, each of these five factors is unable to reliably distinguish between the injury rate groups, but when they are considered in combination, the predictive power increased tenfold.

49.3.5 Validation and Predictive Ability

The iLMM's LBD risk model is one of a few tools for assessing injury risk that have been tested for their predictive abilities. It was validated in a prospective study (Marras et al., 2000) to determine how well it reflects a job's LBD risk. A total of 36 jobs from 16 different companies were tracked in a prospective cohort study, in which 142 employees performing these jobs were studied both before and after workplace interventions were made. Assessments of LBD risk were made using the iLMM and associated model, and the jobs' LBD incidence rates also were computed pre- and postintervention. The results from this study indicated that a statistically significant correlation existed between changes in the jobs' estimated LBD risk assessment values (using the iLMM) and changes in their actual LBD injury rates during the observation period. This relationship is graphically depicted in Figure 49.4. As this figure shows, jobs whose interventions produced LBD risk values still considered "high" resulted in small changes, on average, in their actual low-back injury rates. For those jobs where the intervention produced a "medium" level of LBD risk, the average injury rate drop was over 50%. Finally, the

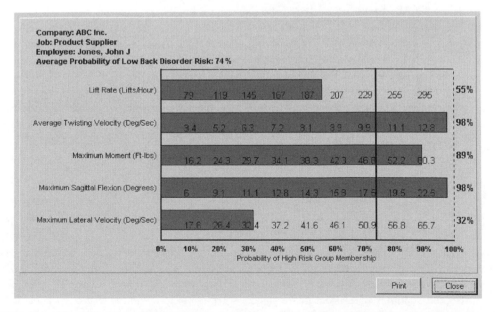

FIGURE 49.3 The LBD risk model, showing the five factors scaled relative to risk. The vertical line indicates the overall probability of high-risk group membership, or LBD risk, for a particular job.

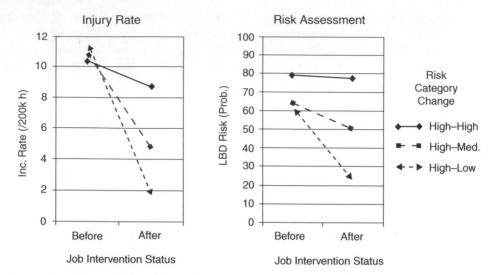

FIGURE 49.4   The impact of job interventions on LBD risk assessments and LBD injury rates, using the iLMM, for three groups. The "high–high" group indicates those jobs where LBD risk remained high following a job change. The group of "high–medium" jobs were those whose interventions produced a moderate drop in LBD risk. The "high–low" category included those jobs where the intervention resulted in a large reduction in LBD risk. These charts show that, on average, changes in a job's resulting risk assessment level following an intervention were positively correlated to the job's LBD injury rate.

groups of jobs whose interventions produced an LBD risk assessed as "low" resulted in an average injury rate drop of approximately 85%.

A concern in the assessment of jobs having a high LBD risk is the possibility that employees having had a back injury, or currently experiencing low back pain, may move differently than those with healthy spines. This could influence their trunk kinematics and the resulting risk assessment for a job using the iLMM. However, Ferguson et al. (in review) found this concern to be unfounded. They studied 200 employees who had returned to their jobs following a low back injury, as well as 200 controls having no reported low back pain who were experienced performing the same jobs. These individuals all were monitored at their worksites while wearing the iLMM. The data showed no statistically significant differences between those two groups of employees on any trunk motion measures. Further, the LBD risk assessments using the iLMM also did not differ between these two groups.

These results support the use of the iLMM and the LBD risk model as an accurate method for determining a job's potential to produce low-back injuries in employees performing the work. It also has been found to be predictive, in that risk level changes arising from job interventions were positively correlated with actual changes in LBD injury rates.

49.4   Benefits of the iLMM and the LBD Risk Model

Use of the iLMM and its computation of LBD risk provides several advantages to those assessing injury risk for material handling activities. First, the iLMM can determine the instantaneous three-dimensional position, velocity, and acceleration of the trunk while individuals perform their actual job tasks, not work simulated instead in a laboratory or artificial work setting. This ability eliminates any question of differences between trunk motions gathered in simulated work environments from those actually obtained on the job. Also, the quantitative iLMM data are objective. Trunk motions and the resulting calculations of LBD risk are determined irrespective of an investigator's (perhaps unintentionally biased) view of the work.

A second advantage of the iLMM system is that materials handling job risk is assessed relative to a large database of jobs from diverse manufacturing environments. These jobs encompass a wide range of LBD risk levels and actual low back injury rates. This enables the investigator to determine "how much is too much." That is, the model can assess how similar a job is to others known to have high rates of low back injury associated with them. It also allows one to rank several jobs, based on their LBD risk values, and to study solutions for those having the greatest likelihood of producing injury.

Third, the model demonstrates that it is the *combination* of several factors that determines a job's level of LBD risk. In other words, there is no one factor that is responsible for risk; rather, one evaluates risk based on a mix of five inputs related to a job's work requirements. An example of this model's benefit would be a job where a heavy load must be handled, resulting in a large maximum moment value on the risk chart. If this load weight cannot be reduced using mechanical means, the job's overall risk level *can* be greatly reduced if an intervention produces significant reductions in the magnitude of the model's remaining four factors.

Fourth, the risk model enables the investigator to quantitatively assess and compare each task within a job. Specific factors that contribute to a task's risk can be identified, as are the tasks that most contribute to the job's overall LBD risk. This information pinpoints the specific tasks and the factors therein that must be addressed during job redesign to reduce the job's injury risk potential.

A fifth benefit of this LBD risk model is the assistance it can provide to the ergonomics intervention process. Modified jobs can be remonitored using the iLMM, and the effects of those changes can be quantified and compared with those values determined prior to the intervention. Traditionally, the effects of job changes on the numbers of related musculoskeletal strains (the job's incident rate) may take several years to appear. The iLMM can produce more timely feedback to the investigator regarding anticipated returns on the redesign investment (i.e., just the time needed to analyze the data). Thus, for jobs that produce only minimal reductions in LBD risk due to planned redesign efforts, further (and perhaps different) improvements can be attempted.

49.5 Applications: How to Use the iLMM and LBD Risk Model

49.5.1 Recommended Equipment

The iLMM system comprises the equipment listed next. It is important to make sure that all components are available at any data collection site.

49.5.2 Adjustable Industrial Lumbar Motion Monitor

The iLMM is battery powered and contains a wireless, digital telemetry transmitter device. These features allow data gathering for several hours and at distances of up to 100 ft or more from the data collection computer. A single iLMM can be set to four different sizes — extra small, small, medium, and large — by changing the monitor's linkage system. The selection of the appropriate size worn by individuals depends both on their spine lengths and work activity. For example, an iLMM adjusted to the medium size may be appropriate for most males of average height, unless their job requires extreme forward flexion. In that case, they may be better suited for the large-size iLMM. Similarly, the small iLMM size is more appropriate for shorter individuals or those of moderate height who do little forward bending.

49.5.3 iLMM Harnesses

The iLMM system consists of a shoulder harness and a waist belt. Each is adjustable and designed to fit most males and females who may need to wear the device. The harness and belt have adjustable straps to allow for individual differences. A slide plate is attached to the shoulder harness, into which the iLMM moves. The iLMM is attached to the waist belt before it is placed on an individual.

49.5.4   Data Collection System

The iLMM data gathering and analysis software requires a computer having a PC-based operating system. (The software used in conjunction with the iLMM will be discussed in a later section.) For portability purposes, it is recommended that a notebook computer be used, having a minimum 133 MHz processor and 32 MB of RAM. Using a notebook on battery power will ease data collection. Signals from the digital telemetry transmitter on the iLMM are sent to the receiver unit attached to the computer. The system can use battery power if data collection takes place in a remote location. However, cabling also is available for power to be supplied from a 110-V source. In the event of a batter-power outage or electrical disturbance in the environment, the iLMM system is also supplied with a hardwire that connects the iLMM to the data collection computer.

These iLMM system's components all fit into a single, rugged case. It is designed to securely house the aforementioned equipment during both use and transport. The case also doubles as a work surface, to simplify use during data collection. The system's carrying case also contains an extendable handle, and the entire unit is on wheels, to maximize portability around a work site.

49.5.5   Other Useful Equipment

The iLMM provides data regarding the trunk motions required of a job. However, other information is important and needs to be collected at the job site to fully assess the job's risk. Listed next are additional equipment that should be part of the iLMM gear.

49.5.5.1   Scale or Force Gauge

To use the iLMM's LBD risk model, it is critical to record the weights of items handled by individuals during data collection. A scale or force gauge is needed, and it is recommended that one be used that is capable of reading object weights and forces from 1 to at least 100 lb.

49.5.5.2   Tape Measure

A heavy-duty tape measure is needed to record the distances from which objects are handled out from the body.

49.5.5.3   Extension Cord/Outlet Strip

Power outlets are not always conveniently located near where data collection takes place. In situations where battery life may be an issue during data gathering, an extension cord with multiple plugs provides more flexibility in locating the computer closer to the job site and in full view of the individual monitored.

49.6   Selecting the Job(s) to Monitor

The iLMM can be used to monitor the positions, velocities, and accelerations of the trunk in the three cardinal planes for any job in which trunk movements are required. This type of information can be important when one needs to describe the motion characteristics required of a job or activity. The LBD risk model is composed of five workplace factors that, when combined, assess a job's probability of having a high number of LBDs. This model is also useful for establishing relative comparisons between several tasks that comprise a job.

The industrial jobs used to develop the iLMM risk model had MMH frequencies ranging from approximately six to 1500 lifting tasks required per hour. Thus, the tasks to be assessed using this model should fall within this scope of repetition. Most MMH jobs fit this profile. For example, in automobile assembly, job cycles are repeated every 1 to 2 min, and the parts themselves often are standard in size and weight. Palletizing jobs may require very different types of objects to be handled, but the task of continual lifting from one storage area to another remains the same. These types of jobs are very consistent with those used to develop the risk model database, and the job's LBD risk can be easily assessed with the iLMM's software.

Some jobs are less repetitive or have more job variation than assembly or palletizing tasks. However, the LBD risk model still can be used to make relative comparisons between the tasks that comprise the job. For jobs that require a large number of tasks, the risk model is helpful in comparing the factors that make up the model. This will allow for the ergonomist to assess trade-offs between such factors as lifting frequency, object weights handled, and the trunk motions required for the different tasks. It should be noted that there is inherent variability in the way an individual may perform the same task repeatedly. Because of this reality, one study (Allread et al., 2000) suggests that at least three repetitions of a task cycle be collected for each of three individuals familiar with the job. This will best reduce this inherent variability found in MMH work.

Another issue to be considered is job rotation. Employers often use a variety of rotation schemes for job processes. If the job to be monitored requires no rotation (employees perform the same job every day/week/month), then the risk assessment can be directly related to the tasks observed. Jobs in which individuals rotate regularly between a few work areas also may be used in assessing LBD risk, if this rotation schedule is fixed. When the job rotation requires employees to do completely different jobs on an hourly or weekly basis, it becomes difficult to relate a task's risk values to the overall job risk, since many tasks could contribute to the risk assessment. This issue is key to determining a job's suitability for LBD risk assessment. That is, does the job's work structure enable one to define the job in terms of a few repeatable, consistently performed tasks?

There may be some jobs that fit within the LBD risk model profile but still should not be monitored. Seated jobs may require repetitive activities, but they usually are not ones that require significant material handling. In any event, the iLMM may rub against a chair's back or the waist belt will shift from its position on the hips during seated work, and erroneous iLMM output will result. Also, jobs that require close contact of the iLMM with a finished product could produce scratches on the product, and the employer may not want to risk product damage. Finally, exposure to water or other liquids may damage the iLMM or its components.

49.7 Defining the Major Components of the Job through a Task Analysis

It is very important to properly define all relevant tasks that make up the job under investigation. These tasks should encompass the range of materials handling work that is required of the job — especially those tasks that may present a risk of LBD. The tasks also should be defined so they are meaningful to those who will be interested in the iLMM results. One way to define these job tasks is through a task analysis. A task analysis defines the discrete events of a job.

Take, for example, a job performed in a food processing plant (shown in Figure 49.5). An employee places 12 frozen food packages, two at a time, into a box, records the date and time of the packaging, then loads the packed box onto a pallet. A simple task analysis of this job is shown in Table 49.3. Two of the three tasks identified would qualify as a materials handling task for this job — loading the individual packages (Figure 49.5a) and placing the box onto the pallet (Figure 49.5c). Recording the date and time of the packaging (Figure 49.5b) would not be considered relevant material handling since low forces and exertions are required. Any trunk movement related to this task would not be considered in further analyses. However, trunk kinematics and the risk probability could be determined for the two relevant MMH tasks identified.

It may be convenient to further subdivide and redefine a task that is similar in all work dimensions except for one. For instance, in the example given, a fully loaded pallet may contain many boxes stacked several layers high. Tasks could be defined separately as "place box on layer 1," "place box on layer 2," etc. This may assist in data interpretation. That is, differences in trunk kinematics could be

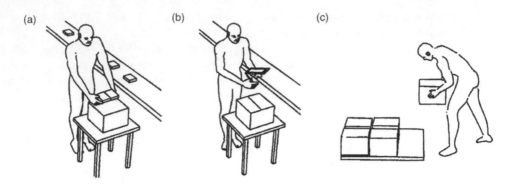

FIGURE 49.5 The three tasks of the frozen food packaging job: (a) place two packages in box; (b) record date/time of the packaging and (c) load full box onto pallet.

interpreted as a function of the defined job tasks, as well as in terms of workplace factors such as a box's location on the pallet.

49.8 Collecting and Recording Workplace Data for Risk Assessment

The trunk motion information that is used in the LBD risk model is automatically stored in the data collection software. Trunk motions include the position, the velocity of movement, and the related acceleration of the trunk in all three directions of motion during a task — the sagittal (forward bending) plane, the lateral (side bending) plane, and the rotational (twisting) plane. Two other components, lift rate and maximum moment, must be determined and input manually.

49.8.1 Lift Rate

Lift rate is defined as the *total* number of MMH actions that a job requires per hour. This number is related to the lifts for all tasks combined; it does not change from one task to another within a job. As the task analysis in Table 49.2 found, on average, one box is fully prepared and loaded onto the pallet per minute. Assuming this rate represents an average job cycle across the work shift, the packaging task alone would require 360 lifts per hour, since the product is placed in each of the 60 boxes six times (two packages per lift). For the palletizing task, 60 lifts are required per hour, for each fully packed box. Both tasks combine for a total of 420 lifts required per hour for this job. The non-MMH date/time recording task would not influence the lift frequency of the job.

Because the lift rate value is directly input into the LBD risk model, it is very important to get an accurate estimate of the lifting frequency. It may be necessary to confirm the task analysis results by questioning the employees familiar with the job or the job supervisor.

TABLE 49.3 Task Analysis of One Cycle of the Frozen Food Packaging Job

Task	Length of Task	Notes
Place two food packages in box	42 sec	Task time is 7 sec per two boxes packaged; 12 packages fit into each box
Record date/time of packaging	10 sec	No manual materials handling required
Load full box onto pallet	8 sec	Full pallet contains seven layers of boxes
Total time of job tasks	60 sec	

49.8.2   Maximum Moment

Maximum moment is defined as the external horizontal moment generated about the spine. A moment is composed of two factors, the weight of the object being lifted and the horizontal distance from the spine at which it is handled.

49.8.2.1   Weight

Each object that is handled on a job must be weighed and recorded in the software or noted manually and input during data analysis. If objects of varying weights are handled, then each must be weighed individually and recorded. This often occurs, for example, in mail and freight delivery operations. In the aforementioned food processing example, the weights are constant for each task. The combined weight of the two food packages lifted together needs to be recorded, as does the weight of a fully packed box that is palletized.

49.8.2.2   Horizontal Distance

A tape measure is needed to determine how far from the spine objects are being handled for each task. With the tape measure held *horizontally*, one must measure the distance from the spine at the lumbosacral joint (near the top of the hips) to the center of the hands when the task is being performed. Obviously, as an individual handles objects, this distance changes as the object is moved. It is important to determine at what point during the task the distance is the greatest (i.e., generating the greatest external moment about the spine) and to record this length. An ergonomist correctly measuring the horizontal distance is shown in Figure 49.6(a). Here, the tape measure is kept level to determine the length from the individual's lumbosacral joint to the center of the hands. The *incorrect* approach is being used in Figure 49.6(b), since the distance being measured is not horizontal.

For some MMH jobs, such as those on an assembly line, jobs can be very repetitive and the employees' actions and movements rather consistent. In these cases, the horizontal distances at which items are held may not vary much as identical objects are handled during each work cycle. For other jobs, such as when pallets are loaded or unloaded, each cycle can produce very different trunk motions, since objects are being handled to/from different areas. This will likely change the horizontal distance at which an employee handles the load. Because the maximum moment value is directly input into the LBD risk model, it is important that these distances be accurately measured and that changes in the distances for each task cycle being monitored by the iLMM are recorded.

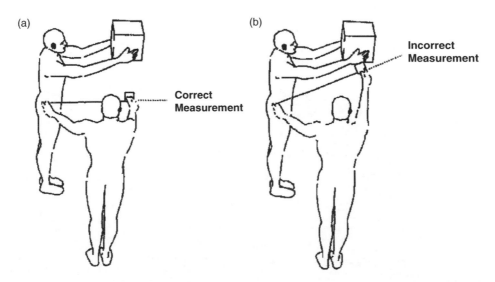

FIGURE 49.6   Correct (a) and incorrect (b) methods of measuring the distance a load is held from the lumbosacral joint.

The measurement of these horizontal distances should not interfere with the work being done by the employee. The ergonomist who measures these distances should stand close enough to the individual to get accurate readings, but far enough away to not disturb the work. It is important that the individual be able to move naturally at the job site. A monitored individual being crowded by an investigator will move differently, change his or her trunk motions, and give erroneous information via the iLMM concerning the required activities of the task.

49.9   Setting Up the iLMM for Data Collection

After the job has been selected, the tasks identified, and basic work place information has been gathered, the data collection process can begin. The proper iLMM size needs to be chosen and prepared for use. This process is depicted in a flow chart (Figure 49.7) and is described next.

49.9.1   Selecting the Correct iLMM Size

The standing height of the individual to be monitored is usually the best indicator of which iLMM size to use. It is important that the top of the iLMM moves easily within the slide plate that is attached to the shoulder harness. The iLMM should not separate from the slide plate when the individual performs the activity to be monitored (indicating that too small of a size is being used), nor should the T-sections of the iLMM move within the slide plate (signifying that too large of a size has been selected). Thus, the type of trunk movement required of the job also must be considered when choosing the appropriate iLMM size, especially that involving extreme forward flexion.

49.9.2   Adjusting the iLMM

Each iLMM is individually calibrated during manufacturing, and there is a specific position designated as its "neutral" position. During normal usage, this position may change slightly and must be readjusted (using the software) before every subsequent use. After sizing the iLMM, it must be secured in its zero-plate, and its voltage offsets adjusted using the supplied iLMM software. The three traces shown in the software (representing the three cardinal planes of motion) should show realistic position values. Every iLMM must be adjusted in this manner before it is placed on an individual. This readies the iLMM for data collection.

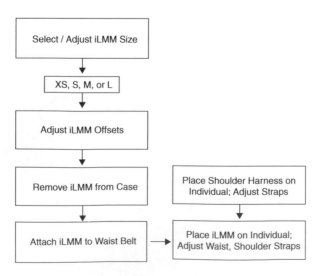

FIGURE 49.7   Flow chart of the iLMM setup procedure.

49.9.3 Putting the iLMM on the Employee

After the iLMM has been checked to be working properly, it is ready to be placed on an employee. When handling the iLMM, it should never be bent into extreme angles, as damage to the unit could result. The following steps will ensure proper placement.

49.9.3.1 Attach Shoulder Harness

The adjustable shoulder harness should fit snugly when worn, so that it does not move during data collection; however, it should enable the individual to breath normally. The slide plate should be centered on the employee's back between the shoulder blades.

49.9.3.2 Put iLMM on Waist Belt

Before being placed on the employee, the iLMM must first be placed on the waist belt. This is most easily done by laying the waist belt out on a flat surface. Then the unit can be positioned on the belt's bolts and secured with the nuts attached to the base of the iLMM.

49.9.3.3 Place Waist Belt/iLMM on Individual

This is done by first positioning the top of the iLMM into the slide plate on the shoulder harness. Then, the waist belt is to be fit over the employee's hips and momentarily secured with the belt's Velcro® strap. Next, locate the tops of the hips (the iliac crest) and align the top edge of the waist belt with this ridge. This will place the base of the iLMM (at the lowest T-section of the unit) at the individual's lumbosacral joint. Finally, ensure that the iLMM is centered on the back, so that the monitor's T-sections are aligned with the slide plate on the shoulder harness. If not, one or both of the following adjustments can be made:

1. The waist belt can be moved to the left or right
2. The shoulder harness can be loosened and shifted

A properly fitted iLMM is shown in Figure 49.8. Once the iLMM is positioned correctly, securely tighten the Velcro strap on the waist belt, and secure the leg straps attached to the belt. It is important

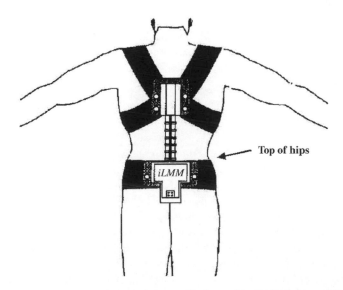

FIGURE 49.8 Proper placement of the iLMM on the torso. The base of the iLMM is located slightly below the top of the hips, and the iLMM T-sections are aligned vertically.

to always use these leg straps. They prevent the iLMM from moving about the hips, which otherwise would move the base of the iLMM from its proper position and result in erroneous data.

49.10   Collecting the Data

Once the iLMM is properly placed on an individual and its signal is being read by the computer, the investigator is ready to collect job data. The iLMM may feel somewhat unnatural to the individual at first, so it is suggested that she or he be given some time (a few job cycles, for example) to become accustomed to it before collecting data.

It not already done, the company, job, task, and employee data need to be entered into the iLMM software. Follow the software instructions for this procedure. An example of the job profile selection screen using the Ballet 2.0 data collection and analysis software is shown in Appendix A-1 (Figure 49.A1). After the task data are input, and the data collection screen is open and showing the three traces (for the sagittal, lateral, and twisting planes of motion), data gathering can begin.

The main goal of data collection is to gather information on trunk motions that are representative of all the work required of the job. For example, if the job requires handling objects of widely varying weights or from different locations (on a pallet, for example), then the data should be gathered on these tasks. The more data that are collected, the more it is likely to represent the requirements of the job. An example iLMM data collection screen using the Ballet 2.0 software is presented in Appendix A-2. Notice that only the motions associated with the designated job tasks are marked, using the computer keyboard's function keys.

It is extremely important to keep a record of the weights handled and the measured horizontal lifting distances for each task cycle collected, so that these data are matched with the corresponding trunk motion data for that cycle. The investigator may wish to develop a data collection form beforehand to keep track of this information.

49.11   Analyzing and Interpreting the iLMM Data

By following the guidelines in the previous sections of this chapter, data analysis and interpretation will be made easier. Tasks of a job that are carefully chosen will assist in the job's interpretation of its risk probability and lead to results having more practical significance. The software used to collect and analyze iLMM data provides trunk kinematic information and risk probability charts for each task defined for a job and for each individual who was monitored. It is beyond the scope of this chapter to provide iLMM software documentation. Instead, explanations of how these software outputs can be interpreted will be discussed.

49.11.1   Trunk Kinematic Information

Information on the specific trunk motions produced by individuals during each task can be obtained from the software. This includes the trunk positions and the velocities and accelerations produced by the trunk for each plane of motion for those cycles of a task chosen by the ergonomist. This information can be useful for general descriptions of the material handling or for comparisons with other tasks or jobs. Use of these data can be valuable for investigators who have formed hypotheses about, for example, what tasks require more trunk motions than others.

Output from the frozen foods example job is shown in Table 49.4. As the header shows, these data include trials of data collected for the food packaging task of this job. Information includes minimum and maximum trunk positions, average and maximum velocities, and maximum accelerations for each plane of motion. From these data, average motions over the six trials could be computed, as could the range of movement required and the amount of variation for each kinematic parameter.

TABLE 49.4 Output Data Showing Low-Back Kinematics for Six Trials of One Job Task

Company: Frosty Foods
Job: Frozen foods packager
Employee: Jim Hayes
Task: Place two packages in box
Date: 12/11/03
Run: 1

Trial	Time	Sagittal (forward bending)					Lateral (side bending)					Twisting (rotational)				
		Min. pos.	Max. pos.	Avg. vel.	Max. vel.	Max. acc.	Min. pos.	Max. pos.	Avg. vel.	Max. vel.	Max. acc.	Min. pos.	Max. pos.	Avg. vel.	Max. vel.	Max. acc.
1	10:46:17	−3.6	4.1	2.1	16.4	96.5	−15.2	−1.2	3.0	41.3	227.6	−16.7	−0.1	4.7	32.7	279.7
2	10:47:32	−6.5	5.4	3.4	24.6	115.3	−13.2	−0.9	3.9	33.5	181.0	−16.7	1.6	5.8	30.6	241.9
3	10:51:02	−5.1	6.3	3.3	25.4	123.8	−14.7	−0.7	4.0	37.7	222.2	−16.2	0.9	6.1	41.6	258.4
4	10:54:12	−18.0	6.4	5.6	17.7	94.8	−2.1	9.4	2.8	8.8	46.7	−35.0	−17.7	3.2	22.6	215.3
5	10:58:56	−11.1	10.3	4.7	27.3	116.7	−4.6	8.0	2.9	18.4	70.4	−20.2	−7.6	3.8	23.1	145.7
6	11:01:29	−4.1	10.4	3.6	21.2	89.1	−3.2	8.0	2.5	12.4	69.2	−25.0	−4.8	4.0	23.6	163.7

49.11.2 Probability of High-Risk Group Membership (LBD Risk)

The software will produce information that compares the job tasks of interest with a database of jobs previously determined to have high or low injury rates to the low back associated with them. The interpretation of data is best described through use of the food packager example.

In the food packaging job, two tasks were defined. The first task involved placing individual frozen food packages into a box (12 in all), and the second task involved loading the filled box of packages onto a pallet. Risk probability charts are shown for each task in Figure 49.9. In Figure 49.9(a), the package loading task was found to have a risk probability of 42%. (Probability values are calculated by averaging the individual logits from each of the five risk factors—in Figure 49.8(a), $(98\% + 31\% + 1\% + 29\% + 49\%)/5 = 42\%$). The risk charts for this first task that were produced using the iLMM software are shown in the appendix. The probability value for the palletizing task (Figure 49.9b) was calculated to be 62%, clearly the task more similar to those considered "high risk." These two probability values are to be used only for comparison purposes. It is the chart in Figure 49.9(c) that reflects the true risk value for the entire job. This chart summarizes the largest values for each risk factor across both tasks making up the job. The value shown on this chart, and thus the risk for this material handling job, is 66%.

A closer examination of the top two charts in Figure 49.9 shows which factors most contributed to the job summary values in Figure 49.9(c). Each of the five factors is discussed separately.

- *Lift rate.* The lifting frequency for the entire job was 420 lifts/h. Because this variable is composed of the total number of lifts from both tasks, this value is shown to be the same on all charts in Figure 49.9. As indicated by the length of the lift rate bar on the charts, this rate is very rapid and is comparable to some of the highest frequency material handling jobs found in industry.
- *Average twisting velocity.* The amount of twisting velocity required for the package loading task was fairly low, but it was moderately high for the palletizing task, as indicated by the length of the bars on these charts in Figure 49.9. The greater value of the two is used in the job summary chart in Figure 49.9(c), which was taken from the palletizing task.
- *Maximum moment.* As shown on the charts in Figure 49.9, the external moment values were low for both tasks comprising this job. The low weight of the individual packages (each at one pound) and the fully packed box being palletized (12 lb) generated low maximum external moment values. The greater moment value from the palletizing task was used in the job summary chart.
- *Maximum sagittal flexion.* The package loading task was performed while employees were in relatively upright postures. This is reflected in Figure 49.9(a) by a short sagittal flexion bar on the chart. However, during box palletizing, those boxes placed on the lower layers of the stack required much forward trunk flexion. Figure 49.9(b) depicts these higher angles by the long bar for this factor. Subsequently, this higher value of the two tasks resulted in it being used in the job summary chart in Figure 49.9(c).
- *Maximum lateral velocity.* The iLMM determined that lateral velocities generated during the package loading task were higher than those found during box palletizing. The values for the previous four factors all were larger during handing of the full box and, thus, were used in the job summary chart. However, it is the lateral velocity value from the package loading task that must be used in the job summary, since it is the greater of the two tasks analyzed.

It is the job summary value of 66% (taken from the chart in Figure 49.9c) that represents the probability of LBD risk for this example food processing job. The value indicates that, on the continuum of low-risk jobs (0%) to high-risk jobs (100%), this particular job has a 66% likelihood of being considered "high risk." As stated earlier in this chapter, a high-risk job was defined as one having 12 or more (with an average of 26.4) low back strains per 200,000 h (or 100 workers/yr) of employee exposure. Results here could be interpreted as indicating that this particular job has a relatively high chance of producing a large number of low back strain injuries among individuals who do this job.

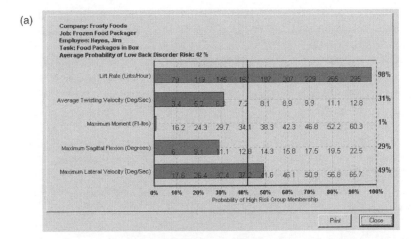

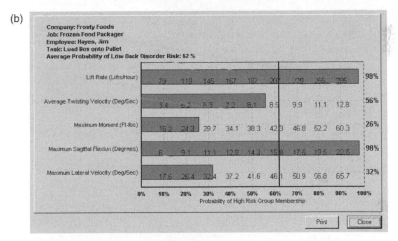

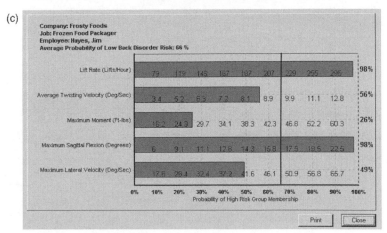

FIGURE 49.9   Risk value charts for the frozen food packaging job: (a) "place two food packages in box" task; (b) "load full box onto pallet" task; and (c) summary of all job tasks.

The primary goal of the iLMM analysis could be simply to determine whether or not a job presents a risk of low back strain to employees doing its work. This can be determined by calculating the LBD risk value from a job summary. As Figure 49.10 shows, there is some overlap in risk values among jobs defined as low and high risk. However, our analyses (Marras et al., 1993) found that those jobs with fairly low probability values (below 30%) were much more likely to be low-risk jobs. Similarly, very few jobs having a risk value over 60% were defined as "low risk," and no low-risk jobs had risk values over 70%. That is, jobs with probability values above 60% are virtually assured to have some low back injury risk associated with them.

A second goal of the analysis may be to compare one job task with another, to determine which ones require more trunk motions and external moments about the spine. This exercise can assist in learning which tasks should be the focus of redesign efforts. Introducing ergonomics modifications to tasks already found to have low-risk values probably will have little real impact on improving the job overall. However, making changes to tasks whose individual factors contribute to the job's summary risk probability will reduce the probability for the entire job.

A third goal of the analysis may be to determine, for specific tasks, which individual components are most responsible for its composite probability value. This type of analysis provides direct information regarding how the job's requirements may affect those factors used in this probability model. Examples of ergonomics improvements that can be made on a job to affect each component are listed next. These examples should not be considered an exhaustive list of possible changes that can be implemented to improve working conditions.

49.11.2.1 Lift Rate

The total number of material handling actions required of the job affects this risk factor. Thus, reducing this rate will reduce the overall risk value. Management typically does not favor this type of change, as it is believed to reduce productivity. However, there are ways to redesign jobs to reduce lifting frequency:

1. *Rearrange job tasks.* For jobs in which a number of tasks comprise the job, one method is to rearrange tasks with those of other jobs. This may more evenly distribute the lifting frequencies of several jobs, some of which may be considerably lower than the job monitored.
2. *Rotate jobs.* Rotating employees between a job having a high lifting frequency with one having a much lower frequency also will reduce the rate for the job of interest. The effects of the jobs into

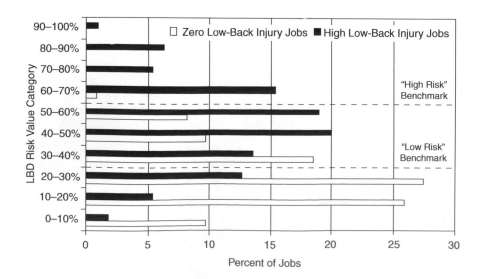

FIGURE 49.10   Distribution of LBD risk values for jobs having high and low rates of low back disorders.

which employees are rotated must also be considered, however. It should be cautioned that this and the previous approach are most beneficial if the jobs that are rearranged or included in the rotation allow employees to use different muscle groups to perform the job. Rotating individuals into jobs that require the same muscles be used likely will have either no benefit or could produce greater musculoskeletal stress.

3. *Add employees.* Dividing the job so that added personnel are available to perform the job will distribute the work across more people and lower the job's required lift rate. Of course, the cost of the additional employees must be compared with the benefits of reduced low back strains and their related costs.

4. *Automate.* It may be possible that some job tasks can be automated through new equipment or the use of robotics. This method of assisting the material handler will undoubtedly reduce the lifting frequency of the job and the overall job requirements.

49.11.2.2   Average Twisting Velocity

Rapid twisting of the trunk can result from a number of situations. If a work area is designed such that material transfer from one location to another is difficult or if these two locations are not convenient to one another, a high twisting velocity may result. High velocities often result because work areas do not allow employees to move their feet to handle goods. In such a situation, a turning action normally done that included movement of the trunk, hips, legs, and ankles would be more concentrated in the trunk, and higher twisting velocities may occur. It would be difficult to reduce the speed at which individuals twist simply by instructing them to "slow down." However, engineering controls that can reduce twisting velocity include:

1. *Place work in front of material handler.* Move the locations of the lifting origin and destination, so that they are more convenient to one another. In other words, create a workplace in which the material transfer requires moving in as few planes of motion as possible, thus allowing the employee to remain is a more neutral posture.

2. *Spread out congested work areas.* Work areas that allow employees to walk or take at least one step during handling often reduces the twisting velocity of the job. This occurs because the added movement allows one to get into a position in which the entire body can assist in the transfer, rather than just the trunk.

3. *Raise working heights.* Depending on the location of goods to be handled, lower levels of work that require a great deal of sagittal flexion also may require additional trunk twisting. This is often the case when the work requires asymmetric lifting. By raising the work heights, not only will forward flexion be reduced, but twisting velocity can decrease because the handling distances are more appropriately located.

49.11.2.3   Maximum Moment

Because an external moment is the product of an object's weight and the distance from the body at which it is handled, reducing either of these two factors will reduce external moment. Various examples are described next:

1. *Reduce weight requirements.* Some work situations allow employees to handle as many units or as much weight as they feel is acceptable. However, to work faster, some employees may handle more goods at one time than is physically safe. A limit on the numbers/weights of objects that can be handled at any given time can reduce moment values. In some environments, raw materials handled are received from a supplier in bulk quantities. The weights of these materials may produce excessive moment values. By working with these suppliers, an arrangement may be possible to package materials in smaller, lower-weight containers. The weight changes just described likely will *increase* the lifting frequency of the job, however, so this trade-off in the risk model should be examined.

2. *Install material handing aids.* For goods that are of uniform shape or size, several types of lifting aids are commercially available to provide handling assistance. These can be adapted to a wide range of work environments. Handling aids, such as lifting hoists, when incorporated successfully, greatly reduce the loading forces on the spine and result in much lower moment values. The device should be considered carefully, since a handling aid that is difficult to use or greatly slows the job process likely will be abandoned by employees. In addition, some material handling aids can reduce the distance at which objects are handled. For example, lift/tilt tables are available commercially and are able to be adapted to specific needs. These devices can raise or angle objects or bins of goods so that they can be more easily accessed. This can result in reduced horizontal reach distances.

3. *Evaluate the transfer locations.* The distance from the body at which individuals handle goods often is greatest during the initial or final contact with the product. This is often true during palletizing operations, in which cases need to be placed properly on a skid to ensure the load's stability. During carrying, for example, people tend to bring the product closer to their bodies. An evaluation of these locations may detect workplace arrangements that cause individuals to reach further than is necessary to handle objects.

49.11.2.4 Maximum Sagittal Flexion

The more individuals work in upright positions during material handling, the lower their trunk flexion will be. This reduces subsequent risk probabilities for jobs. Reducing sagittal flexion can be accomplished by eliminating tasks that, for example, require loads below knee level to be handled. Several interventions can prevent these situations:

1. *Raise the heights of loads placed near the floor.* Objects can be raised from the floor during material handling work by a number of different ways. For palletizing tasks, stacking a skid underneath the pallet will raise the height of the bottom-most objects. Lift tables and self-raising devices placed underneath objects also will raise them higher off the floor and reduce sagittal flexion. These changes should be evaluated for safety considerations and to ensure that objects at the top of the pallet can still be accessed. Other tables are available commercially that tilt or swivel objects, such as those on pallets, which bring the loads closer to the material handlers.

2. *Adjust working heights relative to an individual's standing height.* Work areas that are adjustable to accommodate those performing the work also can reduce trunk flexion. Work tables can be constructed or purchased that move vertically to a position most comfortable to the user. On assembly lines, conveyors can be adjusted that raise or lower objects depending on the work being performed at a specific location. Alternately, work areas alongside the lines, under the employees themselves, can be raised or lowered to produce the correct working height.

3. *Train employees on proper lifting techniques.* For some work situations, vertical adjustability may not be technically or economically feasible. In these cases, employees can be educated in proper lifting techniques aimed to reduce back strain and reduce the amount of sagittal flexion required.

49.11.2.5 Maximum Lateral Velocity

High lateral velocity values on the risk charts indicate that the material handling work requires rapid sideways bending. This motion may be difficult to visualize, but it usually indicates that work is not being performed in front of one's body, but asymmetrically instead. Workplace modifications that more conveniently locate or raise the work relative to the material handler (as already noted earlier) can assist in reducing this factor on the probability charts. A case study conducted by Stuart-Buttle (1995) found that reductions in lateral velocities and sagittal flexion can be achieved through workplace interventions in MMH tasks using lift tables. This paper also illustrated the importance of testing the impact of workplace modifications. The initial installation of this table actually produced higher iLMM risk values and more employee dissatisfaction. The iLMM analysis identified the problems with the new system and provided feedback about how the workplace needed to be further changed to produce actual risk reduction.

From the example job modifications just discussed, it is important to understand that these five work place factors are interrelated. None of these factors responds independently from the others. For instance, adding a lift table to palletizing work may reduce sagittal flexion, because the load is being raised. However, it also can lower the external moment and twisting velocity values, because the load may be held closer to the body during handling and be more easily accessed. If the work is self-paced, the lifting rate actually could *increase* since the work may be less physically demanding, and those affected may be capable and willing to handle more material. This example illustrates the trade-offs that must be considered when evaluating the probability of risk for a job and implementing ergonomic interventions.

49.11.3 Interpreting Results from Several Individuals

Casual observation of any material handling activity will show that people usually perform the same job differently. Employees inherently vary in how they do manual work, and this will produce different trunk motions across individuals performing a task. To account for these differences, the investigator may wish to use the iLMM to monitor several employees who perform the same job and then analyze and interpret the results in terms of LBD risk averaged across these individuals. In a study of trunk motion variability, Allread et al. (2000) reviewed iLMM data for tasks in which load weights and the origin and destination heights of the objects handled did not vary. Statistical analyses found that no additional reduction in trunk kinematic variation was achieved for data in which three individuals performed three cycles of the job task. These results suggest the minimum amount of iLMM data that should be gathered to capture the trunk motions reflective of an industrial task. This finding also implies that additional data, particularly the number of task cycles, should be collected if workplace requirements (e.g., load weights, lifting heights) vary within a task.

Appendix A1

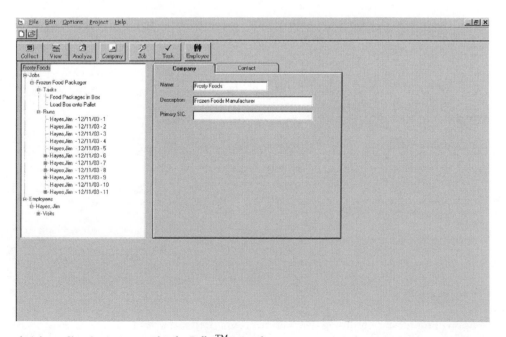

Example job profile selection screen for the Ballet™ 2.0 software.

Appendix A2

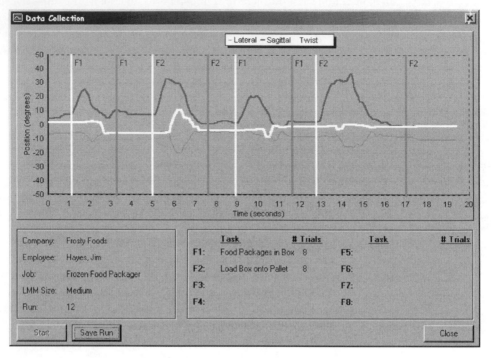

Example data collection screen for the Ballet™ 2.0 software.

References

Allread WG, Marras WS, and Burr DL, Measuring trunk motions in industry: variability due to task factors, individual differences, and the amount of data collected, *Ergonomics*, 43(6): 691–701, 2000.

Andersson GBJ, *The Adult Spine. Principles and Practice*, 2nd ed. Raven, New York, 1997.

Bigos SJ, Spengler DM, Martin NA, Zeh J, Fisher L, Nachemson A, and Wang MH, Back injuries in industry: a retrospective study. II. Injury factors, *Spine*, 11(3): 246–251, 1986.

Cats-Baril W, Cost of low back pain prevention. Paper presented at the "Low Back Pain Prevention, Control, and Treatment Symposium," St. Louis, MO, March 11–13, 1996.

Ferguson SA, Marras WS, and Burr D, The influence of individual low back health status on workplace trunk kinematics and risk of low back disorder, *Ergonomics*, 47(11): 1226–1237, 2004.

Gill KP and Callaghan MJ, Intratester and intertester reproducibility of the lumbar motion monitor as a measure of range, velocity and acceleration of the thoracolumbar spine, *Clinical Biomechanics*, 11(7): 418–421, 1996.

Guo HR, Tanaka S, Halperin WE, and Cameron LL, Back pain prevalence in U.S. industry and estimates of lost workdays, *American Journal of Public Health*, 89(7): 109–1035, 1999.

Magora A, Investigation of the relation between low back pain and occupation — IV, physical requirements: bending, rotation, reaching and sudden maximal effort, *Scandinavian Journal of Rehabilitation Medicine*, 5: 191–196, 1973.

Marras WS, Fathallah FA, Miller RJ, Davis SW, and Mirka GA, Accuracy of a three-dimensional lumbar motion monitor for recording dynamic trunk motion characteristics, *International Journal of Industrial Ergonomics*, 9(1): 75–87, 1992.

Marras WS, Lavender SA, Leurgans S, Rajulu S, Allread WG, Fathallah F, and Ferguson SA, The role of dynamic three dimensional trunk motion in occupationally-related low back disorders: the effects of workplace factors, trunk position and trunk motion characteristics on injury, *Spine*, 18(5): 617–628, 1993.

Marras WS, Allread WG, Burr DL, and Fathallah FA, Prospective validation of a low-back disorder risk model and assessment of ergonomic interventions associated with manual materials handling tasks, *Ergonomics*, 43(11): 1866–1886, 2000.

National Center for Health Statistics, *Prevalence of Selected Impairments*. United States Government Printing Office, Series 10, No. 134, 1977.

National Institute for Occupational Safety and Health (NIOSH), *Work Practices Guide for Manual Lifting*. Department of Health and Human Services (DHHS), National Institute for Occupational Safety and Health, Publication No. 81–12, 1981.

Punnet L, Fine LJ, Keyserling WM, Herrin GD, and Chaffin DB, Back disorders and nonneutral trunk postures of automobile assembly workers. *Scandinavian Journal of Work, Environment, and Health*, 17: 337–346, 1991.

Putz-Anderson V and Waters TR, Revisions in NIOSH guide to manual lifting. Paper presented at a national conference entitled, "A National Strategy for Occupational Musculoskeletal Injury Prevention—Implementation Issues and Research Needs," Ann Arbor, MI, University of Michigan, 1991.

Reisbord LS and Greenland S, Factors associated with self-reported back pain prevalence: a population-based study, *Journal of Chronic Diseases*, 38: 691–702, 1985.

Stuart-Buttle C, A case study of factors influencing the effectiveness of scissor lifts for box palletizing, *American Industrial Hygiene Association Journal*, 56(11): 1127–1132, 1995.

50

The ACGIH TLV® for Low Back Risk

William S. Marras
The Ohio State University

Chris Hamrick
Ohio Bureau of Worker's Compensation

50.1 Overview

The Lifting TLV (threshold limit value) was the product of a team of scientists whose goal was to develop a lifting guideline that was accurate, used the latest scientific information, and easy to use. This group of scientists consisted of Lawrence J. Fine, Christopher Hamrick, W. Monroe Keyserling, William S. Marras, Robert Norman, Barbara Silverstein, and Thomas Waters with correspondence members consisting of Peter Buckle and John W. Frank. The product of this group's work has been presented as an ACGIH 2005 TLV. This chapter is based upon the work of the committee establishing this NIE (notice of intent to establish) of which the two authors of this chapter were members.

The scientific rationale for this TLV is based on the most recent biomechanical, psychophysical, and epidemiological studies, which together demonstrated a causal association between lifting activities and increase risk of low back disorders characterized by pain and the temporary or prolonged inability to perform normal occupational and nonoccupational activities. The model for the structure of this TLV was based upon the structure of the Finnish lifting guidelines and the 1995 Occupational Safety and Health Administration (OSHA) draft standard. These approaches were latter incorporated in the State of Washington Ergonomic Rule. The approach for these efforts was to define the vertical and horizontal space relative to the base of the spine (L5/S1) of the material handler. Surveillance studies have shown that the single strongest indicator of risk for a lifting task was the load moment (weight of the object lifted times the distance from the spine) relative to L5/S1 (Marras et al., 1993). In addition, biomechanical studies have indicated that the vertical location of a load relative spine has profound biomechanical implications for spine loading and tolerance (Marras, Granata, Davis, Allread & Jorgensen, 1999). Therefore, a tool that identifies the origin location of the load to be lifted from a horizontal and vertical location perspective was a reasonable approach.

Lift origins were divided into 12 horizontal and vertical zones relative within the sagittal plane of the body (Table 50.1). The lift height zones consisted of four regions: (1) the region from 30 cm above to 8 cm below shoulder height (reach limit), (2) knuckle height to below shoulder height, (3) middle shin height to knuckle height, and (4) floor to middle shin height. Horizontal location origins were divided

TABLE 50.1   Table with Cell Names

Anatomical Landmarks	Start Height of Lift (Use of anatomical landmarks from column one is preferred)	Close Lifts: origin < 30 cm from mid-point between inner ankle bones	Near Lifts: origin 30–60 cm from mid-point between inner ankle bones	Extend Lifts: origin ≥ 60–80 cm from mid-point between inner ankle bones
From overhead reach limit to 8 cm below shoulder height [Ht] (Axilla)	132–183 cm	Cell A	Cell B	No known safe limit for repetitive lifting
Knuckle Ht to below shoulder Ht	>82 to 132 cm	Cell C	Cell D	Cell E
Middle shin Ht to knuckle Ht	>30 to 82 cm	Cell F	Cell G	Cell H
Floor to middle shin Ht	0–30 cm	Cell I	No known safe limit for repetitive lifting	No known safe limit for repetitive lifting

into three zones and were defined relative to the horizontal distance of the object lifted from L5/S1. These horizontal zones were defined as: (1) close lifts — less than 30 cm from the spine, (2) intermediate lifts — 30–60 cm from the spine, and (3) extended lifts whose origins were between 60 and 80 cm from the spine.

Once the lift origin was defined, the maximum weight of lift was determined when lifting within each of these zones. In order to arrive at an acceptable weight of lift from each of these 12 zones information based upon biomechanical, psychophysical, epidemiological, and historical surveillance data were compared so that patterns of risk could be established. Although previous National Institute of Occupational Safety and Health (NIOSH) efforts have attempted to look at risk consistency between the different approaches, none have assessed the patterns of risk using modern approaches.

The logic behind this effort assumes that several workplace factors influence the risk of low back disorder (LBD). These factors include: (1) exposure to high moments, (2) lifting from extreme postures, and (3) repetitive lifting. Biomechanical reasoning dictates that risk occurs when an imposed load on a structure exceeds the tolerance level of the structure (McGill, 1997). The workplace factors acknowledged by the TLV, such as moments, posture, and repetition impose loads on the spine. These loads are compared to biomechanical, physiological, and psychophysical tolerances. A recent review of the literature has indicated that the better defined the conditions of a lifting task, the stronger the association with risk (NRC, 2001). The TLV assessment incorporates the most powerful of these defining factors.

50.2   TLV Development Background

50.2.1   Spine Load Estimates

Accurate estimation of spinal loads depends on predicting the internal loads (muscle activities and passive tissue contributions) that are needed to support or counterbalance the external loads (load that is lifted or moved). The sum of the internal and external forces define spinal loading. Early approaches paved the way for current methods (Chaffin, 1975; Garg, Chaffin & Frievalds, 1982; Schultz, Andersson, Ortengren, Haderspeck & Nachemson, 1982a; Schultz, Andersson, Ortengren, Bjork & Nordin, 1982b; Schultz, Haderspeck, Warwick & Portillo, 1983; Schultz & Andersson, 1981). These early models predicted internal forces based upon the minimum required activity to balance the external load. However, these methods often found it difficult to explain the coactivity of the muscles often observed during realistic dynamic exertions. A method to measure indirectly spinal loads was to employ biologically assisted models. With these models, one monitors the biological output from many muscles to directly assess which muscles are active in response to an external load. Current methods use the real time recording of electromyographic (EMG) activity from trunk muscles and three dimensional geometric model of the trunk to predict the three-dimensional loading of the spine under dynamic lifting conditions over time (Granata & Marras, 1993, 1995; Marras & Granata, 1995, 1997a, b; Marras & Sommerich, 1991a, b; McGill & Norman, 1985, 1986; van Dieen, Hoozemans, van der Beek & Mullender, 2002; van Dieen, JJ, Groen, Toussaint & Meijer, 2001). Applications of these models have demonstrated that spine loading varies as a function of repetition (Granata, Marras & Davis, 1999), forward bending (Granata & Marras, 1995; Marras & Sommerich, 1991b), and trunk moment (Granata & Marras, 1993, 1995; Marras & Sommerich, 1991b). Loading can occur in compression, shear, or torsion. To date, these models are the most accurate models available for the assessment of realistic work conditions. When the results of these studies are combined with the epidemiological studies that lifting below knuckle height, or at a greater horizontal distance from the trunk are more hazardous than lifting done between knuckle height and mid-chest height close to the body, a strong rationale is present for assessing LBD risk as a function of load location during lifting.

50.2.2 Load Tolerance

While the tolerance limits for spine damage is not completely understood, for the vertebral end plate, the range of *in vitro* tolerances is known from laboratory studies of the relationship of compression forces. All direct tolerance data has been derived from cadaveric tissue damage to the disc or the vertebral end plates. Several possible mechanisms of injury are thought to exist. One of the more plausible mechanisms for LBDs involves microfracture of the vertebral end plates. As healing occurs, scar tissue develops at the endplate that interferes with nutrient delivery to the disc. This loss of nutrient results in atrophy of the disc fibers, which may initiate chronic damage to the disc and may result in disc degeneration or herniation. There is some scientific evidence that the *in vivo* and *in vitro* tolerance levels do not differ greatly (Waters, Putz-Anderson, Garg & Fine, 1993; Yoganandan, 1986). Increasing levels of disc compression initiate a number of other harmful disc responses at the cellular level, thus, providing further evidence of a cumulative damage to the spine (Lotz & Chin, 2000; Lotz, Colliou, Chin, Duncan & Liebenberg, 1998).

Jager and coworkers (Jager & Luttmann, 1999; Jager, Luttmann & Laurig, 1991) have shown that lumbar vertebra tolerances vary as a function of gender and age. Their data suggests that approximately 30% of lumbar segments have a tolerance of 3.4 kN or less (Waters et al., 1993). This value of 3.4 kN was selected as a tolerance criterion in developing the NIOSH lifting equation. While there is uncertainty about whether this is a reliable predictor of risk for low back disorders, there is epidemiological data to suggest that it is a reasonable one (Chaffin & Park, 1973; Herrin, Jaraiedi & Anderson, 1986). Chaffin and Park found that the incidence for jobs with less than 2.5 kN of spine compression was less than 5%, while jobs with more than 4.5 kN of compression had an incidence of more than 10%. Andersson, Svensson, and Oden (1983) reported that when males performed lifting tasks resulting in spine compression forces greater than 3.4 kN, they had a 40% higher incidence rate of low back pain than did males employed in jobs with lower predicted forces. Herrin et al. (1986) found that jobs with compression forces between 4.5 and 6.8 kN had an incidence rate of 1.5 times greater than jobs with compression forces that were less than 4.5 kN.

Tolerance to spine loading is reduced in highly repetitive tasks or when there is substantial flexion of the spine. Vertebral strength is reduced by 30% with 10 loading cycles and by 50% with 5000 loading cycles (Brinkmann, Biggermann & Hilweg, 1988). Solomonow, Zhou, Baratta, Lu, and Harris (1999) demonstrated that cyclical loading induces creep into the viscoelastic tissues of the spinal tissues which desensitizes the mechanoreceptors, possibly increasing exposure of the tissues to instability and risk of injury even before the muscles fatigue. In addition, the posture of the spine at which point the load is applied appears to be of great significance to the tolerance of the spine as well as to the ability of the spine to receive nutrients. A flexed spine may be as much as 40% weaker than during an upright posture (Gunning, Callaghan & McGill, 2001). Such observations may explain why in some epidemiological studies the highest risks are seen in highly repetitive lifting tasks (Marras et al., 1993) or when the spine is flexed [(Punnett, Fine, Keyserling, Herrin & Chaffin, 1991; Vingard et al., 2000).]

Lifting tasks not only generate compressive forces on the spine, but also generate three-dimensional loading of the spine. As a result damage can occur not only from compressive forces but also from shear loading. Recently tolerances have been estimated for shear loading of the spine. These are expected to occur between 750 and 1000 N (McGill, 1997). Quantitative workplace measures by Marras et al. (1993, 1995) and Norman et al. (1998) have evaluated the biomechanical factors associated with jobs that put the worker at a high risk of LBD. Both of these studies have evaluated the three-dimensional factors that are associated with risk. The results of these studies agree well with the load-tolerance model as well as the increase risk with substantial spinal flexion.

50.2.3 Studies of Lift Location

Several studies have employed EMG-assisted models to assess spine load in compression, lateral shear, and anterior–posterior shear relative to spine tolerance limits (Marras, Granata, Davis, Allread & Jorgensen, 1997; Marras & Davis, 1998; W. S. Marras et al., 1999). In these studies, workers were asked to lift from various horizontal and vertical locations on a pallet. Spine compression and shear as well as back injury risk (Marras et al., 1993) were evaluated as a function of load weight magnitude and location of the load on the pallet. This analysis indicated that the location of the load in space had a far more dramatic effect on spinal loading than did the weight of the object lifted. Table 50.2 summarizes the spine compression relative to the spine tolerances for the load lifted from various locations (Figure 50.1) as well as for loads of various weights and with different handle conditions. This analysis has shown quantitatively that the greatest influence on spine loading is the location from which the case is lifted from a pallet. Spine loading measures have shown that the lower regions of the pallet (regions E and F) contribute the most to spine loading. These regions also represent the conditions where most of the spine compression distribution and a large portion of the A/P shear forces exceed the spine tolerance limits (S. McGill, 2002; S. M. McGill, 1997; NIOSH, 1981). The compressive forces exceeded the tolerance limits of the spine more often (a greater portion of the distribution) than did the A/P shear forces. In fact, even the

TABLE 50.2 Summary of the Percentage of Data within the Benchmark Zones for Spine Compression

| Region on the Pallet | Benchmarks | Case Weight | | | | | |
| | | 40 lbs (18.2 kg) | | 50 lbs (22.7 kg) | | 60 lbs (27.3 kg) | |
		Handles	No Handles	Handles	No Handles	Handles	No Handles
A	<3400 N	100.0	100.0	100.0	99.2	99.2	100.0
	3400–6400 N	0.0	0.0	0.0	0.8	0.8	0.0
	>6400 N	0.0	0.0	0.0	0.0	0.0	0.0
B	<3400 N	98.2	89.1	84.5	76.4	83.6	67.3
	3400–6400 N	1.8	10.9	15.5	23.6	16.4	32.7
	>6400 N	0.0	0.0	0.0	0.0	0.0	0.0
C	<3400 N	98.7	91.3	94.7	82.7	92.6	76.0
	3400–6400 N	1.3	8.7	5.3	17.3	7.4	23.3
	>6400 N	0.0	0.0	0.0	0.0	0.0	0.7
D	<3400 N	88.7	82.0	80.7	75.3	76.7	64.7
	3400–6400 N	11.3	18.0	19.3	24.7	23.3	34.6
	>6400 N	0.0	0.0	0.0	0.0	0.0	0.7
E	<3400 N	45.3	30.0	29.3	14.0	16.0	3.3
	3400–6400 N	52.0	62.0	62.7	65.3	72.0	66.0
	>6400 N	2.7	8.0	8.0	20.7	12.0	30.7
F	<3400 N	35.3	24.0	30.0	10.7	9.3	2.0
	3400–6400 N	60.7	67.3	56.7	65.3	71.3	62.0
	>6400 N	4.0	8.7	13.3	24.0	19.3	36.0

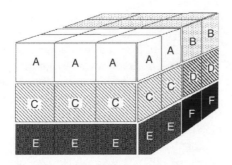

FIGURE 50.1 Load locations when lifting from a pallet. Refers to region locations identified in Table 50.2.

relatively light 18 kg case, without handles, resulted in over 70% of the lifts exceeding the 3400 N spine compression tolerance limit.

This effort has also indicated that this goal of elevating the load can sometimes be partially accomplished by including handles in the case to be lifted. Handles interact strongly with load location to mediate spine loading and risk. In order to appreciate how spine compression loading changed as a function of these features as well as the region factor, Table 50.2 has been included to indicate the percentage of observations that exceed the two compression force "bench-marks." This table indicates how the inclusion of handles and cases of different weights interact with pallet region to define loading, and therefore, risk of LBD. Similar analyses for the shear forces indicated that, for the most part, spine compression was more problematic than the shear forces for the weight factor alone. Table 50.1 indicates that the practical benefit of handles is realized for the heavier cases located at the lower levels of the pallet. Specifically, regardless of case weight, the inclusion of handles can reduce the percentage of observations that exceed the 6400 N limit by approximately 50%. The effect of handles also interacts with the region factor. Compression forces often exceed the tolerance limit for all case weights at the lowest layer of the pallet. The decrease in compression force by the use of handles most likely resulted from the decrease in the external moments. This would be expected since the handles were located at the top of the cases and subjects were not required to bend the torso to the same extent. This would reduce the external moment generated by the torso mass. Further evaluation of the model results confirmed that significant differences in maximum L_5/S_1 compression force were associated with differences in three of the four maximum moment measurements (i.e., sagittal, twisting, and resultant moment). The external moment associated with cases containing handles was significantly less than the moment for cases without handles. The most pronounced reduction of maximum moment was for the sagittal moment, which reduced to 151.8 N m from 176.4 N m (a 13.9% decrease) by the inclusion of handles on the cases. Examination of the results also shows that the effect of including handles is to raise the lift point of the case. This affects the length–strength relationship of the trunk muscles moving the load position to a location where the length–strength relationship of the muscle results the most power. Thus, including handles is analogous to slightly changing the region of the case.

A large portion of the A/P shear distribution also exceeded the 1000 N tolerance at regions E and F when no handles were present on the case. Specifically, A/P shear was found to be problematic for the 27.3 kg (60 lb) cases in regions E and F. Handles were also found to reduce this average peak shear to well below the 1000 N tolerance. Observations of the subjects indicated that this reduction in shear associated with handles was due to the fact that subjects could grab the handle and slide the load toward them as opposed to reaching to the far end of the box to lift as was observed without handles.

These results based upon spine loading were also compared to an analysis of the same features evaluated with a LBD risk model (Marras et al., 1993). This analysis yielded very similar results. Thus, the spinal loading profiles were for the most part similar to trunk motion and workplace factor measures that match trunk motions and workplace factors historically associated with jobs placing workers at an increased risk of LBD.

These studies form the basis for the limits of lift in the various lifting zones defined in Table 50.1. Collectively, these biomechanical studies clearly indicated the need for lower weights in lifts at the greatest distance from the body and at the lower lifting heights. The selected weights in cells E and F, and G are based in part on the implications of the biomechanical studies.

The epidemiological literature provides additional scientific justification that lifting below knuckle height, or at a greater horizontal distance from the trunk are more hazardous than lifting done between knuckle height and mid-chest height close to the body. Kerr et al. (2001) had reported that more than 20° of trunk flexion for more than 1.6 h was associated with an increased risk of LBDs (odds ratio, OR = 1.4). With more than 50° of peak flexion the risk increased even further (OR = 2.4). Punnett et al. (1991) reported an odds ratio of 4.2 for work with more than 20° of trunk flexion for less than 10% of the cycle or shift (48 min) and a higher risk of 6.1 for similar work for more than 10% of the cycle. With trunk flexion of more than 45° for more than 10% of the cycle the risk increased even further (OR = 8.9). Four other studies reported related findings. Vingard et al. (2000) reported an odds ratio of 1.8 for trunk flexion more than one hour a day for men, but not for women. Holmstrom, Lindell, and Moritz (1992) reported an odds ratio of 1.3 for stooping more than four hours per day. The Marras study (Marras et al., 1993, 1995) quantitatively monitored 114 different workplace variables in over 500 jobs that were classified according to historical risk of LBD. These analyses were able to show that many, biomechanical workplace factors (such as trunk velocities) were associated with risk. However, when a multivariate logistic model of risk was considered, five factors in combination (lift frequency, sagittal torso bending angle, lateral velocity, twisting velocity, and external load moment), described the relationship with risk of reporting a LBD incidence (OR = 10.7) and LBD lost or restricted time (OR = 10.6) very well. Two of these factors (sagittal torso bending angle, and external load moment) are most important in lifts that occur at some distance from the body and at lower heights. Further analysis of this database strongly provides two important pieces of information relative to load origin location. First, this epidemiologic information indicates that it is highly unusual for industry to design jobs requiring routine lifting from floor level. Second, there is a dramatic change in the proportion of high-risk jobs vs. low-risk jobs associated with this database as a function of region. Using this historical risk (surveillance) database, very few low-risk jobs have been observed where lifting originates from low vertical heights compared to high-risk jobs.

Analysis of The Ohio State University (OSU) database indicated that six of the highly repetitive lifting tasks performed in cell B (Table 50.6) involved lifting of weights less than 15 lbs (6.8 kg) while five of the six high-risk jobs involved lifting more than 15 lbs (6.8 kg). Additional supportive evidence for the consequences of load during lifting at shoulder height and above the shoulders has been published in the biomechanical literature by Lindbeck et al. (1997) and Sporrong, Sandsjo, Kadefors, and Herberts (1999). Sporrong and associates found that the supraspinatus muscle is heavily loaded in arm positions with substantial abduction, whereas Lindbeck et al. found that work in overhead positions results in high pressure in the supraspinatus muscle, thus increasing the shoulder load. There is substantial evidence from epidemiological studies that overhead work is associated with increase of shoulder musculoskeletal disorders.

This evidence, when considered collectively, demonstrated increasing risk with increasing exposure for forward flexion of the back. In addition to biomechanical and epidemiological studies, the psychophysical studies indicate that for a considerable fraction of the working population TLV weights would be unacceptable. Eventually, even lower weight recommendations may be needed for a strongly protective TLV for repetitive lifting tasks.

50.2.4 Repetition Modifiers

Examination of biomechanical modeling results has demonstrated that spine loading varies as a function of repetition (Granata et al., 1999; Marras et al., 2004) and trunk moment (Granata & Marras, 1993, 1995; Marras & Granata, 1997a; Marras et al., 1993, 1995, 2004; Marras & Sommerich, 1991b). Often spine loading increases after the first 2 h of exposure (Marras et al., 2004). Compression tolerance to spine loading appears reduced in highly repetitive tasks or when there is substantial flexion of the spine. Compressive strength of the vertebrae is reduced by 30% with 10 loading cycles and by 50% with 5000 loading cycles (Brinkmann et al., 1988). Solomonow et al. (1999) showed that cyclical loading induces creep into the viscoelastic tissues of the spinal tissues which desensitizes the mechanoreceptors, possibly increasing exposure of the tissues to instability and risk of injury even before the muscles fatigue. Such observations may explain why in some epidemiological studies the highest risks are seen in highly repetitive lifting tasks (Marras et al., 1993, 1995).

There are several epidemiological studies which have focused on highly repetitive lifting and complement the findings of biomechanical studies (Chaffin & Park, 1973; Kelsey et al., 1984; Kerr et al., 2001; Liira, Shannon, Chambers & Haines, 1996; Liles, Deivanayagam, Ayoub & Mahajan, 1984; Marras et al., 1995; Waters et al., 1999). In a case referent study of hospitalization for herniated lumbar discs, Kelsey et al. (1984) found increased risks associated with lifting more than 25 lbs 25 times per day (RR = 3.5); lifting more than 25 lbs while twisting more than five times per day (RR = 3.1); and for lifting more than 25 lbs per day while twisting and having the knee straight (RR = 6.1). Similarly, Punnett et al. (1991) found an increased risk of low back disorders with frequent lifts (more than 10 lbs more than once per minute). Liira et al. (1996) found increased odds of long-term low back problems for those bending/lifting more than 50 times per day (OR = 1.7), for frequent lifts of less than 50 lbs (OR = 1.5), awkward trunk postures (OR = 2.3) and whole body vibration (OR = 1.8). As the number of risk factors increased, so did the risk. The Marras studies (1993, 1995) quantitatively monitored 114 different workplace variables in over 500 jobs that were classified according to historical risk of LBD. These analyses were able to show that many, biomechanical workplace factors (such as trunk velocities) were associated with risk. However, when a multivariate logistic model of risk was considered, five factors in combination (lift frequency, sagittal torso bending angle, lateral velocity, twisting velocity, and external load moment), described the relationship with risk of reporting a LBD incidence (OR = 10.7) and LBD lost or restricted time (OR = 10.6) very well. The high-risk jobs (OR = 3.3) averaged 3.77 lifts per minute and 20-lb loads. All studies provide some evidence that repetitive lifting is hazardous, and most provide substantial evidence. The Marras et al. (1993, 1995) studies provided the most detailed information on the most highly repetitive lifts and was used to select the weights for the various cells in the TLV tables. The weights for cells B, D, and G primarily determined by identifying weights below which there were a large proportion of low risk jobs (i.e., with incidence rates of zero cases) and a small proportion of the high-risk jobs (i.e., with incidence rates of 12 cases per 100 FTEs (full time employees)). For example, the weight in cell D is 20 lbs (9.1 kg). In the OSU database on low- and high-risk jobs, number of low jobs below 20 lbs (9.1 kg) and above 20 lbs (9.1 kg) was 49/6 for low-risk jobs and corresponding number was 15/30 for high-risk jobs. The weight for lifts in cell G is 15 lbs (6.8 kg). In the OSU database on low- and high-risk jobs, number of low jobs below 15 lbs (6.8 kg) and above 15 lbs (6.8 kg) was 20/4 for low-risk jobs and 2/21 for high-risk jobs. The data from the OSU studies are based on experienced workers performing repetitive lifting tasks and thus is more predictive of observed elevated risk for LBD.

50.3 Using the Lifting TLV

50.3.1 About the Lifting TLV

The Lifting TLVs recommend workplace lifting conditions under which it is believed nearly all workers may be repeatedly exposed, day after day, without developing work-related low back and shoulder disorders associated with repetitive lifting tasks (ACGIH, 2004). Appropriate control measures should be implemented any time the Lifting TLVs are exceeded or lifting-related musculoskeletal disorders are detected.

The tool consists of three charts; the chart used to determine the TLV is a function of the lifting duration and lifting frequency. Then, based upon the one of four categories for height of the lift (floor to mid-shin height, mid-shin to knuckle height, knuckle height to below shoulder, and below shoulder to 30 cm above shoulder) and one of three categories of horizontal lift distance (close, intermediate, and extended), a TLV for weight is given. In some cases, the tables indicate that there is no known safe limit for repetitive lifting under those conditions.

The TLV is limited to two-handed mono-lifting tasks performed within 30° of the sagittal plane, so tasks requiring a large amount of trunk twisting should not be analyzed with this tool. Additionally, the ACGIH states that the guide does not apply to tasks where there are >360 lifts per hour, lifting more than 8 h per day, constrained body posture, high heat or humidity, lifting unstable objects, poor hand couplings, or unstable footing. Professional judgment should be used to determine weight limits under these conditions.

If any of the conditions in the following list are present, then professional judgment should be used to reduce weight limits below those recommended in the TLVs:

- Lifting at a frequency higher than 360 lifts per hour
- Extended work shifts: lifting performed for longer than 8 h per day
- High asymmetry: lifting more than 30° away from the sagittal plane
- One-handed lifting
- Lifting while seated or kneeling
- High heat and humidity
- Lifting unstable objects (e.g., liquids with shifting center of mass)
- Poor hand coupling: lack of handles, cut-outs, or other grasping points
- Unstable footing (e.g., inability to support the body with both feet while standing)

The lifting TLV does incorporate relatively complex data into a format that is quick, easy to use, and easy to interpret; hence, it is a very useful tool to quickly assess many lifting tasks. The results can also direct the user to job redesign strategies. For example, if the lifting conditions exceed the TLV, the user can then find cells in the table that would not exceed the TLV, and then redesign the job accordingly. Since the results are presented in a straightforward, intuitive format, the TLV also can be useful when requesting support from management for resources to institute ergonomic interventions.

50.3.2   Instructions for Determining the Lifting TLV

1. Understand the limitations and basis of the TLVs
2. Determine if the task duration is less than or equal to 2 h per day or greater than 2 h per day
3. Determine the lifting frequency as the number of lifts a worker performs per hour
4. Use the TLV table that corresponds to the duration and lifting frequency of the task (Table 50.3 can be used to determine the appropriate TLV table given the frequency and the duration.)
5. Determine the lifting zone height based on the location of the hands at the beginning of the lift (Figure 50.2.)
6. Determine the horizontal location of the lift by measuring the horizontal distance from the midpoint between the inner ankle bones to the midpoint between the hands at the beginning of the lift
7. Determine the TLV for the lifting task, as displayed in the table cell that corresponds to the lifting zone and horizontal distance in the appropriate table, based upon frequency and duration

50.3.3   If a Lifting Task Exceeds the TLV

If the weight being lifted in the task exceeds the TLV, then changes must be made to the task to ensure that the weight is within the limit. Factors such as the characteristics of the task, type of industry, and

TABLE 50.3   TLV Table Corresponding to Specified Levels of Lifting Frequency and Lifting Duration

Lifts per hour	Duration of Task per day	
	≤2 h	>2 h
≤60	Table 50.4	
≤12		Table 50.4
>12 and ≤30		Table 50.5
>60 and ≤360	Table 50.6	
>30 and ≤360		Table 50.6

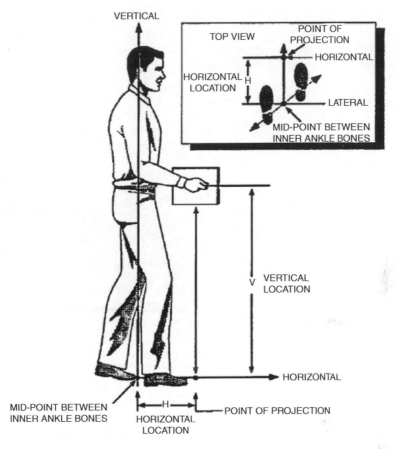

FIGURE 50.2   Graphic representation of hand location.

economics will dictate the appropriate task redesign strategy. The following hierarchy of controls is suggested when redesigning manual material handling tasks.

1. *Eliminate unnecessary lifting.* Whenever possible, eliminate manual materials handling by combining operations or shortening the distances that material must be moved. Look at material flow through the facility, and eliminate any unnecessary lifts. By doing so, we eliminate worker exposure to the musculoskeletal disorder risk factors. In addition, the overall efficiency of a facility is generally improved as time previously required to manually handle materials can be used for other productive tasks.

2. *Automate or mechanize lifting.* If it is not possible to eliminate the lift, consider automating the lifting task or using a mechanical lifting device. Devices such as hoists, cranes and manipulators can eliminate the forces on the spine associated with manual materials handling. Therefore, the likelihood of back injuries is also reduced.

3. *Modify the job to fit within worker capabilities.* If material must still be handled manually (or until one of the above approaches can be implemented), design the task to reduce the stress on the body as much as possible, with emphasis on ensuring that the weight lifted is below the Lifting TLV. Some strategies for job design include:
   - Allow for lifting loads as close to the body as possible. Some techniques to reduce reaching distances are: (a) eliminate any barriers such as the sides of bins or boxes, (b) use a turn table for loads on pallets, and (c) use a tilt table to allow for better access into bins.

- Place the load as close to waist height as possible. This may be accomplished by using adjustable lift tables or inclined conveyors to locate the object to be handled at waist height.
- Reduce the need to twist the trunk by reorienting the lifting origins and destinations.
- Reduce the weight of the load being lifted so that the weights are within the Lifting TLV.

50.3.4   Example 1

As an example of how the TLV can identify lifting tasks which put workers at risk of injury, consider the following scenario, as seen in a manufacturing facility.

A producer of refractory products for use in metal melting and high-temperature industrial applications performs a mixing task which requires the operator to make the mix for the forming of parts (Figure 50.3). The mix is made in batches and each batch requires loading a bag of filler weighing 18 kg. The operator mixes 80 batches per 8-h shift, or 10 per hour. The hopper opening is between the operator and the staging area for the bags, and a dust-control ventilation duct is also between the operator and the bag to

FIGURE 50.3   Lifting filler bag.

be lifted. Therefore, the operator is required to extend the hands away from the trunk to lift the bags. At the point where the bag is lifted, the hands are 75 cm from the midpoint between the ankles, and the bag is between knuckle height to below shoulder height.

Because the task is performed for more than 2 h per day (actual task duration is 8 h per day), and the frequency of lift is ≤12 lifts per hour (actual lifting frequency is ten lifts per hour), the appropriate TLV table for this task is Table 50.4. Using the cell which corresponds to a lifting height zone between "Knuckle height to below shoulder" and a horizontal location of lift of "Extended Lifts," the TLV for this task is 9 kg. The actual weight being lifted is 18 kg, which exceeds the TLV.

Given that the weight lifted in this task is twice the Lifting TLV, the company realized that the task must be redesigned. By considering the hierarchy of controls, they determined that the best method of control was to eliminate the lift. The company installed a "super sack" system whereby the filler is purchased in bulk, and screw feeders controlled by the operator are used to deliver the appropriate amount of filler to the mix. The operator could then make the mix with minimal manual handling of materials, and the relatively high risk of injury posed by the lifting task was eliminated.

50.3.5   Example 2

Another example of using the TLV to identify risk of injury is seen in the following situation from a machine and welding shop.

One of the parts produced by the shop is a "shell" used in air compressors. It must be lifted from floor level to a press, located at about chest level (Figure 50.4). The shell weighs 23 kg. At the current production rate, the machine operator must perform a lift approximately every 3.5 mins, or about 17 lifts per hour. There are no obstacles, and the load is relatively compact (51 cm long and 33 cm in diameter), so the operator can get close to the load.

Because this task is performed for more than 2 h per day (actual task duration is 8 h per day), and the frequency of lift is >12 lifts per hour but ≤30 lifts per hour (actual lifting frequency is 17 lifts per hour), Table 50.5 is the appropriate table to determine the TLV for this task. Using the cell which corresponds to a lifting height zone of "Floor to middle shin height" and a horizontal location of lift of "Close Lifts," the TLV for this task is 9 kg. The actual weight being lifted (23 kg) is clearly greater than the TLV.

TABLE 50.4 TLVs for Lifting Tasks ≤2 h per day with ≤60 Lifts per hour or >2 h per day with ≤12 Lifts per hour

Lifting Height Zone	Horizontal Location of Lift		
	Close Lifts: Origin <30 cm from Midpoint between Inner Ankle Bones	Intermediate Lifts: Origin 30–60 cm from Midpoint between Inner Ankle Bones	Extended Lifts: Origin >60 to 80 cm from Midpoint between Inner Ankle Bones[a]
Reach limit[b] from 30 cm above to 8 cm below shoulder height	16 kg	7 kg	No known safe limit for repetitive lifing[c]
Knuckle height[d] to below shoulder	32 kg	16 kg	9 kg
Middle shin height to knuckle height[d]	18 kg	14 kg	7 kg
Floor to middle shin height	14 kg	No known safe limit for repetitive lifing[c]	No known safe limit for repetitive lifing[c]

[a]Lifting tasks should not be started at a horizontal reach distance more than 80 cm from the midpoint between the inner angle bones (Figure 50.2)

[b]Routine lifting tasks should not be conducted from starting heights greater than 30 cm above the shoulder or more than 180 cm above floor level (Figure 50.2)

[c]Routine lifting tasks should not be performed for shaded table entries marked "No known safe limit for repetitive lifting." While the available evidence does not permit identification of safe weight limits in the shaded regions, professional judgment may be used to determine if infrequent lifts of lifht weights may be safe.

[d]Anatomical landmark for knuckle height assumes the worker is standing erect with arms hanging at the side.

Source: Reprinted from ACGIH, 2004 Threshold Limit Values for Chemical Substance and Physical Agents & Biological Exposure Indices, 2004. Cincinnati, OH: ACGIH. With permission.

The company redesigned the lifting task by installing a manipulator attached to an overhead crane so that the shell could be lifted mechanically, thereby nearly eliminating the muscular effort required by the machine operator (Figure 50.5). By reducing the muscular effort, the mechanical stress on the spine is also reduced, thereby substantially reducing the risk of injury to the machine operator.

In the year prior to installing the manipulator, the company reported nine injuries associated with this task: two back sprains, six neck sprains, and one wrist sprain. In the year after the change, no injuries have been reported to the population performing this task. The company also indicated that productivity has increased by nearly 70%, and scrap rate has decreased from 2 to 0.8%.

As these examples illustrate, the Lifting TLV is a practical, easy to use tool that can be used in the workplace to assess risk. The results can also be used to determine jobs which are candidates for redesign and as an assessment tool after making job modifications to determine if the risk of injury has been reduced to acceptable levels.

FIGURE 50.4 Lifting shell to press.

TABLE 50.5   TLVs for Lifting Tasks >2 h per day with >12 and ≤ 30 Lifts per hour or ≤2 h per day with >60 and ≤360 Lifts per hour

Lifting Height Zone	Horizontal Location of Lift		
	Close Lifts: Origin <30 cm from Midpoint between Inner Ankle Bones	Intermediate Lifts: Origin 30–60 cm from Midpoint between Inner Ankle Bones	Extended Lifts: Origin >60 to 80 cm from Midpoint between Inner Ankle Bones[a]
Reach limit[b] from 30 cm above to 8 cm below shoulder height	14 kg	5 kg	No known safe limit for repetitive lifing[c]
Knuckle height[d] to below shoulder	27 kg	14 kg	7 kg
Middle shin height to knuckle height[d]	16 kg	11 kg	5 kg
Floor to middle shin height	9 kg	No known safe limit for repetitive lifing[c]	No known safe limit for repetitive lifing[c]

[a]Lifting tasks should not be started at a horizontal reach distance more than 80 cm from the midpoint between the inner angle bones (Figure 50.2)

[b]Routine lifting tasks should not be conducted from starting heights greater than 30 cm above the shoulder or more than 180 cm above floor level (Figure 50.2.)

[c]Routine lifting tasks should not be performed for shaded table entries marked "No known safe limit for repetitive lifting." While the available evidence does not permit identification of safe weight limits in the shaded regions, professional judgment may be used to determine if infrequent lifts of lifht weights may be safe.

[d]Anatomical landmark for knuckle height assumes the worker is standing erect with arms hanging at the side.

Source: Reprinted from ACGIH, 2004 Threshold Limit Values for Chemical Substance and Physical Agents & Biological Exposure Indices, 2004. Cincinnati, OH: ACGIH. With permission.

TABLE 50.6   TLVs for Lifting Tasks >2 h per day with >30 and ≤360 Lifts per hour.

Lifting Height Zone	Horizontal Location of Lift		
	Close Lifts: Origin < 30 cm from Midpoint between Inner Ankle Bones	Intermediate Lifts: Origin 30–60 cm from Midpoint between Inner Ankle Bones	Extended Lifts: Origin > 60 to 80 cm from Midpoint between Inner Ankle Bones[a]
Reach limit[b] from 30 cm above to 8 cm below shoulder height	11 kg	No known safe limit for repetitive lifing[c]	No known safe limit for repetitive lifing[c]
Knuckle height[d] to below shoulder	14 kg	9 kg	5 kg
Middle shin height to knuckle height[d]	9 kg	7 kg	2 kg
Floor to middle shin height	No known safe limit for repetitive lifing[c]	No known safe limit for repetitive lifing[c]	No known safe limit for repetitive lifing[c]

[a]Lifting tasks should not be started at a horizontal reach distance more than 80 cm from the midpoint between the inner angle bones (Figure 50.2)

[b]Routine lifting tasks should not be conducted from starting heights greater than 30 cm above the shoulder or more than 180 cm above floor level (Figure 50.2)

[c]Routine lifting tasks should not be performed for shaded table entries marked "No known safe limit for repetitive lifting." While the available evidence does not permit identification of safe weight limits in the shaded regions, professional judgment may be used to determine if infrequent lifts of lifht weights may be safe.

[d]Anatomical landmark for knuckle height assumes the worker is standing erect with arms hanging at the side.

Source: Reprinted from ACGIH, 2004 Threshold Limit Values for Chemical Substance and Physical Agents & Biological Exposure Indices, 2004. Cincinnati, OH: ACGIH. With permission.

FIGURE 50.5   Using a manipulator to lift shell onto machine.

References

American Conference of Governmental Industrial Hygienists (ACGIH) (2004). *2004 Threshold Limit Values for Chemical Substances and Physical Agents & Biological Exposure Indices.* Cincinnati, OH: ACGIH.

Andersson, G. B., Svensson, H. O., & Oden, A. (1983). The intensity of work recovery in low back pain. *Spine*, 8: 880–884.

Brinkmann, P., Biggermann, M., & Hilweg, D. (1988). Fatigue fracture of human lumbar vertebrae. *Clin Biomech (Bristol, Avon)* 3: S1–S23.

Chaffin, D. B. (1975). A computerized biomechanical model: development of and use in studying gross body actions. *J. Biomech.*, 2: 429–441.

Chaffin, D. B. & Park, K. S. (1973). A longitudinal study of low-back pain as associated with occupational weight lifting factors. *Am. Ind. Hyg. Assoc. J.*, 34: 513–525.

Garg, A., Chaffin, D. B., & Frievalds, A. (1982). Biomechanical stresses from manual load lifting: a static vs dynamic evaluation. *IIE Trans*, 14: 272–281.

Granata, K. P. & Marras, W. S. (1993). An EMG-assisted model of loads on the lumbar spine during asymmetric trunk extensions. *J. Biomech.*, 26: 1429–1438.

Granata, K. P. & Marras, W. S. (1995). An EMG-assisted model of trunk loading during free-dynamic lifting. *J. Biomech.*, 28: 1309–1317.

Granata, K. P., Marras, W. S., & Davis, K. G. (1999). Variation in spinal load and trunk dynamics during repeated lifting exertions. *Clin. Biomech. (Bristol, Avon)*, 14: 367–375.

Gunning, J. L., Callaghan, J. P., & McGill, S. M. (2001). Spinal posture and prior loading history modulate compressive strength and type of failure in the spine: a biomechanical study using a porcine cervical spine model. *Clin. Biomech. (Bristol, Avon)*, 16: 471–480.

Herrin, G. D., Jaraiedi, M., & Anderson, C. K. (1986). Prediction of overexertion injuries using biomechanical and psychophysical models. *Am. Ind. Hyg. Assoc. J.*, 47: 322–330.

Holmstrom, E. B., Lindell, J., & Moritz, U. (1992). Low back and neck/shoulder pain in construction workers: occupational workload and psychosocial risk factors. Part 2. Relationship to neck and shoulder pain. *Spine*, 17: 672–677.

Jager, M., & Luttmann, A. (1999). Critical survey on the biomechanical criterion in the NIOSH method for the design and evaluation of manual lifting tasks. *Int. J. Indu. Ergon.*, 23: 331–337.

Jager, M., Luttmann, A., & Laurig, W. (1991). Lumbar load during one-hand bricklaying. *Int. J. Ind. Ergon.*, 8: 261–277.

Kelsey, J. L., Githens, P. B., White, A. A. I., Holford, T. R., Walters, S. D., O'Conner, T., Ostfeld, A. M., Weil, U., Southwick, W. O., & Calogero, J. A. (1984). An epidemiologic study of lifting and twisting on the job and risk for acute prolapsed lumbar intervertebral disc. *J. Ortho. Res.*, 2: 61–66.

Kerr, M. S., Frank, J. W., Shannon, H. S., Norman, R. W., Wells, R. P., Neumann, W. P., & Bombardier, C. (2001). Biomechanical and psychosocial risk factors for low back pain at work. *Am. J. Public Health*, 91: 1069–1075.

Liira, J. P., Shannon, H. S., Chambers, L. W., & Haines, T. A. (1996). Long–term back problems and physical work exposures in the 1990 Ontario Health Survey. *Am. J. Public Health*, 86: 382–387.

Liles, D. H., Deivanayagam, S., Ayoub, M. M., & Mahajan, P. (1984). A job severity index for the evaluation and control of lifting injury. *Hum. Factors*, 26: 683–693.

Lindbeck, L., Karlsson, D., Kihlberg, S., Kjellberg, K., Rabenius, K., Stenlund, B., & Tollqvist, J. (1997). A method to determine joint moments and force distributions in the shoulders during ceiling work — a study on house painters. *Clin. Biomech (Bristol, Avon)*, 12: 452–460.

Lotz, J. C. & Chin, J. R. (2000). Intervertebral disc cell death is dependent on the magnitude and duration of spinal loading. *Spine*, 25: 1477–1483.

Lotz, J. C., Colliou, O. K., Chin, J. R., Duncan, N. A., & Liebenberg, E. (1998). Compression-induced degeneration of the intervertebral disc: an in vivo mouse model and finite-element study. *Spine*, 23: 2493–2506.

Marras, W., Granata, K., Davis, K., Allread, W., & Jorgensen, M. (1997). Spine loading and probability of low back disorder risk as a function of box location on a pallet. *Int. J. Hum. Factors Manuf.*, 7: 323–336.

Marras, W. S. & Davis, K. G. (1998). Spine loading during asymmetric lifting using one versus two hands. *Ergonomics*, 41: 817–834.

Marras, W. S., Ferguson, S. A., Lavender, S. A., Burr, D., Chany, A. M., Parakkat, J., & Yang, G. (2004). Impact of experience and workday length on spine loading. Paper presented at the International Society for the Study of the Lumbar Spine, Porto, Portugal.

Marras, W. S. & Granata, K. P. (1995). A biomechanical assessment and model of axial twisting in the thoracolumbar spine. *Spine*, 20: 1440–1451.

Marras, W. S. & Granata, K. P. (1997a). The development of an EMG-assisted model to assess spine loading during whole-body free-dynamic lifting. *J. Electromyogr. Kinesiol.*, 7: 259–268.

Marras, W. S. & Granata, K. P. (1997b). Spine loading during trunk lateral bending motions. *J. Biomech.*, 30: 697–703.

Marras, W. S., Granata, K. P., Davis, K. G., Allread, W. G., & Jorgensen, M. J. (1999). Effects of box features on spine loading during warehouse order selecting. *Ergonomics*, 42: 980–996.

Marras, W. S., Lavender, S. A., Leurgans, S. E., Fathallah, F. A., Ferguson, S. A., Allread, W. G., & Rajulu, S. L. (1995). Biomechanical risk factors for occupationally related low back disorders. *Ergonomics*, 38: 377–410.

Marras, W. S., Lavender, S. A., Leurgans, S. E., Rajulu, S. L., Allread, W. G., Fathallah, F. A., & Ferguson, S. A. (1993). The role of dynamic three-dimensional trunk motion in occupationally-related low back disorders. The effects of workplace factors, trunk position, and trunk motion characteristics on risk of injury. *Spine*, 18: 617–628.

Marras, W. S. & Sommerich, C. M. (1991a). A three-dimensional motion model of loads on the lumbar spine: I. Model structure. *Hum. Factors*, 33: 123–137.

Marras, W. S. & Sommerich, C. M. (1991b). A three-dimensional motion model of loads on the lumbar spine: II. Model validation. *Hum. Factors*, 33: 139–149.

McGill, S. (2002). *Low Back Disorders: Evidence-Based Prevention and Rehabilitation*. Champaign, IL: Human Kinetics.

McGill, S. M. (1997). The biomechanics of low back injury: implications on current practice in industry and the clinic. *J. Biomech.*, 30: 465–475.

McGill, S. M. & Norman, R. W. (1985). Dynamically and statically determined low back moments during lifting. *J. Biomech.*, 18: 877–885.

McGill, S. M. & Norman, R. W. (1986). Partitioning of the L4–L5 dynamic moment into disc, ligamentous, and muscular components during lifting [see comments]. *Spine*, 11: 666–678.

National Institute of Occupational Safety and Health (NIOSH), (1981). *Work Practices Guide For Manual Lifting*. Cincinnati, OH: Department of Health and Human Services (DHHS), NIOSH.

Norman, R., Wells, R., Neumann, P., Frank, J., Shannon, H., & Kerr, M. (1998). A comparison of peak vs. cumulative physical work exposure risk factors for the reporting of low back pain in the automotive industry. *Clin. Biomech. (Bristol, Avon)*, 13: 561–573.

National Research Council (NRC) (2001). *Musculoskeletal Disorders and the Workplace: Low Back and Upper Extremity*. Washington, DC: National Academy Press.

Punnett, L., Fine, L. J., Keyserling, W. M., Herrin, G. D., & Chaffin, D. B. (1991). Back disorders and nonneutral trunk postures of automobile assembly workers. *Scand. J. Work. Environ. Health.*, 17: 337–346.

Schultz, A., Andersson, G., Ortengren, R., Haderspeck, K., & Nachemson, A. (1982a). Loads on the lumbar spine. Validation of a biomechanical analysis by measurements of intradiscal pressures and myoelectric signals. *J. Bone. Joint. Surg. [Am].*, 64: 713–720.

Schultz, A., Andersson, G. B., Ortengren, R., Bjork, R., & Nordin, M. (1982b). Analysis and quantitative myoelectric measurements of loads on the lumbar spine when holding weights in standing postures. *Spine*, 7: 390–397.

Schultz, A., Haderspeck, K., Warwick, D., & Portillo, D. (1983). Use of lumbar trunk muscles in isometric performance of mechanically complex standing tasks. *J. Orthop. Res.*, 1: 77–91.

Schultz, A. B. & Andersson, G. B. (1981). Analysis of loads on the lumbar spine. *Spine*, 6: 76–82.

Solomonow, M., Zhou, B. H., Baratta, R. V., Lu, Y., & Harris, M. (1999). Biomechanics of increased exposure to lumbar injury caused by cyclic loading: Part 1. Loss of reflexive muscular stabilization. *Spine*, 24: 2426–2434.

Sporrong, H., Sandsjo, L., Kadefors, R., & Herberts, P. (1999). Assessment of workload and arm position during different work sequences: a study with portable devices on construction workers. *Appl. Ergon.*, 30: 495–503.

van Dieen, J. H., Hoozemans, M. J., van der Beek, A. J., & Mullender, M. (2002). Precision of estimates of mean and peak spinal loads in lifting. *J. Biomech.*, 35: 979–982.

van Dieen, J. H., JJ, M. D., Groen, V., Toussaint, H. M., & Meijer, O. G. (2001). Within-subject variability in low back load in a repetitively performed, mildly constrained lifting task. *Spine*, 26: 1799–1804.

Vingard, E., Alfredsson, L., Hagberg, M., Kilbom, A., Theorell, T., Waldenstrom, M., Hjelm, E. W., Wiktorin, C., & Hogstedt, C. (2000). To what extent do current and past physical and psychosocial occupational factors explain care-seeking for low back pain in a working population? Results from the Musculoskeletal Intervention Center-Norrtalje Study. *Spine*, 25: 493–500.

Waters, T. R., Baron, S. L., Piacitelli, L. A., Anderson, V. P., Skov, T., Haring-Sweeney, M., Wall, D. K., & Fine, L. J. (1999). Evaluation of the revised NIOSH lifting equation. A cross-sectional epidemiologic study. *Spine*, 24: 386–394; discussion 395.

Waters, T. R., Putz-Anderson, V., Garg, A., & Fine, L. J. (1993). Revised NIOSH equation for the design and evaluation of manual lifting tasks. *Ergonomics*, 36: 749–776.

Yoganandan, N. (1986). Biomechanical identification of injury to an intervetebral joint. *Clin. Biomech.*, 1: 149.

Index

U

Z

Printed and bound by CPI Group (UK) Ltd, Croydon, CR0 4YY

01/11/2024

01782638-0001